Entomopathogenic Nematodes as Biological Control Agents

We dedicate this book to Professor Randy Gaugler and Professor Itamar Glazer. Both scientists have contributed immensely to the field of entomopathogenic nematology.

Professor Gaugler holds degrees from North Dakota State, North Carolina State, and the University of Wisconsin. Randy joined Rutgers University in 1982. He reached the rank of Distinguished Professor in 1995 and Distinguished Professor Emeritus in 2020. Randy is clearly a foremost pioneer in the discipline. The Gaugler laboratory conducted foundational research in several key areas of entomopathogenic nematology, including behavioral ecology (foraging dynamics), population ecology, genetics, and production technology. His 300 research papers and four books (including two keystone books on entomopathogenic nematodes) have earned more than 22,000 citations. Randy has received more than 40 honors and awards and has been elected a Fellow of four scientific societies including the American Association for the Advancement of Science.

Professor Glazer obtained degrees from Hebrew University. Also a pioneer in entomopathogenic nematology, he was a Senior Scientist at Biosys Inc. (a founding company for entomopathogenic nematode production) from 1985 to 1987. Itamar joined the Department of Entomology and Nematology, ARO, Volcani Center, in 1987 where he held various leadership positions including Head of the Nematology Department, Director of Technology Transfer and Head of the Plant Science Institute; he retired from the Agricultural Research Organization in 2021. Itamar established himself as a leader in entomopathogenic nematode research, particularly in the biology and genetics of entomopathogenic nematodes under environmental stress. Itamar has produced over 300 publications, which include 132 reviewed publications, 25 invited reviews (including editing three books), and 145 presentations at scientific meetings.

In addition to dedicating the book to Professors Gaugler and Glazer based on their groundbreaking contributions to the discipline, we also want to honor and thank these two scientists for serving as our most excellent mentors. They were our first postdoctoral advisors (Randy for Ed Lewis and Itamar for David Shapiro-Ilan) and thus introduced us in earnest to this fascinating field of entomopathogenic nematology. We are extremely grateful to Randy and Itamar for their friendship and support, as they were instrumental in launching our own careers.

Randy Gaugler

Itamar Glazer

Entomopathogenic Nematodes as Biological Control Agents

Edited by

David I. Shapiro-Ilan

USDA-ARS, Southeastern Fruit and Tree Nut Research Station, USA

and

Edwin E. Lewis

Department of Entomology, Plant Pathology and Nematology, University of Idaho, USA

CABI

CABI is a trading name of CAB International

CABI
Nosworthy Way
Wallingford
Oxfordshire OX10 8DE
UK

Tel: +44 (0)1491 832111
E-mail: info@cabi.org
Website: www.cabi.org

CABI
200 Portland Street
Boston
MA 02114
USA

Tel: +1 (617)682-9015
E-mail: cabi-nao@cabi.org

A catalogue record for this book is available from the British Library, London, UK.

Library of Congress Cataloging-in-Publication Data

Names: Shapiro-Ilan, David I., editor. | Lewis, Edwin E, editor.
Title: Entomopathogenic nematodes as biological control agents / edited by David Shapiro-Ilan and Edwin E. Lewis
Description: Boston, MA : CAB International, [2024] | Includes bibliographical references and index. | Summary: "Entomopathogenic nematodes (EPNs) are biocontrol agents that are used to control a wide variety of insect pests within agriculture and forestry. In addition to their use as bio-pesticides, EPNs have a fascinating biology and are thus considered model organisms in ecology, symbiosis and pathogenesis. This book presents basic knowledge and diverse applications to illustrate how EPNs play an important role as potent biocontrol solutions. It describes: Fundamental concepts such as biology, taxonomy, symbiosis genomics and behavioural ecology. Aspects of commercialization, including mass production, formulation, safety and regulation, and marketing. Diverse cropping systems e.g. maize, wheat and grains, citrus, orchard systems, berries, vine crops, vegetables and turf. Other applications including urban, nursery, forestry, greenhouse, veterinary and medical. Ecological considerations and applications in conservation biocontrol. This book is a must have for all pest management professionals including those practicing integrated pest management strategies"-- Provided by publisher
Identifiers: LCCN 2024014498 (print) | LCCN 2024014499 (ebook) | ISBN 9781800620308 (hardback) | ISBN 9781800620315 (ebook) | ISBN 9781800620322 (epub)
Subjects: LCSH: Nematodes as biological pest control agents. | Agricultural pests--Biological control. | Microbial pesticides.
Classification: LCC SB976.N46 E68 2024 (print) | LCC SB976.N46 (ebook) | DDC 632/.96--dc23/eng/20240715
LC record available at https://lccn.loc.gov/2024014498
LC ebook record available at https://lccn.loc.gov/2024014499

ISBN-13: 9781800620308 (hardback)
9781800620315 (ePDF)
9781800620322 (ePub)

DOI: 10.1079/9781800620322.0000

Commissioning Editor: Rebecca Stubbs
Editorial Assistant: Helen Elliott
Production Editor: Shankari Wilford

Typeset by Straive, Pondicherry, India
Printed and bound in the UK by CPI Group (UK) Ltd, Croydon, CR0 4YY

Contents

Contributors

Steven P. Arthurs, BioBee USA, Salisbury, MD 21804, USA. Email: steven.arthurs@biobee.us

Anil Baniya, Department of Nematology, University of California, Riverside, CA 92507, USA. Email: abaniya@ucr.edu

Rubén Blanco-Pérez, Soils, Biosystems, and Agroforestry Ecology Department, Misión Biológica de Galicia (BMG-CSIC), Pazo de Salcedo. Carballeira, 8 Salcedo, 36143 Pontevedra, Spain. Email: ruben.blanco@mbg.csic.es

Mary Barbercheck, Department of Entomology, The Pennsylvania State University, University Park, PA 16802, USA. Email: meb34@psu.edu

Helge B. Bode, Department of Natural Products in Organismic Interactions, Max-Planck-Institute for Terrestrial Microbiology, 35043 Marburg, Germany; Chemical Biology, Department of Chemistry, Philipps University Marburg, 35043 Marburg, Germany; Molecular Biotechnology, Department of Biosciences, Goethe University Frankfurt, 60438 Frankfurt am Main, Germany; Senckenberg Gesellschaft für Naturforschung, 60325 Frankfurt am Main, Germany. Email: Helge.Bode@mpi-marburg.mpg.de

Raquel Campos Herrera, Instituto de Ciencias de la Vid y del Vino (ICVV), Gobierno de La Rioja, CSIC, Universidad de La Rioja, Logroño, 26007, Spain. Email: raquel.campos@icvv.es

Julie G. Chacon-Orozco, Laboratório de Controle Biológico, Instituto Biológico, Campinas, São Paulo, Brazil, Cep 13101-680. Email: jchaconorozco@gmail.com

Harun Cimen, Recombinant DNA and Recombinant Protein Application and Research Center, Aydin Adnan Menderes University, 09100 Aydin, Turkey. Email: hcimen@adu.edu.tr

Regina K. Cruzado-Gutiérrez, Institute for Health in the Human Ecosystem, University of Idaho, Moscow, ID 83844, USA. Email: karin@uidaho.edu

Surendra K. Dara, University of California Division of Agriculture and Natural Resources, San Luis Obispo, CA 93401, USA. Current address: Department of Horticulture, Oregon State University, Corvallis, Oregon, USA. Email: surendra_d@hotmail.com

Adler R. Dillman, Department of Nematology, University of California, Riverside, CA 92507, USA. Email: adlerd@ucr.edu

Larry Duncan, University of Florida, IFAS, Citrus Research and Education Center, 700 Experiment Station Road, Lake Alfred, FL33850, USA. Email: lwduncan@ufl.edu

Ioannis Eleftherianos, Infection and Innate Immunity Laboratory, Department of Biological Sciences, The George Washington University, Science and Engineering Hall, 800 22nd Street NW, Washington DC 20052, USA. Email: ioannise@gwu.edu

Shane Foye, Department of Entomology, Rutgers University, New Brunswick, NJ 08901, USA. Current address: Department of Entomology, University of Wisconsin, Madison, WI 53706, USA. Email: shanefoye1989@gmail.com

Patrick Frettinger, BASF France SAS, 21 Chemin de la Sauvegarde, 69130 Ecully, France. Email: patrick.frettinger@basf.com

Fernando Garcia del Pino, Department of Animal Biology, Vegetal Biology and Ecology, Faculty of Bioscience, Universitat Autònoma de Barcelona, 08193 Bellaterra, Barcelona, Spain. Email: Fernando.Garcia@uab.cat

Itamar Glazer, Institute of Plant Protection, ARO, Volcani Center, Rishon LeZion, Israel. Email: glazerit@volcani.agri.gov.il

John A. Goolsby, US Department of Agriculture, Agricultural Research Service, Cattle Fever Tick Research Unit, 22675 N. Moorefield Rd., Moore Airbase, Bldg. 6419, Edinburg, TX 78541, USA. Email: goolsby1600@gmail.com

Dawn H. Gouge, University of Arizona, Maricopa Agricultural Center, Maricopa, AZ 85138, USA, Email: dhgouge@arizona.edu

Richou Han, Institute of Zoology, Guangdong Academy of Sciences, Guangzhou, China, Email: hanrc@giz.gd.cn

Sebnem Hazal Gulsen, Kocarli Vocational School, Aydin Adnan Menderes University, 09100 Aydın, Turkey. Email: Shgulsen@adu.edu.tr

Canan Hazir, Aydin Health Services Vocational School, Aydin Adnan Menderes University, 09100 Aydin, Turkey. Email: canan.hazir@adu.edu.tr

Selcuk Hazir, Department of Biology, Faculty of Science, Aydin Adnan Menderes University, 09100 Aydin, Turkey. Email: shazir@adu.edu.tr

Ivan Hiltpold, Entomology and Nematology, Division of Plant Protection, Agroscope, 1260 Nyon, Switzerland. Email: ivan.hiltpold@agroscope.admin.ch

Ganpati B. Jagdale, Extension Nematology Laboratory, Department of Plant Pathology, University of Georgia, 2350 College Station Road, Athens, GA 30605, USA, email. Email: gbjagdal@uga.edu

Rinus Knoetze, Department of Conservation Ecology and Entomology, Faculty of AgriSciences, Stellenbosch University, Private Bag X1, Matieland 7602, Stellenbosch; Plant Protection Division, Agricultural Research Council (ARC), Infruitec-Nietvoorbij, Private Bag X5026, Stellenbosch 7599, South Africa. Email: knoetzer@arc.agric.za

Albrecht Koppenhöfer, Department of Entomology, Rutgers University, New Brunswick, NJ 08901, USA. Email: a.koppenhofer@rutgers.edu

Gabriela Lankin, Department of Plant Protection, Faculty of Agricultural Sciences, Universidad de Chile, Chile. Email: glankin@uchile.cl

Luis G. Leite, Laboratório de Controle Biológico, Instituto Biológico, Campinas, São Paulo, Brazil, Cep 13101-680. Email: luis.leite@sp.gov.br

Edwin E. Lewis, Department of Entomology, Plant Pathology and Nematology, University of Idaho, Moscow, ID 83844-2329, USA. Email: eelewis@uidaho.edu

Diana K. Londono, BASF Corporation, 26 Davis Drive, Research Triangle Park, North Carolina, NC 27709-3528, USA, Email: diana.londono@basf.com

Antoinette P. Malan, Department of Conservation Ecology and Entomology, Faculty of AgriSciences, Stellenbosch University, Private Bag X1, Matieland 7602, Stellenbosch, South Africa. Email: apm@sun.ac.za

Dana Ment, Institute of Plant Protection, ARO, The Volcani Center, Rishon LeZion, Israel. Email: danam@volcani.agri.gov.il

Jayashree Ramakrishnan, Institute of Plant Protection, ARO, The Volcani Center, Rishon LeZion, Israel; The Robert H. Smith Faculty of Agriculture, Food & Environment, The Hebrew University of Jerusalem, Rehovot, Israel. Email: jramakrishnan28@gmail.com

Gadi V.P. Reddy, USDA-ARS, Southern Insect Management Research Unit, 141 Experiment Station Road, PO Box 346, Stoneville, MS 38776, USA. Email: Gadi.Reddy@usda.gov

Cesar Rodriguez-Saona, PE Marucci Center for Blueberry and Cranberry Research and Extension, Rutgers University, 125A Lake Oswego Rd., Chatsworth, NJ 08019, USA. Email: crodriguez@njaes.rutgers.edu

Weibin Ruan, College of Life Sciences, Nankai University, Tianjin, 30071, China, Email: ruanweibin@nankai.edu.cn

Michael Samish, Kimron Veterinary Institute, Bet-Dagan, PO Box 12, 50250, Israel. Email: samishmichael@gmail.com

Ernesto San-Blas, Laboratory of Nematology, Institute of Agri-food, Animal and Environmental Sciences (ICA3), Universidad de O'Higgins, Campus Colchagua, Chile. Email: esanblas@yahoo.com or esanblas@uoh.cl

Ramandeep Kaur Sandhi, FMC Corporation, Stine Research Center, 1090 Elkton Road, Newark, DE 19711, USA. Email: Raman.Sandhi@fmc.com

David Shapiro-Ilan, USDA-ARS, Southeastern Fruit and Tree Nut Research Station, 21 Dunbar Road, Byron, GA 31008, USA. Email: David.Shapiro@usda.gov

Ana Luiza Sousa, Department of Entomology, Rutgers University, New Brunswick, New Jersey, USA. Current address: Syngenta Seeds Development—R&D Innovation Center, Malta, Illinois 60150, USA. Email: ana_luiza.sousa@syngenta.com

Tatyana Stefanovska, Department of Entomology, Integrated Pest Management and Plant Quarantine, National University of Life and Environmental Science, Kiev, Ukraine. Email: steftat@hotmail.com.

Glen Stevens, Department of Entomology, Plant Pathology and Nematology, University of Idaho, Moscow, ID 83844, USA. Email: glens@uidaho.edu

S. Patricia Stock, School of Animal and Comparative Biomedical Sciences, University of Arizona, 1117 Lowell Street, Tucson, AZ 85721, USA; College of Agriculture, California State University, 321 Plumas Hall, 400 1st Street, Chico, CA 95929, USA. Email: spstock@arizona.edu

Yoelvis Sulbaran, Institute of Agri-food, Animal and Environmental Sciences (ICA3), Universidad de O'Higgins, Chile. Email: yoelvis.sulbaran@uoh.cl

Ghada Tafesh-Edwards, Infection and Innate Immunity Lab, Department of Biological Sciences, The George Washington University, Science and Engineering Hall, 800 22nd Street NW, Washington DC 20052, USA. Email: gtafesh@gwmail.gwu.edu

Stefan Toepfer, CAB International, Delemont, Switzerland. Email: s.toepfer@cabi.org

Mustapha Touray, Department of Biology, Faculty of Science, Aydin Adnan Menderes University, 09100 Aydin, Turkey. Email: mtpha.touray@gmail.com

Derya Ulug, Department of Biology, Faculty of Science, Aydin Adnan Menderes University, 09100 Aydin, Turkey. Email: deryaasici@adu.edu.tr

Bart Vandenbossche, e-nema Ltd., Schwentinental, Germany. Email: b.vandenbossche@e-nema.de

Christopher D. Williams, School of Biological and Environmental Sciences, Liverpool John Moores University, Liverpool L3 3AF, UK. Email: C.D.Williams@ljmu.ac.uk

Sheng-Yen Wu, State Key Laboratory of Ecological Pest Control for Fujian and Taiwan Crops, Ministerial and Provincial Joint Innovation Centre for Safety Production of Cross-Strait Crops, Fujian Agriculture and Forestry University, Fuzhou, 350002, China. Email: sywu@fafu.edu.cn

Xun Yan, Zhongkai University of Agriculture and Engineering, Guangzhou, PR China. Email: yanxun@zhku.edu.cn

Preface

Interest in biological control, defined as the use of living organisms for pest control, continues to rise due to increasing public awareness of the negative impact of chemical pesticides. Large-scale commercial interest in the development and implementation of biological control methods particularly surged after the publication of Rachael Carson's seminal book, *Silent Spring*, in 1962, which documented the environmental impacts of indiscriminate use of chemical pesticides. Synthetic chemical pesticides harm natural enemies that keep pest populations under check, but their indiscriminate use can induce pesticide resistance in target pests, requiring even higher pesticide application rates. Chemical pesticide use can also lead to the resurgence of otherwise minor pests, requiring new chemicals or methods of control. Risks to human health posed by widespread application of chemical pesticides, particularly in developing countries, continues to remain high due to illiteracy and lack of training. Persistent chemical pesticides can enter the food chain through air and water, harming non-target organisms including humans. Use of biological control agents has been increasing over the past several decades and is expected to rise further over the next decade due to regulations-driven replacement of synthetic pesticides and rapidly growing organic farming, particularly in Europe. The market share of biological control agents exceeded US$3.0 billion in 2018 and is projected to grow at 15% compounded annual growth rate up to 2025.

The use of predatory and parasitic insects in biological control is over a century old, but serious interest in nematodes as biological control agents only started in the 1980s. Since then, tremendous new knowledge has been generated that has helped in the development of nematodes as practical biological control agents. Among the different types of nematodes, entomopathogenic nematodes in the families Heterorhabditidae and Steinernematidae have received most interest as biological control agents of insect pests because of their mutualistic association with *Photorhabdus* and *Xenorhabdus* bacteria, respectively. This association enables entomopathogenic nematodes to exploit a wide range of insect species and facilitates their large-scale mass production through fermentation.

The editors of this book are world renowned for their contributions to entomopathogenic nematology and for creating simple and innovative application techniques for using nematodes to control insect pests. They have assembled a team of researchers from around the world to contribute chapters on basic and applied aspects of biocontrol nematodes. This book uses the basic organizational framework used in a previous volume, *Nematodes As Biocontrol Agents*, edited by Parwinder S. Grewal, Ralph-Udo Ehlers, and David I. Shapiro-Ilan, published by CABI in 2005. Most authors have thoroughly updated their chapters and have beautifully summarized the new information that has accumulated over the past 20 or so years. This volume will undoubtedly serve as the authoritative volume on entomopathogenic nematodes for another decade or more before significant new information accumulates again. This book is a must-have for all pest management professionals including those practicing integrated pest management strategies.

Parwinder S. Grewal
University of Hawaii at Manoa

Glossary

Amphimixis: Sexual reproduction resulting from the fusion of gametes from two separate organisms, which are usually referred to as male and female.
Axenic: Free from associated organisms.
Biocontrol: The introduction of natural enemies (parasites, parasitoids, predators, or pathogens) to suppress pest populations; some include certain by-products of natural enemies in the definition.
Commensalism: A symbiotic relationship between two species in which one of the organisms benefits and the other is not apparently affected.
Dauer stage or dauer larva: A developmentally arrested dispersal stage in certain nematodes; in entomopathogenic nematodes, it is the only free-living stage (also known as an infective juvenile).
Entomogenous: Refers to organisms growing in or on the bodies of insects; denotes a parasitic or other intimate symbiotic relationship.
Entomoparasitic: Parasitic to insects; a relationship between an organism (e.g. a nematode) and an insect, in which the organism benefits at the insect host's expense. Host mortality is not necessarily a requirement for the parasite's development. Nematode examples include Mermithidae, Allantonematidae, Parasitylenchidae, Phaenopsitylenchidae, Iotonchidae, Acugutturidae, Parasitaphelenchidae, Entaphelenchidae, and Thelastomatidae.
Entomopathogenic: A microorganism or nematode capable of causing disease in insects; in insect nematology, the term is specifically used to refer to parasitic nematodes that are mutualistically associated with bacterial symbionts. All life stages of the nematode, except for the free-living third-stage infective juvenile or dauer stage, are found inside the insect host. Nematode examples are Steinernematidae and Heterorhabditidae.
Entomophilic: Having an affinity for insects ('insect loving'); for nematodes, can refer to any association with insects (parasitic or non-parasitic).
Epizootic: An outbreak of disease in which there is an unusually large number of cases.
Hermaphrodite: An individual that has characteristics of both males (often but not limited to sperm) and females (often but not limited to eggs). Some species of hermaphrodites can self-fertilize (e.g. entomopathogenic nematodes in the genus *Heterorhabditis*), which would be sexual reproduction resulting from the fusion of two gametes that originate from the same individual organism. Other species cannot self-fertilize (e.g. most earthworms).
Incidence: The number of new cases of a particular disease within a given time period.
Infectivity: The ability of an organism to enter a susceptible host, resulting in presence of the organism within the host (whether or not this causes detectable pathological effects); the ability to produce infection.

In vitro: Outside the living organism, in an artificial environment.
In vivo: Within the living organism.
Mutualism: A symbiotic relationship between two different species in which both jointly benefit.
Patent infection: An overt infection with distinct signs and symptoms of the disease.
Pathogenicity: The quality or state of being pathogenic; the potential ability to produce disease (an 'all-or-none' concept).
Phoretic: A symbiotic relationship in which one organism associates with another in order to obtain transportation, causing little or no detectable pathology to the host; examples of nematodes having a phoretic association with insects include certain members of the Rhabditidae, Diplogastridae and Aphelenchidae.
Prevalence: The total number of cases of a particular disease at a given moment in time.
Sign: An objective manifestation of disease indicated by an alteration in structure.
Symbiosis: The living together of individuals of two different species, particularly the living together of two dissimilar species in an intimate association (e.g. mutualism, commensalism, parasitism).
Symptom: Any objective aberration in behavior or function indicating disease.
Virulence: The disease-producing power of an organism; the degree of pathogenicity within a group or species.

Sources

Lacey, L.A. and Brooks, W.M. (1997) Initial handling and diagnosis of diseased insects. In: Lacey, L.A. (ed.) *Manual of Techniques in Insect Pathology*. Academic Press, San Diego, California, pp. 1–15.
Poinar, G.O. Jr. (1975) *Entomogenous Nematodes: A Manual and Host List of Insect–Nematode Associations*. Brill, Leiden, Netherlands.
Shapiro-Ilan, D.I., Fuxa, J.R., Lacey, L.A., Onstad, D.W., and Kaya, H.K. (2005) Definitions of pathogenicity and virulence in invertebrate pathology. *Journal of Invertebrate Pathology* 88, 1–7.
Steinhaus, E.A. and Martignoni, M.E. (1970) *An Abridged Glossary of Terms Used in Invertebrate Pathology*, 2nd edn. USDA Forest Service, Pacific Northwest Forest and Range Experiment Station, Portland, Oregon.
Stock, S.P. (2002) Glossary of terms used in insect nematology. *Nematology Newsletter* 2002 (Issue 3), 17.

1 Diversity and Systematics of Nematodes Used in Biocontrol

S. Patricia Stock[1,2*]

[1]*School of Animal and Comparative Biomedical Sciences, University of Arizona, Tucson, Arizona, USA;* [2]*College of Agriculture, California State University, Chico, California, USA*

Abstract

This chapter provides an updated revision on the diversity and systematics of entomopathogenic nematodes (EPNs) (Steinernematidae and Heterorhabditidae). It also addresses other nematode families considered in biocontrol, thus placing EPNs within the larger phylogenetic picture of the role of nematodes in biocontrol. A historical summary of the different molecular methods and approaches considered to supplement taxa descriptions and their evolutionary relationships is presented. Next-generation sequencing technologies are summarized in relation to their role in the resolution of taxonomic conflicts and in shedding light on the continued understanding of intra- and interspecific variation among EPNs. The last section introduces the concept of integrative taxonomy and reflects on the need to avoid taxonomic vandalism when making nomenclatural changes and decisions.

1.1 Introduction

Among the numerous beneficial organisms considered in biocontrol are nematodes that associate with pests of potential importance in agriculture, forestry, or health (Poinar, 1983; Petersen, 1985; Gaugler and Kaya, 1990; Bedding, 1993; Wilson and Gaugler, 2000; Grewal *et al.*, 2003; Askary, 2010; Belien, 2018). Their associations range from "casual" (i.e. phoretic, commensal) to obligate parasitism and pathogenesis. The number of newly discovered species with biocontrol potential has increased significantly over the past two decades. Over 12% of the newly described species between 2011 and 2019 are associated with invertebrates including insects, and many of them have biocontrol potential (Hodda, 2022).

Their prompt and accurate identification is essential for the acquisition of permits and their successful implementation in control programs (Lacey *et al.*, 2001; MacLeod *et al.*, 2010; Frewin *et al.*, 2013; Ashfaq and Hebert, 2016). Morphological/morphometric methods have been the cornerstone in the identification of nematodes; however, over the past two decades, multidisciplinary approaches encompassing morphological, molecular, distribution, and bionomic data have been widely adopted. A variety of molecular methods and techniques have been implemented to complement traditional methods. Many unculturable taxa or those with no obvious morphological differences could not have been described without the aid of molecular methods (Bogale *et al.*, 2020). Additionally, the lack of properly trained taxonomists continues

*Email: spstock@arizona.edu

 Entomopathogenic Nematodes as Biological Control Agents (D.I. Shapiro-Ilan and E.E. Lewis eds)
DOI: 10.1079/9781800620322.0001

to challenge the expedited and accurate identification of taxa.

Some early molecular approaches included isoenzyme patterns, restriction fragment length polymorphisms (RFLPs), random amplified polymorphic DNA (RAPD) analysis and the polymerase chain reaction (PCR). More recently, the latest generation technologies (i.e. quantitative PCR (qPCR), DNA sequencing, and high-throughput sequencing), together with bioinformatic tools have proven useful not only in systematics but also for studying the diversity of nematodes (Gasser *et al.*, 2008; Porazinska *et al.*, 2009).

In this chapter, the focus is on the systematics and diversity of the most prominent nematode groups considered in biocontrol of economically important pests. A summary of the molecular methods and markers used is also presented. Although the rest of this book is focused solely on entomopathogenic nematodes (EPNs) in the families Heterorhabditidae and Steinernematidae, this chapter also includes the systematics of other nematodes used in biocontrol, thereby placing EPNs within the larger phylogenetic picture of nematodes used in biocontrol.

1.2 Classification

More than 30 nematode families include species that parasitize and/or are associated with insects (Poinar, 1983, 1990; Kaya and Stock, 1997; de Ley and Blaxter, 2004; Stock and Goodrich-Blair, 2012; Stock, 2015). Because of their biocontrol potential, research has concentrated on five families: Mermithidae, Neotylenchidae, Rhabditidae, Steinernematidae and Heterorhabditidae. The latter two have received the most attention as control agents of soil insect pests (Lacey *et al.*, 2001; Campos-Herrera, 2015). More recently, soil-dwelling facultative scavengers, *Oscheius* spp. (Rhabditidae), have been investigated as promising alternatives for insect pest management (Zhang *et al.*, 2008, 2019, 2022).

The biocontrol potential of nematodes is not restricted to insects. For example, *Phasmarhabditis hermaphrodita* (Schneider) (Rhabditidae) is known to suppress several slug species and is currently used as a biological molluscicide (Wilson *et al.*, 1993, 1994; Glen and Wilson, 1997; Wilson and Gaugler, 2000; Rae *et al.*, 2009; Wilson *et al.*, 2012; Pieterse *et al.*, 2017). Moreover, several predatory mononchids, dorylaimids, nygolaimids, diplogasterids, and the fungal-feeding nematode *Aphelenchus avenae* Bastian have been studied as potential biological control agents of plant-parasitic nematodes and other plant pathogens (Kasab and Abdel-Kader, 1996; Matsunaga *et al.*, 1996; Lootsma and Scholte, 1997; Choudhury and Sivakumar, 2000; Haraguchi and Yoshiga, 2020; Zhang *et al.*, 2022).

In this chapter, the classification scheme developed by de Ley and Blaxter (2002, 2004) is followed to list the most prominent groups with biocontrol potential (Table 1.1).

1.3 Diagnosis of Major Groups

1.3.1 Family Steinernematidae Chitwood & Chitwood, 1937 (Fig. 1.1)

1.3.1.1 Diagnostic characters

Adults with truncated to slightly rounded head. Six fused lips, but tips distinct, with one labial papilla each. Four cephalic papillae present. Amphids small. Stoma reduced, short and wide, with inconspicuous sclerotized walls. Esophagus rhabditoid, set off from intestine. Nerve ring surrounding isthmus or anterior part of basal bulb. Excretory pore opening distinct. Females with paired opposed ovaries. Vagina short, muscular. Vulva located near middle of body, with or without protruding lips. Epiptygma present or absent. Males monorchic, testis reflexed. Spicules paired, symmetrical. Gubernaculum present. One single midventral and 10–14 pairs of genital papillae present of which 7–10 pairs are precloacal. Tail rounded, digitated or mucronated. Third-stage infective juvenile (IJ) with collapsed stoma. Cuticle annulated, lateral field with six to eight ridges in middle of body. Esophagus and intestine collapsed. Specialized bacterial pouch located at beginning of intestine with variable shape. Excretory pore distinct, anterior to nerve ring. Tail conoid or filiform, with variable hyaline portion. Phasmids present, prominent or inconspicuous.

The family Steinernematidae currently comprises two genera, *Steinernema* Travassos, 1927 and *Neosteinernema* Nguyen & Smart, 1994. There

Table 1.1. Major groups in the phylum Nematoda considered as biological control agents. Classification based on de Ley and Blaxter, 2002.

CLASS CHROMADOREA INGLIS, 1983
Subclass Chromadoria Pearse, 1942
ORDER RHABDITIDA CHITWOOD, 1933
Suborder Tylenchina Thorne, 1949
Infraorder Panagrolaimomorpha de Ley & Blaxter, 2002
Superfamily Strongyloidoidea Chitwood & McIntosh, 1934
Family Steinernematidae Travassos, 1927
Infraorder Tylenchomorpha de Ley & Blaxter, 2002
Family Neotylenchidae Thorne, 1941
Suborder Rhabditina Chitwood, 1933
Infraorder Rhabditomorpha de Ley & Blaxter, 2002
Superfamily Rhabditoidea Örley, 1880
Family Rhabditidae Örley, 1880
Superfamily Strongyloidea Baird, 1853
Family Heterorhabditidae Poinar, 1975
CLASS ENOPLEA INGLIS, 1983
Subclass Dorylaimia Inglis, 1983
ORDER MERMITHIDA HYMAN, 1951
Suborder Mermithina Andrássy, 1974
Superfamily Mermithoidea Braun, 1883
Family Mermithidae Braun, 1883

are currently over 100 species described in the genus *Steinernema*, while only one species is known for the genus *Neosteinernema* (*N. longicurvicauda*) (Table 1.2).

1.3.1.2 Bionomics

Steinernematids are generically known as EPNs. They are mutualistically associated with *Xenorhabdus* (Gamma-Proteobacteria, Enterobacteriaceae) (Boemare, 2002). This symbiotic partnership is antagonistic to a wide range of insects (Poinar, 1990; Stock and Hunt, 2005; Georgis *et al.*, 2006; see Chapter 9, this volume).

1.3.1.3 Phylogenetic relationships

Steinernema spp. are more closely related to *Strongyloides stercoralis* than to *Ascaris suum*, *Caenorhabditis elegans*, *Ancylostoma duodenale* or *Necator americanus* and are members of clade V (Blaxter *et al.*, 1998; de Ley and Blaxter, 2004; Blaxter *et al.*, 2016). According to Holterman *et al.* (2006), *Steinernema* are placed as closely related to *Bursaphelenchus xylophilus*, a plant parasite, and the free-living *Panagrellus redivivus*.

The first explicit hypothesis for evolutionary relationships among *Steinernema* spp. was based on genetic distances calculated from ribosomal DNA (rDNA) restriction sites for 12 species (Reid, 1994). Subsequently, relationships were considered from RFLP patterns of the internal transcribed spacer (ITS) region of rDNA (Reid *et al.*, 1997).

Liu *et al.* (1997) investigated partial small subunit (18S) rDNA sequences to study evolutionary relationships; however, interpretation of the results was hampered by a number of factors including reduced taxon sampling, uncertainties in character homology, insufficient number of phylogenetically informative characters and tree-building methods (Stock *et al.*, 2001; Nadler *et al.*, 2006).

In the early 2000s, single-gene sequences of nuclear and mitochondrial genes were investigated not only for diagnostics but also to assess their phylogenetic relationships (Stock *et al.*, 2001; Spiridonov *et al.*, 2004b). The first comprehensive testable hypothesis of phylogenetic relationships for species in *Steinernema* considered the combination of 28S rDNA sequences and morphological datasets (Stock *et al.*, 2001). Nadler *et al.* (2006) contemplated a three-gene (28S, 12S and *cox1*) approach together with morphological data to help overcome the distorting effect of stochastic sequence variation of single-gene approaches. This study generated

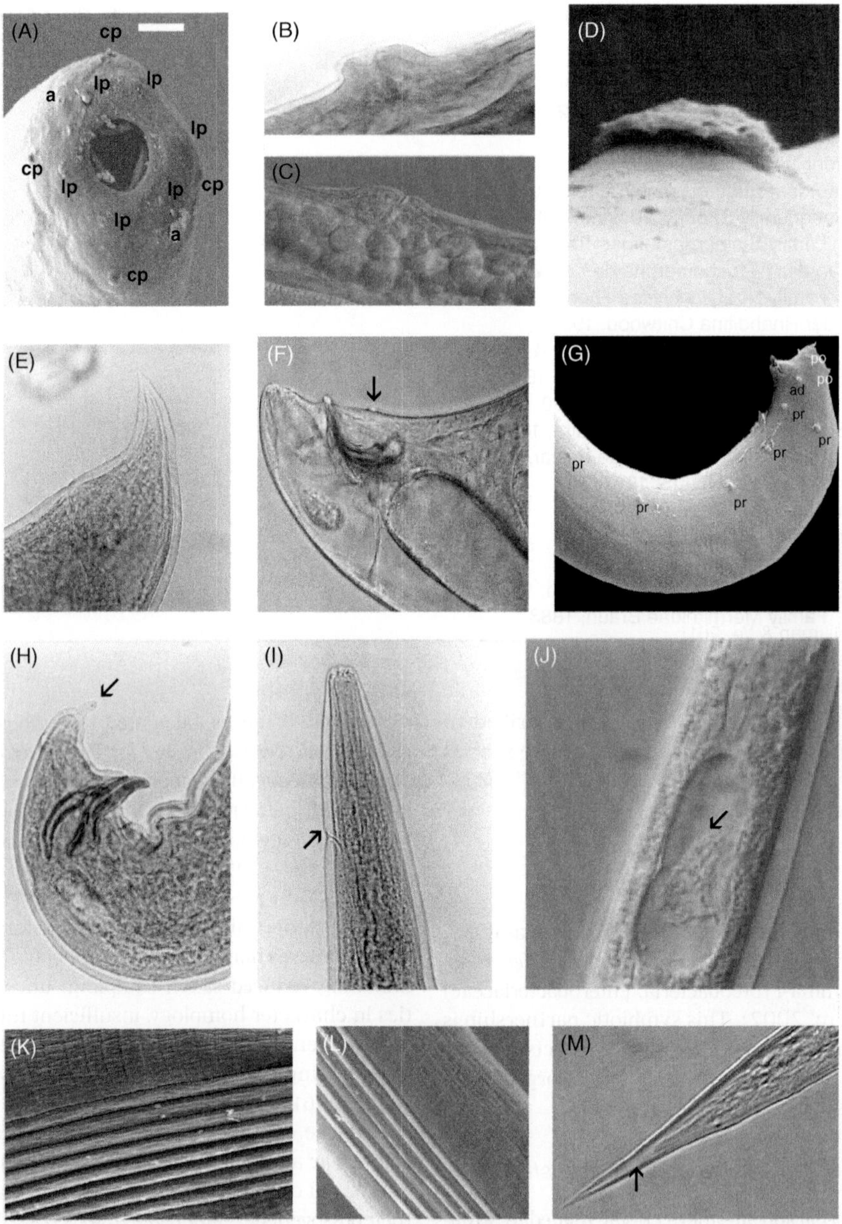

Fig. 1.1. Family Steinernematidae. (A–E) First-generation female. (A) Scanning electron micrograph (SEM) micrograph showing stomatal opening and labial (lp) and cephalic (cp) papillae and amphids (a). (B) Protruding vulval lips (lateral view). (C) Slightly protruding vulval lips (lateral view). (D) Epiptygma. (E) Tail (lateral view). (F–H) First-generation male. (F) Tail showing single ventral papillae (arrow; lateral view). (G) SEM of tail showing precloacal (pr), adcloacal (ad) and postcloacal (po) papillae (lateral view). (H) Tail (lateral view) showing mucro (arrow). (I–N) Third-stage infective juvenile. (I) Anterior end showing excretory pore (arrow). (J) Bacterial pouch (lateral view) showing clump of bacterial cells (arrow). (K, L) SEMs of lateral field pattern with eight (K) and six (L) ridges. (M) Tail (lateral view), showing hyaline portion (arrow). Scale: (A) and (L) 5.5 µm; (B), (C), (E) and (F) 25 µm; (D) 35 µm; (G) 40 µm; (H) 23.5 µm; (I) and (J) 16 µm; (K) 4 µm; (M) 10 µm. All scales are based on the scale bar in (A).

Table 1.2. Taxonomic summary of the family Steinernematidae Chitwood and Chitwood, 1937 (syn. Neoaplectanidae Sobolev, 1953)

Taxa	Origin[a]	GenBank sequence data (accession no.)
Genus: Steinernema Travassos, 1927		
Type species		
S. kraussei (Steiner, 1923) Travassos, 1927	Europe (Germany), North America	28S (AF331896)
Other species		
S. abbasi Elawad, Ahmad & Reid, 1997	Asia (Oman)	18S (AY035764), 28S (AF331890), ITS1, -2 (AY248749)
S. aciari Qiu, Yan, Zhou, Nguyen & Pang, 2005	Asia (China)	28S (AY787660), ITS1, 2 (AY787861, GU395637)
S. affine (Bovien, 1937) Wouts, Mráček, Gerdin & Bedding, 1982	Europe (Denmark)	18S (AY035765), 28S (AF331899), ITS1, -2 (AF331912)
S. anatoliense Hazir, Stock & Keskin, 2003	Asia (Turkey)	NA
S. akhursti Qiu, Hu, Zhou, Mei, Nguyen & Pang, 2005	Asia (China)	18S partial (KT878310), ITS1, -2 (DQ375757), 28S partial (KF289902)
S. apuliae Triggiani, Mráček & Reid, 2004	Italy	28S (KU194621, GU569044), ITS1, -2 (GU569044), CO I (GU569063), CS3 (GU599901)
S. arasbaranense Nikdel, Nikman & Ye, 2011	Asia (Iran)	ITS1, ITS2 (FJ860039)
S. arenarium (Artyukhovsky, 1967) Wouts, Mráček, Gerdin & Bedding, 1982	Asia (Central Russia)	18S (U70639), 28S (AF331892), ITS1 (AF192985), CO II-16S (AF192992)
S. ashiuense Phan, Takemoto & Futai, 2006	Asia (China)	28S (FJ235126), ITS1, -2 (FJ165550, DQ354694)
S. asiaticum Anis, Shahina, Reid & Rowe, 2002	Asia (Pakistan)	NA
S. australe Edgington, Buddie, Tymo, Hunt, Nguyen, Franc, Merino & Moore, 2009	South America (Chile)	28S (FJ235126), ITS1, -2 (FJ235125, MH45388)
S. backanense Phan, Spiridonov, Subbotin & Moens, 2006	Asia (Vietnam)	ITS1, -2 (AY487918)
S. balochiense Fayyaz, Khanum, Ali, Solangi, Gulsher & Javed, 2015	Asia (Pakistan)	ITS (JX135547), 28S (JX068821), 12S (JN571100)
S. batswanae Didiza, Lephoto & Gray, 2021	Africa (South Africa)	ITS1, -2 (MH697401)
S. beddingi Qiu, Hu, Zhou, Pang & Nguyen, 2005	Asia (China)	28S (AY603396), ITS (AY603397)
S. bertusi Katumanyane, Malan, Tiedt & Hurley, 2020	Africa (South Africa)	ITS1, -2 (MW618707)
S. beitlechemi Çimen, Pûža, Nermuť, Hatting, Ramakuwela, Faktorová & Hazir, 2016	Africa (South Africa)	28S (KT580949), ITS1, -2 (KT373856, KT3855)
S. bicornutum Tallosi, Peters & Ehlers, 1995	Europe (Yugoslavia)	28S (AF331904), ITS1, -2 (AF121048)
S. biddulphi Çimen, Pûža, Nermuť, Hatting, Ramakuwela, Faktorová & Hazir, 2016	Africa (South Africa)	28S (KT580950), ITS1, -2 (KT373857)
S. boemarei Lee, Sicard, Skeie & Stock, 2009	Europe (France)	28S (FJ152415), ITS1, -2 (FJ152414), 12S (GU569027), CO I (GU569065)
S. brazilense Nguyen, Ginarte, Leite, dos Santos & Harakava, 2010	South America (Brazil)	28S (FJ410326), ITS1, -2 (FJ410325)

Continued

Table 1.2. Continued.

Taxa	Origin[a]	GenBank sequence data (accession no.)
S. cameroonense Kanga, Trinh, Waeyenberge, Spiridonov, Hauser & Moens, 2012	Africa (Cameroon)	28S (JX985265), ITS1, -2 (JX985267, OD921260)
S. carpocapsae (Weiser, 1955) Wouts, Mráček, Gerdin & Bedding, 1982	Asia, Europe (Czechoslovakia), North America, South America	18S (U70633, AF36604), 28S (AF331900), ITS1 (AF192987, AF036947), ITS1, -2 (AF331913, AF121049), COII-16S (AF192995), SAT (U12680)
S. caudatum Xu, Wang & Li, 1991	Asia (China)	NA
S. ceratophorum Jian, Reid & Hunt, 1997	Asia (China)	28S (AF331888), ITS1, -2, (AF440765)
S. changbaiense Ma, Chen, de Clercq, Han & Moens, 2012	Asia (China)	28S (JN830950) ITS1, -2 (JN865168)
S. cholashanense Nguyen, Pûža & Mráček, 2008	Asia (China)	28S (EF520284), ITS1, -2 (EF431959)
S. citrae Stokwe, Malan, Nguyen, Knoetze & Tiedt, 2011	Africa (South Africa)	28S (GU004534), ITS1, -2 (EU740970)
S. colombiense López-Nunez, Plichta, Góngora-Botero & Stock, 2008	South America (Colombia)	28S (EU345420), ITS1, -2 (345421)
S. costaricense Uribe-Lorío, Mora & Stock, 2007	Central America (Costa Rica)	28S (GU569047, EF187017, JX493015), ITS1, -2 (OK305941, OK319047), CO I (GU569066)
S. cubanum Mráček, Hernandez & Boemare, 1994	Central America (Cuba)	28S (AF331889)
S. cumgarense Phan, Spiridonov, Subbotin & Moens, 2006	Asia, Vietnam	ITS1, -2 (AY170339, AY487920)
S. diaprepesi Nguyen & Duncan, 2002	North America (USA)	ITS1, -2 (AF440764)
S. eapokense Tamiru, Waeyenberge, Hailu, Ehlers, Pûža & Mráček, 2006	Asia, Vietnam	ITS1, -2 (AY4879171)
S. ethiopiense Phan, Spiridonov, Subbotin & Moens, 2006	Africa (Ethiopia)	ITS1, -2, (JN651414)
S. fabii Abate, Malan, Tiedt, Wingfield, Slippers & Hurley, 2016	Africa (south Africa)	28S (KR527217), ITS1, -2 (KR527216)
S. feltiae (Filipjev, 1934) Wouts, Mráček, Gerdin & Bedding, 1982	Europe (Russia), North America, South America	18S (U70634, AY035766), 28S (AF331906), ITS1 (AF92983, AF92982), ITS1, -2 (AF121050), mRNA-gsy-1 (AF241845), CO II-16S (AF192991, AF192990)
S. glaseri (Steiner, 1929) Wouts, Mráček, Gerdin & Bedding, 1982	Asia, North America (USA), South America	18S (U70640), 28S (AF331908), ITS1 (AF192986), ITS1, -2 (AF122015), CO II-16S (AF192993), SAT (U19929)
S. goweni San-Blas, Morales-Montero, Portillo, Nermuť & Pûža 2016	South America (Venezuela)	28S (KR781039), ITS1, -2 (KR781038)
S. guangdongense Qiu, Fang, Zhou, Pang & Nguyen, 2004	Asia (China)	28S (AY169558), ITS1, -2 (AY170341)
S. hebeiense Chen, Li, Yan, Spiridonov & Moens,2006	Asia (China)	28S (DQ399664), ITS1, -2 (DQ105794)
S. hermaphroditum Stock, Griffin & Chaenari, 2004	Asia (Indonesia)	28S (AY598358, MF693228), ITS1, -2 (JQ687355, MW131973), 12S (AY944013), CO I (AY943987)

Continued

Table 1.2. Continued.

Taxa	Origin[a]	GenBank sequence data (accession no.)
S. huense Phan, Mráček, Pûža, Nermuť & Jarošová, 2014	Asia (Vietnam)	28S (KF857582), ITS1, -2, (KF857581)
S. ichnusae Tarasco, Mráček, Nguyen & Triggiani, 2008	Europe (Italy)	28S (421130), ITS1, -2, (EU421129)
S. innovationi Çimen, Lee, Hatting, Hazir & Stock, 2015	Africa (South Africa)	28S (KJ578794), ITS 1, -2 (KJ578793)
S. intermedium (Poinar, 1985) Mamiya, 1988	North America (USA), Europe	18S (U70636), 28S (AF331909), ITS1 (AF192989), ITS1, -2 (AF33916, AF122016)
S. jeffreyense Malan, Knoetze & Tiedt, 2015	Africa (South Africa)	28S (KF857582), ITS1, -2 (KF857581)
S. jollieti Spiridonov, Krasomil-Osterfeld & Moens, 2004	North America (USA)	28S (GU569051), ITS 1, -2 (AY171265)
S. karii Waturu, Hunt & Reid, 1997	Africa (Kenya)	18S (AJ417021), 28S (AF331902)
S. khoisanae Nguyen, Malan & Gozel, 2006	Africa (South Africa)	28S (DQ314289), ITS1, -2 (DQ314287)
S. khuongi Stock, Campos-Herrera, El-Borai & Duncan, 2019	North America (USA)	28S (GU177835), ITS1, -2 (GU174002)
S. kushidai Mamiya, 1988	Asia (Japan)	28S (AF331897), ITS1, -2 (AF192984),
S. lamjungense Khatri-Chhetri, Waeyenberge, Spiridonov, Manandhar & Moes, 2011	Asia (Nepal)	28S (HM000102), ITS1, -2 (HM000101)
S. leizhouense Nguyen, Qiu, Zhou & Pang, 2006	Asia (China)	28S (AY169557), ITS1, -2 (AY170340)
S. litchii Steyn, Knoetze, Tiedt & Malan, 2017	Africa (South Africa)	28S (KY569031), ITS1, -2 (KP325086)
S. littorale Yoshida, 2004	Asia (Japan)	ITS1, -2, (LC157428, AB243441, KY807712)
S. loci Phan, Nguyen & Moens, 2001	Asia (Vietnam)	ITS1, -2 (AY355443)
S. longicaudum Shen & Wang, 1992	Asia (China), North America	18S (AY035767), 28S (AF331894)
S. minutum Maneesakorn, Grewal & Chandrapatya, 2010	Asia (Thailand)	ITS1, -2, (GU647156, KY807711, KY807715, KY807717), CO I (MW287378)
S. monticolum Stock, Choo & Kaya, 1997	Asia (Korea)	28S (AF331895), ITS1, -2 (AF331914, AF122017)
S. nguyeni Malan, Knoetze & Tiedt, 2016	Africa (South Africa)	28S (KR815816), ITS1, -2 (KP325084, MW915805, MT680196, MT940450, MT702997, MT702994)
S. neocurtillae Nguyen & Smart, 1992	North America (USA)	ITS1, -2 (AF122018)
S. nepalense Khatri-Chhetri, Waeyenberge, Spiridonov, Manandhar & Moes, 2011	Asia (Nepal)	ITS1, -2(HQ190044)
S. nyetense Kanga, Trinh, Waeyenberger, Spiridonov, Hauser & Moens, 2012	Africa (Cameroon)	28S (JX985264), ITS1, -2 (JX985266)
S. oregonense Liu & Berry, 1996	North America (USA)	18S (U70637), 28S (AF331891), ITS1, -2 (AF122019),
S. pakistanense Shahina, Anis, Reid, Rowe & Maqbool, 2001	Asia (Pakistan)	28S (MF289982, MF289983) ITS1, -2 (MF289981, MF116307, MK491793-98, JN157771, JN157770), 12S (JX144739)
S. papillatum San-Blas, Portillo, Nermuť, Pûža & Morales-Montero, 2015	South America (Venezuela)	ITS1, -2 (KJ913707, KM2111574)

Continued

Table 1.2. Continued.

Taxa	Origin[a]	GenBank sequence data (accession no.)
S. phyllophagae Nguyen & Buss, 2011	North America (USA)	28S (FJ666054), ITS1, -2, (FJ410327)
S. poinari Mráček, Půža & Nermuť, 2014	Europe (Czech Republic)	28S (KF145174), ITS1, -2, (KF145173)
S. populi Tian, Hu, Li, Zhang, Půža, Shapiro-Ilan, Zhao, Liu, Zhou, Ding & Wang, 2022	Asia (China)	28S (MZ367685), ITS1, -2 (MZ367621)
S. puertoricense Román & Figueroa, 1994	Central America (Puerto Rico)	28S (AF331903)
S. pui Qiu, Zhao, Wu, Lv & Pang, 2011	Asia (China)	28S (GU395618) ITS1, -2 (GU395636)
S. puntauvense Uribe-Lorío, Mora & Stock, 2007	Central America (Costa Rica)	28S (EF187018, GU569056), ITS1, -2 (FJ381665)
S. pwaniense Půža, Nermuť, Mráček, Gengler & Haukeland, 2017		28S (KT358813), ITS1, -2 (KT358811)
S. rarum (de Doucet, 1986) Mamiya, 1988	South America (Argentina), North America	28S (AY253296, AF331905)
S. riobrave Cabanillas, Poinar & Raulston, 1994	North America (USA)	18S (U70635), 28S (AF331893), COII-16S (AF192994)
S. ritteri de Doucet & Doucet, 1990	South America (Argentina)	NA
S. robustispiculum Phan, Subbotin, Waeyenberge & Moens, 2005	Asia (China)	ITS1, -2 (AY355442)
S. sacchari Nthenga, Knoetze, Berry, Tiedt & Malan, 2014		28S (KC633096), ITS1, -2 (KC633095)
S. sangi Phan, Nguyen & Moens, 2001	Asia (Vietnam)	ITS1, -2, (AY355441)
S. sasonense Phan, Spiridonov, Subbotin & Moens, 2006	Asia (Vietnam)	ITS1, 2 (AY487919)
S. scapterisci Nguyen & Smart, 1990	South America (Uruguay)	28S (AF331898), ITS1, -2 (AF122020, AF331915)
S. scarabaei Stock & Koppenhöfer, 2003	North America (USA)	28S (AY172023)
S. schliemanni Spiridonov, Waeyenberge & Moens, 2010	Europe (Germany)	28S (HM78113), ITS1, -2 (HM778112)
S. siamkayai Stock, Somsook & Kaya, 1998	Asia (Thailand)	28S (AF331907), ITS1, -2 (AF331917)
S. sichuanense Mráček, Nguyen, Tailliez, Boemare & Chen, 2006	Asia (China)	28S (DQ884966), ITS1, -2 (DQ884965)
S. silvaticum Sturhan, Spiridonov & Mráček, 2005	Europe (Germany)	28S (KC631428, MG547576-79), ITS1, -2 (KC631432-34, KU736743, KU736744-46, MG543845-48), CO I (MG547572-75)
S. surketense Khatri-Chhetri, Waeyenberge, Spiridonov, Manandhar & Moes, 2011	Asia (Nepal)	ITS1, -2 (HQ190042)
S. tami Luc, Nguyen, Reid & Spiridonov, 2000	Asia (Vietnam)	18S (AY035768)
S. thanhi Phan, Nguyen & Moens, 2001	Asia (Vietnam)	ITS1, -2 (AY355444)
S. tbilisiense Gorgadze, Lortkhipanidze, Ogier, Tailliez & Burjanadze, 2015	Europe (Georgia)	ITS1, -2 (KC287225)
S. tielingense Ma, Chen, Li, Han, Khatri-Chhetri, DeClercq & Moens, 2012	Asia (China)	28S (GU994202, LC571881), ITS1, -2 (GU994201, LC571880)

Continued

Table 1.2. Continued.

Taxa	Origin[a]	GenBank sequence data (accession no.)
S. tophus Çimen, Lee, Hatting, Hazir & Stock, 2014	Africa (South Africa)	28S (KJ701240), ITS1, -2 (KJ701241)
S. unicornum Edgington, Buddie, Tymo, France, Merino & Hunt, 2009	South America (Chile)	28S (GU191462, GU191463, MT586275), ITS1, -2 (GQ497167, GU19146, MT586276)
S. vulcanicum Clausi, Longo, Rappazzo, Tarasco & Vinciguerra, 2011	Europe (Italy)	28S (MF491515), ITS1 (GU929442), ITS2 (MF491477), 12S (GU929443)
S. websteri Cutler & Stock, 2003	Asia (China)	28S (JF503100, AY841762), 18S (FJ381666.1), CO I (GU569074)
S. weiseri Mráček, Sturhan & Reid, 2003	Europe (Germany)	28S (DQ400854, FJ165549, GU569059), ITS1, -2 (KJ696685, LR745522-23, MN861053-56), 12S (GU569040), CO I (GU56907)
S. xinbinense Ma, Chen, de Clercq, Waeyenberge, Han & Moens, 2012	Asia (China)	ITS1, -2 (JN171593)
S. xueshanense Mráček, Liu & Nguyen, 2009	Asia (China)	28S (FJ666053), ITS1, -2 (FJ666052)
S. yirgalemense Nguyen, Tesfamariam, Gozel, Gaugler & Adams, 2004	Africa (Ethiopia)	28S (AY748451), ITS1, -2 (AY748450, EU625295)
Genus: *Neosteinernema* Nguyen & Smart, 1994		
Type and only species		
Neosteinernema longicurvicauda Nguyen & Smart, 1994	North America (USA)	NA

NA, no sequences available.
[a]Country of original isolation in parenthesis.

well-resolved phylogenetic trees that provided stronger and further support for the recognition of five evolutionary clades in this genus: "Carpocapsae," "Feltiae," "Glaseri," "Intermedium" and "Bicornutum," which continue to be recognized as new taxa are described.

More recently, next-generation sequencing coupled with powerful bioinformatics tools have made possible the *de novo* acquisition of several genomes, providing a new platform for interpretation of the evolutionary relationships in this family. For example, Dillman *et al.* (2015) considered the genomes of five *Steinernema* spp. to investigate gene conservation across species, and identified genes uniquely expanded in insect parasites. In this study, *Steinernema monticolum*, a species previously hypothesized to be closer to *Steinernema carpocapsae* and *Steinernema scapterisci* (Stock *et al.*, 2001; Nadler *et al.*, 2006), was found to be more closely related to *Steinernema feltiae* than to any of the other species. Certainly, as more genomes become available, a more robust interpretation of the evolutionary relationships among *Steinernema* will become accessible and will also help define species boundaries.

1.3.2 Family Neotylenchidae Thorne, 1941

1.3.2.1 Diagnostic characters (modified from Siddiqi, 2000)

Free-living stages with smooth or finely striated cuticle. Stylet well developed, less than 20 mm long; basal knobs may be bifid. Esophagus fusiform, basal bulb absent. Esophageal glands free in body cavity, extending over intestine. Dorsal gland orifice close to stylet base. Nerve ring generally circumintestinal, posterior to, or at level of, esophago-intestinal junction. Excretory pore anterior or posterior to nerve ring. Females monodelphic or prodelphic. Vulva in posterior region, postvulval sac present or absent. Tail conoid, subcylindroid or cylindroid. Males monorchic, testis outstretched. Bursa present or absent. Spicules paired, small, cephalated or arcuate, distally pointed. Gubernaculum present or absent. Preadult females (free-living) with hypertrophied stylet and esophagus. Ovary immature. Uterus long. Mature parasitic females obese, sausage-shaped or elongate tuboid. Stylet and esophagus non-functional. Uterus hypertrophied but not everted.

1.3.2.2 Bionomics

Members of this family have a free-living generation alternating with an insect-parasitic generation. The genus *Deladenus* (= *Beddingia*) Thorne, 1941 currently comprises 27 nominal species with *Deladenus siricidicola* Bedding, 1968, a parasite of the wood wasp *Sirex noctilio*, being the only taxon currently used in biocontrol.

1.3.2.3 Deladenus *(=* Beddingia*) (Fig. 1.2)*

DIAGNOSTIC CHARACTERS (MODIFIED AFTER SIDDIQI, 2000). Free-living adult stages (mycetophagous) straight or slightly ventrally curved. Body cylindrical, tapering anteriorly and posteriorly to vulva; slender in young females but obese or swollen in mature females. Cuticle with fine transverse striae. Stylet small, basal knobs weak to moderately developed and rounded. Esophagus cylindroid. Esophago-intestinal junction at, or anterior to, nerve ring. Dorsal gland large, subventral glands reduced. Nerve ring surrounding isthmus. Excretory pore location variable. Hemizonid anterior or posterior to excretory pore. Female reproductive system monovarial, amphidelphic. Ovary outstretched and flexed. Spermatheca elongate. Vulva protuberant or not and extremely posterior. Vulval sac present or absent. Males monorchic, testis outstretched. Spicules paired, moderately robust and arcuate. Gubernaculum present. Tail conical or elongate conoid. Bursa present. Parasitic females obese, with body elongate. Cephalic region overgrown by body enlargement. Stylet present, hypertrophied, stout. Esophagus and esophageal glands hypertrophied in young females but degenerate in mature females. Vulva a transverse slit, lips not protuberant. Short post-vulval sac secondarily formed in impregnated young females.

PHYLOGENETIC RELATIONSHIPS *Deladenus* is considered paraphyletic and more closely related to either ancestral forms of the Hexatylina or forms typical to the Anguinata (Holterman *et al.*, 2006). *Deladenus* spp. are grouped into two clades, one of them uniting members of the superfamilies Iotonchioidea and the remainder species being included in Sphaerularioidea.

Morris *et al.* (2013) investigated the evolutionary relationships of *Deladenus* spp. for North America and studied their cophylogeny with *Sirex* spp. wasp hosts considering three genes (mtCO I, LSU and ITS). The study revealed cospeciation between all four North American *Sirex* spp. and their associated *Deladenus* parasites.

1.3.3 Family Rhabditidae Oerley, 1880

1.3.3.1 Diagnostic characters

Stoma commonly cylindrical without distinct separation of cheilo-, gymno- and stegostom. Stoma two or more times as long as wide. Usually with six distinct lips, each with one cephalic papilla. Amphids pore-like. Esophagus clearly divided into corpus (procorpus and metacorpus) and postcorpus. Male spicules separate or fused distally. Gubernaculum present. Bursa mostly well developed, peloderan or leptoderan, occasionally small or rudimentary. Nine or ten pairs of genital papillae (bursal rays). Females with one or two ovaries.

1.3.3.2 Bionomics

Most members of this family are free-living bacterivores, although several *Phasmarhabditis* spp. have been found to have parasitic associations with terrestrial slugs and snails. At present, 20 species have been described around the world. *P. hermaphrodita* (Schneider, 1859) was the first species found to be capable of killing several slug and snail pests. This species is widely used as a biological control agent, and is mass produced and commercialized as a molluscicide (Wilson *et al.*, 1994; Glen and Wilson, 1997). Currently, there are many other species being tested for their biocontrol potential against terrestrial gastropods. In particular, three European species, *Phasmarhabditis bohemica*, *P. bonaquaense* and *P. apuliae* have been demonstrated to grow in liquid and solid media and with different bacterial species (Nermuť *et al.*, 2022).

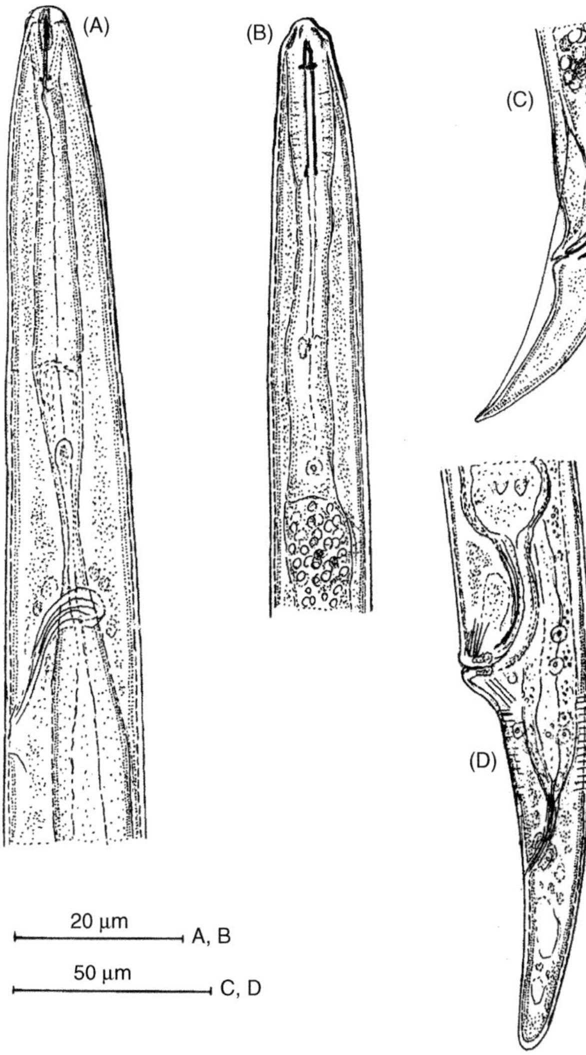

Fig. 1.2. *Deladenus siricidicola* Bedding. (A) Esophageal region of fungus-feeding female. (B) Esophageal region of entomoparasitic preadult female *Deladenus* sp. (C) Male tail region. (D) Posterior region of fungus-feeding female. After Siddiqi (2000), courtesy of CAB International.

1.3.3.3 Phasmarhabditis *Andrássy, 1976 (Fig. 1.3)*

DIAGNOSTIC CHARACTERS Body straight when heat killed, robust, elongate and tapering gradually to bluntly rounded head end. Cuticle with fine transverse and longitudinal striations. Lips rounded, arranged in three pairs each bearing a prominent labial papilla. Stoma rounded, triangular in cross-section. Stegostom well developed and with minute tubercules. Esophageal collar present. Esophagus with well-developed, cylindrical corpus. Basal bulb with prominent valve plates. Excretory pore usually anterior to basal bulb. Nerve ring surrounding isthmus. Deirids prominent. Females didelphic, amphidelphic. Vulva located at mid-body level. Males (when present) monorchic. Spicules separate. Bursa peloderan, open, with nine pairs of genital papillae. Tail

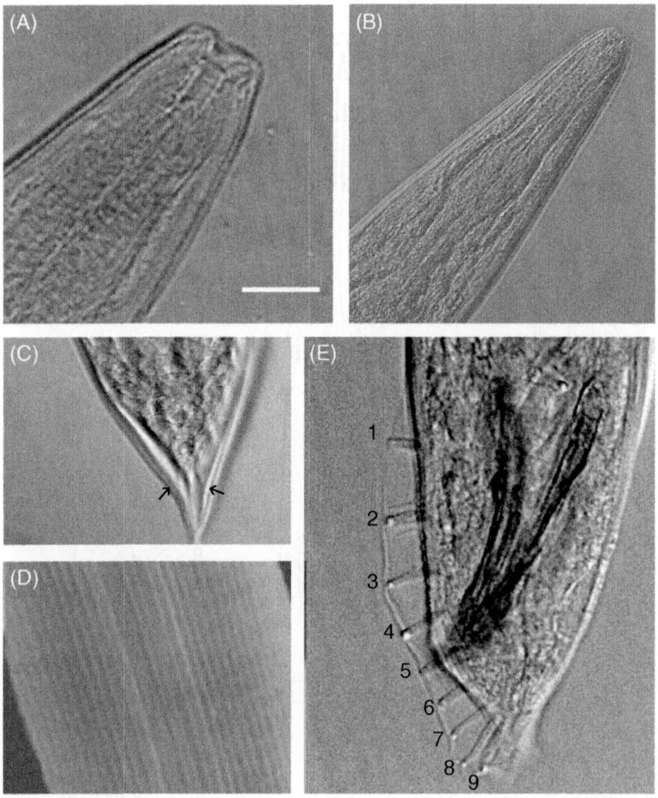

Fig. 1.3. *Phasmarhabditis* Andrássy. (A) Female stoma (dorsal view) of *P. hermaphrodita*. (B) Esophageal region (lateral view) of *P. hermaphrodita*. (C) Female tail of *P. neopapillosa* showing phasmids (arrows). (D) Lateral field of *P. hermaphrodita*. (E) Male tail of *P. neopapillosa* indicating the number of pairs of papillae (1–9). Scale: (A) and (E) 10 µm; (B) and (C) 25 µm; (D) 12 µm. All scales are based on the scale bar in (A).

conical, spicate or cupola-shaped. Phasmids prominent and sometimes protruding (Table 1.3).

PHYLOGENETIC RELATIONSHIPS Howe *et al.* (2020) used a segment of mitochondrial DNA (mtDNA) sequences including various genes (*rrnL, nad3, nad5, tRNA-Ala, tRNA-Pro, tRNA-Val, nad6, nad4L, tRNA-Trp, tRNA-Glu* and *rrnS*) to investigate the phylogenetic relationships of *Phasmarhabditis* spp. and demonstrated that the *Phasmarhabditis* spp. form a monophyletic group, with *P. hermaphrodita* being most closely related to *Phasmarhabditis neopapillosa* but more distantly related to other species. Furthermore, the authors showed *P. hermaphrodita* strains from New Zealand and the USA had mtDNA haplotypes nearly identical to the commercially available strain (Nemaslug®), and thus these strains were placed together in an intraspecific monophyletic clade.

1.3.4 Family Heterorhabditidae Poinar, 1976 (Fig. 1.4)

1.3.4.1 Diagnostic characters

Adults with six distinct protruding pointed lips surrounding oral aperture, each lip bearing one labial papilla. Stoma short and wide. Esophagus rhabditoid. Corpus cylindrical, metacorpus not differentiated. Isthmus short. Basal bulb pyriform with reduced valve. Excretory pore usually located at level of basal bulb. Hermaphrodite (first generation) with an ovotestis. Vulva located near middle of body. Post-anal swelling present or absent. Tail terminus blunt, with or without a mucro. Females (second generation) amphidelphic, ovaries with reflexed portions often extending past vulval opening. Vulva located near middle of body, with or without protruding lips. Tail conoid;

Table 1.3. Key diagnostic features of *Phasmarhabditis* spp. All measurements are in microns.

Species	Female			Male		
	Total body length (range)	Tail		Total body length (range)	Bursa shape	Spicule length
		Shape	Length			
P. hermaphrodita[a] (Schneider, 1859) Andrássy, 1983	1799 (1509–2372)	Elongate, conoid	3–4 anal body widths long	Males are extraordinarily rare		
P. apuliae Nermuť, Půža & Mráček 2016	2623 (2262–2848)	Tail long tapering, prominent papilla-like phasmids present	>3 anal body widths long	2096 (1898–2363)	Peloderan	1.3 times longer than tail; spicules very slightly arcuate, almost straight, with sharp tip
P. bohemica Nermuť, Půža, Mekete & Mráček, 2017	2079 (1777–2222)	Tail tapering to tip, with prominent papilliform phasmids located laterally in mid-tail region	1.5–2 anal body widths long	1686 (1515–1818)	Peloderan	Spicules paired, very slightly arcuate, almost straight, with sharp tip
P. bonaquaense Nermuť, Půža, Mekete & Mráček, 2016	2349 (1878–2626)	Tail cupola-shaped with long hyaline filiform terminus; very prominent papilla-like phasmids present, located at base of terminus	>5 times anal body widths long	1829 (1414–2121)	Peloderan	~1.5 anal body widths long
P. californica de Ley, Holovachov, McDonnell, Bert, Paine & de Ley, 2016	1501 (1298–1757)	Tail conoid, wider part forming 38% (28–54%) of tail length, hyaline tail tip longer	2.5–3 times anal body widths long	No males found	No males found	No males found
P. circassica Ivanova, Geraskina & Spiridonov, 2020	1830 (1522–2510)	Tail end initially widely conoid, followed by sturdy spike corresponding to ~55% of total tail length	55 (33–57)	1200 (916–1607)	Pelodera	Spicules nearly straight, uniformly thick for most of spicule length, tapering to tips and about twice as long as body diameter

Continued

Table 1.3. Continued.

	Female			Male		
		Tail				
Species	Total body length (range)	Shape	Length	Total body length (range)	Bursa shape	Spicule length
P. clausiliiae Ivanova, Geraskina & Spiridonov, 2020	1276 (1050–1730)	Tail end initially widely conoid, followed by sturdy spike corresponding to ~58% of total tail length	82 (54–125)	1183 (905–1540)	Peloderan	Spicules about twice as long as corresponding body diameter, nearly straight, tapering to distal tips, manubrium small
P. kenyaensis Pieterse, Rowson, Tiedt, Malan, Haukeland & Ross, 2020	2151 (1918–2340)	Tail conoid, gradually tapering to a sharp tip, with two phasmids located at ~40% of tail length	~3.5 times as long as tail	1989 (1835–2034)	Peloderan and narrow	Spicules cephalate, paired and separate, slightly curved, gradually tapering to a pointed terminus
P. meridionalis Ivanova & Spiridonov, 2017	1612 (1057–1931)	Tail end cupola-shaped with filamentous terminus	Tail long 68 (44–97), with filamentous terminus ~67% of total tail length	1317 (1159–1526)	Open, peloderan	Nearly straight, twice as long as body diameter
P. neopapillosa (Mengert in Osche, 1952) Andrássy, 1983[a]	2227 (1817–2449)	Elongate, conoid	3–4 anal body widths long	1585 (1432–1771)	Well developed	1.5 times as long as tail
P. nidrosiensis (Allgén, 1933) Andrássy, 1983[b]	1000–1750	Tail cupola-shaped with pointed tip	1.5–2 anal body widths long	900–1720	Small and narrow	Twice as long as tail
P. papillosa (Schneider, 1866) Andrássy, 1976[b,c]	1600–3400	Tail cupola-shaped with pointed tip	1.5–2 anal body widths long	1200–2400	Well developed	1–1.5 times as long as tail
P. quinamensis Ivanova & Spiridonov, 2021	134 (110–1610)	Tail initially widely conoid then narrowing into thread-like terminal hyaline part nearly twice as long as wider part	89 (65–127)	1052 (920–1250)	Bursa wide, open, fan-like	Spicules straight except for the distal tips, which are slightly bent ventrally; incision ~4 µm long in distal part of each spicule; distal tips with holes

P. safricana Ross, Pieterse, Malan & Ivanova, 2018	1598 (1160–2071)	Tail end "dome-shaped with spike about 50% tail length"	61 (46–87)	1477 (1220–1702)	Bursa open, wide peloderan	Spicules nearly straight, at least twice as long as body diameter
P. valida (Sudhaus, 1974) Andrássy, 1983[b]	NA	Cupola-shaped with pointed tip	1.5–2 anal body widths long	NA	Well developed	NA
P. villasmundi Ivanova, Clausi, Leoni & Spiridonov, 2022	2107 (1150–2740)	Tail elongated, conical	122 (82–176)	1510 (1090–1990)	Bursa wide, open, peloderan	Spicules straight, 82 (67–96), except for distal tips slightly bent in ventral direction
P. zhejiangensis Zhang & Liu, 2020	1911 (1440–2355)	Tail pointed with filiform tip	73.5 (59.9–83.5)	1486 (1014–1842)	Peloderan, open bursa	Spicules are 58 (48–68) long on average; anterior end of spicules slightly enlarged

NA, not available.
[a]After Hooper *et al.* (1999).
[b]After Andrássy (1983).
[c]Type species.

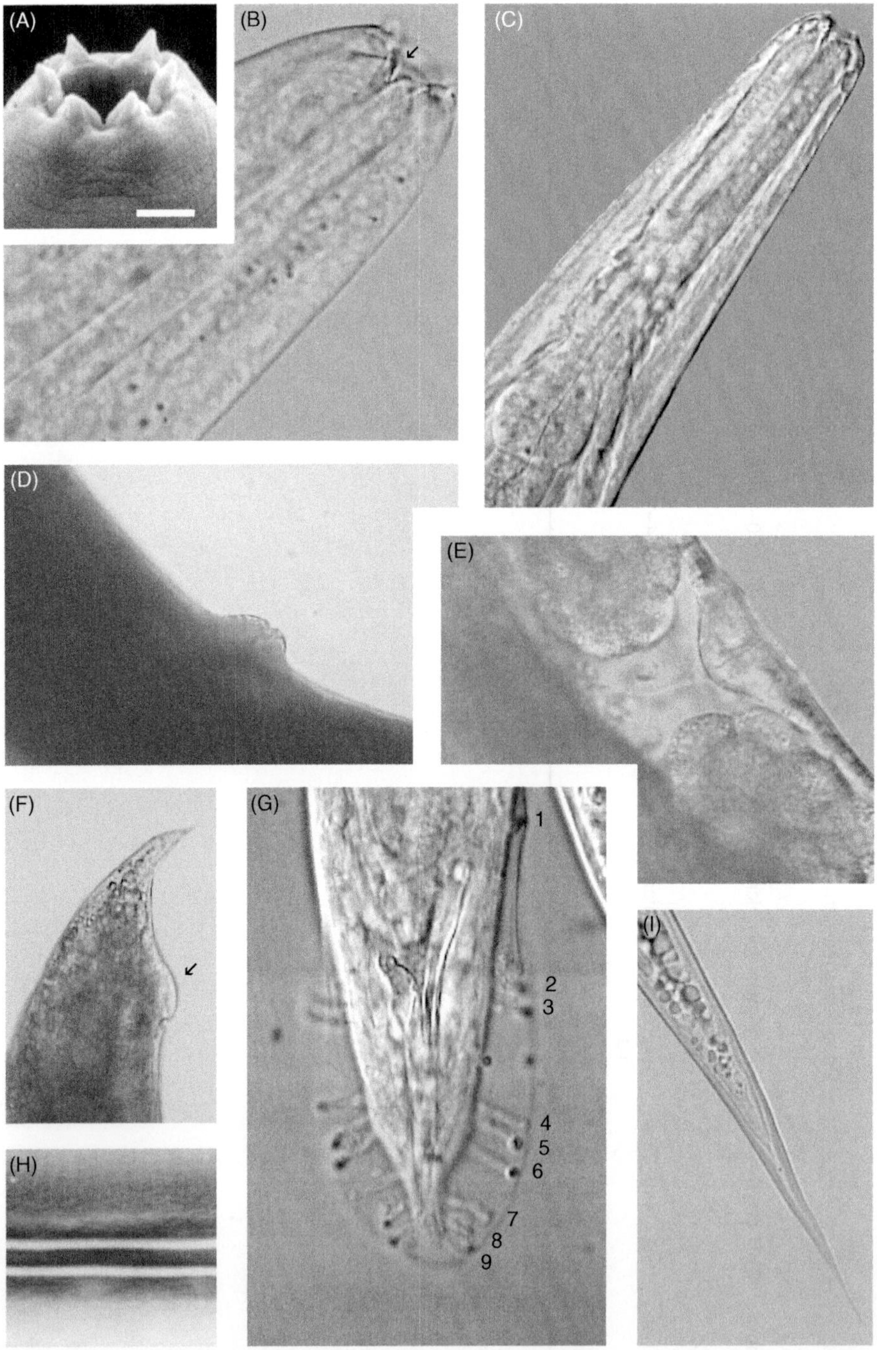

Fig. 1.4. Family Heterorhabditidae. (A–E) First-generation hermaphrodite. (A) Scanning electron micrograph of the anterior end. (B) Anterior end (lateral view) showing stoma (arrow). (C) Esophagus (lateral view). (D) Protruding vulval lips (lateral view). (E) Non-protruding vulva (lateral view). (F) Tail (lateral view) showing post-anal swelling (arrow). (G) Tail of second-generation male (ventral view) showing arrangement of genital papillae. (H, I) Third-stage infective juvenile. (H) Tail (lateral view). (I) Lateral field pattern. Scale: (A) 4.5 μm; (B) 12 μm; (C) and(D) 25 μm; (E) 20 μm; (F) 15 μm; (G) and (I) 6.5 μm; (H) 3.5 μm. All scales are based on the scale bar in (A).

post-anal swelling present or absent. Males (second generation) monorchic. Spicules paired, symmetrical, straight or arcuate, with pointed tips. Gubernaculum slender, about half the length of spicules. Bursa open, peloderan, attended by a complement of nine pairs of bursal rays (papillae). IJ ensheathed in cuticle of second-stage juvenile. Cuticle of second-stage juvenile with longitudinal ridges throughout most of body length, and a tessellate pattern in anteriormost region. Lateral field with two ridges. Prominent cuticular dorsal tooth present. Excretory pore located posterior to basal bulb. Tail short, conoid, tapering to a small spike-like tip.

1.3.4.2 Bionomics

Heterorhabditids have a similar life cycle to steinernematids, and are also known as EPNs. Unlike *Steinernema* spp., *Heterorhabditis* adults resulting from IJs are hermaphroditic. Eggs laid by the hermaphrodites produce juveniles that develop into males and females. Males and females mate and produce eggs that develop into IJs. The family Heterorhabditidae consists of one genus, *Heterorhabditis* Poinar, 1976, with *Heterorhabditis bacteriophora* as the described type species and more than 20 other described species (Table 1.4).

1.3.4.3 Phylogenetic relationships

Various studies have explored the evolutionary relationships among *Heterorhabditis* spp. considering nucleotide sequence data from nuclear (28S, 18S and ITS1) and mitochondrial (ND4) genes (Curran and Driver, 1994; Reid, 1994; Liu *et al.*, 1997, 1999; Adams *et al.*, 1998). Curran and Driver (1994) presented the first evolutionary hypothesis using a combination of RFLP analysis and partial sequences of 28S rDNA. This study recognized species identity based on morphological and cross-hybridization tests, but the limited taxon sampling did not provide a suitable framework to interpret their phylogenetic relationships.

Reid (1994) assessed evolutionary relationships among several *Steinernema* and *Heterorhabditis* spp. considering RFLP analysis of ITS rDNA, demonstrating a close relationship between *H. megidis* and *Heterorhabditis* spp. of the Northwest European group, now considered to be conspecific. In another study, Liu *et al.* (1997) inferred phylogenetic relationships for both EPN families using partial 18S rDNA sequences, concluding that this marker is too conserved to resolve relationships among *Heterorhabditis* spp.

Adams *et al.* (1998) investigated the evolutionary relationships among *Heterorhabditis* spp. using sequences of the ITS1 of rDNA, reinforcing the relationship among closely related species but not for distantly related species. Phan *et al.* (2003) compared *Heterorhabditis* strains and species showing that tropical and subtropical species such as *Heterorhabditis indica* and *Heterorhabditis baujardi* formed a distinct clade that was distinct from those from temperate regions. Later, Nguyen *et al.* (2008a,b) proposed three distinct monophyletic clades: "Indica", "Bacteriophora" and "Megidis" groups.

Mitochondrial genes have also been explored to study the evolutionary history of *Heterorhabditis* spp. (Liu *et al.*, 1999; Saeb and Grewal, 2014). Overall, these studies are in congruence with those reflected by ITS sequence analyses. A more recent phylogenetic study by Dhakal *et al.* (2020) considered a multilocus approach that included the D2–D3 expansion segments of 28S rDNA, ITS rDNA, *COI* mtDNA, and the thin filament (F-actin)-associated protein (*unc-87*) and calmodulin (*cmd-1*) genes, recognizing the previously depicted three evolutionary clades.

1.3.5 Family Mermithidae Braun, 1883 (Fig. 1.5)

1.3.5.1 Diagnostic characters

Long slender nematodes sometimes reaching a length of 50 cm, but usually between 1 cm and 10 cm. Cuticle smooth or with crisscross fibers. Anterior end containing two, four or six cephalic papillae and rarely a pair of lateral mouth papillae. Amphids tube-like or modified pouch-like. Esophagus modified into a slender tube surrounded posteriorly by stichosomal tissue. Intestine modified into a trophosome or food-storage organ forming a blind sac soon after the nematodes enter a host. Preparasitic juveniles with a functional stylet and a pair of penetration glands that degenerate after host invasion. Ovaries paired; muscular vagina straight or curved. Males with a single fused or paired spicules. Gubernaculum

Table 1.4. Taxonomic summary of described *Heterorhabditis* spp.

Taxa	Biogeography[a]	GenBank sequence data (accession no.)
Type and only genus: *Heterorhabditis* Poinar, 1976 (syn. *Chromonema* Khan, Brooks and Hirschmann, 1976)		
Type species		
H. bacteriophora Poinar, 1976 (syn. *Chromonema heliothidis* Khan, Brooks & Hirschmann, 1976) *H. heliothidis* (Khan, Brooks & Hirschmann, 1976) Poinar, Thomas & Hess, 1977 *H. argentinensis* Stock, 1993[b]	Africa, Asia, Australia, Central America, Europe, North America (USA), South America	18S (AF036593), 5.8S (U65497), ITS1 (AF029708, AF029706), ITS1, -2 (MT372491), 28S (D3) (U47560, MT372509), ND4 (AF066890, AF066888), SAT (U19928), CO I (MT373729), unc-87 (MT375796), cmd-1 (MT406138)
Other species		
H. amazonensis Andaló, Nguyen & Moino, 2006	South America (Brazil)	ITS1, -2 (DQ665222, MK262740, MT372499, KU870321, OP727808), CO I (OP727808, OP727564, OP727562), cmd-1 (MT406142), umc-87 (MT375792), 28S (MT373738)
H. atacamensis Edgington, Buddie, Moore, France, Merino & Hunt, 2010	South America (Chile)	
H. baujardi Phan, Subbotin, Nguyen & Moens, 2003	Asia (Vietnam)	28S (MT372503), ITS1 (AF548768), ITS1, -2, (MT372503), CO I (MT373736) umc-87 (MT375791), cmd-1 (MT406139)
H. brevicaudis Liu, 1994	Asia (China)	ITS1, -2 (AF548768)
H. beicherriana Li, Liu, Nermuť, Puza & Mráček, 2012	Asia (China)	28S (MT372511), ITS1, -2, (MT372490), CO I (MT373730) unc-87 (MT375797)
H. downesi Stock, Griffin & Burnell, 2002	Europe (Ireland)	ITS1 (AF029713)
H. floridensis Nguyen, Gozel, Koppenhöfer & Adams, 2006	North America (USA)	28S (MT372504), ITS1, -2 (MT372501), CO I (MT373737), unc-87 (MT375793), cmd-1 (MT406141)
H. georgiana Nguyen, Shapiro-Ilan & Mbata, 2008	North America (USA)	28S (MT372510), ITS1, -2 (MT372492), CO I (MT373731), unc-87(MT375798, MT375799)
H. gerrardi Plichta, Joyce, Clarke, Waterfield & Stock, 2009	Oceania (Australia)	ITS1, -2 (FJ152545)
H. indica Poinar, Karunakar & David, 1992 (syn. *H. hawaiiensis* Gardner, Stock & Kaya, 1994)	Asia (India), Central America, North America	18S (U70628), ITS1 (AF029710, AF029707), ND4 (AF066879, AF066878), SAT (U68112)
H. marelata Liu & Berry, 1996 (syn. *H. hepialius* Stock, Strong & Gardner, 1996)	North America (USA)	18S (AF083004, U70630), ITS1 (AF029713, AF029709), ND4 (AF06881, AF066880)

Continued

Table 1.4. Continued.

Taxa	Biogeography[a]	GenBank sequence data (accession no.)
H. megidis Poinar, Jackson & Klein, 1987	North America (USA), Europe	18S (AF70631), ITS1 (AF029711), ITS1, -2 (AY293284), ND4 (AF066885)
H. mexicana Nguyen, Shapiro-Ilan, Stuart, McCoy, James & Adams, 2004	North America (Mexico)	ITS1, -2 (AY321478, ON489248, ON489249, ON482174, MK421512)
H. noenieputensis Malan, Knoetze & Tiedt, 2014	Africa (South Africa)	28S (MT372505), ITS1, -2 (MT372497), CO I (MT373728)
H. poinari Kakulia & Mikaia, 1997[b]	Asia (Georgia)	NA
H. safricana Malan, Nguyen, de Waal & Tiedt, 2008	Africa (South Africa)	28S (MT889743), ITS1, -2 (EF488006, MW915609, MW915610, MW09290)
H. sonorensis Stock, Rivera-Orduno & Flores-Lara, 2009	North America (Mexico, USA)	ITS1, -2 (FJ477730, FJ477731)
H. taysearae Shamseldean, Abou El-Sooud, Abd-Elgawad & Saleh, 1996	Asia (Egypt)	NA
H. zealandica Poinar, 1990	Australia (New Zealand)	28 S(MT372508), ITS1, -2 (MT372493), ITS1 (AF029705), CO I (MT373734)

NA, no sequences available.
[a]Country of original isolation shown in parentheses.
[c]Species inquirenda.

and bursa absent. Several rows of genital papillae usually present (Table 1.5).

1.3.5.2 Bionomics

All known species are obligate parasites of terrestrial and aquatic arthropods and insect groups, including Orthoptera, Dermaptera, Hemiptera, Lepidoptera, Diptera, Coleoptera and Hymenoptera. Mermithids with significant biocontrol potential include *Romanomermis culicivorax* (a parasite of mosquito larvae) (Petersen, 1985), *Oesophagomermis* (= *Filipjevimermis*) *leipsandra* (a parasite of larval banded cucumber beetle, *Diabrotica balteata*) (Creighton and Fassuliotis, 1983), *Mermis nigrescens* (a parasite of grasshoppers) (Webster and Thong, 1984) and *Agamermis unka* (a parasite of white and brown planthoppers) (Choo *et al.*, 1989; Choo and Kaya, 1994).

1.3.5.3 Phylogenetic relationships

According to Ahmed and Holovachov (2021), the Mermithidae are monophyletic based on available 18S rDNA-based sequences. The evolutionary relationships in this family have not been investigated extensively. An early study by Gafurov (1996), based on morphological traits, established seven subfamilies. More recently, Tobias (2017) considered partial 18S rRNA gene, partial 28S rRNA gene and partial CO I gene markers showing a trifurcation at the most basal node within the family, and determined three clades: clade 1 contained aquatic and terrestrial genera such as *Allomermis*, *Mermis*, *Heleidomermis*, *Limnomermis*, *Isomermis*, *Gastromermis*, *Strelkovimermis*, *Amphimermis* and *Romanomermis*; clade 2 contained *Octomyomermis*, *Hexamermis*, *Agamermis* and *Ovomermis*; and clade 3 encompasses *Tautomermis* taxa.

1.4 Molecular Approaches and Their Application in Nematode Taxonomy and Phylogenetics

Molecular methods are widely recognized as a powerful tool for the diagnostics of natural enemies. Accuracy of identification is also critical

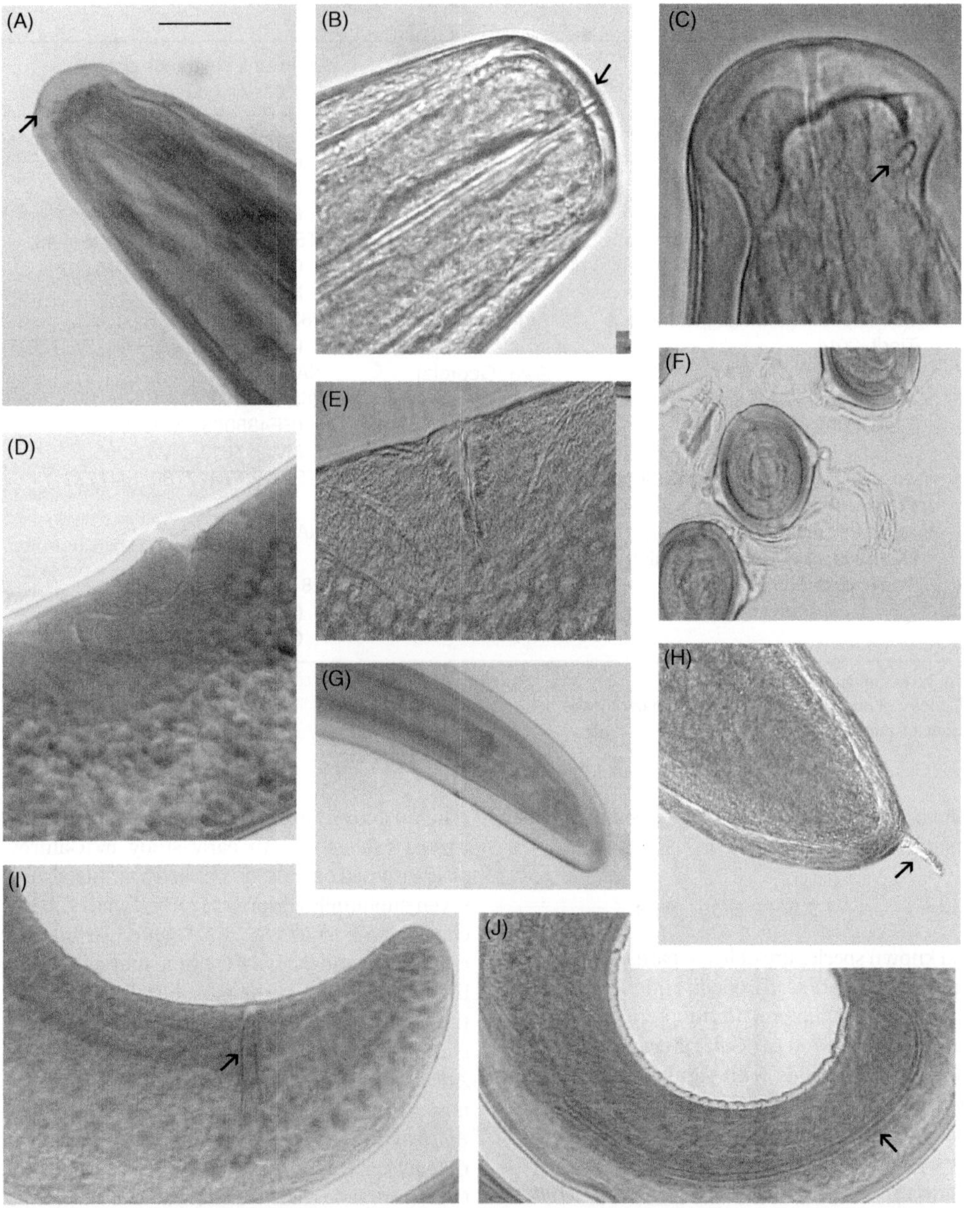

Fig. 1.5. Family Mermithidae. (A) Anterior end of female showing slightly shifted oral aperture (arrow). (B) Anterior end of post-parasitic juvenile (J4) showing terminal oral aperture (arrow; dorsal view). (C) Anterior end (dorsolateral view) showing amphid position (arrow). (D) S-shaped vagina. (E) Pear-shaped vagina (lateral view). (F) Eggs with byssi. (G, H) Tail of post-parasitic juvenile (J4) without digitate appendage (G) and with digitate appendage (H). (I, J) Male tail (lateral view) showing short spicules (I) and long curved spicules (J). Scale: (A) 18 µm; (B) and (C) 12 µm; (D) and (E) 45 µm; (F) and (I) 40 µm; (G) and (J) 20 µm; (H) 25 µm. All scales are based on the scale bar in (A).

to track the establishment, deduce population structure and monitor impacts on non-target organisms (Gariepy *et al.*, 2014). Molecular identification and detection techniques have also helped manage the decline in specialist skills, particularly in nematology, where the scarcity of expert taxonomists cannot keep up with the needs and demands for description of unknown taxa.

Table 1.5. Key diagnostic features of mermithid genera considered in biological control.

Diagnostic feature	*Agamermis* Cobb, Steiner and Christie, 1923	*Mermis* Dujardin, 1842	*Oesophagomermis* Artyukhovsky, 1969	*Romanomermis* Coman, 1961	*Strelkovimermis* Rubzov, 1969
Cephalic papillae	6	4	6	6	6
Labial papillae	Absent	Present (2)	Absent	Absent	Absent
Oral opening	Terminal	Absent	Terminal or slightly shifted to ventral side	Terminal	Terminal or slightly shifted to ventral side
Hypodermal cords	6	6	6	8	6
Vagina shape	S-shaped	S-shaped	S-shaped	Pear-shaped	S-shaped
Bursal sleeve	Absent	Absent	Absent	Absent	May be present
Parasitic and post-parasitic tail	With crater-like terminus	With tail appendage	With small tail appendage	With tail appendage	With tail appendage
Eggs	Without byssi	With byssi	Without byssi	Without byssi	Without byssi

A variety of molecular methods and markers have been implemented for the diagnosis, identification and assessment of evolutionary relationships of nematode species with potential consideration in biological control. Each of these methods has its own advantages and/or disadvantages. Nucleotide sequencing has greatly transformed our understanding of the evolutionary relationships of nematode taxa and has also been shown to be a complementary tool for taxa diagnoses. A historical overview is provided below, considering the most common methods and markers used for the diagnostics and identification of nematode taxa with biocontrol potential.

1.4.1 Molecular methods

1.4.1.1 PCR

Since the early 1990s, the PCR technique has been the most widely used tool for nematode identification and for studying their genetic diversity. Once the target DNA is generated by PCR amplification, it can be further characterized by various analyses including RFLP, RAPD and sequencing.

1.4.1.2 RAPD

The RAPD-PCR approach uses a single random primer of about ten nucleotides to create genomic fingerprints. This technique was often used for estimating genetic diversity among individual specimens, populations and closely related species. It was also applied to identify taxa in three families with biocontrol potential including Heterorhabditidae and Steinernematidae. Various studies considered this approach as a complementary tool to morphology in the identification of *Heterorhabditis* and *Steinernema* spp. (Gardner *et al.*, 1994; Liu and Berry, 1996a). It was also considered to measure the genetic variability among isolates of these genera (Liu and Berry, 1995, 1996a; Hashmi *et al.*, 1996). A study by Liu and Berry (1996a) also used this approach to assess the phylogenetic relationships of these taxa.

1.4.1.3 RFLP

Restriction enzymes and PCR-RFLP can reveal variations in sequences in PCR products. The PCR products obtained from different species or populations are digested by restriction enzymes, and the resulting fragments separated by electrophoresis may yield differences in length that occur within restriction sites, creating different RFLP profiles. PCR-RFLP of ITS rDNA were used for many years for the diagnosis and identification of unknown *Steinernema* and *Heterorhabditis* spp. (Reid and Hominick, 1992; Reid *et al.*, 1997; Stock *et al.*, 1998; Anis *et al.*, 2000; Stack *et al.*, 2000; Hussaini *et al.*, 2001; Phan *et al.*, 2001a,b). Furthermore, the PCR-RFLP approach was considered to interpret evolutionary relationships among EPNs (Reid, 1994; Reid *et al.*, 1997). However, many restriction enzymes were needed to generate profiles to diagnose *Steinernema* spp., resulting in a very expensive method. Moreover, it was also recognized that restriction enzymes vary in their efficiency for generating RFLPs (Whitkus *et al.*, 1994), and that without restriction site maps, fragment patterns cannot reliably produce homologous characters that are required to infer phylogenetic relationships or delimit species.

1.4.1.4 Amplified fragment length polymorphisms (AFLPs)

The AFLP technique creates high-density linkage maps. Restriction fragments are amplified through PCR from complete digestion of genomic DNA, using a combination of two restriction enzymes. AFLP and RAPD techniques are similar, but AFLP results are more reliable under strict experimental parameters.

AFLPs have been used to assess the genetic variation of EPN populations. For example, Dillon *et al.* (2008) used AFLP analysis to assess the persistence of exotic and native populations of *S. feltiae*.

In a more recent study, Wang *et al.* (2013) used single-enzyme AFLP analysis to assess the genetic diversity among *S. monticolum*, revealing a high level of genetic diversity among several Korean populations. However, this study revealed that there was no correlation between genetic similarity and geographical proximity.

Peteira *et al.* (2015) used AFLPs and RAPDs to study the genetic variation of Cuban *Heterorhabditis* populations as well as other biological control agents with the goal of establishing accurate molecular markers for their identification but also to monitor their prevalence in the field.

1.4.1.5 Sequence-based analysis

The Sanger sequencing, also known as the "chain-termination method" was developed almost 40 years ago (Sanger *et al.*, 1977). For more than three decades, sequence-based methods have been widely used in nematode systematics. This is mostly because of their inexpensiveness and demonstrated variation yield (within and among nematode species) when compared with other methods previously considered.

Sequencing analyses have involved nucleotide data from nuclear RNA, mtDNA and full genomes, and have proven to be a suitable tool not only in taxonomy but also in assessing phylogenetic relationships (Powers *et al.*, 1994; Hyman and Azevedo, 1996; Adams *et al.*, 1998; Szalanski *et al.*, 2000; Stock *et al.*, 2001; Nadler *et al.*, 2006; Zeng *et al.*, 2015; Mazza *et al.*, 2017; Morris *et al.*, 2018; Watanabe *et al.*, 2021). Sequence analyses have also been considered for species delimitation (Adams *et al.*, 1998; Nguyen *et al.*, 2001; Stock *et al.*, 2001; van Megen *et al.*, 2009; Dillman *et al.*, 2015; Stock, 2015; Morris *et al.*, 2018; Didiza *et al.*, 2021).

Currently, databases such as GenBank (www.ncbi.nlm.nih.gov; accessed 1 April 2024) and NEMBASE (http://nematodes.org/; accessed 1 April 2024) are public depositories for sequence data, which facilitate nematode identification through comparison. However, a critical aspect to consider is the quality and authenticity of sequences deposited in the databases, which is essential for accurate identification.

A key element in the process of validating taxa through the comparison of sequences is the assessment of their alignment. Although alignments are generated with specific software and sequences are deposited into a database, there is generally a lack of explicit information about the parameters used for their editing (usually done through specific software packages or manually). Manual editing and trimming of sequences are particularly common in larger datasets containing ambiguous sites (Nadler *et al.*, 2006; Stock, 2009). These actions introduce variations in alignments that may have implications in the interpretation of data and subsequent taxonomic decisions made. Although most journals require sequences to be deposited in open-access databases as part of the process of publication and acceptance of a manuscript, there are currently no requirements for deposition of sequence alignments (Bogale *et al.*, 2020). This is a process that should be considered and implemented for a more transparent process and validation of data.

1.4.1.6 Next-generation sequencing technologies

Second-generation sequencing technologies have allowed the sequencing of genomes at an affordable time–cost scale (Porazinska *et al.*, 2009; Murfin *et al.*, 2012). Different commercial platforms have been used including: (i) emulsion PCR (emPCR; Roche-454 Life Sciences, Basel, Switzerland); (ii) reversible terminator (RT; Illumina, San Diego, California); (iii) sequencing by oligonucleotide ligation and detection (SOLiD; Thermo Fisher Scientific-Life Technologies; Waltham, Massachusetts); and (iv) ion torrent chip (IonT; Thermo Fisher Scientific-Life Technologies, Waltham, Massachusetts). Among these, the 454 sequencing platform or "pyrosequencing" has enabled the rapid and accurate quantification of sequence variation and has been demonstrated to be a highly adaptable tool for addressing ecological and biological questions in many fields including nematology. For example, 454 pyrosequencing was used to analyze transcriptome expression in *Steinernema* spp. revealing differences in gene expression due to temperature stress (Rougon-Cardoso *et al.*, 2016; Yaari *et al.*, 2016).

With the advent of third-generation sequencing technologies and enhanced infrastructure in computing algorithms, bioinformatics and genomics, many new nematode genome projects are currently under way. For example, the "959 nematode genomes" initiative (www.nematodegenomes.org; accessed 1 April 2024) has established a wonderful platform to promote, stimulate and coordinate genome projects within the Nematoda (Kumar *et al.*, 2012).

Access to full genomes will not only allow a better understanding of the phylogeny of nematodes but will also contribute to a better interpretation of the dynamics of important traits of taxa in this phylum and their evolutionary relationships (Kumar *et al.*, 2012; Baniya *et al.*,

2020; Regeai *et al.*, 2021). Furthermore, the combination of multi-omics techniques, complemented with morphological and physiological ones, is permitting a holistic approach to deciphering diverse biological systems including those of nematodes.

One of the most relevant breakthroughs from these technologies is the ability for direct sequencing of RNA, without retrotranscription or amplification steps, avoiding the biases related to these issues. Like structural genomics described above, functional genomics or transcriptomics is being used in different fields, such as agronomy and medicine (Dorado *et al.*, 2021). For example, Dillman *et al.* (2015) identified genes uniquely expanded in insect parasites. Specifically, protein domain analysis in five *Steinernema* genomes revealed a striking expansion of numerous putative parasitism genes. Furthermore, stage-specific gene expression of some of these expanded families provided support to the notion of these genes in the insect parasitism and revealed sets of conserved, non-coding, regulatory motifs associated with orthologous genes shared between *Steinernema* and *C. elegans*.

More recently Lefoulon *et al.* (2022) performed a comparative dual RNA-sequencing analysis of IJs of two *Steinernema–Xenorhabdus* partnerships. Study focused on the fitness costs of their symbiotic relationship considering the presence/absence of the bacterial symbionts and under three different rearing scenarios.

1.4.2 Targeted genes

1.4.2.1 Nuclear ribosomal DNA genes

Nuclear rDNA has been a common marker involved in the delimitation of nematodes at different taxonomic levels (e.g. Curran and Driver, 1994; Blaxter *et al.*, 1998; Stock, 2005; Subbotin, 2021). In particular, three loci have been extensively used: (i) the small subunit (18S); (ii) the ITS; and (iii) the large subunit (28S).

18S OR SMALL SUBUNIT (SSU) GENE OF RDNA. Sequences of the 18S rDNA gene have been used to assess EPN evolutionary relationships, depicting Steinernematidae and Heterorhabditidae as representing distinct, unrelated lineages. While steinernematids are recognized as members of clade IV, heterorhabditids are placed within clade V (Blaxter *et al.*, 1998; de Ley and Blaxter, 2004). However, at the species level, it was demonstrated early on that the 18S gene is too conserved to resolve relationships at the species level (Liu *et al.*, 1997; Stock *et al.*, 2001).

This gene has also been used to investigate the phylogeny of the Mermithidae and also to create a barcode system for diagnostic purposes (St-Onge *et al.*, 2008; Crainey *et al.*, 2009; Yeates and Buckley, 2009; Kamminga *et al.*, 2012; Kobylinski *et al.*, 2012; Kubo *et al.*, 2016; Tripodi and Strange, 2018). For example, Crainey *et al.* (2009) used the 18S rDNA gene to develop a barcode for the diagnosis of *Isomermis lairdi*, a species known to have a major impact on populations of *Simulium damnosum* (Diptera: Simuliidae). This study also revealed that this mermithid species forms a monophyletic group with two other taxa, *Gastromermis viridis* and *Isomermis wisconsinensis*.

ITS REGION OF RDNA. The ITS gene has also been widely used in EPN systematics (Powers *et al.*, 1997). For example, in the Heterorhabditidae, the ITS1 region has been demonstrated to exhibit sufficient genetic variation for differentiating *Heterorhabditis* spp. Additionally, it was proven valuable for delimitation and interpretation of evolutionary relationships among species (Adams *et al.*, 1998). The ITS1 and -2 regions, including the 5.8S gene of rDNA, were investigated early on to assess phylogenetic relationships and delimit species of *Steinernema* (Nguyen *et al.*, 2001). However, because of its conserved nature, in particular the 5.8S locus, it is uninformative to resolve phylogenetic relationships or delimit of terminal taxa (Nguyen *et al.*, 2001).

The ITS has also been used to explore phylogenetic relationships of *Deladenus* spp., in particular, the four different types of adults (male, mycophagous female, infective female, and parasitic female) and the similarity of morphological characters among species. The taxonomy of this genus has become more dependent on molecular markers for the diagnosis and identification of species (Kanzaki *et al.*, 2016; Morris *et al.*, 2018; Heydari *et al.*, 2020).

ITS1 has also been explored to assess phylogenetic relationships of the family Mermithidae, in relation to the genus *Mermis* (Presswell *et al.*, 2015).

Phasmarhabditis taxa have also been diagnosed using ITS rDNA genes, as well exploring the evolutionary relationships of species within the genus (Nermuť *et al.*, 2016a,b; Ross *et al.*, 2018; Pieterse *et al.*, 2020; Ivanova and Spiridonov, 2021).

28S OR LARGE SUBUNIT (LSU) OF RDNA. As previously mentioned, LSU sequence data has been widely used to assess phylogenetic relationships among *Steinernema* and *Heterorhabitis* spp. (Stock *et al.*, 2001; Nadler *et al.*, 2006; Stock, 2009; Lulamba and Serepa-Dlamini, 2020; Ramos *et al.*, 2022).

Stock *et al.* (2001) showed that the D2/D3 domains of 28S rDNA are suitable and informative for interpreting evolutionary relationships among *Steinernema* spp. (see section 1.3.4.3). This region is also considered to be an effective and reliable approach for terminal taxa delimitation and for diagnostic purposes (Stock *et al.*, 2001; Stock, 2009; Spiridonov and Subbotin, 2016). Diagnoses of *Heterorhabditis* spp. have also been done considering the D2/D3 domains (Stock, 2009; Hunt and Nguyen, 2016) and has also been used to assess phylogenetic relationships among species in this genus.

Sequences of the D2/D3 domain have also been used to identify new taxa in the Mermithidae (Kobylinski *et al.*, 2012; Presswell *et al.*, 2015; Tong *et al.*, 2021). However, due to the low number of orthologous sequences in the available databases, they have not been widely used to reconstruct evolutionary relationships in this family (Mazza *et al.*, 2017).

In a more recent study, Yoshino and Waki (2021) utilized 28S sequences to develop a phylogenic framework of mermithids isolated from Isopoda. This study revealed two distinct clades. One of the groups included terrestrial taxa such as *Hexamermis* and *Thaumamermis*, a taxon also known to parasitize Crustacea. The second clade encompassed aquatic genera including *Romanomermis*, *Strelkovimermis* and other genera.

LSU sequences have also been used to assess the validity of new *Phasmarhabditis* spp. They have also been used to investigate their phylogenetic relationships (de Ley *et al.* 2016; Ivanova and Spiridonov, 2021; Gorgadze *et al.*, 2022).

1.4.2.2 Mitochondrial genes

A few mitochondrial genes have been investigated to assess the genetic variation within and among nematodes with potential as biological control agents. For example, Powers *et al.* (1986) studied the molecular structure of nematode mtDNA using the mermithid *R. culicivorax*. In a later study, Powers *et al.* (1994) compared several mtDNA genes (e.g. NADH dehydrogenase subunit 3 (ND3), large rDNA and cytochrome *b* genes) to measure the genetic divergence from several nematode species.

Blouin *et al.* (1999) and Liu *et al.* (1999) studied the genetic variation among several *Heterorhabditis marelata* populations using the ND4 gene of mtDNA and found limited intraspecific variation. Other mtDNA genes studied include CO II and 18S rDNA (Szalanski *et al.*, 2000). These loci showed variation at the species level and proved useful for discrimination among a selection of *Steinernema* spp. However, they failed to show variation at the intraspecific level when tested with several *S. feltiae* populations.

Hyman *et al.* (2011) investigated full mitochondrial sequences of three mosquito parasites, *Romanomermis culicivorax*, *R. nielseni* and *R. iyengari*. In this analysis, the authors found duplicated expanses encompassing different regions of the mitochondrial genomes in each of the congeners. Moreover, the study revealed that these mtDNA sequences shared few rDNA and protein gene junctions, indicating extensive gene rearrangement within the *Romanomermis* lineage (Hyman *et al.*, 2011). Rapid structural changes were also observed at the conspecific level where no two individual nematodes carry the same haplotype (Hyman *et al.*, 2011).

Tobias (2017) also considered CO I to characterize the evolutionary genetics of *Thaumamermis zealandica* and its host, the sandhopper *Bellorchestia quoyana* (Talitridae: Amphipoda), revealing an absence of intraspecific variation among the studied populations. The combined sequences of 18S, 28S and CO I were used to assess the phylogenetic relationships of Mermithidae, revealing three major clades and determined that 18S and CO I genes are not informative in inferring evolutionary relationships in this family.

CO I has also been used in the systematics and taxonomy of *Deladenus* spp. in combination with ITS genes, in particular in relation to the identification of cryptic species within *D. siricidicola* and *D. canii* clades (Kanzaki *et al.*, 2016; Morris *et al.*, 2018).

1.5 Integrative Taxonomy and a Reflection on Nomenclatural Actions

For many decades, morphology-based classification has been the foundation of nematode taxonomy. However, with the scarcity of trained taxonomists and the need for fast and accurate diagnoses of taxa, molecular approaches have been incorporated as a routine practice. Furthermore, many unculturable and cryptic taxa have been described with the aid of specific genetic markers. Despite these efforts, congruence in the identification of taxa using different molecular markers has not always been achievable. Analogously, taxa described based on morphological features do not always correspond with those based on molecular information. No single method is self-sufficient, and the choice of method(s) is usually based on the question asked, the nature of the samples and the availability of resources (Tixier *et al.*, 2011; Bogale *et al.*, 2020). Thus, resolving specific taxonomic matters requires the implementation of a multidisciplinary approach that includes morphological, molecular, geographic and bionomic data to avoid creating "taxonomic noise".

Although lumping and splitting of taxa are valid nomenclatural actions, proper and conventional steps such as the examination of type specimens, detailed comparison of sequence data and alignments, and validation of biological data, among others, should be followed to avoid generating further confusion. The lack of adherence to these steps was coined by Wood *et al.* (2020) as "taxonomic vandalism". Specifically, the term is used to address the current challenges with the synonymy of taxa without taking proper steps such as examining voucher specimens (which should be deposited in internationally accessible biodiversity repositories and/or museums), and the provision of data and rigorous analyses that justify and support any nomenclatural change. Furthermore, no naming priority or recognition over any existing proper scientific names should be acknowledged or given without a peer-reviewed taxonomic or revisionary classification study and the provision of accompanying supporting data (Wood *et al.*, 2020).

In this respect, a few EPN species recently synonymized have not followed proper taxonomic steps and thus should be disregarded. For example, several *Steinernema* spp. in the "Carpocapsae" group were synonymized by Hunt and Subbotin (2016) based on the justification that they had similar sequences (without providing supporting data). The rationale was based on the paucity of phylogenetically informative characters in the "Carpocapsae" clade. However, this issue was recognized by Stock *et al.* (2001) when presenting the first combined molecular and morphological phylogenetic framework for *Steinernema*. Furthermore, multilocus sequence analyses and morphological traits study by Nadler *et al.* (2006) provided further support for validity for the species in the "Carpocapsae" clade with a robust evolutionary hypothesis that maximized the explanatory power of available character datasets for *Steinernema* spp. and in relation to closely related species. Therefore, the synonymy proposed by Hunt and Subbotin (2016) is unsustained based on two premises: (i) lack of a complete taxonomic study and supporting data that validate their taxonomic action; and (ii) currently studied loci and molecular markers do not provide robust evidence to interpret the overlap between intraspecies variation and insufficient interspecies divergence.

Similarly, as described earlier, Dhakal *et al.* (2020) synonymized several *Heterorhabditis* spp. based on the examination of character states of sequences of a selection of genes, including the segments of 28S (D2/D3) and ITS of rDNA, and the genes CO I, *unc-87* (encoding thin filament (F-actin)-associated protein) and *cmd-1* (encoding calmodulin). In this study, many of the sequences utilized were generated from strains that were not the "type strains" used in the original species description, thus representing another challenge from a comparative standpoint and when considering making any taxonomic changes.

Similarly, the synonymy of *Heterorhabditis sonorensis* and *H. taysarae* by Dhakal *et al.* (2020) based on the exclusive interpretation of sequences data, and without the provisioning of supporting data (i.e. sequence analyses and alignments) should be discounted. *Heterorhabditis* spp. have been described for many decades using a multidimensional approach including morphological/morphometric data, cross-hybridization assays and analyses of evolutionary relationships (considering nuclear and mitochondrial genes). Thus, all these efforts and this body of knowledge should not be disregarded to be replaced with a minimalist approach.

One of the major challenges of modern biology is to develop accurate and reliable technologies for a rapid screening of DNA sequence variation. Moreover, it is also critical to have a thorough understanding of the genetic composition of all species. If possible, a representative sample of individuals should be genotyped, preferentially from different geographic locations or from different hosts, for parasitic taxa. In addition, voucher specimens (frozen tissues, nucleic acids and live cultures) should be preserved and curated for future studies.

There is no perfect DNA-typing method, and the choice of a particular technique is often a compromise that depends on several factors, including available expertise, funding, time limitations and the research question being asked; all points should be considered to avoid making erroneous interpretations and creating taxonomic noise. Furthermore, the rampant growth of nematode full-genome sequences should shed light on the resolution of taxonomic conflicts and provide a better understanding of intra- and interspecific variation. Until further information is available, the recommendation is to follow a balanced approach in making any taxonomic decisions.

1.6 References

Abate, B.A., Malan, A.P., Tiedt, L.R., Wingfield, M.J., Slippers, B. and Hurley, B.P. (2016) *Steinernema fabii* n. sp. (Rhabditida: Steinernematidae), a new entomopathogenic nematode from South Africa. *Nematology* 18, 235–255.

Adams, B.J., Burnell, A.M. and Powers, T.O. (1998) A phylogenetic analysis of *Heterorhabditis* (Nemata: Rhabditidae) based on internal transcribed spacer 1 DNA sequence data. *Journal of Nematology* 30, 22–39.

Ahmed, M. and Holovachov, O. (2021) Twenty years after De Ley and Blaxter-How far did we progress in understanding the phylogeny of the phylum Nematoda? *Animals* 11: 3479.

Andaló, V., Nguyen, K.B. and Moino, A. Jr. (2006) *Heterorhabditis amazonensis* n. sp. (Rhabditida: Heterorhabditidae) from Amazonas, Brazil. *Nematology* 8, 853–867.

Andrássy, I. (1983) *A Taxonomic Review of the Suborder Rhabditina (Nematoda: Secernentia)*. Office of Scientific and Technical Research Overseas, France.

Anis, M., Shahina, F., Reid, A.P. and Maqbool, M.A. (2000) Redescription of *Heterorhabditis indica* Poinar *et al.*, 1992 (Rhabditida: Heterorhabditidae) from Pakistan. *Pakistan Journal of Nematology* 18, 11–27.

Anis, M., Shahina, F., Reid, A.P. and Rowe, J. (2002) *Steinernema asiaticum* sp. n. (Rhabditida: Steinernematidae) from Pakistan. *International Journal of Nematology* 12, 220–231.

Artyukhovsky, A.K. (1967) [*Neoaplectana arenaria* nov. sp. (Steinernematidae: Nematoda) inducing nematode disease in chafers of the Voronezh region]. *Trudy Voroneszhskogo Gosudarstvennogo Zapovednika* 15, 94–100 (in Russian).

Artyukhovski, A.K. (1969) New genera of the family Mermithidae (Enoplida, Nematoda). *Zoologicheskii Zhurnal* 48, 1309–1319.

Ashfaq, M. and Hebert, P.D. (2016) DNA barcodes for bio-surveillance: regulated and economically important arthropod plant pests. *Genome* 59, 933–945.

Askary, T.H. (2010) Nematodes as biocontrol agents. In: Lichtfouse, E. (ed.) *Sociology, Organic Farming, Climate Change and Soil Science*. Springer, Dordrecht, Netherlands, pp. 347–378.

Baniya, A., Huguet-Tapia, J.C. and DiGennaro, P. (2020) A draft genome of *Steinernema diaprepesi*. *Journal of Nematology* 52: e2020-69.

Bedding, R.A. (1968) *Deladenus wilsoni* n. sp. and *Deladenus siricidicola* n. sp. (Neotylenchidae) entomophagous–mycetophagous nematodes parasitic of siricid woodwasps. *Nematologica* 18, 482–493.

Bedding, R.A. (1993) Biological control of *Sirex noctilio* using the nematode *Beddingia siricidicola*. In: Bedding, R.A., Akhurst, R. and Kaya, H.K. (eds.) *Nematodes and the Biological Control of Insect Pests*. CSIRO Publications, East Melbourne, Australia, pp. 11–20.

Belien, T. (2018) Entomopathogenic nematodes as biocontrol agents of insect pests in orchards. *CAB Reviews* 2018, 1–11.

Blaxter, M.L., de Ley, P., Garey, J.R., Liu, L.X., Scheldeman, P. *et al.* (1998) A molecular evolutionary framework for the phylum Nematoda. *Nature* 392, 71–75.

Blaxter, M., Koutsovoulos, G., Jones, M., Kumar, S., Elsworth, B. *et al.* (2016) Phylogenomics of Nematoda. In: Olson, P.D., Hughes, J. and Cotton, JAE (eds.) *Next Generation Systematics*. Cambridge University Press, Cambridge, UK, pp. 62–83.

Blouin, M.S., Liu, J. and Berry, R.E. (1999) Life cycle variation and the genetic structure of nematode populations. *Heredity* 83, 253–259.

Boemare, N. (2002) Biology, taxonomy and systematics of *Photorhabdus* and *Xenorhabdus*. In: Gaugler, R. (ed.) *Entomopathogenic Nematology*. CAB International, Wallingford, UK, pp. 35–56.

Bogale, M., Baniya, A. and DiGennaro, P. (2020) Nematode identification techniques and recent advances. *Plants* 9: 1260.

Bovien, P. (1937) *Some Types of Association between Nematodes and Insects*. Videnskabelige Meddelelser fra Dansk Naturhistorisk Forening, Copenhagen.

Cabanillas, H.E., Poinar, G.O. Jr. and Raulston, J.R. (1994) *Steinernema riobravis* n. sp. (Rhabditida: Steinernematidae) from Texas. *Fundamental and Applied Nematology* 17, 123–131.

Campos-Herrera, R. (ed.) (2015) *Nematode Pathogenesis of Insects and Other Pests: Ecology and Applied Technologies for Sustainable Plant and Crop Protection*. Springer, Cham, Switzerland.

Chen, S., Li, X., Yan, A., Spiridonov, S.E. and Moens, M. (2006) A new entomopathogenic nematode, *Steinernema hebeiense* sp. n. (Rhabditida: Steinernematidae), from North China. *Nematology* 8, 563–574.

Choo, H.Y. and Kaya, H.K. (1994) Biological control of the brown planthopper by a mermithid nematode. *Korean Journal of Applied Entomology* 33, 207–215.

Choo, H.Y., Kaya, H.K. and Kim, J.B. (1989) *Agamermis unka* (Mermithidae) parasitism of *Nilaparvata lugens* in rice fields in Korea. *Journal of Nematology* 21, 254–259.

Çimen, H., Lee, M.M., Hatting, J., Hazir, S. and Stock, S.P. (2014) *Steinernema tophus* sp. n. (Nematoda: Steinernematidae), a new entomopathogenic nematode from South Africa. *Zootaxa* 3821: 337–353.

Çimen, H., Lee, M.M., Hatting, J., Hazir, S. and Stock, S.P. (2015) *Steinernema innovationi* n. sp. (Panagrolaimomorpha: Steinernematidae), a new entomopathogenic nematode species from South Africa. *Journal of Helminthology* 89, 415–427.

Çimen, H., Půža, V., Nermuť, J., Hatting, J., Ramakuwela, T., Faktorova, L. and Hazir, S. (2016) *Steinernema beitlechemi* n. sp., a new entomopathogenic nematode (Nematoda: Steinernematidae) from South Africa. *Journal of Nematology* 48, 148–158.

Choudhary, B.N and Sivakumar, C.V. (2000) Biocontrol potential of *Mylonchulus minor* against some plant parasitic nematodes. *Annals of Plant Protection Sciences* 8, 53–57.

Clausi, M., Longo, A., Rappazzo, G., Tarasco, E. and Vinciguerra, M.T. (2011) *Steinernema vulcanicum* n. sp. (Rhabditida: Steinernematidae), a new entomopathogenic nematode species from Sicily (Italy). *Nematology* 13, 409–423.

Cobb, N.A., Steiner, G. and Christie, J.R. (1923) *Agametmis decaudata* Cobb, Steiner and Christie; a nema parasite of grasshoppers and other insects. *Journal of Agricultural Research* 23, 921–926.

Coman, D. (1961) Mermithidae. *Fauna Republicii Socialiste Romania* 2: 1–61.

Crainey, J.L., Wilson, M.D. and Post, R.J. (2009) An 18S ribosomal DNA barcode for the study of *Isomermis lairdi*, a parasite of the blackfly *Simulium damnosum* sl. *Medical and Veterinary Entomology* 23, 238–244.

Creighton, C.S. and Fassuliotis, G. (1983) Infectivity and suppression of the banded cucumber beetle (Coleoptera: Chrysomelidae) by the mermithid nematode *Filipjevimermis leipsandra* (Mermithida: Mermithidae). *Journal of Economic Entomology* 76, 615–618.

Curran, J. and Driver, F. (1994) Molecular taxonomy of *Heterorhabditis*. In: Burnell A.M., Ehlers, R.-U. and Masson J.-P. (eds.) *Genetics of Entomopathogenic Nematode–Bacterium Complexes*. European Commission Publication, Luxembourg, pp. 41–48.

Cutler, C.G. and Stock, S.P. (2003) *Steinernema websteri* n. sp. (Rhabditida: Steinermeatidae), a new entomopathogenic nematode from China. *Nematologia Mediterranea* 31, 215–224.

de Doucet, M.M.A. (1986) A new species of *Neoaplectana* Steiner, 1929 (Nematoda: Steinernematidae) from Cordoba, Argentina. *Revue Nematologie* 9, 317–323.

de Doucet, M.M.A. and Doucet, M.E. (1990) *Steinernema ritteri* n. sp. (Nematoda: Steinernematidae) with a key to the species of the genus. *Nematologica* 36, 257–265.

de Ley, P. and Blaxter, M. (2002) Systematic position and phylogeny. In: Lee, D.L. (ed.) *The Biology of Nematodes.* CRC Press, London, pp. 1–30.

de Ley, P. and Blaxter, M.L. (2004) A new system for Nematoda: combining morphological characters with molecular trees and translating clades into ranks and taxa. In: *Proceedings of the Fourth International Congress of Nematology, 8–13 June 2002, Tenerife, Spain*. Brill, Leiden, Netherlands, pp. 633–653.

de Ley, I.T., Holovachov, O., McDonnell, R.J., Bert, W., Paine, T.D. and de Ley, P. (2016) Description of *Phasmarhabditis californica* n. sp. and first report of *P. papillosa* (Nematoda: Rhabditidae) from invasive slugs in the USA. *Nematology* 18, 175–193.

Dhakal, M., Nguyen, K.B., Hunt, D.J., Ehlers, R.U., Spiridonov, S.E. and Subbotin, S.A. (2020) Molecular identification, phylogeny and phylogeography of the entomopathogenic nematodes of the genus *Heterorhabditis* Poinar, 1976: a multigene approach. *Nematology* 23, 451–466.

Didiza, L., Lephoto, T.E. and Gray, V.M. (2021) Morphological and molecular phylogenetic description of *Steinernema batswanae* n. sp. (Rhabditida: Steinernematidae): a new species of an entomopathogenic nematode from South Africa. *Archives of Phytopathology and Plant Protection* 54, 1603–1616.

Dillman, A.R., Macchietto, M., Porter, C.F., Rogers, A., Williams, B. *et al.* (2015) Comparative genomics of *Steinernema* reveals deeply conserved gene regulatory networks. *Genome Biology* 16: 200.

Dillon, A.B., Rolston, A.N., Meade, C.V., Downes, M.J. and Griffin, C.T. (2008) Establishment, persistence, and introgression of entomopathogenic nematodes in a forest ecosystem. *Ecological Applications* 18, 735–747.

Dorado, G., Gálvez, S., Rosales, T.E., Vásquez, V.F. and Hernández, P. (2021) Analyzing modern biomolecules: the revolution of nucleic-acid sequencing – review. *Biomolecules* 11: 1111.

Dujardin, F. (1842) Memoire sur la structure anatomique des Gordius et d'un autre helminthe, le Mermis, qu'on a con fondu avec eux. *Annales des Sciences Naturelles* 18, 129–151.

Edgington, S., Buddie, A.G., Tymo, L., Hunt, D.J., Nguyen, K.B. *et al.* (2009a) *Steinernema australe* n. sp. (Panagrolaimomorpha: Steinernematidae), a new entomopathogenic nematode from Isla Magdalena, Chile. *Nematology* 11, 699–717.

Edgington, S., Buddie, A.G., Tymo, L.M., France, A., Merino, L. and Hunt, D.J. (2009b) *Steinernema unicornum* sp. n. (Panagrolaimomorpha: Steinernematidae), a new entomopathogenic nematode from Tierra del Fuego, Chile. *Journal of Nematode Morphology and Systematics* 12, 113–131.

Edgington, S., Buddie, A.G., Moore, D., France, A., Merino, L. and Hunt, D.J. (2011) *Heterorhabditis atacamensis* n. sp. (Nematoda: Heterorhabditidae), a new entomopathogenic nematode from the Atacama Desert, Chile. *Journal of Helminthology* 85, 381–394.

Elawad, S., Ahmad, W. and Reid, A.P. (1997) *Steinernema abbasi* sp. n. (Nematoda: Steinernematidae) from the Sultanate of Oman. *Fundamental and Applied Nematology* 20, 435–442.

Fayyaz, S., Khanum, T.A., Ali, S., Solangi, G.S., Gulsher, M. and Javed, S. (2015) *Steinernema balochiense* n. sp. (Rhabditida: Steinernematidae) a new entomopathogenic nematode from Pakistan. *Zootaxa* 3904, 387–402.

Filipjev, I. (1934) Miscellanea Nematologica. I. Eine neue Art der Gattung *Neoaplectana* Steiner nebst Bemerkunge über die systematische Stellung der letzteren. *Magazin de Parasitologie de l'Institut Zoologique de l'Academie de l'USSR* 4, 229–240.

Frewin, A.J., Scott-Dupree, C. and Hanner, R. (2013) DNA barcoding for plant protection: applications and summary of available data for arthropod pests. *CAB Reviews* 2013, 1–13.

Gafurov, A.K. (1996) *Mermitidy: Sistematika, Biologiya, Filogeniya, Prakticheskoe Znachenie.* Donish, Dushanbe.

Gardner, S.L., Stock, S.P. and Kaya, H.K. (1994) A new species of *Heterorhabditis* from the Hawaiian Islands. *Journal of Parasitology* 80, 100–106.

Gariepy, T.D., Haye, T. and Zhang, J. (2014) A molecular diagnostic tool for the preliminary assessment of host–parasitoid associations in biological control programmes for a new invasive pest. *Molecular Ecology* 23, 3912–3924.

Gasser, R.B., Bott, N.J., Chilton, N.B., Hunt, P. and Beveridge, I. (2008) Toward practical, DNA-based diagnostic methods for parasitic nematodes of livestock – bionomic and biotechnological implications. *Biotechnology Advances* 26, 325–334.

Gaugler, R. and Kaya, H.K. (eds.) (1990) *Entomopathogenic Nematodes in Biological Control.* CRC Press, Boca Raton, Florida.

Georgis, R., Koppenhöfer, A.M., Lacey, L.A., Bélair, G., Duncan, L.W. *et al.* (2006). Successes and failures in the use of parasitic nematodes for pest control. *Biological Control* 38, 103–123.

Glen, D.M. and Wilson, M. (1997) Slug-parasitic nematodes as biocontrol agents for slugs. *Agro Food Industry Hi Technology* 8, 23–27.

Gorgadze, O., Lortkhipanidze, M., Ogier, J.C., Tailliez, P. and Burjanadze, M. (2015) *Steinernema tbilisiensis* sp. n. (Nematoda: Steinernematidae) – a new species of entomopathogenic nematode from Georgia. *Journal of Agricultural Science and Technology* A5, 13–18.

Gorgadze, O., Troccoli, A., Fanelli, E., Tarasco, E. and de Luca, F. (2022) *Phasmarhabditis thesamica* n. sp. (Nematoda: Rhabditidae), a new slug nematode from southern slope of Caucasus, Georgia. *Nematology* 24, 617–629.

Grewal, P.S., Grewal, S.K., Tan, L. and Adams, B. (2003) Parasitism of molluscs by nematodes: types of associations and evolutionary trends .*Journal of Nematology* 35, 146–156.

Haraguchi, S. and Yoshiga, T. (2020) Potential of the fungal feeding nematode *Aphelenchus avenae* to control fungi and the plant parasitic nematode *Ditylenchus destructor* associated with garlic. *Biological Control* 143: 104203.

Hashmi, G., Glazer, I. and Gaugler, R. (1996) Molecular comparisons of entomopathogenic nematodes using randomly amplified polymorphic DNA (RAPD) markers. *Fundamental and Applied Nematology* 19, 399–406.

Hazir, S., Stock, S.P. and Keskin, N. (2003) A new entomopathogenic nematode, *Steinernema anatoliense* n. sp. (Rhabditida: Steinernematidae), from Turkey. *Systematic Parasitology* 55, 211–220.

Heydari, F., Abolafia, J. and Pedram, M. (2020) Description of *Deladenus brevis* n. sp. (Sphaerularioidea: Neotylenchidae) from Iran: a morphological and molecular phylogenetic study. *Journal of Nematology* 52, 1–13.

Hodda, M. (2022) Phylum Nematoda: trends in species descriptions, the documentation of diversity, systematics, and the species concept. *Zootaxa* 5114, 290–317.

Holterman, M., van der Wurff, A., van den Elsen, S., van Megen, H., Bongers, T. *et al.* (2006) Phylum-wide analysis of SSU rDNA reveals deep phylogenetic relationships among nematodes and accelerated evolution toward crown clades. *Molecular Biology and Evolution* 23, 1792–1800.

Hooper, D.J., Wilson, M.J., Rowe, J.A. and Glen, D.M. (1999) Some observations on the morphology and protein profiles of the slug-parasitic nematodes *Phasmarhabditis hermaphrodita* and *P. neopapillosa* (Nematoda: Rhabditidae). *Nematology* 1, 173–182.

Howe, D.K., Ha, A.D., Colton, A., de Ley, I.T., Rae, R.G. *et al.* (2020) Phylogenetic evidence for the invasion of a commercialized European *Phasmarhabditis hermaphrodita* lineage into North America and New Zealand. *PLOS ONE* 15: e0237249.

Hunt, D.J. and Nguyen, K.B. (eds.) (2016) *Advances in Entomopathogenic Nematode Taxonomy and Phylogeny*. Nematology Monographs and Perspectives, Vol. 12. Brill, Leiden, Netherlands.

Hunt, D.J. and Subbotin, S.A. (2016) Taxonomy and systematics. In: Hunt, D.J. and Nguyen, K.B. (eds.) *Advances in Entomopathogenic Nematode Taxonomy and Phylogeny. Nematology Monographs and Perspectives*, Vol. 12. Brill, Leiden, Netherlands, pp. 13–58.

Hussaini, S.S., Ansari, M.A., Ahmad, W. and Subbotin, S.A. (2001) Identification of some Indian populations of *Steinernema* species (Nematoda) by RFLP analysis of the ITS region of rDNA. *International Journal of Nematology* 11, 73–76.

Hyman, B.C. and Azevedo, J.L. (1996) Similar evolutionary patterning among repeated and single copy nematode mitochondrial genes. *Molecular Biology and Evolution* 13, 221–232.

Hyman, B.C., Lewis, S.C., Tang, S. and Wu, Z. (2011) Rampant gene rearrangement and haplotype hypervariation among nematode mitochondrial genomes. *Genetica* 139, 611–615.

Ivanova, E.S. and Spiridonov, S.E. (2017) *Phasmarhabditis meridionalis* sp. n. (Nematoda: Rhabditidae) from a land snail *Quantula striata* (Gastropoda: Dyakiidae) from south of Vietnam. *Russian Journal of Nematology* 25, 129–140.

Ivanova, E.S. and Spiridonov, S.E. (2021) *Phasmarhabditis quinamensis* sp. n. (Nematoda: Rhabditidae) from tropical terrestrial gastropods in southern Vietnam. *Nematology* 24, 225–239.

Ivanova, E.S., Geraskina, A.P. and Spiridonov, S.E. (2020) Two new species of *Phasmarhabditis* Andrássy, 1976 (Nematoda: Rhabditidae) associated with land snails in Northwest Caucasus, Russian Federation: description and molecular affiliation. *Nematology* 22, 179–197.

Ivanova, E.S., Gorgadze, O.A., Lortkhipanidze, M.A. and Spiridonov, S.E. (2021) *Phasmarhabditis akhaldaba* sp. n. associated with a slug *Deroceras reticulatum* in Lesser Caucasus mountains in Republic of Georgia. *Russian Journal of Nematology* 29, 75–88.

Ivanova, E.S., Clausi, M., Leone, D. and Spiridonov, S.E. (2022) *Phasmarhabditis villasmundi* sp. n. infecting land gastropods in the Nature Reserve 'Speleological Complex Villasmundo – S. Alfio' in Syracuse Province, Sicily. *Nematology* 25, 169–180.

Jian, H., Reid, A.P. and Hunt, D.J. (1997) *Steinernema ceratophorum* (Nematoda: Steinernematidae), a new entomopathogenic nematode from north-east China. *Systematic Parasitology* 37, 115–125.

Kakulia, G. and Mikaia, N. (1997) New species of the nematode *Heterorhabditis poinari* sp. nov. (Rhabditida: Heterorhabditidae). *Parasitology and Helminthology* 155, 457–459.

Kamminga, K.L., Davis, J.A., Stock, S.P. and Richter, A.R. (2012) First report of a mermithid nematode infecting *Piezodorus guildinii* and *Acrosternum hilare* (Hemiptera: Pentatomidae) in the United States. *Florida Entomologist* 95, 214–217.

Kanga, F.N., Trinh, P.Q., Waeyenberge, L., Spiridonov, S.E., Hauser, S. and Moens, M. (2012) Two new species of Steinernema Travassos, 1927 from the humid forest of southern Cameroon. *Russian Journal of Nematology* 20, 15–36.

Kanzaki, N., Tanaka, S.E., Fitza, K., Kosaka, H., Slippers, B. *et al.* (2016) *Deladenus nitobei* n. sp. (Tylenchomorpha: Allantonematidae) isolated from *Sirex nitobei* (Hymenoptera: Siricidae) from Aomori, Japan, a new member of the *Siricidicola* superspecies. *Nematology* 18, 1199–1217.

Kasab, A.S. and Abdel-Kader, K. (1996) Suppressing root-knot, citrus and reniform nematodes by use of less irrigation frequency and predacious nematode in olive. *Annals of Agricultural Science of Cairo* 41, 511–520.

Katumanyane, A., Malan, A.P., Tiedt, L.R. and Hurley, B.P. (2020) *Steinernema bertusi* n. sp. (Rhabditida: Steinernematidae), a new entomopathogenic nematode from South Africa. *Nematology* 22, 343–360.

Kaya, H.K. and Stock, S.P. (1997) Techniques in insect nematology. In: Lacey, L.A. (ed.) *Manual of Techniques in Insect Pathology*. Biological Techniques Series, Academic Press, San Diego, California, pp. 281–324.

Khan, A., Brooks, W.M. and Hirschmann, H. (1976) *Chromonema heliothidis* n. gen., n. sp. (Steinernematidae, Nematoda), a parasite of *Heliothis zea* (Noctuidae, Lepidoptera), and other insects. *Journal of Nematology* 8, 159–168.

Khatri-Chhetri, H.B., Waeyenberge, L., Spiridonov, S., Manandhar, H.K. and Moens, M. (2011a) *Steinernema lamjungense* n. sp. (Rhabditida: Steinernematidae), a new species of entomopathogenic nematode from Lamjung district, Nepal. *Nematology* 13, 589–605.

Khatri-Chhetri, H.B., Waeyenberge, L., Spiridonov, S., Manandhar, H.K. and Moens, M. (2011b) Two new species of *Steinernema* Travassos, 1927 with short infective juveniles from Nepal. *Russian Journal of Nematology* 19, 53–74.

Kobylinski, K.C., Sylla, M., Black, W. and Foy, B.D. (2012) Mermithid nematodes found in adult *Anopheles* from southeastern Senegal. *Parasites & Vectors* 5: 131.

Kubo, R., Ugajin, A. and Ono, M. (2016) Molecular phylogenetic analysis of mermithid nematodes (Mermithida: Mermithidae) discovered from Japanese bumblebee (Hymenoptera: Bombinae) and behavioral observation of an infected bumblebee. *Applied Entomology and Zoology* 51, 549–554.

Kumar, S., Koutsovoulos, G., Kaur, G. and Blaxter, M. (2012) Toward 959 nematode genomes. *Worm* 1, 42–50.

Lacey, L.A., Frutos, R., Kaya, H.K. and Vails, P. (2001) Insect pathogens as biological control agents: do they have a future? *Biological Control* 21, 230–248.

Lee, M.M., Sicard, M., Skeie, M. and Stock, S.P. (2009) *Steinernema boemarei* n. sp. (Nematoda: Steinernematidae), a new entomopathogenic nematode from southern France. *Systematic Parasitology* 72, 127–141.

Lefoulon, E., McMullen, J.G. and Stock, S.P. (2022) Transcriptomic analysis of *Steinernema* nematodes highlights metabolic costs associated to *Xenorhabdus* endosymbiont association and rearing conditions. *Frontiers in Physiology* 13: 821845.

Li, X.-Y., L., Liu, Q.-Z., Nermuť, J., Půža, V. and Mráček, Z. (2012) *Heterorhabditis beicherriana* n. sp. (Nematoda: Heterorhabditidae), a new entomopathogenic nematode from the Shunyi district of Beijing, China. *Zootaxa* 3569, 25–40.

Liu, J. (1994) A new species in the genus *Heterorhabditis* from China (Rhabditida: Heterorhabditidae)]. *Acta Zootaxonomica Sinica* 19, 268–272.

Liu, J. and Berry, R.E. (1995) Differentiation of isolates in the genus *Steinernema* (Nematoda: Steinernematidae) by random amplified polymorphic DNA fragments and morphological characters. *Parasitology* 111, 119–125.

Liu, J. and Berry, R.E. (1996a) Phylogenetic analysis of the genus *Steinernema* by morphological characters and randomly amplified polymorphic DNA fragments. *Fundamental and Applied Nematology* 19, 463–469.

Liu, J. and Berry, R.E. (1996b) *Steinernema oregonensis* n. sp. (Rhabditida: Steinernematidae) from Oregon, USA. *Fundamental and Applied Nematology* 19, 375–380.

Liu, J. and Berry, R.E. (1996c) *Heterorhabditis marelatus* n. sp. (Rhabditida: Heterorhabditidae) from Oregon. *Journal of Invertebrate Pathology* 67, 48–54.

Liu, J., Berry, R.E. and Moldenke, A.F. (1997) Phylogenetic relationships of entomopathogenic nematodes (Heterorhabditdae and Steinernematidae) inferred from partial 18S rRNA gene sequences. *Journal of Invertebrate Pathology* 69, 246–252.

Liu, J., Berry, R.E. and Blouin, M.S. (1999) Molecular differentiation and phylogeny of entomopathogenic nematodes (Rhabditida: Heterorhabditidae) based on ND4 gene sequences of mitochondrial DNA. *Journal of Parasitology* 85, 709–715.

Lootsma, M. and Scholte, K. (1997) Effects of the springtail *Folsomia fimetaria* and the nematode *Aphelenchus avenae* on *Rhizoctonia solani* stem infection of potato at temperatures of 10 and 15°C. *Plant Pathology* 46, 203–208.

López-Nunez, J.C., Plichta, K., Gongora-Botero, C.E. and Stock, S.P. (2008) A new entomopathogenic nematode, *Steinernema colombiense* n. sp. (Nematoda: Steinernematidae), from Colombia. *Nematology* 10, 561–574.

Luc, V.P., Nguyen, K.B., Reid, A.P. and Spiridonov, S.E. (2000) *Steinernema tami* sp. n. (Rhabditida: Steinernematidae) from Cat Tien Forest, Vietnam. *Russian Journal of Nematology* 8, 33–43.

Lulamba, T.E. and Serepa-Dlamini, M.H. (2020) Molecular identification of a *Heterorhabditis* entomopathogenic nematode isolated from the northernmost region of South Africa. *Egyptian Journal of Biological Pest Control* 30, 1–9.

Ma, J., Chen, S., de Clercq, P., Han, R. and Moens, M. (2012a) *Steinernema changbaiense* sp. n. (Rhabditida: Steinernematidae), a new species of entomopathogenic nematodes from Northeast China. *Russian Journal of Nematology* 20, 97–112.

Ma, J., Chen, S., Li, X., Han, R., Khatri-Chhetri, H.B. *et al.* (2012b) A new entomopathogenic nematode, *Steinernema tielingense* n. sp. (Rhabditida: Steinernematidae), from north China. *Nematology* 14, 321–338.

Ma, J., Chen, S., de Clercq, P., Waeyenberge, L., Han, R. and Moens, M. (2012c) A new entomopathogenic nematode, *Steinernema xinbinense* n. sp. (Nematoda: Steinernematidae), from north China. *Nematology* 14, 723–739.

MacLeod, A., Pautasso, M., Jeger, M.J. and Haines-Young, R. (2010) Evolution of the international regulation of plant pests and challenges for future plant health. *Food Security* 2, 49–70.

Malan, A.P., Nguyen, K.B., de Waal, J.Y. and Tiedt, L. (2008) *Heterorhabditis safricana* n. sp. (Rhabditida: Heterorhabditidae), a new entomopathogenic nematode from South Africa. *Nematology* 10, 381–396.

Malan, A.P., Knoetze, R. and Tiedt, L. (2014) *Heterorhabditis noenieputensis* n. sp. (Rhabditida: Heterorhabditidae), a new entomopathogenic nematode from South Africa. *Journal of Helminthology* 88, 139–151.

Malan, A., Knoetze, R. and Tiedt, L. (2015) *Steinernema jeffreyense* n. sp. (Rhabditida: Steinernematidae), a new entomopathogenic nematode from South Africa. *Journal of Helminthology* 90, 262–278.

Malan, A.P., Knoetze, R. and Tiedt, L.R. (2016) *Steinernema nguyeni* n. sp. (Rhabditida: Steinernematidae), a new entomopathogenic nematode from South Africa. *Nematology* 18, 571–590.

Mamiya, Y. (1988) *Steinernema kushidai* n. sp. (Nematoda: Steinernematidae) associated with scarabaeid beetle larvae from Shizuoka, Japan. *Applied Entomology and Zoolology* 23, 313–320.

Maneesakorn, P., Grewal, P.S. and Chandrapatya, A. (2010) *Steinernema minutum* sp. nov. (Rhabditida: Steinernematidae): a new entomopathogenic nematode from Thailand. *International Journal of Nematology* 20, 27–42.

Matsunaga, T., Wang, X.D. and Ishibashi, N. (1996) Interference by entomopathogenic and fungivorous nematodes of root invasion by plant-parasitic nematodes. *Japanese Journal of Nematology* 26, 1–11.

Mazza, G., Paoli, F., Strangi, A., Torrini, G., Marianelli, L. *et al.* (2017) *Hexamermis popilliae* n. sp. (Nematoda: Mermithidae) parasitizing the Japanese beetle *Popillia japonica* Newman (Coleoptera: Scarabaeidae) in Italy. *Systematic Parasitology* 94, 915–926.

Morris, E.E., Kepler, R.M., Long, S.J., Williams, D.W. and Hajek, A.E. (2013) Phylogenetic analysis of *Deladenus* nematodes parasitizing northeastern North American *Sirex* species. *Journal of Invertebrate Pathology* 113, 177–183.

Morris, E.E., Stock, S.P., Castrillo, L.A., Williams, D.W. and Hajek, A.E. (2018) Characterisation of the dimorphic *Deladenus beddingi* n. sp. and its associated woodwasp and fungus. *Nematology* 20, 939–955.

Mracek, Z., Hernandez, E.A. and Boemare, N.E. (1994) *Steinernema cubana* sp. n. (Nematoda: Rhabditida: Steinernematidae) and the preliminary characterization of its associated bacterium. *Journal of Invertebrate Pathology* 64, 123–129.

Mráček, Z., Sturhan, D., Reid, A.P. (2003) *Steinernema weiseri* n. sp. (Rhabditida, Steirnematidae), a new entomopathogenic nematode from Europe. *Systematic Parasitology* 56, 37–47.

Mráček, Z., Nguyen, K.B., Tailliez, P., Boemare, N.L. and Chen, S. (2006) *Steinernema sichuanense* n. sp. (Rhabditida, Steinernematidae), a new species of entomopathogenic nematode from the province of Sichuan, east Tibetan Mts., China. *Journal of Invertebrate Pathology* 93, 157–169.

Mráček, Z., Liu, Q.-Z. and Nguyen, K.B. (2009) *Steinernema xueshanense* n. sp. (Rhabditida, Steinernematidae), a new species of entomopathogenic nematode from the province of Yunnan, southeast Tibetan Mts., China. *Journal of Invertebrate Pathology* 102, 69–78.

Mráček, Z., Půža, V. and Nermuť, J. (2014) *Steinernema poinari* sp. n. (Nematoda: Steinernematidae) a new entomopathogenic nematode from the Czech Republic. *Zootaxa* 3760, 336–350.
Murfin, K.E., Dillman, A.R., Foster, J.M., Bulgheresi, S., Slatko, B.E. *et al.* (2012) Nematode–bacterium symbioses – cooperation and conflict revealed in the "omics" age. *Biological Bulletin* 223, 85–102.
Nadler, S.A., Bolotin, E. and Stock, S.P. (2006) Phylogenetic relationships of *Steinernema* Travassos, 1927 (Nematoda: Cephalobina: Steinernematidae) based on nuclear, mitochondrial, and morphological data. *Systematic Parasitology* 63, 159–179.
Nermuť, J., Pûža, V. and Mráček, Z. (2016a) *Phasmarhabditis apuliae* n. sp. (Nematoda: Rhabditidae), a new rhabditid nematode from milacid slugs. *Nematology* 18, 1095–1112.
Nermuť, J., Půža, V., Mekete, T. and Mráček, Z. (2016b) *Phasmarhabditis bonaquaense* n. sp. (Nematoda: Rhabditidae), a new slug-parasitic nematode from the Czech Republic. *Zootaxa* 4179, 30–546.
Nermuť, J., Půža, V., Mekete, T. and Mráček, Z. (2017) *Phasmarhabditis bohemica* n. sp. (Nematoda: Rhabditidae), a slug-parasitic nematode from the Czech Republic. *Nematology* 19, 93–107.
Nermuť, J., Holley, M., Ortmayer, L.F. and Půža, V. (2022) New players in the field: ecology and biocontrol potential of three species of mollusk-parasitic nematodes of the genus *Phasmarhabditis*. *BioControl* 67, 543–553.
Nguyen, K.B. and Duncan, L.W. (2002) *Steinernema diaprepesi* n. sp. (Rhabditida: Steinernematidae), a parasite of the citrus root weevil *Diaprepes abbreviatus* (L) (Coleoptera: Curculionidae). *Journal of Nematology* 34, 159–170.
Nguyen, K.B. and Buss, E. (2011) *Steinernema phyllophagae* n. sp. (Rhabditida: Steinernematidae), a new entomopathogenic nematode from Florida, USA. *Nematology* 13, 425–442.
Nguyen, K.B. and Smart, G.C. Jr. (1990) *Steinernema scapterisci* n. sp. (Rhabditida: Steinernematidae). *Journal of Nematology* 22, 187–199.
Nguyen, K.B. and Smart, G.C. Jr. (1992) *Steinernema neocurtillis* n. sp. (Rhabditida: Steinernematidae) and a key to species of the genus *Steinernema*. *Journal of Nematology* 4, 321–462
Nguyen, K.B. and Smart, G.C. (1994) *Neosteinernema longicurvicauda* n. gen., n. sp. (Rhabditida: Steinernematidae), a parasite of the termite *Reticuldermes flavipes* (Koller). *Journal of Nematology* 26, 162–174.
Nguyen, K.B., Maruniak, J. and Adams, J.B. (2001) Diagnostic and phylogenetic utility of the rDNA internal transcribed spacer sequences of *Steinernema*. *Journal of Nematology* 33, 73–82.
Nguyen, K., Sharpiro-Ilan, D., Stuart, R., McCoy, C., James, R. and Adams, B. (2004a) *Heterorhabditis mexicana* n. sp. (Rhabditida: Heterorhabditidae) from Tamaulipas, Mexico, and morphological studies of the bursa of *Heterorhabditis* spp. *Nematology* 6, 231–244.
Nguyen, K.B., Tesfamariam, M., Gozel, U., Gaugler, R. and Adams, B.J. (2004b) *Steinernema yirgalemense* n. sp. (Rhabditida: Steinernematidae) from Ethiopia. *Nematology*, 6, 839–856.
Nguyen, K.B., Gozel, U., Koppenhoefer, H.S. and Adams, B.J. (2006a) *Heterorhabditis floridensis* n. sp. (Rhabditida: Heterorhabditidae) from Florida. *Zootaxa* 1177, 1–19.
Nguyen, K.B., Qiu, L., Zhou, Y. and Pang, Y. (2006b) *Steinernema leizhouense* sp. n. (Nematoda: Steinernematidae), a new entomopathogenic nematode from southern China. *Russian Journal of Nematology* 14, 101–118.
Nguyen, K.B., Malan, A.P. and Gozel, U. (2006c) *Steinernema khoisanae* n. sp. (Rhabditida: Steinernematidae), a new entomopathogenic nematode from South Africa. *Nematology* 8, 157–175.
Nguyen, K.B., Půža, V. and Mráček, Z. (2008a) *Steinernema cholashanense* n. sp. (Rhabditida, Steinernematidae) a new species of entomopathogenic nematode from the province of Sichuan, Chola Shan Mountains, China. *Journal of Invertebrate Pathology* 97, 251–264.
Nguyen, K.B., Shapiro-Ilan, D.I. and Mbata, G.N. (2008b) *Heterorhabditis georgiana* n. sp. (Rhabditida: Heterorhabditidae) from Georgia, USA. *Nematology* 10, 433–448.
Nguyen, K.B., Ginarte, C.M.A., Leite, L.G., dos Santos, J.M. and Harakava, R. (2010) *Steinernema brazilense* n. sp. (Rhabditida: Steinernematidae), a new entomopathogenic nematode from Mato Grosso, Brazil. *Journal of Invertebrate Pathology* 103, 8–20.
Nikdel, M., Niknam, G. and Ye, W. (2011) *Steinernema arasbaranense* sp. n. (Nematoda: Steinernematidae), a new entomopathogenic nematode from Arasbaran forests, Iran. *Nematologia Mediterranea* 39, 17–28
Nthenga, I., Knoetze, R., Berry, S., Tiedt, L.R. and Malan, A.P. (2014) *Steinernema sacchari* n. sp. (Rhabditida: Steinernematidae), a new entomopathogenic nematode from South Africa. *Nematology* 16, 475–494.
Peteira, B., Martínez, B., Rodríguez, M.G., González, I., Infante, D. *et al.* (2015) Marcadores moleculares en la investigación y desarrollo de agentes de control biológico. *Revista de Protección Vegetal* 30, 71–71.
Petersen, J.J. (1985) Nematode parasites. *American Mosquito Control Association Bulletin* 6, 110–122.

Phan, K.L., Nguyen, N.C. and Moens, M. (2001a) *Steinernema loci* sp. n. and *Steinernema thanhi* sp. n. (Rhabditida: Steinernematidae) from Vietnam. *Nematology* 3, 503–514.

Phan, K.L., Nguyen, N.C. and Moens, M. (2001b) *Steinernema sangi* sp. n. (Rhabditida: Steinernematidae) from Vietnam. *Russian Journal of Nematology* 9, 1–7.

Phan, K.L., Subbotin, S.A., Nguyen, N.C. and Moens, M. (2003) *Heterorhabditis baujardi* sp. n. (Rhabditida: Heterorhabditidae) from Vietnam and morphometric data for *H. indica* populations. *Nematology* 5, 367–382.

Phan, L.K., Subbotin, S.A., Waeyenberge, L. and Moens, M. (2005) A new entomopathogenic nematode, *Steinernema robustispiculum* n. sp. (Rhabditida: Steinernematidae), from Chumomray National Park in Vietnam. *Systematic Parasitology* 60, 23–32.

Phan, K.L., Spiridonov, S.E., Subbotin, S.A. and Moens, M. (2006a) Four new species of *Steinernema* Travassos, 1928 with short infective juveniles from Vietnam. *Russian Journal of Nematology* 14, 11–29.

Phan, L.K., Takemoto, S. and Futai, K. (2006b) *Steinernema ashiuense* sp. n. (Nematoda: Steinernematidae), a new entomopathogenic nematode from Japan. *Nematology* 8, 681–690.

Phan, K.L., Mráček, Z., Půža, V., Nermuť, J. and Jarošová, A. (2014) *Steinernema huense* sp. n., a new entomopathogenic nematode (Nematoda: Steinernematidae) from Vietnam. *Nematology* 16, 761–775.

Pieterse, A., Malan, A.P. and Ross, J.L. (2017) Nematodes that associate with terrestrial mollusks as definitive hosts, including *Phasmarhabditis hermaphrodita* (Rhabditida: Rhabditidae) and its development as a biological molluscicide. *Journal of Helminthology* 91, 517–527.

Pieterse, A., Rowson, B., Tiedt, L., Malan, A.P., Haukeland, S. and Ross, J.L. (2020) *Phasmarhabditis kenyaensis* n. sp. (Nematoda: Rhabditidae) from the slug, *Polytoxon robustum*, in Kenya. *Nematology* 23, 229–245.

Plichta, K.L., Joyce, S.A., Clarke, D., Waterfield, N. and Stock, S.P. (2009) *Heterorhabditis gerrardi* n. sp. (Nematoda: Heterorhabditidae): the hidden host of *Photorhabdus asymbiotica* (Enterobacteriaceae: γ-Proteobacteria). *Journal of Helminthology* 83, 309–320.

Poinar, G.O. Jr. (1976) Description and biology of a new insect parasitic rhabditoid, *Heterorhabditis bacteriophora* n. gen. n. sp. (Rhabditida, Heterorhabditidae n. fam.). *Nematologica* 21, 463–470.

Poinar, G.O. Jr. (1983) *The Natural History of Nematodes.* Prentice-Hall, Englewood Cliffs, New Jersey.

Poinar, G.O. Jr. (1985) *Neoaplectana intermedia* n. sp. (Steinernematidae: Nematoda) from South Carolina. *Revue Nematologie* 8, 321–327.

Poinar, G.O. Jr. (1990) Entomopathogenic nematodes in biological control. In: Gaugler, R. and Kaya, K.H. (eds) *Taxonomy and Biology of Steinernematidae and Heterorhabditidae.* CRC Press, Boca Raton, Florida, pp. 23–74.

Poinar, G.O. Jr., Thomas, G.M., and Hess, R. (1977) Characteristics of the specific bacterium associated with *Heterorhabditis bacteriophora* (Heterorhabditidae: Rhabditida). *Nematologica* 23, 97–102.

Poinar, G.O. Jr., Jackson, T. and Klein, M. (1987) *Heterorhabditis megidis* sp. n. (Heterorhabditidae: Rhabditida), parasitic in the Japanese beetle, *Popillia japonica* (Scarabaeidae: Coleoptera), in Ohio. *Proceedings of the Helminthological Society of Washington* 54, 53–59.

Poinar, G.O. Jr., Karunakar, G.K. and David, H. (1992) *Heterorhabditis indicus* n. sp. (Rhabditida: Nematoda) from India: separation of *Heterorhabditis* spp. by infective juveniles. *Fundamental and Applied Nematology* 15, 467–472.

Porazinska, D.L., Giblin-Davis, R.M., Faller, L., Farmerie, W., Kanzaki, N. *et al.* (2009) Evaluating high-throughput sequencing as a method for metagenomic analysis of nematode diversity. *Molecular Ecology Resources* 9, 1439–1450.

Powers, T.O., Platzer, E.G. and Hyman, B.C. (1986) Large mitochondrial genome and mitochondrial DNA size polymorphism in the mosquito parasite, *Romanomermis culicivorax. Current Genetics* 11, 71–77.

Powers, T.O., Harris, T.S. and Hyman, B.C. (1994) Mitochondrial DNA sequence divergence among *Meloidogyne incognita, Romanomermis culicivorax, Ascaris suum* and *Caenorhabditis elegans*. *Journal of Nematology* 25, 564–572.

Powers, T.O., Todd, T.C., Burnell, A.M., Muray, P.C.B., Fleming, C.C. *et al.* (1997) The rDNA internal transcribed spacer region as a taxonomic marker for nematodes. *Journal of Nematology* 29, 441–450.

Presswell, B., Evans, S., Poulin, R. and Jorge, F. (2015) Morphological and molecular characterization of *Mermis nigrescens* Dujardin, 1842 (Nematoda: Mermithidae) parasitizing the introduced European earwig (Dermaptera: Forficulidae) in New Zealand. *Journal of Helminthology* 89, 267–276.

Prabhuraj, A., Viraktamath, C.A. and Kumar, A.R.V. (2002) Pathogenicity of two entomopathogenic nematodes against *Holotrichia serrata* (Fabricius) (Coleoptera: Scarabaeidae) and *Spodoptera litura* (Fabricius) (Lepidoptera: Noctuidae). In: *Proceedings of the Symposium of Biological Control of Lepidopteran Pests*. Society for Biocontrol Advancement, Bangalore, pp. 205–209.

Půža, V., Nermut, J., Mráček, Z., Gengler, S. and Haukeland, S. (2017) *Steinernema pwaniensis* n. sp., a new entomopathogenic nematode (Nematoda: Steinernematidae) from Tanzania. *Journal of Helminthology* 91, 20–34.

Qiu, L., Fang, Y., Zhou, Y., Pang, Y.I. and Nguyen, K.B. (2004) *Steinernema guangdongense* sp. n. (Nematoda: Steinernematidae), a new entomopathogenic nematode from southern China with a note on *S. serratum* (*nomen nudum*). *Zootaxa* 704, 1–20.

Qiu, L., Hu, X., Zhou, Y., Pang, Y. and Nguyen, K.B. (2005a) *Steinernema beddingi* n. sp. (Nematoda: Steinernematidae), a new entomopathogenic nematode from Yunnan, China. *Nematology* 7, 737–749.

Qiu, L., Hu, X., Zhou, Y., Mei, S., Nguyen, K.B., and Pang, Y. (2005b) *Steinernema akhursti* sp. n. (Nematoda: Steinernematidae) from Yunnan, China. *Journal of Invertebrate Pathology* 90, 151–160.

Qiu, L., Yan, X., Zhou, Y., Nguyen, K.B. and Pang, Y. (2005c) *Steinernema aciari* sp. n. (Nematoda: Steinernematidae), a new entomopathogenic nematode from Guangdong, China. *Journal of Invertebrate Pathology* 88, 58–69.

Qiu, L., Zhao, J., Wu, Z., Lv, Z. and Pang, Y. (2011) *Steinernema pui* sp. n. (Rhabditida, Steinernematidae), a new entomopathogenic nematode from Yunnan, China. *Zootaxa* 2767, 1–13.

Rae, R.G., Robertson, J.F. and Wilson, M.J. (2009) Chemoattraction and host preference of the gastropod parasitic nematode *Phasmarhabditis hermaphrodita*. *Journal of Parasitology* 95, 517–526.

Ramos, R.J., Dichusa, C.A., Aryal, S., Crispolon, E.J., Sumaya, N.P.D. and Sumaya, N.H. (2022) Isolation, identification and virulence of the local entomopathogenic nematode *Heterorhabditis indica* in the southern Philippines against the sugarcane white grub, *Leucopholis irrorata* (Chevrolat) (Coleoptera: Scarabaeidae). *International Journal of Tropical Insect Science* 42, 2739–2750.

Regeai, S.O., Fitzpatrick, D.A., Burnell, A.M. and Kakouli-Duarte, T. (2021) Characterisation of the complete mitochondrial genome of the insect-parasitic nematode *Heterorhabditis bacteriophora*: an idiosyncratic gene order and the presence of multiple long non-coding regions. *Nematology* 24, 37–54.

Reid, A.P. (1994) Molecular taxonomy of *Steinernema*. In: Burnell, A.M., Ehlers, R.-U. and Masson, J.P. (eds.) *Genetics of Entomopathogenic Nematode-Bacterium Complexes*. European Commission Publications, Luxembourg, pp. 49–58.

Reid, A.P. and Hominick, W.M. (1992) Restriction fragment length polymorphisms within ribosomal DNA repeat unit of British entomopathogenic nematodes (Rhabditida: Steinernematidae). *Parasitology* 105, 317–323.

Reid, A.P., Hominick, W.M. and Briscoe, B.R. (1997) Molecular taxonomy and phylogeny of entomopathogenic nematode species (Rhabditida: Steinernematidae) by RFLP analysis of the ITS region of rDNA repeat unit. *Systematic Parasitology* 37, 187–193.

Román, J. and Figueroa, W. (1994) *Steinernema puertoricensis* n. sp. (Rhabditida: Steinernematidae), a new entomopathogenic nematode from Puerto Rico. *Journal of Agriculture of the University of Puerto Rico* 78, 168–175.

Ross, J.L., Pieterse, A., Malan, A.P. and Ivanova, E. (2018) *Phasmarhabditis safricana* n. sp. (Nematoda: Rhabditidae), a parasite of the slug *Deroceras reticulatum* from South Africa. *Zootaxa* 4420, 391–404.

Rougon-Cardoso, A., Flores-Ponce, M., Ramos-Aboites, H.E., Martínez-Guerrero, C.E., Hao, Y.J. *et al.* (2016) The genome, transcriptome, and proteome of the nematode *Steinernema carpocapsae*: evolutionary signatures of a pathogenic lifestyle. *Scientific Reports* 6: 37536.

Rubzov, I.A. (1969) [New and little-known species of nematodes from the Danube.] *Gidrobiologicheskii Zhurnal* 5, 49–59 (in Russian).

Saeb, A.T. and Grewal, P.S. (2014) NADH dehydrogenase subunit 4 is not an ideal target for phylogeny reconstruction or genetic differentiation among entomopathognic nematodes. *SOP Transactions on Inheritance and Genetic Engineering.* Available at: www.academia.edu/33832552/NADH_Dehydrogenase_Subunit_4_Is_Not_an_Ideal_Target_for_Phylogeny_Reconstruction_or_Genetic_Differentiation_Among_Entomopathognic_Nematodes_RHABDITIDA_HETERORHABDITIDAE (accessed 24 April 2024).

San-Blas, E., Portillo, E., Nermuť, J., Půža, V. and Morales-Montero, P. (2015) *Steinernema papillatum* n. sp. (Rhabditida: Steinernematidae), a new entomopathogenic nematode from Venezuela. *Nematology* 17, 1081–1097.

San-Blas, E., Morales-Montero, P., Portillo, E., Nermuť, J. and Puza, V. (2016) *Steinernema goweni* n. sp. (Rhabditida: Steinernematidae), a new entomopathogenic nematode from Zulia State, Venezuela. *Zootaxa* 4067, 200–214.

Sanger, F., Nicklen, S. and Coulson, A.R. (1977) DNA sequencing with chain-terminating inhibitors. *Proceedings of the National Academy of Sciences USA* 74, 5463–5467.

Schneider, A.F. (1859) Über eine Nematodenlarvae und gewisse Verschiedenheiten in den Geschlechtsorganen der Nematoden. *Zeistschrift für Wissenschaftliche Zoologie* 10, 176–178.

Schneider, A.F. (1866) *Monographie der Nematoden*. G. Reimer, Berlin.

Shahina, F., Anis, M., Reid, A.P., Rowe, J. and Maqbool, M.A. (2001) *Steinernema pakistanense* sp. n. (Rhabditidae: Steinernematodae) from Pakistan. *International Journal of Nematology* 11, 124–133.

Shamseldean, M.M., Abou El-Sooud, A.B., Abd-Elgawad, M.M. and Saleh, M.M. (1996) Identification of a new *Heterorhabditis* species from Egypt, *Heterorhabditis taysearae* n. sp. (Rhabditida: Heterorhabditidae). *Egyptian Journal of Biological Control* 6, 129–138.

Shen, C.P. and Wang, G.H. (1992) [Description of an entomopathogenic nematode, *Steinernema longicaudum* sp. nov. and its application]. In: *Proceedings of the XIX International Congress of Entomology, Beijing, China, 28 June–4 July*. Entomological Society of China, Beijing, pp. 220–231 (in Chinese).

Siddiqi, M.R. (2000) *Tylenchida Parasites of Plants and Insects*. CAB International, Wallingford, UK, pp. 633–751.

Spiridonov, S.E. and Subbotin, S.A. (2016) Phylogeny and phylogeography of Heterorhabditis and Steinernema. In: Hunt, D.J. and Nguyen, K.B. (eds.) *Advances in Entomopathogenic Nematode Taxonomy and Phylogeny*. Nematology Monographs and Perspectives, Vol. 12. Brill, Leiden, Netherlands, pp. 413–427.

Spiridonov, S.E., Krasomil-Osterfeld, K. and Moens, M. (2004a) *Steinernema jollieti* sp n. (Rhabditida: Steinernematidae), a new entomopathogenic nematode from the American midwest. *Russian Journal of Nematology* 12, 85–95.

Spiridonov, S.E., Reid, A.P., Podrucka, K., Subbotin, S.A. and Moens, M. (2004b) Phylogenetic relationships within the genus *Steinernema* (Nematoda: Rhabditida) as inferred from analyses of sequences of the ITS1–5.8S–ITS2 region of rDNA and morphological features. *Nematology* 6, 547–566.

Spiridonov, S.E., Waeyenberge, L. and Moens, M. (2010) *Steinernema schliemanni* sp. n. (Steinernematidae; Rhabditida) – a new species of steinernematids of the 'monticolum'group from Europe. *Russian Journal of Nematology* 18, 175–190.

Stack, C.M., Easwaramoorthy, S.G., Metha, U.K., Downes, M.J., Griffin, C.T. and Burnell, A.M. (2000) Molecular characterisation of *Heterorhabditis indica* isolates from India, Kenya, Indonesia and Cuba. *Nematology* 2, 477–487.

Steiner, G. (1923) *Aplectana kraussei* n. sp. eine in der Blattwespe *Lyda* sp. Parasitierende Nematodenform, nebst Bemerkungen über das Seitenorgan der parasitischen Nematoden. *Zentralbatt fur Bakteriologiel Parasitenkunde Infektionskrankheiten und Hygiene Abteilung* 2, 14–18.

Steiner, G. (1929) *Neoaplectana glaseri*, n. g. n. sp. (Oxyuridae), a new nemic parasite of the japanese beetle (*Popillia japonica* Newm.). *Journal of the Washington Academy of Science* 19. 436–440.

Steyn, W.P., Knoetze, R., Tiedt, L.R. and Malan, A.P. (2017) *Steinernema litchii* n. sp. (Rhabditida: Steinernematidae), a new entomopathogenic nematode from South Africa. *Nematology* 19, 1157–1177.

Stock, S.P. (1993) A new species of the genus Poinar, 1976 (Nematoda: Heterorhabditidae) parasitizing *Heterorhabditis graphognathus* sp. larva (Coleoptera: Curculionidae) from Argentina. *Research and Reviews in Parasitology* 53, 103–107.

Stock, S.P. (2009) Molecular approaches and the taxonomy of insect-parasitic and pathogenic nematodes. In: Stock, S.P., Vandenberg, J., Glazer I. and Boemare, N. (eds) *Insect Pathogens: Molecular Approaches and Techniques*. CAB International, Wallingford, UK, pp. 71–94.

Stock, S.P. (2015) Diversity, biology and evolutionary relationships. In: Campos-Herrera, R. (ed.) *Nematode Pathogenesis of Insects and Other Pests: Ecology and Applied Technologies for Sustainable Plant and Crop Protection*. Springer, Cham, Switzerland, pp. 3–27.

Stock, S.P. and Goodrich-Blair, H. (2012) Nematode parasites, pathogens and associates of insects and invertebrates of economic importance. In: Lacey, L. (ed.) *Manual of Techniques in Invertebrate Pathology*, 2nd edn. Academic Press, London, pp. 373–426.

Stock, S.P. and Hunt, D.J. (2005) Morphology and systematics of nematodes used in biocontrol. In: Grewal, P.S., Ehlers, R.U. and Shapiro-Ilan, D. (eds) *Nematodes as Biocontrol Agents*. CAB International, Wallingford, UK, pp. 3–43.

Stock, S.P. and Koppenhöfer, A.M. (2003) *Steinernema scarabaei* n. sp. (Rhabditida: Steinernematidae), a natural pathogen of scarab beetle larvae (Coleoptera: Scarabaeidae) from New Jersey, USA. *Nematology* 5, 191–204.

Stock, S., Strong, D. and Gardner, S. (1996) Identification of *Heterorhabditis* (Nematoda: Heterorhabditidae) from California with a new species isolated from the larvae of the ghost moth *Hepialis californicus* (Lepidoptera: Hepialidae) from the Bodega Bay Natural Reserve. *Fundamental and Applied Nematology* 19, 585–592.

Stock, S.P., Choo, H.Y. and Kaya, H.K. (1997) An entomopathogenic nematode, *Steinernema monticolum* sp. n. (Rhabditida: Steinernematidae) from Korea with a key to other species. *Nematologica* 43, 15–29.
Stock, S.P., Somsook, V. and Reid, A.P. (1998) *Steinernema siamkayai* n. sp. (Rhabditida: Steinernematidae), an entomopathogenic nematode from Thailand. *Systematic Parasitology* 41, 105–113.
Stock, S.P., Campbell, J.F. and Nadler, S.A. (2001) Phylogeny of *Steinernema* Travassos, 1927 (Cephalobina: Steinernematidae) inferred from ribosomal DNA sequences and morphological characters. *Journal of Parasitology* 87, 877–889.
Stock, S.P., Griffin, C.T. and Burnell A.M. (2002) Morphological characterization of three isolates of *Heterorhabditis* Poinar, 1976 from the 'Irish group' (Nematoda: Rhabditida: Heterorhabditidae) and additional evidence supporting their recognition as a distinct species, *H. downesi n. sp. Systematic Parasitology* 51, 95–106.
Stock, S.P., Griffin, C.T. and Chaerani, R. (2004) Morphological and molecular characterisation of *Steinernema hermaphroditum* n. sp. (Nematoda: Steinernematidae), an entomopathogenic nematode from Indonesia, and its phylogenetic relationships with other members of the genus. *Nematology* 6, 401–412.
Stock, S.P., Rivera-Orduño, B. and Flores-Lara, Y. (2009) *Heterorhabditis sonorensis* n. sp. (Nematoda: Heterorhabditidae), a natural pathogen of the seasonal cicada *Diceroprocta ornea* (Walker) (Homoptera: Cicadidae) in the Sonoran Desert. *Journal of Invertebrate Pathology* 100, 175–184.
Stock, S.P., Campos-Herrera, R., El-Borai, F.E. and Duncan, L.W. (2019) *Steinernema khuongi* n. sp. (Panagrolaimomorpha, Steinernematidae), a new entomopathogenic nematode species from Florida, USA. *Journal of Helminthology* 93, 226–241.
Stokwe, N.F., Malan, A.P., Nguyen, K.B., Knoetze, R. and Tiedt, L. (2011) *Steinernema citrae* n. sp. (Rhabditida: Steinernematidae), a new entomopathogenic nematode from South Africa. *Nematology* 13, 569–587.
St-Onge, M., LaRue, B. and Charpentier, G. (2008) A molecular revision of the taxonomic status of mermithid parasites of black flies from Quebec (Canada). *Journal of Invertebrate Pathology* 98, 299–306.
Sturhan, D., Spiridonov, S. and Mráček, Z. (2005) *Steinernema silvaticum* sp. n. (Rhabditida: Steinernematidae), a new entomopathogenic nematode from Europe. *Nematology* 7, 227–241.
Subbotin, S.A. (2021) Molecular identification of nematodes using polymerase chain reaction (PCR). In: *Techniques for Work with Plant and Soil Nematodes*. CAB International, Wallingford, UK, pp. 218–239.
Sudhaus, W. (1974) Zur Systematik, Verbreitung, Ökologie und Biologie neuer und wenig bekannter Rhabditiden (Nematoda). 1. Teil. *Zoologische Jahrbucher, Abteilung-fur Systematik, Ökologie und Geographie der Tiere.* 101, 417–465.
Szalanski, A.L., Taylor, D.B. and Mullin, P.G. (2000) Assessing nuclear and mitochondrial DNA sequence variation within *Steinernema* (Rhabditida: Steinernematidae). *Journal of Nematology* 32, 229–233.
Tallosi, B., Peters, A. and Ehlers, R.-U. (1995) *Steinermea bicornutum* sp. n. (Rhabditida: Steinernematidae) from Vojvodina, Yugoslavia. *Russian Journal of Nematology* 3, 71–80.
Tamiru, T., Waeyenberge, L., Hailu, T., Ehlers, R.U., Půža, V. and Mráček, Z. (2012) *Steinernema ethiopiense* sp. n. (Rhabditida: Steinernematidae), a new entomopathogenic nematode from Ethiopia. *Nematology* 14, 741–757.
Tarasco, E., Mráček, Z., Nguyen, K.B. and Triggiani, O. (2008) *Steinernema ichnusae* sp. n. (Nematoda: Steinernematidae) a new entomopathogenic nematode from Sardinia Island (Italy). *Journal of Invertebrate Pathology* 99, 173–185.
Tian, C.L., Zhu, F., Li, X.Y., Zhang, J.H., Půža, V., *et al.* (2022) *Steinernema populi* n. sp. (Panagrolaimomorpha, Steinernematidae), a new entomopathogenic nematode species from China. *Journal of Helminthology* 96: e57.
Tixier, M.S., Tsolakis, H., Ragusa, S., Poinso, A., Ferrero, M., *et al.* (2011) Integrative taxonomy demonstrates the unexpected synonymy between two predatory mite species: *Cydnodromus idaeus* and *C. picanus* (Acari: Phytoseiidae). *Invertebrate Systematics* 25, 273–281.
Tobias, Z.J.C. (2017) The evolutionary genetics of two host–parasite systems on New Zealand's South Island. PhD thesis, University of Otago, Dunedin, New Zealand.
Tong, X., Kanzaki, N. and Akimoto, S.I. (2021) First record of a mermithid nematode (Nematoda: Mermithidae) parasitizing winged females of gall-forming aphids (Hemiptera: Aphididae: Eriosomatinae). *Entomological Science* 24, 354–360.
Travassos, L. (1927) Sobre o genero Oxysomatium. *Boletim Biologico, Sao Paulo* 5, 20–21.
Triggiani, O., Mráček, Z. and Reid, A. (2004): *Steinernema apuliae* sp. n. (Rhabditida: Steinernematidae): a new entomopathogenic nematode from southern Italy. *Zootaxa* 460, 1–12.

Tropdi, A.D. and Strange, J.P. (2018) Rarely reported, widely distributed, and unexpectedly diverse: molecular characterization of mermithid nematodes (Nematoda: Mermithidae) infecting bumble bees (Hymenoptera: Apidae: Bombus) in the USA. *Parasitology* 145, 1558–1563.

Uribe-Lorío, L., Mora, M. and Stock, S.P. (2007) *Steinernema costaricense* n. sp. and *S. puntauvense* n. sp. (Rhabditida: Steinernematidae), two new entomopathogenic nematodes from Costa Rica. *Systematic Parasitology* 68, 67–182.

van Megen, H., van den Elsen, S., Holterman, M., Karssen, G., Mooyman, P., *et al.* (2009) A phylogenetic tree of nematodes based on about 1200 full-length small subunit ribosomal DNA sequences. *Nematology* 11, 927–950.

Wang, H., Jung, Y.H., Son, D. and Choo, H.Y. (2013) High level of genetic diversity among *Steinernema monticolum* in Korea revealed by single enzyme amplified fragment length polymorphism. *Journal of Invertebrate Pathology* 113, 146–151.

Watanabe, S., Tsunashima, A., Itoyama, K. and Shinya, R. (2021) Survey of mermithid nematodes (Mermithida: Mermithidae) infecting fruit-piercing stink bugs (Hemiptera: Pentatomidae) in Japan. *Applied Entomology and Zoology* 56, 27–39.

Waturu, C.N., Hunt, D.J. and Reid, A.P. (1997) *Steinernema karii* sp. n. (Nematoda: Steinernematidae) a new entomopathogenic nematode from Kenya. *International Journal of Nematology* 7, 68–75.

Webster, J.M. and Thong, C.H.S. (1984) Nematode parasites of orthopterans. In: Nickle, W.R. (ed.) *Plant and Insect Nematodes*. Marcel Dekker, New York, pp. 697–726.

Weiser, J. (1955) *Neoaplectana carpocapsae* n. sp. (Anguillulata, Steinernematidae) novy cizopasnik housenek obalace jablecnebo, *Carpocapsae pomonella* L. *Vestnik Česká Společnost Zoologicka* 19, 44–52.

Whitkus, R., Doebley, J. and Wendel, J.F. (1994) Nuclear DNA markers in systematics and evolution. In: Phillips, L. and Vasil, I.K. (eds.) *DNA-based Markers in Plants.* Kluwer Academic Publishers, Dordrecht, Netherlands, pp. 116–141.

Wilson, M.J. and Gaugler, R. (2000) Terrestrial mollusk pests. In: Lacey, L. and Kaya, H.K. (eds.) *Field Manual of Techniques in Invertebrate Pathology*. Springer, Dordrecht, Netherlands, pp. 787–804.

Wilson, M.J., Glen, D.M. and George, S.K. (1993) The rhabditid nematode *Phasmarhabditis hermaphrodita* as a potential biological control agent for slugs. *Biocontrol Science and Technology* 3, 503–511.

Wilson, M.J., Glen, D.M., George, S.K., Pearce, J.D. and Wiltshire, C.W. (1994) Biological control of slugs in winter wheat using the rhabditid nematode *Phasmarhabditis hermaphrodita. Annals of Applied Biology* 125, 377–390.

Wilson, M.J., Burch, G., Tourna, M., Aalders, L.T. and Barker, G.M. (2012) The potential of a New Zealand strain of *Phasmarhabditis hermaphrodita* for biological control of slugs. *New Zealand Plant Protection* 65, 161–165.

Wood, P.L. Jr., Guo, X., Travers, S.L., Su, Y.-C., Olson, K.V. *et al.* (2020) Parachute geckos free fall into synonymy: gekko phylogeny, and a new subgeneric classification, inferred from thousands of ultraconserved elements. *Molecular Phylogenetics and Evolution* 146, 1–11.

Wouts, W.M., Mráček, Z., Gerdin, S. and Bedding, R.A. (1982) *Neoaplectana* Steiner, 1929 a junior synonym of *Steinernema* Travassos, 1927 (Nematoda; Rhabditida). *Systematic Parasitology* 4, 147–154.

Xu, Z., Wang, G. and Li, X. (1991) [A new species of the genus *Steinernema* (Rhabditida: Steinernematidae).] *Zoological Research* 12, 17–20 (in Chinese).

Yaari, M., Doron-Faigenboim, A., Koltai, H., Salame, L. and Glazer, I. (2016) Transcriptome analysis of stress tolerance in entomopathogenic nematodes of the genus *Steinernema. International Journal for Parasitology* 46, 83–95.

Yeates, G.W. and Buckley, T.R. (2009) First records of mermithid nematodes (Nematoda: Mermithidae) parasitising stick insects (Insecta: Phasmatodea). *New Zealand Journal of Zoology* 36, 35–39.

Yoshida, M. (2004) *Steinernema litorale* n. sp. (Rhabditida: Steinernematidae), a new entomopathogenic nematode from Japan. *Nematology* 6, 819–838.

Yoshino, H. and Waki, T. (2021) First report on Mermithidae (Mermithida) infection in *Ligidium* sp. (Isopoda, Ligiidae). *Parasitology International* 82: 102304.

Zeng, Y., Ye, W., Kerns, J., Tredway, L., Martin, S. and Martin, M. (2015) Molecular characterization and phylogenetic relationships of plant-parasitic nematodes associated with turfgrasses in North Carolina and South Carolina, United States. *Plant Disease* 99, 982–993.

Zhang, C.N. and Liu, Q.Z. (2020) *Phasmarhabditis zhejiangensis* sp. nov. (Nematoda: Rhabditidae), a new rhabditid nematode from Zhejiang, China. *PLoS One* 15: e0241413.

Zhang, C., Liu, J., Xu, M., Sun, J., Yang, S., *et al.* (2008) *Heterorhabditidoides chongmingensis* gen. nov., sp. nov. (Rhabditida: Rhabditidae), a novel member of the entomopathogenic nematodes. *Journal of Invertebrate Pathology* 98, 153–168.

Zhang, K., Baiocchi, T., Lu, D., Chang, D.Z. and Dillman, A.R. (2019) Differentiating between scavengers and entomopathogenic nematodes: which is *Oscheius chongmingensis*? *Journal of Invertebrate Pathology* 167: 107245.

Zhang, P., Zhang, W. and Hu, S. (2022) Fungivorous nematode *Aphelenchus avenae* and collembola *Hypogastrura perplexa* alleviate damping-off disease caused by *Pythium ultimum* in tomato. *Plant and Soil* 482, 175–189.

2 The Biology of Entomopathogenic Nematodes

Ernesto San-Blas[1]*, Yoelvis Sulbaran[1] and Gabriela Lankin[2]
[1]*Institute of Agri-food, Animal and Environmental Sciences (ICA3), Universidad de O'Higgins, Campus Colchagua, Chile;* [2]*Department of Plant Protection, Faculty of Agricultural Sciences, Universidad de Chile, Chile*

Abstract

Despite the recent knowledge on entomopathogenic nematodes (EPNs), there are still biological aspects that should be considered to fully understand these peculiar organisms, which live in diverse and changing environments. EPNs and their symbiont bacteria should be taken as holobionts, where the roles and functions that each symbiont plays are fundamental to a good performance as a biological control agent. We propose that EPNs must face three different and complex environments during their life cycle: the soil, the living host and the dead host. We present detailed aspects of their life cycle. Their biological and ecological success in relation to the symbiotic interaction with the bacteria and adapting to the challenges imposed by each environment indicates the enormous plasticity and adaptability of these organisms. Thus, considering these important features in future studies will allow us to optimize EPN efficacy as biological control agents under field conditions.

2.1 Introduction

The biology of entomopathogenic nematodes (EPNs) may seem quite simple at first glance: they are soil-living organisms that invade and kill an insect host, reproducing inside the cadaver to produce a new generation that abandon the cadaver searching for the next host to invade. In a strict sense, EPNs do exactly this in nature, in laboratory conditions and even on the International Space Station (Kaplan *et al.*, 2020b). However, this oversimplified life cycle is a complex process that combines a plethora of biological and environmental factors.

Traditionally, EPNs have been considered within the families Steinernematidae and Heterorhabditidae, with three representative genera: *Steinernema* (102 species), *Neosteinernema* (one species) and *Heterorhabditis* (22 species) (Hunt and Nguyen, 2017; Koppenhöfer *et al.*, 2020; Hazir *et al.*, 2022). Despite their common ability to kill insects with the help of symbiotic bacteria, these two families evolved independently from different ancestors (Blaxter *et al.*, 1998). This obligated symbiosis gives the name "entomopathogenic" to this group of nematodes. The most important stage for biocontrol is the infective juvenile (IJ), a specialized stage capable of surviving and persisting in the soil under harsh conditions. This is the only stage capable of surviving in nature outside an insect cadaver, locating, invading and killing new hosts to continue the life cycle (Fig. 2.1).

*Email: esanblas@uoh.cl or esanblas@yahoo.com

 Entomopathogenic Nematodes as Biological Control Agents (D.I. Shapiro-Ilan and E.E. Lewis eds)
DOI: 10.1079/9781800620322.0002

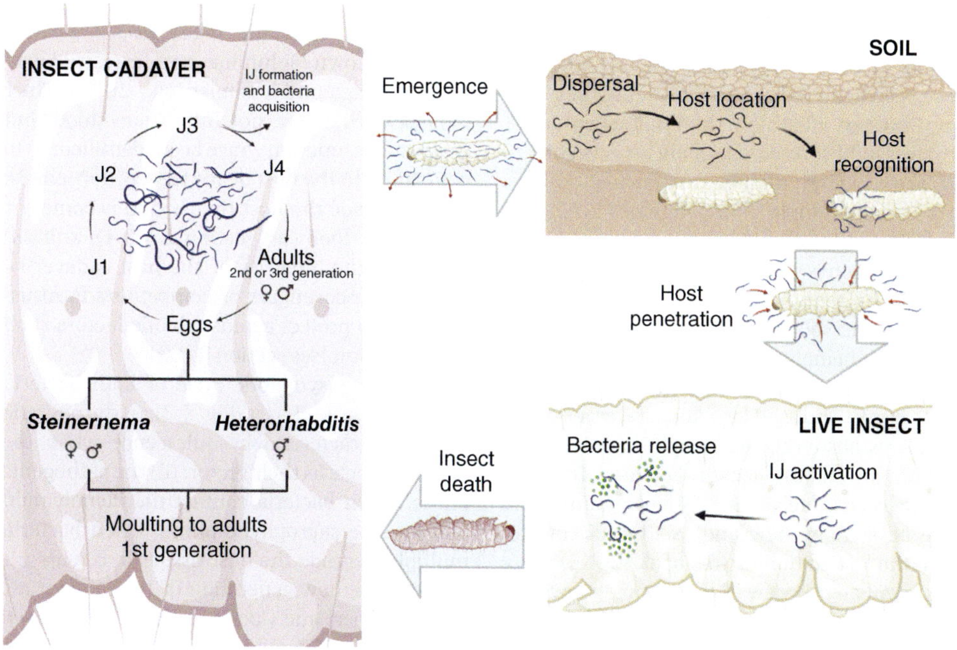

Fig. 2.1. Life cycle of entomopathogenic nematodes (EPNs). During their life cycle, EPNs move across three interfaces (arrows) to three environments. (i) In the soil, infective juveniles (IJs) must follow physical and chemical cues to find and recognize an adequate host. (ii) After penetration into the live host, IJs activate and release the symbiotic bacteria carried in their intestine. In this environment, among other functions, the bacteria reproduce, becoming food for the nematodes and killing the host. (3) Within the dead host, IJs molt to first-generation adults and reproduce during one, two or three generations (depending on the host size) until the resources are depleted and the cadaver is overcrowded. Then they reassociate with the bacteria, emerge from the cadaver and disperse to start the search for new hosts. Figure created using Biorender.com.

A recent debate has arisen to consider other rhabditids that also kill insects, such as *Oscheius* sp. and *Caenorhabditis briggsae* (e.g. Abebe *et al.*, 2010; Pervez *et al.*, 2013; Serepa-Dlamini and Gray, 2018). Dillman *et al.* (2012a) revised the definition of "entomopathogeny", defining six essential parameters: (i) the IJ must carry pathogenic bacteria; (ii) it must locate their host actively and infect them by entering into the hemocoel; (iii) it must efficiently release the bacteria into the insect hemolymph; (iv) this must result in the rapid death of the insect (i.e. 24–120 h) followed by bacterial proliferation and nematode reproduction driven by cadaver-nutrient utilization; (v) new generations of IJs must reassociate with pathogenic bacteria; and (vi) they emerge from the nutrient-depleted cadaver to search for new hosts. While the discussion continues, this chapter is focused on the biology of *Steinernema* and *Heterorhabditis* nematodes.

Some EPNs are relatively host specific, whereas others have a broader host range. For example, *Steinernema feltiae*, *Steinernema carpocapsae* and *Heterorhabditis indica* infect many insect species from several families and orders (Peters, 1996; San-Blas *et al.*, 2019), while *Steinernema scapterisci* primarily infects orthopterans, and *Steinernema scarabaei* and *Steinernema kushidai* prefer scarab larvae (Mamiya, 1988; Nguyen and Smart, 1991). Nevertheless, almost all described EPNs have been isolated from the soil, using the highly susceptible insect *Galleria mellonella* (Lepidoptera: Pyralidae) as bait, and it is likely that many other host-specific undescribed nematode species will be isolated from soil.

Due to their remarkable aptitudes, EPNs have been produced commercially since 1981 (Poinar and Grewal, 2012), and many advances have been achieved by understanding their biology. However, after more than half a century of progress in EPN research, many aspects governing their biology remain unknown (San-Blas, 2013). For example, how long does an EPN IJ live? The answer to this very basic and even naive question is still uncertain. IJs often survive for years in laboratories, and for months, maybe years, in the soil. However, after reaching the host insect hemolymph, the EPNs lifespan is reduced to hours or days. In fact, the life cycle from egg to egg, excluding IJ formation, takes just a few days. Thus, answering this "simple" question could become an endless debate. Fortunately, every day we learn more about EPN biology, and there have been many new and exciting discoveries since the last edition of this book (2005). Here we present an overview of the current knowledge of the biology of EPNs.

2.2 The EPN–Bacteria Complex: a Holobiont

One of the fundamental features of EPNs is that these "beings" cannot be considered as separate individuals but rather as holobionts (Aldana and Robeva, 2021). They harbor and vector obligate symbiotic bacteria, which, in turn, help the nematode to carry out many different functions (Mucci *et al.*, 2022). EPNs without bacteria do not occur naturally, and the bacteria are not capable of surviving in the soil outside either the IJ or the insect host. This interaction, with multiple and complex emerging traits, is exactly what makes them such successful entomopathogenic organisms.

2.2.1 The symbiotic bacteria

Photorhabdus and *Xenorhabdus* spp., from the class γ-Proteobacteria, family Morganellaceae (Adeolu *et al.*, 2016) (Enterobacteriales), are Gram-negative, non-spore-forming, facultative anaerobes, and are rod-shaped with peritrichous flagella (Fischer-Le Saux *et al.*, 1999). *Photorhabdus* and *Xenorhabdus* spp. associate with *Heterorhabditis* and *Steinernema* spp., respectively. Additionally, *Photorhabdus* spp. are the only known bioluminescent terrestrial bacteria (Sajnaga and Kazimierczak, 2020). Their biological roles are not fully understood, but three major functions have been identified: (i) to contribute to the host death (see section 2.3.3); (ii) to degrade the host tissues and become the nematodes' food (see section 2.3.2.4); and (iii) to protect the integrity of the host cadaver by preventing scavengers or competitors from using it, or to protect against abiotic factors such as desiccation (see section 2.3.3).

The endosymbiotic cycle of both bacteria comprises three phases (Fig. 2.2): (i) the phoretic phase, characterized by quiescence when the nematode acts as their vector; (ii) the pathogenic phase, when bacteria contact the hemolymph; and (iii) the saprophytic phase, when bacteria multiply, degrade the host tissues and reassociate with the new generation of IJs (Ciche *et al.*, 2006; Herbert and Goodrich-Blair, 2007; Stock and Blair, 2008). Until recently, the EPN–bacteria interaction was thought to be very specific. Now we know that, while *Steinernema* spp. associate with only one species of *Xenorhabdus* (although a bacterial species can associate with more than one EPN species) (Lee and Stock, 2010), some *Heterorhabditis* spp. can associate with more than one species (or subspecies) of *Photorhabdus* spp. (Maher *et al.*, 2021).

2.3 The EPN Life Cycle

During their life cycle, EPNs experience dramatic changes in their bodies: their metabolism activates, they become adults, mate, lay eggs, etc. These changes respond to the need to produce offspring, and also to survive three very different environments (Fig. 2.1): (i) the IJs must move through the soil, locate and infect a potential host, and recognize and penetrate it; (ii) within the living insect, IJs must reach the hemolymph, release the bacterial symbiont and overcome the insect's immune response to finally kill it; and (iii) in the host cadaver, the tissues are degraded by the bacteria (which become the food of the EPNs), the nematodes molt to adults, reproduction begins and new generations of nematodes occur.

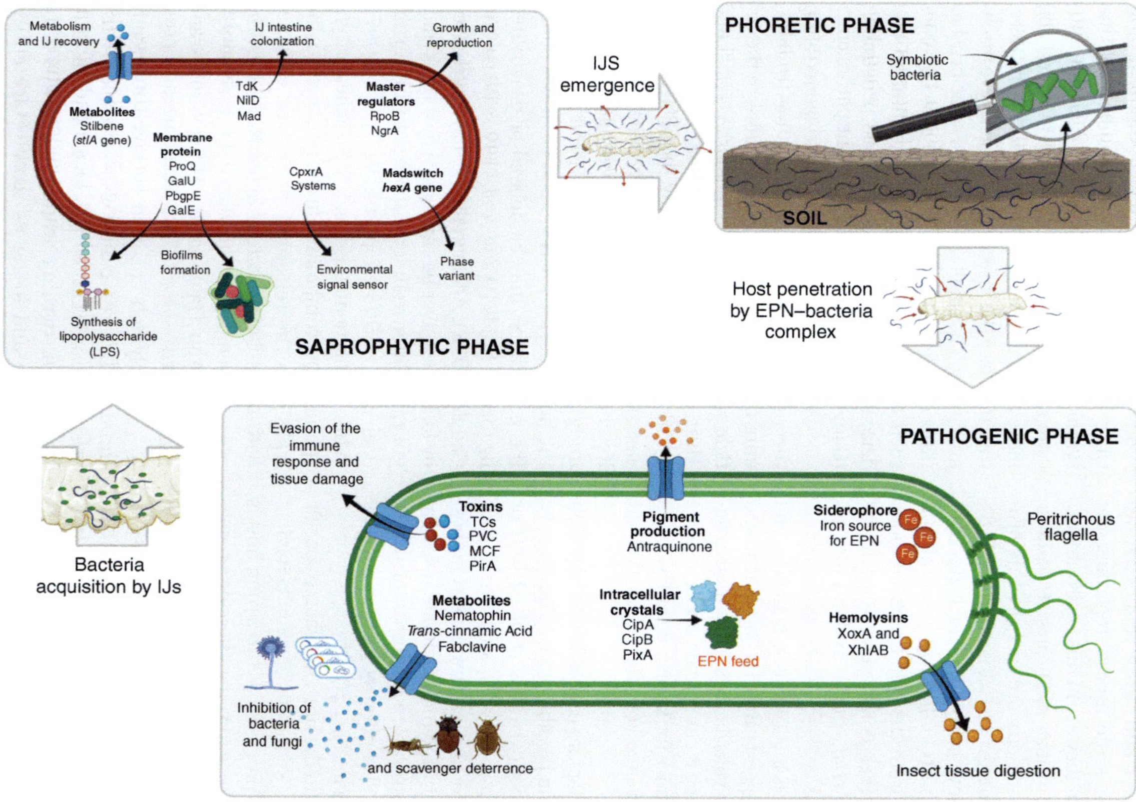

Fig. 2.2. Life cycle of the symbiotic bacteria. The symbiotic bacteria go through three phases during the infective cycle. (i) During the phoretic phase, they are vectored in the intestine of the infective juveniles (IJs) from the soil to the hemocoel of the insect host. (ii) Once released in the hemocoel, bacteria enter the pathogenic phase, where they serve several important functions for the entomopathogenic nematode (EPN)–bacteria holobiont: production of food for the EPN (iron-containing siderophores, intracellular crystals), toxins to evade the host immune response and help with the process of host tissue degradation, metabolites to inhibit the growth of saprophytic bacteria and fungi and deter scavenger microorganisms, hemolysins for insect tissue digestion, and pigments for an aposematic coloration of the insect cadaver. (iii) During the saprophytic phase, bacteria have a whole set of new functions, such as attaching to the intestine, production of biofilms and establishment of symbiosis. Figure created using Biorender.com.

2.3.1 Key processes in the soil

To achieve successful infection, IJs must disperse from the cadaver, search and forage to locate a potential host, and recognize it once they have made contact with it. All of these concepts are well described in Chapters 6 and 7 (this volume), so we will briefly discuss some aspects regarding the biology of IJs when they are in this environment.

IJs abandon their natal insect cadaver when certain conditions are met. In this process, known as emergence (see section 2.3.3.8), the IJs have to move away from the cadaver to search for new hosts, following physical and chemical stimuli from the soil, the insect host, the plants and other conspecifics (largely explained by Stuart *et al.*, 2015, and in Chapters 5–7, this volume).

Host seeking is a very plastic behavior, and some evidence suggests that IJs can learn, and memorize specific volatile cues in their environment. Their learning is cue specific, and it appears to increase preference for known volatiles, while decreasing preference for new ones (Willett *et al.*, 2015).

A myriad of factors may affect the decision of IJs to attack the possible hosts. They can be attracted to insects (Kruitbos *et al.*, 2009) or other non-host invertebrates, such as isopods (Eng *et al.*, 2005) and earthworms (Campos-Herrera *et al.*, 2006), which facilitate dispersal to new and more favorable environmental niches, a phenomenon called phoresy, gaining survival advantages (Dillman *et al.*, 2012b).

When closer to the host, IJs react to specific host signals that stimulate certain behaviors and physiological processes. If the wrong host is chosen, the nematode may die; therefore, recognition is key for success in penetrating the host's body (Lewis *et al.*, 2006). These physical and chemical stimuli may come from the host cuticle, gut content, feces or other volatile by-products of the insect's metabolism (Grewal *et al.*, 1993a,b, 1997).

Some nematodes can recognize hosts that have already been infected by conspecifics, and a massive attack may facilitate overcoming host defenses (Peters and Ehlers, 1997). However, overcrowded hosts may deter IJs by avoiding competition for resources and inbreeding (Koppenhöfer and Kaya, 1995; Ryder and Griffin, 2002).

2.3.2 Key processes inside the living host

2.3.2.1 Penetration and reaching hemolymph

After recognition, the attack by the IJs begins. Once the nematodes enter the hemocoel, the symbiotic bacteria are released, resulting in septicemia and toxemia. First, the IJ attaches and moves along the insect cuticle to find entry. Here, they may face the insect's passive defensive mechanisms such as small spiracles and setae brushes (Koppenhöfer *et al.*, 2007, 2011) or active ones such as grooming, chewing, regurgitation and defecation, among others, to avoid penetration (Gaugler, 1988; Gaugler *et al.*, 1994).

As the reader may conclude, penetration through the host's natural openings (mouth, anus or spiracles) or by breaking through the cuticle at the intersegmental membranes (Eidt and Thurston, 1995; Simões and Rosa, 1996) is not a risk-free process. IJs that enter the host's gastrointestinal system must be able to tolerate its chemical environment and penetrate the midgut or the Malpighian tubules toward the hemocoel (Sicard *et al.*, 2004). At this point, the penetration may be delayed or deactivated by the gut fluids (Wang *et al.*, 1995; Wang and Gaugler, 1998). Additionally, IJs may be removed by food passage or be challenged by a thick midgut epithelium, as in some species of white grubs (Cui *et al.*, 1993; Koppenhöfer *et al.*, 2007). While steinernematids break the tracheal or gut epithelia by combining direct pressure to the walls with the release of proteolytic enzymes with a histolytic effect, in heterorhabditids, the process seems to be mediated by mechanical rupture (Abu Hatab *et al.*, 1995; Simões and Rosa, 1996).

Hemolymph can also be reached directly through the host cuticle. For example, *Heterorhabditis bacteriophora* penetrates with a drill-like motion, twisting around its focal point of anchorage and losing its second-stage cuticle (becoming exsheathed) (Dziedziech *et al.*, 2020). In cuticles of *Tipula* spp. exposed to *S. feltiae*, some small rounded holes have been observed (Peters and Ehlers, 1994), so it is probable that *Steinernema* spp. excrete the same proteolytic enzymes used to break the midgut.

2.3.2.2 IJ activation

The IJ, a diapause-like stage, is in an arrested development stage, which irreversibly changes by specific stimuli. After reaching the hemolymph, IJs open their digestive system to release the symbiotic bacteria (Strauch and Ehlers, 1998), resuming their life cycles. In this dramatic process, many morphological and physiological changes occur (Lu *et al.*, 2017a), including the stoma (mouth) opening, exsheathment (*Heterorhabditis* spp.), and a strong activity of the pharyngeal gland cells and the genital primordia (Dolan *et al.*, 2002). For this reason, the term "IJ activation" has been suggested instead of "recovery," which is based on *Caenorhabditis elegans* dauer larvae (Golden and Riddle, 1982; Lu *et al.*, 2017a). In this process, three activation stages can be recognized in the IJs: (i) non-activated, with no observable pharyngeal bulb (generally invisible) and closed stoma; (ii) partially activated, with open stoma and partially expanded pharyngeal bulb, but with no pumping; and (iii) fully activated, with a fully expanded pharyngeal bulb and discernible pumping (Lu *et al.*, 2017a).

The mechanisms governing this process are not fully understood, but it may be host specific (Alonso *et al.*, 2018). For example, *S. scapterisci* more readily activates in cricket tissue homogenate compared with the tissues of other host species (Lu *et al.*, 2017b). Additionally, *S. carpocapsae* and *S. feltiae* activate differentially in silkworm (*Bombyx mori*) and mealworm (*Tenebrio molitor*) tissues but similarly in superworms (*Zophobas morio*) (Alonso *et al.*, 2018).

Other stimuli can have major (factors) or minor (cofactors) roles in IJ activation (Kikuta *et al.*, 2009) or food signals (Strauch and Ehlers, 1998). EPNs activate within hours of reaching the hemolymph (Strauch and Ehlers, 1998; Nagata *et al.*, 2008). Kikuta *et al.* (2008, 2009) demonstrated that *S. carpocapsae* was activated in *Spodoptera frugiperda* cell cultures but only in the presence of their bacterial symbionts (Strauch and Ehlers, 1998). However, activation was more effective when the IJs were exposed to *B. mori* hemolymph, suggesting the existence of additional activation factors (Kikuta *et al.*, 2009). The nature of these signaling compounds remains unknown; an electronegative compound of <700 Da was isolated from bacterial exudates (by an *in vitro* process) of *Photorhabdus luminescens* and proved to be effective as an IJ activation factor for *H. bacteriophora*, but its identification remains unsolved (Aumann and Ehlers, 2001). Other cofactors that prompt IJ activation include carbon dioxide (Jessen *et al.*, 2000) and ammonia (San-Blas *et al.*, 2014).

2.3.2.3 Bacterial release

Later in the IJ activation process, the symbiotic bacteria are released. In *Steinernema* spp., bacteria are released by defecation (Goodrich-Blair, 2007; Snyder *et al.*, 2007) and in *Heterorhabditis* spp. by regurgitation (Ciche and Ensign, 2003).

In *Steinernema* spp., the release of bacteria is an active process initiated by the nematode's pharynx. When IJs of *S. carpocapsae* contact the hemolymph, their esophagus pumping activates, helping the bacteria to depart from the receptacle, the distal region of which expands, allowing the bacteria to pass toward the intestine. The peristaltic movements of the nematode's intestine then help release the bacteria (Snyder *et al.*, 2007). In contrast, in *Heterorhabditis* spp., the bacteria gradually migrate to the pharynx to exit through the mouth after 60 min. Apparently, a low-molecular-weight, heat- and protease-stable, anionic component in the hemolymph induces regurgitation (Ciche and Ensign, 2003).

2.3.2.4 Killing the host

Both the nematode and the bacteria participate in killing the host. Host immune suppression is described in Chapter 3 (this volume), but here we describe the biological aspects of this process. The exact role of bacteria and nematodes in overcoming the immune response of the host is not fully understood (Koppenhöfer *et al.*, 2020), as only a small number of nematode–bacteria combinations in a few hosts have been evaluated. In *Photorhabdus–Heterorhabditis* combinations, the bacteria play the major role in killing the host, while in *Steinernema–Xenorhabdus* combinations the nematodes play a more active role in the virulence (Lu *et al.*, 2017a).

THE KILLING POTENTIAL OF NEMATODES. The general concept is that the symbiotic bacteria are solely responsible for producing septicemia and/or toxemia to kill the host. However, it is also known that axenic steinernematid IJs (nematodes with no symbionts) kill insects (Boemare *et al.*, 1982; Burman, 1982; Ehlers *et al.*, 1997). Recently, their role in the killing machine was demonstrated by transcriptome analysis (Lu *et al.*, 2017a). Activated *S. carpocapsae* secreted lethal venom proteins that were highly toxic to insect larvae, paralyzing and eventually killing them. Proteins isolated from axenic IJs corroborated their origin and were only produced after IJ activation; therefore, these are different from the histolytic proteins secreted by the nematodes to enter the hemocoel (Lu *et al.*, 2017a).

In the case of *Heterorhabditis* spp., their ability to kill the host without their symbionts is unclear. Some attempts to obtain heterorhabditid IJs with non-specific *P. luminescens* as the symbiont resulted in IJs incapable of retaining the bacteria and of killing *G. mellonella* larvae, suggesting an inability to produce entomotoxins (Han *et al.*, 1991; Gerritsen and Smits, 1993; Han and Ehlers, 2000). However, more recently, axenic heterorhabditid IJs capable of killing *Drosophila melanogaster* larvae were reported (Hallem *et al.*, 2007).

The release of bacteria by the IJs can take from 30 min to 6 h after contacting the hemolymph (Wang *et al.*, 1995; Bowen *et al.*, 1998). Thus, they must evade or suppress the host immune response until the bacteria are released (Brivio *et al.*, 2002). For this, IJs of both genera can produce cuticular lipids, proteins and other unknown compounds that enable them to evade the host immune system or to escape encapsulation or melanization by hemocytes (Mastore and Brivio, 2008; Brivio *et al.*, 2018; see Chapter 3, this volume).

THE ROLE OF BACTERIA. A fundamental role is played by *Xenorhabdus* and *Photorhabdus* spp. in evading the host's immune response and generating sepsis and/or toxemia leading to its death. During infection, the bacteria activate their genetic machinery, a repertoire of cellular mechanisms and virulence factors necessary for colonization. At the start of infection, *Xenorhabdus* spp. occupy two specific places inside the host. In the larvae of *Spodoptera littoralis*, they have been found in hemolymph plasma and connective tissues (among muscle fibers) of their anterior midgut. In *Locusta migratoria*, after the injection of *Xenorhabdus* spp., there was no colonization of the midgut, but rather within the hematopoietic organ (the site of hemocyte production) and hemolymph (Sicard *et al.*, 2004).

The insecticidal activity of both bacteria relates to the production of specific proteins and secondary metabolites (Bode, 2009; Sheets *et al.*, 2011; Rosa da Silva *et al.*, 2017), such as oxindole and benzylideneacetone, toxins that induce immunosuppression and destroy the epithelial cells of the midgut of the hosts (Park and Kim, 2000; Seo *et al.*, 2012). In the case of *Xenorhabdus* spp., the bacteria produce metabolites with insecticidal activity (Akhurst, 1982; Sergeant *et al.*, 2006; Bode, 2009; Fuchs *et al.*, 2014). Additionally, *Photorhabdus* spp. produce toxins such as MCF (Makes Caterpillar Floppy) and PVC (*Photorhabdus* Virulence Cassettes) that promote apoptosis of hemocytes (Rodou *et al.*, 2010), dimeric PirAB (insect-related protein) toxin with cytotoxic effects (Rodou *et al.*, 2010) and mono-ADP-ribosyltransferase (mARTs) that destabilize the cytoskeleton of the insect cells (Rodou *et al.*, 2010; Aktories *et al.*, 2014).

2.3.3 Key processes inside the dead host

2.3.3.1 The bacteria as food source

Both bacteria secrete proteases, lipases, hemolysins and chitinases that change the insect cadaver into a usable substrate (Forst *et al.*, 1997; Richards and Goodrich-Blair, 2009). EPNs feed directly on the bacterial biomass growing in the host cadaver, indicating that the symbionts provide specific nutrients to nematodes during their development (Akhurst, 1982; Liu *et al.*, 2000; Adams *et al.*, 2006). However, their complete nutrition cycle has not yet been elucidated, and few mechanisms have been described.

Among those mechanisms described, bacterial siderophores (which facilitate the uptake of iron) are essential for the development of both the bacteria and the nematode (Postle and Larsen, 2007; Watson *et al.*, 2010). However, in the stationary phase, both bacteria produce intracellular protein crystals with high levels of essential amino acids to nourish the nematode (Bintrim and Ensign, 1998; Bowen and Ensign, 2001).

2.3.3.2 Molting to adults and reproduction

As the nematodes start feeding on the bacteria, the development of the fourth larval stage (also called preadults) and the adults occurs. *Xenorhabdus nematophila* feeds on the insect tissues and then the *S. carpocapsae* nematodes feed solely on the bacteria or by endotokia matricida (EM; see section 2.3.3.4) (Mucci *et al.*, 2022). The development of preadult gonads starts, and reproduction takes place, and while first-generation IJs in heterorhabditids become self-fertile hermaphrodite adults, in *Steinernema* spp. they develop into males or females (Poinar, 1990), except for *Steinernema hermaphroditum*, which reproduces by hemaphroditism in its first generation (Griffin *et al.*, 2001). In successive generations, both genera produce males and females (although *Heterorhabditis* spp. also keep producing hermaphrodites) (Koltai *et al.*, 1995; Wang and Bedding, 1996; Johnigk and Ehlers, 1999; Griffin, 2012).

Thus, *Heterorhabditis* spp. reproduce with a single IJ, while *Steinernema* spp., except for *S. hermaphroditum*, need at least one male and one female (Downes and Griffin, 1996); moreover, their sex ratio is not always 1:1. Sex in EPNs is determined genetically (Bohan and Hominick, 1997), and several species such as *Steinernema affine* are markedly female biased (about 90%) (Bastidas *et al.*, 2014), whereas others, such as *S. feltiae*, *S. riobrave* and *S. glaseri*, tend to be moderately biased to females or relatively balanced (Stuart *et al.*, 1998; Alsaiyah *et al.*, 2009; Bastidas *et al.*, 2014). To ensure that both sexes are present in the dead host, the general "male colonizing hypothesis" was proposed by Grewal *et al.* (1993c). Although controversial, because colonization patterns vary among species (e.g. Alsaiyah *et al.*, 2009; Griffin, 2012; Bastidas *et al.*, 2014), we now know that this occurs only in highly female-biased species/strains.

2.3.3.3 Mating

Copulation also differs among the genera. *Steinernema* males coil around the female at the vulva region, whereas in *Heterorhabditis* spp., the male and female align their bodies parallel, with their heads pointing in opposite directions. The male–female encounters are not driven randomly, and sexual maturity of males may depend on female presence (Ebssa *et al.*, 2008). In *S. carpocapsae*, the production of attractive pheromones by virgin females mediates the male's approach (Neves *et al.*, 1998). In fact, fecundated females are less attractive, probably due to a reduction in pheromone production (Lewis *et al.*, 2002). These chemical diffusates also have a strong effect on male development. For instance, *Steinernema longicaudum* growing alone is incapable of maturing sexually and producing sperm, which does happen when a female is present, even when separated by a permeable membrane (Ebssa *et al.*, 2008).

Fierce fighting between males competing for sexual partners can result in death of the rivals, especially among first-generation males, as the small invading population allows monopolizing of the females (very unusual in second and/or third generations) (Zenner *et al.*, 2014). Additionally, in *S. longicaudum*, the males are able to recognize their kin, and the fighting rate decreases among relatives (Kapranas *et al.*, 2016). In contrast, in mixed infections by two *Steinernema* spp., the males of one species kill females and males of the other (O'Callaghan *et al.*, 2014).

2.3.3.4 Endotokia matricida

EM is the intrauterine development of juveniles, which concludes with the death of the mother nematode. Although more frequent in heterorhabditids than in steinernematids, this adaptive strategy is common in both genera and in other rhabditids. Furthermore, heterorhabditid IJs develop exclusively via EM (Ciche *et al.*, 2008). The first- and second-generation hermaphrodites and females, respectively, produce eggs. Some of these eggs become next-generation females and hermaphrodites, and the rest remain within the uterus and EM occurs. In the third generation, hermaphrodites and females only reproduce by EM (Wang and Bedding, 1996; Johnigk and Ehlers, 1999). EM occurs as follows: newly hatched first juveniles feed on embryos held in the ovaries. The mother eventually dies, and the juveniles become IJs soon after, emerging from her cadaver. Steinernematid IJs emerge mainly through the natural openings of the maternal carcass (mouth, vulva and anus), while heterorhabditids mostly emerge transcuticularly using a tooth-like plaque located in their head (Baliadi *et al.*, 2001).

2.3.3.5 IJ formation and bacterial acquisition

IJ FORMATION. IJs are modified third-stage juveniles. When some conditions are met, such as EPN overcrowding or lack of quality food (Popiel *et al.*, 1989), the second-stage juveniles become IJs instead of "normal" third-stage juveniles (Poinar and Leutenegger, 1968), through a series of changes: molting begins, the bacterial symbiosis is established, the digestive tract collapses closing the anus and mouth, and the second-stage cuticle (sheath) is retained (Poinar and Leutenegger, 1968).

ROLE OF THE SECOND CUTICLE. The sheath, an extra protective layer, is lost when some parasitic species pass from the free-living stage to parasitism (Rogers and Sommerville, 1957). While in heterorhabditids it is strongly fitted around their body, in steinernematids it is quite loose and is not retained when moving through the soil or any other substrate (Timper and Kaya, 1989; Campbell and Gaugler, 1991a). In fact, the sheath of *Steinernema* spp. reduces motility and the sensitivity of the sensory receptors enclosed by it (Campbell and Gaugler, 1992). Losing the sheath does not affect the pathogenicity of steinernematids, but in some heterorhabditids, there could be a reduction in their ability to infect larvae (Campbell and Gaugler, 1991a, 1992; Dempsey and Griffin, 2003), suggesting a potentially more important role in the survival of IJs of *Heterorhabditis* spp. than those of *Steinernema* spp. (O'Leary *et al.*, 1998; Dempsey and Griffin, 2003).

The protective role of the sheath against abiotic and biotic factors is evident. In several heterorhabditids, it protects against desiccation (Campbell and Gaugler, 1991b; Menti *et al.*, 1997; O'Leary *et al.*, 1998) and freezing (Wharton and Surrey, 1994). In *Steinernema* spp. it protects against nematophagous fungi such as *Hirsutella rhossiliensis* or *Drechmeria coniospora* (Timper and Kaya, 1989). Additionally, the sheath reduces IJ predation by mesostigmatid mites, which are able to puncture the sheath but not the IJ cuticle (Epsky *et al.*, 1988).

The loss of the sheath in *Heterorhabditis* spp. is generally associated with penetration into the insect host, as they force their tooth-like structure into the insects' cuticle or midgut (Bedding and Molyneux, 1982). In many insects the sheaths have been found partially embedded in the cuticle or outside the host after penetration (Bedding and Molyneux, 1982; Dziedziech *et al.*, 2020).

2.3.3.6 Bacterial acquisition

Bacterial acquisition differs between both genera. *Steinernema* spp. produce an intestinal receptacle anteriorly (also called bacterial vesicle or pouch) (Bird and Akhurst, 1983; Goodrich-Blair and Clarke, 2007; Snyder *et al.*, 2007), whereas *Heterorhabditis* spp. harbors the bacteria all along their intestines (Goodrich-Blair and Clarke, 2007).

Steinernematids acquire the bacteria horizontally (within the host environment) through three main sequential events: (i) one or a few bacterial cells colonize the nematode pharyngeal–intestinal valve; (ii) the intestine is contracted as the bacteria remain in the pharyngeal–intestinal valve; and (iii) as the nematodes are forming the IJ cuticle, the intestine is decontracted forming the receptacle and the bacteria occupy it (Chaston *et al.*, 2013). The receptacle is formed by two specialized intestinal cells (Martens and Goodrich-Blair, 2005), and once the bacterial cells have established within it, they proliferate until they reach the maximum population density (20–200 cells) (Martens *et al.*, 2003).

Heterorhabditis spp. acquire the bacteria vertically (maternally) via EM (Ciche *et al.*, 2008), through seven main sequential events: (i) bacteria adhere to the maternal posterior intestine; (ii) they grow within the intestinal lumen; (iii) they invade the rectal gland cells; (iv) they are released into the maternal body cavity; (v) they adhere to the pharyngeal–intestinal valve cells of the new hatched juveniles; (vi) they invade the pharyngeal–intestinal valve cells; and (vii) they colonize the IJ intestinal lumen. As nematodes develop into adults, they feed on the bacteria again, but a number of viable cells remain in the intestines of the maternal nematodes. The symbiont transmission initiates with the adhesion of *Photorhabdus* cells to the two most posterior intestinal cells producing a persisting biofilm. The bacteria then invade the rectal gland cells and grow in their vacuoles for 4–5 days, while the embryos hatch in the mother's uterus and start developing. The rectal gland cells lyse, releasing the bacteria into the maternal cavity where the new juveniles are feeding and the bacteria adhere to the intestine when the IJ stage is forming to finally establish the symbiosis (Ciche *et al.*, 2008; Somvanshi *et al.*, 2010).

2.3.3.7 The "other job" of the symbiotic bacteria

Throughout the infection cycle, EPNs are exposed to many natural enemies (Chapter 8, this volume) and competitors (micro- and macroorganisms) that try to take advantage of the resources provided by the insect cadaver (Karthik Raja *et al.*, 2021). Besides all the functions of bacteria, they also protect the cadavers by producing additional secondary metabolites of two types: (i) those that inhibit the growth of opportunistic microorganisms such as bacteria and fungi (e.g. Fuchs *et al.*, 2014; Muangpat *et al.*, 2020; Yimthin *et al.*, 2021; Chapter 9, this volume) and, those that deter macroorganisms from scavenging the insect cadavers (Kaya *et al.*, 1998; Gulcu *et al.*, 2012; Chapter 8, this volume).

Bacteria also act as mediators of ecologically important traits for their hosts, and the capacity of EPNs to form symbiosis with two different bacterial species allows them to extend their ecological niche. For example, in a dune habitat, the *Photorhabdus temperata* subsp. *cinerea* in association with *Heterorhabditis downesi* was consistently more frequent than the *P. temperata* subsp. *temperata* in the front parts of the dunes, where harsh conditions, including drought, prevail (Maher *et al.*, 2017); however, both subspecies populations were equally frequent in the dune grassland further from the sea (a softer environment). In laboratory trials, the survival rates of nematodes exposed to dry conditions increased when the insect cadaver was colonized by *P. temperata* subsp. *cinerea* (Maher *et al.*, 2017).

2.3.3.8 Emergence

To finish their life cycle inside the insect cadaver, IJs actively emerge into the soil to find new hosts, triggered by a combination of physical and chemical stimuli such as food depletion or overcrowding (Nguyen and Smart, 1992; Ryder and Griffin, 2002), or moisture and temperature (Gaugler and Brown, 1997; Koppenhöfer *et al.*, 1997; Lewis and Shapiro-Ilan, 2002). However, under inadequate environmental conditions, such as dry or freezing conditions, IJs can survive for up to 50 days inside the host cadaver (with no emergence) (Gaugler and Brown, 1997; Koppenhöfer *et al.*, 1997; Lewis and Shapiro-Ilan, 2002). EPNs are hygroscopic-positive organisms (they move toward water), so the emergence is delayed with low moisture (Gaugler and Brown, 1997; Koppenhöfer *et al.*, 1997).

The insect cadaver becomes overcrowded when thousands of the new hatched nematodes start growing, making the food source scarce at some point. Simultaneously, the growing nematodes will excrete metabolites such as pheromones (ascarosides) and other small molecules such as prenol or acetyl methyl carbinol or by-products of their digestion (especially ammonia in *Steinernema* spp.), which accumulate and stimulate IJ emergence (Shapiro *et al.*, 2000; San-Blas *et al.*, 2008a,b, 2014; Choe *et al.*, 2012; Kaplan *et al.*, 2012, 2020a; Noguez *et al.*, 2012; Baiocchi *et al.*, 2017; Hartley *et al.*, 2019; Kin *et al.*, 2019).

2.4 Entomopathogenic Nematodes as Scavengers

When it comes to adaptability and niche exploitation, EPNs display an enormous plasticity. Facultative scavenging of EPNs was described by San-Blas and Gowen (2008) as an alternative to complete their life cycles in insects already killed by other causes, such as extreme heat, cold or pesticides (Půža and Mráček, 2010; San-Blas *et al.*, 2012; Nalinci *et al.*, 2021).

While easy to observe under laboratory conditions, this behavior is not fully understood either in the laboratory or in nature (Blanco-Pérez *et al.*, 2019). In olfactory assays, nematodes follow unidentified cues produced by insect cadavers and reach them in hours (San-Blas and Gowen, 2008), so this process might be underestimated in nature. In fact, scavenging would broaden their host range to species that they cannot kill (at least in theory). For instance, wireworms (Coleoptera: Elateridae) have low susceptibility to EPNs (Williams *et al.*, 2022), but when dead larvae were offered to EPNs as a scavenging resource, there was a higher level of colonization (Půža and Mráček, 2010; San-Blas *et al.*, 2012).

Most studies have considered only EPNs and the cadaver, so they can scavenge as they please. For instance, *S. feltiae* and *S. affine* colonized more dead than live *G. mellonella*, probably due to the lack of the insect immune responses (San-Blas *et al.*, 2008b). Scavenging is more frequent in steinernematids and depends on a number of factors, such as nematode and insect species, number of nematodes in the inoculum, cadaver age, competition with other nematodes or microorganisms, and the cause of death (San-Blas and Gowen, 2008; San-Blas *et al.*, 2008b; Blanco-Pérez *et al.*, 2019). Freeze-killed, non-sterilized *G. mellonella* larvae are decomposed by their endogenous or environmental microflora within days, and some species of *Steinernema* spp. (*S. glaseri* and *S. carpocapsae*) can successfully reproduce in these cadavers 8 days after death. However, when scavenging is done by EPNs and free-living bacterivorous nematodes simultaneously, the latter can also facilitate EPN reproduction by feeding on the endogenous bacteria (Blanco-Pérez *et al.*, 2017).

An extreme example of "using any available resource" that has been observed is scavenging on insecticide-killed larvae. One may think that pesticides negatively affect EPNs, both by direct exposure or by the nematodes trying to settle in pesticide-contaminated cadavers. Interestingly, negative effects occur only with some insecticides (Krishnayyaand and Grewal, 2002; Mbata and Shapiro-Ilan, 2013). For example, *S. carpocapsae* significantly invaded more insecticide-killed than freeze-killed *G. mellonella* and *Hyphantria cunea* (Lepidoptera: Arctiidae) larvae, and more progeny IJs emerged from spinosad-killed insects than from the freeze-killed control (Nalinci *et al.*, 2021). Although still enigmatic, this choice reconfirms the plasticity of EPN biology when extending their host range is a matter of surviving.

2.5 Future Perspectives

Our understanding of EPN biology has evolved rapidly in the last decades, resulting in improvements in many areas, such as liquid mass production, field application, long-term survival and promoting their use as biological control agents (Griffin, 2012; Koppenhöfer *et al.*, 2020; Hazir *et al.*, 2022). However, inconsistency in EPN efficacy under field conditions is still one of the biggest challenges for their wider adoption and should be partially attributed to many aspects of their biology that are still unresolved (San-Blas, 2013).

Although there are more than 120 described EPNs, only relatively few of them are commercialized as biological control agents and the vast majority remain in different laboratories and collections waiting for their biological characterization. There is a hidden treasure in these EPN "banks" regarding new discoveries and understanding of key and even unusual processes and behaviors that would allow us to have a broader picture in relation to their general biology and application.

One of the concepts introduced in this chapter is that the complex EPN–bacteria is a holobiont and not two separate organisms "living together." The forthcoming studies on their interaction should observe the holobiont itself, in all steps on their life cycle and in those related to the biological importance of shifting the symbiosis with one or another species of bacteria (Maher *et al.*, 2017).

Another key aspect to consider going forward is that EPNs pass through three different environments during its life cycle: the soil, the living host and the dead host. The current vision does not take into account the important challenges faced in each step and even the existence of a possible alternative environment: an already-dead insect that would allow scavenging (San-Blas and Gowen, 2008). Advances in studying the changes suffered by the holobiont in such interfaces would allow us to update the model nematode–insect association evolution (Dillman *et al.*, 2012a) and the EPN life cycle in the future.

How EPNs live in nature and how the processes involved in their life cycles explain their long-term survival is still a challenge. Many important biological discoveries since the last version of this book, such as male fighting (Zenner *et al.*, 2014), memory acquisition (Willett *et al.*, 2015) and ascaroside production (Kaplan *et al.*, 2012), among many exciting others, have elucidated new aspects of the EPN life cycle, but the whole picture of the EPNs living in nature remains blurry. The plasticity of EPNs is well known in laboratory conditions, so many of those aspects driving EPN survival, long-term persistence and recycling should be answered by designing new ways to observe them in natural conditions.

We are starting to understand which molecules trigger or block EPN biological processes, but many remain unknown. With the advent and development of new technologies, the identity of these molecules has been revealed, so we expect to discover many others in the coming years.

The world is changing and new challenges in food production are constantly appearing. Crops are moving to new environments followed by their insect pests. How could the biological aspects of EPNs be optimized in a changing environment, where soil temperature and humidity will change season after season (as well as the soil fauna)? The conception of sustainable agriculture is gaining ground by markets, politicians and the general public, and EPNs represent a fundamental tool for integrated pest management programs, so the importance of digging deeper in the biology of these organisms is beyond doubt. In this way, the production and use of EPNs will be substantially improved and they will continue to be a reliable, safe and effective tool for agriculture in a changing world.

2.6 References

Abebe, E., Jumba, M., Bonner, K., Gray, V., Morris, K. and Thomas, W.K. (2010) An entomopathogenic *Caenorhabditis briggsae*. *Journal of Experimental Biology* 213, 3223–3229.

Abu Hatab, M., Selvan, S. and Gaugler, R. (1995) Role of proteases in penetration of insect gut by the entomopathogenic nematode *Steinernema glaseri* (Nematoda: Steinernematidae). *Journal of Invertebrate Pathology* 66, 125–130.

Adams, B.J., Fodor, A., Koppenhöfer, H.S., Stackebrandt, E., Stock, S.P. and Klein, M.G. (2006) Biodiversity and systematics of nematode–bacterium entomopathogens. *Biological Control* 37, 32–49.

Adeolu, M., Alnajar, S., Naushad, S.S. and Gupta, R. (2016) Genome-based phylogeny and taxonomy of the 'Enterobacteriales': proposal for Enterobacterales ord. nov. divided into the families Enterobacteriaceae, Erwiniaceae fam. nov., Pectobacteriaceae fam. nov., Yersiniaceae fam. nov., Hafniaceae fam. nov., Morganellaceae fam. nov., and Budviciaceae fam. nov. *International Journal of Systematic and Evolutionary Microbiology* 66, 5575–5599.

Akhurst, R.J. (1982) Antibiotic activity of *Xenorhabdus* spp., bacteria symbiotically associated with insect pathogenic nematodes of the families Heterorhabditidae and Steinernematidae. *Microbiology* 128, 3061–3065.

Aktories, K., Schmidt, G. and Lang, A.E. (2014) *Photorhabdus luminescens* toxins TccC3 and TccC5: insecticidal ADP-ribosyltransferases that modify threonine and glutamine. *Current Topics in Microbiology and Immunology* 384, 53–67.

Aldana, M. and Robeva, R. (2021) New challenges in systems biology: understanding the holobiont. *Frontiers in Physiology* 12: 662878.

Alonso, V., Nasrolahi, S. and Dillman, A. (2018) Host-specific activation of entomopathogenic nematode infective juveniles. *Insects* 9: 59.

Alsaiyah, M.A.M., Ebssa, L., Zenner, A., O'Callaghan, K.M. and Griffin, C.T. (2009) Sex ratios and sex-biased infection behaviour in the entomopathogenic nematode genus *Steinernema*. *International Journal for Parasitology* 39, 725–734.

Aumann, J. and Ehlers, R.-U. (2001) Physico-chemical properties and mode of action of a signal from the symbiotic bacterium *Photorhabdus luminescens* inducing dauer juvenile recovery in the entomopathogenic nematode *Heterorhabditis bacteriophora*. *Nematology* 3, 849–853.

Baiocchi, T., Lee, G., Choe, D.-H. and Dillman, A.R. (2017) Host seeking parasitic nematodes use specific odors to assess host resources. *Scientific Reports* 7: 6270.

Baliadi, Y., Yoshiga, T. and Kondo, E. (2001) Development of *Endotokia matricida* and emergence of originating infective juveniles of steinernematid and heterorhabditid nematodes. *Japanese Journal of Nematology* 31, 26–36.

Bastidas, B., Portillo, E. and San-Blas, E. (2014) Size does matter: the life cycle of *Steinernema* spp. in micro-insect hosts. *Journal of Invertebrate Pathology* 121, 46–55.

Bedding, R.A. and Molyneux, A.S. (1982) Penetration of insect cuticle by infective juveniles of *Heterorhabditis* spp. (Heterorhabditidae: Nematoda). *Nematologica* 28, 354–359.

Bintrim, S.B. and Ensign, J.C. (1998) Insertional inactivation of genes encoding the crystalline inclusion proteins of *Photorhabdus luminescens* results in mutants with pleiotropic phenotypes. *Journal of Bacteriology* 180, 1261–1269.

Bird, A.F. and Akhurst, R.J. (1983) The nature of the intestinal vesicle in nematodes of the family Steinernematidae. *International Journal for Parasitology* 13, 599–606.

Blanco-Pérez, R., Bueno-Pallero, F.Á., Neto, L. and Campos-Herrera, R. (2017) Reproductive efficiency of entomopathogenic nematodes as scavengers. Are they able to fight for insect's cadavers? *Journal of Invertebrate Pathology* 148, 1–9.

Blanco-Pérez, R., Bueno-Pallero, F.Á., Vicente-Díez, I., Marco-Mancebón, V.S., Pérez-Moreno, I. and Campos-Herrera, R. (2019) Scavenging behavior and interspecific competition decrease offspring fitness of the entomopathogenic nematode *Steinernema feltiae*. *Journal of Invertebrate Pathology* 164, 5–15.

Blaxter, M.L., de Ley, P., Garey, J.R., Liu, L.X., Scheldeman, P. *et al.* (1998) A molecular evolutionary framework for the phylum Nematoda. *Nature* 392, 71–75.

Bode, H.B. (2009) Entomopathogenic bacteria as a source of secondary metabolites. *Current Opinion in Chemical Biology* 13, 224–230.

Boemare, N.E., Laumond, C. and Luciani, J. (1982) Mise en évidence d'une toxicogènèse provoquée du nèmatode entomophage *Neoaplectana carpocapsae* Weiser chez l'insecte *Galleria mellonella*. *Comptes Rendus des Seances de l'Academie des Sciences Serie III Sciences de la Vie* 295, 543–546.

Bohan, D.A. and Hominick, W.M. (1997) Sex and the dynamics of infection in the entomopathogenic nematode *Steinernema feltiae*. *Journal of Helminthology* 71, 197–202.

Bowen, D.J. and Ensign, J.C. (2001) Isolation and characterization of intracellular protein inclusions produced by the entomopathogenic bacterium *Photorhabdus luminescens*. *Applied and Environmental Microbiology* 67, 4834–4841.

Bowen, D., Rocheleau, T.A., Blackburn, M., Andreev, O., Golubeva, E. *et al.* (1998) Insecticidal toxins from the bacterium *Photorhabdus luminescens*. *Science* 280, 2129–2132.

Brivio, M.F., Pagani, M. and Restelli, S. (2002) Immune suppression of *Galleria mellonella* (Insecta, Lepidoptera) humoral defenses induced by *Steinernema feltiae* (Nematoda, Rhabditida): involvement of the parasite cuticle. *Experimental Parasitology* 101, 149–156.

Brivio, M.F., Toscano, A., de Pasquale, S.M., de Lerma Barbaro, A., Giovannardi, S. *et al* (2018) Surface protein components from entomopathogenic nematodes and their symbiotic bacteria: effects on immune responses of the greater wax moth, *Galleria mellonella* (Lepidoptera: Pyralidae). *Pest Management Science* 74, 2089–2099.

Burman, M. (1982) *Neoaplectana carpocapsae*: Toxin production by axenic insect parasitic nematodes. *Nematologica* 28, 62–70.

Campbell, L.R. and Gaugler, R. (1991a) Mechanisms for exsheathment of entomopathogenic nematodes. *International Journal for Parasitology* 21, 219–224.

Campbell, L.R. and Gaugler, R. (1991b) Role of the sheath in desiccation tolerance of two entomopathogenic nematodes. *Nematologica* 37, 324–332.

Campbell, L.R. and Gaugler, R. (1992) Effect of exsheathment on motility and pathogenicity of two entomopathogenic nematode species. *Journal of Nematology* 24, 365–370.

Campos-Herrera, R., Trigo, D. and Gutiérrez, C. (2006) Phoresy of the entomopathogenic nematode *Steinernema feltiae* by the earthworm *Eisenia fetida*. *Journal of Invertebrate Pathology* 92, 50–54.

Chaston, J.M., Murfin, K.E., Heath-Heckman, E.A. and Goodrich-Blair, H. (2013) Previously unrecognized stages of species-specific colonization in the mutualism between *Xenorhabdus* bacteria and *Steinernema* nematodes. *Cellular Microbiology* 15, 1545–1559.

Choe, A., von Reuss, S.H., Kogan, D., Gasser, R.B., Platzer, E.G. *et al.* (2012) Ascaroside signaling is widely conserved among nematodes. *Current Biology* 22, 772–780.

Ciche, T.A. and Ensign, J.C. (2003) For the insect pathogen *Photorhabdus luminescens*, which end of a nematode is out? *Applied and Environmental Microbiology* 69, 1890–1897.

Ciche, T.A., Darby, C., Ehlers, R.-U., Forst, S. and Goodrich-Blair, H. (2006) Dangerous liaisons: the symbiosis of entomopathogenic nematodes and bacteria. *Biological Control* 38, 22–46.

Ciche, T.A., Kim, K., Kaufmann-Daszczuk, B., Nguyen, K.C.Q. and Hall, D.H. (2008) Cell invasion and matricide during *Photorhabdus luminescens* transmission by *Heterorhabditis bacteriophora* nematodes. *Applied and Environmental Microbiology* 74, 2275–2287.

Cui, L., Gaugler, R. and Wang, Y. (1993) Penetration of steinernematid nematodes (Nematoda: Steinernematidae) into Japanese beetle larvae, *Popillia japonica* (Coleoptera: Scarabaeidae). *Journal of Invertebrate Pathology* 62, 73–78.

Dempsey, C. and Griffin, C. (2003) The infectivity and behaviour of exsheathed and ensheathed *Heterorhabditis megidis* infective juveniles. *Nematology* 5, 49–53.
Dillman, A.R., Chaston, J.M., Adams, B.J., Ciche, T.A., Goodrich-Blair, H. *et al.* (2012a) An entomopathogenic nematode by any other name. *PLOS Pathogens* 8: e1002527.
Dillman, A.R., Guillermin, M.L., Lee, J.H., Kim, B., Sternberg, P.W. and Hallem, E.A. (2012b) Olfaction shapes host–parasite interactions in parasitic nematodes. *Proceedings of the National Academy of Sciences USA* 109, E2324–E2333.
Dolan, K.M., Jones, J.T. and Burnell, A.M. (2002) Detection of changes occurring during recovery from the dauer stage in *Heterorhabditis bacteriophora*. *Parasitology* 125, 71–81.
Downes, M.J. and Griffin, C.T. (1996) Dispersal behaviour and transmission strategies of the entomopathogenic nematodes *Heterorhabditis* and *Steinernema*. *Biocontrol Science and Technology* 6, 347–356.
Dziedziech, A., Shivankar, S. and Theopold, U. (2020) High-resolution infection kinetics of entomopathogenic nematodes entering *Drosophila melanogaster*. *Insects* 11: 60.
Ebssa, L., Dix, I. and Griffin, C.T. (2008) Female presence is required for male sexual maturity in the nematode *Steinernema longicaudum*. *Current Biology* 18, R997–R998.
Ehlers, R.-U., Wulff, A. and Peters, A. (1997) Pathogenicity of axenic *Steinernema feltiae*, *Xenorhabdus bovienii*, and the bacto-helminthic complex to larvae of *Tipula oleracea* (Diptera) and *Galleria mellonella* (Lepidoptera). *Journal of Invertebrate Pathology* 69, 212–217.
Eidt, D.C. and Thurston, G.S. (1995) Physical deterrents to infection by entomopathogenic nematodes in wireworms (Coleoptera: Elateridae) and other soil insects. *Canadian Entomologist* 127, 423–429.
Eng, M.S., Preisser, E.L. and Strong, D.R. (2005) Phoresy of the entomopathogenic nematode *Heterorhabditis marelatus* by a non-host organism, the isopod *Porcellio scaber*. *Journal of Invertebrate Pathology* 88, 173–176.
Epsky, N.D., Walter, D.E. and Capinera, J.L. (1988) Potential role of nematophagous microarthropods as biotic mortality factors of entomogenous nematodes (Rhabditida: Steinernematidae, Heterorhabditidae). *Journal of Economic Entomology* 81, 821–825.
Fischer-Le Saux, M., Viallard, V., Brunel, B., Normand, P. and Boemare, N.E. (1999) Polyphasic classification of the genus *Photorhabdus* and proposal of new taxa: *P. luminescens* subsp. *luminescens* subsp. nov., *P. luminescens* subsp. *akhurstii* subsp. nov., *P. luminescens* subsp. *laumondii* subsp. nov., *P. temperata* sp. nov., *P. temperata* subsp. *temperata* subsp. nov. and *P. asymbiotica* sp. nov. *International Journal of Systematic Bacteriology* 49, 1645–1656.
Forst, S., Dowds, B., Boemare, N. and Stackebrandt, E. (1997) *Xenorhabdus* and *Photorhabdus* spp.: bugs that kill bugs. *Annual Review of Microbiology* 51, 47–72.
Fuchs, S.W., Grundmann, F., Kurz, M., Kaiser, M. and Bode, H.B. (2014) Fabclavines: bioactive peptide–polyketide–polyamino hybrids from *Xenorhabdus*. *ChemBioChem* 15, 512–516.
Gaugler, R. (1988) Ecological considerations in the biological control of soil-inhabiting insects with entomopathogenic nematodes. *Agriculture, Ecosystems & Environment* 24, 351–360.
Gaugler, R. and Brown, I.M. (1997) Temperature and humidity influence emergence and survival of entomopathogenic nematodes. *Nematologica* 43, 363–375.
Gaugler, R., Wang, Y.I. and Campbell, J.F. (1994) Aggressive and evasive behaviors in *Popillia japonica* (Coleoptera: Scarabaeidae) larvae: defenses against entomopathogenic nematode attack. *Journal of Invertebrate Pathology* 64, 193–199.
Gerritsen, L.J.M. and Smits, P.H. (1993) Variation in pathogenicity of recombinations of *Heterorhabditis* and *Xenorhabdus luminescens* strains. *Fundamental and Applied Nematology* 16, 367–373.
Golden, J.W. and Riddle, D.L. (1982) A pheromone influences larval development in the nematode *Caenorhabditis elegans*. *Science* 218, 578–580.
Goodrich-Blair, H. (2007) They've got a ticket to ride: *Xenorhabdus nematophila–Steinernema carpocapsae* symbiosis. *Current Opinion in Microbiology* 10, 225–230.
Goodrich-Blair, H. and Clarke, D.J. (2007) Mutualism and pathogenesis in *Xenorhabdus* and *Photorhabdus*: two roads to the same destination. *Molecular Microbiology* 64, 260–268.
Grewal, P.S., Gaugler, R. and Lewis, E.E. (1993a) Host recognition behavior by entomopathogenic nematodes during contact with insect gut contents. *Journal of Parasitology* 79, 495–503.
Grewal, P.S., Gaugler, R. and Selvan, S. (1993b) Host recognition by entomopathogenic nematodes: behavioral response to contact with host feces. *Journal of Chemical Ecology* 19, 1219–1231.
Grewal, P.S., Selvan, S., Lewis, E.E. and Gaugler, R. (1993c) Male insect-parasitic nematodes: a colonizing sex. *Experientia* 49, 605–608.

Grewal, P.S., Gaugler, R. and Selvan, S. (1997) Host recognition by entomopathogenic nematodes: behavioral response to contact with host feces. *Journal of Chemical Ecology* 19, 1219–1231.
Griffin, C.T. (2012) Perspectives on the behavior of entomopathogenic nematodes from dispersal to reproduction: traits contributing to nematode fitness and biocontrol efficacy. *Journal of Nematology* 44, 177–184.
Griffin, C.T., O'Callaghan, K.M. and Dix, I. (2001) A self-fertile species of *Steinernema* from Indonesia: further evidence of convergent evolution amongst entomopathogenic nematodes? *Parasitology* 122, 181–186.
Gulcu, B., Hazir, S. and Kaya, H.K. (2012) Scavenger deterrent factor (SDF) from symbiotic bacteria of entomopathogenic nematodes. *Journal of Invertebrate Pathology* 110, 326–333.
Hallem, E.A., Rengarajan, M., Ciche, T.A. and Sternberg, P.W. (2007) Nematodes, bacteria, and flies: a tripartite model for nematode parasitism. *Current Biology* 17, 898–904.
Han, R. and Ehlers, R.-U. (2000) Pathogenicity, development, and reproduction of *Heterorhabditis bacteriophora* and *Steinernema carpocapsae* under axenic *in vivo* conditions. *Journal of Invertebrate Pathology* 75, 55–58.
Han, R., Wouts, W.M. and Li, L. (1991) Development and virulence of *Heterorhabditis* spp. strains associated with different *Xenorhabdus luminescens* isolates. *Journal of Invertebrate Pathology* 58, 27–32.
Hartley, C.J., Lillis, P.E., Owens, R.A. and Griffin, C.T. (2019) Infective juveniles of entomopathogenic nematodes (*Steinernema* and *Heterorhabditis*) secrete ascarosides and respond to interspecific dispersal signals. *Journal of Invertebrate Pathology* 168: 107257.
Hazir, S., Kaya, H., Touray, M., Cimen, H. and Shapiro-Ilan, D. (2022) Basic laboratory and field manual for conducting research with the entomopathogenic nematodes, *Steinernema* and *Heterorhabditis*, and their bacterial symbionts. *Turkish Journal of Zoology* 46, 305–350.
Herbert, E.E. and Goodrich-Blair, H. (2007) Friend and foe: the two faces of *Xenorhabdus nematophila*. *Nature Reviews Microbiology* 5, 634–646.
Hunt, D.J. and Nguyen, K.B. (eds.) (2017) *Advances in Entomopathogenic Nematode Taxonomy and Phylogeny*. Nematology Monographs and Perspectives, Vol. 12. Brill, Leiden, Netherlands.
Jessen, P., Luttmann, R., Ehlers, R.-U., Strauch, O. and Wyss, U. (2000) Carbon dioxide triggers recovery from dauer juvenile stage in entomopathogenic nematodes (*Heterorhabditis* spp.). *Nematology* 2, 319–324.
Johnigk, S.-A. and Ehlers, R.-U. (1999) Juvenile development and life cycle of *Heterorhabditis bacteriophora* and *H. indica* (Nematoda: Heterorhabditidae). *Nematology* 1, 251–260.
Kaplan, F., Alborn, H.T., von Reuss, S.H., Ajredini, R., Ali, J.G. *et al.* (2012) Interspecific nematode signals regulate dispersal behavior. *PLOS ONE* 7: e38735.
Kaplan, F., Perret-Gentil, A., Giurintano, J., Stevens, G., Erdogan, H. *et al.* (2020a) Conspecific and heterospecific pheromones stimulate dispersal of entomopathogenic nematodes during quiescence. *Scientific Reports* 10: 5738.
Kaplan, F., Shapiro-Ilan, D. and Schiller, K.C. (2020b) Dynamics of entomopathogenic nematode foraging and infectivity in microgravity. *npj Microgravity* 6: 20.
Kapranas, A., Maher, A.M.D. and Griffin, C.T. (2016) Higher relatedness mitigates mortality in a nematode with lethal male fighting. *Journal of Evolutionary Biology* 29, 344–351.
Karthik Raja, R., Arun, A., Touray, M., Hazal Gulsen, S., Cimen, H. *et al.* (2021) Antagonists and defense mechanisms of entomopathogenic nematodes and their mutualistic bacteria. *Biological Control* 152: 104452.
Kaya, H.K., Koppenhöfer, A.M. and Johnson, M. (1998) Natural enemies of entomopathogenic nematodes. *Nematological Research* 28, 13–21.
Kikuta, S., Kiuchi, T., Aoki, F. and Nagata, M. (2008) Development of an entomopathogenic nematode, *Steinernema carpocapsae*, in cultivated insect cells under axenic condition. *Nematology* 10, 84–851.
Kikuta, S., Kiuchi, T., Aoki, F. and Nagata, M. (2009) Recovery of infective juveniles of the entomopathogenic nematode *Steinernema carpocapsae* via factors produced by insect cells and symbiotic bacteria. *Nematology* 11, 611–618.
Kin, K., Baiocchi, T. and Dillman, A.R. (2019) Dispersal and repulsion of entomopathogenic nematodes to prenol. *Biology (Basel)* 8: 58.
Koltai, H., Glazer, I. and Segal, D. (1995) Reproduction of the entomopathogenic nematode *Heterorhabditis bacteriophora* Poinar, 1976: hermaphroditism vs amphimixis. *Fundamental and Applied Nematology* 18, 55–61.
Koppenhöfer, A.M. and Kaya, H.K. (1995) Density-dependent effects on *Steinernema glaseri* (Nematoda: Steinernematidae) within an insect host. *Journal of Parasitology* 81, 797–799.

Koppenhöfer, A.M., Baur, M.E., Stock, S.P., Choo, H.Y., Chinnasri, B. and Kaya, H.K. (1997) Survival of entomopathogenic nematodes within host cadavers in dry soil. *Applied Soil Ecology* 6, 231–240.

Koppenhöfer, A.M., Grewal, P.S. and Fuzy, E.M. (2007) Differences in penetration routes and establishment rates of four entomopathogenic nematode species into four white grub species. *Journal of Invertebrate Pathology* 94, 184–195.

Koppenhöfer, A.M., Grewal, P.S. and Kaya, H.K. (2011) Synergism of imidacloprid and entomopathogenic nematodes against white grubs: the mechanism. *Entomologia Experimentalis et Applicata* 94, 283–293.

Koppenhöfer, A.M., Shapiro-Ilan, D.I. and Hiltpold, I. (2020) Entomopathogenic nematodes in sustainable food production. *Frontiers in Sustainable Food Systems* 4: 125.

Krishnayyaand, P.V. and Grewal, P.S. (2002) Effect of neem and selected fungicides on viability and virulence of the entomopathogenic nematode *Steinernema feltiae*. *Biocontrol Science and Technology* 12, 259–266.

Kruitbos, L., Heritage, S. and Wilson, M. (2009) Phoretic dispersal of entomopathogenic nematodes by *Hylobius abietis*. *Nematology* 11, 419–427.

Lee, M.-M. and Stock, S.P. (2010) A multilocus approach to assessing co-evolutionary relationships between *Steinernema* spp. (Nematoda: Steinernematidae) and their bacterial symbionts *Xenorhabdus* spp. (γ-Proteobacteria: Enterobacteriaceae). *Systematic Parasitology* 77, 1–12.

Lewis, E.E. and Shapiro-Ilan, D.I. (2002) Host cadavers protect entomopathogenic nematodes during freezing. *Journal of Invertebrate Pathology* 81, 25–32.

Lewis, E.E., Barbarosa, B. and Gaugler, R. (2002) Mating and sexual communication by *Steinernema carpocapsae* (Nemata: Steinernematidae). *Journal of Nematology* 34, 328–231.

Lewis, E.E., Campbell, J., Griffin, C., Kaya, H. and Peters, A. (2006) Behavioral ecology of entomopathogenic nematodes. *Biological Control* 38, 66–79.

Liu, J., Poinar, G.O. and Berry, R.E. (2000) Control of insect pests with entomopathogenic nematodes: the impact of molecular biology and phylogenetic reconstruction. *Annual Review of Entomology* 45, 287–306.

Lu, D., Macchietto, M., Chang, D., Barros, M.M., Baldwin, J. *et al.* (2017a) Activated entomopathogenic nematode infective juveniles release lethal venom proteins. *PLOS Pathogens* 13: e1006302.

Lu, D., Sepulveda, C. and Dillman, A.R. (2017b) Infective juveniles of the entomopathogenic nematode *Steinernema scapterisci* are preferentially activated by cricket tissue. *PLOS ONE* 12: e0169410.

Maher, A.M.D., Asaiyah, M.A.M., Brophy, C. and Griffin, C.T. (2017) An entomopathogenic nematode extends its niche by associating with different symbionts. *Microbial Ecology* 73, 211–223.

Maher, A.M.D., Asaiyah, M., Quinn, S., Burke, R., Wolff, H. *et al.* (2021) Competition and co-existence of two *Photorhabdus* symbionts with a nematode host. *Microbial Ecology* 81, 223–239.

Mamiya, Y. (1988) *Steinernema kushidai* n. sp. (Nematoda : Steinernematidae) associated with scarabaeid beetle larvae from Shizuoka, Japan. *Applied Entomology and Zoology* 23, 313–320.

Martens, E.C. and Goodrich-Blair, H. (2005) The *Steinernema carpocapsae* intestinal vesicle contains a subcellular structure with which *Xenorhabdus nematophila* associates during colonization initiation. *Cellular Microbiology* 7, 1723–1735.

Martens, E.C., Heungens, K. and Goodrich-Blair, H. (2003) Early colonization events in the mutualistic association between *Steinernema carpocapsae* nematodes and *Xenorhabdus nematophila* bacteria. *Journal of Bacteriology* 185, 3147–3154.

Mastore, M. and Brivio, M.F. (2008) Cuticular surface lipids are responsible for disguise properties of an entomoparasite against host cellular responses. *Developmental & Comparative Immunology* 32, 1050–1062.

Mbata, G.N. and Shapiro-Ilan, D. (2013) The potential for controlling *Pangaeus bilineatus* (Heteroptera: Cydnidae) using a combination of entomopathogens and an insecticide. *Journal of Economic Entomology* 106, 2072–2076.

Menti, H., Wright, D.J. and Perry, R.N. (1997) Desiccation survival of populations of the entomopathogenic nematodes *Steinernema feltiae* and *Heterorhabditis megidis* from Greece and the UK. *Journal of Helminthology* 71, 41–46.

Muangpat, P., Suwannaroj, M., Yimthin, T., Fukruksa, C., Sitthisak, S. *et al.* (2020) Antibacterial activity of *Xenorhabdus* and *Photorhabdus* isolated from entomopathogenic nematodes against antibiotic-resistant bacteria. *PLOS ONE* 15: e0234129.

Mucci, N.C., Jones, K.A., Cao, M., Wyatt, M.R., Foye, S. *et al.* (2022) Apex predator nematodes and meso-predator bacteria consume their basal insect prey through discrete stages of chemical transformations. *mSystems* 7: e00312-22. doi:10.1101/2021.02.13.430781

Nagata, M., Kiuchi, T., Kikuta, S. and Aoki, F. (2008) Development of an entomopathogenic nematode, *Steinernema carpocapsae*, in cultured insect cells under axenic conditions. *Nematology* 10, 845–851.

Nalinci, E., Karagoz, M., Gulcu, B., Ulug, D., Hazal Gulsen, S. *et al.* (2021) The effect of chemical insecticides on the scavenging performance of *Steinernema carpocapsae*: direct effects and exposure to insects killed by chemical insecticides. *Journal of Invertebrate Pathology* 184: 107641.
Neves, J.M., Simoes, N. and Mota, M. (1998). Evidence for a sex pheromone in *Steinenema carpocapsae*. *Nematologica* 44, 95–98.
Nguyen, K.B. and Smart, G.C. (1991) Pathogenicity of *Steinernema scapterisci* to selected invertebrates. *Journal of Nematology* 23, 7–11.
Nguyen, K.B. and Smart, G.C. (1992) Life cycle of *Steinernema scapterisci* Nguyen & Smart, 1990. *Journal of Nematology* 24, 160–169.
Noguez, J.H., Conner, E.S., Zhou, Y., Ciche, T.A., Ragains, J.R. and Butcher, R.A. (2012) A novel ascaroside controls the parasitic life cycle of the entomopathogenic nematode *Heterorhabditis bacteriophora*. *ACS Chemical Biology* 7, 961–966.
O'Callaghan, K.M., Zenner, A.N.R.L., Hartley, C.J. and Griffin, C.T. (2014) Interference competition in entomopathogenic nematodes: male *Steinernema* kill members of their own and other species. *International Journal for Parasitology* 44, 1009–1017.
O'Leary, S., Burnell, A. and Kusel, J. (1998) Biophysical properties of the surface of desiccation-tolerant mutants and parental strain of the entomopathogenic nematode *Heterorhabditis megidis* (strain UK211). *Parasitology* 117, 337–345.
Park, Y. and Kim, Y. (2000) Eicosanoids rescue *Spodoptera exigua* infected with *Xenorhabdus nematophilus*, the symbiotic bacteria to the entomopathogenic nematode *Steinernema carpocapsae*. *Journal of Insect Physiology* 46, 1469–1476.
Pervez, R., Eapen, S.J., Devasahayam, S. and Jacob, T.K. (2013) A new species of entomopathogenic nematode *Oscheius gingeri* sp. n. (Nematoda: Rhabditidae) from ginger rhizosphere. *Archives of Phytopathology and Plant Protection* 46, 526–535.
Peters, A. (1996) The natural host range of *Steinernema* and *Heterorhabditis* spp. and their impact on insect populations. *Biocontrol Science and Technology* 6, 389–402.
Peters, A. and Ehlers, R.-U. (1994) Susceptibility of leatherjackets (*Tipula paludosa* and *Tipula oleracea*; Tipulidae; Nematocera) to the entomopathogenic nematode *Steinernema feltiae*. *Journal of Invertebrate Pathology* 63, 163–171.
Peters, A. and Ehlers, R. (1997) Encapsulation of the entomopathogenic nematode *Steinernema feltiae* in *Tipula oleracea*. *Journal of Invertebrate Pathology* 69, 218–222.
Poinar, G.O. Jr. (1990) Taxonomy and biology of Steinernematidae and Heterorhabditidae. In: Gaugler, R. and Kaya, H.K. (eds.) *Entomopathogenic Nematodes in Biological Control*. CRC Press, Boca Raton, p. 42.
Poinar, G.O. Jr. and Grewal, P.S. (2012) History of entomopathogenic nematology. *Journal of Nematology* 44, 153–161.
Poinar, G.O. Jr. and Leutenegger, R. (1968) Anatomy of the infective and normal third-stage juveniles of *Neoaplectana carpocapsae* Weiser (Steinernematidae: Nematoda). *Journal of Parasitology* 54, 340–350.
Popiel, I., Grove, D.L. and Friedman, M.J. (1989) Infective juvenile formation in the insect parasitic nematode *Steinernema feltiae*. *Parasitology* 99, 77–81.
Postle, K. and Larsen, R.A. (2007) TonB-dependent energy transduction between outer and cytoplasmic membranes. *BioMetals* 20, 453–465.
Půža, V. and Mráček, Z. (2010) Does scavenging extend the host range of entomopathogenic nematodes (Nematoda: Steinernematidae)? *Journal of Invertebrate Pathology* 104, 1–3.
Richards, G.R. and Goodrich-Blair, H. (2009) Masters of conquest and pillage: *Xenorhabdus nematophila* global regulators control transitions from virulence to nutrient acquisition. *Cellular Microbiology* 11, 1025–1033.
Rodou, A., Ankrah, D.O. and Stathopoulos, C. (2010) Toxins and secretion systems of *Photorhabdus luminescens*. *Toxins (Basel)* 2, 1250–1264
Rogers, W.P. and Sommerville, R.I. (1957) Physiology of exsheathment in nematodes and its relation to parasitism. *Nature* 179, 619–621.
Rosa da Silva, J.L., Undurraga Schwalm, F., Eugênio Silva, C., da Costa, M., Heermann, R. and Santos da Silva, O. (2017) Larvicidal and growth-inhibitory activity of entomopathogenic bacteria culture fluids against *Aedes aegypti* (Diptera: Culicidae). *Journal of Economic Entomology* 110, 378–385.
Ryder, J.J. and Griffin, C.T. (2002) Density-dependent fecundity and infective juvenile production in the entomopathogenic nematode, *Heterorhabditis megidis*. *Parasitology* 125, 83–92.
Sajnaga, E. and Kazimierczak, W. (2020) Evolution and taxonomy of nematode-associated entomopathogenic bacteria of the genera *Xenorhabdus* and *Photorhabdus*: an overview. *Symbiosis* 80, 1–13.

San-Blas, E. (2013) Progress on entomopathogenic nematology research: a bibliometric study of the last three decades: 1980–2010. *Biological Control* 66, 102–124.

San-Blas, E. and Gowen, S.R. (2008) Facultative scavenging as a survival strategy of entomopathogenic nematodes. *International Journal for Parasitology* 38, 85–91.

San-Blas, E., Gowen, S.R. and Pembroke, B. (2008a) *Steinernema feltiae*: ammonia triggers the emergence of their infective juveniles. *Experimental Parasitology* 119, 180–185.

San-Blas, E., Pembroke, B. and Gowen, S. (2008b) Scavenging or infection? Possible host choosing by entomopathogenic nematodes. *Nematology* 10, 251–259.

San-Blas, E., Pembroke, B. and Gowen, S.R. (2012) Scavenging and infection of different hosts by *Steinernema carpocapsae*. *Nematropica* 42, 123–130.

San-Blas, E., Pirela, D., García, D. and Portillo, E. (2014) Ammonia concentration at emergence and its effects on the recovery of different species of entomopathogenic nematodes. *Experimental Parasitology* 144, 1–5.

San-Blas, E., Campos-Herrera, R., Dolinski, C., Monteiro, C., Andaló, V. *et al.* (2019) Entomopathogenic nematology in Latin America: a brief history, current research and future prospects. *Journal of Invertebrate Pathology* 165, 22–45.

Seo, S., Lee, S., Hong, Y. and Kim, Y. (2012) Phospholipase A2 inhibitors synthesized by two entomopathogenic bacteria, *Xenorhabdus nematophila* and *Photorhabdus temperata* subsp. *temperata*. *Applied and Environmental Microbiology* 78, 3816–3823.

Serepa-Dlamini, M.H. and Gray, V.M. (2018) A new species of entomopathogenic nematode *Oscheius safricana* n. sp. (Nematoda: Rhabditidae) from South Africa. *Archives of Phytopathology and Plant Protection* 51, 309–321.

Sergeant, M., Baxter, L., Jarrett, P., Shaw, E., Ousley, M. *et al.* (2006) Identification, typing, and insecticidal activity of *Xenorhabdus* isolates from entomopathogenic nematodes in United Kingdom soil and characterization of the *xpt* toxin loci. *Applied and Environmental Microbiology* 72, 5895–5907.

Shapiro, D.I., Lewis, E.E., Paramasivam, S. and McCoy, C.W. (2000) Nitrogen partitioning in *Heterorhabditis bacteriophora*-infected hosts and the effects of nitrogen on attraction/repulsion. *Journal of Invertebrate Pathology* 76, 43–48.

Sheets, J.J., Hey, T.D., Fencil, K.J., Burton, S.L., Ni, W. *et al.* (2011) Insecticidal toxin complex proteins from *Xenorhabdus nematophilus*: structure and pore formation. *Journal of Biological Chemistry* 286, 22742–22749.

Sicard, M., Brugirard-Ricaud, K., Pagès, S., Lanois, A., Boemare, N.E. *et al.* (2004) Stages of infection during the tripartite interaction between *Xenorhabdus nematophila*, its nematode vector, and insect hosts. *Applied and Environmental Microbiology* 70, 6473–6480.

Simões, N. and Rosa, J.S. (1996) Pathogenicity and host specificity of entomopathogenic nematodes. *Biocontrol Science and Technology* 6, 403–412.

Snyder, H., Stock, S.P., Kim, S.-K., Flores-Lara, Y. and Forst, S. (2007) New insights into the colonization and release processes of *Xenorhabdus nematophila* and the morphology and ultrastructure of the bacterial receptacle of its nematode host, *Steinernema carpocapsae*. *Applied and Environmental Microbiology* 73, 5338–5346.

Somvanshi, V.S., Kaufmann-Daszczuk, B., Kim, K., Mallon, S. and Ciche, T.A. (2010) *Photorhabdus* phase variants express a novel fimbrial locus, mad, essential for symbiosis. *Molecular Microbiology* 77, 1021–1038.

Stock, S.P. and Blair, H.G. (2008) Entomopathogenic nematodes and their bacterial symbionts: the inside out of a mutualistic association. *Symbiosis* 46, 65–75.

Strauch, O. and Ehlers, R.-U. (1998) Food signal production of *Photorhabdus luminescens* inducing the recovery of entomopathogenic nematodes *Heterorhabditis* spp. in liquid culture. *Applied Microbiology and Biotechnology* 50, 369–374.

Stuart, R.J., Hatab, M.A. and Gaugler, R. (1998) Sex ratio and the infection process in entomopathogenic nematodes: are males the colonizing sex? *Journal of Invertebrate Pathology* 72, 288–295.

Stuart, R.J., Barbercheck, M.E. and Grewal, P.S. (2015) Entomopathogenic nematodes in the soil environment: distributions, interactions and the influence of biotic and abiotic factors. In: Campos-Herrera, R. (ed.) *Nematode Pathogenesis of Insects and Other Pests: Ecology and Applied Technologies for Sustainable Plant and Crop Protection*. Springer, Cham, Switzerland, pp. 97–137.

Timper, P. and Kaya, H.K. (1989) Role of the second-stage cuticle of entomogenous nematodes in preventing infection by nematophagous fungi. *Journal of Invertebrate Pathology* 54, 314–321.

Wang, J. and Bedding, R.A. (1996) Population development of *Heterorhabditis bacteriophora* and *Steinernema carpocapsae* in the larvae of *Galleria mellonella*. *Fundamental and Applied Nematology* 19, 363–367.
Wang, Y. and Gaugler, R. (1998) Host and penetration site location by entomopathogenic nematodes against Japanese beetle larvae. *Journal of Invertebrate Pathology* 72, 313–318.
Wang, Y., Campbell, J.F. and Gaugler, R. (1995) Infection of entomopathogenic nematodes *Steinernema glaseri* and *Heterorhabditis bacteriophora* against *Popillia japonica* (Coleoptera: Scarabaeidae) larvae. *Journal of Invertebrate Pathology* 66, 178–184.
Watson, R.J., Millichap, P., Joyce, S.A., Reynolds, S. and Clarke, D.J. (2010) The role of iron uptake in pathogenicity and symbiosis in *Photorhabdus luminescens* TT01. *BMC Microbiology* 10: 177.
Wharton, D.A. and Surrey, M.R. (1994) Cold tolerance mechanisms of the infective larvae of the insect parasitic nematode, *Heterorhabditis zealandica* Poinar. *Cryo-Letters* 15, 353–360.
Willett, D.S., Alborn, H.T., Duncan, L.W. and Stelinski, L.L. (2015) Social networks of educated nematodes. *Scientific Reports* 5: 14388.
Williams, L., Cherry, R. and Shapiro-Ilan, D. (2022) Effect of host size on susceptibility of *Melanotus communis* (Coleoptera: Elateridae) wireworms to entomopathogens. *Journal of Nematology* 54: 20220033.
Yimthin, T., Fukruksa, C., Muangpat, P., Dumidae, A., Wattanachaiyingcharoen, W. *et al.* (2021) A study on *Xenorhabdus* and *Photorhabdus* isolates from Northeastern Thailand: identification, antibacterial activity, and association with entomopathogenic nematode hosts. *PLOS ONE* 16: e0255943.
Zenner, A.N.R.L., O'Callaghan, K.M. and Griffin, C.T. (2014) Lethal fighting in nematodes is dependent on developmental pathway: male–male fighting in the entomopathogenic nematode *Steinernema longicaudum*. *PLOS ONE* 9: e89385.

3 Insect Innate Immunity against Entomopathogenic Nematodes

Ioannis Eleftherianos* and Ghada Tafesh-Edwards
Infection and Innate Immunity Laboratory, Department of Biological Sciences, The George Washington University, Washington, DC, USA

Abstract
Entomopathogenic nematodes (EPNs) form mutualistic species-specific relationships with Gram-negative bacteria. These insect pathogens are widely used in agricultural practices for the efficient management of noxious insect pests and disease vectors. In addition, the use of EPNs together with natural insect hosts or insect models generates significant information for the role of insect immune signaling and function against these potent insect parasites and the effector molecules they produce during infection. This research has started to uncover the molecular mechanisms that regulate the activation of humoral, cellular and coagulation responses, and the EPN immunomodulatory strategies that impair the host immune system. Here, we aim to review the recent advances in the interaction of EPNs with insect host defenses and address their contribution to the development of novel means of biological control through undermining key insect anti-nematode immune reactions.

3.1 Introduction

Nematodes or roundworms are multicellular animals that are found in diverse habitats on Earth (Dillman and Sternberg, 2012; Eleftherianos *et al.*, 2017). Many species of nematodes are "free-living," living in soil, marine and freshwater habitats. Among the parasitic nematodes, entomopathogenic nematodes (EPNs) have developed a mutualistic relationship with Gram-negative bacteria and possess the ability to locate and invade insects. The lethal partnership between an EPN and its mutualistic bacteria leads to rapid insect death and proliferation of the pathogens (Dillman and Sternberg, 2012). The extraordinary efficiency of EPNs to eliminate insects has attracted the attention of the relevant research community to explore their insect-pathogenic properties and convert them into efficient biological control agents for the integrated management of noxious insect pests (Piedra Buena *et al.*, 2015). This research has led to the commercialization of EPNs as biopesticides, their mass production and their incorporation into crop protection schemes (Jess *et al.*, 2005; Shapiro-Ilan *et al.*, 2005; Dara, 2017; Wraight *et al.*, 2017).

The most well-studied EPNs are members of the families Steinernematidae and Heterorhabditidae, and several *Steinernema* and *Heterorhabditis* spp. have been commercialized and applied to agricultural settings (Ehlers, 2001). The selection of EPNs for the management of insect pests is based on various factors that relate to their host range and the foraging strategy of

*Email: ioannise@gwu.edu

DOI: 10.1079/9781800620322.0003

the nematode parasite. In addition, environmental factors and their effects on nematode survival and infection efficiency such as temperature, moisture, soil type, exposure to ultraviolet light, salinity and organic content of soil are also considered (Lacey and Georgis, 2012).

Understanding the details of the interaction between EPNs, their associated bacterial partners and the physiological response of the infected insects, including the insect immune response to EPN–bacteria complexes, is considered an imperative step for improving the efficacy of these beneficial pathogens. Expanding this knowledge will provide superior insect pest control and therefore will limit crop yield losses (Kenney and Eleftherianos, 2016). This chapter describes the recent developments and progress in the novel field of insect immunity against EPNs. In addition, elucidating the different types of EPN immunomodulatory strategies will provide a robust platform for identifying and characterizing conserved mechanisms employed by vertebrate nematode parasites to disrupt the host anti-nematode innate immune response in humans.

3.2 The Life Cycle of Entomopathogenic Nematodes

EPNs complete their life cycle and multiply in the insect cadaver before they exit and spread to look for new hosts. The EPN life cycle comprises the egg stage, four juvenile stages and the adult stage (Fig. 3.1) (see Chapter 2, this volume, for a detailed description of EPN basic biology). The central characteristic of the EPN life cycle is the mutualistic symbiosis with Gram-negative bacteria that participate in the insect infection process. The bacteria are localized in the gut of the third-stage infective juveniles (IJs). IJs are considered equivalent to the dauer juvenile stage of the model nematode *Caenorhabditis elegans* (Shapiro-Ilan *et al.*, 2017). The IJ, the only non-feeding, free-living stage, can survive for several months in the absence of a suitable host (Hominick, 1990). IJs occur naturally in the soil and use environmental and host cues such as carbon dioxide, vibration and other chemicals produced in waste products to locate and recognize suitable insect hosts before they invade them (Shapiro-Ilan *et al.*, 2017).

Heterorhabditis and *Steinernema* spp. are associated with bacteria of the genera *Photorhabdus* and *Xenorhabdus*, respectively, and the relationship between nematodes and bacteria is highly specific (Ferreira and Malan, 2014). Both the nematodes and their mutualistic bacteria are highly pathogenic, and they are able to kill the infected insect host within a few days of invasion (Shapiro-Ilan *et al.*, 2017). The contribution of the EPNs and their associated bacteria to the overall pathogenicity of the nematode–bacteria complex is currently under investigation. The role of virulence in the bacteria was recognized early, yet the role of the nematodes contributing to pathogenicity has only been progressively elucidated in the past couple of decades (Han and Ehlers, 2000; Sicard *et al.*, 2003; Dillman and Sternberg, 2012; Lu *et al.*, 2017). Further studies with *Heterorhabditis–Photorhabdus* and *Steinernema–Xenorhabdus* nematode–bacteria complexes together with natural hosts will provide additional evidence on the synergism of the mutualistic partners and their specific strategies to defeat the insect host.

The initial stage of EPN infection involves gaining access to the insect host by penetrating the intersegmental membranes of the cuticle or entering via natural openings (mouth, anus or spiracles). Following invasion of the insect body, IJs deliver their mutualistic bacteria into the host's hemolymph by regurgitation or defecation (Eleftherianos *et al.*, 2017). The timing of bacterial release differs between *Heterorhabditis* and *Steinernema* spp. *Heterorhabditis* spp. nematodes discharge *Photorhabdus* spp. bacteria within 30 min of invasion, but it takes 4–6 h for *Steinernema* spp. nematodes to eject their *Xenorhabdus* spp. bacteria (Li *et al.*, 2007). When the insect host succumbs due to toxemia or septicemia, the nematodes continue to feed on the bacteria, which have multiplied exponentially and spread to several tissues (ffrench-Constant *et al.*, 2003). As a result, EPNs grow and molt to the fourth stage and then to the adult stage, eventually completing one to three generations in the dead host.

Heterorhabditid IJs develop into self-fertilizing hermaphroditic adults in the first generation and subsequent generations consist of males, females and hermaphroditic individuals. When space and food resources are consumed, a new generation of non-feeding IJs exits the host cadaver and disseminates into the soil, hunting for a new host to extend their life cycle (Hominick, 1990). In contrast, *Steinernema* IJs, with the exception of *Steinernema*

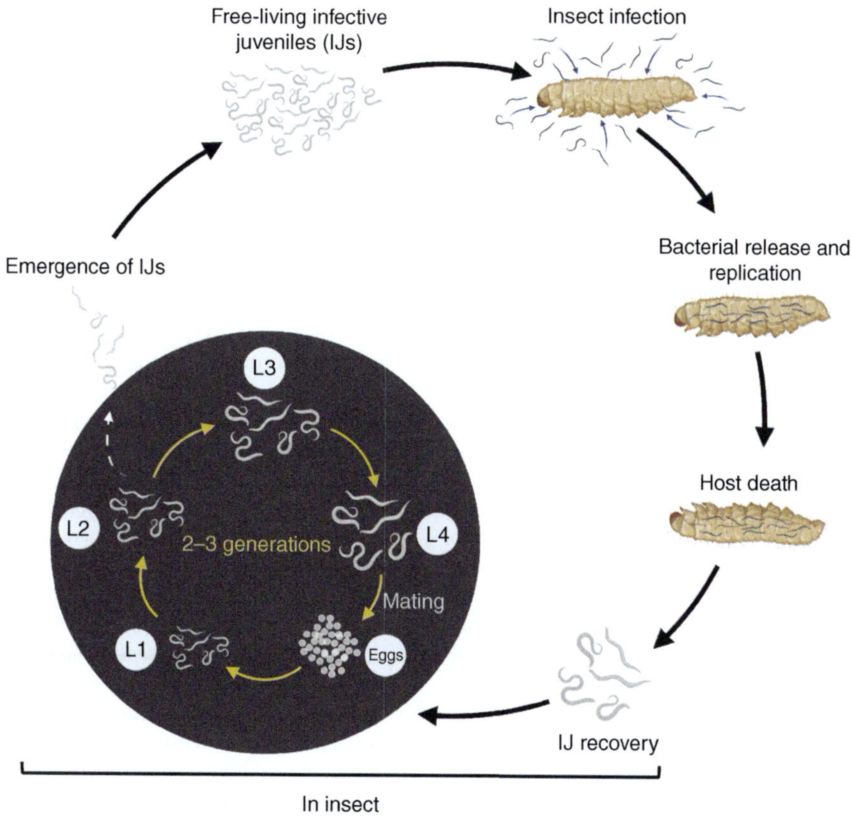

Fig. 3.1. When the infected insect host dies, the entomopathogenic nematode non-feeding and free-living stage, called infective juveniles (IJs), exits the insect cadaver to search for new insect hosts to invade and parasitize. When the nematode IJs enter the insect by penetrating the cuticle or gaining access to the insect body cavity through natural openings, they expel their mutualistic bacterial partners into the insect hemolymph. The released bacteria replicate at high rates and secrete toxins and virulence factors that target various insect tissues. Following insect death, the nematodes reacquire the bacteria by feeding on the degraded tissues. In this way, the nematodes complete two or three generations in the infected insect host. Each generation consists of the egg stage, four larval stages (L1–L4), and the adult stage. *Steinernema* spp. IJs may become males or females, whereas *Heterorhabditis* spp. develop into hermaphrodites that later produce males and females within the host.

hermaphroditum, only produce adult males and females (no hermaphrodites).

3.3 The Insect Innate Immune System

Insects have developed an innate immune system that responds to microbial invaders and eukaryotic parasites and is controlled by a complex and diverse array of signaling pathways (Castillo *et al.*, 2011; Kenney and Eleftherianos, 2016; Riera *et al.*, 2016). The insect innate immune response exhibits physical and chemical barriers, and humoral and cellular reactions, which are also found in all jawed vertebrates (Riera *et al.*, 2016). Physical barriers such as the larval cuticle, adult exoskeleton, chitinous lining of the tracheae and peritrophic membrane of the midgut are the first lines of defense against infectious agents (Lemaitre and Hoffmann, 2007; Davis and Engström, 2012).

Previous work with model insects has revealed the molecular factors that EPNs and their mutualistic bacteria employ to interfere with the insect immune system, as well as the hosts' immune strategies to promote their survival and propagation (Castillo *et al.*, 2011; Kenney and Eleftherianos, 2016; Eleftherianos *et al.*, 2017). The conservation of primary components of innate immunity between vertebrates and invertebrates renders insects as powerful models for clarifying innate immune functions in more complex organisms (Riera Romo *et al.*, 2016). For example, genetic and molecular studies in the fruit fly (*Drosophila melanogaster*) model have uncovered the innate immune mechanisms regulating the signaling pathways mediating antimicrobial peptide (AMP) gene expression in animals (Lemaitre and Hoffmann, 2007). Therefore, investigation and characterization of anti-nematode mechanisms that regulate the insect immune response to EPNs will contribute substantially to the development of better strategies to control insect pests and potentially nematode parasites that infect mammals (Campos-Herrera *et al.*, 2012).

When parasitic nematodes overcome the initial barrier of defense, they gain access into the insect body cavity where they are detected through currently unknown mechanisms by the insect immune system (Stokes *et al.*, 2015). Recognition of the parasites leads to the activation of molecular signaling pathways that regulate the synthesis of humoral effector molecules including AMPs and reactive oxygen species, as well as cellular immune mechanisms such as phagocytosis, melanization and encapsulation (Fig. 3.2) (Lemaitre and Hoffmann, 2007; Vlisidou and Wood, 2015). The insect humoral and cellular anti-nematode immune responses will be presented in detail below.

3.3.1 Insect humoral immune responses

The most studied insect humoral responses, especially in *Drosophila* spp., are predominantly regulated by two distinct nuclear factor (NF-κB) signaling pathways, the Toll and Imd (immune deficiency) pathways (Lemaitre and Hoffmann, 2007; Castillo *et al.*, 2011; Kenney and Eleftherianos, 2016). Induction of these signaling cascades ultimately leads to the transient production of AMPs by the fat body, the equivalent to the mammalian liver. Seven classes of immune-inducible AMPs have been thoroughly investigated in *Drosophila* spp.: Diptericin, Attacin, Drosocin and Cecropin are mainly active against Gram-negative bacteria; Defensin acts against Gram-positive bacteria; and Drosomycin and Metchnikowin have antifungal properties (Lemaitre and Hoffmann, 2007). Interestingly, Attacin, Drosocin, Cecropin and Defensin are also present in other insects, whereas Drosomycin, Diptericin and Metchnikowin have been found so far only in *Drosophila* spp. Other immune molecules with anti-pathogen properties have been found in lepidopteran insects (moths and butterflies) and include C-type lectins, hemolin, serpins and hemolymph proteases (Casanova-Torres and Goodrich-Blair, 2013).

The Toll pathway is homologous to mammalian Toll/interleukin-1 (IL-1) receptor signaling. It is an evolutionarily conserved signaling pathway that controls the establishment of the dorso-ventral axis of *Drosophila* spp. embryos (Moussian and Roth, 2005). Toll shares the cytoplasmic TIR (Toll/IL-1 receptor) domain and leucine-rich repeats with mammalian Toll-like receptors. In addition, the Toll pathway regulates the systemic expression of AMPs against Gram-positive bacterial and fungal infections (Lindsay and Wasserman, 2014). Toll is induced on detection of lysine-type peptidoglycan (PGN) by the PGN recognition proteins PGRP-SA and PGRP-SD, which are circulating pattern-recognition receptors. Recognition of microbial patterns activates a common proteolytic cascade downstream of the pattern-recognition receptors and leads to the induction of the cytokine Spätzle, which binds to the cell transmembrane receptor Toll and causes the receptor's dimerization (Lindsay and Wasserman, 2014). Activation of the Toll pathway terminates with the transcription of genes encoding the AMPs Drosomycin, Metchnikowin and Defensin. Curiously, previous transcriptomic studies in *D. melanogaster* larvae have shown that infection by *Steinernema carpocapsae* strongly upregulates the Toll receptor-associated adaptor MyD88, suggesting the participation of this pathway in the regulation of the insect anti-nematode immune response (Yadav *et al.*, 2018).

The Imd signaling pathway is homologous to the mammalian tumor necrosis factor (TNF)

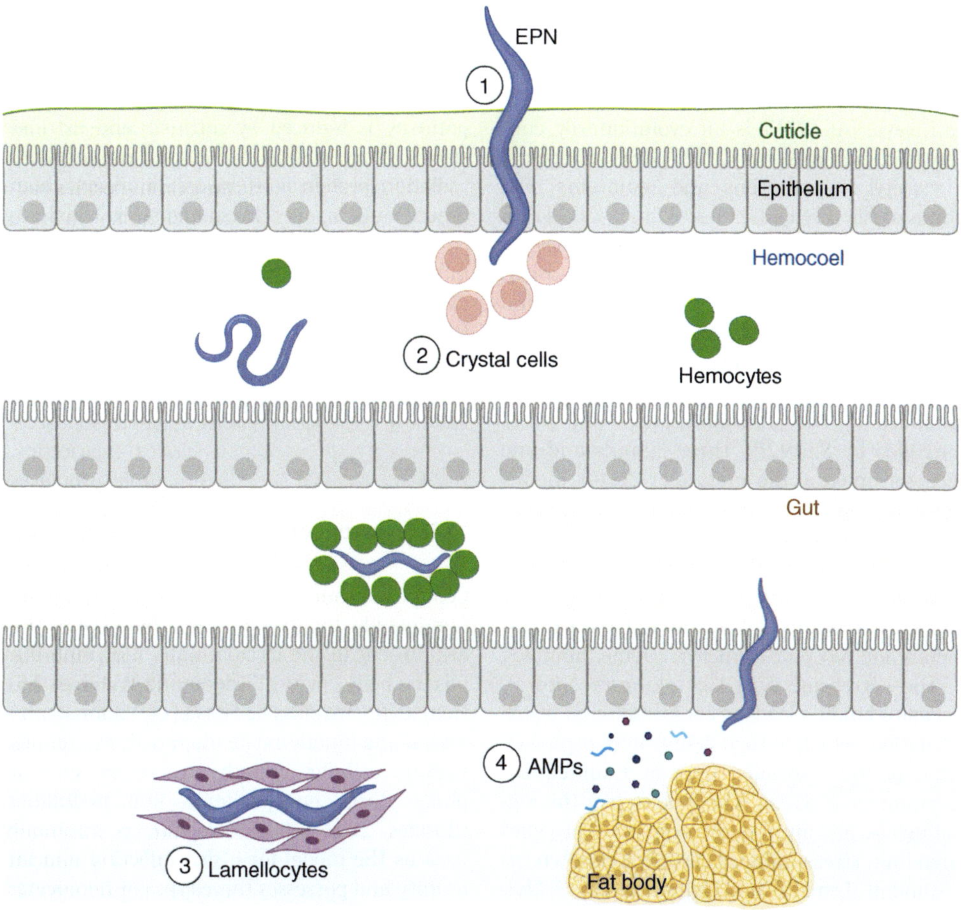

Fig. 3.2. Insect immune responses against entomopathogenic nematodes (EPNs). (1) Insect larvae prevent the entry of EPNs by physical barriers consisting of the larval cuticle and the chitinous lining of the tracheae. (2) On EPN invasion, crystal cells move rapidly to the penetration site to secrete melanin to seal the wound and restrict the access and spread of the parasites. Melanin production is regulated through the activation of the prophenoloxidase cascade. (3) EPNs migration in the insect body cavity induces cellular immune reactions, which mainly involve the aggregation of hemocyte populations and the encapsulation response by lamellocytes. (4) The insect humoral immune response occurs primarily in the fat body and involves the expression of antimicrobial peptides (AMPs) and other stress-related molecules.

receptor 1 (TNFR1) signaling pathway (Kleino and Silverman, 2014). Imd signaling responds to Gram-negative bacterial infections and controls both local and systemic expression of AMPs by the gut epithelial cells and the fat body cells, respectively. Previous work has indicated that the expression of certain components of the Imd pathway is influenced following infection with *Photorhabdus luminescens* and *Xenorhabdus nematophila* bacteria (Aymeric *et al*., 2010). Initiation of the Imd pathway involves the detection of Dap-type PGN by the recognition receptors PGRP-LC and PGRP-LE, a process that leads to the activation of the intracellular adaptor molecule Imd (homolog of receptor-interacting protein, RIP) and the downstream protein complexes of the signaling cascade (Kleino and Silverman, 2014). Transduction of the signal to the nucleus triggers the expression of AMP genes, including *diptericin, attacin, drosocin* and *cecropin* (Lemaitre and Hoffmann, 2007; Kleino and Silverman, 2014).

Anti-pathogen properties in insects are also modulated by the janus kinase/signal transducer and activator of transcription (JAK/STAT) and the Jun N-terminal kinase (JNK) signaling pathways. JAK/STAT is an evolutionarily conserved immune pathway that is directed against a variety of pathogens and maintains host homeostasis (Myllymäki and Rämet, 2014). In addition, the JAK/STAT pathway is involved in the regulation of cell growth, differentiation, apoptosis and inflammation (Myllymäki and Rämet, 2014; Bang, 2019). In contrast to the multiple JAK/STAT pathway components in vertebrates, *Drosophila* spp. possess only one JAK (encoded by the gene *hopscotch*), one STAT (encoded by *Stat92E*), three ligands and one receptor, making the molecular events in this signaling pathway less redundant and simpler to analyze (Myllymäki and Rämet, 2014; Bang, 2019). Initiation of the JAK/STAT pathway in *Drosophila* spp. is triggered by the extracellular binding of the unpaired ligand Upd to the transmembrane receptor Domeless (Dome, homolog of the vertebrate type I cytokine receptors). Receptor binding leads to conformational transformation, which in turn results in phosphorylation of the associated JAKs and subsequent recruitment of STATs. The final step of this signaling cascade involves the expression of genes encoding stress-induced proteins known as Turandots (Lemaitre and Hoffmann, 2007; Myllymäki and Rämet, 2014). Experiments with loss-of-function mutants and employment of RNA interference-based gene silencing in *Drosophila* and mosquito species have produced valuable evidence for the functional significance of JAK/STAT signaling members and regulation of immune responses (Bang, 2019). Interestingly, infection of *D. melanogaster* larvae with *S. carpocapsae* nematodes deficient of their *X. nematophila* bacteria upregulates the receptor Dome. This illustrates the potential involvement of the JAK/STAT pathway in the recognition of EPN molecular patterns (Yadav *et al.*, 2018).

JNK is an evolutionarily conserved mitogen-activated protein kinase (MAPK) that diverges from the Imd pathway and participates in several biological processes including cell-cycle regulation, apoptosis and phagocytosis (Tafesh-Edwards and Eleftherianos, 2020). Many of the JNK pathway components in *Drosophila* spp. are also present in vertebrates. These include Hemipterous (Hep), an orthologue of human MAP2K7, Bsk (DJNK) and DJun/DFos (Fos proto-oncogene), which control dorsoventral patterning and immunity (Sluss *et al.*, 1996). The JNK pathway is induced by intrinsic and extrinsic stress factors, such as oxidative stress, ultraviolet radiation, protein synthesis inhibitors and chemotherapeutic agents. In addition, JNK signaling can be activated by cell death ligands and pro-inflammatory cytokines such as TNF-α and IL-1 (Sluss *et al.*, 1996; Shen and Liu, 2006). Importantly, the JNK pathway has previously been shown to provide a survival advantage to *D. melanogaster* larvae following infection with axenic *S. carpocapsae* (Yadav *et al.*, 2018).

3.3.2 Insect cellular immune responses

Cellular immune mechanisms in insects are regulated by hemocytes (insect blood cells) that are present in the circulating hemolymph that fills the body cavity (Vlisidou and Wood, 2015). Hemocytes are classified based on their morphological and functional characteristics to perform various cellular immune responses such as phagocytosis, encapsulation and nodulation (Rosales, 2017). *D. melanogaster* is commonly used as the model for insect infection and immunity and possesses three types of hemocytes: plasmatocytes, crystal cells and lamellocytes (Vlisidou and Wood, 2015).

Plasmatocytes are macrophage-like cells that comprise the majority of the hemocytes in the circulating hemolymph. Their main function is to perform phagocytosis involving the engulfment and degradation of microbes, foreign particles or apoptotic cells (Gold and Brückner, 2015). The recognition process occurs through cell-surface proteins including scavenger receptors CD36 homolog Croquemort (Crq), Draper and NimC4/Simu, and the immunoglobulin superfamily protein DSCAM (Down syndrome cell-adhesion molecule). However, in *D. melanogaster* larvae and adult flies, recognition and engulfment of invading microbes depend mainly on the cell-surface receptors NimC1 and Eater (Chung and Kocks, 2011). Nimc1 detects Gram-negative bacteria, and Eater is the main phagocytic receptor for Gram-positive bacterial cells; Eater also facilitates the binding of microbes to hemocytes. Interestingly, both receptors synergize to perform the phagocytosis of bacteria, and Eater can bypass the requirement for NimC1 (Chung and Kocks, 2011).

Crystal cells are relatively large cells with crystalline inclusions. These cells are not capable of performing phagocytosis of microbes, but they regulate the melanization response that connects humoral and cellular immunity (Dudzic *et al.*, 2015). Crystal cells are responsible for the rapid synthesis of melanin at the site of infection in order to restrict the dissemination of microbial invaders and accelerate wound healing by forming a scab at the epithelial site of injury. Melanin synthesis is mediated through the function of the phenoloxidase (PO) enzyme. The PO enzyme is synthesized as an inactive precursor called prophenoloxidase (PPO) that is cleaved to generate active PO as a result of proteolytic cascade activation. Active PO catalyzes the oxidation of phenols to quinones, which eventually polymerize into melanin. The *D. melanogaster* genome encodes three *PPO* genes. *PPO1* and *PPO2* are expressed mainly in crystal cells; the cells are easily disrupted to release their contents into the hemolymph upon activation. Melanin formation has been associated with several immune roles in *D. melanogaster* including the encapsulation of parasites (Dudzic *et al.*, 2015). Specifically, deposition of melanin on the parasite forms a physical barrier, allowing the localized and confined production of toxic compounds while ensuring protection of the host. In addition, by-products of PO activity include reactive oxygen species, which are thought to contribute to the killing of pathogens.

Lamellocytes are large, flat, adherent cells that are not found in embryos or adults and are rarely present in healthy larvae. The main function of lamellocytes is the encapsulation and neutralization of objects too large to be phagocytosed by plasmatocytes, such as parasitoid eggs (Lemaitre and Hoffmann, 2007; Vlisidou and Wood, 2015). Lamellocytes are produced primarily in the hematopoietic organ (lymph gland), which is activated on recognition of foreign invaders. This process leads to the rapid proliferation of lamellocytes within the *D. melanogaster* hemocoel (Honti *et al.*, 2014; Satyavathi *et al.*, 2014). Expansion of lamellocytes aids in the formation of large nodules consisting of hemocyte aggregates (nodulation response) and the subsequent entrapment of the foreign object (encapsulation response). Capsule formation around the invading organism results in the production of reactive cytotoxic molecules, which have lethal properties (Carton *et al.*, 2009).

3.3.3 Insect coagulation response

The coagulation response in insects ensures fast and efficient closure of wounds, prevents hemolymph loss and opposes systemic infections (Loof *et al.*, 2011). Hemolymph coagulation in insects improves homeostasis and participates in the host immune response against EPNs (Loof *et al.*, 2011; Dziedziech *et al.*, 2020). In particular, insect coagulation can be viewed as a primary link between humoral and cellular responses that expands the overall systemic immune response against various pathogens, activating successive signaling cascades in cells and regulating their transcriptional responses accordingly.

Most of the current information on hemolymph coagulation in insects derives from insect models such as *D. melanogaster* (Theopold *et al.*, 2014; Dziedziech *et al.*, 2020) and natural insect pests such as the fall armyworm, *Spodoptera frugiperda* (Huot *et al.*, 2019). A common feature of coagulation in insects is found in the family of transglutaminases (homologs of the vertebrate clotting factor XIIIa) (Theopold *et al.*, 2014). Insect transglutaminase activity accumulates on microbial surfaces, leading to their sequestration into the clot (Wang *et al.*, 2010). As a consequence, pathogens can be either destroyed extracellularly through the action of AMPs or eliminated intracellularly through phagocytosis. Subsequent steps also include melanization events and epithelial movements (Galko and Krasnow, 2004).

Analysis of hemolymph clots in *D. melanogaster* has shown that hemolectin (homolog of the vertebrate clotting factor von Willebrand factor) is expressed by plasmatocytes and is the most abundant protein in a clot (Goto *et al.*, 2001). Hemolymph coagulation is independent of both melanin production and PO activity. Instead, melanin production and PO activity are regulated by plasmatocytes and several clotting factors (Goto *et al.*, 2003). These clotting factors include transglutaminase (TG) and Fondue (Galko and Krasnow, 2004; Wang *et al.*, 2010;

Theopold *et al.*, 2014). TG is expressed in hemocytes while Fondue is found mainly in the fat body and is regulated through the Toll pathway. This indicates the participation of both cellular and humoral factors in blood coagulation with indispensable contributions from hemocytes. Interestingly, TG is the only evolutionarily conserved component of clotting cascades, and synergizes with Fondue to crosslink clot fibers formed by hemolectin to respond rapidly to injury and microbial infection (Johansson *et al.*, 2005; Scherfer *et al.*, 2006).

3.4 Insect Immune Response to Entomopathogenic Nematodes

Insects can activate a robust systemic immune response against EPNs that invade their hemocoel. Insect anti-nematode immune responses mount distinct signaling pathways and effector molecules. The specific immune responses include activation of hemolymph clotting, triggering the melanization reaction and ultimately encapsulating the nematode in layers of hemocytes (Castillo *et al.*, 2011). However, EPNs employ distinct tactics to interfere with the insect immune system, especially during the early stages of infection. EPN immune system suppression strategies aim to facilitate the survival and replication of both the nematode parasites and their mutualistic bacteria (Toubarro *et al.*, 2013b). In particular, the insect immune response is sequentially challenged by the two mutualistic partners. The sequential response is due to the delayed release of the bacteria by their nematode host once the parasites reach the hemolymph (Li *et al.*, 2007). However, the exact insect immune response against EPNs depends strongly on host and nematode specificity.

Comparative analysis of the infectivity of *S. carpocapsae*, *Steinernema glaseri*, *Steinernema scapterisci* and *Heterorhabditis bacteriophora* to Japanese beetle (*Popillia japonica*) larvae suggests that *P. japonica* mounts strong encapsulation and melanization responses to all species except *S. glaseri* (Wang *et al.*, 1994). Although *H. bacteriophora* and *S. carpocapsae* can overcome the immune response, *S. scapterisci* is neutralized by the host immune system. In addition, infection of *S. scapterisci* and *H. bacteriophora* in the adult stage of the house cricket, *Acheta domesticus*, results in successful removal of the parasites without affecting the mortality of the host (Wang *et al.*, 1994). EPN inhibition via encapsulation is also noticeable in infection of the Colorado potato beetle, *Leptinotarsa decemlineata*, with *Heterorhabditis marelata* nematodes (Armer *et al.*, 2004). Similar research suggests that *S. glaseri* infection in *P. japonica* larvae provokes reduced encapsulation and melanization responses in comparison with *H. bacteriophora* infection, thus strengthening parallel reports of *S. glaseri* persistence in this particular host (Li *et al.*, 2007). Despite the success of *S. glaseri* in *P. japonica*, this nematode fails to establish an infection in *A. domesticus*. Although *S. glaseri* FL strain, *Steinernema scarabaei*, *Steinernema feltiae* and *H. bacteriophora* are efficiently melanized by hemocytes from the oriental beetle, *Exomala orientalis*, tobacco hornworm, *Manduca sexta*, and *P. japonica*, *S. glaseri* NC strain can escape from hemocyte encapsulation and suppress the immune responses in these hosts (Li *et al.*, 2007).

The findings depicted above provide strong evidence not only that the potency of immune responses varies between insect hosts, but also that defense reactions can be specific to the nematode species and strain. It is worth noting that variations in EPN infectivity and especially the ability of the parasites to evade the host immune system can further elicit distinct differences in the activation of humoral and cellular responses as well as the melanization reaction. It is possible that differences in EPN recognition between insect hosts could reflect changes in the regulation of immune mechanisms against nematodes and their related bacteria, which could render a particular insect species resistant or susceptible to EPN infection.

3.5 Nematode Immunomodulatory Strategies

The success of EPNs as potent insect pathogens is due to their ability to produce and release an assortment of molecules with pathogenic properties and molecules that participate in suppression of the host immune system. This strategy is known as immunomodulation, and allows EPNs not only to survive and migrate within the host but also to cause a powerful infection that is characterized by extensive damage on host tissues (Cooper and Eleftherianos, 2016; Bobardt *et al.*, 2020). This is achieved mainly though suppression of the host defense during EPN infection by modifying the activation of the host immune response and involves three possible strategies: (i) EPNs are recognized as self and thus prevent the activation of immune responses from the host; (ii) EPNs are not recognized as non-self by the host immune system and thus avoid detection; and (iii) EPNs are recognized as non-self by the host immune system but they use tactics to inhibit host immune signaling and function (Eleftherianos *et al.*, 2017). The different immunomodulation strategies employed by EPNs to persist in the host comprise processes and components that differ with the nematode species and the infected insect host (Klei, 1997; Hartmann and Lucius, 2003; McSorley *et al.*, 2013). For instance, EPNs can move to distant tissues where they become undetected by the host immune system (anatomical seclusion), they can sequester hemolymph proteins to avoid immune recognition (camouflage or mimicry), and they release factors that actively reduce host immune activities (interference) (Castillo *et al.*, 2011; Cooper and Eleftherianos, 2016; Eleftherianos *et al.*, 2017; Bobardt *et al.*, 2020). Expanding the current knowledge on EPN immunomodulation strategies will undoubtedly contribute toward designing improved biological control procedures to combat insect pests and obtaining crucial information for preventing parasitic nematode infections in humans.

EPN immunomodulation is predominantly associated with nematode body surface compounds and the excretory/secretory (ES) effector molecules they produce during insect infection (Castillo *et al.*, 2011; Cooper and Eleftherianos, 2016; Eleftherianos *et al.*, 2017; Bobardt *et al.*, 2020). More precisely, epicuticular lipids from *S. feltiae* can reduce antibacterial peptide synthesis in *Galleria mellonella*. Following bacterial infection, *G. mellonella* larvae previously challenged with purified *S. feltiae* cuticles exhibit a decreased AMP response and a lack of bacterial clearance from the hemocoel, which is due to the binding of certain hemolymph proteins by *S. feltiae* cuticular lipids (Brivio *et al.*, 2002). Interestingly, degradation of the EPN cuticular lipids prior to injection into insect larvae inhibits AMP inhibitory effects. This further supports rescue experiments in which simultaneous injection of purified hemolymph proteins with bacterial cells into insects that have been previously infected results in the recovery of the host's antimicrobial response (Brivio *et al.*, 2002; Eleftherianos *et al.*, 2017).

Regarding the ES effectors that are responsible for EPN immunomodulation, recent studies in *H. bacteriophora* have revealed that their ES products are lethal to their insect hosts at high concentrations (Kenney *et al.*, 2019, 2021). Proteases have been identified as major components that *Heterorhabditis* and *Steinernema* spp. employ to dampen the host immune system and to facilitate nematode invasion through the digestion of proteins within the infected host. In particular, the well-conserved AMP Cecropin can be suppressed in *G. mellonella* and other lepidopteran and hymenopteran insects by *H. bacteriophora* nematodes and their related *Photorhabdus luminescens* bacteria through the production and release of a specific protease (Jarosz, 1998). Furthermore, application of proteins extracted from activated *H. bacteriophora* (nematodes that had previously been exposed to insect hemolymph) to *D. melanogaster* adults, larvae and isolated larval fat bodies substantially suppresses expression of the AMP-encoding gene *diptericin*. The *diptericin* gene is regulated by the Imd pathway and targets Gram-negative bacteria (Kenney *et al.*, 2021). This forms an immunomodulatory mechanism that promotes the infection of both the EPN and its associated *Photorhabdus luminescens* bacteria (the bacteria would otherwise be eliminated by the insect host). Therefore, the activated nematode ES products induce a more rapid propagation of *Photorhabdus luminescens* by reducing the expression of Diptericin. The EPN immunomodulatory strategy leads to faster insect host mortality. Recently, a putative uridine diphosphate glycosyltransferase (*Heterorhabditis bacteriophora* (Hb)-UGT-1), invertebrate-type lysozyme (Hb-ILYS-1) and serine carboxypeptidase (Hb-SC-1) have been characterized and each has demonstrated a distinct set of immunomodulatory properties (Kenney *et al.*, 2020a,b). In particular, injection of the recombinant form of Hb-SC-1 was shown to be lethal to adult flies, which is due to its ability to prevent the full upregulation of AMPs, to limit PO enzyme activity and to suppress phagocytosis (Fig. 3.3) (Kenney *et al.*, 2020b).

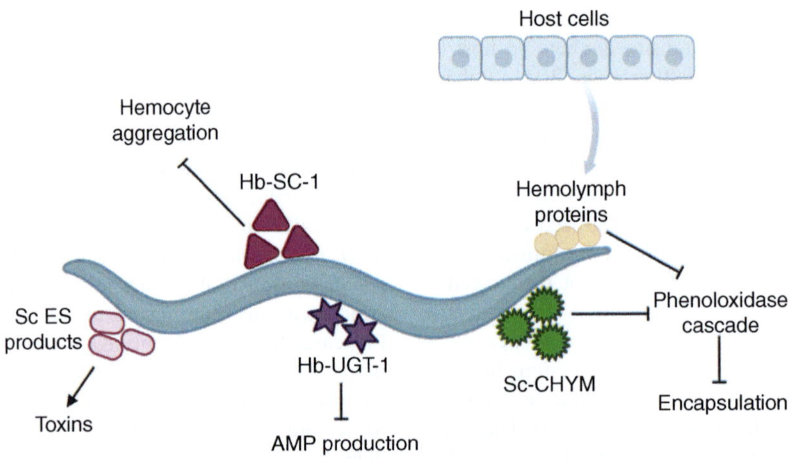

Fig. 3.3. Interaction of entomopathogenic nematodes (EPNs) with the insect immune system. EPNs are capable of using their cuticular lipids to recruit host hemolymph proteins, a process that results in reduced phenoloxidase enzyme activity and melanin production. They can also interfere with the prophenoloxidase cascade through the production of immunomodulatory molecules (e.g. *Heterorhabditis bacteriophora* (Hb) excretory/secretory (ES) products, Sc-CHYM and Hb-SC-1). EPNs can use their cuticular lipids and secreted molecules (e.g. *Steinernema carpocapsae* (Sc) ES products, Sc-CHYM and Hb-SC-1) to inhibit hemocytic functions. EPNs can further employ their cuticular lipids and various secreted components (e.g. Sc ES products, Hb ES products, Hb-UGT-1 and Hb-SC-1) to suppress the synthesis of antimicrobial peptides during infection of the insect host. Finally, some EPN secreted effectors (e.g. Sc ES products, Hb ES products, Hb-SC-1) are directly toxic and cause lethality to insect hosts.

The ES molecules of EPNs are also able to suppress the insect cellular response in order to facilitate their dissemination of nematodes in the host and the spread of their bacteria. Research with *S. carpocapsae* and *X. nematophila* have shown that infection with nematodes at early stages and before they release their endosymbiotic bacteria causes a sharp decrease in the total number of hemocytes (the central modulators of the insect cellular immune response). These findings support the idea that the EPNs attack the host cellular immune response to accelerate the propagation of their bacterial partners (Garriga *et al.*, 2020). Similar experiments with *S. carpocapsae* IJs have found that ES products from this EPN can disrupt the adhesion of plasmatocytes and granular cells during infection of the lepidopteran *G. mellonella* and the forest tent caterpillar moth, *Malacosoma disstria* (Walter *et al.*, 2008). This immunomodulatory effect also limits the PO response in the hemolymph and negatively impacts the encapsulation of the parasites (Toubarro *et al.*, 2009). Early studies in the armyworm moth, *Mythimna unipuncta*, infected with *S. carpocapsae* indicated that the production of EPN factors reduced the number of hemocytes and interfered with hemocyte activity (Ribeiro *et al.*, 1999). Isolation and characterization of these factors revealed their non-sticky and cytotoxic properties toward plasmatocytes and granular cells, respectively; these properties probably restrict nodulation and encapsulation responses in the infected host.

Other studies using the *G. mellonella* model reveal that the cuticle of *S. feltiae* has the ability to sequester key insect hemolymph proteins and use them as a shield to promote the suppression of hemocyte aggregation and encapsulation, and therefore the persistence of the nematode parasites in the host (Mastore and Brivio, 2008). In addition, *S. carpocapsae* IJs produce the chymotrypsin-like serine protease Sc-SP-3, which is expressed only in this EPN stage (Toubarro *et al.*, 2009). Interestingly, Sc-SP-3 is expressed at higher levels when the nematodes are present in the midgut compared with those that enter the hemocoel. Sc-SP-3 has multifunctional activity because it causes histolysis in the insect midgut, digests extracellular proteins and induces apoptosis in insect cells. The multi-tasking capacity of this chymotrypsin-like protease indicates its central role in *S. carpocapsae* pathogenesis (Toubarro *et al.*, 2009). In the *G. mellonella* model, another *S. carpocapsae* secreted trypsin-like serine protease affects the host cellular immune response by inducing severe morphological changes to hemocytes and preventing their spread (Balasubramanian *et al.*, 2010). Likewise, the BPTI–Kunitz family of inhibitors (Sc-KU-4), produced by *S. carpocapsae*, not only inhibits the encapsulation response but also blocks hemocyte aggregation (Toubarro *et al.*, 2013a).

In order to evade the host immune response, EPNs utilize their cuticle or ES products to interfere with the insect PO response and suppress the melanization cascade. Infection of *G. mellonella* larvae with *S. feltiae* nematodes drastically reduces the enzymatic activity of the PPO cascade in the host hemolymph (Brivio *et al.*, 2002). The mechanism behind this immunomodulatory effect involves *S. feltiae* cuticle lipidic molecules, which interact with the host humoral factors and remove them from the insect hemolymph (Brivio *et al.*, 2004). Sequestering host immune components possibly leads to the substantial reduction in PPO cascade activity and lower encapsulation response. Biochemical and molecular characterization of an effector molecule (chymotrypsin protease, Sc-CHYM), which is secreted by the IJs of *S. carpocapsae*, has revealed its role as an EPN virulence factor due to its ability to prevent activation of the PPO system in the hemolymph of *G. mellonella* larvae and subsequently inhibit the melanization and encapsulation responses against these nematodes (Balasubramanian *et al.*, 2009). Remarkably, sequencing reveals that Sc-CHYM is clustered with several insect chymotrypsins, forming an ancestral branch in the phylogenetic tree; the result suggests that Sc-CHYM branches off at an early stage of a cluster divergence (Balasubramanian *et al.*, 2009). More recently, isolation and purification of ES products produced by *H. bacteriophora* has further displayed a number of peptides that can suppress the PO enzyme activity in the hemolymph of *G. mellonella* (Eliáš *et al.*, 2020). Six of the identified peptide sequences include molecules that coordinate the insect immune system through modulating the ubiquitination and Toll signaling pathway activity.

The use of transcriptomic and functional approaches has deciphered exciting new knowledge on the pathogenesis strategies employed by EPNs. This knowledge is crucial for identifying the host immune components that interact with EPN invasion and migration in the host during infection. EPN immunomodulatory strategies are highly specific to the nematode species and the infected insect host. They are launched to weaken the host immune system and create suitable conditions for the nematodes and their mutualistic bacteria. For example, hemocytes in *G. mellonella* can detect the presence of *H. bacteriophora* but not *S. carpocapsae* or *S. glaseri* IJs (Ebrahimi *et al.*, 2011). Suppression of humoral and cellular immune mechanisms promotes nematode migration in the host and enhances EPN pathogenicity (Ebrahimi *et al.*, 2011). More recent studies have confirmed that *H. bacteriophora* can produce ES products that disrupt host immune functions in *D. melanogaster* (Kenney *et al.*, 2019, 2021). However, little is known about the effect of EPN ES products on the innate immune system of natural insect hosts. These effects could be the result of a collection of different ES components accomplishing a variety of roles, each of which might be important for disrupting specific innate immune signaling pathways and host defenses. Further identification and characterization of the specific EPN-derived molecules that target the insect immune system and determination of their conservation in mammalian parasitic nematodes will potentially lead to the development of new treatments for infectious diseases.

3.6 Conclusions and Perspectives

A unique feature of EPNs as biological control agents involves their species-specific mutualism with Gram-negative bacteria. EPN–bacterial complexes are capable of entering suitable insect hosts and establishing infection. This extraordinary natural system of mutualism and synergistic infection is widely used in biological control practices as ecologically friendly alternatives to chemical insecticides for the efficient control of agricultural insect pests and disease vectors. In addition, EPNs constitute suitable experimental models for understanding the molecular basis of nematode parasitism and bacterial pathogenicity in relation to the host innate immune system (Tafesh-Edwards and Eleftherianos, 2022). Although significant discoveries have been made recently to elucidate the contribution of the EPN and their mutualistic bacteria in overcoming the host innate immune response, future studies will focus on elucidating the precise nature of the co-operation between the immunomodulatory properties of the ES products secreted by the EPNs and the tissue-degrading properties of virulence factors produced by the bacterial partners during the different stages of the infection. Further investigation will be greatly facilitated by the successful employment of targeted knockout, or alternatively targeted knockdown approaches, together with omic studies in the EPN IJ stage. The use of innovative research strategies will lead to the identification of original effector molecules from the EPNs, which will in turn provide novel information on the interaction of the EPN–bacteria complexes with host innate immune signaling and the regulated humoral and cellular immune function. Performing experiments with genetically amenable insect models will further promote detection of the host immune components that are activated or suppressed in response to EPN assault. Clarifying the events that define EPN–insect host interactions at the molecular and mechanistic levels will prove crucial for developing and integrating advanced pest management tactics in the field.

3.7 References

Armer, C.A., Rao, S. and Berry, R.E. (2004) Insect cellular and chemical limitations to pathogen development: the Colorado potato beetle, the nematode *Heterorhabditis marelatus*, and its symbiotic bacteria. *Journal of Invertebrate Pathology* 87, 114–122.

Aymeric, J.L., Givaudan, A. and Duvic, B. (2010) *Imd* pathway is involved in the interaction of *Drosophila melanogaster* with the entomopathogenic bacteria, *Xenorhabdus nematophila* and *Photorhabdus luminescens*. *Molecular Immunology* 47, 2342–2348.

Balasubramanian, N., Hao, Y.J., Toubarro, D., Nascimento, G. and Simões, N. (2009) Purification, biochemical and molecular analysis of a chymotrypsin protease with prophenoloxidase suppression activity from the entomopathogenic nematode *Steinernema carpocapsae*. *International Journal for Parasitology* 39, 975–984.

Balasubramanian, N., Toubarro, D. and Simões, N. (2010) Biochemical study and *in vitro* insect immune suppression by a trypsin-like secreted protease from the nematode *Steinernema carpocapsae*. *Parasite Immunology* 32, 165–175.

Bang, I.S. (2019) JAK/STAT signaling in insect innate immunity. *Entomological Research* 49, 339–353.

Bobardt, S.D., Dillman, A.R. and Nair, M.G. (2020) The two faces of nematode infection: virulence and immunomodulatory molecules from nematode parasites of mammals, insects and plants. *Frontiers in Microbiology* 11: 577846.

Brivio, M.F., Pagani, M. and Restelli, S. (2002) Immune suppression of *Galleria mellonella* (Insecta, Lepidoptera) humoral defenses induced by *Steinernema feltiae* (Nematoda, Rhabditida): involvement of the parasite cuticle. *Experimental Parasitology* 101, 149–156.

Brivio, M.F., Mastore, M. and Moro, M. (2004) The role of *Steinernema feltiae* body-surface lipids in host–parasite immunological interactions. *Molecular and Biochemical Parasitology* 135, 111–121.

Campos-Herrera, R., Barbercheck, M., Hoy, C.W. and Stock, S.P. (2012) Entomopathogenic nematodes as a model system for advancing the frontiers of ecology. *Journal of Nematology* 44, 162–176.

Carton, Y., Frey, F. and Nappi, A.J. (2009) Parasite-induced changes in nitric oxide levels in *Drosophila paramelanica*. *Journal of Parasitology* 95, 1134–1141.

Casanova-Torres, Á.M. and Goodrich-Blair, H. (2013) Immune signaling and antimicrobial peptide expression in Lepidoptera. *Insects* 4, 320–338.

Castillo, J.C., Reynolds, S.E. and Eleftherianos, I. (2011) Insect immune responses to nematode parasites. *Trends in Parasitology* 27, 537–547.

Chung, Y.S. and Kocks, C. (2011) Recognition of pathogenic microbes by the *Drosophila* phagocytic pattern recognition receptor Eater. *Journal of Biological Chemistry* 286, 26524–26532.

Cooper, D. and Eleftherianos, I. (2016) Parasitic nematode immunomodulatory strategies: recent advances and perspectives. *Pathogens (Basel)* 5: 58.

Dara, S.K. (2017) Microbial control of arthropod pests in small fruit and vegetables in temperate climate. In: Lacey, L.A. (ed.) *Microbial Control of Insect and Mite Pests: From Theory to Practice*. Academic Press, London, pp. 209–221.
Davis, M.M. and Engström, Y. (2012) Immune response in the barrier epithelia: lessons from the fruit fly *Drosophila melanogaster*. *Journal of Innate Immunity* 4, 273–283.
Dillman, A.R. and Sternberg, P.W. (2012) Entomopathogenic nematodes. *Current Biology* 22, R430–R431.
Dudzic, J.P., Kondo, S., Ueda, R., Bergman, C.M. and Lemaitre, B. (2015) *Drosophila* innate immunity: regional and functional specialization of prophenoloxidases. *BMC Biology* 13: 81.
Dziedziech, A., Shivankar, S. and Theopold, U. (2020) *Drosophila melanogaster* responses against entomopathogenic nematodes: focus on hemolymph clots. *Insects* 11: 62.
Ebrahimi, L., Niknam, G. and Dunphy, G.B. (2011) Hemocyte responses of the Colorado potato beetle, *Leptinotarsa decemlineata*, and the greater wax moth, *Galleria mellonella*, to the entomopathogenic nematodes, *Steinernema feltiae* and *Heterorhabditis bacteriophora*. *Journal of Insect Science* 11: 75.
Ehlers, R.U. (2001) Mass production of entomopathogenic nematodes for plant protection. *Applied Microbiology and Biotechnology* 56, 623–633.
Eleftherianos, I., Shokal, U., Yadav, S., Kenney, E. and Maldonado, T. (2017) Insect immunity to entomopathogenic nematodes and their mutualistic bacteria. *Current Topics in Microbiology and Immunology* 402, 123–156.
Eliáš, S., Hurychová, J., Toubarro, D., Frias, J., Kunc, M. *et al.* (2020) Bioactive excreted/secreted products of entomopathogenic nematode *Heterorhabditis bacteriophora* inhibit the phenoloxidase activity during the infection. *Insects* 11: 353.
Ferreira, T. and Malan, A.P. (2014) *Xenorhabdus* and *Photorhabdus*, bacterial symbionts of the entomopathogenic nematodes *Steinernema* and *Heterorhabditis* and their *in vitro* liquid mass culture: a review. *African Entomology* 22, 1–14.
ffrench-Constant, R., Waterfield, N., Daborn, P., Joyce, S., Bennett, H., *et al.* (2003) *Photorhabdus*: towards a functional genomic analysis of a symbiont and pathogen. *FEMS Microbiology Reviews* 26, 433–456.
Galko, M.J. and Krasnow, M.A. (2004) Cellular and genetic analysis of wound healing in *Drosophila* larvae. *PLOS Biology* 2: e239.
Garriga, A., Mastore, M., Morton, A., Pino, F.G.D. and Brivio, M.F. (2020) Immune response of *Drosophila suzukii* larvae to infection with the nematobacterial complex *Steinernema carpocapsae–Xenorhabdus nematophila*. *Insects* 11: 210.
Gold, K.S. and Brückner, K. (2015) Macrophages and cellular immunity in *Drosophila melanogaster*. *Seminars in Immunology* 27, 357–368.
Goto, A., Kumagai, T., Kumagai, C., Hirose, J., Narita, H. *et al.* (2001) A *Drosophila* haemocyte-specific protein, hemolectin, similar to human von Willebrand factor. *Biochemical Journal* 359, 99–108.
Goto, A., Kadowaki, T. and Kitagawa, Y. (2003) *Drosophila* hemolectin gene is expressed in embryonic and larval hemocytes and its knock down causes bleeding defects. *Developmental Biology* 264, 582–591.
Han, R. and Ehlers, R.U. (2000) Pathogenicity, development, and reproduction of *Heterorhabditis bacteriophora* and *Steinernema carpocapsae* under axenic *in vivo* conditions. *Journal of Invertebrate Pathology* 75, 55–58.
Hartmann, S. and Lucius, R. (2003) Modulation of host immune responses by nematode cystatins. *International Journal for Parasitology* 33, 1291–1302.
Hominick, W.M. (1990) Entomopathogenic rhabditid nematodes and pest control. *Parasitology Today* 6, 148–152.
Honti, V., Csordás, G., Kurucz, É., Márkus, R. and Andó, I. (2014) The cell-mediated immunity of *Drosophila melanogaster*: hemocyte lineages, immune compartments, microanatomy and regulation. *Developmental and Comparative Immunology* 42, 47–56.
Huot, L., George, S., Girard, P.A., Severac, D., Nègre, A. *et al.* (2019) *Spodoptera frugiperda* transcriptional response to infestation by *Steinernema carpocapsae*. *Scientific Reports* 9: 12879.
Jarosz, J. (1998) Active resistance of entomophagous rhabditid *Heterorhabditis bacteriophora* to insect immunity. *Parasitology* 117, 201–208.
Jess, S., Schweizer, H. and Kilpatrick, M. (2005) Mushroom applications. In: Grewal, P.S., Ehlers, R.-U. and Shapiro-Ilan, D.I. (eds.) *Nematodes as Biocontrol Agents*. CAB International, Wallingford, UK, pp. 191–213.
Johansson, K.C., Metzendorf, C. and Söderhäll, K. (2005) Microarray analysis of immune challenged *Drosophila* hemocytes. *Experimental Cell Research* 305, 145–155.
Kenney, E. and Eleftherianos, I. (2016) Entomopathogenic and plant pathogenic nematodes as opposing forces in agriculture. *International Journal for Parasitology* 46, 13–19.

Kenney, E., Hawdon, J.M., O'Halloran, D. and Eleftherianos, I. (2019) *Heterorhabditis bacteriophora* excreted-secreted products enable infection by *Photorhabdus luminescens* through suppression of the Imd pathway. *Frontiers in Immunology* 10: 2372.

Kenney, E., Yaparla, A., Hawdon, J.M., O'Halloran, D.M., Grayfer, L. and Eleftherianos, I. (2020a) A putative UDP-glycosyltransferase from *Heterorhabditis bacteriophora* suppresses antimicrobial peptide gene expression and factors related to ecdysone signaling. *Scientific Reports* 10: 12312.

Kenney, E., Yaparla, A., Hawdon, J.M., O'Halloran, D.M., Grayfer, L. and Eleftherianos, I. (2020b) A putative lysozyme and serine carboxypeptidase from *Heterorhabditis bacteriophora* show differential virulence capacities in *Drosophila melanogaster*. *Developmental and Comparative Immunology* 114: 103820.

Kenney, E., Hawdon, J.M., O'Halloran, D.M. and Eleftherianos, I. (2021) Secreted virulence factors from *Heterorhabditis bacteriophora* highlight its utility as a model parasite among Clade V nematodes. *International Journal for Parasitology* 51, 321–325.

Klei, T.R. (1997) Immunological control of gastrointestinal nematode infections. *Veterinary Parasitology* 72, 507–523.

Kleino, A. and Silverman, N. (2014) The *Drosophila* IMD pathway in the activation of the humoral immune response. *Developmental and Comparative Immunology* 42, 25–35.

Lacey, L.A. and Georgis, R. (2012) Entomopathogenic nematodes for control of insect pests above and below ground with comments on commercial production. *Journal of Nematology* 44, 218–225.

Lemaitre, B. and Hoffmann, J. (2007) The host defense of *Drosophila melanogaster*. *Annual Review of Immunology* 25, 697–743.

Li, X.Y., Cowles, R.S., Cowles, E.A., Gaugler, R. and Cox-Foster, D.L. (2007) Relationship between the successful infection by entomopathogenic nematodes and the host immune response. *International Journal for Parasitology* 37, 365–374.

Lindsay, S.A. and Wasserman, S.A. (2014) Conventional and non-conventional *Drosophila* Toll signaling. *Developmental and Comparative Immunology* 42, 16–24.

Loof, T.G., Schmidt, O., Herwald, H. and Theopold, U. (2011) Coagulation systems of invertebrates and vertebrates and their roles in innate immunity: the same side of two coins? *Journal of Innate Immunity* 3, 34–40.

Lu, D., Macchietto, M., Chang, D., Barros, M.M., Baldwin, J. *et al.* (2017) Activated entomopathogenic nematode infective juveniles release lethal venom proteins. *PLOS Pathogens* 13: e1006302.

Mastore, M. and Brivio, M.F. (2008) Cuticular surface lipids are responsible for disguise properties of an entomoparasite against host cellular responses. *Developmental and Comparative Immunology* 32, 1050–1062.

McSorley, H.J., Hewitson, J.P. and Maizels, R.M. (2013) Immunomodulation by helminth parasites: defining mechanisms and mediators. *International Journal for Parasitology* 43, 301–310.

Moussian, B. and Roth, S. (2005) Dorsoventral axis formation in the *Drosophila* embryo – shaping and transducing a morphogen gradient. *Current Biology* 15, R887–R899.

Myllymäki, H. and Rämet, M. (2014) JAK/STAT pathway in *Drosophila* immunity. *Scandinavian Journal of Immunology* 79, 377–385.

Piedra Buena, A., López-Cepero, J. and Campos-Herrera, R. (2015) Entomopathogenic nematode production and application: regulation, ecological impact and non-target effects. In: Campos-Herrera, R. (ed.) *Nematode Pathogenesis of Insects and Other Pests: Ecology and Applied Technologies for Sustainable Plant and Crop Protection*. Springer, Cham, Switzerland, pp. 253–280.

Ribeiro, C., Duvic, B., Oliveira, P., Givaudan, A., Palha, F. *et al.* (1999) Insect immunity-effects of factors produced by a nematobacterial complex on immunocompetent cells. *Journal of Insect Physiology* 45, 677–685.

Riera Romo, M., Pérez-Martínez, D. and Castillo Ferrer, C. (2016) Innate immunity in vertebrates: an overview. *Immunology* 148, 125–139.

Rosales, C. (2017) Cellular and molecular mechanisms of insect immunity. In: Shields, V.D.C. (ed.) *Insect Physiology and Ecology*. IntechOpen. Available at: www.intechopen.com/books/insect-physiology-and-ecology/cellular-and-molecular-mechanisms-of-insect-immunity (accessed 13 March 2024).

Satyavathi, V.V., Minz, A. and Nagaraju, J. (2014) Nodulation: an unexplored cellular defense mechanism in insects. *Cell Signalling* 26, 1753–1763.

Scherfer, C., Qazi, M.R., Takahashi, K., Ueda, R., Dushay, M.S. *et al.* (2006) The Toll immune-regulated *Drosophila* protein Fondue is involved in hemolymph clotting and puparium formation. *Developmental Biology* 295, 156–163.

Shapiro-Ilan, D.I., Duncan, L.W., Lacey, L.A. and Han, R. (2005) Orchard applications. In: Grewal, P.S., Ehlers, R.-U. and Shapiro-Ilan, D.I. (eds.) *Nematodes as Biocontrol Agents*. CAB International, Wallingford, UK, pp. 215–229.

Shapiro-Ilan, D.I., Hazir, S. and Glazer, I. (2017) Basic and applied research: Entomopathogenic nematodes. In: Lacey, L.A. (ed.) *Microbial Agents for Control of Insect Pests: From Discovery to Commercial Development and Use*. Academic Press, Amsterdam, pp. 91–105.

Shen, H.M. and Liu, Z.G. (2006) JNK signaling pathway is a key modulator in cell death mediated by reactive oxygen and nitrogen species. *Free Radical Biology and Medicine* 40, 928–939.

Sicard, M., Le Brun, N., Pages, S., Godelle, B., Boemare, N. *et al.* (2003) Effect of native *Xenorhabdus* on the fitness of their *Steinernema* hosts: contrasting types of interaction. *Parasitology Research* 91, 520–524.

Sluss, H.K., Han, Z., Barrett, T., Goberdhan, D.C., Wilson, C. *et al.* (1996) A JNK signal transduction pathway that mediates morphogenesis and an immune response in *Drosophila*. *Genes and Development* 10, 2745–2758.

Stokes, B.A., Yadav, S., Shokal, U., Smith, L.C. and Eleftherianos, I. (2015) Bacterial and fungal pattern recognition receptors in homologous innate signaling pathways of insects and mammals. *Frontiers in Microbiology* 6: 19.

Tafesh-Edwards, G. and Eleftherianos, I. (2020) JNK signaling in *Drosophila* immunity and homeostasis. *Immunology Letters* 226, 7–11.

Tafesh-Edwards, G. and Eleftherianos, I. (2022) Entomopathogenic nematodes as a model for probing the insect immune system. In: Glazer, I., Shapiro-Ilan, D.I. and Sternberg, P.W. (eds.) *Nematodes as Model Organisms*. CAB International, Wallingford, UK, pp. 42–57.

Theopold, U., Krautz, R. and Dushay, M.S. (2014) The *Drosophila* clotting system and its messages for mammals. *Developmental and Comparative Immunology* 42, 42–46.

Toubarro, D., Lucena-Robles, M., Nascimento, G., Costa, G., Montiel, R. *et al.* (2009) An apoptosis-inducing serine protease secreted by the entomopathogenic nematode *Steinernema carpocapsae*. *International Journal for Parasitology* 39, 1319–1330.

Toubarro, D., Avila, M.M., Hao, Y., Balasubramanian, N., Jing, Y. *et al.* (2013a) A serpin released by an entomopathogen impairs clot formation in insect defense system. *PLOS ONE* 8: e69161.

Toubarro, D., Avila, M.M., Montiel, R., and Simões, N. (2013b) A pathogenic nematode targets recognition proteins to avoid insect defenses. *PLOS ONE* 8: e75691.

Vlisidou, I. and Wood, W. (2015) *Drosophila* blood cells and their role in immune responses. *FEBS Journal* 282, 1368–1382.

Walter, T.N., Dunphy, G.B. and Mandato, C.A. (2008) *Steinernema carpocapsae* DD136: metabolites limit the non-self adhesion responses of haemocytes of two lepidopteran larvae, *Galleria mellonella* (F. Pyralidae) and *Malacosoma disstria* (F. Lasiocampidae). *Experimental Parasitology* 120, 161–174.

Wang, Y., Gaugler, R. and Cui, L. (1994) Variations in immune response of *Popillia japonica* and *Acheta domesticus* to *Heterorhabditis bacteriophora* and *Steinernema* species. *Journal of Nematology* 26, 11–18.

Wang, Z., Wilhelmsson, C., Hyrsl, P., Loof, T.G., Dobes, P. *et al.* (2010) Pathogen entrapment by transglutaminase – a conserved early innate immune mechanism. *PLOS Pathogens* 6: e1000763.

Wraight, S.P., Lopes, R.B. and Faria, M. (2017) Microbial control of mite and insect pests of greenhouse crops. In: Lacey, L.A. (ed.) *Microbial Control of Insect and Mite Pests: From Theory to Practice*. Academic Press, London, pp. 237–252.

Yadav, S., Gupta, S. and Eleftherianos, I. (2018) Differential regulation of immune signaling and survival response in *Drosophila melanogaster* larvae upon *Steinernema carpocapsae* nematode infection. *Insects* 9: 17.

4 Genomics of Entomopathogenic Nematodes

Anil Baniya and Adler R. Dillman*
Department of Nematology, University of California, Riverside, CA, USA

Abstract

Nematodes have been a focus of genomics since the beginning of large-scale sequencing initiatives in the early 1990s. The use of genome sequences and post-genomic tools has had a significant impact on different fields of biology. Several entomopathogenic nematode (EPN) genomes have been sequenced, with many more on the way. Genomes of EPNs are invaluable to understanding the molecular mechanisms of parasitism, symbiosis and host–parasite interactions. With improvements in genome sequencing technologies and a refined annotation pipeline, more complete nematode genomes are now available. The increased number of genomes allows more detailed comparative studies, particularly when combined with post-genomic tools such as transcriptomics and proteomics. Genome information will make EPNs better model systems and will provide a foundation for their improvement as biological control agents. This chapter provides an overview of current EPN genomics and future direction.

4.1 Introduction

Nematodes are deceptively simple thread-like worms with an elongated stomach and a reproductive system enclosed in a tough outer cuticle. They are ubiquitous organisms renowned for their small size and capacity to adapt to harsh and shifting environmental conditions. It is estimated that there are as many as a million species of nematodes, although only around 26,000 species have been described. Based on trophic ecology, nematodes can be categorized as free living (bacteriovorus or fungivorous), predacious or parasitic in nature. Free-living nematodes are vital to environmental health and are crucial to nutrient cycling. In contrast, parasitic nematodes present a major challenge to human health and agriculture. Important groups of parasitic nematodes include plant-parasitic nematodes, hookworms, *Ascaris* spp., whipworm and filarial nematodes. These nematodes inflict terrible suffering on human and animal populations, and cause more than US$150 billion dollars of loss annually in crop damage (Bogale *et al.*, 2020). Among the parasitic nematodes, entomopathogenic nematodes (EPNs) are a unique group of nematodes that are lethal parasites of insects but are not harmful to humans or plants. Studies of the molecular and biological mechanisms by which EPNs kill insects are of great importance because of their ability to control insect pests of agriculture and the possibility of having toxins that are conserved in animal/human-parasitic nematodes (Shapiro-Ilan *et al.*, 2018).

*Email: adlerd@ucr.edu

 Entomopathogenic Nematodes as Biological Control Agents (D.I. Shapiro-Ilan and E.E. Lewis eds)
DOI: 10.1079/9781800620322.0004

Genome sequencing and annotation of the genes within a genome are regarded as the first step of an in-depth understanding of the biological processes within an organism. The huge genetic content of species within a phylum can be indexed and organized naturally through genome sequencing. The most important and long-lasting piece of information we can currently acquire for each life form is its genetic blueprint, which is represented by its genome (The *C. elegans* Sequencing Consortium, 1998). With advancements in genome sequencing technologies and bioinformatic pipelines, genome assembly and annotation are constantly improving. As a result, there will be more improved genomes available in the future, and preliminary estimates of genome sequences and descriptions of the genes can be improved. Since the start of large-scale sequencing initiatives in the early 1990s, nematodes have been a focus of genomics. *Caenorhabditis elegans* was the first multicellular organism whose entire genome was sequenced, serving as a foundation for subsequent nematode genomics (The *C. elegans* Sequencing Consortium, 1998). The nematode *C. elegans* has proven invaluable for genomic analysis in different fields of biology. Thousands of researchers around the world have contributed to our understanding of the function of the 20,431 protein-coding genes (The *C. elegans* Sequencing Consortium, 1998; Harris *et al.*, 2010). This is probably due to the same factors that make this nematode so invaluable in genetics: (i) it is easy to culture in the laboratory; (ii) it has simple, rapid and invariant development; (iii) many of its biological principles are universal, even if the specific details are not; and (iv) the more we understand a certain biological concept, the more interesting it tends to become (Horvitz and Sternberg, 1982). The genome sequencing of *Caenorhabditis briggsae* in 2003 paved the way for comparative genomics by comparing the genomes of *C. elegans* and *C. briggsae*. It provided a powerful tool to understand the genetic basis for the evolution of these relatively simple metazoans. These two *Caenorhabditis* spp. are nearly indistinguishable morphologically at the microscopic level. Comparative genomics predicts that these species diverged between 80 and 110 million years ago, suggesting that they are more evolutionary distant than humans are from mice, which are estimated to have diverged around 75 million years ago (Stein *et al.*, 2003). Following the availability of genome sequences from other *Caenorhabditis* spp. such as *C. remanei*, *C. brenneri* and *C. japonica*, the annotation of the *C. elegans* genome has been improved. This has aided our understanding of the evolutionary mechanisms that shape the genomes of nematodes. The genome of *Pristionchus pacificus* became available in 2008, providing a unique perspective on its lifestyle (Dieterich *et al.*, 2008). Despite the availability of comprehensive genomic information for free-living nematodes, little information about most of the other important members of the phylum Nematoda was available.

The genomics of parasitic nematodes, which account for roughly half of all described nematode species (~44% of described species are parasites of vertebrates and invertebrates, while ~15% are parasites of plants) and have significant implications for human health and nutrition, are largely unknown (Lambert and Bekal, 2002). Comparative genome analysis of the human filarial nematode *Brugia malayi* and *Caenorhabditis* spp. revealed notable differences in the predicted proteome. The predicted proteome laid the groundwork for effective drug design and demonstrated how *Brugia malayi* adapts to niches in its human and vector hosts. This study was remarkable in understanding the molecular basis of a mutualistic relationship between *Brugia malayi* and its endosymbiont *Wolbachia* (Ghedin *et al.*, 2007). This example demonstrates the value of obtaining additional genome data from representative species across the phylum for comparative genomic analyses.

Sequencing technology has evolved significantly over the past 20 years, moving from hierarchical sequencing and assembly of cloned DNA segments (as utilized in the *C. elegans* experiment) to shotgun, high-throughput 500 bp reads produced by 454 Roche sequencing and the much more affordable 150 bp reads produced by Illumina sequencing (Mardis 2008; Jex *et al.*, 2011; Dillman *et al.*, 2012b; Coghlan *et al.*, 2018). It is labor intensive and error prone to use short-read sequencing, as homopolymeric regions and large repeats cannot be effectively sequenced with this technology. An alternative method for the *de novo* assembly of complicated novel genomes is offered by long-read sequencing techniques. In comparison with short-read platforms, long-read, single-molecule sequencing technologies have extended read lengths by a factor of 100–1000 and may cover much bigger repetitive regions. Until recently, the Pacific Biosciences Single-Molecule Real-Time sequencing system was the leading platform of this type. Pacific Biosciences (PacBio) sequencing currently averages 10 kb read lengths, with consensus sequencing error rates comparable to short-read sequencing. In 2014, Oxford Nanopore Technologies introduced the MinION nanopore sequencer. MinION sequencing is a viable option for whole-genome sequencing of complex metazoan genomes due to its long read lengths, recent improvements in performance and the development of software to assemble genomes (Koren *et al.*, 2017). Since the publication of the first nematode genome in 1998, hundreds of nematode genomes have been sequenced (Table 4.1), with many more in preparation.

Because nematodes are ecologically diverse and have a high species richness, the genome sequencing effort should be guided by phylogenetic relationships as well as human health and agricultural concerns (Dillman *et al.*, 2012b). Nematodes within the phylum Nematoda are divided into three major clades: Enoplia, Dorylaimia, and Chromadoria (de Ley and Blaxter, 2002, 2004; Hodda, 2011). Chromadoria is further subdivided into ten clades, which, along with Enoplia and Dorylaimia, forms 12 major monophyletic branches within the phylum (Fig. 4.1) (Holterman *et al.*, 2006; van Megen *et al.*, 2009; Blaxter, 2011). Until now, most of the genome sequencing efforts have focused on the Chromadoria clade. This group includes *C. elegans* as well as other economically significant nematodes related to humans, animals, and plants (Fig. 4.1). A systematic genomic analysis of nematodes would aid in a better understanding of nematode evolution, improve comparative studies, and could reveal striking differences among species across the phylum.

EPNs within the genera *Heterorhabditis* and *Steinernema* are insect parasites. They have symbiotic bacteria in their intestines that are mutually associated with the nematode (*Photorhabdus* and *Xenorhabdus* spp., respectively). The biology of EPNs allows researchers to investigate both mutualistic and parasitic symbiosis. *Steinernema* spp. (clade IV by Blaxter *et al.*, 1998, and clade 10 by van Megen *et al.*, 2009) and *Heterorhabditis* spp. (clade V by Blaxter *et al.*, 1998, and clade 10 by van Megen *et al.*, 2009) are two groups of EPNs that are phylogenetically unrelated (Fig. 4.1). These EPNs are a fascinating system for studying both convergent and parallel evolution. They are unusual intermediates between predators and parasitoids, and there have been numerous studies on their host-seeking behavior (Dillman *et al.*, 2012a). They seek hosts primarily through chemoreception, and some are capable of jumping, which is an unusual behavior in nematodes and is unique to some *Steinernema* spp. (Backhaus *et al.*, 1999; Campbell *et al.*, 2003; Baiocchi *et al.*, 2019). Furthermore, EPNs are used for biological pest control in agriculture; approximately 12 species have been commercialized for biocontrol applications (Shapiro-Ilan *et al.*, 2020). The genomes of EPNs can be used for artificial selection and genomics-aided breeding to discover the diversity of wild gene variants and to select candidate genes that influence desirable features (Lu *et al.*, 2016).

Genomics technology is now being used in entomopathogenic nematology. An effort has been made to sequence the genomes of both *Heterorhabditis* and *Steinernema* spp. nematodes. The *Heterorhabditis bacteriophora* TTO1 strain was the first EPN studied to generate a draft sequence, expressed sequence tags (ESTs) and a physical map (Ciche, 2007). Transcriptomic analysis of *H. bacteriophora* was published in 2009 (Bai *et al.*, 2009). The draft genome sequence of this nematode was published in 2013 (Bai *et al.*, 2013). An improved annotation of *H. bacteriophora* was made publicly available in 2018 (McLean *et al.*, 2018; Vadnal *et al.*, 2018). In 2021, a draft genome of *Heterorhabditis indica* became available (Bhat *et al.*, 2021). Similarly, an effort to uncover the genomics of the genus *Steinernema* began with the genome announcement in 2015 of five *Steinernema* genomes: *S. carpocapsae*, *S. scapterisci*, *S. monticolum*, *S. feltiae* and *S. glaseri* (Dillman *et al.*, 2015). The genome and transcriptome of an additional strain of *S. carpocapsae* was published in 2016 (Rougon-Cardoso *et al.*, 2016). In 2019, an effort to create the complete genome of *S. carpocapsae* using hybrid assembly (a combination of both short Illumina and long PacBio reads) was able to generate a high-quality genome with a complete X chromosome (Serra *et al.*, 2019). There is also a draft genome assembly of nematodes from their native habitat for *S. feltiae*, *S. diaprepesi* and *S. khuongi* (Baniya *et al.*, 2020; Fu *et al.*, 2020; Baniya and Digennaro, 2021). In addition to the existing EPN genomes, improved genomes of other EPNs are in progress. These sequences will

Table 4.1. Genome statistics of 171 published and publicly available genomes of nematodes as viewed in https://parasite.wormbase.org/species.html and from the NCBI genome database (www.ncbi.nlm.nih.gov/genome/) (both accessed 4 April 2024).

Species name	Provider	BioProject ID	Clade	N50[a]	Genome size	No. of scaffolds	No. of coding genes
Acanthocheilonema viteae	University of Edinburgh	PRJEB1697	Clade III	25,752	77,350,906	6,796	10,397
Acrobeloides nanus	University College London	PRJEB26554	Clade IV	19,572	248,054,631	30,759	32,984
Ancylostoma caninum	McDonnell Genome Institute	PRJNA72585	Clade V	255,299	465,750,606	25,339	30,198
Ancylostoma ceylanicum	Cornell University	PRJNA231479	Clade V	668,412	313,092,710	1,736	36,687
Ancylostoma ceylanicum	McDonnell Genome Institute	PRJNA72583	Clade V	372,147	348,994,891	8,098	17,509
Ancylostoma duodenale	McDonnell Genome Institute	PRJNA72581	Clade V	10,112	332,878,295	69,984	27,485
Angiostrongylus cantonensis	Wellcome Sanger Institute	PRJEB493	Clade V	43,712	253,242,791	18,635	14,520
Angiostrongylus cantonensis	Sun Yat-sen University	PRJNA350391	Clade V	860,760	293,306,699	26,577	10,314
Angiostrongylus costaricensis	Wellcome Sanger Institute	PRJEB494	Clade V	111,998	262,782,696	6,384	13,417
Angiostrongylus vasorum	University of Zurich	PRJNA663250	Clade V	1,673,247	279,948,744	468	13,766
Anisakis simplex	Wellcome Sanger Institute	PRJEB496	Clade III	9,229	126,869,778	42,005	20,971
Ascaris lumbricoides	Wellcome Sanger Institute	PRJEB4950	Clade III	55,909	316,975,410	31,710	23,604
Ascaris suum	University of Colorado Denver	PRJNA62057	Clade III	4,646,302	298,028,455	415	16,778
Ascaris suum	University of Melbourne	PRJNA80881	Clade III	407,675	272,782,664	29,831	18,542
Brugia malayi	WormBase	PRJNA10729	Clade III	14,214,749	88,235,797	197	10,878
Brugia pahangi	Wellcome Sanger Institute	PRJEB497	Clade III	65,530	90,545,113	14,029	14,674
Brugia timori	Wellcome Sanger Institute	PRJEB4663	Clade III	4,903	64,930,714	23,963	16,203
Bursaphelenchus okinawaensis	University of Miyazaki	PRJEB40023	Clade IV	11,641,800	69,956,301	7	14,593
Bursaphelenchus xylophilus	Wellcome Sanger Institute	PRJEA64437	Clade IV	949,830	74,561,461	5,527	17,704
Bursaphelenchus xylophilus	University of Miyazaki	PRJEB40022	Clade IV	12,794,883	78,274,012	11	15,884
Caenorhabditis angaria	California Institute of Technology	PRJNA51225	Clade V	79,665	105,997,628	34,621	27,970
Caenorhabditis becei	BANG	PRJEB28243	Clade V	767,505	87,867,412	1,565	23,926
Caenorhabditis bovis	BANG	PRJEB34497	Clade V	7,557,458	62,726,283	35	13,128
Caenorhabditis brenneri	WormBase	PRJNA20035	Clade V	377,250	190,369,721	3,305	30,705
Caenorhabditis briggsae	WormBase	PRJNA10731	Clade V	17,485,439	108,384,165	367	20,821
Caenorhabditis elegans	WormBase	PRJNA13758	Clade V	17,493,829	100,286,401	7	19,985
Caenorhabditis inopinata	University of Miyazaki	PRJDB5687	Clade V	20,594,552	122,994,328	6	21,443
Caenorhabditis japonica	WormBase	PRJNA12591	Clade V	94,114	166,256,191	18,817	29,935
Caenorhabditis latens	University of Oregon	PRJNA248912	Clade V	366,678	117,124,750	1,858	24,855
Caenorhabditis nigoni	Cornell University	PRJNA384657	Clade V	20,390,332	129,488,540	155	29,167
Caenorhabditis panamensis	BANG	PRJEB28259	Clade V	487,150	79,045,575	985	22,312

Continued

Table 4.1. Continued.

Species name	Provider	BioProject ID	Clade	N50[a]	Genome size	No. of scaffolds	No. of coding genes
Caenorhabditis parvicauda	BANG	PRJEB12595	Clade V	44,393	93,733,926	5,716	16,412
Caenorhabditis quiockensis	BANG	PRJEB11354	Clade V	139,266	100,371,645	4,890	22,238
Caenorhabditis remanei	University of Oregon	PRJNA248909	Clade V	1,522,088	118,549,266	1,591	24,977
Caenorhabditis remanei	University of Oregon	PRJNA248911	Clade V	1,765,890	124,541,912	912	24,867
Caenorhabditis remanei	WormBase	PRJNA53967	Clade V	435,512	145,442,736	3,670	31,438
Caenorhabditis remanei	University of Oregon	PRJNA577507	Clade V	21,501,900	130,480,874	187	26,308
Caenorhabditis sinica	University of Edinburgh	PRJNA194557	Clade V	25,151	131,797,386	15,261	34,696
Caenorhabditis sulstoni	BANG	PRJEB12601	Clade V	136,669	65,098,870	2,044	18,192
Caenorhabditis tribulationis	BANG	PRJEB12608	Clade V	224,519	101,228,815	3,276	24,760
Caenorhabditis tropicalis	McDonnell Genome Institute	PRJNA53597	Clade V	20,921,866	79,321,433	665	22,326
Caenorhabditis uteleia	BANG	PRJEB12600	Clade V	175,547	104,046,868	3,222	27,614
Caenorhabditis waitukubuli	BANG	PRJEB12602	Clade V	15,045	91,416,478	21,203	30,089
Caenorhabditis zanzibari	BANG	PRJEB12596	Clade V	91,045	101,086,038	3,127	22,198
Cercopithifilaria johnstoni	Wellcome Sanger Institute	PRJEB47283	Clade III	98,884	76,938,708	2,092	10,438
Cylicostephanus goldi	Wellcome Sanger Institute	PRJEB498	Clade V	1,235	173,362,437	154,509	13,906
Dictyocaulus viviparus	University of Edinburgh	PRJEB5116	Clade V	22,491	169,388,535	17,715	14,306
Dictyocaulus viviparus	McDonnell Genome Institute	PRJNA72587	Clade V	225,234	161,038,511	7,175	13,514
Diploscapter coronatus	National Institute of Genetics, Japan	PRJDB3143	Clade V	1,007,652	170,470,382	511	34,063
Diploscapter pachys	New York University	PRJNA280107	Clade V	124,169	157,719,873	10,775	27,341
Dirofilaria immitis	University of Edinburgh	PRJEB1797	Clade III	71,050	88,309,529	16,061	12,872
Ditylenchus destructor	Huazhong Agricultural University	PRJNA312427	Clade IV	555,026	111,138,200	1,761	13,931
Ditylenchus dipsaci	Agriculture and Agri-Food Canada	PRJNA498219	Clade IV	287,009	227,234,012	1,394	26,428
Dracunculus medinensis	Wellcome Sanger Institute	PRJEB500	Clade III	643,251	103,750,625	1,347	10,969
Elaeophora elaphi	Wellcome Sanger Institute	PRJEB502	Clade III	25,422	82,568,297	8,078	10,410
Enterobius vermicularis	Wellcome Sanger Institute	PRJEB503	Clade III	20,546	150,050,672	19,832	12,895
Globodera pallida	Wellcome Sanger Institute	PRJEB123	Clade IV	120,163	123,625,196	6,873	16,403
Globodera rostochiensis	Wageningen University and Research	PRJNA695196	Clade IV	3,274,370	92,682,755	88	17,856
Globodera rostochiensis	Wageningen University and Research	PRJNA695196	Clade IV	2,102,824	101,587,200	135	19,459
Globodera rostochiensis	University of Edinburgh	PRJEB13504	Clade IV	88,495	95,876,286	4,281	14,308
Gongylonema pulchrum	Wellcome Sanger Institute	PRJEB505	Clade III	4,421	322,334,493	128,401	27,158
Haemonchus contortus	Wellcome Sanger Institute	PRJEB506	Clade V	47,382,676	283,439,308	7	19,623

Haemonchus contortus	University of Melbourne	PRJNA205202	Clade V	56,143	319,640,208	14,419	23,610
Haemonchus placei	Wellcome Sanger Institute	PRJEB509	Clade V	37,562	259,125,910	24,923	21,928
Halicephalobus mephisto	American University, Washington, DC	PRJNA528747	Clade IV	313,311	61,428,783	880	12,546
Halicephalobus nkz332	Indiana University	PRJNA555616	Clade IV	60,483	47,438,423	5,084	10,997
Heligmosomoides polygyrus	Wellcome Sanger Institute	PRJEB1203	Clade V	35,732	560,662,278	44,726	27,459
Heligmosomoides polygyrus	University of Edinburgh	PRJEB15396	Clade V	179,069	696,954,138	23,647	23,471
Heterodera glycines	Iowa State University	PRJNA381081	Clade IV	17,907,690	157,978,452	9	22,465
Heterodera schachtii	The *H. schachtii* genome consortium	PRJNA722882	Clade IV	1,273,070	179,246,932	395	26,739
Heterorhabditis bacteriophora	McDonnell Genome Institute	PRJNA13977	Clade V	312,328	77,007,652	1,263	20,964
Heterorhabditis indica	IARI	PRJNA720543	Clade V	78,124	91,271,500	3,550	10,494
Litomosoides sigmodontis	University of Edinburgh	PRJEB3075	Clade III	45,733	64,813,410	3,165	10,246
Loa loa	Institute for Genome Sciences	PRJNA246086	Clade III	180,288	96,405,388	2,250	12,473
Loa loa	Broad Institute	PRJNA37757	Clade III	172,043	91,373,458	5,773	14,908
Meloidogyne arenaria	INRA	PRJEB8714	Clade IV	16,417	258,067,405	26,196	101,269
Meloidogyne arenaria	University of Hull	PRJNA340324	Clade IV	10,460	163,746,389	46,436	27,548
Meloidogyne arenaria	RIKEN	PRJNA438575	Clade IV	204,341	284,032,373	2,224	76,700
Meloidogyne enterolobii	University of Hull	PRJNA340324	Clade IV	10,529	162,967,259	42,008	29,473
Meloidogyne enterolobii	INRAE	PRJEB36431	Clade IV	143,476	240,054,310	4,437	63,841
Meloidogyne exigua	IRD	PRJNA725818	Clade IV	1,882,513	42,101,100	206	NA
Meloidogyne floridensis	University of Edinburgh	PRJEB6016	Clade IV	3,688	96,673,063	58,696	49,938
Meloidogyne floridensis	University of Hull	PRJNA340324	Clade IV	13,230	74,846,012	8,887	12,762
Meloidogyne graminicola	IARI	PRJNA411966	Clade IV	20,427	38,184,958	4,304	10,895
Meloidogyne graminicola	IARI	PRJNA615787	Clade IV	294,907	41,550,000	283	10,283
Meloidogyne hapla	North Carolina State University	PRJNA29083	Clade IV	37,501	53,017,507	3,452	14,419
Meloidogyne incognita	INRA	PRJEB8714	Clade IV	38,543	183,531,997	12,091	43,718
Meloidogyne incognita	University of Hull	PRJNA340324	Clade IV	16,480	121,964,221	33,351	21,830
Meloidogyne javanica	INRA	PRJEB8714	Clade IV	10,373	235,798,407	31,341	97,208
Meloidogyne javanica	University of Hull	PRJNA340324	Clade IV	14,103	150,345,270	34,316	24,679
Meloidogyne luci	Wellcome Sanger Institute	PRJEB27977	Clade IV	1,711,905	209,155,000	327	NA
Meloidogyne oryzae	ICAR-IARI	PRJNA411966	Clade IV	20,400	38,180,000	4,304	10,196
Mesorhabditis belari	ENS Lyon	PRJEB30104	Clade V	462,517	163,453,184	1,721	27,588
Micoletzkya japonica	MPI for Developmental Biology	PRJEB27334	Clade V	188,443	201,900,956	137,965	19,855
Necator americanus	McDonnell Genome Institute	PRJNA72135	Clade V	211,860	244,075,060	11,864	15,728
Nippostrongylus brasiliensis	Wellcome Sanger Institute	PRJEB511	Clade V	33,446	294,400,206	29,375	22,796

Continued

Table 4.1. Continued.

Species name	Provider	BioProject ID	Clade	N50[a]	Genome size	No. of scaffolds	No. of coding genes
Oesophagostomum dentatum	McDonnell Genome Institute	PRJNA72579	Clade V	19,257	489,973,581	99,770	25,291
Onchocerca flexuosa	Wellcome Sanger Institute	PRJEB512	Clade III	2,932	86,175,476	45,472	16,119
Onchocerca flexuosa	McDonnell Genome Institute	PRJNA230512	Clade III	540,294	67,740,367	1,604	8,140
Onchocerca ochengi	Wellcome Sanger Institute	PRJEB1204	Clade III	16,140	91,660,559	20,243	12,816
Onchocerca ochengi	University of Edinburgh	PRJEB1465	Clade III	12,275	95,513,350	24,057	13,990
Onchocerca volvulus	WormBase	PRJEB513	Clade III	25,485,961	96,427,137	708	12,109
Oscheius tipulae	University of Edinburgh	PRJEB15512	Clade V	1,203,411	59,468,623	191	14,626
Oscheius tipulae	Wellcome Sanger Institute	PRJNA644888	Clade V	10,817,478	60,923,166	7	16,356
Panagrellus redivivus	California Institute of Technology	PRJNA186477	Clade IV	262,414	65,093,147	940	24,249
Panagrolaimus davidi	University College London	PRJEB32708	Clade IV	4,441	118,052,509	39,598	25,806
Panagrolaimus es5	University College London	PRJEB32708	Clade IV	6,292	90,240,976	25,284	24,678
Panagrolaimus ps1159	University College London	PRJEB32708	Clade IV	9,884	84,957,092	17,628	24,802
Panagrolaimus superbus	University College London	PRJEB32708	Clade IV	1,889	76,664,811	53,192	19,887
Parapristionchus giblindavisi	MPI for Developmental Biology	PRJEB27334	Clade V	111,900	200,833,897	7,303	22,872
Parascaris equorum	Wellcome Sanger Institute	PRJEB514	Clade III	5,333	185,376,996	60,399	14,472
Parascaris univalens	University of Colorado Denver	PRJNA386823	Clade III	1,825,986	253,353,821	1,274	14,323
Parastrongyloides trichosuri	Wellcome Sanger Institute	PRJEB515	Clade IV	836,942	42,486,966	1,810	15,010
Plectus sambesii	MRC LMS	PRJNA390260	Clade C	23,404	186,666,232	18,974	39,067
Pristionchus arcanus	MPI for Developmental Biology	PRJEB27334	Clade V	269,953	202,757,162	4,263	39,331
Pristionchus entomophagus	MPI for Developmental Biology	PRJEB27334	Clade V	367,513	264,161,588	72,721	31,150
Pristionchus exspectatus	MPI for Developmental Biology	PRJEB24288	Clade V	142,175	177,635,997	4,412	35,595
Pristionchus fissidentatus	MPI for Developmental Biology	PRJEB27334	Clade V	441,872	247,442,930	56,870	28,283
Pristionchus japonicus	MPI for Developmental Biology	PRJEB27334	Clade V	447,930	223,093,336	33,291	34,293
Pristionchus maxplancki	MPI for Developmental Biology	PRJEB27334	Clade V	308,537	265,592,641	69,506	37,393
Pristionchus mayeri	MPI for Developmental Biology	PRJEB27334	Clade V	234,324	297,448,349	84,599	32,719
Pristionchus pacificus	WormBase	PRJNA12644	Clade V	23,915,096	158,500,396	47	26,342
Propanagrolaimus ju765	University College London	PRJEB32708	Clade IV	10,830	64,262,255	13,268	20,756
Rhabditophanes sp. KR3021	Wellcome Sanger Institute	PRJEB1297	Clade IV	537,195	47,267,908	471	13,496
Romanomermis culicivorax	University of Cologne	PRJEB1358	Clade I	17,583	322,765,761	62,537	48,376
Setaria digitata	University of Colombo	PRJNA479729	Clade III	120,860	78,770,088	1,879	10,531
Soboliphyme baturini	Wellcome Sanger Institute	PRJEB516	Clade I	19,742	218,332,512	22,336	13,193
Steinernema carpocapsae	University of California, Irvine	PRJNA202318	Clade IV	7,362,381	84,512,609	16	30,931

Steinernema carpocapsae	University of California, Irvine	PRJNA202318	Clade IV	299,566	86,127,942	1,578	28,307
Steinernema diaprepesi	University of Florida	PRJNA605202	Clade IV	11,484	118,236,000	35,417	15,094
Steinernema feltiae	California Institute of Technology	PRJNA204661	Clade IV	47,472	82,627,153	5,839	33,457
Steinernema feltiae	Washington State University	PRJNA353610	Clade IV	60,158	121,603,157	4,678	32,303
Steinernema glaseri	California Institute of Technology	PRJNA204943	Clade IV	37,382	92,932,499	7,514	34,137
Steinernema monticolum	California Institute of Technology	PRJNA205067	Clade IV	11,533	89,356,173	14,332	36,005
Steinernema khuongi	University of Florida	PRJNA670677	Clade IV	46,074	81,760,000	8,783	NA
Steinernema scapterisci	California Institute of Technology	PRJNA204942	Clade IV	90,783	79,776,474	2,877	31,376
Strongyloides papillosus	Wellcome Sanger Institute	PRJEB525	Clade IV	86,067	60,447,969	4,703	18,456
Strongyloides ratti	Wellcome Sanger Institute	PRJEB125	Clade IV	11,693,564	43,166,851	136	12,464
Strongyloides stercoralis	Wellcome Sanger Institute	PRJEB528	Clade IV	431,128	42,674,647	800	13,123
Strongyloides venezuelensis	Wellcome Sanger Institute	PRJEB530	Clade IV	715,404	52,178,999	587	16,904
Strongylus vulgaris	Wellcome Sanger Institute	PRJEB531	Clade V	2,377	291,096,078	167,310	21,071
Syphacia muris	Wellcome Sanger Institute	PRJEB524	Clade III	60,614	99,093,317	3,879	11,117
Teladorsagia circumcincta	McDonnell Genome Institute	PRJNA72569	Clade V	46,814	700,614,188	81,734	25,572
Thelazia callipaeda	Wellcome Sanger Institute	PRJEB1205	Clade III	50,894	75,392,092	5,778	10,912
Toxocara canis	Wellcome Sanger Institute	PRJEB533	Clade III	31,084	299,788,016	29,185	20,264
Toxocara canis	University of Melbourne	PRJNA248777	Clade III	374,962	317,115,901	22,857	18,596
Trichinella britovi	University of Melbourne	PRJNA257433	Clade I	147,150	51,504,795	7,991	16,058
Trichinella murrelli	University of Melbourne	PRJNA257433	Clade I	106,269	49,016,728	5,242	14,852
Trichinella nativa	McDonnell Genome Institute	PRJNA179527	Clade I	12,611	49,070,491	21,024	10,175
Trichinella nativa	University of Melbourne	PRJNA257433	Clade I	141,195	47,933,238	4,443	13,545
Trichinella nelsoni	University of Melbourne	PRJNA257433	Clade I	293,867	47,436,005	2,969	13,229
Trichinella papuae	University of Melbourne	PRJNA257433	Clade I	222,396	46,844,510	2,542	11,837
Trichinella patagoniensis	University of Melbourne	PRJNA257433	Clade I	154,103	49,773,209	6,623	15,319
Trichinella pseudospiralis	University of Melbourne	PRJNA257433	Clade I	235,426	49,162,916	7,221	12,659
Trichinella pseudospiralis	University of Melbourne	PRJNA257433	Clade I	167,180	46,055,013	1,379	11,005
Trichinella pseudospiralis	University of Melbourne	PRJNA257433	Clade I	287,133	49,152,658	6,594	12,455
Trichinella pseudospiralis	University of Melbourne	PRJNA257433	Clade I	234,172	48,398,615	5,956	14,621
Trichinella pseudospiralis	University of Melbourne	PRJNA257433	Clade I	112,255	48,106,863	7,462	13,717
Trichinella spiralis	McDonnell Genome Institute	PRJNA12603	Clade I	6,373,445	63,525,422	6,863	16,380
Trichinella spiralis	University of Melbourne	PRJNA257433	Clade I	212,546	50,025,957	7,652	14,737
Trichinella sp. T6	University of Melbourne	PRJNA257433	Clade I	158,103	50,853,738	5,951	15,241
Trichinella sp. T8	University of Melbourne	PRJNA257433	Clade I	239,129	49,331,761	4,124	14,919

Continued

Table 4.1. Continued.

Species name	Provider	BioProject ID	Clade	N50[a]	Genome size	No. of scaffolds	No. of coding genes
Trichinella sp. T9	University of Melbourne	PRJNA257433	Clade I	212,690	49,075,249	5,752	13,111
Trichinella zimbabwensis	University of Melbourne	PRJNA257433	Clade I	205,645	50,795,114	10,839	14,764
Trichuris muris	WormBase	PRJEB126	Clade I	28,941,788	111,837,642	803	14,995
Trichuris suis	McDonnell Genome Institute	PRJNA179528	Clade I	1,322,386	63,839,715	306	9,831
Trichuris suis	University of Melbourne	PRJNA208415	Clade I	503,034	74,234,559	4,292	14,436
Trichuris suis	University of Melbourne	PRJNA208416	Clade I	443,734	71,056,402	3,284	14,261
Trichuris trichiura	Wellcome Sanger Institute	PRJEB535	Clade I	69,794	75,496,394	4,156	9,650
Wuchereria bancrofti	Wellcome Sanger Institute	PRJEB536	Clade III	9,872	76,991,470	13,520	13,058
Wuchereria bancrofti	Case Western Reserve University	PRJNA275548	Clade III	56,656	90,325,107	5,105	11,068

[a]N50, sequence length of the shortest contig at 50% of the total assembly length.
BANG, Nematode and Neglected Genomics, IEB, University of Edinburgh; ENS Lyon, École Normale Supérieure de Lyon; IARI, Indian Agricultural Research Institute; ICAR, Indian Council of Agricultural Research; INRA, Institut National de la Recherche Agronomique; INRAE, National Research Institute for Agriculture, Food and the Environment; IRD Institut de Recherche pour le Developpement; MPI, Max Planck Institute; MRC LMS, Medical Research Council Laboratory of Medical Sciences; RIKEN, Institute of Physical and Chemical Research.

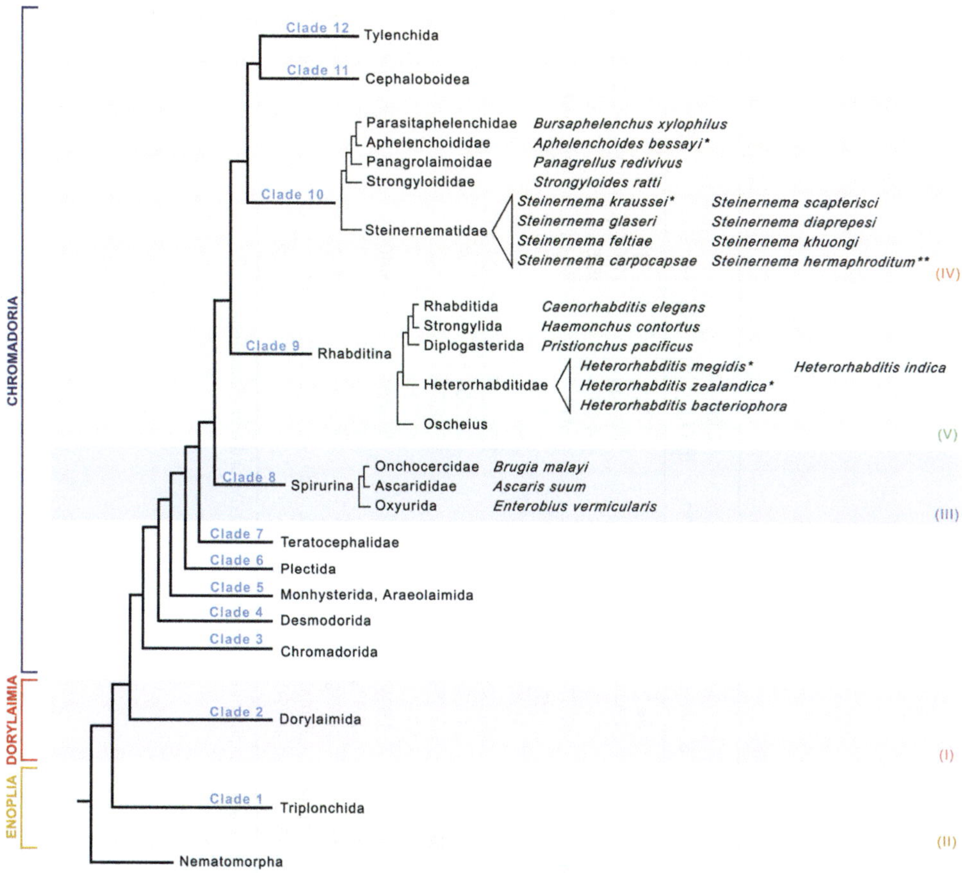

Fig. 4.1. Schematic phylogeny of nematodes. Clades 1–12 based on van Megen *et al.* (2009), and Roman numerals in parentheses on the right side correspond to clades as defined by Blaxter *et al.* (1998). Representative species in clades 8–10 are shown on the right side of the tree. *, Species whose genome sequences are not available; **, species whose genome sequences are in progress.

serve as a foundation for studying the biology of these nematodes and should aid toward developing sustainable strategies to improve the effectiveness of EPNs as biocontrol agents of important agricultural pests. These genomes will also aid in the development of these nematodes as a model system for studying host–parasite interactions, mutualism and symbiosis.

4.2 Genomics of *Steinernema* spp.

Currently, seven genomes of *Steinernema* spp. have been sequenced and made available to the public (Table 4.2). Sequencing and annotation of *S. carpocapsae* All strain and its other four congeners, *S. scapterisci*, *S. glaseri*, *S. feltiae* and *S. monticolum*, provided a foundation for comparative genomics within the genus *Steinernema*. The genome size of these nematodes ranged from 80 to 90 Mb and predicted protein-coding genes ranged from 28,000 to 36,000 genes, although these estimates may be high due to the draft nature of the assembly. The highest N50 (the sequence length of the shortest contig at 50% of the total assembly length) of a scaffolded assembly was initially observed in the *S. carpocapsae* scaffold as N50 = 299 kb. The availability of the genome sequence allowed a better understanding of the phylogenetic relationship between these congeners. A supermatrix of 3885 strictly

Table 4.2. Genome statistics and features of published *Steinernema* spp. genomes.

Feature	*S. carpocapsae*	*S. scapterisci*	*S. feltiae* (SN strain)	*S. glaseri*	*S. monticolum*	*S. feltiae* (NW strain)	*S. diaprepesi*	*S. khuongi*
Estimated genome size (Mb)	84.5	79.4	82.4	92.9	89.3	121.6	118	82
N50 (bp)	7,362,381	90,783	47,472	37,444	11,556	60,433	11,474	42,000
Number of scaffolds/contigs	16	2,877	5,839	7,515	14,331	4,678	35,545	8,794
GC content (%)	45.7	47.98	46.99	47.63	42.01	NA	45.01	47.88
N content (Mb)	NA	0.76	2.76	3.37	4.34	0	0	0
N content (%)	0.54	0.96	3.36	3.64	4.87	0	0	
Maximum scaffold size (bp)	20,922,283	1,149,164	1,470,990	339,094	110,081	1,315,981	1,706,490	428,226
Predicted genes	30,957	31,378	36,434	34,143	36,007	32,304	NA	NA
Complete BUSCO	87%	84.5%	84.32%	59.4%	69.2%	87.27%	85%	86.6% (850)[a]
Single-copy BUSCOs	NA	79.3%	80.55%	57.8%	65.4%	76.68%	79.60%	81.3% (798)[a]
Duplicated BUSCOs	NA	5.2%	3.77%	1.6%	3.8%	10.59%	5.40%	5.3% (52)[a]
Fragmented BUSCOs	7%	8.1%	8.15%	12.9%	12.5%	7.33%	7.10%	6.5% (64)[a]
Missing BUSCOs	6%	7.4%	7.54%	27.7%	18.3%	5.40%	7.90%	6.9% (68)[a]
Reference	Serra *et al.* (2019)	Dillman *et al.* (2015)	Dillman *et al.* (2015)	Dillman *et al.* (2015)	Dillman *et al.* (2015)	Fu *et al.* (2020)	Baniya *et al.* (2020)	Baniya and Digennaro (2021)

BUSCO, benchmarking universal single-copy orthologs; NA, not available; N50, sequence length of the shortest contig at 50% of the total assembly length.
[a]Number of genes shown in parentheses.

orthologous genes (1:1:1:1:1:1) of these nematodes provided better insight into the phylogenetic relationship, placing *S. monticolum* closer to *S. feltiae*, which was otherwise believed to be closer to *S. carpocapsae* and *S. scapterisci* (Dillman *et al.*, 2015). A total of 28,313 predicted proteins from *S. carpocapsae* was compared with ten species among different groups of nematodes including free-living, necromenic, plant-parasitic and vertebrate-parasitic species. Comparative analysis identified 17,653 (62.3%) *S. carpocapsae* proteins among 10,350 orthologous clusters homologous to at least one of the species used in the comparison. This work also led to the discovery of 486 highly conserved genes that will be invaluable for studying phylogenetic resolution among nematodes in future work. Approximately one-third (37.7%) of the proteome (10,660 proteins) were orphans, implying that they are either unique to *S. carpocapsae* or the primary sequences are so diversely evolved that they could not be assigned any orthology based on sequence similarity alone. Future transgenic efforts in EPNs may benefit from learning more about the function of these orphan proteins, which may identify genes crucial for survival, persistence or infection (Dillman *et al.*, 2015).

The high-quality genome of *S. carpocapsae* made it possible to predict a total of 17,518 Pfam domains from 3256 different Pfam accession categories. Analysis of these Pfam domains led to the identification of gene families in *Steinernema* that may have evolved because of adaptation to a parasitic lifestyle. The relative abundance of the Pfam domains in the genomes of *S. carpocapsae* and other parasitic nematodes such as *Bursaphelenchus xylophilus*, *Brugia malayi*, *Ascaris suum* and free-living *C. elegans* revealed that most Pfam domains were present in similar levels in all genomes with a few notable differences in certain families of G protein-coupled receptors (GPCRs), proteases, protease inhibitors, and fatty acid- and retinol-binding (FAR) proteins. Pfam domains among sequenced *Steinernema* congeners were compared and revealed that some protease and protease inhibitor families and FAR proteins appeared to be conserved in all species; however, there was wide variation in the domain richness among certain families of GPCRs, proteases and protease inhibitors (Dillman *et al.*, 2015). *Steinernema*-specific expansions of serine and metalloproteases were discovered by multi-genome comparison. The importance of the proteases and protease inhibitor group of proteins for future selection and recombinant studies has been highlighted by comparative genomics because they are known to be important in *Steinernema* invasion, suppressing insect host immunity, tissue degradation and host-killing (Abuhatab *et al.*, 1995; Balasubramanian *et al.*, 2009, 2010; Toubarro *et al.*, 2009, 2010). Functional studies have demonstrated the role of protease inhibitors in nematode evasion of host immunity, and genome analyses have revealed the expansion of several protease inhibitor families in steinernematids. One intriguing idea is that the secretome repertoire of EPNs may influence their host range and specificity. Using genetic transformation, the host range and specificity of the infection may be able to be altered by adding or removing specific secreted products from the secretome (Dillman *et al.*, 2012b).

Another incredibly interesting gene family that has been expanded in the genomes of steinernematids is the FAR protein family. Although their precise physiological role was not fully understood at that time, FAR proteins were assumed to play a significant role in parasitism by helping to sequester host fatty acids and/or retinoids to facilitate immune evasion or suppression. FAR proteins have been demonstrated to have an immunomodulatory effect in plant and animal model systems. This effect is most likely caused by sequestering lipids essential for immune signaling (Garofalo *et al.*, 2002; Kennedy *et al.*, 2013; Parks *et al.*, 2021). Understanding how proteins act mechanistically is crucial for biocontrol, disease treatment and prevention. The availability of genomic sequences from multiple *Steinernema* spp. provides several candidate genes and gene families that could be employed to improve EPN performance (Lu *et al.*, 2016).

The genome sequence also aided in identifying gene family expansions important to the parasitic nature of steinernematids. The detection of the genes and pathways allowed researchers to formulate new hypotheses about genes that are likely to be important in parasitism and to delve deeper into their genealogy. For example, the discovery of *S. carpocapsae* venom proteins and the dynamics of the transcriptome during infective juvenile activation were both made possible by this research. It established a novel paradigm that EPNs not only operate as carriers for entomopathogenic bacteria but also significantly contribute to toxicity in the host. The project to investigate EPN secretions was inspired by the genome's predicted proteome. These secretions are rich in proteins that are conserved in other parasitic nematodes, offering fresh evidence for the possibility that EPNs and vertebrate-parasitic nematodes use comparable molecules to facilitate parasitism. It thus demonstrated that vertebrate-parasitic nematodes can be studied using EPNs as a model system, notably for the bioactive substances they release to regulate their hosts (Dillman *et al.*, 2015; Lu *et al.*, 2017; Chang *et al.*, 2019). *S. carpocapsae* and *S. feltiae* were also used to investigate the extent of expression conservation in orthologous genes associated with embryonic development across nematode families. Based on the high conservation of orthologous genes between *Caenorhabditis* and *Steinernema* spp., this study discovered a funnel-shaped model of embryonic development (Macchietto *et al.*, 2017).

An initial understanding of EPN biology suggested that bacteria, as opposed to nematodes, played an active role in pathogenicity to insects (Li *et al.*, 2009; Mastore *et al.*, 2015; Brivio and Mastore, 2018). None the less, the increase in nematode genome sequences has provided evidence indicating a more active role of the nematode in the pathogenic process. The entomopathogenic nematodes in Steinernematidae are phylogenetically related to the mammalian-parasitic nematodes in Strongyloididae (Tylenchina; Panagrolaimomorpha), implying a transition to vertebrate parasitism via host switching (Blaxter and Koutsovoulos, 2015).

In 2016, a high-quality genome of *S. carpocapsae* Breton strain was published. The goal was to investigate the genome and transcriptome of *S. carpocapsae* to understand the origin and mechanisms of strongylid parasites, which could have implications for human health (Rougon-Cardoso *et al.*, 2016). The final draft genome was 84.6 Mb in 347 scaffolds, with an N50 of 1.24 Mb and a largest scaffold of 8.7 Mb. The quality of this genome was significantly higher than that of *S. carpocapsae* All strain, which had an N50 of 0.3 Mb and a largest scaffold of 1.7 Mb. In the Breton strain, genomic heterozygosity was reduced by making an isogenic line before sequencing it, a step that was skipped for the All strain. Genome annotation of the Breton strain identified 16,333 protein-coding genes, significantly fewer than the predicted proteins from the All strain (28,313 genes).

Gene prediction in the Breton strain was done using a hidden Markov model trained on *S. carpocapsae* gene structures with *ab initio* predictions using genomic information from *C. elegans* genes and *B. malayi* genes. In All strain gene prediction, the Augustus software package was used, with parameters optimized only for *C. elegans*. Furthermore, mass spectrometry verified the expression of 3773 predicted genes in the Breton strain. The gene ontology (GO) term analysis of the predicted genes of the Breton strain revealed 135 enriched GO terms when compared with *C. elegans*. These enriched GO terms are related to parasitism. The high quality of the genome also enabled the prediction of non-coding RNA (ncRNA). They identified 1097 transfer RNAS (tRNAs), 40 ribosomal RNAs (rRNAs; 15 5S rRNAs, one 5.8S rRNA and 24 8S rRNAs), 38 microRNA (miRNA) hairpins and 146 other ncRNAs. Parasitic nematodes (*Meloidogyne hapla*, *Bursaphelenchus xylophilus*, *Meloidogyne incognita*, *S. carpocapsae*, *A. suum*, *Brugia malayi* and *H. bacteriophora*) and free-living nematodes (*C. elegans*, *Pristionchus pacificus*, *Panagrellus redivivus* and *Caenorhabditis remanei*) were examined for ncRNA family abundance, and the parasitic nematodes had at least twice the amounts of the following ncRNAs: mir-2444, SeC (a tRNA), and miR-227, miR-100/miR-10, miR-2b, ACEA_U3 (a small nucleolar RNA) and miR-4455 (miRNAs). These findings suggest that ncRNAs may have a functional role in the pathogenic lifestyle, but further research is needed to validate the correlation between these families and parasitism.

RNA sequencing experiments to study miRNA expression of nematodes in response to insect hemolymph exposure identified five miRNAs in *S. carpocapsae* (miR-84-5p, miR-84-3p, let-7-5p, miR-31-3p and Cluster_21397_3p) whose expression increased after hemolymph induction. In *C. elegans*, double mutants of miR-84 and miR-48 have a delayed molting phenotype and accumulate double cuticles. This finding suggests that miR-84 and let-7 upregulation in *S. carpocapsae* may be involved in the molting process triggered by insect hemolymph contact. Distribution of proteases such as aspartic, cysteine and threonine from *S. carpocapsae* was in similar proportion to that of other parasitic nematodes such as *Brugia malayi*, *Strongyloides ratti*, the free-living nematode *C. elegans* and the necromanic nematode *Pristionchus pacificus*. The number of serine proteases in *S. carpocapsae* and *Pristionchus pacificus* genomes was higher.

Analysis of the genome of *S. carpocapsae* predicted 1421 putative excreted proteins. These predicted proteins are involved in immune evasion and suppression of host immune responses. Predicted proteins of *S. carpocapsae* are shared among different parasitic nematodes. Proteins such as lipases, saposins and transthyretin-like proteins are expressed in an early stage of *Haemonchus contortus*. Some of the excretory/secretory products predicted in *S. carpocapsae* involved in parasitism are serine proteases, astacin metalloprotease, papain-type aspartic and cysteine proteases. A phylogenetic analysis of *S. carpocapsae* using 245 proteins from strictly one-to-one orthologous genes from nine nematode species indicated that *Steinernema* is more closely related to *Strongyloides* spp. than to *Caenorhabditis* spp. The basal position of *Steinernema* to *Strongyloides* supports the hypothesis that this vertebrate parasite evolved from an entomopathogenic ancestor (Blaxter and Koutsovoulos, 2015). Identification of the protein ATAD-3 by a thorough examination of gene functions enriched at various levels raises the possibility that mitochondria might have a role in the evolution of nematode parasitism. However, more studies are required to confirm this. The high-quality genome of the *S. carpocapsae* Breton strain is a valuable resource for improving our understanding of host–parasite interactions and genetically improving entomopathogens as biological control agents (Rougon-Cardoso *et al.*, 2016).

Most of the genome sequencing for the *Steinernema* genome was done using short-read sequencing techniques such as Illumina. These genomes were invaluable to understanding the biology of nematodes but were highly fragmented. These fragments deter the analysis of conserved gene sequences, which could play a significantly larger role in the evolution of this phylum. To better understand the mechanism of *Steinernema* genome evolution, the genome of *S. carpocapsae* was reassembled using PacBio long-read sequencing and optical mapping of the assembled sequences using the BioNano Irys system.

The new genome of *S. carpocapsae* has 16 scaffolds, approximately 31,000 predicted genes and an identified X chromosome. The genome revealed several protease families involved in parasitism. Gene expression differences in catalytic activities between males and females were also observed (Serra *et al.*, 2019). DNA extraction was done from unmated males and females individually for Illumina sequencing. When DNA from individual males and females was mapped with the assembled scaffold, a twofold difference in coverage was observed between males and females in the largest scaffold, which was named the X chromosome. Based on one-to-one orthologs, a macrosyntenic analysis of the *S. carpocapsae* and *C. elegans* genomes revealed that the *S. carpocapsae* X chromosome is homologous to a different section of *C. elegans* and *Pristionchus pacificus* chromosomes II, X and IV.

The *S. carpocapsae* genome contained 254 serine protease genes and 228 unique metalloprotease genes involved in parasitism. When these genes were mapped to scaffolds, they aggregated in a cluster. The majority of the venomous metalloprotease genes were found on scaffold 7, while the serine protein gene clusters were found on three different scaffolds, 5, 8 and 14. Many of the metalloproteins have *C. elegans* orthologs, with a few duplications; however, serine proteins appear to have expanded rapidly and extensively in *S. carpocapsae*.

Comparative gene expression analysis of the adult males and females found 1026 genes upregulated in females and 2328 genes downregulated in males. The GO term analysis of the upregulated genes is related to sperm motility, male gamete formation, catalytic activity and biological regulation. Similarly, the downregulated genes are enriched for immune response, embryo development, catalytic activity and lipid metabolic process. The catalytic activity was the only GO term that was similar between males and females. Further investigation into the female catalytic activity GO terms found that they are enriched for serine-type peptidase activity, amine-lyase activity, serine hydrolase activity and peptidase activity. The catalytic activity GO terms of males, in contrast, are enriched for phosphatase activity, kinase activity, phosphotransferase activity and phosphoric ester hydrolase activity. Conservation of the *C. elegans* sex-determination pathway was compared with that of *S. carpocapsae*. The genome of *S. carpocapsae* contained 19 of the 27 genes in the *C. elegans* sex-determination pathway. These genes could be the key players in the nematode sex-determination pathway (Serra *et al.*, 2019).

Locally adapted EPNs are sometimes more effective than exotic nematodes at controlling agricultural pests (Gaugler, 1988; Duncan *et al.*, 2003, 2013; Stuart *et al.*, 2008; Hiltpold, 2015; Shields *et al.*, 2022). Examining these wild populations and species may reveal genetic changes that occurred during domestication and natural selection, allowing better biodiversity preservation, trait selection and targeted breeding (Adhikari *et al.*, 2009; Wu *et al.*, 2018). Efforts to evaluate the genome of native *Steinernema* spp. began with the genome sequence of a field-collected *S. feltiae* NW strain from Washington, USA (Fu *et al.*, 2020). The PacBio sequencing platform was used to sequence the genome. The genome size of the field-collected isolates of the NW strain is 121.6 Mb, which is 47% larger than the previously published laboratory-cultured *S. feltiae* SN strain genome, with a smaller number of contigs and improved genome completeness (Table 5.2). A whole-genome alignment approach was used to conduct a comparative genomic analysis of the NW assembly with the earlier SN assembly. More than 90% of the contigs/scaffolds in the two genomes were aligned. A total of 29 Mb of NW sequences failed to align with the SN assembly, indicating strain divergence. Despite the larger assembled genome size, genome annotation predicted 12% fewer genes/proteins than the previous annotation of the SN genome. The apparent intraspecific variation in genome size could result from several factors, including different environmental conditions, unrecognized speciation events and sequencing technology errors. A study of different *S. feltiae* strains using flow cytometry could therefore provide a more accurate estimate of genome size or confirm the apparent changes in genome size reported throughout the two genome sequencing projects (Fu *et al.*, 2020).

Similarly, the genome of *S. diaprepesi*, a native nematode from the central region of Florida citrus orchards, was published in 2020. The Illumina platform was used for genome sequencing. The completed genome had a BUSCO (benchmarking universal single-copy orthologs) score of 85% and was approximately 118 Mb in size (Table 5.2) (Baniya *et al.*, 2020). Another native genome from the Florida Flatwood region, *S. khuongi*, was published in 2021. The genome of *S. khuongi* is 82 Mb in size and has a BUSCO score of 86.6% (Table 5.2). The nematode's symbiotic bacteria were previously unknown, but simultaneous genome sequencing and comparative genomics of the nematode and bacteria identified the bacteria as *Xenorhabdus poinarii* (Baniya and Digennaro, 2021).

4.3 Genomics of *Heterorhabditis* spp.

The National Human Genome Research Institute (NHGRI) announced in June 2005 that it would focus on *H. bacteriophora* TTO1 strain for a high-quality draft genome. This nematode was inbred for 13 generations and the genome size was determined to be 111.4 ± 1.1 Mb using a flow cytometry technique (Bennett *et al.*, 2003). In 2006, random sequencing of clones from an infective juvenile stage cDNA library yielded 1246 ESTs (Sandhu *et al.*, 2006). In 2009, the adult *H. bacteriophora* TTO1 strain's cDNA libraries produced 31,485 high-quality ESTs (Bai *et al.*, 2009). A total of 10,886 distinct EST sequences were identified from adult *H. bacteriophora* cDNA libraries. Analysis of the ESTs using BLAST (basic local alignment search tool) revealed that these ESTs have a role in the defense response, parasitism, RNA interference (RNAi), the stress response and dauer-related processes. This work laid the foundation for future in-depth gene function analysis, genome annotation and microarray development (Bai *et al.*, 2007, 2009). In 2013, a draft genome of *H. bacteriophora* was published. The genome was sequenced using short-read sequencing technology. The final draft genome consists of 1263 scaffolds with a total genome length of 77,007,652 bp. The genome has a GC content of 32.2%, similar to the plant-parasitic nematodes *M. incognita* and *M. hapla*, the human-parasitic nematode *Brugia malayi* and the free-living nematode *C. elegans*. The assembled genome's N50 is 312,328 bp. A total of 21,250 protein-coding genes were predicted from the *H. bacteriophora* genome using *ab initio* gene prediction parameters optimized for *C. elegans*. Insulin/insulin-like growth factor 1 (IGF-1) signaling pathways regulate growth and development in *C. elegans* dauers (Murphy and Hu, 2013). These pathways are critical for the development of the dauer and adult stages, as well as for longevity, stress resistance and innate immunity. The genome analysis of *H. bacteriophora* revealed that 19 of the 23 genes in the *C. elegans* insulin/IGF-1 signaling pathway are conserved in *H. bacteriophora*. These genes could be candidates for genetic manipulation to improve longevity and/or stress resistance, resulting in improved EPN field efficacy (Bai *et al.*, 2013). Based on the sequence similarity to *C. elegans* gene products, RNAi genes in *H. bacteriophora*, *Brugia malayi* and *M. hapla* were identified. All four species shared four genes: *drsh-1*, *ego-1*, *rsd-3* and *smg-2*. The *sid-1* gene, essential for systemic RNAi in *C. elegans*, was detected only in *C. elegans* and *H. bacteriophora*.

The protein family domain (Pfam) analysis predicted 7957 Pfam domains with 4144 different Pfam accessions. One highly enriched protein family in the genome is the GPCR family. Some of the highly enriched predicted protein families in the genome of *H. bacteriophora* are 82 GPCRs and 24 members of the nuclear hormone receptor (NHR) gene family. Comparative genomics of the Pfam domains of the *H. bacteriophora* genome with ten other species of nematodes identified 56 different protein domains that are significantly different from other nematodes in comparison. The Mariner transposase (PF01359.11) was the most common protein domain index, accounting for 138 in *H. bacteriophora*, 65 in *C. japonica*, one in *M. incognita* and *M. hapla*, and none in *C. elegans* or *Brugia malayi*. In contrast, only nine C-type lectin domain-containing proteins were found in *H. bacteriophora* compared with 133 in *C. elegans*. The decrease in C-lectin domain proteins in *H. bacteriophora* compared with *C. elegans* could be attributed to its mutualistic relationship with *Photorhabdus luminescens* bacteria (C.G. Bhat *et al.*, 2019). In *H. bacteriophora*, 134 potential miRNA genes representing 26 different animal miRNA species, 254 tRNA genes and one tRNA pseudogene were predicted. There are 3794 microsatellite loci predicted in 506 contigs of the *H. bacteriophora* genome. These microsatellite markers are extremely useful for determining the phylogeographic, demographic and genetic structure of *H. bacteriophora* populations. EPN-secreted proteases are known to influence nematode hemolymph penetration, insect host tissue degradation and immune suppression. *H. bacteriophora* secretome analysis predicted 603 secreted proteins. There were only about 30 predicted protease and protease inhibitors among these proteins. This could be due to the nematode's reliance on *Photorhabdus luminescens* for immune suppression and tissue degradation of the insect host (Bai *et al.*, 2013). This genome information will help us understand the fundamental processes of gut mutualism and insect parasitism in nematodes. The genome sequence, combined with RNAi gene-silencing targets, provides a powerful reverse genetic approach for investigating the functions of signaling pathways and transcription factors in insect parasitism and symbiosis (Bai *et al.*, 2013).

The genome sequence of *H. bacteriophora* determined by Bai *et al.* (2013) has a high rate of completeness of 99.6% when assessed using CEGMA (core eukaryote gene mapping approach), and analysis of the BUSCO version 2.0.1 revealed that it is 80.9% complete with 5.6% fragmented hits for the BUSCO Eukaryota gene set. However, the BUSCO score for the predicted proteome was relatively low, with 47.8% complete and 34.7% fragmented (McLean *et al.*, 2018). The genome of *H. bacteriophora* was predicted to contain 21,250 proteins. When these putative proteins were examined further, nearly half of them had no matches in GenBank's non-redundant protein database, indicating the genome's novelty. Furthermore, when compared with the proteomes of other clade V nematodes such as *C. elegans* and animal parasitic nematodes, *H. bacteriophora* had fewer shared orthologous clusters and KEGG (Kyoto Encyclopedia of Genes and Genomes) loci in many metabolic categories (Bai *et al.*, 2013). The *H. bacteriophora* genome and gene predictions were repredicted using the most recent tools (BRAKER1) and methods to see whether these unusual features reflect a truly divergent proteome (McLean *et al.*, 2018). The repredicted proteome now has 15,747 predicted protein-coding genes, which is fewer than the previously predicted 20,964 genes. However, the newly predicted proteome has improved BUSCO scores of 94% complete and 4.3% fragmented, which is significantly higher than the previous 47.8% complete and 34.7% fragmented BUSCO. The improved genome annotation of *H. bacteriophora* is now available, and this information will aid in the exploration of genes involved in its parasitic and symbiotic interactions, as well as supporting genetic manipulation and enhancing its utility as a biological control agent (McLean *et al.*, 2018; Vadnal *et al.*, 2018).

The genome sequence of *H. indica* strain IARI-EPN-Hms1 was also made public in 2021. The genome was sequenced using the Illumina HiSeq 4000 platform for short-read sequencing. The final draft genome is 91.26 Mb in length and contains 3538 scaffolds. The genome's N50 is 587 kb, with an average scaffold length of 25.79 kb and a GC content of 35.31%. BUSCO version 5.2.2 was used to assess genome completeness against the 3131 BUSCOs in the nematoda_odb10 database. The *H. indica* genome BUSCO score was 84% (2617) (complete and single-copy: 2617 (83.6%); complete and duplicated: 13 (0.4%)), whereas 205 (6.5%) BUSCOs were fragmented and 296 (9.5%) were missing. Functional annotation of the genome resulted in 9596 genes. Comparative genomics using *H. indica* and four other nematode genomes to identify orthologous genes revealed that *H. indica* shared 4526 with *C. elegans*, 4059 with S. *carpocapsae*, 2848 groups with *H. bacteriophora* and 3396 with *Oscheius tipulae* (Bhat *et al.*, 2021).

4.4 The Future of EPN Genomics

Entomopathogenic nematology is entering a new era of comparative and functional genomics, which will aid in the advancement of EPNs as a biological control agent and model system for studying host–parasite interactions, mutualism and symbiosis. The genome sequences of seven *Steinernema* and two *Heterorhabditis* spp. are now available. The sequencing of additional nematode genomes for *Steinernema* and *Heterorhabditis* spp. is currently under way. *C. elegans* is a highly effective organism in the laboratory for studying the genetic underpinnings of a wide range of biological topics, such as how genes determine complex structures and behaviors (Brenner, 1974). The mode of reproduction of *C. elegans* is a highly tractable system comprised of self-fertilizing hermaphrodites and males. This facilitates genetic screening, and mutagenesis can produce many mutants (Brenner, 1974). However, *C. elegans* lacks a highly specific association with its microbes, such as the nematode–bacteria relationship found among EPNs. A tractable nematode system with a naturally occurring species-specific microbial symbiosis could also be useful for studying symbiotic interactions. The discovery of *Steinernema hermaphroditum*, a self-fertilizing hermaphrodite EPN, has led to mutagenesis screens with visible and heritable phenotypes. The availability of a complete and annotated *S. hermaphroditum* genome would help to target genes and identify genes for forward and reverse genetics, making this nematode a good model for studying mutualism, symbiosis and host–parasite interaction (A.H. Bhat *et al.*, 2019; Cao *et al.*, 2022). A genome project for *S. hermaphroditum* is currently in progress. Current sequencing technology approaches from Illumina, Roche 454 and Applied Biosystems (SOLiD) enables an exponential increase in the number of sequences per unit time or cost compared with classic Sanger dideoxy chain termination (Morozova and Marra, 2008). Despite the massive amounts of data they can generate, one of the drawbacks of these technologies is the short read length, which limits our ability to produce complete *de novo* genome assembly. Third-generation sequencing approaches are surpassing short-read sequencing. These new methods are based on single DNA molecule monitoring and combine higher-density parallel sequencing with longer read lengths. The PacBio sequencing approach, which depends on real-time DNA polymerase monitoring, is currently widely used (Eid *et al.*, 2009). Similarly, Oxford Nanopore technology combines exonuclease activity with the transit of released nucleotides through a nanopore (Clarke *et al.*, 2009). Both technologies aim to reduce the cost of producing a human genome to less than US$1,000. The anticipation of these technologies has resulted in an unprecedented rate of organism sequencing. Current best practices for *de novo* nematode genome sequencing are to use the hybrid sequencing method (long-read sequencing using a MinION or PacBio sequencer and assembling the genome, error correction using the short-read fragments, and arranging the scaffolds into chromosomes using high-throughput chromatin conformation capture (Hi-C) genome scaffolding techniques) (Lian *et al.*, 2019; Serra *et al.*, 2019; Foster *et al.*, 2020). Due to the rapidly declining cost of sequencing, the acquisition of assembled genomes from any species of interest will become commonplace in the coming years. Genome and comparative analysis will provide a better understanding of EPN biology, allowing improved efficacy in biological control and enhancing the utility of EPNs as a model system for mutualism, symbiosis and host–parasite interactions.

4.5 References

Abuhatab, M., Selvan, S. and Gaugler, R. (1995) Role of proteases in penetration of insect gut by the entomopathogenic nematode *Steinernema glaseri* (Nematoda: Steinernematidae). *Journal of Invertebrate Pathology* 66, 125–130.

Adhikari, B.N., Chin-Yo, L., Xiaodong, B., Ciche, T.A., Grewal, P.S. *et al.* (2009) Transcriptional profiling of trait deterioration in the insect pathogenic nematode *Heterorhabditis bacteriophora. BMC Genomics* 10: 609.

Backhaus, S., Simmonds, R.W., Loshak, A., Davis, J.C. and Packard, R.E. (1999) How and why a parasitic nematode jumps. *Nature* 397, 485–486.

Bai, X., Adams, B.J., Ciche, T.A., Clifton, S., Gaugler, R. *et al.* (2009) Transcriptomic analysis of the entomopathogenic nematode *Heterorhabditis bacteriophora* TTO1. *BMC Genomics* 10: 205.

Bai, X., Adams, B.J., Ciche, T.A., Clifton, S., Gaugler, R. *et al.* (2013) A lover and a fighter: the genome sequence of an entomopathogenic nematode *Heterorhabditis bacteriophora*. *PLOS ONE* 8: e69618.

Bai, X., Grewal, P.S., Hogenhout, S.A., Adams, B.J., Ciche, T.A. *et al.* (2007) Expressed sequence tag analysis of gene representation in insect parasitic nematode *Heterorhabditis bacteriophora*. *Journal of Parasitology* 93, 1343–1349.

Baiocchi, T., Braun, L. and Dillman, A.R. (2019) Touch-stimulation increases host-seeking behavior in *Steinernema carpocapsae*. *Journal of Nematology* 51: e2019-67.

Balasubramanian, N., Hao, Y.J., Toubarro, D., Nascimento, G. and Simões, N. (2009) Purification, biochemical and molecular analysis of a chymotrypsin protease with prophenoloxidase suppression activity from the entomopathogenic nematode *Steinernema carpocapsae*. *International Journal for Parasitology* 39, 975–984.

Balasubramanian, N., Toubarro, D. and Simões, N. (2010) Biochemical study and *in vitro* insect immune suppression by a trypsin-like secreted protease from the nematode *Steinernema carpocapsae*. *Parasite Immunology* 32, 165–175.

Baniya, A. and Digennaro, P. (2021) Genome announcement of *Steinernema khuongi* and its associated symbiont from Florida. *G3 (Bethesda)* 11: jkab053.

Baniya, A., Huguet-Tapia, J.C. and DiGennaro, P. (2020) A draft genome of *Steinernema diaprepesi*. *Journal of Nematology* 52: e2020-69.

Bennett, M.D., Leitch, I.J., Price, H.J. and Johnston, J.S. (2003) Comparisons with *Caenorhabditis* (approximately 100 Mb) and *Drosophila* (approximately 175 Mb) using flow cytometry show genome size in *Arabidopsis* to be approximately 157 Mb and thus approximately 25% larger than the *Arabidopsis* genome initiative estimate of approximately 125 Mb. *Annals of Botany* 91, 547–557.

Bhat, A.H., Chaubey, A.K., Shokoohi, E. and Mashela, P.W. (2019) Study of *Steinernema hermaphroditum* (Nematoda, Rhabditida), from the West Uttar Pradesh, India. *Acta Parasitologica* 64, 720–737.

Bhat, C.G., Chauhan, K., Phani, V., Papolu, P.K., Rao, U. and Somvanshi, V.S. (2019) Expression of *Heterorhabditis bacteriophora* C-type lectins, *Hb-clec-1* and *Hb-clec-78*, in context of symbiosis with *Photorhabdus* bacteria. *Symbiosis* 77, 49–58.

Bhat, C.G., Somvanshi, V.S., Budhwar, R., Godwin, J. and Rao, U. (2021) Nematode genome announcement: the draft genome sequence of entomopathogenic nematode *Heterorhabditis indica*. *Journal of Nematology* 53: e2021-101.

Blaxter, M. (2011) Nematodes: the worm and its relatives. *PLOS Biology* 9: e1001050.

Blaxter, M. and Koutsovoulos, G. (2015) The evolution of parasitism in Nematoda. *Parasitology* 142 (Suppl. 1), S26.

Blaxter, M.L., de Ley, P., Garey, J.R., Llu, L.X., Scheldeman, P., Vierstraete, A. *et al.* (1998) A molecular evolutionary framework for the phylum Nematoda. *Nature* 392, 71–75.

Bogale, M., Baniya, A. and Digennaro, P. (2020) Nematode identification techniques and recent advances. *Plants* 9: 1260.

Brenner, S. (1974) The genetics of *Caenorhabditis elegans*. *Genetics* 77, 71–94.

Brivio, M.F. and Mastore, M. (2018) Nematobacterial complexes and insect hosts: different weapons for the same war. *Insects* 9: 117.

Campbell, J.F., Lewis, E.E., Stock, S.P., Nadler, S. and Kaya, H.K. (2003) Evolution of host search strategies in entomopathogenic nematodes. *Journal of Nematology* 35, 142–145.

Cao, M., Schwartz, H.T., Tan, C.H. and Sternberg, P.W. (2022) The entomopathogenic nematode *Steinernema hermaphroditum* is a self-fertilizing hermaphrodite and a genetically tractable system for the study of parasitic and mutualistic symbiosis. *Genetics* 220: iyab170.

Chang, D.Z., Serra, L., Lu, D., Mortazavi, A. and Dillman, A.R. (2019) A core set of venom proteins is released by entomopathogenic nematodes in the genus *Steinernema*. *PLOS Pathogens* 15: e1007626.

Ciche, T. (2007) The biology and genome of *Heterorhabditis bacteriophora*. In: *WormBook: The Online Review of C. elegans Biology*. Wormbook, Pasadena, California. Available at: https://www.ncbi.nlm.nih.gov/books/NBK19785/ (accessed 13 March 2024).

Clarke, J., Wu, H.C., Jayasinghe, L., Patel, A., Reid, S. and Bayley, H. (2009) Continuous base identification for single-molecule nanopore DNA sequencing. *Nature Nanotechnology* 4, 265–270.

Coghlan, A., Tyagi, R., Cotton, J.A., Holroyd, N., Rosa, B.A. *et al.* (2018) Comparative genomics of the major parasitic worms. *Nature Genetics* 51, 163–174.

de Ley, P. and Blaxter, M. (2002) Systematic position and phylogeny. In: Lee, D.L. (ed.) *The Biology of Nematodes*. CRC Press, London, pp. 1–30.

de Ley, P. and Blaxter, M.L. (2004) A new system for Nematoda: combining morphological characters with molecular trees, and translating clades into ranks and taxa. In: Cook, R. and Hunt, D.J. (eds.) *Proceedings of the Fourth International Congress of Nematology*, Vol. 2. Brill, Leiden, Netherlands, pp. 633–653.

Dieterich, C., Clifton, S.W., Schuster, L.N., Chinwalla, A., Delehaunty, K. *et al.* (2008) The *Pristionchus pacificus* genome provides a unique perspective on nematode lifestyle and parasitism. *Nature Genetics* 40, 1193–1198.

Dillman, A.R., Chaston, J.M., Adams, B.J., Ciche, T.A., Goodrich-Blair, H. *et al.* (2012a) An entomopathogenic nematode by any other name. *PLOS Pathogens* 8: e1002527.

Dillman, A.R., Mortazavi, A. and Sternberg, P.W. (2012b) Incorporating genomics into the toolkit of nematology. *Journal of Nematology* 44, 191–205.

Dillman, A.R., Macchietto, M., Porter, C.F., Rogers, A., Williams, B. *et al.* (2015) Comparative genomics of *Steinernema* reveals deeply conserved gene regulatory networks. *Genome Biology* 16: 200.

Duncan, L.W., Graham, J.H., Dunn, D.C., Zellers, J., McCoy, C.W. and Nguyen, K. (2003) Incidence of endemic entomopathogenic nematodes following application of *Steinernema riobrave* for control of *Diaprepes abbreviatus*. *Journal of Nematology* 35, 178–186.

Duncan, L.W., Stuart, R.J., El-Borai, F.E., Campos-Herrera, R., Pathak, E. *et al.* (2013) Modifying orchard planting sites conserves entomopathogenic nematodes, reduces weevil herbivory and increases citrus tree growth, survival and fruit yield. *Biological Control* 64, 26–36.

Eid, J., Fehr, A., Gray, J., Luong, K., Lyle, J. *et al.* (2009) Real-time DNA sequencing from single polymerase molecules. *Science* 323, 133–138.

Foster, J.M., Grote, A., Mattick, J., Tracey, A., Tsai, Y.C. *et al.* (2020) Sex chromosome evolution in parasitic nematodes of humans. *Nature Communications* 11: 1964.

Fu, Z., Li, Y., Elling, A.A. and Snyder, W.E. (2020) A draft genome of a field-collected *Steinernema feltiae* strain NW. *Journal of Nematology* 52, e2020-03.

Garofalo, A., Kläger, S.L., Rowlinson, M.C., Nirmalan, N., Klion, A. *et al.* (2002) The FAR proteins of filarial nematodes: secretion, glycosylation and lipid binding characteristics. *Molecular and Biochemical Parasitology* 122, 161–170.

Gaugler, R. (1988) Ecological considerations in the biological control of soil-inhabiting insects with entomopathogenic nematodes. *Agriculture, Ecosystems and Environment* 24, 351–360.

Ghedin, E., Wang, S., Spiro, D., Caler, E., Zhao, Q. *et al.* (2007) Draft genome of the filarial nematode parasite *Brugia malayi*. *Science* 317, 1756–1760.

Harris, T.W., Antoshechkin, I., Bieri, T., Blasiar, D., Chan, J. *et al.* (2010) WormBase: a comprehensive resource for nematode research. *Nucleic Acids Research* 38 (Suppl. 1), D463–D467.

Hiltpold, I. (2015) Prospects in the application technology and formulation of entomopathogenic nematodes for biological control of insect pests. In: *Nematode Pathogenesis of Insects and Other Pests: Ecology and Applied Technologies for Sustainable Plant and Crop Protection*. Springer, Cham, Switzerland, pp. 187–205.

Hodda, M. (2011) Phylum Nematoda Cobb 1932. *Zootaxa* 3148, 63–95.

Holterman, M., van der Wurff, A., van den Elsen, S., van Megen, H., Bongers, T. *et al.* (2006) Phylum-wide analysis of SSU rDNA reveals deep phylogenetic relationships among nematodes and accelerated evolution toward crown clades. *Molecular Biology and Evolution* 23, 1792–1800.

Horvitz, H.R. and Sternberg, P.W. (1982) Nematode postembryonic cell lineages. *Journal of Nematology* 14, 240–248.

Jex, A.R., Liu, S., Li, B., Young, N.D., Hall, R.S. *et al.* (2011) *Ascaris suum* draft genome. *Nature* 479, 529–533.

Kennedy, M.W., Corsico, B., Cooper, A. and Smith, B.O. (2013) The unusual lipid-binding proteins of nematodes: NPAs, nemFABPs and FARs. In: Kennedy, M.W. and Harnett, W. (eds.) *Parasitic Nematodes: Molecular Biology, Biochemistry and Immunology*, Vol 2. CABI International, Wallingford, UK, pp. 397–412.

Koren, S., Walenz, B.P., Berlin, K., Miller, J.R., Bergman, N.H. and Phillippy, A.M. (2017) Canu: scalable and accurate long-read assembly via adaptive *k*-mer weighting and repeat separation. *Genome Research* 27, 722–736.

Lambert, K. and Bekal, S. (2002) Introduction to plant-parasitic nematodes. Plant Health Instructor. Available at: www.apsnet.org/edcenter/disandpath/nematode/intro/Pages/IntroNematodes.aspx (accessed 13 March 2024).

Li, X., Cowles, E.A., Cowles, R.S., Gaugler, R. and Cox-Foster, D.L. (2009) Characterization of immunosuppressive surface coat proteins from *Steinernema glaseri* that selectively kill blood cells in susceptible hosts. *Molecular and Biochemical Parasitology* 165, 162–169.

Lian, Y., Wei, H., Wang, J., Lei, C., Li, H. *et al.* (2019) Chromosome-level reference genome of X12, a highly virulent race of the soybean cyst nematode *Heterodera glycines*. *Molecular Ecology Resources* 19, 1637–1646.

Lu, D., Baiocchi, T. and Dillman, A.R. (2016) Genomics of entomopathogenic nematodes and implications for pest control. *Trends in Parasitology* 32, 588–598.

Lu, D., Macchietto, M., Chang, D., Barros, M.M., Baldwin, J. *et al.* (2017) Activated entomopathogenic nematode infective juveniles release lethal venom proteins. *PLOS Pathogens* 13: e1006302.

Macchietto, M., Angdembey, D., Heidarpour, N., Serra, L., Rodriguez, B. *et al.* (2017) Comparative transcriptomics of *Steinernema* and *Caenorhabditis* single embryos reveals orthologous gene expression convergence during late embryogenesis. *Genome Biology and Evolution* 9, 2681–2696.

Mardis, E.R. (2008) Next-generation DNA sequencing methods. *Annual Review of Genomics and Human Genetics* 9, 387–402.

Mastore, M., Arizza, V., Manachini, B. and Brivio, M.F. (2015) Modulation of immune responses of *Rhynchophorus ferrugineus* (Insecta: Coleoptera) induced by the entomopathogenic nematode *Steinernema carpocapsae* (Nematoda: Rhabditida). *Insect Science* 22, 748–760.

McLean, F., Berger, D., Laetsch, D.R., Schwartz, H.T. and Blaxter, M. (2018) Improving the annotation of the *Heterorhabditis bacteriophora* genome. *GigaScience* 7: giy034.

Morozova, O. and Marra, M.A. (2008) Applications of next-generation sequencing technologies in functional genomics. *Genomics* 92, 255–264.

Murphy, C.T. and Hu, P.J. (2013) Insulin/insulin-like growth factor signaling in *C. elegans*. In: *WormBook: The Online Review of* C. elegans *Biology*. WormBook, Pasadena, California. Available at: www.ncbi.nlm.nih.gov/books/NBK179230/ (accessed 13 March 2024).

Parks, S.C., Nguyen, S., Nasrolahi, S., Bhat, C., Juncaj, D. *et al.* (2021) Parasitic nematode fatty acid- and retinol-binding proteins compromise host immunity by interfering with host lipid signaling pathways. *PLOS Pathogens* 17: e1010027.

Rougon-Cardoso, A., Flores-Ponce, M., Ramos-Aboites, H.E., Martínez-Guerrero, C.E., Hao, Y.J. *et al.* (2016) The genome, transcriptome, and proteome of the nematode *Steinernema carpocapsae*: evolutionary signatures of a pathogenic lifestyle. *Scientific Reports* 6: 37536.

Sandhu, S.K., Jagdale, G.B., Hogenhout, S.A. and Grewal, P.S. (2006) Comparative analysis of the expressed genome of the infective juvenile entomopathogenic nematode, *Heterorhabditis bacteriophora*. *Molecular and Biochemical Parasitology* 145, 239–244.

Serra, L., Macchietto, M., Macias-Muñoz, A., McGill, C.J., Rodriguez, I.M. *et al.* (2019) Hybrid assembly of the genome of the entomopathogenic nematode *Steinernema carpocapsae* identifies the X-chromosome. *G3 (Bethesda)* 9, 2687–2697.

Shapiro-Ilan, D.I., Hiltpold, I. and Lewis, E.E. (2018) Nematodes. In: Hajek, A.E. and Shapiro-Ilan D.I. (eds.) *Ecology of Invertebrate Diseases*. Wiley, Oxford, UK, pp. 415–440.

Shapiro-Ilan, D.I., Hazir, S. and Glazer, I. (2020) Advances in use of entomopathogenic nematodes in IPM. In: Kogan, M. and Heinrichs, E.A. (eds.) *Integrated Management of Insect Pests: Current and Future Developments*. Burleigh Dodds Science Publishing, Cambridge, UK, pp. 649–678.

Shields, E., Testa, A., Rusinek, T. and Bornt, C. (2022) Management of wireworms in sweet potatoes with persistent NY entomopathogenic nematodes. *Great Lakes Entomologist* 54, 73–80.

Stein, L.D., Bao, Z., Blasiar, D., Blumenthal, T., Brent, M.R. *et al.* (2003) The genome sequence of *Caenorhabditis briggsae*: a platform for comparative genomics. *PLOS Biology* 1: e45.

Stuart, R.J., Borai, F.E. and Duncan, L.W. (2008) From augmentation to conservation of entomopathogenic nematodes: trophic cascades, habitat manipulation and enhanced biological control of *Diaprepes abbreviatus* root weevils in Florida citrus groves. *Journal of Nematology* 40, 73–84.

The *C. elegans* Sequencing Consortium (1998) Genome sequence of the nematode *C. elegans*: a platform for investigating biology. *Science* 282, 2012–2018.

Toubarro, D., Lucena-Robles, M., Nascimento, G., Costa, G., Montiel, R. *et al.* (2009) An apoptosis-inducing serine protease secreted by the entomopathogenic nematode *Steinernema carpocapsae*. *International Journal for Parasitology* 39, 1319–1330.

Toubarro, D., Lucena-Robles, M., Nascimento, G., Santos, R., Montiel, R. *et al.* (2010) Serine protease-mediated host invasion by the parasitic nematode *Steinernema carpocapsae*. *Journal of Biological Chemistry* 285, 30666–30675.

Vadnal, J., Granger, O.G., Ratnappan, R., Eleftherianos, I., O'Halloran, D.M. and Hawdon, J.M. (2018) Refined *ab initio* gene predictions of *Heterorhabditis bacteriophora* using RNA-seq. *International Journal for Parasitology* 48, 585–590.

van Megen, H., van den Elsen, S., Holterman, M., Karssen, G., Mooyman, P. *et al.* (2009) A phylogenetic tree of nematodes based on about 1200 full-length small subunit ribosomal DNA sequences. *Nematology* 11, 927–950.

Wu, S., Lau, K.H., Cao, Q., Hamilton, J.P., Sun, H. *et al.* (2018) Genome sequences of two diploid wild relatives of cultivated sweetpotato reveal targets for genetic improvement. *Nature Communications* 9: 4580.

5 Survival Mechanisms of Entomopathogenic Nematodes: Current Knowledge and its Integration toward a Better Performing Biological Control Agent

Jayashree Ramakrishnan[1,2], Dana Ment[1] and Itamar Glazer[1*]
[1]*Institute of Plant Protection, ARO, The Volcani Center, Rishon LeZion, Israel;*
[2]*The Robert H. Smith Faculty of Agriculture, Food & Environment, The Hebrew University of Jerusalem, Rehovot, Israel*

Abstract

As entomopathogenic nematodes (EPNs) are used as biological control agents, their survival and persistence is crucial to ensure success in application against insect pests. The survival of EPNs is dependent on abiotic and biotic factors in the environment. Abiotic stress environments such as desiccation, temperature and ultraviolet radiation severely impact their performance in the field. Hence, a comprehensive understanding of the underlying survival mechanisms enabling protection and tolerance is imperative to realistically enhance their performance in the field. Thus, identifying key players among each stress could invariably contribute toward developing more robust, reliable solutions for application strategies including genetic tools and formulation technologies.

5.1 Introduction

Nematodes are among the most abundant organisms on earth, filling all trophic levels in the soil food web, as they exist in almost every possible habitat and ecosystem (Bernard, 1992; Ettema, 1998; de Ley, 2006; Powers *et al.*, 2009; van den Hoogen *et al.*, 2019). Indeed, these organisms can be found in aquatic, marine and terrestrial ecosystems ranging from the tropics to the poles and from the highest to the lowest of elevations (van den Hoogen *et al.*, 2019).

The natural habitat of entomopathogenic nematodes (EPNs) in the families Steinernematidae and Heterorhabditidae is soil, which brings certain challenges. Soil is a difficult environment for persistence of any organism due to its complexity in physical, chemical and biological interfaces. Nevertheless, EPNs have been identified from soils throughout the world in ecosystems ranging from subarctic to arid and temperate to tropical climates (Poinar, 1990; Hominick *et al.*, 1996). It is assumed that along the course of evolution, EPNs, just like other terrestrial organisms, adopted unique survival mechanisms to resist environmental extremes.

As EPNs are used as biological control agents against insect pests, their survival and persistence is crucial to ensure success in application (Lacey and Georgis, 2012; Shapiro-Ilan *et al.*, 2019; Glazer, 2022). It is, for example, imperative for EPNs to possess survival mechanisms when they are

*Email: glazerit@volcani.agri.gov.il

DOI: 10.1079/9781800620322.0005

applied at extreme conditions resulting in rapid desiccation and/or exposure to ultraviolet (UV) radiation on the soil surface or aboveground (e.g. plant foliage, tree trunks). These two distinct and diverse ecological niches are characterized by temporal variation of temperatures, salinity, moisture and oxygen concentrations. In this chapter, the survival mechanisms of EPNs and implications for their use as beneficial biological control agents will be described. As the molecular and physiological mechanisms of environmental tolerance were also studied in certain model nematodes species such as *Caenorhabditis elegans* and members of the genus *Panagrolaimus*, we will review the knowledge relevant to EPNs.

5.2 The Resilient Developmental Stage of EPNs: Dauer Juveniles

In many parasitic nematodes, which spend part of their life cycle in a protected and favorable environment (i.e. the host), only free-living stages have the capability to survive the harsh environment outside the host. Among several Rhabditida families, environmental stress (e.g. lack of food, high population density, heat) induce the development of a unique juvenile stage, the "dauer juvenile" (Riddle, 1988). This stage is adapted morphologically and physiologically to remain in the environment without feeding while searching for a new food source. In EPNs from the families Steinernematidae and Heterorhabditidae, the dauer stage is also the infective stage, termed the infective juvenile (IJ). IJs locate and invade insect hosts (Lewis *et al.*, 1992) and also provide protection to symbiotic bacteria carried in their intestine (Akhurst, 1993). The IJ integument is comprised of two layers of external membrane, the cuticles of the third stage and of the second molt, which are retained to provide additional protection (Timper and Kaya, 1989; Campbell and Gaugler, 1991).

IJs do not feed and so are dependent on internal energy sources to survive until a host is invaded. Quantitative and qualitative analyses of biochemical reserves of EPNs have shown that lipid content and fatty acid composition is highly dependent on media components (Selvan *et al.*, 1993b; Lewis *et al.*, 1995; Patel *et al.*, 1997; Abu Hatab *et al.*, 1998).

The other significant energy reserve found in nematodes is glycogen (Barrett and Wright, 1998; Wright, 1998). Glycogen as a storage compound occurs in appreciable amounts in some EPNs, but its functional significance has been less clear (Selvan *et al.* 1993a,b). Studies by Lewis *et al.* (1995) and Patel *et al.* (1997) indicated that glycogen may play a significant, even if secondary, role to neutral lipids in the maintenance of infectivity in several species.

From the above studies on energy stores, it is clear that survival of IJs is a function of their metabolic rates and their initial levels of energy reserves. Availability of energy reserves is essential to support the physiological and behavioral processes accompanying acclimatization to environmental stress. The importance of energy sources for stress tolerance will be discussed further.

Although IJs have moderate tolerance to environmental extremes, extensive attention has been devoted to their survival capabilities in storage and under field conditions (Glazer, 2002; Burnell and Tunnacliffe, 2011). Specifically, these studies include evaluation of desiccation and osmotic stress tolerance (Glazer, 2002, 2015) and sensitivity to extreme temperatures (Grewal *et al.*, 1994, 2011). The key reason for characterizing these tolerances was to promote understanding of how well IJs survive in order to increase persistence and efficacy in the field, in storage and for shelf life of these biological control agents (Glazer, 2015; Sumaya *et al.*, 2017).

5.2.1 The role of the cuticle in dauer juvenile survival in the environment

The nematode cuticle or integument is a highly ordered extracellular multi-layered structure inclosing epicuticle, cortical, median and basal layers of varying osmiophilic nature. Patel and Wright (1998) emphasized the important functional role of the cuticle in EPN survival under desiccating conditions. Based on ultrastructural studies on *Steinernema* spp., the epicuticle layer acts as a semi-permeable barrier and regulates cuticle permeability, while its thickness could confer protection during rapid desiccation (Kondo and Ishibashi, 1989; Bird and Bird, 1991). However, Patel and Wright (1998) compared cuticles of Steinernematidae (*Steinernema carpocapsae*,

S. feltiae and *S. riobrave*) and suggested that the chemical composition of the epicuticle layer regulates the rate of water loss from the nematode body, meaning that lipid components in the epicuticle could retain water owing to its less osmiophilic nature (Dunphy and Webster, 1987). O'Leary *et al.* isolated a mutant in *Heterorhabditis megidis* with enhanced survival at rapid desiccation (57% relative humidity (RH)), owing to negative charge on the sheath surface in comparison to wild type. The study concluded that negative charge of the sheath influenced water retention over the cuticle, while removal of the sheath resulted in loss of this phenomenon (O'Leary *et al.*, 1997, 1998, 2001). Hence, cuticle permeability is a key factor determining the ability of nematodes to survive evaporative or osmotic water loss (Piggott *et al.*, 2002). In addition, an increase in cuticle permeability was correlated to the age of IJs (75 days since emergence) leading to greater water loss and reduced desiccation tolerance (Patel *et al.*, 1998).

5.3 How EPNs Respond to Extreme Conditions

Quiescence is a form of dormancy involving physiological adaptations that enhance survival (Perry and Moens, 2011). Quiescence is a facultative response, involving a lowered metabolism in response to unfavorable environmental conditions, and is readily reversible when favorable conditions return. If the stress persists, some organisms can enter a state of cryptobiosis where there is no measurable metabolism. The type of quiescence induced in organisms by unfavorable environmental conditions such as absence of water, osmotic stress, extreme temperatures (heat or freezing) and lack of oxygen are termed anhydrobiosis, osmobiosis, thermobiosis, cryobiosis and anoxybiosis, respectively (Barrett, 1991). Under natural conditions these situations may intertwine, and the organisms may face multi-stress situations, as will be discussed further. At this point, we will outline the most important stress factors and characterize the response of certain nematodes to these conditions. In addition, we will summarize the current knowledge on behavioral, physiological and genetic adaptations to these stresses and the formulation technologies designed in order to protect EPNs from stress factors.

5.3.1 Desiccation tolerance

The rate of water loss from the body of the nematode is a key factor in its ability to survive extended periods without water in an anhydrobiosis state (Crowe and Madin, 1975). Herein, we will refer to two schematic situations in which water is removed from the body of the IJs of *Steinernema* and *Heterorhabditis* spp.: (i) gradual desiccation, whereby water is removed slowly (over a period of ≥72–96 h); this usually occurs in soil or cryptic habitats; and (ii) rapid desiccation, whereby water is evaporated within minutes or several hours; this is common when EPNs are applied to the soil surface or to plant foliage as biological control agents. The response and possible mechanisms for protection may be different under the two situations, as we will further discuss.

5.3.1.1 Gradual desiccation

Plants can remain viable under dry conditions, mainly in the form of seeds. Among invertebrates, a handful of animals such as tardigrades, brine shrimps and nematodes can overcome dry conditions (Cooper and van Gundy, 1971; Wharton, 1986; Perry and Moens, 2011), even though some are aquatic organisms and need a film of water surrounding their body in order to move (Norton, 1979). Some nematodes can survive in a desiccated state for many years, such as *Anguina tritici* and *Ditylenchus dipsaci* (L2 and L4 stages, respectively) (Cooper and van Gundy, 1971; Perry, 1977; Wharton, 1986).

Survival of EPNs in soil is attributed to entering a dormant state (anhydrobiosis) as soil moisture decreases gradually, but EPNs with poor desiccation tolerance and low anhydrobiotic capabilities may practice desiccation-avoidance behavior. For example, downward movement in response to a soil moisture gradient was observed in the highly desiccation-tolerant *S. carpocapsae* and the poorly desiccation-tolerant *Heterorhabditis bacteriophora* (Salame and Glazer, 2015). *H. bacteriophora* IJs abandoned the upper soil layers as dryness intensified, whereas over 70% of *S. carpocapsae* IJs remained in the upper layer (Salame and Glazer, 2015).

EPNs have been subjected to intensive study for their ability to withstand gradual desiccation. They can persist for 2–3 weeks in dry soil (Kaya, 1990; Kung *et al.*, 1990). Most studies investigating the ability of *Steinernema* spp. to survive desiccation have focused on *S. carpocapsae* (Simons and Poinar, 1973; Ishibashi *et al.*, 1987; Womersley, 1990; Glazer, 1992). The general finding has been that various strains of *S. carpocapsae* can survive for appreciable lengths of time under slow drying conditions. However, differences in survival capabilities were observed among *Steinernema* spp. (Kung *et al.*, 1990; Baur *et al.*, 1995; Solomon *et al.*, 2000).

For heterorhabditids, Surrey and Wharton (1995) tested the desiccation survival of *Heterorhabditis zealandica*. Their experiments indicated that survival was poor once water had been lost from the substrate. Poor survival of *H. megidis* strain UK211 was also reported by O'Leary *et al.* (1998) who exposed the nematodes directly to 57% RH for 3 h without preconditioning at higher RH levels. A substantial difference was observed in survival ability among IJs of 12 new heterorhabditid populations, isolated from different climatic regions in Israel. The nematodes were preconditioned at 97% RH for 72 h followed by exposure for an additional 72 h at 93% RH (Liu and Glazer, 2000).

In a broad comparison of EPN strains and species (ten species and 21 strains), *S. carpocapsae* exhibited the highest level of desiccation tolerance among species followed by *S. feltiae* and *S. rarum*. The heterorhabditid species exhibited the least desiccation tolerance, and *S. riobrave* and *S. glaseri* were intermediate. Intraspecific variation was not pronounced (unlike interspecific variation) (Shapiro-Ilan *et al.*, 2014b).

EPNs require preconditioning for a slow removal of water content from the organism prior to exposure to extreme desiccating conditions. This preconditioning period enables the nematodes to induce the necessary biochemical changes required to survive in an anhydrobiotic state. Such nematodes are found in soil habitats that experience slow rates of water loss. The majority of anhydrobiotic nematodes are believed to fall into this latter group (Womersley, 1987; Burnell and Tunacliffe, 2011; Erkut *et al.*, 2011).

Coiling is a behavioral response to dehydration that reduces the surface area of the nematode exposed to drying conditions. Behavioral adaptations such as loose coiling has been induced in *S. glaseri* exposed to elevated RHs (97%), and in *S. carpocapsae* on slow-drying model substrates (Womersley, 1990). However, in general, coiling has not been consistently observed during desiccation and does not correlate with survival. In addition, positive correlation was observed between clump size and survival (Solomon and Glazer, 1999). Clumping aids the survival of *S. carpocapsae* IJs when they are desiccated in large aggregates (>1000 individuals) (Simons and Poinar, 1973).

Successful entry into long-term anhydrobiosis depends on subsequent biochemical and molecular adaptations. The biochemical mechanisms involved in the induction of anhydrobiosis are not fully understood. Previous work on the biochemistry and molecular genetics of anhydrobiosis in nematodes has been focused primarily on the fungal-feeding soil nematode *Aphelenchus avenae*, which is a slow-dehydration strategist (Crowe and Madin, 1975). However, recent studies have expanded this work on organisms including *C. elegans*, *Panagrolaimus* spp., *Plectus murrayi* and others, revealing several key players in the mechanisms involved (Shannon *et al.*, 2005; Adhikari *et al.*, 2009; Erkut *et al.*, 2011, 2013; Tyson *et al.*, 2012).

Yaari *et al.* (2016) used 454 pyrosequencing to analyze transcriptome expression in *Steinernema* species that differ their tolerance to desiccation and heat stress. Analyzing gene expression patterns of the stress response showed a large fraction of downregulated genes in the desiccation-tolerant nematode *S. riobrave*, whereas larger fractions of the genes in the susceptible *S. feltiae* strains were upregulated. This phenomenon was further observed in metabolic pathways and the expression of specific stress-related genes. Somvanshi *et al.* (2008) studied the regulation of four genes associated with stress tolerance in five EPNs with varying resistance to desiccation. Interestingly, all four genes showed the greatest degree of upregulation in the least desiccation-resistant nematode species; an inverse correlation between gene expression and the phenotype of desiccation tolerance in the examined EPNs was also observed.

BIOCHEMICAL CHANGES. One biochemical change that has been reported in anhydrobiotic nematodes is the accumulation of polyols such as glycerol and sugars (trehalose), which are believed to act as protectants of biological membranes and intracellular proteins during dehydration (Womersley, 1990; Barrett, 1991; Behm,

1997). Solomon *et al.* (2000) recorded increased trehalose levels in the EPN *S. feltiae* as reported for other anhydrobiotic nematodes during the dehydration process (Womersley, 1987; Crowe and Crowe, 1992; Solomon and Glazer, 1999). Trehalose accumulation has also been associated with the onset of desiccation in several anhydrobiotic nematodes including *Aphelenchus avenae*, *Anguina tritici*, *Ditylenchus dipsaci* and several *Panagrolaimus* spp. (Shannon *et al.*, 2005). The mode of action of these non-reducing disaccharides, trehalose (in animals and fungi) and sucrose (in plants), is thought to be either as water-replacement molecules or as vitrification agents, and their production through the glyoxylate shunt/gluconeogenesis pathway was discovered recently (Erkut *et al.*, 2016). Recently, several novel pathways were identified to be involved during induction of anhydrobiosis in the protection mechanism of *Heterorhabditis indica* such as the shikimate and saccharopine pathways (Balakumaran *et al.*, 2022). However, all changes have been characterized during the preconditioning process. Levels of trehalose were quantified in rapidly desiccated *Heterorhabditis* spp. and preconditioned *S. carpocapsae* and it was found that *Heterorhabditis* spp. and *S. feltiae* favored the synthesis of glycerol for the protective effect (O'Leary *et al.*, 2001; Gal *et al.*, 2003). However, preconditioned *S. carpocapsae* favored the synthesis of trehalose for protection from desiccation.

Late embryogenesis abundant (LEA) proteins have been associated with tolerance to desiccation in seeds, pollen and desiccation-resistant plants. Solomon *et al.* (2000) were the first to identify the increase of this protein in the *S. feltiae*. The genes encoding this desiccation-related protein or homologs were then reported in *Aphelenchus avenae*, *C. elegans* and *Bursaphelenchus xylophilus* (Browne *et al.*, 2004; Gal *et al.*, 2005). This finding led to further discovery of this group of proteins in other anhydrobiotic animals. In nematodes and other invertebrates, almost all LEA proteins characterized belong to group 3 (Pfam LEA_4) (Sharon *et al.*, 2009). Expression of LEA proteins has been associated with tolerance to water stress caused by desiccation, salinity or low temperature (Burnell and Tunnacliffe, 2011). LEA proteins may protect cellular components against the effects of desiccation (Goyal *et al.*, 2005). Furthermore, reduction of expression of LEA genes by RNA interference in *C. elegans* (Gal *et al.*, 2004) or by mutation in bacteria (Battista *et al.*, 2001), has been shown to decrease resistance to desiccation.

Membrane fluidity is affected by phospholipid composition and the degree of fatty acid desaturation. The introduction of unsaturated bonds into the fatty acids of membrane phospholipids decreases the temperature for the transition from gel to liquid crystalline state (Los and Murata, 1998). Recently, phospholipid remodeling was also suggested to retain the liquid crystalline state (Abusharakh *et al.*, 2014). Synthesis of transcripts encoding the fatty acid desaturase 2 gene (*fat-2*) was strongly upregulated in response to desiccation in *S. carpocapsae* (Tyson *et al.*, 2007) and *S. feltiae* (Yaari *et al.*, 2016).

Another group of molecules known to have a protective role under stress conditions is the heat-shock proteins (Hsps), which are among the largest groups of molecular chaperones. Molecular chaperones and unfolded-protein-response Hsps are essential for the correct folding and maturation of a wide diversity of client proteins and for protection of proteins from stress-induced unfolding and aggregation (Morimoto, 2008; Richter *et al.*, 2010). The *Steinernema* data set contains representatives of all Hsp classes characteristic of nematodes, including four distinct small Hsps. These are the major "holding" chaperones, retaining unfolded proteins in a conformation suitable for subsequent refolding and thus preventing their irreversible aggregation (Eyles and Gierasch, 2010; Stengel *et al.*, 2010). In a proteomic analysis of desiccated *S. feltiae*, Chen *et al.* (2006) reported more than 400 protein spots identified by two-dimensional gel electrophoresis. Ten spots showed changes in abundance on desiccation, one of which was the chaperonin Hsp60. The evidence for involvement of classical protein homeostasis systems is also rather tentative in other anhydrobiotes. For example, tardigrades upregulate Hsp70 during anhydrobiosis, although it seems to be more highly expressed during recovery from the dry state (Schill *et al.*, 2004).

5.3.1.2 Rapid desiccation

On exposed surfaces such as foliar applications, EPNs are typically unprotected in the field, representing a challenging environment for their application (Belien, 2018). In addition, nematodes require a film of water surrounding them to maintain motility and activity (Perry *et al.*, 2012). Thus, on exposed surfaces, rapid desiccation is the major limiting factor for the survival of EPNs (Glazer, 1992; Shapiro-Ilan *et al.*, 2006).

Glazer (1992) estimated the survival of EPNs to be minutes or hours due to exposure to reduced humidity levels (Glazer, 1992). RH level was identified as a major factor that directly influenced the low survival of EPNs by rapid rate of water loss (Glazer and Navon, 1990; Glazer, 1992). Patel *et al.* (1997) characterized the rate of water loss from EPNs under rapid desiccation conditions and identified that *S. carpocapsae* had the slowest water loss rate in comparison with *S. feltiae* and *S. glaseri*. In addition, the survival capability of *S. carpocapsae* was attributed to the cuticle permeability as well as the physical and behavioral adaptations (e.g. coiling, aggregation). Recently, *in vitro* studies indicated water loss of 75% and 82% at 75% RH and 64% RH within 2 h of rapid desiccation, respectively (Ramakrishnan *et al.*, 2022). Such drastic water loss rates affect the structures and functionality of the nematodes' organelles, putting enormous pressure on the coping and survival mechanisms.

To avoid these stressful situations, specific formulations can assist in protecting EPNs (see Chapter 11, this volume). For example, formulations based on Pickering emulsions have recently been studied. Specifically, silica Pickering emulsion gel (SPEG) and titanium Pickering emulsion (TPE) were screened for their protection capabilities (survival and efficacy) in comparison with control (nematodes in water) when applied to cotton plants at controlled conditions of 25°C and average RH of 77.5±6.1%. The TPE and SPEG formulations are based on oil in water and inverse Pickering emulsion, respectively. The results indicated extension of survival at 96 h by SPEG (>45%) and TPE (>25%) in comparison with the control (0%). At 48–96 h, an increase in efficacy 3.5–4-fold was observed for EPNs protected in SPEG in comparison with water (Ramakrishnan *et al.*, 2023). Beck *et al.* (2013) showed that EPN deposition and spread on the leaf surface is important to improve the efficacy of foliar application of EPNs. It is also imperative to maintain a stable number of live IJs cm^{-2} of the leaf over time to increase the chances of contact between the host and pathogen. The number of live IJs retained in SPEG and TPE was 1.2 and 0.5 cm^{-2} in comparison with only 0.18 IJ cm^{-2} in the control 48 h after application. SPEG accounted for the slowest decrease of live IJs per surface area compared with TPE and control samples over time, exhibiting a sixfold increase at 48 h. With regard to IJ retention, this suggests an even higher increase in probability of successful nematode–host interactions when accounting for both time and nematode survival (Ramakrishnan *et al.*, 2023).

5.3.2 Osmotic stress

Nematodes, as terrestrial organisms, may have to tolerate considerable variation in the concentrations of salts in the environment and thus are subject to osmotic stress (Evans, 1976; Wharton, 1986). The degree to which this is necessary depends on the habitat of the nematode. Marine nematodes are thought to be isosmotic to the surrounding seawater (Wright, 2004) and do not have to face much variation in the osmotic and ionic composition of their surroundings.

The ability to enter a metabolic state of osmobiosis due to osmotic pressure (Keilin, 1959) may be considered to be similar to anhydrobiosis induced by water evaporation (Cooper and van Gundy, 1971; Perry and Moens, 2011). The mechanisms of osmoregulation among the steinernematid and heterorhabditid EPNs are not known, although studies have demonstrated their ability to withstand osmotic stress (Thurston *et al.*, 1994; Finnegan *et al.*, 1999). Glazer and Salame (2000) evaluated the effect of different osmolytes on the viability of *S. carpocapsae*. Exposing fresh IJs to a high temperature (45°C) resulted in a rapid reduction in viability, yet under the same assay conditions, the mortality of evaporatively and osmotically desiccated nematodes showed enhanced ability to withstand temperature stress. When IJs of *S. carpocapsae* were exposed to different salt solutions, high mortality was recorded among the nematodes within the first 24 h of exposure. In contrast, non-ionic solutes such as glycerol at concentrations of 2.2–3.8 M, PEG 300 (polyethelyne glycol with molecular weight 300) at concentrations of 1.2–1.6 M and PEG 600 at 0.8 M did not reduce IJ viability, although all nematodes were found to have shrunk after 48 h. Furthermore, 72 h exposure to solutions of non-ionic solutes resulted in an increase in heat tolerance similar to that of evaporatively dehydrated nematodes. High viability was also recorded among nematodes that were stored for 72 days following a gradual increase in glycerol concentration.

Using PEG 600, Strauch *et al.* (2004) demonstrated induction of osmo/anhydrobiosis in IJs of *H. bacteriophora* following exposure to a solution with a water activity (a_w) of 0.97 for 72 h. The treated IJs were then able to tolerate further exposure to a_w of 0.90. This led to further use of osmotic solutions for genetic enhancement of desiccation tolerance of inbred lines of *H. bacteriophora* (Anbesse *et al.*, 2012, 2013).

Recently, Levy *et al.* (2020) utilized a similar approach of osmotic desiccation with PEG 600 to screen natural populations of *H. bacteriophora* for genetic diversity. Gene expression patterns and genomic variation were studied following osmotic desiccation stress. Transcriptomic studies enable identification of genes and molecular markers as genetic selection tools for enhancement of EPN ability to withstand environmental stress conditions.

When a nematode is immersed in distilled water or another hypo-osmotic media, water will enter and if not removed the nematode will swell and die. The behavior and structure of the excretory system in some nematodes suggests that it is involved in the removal of excess water. A variety of nematode structures have been suggested to have an excretory function (Wright and Newall, 1976; Wright, 2004), but water removal has been most closely associated with the tubular type of excretory system found in Rhabditida nematodes. This excretory system often has associated glands and is referred to as a secretory/excretory system (Wright, 2004).

The cuticle is likely to be a major route of water flux under osmotic stress. Marine nematodes are usually isosmotic to seawater and have high cuticular water permeability. Soil nematodes, some plant-parasitic nematodes and the free-living stages of animal-parasitic nematodes are more likely to be exposed to osmotic stress and have much lower cuticular permeability (Wright and Newall, 1976; Wharton *et al.*, 1988). Slowing down the rate of water flux allows time for osmoregulatory mechanisms to operate. Low cuticular permeability is associated with the ability to survive other environmental stress, such as desiccation (Wharton *et al.*, 1988; Perry and Moens, 2011). The decreased ability of *Caenorhabditis briggsae* to withstand osmotic stress with age has been associated with changes in the epicuticle and an increase in permeability (Searcy *et al.*, 1976).

The IJs of the EPN *S. carpocapsae* synthesize glycerol and trehalose during osmotic dehydration (Qiu and Bedding, 2002). The patterns of protein synthesis of osmotically dehydrated *S. feltiae* IJs (exposed to 24% glycerol) and unstressed controls were compared using two-dimensional electrophoresis. Ten novel proteins were produced and a further ten upregulated in response to osmotic dehydration. The proteins that could be identified by mass spectrometry included actin (filament formation and muscle fibers), proteasome regulatory particle (involved in protein degradation), and GroEL and GroES (members of the Hsp60 family of chaperones involved in protein folding) (Chen *et al.*, 2005). A further study compared the response to osmotic dehydration (24% glycerol) and evaporative dehydration (97% followed by 85% RH) in *S. feltiae* (Chen *et al.*, 2006). Many of the proteins identified were expressed in response to both osmotic and evaporative dehydration, including proteins known to be involved in stress responses, such as Hsp60, coenzyme Q (ubiquinone), biosynthesis protein 4, inositol monophosphatase and fumarate lyase, and also proteins involved in the regulation of the cell cycle, gene transcription and the organization of macromolecular structures.

5.3.3 Extreme temperatures

5.3.3.1 Cold tolerance

Nematodes, including EPNs, inhabiting the soil are exposed to freezing temperatures in temperate, arctic and subarctic regions as well as at high altitudes. Nematodes are essentially aquatic organisms, and unless they are desiccated, face the risk of inoculative freezing by the ice in their surroundings seeding, via body orifices, and freezing their body fluids. Wharton (2011) comprehensively reviewed the cold-tolerance strategies of nematodes. There are a number of strategies by which nematodes may survive subzero temperatures. The two main cold-tolerance strategies are freeze avoidance and freezing tolerance.

Interest in cold tolerance of EPNs is related to their overwintering survival as well as their efficacy at lower temperatures (>15°C). Steinernematids have been isolated in many sites in northern Europe (Burman *et al.*, 1986; Husberg *et al.*, 1988) and Canada (Mráček and Webster, 1993), while heterorhabditids have been isolated from northern parts of Europe (Griffin and Downes, 1991). Isolation of EPNs from such cold regions indicates that they are able to withstand freezing conditions. Brown and Gaugler (1996) demonstrated that *S. feltiae*, *Steinernema anomali* and *H. bacteriophora* were all freezing tolerant with a lower lethal temperature of −22, −14 and −19°C, respectively. In a broad comparison of freezing tolerance among ten EPN species, poor freezing tolerance was observed in *H. indica*, *S. glaseri*, *Steinernema rarum* and *S. riobrave*, whereas *Heterorhabditis georgiana* and *S. feltiae* exhibited the highest freezing tolerance; significant intraspecies variation among strains was also observed (Shapiro-Ilan *et al.*, 2014b). Surrey and Wharton (1995) showed that *H. zealandica* is freeze avoiding. Cryopreservation studies have demonstrated that EPNs can be stored indefinitely in liquid nitrogen (Popiel and Vasquez, 1989; Curran *et al.*, 1992).

Knowledge about the biochemical and molecular changes involved in cold tolerance has accumulated in recent years. Trehalose appears to be the main cryoprotectant/antifreeze used by nematodes. An increase in trehalose concentration during low-temperature acclimation has been shown in *Panagrolaimus davidi* (Wharton *et al.*, 2000), *Nematodirus battus* eggs (Ash and Atkinson, 1986), *Steinernema kushidai* (Ogura and Nakashima, 1997), *S. carpocapsae* (Qiu and Bedding, 2000), *S. feltiae* and *S. riobrave* (Grewal and Jagdale, 2002; Jagdale and Grewal, 2003) and *Heterodera glycines* (Yen *et al.*, 1996), and as a cold shock response in *Heterorhabditis bacteriophora* (Jagdale *et al.*, 2005). Increased trehalose concentration is associated with increased freezing survival or cold tolerance in *Panagrolaimus davidi*, *S. kushidai* and *S. carpocapsae* but not in *S. feltiae* and *S. riobrave* (Ogura and Nakashima, 1997; Wharton *et al.*, 2000; Jagdale and Grewal, 2003).

The concentration of trehalose is mediated by trehalose-6-phosphate synthase (gene: *tps*). The presence of these enzymes has been demonstrated in a variety of nematodes (Behm, 1997). Cold shock increases the activity of trehalose 6-phosphate synthase and decreases the activity of trehalase in *H. bacteriophora*, resulting in an increase in trehalose concentration (Jagdale *et al.*, 2005). There is less evidence for the role of other organic solutes in nematode cold tolerance. Nematodes synthesize a variety of polyols, including glycerol, inositol, ribitol and sorbitol (Womersley, 1981; Barrett, 1991).

Unsaturated fatty acids (UFAs) are more fluid than saturated fatty acids, and hence an increase in the proportion of UFAs enables the organism's membranes to resist solidification at lower temperatures (Margesin *et al.*, 2007). The proportion of UFAs is increased at lower temperatures in *S. feltiae*, *S. carpocapsae* and *S. riobrave* (Jagdale and Gordon, 1997) and in *C. elegans* (Tanaka *et al.*, 1996). Chilling-resistant *C. elegans* can be produced by culturing at 10°C. These organisms will survive exposure to 0°C, while those cultured at 25°C do not survive (Murray *et al.*, 2007). Chilling-resistant worms have an increased proportion of UFAs compared with chilling-sensitive worms.

5.3.3.2 Heat tolerance

The optimal temperature, referred to as the "temperature niche breadth," for development of species or genera varies and depends on the normal growth habitat of the organism (Anderson and Coleman, 1982; Stiernagle, 2006). Hence, high temperatures (>32°C) have an adverse effect on reproduction, growth and survival of many organisms (Wharton, 1986), including nematodes (Zervos *et al.*, 1991; Grewal *et al.*, 1994). Our understanding of the nematode's response to heat comes largely from genetic and molecular studies using the free-living model nematode *C. elegans* (Devaney, 2011).

High temperatures can adversely affect EPN efficacy (Hiltpold, 2015). Intensive studies had been devoted to understanding EPN activity and persistence under high temperatures as well as their survival in storage and shipment at elevated temperatures. New heterorhabditid isolates from the arid region of Israel (Glazer *et al.*, 1991) and Egypt (Shamseldean and Abd-Elgawad, 1994) or the tropic climate of Sri Lanka (Amarasinghe *et al.*, 1994) were shown to be heat tolerant. Shapiro *et al.* (1996) reported that the heat tolerance trait in the IS-5 strain of

Heterorhabditis sp., one of the isolates from Israel, was stable and genetically based.

Grewal *et al.* (2004, 2006) reported that temperature adaptation of EPN species appears to be a continuum, with species representing both the cold and warm extremes. Measuring the thermal niche breadths of infection (host penetration and host death), establishment (development of the IJs to adults in the host following infection) and reproduction of eight species, they concluded that *S. feltiae* represented tolerance to the coldest extreme and *S. riobrave* to the warmest extreme. The EPN thermal niche breadths were progressively narrower for movement (4–40°C), host penetration (5–39°C), host death (8–39°C) and reproduction (10–35°C) (Grewal *et al.*, 1994, 2004, 2006). *Heterorhabditis floridensis* was found to have an extremely broad temperature niche breadth for infectivity with activity from 17°C to 39°C (Shapiro-Ilan *et al.*, 2014a).

Grewal (2000) demonstrated that temperature has a strong influence on IJ survival in storage and this influence varies with EPN species. For example, 50% of *S. feltiae* IJs survived in tap water for 42–44 and 27–30 weeks at 5°C and 25°C, respectively. The survival of *S. riobrave* IJs was similar between 25°C and 5°C, perhaps due to its lack of cold adaptation. Heat tolerance also varies substantially among species and strains of EPNs. The survival of IJs exposed to 40°C for 2 h varied between 13% and 90% among 15 *H. bacteriophora* strains (Grewal *et al.*, 2002) and between 37% and 82% in 15 *S. carpocapsae* strains (Somasekhar *et al.*, 2002).

Mukuka *et al.* (2010) screened 36 natural populations and 18 hybrid or inbred strains of *Heterorhabditis* for their response to high temperature. The nematodes were tested with or without prior adaptation to heat at 35°C for 3 h. The mean tolerated temperature ranged from 33.3°C to 40.1°C for non-adapted and from 34.8°C to 39.2°C for adapted strain populations. *H. indica* was the most tolerant, followed by *H. bacteriophora* and *H. megidis*. Correlation between heat tolerance and mean annual temperature at the place of origin of the strains was weak.

Heat tolerance in EPN IJs also appears to be facilitated by trehalose accumulation (Grewal *et al.*, 2006). Jagdale and Grewal (2003) showed that *S. feltiae*, *S. carpocapsae* and *S. riobrave* accumulated trehalose when acclimatized at either 5°C or 35°C, but the amount of trehalose accumulation differed by species and temperature. Enhanced heat tolerance (at 40°C for 8 h) was positively correlated with the increased trehalose levels in warm- and cold-acclimatized *S. carpocapsae* and *S. feltiae* but not in *S. riobrave*.

Production of *Hsp70* was also detected in *H. bacteriophora* HP88 strain (Selvan *et al.*, 1996), even though temperatures above 32°C hamper reproduction, activity and viability of this strain (Grewal *et al.*, 1994). Hashmi *et al.* (1997) analyzed the molecular organization of *Hsp70* genes in five species and isolates of heterorhabditids and one steinernematid species using the polymerase chain reaction and restriction fragment length polymorphism (RFLP). They found homology with *Hsp70* genes of *C. elegans*. In addition, RFLPs with an *Hsp70* probe revealed different banding patterns for *Heterorhabditis* spp. and isolates. Analysis of the *H. bacteriophora* genome revealed large numbers of Hsps, including Hsp1, Hsp3, Hsp6, Hsp12.2, Hsp16.2, Hsp17, Hsp25, Hsp43, Hsp60, Hsp70 and other Hsps such as STI-1, STC-1, DAF-21, DNJ-12 and DNJ-13 (Grewal *et al.*, 2011).

In yeasts, deletion of either the *tps1* or *tps2* gene, which encode the T6PS complex, causes an inability to accumulate trehalose during heat shock, which, in turn, significantly reduces thermotolerance in yeast (Bell *et al.*, 1992; Virgilio *et al.*, 1993, 1994; Hottiger *et al.*, 1994). As *H. bacteriophora* accumulates trehalose and increases T6PS activity after heat shock (Jagdale *et al.*, 2005), it is possible that this enzyme in nematodes, as in yeast, may also be acting as an Hsp and performing a role in thermotolerance.

In the *H. bacteriophora* genome, a number of transcription factors/regulators involved in stress responses have also been found, including HSF1, a heat-shock transcription factor functioning as a transcriptional regulator of stress-induced gene expression and involved in nematode larval development, innate immunity and regulation of the adult lifespan (Baugh and Sternberg, 2006; Singh and Aballay, 2006). Skinhead transcription factor (SKN-1) functions in the p38 mitogen-activated protein kinase pathway to regulate the oxidative stress response and in parallel to DAF-16 in the DAF-2-mediated insulin/IGF-1-like signaling pathway to regulate the adult lifespan (Tullet *et al.*, 2008). Other transcriptional factors/regulators identified in *H. bacteriophora* include GFL-1, CUL-4, LAG-1, DIN-1 and DJR-1.1.

5.3.4 Oxidative stress

Oxygen is an essential component for maintaining life for most organisms, including nematodes. Oxygen stress and the oxidative effect have been studied intensely with EPNs due to the deleterious effect of lack of oxygen at fermentation, storage and even under certain field conditions. There are large differences in the ability of EPN species and strains to withstand hypoxic conditions (summarized by Grewal *et al.*, 2011). Burman and Pye (1980) reported that *S. carpocapsae* can withstand oxygen tensions as low as 0.5% saturation at 20°C. In sandy soil, survival of *S. carpocapsae* and *S. glaseri* decreased as the oxygen concentration decreased from 20% to 1% (Kung *et al.*, 1990). In *H. bacteriophora*, Grewal *et al.* (2002) found a strong correlation of IJ longevity with heat, UV radiation and hypoxia tolerance. While *H. bacteriophora* can withstand anoxic conditions in water at 25°C for only 4 days, *S. carpocapsae* can tolerate such conditions for up to 10 days (Somasekhar *et al.*, 2002). There also appears to be more variation in hypoxia tolerance among strains of *H. bacteriophora* (Grewal *et al.*, 2002) than those of *S. carpocapsae* (Somasekhar *et al.*, 2002). Qiu and Bedding (2000) reported that *S. carpocapsae* IJs incubated in M9 buffer at 23°C under absolute anaerobic conditions were fully inactivated in 16 h but could be revived when returned to aerobic conditions if exposure to anaerobic conditions was for not more than 7 days. The survival time under anaerobic conditions was significantly affected by temperature, with 90% survival times being 20, 7 and 5 days at 5°C, 23°C and 28°C, respectively.

In addition to this finding, a strong correlation was found between oxidative stress tolerance and IJ longevity in *H. bacteriophora* (Sumaya *et al.*, 2017). The same strains that expressed increased oxidative stress resistance also elicited a high longevity and soil persistence. Sumaya *et al.* (2018) applied oxidative stress-tolerance assays as a selection method to improve *H. bacteriophora* IJ longevity following hybridization and mutagenesis.

Many enzymatic antioxidant defense systems have been reported to be upregulated in response to desiccation or osmotic stress in microorganisms, plants and animals, and in both desiccation-sensitive and anhydrobiotic nematodes. Some examples in nematodes include: aldehyde dehydrogenase in *S. feltiae* (Gal *et al.*, 2003; Yaari *et al.*, 2016), glutaredoxin and mitochondrial manganese superoxide dismutase in *S. carpocapsae* (Tyson *et al.*, 2007), glutaredoxin (Browne *et al.*, 2004) and glutathione peroxidase (Reardon *et al.*, 2010; Yaari *et al.*, 2016) in *Aphelenchus avenae*, and aldehyde dehydrogenase, glutathione *S*-transferase (GST) and glutathione peroxidase in *Plectus murrayi* (Adhikari *et al.*, 2009). A number of detoxification strategies seem to be similar in response to desiccation and oxidative stress such as reactive oxygen species like superoxide dismutase, Ras-related protein and GST. GSTs are a diverse superfamily of multi-functional proteins that play prominent roles in detoxification metabolism in nematodes and are known to be involved in several functions, including xenobiotic detoxification and oxidative stress tolerance (Rossum *et al.*, 2001; Leiers *et al.*, 2003; Lindblom and Dodd, 2006).

5.4 Conclusions and Future Directions

Survival in the environment is a key factor in the development of EPNs as biological control agents. In the past three decades, much research has focused on physiological mechanisms of stress tolerance in IJs. Studies on molecular mechanisms controlling IJ longevity and stress tolerance have accelerated in the last decade. The availability of genomic sequence data and putative proteomes provides a large number of genes that could be useful to increase IJ persistence and survival in the soil. Several known stress-tolerance genes such as Hsps, trehalose-related molecules and pathways, as well as their orthologs and paralogs that have been expanded in EPN genomes, remain to be functionally tested. The link between survival mechanisms, particularly the response to heat stress, and nematode longevity, as well as the fascinating ability to survive long periods under extreme conditions (low RH or subzero temperatures), provide the foundation to explore these mechanisms in other organisms. Kaczmarek *et al.* (2019) recently reviewed our knowledge about the impact of anhydrobiosis on longevity and fitness of nematodes, rotifers and tardigrades, indicating that anhydrobiosis has been proposed to eliminate or slow down ageing symptoms. EPN research is burgeoning with possibilities, but much remains intractable without the application of more molecular tools (Lu *et al.*, 2016). However, molecular advances alone will not lead to actual improvements in the efficacy of EPNs as biological control agents. It is also imperative to improve application and delivery systems to increase the survival and activity of EPNs under a wider range of ambient temperatures and RHs. This, in turn, will extend the range of target pests and agricultural habitats where EPNs can be applied efficaciously. In particular, their use on exposed surfaces such as soil, bark and foliage will be enhanced.

5.5 References

Abu Hatab, M.A., Gaugler, R. and Ehlers, R.U. (1998) Influence of culture method on *Steinernema glaseri* lipids. *Journal of Parasitology* 84, 215–221.

Abusharkh, S.E., Erkut, C., Oertel, J., Kurzchalia, T.V. and Fahmy, K. (2014) The role of phospholipid headgroup composition and trehalose in the desiccation tolerance of *Caenorhabditis elegans*. *Langmuir* 30, 12897–12906.

Adhikari, B.N., Lin, C.-Y., Bai, X., Ciche, T.A., Grewal, P.S. *et al.* (2009) Transcriptional profiling of trait deterioration in the insect pathogenic nematode *Heterorhabditis bacteriophora*. *BMC Genomics* 10: 609.

Akhurst, R. (1993) Bacterial symbionts of entomopathogenic nematodes – the power behind the throne. In: Bedding, R.A., Akhurst, R. and Kaya, H.K. (eds.) *Nematodes and the Biological Control of Insect Pests*. CSIRO Publications, East Melbourne, Australia, pp. 127–136.

Amarasinghe, L.D., Hominick, W.M., Briscoe, B.R. and Reid, A.P. (1994) Occurrence and distribution of entomopathogenic nematodes in Sri Lanka. *Journal of Helminthology* 68, 277–286.

Anbesse, S.A., Strauch, O. and Ehlers, R.-U. (2012) Genetic improvement of the biological control nematode *Heterorhabditis bacteriophora* (Rhabditidomorpha: Heterorhabditidae): heterosis effect enhances desiccation but not heat tolerance. *Biocontrol Science and Technology* 22, 1035–1045.

Anbesse, S., Sumaya, N.H., Dörfler, A.V., Strauch, O. and Ehlers, R.-U. (2013) Selective breeding for desiccation tolerance in liquid culture provides genetically stable inbred lines of the entomopathogenic nematode *Heterorhabditis bacteriophora*. *Applied Microbiology and Biotechnology* 97, 731–739.

Anderson, R.V. and Coleman, D.C. (1982) Nematode temperature responses: a niche dimension in populations of bacterial-feeding nematodes. *Journal of Nematology* 14, 69–76.

Ash, C. and Atkinson, H.J. (1986) *Nematodirus battus*: development of cold hardiness in dormant eggs. *Experimental Parasitology* 62, 24–28.

Balakumaran, M., Chidambaranathan, P., Tej Kumar, J.P., Sirohi, A., Jain, P.K. *et al.* (2022) Deciphering the mechanism of anhydrobiosis in the entomopathogenic nematode *Heterorhabditis indica* through comparative transcriptomics. *PLOS ONE* 17: e0275342.

Barrett, J. (1991) Anhydrobiotic nematodes. *Agricultural Zoology Reviews* 4, 161–176.

Barrett, J. and Wright, D. (1998) Intermediary metabolism. In: Perry, R.N. and Wright, D.J. (eds.) *The Physiology and Biochemistry of Free-living and Plant-parasitic Nematodes*. CAB International, Wallingford, UK, pp. 331–353.

Battista, J.R., Park, M.-J. and McLemore, A.E. (2001) Inactivation of two homologues of proteins presumed to be involved in the desiccation tolerance of plants sensitizes *Deinococcus radiodurans* R1 to desiccation. *Cryobiology* 43, 133–139.

Baugh, L.R. and Sternberg, P.W. (2006) DAF-16/FOXO regulates transcription of cki-1/Cip/Kip and repression of *lin-4* during *C. elegans* L1 arrest. *Current Biology* 16, 780–785.

Baur, M., Kaya, H. and Thurston, G. (1995) Factors affecting entomopathogenic nematode infection of *Plutella xylostella* on a leaf surface. *Entomologia Experimentalis et Applicata* 77, 239–250.

Beck, B., Brusselman, E., Nuyttens, D., Moens, M., Pollet, S. *et al.* (2013). Improving foliar applications of entomopathogenic nematodes by selecting adjuvants and spray nozzles. *Biocontrol Science and Technology* 23, 507–520.

Behm, C.A. (1997) The role of trehalose in the physiology of nematodes. *International Journal for Parasitology* 27, 215–229.

Belien, T. (2018) Entomopathogenic nematodes as biocontrol agents of insect pests in orchards. *CAB Reviews* 2018, 1–11.

Bell, W., Klaassen, P., Ohnacker, M., Boller, T., Herweijer, M. *et al.* (1992) Characterization of the 56-kDa subunit of yeast trehalose-6-phosphate synthase and cloning of its gene reveal its identity with the product of CIF1, a regulator of carbon catabolite inactivation. *European Journal of Biochemistry* 209, 951–959.

Bernard, E.C. (1992) Soil nematode biodiversity. *Biology and Fertility of Soils* 14, 99–103.

Bird, A.F. and Bird, J. (1991) The exoskeleton. In: Bird, A.F. and Bird, J. (eds.) *The Structure of Nematodes*, 2nd edn. Academic Press, San Diego, California, pp. 44–74.

Brown, I.M. and Gaugler, R. (1996) Cold tolerance of steinernematid and heterorhabditid nematodes. *Journal of Thermal Biology* 21, 115–121.

Browne, J.A., Dolan, K.M., Tyson, T., Goyal, K., Tunnacliffe, A. and Burnell, A.M. (2004) Dehydration-specific induction of hydrophilic protein genes in the anhydrobiotic nematode *Aphelenchus avenae*. *Eukaryotic Cell* 3, 966–975.

Burman, M. and Pye, A.E. (1980) *Neoaplectana carpocapsae*: respiration of infective juveniles. *Nematologica* 26, 214–219.

Burman, M., Abrahamsson, K., Ascard, J., Sjoberg, A. and Erikson, B. (1986) Distribution of insect parasitic nematodes in Sweden. In: Samson, R.A., Vlak, J.M. and Peters, D. (eds.) *Proceedings of the 4th International Colloquium of Invertebrate Pathology, 18–22 August, Veldhoven, The Netherlands*, p. 312.

Burnell, A. and Tunnacliffe, A. (2011) Gene induction and desiccation stress in nematodes. In: Perry, R.N. and Wharton, D.A. (eds.) *Molecular and Physiological Basis of Nematode Survival*. CAB International, Wallingford, UK, pp. 126–156.

Campbell, L.R. and Gaugler, R. (1991) Role of the sheath in desiccation tolerance of two entomopathogenic nematodes. *Nematologica*, 37, 324–332.

Chen, S., Glazer, I., Gollop, N., Cash, P., Argo, E. *et al.* (2006) Proteomic analysis of the entomopathogenic nematode *Steinernema feltiae* IS-6 IJs under evaporative and osmotic stresses. *Molecular and Biochemical Parasitology* 145, 195–204.

Chen, S., Gollop, N. and Glazer, I. (2005) Cross-stress tolerance and expression of stress-related proteins in osmotically desiccated entomopathogenic *Steinernema feltiae* IS-6. *Parasitology* 131, 695–703.

Cooper, A. Jr. and van Gundy, S. (1971) Senescence, quiescence, and cryptobiosis. In: Zuckerman, B.M., Mai, W.F. and Rohde, R.A. (eds.) *Plant Parasitic Nematodes, Vol. 2. Cytogenetics, Host–Parasite Interactions, and Physiology*. Academic Press, New York, pp. 297–318.

Crowe, J. and Crowe, L. (1992) Membrane integrity in anhydrobiotic organisms: toward a mechanism for stabilizing dry cells. In: Somero, G.N., Osmond, C.B. and Bolis, C.L. (eds.) *Water and Life: Comparative Analysis of Water Relationships at the Organismic, Cellular, and Molecular Levels*. Springer, Berlin, pp. 87–103.

Crowe, J.H. and Madin, K. (1975) Anhydrobiosis in nematodes: evaporative water loss and survival. *Journal of Experimental Zoology* 193, 323–333.

Curran, J., Gilbert, C. and Butler, K. (1992) Routine cryopreservation of isolates of *Steinernema* and *Heterorhabditis* spp. *Journal of Nematology* 24, 269–270.

de Ley, P. (2006) A quick tour of nematode diversity and the backbone of nematode phylogeny. In: *WormBook: The Online Review of* C. elegans *Biology*. Wormbook, Pasadena, California. Available at: https://doi.org/10.1895/wormbook.1.41.1 (accessed 14 March 2024).

Devaney, E. (2011) Thermobiotic survival. In: Perry, R.N. and Wharton, D.A. (eds.) *Molecular and Physiological Basis of Nematode Survival*. CAB International, Wallingford, UK, pp. 233–255.

Dunphy, G.B. and Webster, J.M. (1987) Partially characterized components of the epicuticle of dauer juvenile *Steinernema feltiae* and their influence on hemocyte activity in *Galleria mellonella*. *Journal of Parasitology* 73, 584–588.

Erkut, C., Penkov, S., Khesbak, H., Vorkel, D., Verbavatz, J.-M. *et al.* (2011) Trehalose renders the dauer larva of *Caenorhabditis elegans* resistant to extreme desiccation. *Current Biology* 21, 1331–1336.

Erkut, C., Vasilj, A., Boland, S., Habermann, B., Shevchenko, A. and Kurzchalia, T.V. (2013) Molecular strategies of the *Caenorhabditis elegans* dauer larva to survive extreme desiccation. *PLOS ONE* 8: e82473.

Erkut, C., Gade, V.R., Laxman, S. and Kurzchalia, T.V. (2016) The glyoxylate shunt is essential for desiccation tolerance in *C. elegans* and budding yeast. *Elife* 5: e13614.

Ettema, C.H. (1998) Soil nematode diversity: species coexistence and ecosystem function. *Journal of Nematology* 30, 159–169.

Evans, A. (1976) Survival strategies in nematodes. In: Croll, N.A. (ed.) *The Organization of Nematodes*. Academic Press, London, pp. 383–424.

Eyles, S.J. and Gierasch, L.M. (2010) Nature's molecular sponges: small heat shock proteins grow into their chaperone roles. *Proceedings of the National Academy of Sciences USA* 107, 2727–2728.

Finnegan, M.M., Downes, M.J., O'Regan, M. and Griffin, C.T. (1999) Effect of salt and temperature stresses on survival and infectivity of *Heterorhabditis* spp. IJs. *Nematology* 1, 69–78.

Gal, T.Z., Glazer, I. and Koltai, H. (2003) Differential gene expression during desiccation stress in the insect-killing nematode *Steinernema feltiae* IS-6. *Journal of Parasitology* 89, 761–766.

Gal, T., Glazer, I. and Koltai, H. (2004) A LEA3 family member is involved in survival of *C. elegans* during exposure to stress. *FEBS Letters* 577, 21–26.

Gal, T.Z., Glazer, I. and Koltai, H. (2005) Stressed worms: responding to the post-genomics era. *Molecular and Biochemical Parasitology* 143, 1–5.

Glazer, I. (1992) Survival and efficacy of *Steinernema carpocapsae* in an exposed environment. *Biocontrol Science and Technology* 2, 101–107.

Glazer, I. (2002) Survival biology. In: Gaugler, R. (ed.) *Entomopathogenic Nematology*. CAB International, Wallingford, UK, pp. 169–187.

Glazer, I. (2015) Improvement of entomopathogenic nematodes: a genetic approach. In: Campos-Herrera, R. (ed.) *Nematode Pathogenesis of Insects and Other Pests: Ecology and Applied Technologies for Sustainable Plant and Crop Protection*. Springer, Cham, Switzerland, pp. 29–55.

Glazer, I. (2022) Stress and survival mechanisms. In: Glazer, I., Shapiro-Ilan, D.I. and Sternberg, P.W. (eds.) *Nematodes as Model Organisms*. CAB International, Wallingford, UK, pp. 215–243.

Glazer, I. and Navon, A. (1990) Activity and persistence of entomoparasitic nematodes tested against *Heliothis armigera* (Lepidoptera: Noctuidae). *Journal of Economic Entomology* 83, 1795–1800.

Glazer, I. and Salame, L. (2000) Osmotic survival of the entomopathogenic nematode *Steinernema carpocapsae*. *Biological Control* 18, 251–257.

Glazer, I., Liran, N. and Steinberger, Y. (1991) A survey of entomopathogenic nematodes (Rhabditida) in the Negev Desert. *Phytoparasitica* 19, 291–300.

Goyal, K., Browne, J., Burnell, A. and Tunnacliffe, A. (2005) Dehydration-induced *tps* gene transcripts from an anhydrobiotic nematode contain novel spliced leaders and encode atypical GT-20 family proteins. *Biochimie* 87, 565–574.

Grewal, P.S. (2000) Enhanced ambient storage stability of an entomopathogenic nematode through anhydrobiosis. *Pest Management Science* 56, 401–406.

Grewal, P.S. and Jagdale, G.B. (2002) Enhanced trehalose accumulation and desiccation survival of entomopathogenic nematodes through cold preacclimation. *Biocontrol Science and Technology* 12, 533–545.

Grewal, P., Selvan, S. and Gaugler, R. (1994) Thermal adaptation of entomopathogenic nematodes: niche breadth for infection, establishment, and reproduction. *Journal of Thermal Biology* 19, 245–253.

Grewal, P.S., Wang, X. and Taylor, R.A.J. (2002) Dauer juvenile longevity and stress tolerance in natural populations of entomopathogenic nematodes: is there a relationship? *International Journal for Parasitology* 32, 717–725.

Grewal, P., Power, K., Grewal, S., Suggars, A. and Haupricht, S. (2004) Enhanced consistency in biological control of white grubs (Coleoptera: Scarabaeidae) with new strains of entomopathogenic nematodes. *Biological Control* 30, 73–82.

Grewal, P.S., Bornstein-Forst, S., Burnell, A.M., Glazer, I. and Jagdale, G.B. (2006) Physiological, genetic, and molecular mechanisms of chemoreception, thermobiosis, and anhydrobiosis in entomopathogenic nematodes. *Biological Control* 38, 54–65.

Grewal, P.S., Bai, X. and Jagdale, G.B. (2011) Longevity and stress tolerance of entomopathogenic nematodes. In: Perry, R.N. and Wharton, D.A. (eds.) *Molecular and Physiological Basis of Nematode Survival*. CAB International, Wallingford, UK, pp. 157–181.

Griffin, C.T. and Downes, M.J. (1991) Low temperature activity in *Heterorhabditis* sp. (Nematoda: Heterorhabditidae). *Nematologica* 37, 83–91.

Hashmi, G., Hashmi, S., Selvan, S., Grewal, P. and Gaugler, R. (1997) Polymorphism in heat shock protein gene (*hsp70*) in entomopathogenic nematodes (Rhabditida). *Journal of Thermal Biology* 22, 143–149.

Hiltpold, I. (2015) Prospects in the application technology and formulation of entomopathogenic nematodes for biological control of insect pests. In: Campos-Herrera, R. (ed.) *Nematode Pathogenesis of Insects and Other Pests: Ecology and Applied Technologies for Sustainable Plant and Crop Protection*. Springer, Cham, Switzerland, pp. 187–205.

Hominick, W.M., Reid, A.P., Bohan, D.A. and Briscoe, B.R. (1996) Entomopathogenic nematodes: biodiversity, geographical distribution and the convention on biological diversity. *Biocontrol Science and Technology* 6, 317–332.

Hottiger, T., Virgilio, C., Hall, M.N., Boller, T. and Wiemken, A. (1994) The role of trehalose synthesis for the acquisition of thermotolerance in yeast. II. Physiological concentrations of trehalose increase the thermal stability of proteins *in vitro*. *European Journal of Biochemistry* 219, 187–193.

Husberg, G., Vänninen, I. and Hokkanen, H. (1988) Insect pathogenic fungi and nematodes in fields in Finland. *Växtskyddsnotiser* 52, 38–42.

Ishibashi, N., Tojo, S. and Hatate, H. (1987) Desiccation survival of *Steinernema feltiae* str. DD-136 and possible desiccation protectants for foliage application of nematodes. In: Ishibashi, N. (ed.) *Recent Advances in Biological Control of Insect Pests by Entomogenous Nematodes in Japan*. Saga University, Saga, Japan, pp. 139–144.

Jagdale, G.B. and Gordon, R. (1997) Effect of temperature on the composition of fatty acids in total lipids and phospholipids of entomopathogenic nematodes. *Journal of Thermal Biology* 22, 245–251.

Jagdale, G.B. and Grewal, P.S. (2003) Acclimation of entomopathogenic nematodes to novel temperatures: trehalose accumulation and the acquisition of thermotolerance. *International Journal for Parasitology* 33, 145–152.

Jagdale, G.B., Grewal, P.S. and Salminen, S.O. (2005) Both heat-shock and cold-shock influence trehalose metabolism in an entomopathogenic nematode. *Journal of Parasitology* 91, 988–994.

Kaczmarek, Ł., Roszkowska, M., Fontaneto, D., Jezierska, M., Pietrzak, B. *et al.* (2019) Staying young and fit? Ontogenetic and phylogenetic consequences of animal anhydrobiosis. *Journal of Zoology* 309, 1–11.

Kaya, H.K. (1990) Soil ecology. In: Gaugler, R. and Kaya, H.K. (eds) *Entomopathogenic Nematodes in Biological Control*. CRC Press, Boca Raton, FL, pp. 93–116.

Keilin (1959) The Leeuwenhoek Lecture: the problem of anabiosis or latent life: history and current concept. *Proceedings of the Royal Society of London B: Biological Sciences* 150, 149–191.

Kondo, E. and Ishibashi, N. (1989) Ultrastructural characteristics of the infective juveniles of *Steinernema* spp. (Rhabditida: Steinernematidae) with reference to their motility and survival. *Applied Entomology and Zoology* 24, 103–111.

Kung, S.-P., Gaugler, R. and Kaya, H.K. (1990) Soil type and entomopathogenic nematode persistence. *Journal of Invertebrate Pathology* 55, 401–406.

Lacey, L.A. and Georgis, R. (2012) Entomopathogenic nematodes for control of insect pests above and below ground with comments on commercial production. *Journal of Nematology* 44, 218–225.

Leiers, B., Kampkötter, A., Grevelding, C.G., Link, C.D., Johnson, T.E. and Henkle-Dührsen, K. (2003) A stress-responsive glutathione *S*-transferase confers resistance to oxidative stress in *Caenorhabditis elegans*. *Free Radical Biology and Medicine* 34, 1405–1415.

Levy, N., Faigenboim, A., Salame, L., Molina, C., Ehlers, R.-U. *et al.* (2020) Characterization of the phenotypic and genotypic tolerance to abiotic stresses of natural populations of *Heterorhabditis bacteriophora*. *Scientific Reports* 10: 10500.

Lewis, E.E., Gaugler, R. and Harrison, R. (1992) Entomopathogenic nematode host finding: response to host contact cues by cruise and ambush foragers. *Parasitology* 105, 309–315.

Lewis, E.E., Selvan, S., Campbell, J.F. and Gaugler, R. (1995) Changes in foraging behaviour during the infective stage of entomopathogenic nematodes. *Parasitology* 110, 583–590.

Lindblom, T.H. and Dodd, A.K. (2006) Xenobiotic detoxification in the nematode *Caenorhabditis elegans*. *Journal of Experimental Zoology A: Comparative Experimental Biology* 305A, 720–730.

Liu, Q.Z. and Glazer, I. (2000) Desiccation survival of entomopathogenic nematodes of the genus *Heterorhabditis*. *Phytoparasitica* 28, 331–340.

Los, D.A. and Murata, N. (1998) Structure and expression of fatty acid desaturases. *Biochimica et Biophysica Acta* 1394, 3–15.

Lu, D., Baiocchi, T. and Dillman, A.R. (2016) Genomics of entomopathogenic nematodes and implications for pest control. *Trends in Parasitology* 32, 588–598.

Margesin, R., Neuner, G. and Storey, K.B. (2007) Cold-loving microbes, plants, and animals – fundamental and applied aspects. *Naturwissenschaften* 94, 77–99.

Morimoto, R.I. (2008) Proteotoxic stress and inducible chaperone networks in neurodegenerative disease and aging. *Genes & Development* 22, 1427–1438.

Mráček, Z. and Webster, J. (1993) Survey of Steinernematidae and Heterorhabditidae (Rhabditida, Nematoda) in Western Canada. *Journal of Nematology* 25, 710–717.

Mukuka, J., Strauch, O., Waeyenberge, L., Viaene, N., Moens, M. and Ehlers, R.-U. (2010) Heat tolerance among different strains of the entomopathogenic nematode *Heterorhabditis bacteriophora*. *BioControl* 55, 423–434.

Murray, P., Hayward, S.A.L., Govan, G.G., Gracey, A.Y. and Cossins, A.R. (2007) An explicit test of the phospholipid saturation hypothesis of acquired cold tolerance in *Caenorhabditis elegans*. *Proceedings of the National Academy of Sciences USA* 104, 5489–5494.

Norton, D.C. (1979) Ecology of plant-parasitic nematodes. 1978. *Soil Science* 127, 63.

O'Leary, S.A. and Burnell, A.M. (1997) The isolation of mutants of *Heterorhabditis megidis* (strain UK211) with increased desiccation tolerance. *Fundamental and Applied Nematology* 20, 197–205.

O'Leary, S.A., Burnell, A.M. and Kusel, J.R. (1998) Biophysical properties of the surface of desiccation-tolerant mutants and parental strain of the entomopathogenic nematode *Heterorhabditis megidis* (strain UK211). *Parasitology* 117, 337–345.

O'Leary, S.A., Power, A.P., Stack, C.M. and Burnell, A.M. (2001) Behavioural and physiological responses of infective juveniles of the entomopathogenic nematode *Heterorhabditis* to desiccation. *Biocontrol* 46, 345–362.

Ogura, N. and Nakashima, T. (1997) Cold tolerance and preconditioning of infective juveniles of *Steinernema kushidai* (Nematoda: Steinernematidae). *Nematologica* 43, 107–115.

Patel, M.N. and Wright, D.J. (1998) The ultrastructure of the cuticle and sheath of infective juveniles of entomopathogenic steinernematid nematodes. *Journal of Helminthology* 72, 257–266.

Patel, M.N., Perry, R.N. and Wright, D.J. (1997) Desiccation survival and water contents of entomopathogenic nematodes, *Steinernema* spp. (Rhabditida: Steinernematidae). *International Journal for Parasitology* 27, 61–70.

Perry, R.N. (1977) Desiccation survival of larval and abult stages of the plant parasitic nematodes, *Ditylenchus dipsaci and D. myceliophagus. Parasitology* 74, 139–148.

Perry, R.N. and Moens, M. (2011) Survival of parasitic nematodes outside the host. In: Perry, R.N. and Wharton, D.A. (eds.) *Molecular and Physiological Basis of Nematode Survival*. CAB International, Wallingford, UK, pp. 1–27.

Perry, R.N., Ehlers, R.-U. and Glazer, I. (2012) A realistic appraisal of methods to enhance desiccation tolerance of entomopathogenic nematodes. *Journal of Nematology* 44, 185–190.

Piggott, S., Liu, Q.-Z., Glazer, I. and Wright, D.J. (2002) Does osmoregulatory behaviour in entomopathogenic nematodes predispose desiccation tolerance? *Nematology* 4, 483–487.

Poinar, G.O. (1990) Taxonomy and biology of Steinernematidae and Heterorhabditidae. In: Gaugler, R. and Kaya, H.K. (eds.) (1990) *Entomopathogenic Nematodes in Biological Control*. CRC Press, Boca Raton, Florida, pp. 21–62.

Popiel, I. and Vasquez, E. (1989) Cryoprotection of *Steinernema carpocapsae* and *Heterorhabditis bacteriophora*. *Journal of Nematology* 23, 423–437.

Powers, T.O., Neher, D.A., Mullin, P., Esquivel, A., Giblin-Davis, R.M. *et al.* (2009) Tropical nematode diversity: vertical stratification of nematode communities in a Costa Rican humid lowland rainforest. *Molecular Ecology* 18, 985–996.

Qiu, L. and Bedding, R.A. (2000) Energy metabolism and survival of the infective juveniles of *Steinernema carpocapsae* under oxygen-deficient conditions. *Journal of Nematology* 32, 271–280.

Qiu, L. and Bedding, R.A. (2002) Characteristics of protectant synthesis of infective juveniles of *Steinernema carpocapsae* and importance of glycerol as a protectant for survival of the nematodes during osmotic dehydration. *Comparative Biochemistry and Physiology B: Biochemistry and Molecular Biology* 131, 757–765.

Ramakrishnan, J., Salame, L., Nasser, A., Glazer, I. and Ment, D. (2022) Survival and efficacy of entomopathogenic nematodes on exposed surfaces. *Scientific Reports* 12: 4629.

Ramakrishnan, J., Salame, L., Mani, K.A., Feldbaum, R., Karavani, E. *et al.* (2023) Increasing the survival and efficacy of entomopathogenic nematodes on exposed surfaces by Pickering emulsion formulations offers new venue for foliar pest management. *Journal of Invertebrate Pathology* 199: 107938.

Reardon, W., Chakrabortee, S., Pereira, T.C., Tyson, T., Banton, M.C. *et al.* (2010) Expression profiling and cross-species RNA interference (RNAi) of desiccation-induced transcripts in the anhydrobiotic nematode *Aphelenchus avenae*. *BMC Molecular Biology* 11: 6.

Richter, K., Haslbeck, M. and Buchner, J. (2010) The heat shock response: life on the verge of death. *Molecular Cell* 40, 253–266.

Riddle, D.L. (1988) 12 The dauer larva. *Cold Spring Harbor Monograph Archive* 17, 393–412.

Rossum, A.J. van, Brophy, P.M., Tait, A., Barrett, J. and Jefferies, J.R. (2001) Proteomic identification of glutathione *S*-transferases from the model nematode *Caenorhabditis elegans*. *Proteomics* 1, 1463–1468.

Salame, L. and Glazer, I. (2015) Stress avoidance: vertical movement of entomopathogenic nematodes in response to soil moisture gradient. *Phytoparasitica* 43, 647–655.

Schill, R.O., Steinbrück, G.H.B. and Köhler, H.-R. (2004) Stress gene (*hsp70*) sequences and quantitative expression in *Milnesium tardigradum* (Tardigrada) during active and cryptobiotic stages. *Journal of Experimental Biology* 207, 1607–1613.

Searcy, D.G., Kisiel, M.J. and Zuckerman, B.M. (1976) Age-related increase of cuticle permeability in the nematode *Caenorhabditis briggsae*. *Experimental Aging Research* 2, 293–301.

Selvan, S., Gaugler, R. and Lewis, E.E. (1993a) Biochemical energy reserves of entomopathogenic nematodes. *Journal of Parasitology* 79, 167–172.

Selvan, S., Gaugler, R. and Grewal, P.S. (1993b) Water content and fatty acid composition of infective juvenile entomopathogenic nematodes during storage. *Journal of Parasitology* 79, 510–516.

Selvan, S., Grewal, P.S., Leustek, T. and Gaugler, R. (1996) Heat shock enhances thermotolerance of infective juvenile insect-parasitic nematodes *Heterorhabditis bacteriophora* (Rhabditida: Heterorhabditidae). *Experientia* 52, 727–730.

Shamseldean, M.M. and Abd-Elgawad, M.M. (1994) Natural occurrences of insect pathogenic nematodes (Rhabditida: Heterorhabditidae) in Egyptian soils. *Afro-Asian Journal of Nematology* 4, 151–154.

Shannon, A.J., Browne, J.A., Boyd, J., Fitzpatrick, D.A. and Burnell, A.M. (2005) The anhydrobiotic potential and molecular phylogenetics of species and strains of *Panagrolaimus* (Nematoda, Panagrolaimidae). *Journal of Experimental Biology* 208, 2433–2445.

Shapiro, D.I., Glazer, I. and Segal, D. (1996) Trait stability and fitness of the heat tolerant entomopathogenic nematode *Heterorhabditis bacteriophora* IS5 strain. *Biological Control* 6, 238–244.

Shapiro-Ilan, D.I., Gouge, D.H., Piggott, S.J. and Fife, J.P. (2006) Application technology and environmental considerations for use of entomopathogenic nematodes in biological control. *Biological Control* 38, 124–133.

Shapiro-Ilan, D.I., Blackburn, D., Duncan, L., El-Borai, F.E., Koppenhöfer, H. *et al.* (2014a) Characterization of biocontrol traits in *Heterorhabditis floridensis*: a species with broad temperature tolerance. *Journal of Nematology* 46, 336–345.

Shapiro-Ilan, D.I., Brown, I. and Lewis, E.E. (2014b) Freezing and desiccation tolerance in entomopathogenic nematodes: diversity and correlation of traits. *Journal of Nematology* 46, 27–34.

Shapiro-Ilan, D., Hazir, S. and Glazer, I. (2019) Integrated management of insect pests: current and future developments, In: Kogan, M. and Heinrichs, E.A. (eds.) *Advances in Use of Entomopathogenic Nematodes in Integrated Pest Management*. Burleigh Dodds Science Publishing, Cambridge, UK, pp. 649–678.

Sharon, M.A., Kozarova, A., Clegg, J.S., Vacratsis, P.O. and Warner, A.H. (2009) Characterization of a group 1 late embryogenesis abundant protein in encysted embryos of the brine shrimp *Artemia franciscana*. *Biochemistry and Cell Biology* 87, 415–430.

Simons, W.R. and Poinar, G.O. (1973) The ability of *Neoaplectana carpocapsae* (Steinernematidae: Nematodea) to survive extended periods of desiccation. *Journal of Invertebrate Pathology* 22, 228–230.

Singh, V. and Aballay, A. (2006) Heat-shock transcription factor (HSF)-1 pathway required for *Caenorhabditis elegans* immunity. *Proceedings of the National Academy of Sciences USA* 103, 13092–13097.

Solomon, A. and Glazer, I. (1999) Desiccation survival of the entomopathogenic nematode *Steinernema feltiae*: induction of anhydrobiosis. *Nematology* 1, 61–68.

Solomon, A., Salomon, R., Paperna, I. and Glazer, I. (2000) Desiccation stress of entomopathogenic nematodes induces the accumulation of a novel heat-stable protein. *Parasitology* 121, 409–416.

Somasekhar, N., Grewal, P.S. and Klein, M.G. (2002) Genetic variability in stress tolerance and fitness among natural populations of *Steinernema carpocapsae*. *Biological Control* 23, 303–310.

Somvanshi, V.S., Koltai, H. and Glazer, I. (2008) Expression of different desiccation-tolerance related genes in various species of entomopathogenic nematodes. *Molecular and Biochemical Parasitology* 158, 65–71.

Stengel, F., Baldwin, A.J., Painter, A.J., Jaya, N., Basha, E. *et al.* (2010) Quaternary dynamics and plasticity underlie small heat shock protein chaperone function. *Proceedings of the National Academy of Sciences USA* 107, 2007–2012.

Stiernagle, T. (2006) Maintenance *of C. elegans*. *WormBook: The Online Review of* C. elegans *Biology*. Wormbook, Pasadena, California. Available at: https://doi.org/10.1895/wormbook.1.101.1 (accessed 14 March 2024).

Strauch, O., Oestergaard, J., Hollmer, S. and Ehlers, R.-U. (2004) Genetic improvement of the desiccation tolerance of the entomopathogenic nematode *Heterorhabditis bacteriophora* through selective breeding. *Biological Control* 31, 218–226.

Sumaya, N.H., Aryal, S., Vandenbossche, B., Barg, M., Doerfler, V. *et al.* (2017) Phenotyping dauer juvenile oxidative stress tolerance, longevity and persistence within wild type and inbred lines of the entomopathogenic nematode *Heterorhabditis bacteriophora*. *Nematology* 19, 971–986.

Sumaya, N.H., Gohil, R., Okolo, C., Addis, T., Doerfler, V. *et al.* (2018) Applying inbreeding, hybridization and mutagenesis to improve oxidative stress tolerance and longevity of the entomopathogenic nematode *Heterorhabditis bacteriophora*. *Journal of Invertebrate Pathology* 151, 50–58.

Surrey, M. and Wharton, D. (1995) Desiccation survival of the infective larvae of the insect parasitic nematode, *Heterorhabditis zealandica* Poinar. *International Journal for Parasitology* 25, 749–752.

Tanaka, T., Ikita, K., Ashida, T., Motoyama, Y., Yamaguchi, Y. and Satouchi, K. (1996) Effects of growth temperature on the fatty acid composition of the free-living nematode *Caenorhabditis elegans*. *Lipids* 31, 1173–1178.

Thurston, G.S., Ni, Y. and Kaya, H.K. (1994) Influence of salinity on survival and infectivity of entomopathogenic: nematodes. *Journal of Nematology* 26, 345–351.

Timper, P. and Kaya, H.K. (1989) Role of the second-stage cuticle of entomogenous nematodes in preventing infection by nematophagous fungi. *Journal of Invertebrate Pathology* 54, 314–321.

Tullet, J.M.A., Hertweck, M., An, J.H., Baker, J., Hwang, J.Y. *et al.* (2008) Direct inhibition of the longevity-promoting factor SKN-1 by insulin-like signaling in *C. elegans*. *Cell* 132, 1025–1038.

Tyson, T., O'Mahony Zamora, G., Wong, S., Skelton, M., Daly, B. *et al.* (2012) A molecular analysis of desiccation tolerance mechanisms in the anhydrobiotic nematode *Panagrolaimus superbus* using expressed sequenced tags. *BMC Research Notes* 5: 68.

Tyson, T., Reardon, W., Browne, J.A. and Burnell, A.M. (2007) Gene induction by desiccation stress in the entomopathogenic nematode *Steinernema carpocapsae* reveals parallels with drought tolerance mechanisms in plants. *International Journal for Parasitology* 37, 763–776.

van den Hoogen, J., Geisen, S., Routh, D., Ferris, H., Traunspurger, W. *et al.* (2019) Soil nematode abundance and functional group composition at a global scale. *Nature* 572, 194–198.

Virgilio, C., Burckert, N., Bell, W., Jeno, P., Boller, T. and Wiemken, A. (1993) Disruption of *TPS2*, the gene encoding the 100-kDa subunit of the trehalose-6-phosphate synthase/phosphatase complex in *Saccharomyces cerevisiae*, causes accumulation of trehalose-6-phosphate and loss of trehalose-6-phosphate phosphatase activity. *European Journal of Biochemistry* 212, 315–323.

Virgilio, C., Hottiger, T., Dominguez, J., Boller, T. and Wiemken, A. (1994) The role of trehalose synthesis for the acquisition of thermotolerance in yeast. I. Genetic evidence that trehalose is a thermoprotectant. *European Journal of Biochemistry* 219, 179–186.

Wharton, D.A. (1986) *A Functional Biology of Nematodes*. Croom Helm, London.

Wharton, D.A. (2011) Cold tolerance. In: Perry, R.N. and Wharton, D.A. (eds.) *Molecular and Physiological Basis of Nematode Survival*. CAB International, Wallingford, UK, pp. 182–204.

Wharton, D.A., Preston, C.M., Barrett, J. and Perry, R.N. (1988) Changes in cuticular permeability associated with recovery from anhydrobiosis in the plant parasitic nematode, *Ditylenchus dipsaci*. *Parasitology* 97, 317–330.

Wharton, D.A., Judge, K.F. and Worland, M.R. (2000) Cold acclimation and cryoprotectants in a freeze-tolerant Antarctic nematode, *Panagrolaimus davidi*. *Journal of Comparative Physiology B* 170, 321–327.

Womersley, C. (1981) Biochemical and physiological aspects of anhydrobiosis. *Comparative Biochemistry and Physiology Part B: Comparative Biochemistry* 70, 669–678.

Womersley, C. (1987) A reevaluation of strategies employed by nematode anhydrobites in relation to their natural environment. *Vistas on Nematology*, pp. 165–173.

Womersley, C.Z. (1990) Dehydration survival and anhydrobiotic potential. In: Veech, J.A. and Dickson, D.W. (eds.) *Vistas on Nematology*. Society of Nematologists, Hyattsville, Maryland, pp. 117–137.

Wright, D.J. (2004) Osmoregulatory and excretory behaviour. In: Gaugler, R. and Bilrami, A.L. (eds.) *Nematode Behaviour*. CAB International, Wallingford, UK, pp. 177–196.

Wright, D. and Newall, D. (1976) Nitrogen excretion, osmotic and ionic regulation in nematodes. In: Croll, N.A. (ed.) *The Organization of Nematodes*. Academic Press, London, pp. 163–210.

Wright, D. (1998) Respiratory physiology, nitrogen excretion and osmotic and ionic regulation. In: Perry, R.N. and Wright, D.J. (eds.) *The Physiology and Biochemistry of Free-living and Plant-parasitic Nematodes*. CAB International, Wallingford, UK, pp. 103–131.

Yaari, M., Doron-Faigenboim, A., Koltai, H., Salame, L. and Glazer, I. (2016) Transcriptome analysis of stress tolerance in entomopathogenic nematodes of the genus *Steinernema*. *International Journal for Parasitology* 46, 83–95.

Yen, J.H., Niblack, T.L., Karr, A.L. and Wiebold, W.J. (1996) Seasonal biochemical changes in eggs of *Heterodera glycines* in Missouri. *Journal of Nematology* 28, 442–450.

Zervos, S., Johnson, S.C. and Webster, J.M. (1991) Effect of temperature and inoculum size on reproduction and development of *Heterorhabditis heliothidis* and *Steinernema glaseri* (Nematoda: Rhabditoidea) in *Galleria mellonella*. *Canadian Journal of Zoology* 69, 1261–1264.

6 Behavioral Ecology of Entomopathogenic Nematodes

Edwin E. Lewis[1]*, Glen Stevens[1], Ivan Hiltpold[2] and David Shapiro-Ilan[3]
[1]*Department of Entomology, Plant Pathology and Nematology, University of Idaho, Moscow, Idaho, USA;* [2]*Entomology and Nematology, Division of Plant Protection, Agroscope, Nyon, Switzerland;* [3]*USDA-ARS, Southeastern Fruit and Tree Nut Research Station, Byron, Georgia, USA*

Abstract
The behavioral ecology of entomopathogenic nematodes (EPNs) has been studied in several different contexts. Starting in the 1980s, studies of foraging behavior and infection behavior were conducted with the objective of determining how to use EPNs effectively and concentrated on narrowing the gap between laboratory results and field results. This emphasis remains. But as research questions grew more complicated and methods more sophisticated, EPNs began to emerge as model systems to study various theoretical underpinnings of animal behavior. Here, we address primarily the results of studies that can be used to improve the biocontrol potential of EPNs, for example studies of how EPNs behave when in contact with different substrates, or infection dynamics. We also attempt to link infective juvenile (IJ) behavior with field distributions. This work is applicable to explain what happens to IJs after application, for example. We also touch on studies that do not address biocontrol directly, such as male–male conflict within infected cadavers. Finally, we review recent work that goes past documenting and understanding EPN behavioral ecology and seeks to use what has been learned to manipulate these behaviors to benefit biological control outcomes. Early studies on EPN behavior sought to discover what IJs did in response to putative hosts. The next step was to build on these studies to improve our predictive capabilities about when and where EPNs would be successful biological control agents. Most recently, the work transcends understanding and predicting behaviors to actually changing EPN behavior to improve their effectiveness.

6.1 Introduction: Why Does Behavioral Ecology Matter for Biological Control with Entomopathogenic Nematodes?

The behavior and ecology of a biological control agent can affect its efficacy profoundly. As stated by Georgis and Gaugler (1991), a major challenge in using entomopathogenic nematodes (EPNs) for biological control is improving their predictability. Although this statement was made decades ago, it stands as a goal of current research today and will remain so for the foreseeable future. Of course, having predictably positive outcomes of EPN applications (positive meaning reliable levels of reduction of pest populations) is extremely important, especially considering that EPNs are more expensive to acquire and demand more expertise from the applicator than do most synthetic chemical pesticides or even most pest

*Email: eelewis@uidaho.edu

DOI: 10.1079/9781800620322.0006

management products based on other insect pathogens. Failures of EPN applications have been attributed to several causes that include improper application (Georgis *et al.*, 2006), application to areas with soil types not amenable to EPN use, poor product quality, incompatibility with other concurrent practices, targeting an insect pest not susceptible to the EPN species applied, and a long list of other conditions and practices (Georgis *et al.*, 2006). Not all of these documented instances of disappointing performance can be attributed to a lack of understanding of EPN behavior, but behavior does play a role in many of them (Lewis *et al.*, 2006).

Studies of the behavioral ecology of EPNs are focused almost entirely on the infective juvenile (IJ) stage, which is the life stage comprising most products, the stage that searches for and infects new hosts, and the only life stage that exists outside the host. Thus, most aspects of predictability depend on knowing what the IJs will do after application, in terms of searching behavior, longevity and infection behaviors. In this chapter, we discuss two broad categories of EPN behavioral ecology: (i) direct influences, which include whether they can find, infect and kill a target insect pest; and (ii) indirect effects, which include how long they will survive in the field, how they will be distributed and whether they can survive exposure to various environmental stressors.

6.2 The Entomopathogenic Nematode Foraging Strategy Paradigm as a Behavioral Model

The paradigm of EPN foraging strategies ranging from a stationary mode of foraging, termed "ambushing," and a highly mobile foraging mode, termed "cruising" (Lewis *et al.*, 1992; Lewis *et al.*, 1993), has been a subject of debate since the 1990s. While there have been studies showing exceptions to the idea that species exhibit a consistent array of behaviors related to foraging (e.g. Wilson *et al.*, 2012), the utility of this paradigm has been to provide a structure of related behaviors that enable the construction of hypotheses about nematode behavior and ecology. The suite of correlated behaviors that are expressed by nematodes employing a particular foraging strategy can be termed behavioral syndromes (Sih *et al.*, 2004). As defined by Sih *et al.* (2004), these syndromes can be expressed either within a given behavioral context (e.g. how members of a species would express movement behaviors on different substrates) or in different contexts (e.g. how members of a species with a particular foraging strategy might react to predators, conspecific individuals or competitors from different species). The impact of foraging strategies on biological control success is generally attributed to how an IJ forages in relation to the behavior of the target pest, in other words, whether they will be in the same place at the same time. Thus: (i) what IJs find depends on how they search for it; and (ii) distribution and dispersion are determined by IJ behavior. The responses to stress are also partly behavioral, so these are discussed here as well.

6.3 Matching Entomopathogenic Nematode Species with Target Hosts

What you find depends on how you look for it. When considering foraging strategies, EPN species that forage by spending most of their time standing on their tails near the surface of the soil are most likely to encounter mobile hosts near the soil surface, while EPN species that forage by moving through the soil matrix are most likely to encounter relatively sedentary hosts below the soil surface. However, the interplay between various species of IJs and their potential hosts can be more complicated and includes host recognition based on the characteristics of the host (Lewis *et al.*, 1996; Baiocchi *et al.*, 2019) or their byproducts (Grewal *et al.*, 1993a; Baiocchi *et al.*, 2017; Kin *et al.*, 2019), assessment of several aspects of host quality (Grewal *et al.*, 1997; Ramos-Rodriguez *et al.*, 2007; Fushing *et al.*, 2008; Hallem *et al.*, 2011) and the ability of the infection to produce progeny.

6.3.1 Host behavior and IJ behavior

Arthropods and EPNs vary in their movement-associated behaviors and their location within the habitat. Because EPNs are generally not tolerant of desiccation, temperature extremes or ultraviolet (UV) light exposure, they are conventionally relegated to cryptic or subterranean habitats. For EPNs to be successful, they need their foraging behavior to align with the behavior and distribution of the target host. Once EPNs are applied to an area, they engage in searching behaviors that can differ remarkably among species (Lewis *et al.*, 2006). Campbell *et al.* (2003) described the phylogenetic underpinnings of foraging behaviors among EPNs and showed that evolutionary relationships among EPN species in the genus *Steinernema* can be useful in predicting how a new and unstudied species may forage. Considering how many newly isolated EPNs are largely uncharacterized, and often even unidentified to the species level, this may be a helpful tool when deciding whether further research is warranted for a particular isolate.

The importance of matching EPN behavior to insect targets has been shown repeatedly to be one aspect of improving EPN efficacy. In the 1980s, there were several tests of host range that were completed in Petri dishes where the host ranges of EPNs were thought to include insects they would be unlikely to encounter in field conditions. For example, *Steinernema carpocapsae* IJs were found to be able to infect, kill and reproduce in larvae of the Japanese beetle, *Popillia japonica*, but extensive field trials revealed that, despite their ability to kill *Popillia japonica* larvae in laboratory conditions, *S. carpocapsae* IJs were not effective control agents of these turf pests when applied in the field (Forschler and Gardner, 1991). However, several other EPN species were found to be very effective (Simoes *et al.*, 1993). The reason for the lack of efficacy was twofold: (i) *S. carpocapsae* IJs forage mostly by standing on their tails near the soil surface waiting to attach to passing insects (which would be unlikely to include scarab larvae); and (ii) the *Popillia japonica* immune system has been shown to effectively recognize and mount humoral and cellular immune responses in response to *S. carpocapsae* IJs in the hemocoel (Wang *et al.*, 1994). There are other examples of mismatches between EPN species and target pests. There are also scenarios where EPNs that are classified as ambushers, such as *S. carpocapsae*, do control insect species found fairly deep in the soil or other cryptic habits, as demonstrated by Harvey and Griffin (2016) with *S. carpocapsae* being used successfully to control the large pine weevil, *Hylobius abietis*, and also in the control of the peachtree borer, *Synanthedon exitiosa* (Wong *et al.*, 2022) or the western corn rootworm, *Diabrotica virgifera virgifera* (Hiltpold and Hibbard, 2018). While the foraging strategy paradigm is a useful tool in developing hypotheses about other related behaviors of EPN IJs and may provide a place to start when studying new EPN isolates, care needs to be taken in overstating the relationships between host behavior and EPN IJ behavior in terms of the nematodes' ability to find hosts.

Potential hosts are not always passive and often respond behaviorally to the presence of EPNs in their proximity, and these behaviors could potentially impact efficacy. Potential hosts that have been incapacitated by physical restraint, sublethal intoxication by pesticides or parasitoid venom do not engage in defensive behaviors and are subjected to higher rates of infection by EPNs. Gaugler *et al.* (1994) demonstrated the efficacy of grooming behavior by *Popillia japonica* larvae as a defense against EPN attack. When larvae were put in contact with even one IJ of *Heterorhabditis bacteriophora*, they responded by brushing with their legs and rubbing with the raster (the last abdominal segment, which is covered with stout bristles), which proved an effective defense, removing more than 60% of the IJs. However, when the larvae were restrained with a wire cage, IJ penetration into the hemocoel increased significantly. Other ways to reduce host grooming behaviors have been explored; intoxicating scarab larvae with a sublethal dose of imidacloprid also reduces this behavior and can increase efficacy of the EPNs (Koppenhofer *et al.*, 2000). Chapter 27 (this volume) describes interactions between EPNs and other inputs for plant production; EPNs are generally found to be compatible with insecticides. Combining insecticides or other insect pathogens with EPNs has been effective in some systems, thus broadening the idea of exploiting behavioral alterations of hosts; for example, *S. carpocapsae* control improves when combined with insecticides against *Bradysia odoriphaga* larvae in chive, onion and garlic production (Chang *et al.*, 2023), applications of imidacloprid combined with EPNs were more effective than either material alone in reducing the peanut burrower bug, *Pangaeus bilineatus* (Hemiptera: Cydnidae) (Mbata *et al.*, 2022), and combining *Heterorhabditis amazonensis* JPM4 with several insecticides provided superior control of the tomato leafminer, *Tuta absoluta* (Lepidoptera: Gelechiidae), than either the insecticides or EPNs alone (Sabino *et al.*, 2019).

6.3.2 Biotic and abiotic substrate conditions that affect IJ behavior

EPNs respond directionally to a variety of stimuli that are thought to be involved in host location and evaluation, such as carbon dioxide (Lewis *et al.*, 1993; Banerjee and Hallem, 2020), vibration (Torr *et al.*, 2004), temperature (Hang *et al.*, 2007), nitrogen-based compounds (Grewal *et al.*, 1993a), and electromagnetic stimuli (Shapiro-Ilan *et al.*, 2012). Directional movement has also been observed when EPNs were exposed to volatile organic compounds emitted by their insect host (Hallem *et al.*, 2011) or by host-damaged roots (Rasmann *et al.*, 2005). Movement and distribution of EPNs in the field also result from passive means (e.g. water movement), phoretic associations (Shapiro *et al.*, 1995; Eng *et al.*, 2005; Campos-Herrera *et al.*, 2006) and infected insects that remain mobile for a short time after infection. The fact that EPNs move throughout the environment, albeit with some limitations, has been touted as one of their attributes, making them sought-after biological control agents. Here, we will address EPN behavior and soil characteristics, application of EPNs to cryptic habitats that exploit foraging behavior and special application methods that expand their normal range of conditions. EPNs are sensitive to dry conditions, UV light exposure, temperature extremes, etc., all of which contribute to their relegation largely to soil environments. How EPNs respond to various stressors in their environment (heat or drought conditions) is addressed elsewhere (see Chapter 5, this volume); here, we discuss briefly how aspects of the environment can impact EPN behavior. EPNs move through a water film that coats the soil particles. Thus, ample pore space (at least the diameter of the IJ) must exist in the soil, as well as adequate moisture. In heavy clay soils, where particle size and thus pore size is small, EPNs are not effective if host finding requires movement through the soil matrix (Choo and Kaya, 1991). Kaspi *et al.* (2010) proposed that grouping soils into categories that reflect characteristics essential to EPN movement and survival could be useful in determining soil conditions that would be amenable to EPN application. They found that the single most important soil factor affecting EPN movement was soil particle size distribution, but second in rank was soil salinity. Other factors influencing EPN behavior and efficacy in different soil conditions include temperature extremes (Ali and Wharton, 2013), salinity (Nielsen *et al.*, 2011), moisture (El-Borai *et al.*, 2016) and electrical fields (Shapiro-Ilan *et al.*, 2012). EPNs have been shown to avoid potentially damaging conditions by moving away in space (Nielsen *et al.*, 2011) or time (Lewis and Shapiro-Ilan, 2002; Spence *et al.*, 2010).

Biotic antagonists also influence the behavior of EPN IJs and other nematodes in soil. The significant literature concerning the biological control of nematodes that are parasites of plants, livestock and humans can provide a guide to the challenges that EPNs face when applied to soil. The natural enemies associated with EPNs include predatory nematodes (Kanwar *et al.*, 2021), predatory arthropods (Karagoz *et al.*, 2007; Ulug *et al.*, 2014), fungal pathogens and bacterial pathogens (see Chapter 8, this volume). The behavior of EPN IJs affects the likelihood of encountering the sedentary pathogens in two ways. The first is the relative degrees of movement between the EPN and natural enemy; mobile natural enemies are most likely to encounter relatively sedentary EPNs and highly mobile EPNs are most likely to encounter more sedentary natural enemies, such as pathogens. In addition to the influence of relative mobility, EPNs have been shown to be repelled by the nematode-trapping fungi in the genus *Arthrobotrys*. However, EPNs are not repelled by endoparasitic fungi of nematodes. Much of what we know about EPN movement is based on tests done in very controlled conditions. Indeed, there have been precious few movement trials conducted in the presence of EPN natural enemies; these have highlighted the challenges of understanding EPN dispersal and host-finding behaviors when trials are conducted in conditions of plain sand, sterilized soil, pleuronic gels or agar surfaces.

Soil is not the only medium to which EPNs are applied for biological control purposes, and EPN movement can be facilitated by specialized application strategies. EPN movement has been exploited to control several species of boring insects, as the IJs can move along galleries formed in woody plants. One example of exploiting EPN mobility against a cryptic insect is the use of EPNs against the lesser peach tree borer, *Synanthedon pictipes* (Lepidoptera: Sesiidae). Borers are difficult to reach using conventional pesticides because they are protected by their location within the plant. *S. carpocapsae* has been shown to readily infect and kill this insect species (Shapiro-Ilan and Cottrell, 2006), and they can move through galleries. The remaining challenge was to deliver the IJs to the gallery opening, and this was achieved by applying the EPNs in a mix of the fire retardant Barricade® (Shapiro-Ilan *et al.*, 2016). Use of Barricade® is discussed in other chapters in this volume for use in greenhouse (Chapter 25, this volume) and in vegetable production (Chapter 20, this volume). Using specialized formulation or application methods may be a way to exploit EPN movement and host finding in situations where they normally would not be effective due to their inability to move on dry surfaces. Similar efforts have been made in controlling other borers, leaf miners and foliar pests such as the diamondback moth, *Plutella xylostella* (Chapter 20, this volume).

6.4 Entomopathogenic Nematode Distribution at the Microscale: Patchiness and Social Behaviors

It is well established that EPNs exhibit patchy distribution throughout spatial scales, a trait they share with all nematodes (e.g. Liu *et al.*, 2019). Many factors likely underlie the irregular distribution of EPNs at meter scales, including abiotic factors such as temperature, moisture, or even electrical fields and non-EPN biotic factors (see Chapter 7, this volume).

When one considers EPNs independently, interacting within a guild of multiple potential EPN species at the meter scale, there are exciting interactions within and among EPN populations. Phenomena such as mass emergence, dispersal and associated sex biases have been well researched. While other social phenomena such as competition within the host and interspecific interactions among EPNs continue to be the focus of research, recent experiments focused on social phenomena such as group movement, following behavior and joining behavior within and among EPN species is adding complexity to our understanding. The meter scale is likely the spatial framework that is most appropriate for these considerations, as it matches with our general understanding of the limit of field dispersal of a given population (e.g. Downes and Griffin, 1996; Bal *et al.*, 2017), excepting large-scale commercial applications of formulated EPNs in managed systems for inundative biological control.

6.4.1 Mass emergence and dispersal

Mass emergence seems more relevant to IJs from insect cadavers than inundative applications. Late-stage EPN-infected insect cadavers can produce tens to hundreds of thousands of IJs, even if some are nematodes of other trophic groups (e.g. free-living nematodes living within the infection) (Rolston *et al.*, 2006; Blanco-Pérez *et al.*, 2017). Earlier studies demonstrated greater dispersal (Shapiro and Glazer, 1996) and infectivity (Shapiro and Lewis, 1999) when IJs emerge from hosts rather than being applied via aqueous suspension, so cadaver applications have been used to control a wide range of pests (Shapiro-Ilan *et al.*, 2003; Bruck *et al.*, 2005; Del Valle *et al.*, 2008). The extent of dispersal may vary between IJs emerging from their cadaver early versus late in the overall emergence; in the EPN *S. carpocapsae*, IJs that emerged within the first 2 days of the mass emergence showed three times higher dispersal rates (as a proportion of the population that dispersed) than on day 7 of emergence (Fujimoto *et al.*, 2007). Many of the emerging IJs leave the area of the host (Fujimoto *et al.*, 2007), although some *Steinernema feltiae* IJs remain near the cadaver for 9 days or more, even in the presence of new potential hosts (Rolston *et al.*, 2006).

The presence of root analogs facilitated the dispersal of EPN populations in laboratory trials. For example, Ennis *et al.* (2010) used branches in arenas with *S. carpocapsae* and pine weevil (*Hylobius abietis*) larvae as hosts, while Demarta *et al.* (2014) used Teflon-wrapped wire root analogs to examine interactions between *Heterorhabditis megidis* and *Galleria mellonella* larvae. Hence, natural roots and presumably root channels, both more common in natural habitats than in laboratory trials, should promote dispersal of EPNs in natural environments. This demonstrates another instance where laboratory-based trials may not predict the outcomes of field tests, some of which have been described above. Interestingly, increasing the complexity of artificial root systems reduced host finding in the EPN *H. megidis*, although the addition of synthetic root volatiles where the larvae were placed reversed this pattern (Demarta *et al.*, 2014). Similar behavior was observed in presence of actual plant roots (Hiltpold *et al.*, 2020).

Sex ratios can vary among EPN species and even populations, both for patterns of emergence and dispersal and for development within the host. For instance, Lewis and Gaugler (1994) showed that *Steinernema glaseri* males emerge earlier and are more sensitive to volatile host cues than females, whereas IJs of *S. carpocapsae* were biased toward early female emergence. Furthermore, a comparison across five steinernematid species revealed early male emergence in *S. glaseri*, *S. longicaudum* and *S. kraussei*, but later male emergence in *S. carpocapsae* and *S. feltiae* (Alsaiyah *et al.*, 2009). Continuing with differences between the sexes in EPN IJ behavior, males were proposed to colonize hosts before females by Grewal *et al.* (1993b). However, this was reconsidered by Stuart *et al.* (1998), where they found that males may respond more strongly to host volatiles, but there was no consistent difference in who invaded first. Considering subsequent results, the male colonizer hypothesis appears to be true for some species but not all. Further work with *S. carpocapsae* showed no sex ratio differences in dispersing IJs, but for those that stayed on hosts, early emergers were male biased and later-emerging IJs were female biased (Fujimoto *et al.*, 2007). However, developing EPNs inside the host appear to be female biased in *S. feltiae* (Campos-Herrera and Gutiérrez, 2014) and *S. longicaudum* (Alsaiyah *et al.*, 2009), and could show intraspecific variability (Campos-Herrera and Gutiérrez, 2014). Clearly, how the sex ratio relates to EPN biology and behavior remains enigmatic.

6.5 Intra- and Interspecific Interactions Inside the Host

In a rare examination of EPN behavior in stages other than the IJ, a fascinating series of experiments conducted primarily in hanging drops of insect hemolymph revealed behavioral interactions within and among EPN populations in the genus *Steinernema*. Assessments of interference competition among four species from three of the five clades of the genus *Steinernema* (*S. longicaudum*, *S. carpocapsae*, *S. kraussei* and *S. feltiae*) revealed both intra- and interspecific aggressions: male nematodes in the host killed both conspecific and heterospecific males and heterospecific females (O'Callaghan *et al.*, 2014). Differences existed among species in competitive strength, with *S. longicaudum* causing the highest mortality rates. Perhaps even more interesting, marked differences in aggression were observed between two populations of *S. feltiae*: strain 4CFMO was a killer, while strain UK76 was not. Follow-on work identified the role of developmental pathways in aggressive behavior: *S. longicaudum* males that had developed from IJs (post-IJ males that originally invaded the host) were more aggressive in encounters than males that had not passed through the dauer stage (generations subsequent to the original invaders) (Zenner *et al.*, 2014). Furthermore, comparisons among groups of *S. longicaudum* IJs with different levels of kinship (single line versus three-line mixes) showed lower mortality rates between closely related individuals, suggesting that kin selection is working to reduce aggression among EPNs within the host (Kapranas *et al.*, 2016). Nevertheless, *S. longicaudum* males also appear to discriminate according to the perceived resource value of females: they fought more often and for longer for virgin females than for mated females (Kapranas *et al.*, 2020). In this set of experiments, mated males also tended to defeat virgin males, suggesting either a superior competitive ability or an increased tendency to attack.

6.6 Group and Individual Movement: Aggregation, Following and Joining

While mass emergence from the insect cadaver certainly influences the patchy dynamics exhibited by EPN populations in the field (e.g. Stuart and Gaugler, 1994; Campbell *et al.*, 1996), questions about whether and how this aggregate behavior is maintained or even reinforced after the IJs leave the immediate area near the host are an increasing focus of research. Studies such as Rolston *et al.* (2006) illustrate that freshly emerged IJs rapidly attack and colonize new hosts if available, which contrasts with the "phased infectivity hypothesis" originally proposed by Bohan and Hominick (1997), who suggested that there existed a population of non-infectious *S. feltiae* in a cohort of IJs from a single cadaver. Whether or not all IJs immediately infect available suitable hosts, it is unclear how these social dynamics apply when there are no hosts to immediately infect; in natural conditions, it may easily take months for IJs to find new resources. These questions suggest a focus on group dispersal, following and joining, which are areas of recent interest.

EPNs travel and infect in groups. In experimental arenas without resources (boxes filled with damp sand), *H. bacteriophora*, *H. indica*, *S. carpocapsae*, *S. feltiae*, *S. glaseri* and *S. riobrave* IJs showed group movement when initial assessment was made at four points around the central introduction point (Shapiro-Ilan *et al.*, 2014). A follow-on assessment maintained the same no-resource environment but increased the complexity of the arena and of sampling: IJs assessed at three distances from hosts at 48 and 72 h post-addition. In both cases, group movement was maintained across space and time, suggesting that nematodes move in an aggregative fashion *en masse* through the soil environment (Ruan *et al.*, 2018). Although IJs applied to soil via infected cadaver versus those applied in water appear to differ in terms of dispersal and infectivity (see above), as regards their group movement dynamics, the IJs demonstrated continuous aggregation regardless of the application method (Ruan *et al.*, 2018). Recent work assessed aggregation behaviors among three different *Steinernema* spp. (S. *carpocapsae*, *S. feltiae* and *S. glaseri*), and focused on their aggregation behavior when either mixed together and co-applied or when each species was added to opposite corners of a sand arena (Stevens *et al.*, 2023). These trials showed that all three species demonstrated aggregative group movement when alone. When in an arena with another EPN species, significant differences in aggregation behavior were observed between co-applied and separate-corner applications: IJs in separate-corner applications aggregated more than they did when alone, while co-applied species aggregated less than when alone. This suggests that there are species-specific signals and cues that IJs use to promote conspecific aggregation, and that mixing species of EPN together can confuse these signals.

How do IJs maintain group membership? Recent investigations showed that IJs exhibit trail-following behaviors (Erdogan *et al.*, 2021b). In these experiments, IJs (followers) introduced individually into an agar channel were monitored to assess whether they followed the path of earlier single IJs (leaders) removed from the system immediately before the followers were introduced. Intra- and interspecific outcomes were complex to interpret and were heavily influenced by whether or not lead IJs had experienced previous contacts with *G. mellonella* cuticle. In conspecific trials, 80% of *S. glaseri* following IJs moved in the same direction as lead IJs, regardless of whether or not the lead IJ had been exposed to *G. mellonella* cuticle, whereas the other two species (*S. carpocapsae* and *S. feltiae*) reached similar percentages only if the lead IJs had been exposed to host contact. In heterospecific trials (lead and following IJs of different species), *S. glaseri* was willing to follow either species if the lead IJ had experienced host contact but not "naïve" IJs. In contrast, following *S. carpocapsae* IJs tracked experienced *S. glaseri* IJs, and following *S. feltiae* IJs avoided *S. carpocapsae* IJs that had experienced host contact. None of these interactions appeared to be linked to competitive behaviors: the most competitive species in side-by-side infections were not more likely to follow, and the species that exhibited the following behavior the most (*S. glaseri*) always suffered in terms of IJ production when exposed to waxworms combined with any of the other EPN species. Comprehensively, IJs leave a trail to follow that allows EPNs to distinguish the species of a lead IJ and some elements related to its experience. Moreover, it seems that several factors influence this behavior. Given that IJs undergo behavioral changes when exposed to contact with hosts (Lewis *et al.*, 1992, Lewis *et al.*, 1995), it is perhaps not surprising that nematodes can both follow a leader and discriminate between various qualities of a potential leader.

Similar to trail following, IJs appear to join existing groups of IJs under some conditions. Using a six-arm sand-filled olfactometer, Stevens *et al.* (2023) showed that dispersal of *S. glaseri* IJs was non-random, with IJs aggregating in particular arms in both the absence of cues and when there were hosts present in the cue pots of all six arms. When comparing IJ attraction to cue pots containing either IJs or sand only, dispersing *S. glaseri* IJs demonstrated an inclination to move toward the arms containing conspecific nematodes, in other words to join conspecific groups. This response was particularly strong when those conspecific IJs had experienced host contact. This group-joining behavior by *S. glaseri* also extended to heterospecific IJ groups but only in the case of one species (*S. feltiae* was attractive, while *S. carpocapsae* was not), and only when the *S. feltiae* IJs had experienced host contact.

The idea that EPNs demonstrate these group behaviors may give clues to their behaviors after application and how these behaviors can influence efficacy. Thus, three aspects of their behavior could lead to aggregated population structure in the field, even after application. First, EPNs have limited ability to move on their own; they respond to various cues and signals but only on the scale of a few centimeters per day. When IJs emerge from their host at the end of an infection, they readily disperse but still only several centimeters. Second, they not only remain in groups but also form groups of conspecific individuals. Third, infected hosts are highly attractive to conspecific IJs outside the host, reinforcing their aggregated distribution. Being able to hypothesize what distribution IJs will form after an application is somewhat useful, but can this knowledge be used to manipulate behavior in terms of dispersal and infection?

6.7 Manipulating Searching and Infection Behaviors

Understanding and predicting EPN behaviors associated with their success or failure in controlling insects has been a goal for decades. But can their behavior be manipulated to improve efficacy? This would represent a "next step" in the aims of studying EPN behavior. One could make the argument that efforts should be made to isolate and identify new EPN species and strains that have characteristics desirable for control. While some efforts have been made in this direction, they are limited by the presence or absence of characteristics that are not directly related to performance in field conditions, such as their ease of culturing. If a new EPN isolate has characteristics that make it tolerant to some environmental extreme but it is not readily culturable, there will be limited utility of this isolate. Thus, efforts that focus on improving the performance of strains that are already commercially available may yield positive results in the shortest time frame.

The production of EPNs commercially is covered in Chapter 10 (this volume). There are several methods of *in vitro* and *in vivo* production, but what most of these methods have in common is that they separate the IJ from any materials that are associated with the infection or the cadaver. The impact of typical EPN production on infection and dispersal behaviors was first documented by Shapiro and Glazer (1996) and Shapiro and Lewis (1999); these studies showed that EPN dispersal behaviors and infection behaviors were negatively impacted by their being harvested in water using White traps. In addition to showing that IJs emerging directly from their host cadavers to the substrate had increased dispersal and infectivity, and later tolerance to head desiccating conditions (Perez *et al.*, 2003; Gulzar *et al.*, 2020), these studies showed that when IJs harvested from White traps were exposed to macerated cadavers of conspecific EPN-infected insects, the original capabilities and tolerances were recovered. Of course, commercial production is vastly more sophisticated than harvesting EPNs in White traps, but all of these methods remove the IJs from the influence of ascaroside pheromones that are associated with the host cadaver. We described above how EPN IJs show better efficacy in some situations when they are applied as infected cadavers (i.e. they are not removed from association with the cadaver), but there are logistical challenges to this method. The question remains: how can these infection and dispersal behaviors be recovered when IJs are produced on a commercial scale?

Nematode pheromones, known as ascarosides, play key roles in development, dispersal and infection behaviors of most nematodes, including EPNs (Srinivasan *et al.*, 2008; Kaplan *et al.*, 2012). Indeed, crude pheromone extracts isolated from late-stage infected cadavers have been tested repeatedly for their effects on EPN dispersal and infection behaviors (Kaplan *et al.*, 2020; Erdogan *et al.*, 2021a). This pheromone mixture has been proposed as an adjuvant to expose to IJs prior to tank mix and application (Oliveira-Hofman *et al.*, 2019). The intent is to increase both dispersal and infectivity (Shapiro-Ilan *et al.*, 2019).

Ascarosides also have the surprising effect on plants of inducing defensive reactions. Considering the vast array of plant-parasitic nematode species that infect plants on a regular basis, it would not be surprising that exposure to ascarosides secreted by these nematodes would induce plant defensive reactions (Manosalva *et al.*, 2015). However, plant defences are also induced by ascarosides associated with EPNs, due to the highly conserved nature of ascarosides; that is, EPNs and plant-parasitic nematodes share ascr#18, which is the nematode-associated molecular pattern (NAMP) responsible for increasing resistance in *Arabidopsis*, tomato, potato and barley to viral, bacterial, oomycete, fungal and nematode infections. Ascarosides associated with the pinewood nematode, *Bursaphelenchus xylophilus*, also have been shown to promote early pupation by their vector beetle, *Monochamus alternatus* (Coleoptera: Cerambycidae) (Zhao *et al.*, 2016). Thus, pheromones associated with EPNs and other nematode species alter the physiology of plants and insects; how might these chemicals be used in IPM systems in the future?

In addition to pheromones, EPNs respond to other environmental cues and signals. IJs exhibit positive or negative chemotaxis toward odorants (blended or single) typically emitted by their insect host (Hallem *et al.*, 2011). Similarly, host cues trigger jumping in more sessile species of EPN IJs (Hallem *et al.*, 2011), a behavior allowing them to reach moving insects in the soil matrix. EPN-infected insect hosts also emit particular odorants that reveal their status as potential hosts to surrounding IJs (Baiocchi *et al.*, 2017); lower concentrations of the cues (typically emitted early after infection) attract IJs toward the insect cadavers, whereas higher concentrations (typically emitted longer after infection) repel them. IJs also respond to odorants emitted by insect-damaged roots, therefore signaling the presence of potential insect hosts (Rasmann *et al.*, 2005; Hiltpold *et al.*, 2010c; Ali *et al.*, 2010). Whether emitted by their host directly or induced by their host damage to a plant, these environmental cues allow EPN IJs to efficiently navigate, in groups, the soil matrix and increase their chances of encountering a host. It is therefore tempting to imagine manipulations of such chemically mediated interactions to improve the effectiveness of EPNs at managing economically important pests. Successful attempts of such manipulations have been taken either by improving the responsiveness of IJs to certain plant odorants (Hiltpold *et al.*, 2010a,b) or by modifying the genome of crop plants to design a plume of odorant highly attractive to IJs (Degenhardt *et al.*, 2009). Under particular circumstances, Hiltpold and Hibbard (2018) have demonstrated that exploiting the ability of EPNs to sense damaged plants may prevent the evolution of resistances against more classical management techniques of soil-dwelling insect pests.

The growing body of literature describing not only EPN behavior but also the mechanisms behind them offers ample options to manipulate the systems to better exploit the potential of these beneficials and increase their competitiveness in the landscape of sustainable pest management techniques.

6.8 Conclusions

Improving predictability has long been a primary goal of EPN-related research; this is true not just in the area of behavioral ecology but also in most applied research efforts. Indeed, most of the chapters of this volume touch on predictability of efficacy being a challenge to end users. Host-finding behaviors, and those behaviors associated with them, remain central to much of the research into IJ behavior. This is not to say that understanding host finding alone will lead to solutions in biological control efforts, but the behavioral syndromes associated with foraging strategies can be used to develop new and insightful hypotheses that will move the discipline forward. Three population-wide characteristics of EPNs that are indirectly related to behavioral ecology are: (i) persistence in the field after application; (ii) distribution in the field after application; and (iii) the stability of distributions in the field. These outcomes of applications are related to group foraging behaviors, as noted above. Patchy distributions of EPNs (and other soil nematodes) have been recognized for decades; understanding the "how" and "why" questions associated with these distributions may allow manipulation of them into distributions favorable to biocontrol efforts.

The isolation, identification and exploitation of ascarosides may allow manipulation of EPNs in ways that will favor biocontrol. Using isolates of these materials from late-stage EPN infections has already shown that dispersal and infection behaviors can be increased. In addition to manipulating EPNs themselves with ascarosides, there may be future efforts to manipulate the insects that are the targets of biocontrol efforts or the physiology of the plants that are to be protected. Because none of these actions happen in a vacuum, care must be taken when deploying such strategies because one effect of an ascaroside on EPNs could be antagonistic to the same ascaroside's impact on the target insect life cycle. As studies continue in this area, the complicated relationships of ascarosides, their intended targets and other members of the system will come to light and may offer new ways to exploit their many different impacts on nematodes, plants and arthropods.

As new species and strains of EPNs are isolated and identified, novel and sometimes useful characteristics will undoubtedly be found. However, as stated above, new isolates must be amenable to laboratory culture, followed by scale-up and formulation before the new characteristics can realistically be exploited for biocontrol. Most research is conducted on a tiny proportion of known EPN species with the most biocontrol potential. As research expands to commercialize more species and strains, new production methods and formulation techniques may allow increased diversity in EPN host ranges, environmental tolerances and soil ecology to enter the market. With these new methods, along with new capabilities to manipulate behavior and physiology, EPN use can diversify and increase.

6.9 References

Ali, F. and Wharton, D.A. (2013) Cold tolerance abilities of two entomopathogenic nematodes, *Steinernema feltiae* and *Heterorhabditis bacteriophora*. *Cryobiology* 66, 24–29.

Ali, J.G., Alborn, H.T. and Stelinski, L.L. (2010) Subterranean herbivore-induced volatiles released by citrus roots upon feeding by *Diaprepes abbreviatus* recruit entomopathogenic nematodes. *Journal of Chemical Ecology* 36, 361–368.

Alsaiyah, M.A.M., Ebssa, L., Zenner, A., O'Callaghan, K.M. and Griffin, C.T. (2009) Sex ratios and sex-biased infection behaviour in the entomopathogenic nematode genus *Steinernema*. *International Journal of Parasitology* 39, 725–734.

Baiocchi, T., Lee, G., Choe, D.-H. and Dillman, A.R. (2017) Host seeking parasitic nematodes use specific odors to assess host resources. *Scientific Reports* 7: 6270.

Baiocchi, T., Braun, L. and Dillman, A.R. (2019) Touch stimulation increases host seeking behavior in *Steinernema carpocapsae*. *Journal of Nematology* 51, 1–5.

Bal, H.K., Acosta, N., Cheng, Z., Grewal, P.S. and Hoy, C.W. (2017) Effect of habitat and soil management on dispersal and distribution patterns of entomopathogenic nematodes. *Applied Soil Ecology* 121, 48–59.

Banerjee, N. and Hallem, E.A. (2020) The role of carbon dioxide in nematode behaviour and physiology. *Parasitology* 147, 841–854.

Blanco-Pérez, R., Bueno-Pallero, F.Á., Neto, L. and Campos-Herrera, R. (2017) Reproductive efficiency of entomopathogenic nematodes as scavengers. Are they able to fight for insect's cadavers? *Journal of Invertebrate Pathology* 148, 1–9.

Bohan, D.A. and Hominick, W.M. (1997) Long-term dynamics of infectiousness within the infective stage pool of the entomopathogenic nematode *Steinernema feltiae* (site 76 strain) Filipjev. *Parasitology* 114, 301–308.

Bruck, D.J., Shapiro-Ilan, D.I. and Lewis, E.E. (2005) Evaluation of application technologies of entomopathogenic nematodes for control of the black vine weevil. *Journal of Economic Entomology* 98, 1884–1889.

Campbell, J.F., Lewis, E.E., Yoder, F. and Gaugler, R. (1996) Entomopathogenic nematode (Heterorhabditidae and Steinernematidae) spatial distribution in turfgrass. *Parasitology* 113, 473–482.

Campbell, J.F., Lewis, E.E., Stock, S.P., Nadler, S. and Kaya, H.K. (2003) Evolution of host search strategies in entomopathogenic nematodes. *Journal of Nematology* 35, 142–145.

Campos-Herrera, R. and Gutiérrez, C. (2014) *Steinernema feltiae* intraspecific variability: infection dynamics and sex-ratio. *Journal of Nematology* 46, 35–43.

Campos-Herrera, R., Trigo, D. and Gutierrez, C. (2006) Phoresy of the entomopathogenic nematode *Steinernema feltiae* by the earthworm *Eisenia fetida*. *Journal of Invertebrate Pathology* 92, 50–54.

Chang, E.E., Huang, M.H., Zhou, X.H., Yu, Y., Wang, C.L. and Li, C.J. (2023) Efficacy of entomopathogenic nematodes combined with insecticides against *Bradysia odoriphaga* larvae. *Nematology* 25, 1–11.

Choo, H.Y. and Kaya, H.K. (1991) Influence of soil texture and presence of roots on host finding by *Heterorhabditis bacteriophora*. *Journal of Invertebrate Pathology* 58, 279–280.

Degenhardt, J., Hiltpold, I., Kollner, T., Frey, M., Gierl, A. *et al.* (2009) Restoring a maize root signal that attracts insect-killing nematodes to control a major pest. *Proceedings of the National Academy of Sciences USA* 106, 13213–13218.

Del Valle, E.E., Dolinski, C., Barreto, E.L.S., Souza, R.M. and Samuels, R.I. (2008) Efficacy of *Heterorhabditis baujardi* LPP7 (Nematoda: Rhabditida) applied in *Galleria mellonella* (Lepidoptera: Pyralidae) insect cadavers to *Conotrachelus psidii* (Coleoptera: Curculionidae) larvae. *Biocontrol Science and Technology* 18, 33–41.

Demarta, L., Hibbard, B.E., Bohn, M.O. and Hiltpold, I. (2014) The role of root architecture in foraging behavior of entomopathogenic nematodes. *Journal of Invertebrate Pathology* 122, 32–39.

Downes, M.J. and Griffin, C.T. (1996) Dispersal behaviour and transmission strategies of the entomopathogenic nematodes *Heterorhabditis* and *Steinernema*. *Biocontrol Science and Technology* 6, 347–356.

El-Borai, F., Killiny, N. and Duncan, L.W. (2016) Consilience in entomopathogenic nematode responses to water potential and their geospatial patterns in Florida. *Frontiers in Microbiology* 7: 356.

Eng, M.S., Preisser, E.L. and Strong, D.R. (2005) Phoresy of the entomopathogenic nematode *Heterorhabditis marelatus* by a non-host organism, the isopod *Porcellio scaber*. *Journal of Invertebrate Pathology* 88, 173–176.

Ennis, D.E., Dillon, A.B. and Griffin, C.T. (2010) Simulated roots and host feeding enhance infection of subterranean insects by the entomopathogenic nematode *Steinernema carpocapsae*. *Journal of Invertebrate Pathology* 103, 140–143.

Erdogan, H., Stevens, G., Stevens, A., Shapiro-Ilan, D., Kaplan, F. *et al.* (2021a) Infected host responses across entomopathogenic nematode phylogeny. *Journal of Nematology* 53: e2021-105.

Erdogan, H., Cruzado-Gutierrez, K., Stevens, G., Shapiro-Ilan, D., Kaplan, F. *et al.* (2021b) Nematodes follow a leader. *Frontiers in Ecology and Evolution* 9. DOI:10.3389/fevo.2021.740351

Forschler, B.T. and Gardner, W.A. (1991) Field efficacy and persistence of enotmogenous nematodes in management of white grubs (Coleoptera, Scarabaeidae) in turn and pasture. *Journal of Economic Entomology* 84, 1454–1459.

Fujimoto, A., Lewis, E.E., Cobanoglu, G. and Kaya, H.K. (2007) Dispersal, infectivity and sex ratio of early- or late-emerging infective juveniles of the entomopathogenic nematode *Steinernema carpocapsae*. *Journal of Nematology* 39, 333–337.

Fushing, H., Zhu, L., Shapiro-Ilan, D.I., Campbell, J.F. and Lewis, E.E. (2008) State-space based mass event-history model I: many decision-making agents with one target. *Annals of Applied Statistics* 2, 1503–1522.

Gaugler, R., Wang, Y. and Campbell, J.F. (1994) Aggressive and evasive behaviors in *Popillia japonica* (Coleoptera: Scarabaeidae) larvae: defenses against entomopathogenic nematode attack. *Journal of Invertebrate Pathology* 64, 193–199.

Georgis, R. and Gaugler, R. (1991) Predictability in biological control using entomopathogenic nematodes. *Journal of Economic Entomology* 84, 713–720.

Georgis, R., Koppenhofer, A.M., Lacey, L.A., Behair, G., Duncan, L.W. *et al.* (2006) Successes and failures in the use of parasitic nematodes for pest control. *Biological Control* 38, 103–123.

Grewal, P.S., Gaugler, R. and Lewis, E.E. (1993a) Host recognition behavior by entomopathogenic nematodes during contact with insect gut contents. *Journal of Parasitology* 79, 495–503.

Grewal, P.S., Selvan, S., Lewis, E.E. and Gaugler, R. (1993b) Male insect-parasitic nematodes: a colonizing sex. *Experientia* 49, 605–608.

Grewal, P.S., Gaugler, R. and Selvan, S. (1997) Host recognition by entomopathogenic nematodes: behavioral response to contact with host feces. *Journal of Chemical Ecology* 19, 1219–1231.

Gulzar, S., Usman, U., Wakil, W., Gulcu, B., Hazir, C. *et al.* (2020) Environmental tolerance of entomopathogenic nematodes differs among nematodes arising from host cadavers versus aqueous suspension. *Journal of Invertebrate Pathology* 175: 107452.

Hallem, E.A., Dillman, A.R., Hong, A.V., Zhang, Y., Yano, J.M. *et al.* (2011) A sensory code for host seeking in parasitic nematodes. *Current Biology* 21, 377–383.

Hang, D.T., Choo, H.Y., Lee, D.W., Lee, S.M., Kaya, H.K. and Park, C.G. (2007) Temperature effects on Korean entomopathogenic nematodes, *Steinernema glaseri* and *S. longicaudum*, and their symbiotic bacteria. *Journal of Microbiology and Biotechnology* 17, 420–427.

Harvey, C.D. and Griffin, C.T. (2016) Local host dependent persistence of the entomopathogenic nematode *Steinernema carpocapsae* used to control the large pine weevil *Hylobius abietis*. *Biocontrol* 61, 185–193.
Hiltpold, I. and Hibbard, B.E. (2018) Indirect root defenses cause induced fitness costs in Bt-resistant western corn rootworm. *Journal of Economic Entomology* 111, 2349–2358.
Hiltpold, I., Baroni, M., Toepfer, S., Kuhlmann, U. and Turlings, T.C. (2010a) Selective breeding of entomopathogenic nematodes for enhanced attraction to a root signal did not reduce their establishment or persistence after field release. *Plant Signaling & Behavior* 5, 1450–1452.
Hiltpold, I., Baroni, M., Toepfer, S., Kuhlmann, U. and Turlings, T.C. (2010b) Selection of entomopathogenic nematodes for enhanced responsiveness to a volatile root signal helps to control a major root pest. *Journal of Experimental Biolology* 213, 2417–2423.
Hiltpold, I., Toepfer, S., Kuhlmann, U. and Turlings, T.C.J. (2010c) How maize root volatiles affect the efficacy of entomopathogenic nematodes in controlling the western corn rootworm? *Chemoecology* 20, 155–162.
Hiltpold, I., Moore, B.D. and Johnson, S.N. (2020) Elevated atmospheric carbon dioxide concentrations alter root morphology and reduce the effectiveness of entomopathogenic nematodes. *Plant Soil* 447, 29–38.
Kanwar, R.S., Patil, J.A. and Yadav, S. (2021) Prospects of using predatory nematodes in biological for plant parasitic nematodes – a review. *Biological Control* 160: 104668.
Kaplan, F., Alborn, H.T., von Reuss, S.H., Ajredini, R., Ali, J.G. *et al.* (2012) Interspecific nematode signals regulate dispersal behavior. *PLOS ONE* 7: e38735.
Kaplan, F., Perret-Gentil, A., Giurintano, J., Stevens, G., Erdogan, H. *et al.* (2020) Conspecific and heterospecific pheromones stimulate dispersal of entomopathogenic nematodes during quiescence. *Scientific Reports* 10: 5738.
Kapranas, A., Maher, A.M.D. and Griffin, C.T. (2016) Higher relatedness mitigates mortality in a nematode with lethal male fighting. *Journal of Evolutionary Biology* 29, 344–351.
Kapranas, A., Zenner, A.N.R.L., Mangan, R. and Griffin, C.T. (2020) Objective and subjective components of resource value in lethal fights between male entomopathogenic nematodes. *Animal Behaviour* 164, 149–154.
Karagoz, M., Gulcu, B., Cakmak, I., Kaya, H.K. and Hazir, S. (2007) Predation of entomopathogenic nematodes by *Sancassania* sp. (Acari: Acaridae). *Entomologica Experimentalis et Applicata* 43, 85–95.
Kaspi, R., Ross, A., Hodson, A.K., Stevens, G.N., Kaya, H.K. and Lewis, E.E. (2010) Foraging efficacy of the entomopathogenic nematode *Steinernema riobrave* in different soil types from California citrus groves. *Applied Soil Ecology* 45, 243–253.
Kin, K., Baiocchi, T. and Dillman, A.R. (2019) Dispersal and repulsion of entomopathogenic nematodes to prenol. *Biology* 8: 58.
Koppenhofer, A.M., Grewal, P.S. and Kaya, H.K. (2000) Synergism of imidacloprid and entomopathogenic nematodes against white grubs: the mechanism. *Entomologica Experimentalis et Applicata* 94, 283–293.
Lewis, E.E. and Gaugler, R. (1994) Entomopathogenic nematode (Rhabdita: Steinernematidae) sex ratio relates to foraging strategy. *Journal of Invertebrate Pathology* 64, 238–242.
Lewis, E.E. and Shapiro-Ilan, D.I. (2002) Host cadavers protect entomopathogenic nematodes during freezing. *Journal of Invertebrate Pathology* 81, 25–32.
Lewis, E.E., Gaugler, R. and Harrison, R. (1992) Entomopathogenic nematode host finding: response to contact cues by cruise and ambush foragers. *Parasitology* 105, 309–315.
Lewis, E.E., Gaugler, R. and Harrison, R. (1993) Response of cruiser and ambusher entomopathogenic nematodes (Steinernematidae) to host volatile cues. *Canadian Journal of Zoology* 71, 765–769.
Lewis, E.E., Grewal, P.S. and Gaugler, R. (1995) Hierarchical order of host cues in parasite foraging strategies. *Parasitology* 110, 207–213.
Lewis, E.E., Ricci, M. and Gaugler, R. (1996) Host recognition behavior reflects host suitability for the entomopathogenic nematode, *Steinernema carpocapsae*. *Parasitology* 113, 573–579.
Lewis, E.E., Campbell, J.F., Griffin, C., Kaya, H.K. and Peters, A. (2006) Behavioral ecology of entomopathogenic nematodes. *Biological Control* 38, 66–79.
Liu, T., Hu, F. and Li, H. (2019) Spatial ecology of soil nematodes: perspectives from global to micro scales. *Soil Biology and Biochemistry* 137: 107565.
Manosalva, P., Manohar, M., von Reuss, S.H., Chen, S.Y., Koch, A. *et al.* (2015) Conserved nematode signalling molecules elicit plant defenses and pathogen resistance. *Nature Communications* 6: 7795.
Mbata, G.N., Li, Y.P. and Shapiro-Ilan, D. (2022) Evaluation of chemical and microbial control options for *Pangaeus bilineatus* (Say) (Hemiptera: Cydnidae) infecting peanut crop. *Pest Management Science* 78, 4719–4727.

Nielsen, A., Spence, K., Nakatani, J. and Lewis, E.E. (2011) Effect of soil salinity on entomopathogenic nematode survival and behaviour. *Nematology* 13, 859–867.

O'Callaghan, K.M., Zenner, A.N.R.L., Hartley, C.J. and Griffin, C.T. (2014) Interference competition in entomopathogenic nematodes: male *Steinernema* kill members of their own and other species. *International Journal of Parasitology* 44, 1009–1017.

Oliveira-Hofman, C., Kaplan, F., Stevens, G., Lewis, E.E., Wu, S.H. *et al.* (2019) Pheromone extracts act as boosters for entomopathogenic nematode efficacy. *Journal of Invertebrate Pathology* 164, 38–42.

Perez, E.E., Lewis, E.E. and Shapiro-Ilan, D.I. (2003) Impact of the host cadaver on survival and infectivity of entomopathogenic nematodes (Rhabditida: Steinernematidae and Heterorhabditidae) under desiccating conditions. *Journal of Invertebrate Pathology* 82, 111–118.

Ramos-Rodriguez, O., Campbell, J.F., Christen, J.M., Shapiro-Ilan, D.I., Lewis, E.E. and Ramaswamy, S.B. (2007) Attraction behaviour of three entomopathogenic nematode species towards infected and uninfected hosts. *Parasitology* 134, 729–738.

Rasmann, S., Köllner, T., Degenhardt, J., Hiltpold, I., Toepfer, S. *et al.* (2005) Recruitment of entomopathogenic nematodes by insect-damaged maize roots. *Nature* 434, 732–737.

Rolston, A.N., Griffin, C.T. and Downes, M.J. (2006) Emergence and dispersal patterns of two isolates of the entomopathogenic nematode *Steinernema feltiae*. *Journal of Nematology* 38, 221–228.

Ruan, W.-B., Shapiro-Ilan, D.I., Lewis, E.E., Kaplan, F., Alborn, H. *et al.* (2018) Movement patterns in entomopathogenic nematodes: continuous vs. temporal. *Journal of Invertebrate Pathology* 151, 137–143.

Sabino, P.H.S., Negrisoli, A.S., Andalo, V., Filgueiras, C.C., Moino, A. and Sales, F.S. (2019) Combined application of entomopathogenic nematodes and insecticides in the control of leaf miner *Tuta absoluta* (Meyrick) (Lepidoptera: Gelechiidae) on tomato. *Neotropical Entomlogy* 48, 314–322.

Shapiro, D.I. and Glazer, I. (1996) Comparison of entomopathogenic nematode infectivity from infected hosts versus aqueous suspension. *Environmental Entomology* 25, 1455–1461.

Shapiro, D.I. and Lewis, E.E. (1999) Comparison of entomopathogenic nematode infectivity from infected hosts versus aqueous suspension. *Environmental Entomology* 28, 907–911.

Shapiro, D.I., Tylka, G.L., Berry, E.C. and Lewis, L.C. (1995) Effects of earthworms on the dispersal of *Steinernema* spp. *Journal of Nematology* 27, 21–28.

Shapiro-Ilan, D.I. and Cottrell, T.E. (2006) Susceptibility of the lesser peachtree borer (Lepidoptera: Sesiidae) to entomopathogenic nematodes under laboratory conditions. *Environmental Entomology* 35, 358–365.

Shapiro-Ilan, D.I., Lewis, E.E., Son, Y. and Tedders, W.L. (2003) Superior efficacy observed in entomopathogenic nematodes applied in infected-host cadavers compared with application in aqueous suspension. *Journal of Invertebrate Pathology* 83, 270–272.

Shapiro-Ilan, D.I., Lewis, E.E., Campbell, J.F. and Kim-Shapiro, D.B. (2012) Directional movement of entomopathogenic nematodes in response to electrical field: effects of species, magnitude of voltage, and infective juvenile age. *Journal of Invertebrate Pathology* 109, 34–40.

Shapiro-Ilan, D.I., Lewis, E.E. and Schliekelman, P. (2014) Aggregative group behavior in insect parasitic nematode dispersal. *International Journal of Parasitology* 44, 49–54.

Shapiro-Ilan, D.I., Cottrell, T.E., Mizell, R.F. and Horton, D.L. (2016) Effiaccy of *Steinernema carpocapsae* plus fire gel applied as a single spray for control of the lesser peachtree borer, *Synanthedon pictipes*. *Biocontrol* 94, 33–36.

Shapiro-Ilan, D.I., Kaplan, F., Oliveira-Hofman, C., Schliekelman, P., Alborn, H.T. and Lewis, E.E. (2019) Conspecific pheromone extracts enhance entomopathogenic activity. *Journal of Nematology* 51, 1–5.

Sih, A., Bell, A.M., Johnson, C. and Ziemba, R.E. (2004) Behavioral syndromes: an integrative overview. *Quarterly Review of Biology* 79, 241–277.

Simoes, N., Laumond, C. and Bonifassi, E. (1993) Effectiveness of *Steinernema* spp. and *Heterorhabditis bacteriophora* against *Popillia japonica* in the Azores. *Journal of Nematology* 25, 480–485.

Spence, K., Stevens, G., Arimoto, H., Ruiz-Vega, J., Kaya, H. and Lewis, E. (2010) Effect of insect cadaver desiccation and soil water potential during rehydration on entomopathogenic nematode (Rhabditida: Steinernematidae and Heterorhabditidae) production and virulence. *Journal of Invertebrate Pathology* 106, 268–273.

Srinivasan, J., Kaplan, F., Ajredini, R., Zachariah, C., Alborn, H.T. *et al.* (2008) A blend of small molecules regulates both mating and development in *Caenorhabditis elegans*. *Nature* 454, 1115–1118.

Stevens, G.N., Erdogan, H., Pimentel, E., Dotson, J., Stevens, A. *et al.* (2023) Group joining behaviors in the entomopathogenic nematode *Steinernema glaseri*. *Biological Control* 181: 105220.

Stuart, R.J. and Gaugler, R. (1994) Patchiness in populations of entomopathogenic nematodes. *Journal of Invertebrate Pathology* 64, 39–45.

Stuart, R.J., Hatab, M.A. and Gaugler, R. (1998) Sex ratio and the infection process in entomopathogenic nematodes: are males the colonizing sex? *Journal of Invertebrate Pathology* 72, 288–295.

Torr, P., Heritage, S. and Wilson, M.J. (2004) Vibrations as a novel signal for host location by parasitic nematodes. *International Journal of Parasitology* 34, 997–999.

Ulug, D., Hazir, S., Kaya, H.K. and Lewis, E.E. (2014) Natural enemies of natural enemies: the potential top-down impact of predators on entomopathogenic nematodes. *Ecological Entomology* 39, 462–469.

Wang, Y., Gaugler, R. and Cui, L.W. (1994) Variations in immune response of *Popillia japonica* and *Acheta domesticus* to *Heterorhabditis bacteriophora* and *Steinernema* species. *Journal of Nematology* 26, 11–18.

Wilson, M.J., Ehlers, R.U. and Glazer, I. (2012) Entomopathogenic nematode foraging strategies – is *Steinernema carpocapsae* really an ambush forager? *Nematology* 14, 389–394.

Wong, C., Oliveira-Hofman, C., Blaauw, B., Chavez, D.J., Jagdale, G. *et al.* (2022) Control of peachtree borer (*Synanthedon exitiosa*) using the nematode *Steinernema carpocapsae*: optimization of application rates and secondary benefits in control of root-feeding weevils. *Agronomy* 12: 2689.

Zenner, A.N.R.L., O'Callaghan, K.M. and Griffin, C.T. (2014) Lethal fighting in nematodes is dependent on developmental pathway: male–male fighting in the entomopathogenic nematode *Steinernema longicaudum*. *PLOS ONE* 9: e89385.

Zhao, L., Zhang, X., Wei, Y., Jhou, J., Zhang, W. *et al.* (2016) Ascarosides coordinate the dispersal of plant parasitic nematode with the metamorphosis of its vector beetle. *Nature Communications* 7: 12341.

7 Population Biology of Entomopathogenic Nematodes: An Ecological and Agronomical Model

Rubén Blanco-Pérez[1], **Glen Stevens**[2] **and Raquel Campos-Herrera**[3]*
[1]*Soils, Biosystems and Agroforestry Ecology Department, Misión Biológica de Galicia (BMG-CSIC), Pontevedra, Spain;* [2]*Department of Entomology, Plant Pathology and Nematology, University of Idaho, Moscow, Idaho, USA;* [3]*Instituto de Ciencias de la Vid y del Vino (ICVV), Gobierno de La Rioja, CSIC, Universidad de La Rioja, Logroño, Spain*

Abstract

Entomopathogenic nematodes (EPNs) are well-known beneficial organisms of crop soils. Expanding our knowledge of the factors determining their distribution in agroecosystems is essential to enhance the ecosystem service that EPNs provide as biological control agents. In this sense, regional surveys have improved the interpretation of habitat preferences and temporal patterns of specific populations, native or applied, in crop soils. Using traditional sampling methods (i.e. insect baits) with the support of molecular tools (e.g. quantitative polymerase chain reaction analysis), field studies have allowed the selection of the main factors driving EPN occurrence worldwide with increasing accuracy. Improved statistical and geostatistical analyses have also helped, for instance to describe the dispersal behavior of EPNs or to connect them with abiotic factors and other soil organisms. All these improvements enable the assembly of predictive ecological models to enrich our knowledge of the distribution and spatiotemporal dynamics of EPN populations within agroecosystems. Such processes facilitate using EPNs as indicators of soil health and as ecological models to select, for example, more sustainable agricultural practices. Future approaches will implement more powerful molecular tools with a finer taxonomic resolution, such as high-throughput sequencing. These advances will serve to optimize the analysis of soil organisms in cryptic environments and to unravel the complexity of the multi-trophic interactions they establish with each other. Ultimately, a more precise understanding of the population biology of biological control agents will facilitate prediction of the efficacy of organisms such as EPNs in agricultural soils.

7.1 Introduction

Entomopathogenic nematodes (EPNs) of the families Steinernematidae and Heterorhabditidae are well-known pathogens of soil-dwelling arthropods (Adams *et al.*, 2006), and their unusual life cycle makes them excellent models for applied ecological and agronomic studies (Campos-Herrera *et al.*, 2012a). The non-feeding infective juvenile (IJ) stage aims to locate and penetrate suitable hosts to complete their life cycle. Pathogenesis is achieved with the assistance of mutualistic enteric γ-proteobacteria in the genera *Xenorhabdus* and *Photorhabdus* (Enterobacterales:

*Email: raquel.campos@icvv.es

 Entomopathogenic Nematodes as Biological Control Agents (D.I. Shapiro-Ilan and E.E. Lewis eds)
DOI: 10.1079/9781800620322.0007

Morganellaceae) for steinernematids and heterorhabditids, respectively (Stock, 2015; Adeolu *et al.*, 2016). IJs must survive hostile abiotic conditions, avoid predators and pathogens, and fight against competitors for the resources within the arthropod cadaver. Likewise, the nematode–bacteria complex must overcome other difficulties within the cadaver, such as fights against inter- and intraspecific EPNs, scavengers and other entomopathogens (Gulcu *et al.*, 2012; Griffin, 2015; Bueno-Pallero *et al.*, 2018). Furthermore, the efficiency of the numerous commercial products based on EPNs available for biological control programs depends on many biotic and abiotic factors, such as crop type, soil characteristics and the availability of target hosts (Lacey *et al.*, 2015).

Approaching the study of EPNs from the perspective of a population is essential for a comprehensive understanding of the ecosystem services they provide as biological control agents. A population is defined most broadly as a group of individuals belonging to the same species that coexist in a specific place and time, with the ability to reproduce or interbreed with each other. How EPN populations persist over time is subject to intrinsic (e.g. genetic diversity, specific searching/migration abilities, relationship with their symbiont) and extrinsic (e.g. environmental constraints, inter- and intraspecific competition, availability of suitable hosts) forces. This chapter summarizes the critical factors contributing to EPN population dynamics at multiple scales and the abiotic and biotic factors affecting their ecology. In addition, we will briefly describe some traditional and molecular techniques employed in studying EPN populations, giving some examples of their application in an agronomical context. As a final remark, we focus on the nematode species in the genera *Steinernema* and *Heterorhabditis*, the so-called traditional EPNs, for which commercial biocontrol formulations are already available (Lacey *et al.*, 2015; Abate *et al.*, 2017). Hence, we will exclude species in the genus *Oscheius* or other genera (Dillman *et al.*, 2012; Zhang *et al.*, 2019).

7.2 Habitat Preference and Temporal Patterns

EPNs exhibit patchy distributions throughout spatial scales, a trait they share with all nematodes (Liu *et al.*, 2019). Many factors likely underlie the irregular distribution of EPNs at microscales, a subject covered in Chapter 6 (this volume). EPNs exist in virtually all terrestrial ecosystems where insects might co-occur, except Antarctica, from natural to agricultural areas, from low to high altitudes, and from healthy to heavy metal-polluted mine soils (Campos-Herrera *et al.*, 2016b; Kergunteuil *et al.*, 2016). Over the decades, numerous surveys have been conducted to expand our understanding of EPN distribution and find new EPN species. In particular, Hominick (2002) compiled and organized the EPN global distribution, mainly from the 1990s to the early 2000s. Since then, other studies have expanded on this global catalog (Abate *et al.*, 2017; San-Blas *et al.*, 2019). The most recent effort recognized 102 *Steinernema* spp. and 22 *Heterorhabditis* spp. (Hazir *et al.*, 2022). However, since then, the new species *Steinernema africanum*, *Steinernema populi*, *Heterorhabditis ruandica* and *Heterorhabditis zacatecana* have been identified (Machado *et al.*, 2021, 2022; Tian *et al.*, 2022). This section summarizes the current EPN distribution, spatially and by species. First, we can discriminate between species of limited geographic distribution, such as *Steinernema ritteri* from Argentina, *Steinernema fabii* from South Africa, *Steinernema huense* from Vietnam and *Steinernema riobrave* from the Rio Grande region of the North America, and ubiquitous species present in almost every continent, particularly *Steinernema feltiae*, *Steinernema carpocapsae*, *Heterorhabditis bacteriophora*, and *Heterorhabditis indica* (Hominick, 2002). Today, geostatistical techniques (e.g. the ecological niche model) can determine whether a particular species may be endemic to a region or have a broader distribution in unexplored areas. For example, Kour (2017) determined probable *H. indica* occurrence on the south coast of the Iberian Peninsula by analyzing its ecoclimatic suitability under the current climate scenario, which was confirmed in subsequent surveys using molecular tools for EPN identification (Campos-Herrera *et al.*, 2016b, 2019a).

Regional surveys contribute to the general knowledge of EPN occurrence in specific regions or habitats. For example, Mráček *et al.* (2005) conducted a comprehensive study in the Czech Republic to assess through insect baiting the effect of many different factors on EPN prevalence, such as ecosystem type, soil properties, season, altitude and predominance of insect hosts. The authors found EPN species in all ecosystems and habitats but in higher abundance in those with high quantities of suitable hosts, mainly sandy soils and arboreal habitats. Based on a similar approach, a study conducted in Southern Cameroon showed that EPN abundance increased with longitude and soil pH, and decreased with altitude (Kanga *et al.*, 2012). Similarly, regional surveys of the two primary ecoregions of the Florida peninsula revealed that soil properties such as soil moisture, clay and organic matter contents contributed to the regulation of native EPNs and their soil food web in citrus orchards and natural areas (Campos-Herrera *et al.*, 2013a, 2016a). In addition, the same abiotic factors explained the differential occurrence of suitable EPN hosts across ecoregions, such as the critical pest the citrus root weevil, *Diaprepes abbreviatus* (Coleoptera: Curculionidae). Furthermore, studies in temperate areas of Switzerland and southern Portugal revealed similar drivers for the EPN community and soil food web, highlighting the influence of the vegetation type, soil pH and clay content (Jaffuel *et al.*, 2018; Campos-Herrera *et al.*, 2019a).

Investigating EPN population fluxes at various scales may help us to identify niche selection and the factors that drive them. For example, vertical migration in the soil profile may help EPNs avoid extreme temperatures or lack of moisture (García Del Pino and Palomo, 1997; Campos-Herrera *et al.*, 2010). Indeed, most regional surveys in temperate climates provide a "snapshot" analysis of samples collected in spring or fall, when EPNs presumably occur in higher numbers (Griffin *et al.*, 1991; Efron *et al.*, 2001). Species-specific seasonal migrations of EPNs through the soil profile have been observed in artificially augmented soil. For example, the dispersion of IJs for inundative EPN applications in alfalfa fields in northern New York depended on the soil type and EPN genera, with steinernematids predominantly present at the soil surface up to 15 cm and heterorhabditids above 35 cm depth (Ferguson *et al.*, 1995). Moreover, differences in the vertical migration of IJs in wintertime can also be related to land uses. For example, *H. bacteriophora* avoided stressful moisture levels by migrating deeper, while desiccation-tolerant species such as *S. carpocapsae* remained at the soil surface (Susurluk, 2009; Salame and Glazer, 2015). EPNs can also move deeper to escape predation and potential competition. For example, some nematophagous fungi commonly inhabit shallower soil horizons, an area often avoided by the EPN species *Steinernema diaprepesi* and *H. indica* in Florida citrus orchards (Fig. 7.1) (Campos-Herrera *et al.*, 2019b). The same study revealed that an ectoparasitic bacterium in the genus *Paenibacillus* exhibited significant phase–space, predator–prey dynamics associated with *S. diaprepesi* occurrence, an EPN species described as a species-specific host for this particular bacterium (El-Borai *et al.*, 2005). Horizontal dispersal may also contribute to EPN coexistence, as the patterns of sympatric EPN species *Steinernema affine* and *Steinernema kraussei* in central European oak woodlands in the Czech Republic resulted from horizontal patchiness rather than vertical distribution (Půža and Mráček, 2010b).

Exogenous disturbances of agroecosystems can modulate the natural distribution of EPNs. Field experiments have shown higher abundances and activity of augmented and native EPNs in organic farming than in conventional agriculture (Ellers-Kirk *et al.*, 2000; Blanco-Pérez *et al.*, 2022a). In contrast, a comprehensive study of different annual cropping systems in Switzerland resulted in no differences between pest management practices, either for EPN populations or for organisms linked to their soil food web (Jaffuel *et al.*, 2016). It seems that a stable rhizosphere can enhance the EPN community, as Campos-Herrera *et al.* (2008, 2010) reported similar EPN activity rates in natural areas (oaks) and organic farming of perennial crops (vineyards and pear orchards) in La Rioja (northern Spain) but much lower activity in conventional agriculture and annual crops. Roots can serve as pathways for EPNs (van Tol *et al.*, 1998; Ennis *et al.*, 2010), and herbivory damage can activate the emission of volatiles from roots that are attractive to IJs in search of hosts (Rasmann *et al.*, 2005; Ali *et al.*, 2010; Hiltpold *et al.*, 2011). However, reduced mechanization and the benefits of soil cover strategies can enrich the EPN community by protecting IJs from harsh conditions and increasing the occurrence of suitable hosts (Shapiro-Ilan *et al.*, 1999; Susurluk and Ehlers, 2008). The results will depend, among other things, on the EPN species, soil properties,

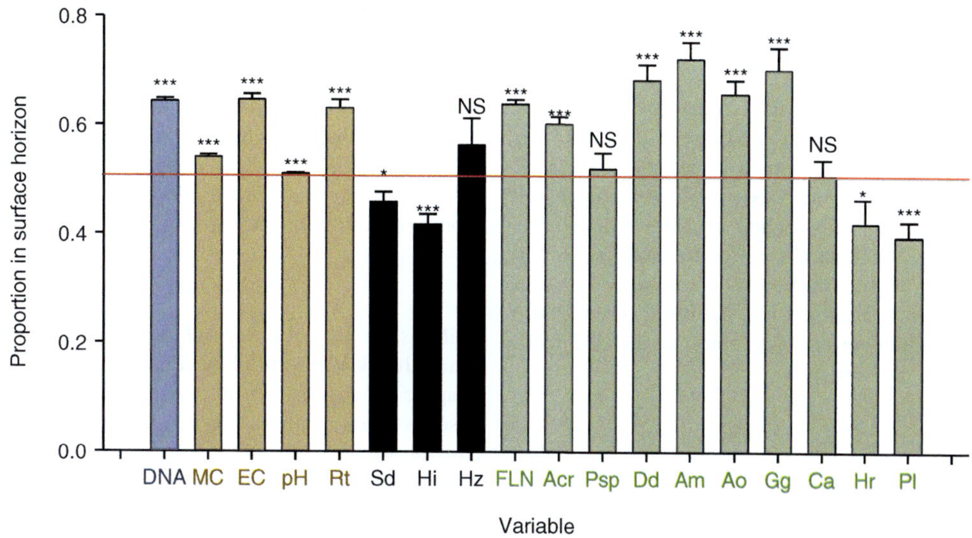

Fig. 7.1. Depth distribution of total DNA (blue bar) and soil properties (yellow bars), entomopathogenic nematodes (EPNs) (black bars) and potential EPN antagonists (green bars). MC, soil moisture content; EC, electrical conductivity; pH, soil pH; Rt, fibrous root mass density, Sd, *Steinernema disprepesi*; Hi, *Heterorhabditis indica*; Hz, *H. zealandica*; FLN, total free living nematodes; Acr, *Acrobeloides* group nematodes; Psp, *Paenibacillus* sp.; Dd, *Drechslerella dactyloides*; Am, *Arthrobotrys musiformis*; Ao, *Arthrobotrys oligospora*; Gg, *Gamsylella gephyropaga*; Ca, *Catenaria anguillulae*; Hr, *Hirsutella rhossiliensis*; Pl, *Purpureocillium lilacinum*. Asterisks indicate proportions significantly larger or smaller than 50%: *, $P = 0.05$; **, $P = 0.01$; ***, $P = 0.001$. Reproduced from Campos-Herrera *et al.* (2019b) with permission.

type and quality of cover crops and amendments, as well as effects on natural enemies and potential competitors (Millar and Barbercheck, 2002; Jaffuel *et al.*, 2017; Blanco-Pérez *et al.*, 2020, 2022b; Khumalo *et al.*, 2021).

Assessing the natural EPN distribution at a regional scale has important practical implications for their regulation and commercialization (Abate *et al.*, 2017). Currently, numerous commercial products based on EPNs are available worldwide with varying levels of demand (reviewed by Lacey *et al.*, 2015; Abate *et al.*, 2017) but involving only a relatively few species. The most common are *S. feltiae*, *S. carpocapsae* and *H. bacteriophora*, with only a few companies commercializing products based on other species (Abate *et al.*, 2017). Furthermore, inundative application, which is the most prominent application approach, typically involves releasing high rates of IJs into the target area with the expectation that reapplication will likely be necessary (Duncan *et al.*, 2003a; Kapranas *et al.*, 2016). The release of commercial EPNs could displace or eventually extinguish native EPNs (Bathon, 1996; Ehlers and Hokkanen, 1996; Millar and Barbercheck, 2001), and also can potentially be spread to other unexpected habitats. For example, *Steinernema scapterisci*, originally from Uruguay, was introduced to Florida to control mole crickets (*Scapteriscus* spp.) in grasslands (Parkman *et al.*, 1993; Parkman and Smart, 1996) and was found two decades later in citrus orchards all over the Florida peninsula (Campos-Herrera *et al.*, 2013a, 2016a). The effects of this practice on native communities and their assemblage with other soil organisms are still largely unknown, making potential changes in the agroecosystem unpredictable.

7.3 Factors Affecting the Natural Distribution of Entomopathogenic Nematodes

7.3.1 Soil abiotic factors

Numerous laboratory or greenhouse experiments have assessed the effect of abiotic factors on the efficacy and prevalence of EPNs. However, as controlled studies tend to exclude unwanted or unknown variables, it is necessary to combine them with field studies to cover, at least in part, soil complexity. Stuart *et al.* (2015) has provided a comprehensive review of the effects of the main abiotic factors on EPN occurrence and distribution. This section provides a brief overview of the impact of environmental factors and soil properties on the EPN community, focusing on their application in agricultural systems, and expanding on specific topics, as needed, with recent research.

7.3.1.1 Temperature

Suboptimal temperature directly (e.g. rates of physiological activity) and indirectly (e.g. evaporation rates and nutrient diffusion) affects all soil organisms. Although the soil acts as a thermal buffer, any variation can affect, for instance, the life cycle of nematodes, slowing or accelerating their development. The optimal thermal niche breadth for EPNs is species specific but is generally more limiting for reproductive success than for virulence (Grewal *et al.*, 1994). Thus, the cold-adapted species *S. feltiae* can only reproduce at 10–25°C, while infection of hosts occurs between 8°C and 30°C. Warm-adapted EPN species such as *S. riobrave* and *Heterorhabditis floridensis* reproduce at 20–35°C and 17–35°C, respectively, but penetration of hosts occurs between 10°C and 39°C (Grewal *et al.*, 1994; Shapiro-Ilan *et al.*, 2014). EPNs can exhibit different mechanisms to survive extreme temperatures (often considered as below 0°C and above 32°C), including trehalose accumulation for cold tolerance (Brown and Gaugler, 1997) or expression of heat-shock proteins for heat tolerance (Selvan *et al.*, 1996; Grewal *et al.*, 2006). Moreover, specific populations isolated from extreme habitats such as deserts have evolved a tolerance for survival (Glazer *et al.*, 1991; Shamseldean and Abd-Elgawad, 1991). Interestingly, traits related to heat tolerance were observed in a hybrid *H. bacteriophora* derived from two populations: a heat-tolerant population isolated in Israel (IS5 strain) mixed with the HP88 strain population (Shapiro-Ilan *et al.*, 1997). Current research has established genetic-based selection tools for beneficial traits related to heat and drought tolerance (Levy *et al.*, 2020). Combining heat and desiccation bioassays with screening of selected genes in various isolates of *H. indica* and *H. bacteriophora* revealed several known stress-related genes and pathways expressed uniquely in these isolates. This line of research provides the basis for further improvement of EPN lines with better survive in heat and desiccation scenarios of particular relevance to climate change and global warming.

7.3.1.2 Moisture

Water is the solvent in the soil for nutrients and other chemicals essential for the metabolism of edaphic organisms, but it also affects the soil gaseous fraction and temperature. Nematodes can only inhabit water-filled pores (Neher, 2010) with adequate surface tension to enable movement (Wallace, 1971). Indeed, Schlüter *et al.* (2022) observed that the most satisfactory soil structure predictor for nematode abundance was the soil porosity (measured with X-ray computed tomography), particularly in the size range of 20–220 μm. Many laboratory studies have shown a species-specific (or even population-specific) capability of EPNs to survive desiccation when exposed to extreme conditions (Kung *et al.*, 1991; Salomon and Glazer, 1999; Koppenhöfer and Fuzy, 2007; Shapiro-Ilan *et al.*, 2014). EPNs can survive unfavorable conditions such as desiccation for 2–3 weeks using different strategies (Kung *et al.*, 1990b). Nevertheless, IJs in dry soils can coil or aggregate (Glazer, 2002) or switch to states of quiescence or anhydrobiosis (Womersley, 1990), increasing trehalose, decreasing glycogen, or synthesizing specific late embryogenesis abundant (LEA) proteins related to water stress (Solomon *et al.*, 2000). In contrast, for aerial EPN applications, nematode survival on exposed surfaces is quite limited, depending on the environmental temperature and relative humidity, and can be restricted to as little as a few hours (Glazer, 1992). Field studies have also provided evidence of the critical importance of soil moisture on EPN distribution and assemblage with other soil members. As an illustration, regional surveys in subtropical (Florida), temperate (Switzerland) and Mediterranean (Portugal) areas showed that the abiotic variables that better explained the distribution of EPNs and associated soil organisms were those related to soil moisture, such as clay content and available water (Campos-Herrera *et al.*, 2013a, 2016a, 2019a; Jaffuel *et al.*, 2018). Similarly,

Preisser and Strong (2004) considered soil moisture a key player in insect suppression by EPNs in a coastal prairie in California, and Williams *et al.* (2013) found it to be a critical factor that modulates EPN efficacy when applied in peaty versus mineral soils in Ireland. In conclusion, soil moisture must be considered to improve the activity and long-term persistence of EPNs for IJ applications in biological control programs (Campos-Herrera *et al.*, 2013a).

7.3.1.3 Texture

Soil texture is one of the most studied factors affecting EPN activity and distribution (Stuart *et al.*, 2015). Defined by the proportion of the particles according to their dimensions, soil texture distinguishes sand, silt and clay particles from larger to smaller sizes. Low porosity is associated with soils with high clay content, which hinders aeration and the movement of soil organisms, whereas sandy soils are more conducive to aeration and nematode movement (Barbercheck, 1992). Indeed, many studies support the concept that heavy soils reduce EPN mobility and virulence (Kung *et al.*, 1990b; Lankin *et al.*, 2020). Similarly, the magnitude of the clay fraction was one of the main factors driving EPN natural occurrence and their assemblage with other soil organisms in surveys conducted in different climate zones, greenhouse and field studies (El-Borai *et al.*, 2012; Campos-Herrera *et al.*, 2013a, 2016a, 2019a; Jaffuel *et al.*, 2018; Blanco-Pérez *et al.*, 2022a). Moreover, soil texture can alter the diffusion patterns of specific volatile organic compounds, such as (E)-β-caryophyllene and pregeijerene, which plants use as cry-for-help signals (Chiriboga *et al.*, 2017; Som *et al.*, 2017). Therefore, the stimuli from a herbivore-damaged plant will be lower and travel less distance in clay soils than in sandy soils, potentially reducing the efficacy of EPNs as biological control agents in these systems. In addition, soil structure, determined by the formation of aggregates (soil particles with variable size, geometry and porosity), provides optimal environments for developing edaphic organisms that affect EPN activity (Portillo-Aguilar *et al.*, 1999). The pores in soil aggregates constitute the habitat of most soil nematodes (with a size of ~2 mm diameter for macroaggregates) and fungi and other multicellular organisms, including smaller nematodes (<1 mm in diameter for microaggregates), whereas most soil bacteria inhabit micropores occurring within soil aggregates (Elliott and Coleman, 1988; Schlüter *et al.*, 2022). Environmental, biological and physical factors, including human activities such as agricultural management, affect soil structure and can consequently alter the nematofauna. For example, soil compaction by agricultural machinery can impact EPN behavior, making different populations more or less suitable for use as biological control agents (Kapranas *et al.*, 2017).

7.3.1.4 Oxygen content

Soil aeration depends on the porosity and water content. The soil oxygen range, for instance, decreases more rapidly with depth in clay or silty soils than in sandy soils (Brady and Weil, 2013). Several studies have shown that EPNs require adequate oxygen levels for optimal activity. For example, Kung *et al.* (1990a) studied the survival and virulence of the EPN species *S. glaseri* and *S. carpocapsae* in sandy loam soil over a range of oxygen concentrations from 1% to 20% for 16 weeks. After 2 weeks, the nematodes that remained at less than 10% oxygen concentration could not penetrate *Galleria mellonella* larvae; at week 16, IJs lost all ability to colonize and kill hosts at 20% oxygen concentration. Moreover, Qiu and Bedding (2000) observed that under anaerobiosis, *S. carpocapsae* maintained regular activity for only 1 week, and the time required for recovery was proportional to the period subject to stress.

7.3.1.5 pH and heavy metals

Soil pH can also affect the effectiveness of EPNs as biological control agents (Barbercheck, 1992). Laboratory studies have shown that EPNs, particularly steinernematids, survive over a broad pH range (from 2 to 11) (Kung *et al.*, 1990a; Khathwayo *et al.*, 2021). Although these laboratory studies are informative, the potential associations that hydrogen (H^+) cations can establish with other ions in field soils will determine EPN survival in nature. For example, clay soils with high levels of exchangeable H^+ ions are considered unstable and potentially toxic due to heavy metals because these cations can dissolve and release aluminum, magnesium and silicon (Brady and

Weil, 2013). Some studies have determined the species-specific toxicity of different heavy metals against specific EPN species, some of them even beneficial for EPN virulence and reproduction (Jaworska and Gospodarek, 2009), which could explain the detection of particular EPN species in areas highly polluted with heavy metals (Campos-Herrera *et al.*, 2016b). It is worth noting, however, that dissolved ions are heterogeneously distributed throughout the soil profile, so individual sites could differ highly from the average pH recorded for a given area. In addition to high levels of pesticide residues and other emerging contaminants that may negatively affect EPN persistence in soils (Campos-Herrera *et al.*, 2008; Gutiérrez *et al.*, 2016), soil applications of EPNs should consider the presence of heavy metals to understand possible limitations in their efficiency. Fortunately, EPN applications are compatible with pH regulators, mineral fertilizers, growth stimulators, vitamins and pigment supplements (Kolombar *et al.*, 2020). Hence, there is an opportunity to formulate these products with EPNs to optimize their use in the agroecosystems.

7.3.2 Biotic factors

Soils are perhaps the most complex and biodiverse of the ecosystems on Earth, comprising organisms ranging from microscopic size (archaea, bacteria, oomycetes and fungi) to meso/macrofauna (e.g. nematodes, mites, collembolans, earthworms, snails and moles), (Orgiazzi *et al.*, 2016). In addition, roots establish chemical pathways that connect soil fauna with the aboveground system (Cavigelli *et al.*, 2012). As part of the soil biota, EPNs interact in complex soil food webs through positive (additive or synergistic), negative (competitive, predatory or antagonistic) or neutral relationships (Preisser, 2003; Duncan *et al.*, 2007; Ram *et al.*, 2008). Other chapters in this volume describe the behavioral ecology of EPNs (Chapter 6) and the antagonistic (Chapter 8) and positive interactions (Chapter 27) they establish with various soil organisms. Here, we briefly review the effects of different soil organisms on EPN occurrence and virulence, emphasizing the agronomic implications highlighted in recent studies. Additional information can be found in the referenced chapters and in the description of intra- and interspecific competition among EPN species, as exemplified by Stuart *et al.* (2015).

7.3.2.1 *Suitable hosts*

The abundance and distribution of suitable hosts are primary factors in regulating EPN population dynamics. Although laboratory studies have suggested a broad range of hosts for EPNs (Peters, 1996), it is more restricted under natural conditions due to differences in spatial or temporal distribution. Field studies have shown a correlation between EPN abundance and host availability (Mráček *et al.*, 2005; Půža and Mráček, 2010b). However, higher EPN numbers are not always associated with a high abundance of suitable hosts. For example, the weevil species *Diaprepes abbreviatus* is more abundant and damaging in citrus groves in the Flatwoods (fine sand and poorly drained soils) than in the Central Ridge (coarse sand and well-drained soils) of Florida (Futch *et al.*, 2005), but Campos-Herrera *et al.* (2013a) found no differences in the total EPN numbers in the two regions.

In contrast, laboratory experiments have demonstrated that EPNs can act as scavengers (San-Blas and Gowen, 2008; Půža and Mráček, 2010a). The extent to which this behavior may help them persist in the agroecosystem requires further research. Recent experiments have reported lower numbers and reduced virulence of IJs emerged from scavenged insect cadavers in the presence of other opportunistic soil organisms such as free-living nematodes (Blanco-Pérez *et al.*, 2019). Finally, there is evidence that EPN infections can have sublethal effects in non-target hosts. For example, the exposure of 13 EPN species/populations to the non-susceptible pupal stage of the onion fly, *Delia antiqua*, showed that the nematode infection altered their developmental periods and increased the risk of mortality (Filgueiras and Willett, 2021). This new venue for EPN applications as life-cycle modulators raises new considerations for studying the impact of hosts on EPN population ecology.

7.3.2.2 *Positive interactions*

EPNs establish positive interactions with many soil organisms whose additive or synergistic effects enhance their long-term persistence in soils (Shapiro-Ilan *et al.*, 2004; Ansari *et al.*, 2010).

Many studies have reported synergistic effects between EPNs and insect-pathogenic viruses, bacteria, fungi and parasitoids, resulting in increased host susceptibility to EPN attack, reducing, for instance, the period of lethal infection (Kaya and Koppenhöfer, 1996; Parkman and Frank, 2002; Ansari *et al.*, 2004; Navarro *et al.*, 2014). An excellent example of synergy is the combination of EPNs and other entomopathogens, such as entomopathogenic fungi (Ansari *et al.*, 2010; Shaurub *et al.*, 2016; Gulzar *et al.*, 2021). This synergy is strongly dependent on the target hosts (species/developmental stage) and the timing and rate of entomopathogen applications (Acevedo *et al.*, 2007; Tarasco *et al.*, 2011; Hajek and Meyling, 2018). Regarding macroinvertebrates, earthworms promote the dispersal of EPNs by soil mixing and phoresy (Shapiro-Ilan *et al.*, 1993, 1995; Shapiro-Ilan and Brown, 2013). Root exudates can also guide IJs to hosts (Rasmann *et al.*, 2005; Ali *et al.*, 2012), and a recent report showed evidence of interaction between symbiotic bacteria of EPNs and roots (Regaiolo *et al.*, 2020). These promising results encourage the exploration of this type of association with plants as a plant growth-promoting association, for example, by regulating belowground multi-trophic interaction cascades to enhance the activity of soil biological control agents while inhibiting potential competitors and plant pests (Stelinski *et al.*, 2019).

7.3.2.3 Negative interactions

Antagonism occurs between EPNs and many soil organisms, but it is particularly relevant when biological control agents (viruses, bacteria, fungi or parasitoids) compete for the same resource. In such interactions, the first biological control agent to colonize the host often compromises the success of subsequent infections. For example, IJs can infect and develop in larvae previously attacked by nucleopolyhedrovirus and granulovirus but it significantly reduces their offspring (Kaya and Brayton, 1978; Bednarek, 1986). Similarly, dual infections of EPNs and bacteria are rare and are strongly dependent on application timing (Kaya and Burlando, 1989; Darissa and Iraki, 2013). Indirect antagonism also arises in combined applications of EPNs and entomopathogenic fungi (Barbercheck and Kaya, 1990; Shapiro-Ilan *et al.*, 2004; Bueno-Pallero *et al.*, 2018). Both entomopathogens release specific bioproducts to avoid competition (Dreyer *et al.*, 2018; Shi and Bode, 2018). Laboratory experiments have shown that EPNs appear to avoid hosts infected with entomopathogenic fungi (Barbercheck and Kaya, 1990; Tarasco *et al.*, 2011). This may reduce antagonistic interactions between the two entomopathogens under natural conditions. However, these interactions may be species specific or even strain/population specific (Bueno-Pallero *et al.*, 2018).

Regarding natural enemies, no viruses have been described that directly attack EPNs (Kaya, 2002; Strong, 2002), despite the isolation of phages from symbiotic bacteria and parasitic protozoa of EPNs (Poinar and Hess, 1988; Boemare *et al.*, 1993). Some ectoparasitic bacteria can colonize the IJ cuticle but show low specificity for EPNs (Chen and Dickson, 1998; Somasekhar and Mehta, 2000; Enright *et al.*, 2003). However, *Paenibacillus* spp. (Bacillales: Paenibacillaceae) could limit migration and encounter suitable hosts of IJs (El-Borai *et al.*, 2005; Enright and Griffin, 2005), and Bajaj and Walia (2005) found evidence of host specificity for isolates of the endospore-forming *Pasteuria* spp. (Bacillales: Pasteuriaceae) and the EPN species *Steinernema pakistanense*. In contrast, certain fungal species have evolved different specialized structures as traps (e.g. constriction rings, adhesive nets, branches) or mature spores, depending on the species, to prey on nematodes (Nordbring-Hertz *et al.*, 2006). Thus, nematophagous fungi fall into three categories: nematode trappers, endoparasites, and egg/cyst parasites, found in all major taxonomic groups (López-Llorca *et al.*, 2007). The susceptibility of EPNs to nematophagous fungi is species specific and environmentally dependent, but the mechanisms behind their interactions are mainly unknown (Koppenhöfer *et al.*, 1996; El-Borai *et al.*, 2009, 2011; Bueno-Pallero *et al.*, 2018). High abundance of nematophagous fungi could alter the EPN community in various crops (Campos-Herrera *et al.*, 2015b; Pathak *et al.*, 2017). Plausibly, soil profile migration helps EPNs to escape from nematophagous fungi, limiting their occurrence in certain areas (Campos-Herrera *et al.*, 2019b).

Several invertebrates (e.g. mites, collembolans, tardigrades) can also reduce EPN activity and occurrence (Poinar, 1979; Ishibashi *et al.*, 1987; Small, 1987). Although molecular tools can now detect predation of certain collembolans and mites on EPNs (Read *et al.*, 2006), evidence for the importance of these interactions under natural conditions has yielded contrasting results (Duncan *et al.*, 2007; Jabbour *et al.*, 2011; Helmberger *et al.*, 2018). Dritsoulas *et al.* (2021) recently reported an inverse correlation between mites in the family Rhodacaridae and *H. indica* abundance using high-throughput sequencing (HTS). Macroinvertebrates can also negatively affect EPN populations. For example, *in vitro* experiments have revealed that the feeding activity of earthworms and their release of cutaneous excreta can compromise EPN fitness in a species-specific manner (Campos-Herrera *et al.*, 2006; Chelkha *et al.*, 2020, 2021). In addition, other nematodes prey on EPNs, such as species of the genera *Clarkus* (Mononchida: Mononchidae) and *Actinolaimus* (Dorylaimida: Actinolaimidae) (Ishibashi *et al.*, 1987). Moreover, different studies have observed that some free-living nematodes can compete with EPNs for the same resource (Duncan *et al.*, 2003a; Campos-Herrera *et al.*, 2012a, 2015a), even affecting the viability of their offspring (Blanco-Pérez *et al.*, 2019).

7.4 Detection Methods and Soil Food Web Modeling

7.4.1 Estimation of activity rates and abundance of EPNs

A better understanding of the natural distribution of EPNs depends on the detection method chosen. In the past, direct isolation of naturally infected hosts collected in the field was commonly used. Nematode emergences from cadavers, separated in the laboratory to avoid contamination, were then evaluated (Bovien, 1937). This technique has facilitated the identification of numerous EPN isolates and their natural hosts (Peters, 1996). However, the most widely used procedure for isolating and studying EPNs is the soil baiting method. Initially developed by Bedding and Akhurst (1975), this methodology recovers EPNs from both *in situ* baits and tests of soil samples retrieved to the laboratory. Although the larval stage of *G. mellonella* was the most common host selected (Dutky and Hough, 1955; Bedding and Akhurst, 1975), the use of specific hosts can be helpful when trying to isolate EPNs with virulence toward a particular target pest, such as white grubs for *Steinernema scarabaei* or mole crickets for *S. scapterisci* (Adams and Nguyen, 2002). In this case, the insect bait can be an initial screening for suitable candidates in biocontrol programs (Duncan *et al.*, 2003b; Shapiro-Ilan *et al.*, 2003). Similarly, to enhance the efficacy of EPN detection in soil samples, strategies such as using two sentinel insect species or baiting at two different temperatures have been explored with promising results (Shapiro-Ilan *et al.*, 2003; Mráček *et al.*, 2005; Campos-Herrera and Gutiérrez, 2009).

Insect baiting also allows EPN quantification from soil samples. Koppenhöfer *et al.* (1998) developed a mathematical model to determine the density of EPN populations by counting the number of insect cadavers recovered per soil sample, repeating each insect bait until no more larval mortality was recorded in successive rounds. However, insect baits have some drawbacks: IJs should be active against hosts during the baiting and outcompete different organisms cohabiting in the same soil sample, including other EPNs, which could lead to underestimation of EPN activities (Campos-Herrera *et al.*, 2015b; Griffin, 2015; Lewis *et al.*, 2015). Some other isolation methods, such as sucrose-gradient centrifugation and Baermann funnels, consist of standard nematode extraction techniques based on densities and active migration of nematodes, respectively (Saunders and All, 1982; Sturhan and Mráček, 2000; Acevedo *et al.*, 2005). Despite some valuable results, these traditional methods are not very useful for assessing EPN abundance because, in a soil sample, a few IJs have to be distinguished from numerous other nematodes, including several rhabditid species of similar morphology. Such approaches make the task so challenging that it only allows short processing numbers of samples by highly specialized nematologists (Hominick, 2002).

The development of molecular tools for identifying and quantifying EPNs from soil samples allows expansion of our knowledge on EPN natural occurrence and their relationships with other soil members (Pathak *et al.*, 2012; Dritsoulas and Duncan, 2020; Dritsoulas *et al.*, 2021). Using species-specific primers and probes is currently the most valuable and informative molecular tool available. The seminal study by Torr *et al.* (2007) demonstrated the positive correlation between EPN abundance detected by quantitative polymerase chain reaction (qPCR) analyses and traditional methods. Subsequent studies have expanded the repertoire of molecular toolkits (e.g. primer/probe sets, functioning conditions) to more species (Table 7.1). The approach resulted in higher sensitivity and efficiency in revealing EPN sympatric distributions than the insect-baiting method (Campos-Herrera *et al.*, 2011a, 2016b). In addition, qPCR analysis allows the evaluation of other soil organisms coexisting with EPNs (only requiring the design of species-specific primer/probe sets for additional target organisms). Thus, screens can be extended to include free-living nematodes (Campos-Herrera *et al.*, 2012a, 2015b, 2019a), nematophagous fungi (Pathak *et al.*, 2012) and ectoparasitic bacteria of the genus *Paenibacillus* (Campos-Herrera *et al.*, 2011b). However, this technique only allows identification and quantification of target organisms, not novel or unexpected species. Another limitation is that primer/probe combinations could amplify DNA sequences of closely related species that may not have been previously detected or known. Therefore, it is strongly recommended to perform appropriate optimizations before analyzing samples in a new system (Campos-Herrera *et al.*, 2015a). Recent molecular techniques may help to overcome some of these limitations. For example, Dritsoulas *et al.* (2020) used HTS with *de novo* universal primers targeting Internal transcribed spacer 1 (ITS1) for nematode detection. Their results were highly congruent with those obtained for EPN soil food web analyses in the Algarve (Portugal) using real-time qPCR (Campos-Herrera *et al.*, 2019a). However, more EPN species were detected through HTS and in higher frequencies (especially for the most dominant species) than in qPCR analysis. Nevertheless, this method also suffers from some limitations. In addition to the high cost, HTS is semi-quantitative and may not detect all EPN species. Moreover, targeting other soil organisms (e.g. fungi and bacteria) would require additional runs for specific primers (Dritsoulas *et al.*, 2021).

In short, different methodologies offer some advantages and limitations for estimating the occurrence and abundance of EPNs and associated organisms within their soil food web. The choice of the most appropriate approach will depend on the objective of the study, access to equipment and funding capacity. Thus, insect baiting will be more appropriate where nematode isolation is essential, for example screening new species/populations in biological control agent selection programs (Bruno *et al.*, 2020; Levy *et al.*, 2020). It is also highly recommended for surveying unexplored sites because novel or unexpected species might be detected. In contrast, qPCR analysis is more suitable for studying multi-trophic interactions of soil members with EPNs (Campos-Herrera *et al.*, 2013a, 2016a, 2019a) or unraveling the impact of soil disturbances such as pollutants or differentiated agronomical management of EPN occurrences (Campos-Herrera *et al.*, 2016b; Blanco-Pérez *et al.*, 2020, 2022a, 2022b). Finally, HTS has proved to be a reliable technique for detecting unexpected EPN species (Dritsoulas *et al.*, 2020), an advantage over qPCR analyses, which are limited to species-specific primers and probes. Hence, moving toward HTS technology is recommended, for example in biodiversity studies and regulation or quarantine programs for commercial importation of EPNs.

7.4.2 Models for the soil food web of EPNs

Sophisticated molecular and statistical tools to identify and quantify soil organisms and relate them to numerous environmental factors enable further studies of EPN population dynamics and assembly of predictive ecological models. For example, direct statistical approaches, such as canonical correspondence analysis, have been valuable in relating nematode occurrence and abundance to abiotic factors (Fiscus and Neher, 2002; Neher *et al.*, 2005; Jaffuel *et al.*, 2018; Campos-Herrera *et al.*, 2019a; Blanco-Pérez *et al.*, 2022a). Furthermore, canonical correspondence analysis allowed the finding that edaphic conditions associated with EPN populations differed for a wide range of free-living nematodes species, arguably due to adapting to entomopathogenic behavior (Hoy *et al.*, 2008). This knowledge can be critical for optimizing soil fertility through farming practices that ensure the ecosystem services provided by native EPN populations. Geostatistical approaches, such as spatial analysis by distance indices (SADIE; Perry and Hewitt, 1991), were also successfully applied to assess rather than infer spatial patterns of EPNs, which help, for instance, to describe their dispersal behavior of emerging IJs (Spiridonov *et al.*, 2007) and to unravel the

Table 7.1. List of available species-specific primers and probes sets designed for entomopathogenic nematode species in the genera *Steinernema* and *Heterorhabditis*.

Species	Sequence (5′→3′)	PCR product size (bp)	Reference
S. affine	F: ACTTAGACGATTGTCTGGCATCT R: CAGGCGCCTATAAAAACCAA P: AGCCACTTCATGCAACGTTTTTGTCA	88	Torr *et al*. (2007)
S. affine group	F: CGATTGCCTGTCATCAAATG R: AAGCAAAAGAGCACCCATCA P: AGGTGCCTGGAGCAGTTGTATGA	130	Blanco-Pérez *et al*. (2022)
S. arenarium	F: CTTTTCGGTGTGAGTGCGTG R: TGTTCCAGTTGCGCCG P: GCGTTGTGGGCTTTGGCGTG	210	Campos-Herrera *et al*. (2019a)
S. bicornutum	F: ACGGAGCAGCTGTATGATCG R: CGAGTCACTGAACCGACCTC P: GCTGTGGTGATATATGCTTGACATTGC	91	Campos-Herrera *et al*. (2015a)
S. carpocapsae	F: CTAACGGCTTTGAAAGGTTTCTAC R: TGCATCAACGAGTGACACAATTATCGA P: GTACATTGTTATCTAAGCGTTTCGTC	203	Campos-Herrera *et al*. (2011a)
S. diaprepesi	F: CTTTCGCTAACGCTTCTGCT R: AGTTCCAGTTGCGTTCGTCA P: TTGCCAGTTGACTTGTACGC	165	Campos-Herrera *et al*. (2011b)
S. feltiae	F: CGTTTTTCTTGCCGACTGATTGGTACA R: TGCGCAATTCCTCGACAAACGA P: ATTTTTCAGAATTTTTCAGAGGCCCTTAC	120	Campos-Herrera *et al*. (2011a)
S. glaseri	F: TTTTTCGACGAGCTATGTACG R: TGATAGCGGACACGACAAAA P: TGTACCTCGTTCGGTGTGAA	108	Campos-Herrera *et al*. (2011b)
S. intermedium	F: GGTTTAGATTTGTTTACGCTTCTCA R: CTCACAGAGCATAACAACGCA P: TCACTTCTAGTGAATGTGCGAATTGCT	99	Campos-Herrera *et al*. (2015a)
S. intermedium group	F: ATTGCTTCTAATATGAGTTGGTTGT R: GCCTTGCTTGAGTTGAGGTC P: TGAATGTGCGAATTGCTGTGCGA	132	Campos-Herrera *et al*. (2015a)
S. khuongi	F: ACGACTATTCGTTGTGTCCGC R: GCGCCGAACCGTAAAG P: CTGACTTGTACGCGATTC	153	Campos-Herrera *et al*. (2011a)[a]
S. kraussei	F: GGCGAGTCGTAACGACTTCTT R: TCGAAGCGATGTATTCTCTCAA P: TTCTAGTTCCAGAAGAACTCAAATGTCCATCACC	137	Torr *et al*. (2007)
S. kraussei-silvaticum	F: TCTGCTGTTTGTTTCGAAGCGA R: TGTCCATCACCACAGTCACG P: ACGGCTACGAAGGGTTTCTGTAGGT	103	Campos-Herrera *et al*. (2015a)
S. neocurtillae	F: TATGACGCGTTTCTTGTCAAATC R: GAGGTCGTAAAACGAAAACATCG P: CTGGTTGACAACTATTTGCTTTGT	161	Campos-Herrera *et al*. (2013)
S. phyllophagae	F: GTTCGGTGTGACGGCATC R: GAAATATAACATTCCAGCGTAGC P: CGCTTGTGTCCGCTATCAT	212	Campos-Herrera *et al*. (2013)
S. poinari	F: TGCTTCTAATGTGAGTTGGCTGT R: ACTCACAGAGCATAATAGTGCAT P: CGCTTCTAGTGAATGTGTGAATTGCC	85	Campos-Herrera *et al*. (2015a)

Continued

Table 7.1. Continued.

Species	Sequence (5′→3′)	PCR product size (bp)	Reference
S. riobrave	F: CGTGCCAACTGGCTTGTAT R: CAAATTGAGGTTGCAACAGA P: TTCGTTAACTAACGTAATCCTTTGC	156	Campos-Herrera *et al.* (2011b)
S. riojaense	F: CGGCTAACTTGTGCGTTGAC R: CGACCCAAAAGGCACAAACG P: AGCTGTCGCGTATGTGCTGTTTCG	86	Blanco-Pérez *et al.* (2020)
S. scapterisci	F: TCGTGCTATGCGGTGAAG R: CTGACAAATTAAGCCAGTTAATCA P: ATTTGGTTAATGGTTTAGCGCGT	222	Campos-Herrera *et al.* (2011a)
S. weiseri	F: TTTCAAAATGTCAGCGGCCC R: TCAGCGGGTAATCTTGCTTGA P: TCGACACAACACGACTCGTTTGT	123	Campos-Herrera *et al.* (2015a)
H. bacteriophora	F: AACGAGAGTTCGGTGATACTGACAAC R: GCCGTGTATACGACACGTCA P: CGTTCAAGTATCTTTATGGGGCAAC	171	Campos-Herrera *et al.* (2011a)
H. floridensis	F: GGCAATACCGTGAACCAAAC R: GAAACACGATATATGTATATAACCACG P: CTCTTAAAGCCGGAGAAATAT	172	Campos-Herrera *et al.* (2011b)
H. indica	F: CATCGATAACTGGTGGCTGA R: CCATCGCTAAGGCTATTTTCC P: CAACCGCCACTATCGGTAAT	165	Campos-Herrera *et al.* (2011b)
H. megidis	F: GCAATGTCGAGTGTCGAACG R: CGCTACACATCCACAGGTACA P: TCCGAATATTGGCAACATGTC	132	Campos-Herrera *et al.* (2015a)
H. zealandica	F: CCGCCTTGCGATATTTGAT R: CGCTGAAGCTAATACATTGTGC P: TCGTTATCTTGTCTCTTGGATACG	93	Campos-Herrera *et al.* (2011b)

F, forward primer; R, reverse primer; P, TaqMan® probe (reporter or quencher not specified).
[a]Described in the original publication as *Steinernema glaseri* group.

interactions they establish with other soil inhabitants (Campos-Herrera *et al.*, 2011b, 2013b). Moreover, ecological models can also predict cropping systems or target pests with high probabilities of success in conservation biocontrol programs. A spatial–temporal study conducted by Campos-Herrera *et al.* (2019b) in Florida citrus groves serves as an illustration. Focused on EPN interactions with the environment, the soil food web and target pests using qPCR analyses, the results revealed a top-down regulation of certain EPN species by antagonistic organisms such as nematophagous fungi and ectoparasitic bacteria. In this latter case, the observations were indeed consistent with classic Lotka–Volterra phase–space, predator–prey dynamics (Fig. 7.2) (Lotka, 1925; Volterra, 1926). Accordingly, this study concluded that *Paenibacillus* sp. records could predict the occurrence of the EPN species *S. diaprepesi* in the following months.

Such efforts are necessary for conservation biological control programs, for example to determine whether EPNs introduced as biological control agents could outcompete native undescribed species (Campos-Herrera *et al.*, 2012b). However, understanding the biogeography of biological control agents often requires finer taxonomic resolution and species identification not limited by specific primer/probe sets than that provided by qPCR analyses (Dritsoulas *et al.*, 2020). As mentioned above, HTS achieves more accurate taxonomic resolution and recognizes rarer taxa than morphological or qPCR analyses (Treonis *et al.*, 2018). Furthermore, Geisen *et al.* (2018), for example, achieved higher taxonomic resolution for

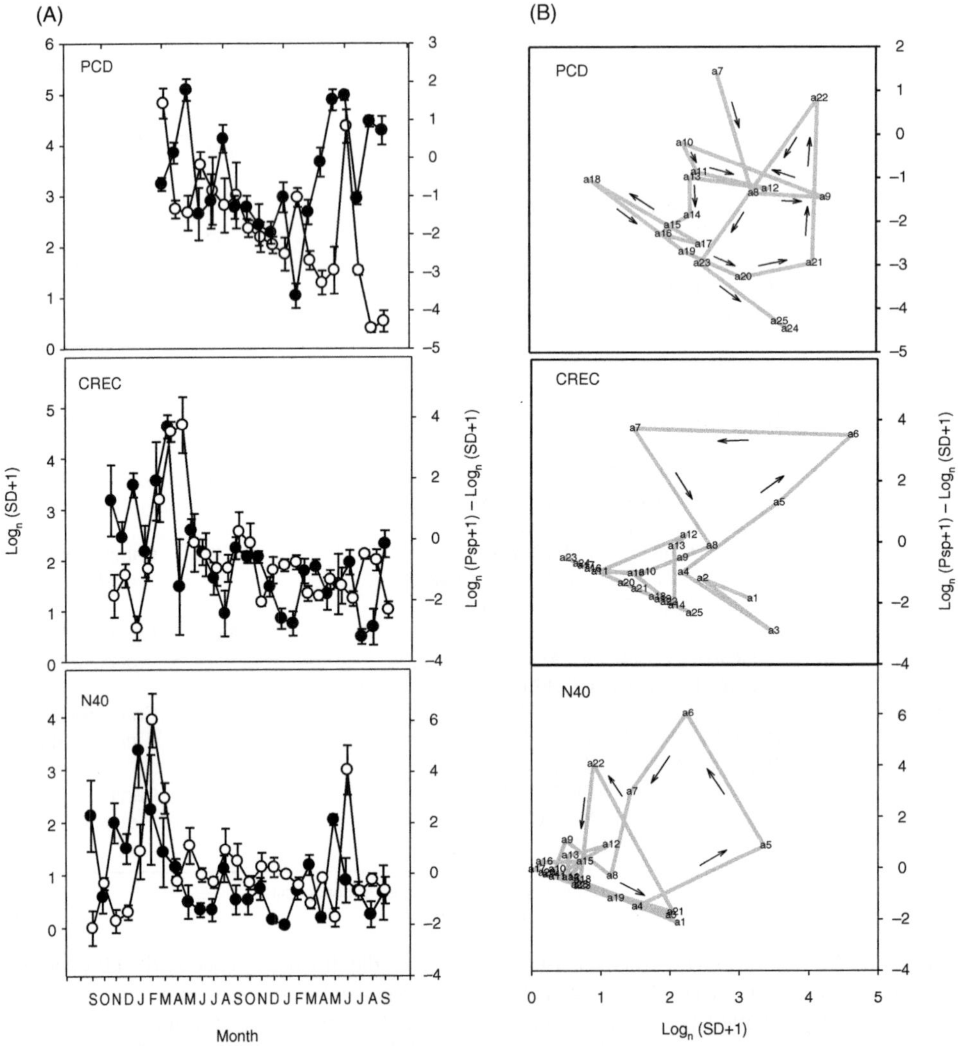

Fig. 7.2. Abundance of *Steinernema diaprepesi* (left panels, black) and *Paenibacillus* sp. endospore infestation rate (left panels, white) at three sites (PCD, CREC and N40) with abundant *S. diaprepesi,* and phase–space diagrams showing the sequential direction of the monthly flux of *Paenibacillus* sp. endospore infestation rate compared with *S. diaprepesi* (Sd) abundance (right panels). Error bars are standard error of the mean. Reproduced from Campos-Herrera *et al.* (2019b) with permission.

native nematode populations at a lower cost than qPCR and morphological approaches. Therefore, as investigated by Dritsoulas *et al.* (2020) by reanalyzing soil samples from the Campos-Herrera *et al.* (2019a) survey, HTS can be a more affordable tool to measure soil organisms in cryptic environments, including unknown organisms, and novel functions of soil communities (He *et al.*, 2010).

7.5 Conclusions and Perspectives

Understanding the distribution, abundance and spatiotemporal dynamics of EPN populations within soil ecosystems is essential to enhance and maintain their service as biological control agents, either from augmentative or conservative approaches. Significant advances in recent years have contributed to expanding our basic and applied knowledge of the natural occurrence of EPNs and their interaction with other soil organisms. Similarly, recent surveys have correlated natural EPN occurrence and the application of alternative managing of agroecosystems with conventional practices of high ecological impact. In this regard, using biological control agents as model organisms and analysis of their population dynamics may be of great interest to ensure the best practices to favor appropriate ecological scenarios, particularly within a climate change context. In the near future, for example, climate change could increase root herbivore damage in crops and favor the spread of agricultural pests while reducing the presence of their natural enemies (Guyer *et al.*, 2021). Hence, regional and temporal surveys become critical not only to establishing the limits of native EPN occurrence or their range of activity in crop soils but also to generate efficient prediction models to assess the sustainability of specific farming practices.

More than ever, a better understanding of the biotic and abiotic factors involved in trophic webs of beneficial organisms is still needed. In the specific case of EPNs, it is crucial to complement traditional sampling and data evaluation tools with novel procedures to achieve successful predictive ecological models. For example, HTS testing, even if it still needs further tuning to improve its accuracy (Waeyenberge *et al.*, 2019), is a promising tool for exploring the biogeography of EPN populations and their soil food web. Such approaches have the potential to provide insights into the multi-trophic interaction that EPNs establish with plants, soil biota and even their own symbionts. Indeed, recent *in vitro* studies have suggested that EPN symbiotic bacteria may be able to survive in the soil associated with roots by changing their phase and interacting with other members of the rhizosphere (Regaiolo *et al.*, 2020), although this phenomena is yet to be demonstrated in natural soil and within plants other than *Arabidopsis thaliana*. In summary, future research should integrate these molecular tools with spatiotemporal dynamics and multi-trophic interaction analyses to comprehend the biogeography of beneficial organisms such as EPNs and identify more sustainable agricultural practices.

7.6 Acknowledgements

This revision is related to the grants awarded to R. Campos-Herrera (TED2021-129169B-I00) and R. Blanco-Pérez (JDC2022-048978-I), funded by MCIN/AEI/10.13039/501100011033 and by "European Union NextGenerationEU/PRTR."

7.7 References

Abate, B.A., Wingfield, M.J., Slippers, B. and Hurley, B.P. (2017) Commercialisation of entomopathogenic nematodes: should import regulations be revised? *Biocontrol Science and Technology* 27, 149–168.

Acevedo, J.P.M., Moino, A.J., Cavalcanti, R.S., Dolinski, C. and Carvalho, F.A. (2005) Sampling and evaluation of techniques for isolation of native entomopathogenic nematodes obtained in Lavras, Manas Gerais. *Nematologia Brasileira* 29, 17–23.

Acevedo, J.P.M., Samuels, R.I., Machado, I.R. and Dolinski, C. (2007) Interactions between isolates of the entomopathogenic fungus *Metarhizium anisopliae* and the entomopathogenic nematode *Heterorhabditis bacteriophora* JPM4 during infection of the sugar cane borer *Diatraea saccharalis* (Lepidoptera: Pyralidae). *Journal of Invertebrate Pathology* 96, 187–192.

Adams, B.J. and Nguyen, K.B. (2002) Taxonomy and systematics. In: Gaugler, R. (ed.) *Entomopathogenic Nematology*. CABI Publishing. Wallingford, UK, pp. 1–33.

Adams, B.J., Fodor, A., Koppenhöfer, H.S., Stackebrandt, E., Stock, S.P. and Klein, M.G. (2006) Biodiversity and systematics of nematode–bacterium entomopathogens. *Biological Control* 37, 32–49.

Adeolu, M., Alnajar, S., Naushad, S. and Gupta, R.S. (2016) Genome-based phylogeny and taxonomy of the '*Enterobacteriales*': proposal for *Enterobacterales* ord. nov. divided into the families *Enterobacteriaceae*, *Erwiniaceae* fam. nov., *Pectobacteriaceae* fam. nov., *Yersiniaceae* fam. nov., *Hafniaceae* fam. nov., *Morganellaceae* fam. nov., and *Budviciaceae* fam. nov. *International Journal of Systematic and Evolutionary Microbiology* 66, 5575–5599.

Ali, J., Alborn, H.T. and Stelinski, L.L. (2010) Subterranean herbivore-induced volatiles released by citrus roots upon feeding by *Diaprepes abbreviatus* recruit entomopathogenic nematodes. *Journal of Chemical Ecology* 36, 361–368.

Ali, J.G., Alborn, H.T., Campos-Herrera, R., Kaplan, F., Duncan, L.W. *et al.* (2012) Subterranean, herbivore-induced plant volatile increases biological control activity of multiple beneficial nematode species in distinct habitats. *PLOS ONE* 7: e38146.

Ansari, M.A., Tirry, L. and Moens, M. (2004) Interaction between *Metarhizium anisopliae* CLO 53 and entomopathogenic nematodes for the control of *Hoplia philanthus*. *Biological Control* 31, 172–180.

Ansari, M.A., Shah, F.A. and Butt, T.M. (2010) The entomopathogenic nematode *Steinernema kraussei* and *Metarhizium anisopliae* work synergistically in controlling overwintering larvae of the black vine weevil, *Otiorhynchus sulcatus*, in strawberry growbags. *Biocontrol Science and Technology* 20, 99–105.

Bajaj, H.K. and Walia, K.K. (2005) Studies on a *Pasteuria* isolate from an entomopathogenic nematode, *Steinernema pakistanense* (Nematoda: Steinernematidae). *Nematology* 7, 637–640.

Barbercheck, M.E. (1992) Effect of soil physical factors on biological control agents of soil insect pests. *Florida Entomologist* 75, 539–548.

Barbercheck, M.E. and Kaya, H.K. (1990) Interactions between *Beauveria bassiana* and the entomogenous nematodes, *Steinernema feltiae* and *Heterorhabditis heliothidis*. *Journal of Invertebrate Pathology* 55, 225–234.

Bathon, H. (1996) Impact of entomopathogenic nematodes on non-target hosts. *Biocontrol Science and Technology* 6, 421–434.

Bedding, R.A. and Akhurst, R.J. (1975) A simple technique for the detection of insect parasitic rhabditid nematodes in soil. *Nematologica* 21, 109–110.

Bednarek, A. (1986) Development of the *Steinernema feltiae* (Filipjev) entomogenous nematodes (Steinernematidae) in the condition of occurrence in the insect body cavity of other pathogens. *Annals Warsaw Agricultural University SGGW-AR: Animal Science* 20, 69–74.

Blanco-Pérez, R., Bueno-Pallero, F.Á., Vicente-Díez, I., Marco-Mancebón, V.S., Pérez-Moreno, I. and Campos-Herrera, R. (2019) Scavenging behavior and interspecific competition decrease offspring fitness of the entomopathogenic nematode *Steinernema feltiae*. *Journal of Invertebrate Pathology* 164, 5–15.

Blanco-Pérez, R., Sáenz-Romo, M.G., Vicente-Díez, I., Ibáñez-Pascual, S., Martínez-Villar, E. *et al.* (2020) Impact of vineyard ground cover management on the occurrence and activity of entomopathogenic nematodes and associated soil organisms. *Agriculture, Ecosystems & Environment* 301: 107028.

Blanco-Pérez, R., Vicente-Díez, I., Ramos-Sáez de Ojer, J.L., Marco-Mancebón, V.S., Pérez-Moreno, I. and Campos-Herrera, R. (2022a) Organic viticulture enhanced the activity of native entomopathogenic nematodes in DOCa Rioja soils (North of Spain). *Agriculture, Ecosystems & Environment* 332: 107931.

Blanco-Pérez, R., Vicente-Díez, I., Pou, A., Pérez-Moreno, I., Marco-Mancebón, V.S. and Campos-Herrera, R. (2022b) Organic mulching modulated native populations of entomopathogenic nematode in vineyard soils differently depending on its potential to control outgrowth of their natural enemies. *Journal of Invertebrate Pathology* 192: 107781.

Boemare, N.E., Boyer-Giglio, M.-H., Thaler, J.-O. and Akhurst, R.J. (1993) The phages and bacteriocins of *Xenorhabdus* sp., symbiont of the nematode *Steinernema* spp. and *Heterorhabditis* spp. In: Bedding, R.A., Akhurst, R. and Kaya, H.K. (eds.) *Nematodes and the Biological Control of Insect Pests*. CSIRO Publications, East Melbourne, Australia, pp. 137–145.

Bovien, P. (1937) *Some Types of Association between Nematodes and Insects. Vol. 1, Videnskabelige Meddelelser fra Dansk Naturhistorisk Forening*. Reitzel, Emdrup, Denmark.

Brady, N.C. and Weil, R.R. (2013) The Nature and Properties of Soils. Pearson, New York.

Brown, I.M. and Gaugler, R. (1997) Temperature and humidity influence emergence and survival of entomopathogenic nematodes. *Nematologica* 43, 363–375.

Bruno, P., Machado, R.A.R., Glauser, G., Köhler, A., Campos-Herrera, R. *et al.* (2020) Entomopathogenic nematodes from Mexico that can overcome the resistance mechanisms of the western corn rootworm. *Scientific Reports* 10: 8257.

Bueno-Pallero, F.Á., Blanco-Pérez, R., Dionísio, L. and Campos-Herrera, R. (2018) Simultaneous exposure of nematophagous fungi, entomopathogenic nematodes and entomopathogenic fungi can modulate belowground insect pest control. *Journal of Invertebrate Pathology* 154, 85–94.

Campos-Herrera, R. and Gutiérrez, C. (2009) Screening Spanish isolates of steinernematid nematodes for use as biological control agents through laboratory and greenhouse microcosm studies. *Journal of Invertebrate Pathology* 100, 100–105.

Campos-Herrera, R., Trigo, D. and Gutiérrez, C. (2006) Phoresy of the entomopathogenic nematode *Steinernema feltiae* by the earthworm *Eisenia fetida*. *Journal of Invertebrate Pathology* 92, 50–54.

Campos-Herrera, R., Gómez-Ros, J.M., Escuer, M., Cuadra, L., Barrios, L. and Gutiérrez, C. (2008) Diversity, occurrence, and life characteristics of natural entomopathogenic nematode populations from La Rioja (Northern Spain) under different agricultural management and their relationships with soil factors. *Soil Biology and Biochemistry* 40, 1474–1484.

Campos-Herrera, R., Piedra-Buena, A., Escuer, M., Montalbán, B. and Gutiérrez, C. (2010) Effect of seasonality and agricultural practices on occurrence of entomopathogenic nematodes and soil characteristics in La Rioja (Northern Spain). *Pedobiologia* 53, 253–258.

Campos-Herrera, R., Johnson, E.G., El-Borai, F.E., Stuart, R.J., Graham, J.H. and Duncan, L.W. (2011a) Long-term stability of entomopathogenic nematode spatial patterns in soil as measured by sentinel insects and real-time PCR assays. *Annals of Applied Biology* 158, 55–68.

Campos-Herrera, R., El-Borai, F.E., Stuart, R.J., Graham, J.H. and Duncan, L.W. (2011b) Entomopathogenic nematodes, phoretic *Paenibacillus* spp., and the use of real time quantitative PCR to explore soil food webs in Florida citrus groves. *Journal of Invertebrate Pathology* 108, 30–39.

Campos-Herrera, R., El-Borai, F.E. and Duncan, L.W. (2012a) Wide interguild relationships among entomopathogenic and free-living nematodes in soil as measured by real time qPCR. *Journal of Invertebrate Pathology* 111, 126–135.

Campos-Herrera, R., Barbercheck, M.E., Hoy, C.W. and Stock, S.P. (2012b) Entomopathogenic nematodes as a model system for advancing the frontiers of ecology. *Journal of Nematology* 44, 162–176.

Campos-Herrera, R., Pathak, E., El-Borai, F.E., Stuart, R.J., Gutiérrez, C. *et al.* (2013a) Geospatial patterns of soil properties and the biological control potential of entomopathogenic nematodes in Florida citrus groves. *Soil Biology and Biochemistry* 66, 163–174.

Campos-Herrera, R., Ali, J.G., Diaz, B.M. and Duncan, L.W. (2013b) Analyzing spatial patterns linked to the ecology of herbivores and their natural enemies in the soil. *Frontiers in Plant Science* 4: 378.

Campos-Herrera, R., Jaffuel, G., Chiriboga, X., Blanco-Pérez, R., Fesselet, M. *et al.* (2015a) Traditional and molecular detection methods reveal intense interguild competition and other multitrophic interactions associated with native entomopathogenic nematodes in Swiss tillage soils. *Plant and Soil* 389, 237–255.

Campos-Herrera, R., Půža, V., Jaffuel, G., Blanco-Pérez, R., Čepulyte-Rakauskiene, R. and Turlings, T.C.J. (2015b) Unraveling the intraguild competition between *Oscheius* spp. nematodes and entomopathogenic nematodes: implications for their natural distribution in Swiss agricultural soils. *Journal of Invertebrate Pathology* 132, 216–227.

Campos-Herrera, R., Rodríguez Martín, J.A., Escuer, M., García-González, M.T., Duncan, L.W. and Gutiérrez, C. (2016a) Entomopathogenic nematode food webs in an ancient, mining pollution gradient in Spain. *Science of the Total Environment* 572, 312–323.

Campos-Herrera, R., El-Borai, F.E., Rodríguez Martín, J.A. and Duncan, L.W. (2016b) Entomopathogenic nematode food web assemblages in Florida natural areas. *Soil Biology and Biochemistry* 93, 105–114.

Campos-Herrera, R., Blanco-Pérez, R., Bueno-Pallero, F.Á., Duarte, A., Nolasco, G. *et al.* (2019a) Vegetation drives assemblages of entomopathogenic nematodes and other soil organisms: evidence from the Algarve, Portugal. *Soil Biology and Biochemistry* 128, 150–163.

Campos-Herrera, R., Stuart, R.J., Pathak, E., El-Borai, F.E. and Duncan, L.W. (2019b) Temporal patterns of entomopathogenic nematodes in Florida citrus orchards: evidence of natural regulation by microorganisms and nematode competitors. *Soil Biology and Biochemistry* 128, 193–204.

Cavigelli, M.A., Maul, J.E. and Szlaveczs, K. (2012) Managing soil biodiversity and ecosystem services. In: Wall, D.H., Bardgett, R.D., Behan-Pelletier, V., Herrick, J.E., Jones, T.H. *et al.* (eds.) *Soil Ecology and Ecosystem Services*. Oxford University Press, Oxford, UK, pp. 337–356.

Chelkha, M., Blanco-Pérez, R., Bueno-Pallero, F.Á., Amghar, S., El Harti, A. and Campos-Herrera, R. (2020) Cutaneous excreta of the earthworm *Eisenia fetida* (Haplotaxida: Lumbricidae) might hinder the biological control performance of entomopathogenic nematodes. *Soil Biology and Biochemistry* 141: 107691.

Chelkha, M., Blanco-Pérez, R., Vicente-Díez, I., Bueno-Pallero, F.Á., Amghar, S. *et al.* (2021) Earthworms and their cutaneous excreta can modify the virulence and reproductive capability of entomopathogenic nematodes and fungi. *Journal of Invertebrate Pathology* 184: 107620.

Chen, Z.X. and Dickson, D.W. (1998) Review of *Pasteuria penetrans*: biology, ecology, and biological control potential. *Journal of Nematology* 30, 313–340.

Chiriboga, X., Campos-Herrera, R., Jaffuel, G., Röder, G. and Turlings, T.C.J. (2017) Diffusion of the maize root signal (E)-β-caryophyllene in soils of different textures and the effects on the migration of the entomopathogenic nematode *Heterorhabditis megidis*. *Rhizosphere* 3, 53–59.

Darissa, O.M. and Iraki, N.M. (2013) *In vivo* and *in vitro* interactions between two entomopathogens: the *bacterium Serratia marcescens* and the nematode *Heterorhabditis indica*. *Russian Journal of Nematology* 21, 83–91.

Dillman, A.R., Chaston, J.M., Adams, B.J., Ciche, T.A., Goodrich-Blair, H. *et al.* (2012) An entomopathogenic nematode by any other name. *PLOS* Pathogens 8: e1002527.

Dreyer, J., Malan, A.P. and Dicks, L.M.T. (2018) Bacteria of the genus *Xenorhabdus*, a novel source of bioactive compounds. *Frontiers in Microbiology* 9: 3177.

Dritsoulas, A. and Duncan, L.W. (2020) Optimizing for taxonomic coverage: a comparison of methods to recover mesofauna from soil. *Journal of Nematology* 52: e2020-104.

Dritsoulas, A., Campos-Herrera, R., Blanco-Pérez, R. and Duncan, L.W. (2020) Comparing high throughput sequencing and real time qPCR for characterizing entomopathogenic nematode biogeography. *Soil Biology and Biochemistry* 145: 107793.

Dritsoulas, A., El-Borai, F.E., Shehata, I.E., Hammam, M.M., El-Ashry, R.M. *et al.* (2021) Reclaimed desert habitats favor entomopathogenic nematode and microarthropod abundance compared to ancient farmlands in the Nile Basin. *Journal of Nematology* 53: e2021-47.

Duncan, L.W., Dunn, D.C., Bague, G. and Nguyen, K. (2003a) Competition between entomopathogenic and free-living bactivorous nematodes in larvae of the weevil *Diaprepes abbreviatus*. *Journal of Nematology* 35, 187–193.

Duncan, L.W., Graham, J.H., Dunn, D.C., Zellers, J., McCoy, C.W. and Nguyen, K.B. (2003b) Incidence of endemic entomopathogenic nematodes following application of *Steinernema riobrave* for control of *Diaprepes abbreviatus*. *Journal of Nematology* 35, 178–186.

Duncan, L.W., Graham, J.H., Zellers, J., Bright, D.B., Dunn, D.C. *et al.* (2007) Food web responses to augmenting the entomopathogenic nematodes in bare and animal manure-mulched soil. *Journal of Nematology* 39, 176–189.

Dutky, S.R. and Hough, W.S. (1955) Note on a parasitic nematode from codling moth larvae, *Carpocapsae pomonella*. *Proceedings of the Entomological Society of Washington* 57, 244.

Efron, D., Nestel, D. and Glazer, I. (2001) Spatial analysis of entomopathogenic nematodes and insect hosts in a citrus grove in a semi-arid region in Israel. *Environmental Entomology* 30, 254–261.

Ehlers, R.U. and Hokkanen, H.M.T. (1996) Insect biocontrol with non-endemic entomopathogenic nematodes (*Steinernema* and *Heterorhabditis* spp.): conclusions and recommendations of a combined OECD and cost workshop on scientific and regulatory policy issues. *Biocontrol Science and Technology* 6, 295–302.

El-Borai, F.E., Duncan, L.W. and Preston, J.F. (2005) Bionomics of a phoretic association between *Paenibacillus* sp. and the entomopathogenic nematode *Steinernema diaprepesi*. *Journal of Nematology* 37, 18–25.

El-Borai, F.E., Bright, D.B., Graham, J.H., Stuart, R.J., Cubero, J. and Duncan, L.W. (2009) Differential susceptibility of entomopathogenic nematodes to nematophagous fungi from Florida citrus orchards. *Nematology* 11, 231–241.

El-Borai, F.E., Campos-Herrera, R., Stuart, R.J. and Duncan, L.W. (2011) Substrate modulation, group effects and the behavioral responses of entomopathogenic nematodes to nematophagous fungi. *Journal of Invertebrate Pathology* 106, 347–356.

El-Borai, F.E., Stuart, R.J., Campos-Herrera, R., Pathak, E. and Duncan, L.W. (2012) Entomopathogenic nematodes, root weevil larvae, and dynamic interactions among soil texture, plant growth, herbivory, and predation. *Journal of Invertebrate Pathology* 109, 134–142.

Ellers-Kirk, C.D., Fleischer, S.J., Snyder, R.H. and Lynch, J.P. (2000) Potential of entomopathogenic nematodes for biological control of *Acalymma vittatum* (Coleoptera: Chrysomelidae) in cucumbers grown in conventional and organic soil management systems. *Journal of Economic Entomology* 93, 605–612.

Elliott, E.T. and Coleman, D.C. (1988) Let the soil work for us. *Ecological Bulletins* 39, 23–32.

Ennis, D.E., Dillon, A.B. and Griffin, C.T. (2010) Simulated roots and host feeding enhance infection of subterranean insects by the entomopathogenic nematode *Steinernema carpocapsae*. *Journal of Invertebrate Pathology* 103, 140–143.

Enright, M.R. and Griffin, C.T. (2005) *Paenibacillus nematophilus* on the entomopathogenic nematode *Heterorhabditis megidis*. *Journal of Invertebrate Pathology* 88, 40–48.

Enright, M.R., McInerney, J.O. and Griffin, C.T. (2003) Characterization of endospore-forming bacteria associated with entomopathogenic nematodes, *Heterorhabditis* spp., and description of *Paenibacillus nematophilus* sp. nov. *International Journal of Systematic and Evolutionary Microbiology* 53, 435–441.

Ferguson, C.S., Schroeder, P.C. and Shields, E.J. (1995) Vertical distribution, persistence, and activity of entomopathogenic nematodes (Nematoda: Heterorhabditidae and Steinernematidae) in alfalfa snout beetle- (Coleoptera: Curculionidae) infested fields. *Environmental Entomology* 24, 149–158.

Filgueiras, C.C. and Willett, D.S. (2021) Non-lethal effects of entomopathogenic nematode infection. *Scientific Reports* 11: 17090.

Fiscus, D.A. and Neher, D.A. (2002) Distinguishing sensitivity of free-living soil nematode genera to physical and chemical disturbances. *Ecological Applications* 12, 565–575.
Futch, S.H., Duncan, L.W. and Zekri, M. (2005) Validation of an area-wide extension program to estimate the seasonal abundance of adult citrus root weevils with unbaited pyramidal traps. *Proceedings of the Florida State Horticultural Society* 117, 143–149.
García Del Pino, F. and Palomo, A. (1997) Temporal study of natural populations of heterorhabditid and steinernematid nematodes in horticultural crop soils. *Fundamental and Applied Nematology* 20, 473–480.
Geisen, S., Snoek, L.B., ten Hooven, F.C., Duyts, H., Kostenko, O. *et al.* (2018) Integrating quantitative morphological and qualitative molecular methods to analyse soil nematode community responses to plant range expansion. *Methods in Ecology and Evolution* 9, 1366–1378.
Glazer, I. (1992) Survival and efficacy of *Steinernema carpocapsae* in an exposed environment. *Biocontrol Science and Technology* 2, 101–107.
Glazer, I. (2002) Survival biology. In: Gaugler, R. (ed.) *Entomopathogenic Nematology*. CAB International, Wallingford, UK, pp. 169–187.
Glazer, I., Gaugler, R. and Segal, D. (1991) Genetics of the nematode *Heterorhabditis bacteriophora* strain HP88: the diversity of beneficial traits. *Journal of Nematology* 23, 324–33.
Grewal, P.S., Selvan, S. and Gaugler, R. (1994) Thermal adaptation of entomopathogenic nematodes: niche breadth for infection, establishment, and reproduction. *Journal of Thermal* Biology 19, 245–253.
Grewal, P.S., Bornstein-Forst, S., Burnell, A.M., Glazer, I. and Jagdale, G.B. (2006) Physiological, genetic, and molecular mechanisms of chemoreception, thermobiosis, and anhydrobiosis in entomopathogenic nematodes. *Biological Control* 38, 54–65.
Griffin, C.T. (2015) Behaviour and population dynamics of entomopathogenic nematodes following application. In: Campos-Herrera, R. (ed.) *Nematode Pathogenesis of Insects and Other Pests: Ecology and Applied Technologies for Sustainable Plant and Crop Protection*. Springer, Cham, Switzerland, pp. 57–95.
Griffin, C.T., Moore, J.F. and Downes, M.J. (1991) Occurrence of insect-parasitic nematodes (Steinernematidae, Heterorhabditidae) in the Republic of Ireland. *Nematologica* 37, 92–100.
Gulcu, B., Hazir, S. and Kaya, H.K. (2012) Scavenger deterrent factor (SDF) from symbiotic bacteria of entomopathogenic nematodes. *Journal of Invertebrate Pathology* 110, 326–333.
Gulzar, S., Wakil, W. and Shapiro-Ilan, D.I. (2021) Combined effect of entomopathogens against *Thrips tabaci* Lindeman (Thysanoptera: Thripidae): laboratory, greenhouse and field trials. *Insects* 12: 456.
Gutiérrez, C., Fernández, C., Escuer, M., Campos-Herrera, R., Beltrán Rodríguez, M.E. *et al.* (2016) Effect of soil properties, heavy metals and emerging contaminants in the soil nematodes diversity. *Environmental Pollution* 213, 184–194.
Guyer, A., van Doan, C., Maurer, C., Machado, R.A.R., Mateo, P. *et al.* (2021) Climate change modulates multitrophic interactions between maize, a root herbivore, and its enemies. *Journal of Chemical Ecology* 47, 889–906.
Hajek, A.E. and Meyling, N.V. (2018) Fungi. In: Hajek, A.E. and Shapiro-Ilan, D.I. (eds.) *Ecology of Invertebrate Diseases*. Wiley, Hoboken, New Jersey, pp. 327–377.
Hazir, S., Kaya, H.K., Touray, M., Çimen, H. and Shapiro-Ilan, D. (2022) Basic laboratory and field manual for conducting research with the entomopathogenic nematodes, *Steinernema* and *Heterorhabditis*, and their bacterial symbionts. *Turkish Journal of Zoology* 46, 305–350.
He, Z., Deng, Y., van Nostrand, J.D., Tu, Q., Xu, M. *et al.* (2010) GeoChip 3.0 as a high-throughput tool for analyzing microbial community composition, structure and functional activity. *ISME Journal* 4, 1167–1179.
Helmberger, M.S., Shields, E.J. and Wickings, K.G. (2018) Soil microarthropod communities reduce *Heterorhabditis bacteriophora* (Nematoda: Heterorhabditidae) host infection. *Agricultural and Forest Entomology* 20, 523–530.
Hiltpold, I., Erb, M., Robert, C.A.M. and Turlings, T.C.J. (2011) Systemic root signalling in a belowground, volatile-mediated tritrophic interaction. *Plant, Cell & Environment* 34, 1267–1275.
Hominick, W.M. (2002) Biogeography. In: Gaugler, R. (ed.) *Entomopathogenic Nematology*. CAB International, Wallingford, UK, pp. 155–143.
Hoy, C.W., Grewal, P.S., Lawrence, J.L., Jagdale, G.B. and Acosta, N. (2008) Canonical correspondence analysis demonstrates unique soil conditions for entomopathogenic nematode species compared with other free-living nematode species. *Biological Control* 46, 371–379.
Ishibashi, N., Young, F., Nakashima, M., Abiru, C. and Haraguchi, N. (1987) Effects of application of DD-136 on silkworm, *Bombyx mori*, predatory insect, *Agriosphodorus dohrni*, parasitoid, *Trichomalus apanteloctenus*, soil mites, and other non-target soil arthropods, with brief notes on feeding behaviour and predatory pressure of soil mites, tardigrades, and predatory nematodes on DD-136 nematodes. In:

Ishibashi, N. (ed.) *Recent Advances in Biological Control of Insect Pests by Entomogenous Nematodes in Japan*. Saga University, Saga, Japan, pp. 158–164.
Jabbour, R., Crowder, D.W., Aultman, E.A. and Snyder, W.E. (2011) Entomopathogen biodiversity increases host mortality. *Biological Control* 59, 277–283.
Jaffuel, G., Mäder, P., Blanco-Pérez, R., Chiriboga, X., Fließbach, A. *et al.* (2016) Prevalence and activity of entomopathogenic nematodes and their antagonists in soils that are subject to different agricultural practices. *Agriculture, Ecosystems & Environment* 230, 329–340.
Jaffuel, G., Blanco-Pérez, R., Büchi, L., Mäder, P., Fließbach, A. *et al.* (2017) Effects of cover crops on the overwintering success of entomopathogenic nematodes and their antagonists. *Applied Soil Ecology* 114, 62–73.
Jaffuel, G., Blanco-Pérez, R., Hug, A.S., Chiriboga, X., Meuli, R.G. *et al.* (2018) The evaluation of entomopathogenic nematode soil food web assemblages across Switzerland reveals major differences among agricultural, grassland and forest ecosystems. *Agriculture, Ecosystems & Environment* 262, 48–57.
Jaworska, M. and Gospodarek, J. (2009) Effect of magnesium on beneficial organisms. *Journal of Elementology* 14, 257–263.
Kanga, F.N., Waeyenberge, L., Hauser, S. and Moens, M. (2012) Distribution of entomopathogenic nematodes in Southern Cameroon. *Journal of Invertebrate Pathology* 109, 41–51.
Kapranas, A., Maher, A.M.D. and Griffin, C.T. (2016) Higher relatedness mitigates mortality in a nematode with lethal male fighting. *Journal of Evolutionary Biology* 29, 344–351.
Kapranas, A., Maher, A.M.D. and Griffin, C.T. (2017) The influence of organic matter content and media compaction on the dispersal of entomopathogenic nematodes with different foraging strategies. *Parasitology* 144, 1956–1963.
Kaya, H.K. (2002) Natural enemies and other antagonists. In: Gaugler, R. (ed.) *Entomopathogenic Nematology*. CAB International, Wallingford, UK, pp. 189–203.
Kaya, H.K. and Brayton, M.A. (1978) Interaction between *Neoaplectana carpocapsae* and a granulosis virus of the armyworm *Pseudaletia unipuncta*. *Journal of Nematology* 10, 350–354.
Kaya, H.K. and Burlando, T.M. (1989) Development of *Steinernema feltiae* (Rhabditida: Steinernematidae) in diseased insect hosts. *Journal of Invertebrate Pathology* 53, 164–168.
Kaya, H.K. and Koppenhöfer, A.M. (1996) Effects of microbial and other antagonistic organism and competition on entomopathogenic nematodes. *Biocontrol Science and Technology* 6, 357–372.
Kergunteuil, A., Campos-Herrera, R., Sánchez-Moreno, S., Vittoz, P. and Rasmann, S. (2016) The abundance, diversity, and metabolic footprint of soil nematodes is highest in high elevation alpine grasslands. *Frontiers in Ecology and Evolution* 4: 84.
Khathwayo, Z., Ramakuwela, T., Hatting, J., Shapiro-Ilan, D.I. and Cochrane, N. (2021) Quantification of pH tolerance levels among entomopathogenic nematodes. *Journal of Nematology* 53: e2021-62.
Khumalo, N.N., Lephoto, T.E. and Gray, V.M. (2021) The effect of organic compost and soil texture on the survival and infectivity of entomopathogenic nematode species. *Archives of Phytopathology and Plant Protection* 54, 1443–1455.
Kolombar, T.M., Gugosyan, Y.A. and Brygadyrenko, V.V. (2020) Impact of mineral fertilizers, growth stimulators, pH regulators, vitamins and pigment supplements on the vitality of entomopathogenic nematodes of Steinernematidae and Heterorhabditidae families. *Regulatory Mechanisms in Biosystems* 11, 323–329.
Koppenhöfer, A.M. and Fuzy, E.M. (2007) Soil moisture effects on infectivity and persistence of the entomopathogenic nematodes *Steinernema scarabaei*, *S. glaseri*, *Heterorhabditis zealandica*, and *H. bacteriophora*. *Applied Soil Ecology* 35, 128–139.
Koppenhöfer, A.M., Jaffee, B.A., Muldoon, A.E., Strong, D.R. and Kaya, H.K. (1996) Effect of nematode-trapping fungi on an entomopathogenic nematode originating from the same field site in California. *Journal of Invertebrate Pathology* 68, 246–252.
Koppenhöfer, A.M., Campbell, J.F., Kaya, H.K. and Gaugler, R. (1998) Estimation of entomopathogenic nematode population density in soil by correlation between bait insect mortality and nematode penetration. *Fundamental and Applied Nematology* 21, 95–102.
Kour, S. (2017) Occurrence and characterization of entomopathogenic nematodes in Fiji Islands and their potential application in biocontrol. PhD thesis, University of the South Pacific, Fiji Islands.
Kung, S.P., Gaugler, R. and Kaya, H.K. (1990a) Soil type and entomopathogenic nematode persistence. *Journal of Invertebrate Pathology* 55, 401–406.
Kung, S.P., Gaugler, R. and Kaya, H.K. (1990b) Influence of soil pH and oxygen on persistence of *Steinernema* spp. *Journal of Nematology* 22, 440–445.

Kung, S.P., Gaugler, R. and Kaya, H.K. (1991) Effects of soil temperature, moisture, and relative humidity on entomopathogenic nematode persistence. *Journal of Invertebrate Pathology* 57, 242–249.

Lacey, L.A., Grzywacz, D., Shapiro-Ilan, D.I., Frutos, R., Brownbridge, M. and Goettel, M.S. (2015) Insect pathogens as biological control agents: back to the future. *Journal of Invertebrate Pathology* 132, 1–41.

Lankin, G., Vidal-Retes, G., Allende, G., Castaneda-Alvarez, C., San-Blas, E. and Aballay, E. (2020) Soil texture, infective juvenile concentration, and soil organic matter influence the efficacy of *Steinernema feltiae* isolate Lican Ray. *Journal of Nematology* 52, 1–11.

Levy, N., Faigenboim, A., Salame, L., Molina, C., Ehlers, R.U. *et al.* (2020) Characterization of the phenotypic and genotypic tolerance to abiotic stresses of natural populations of *Heterorhabditis bacteriophora*. *Scientific Reports* 10:10500.

Lewis, E.E., Hazir, S., Hodson, A. and Gulcu, B. (2015) Trophic relationships of entomopathogenic nematodes in agricultural habitats. In: Campos-Herrera, R. (ed.) *Nematode Pathogenesis of Insects and Other Pests: Ecology and Applied Technologies for Sustainable Plant and Crop Protection*. Springer, Cham, Switzerland, pp. 139–163.

Liu, T., Hu, F. and Li, H. (2019) Spatial ecology of soil nematodes: perspectives from global to micro scales. *Soil Biology and Biochemistry* 137: 107565.

López-Llorca, L.V., Maciá-Vicente, J.G. and Jansson, H.-B. (2007) Mode of action and interactions of nematophagous fungi. In: Ciancio, A. and Mukerji, K.G. (eds.) *Integrated Management of Plant Pests and Diseases, Vol.* 2. Springer Dordrecht, Netherlands, pp. 51–76.

Lotka, A.J. (1925) Elements of physical biology. *Nature* 116, 461.

Machado, R.A.R., Bhat, A.H., Abolafia, J., Muller, A., Bruno, P. *et al.* (2021) Multi-locus phylogenetic analyses uncover species boundaries and reveal the occurrence of two new entomopathogenic nematode species, *Heterorhabditis ruandica* n. sp. and *Heterorhabditis zacatecana* n. sp. *Journal of Nematology* 53: e2021-89.

Machado, R.A.R., Hussain Bhat, A., Abolafia, J., Shokoohi, E., Fallet, P. *et al.* (2022) *Steinernema africanum* n. sp. (Rhabditida, Steinernematidae), a new entomopathogenic nematode species isolated in the Republic of Rwanda. *Journal of Nematology* 54: e2022-1.

Millar, L.C. and Barbercheck, M.E. (2001) Interaction between endemic and introduced entomopathogenic nematodes in conventional-till and no-till corn. *Biological Control* 22, 235–245.

Millar, L.C. and Barbercheck, M.E. (2002) Effects of tillage practices on entomopathogenic nematodes in a corn agroecosystem. *Biological Control* 25, 1–11.

Mráček, Z., Bečvář, S., Kindlmann, P. and Jersáková, J. (2005) Habitat preference for entomopathogenic nematodes, their insect hosts and new faunistic records for the Czech Republic. *Biological Control* 34, 27–37.

Navarro, P.D., McMullen, J.G. and Stock, S.P. (2014) Interactions between the entomopathogenic nematode *Heterorhabditis sonorensis* (Nematoda: Heterorhabditidae) and the saprobic fungus *Fusarium oxysporum* (Ascomycota: Hypocreales). *Journal of Invertebrate Pathology* 115, 41–47.

Neher, D.A. (2010) Ecology of plant and free-living nematodes in natural and agricultural soil. *Annual Review of Phytopathology* 48, 371–394.

Neher, D.A., Wu, J., Barbercheck, M.E. and Anas, O. (2005) Ecosystem type affects interpretation of soil nematode community measures. *Applied Soil Ecology* 30, 47–64.

Nordbring-Hertz, B., Jansson, H.-B. and Tunlid, A. (2006) Nematophagous fungi. In: *eLS*. Wiley, Hoboken, New Jersey. Available at: https://onlinelibrary.wiley.com/doi/book/10.1002/047001590X (accessed 15 March 2024).

Orgiazzi, A., Bardgett, R.D., Barrios, E., Behan-Pelletier, V., Briones, M.J.I. *et al.* (2016) *Global Soil Biodiversity Atlas*. European Commission, Publications Office of the European Union, Luxembourg.

Parkman, J.P. and Frank, J.H. (2002) Interactions between *Ormia depleta* (Diptera: Tachinidae) and *Steinernema scapterisci* (Nematoda: Steinernematidae), natural enemies of pest mole crickets (Orthoptera: Gryllotalpidae). *Biological Control* 31, 1226–1230.

Parkman, J.P. and Smart, G.C. (1996) Entomopathogenic nematodes, a case study: introduction of *Steinernema scapterisci* in Florida. *Biocontrol Science and Technology* 6, 413–420.

Parkman, J.P., Frank, J.H., Nguyen, K.B. and Smart, G.C. (1993) Dispersal of *Steinernema scapterisci* (Rhabditida: Steinernematidae) after inoculative applications for mole cricket (Orthoptera: Gryllotalpidae) control in pastures. *Biological Control* 3, 226–232.

Pathak, E., El-Borai, F.E., Campos-Herrera, R., Johnson, E.G., Stuart, R.J. *et al.* (2012) Use of real-time PCR to discriminate parasitic and saprophagous behaviour by nematophagous fungi. *Fungal Biology* 116, 563–573.

Pathak, E., Campos-Herrera, R., El-Borai, F.E. and Duncan, L.W. (2017) Spatial relationships between entomopathogenic nematodes and nematophagous fungi in Florida citrus orchards. *Journal of Invertebrate Pathology* 144, 37–46.

Perry, J.N. and Hewitt, M. (1991) A new index of aggregation for animal counts. *Biometrics* 47, 1505–1518.

Peters, A. (1996) The natural host range of *Steinernema* and *Heterorhabditis* spp. and their impact on insect populations. *Biocontrol Science and Technology* 6, 389–402.

Poinar, G.O. (1979) *Nematodes for Biological Control of Insects*. CRC Press, Boca Raton, Florida.

Poinar, G.O. and Hess, R.T. (1988) Protozoan diseases. In: Poinar, G.O. and Jansson, H.-B. (eds.) *Diseases of Nematodes*. CRC Press, Boca Raton, Florida, pp. 103–132.

Portillo-Aguilar, C., Villani, M.G., Tauber, M.J., Tauber, C.A. and Nyrop, J.P. (1999) Entomopathogenic nematode (Rhabditida: Heterorhabditidae and Steinernematidae) response to soil texture and bulk density. *Environmental Entomology* 28, 1021–1035.

Preisser, E.L. (2003) Field evidence for a rapidly cascading underground food web. *Ecology* 84, 869–874.

Preisser, E.L. and Strong, D.R. (2004) Climate affects predator control of an herbivore outbreak. *American Naturalist* 163, 754–762.

Půža, V. and Mráček, Z. (2010a) Mechanisms of coexistence of two sympatric entomopathogenic nematodes, *Steinernema affine* and *S. kraussei* (Nematoda: Steinernematidae), in a central European oak woodland soil. *Applied Soil Ecology* 45, 65–70.

Půža, V. and Mráček, Z. (2010b) Does scavenging extend the host range of entomopathogenic nematodes (Nematoda: Steinernematidae)? *Journal of Invertebrate Pathology* 104, 1–3.

Qiu, L. and Bedding, R.A. (2000) Energy metabolism and survival of the infective juveniles of *Steinernema carpocapsae* under oxygen-deficient conditions. *Journal of Nematology* 32, 271–280.

Ram, K., Gruner, D.S., McLaughlin, J.P., Preisser, E.L. and Strong, D.R. (2008) Dynamics of a subterranean trophic cascade in space and time. *Journal of Nematology* 40, 85–92.

Rasmann, S., Kollner, T.G., Degenhardt, J., Hiltpold, I., Toepfer, S. *et al.* (2005) Recruitment of entomopathogenic nematodes by insect-damaged maize roots. *Nature* 434, 732–737.

Read, D.S., Sheppard, S.K., Bruford, M.W., Glen, D.M. and Symondson, W.O. (2006) Molecular detection of predation by soil micro-arthropods on nematodes. *Molecular Ecology* 15, 1963–1972.

Regaiolo, A., Dominelli, N., Andresen, K. and Heermann, R. (2020) The biocontrol agent and insect pathogen *Photorhabdus luminescens* interacts with plant roots. *Applied and Environmental Microbiology* 86: e00891.

Salame, L. and Glazer, I. (2015) Stress avoidance: vertical movement of entomopathogenic nematodes in response to soil moisture gradient. *Phytoparasitica* 43, 647–655.

Salomon, A. and Glazer, I. (1999) Desiccation survival of entompathogenic nematode *Steinernema feltiae*: induction of anhydrobiosis. *Nematology* 1, 61–68.

San-Blas, E. and Gowen, S.R. (2008) Facultative scavenging as a survival strategy of entomopathogenic nematodes. *International Journal for Parasitology* 38, 85–91.

San-Blas, E., Campos-Herrera, R., Dolinski, C., Monteiro, C., Andaló, V. *et al.* (2019) Entomopathogenic nematology in Latin America: a brief history, current research and future prospects. *Journal of Invertebrate Pathology* 165, 22–45.

Saunders, M.C. and All, J.N. (1982) Laboratory extraction methods and field detection of entomophilic Rhabditoid nematodes from soil. *Environmental Entomology* 11, 1164–1165.

Schlüter, S., Gil, E., Doniger, T., Applebaum, I. and Steinberger, Y. (2022) Abundance and community composition of free-living nematodes as a function of soil structure under different vineyard managements. *Applied Soil Ecology* 170: 104291.

Selvan, S., Grewal, P.S., Leustek, T. and Gaugler, R. (1996) Heat shock enhances thermotolerance of infective juvenile insect-parasitic nematodes *Heterorhabditis bacteriophora* (Rhabditida: Heterorhabditidae). *Experientia* 52, 727–730.

Shamseldean, M.M. and Abd-Elgawad, M.M.M. (1991) Natural occurrence of insect pathogenic nematodes (Rhabditida: Heterorhabditidae) in Egyptian soils. *Afro-Asian Journal of Nematology* 4, 151–154.

Shapiro-Ilan, D.I. and Brown, I.M. (2013) Earthworms as phoretic hosts for *Steinernema carpocapsae* and *Beauveria bassiana*: implications for enhanced biological control. *Biological Control* 66, 41–48.

Shapiro-Ilan, D.I., Berry, E.C. and Lewis, E.E. (1993) Interactions between nematodes and earthworms: enhanced dispersal of *Steinernema carpocapsae*. *Journal of Nematology* 25, 189–192.

Shapiro-Ilan, D.I., Tylka, G.L., Berry, E.C. and Lewis, E.E. (1995) Effects of earthworms on the dispersal of *Steinernema* spp. *Journal of Nematology* 27, 21–28.

Shapiro-Ilan, D.I., Glazer, I. and Segal, D. (1997) Genetic improvement of heat tolerance in *Heterorhabditis bacteriophora* through hybridization. *Biological Control* 8, 153–159.
Shapiro-Ilan, D.I., Obrycki, J.J., Lewis, L.C. and Jackson, J.J. (1999) Effects of crop residue on the persistence of *Steinernema carpocapsae*. *Journal of Nematology* 31, 517–519.
Shapiro-Ilan, D.I., Gardner, W.A., Fuxa, J.R., Wood, B.W., Nguyen, K.B. *et al.* (2003) Survey of entomopathogenic nematodes and fungi endemic to pecan orchards of the Southeastern United States and their virulence to the pecan weevil (Coleoptera: Curculionidae). *Environmental Entomology* 32, 187–195.
Shapiro-Ilan, D.I., Jackson, M., Reilly, C.C. and Hotchkiss, M.W. (2004) Effects of combining an entomopathogenic fungi or bacterium with entomopathogenic nematodes on mortality of *Curculio caryae* (Coleoptera: Curculionidae). *Biological Control* 30, 119–126.
Shapiro-Ilan, D.I., Lewis, E.E. and Schliekelman, P. (2014) Aggregative group behavior in insect parasitic nematode dispersal. *International Journal for Parasitology* 44, 49–54.
Shaurub, E.S.H., Reyad, N.F., Abdel-Wahab, H.A. and Ahmed, S.H. (2016) Mortality and nematode production in *Spodoptera littoralis* larvae in relation to dual infection with *Steinernema riobrave*, *Heterorhabditis bacteriophora*, and *Beauveria bassiana*, and the host plant. *Biological Control* 103, 86–94.
Shi, Y.M. and Bode, H.B. (2018) Chemical language and warfare of bacterial natural products in bacteria–nematode–insect interactions. *Natural Product Reports* 35, 309–335.
Small, R.W. (1987) A review of the prey of predatory soil nematodes. *Pedobiologia* 30, 179–206.
Solomon, A., Salomon, R., Paperna, I. and Glazer, I. (2000) Desiccation stress of entomopathogenic nematodes induces the accumulation of a novel heat-stable protein. *Parasitology* 121, 409–416.
Som, S., Willett, D.S. and Alborn, H.T. (2017) Dynamics of belowground volatile diffusion and degradation. *Rhizosphere* 4, 70–74.
Somasekhar, N. and Mehta, U.K. (2000) Infectivity of *Pasteuria penetrans* to entomopathogenic nematodes. *Nematologia Mediterranea* 28, 13–14.
Spiridonov, S.E., Moens, M. and Wilson, M.J. (2007) Fine scale spatial distributions of two entomopathogenic nematodes in a grassland soil. *Applied Soil Ecology* 37, 192–201.
Stelinski, L.L., Willett, D., Rivera, M.J. and Ali, J.G. (2019) 'Tuning' communication among four trophic levels of the root biome to facilitate biological control. *Biological Control* 131, 49–53.
Stock, S.P. (2015) Diversity, biology and evolutionary relationships. I: Campos-Herrera, R. (ed.) *Nematode Pathogenesis of Insects and Other Pests: Ecology and Applied Technologies for Sustainable Plant and Crop Protection*. Springer, Cham, Switzerland, pp. 3–27.
Strong, D.R. (2002) Populations of entomopathogenic nematodes in foodwebs. In: Gaugler, R. (ed.) *Entomopathogenic Nematology*. CAB International, Wallingford, UK, pp. 225–240.
Stuart, R.J., Barbercheck, M.E. and Grewal, P.S. (2015) Entomopathogenic nematodes in the soil environment: distributions, interactions and the influence of biotic and abiotic factors. In: Campos-Herrera, R. (ed.) *Nematode Pathogenesis of Insects and Other Pests: Ecology and Applied Technologies for Sustainable Plant and Crop Protection*. Springer, Cham, Switzerland, pp. 97–137.
Sturhan, D. and Mráček, Z. (2000) Comparison of the *Galleria* baiting technique and a direct extraction method for recovering *Steinernema* (Nematoda: Rhabditida) infective-stage juveniles from soil. *Folia Parasitologica* 47, 315–318.
Susurluk, A.I. (2009) Seasonal and vertical distribution of the entomopathogenic nematodes *Heterorhabditis bacteriophora* (TUR-H2) and *Steinernema feltiae* (TUR-S3) in turf and fallow areas. *Nematology* 11, 321–327.
Susurluk, A. and Ehlers, R.U. (2008) Field persistence of the entomopathogenic nematode *Heterorhabditis bacteriophora* in different crops. *BioControl* 53, 627–641.
Tarasco, E., Santiago Alvarez, C., Triggiani, O. and Moraga, E.Q. (2011) Laboratory studies on the competition for insect haemocoel between *Beauveria bassiana* and *Steinernema ichnusae* recovered in the same ecological niche. *Biocontrol Science and Technology* 21, 693–704.
Tian, C.L., Zhu, F., Li, X.Y., Zhang, J.H., Půža, V. *et al.* (2022) *Steinernema populi* n. sp. (Panagrolaimomorpha, Steinernematidae), a new entomopathogenic nematode species from China. *Journal of Helminthology* 96: e57.
Torr, P., Spiridonov, S.E., Heritage, S. and Wilson, M.J. (2007) Habitat associations of two entomopathogenic nematodes: a quantitative study using real-time quantitative polymerase chain reactions. *Journal of Animal Ecology* 76, 238–245.
Treonis, A.M., Unangst, S.K., Kepler, R.M., Buyer, J.S., Cavigelli, M.A. *et al.* (2018) Characterization of soil nematode communities in three cropping systems through morphological and DNA metabarcoding approaches. *Scientific Reports* 8: 2004.

van Tol, R.W.H.M., van Bexooijen, J. and Ketelaars, T.A.C.M. (1998) Searching behaviour of entomopathogenic nematodes: roots and soil temperature determine success of black vine weevil (*Otiorhynchus sulcatus*) control. *Insect Pathogens and Insect Parasitic Nematodes IOBC/WPRS Bulletin* 21, 187–191.

Volterra, V. (1926) Fluctuations in the abundance of a species considered mathematically. *Nature* 118, 558–560.

Waeyenberge, L., de Sutter, N., Viaene, N. and Haegeman, A. (2019) New insights into nematode DNA-metabarcoding as revealed by the characterization of artificial and spiked nematode communities. *Diversity* 11: 52.

Wallace, H.R. (1971) Abiotic influences in the soil environment. In: Zuckerman, B.M., Mai, W.F. and Rohde, R.A. (eds.) *Plant Parasitic Nematodes, Vol. 1. Morphology, Anatomy, Taxonomy, and ecology*. Academic Press, New York, pp. 257–280.

Williams, C.D., Dillon, A.B., Girling, R.D. and Griffin, C.T. (2013) Organic soils promote the efficacy of entomopathogenic nematodes, with different foraging strategies, in the control of a major forest pest: a meta-analysis of field trial data. *Biological Control* 65, 357–364.

Womersley, C.Z. (1990) Dehydration survival and anhydrobiotic potential. In: Gaugler, R. and Kaya, H.K. (eds.) *Entomopathogenic Nematodes in Biological Control*. CRC Press, Boca Raton, Florida, pp. 116–137.

Zhang, K., Baiocchi, T., Lu, D., Chang, D.Z. and Dillman, A.R. (2019) Differentiating between scavengers and entomopathogenic nematodes: which is *Oscheius chongmingensis*? *Journal of Invertebrate Pathology* 167: 107245.

8 Natural Enemies and Antagonists of Entomopathogenic Nematodes

Selcuk Hazir[1]*, Mustapha Touray[1], Harun Cimen[2], Derya Ulug[1], Canan Hazir[3] and Sebnem Hazal Gulsen[4]

[1]*Department of Biology, Faculty of Science, Aydin Adnan Menderes University, Aydin, Türkiye;* [2]*Recombinant DNA and Recombinant Protein Application and Research Center, Aydin Adnan Menderes University, Aydin, Türkiye;* [3]*Aydin Health Services Vocational School, Aydin Adnan Menderes University, Aydin, Türkiye;* [4]*Kocarli Vocational School, Aydin Adnan Menderes University, Aydin, Türkiye*

Abstract

Steinernema and *Heterorhabditis* spp. nematodes and their respective symbiotic bacteria, *Xenorhabdus* and *Photorhabdus* spp., are effective biological control agents used against various important soil insect pests. These nematodes have little or no negative effects on other non-target organisms and the environment. In their natural habitat (soil), entomopathogenic nematodes share similar niches with several organisms, such as collembolans, nematophagous fungi, mites and other nematodes, that eat or infect them or compete with them for the limited food source inside a shared host. These interactions may directly or indirectly affect nematode survival and persistence, and thus can reduce the efficacy of biological control programs that involve these nematodes.

8.1 Introduction

Soil-dwelling *Steinernema* spp. (family Steinernematidae) and *Heterorhabditis* spp. (family Heterorhabditidae) are obligate entomopathogenic nematodes (EPNs) that are mutually associated with Gram-negative *Xenorhabdus* and *Photorhabdus* spp. bacteria, respectively. These organisms have been isolated from soil environments on all continents except Antarctica and currently at least 119 nematode (102 *Steinernema* and 17 *Heterorhabditis* spp.) and 50 bacterial (27 *Xenorhabdus* and 23 *Photorhabdus* spp.) species have been described (Gulcu *et al.*, 2017; Machado *et al.*, 2018; Bhat *et al.*, 2020; Sajnaga and Kazimierczak, 2020; Hazir *et al.*, 2022). In this nematode–bacteria association, infective juvenile (IJ) forms of the nematodes infect arthropod hosts with the assistance of their vectored bacteria. The bacteria produce a cornucopia of virulence factors in the form of secondary metabolites, toxins and enzymes that aid in suppressing insect immunity and killing the insect host. The bacteria also perform defensive functions against competitors and deter natural enemies (Hominick *et al.*, 1996; Glazer, 2002; Bode, 2009; Gulcu *et al.*, 2012; Ulug *et al.*, 2014). EPNs are one of the most compelling commercialized biocontrol options used to manage insect pests because of a variety of attributes: (i) the environmentally

*Email: selcuk.hazir@gmail.com

 Entomopathogenic Nematodes as Biological Control Agents (D.I. Shapiro-Ilan and E.E. Lewis eds)
DOI: 10.1079/9781800620322.0008

durable and non-feeding IJ stage actively seeks out a broad range of hosts in soil or in cryptic habitats; (ii) the nematodes rarely harm non-target organisms; and (iii) they are amenable to mass production and formulation. In addition, as important safe alternatives to chemical pesticides, these nematodes are not on the compulsory registration list of many countries (Shapiro-Ilan *et al.*, 2012, 2014, 2017).

EPNs share the same highly competitive ecological niche as several groups of organisms. In such environments (i.e. in the soil or inside insect cadavers), all organisms are constantly competing for limited resources using characters or behaviors that maximize their fitness. Both the IJs foraging in soil and the other life stages (which spend 7–20 days or more in infected insect cadavers) are vulnerable to attack from natural enemies that prey on them or compete for food. Some of the main antagonists include collembolans, nematophagous fungi (NFs), tardigrades, mites and other nematodes (Kaya, 2002; Jabbour and Barbercheck, 2011; Stuart *et al.*, 2015; Raja *et al.*, 2021). These organisms are grouped into antagonists of IJs or natural enemies, and antagonists of infected cadavers.

This chapter reviews the body of published research detailing how such competitors, omnivores, scavengers and natural enemies affect populations of EPN–bacteria complexes and the defense mechanisms these organisms employ (Fig. 8.1).

8.2 Enemies and Antagonists of Infective Juveniles

8.2.1 Bacteria

Various bacteria in soil are known pathogens of nematodes; for example, endospore-forming bacteria in the genera *Pasteuria* and *Paenibacillus* have been described as natural antagonists/parasites of plant-parasitic nematodes (PPNs) such as members of the genera *Meloidogyne*, *Heterodera*, *Pratylenchus* and *Xiphinema* (Sayre and Starr, 1988; Bajaj and Walia, 2005; Bakengesa, 2016). Endospores of these bacteria attach to the surface of nematodes and hinder nematode movement and penetration into the plant host; the bacteria can also alter embryogenesis and sterilize the nematode. Such bacteria have potential in biological control of PPNs (Davies *et al.*, 1991, 2011; Vagelas *et al.*, 2012), but they are also potential antagonists of EPNs that occur in similar environments to PPNs. There is limited research on the effects of these bacteria on EPNs. When

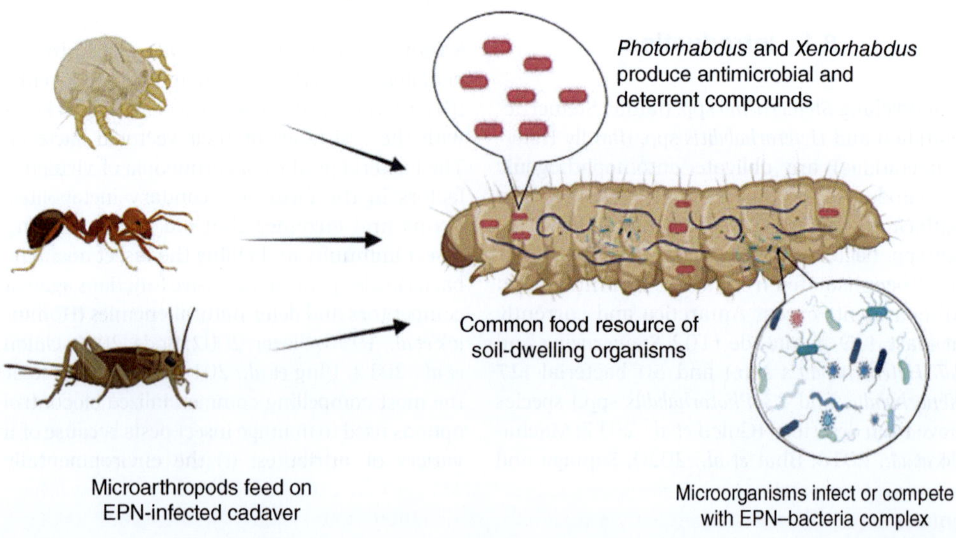

Fig. 8.1. Interaction of entomopathogenic nematode (EPN)–bacteria complexes with other soil-dwelling organisms.

Pasteuria penetrans was mixed with aqueous suspensions of *Steinernema* sp., *Heterorhabditis* sp. and the root-knot nematode, *Meloidogyne incognita*, the bacterial spores attached to the root-knot nematode after 24 h but failed to adhere to EPN IJs after 72 h (Somasekhar and Mehta, 2000). However, the spores of another *Pasteuria* sp. isolate from a soil sample was found inside the pseudocoelom and/or attached to the body cuticle of some *Steinernema pakistanense* IJs, and those IJs with fully matured spores were ineffective against an insect host, indicating a pathogenic interaction between some *Pasteuria* spp. and EPNs (Bajaj and Walia, 2005).

Paenibacillus spp., in contrast, maintain a phoretic and commensal relationship with EPNs. These bacteria have been observed to attach to the cuticular surface of *Steinernema diaprepesi*, *Steinernema glaseri*, *Heterorhabditis indica* and *Heterorhabditis zealandica* IJs and reproduce in the insect host along with the nematodes (Enright *et al.*, 2001, 2003; El-Borai *et al.*, 2003, 2005). Nevertheless, *Paenibacillus nematophilus*, which can suppress *Meloidogyne incognita*, hindered the motility of *H. megidis* in sand and on agar, and reduced its infectivity against *Galleria mellonella* larvae but had no negative effects on IJ survival, size or energy reserve (Enright and Griffin, 2005). Furthermore, assessment of the effects of the secondary metabolites of EPN-symbiotic bacteria on population growth of *Paenibacillus* spp. indicated the spores of *Paenibacillus nematophilus* were resistant to antibiotics produced by *Photorhabdus* spp. (Enright *et al.*, 2003), but growth of a *Paenibacillus* sp. was inhibited by *Xenorhabdus* spp. (El-Borai *et al.*, 2005). These results indicate the detrimental effects of these endospore-forming bacteria on EPN IJs, such as impairing EPN motility and host finding (El-Borai *et al.*, 2005; Campos-Herrera *et al.*, 2013).

Besides the genera mentioned above, studies on the effects of other bacterial species on EPN have determined that, even though *H. indica* killed *Galleria mellonella* larvae pre-fed with *Bacillus subtilis* and *Bacillus licheniformis*, infected cadavers did not develop the characteristic brick-red coloration indicative of heterorhabditid infections, and the number of second-generation females and final nematode progeny was significantly reduced compared with those in the control group (Upadhyay and Mohan, 2021). Similar antagonistic effects were observed for *Pseudomonas fluorescens* on the development of *H. indica* inside *Galleria mellonella* larvae (Kamra and Mohan, 2011).

8.2.2 Nematophagous fungi

NFs are generally saprophytic fungi specialized in trapping and digesting nematodes, which, as the most numerically abundant animals in soil, expediently become prey, especially under unfavorable soil conditions (Nordbring-Hertz, 1988; Soares *et al.*, 2018). These fungi have garnered attention as natural enemies of nematodes, especially against PPNs. Over 700 fungal species described from phyla such as the Ascomycota, Basidiomycota, Chytridiomycota, Oomycota and Zygomycota are predators or endoparasites of adult nematodes and parasites of eggs or cysts (Braga and Araújo, 2014; Li *et al.*, 2015). Predatory forms use specialized hyphae that, on contact, form traps that penetrate and kill the nematodes. Endoparasitic groups use conidia or zoospores that are ingested or adhere to the nematode cuticle and kill the nematodes from inside. Ovicidal groups such as predatory fungi use traps but target nematode eggs and cysts (Braga and Araújo, 2014).

As NFs seem to indiscriminatingly infect different species of nematodes, they may negatively impact EPN populations in soil, especially during biological control programs (Kaya and Koppenhöfer, 1996; El-Borai *et al.*, 2009; Arêdes *et al.*, 2017; Soares *et al.*, 2018). For instance, 14 days after four trapping (*Arthrobotrys oligospora*, *Arthrobotrys dactyloides*, *Arthrobotrys musiformis* and *Gamsylella gephyropaga*) and two endoparasitic (*Catenaria* sp. and *Myzocytium* sp.) NFs were added to soil microcosms with five EPN species (*S. diaprepesi*, *S. glaseri*, *S. riobrave*, *H. indica* and *H. zealandica*), the trapping fungus *Gamsylella gephyropaga* reduced nematode recovery of all species by 56–92% and both endoparasitic NFs reduced the recovery of all EPNs except *H. indica* by at least 82% (El-Borai *et al.*, 2009)

The second-stage cuticle, retained as a sheath by various third-stage IJs, can serve as a protective barrier against conidia of the endoparasitic NFs (Timper and Kaya, 1989). *Hirsutella rhossiliensis* and *Drechmeria coniospora* were less virulent against ensheathed heterorhabditids compared with exsheathed heterorhabditids exposed to dilute sodium hypochlorite, or exsheathed steinernematids moving through soil. Similarly, the predatory fungus *Arthrobotrys oligospora* was observed to be more infective against *S. glaseri* compared with *Heterorhabditis marelata* because, during molting, heterorhabditids retain their second-stage cuticle much longer than steinernematids (Jaffee and Strong, 2005). As well as the second-stage cuticle, foraging behavior and IJ size, there are other crucial factors that protect nematodes from fungal infection. Nematodes with a cruiser strategy such as heterorhabditids (Campbell and Gaugler, 1997) are more responsive to cues and are capable of evading or moving around the fungal traps compared with ambushing nematodes such as *Steinernema carpocapsae* and *S. scapterisci* IJs, which are less mobile (Lewis, 2002; Lewis *et al.*, 2015) and the latter become easier prey for NFs (Arêdes *et al.*, 2017). Additionally, larger IJs, such as *S. glaseri* and *S. diaprepesi*, which adopt a cruiser strategy (Hazir *et al.*, 2003; Del Valle *et al.*, 2014) have the strength to escape capture from predatory fungi (El-Borai *et al.*, 2009). The ability of EPN IJs to overcome such parasitic predation can help maintain a high level of biocontrol efficacy (Raja *et al.*, 2021). Interestingly, certain volatile organic compounds released by plants to recruit natural enemies of insects, such as D-limonene produced by citrus plants, caused EPN IJs to move faster in an *Arthrobotrys dactyloides*-occupied area, thereby reducing their chances of being trapped by the fungus (Willet *et al.*, 2017).

IJs respond differently to NFs depending on the EPN and NF species, and the tested substrate. An extensive study by El-Borai *et al.* (2011) revealed that while *Steinernema* sp., *S. diaprepesi*, *H. indica* and *H. zealandica* IJs were attracted to the activated trapping fungus *Arthrobotrys gephyropaga* and the endoparasitic fungi *Catenaria* sp. and *Myzocytium* sp., *S. riobrave* showed a neutral response on agar plates. In horizontal sand columns, these authors observed that all EPN species were repelled by *Arthrobotrys gephyropaga* and attracted to *Myzocytium* sp. and *Catenaria* sp.; only *H. indica* had a neutral response to *Myzocytium* sp. and *Steinernema* sp. to *Catenaria* sp. These results indicate that there are possible complexities (i.e. mixed responses, aggregation or group movement) and species specificity in the responses of EPNs to NFs, and that experiments using agar can differ markedly from using sand. Thus, the importance of utilizing natural substrates to properly assess the role of fungal semiochemicals in nematode–fungus interactions is emphasized (El-Borai *et al.*, 2011).

Research has assessed the interaction of EPNs and NFs in different soil habitats. For instance, in citrus groves infested with the weevil *Diaprepes abbreviatus* and plant-pathogenic fungi *Phytophthora* spp. in Florida, native communities of EPNs were found to be generally more diverse and more prevalent in well-drained, coarse, sandy soils than in those on the poorly drained, finer-textured soils (Duncan *et al.*, 2003a,b). Weevil abundance in these grooves was directly linked to EPN prevalence, which was affected by factors such as soil type, antagonism by the NF or parasitism by *Paenibacillus* spp. (Duncan *et al.*, 2007; Campos-Herrera *et al.*, 2012). The NFs *Purpureocillium lilacinum* and *Catenaria* sp. have been demonstrated to consistently infect nematodes such as *S. diaprepesi*, *S. riobrave*, *H. indica* and *H. zealandica* at high rates when both of these organisms were applied to sand or to loam soil samples around citrus plant roots (Duncan *et al.*, 2013).

As with endoparasites, these obligate nematode parasites rely mainly on spores (conidia, zoospores) that adhere to the nematode cuticle or are ingested by the nematode in the infected host (Soares *et al.*, 2018). Following infection of infective conidia or spores, the nematode host dies as the fungal hyphae infiltrate and grow inside, usually extending the length of the nematode (Glockling and Holbrook, 2003). Endoparasitic fungi have a restricted host range, even though they are dependent on nematodes as a food resource (Kaya *et al.*, 1998). Several factors such as larger size and greater surface area of the nematode increase the chances of coming in contact with the fungal conidia, as seen in higher mortality rates of *S. diaprepesi* and *S. glaseri* caused by *Hirsutella rhossiliensis* compared with *H. bacteriophora* or *S. carpocapsae* (Kaya *et al.*, 1998; El-Borai *et al.*, 2009).

Overall, NFs are regarded as antagonists that can impact EPN performance in field settings and thus it may be useful to target these organisms in pest management programs to increase EPN efficacy.

8.2.3 Microsporidia

Microsporidia constitute a phylum of obligate intracellular fungal parasites capable of infecting a wide variety of metazoan hosts including EPNs (Kaya, 2002; Troemel *et al.*, 2008). An unidentified microsporidian parasite was found to adversely affect the infectivity and viability of *S. glaseri* (formerly *Neoaplectana glaseri*) (Poinar, 1988). This microsporidian parasite infects tissues in all stages of the nematode while they feed on host insect tissues infected with microsporidia and as a result the nematodes survived poorly at 22°C. The parasite can be passed on to subsequent generations in the nematode population (Poinar, 1988). Vertical transmission has also been reported for *S. carpocapsae* in caterpillars infected with *Nosema mesnili* (Veremchuk and Issi, 1970).

8.2.4 Mites

Macrocheles sp. was the first mite species observed to feed on EPN IJs (Poinar, 1979). Subsequently, *Eugamasus* sp., *Tyrophagus putrescentiae*, *Alicorhagia fragilis*, *Alycus roseus*, *Pergamasus crassipes*, *Gamasellodes vermivorax*, *Cosmolaelaps vacua*, *Antennoseius* sp., *Amblyseius setulus*, *Hypoaspis* sp., *Ascanesoica*, *Ololaelaps veneta* and *Pilogalumna cozadensis* mite species were reported to feed on EPN species including *S. carpocapsae*, *S. feltiae* and *H. bacteriophora* (Ishibashi *et al.*, 1987; Epsky *et al.*, 1988). Certain species such as *Sancassania polyphyllae* (Astigmata) can consume IJs and can also feed on EPN-infected insect cadavers and developing stages of EPNs within the insect host (Figs 8.2 and 8.3) (Karagöz *et al.*, 2007; Ekmen *et al.*, 2010a; Cakmak *et al.*, 2010, 2011, 2013).

Although the above mite species are generally free-living predators or scavengers and can consume EPNs as food, it is not known whether all the mites tested can complete their life cycles

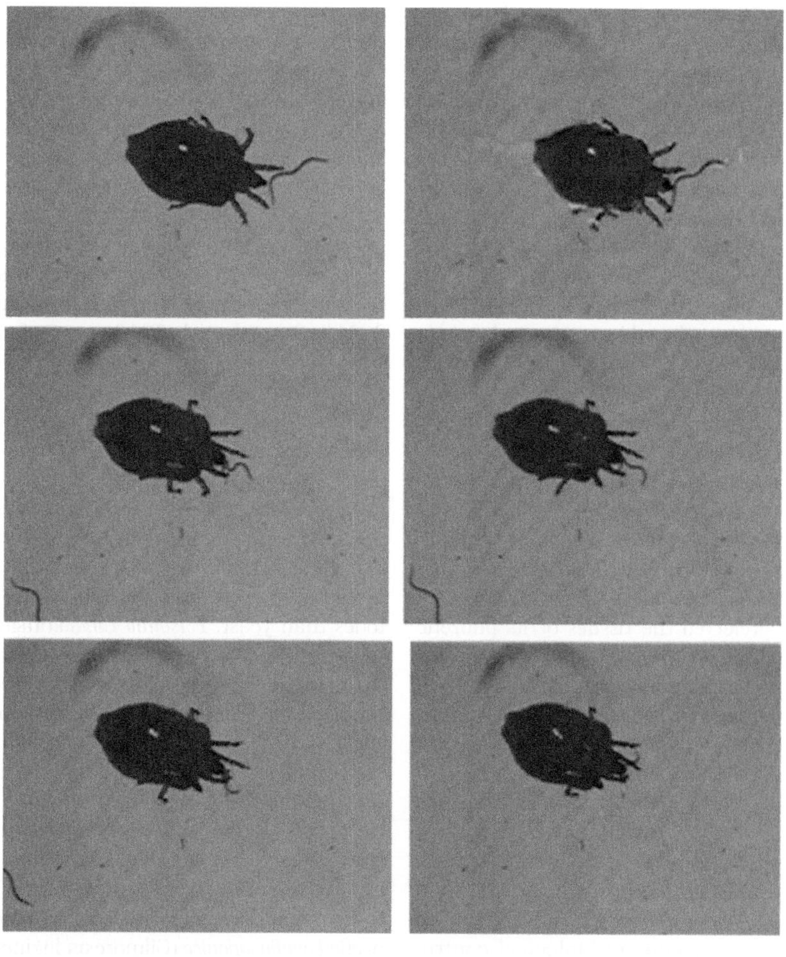

Fig. 8.2. *Sancassania polyphyllae* feeding on a *Steinernema feltiae* infective juvenile.

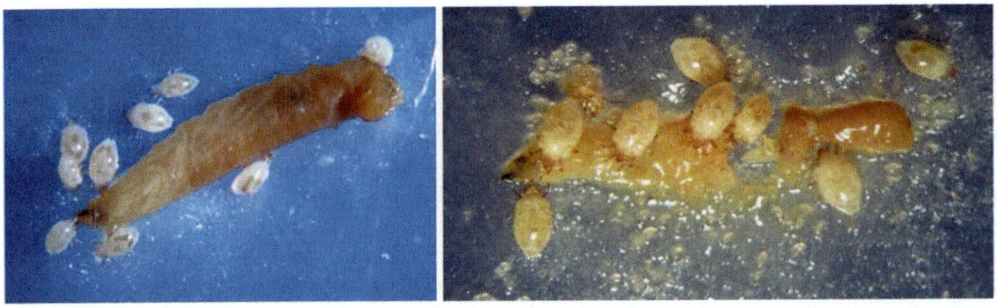

Fig. 8.3. *Sancassania polyphyllae* mites feeding on a nematode-infected insect cadaver.

by feeding on EPNs alone. The larval stages of *Sancassania polyphyllae*, *Tyrophagus putrescentiae* and *Alycus roseus* mites, for instance, cannot develop to the next stage in their life cycle (protonymph) by feeding on EPN IJs, but they can develop into adults by feeding on EPN IJs as nymphs, and female individuals can lay eggs (Epsky *et al.*, 1988; Cakmak *et al.*, 2010). However, the overall development time of *Sancassania polyphyllae* feeding on insect cadavers is significantly shorter than those feeding on IJs (Cakmak *et al.*, 2010). EPNs may rarely infect mites but can sometimes use them as a phoretic host. A phoretic relationship was observed between *S. carpocapsae* and nine mite species, but no phoresy was observed with *H. bacteriophora*. Among the mite species tested, the authors found that only one individual of *Alycus roseus* was infected by EPNs (Epsky *et al.*, 1988). No phoretic relationship existed between some mite species such as *Sancassania polyphyllae* and EPNs, and IJs are not able to infect these mites (Karagöz *et al.*, 2007).

Mites exhibit food choice and foraging preferences in reference to EPN predation. When offered scarab tissues (*Polyphylla fullo*) or waxworm (*Galleria mellonella*) and IJs of *S. feltiae* or *H. bacteriophora* in a food preference study, *Sancassania polyphyllae* preferred the tissues of its phoretic host, *Polyphylla fullo*, over the lepidopteran wax worm tissues or living IJs (Ekmen *et al.*, 2010a). Whereas *Sancassania polyphyllae* was not very successful in finding IJs located at different depths of a soil column, the mites efficiently detected EPN-infected cadavers placed at different soil depths (Ekmen *et al.*, 2010b; Cakmak *et al.*, 2013).

Based on the aforementioned studies, scavenger mites do not appear to be a significant threat to the survival of nematodes when IJs are applied as aqueous suspensions in biological control programs. However, they may feed on insects killed by the nematode before they leave the insect cadaver, negatively affecting the development of nematodes within the cadaver, or they may consume the IJs as they emerge from the cadaver (Fig. 8.3). In addition, EPN-infected cadavers can serve as a food source for various arthropods capable of nematophagy such as predatory mites, staphylinid beetles, acarid mites, and onychiurid and entomobryid collembolans. Microarthropod composition and abundance seems to increase after nematode or cadaver applications (Jabbour and Barbercheck, 2011; Greenwood *et al.*, 2011; Dritsoulas *et al.*, 2022). All these findings suggest that mites may have a negative impact on the biological control of soil insect pests using cadavers containing nematodes (Ekmen *et al.*, 2010b; Cakmak *et al.*, 2013).

8.2.5 Collembolans

Collembolans are primarily fungivores that occasional feed on other animals including nematodes (Gilmore, 1970). Sinellacaeca (Entomobryidae) grew faster when fed with nematodes than yeast. *Folsomia candida* (Isotomidae) preferred the free-living nematode *Caenorhabditis elegans* over the fungus *Cladosporium cladosporioides* (Lee and Widden, 1996). Certain collembolan species such as *Hypogasturas cotti* fed on and could complete their development on *S. carpocapsae* IJs (Ishibashi *et al.*, 1987; Epsky *et al.*, 1988). Other species such as *Folsomia candida* and *Sinella caeca* consumed many *S. feltiae*, *S. carpocapsae* and *S. glaseri* IJs and reduced the effectiveness of these nematodes against larvae of the Japanese beetle *Popillia japonica* (Gilmore and Potter, 1993).

A survey of field soil communities following EPN application revealed increases in the abundance of collembolans (Hodson *et al.*, 2012). Under laboratory conditions, whereas *Folsomia candida* and *Sinella curviseta* have been found to consume *S. feltiae* and *H. bacteriophora* IJs, *Sinella curviseta* did not scavenge on nematode-killed insects (Ulug *et al.*, 2014). PCR techniques and species-specific primers have been used to demonstrate the presence of *Phasmarhabditis hermaphrodita*, *H. megidis* and *S. feltiae* DNA in *Folsomia candida* 24 h after feeding (Read *et al.*, 2006).

The availability of alternative food sources for collembolans in field conditions will affect the predation rates of EPNs. However, some of the studies cited above suggest that some collembola species may adversely affect biological control applications by consuming free-living IJs when applied in aqueous suspensions. Based on a limited number of studies, collembolans do not appear to be a threat to EPN-killed insects that are applied to control insect pests in the greenhouse and field.

8.3 Natural Enemies and Antagonists of Infected Cadavers

EPNs directly or indirectly interact with other organisms in the host. Direct antagonistic interactions encompass predation by another organism on EPNs, whereas indirect antagonism involves competition for the limited food sources inside the shared host.

8.3.1 Entomopathogenic viruses

Entomopathogenic viruses can be significant competitors of EPNs when they coexist in the same host. Although EPNs can develop in insects previously infected by nucleopolyhedroviruses (Bednarek, 1986; Kaya and Burlando, 1989), the integument of the virus-infected insects is extremely weak, especially in the final days before death, and disintegration of the insect's cuticle causes the developing nematodes to become exposed to the external environment, which eventually leads to desiccation. In a study conducted with granulovirus (GV), *S. carpocapsae* developed and reproduced with healthy progeny in GV-infected armyworm larvae (*Spodoptera exigua*), but the number of IJs produced per milligram was significantly lower in GV-infected larvae than in healthy larvae (Kaya and Brayton, 1978). Although combining these two bioagents (viruses and EPNs) may result in synergistic or additive interactions, presumably there will be fewer nutrients due to coinfection of larvae, which adversely affects the nematode reproduction or virus proliferation in the host (Gothama *et al.*, 1995, 1996).

8.3.2 Entomopathogenic bacteria

Entomopathogenic bacteria such as *Bacillus thuringiensis* can coinfect an insect host with EPNs (Bednarek, 1986; Kaya and Burlando, 1989; Poinar *et al.*, 1990). Such dual infections have a negative impact on the growth of the nematodes, even if the developing nematodes can subsist on the reduced level of food in *B. thuringiensis*-infected hosts compared with the controls, and fewer new-generation IJs emerge (Kaya and Burlando, 1989; Poinar *et al.*, 1990). Likewise, *B. thuringiensis* competes with *Xenorhabdus* spp. for food resources when *B. thuringiensis* is inoculated into previously nematode-infected insects; this interaction has a negative impact on both the growth of *Xenorhabdus* spp. and the developmental stages of the nematodes (Poinar *et al.*, 1990). However, dominance in the insect host depends on which is the first organism to infect. For instance, *Steinernema* IJs still sequester their symbiotic bacteria as they emerged from insect cadavers inoculated with *B. thuringiensis* 24 h after nematode infection, but the emerging IJs are malnourished. Nematodes can be affected by the *B. thuringiensis* exotoxin, which has been shown to be toxic to various nematodes (Sayre and Starr, 1988). When a *B. thuringiensis*-resistant strain of *Plutella xylostella* was treated with a combined application of *S. carpocapsae* and *B. thuringiensis* under laboratory conditions, antagonistic interactions were recorded (Baur *et al.*, 1998a). In some cases, additive or synergistic interactions can be observed in combined application of *B. thuringiensis* and certain EPN species, especially against less susceptible insect hosts such as white grubs (Koppenhöfer and Kaya, 1997; Koppenhöfer *et al.*, 1999).

Such positive combinations are relevant to biological control programs.

Limited studies on other entomopathogenic bacteria and EPNs have been conducted. Almost all combinations of *S. carpocapsae* and *H. bacteriophora* with *Serratia marcescens* results in antagonism in *Curculio caryae* larvae control. *Serratia marcescens* may be directly pathogenic to EPNs; this decreases the virulence of nematodes against the host (Shapiro-Ilan *et al.*, 2004).

8.3.3 Entomopathogenic fungi

Dominance of either entomopathogenic fungi or EPNs in coinfected hosts are time and temperature dependent (Barbercheck and Kaya, 1991). The nematodes, their symbiotic bacteria and the fungi all produce compounds to exclude the other; this effect can vary with individual entomopathogenic fungus or symbiont species of the EPN (Ansari *et al.*, 2004; Tarasco *et al.*, 2011). In rare instances of dual infection, the amount of resources in the host available to both organisms will be greatly reduced, even though entomopathogenic fungi colonize different parts of the same cadaver (Tarasco *et al.*, 2011). EPN IJs tend to avoid *Beauveria bassiana*-infected hosts in the soil over non-infected controls. This minimizes any possible antagonistic interactions between EPNs and *Beauveria bassiana* in the soil (Barbercheck and Kaya, 1991). However, *Fusarium solani*, a saprophytic fungus, attracts *S. diaprepesi*, resulting in an increase in the insecticidal efficacy of the EPN (Wu *et al.*, 2018). *Fusarium solani* produces 1-octen-3-ol, a volatile organic compound that attracts *S. diaprepesi* (Wu and Duncan, 2020). Recently, 1-octen-3-ol and 3-octanone, isolated from *Metarhizium brunneum* was found to influence nematode survival and behavior (Hummadi *et al.*, 2021, 2022). The presence of these compounds in the same arena as *S. carpocapsae*, *S. feltiae* and *H. bacteriophora* IJs significantly reduced the penetration efficacy and fecundity of nematodes in insect hosts. Despite the high toxicity of volatile organic compounds toward nematodes, *H. bacteriophora* IJs had a significant preference for areas with these semiochemicals compared with controls in chemotaxis studies (Hummadi *et al.*, 2021).

Interestingly, although EPNs and entomopathogenic fungi differ in their mode of pathogenesis, their common host range allows them to be applied together for the control of various insect pests. Coapplication may produce additive, synergistic or antagonistic interactions on pest mortality (Shapiro-Ilan *et al.*, 2004; Ansari *et al.*, 2006, 2008; Wu *et al.*, 2014; Asan *et al.*, 2017; Wakil *et al.*, 2017). For instance, when *H. indica* or *S. carpocapsae* were simultaneously applied with *Beauveria bassiana* or *Metarhizium anisopliae*, antagonism was observed due to toxins produced by the fungus (Shapiro-Ilan *et al.*, 2004). Nevertheless, positive results can be obtained based on the relative timing of both applications (Abdolmaleki *et al.*, 2017), as well as the rate (Ansari *et al.*, 2004). Understanding the possible interactions between different entomopathogens plays a key role in achieving improved pest control.

8.3.4 Omnivores and scavengers

Several vertebrate and invertebrate animals have been reported to feed on EPN-infected cadavers. These organisms adversely affect EPN populations in soil by directly consuming developing nematodes in the cadavers or by compromising the integrity of the cadavers, which leads to desiccation and death of the nematodes. After the EPN–bacteria complex kills an insect host, within 48 h, the EPN-infected cadavers remain in or on the soil for 7–15 days or more depending on the infecting species and environmental conditions before a new generation of nematodes emerges from the cadaver. During this time, infected cadavers are at risk of being partially or entirely consumed by various scavengers in soil. Interestingly, several ant species (*Linepithema humile*, *Veromessor andrei*, *Pheidole vistana*, *Formica pacifica* and *Monomorium ergatogyna*) were observed to scavenge less on EPN-infected cadavers than on freeze-killed insects and also less on heterorhabditid-infected cadavers compared with steinernematid-infected cadavers (Baur *et al.*, 1998b). The ants also showed a dislike for older cadavers (4 days after nematode infection). *Linepithema humile* workers tried to cover these cadavers with tiny rocks and twigs. They also consumed 50% more of 4-day-old cadavers killed by *Steinernema* spp. compared with those killed by *Heterorhabditis* spp. (Kaya *et al.*,

1998). Injection of the mutualistic bacterium *Photorhabdus luminescens* into the insect host demonstrated that the deterrent was associated with the bacterium (Zhou *et al.*, 2002), and this antifeedant compound was called the "ant deterrent factor(s)." Similar results have been observed by different *Xenorhabdus* spp. against *Linepithema humile* workers (Fig. 8.4). This compound(s) enables the EPN–bacteria complex to survive within the insect cadavers until the new progeny emerge. When supernatants from *Photorhabdus luminescens* cultures were mixed with 5% sucrose, the ants refrained from feeding on the bacterial/sucrose mixture but readily fed on the nutrient broth/sucrose control (Zhou *et al.*, 2002).

Besides ants, other scavengers such as crickets (*Gryllus bimaculatus*) and wasps (*Vespa orientalis*, *Paravespula* sp.) as well as the ant species *Lepisiota frauenfeldi* exhibited similar avoidance behavior and did not feed on 2-day-old or older EPN-infected cadavers (Gulcu *et al.*, 2012). The calliphorid fly, *Chrysomya albiceps*, did not oviposit on meat treated with *Photorhabdus luminescens* supernatant. Based on these findings, "ant deterrent factor(s)" was renamed as "scavenger deterrent factor" (SDF) (Gulcu *et al.*, 2012). In two-choice trials, the carabid beetle *Pterostichus melanarius* preferred freeze-killed *Galleria mellonella* cadavers over those infected with *Steinernema affine* or slugs infected with *Phasmarhabditis hermaphrodita* (Foltan and Puza, 2009). Springtails (*Sinella curviseta*) and *Folsomia candida* were also affected by the SDF (Ulug *et al.*, 2014).

The carabid beetle, *Calosoma granulatum*, also prefers to feed on freeze-killed larvae instead of those infected by EPNs (Mertz *et al.*, 2015). An odor cue from EPN-infected cadavers was suggested to have a repellent effect against the nocturnal predator beetle *Pterostichus madidus*, even in the early stages of infection. As the infection ages, the feeding rate on infected cadavers decreases (Jones *et al.*, 2016).

As well as invertebrate organisms, vertebrates such as the European robin, *Erithacus rubecula* (Fenton *et al.*, 2011), and the great tit, *Parus major* (Jones *et al.*, 2017), have been found to be deterred from consuming *Heterorhabditis* spp.-infected insects. It was believed that these birds, which generally rely on visual cues such as color, texture and shape when selecting food (Gentle, 1975; Forbes, 2010), attributed the aposematic red coloration caused by *Photorhabdus* spp. with the distaste from nematode-infected cadavers (Fenton *et al.*, 2011; Jones *et al.*, 2017). When the omnivorous freshwater cyprinid fish species *Alburnoides bipunctatus*, *Devario aequipinnatus* and *Squalius pursakensis* fed on freeze-killed controls of either *S. feltiae*-, *Steinernema siamkayai*- or *H. bacteriophora*-killed *Galleria mellonella* larvae, they all rejected the EPN-killed insects but readily consumed the freeze-killed insects; they also learned to avoid EPN-killed insects (Raja *et al.*, 2017). Generally, all affected scavengers show a stronger aversion to older cadavers (≥2 days old), indicating that the amount of SDF produced increases with time (Raja *et al.*, 2017).

Fig. 8.4. Scavenger deterrence activity of *Xenorhabdus* spp. against *Linepithema humile* ants. The ants are feeding on *Xenorhabdus doucetiae*-infected and freeze-killed cadavers.

Some omnivores/scavengers do not appear to be affected by SDF. For example, the American cockroach, *Periplaneta americana*, the earwig *Labidura riparia*, the mite *Sancassania polyphyllae*, and the ants *Tetramorium chefketi* and *Pheidole pallidula* were unaffected (Ulug *et al.*, 2014). The ant species *Liometopum occidentale* even consumed more *H. bacteriophora*-infected cadavers than freeze-killed controls (Gulcu *et al.*, 2018). *Sancassania polyphyllae* feeds and reproduces on EPN-infected cadavers as well as consuming IJs emerging from infected larvae (Ekmen *et al.*, 2010a; Ulug *et al.*, 2014). So far, the SDF in supernatants of *Xenorhabdus* and *Photorhabdus* spp. have not been identified. Most recently, two volatile compounds, hexadecanal and 2-heptadecanone, isolated from *S. feltiae*-infected larvae, were confirmed to be highly deterrent to the ant species *Lasius niger*. These compounds are environmentally friendly and can also be a deterrent to other scavengers (Jaffuel *et al.*, 2022).

8.4 Conclusions

Several organisms in soil including bacteria, NFs, mites and collembolans can infect or prey on IJs or infected cadavers containing the developmental stages of nematodes. Some can compete for resources within these insect cadavers. NFs, mites and collembolans can impact IJ populations in soil. Some omnivores and scavengers, especially mites, collembolans, ants and birds, may consume recently infected hosts (<2 days old) and appear to also have negative impacts on incipient nematode stages in the nematode-killed insects. Older cadavers (≥4 days old), however, are protected by the mutualistic bacteria *Xenorhabdus* and *Photorhabdus* spp., which produce an antifeedant/deterrent compound(s) against the omnivores and scavengers as well as antibacterial and antifungal compounds against microbial invaders. Despite these challenges of antagonists, EPNs have successfully spread in various soil habitats around the world except in Antarctica. However, these antagonists may directly or indirectly affect nematode survival and persistence. Antagonists may also reduce the efficacy of biological control programs using these nematodes. Therefore, it is necessary to better understand the relationship of EPNs with these antagonistic species so that improved biological control agents can be achieved. Future research should investigate: (i) the actual impact of antagonists under field conditions, both after biocontrol application and in natural populations, perhaps by using more advanced molecular techniques; (ii) the potential to use SDF or other compounds to enhance control in the field; and (iii) optimization of pathogen combinations for synergy.

8.5 References

Abdolmaleki, A., Rafiee Dastjerdi, H., Tanha Maafi, Z. and Naseri, B. (2017) Virulence of two entomopathogenic nematodes through their interaction with *Beauveria bassiana* and *Bacillus thuringiensis* against *Pieris brassicae* (Lepidoptera: Pieridae). *Journal of Crop Protection* 6, 287–299.

Ansari, M.A., Tirry, L. and Moens, M. (2004) Interaction between *Metarhizium anisopliae* CLO 53 and entomopathogenic nematodes for the control of *Hoplia philanthus*. *Biological Control* 31, 172–180.

Ansari, M.A., Shah, F.A., Tirry, L. and Moens, M. (2006) Field trials against *Hoplia philanthus* (Coleoptera: Scarabaeidae) with a combination of an entomopathogenic nematode and the fungus *Metarhizium anisopliae* CLO 53. *Biological Control* 39, 453–459.

Ansari, M.A., Shah, F.A. and Butt, T.M. (2008) Combined use of entomopathogenic nematodes and *Metarhizium anisopliae* as a new approach for black vine weevil, *Otiorhynchus sulcatus*, control. *Entomologia Experimentalis et Applicata* 129, 340–347.

Arêdes, A., Ribeiro, L.P., Dias, V.M. and Dolinski, C. (2017) Studies on the interaction between tropical nematophagous fungi and entomopathogenic nematodes (Rhabditida) under laboratory conditions. *Nematoda* 4: e112016.

Asan, C., Hazir, S., Cimen, H., Ulug, D., Taylor, J. *et al.* (2017) An innovative strategy for control of the chestnut weevil *Curculio elephas* (Coleoptera: Curculionidae) using *Metarhizium brunneum. Crop Protection* 102, 147–153.

Bajaj, H. and Walia, K.K. (2005) Studies on a *Pasteuria* isolate from an entomopathogenic nematode, *Steinernema pakistanense* (Nematoda: Steinernematidae). *Nematology* 7, 637–640.
Bakengesa, J.A. (2016) Potential of *Paenibacillus* spp. as a biocontrol agent for root-knot nematodes (*Meloidogyne* spp.) PhD thesis, University of Ghent, Ghent, Belgium
Barbercheck, M.E. and Kaya, H.K. (1991) Effect of host condition and soil texture on host finding by entomogenous nematodes *Heterorhabditis bacteriophora* (Rhabditida: Heterorhabditidae) and *Steinernema carpocapsae* (Rhabditida: Steinernematidae). *Environmental Entomology* 20, 582–589.
Baur, M.E., Kaya, H.K., Tabashnik, B.E. and Chilcutt, C.F. (1998a) Suppression of diamondback moth (Lepidoptera: Plutellidae) with an entomopathogenic nematode (Rhabditida: Steinernematidae) and *Bacillus thuringiensis* Berliner. *Journal of Economic Entomology* 91, 1089–1095.
Baur, M.E., Kaya, H.K. and Strong, D.R. (1998b) Foraging ants as scavengers on entomopathogenic nematode-killed insects. *Biological Control* 236, 231–236.
Bednarek, A. (1986) Development of the *Steinernema feltiae* (Filipjev) entomogenous nematode (Steinernematidae) in the conditions of occurrence in the insect's body cavity of other pathogens (preliminary investigation). *Annals of Warsaw Agricultural University Animal Science* 20, 69–74.
Bhat, A.H., Chaubey, A.K. and Askary, T.H. (2020) Global distribution of entomopathogenic nematodes, *Steinernema* and *Heterorhabditis*. *Egyptian Journal of Biological Pest Control* 30: 31.
Bode, H.B. (2009) Entomopathogenic bacteria as a source of secondary metabolites. *Current Opinion in Chemical Biology* 13, 224–230.
Braga, F.R. and Araújo, J.V. (2014) Nematophagous fungi for biological control of gastrointestinal nematodes in domestic animals. *Applied Microbiology and Biotechnology* 98, 71–82.
Cakmak, I., Ekmen, Z.I., Karagoz, M., Hazir, S. and Kaya, H.K. (2010) Development and reproduction of *Sancassania polyphyllae* (Acari: Acaridae) feeding on entomopathogenic nematodes and tissues of insect larvae. *Pedobiologia* 53, 235–240.
Cakmak, I., Karagoz, M., Ekmen, Z.I., Hazir, S. and Kaya, H.K. (2011) Life history of *Sancassania polyphyllae* (Acari: Acaridae) feeding on dissected tissues of its phoretic host, *Polyphylla fullo* (Coleoptera: Scarabaeidae): temperature effects. *Experimental and Applied Acarology* 53, 41–49.
Cakmak, I., Hazir, S., Ulug, D. and Karagoz, M. (2013) Olfactory response of *Sancassania polyphyllae* (Acari: Acaridae) to its phoretic host larva killed by the entomopathogenic nematode, *Steinernema glaseri* (Rhabditida: Steinernematidae). *Biological Control* 65, 212–217.
Campbell, J.F. and Gaugler, R. (1997) Inter-specific variation in entomopathogenic nematode. Foraging strategy: dichotomy or variation along a continuum? *Fundamental and Applied Nematology* 20, 393–398.
Campos-Herrera, R., Barbercheck, M., Hoy, C.W. and Stock, S.P. (2012) Entomopathogenic nematodes as a model system for advancing the frontiers of ecology. *Journal of Nematology* 44, 162–176.
Campos-Herrera, R., Pathak, E., El-Borai, F.E., Schumann, A., Abd-Elgawad, M.M.M. and Duncan, L.W. (2013) New citriculture system suppresses native and augmented entomopathogenic nematodes. *Biological Control* 66, 183–194.
Davies, K., Laird, V. and Kerry, B. (1991) The motility, development and infection of *Meloidogyne incognita* encumbered with spores of the obligate hyperparasite *Pasteuria penetrans*. *Revue de Nématologie* 14, 611–618.
Davies, K.G., Rowe, J., Manzanilla-López, R. and Opperman, C.H. (2011) Re-evaluation of the life-cycle of the nematode-parasitic bacterium *Pasteuria penetrans* in root-knot nematodes, *Meloidogyne* spp. *Nematology* 13, 825–835.
Del Valle, E.E., Balbi, E.I., Lax, P., Rondán Dueñas, J. and Doucet, M.E. (2014) Ecological aspects of an isolate of *Steinernema diaprepesi* (Rhabditida: Steinernematidae) from Argentina. *Biocontrol Science and Technology* 24, 690–704.
Dritsoulas, A., Wu, S.Y., Regmi, H. and Duncan, L.W. (2022) Arthropod community responses reveal potential predators and prey of entomopathogenic nematodes in a citrus orchard. *Agronomy* 12: 2502.
Duncan, L.W., Graham, J.H., Dunn, D.C., Zellers, J., McCoy, C.W. and Nguyen, K. (2003a) Incidence of endemic entomopathogenic nematodes following application of *Steinernema riobrave* for control of *Diaprepes abbreviatus*. *Journal of Nematology* 35, 178–186.
Duncan, L.W., Dunn, D.C., Bague, G. and Nguyen, K. (2003b) Competition between entomopathogenic and free-living bacterivorous nematodes in larvae of the weevil *Diaprepes abbreviatus*. *Journal of Nematology* 35, 187–193.
Duncan, L.W., Graham, J.H., Zellers, J., Bright, D.B., Dunn, D.C. *et al.* (2007) Food web responses to augmenting the entomopathogenic nematodes in bare and animal manure-mulched soil. *Journal of Nematology* 39, 176–189.

Duncan, L.W., Stuart, R.J., El-Borai, F.E., Campos-Herrera, R., Pathak, E. *et al.* (2013) Modifying orchard planting sites conserves entomopathogenic nematodes, reduces weevil herbivory and increases citrus tree growth, survival and fruit yield. *Biological Control* 64, 26–36.

Ekmen, Z.I., Cakmak, I., Karagöz, M., Hazir, S., Ozer, N. and Kaya, H. (2010a) Food preference of *Sancassania polyphyllae* (Acari: Acaridae): living entomopathogenic nematodes or insect tissues? *Biocontrol Science and Technology* 20, 553–566.

Ekmen, Z.I., Hazir, S., Cakmak, I., Ozer, N., Karagöz, M. and Kaya, H.K. (2010b) Potential negative effects on biological control by *Sancassania polyphyllae* (Acari: Acaridae) on an entomopathogenic nematode species. *Biological Control* 54, 166–171.

El-Borai, F.E., Duncan, L.W., Preston, J.F. and Dunn, D. (2003) A phoretic association between a putative *Paenibacillus* sp. and the entomopathogenic nematode *Steinernema diaprepesi*. *Journal of Nematology* 35, 336–337.

El-Borai, F.E., Duncan, L.W. and Preston, J.F. (2005) Bionomics of a phoretic association between *Paenibacillus* sp. and the entomopathogenic nematode *Steinernema diaprepesi*. *Journal of Nematology* 37, 18–25.

El-Borai, F.E., Bright, D.B., Graham, J.H., Stuart, R.J., Cubero, J. and Duncan, L.W. (2009) Differential susceptibility of entomopathogenic nematodes to nematophagous fungi from Florida citrus orchards. *Nematology* 11, 233–243.

El-Borai, F.E., Campos-Herrera, R., Stuart, R.J. and Duncan, L.W. (2011) Substrate modulation, group effects and the behavioral responses of entomopathogenic nematodes to nematophagous fungi. *Journal of Invertebrate Pathology* 106, 347–356.

Enright, M.R. and Griffin, C.T. (2005) Effects of *Paenibacillus nematophilus* on the entomopathogenic nematode *Heterorhabditis megidis*. *Journal of Invertebrate Pathology* 88, 40–48.

Enright, M.R., Finnegan, M., McInerney, J.O., O'Laighleis, M. and Griffin, C. (2001) Specific adherence of sporangia of a *Paenibacillus* sp. bacterium to *Heterorhabditis* spp. nematodes. Hitching a ride to lunch? In: Griffin, C.T., Burnell, A.M., Downes, M.J. and Mulder, R. (eds.) *Developments in Entomopathogenic Nematode/Bacterial Research*. Publications Office of the European Union, Luxembourg, pp. 301–306.

Enright, M.R., McInerney, J.O. and Griffin, C.T. (2003) Characterization of endospore-forming bacteria associated with entomopathogenic nematodes, *Heterorhabditis* spp., and description of *Paenibacillus nematophilus* sp. nov. *International Journal of Systematic and Evolutionary Microbiology* 53, 435–441.

Epsky, N.D., Walter, D.E. and Capinera, J.L. (1988) Potential role of nematophagous micro- arthropods as biotic factors of entomogenous nematodes (Rhabditida: Steinernematidae, Heterorhabditidae). *Journal of Economic Entomology* 81, 821–825.

Fenton, A., Magoolagan, L., Kennedy, Z. and Spencer, K.A. (2011) Parasite-induced warning coloration: a novel form of host manipulation. *Animal Behavior* 81, 417–422.

Foltan, P. and Puza, V. (2009) To complete their life cycle, pathogenic nematode–bacteria complexes deter scavengers from feeding on their host cadaver. *Behavioural Processes* 80, 76–79.

Forbes, S. (2010), Sibling rivalry in the birds and bees. In: Hart, S.L. and Legerstee, M. (eds.) *Handbook of Jealousy: Theory, Research, and Multidisciplinary Approaches*. Wiley, Oxford, UK, pp. 121–143.

Gentle, M.J. (1975) Gustatory behaviour of the chicken and other birds. In: Wright, P., Caryl, P.G. and Vowles, D.M. (eds.) *Neural and Endocrine Aspects of Behaviour in Birds*. Elsevier Science, Amsterdam, Netherlands.

Gilmore, S.K. (1970) Collembola Predation on Nematodes: Extent and Nutritional Significance. Cornell University, Ithaca, New York.

Gilmore, S.K. and Potter, D.A. (1993) Potential role of collembola as biotic mortality agents for entomopathogenic nematodes. *Pedobiologia* 37, 30–38.

Glazer, I. (2002) Survival biology. In: Gaugler, R. (ed.) *Entomopathogenic Nematology*. CAB International, Wallingford, UK, pp. 169–188.

Glockling, S.L. and Holbrook, G.P. (2003) Endoparasites of soil nematodes and rotifers 1: the common and the rare. *Mycologist* 17, 150–154.

Gothama, A.A., Sikorowski, P.P. and Lawrence, G.W. (1995) Interactive effects of *Steinernema carpocapsae* and *Spodoptera exigua* nuclear polyhedrosis virus on *Spodoptera exigua* larvae. *Journal of Invertebrate Pathology* 66, 270–276.

Gothama, A.A., Lawrence, G.W. and Sikorowski, P.P. (1996) Activity and persistence of *Steinernema carpocapsae* and *Spodoptera exigua* nuclear polyhedrosis virus against *S. exigua* larvae on soybean. *Journal of Nematology* 28, 68–74.

Greenwood, C.M., Barbercheck, M.E. and Brownie, C. (2011) Short term response of soil microinvertebrates to application of entomopathogenic nematode-infected insects in two tillage systems. *Pedobiologia* 54, 177–186.
Gulcu, B., Hazir, S. and Kaya, H.K. (2012) Scavenger deterrent factor (SDF) from symbiotic bacteria of entomopathogenic nematodes. *Journal of Invertebrate Pathology* 110, 326–333.
Gulcu, B., Cimen, H., Raja, R.K. and Hazir, S. (2017) Entomopathogenic nematodes and their mutualistic bacteria: their ecology and application as microbial control agents. *Biopesticides International* 13, 79–112.
Gulcu, B., Hazir, S., Lewis, E. and Kaya, H.K. (2018) Evaluation of responses of different ant species (Formicidae) to the scavenger deterrent factor associated with the entomopathogenic nematode-bacterium complex. *European Journal of Entomology* 115, 312–317.
Hazir, S., Kaya, H.K., Stock, P. and Neskin, N. (2003) Entomopathogenic nematodes (Steinernematdiae and Heterorhabditidae) for biological control of soil pests. *Turkish Journal of Biology* 27, 181–202.
Hazir, S., Kaya, H., Touray, M., Cimen, H. and Shapiro-Ilan, D (2022) Basic laboratory and field manual for conducting research with the entomopathogenic nematodes, *Steinernema* and *Heterorhabditis*, and their bacterial symbionts. *Turkish Journal of Zoology* 46, 305–350.
Hodson, A.K., Siegel, J.P. and Lewis, E.E. (2012) Ecological influence of the entomopathogenic nematode, *Steinernema carpocapsae*, on pistachio orchards oil arthropods. *Pedobiologia* 55, 51–58.
Hominick, W.M., Reid, A.P., Bohan, D.A. and Briscoe, B.R. (1996) Entomopathogenic nematodes: biodiversity, geographical distribution and the convention on biological diversity. *Biocontrol Science and Technology* 6, 317–332.
Hummadi, E.H., Dearden, A., Generalovic, T., Clunie, B., Harrott, A. *et al.* (2021) Volatile organic compounds of *Metarhizium brunneum* influence the efficacy of entomopathogenic nematodes in insect control. *Biological Control* 155: 104527.
Hummadi, E.H., Cetin, Y., Demirbek, M., Kardar, N.M., Khan, S. *et al.* (2022) Antimicrobial volatiles of the insect pathogen *Metarhizium brunneum. Journal of Fungi* 8, 326.
Ishibashi, N., Young, F.Z., Nakashima, M., Agabeyru, C. and Haraguchi, N. (1987) Effects of application of DD-136 on silkworm, *Bombyx mori*, predatory insect, *Agriosphodorus dohrni,* parasitoid, *Trichomalusa panteloctenus*, soil mites, and other non-target soil arthropods, with brief notes on feeding behavior and predatory pressure of soil mites, tardigrates, and predatory nematodes on DD-136 nematodes. In: Ishibashi, N. (ed.) *Recent Advances in Biological Control of Insect Pests by Entomogenous Nematodes in Japan*. Saga University, Saga, Japan, pp. 158–164.
Jabbour, R. and Barbercheck, M.E. (2011) Soil microarthropod response to the application of entomopathogenic nematode-killed insects in maize and flower strip habitats. *Pedobiologia* 54, 243–251.
Jaffee, B.A. and Strong, D.R. (2005) Strong bottom-up and weak top-down effects in soil: nematode-parasitized insects and nematode-trapping fungi. *Soil Biology and Biochemistry* 37, 1011–1021.
Jaffuel, G., Krishnamani, S., Machado, R.A., Campos-Herrera, R. and Turlings, T.C. (2022) Potent ant deterrents emitted from nematode-infected insect cadavers. *Journal of Chemical Ecology* 48, 71–78.
Jones, R.S., Fenton, A. and Speed, M.P. (2016) "Parasite-induced aposematism" protects entomopathogenic nematode parasites against invertebrate enemies. *Behavioral Ecology* 27, 645–651.
Jones, R.S., Fenton, A., Speed, M.P. and Mappes, J. (2017) Investment in multiple defences protects a nematode–bacterium symbiosis from predation. *Animal Behaviour* 129, 1–8.
Kamra, A. and Mohan, S. (2011) Antagonistic effect of *Pseudomonas fluorescens* Migula 1895 on the entomopathogenic nematode, *Heterorhabditis indica*. *International Journal of Nematology* 21, 225–229.
Karagöz, M., Gulcu, B., Cakmak, I., Kaya, H.K. and Hazir, S. (2007) Predation of entomopathogenic nematodes by *Sancassania* sp. (Acari: Acaridae). *Experimental and Applied Acarology* 43, 85–95.
Kaya, H.K. (2002) Natural enemies and other antagonists. In: Gaugler, R. (ed.) *Entomopathogenic Nematology*. CAB International, Wallingford, UK, pp. 189–203.
Kaya, H.K. and Brayton, M.A. (1978) Interaction between *Neoaplectana carpocapsae* and a granulosis virus of the armyworm *Pseudaletia unipuncta*. *Journal of Nematology* 10, 350–354.
Kaya, H.K. and Burlando, T.M. (1989) Development of *Steinernema feltiae* (Rhabditida: Steinernematidae) in diseased insect hosts. *Journal of Invertebrate Pathology* 53, 164–168.
Kaya, H.K. and Koppenhöfer, A.M. (1996) Effects of microbial and other antagonistic organism and competition on entomopathogenic nematodes. *Biocontrol Science and Technology* 6, 357–372.
Kaya, H.K., Koppenhöfer, A.M. and Johnson, M. (1998) Natural enemies of entomopathogenic nematodes. *Japan Journal of Nematology* 28, 13–21.
Koppenhöfer, A.M. and Kaya, H.K. (1997) Additive and synergistic interaction between entomopathogenic nematodes and *Bacillus thuringiensis* for scarab grub control. *Biological Control* 8, 131–137.

Koppenhöfer, A.M., Choo, H.Y., Kaya, H.K., Lee, D.W. and Gelernter, W.D. (1999) Increased field and greenhouse efficacy against scarab grubs with a combination of an entomopathogenic nematode and *Bacillus thuringiensis*. *Biological Control* 14, 37–44.

Lee, Q. and Widden, P. (1996) *Folsomia candida*, a 'fungivorous' collembolan, feeds preferentially on nematodes rather than soil fungi. *Soil Biology and Biochemistry* 28, 689–690.

Lewis, E.E. (2002) Behavioral ecology. In Gaugler, R. (ed.) *Entomopathogenic Nematology*. CAB International, Wallingford, UK, pp. 205–224.

Lewis, E.E., Hazir, S., Hodson, A. and Gulcu, B. (2015) Trophic relationships of entomopathogenic nematodes in agricultural habitats. In: Campos-Herrera, R. (ed.) *Nematode Pathogenesis of Insects and Other Pests: Ecology and Applied Technologies for Sustainable Plant and Crop Protection*. Springer, Cham, Switzerland, pp. 139–163.

Li, J., Zou, C., Xu, J., Ji, X., Niu, X., Yang, J. *et al.* (2015) Molecular mechanisms of nematode-nematophagous microbe interactions: basis for biological control of plant-parasitic nematodes. *Annual Review of Phytopathology* 53, 67–95.

Machado, R.A., Wüthrich, D., Kuhnert, P., Arce, C.C., Thönen, L. *et al.* (2018) Whole-genome-based revisit of *Photorhabdus* phylogeny: proposal for the elevation of most *Photorhabdus* subspecies to the species level and description of one novel species *Photorhabdus bodei* sp. nov., and one novel subspecies *Photorhabdus laumondii* subsp. *clarkei* subsp. nov. *International Journal of Systematic and Evolutionary Microbiology* 68, 2664–2681.

Mertz, N.R., Agudelo, E.J.G., Sales, F.S. and Moino, A. Jr. (2015) Effects of entomopathogenic nematodes on the predator *Calosoma granulatum* in the laboratory. *Journal of Insect Behavior* 28, 312–327.

Nordbring-Hertz, B. (1988) Nematophagous fungi: strategies for nematode exploitation and for survival. *Microbiological Sciences* 5, 108–116.

Poinar, G.O. (1979) *Nematodes for Biological Control of Insects*. CRC Press, Boca Raton, Florida.

Poinar, G.O. (1988) A microsporidian parasite of *Neoplectana glaseri* (Steinernematidae: Rhabditida). *Revue de Nématologie* 3, 359–361.

Poinar, G.O., Thomas, G.M. and Lighthart, B. (1990) Bioassay to determine the effect of commercial preparations of *Bacillus thuringiensis* on entomogenous Rhabditoid nematodes. *Agriculture, Ecosystems & Environment* 30, 195–202.

Raja, R.K., Aiswarya, D., Gülcü, B., Raja, M., Perumal, P. *et al.* (2017) Response of three cyprinid fish species to the scavenger deterrent factor produced by the mutualistic bacteria associated with entomopathogenic nematodes. *Journal Invertebrate Pathology* 143, 40–49.

Raja, R.K., Arun, A., Touray, M., Gulsen, S.H., Cimen, H. *et al.* (2021) Antagonists and defense mechanisms of entomopathogenic nematodes and their mutualistic bacteria. *Biological Control* 152: 104452.

Read, D.S., Sheppard, S.K., Bruford, M.W., Glen, D.M. and Symondson, W.O.C. (2006) Molecular detection of predation by soil micro-arthropods on nematodes. *Molecular Ecology* 15, 1963–1972.

Sajnaga, E. and Kazimierczak, W. (2020) Evolution and taxonomy of nematode-associated entomopathogenic bacteria of the genera *Xenorhabdus* and *Photorhabdus*: an overview. *Symbiosis* 80, 1–13.

Sayre, R.M. and Starr, M.P. (1988) Bacterial diseases and antagonists of nematodes. In: Poinar, G.O. and Jansson, H.-B. (eds.) *Diseases of Nematodes, Vol.* 1. CRC Press, Boca Raton, Florida, pp. 3–22.

Shapiro-Ilan, D.I., Jackson, M., Reilly, C.C. and Hotchkiss, M.W. (2004) Effects of combining an entomopathogenic fungi or bacterium with entomopathogenic nematodes on mortality of *Curculio caryae* (Coleoptera: Curculionidae). *Biological Control* 30, 119–126.

Shapiro-Ilan, D.I., Han, R. and Dolinksi, C. (2012) Entomopathogenic nematode production and application technology. *Journal of Nematology* 44, 206–217.

Shapiro-Ilan, D.I., Han, R. and Qiu, X. (2014) Production of entomopathogenic nematodes. In: Morales-Ramos, J., Rojas, G. and Shapiro-Ilan, D.I. (eds.) *Mass Production of Beneficial Organisms: Invertebrates and Entomopathogens*. Academic Press, San Diego, California, pp. 321–355.

Shapiro-Ilan, D., Hazir, S. and Glazer, I. (2017) Basic and applied research: entomopathogenic nematodes: In: Lacey, L.A. (ed.) *Microbial Control of Insect and Mite Pests: From Theory to Practice*. Elsevier, London, pp. 91–105.

Soares, F.E. de F., Sufiate, B.L. and de Queiroz, J.H. (2018) Nematophagous fungi: far beyond the endoparasite, predator and ovicidal groups. *Agriculture and Natural Resources* 52, 1–8.

Somasekhar, N. and Mehta, U.K. (2000) Infectivity of *Pasteuria penetrans* to entomopathogenic nematodes. *Nematologia Mediterranea* 28, 13–14.

Stuart, R.J., Barbercheck, M.E. and Grewal, P.S. (2015) Entomopathogenic nematodes in the soil environment: distributions, interactions and the influence of biotic and abiotic factors. In: Campos-Herrera, R.

(ed.) *Nematode Pathogenesis of Insects and Other Pests: Ecology and Applied Technologies for Sustainable Plant and Crop Protection*. Springer, Cham, Switzerland, pp. 97–137.
Tarasco, E., Alvarez, C.S., Triggiani, O. and Moraga, E.Q. (2011) Laboratory studies on the competition for insect haemocoel between *Beauveria bassiana* and *Steinernema ichnusae* recovered in the same ecological niche. *Biocontrol Science and Technology* 21, 693–704.
Timper, P. and Kaya, H.K. (1989) Role of the second-stage cuticle of entomogenous nematodes in preventing infection by nematophagous fungi. *Journal of Invertebrate Pathology* 54, 314–321.
Troemel, E.R., Felix, M.A., Whiteman, N.K., Barriere, A. and Ausubel, F.M. (2008) Microsporidia are natural intracellular parasites of the nematode *Caenorhabditis elegans*. *PLOS Biology* 6: e309.
Ulug, D., Hazir, S., Kaya, H.K. and Lewis, E. (2014) Natural enemies of natural enemies: the potential top-down impact of predators on entomopathogenic nematode populations. *Ecological Entomology* 39, 462–469.
Upadhyay, A. and Mohan, S. (2021) *Bacillus subtilis* and *B. licheniformis* isolated from *Heterorhabditis indica* infected apple root borer (*Dorysthenes huegelii*) suppresses nematode production in *Galleria mellonella*. *Acta Parasitologica* 66, 989–996.
Vagelas, I.K., Dennett, M.D., Pembroke, B. and Gowen, S.R. (2012) Adhering *Pasteuria penetrans* endospores affect movements of root-knot nematode juveniles. *Phytopathologia Mediterranea* 51, 618–624.
Veremchuk, G.V. and Issi, I.V. (1970) The development of microsporidians of insects in the entomopathogenic nematode *Neoaplectana carpocapsae* (Nematoda: Steinernematodiae). *Parazitologiya* 4, 3–7.
Wakil, W., Yasin, M. and Shapiro-Ilan, D. (2017) Effects of single and combined applications of entomopathogenic fungi and nematodes against *Rhynchophorus ferrugineus* (Olivier). *Scientific Reports* 7: 5971.
Willett, D.S., Alborn, H.T. and Stelinski, L.L. (2017) Multitrophic effects of belowground parasitoid learning. *Scientific Reports* 7: 2067.
Wu, S.Y. and Duncan, L.W. (2020) Recruitment of an insect and its nematode natural enemy by olfactory cues from a saprophytic fungus. *Soil Biology and Biochemistry* 144: 107781.
Wu, S., Youngman, R.R., Kok, L.T., Laub, C.A. and Pfeiffer, D.G. (2014) Interaction between entomopathogenic nematodes and entomopathogenic fungi applied to third instar southern masked chafer white grubs, *Cyclocephala lurida* (Coleoptera: Scarabaeidae), under laboratory and greenhouse conditions. *Biological Control* 76, 65–73.
Wu, S.Y., El-Borai, F.E., Graham, J.H. and Duncan, L.W.. (2018) The saprophytic fungus *Fusarium solani* increases the insecticidal efficacy of the entomopathogenic nematode *Steinernema diaprepesi*. *Journal of Invertebrate Pathology* 159, 87–94.
Zhou, X., Kaya, H.K., Heungens, K. and Goodrich-Blair, H. (2002) Response of ants to a deterrent factor(s) produced by the symbiotic bacteria of entomopathogenic nematodes. *Applied and Environmental Microbiology* 68, 6202–6209.

9 *Xenorhabdus* and *Photorhabdus* Bacterial Metabolites as Tools in Biocontrol

Selcuk Hazir[1]* and Helge B. Bode[2,3,4,5]

[1]*Department of Biology, Faculty of Science, Aydin Adnan Menderes University, Aydin, Türkiye;* [2]*Department of Natural Products in Organismic Interactions, Max-Planck-Institute for Terrestrial Microbiology, Marburg, Germany;* [3]*Chemical Biology, Department of Chemistry, Philipps University Marburg, Marburg, Germany;* [4]*Molecular Biotechnology, Department of Biosciences, Goethe University Frankfurt, Frankfurt am Main, Germany;* [5]*Senckenberg Gesellschaft für Naturforschung, Frankfurt am Main, Germany*

Abstract

Bacteria in the genera *Xenorhabdus* and *Photorhabdus* are symbiotically associated with entomopathogenic nematodes. The genomes of these bacteria harbor 1000 different biosynthetic gene clusters of natural products with specific and complementary ecological functions. Most of these products still await identification. These compounds have various properties ranging from antibacterial, antiprotozoal, acaricidal, antifungal and anticancer to insecticidal activities. This chapter summarizes the bioactive secondary metabolites produced and their potential application in biological control of various pest organisms. Gene regulation by these symbiotic bacteria is also discussed.

9.1 Introduction

As with all living organisms, bacteria produce many compounds during their metabolism. These compounds can typically be categorized as primary and secondary metabolites (also called specialized metabolites or natural products (NPs)). Primary metabolites such as ethanol, lactic acid and certain amino acids are produced during the growth phase of bacteria and are essential for their proper growth, development and reproduction. In contrast, secondary metabolites are organic compounds produced through the modification of primary metabolites, often at the end of the logarithmic or near the stationary phase of growth, and do not play a direct role in growth, development and reproduction, at least not under laboratory conditions. Many of the identified secondary metabolites, however, have an ecological or physiological function such as defense or nutrient acquisition. Secondary metabolites are often antibiotics; for instance, erythromcyin derived from *Saccharopolyspora erythraea* is a classical antibiotic with a wide antimicrobial spectrum. Bacitracin is also a secondary metabolite synthesized as a non-ribosomal peptide synthetase (NRPS) by *Bacillus subtilis* and used in the clinic as a topical antibiotic on cuts and wounds.

Bacteria living in the soil environment and alimentary system of animals must produce various effective secondary metabolites to overwhelm other organisms. Our focus is on bacteria

*Email: selcuk.hazir@gmail.com

 Entomopathogenic Nematodes as Biological Control Agents (D.I. Shapiro-Ilan and E.E. Lewis eds)
DOI: 10.1079/9781800620322.0009

in the genera *Xenorhabdus* and *Photorhabdus* (Enterobacterales: Morganellaceae), which are symbiotically associated with entomopathogenic nematodes (EPNs) belonging to the genera *Steinernema* (Rhabditida: Steinernematidae) and *Heterorhabditis* (Rhabditida: Heterorhabditidae), respectively (Adeolu *et al.*, 2016; Shapiro-Ilan *et al.*, 2020). The third-stage juveniles (also called infective juveniles (IJs)) of these nematodes, which house the bacterial cells in their intestine, survive in the soil environment and search for a suitable insect host. Once IJs enter the insect hemocoel via natural openings (anus, mouth or spiracles) or directly through its thin cuticle, they release the *Xenorhabdus* or *Photorhabdus* spp. bacteria from their intestine into the host hemocoel. The bacteria proliferate and produce various metabolites (i.e. toxins and enzymes) and kill the host within 48 h (Gulcu *et al.*, 2017; Hazir *et al.*, 2003, 2022). The nematodes produce one, two or more generations in the insect hemocoel and feed on bacterial cells and nutrients derived from the insect host. After the nutrients are depleted, the nematodes produce IJs that reacquire the bacteria from the hemocoel and sequester these bacterial cells in their intestine before emerging from the insect cadaver into the soil to search for a new host (Shapiro-Ilan *et al.*, 2017; Gulcu *et al.*, 2018). The life cycle of nematodes within an insect cadaver generally lasts 1–3 weeks before the new generation of IJs emerges from the host cadaver into the soil. During this period, the insect cadavers are a vulnerable food source for various micro- and macroorganisms (Kaya, 2002), and there is a great number of saprophytic microorganism competitors in the insect alimentary system as well. *Xenorhabdus* and *Photorhabdus* spp. produce secondary metabolites to compete with all these organisms to protect the insect cadavers, allowing their nematode hosts to survive and reproduce.

Xenorhabdus or *Photorhabdus* spp. have up to 6.5% of their genomes dedicated to NP production (Chaston *et al.*, 2011). They use NRPSs, polyketide synthetase (PKS) and mixed peptide–polyketide (NRPS-PKS) enzyme systems to produce an enormous repertoire of different NPs to support a complex life cycle involving insect pathogenesis and nematode symbiosis (Brachmann and Bode, 2013; Shi and Bode, 2018; Shi *et al.*, 2022). These compounds have various properties ranging from antibacterial, antiprotozoal, antifungal and anticancer to insecticidal activities (Challinor and Bode, 2015). Although many NPs from *Xenorhabdus* and *Photorhabdus* spp. have been identified, their natural functions remain unknown; overall, these compounds are an untapped reservoir for new therapeutic and chemical entities.

This chapter summarizes the bioactive secondary metabolites produced by *Xenorhabdus* and *Photorhabdus* spp. and their potential application in biological control of various pest organisms. Gene regulation by these symbiotic bacteria is also discussed.

9.2 Arthropod Vector Control

Hematophagous arthropods such as mosquitoes, black flies, sand flies, lice, fleas, ticks and mites are major vectors capable of transmitting a huge number of pathogens to humans and other vertebrates. Mosquitoes belonging to the genera *Aedes*, *Anopheles* and *Culex* pose a significant threat to human health on a global scale due to their role in transmission of malaria, dengue virus, chikungunya virus, Zika virus, West Nile virus, yellow fever virus and several filarial nematode species. These mosquito-borne diseases are a huge problem, especially in resource-limited tropical and subtropical countries and islands; these diseases negatively impact one-third of the world's population. In 2017, malaria caused by *Plasmodium* spp. parasites alone accounted for an estimated 219 million cases and caused 435,000 deaths (WHO, 2018).

Mosquito-borne disease prevention relies mostly on vector control as there are no specific medications and vaccines against most of these disease-causing organisms (Heinz and Stiasny, 2012). The most common method used to reduce and control mosquito populations is the application of chemical pesticides to kill either the adults or the aquatic larvae. Although chemical insecticides can be effective against different life stages of mosquitoes, they are often toxic to humans and other non-target organisms and can contaminate the environment. Moreover, the repeated use of chemicals against insect pests results in pesticide resistance over time. These concerns have led to the emergence of different biological control strategies as an alternative to chemical control of vector-transmitting mosquitoes.

Biolarvicides are the most common biological control agents used in mosquito-control programs. Currently, *Bacillus thuringiensis* var. *israelensis* (*Bti*) and *Bacillus sphaericus* (*Bs*) (syn. *Lysinibacillus sphaericus*) dominate the global market for mosquito larval control (Benelli *et al.*, 2016). These larvicides, however, bring their own share of issues that reduce mosquito control. For example: (i) they tend to sediment quickly and may adsorb to organic matter, reducing their availability and affecting their impact as a larvicide; (ii) *Bti* has side effects; and (iii) perhaps the most concerning and debatable is the reported emergence of resistance in mosquitoes to *Bti* products. These limitations may affect the extensive use of *Bti*/*Bs* biopesticides in the future. Accordingly, we must find alternatives that can augment already-existing mosquito-control agents, such as *Bti*/*Bs* biopesticides or the few remaining effective chemical insecticides. Metabolites of *Xenorhabdus* and *Photorhabdus* spp. are logical resources to target in this endeavor.

Several studies have assessed the efficacy of *Xenorhabdus* and *Photorhabdus* spp. against mosquito larvae (Silva *et al.*, 2013, 2017; Fukruksa *et al.*, 2017; Wagutu *et al.*, 2017; Vitta *et al.*, 2018; Shah *et al.*, 2021). Cell-free supernatants, bacterial pellets or crude fluids have been used in these studies. Silva *et al.* (2013) demonstrated that *Photorhabdus luminescens* and *Xenorhabdus nematophila* cell suspensions were able to kill mosquito larvae when orally ingested and that *P. luminescens* was more pathogenic (73–83%) than *X. nematophila* (42–52%) after 96 h against the third- to fourth-instar *Aedes aegypti*. Besides pellets, culture fluids of the same two bacterial species were highly toxic to mosquito larvae, and the larvicidal activity of fresh, boiled, refrigerated and frozen supernatant of *X. nematophila* was much higher than that of *P. luminescens*. In contrast to *X. nematophila*, the larvicidal activity of *P. luminescens* culture fluid decreased after boiling and freezing. This loss of larvicidal activity in *P. luminescens* indicates that the larvicidal compounds in culture fluid are more likely to be based on protein toxins, which lose their activity when denatured. Moreover, *X. nematophila* secretes secondary metabolites, rather than protein-based toxins, that are highly effective against mosquitoes (Silva *et al.*, 2013). In other studies, bacterial pellets or crude supernatants (broth bacterial culture) of different *Xenorhabdus* and *Photorhabdus* spp./strains were tested against mosquito larvae. *Xenorhabdus ehlersii* (bMH9.2_TH; Fukruksa *et al.*, 2017), *X. stockiae* (bNBP22.2_TH; Vitta *et al.*, 2018), *X. stockiae* (bNN112.3_TH; Yooyangket *et al.*, 2018) and *X. nematophila* (Shah *et al.*, 2021) were observed to be the most effective strains against *A. aegypti*. For *Aedes albopictus*, *X. indica* (bKK26.2_TH; Vitta *et al.*, 2018) and *P. luminescens* subsp. *akhurstii* (bNN121.4_TH; Yooyangket *et al.*, 2018) exhibited the highest toxicity. It seems that secondary metabolite activity varies among different bacterial species/strains. For instance, significant differences were observed among the isolates of *X. stockiae* where larval mortality ranged between 23% and 67%. *P. luminescens* subsp. *akhurstii* strains bNA22.1_TH and bNN121.4_TH caused 21% and 88% larval mortality against *A. aegypti*, respectively (Fukruksa *et al.*, 2017). However, not all *Xenorhabdus* and *Photorhabdus* spp./strains produce effective mosquito larvicidal compounds. *Xenorhabdus miraniensis*, *X. vietnamensis*, *X. japonica* and *P. luminescens* subsp. *hainannensis* showed negligible toxicity to *A. aegypti* larvae (Fukruksa *et al.*, 2017; Yooyangket *et al.*, 2018). Shah *et al.* (2021) indicated that some bacteria such as *X. nematophila* release toxic metabolites with larvicidal activities into the supernatant while others such as *Photorhabdus laumondii* TT01 do not; supernatants of these bacteria were ineffective against the mosquito larvae.

In some cases, disparate results regarding the activity of the same bacterial strains have been reported. For example, Silva *et al.* (2013) reported that the supernatant of *Photorhabdus laumondii* TT01 exhibited 80% larval mortality against *A. aegypti* larvae, whereas Shah *et al.* (2021) indicated that the supernatant of the same strain had less than 10% larvicidal activity after 24 h exposure. However, the bacterial pellet used by Shah *et al.* (2021) caused 75% mortality.

Combined applications of *Xenorhabdus* and *Photorhabdus* spp. with available mosquito-control agents is another control approach. Addition of cultured broth of *X. nematophila*, *X. hominickii* or *Photorhabdus temperata temperata* enhanced *Bti* toxicity against mosquito larvae. The 50% lethal concentration (LC_{50}) values of *Bti* to *A. albopictus* larvae were reduced from 2.9×10^5 to 1.5×10^5 in a *X. nematophila* mixture, 1.7×10^5 in a *X. hominickii* mixture, and 1.9×10^5 in a *Photorhabdus temperata temperata* mixture (Park *et al.*, 2016).

Similar results were obtained for *Culex pipiens pallens*. Subsequently, a new mosquitocidal *Bt* formulation called Dip-Kill, which was a mixture of 80% *X. nematophila*-cultured broth, 10% *Bti* (10^{10} spores ml^{-1}, and 10% preservative was produced. At 1000 ppm, Dip-Kill showed greater efficacy than the commercial *Bti* product at its recommended dose against *Ae albopictus* and *Culex pipiens pallens* (Park *et al.*, 2016). The culture broths of *X. hominickii* and *Photorhabdus temperata temperata* were also able to increase the toxicity of *Bti* against mosquitoes (Shrestha *et al.*, 2011; Park *et al.*, 2016). There was a 10.5-fold enhancement of *Bti* Cry4Ba toxicity with *X. nematophila* and 2.5-fold for *Photorhabdus luminescens* against *A. aegypti* larvae (Park, 2015). Adding benzylideneacetone or oxindole produced from *X. nematophila* and *Photorhabdus temperata temperata* also enhanced *Bti* toxicities to these mosquito larvae (Park *et al.*, 2016). However, none of these studies endeavored to identify which of the array of bioactive larvicidal compounds produced by the EPN symbiotic bacteria are responsible for the activity.

Given that *Xenorhabdus* and *Photorhabdus* spp. produce a wide variety of secondary metabolites (Tobias *et al.*, 2017; Shi and Bode, 2018), it is beneficial to identify bioactive compounds instead of using a conglomeration of compounds with different levels of toxicity, such as crude supernatants or bacterial pellets. Identification of the pure larvicidal compound(s) allows use at lower concentrations with lesser cytotoxicity than the supernatant or pellet. Some studies have been carried out to determine the bioactive larvicidal compound in *Xenorhabdus* and *Photorhabdus* spp. In the genus *Photorhabdus*, PirAB proteins (Ahantarig *et al.*, 2009), tca complex protein (Bowen *et al.*, 1998) and two bioactive anthraquinones (1,3-dimethoxy-8-hydroxy-9,10-anthraquinone and 3-methoxychrysazine; Ahn *et al.*, 2013) were identified as mosquitocidal compounds. Ahantarig *et al.* (2009) found proteins PirA and -B encoded by the loci *plu4093–plu4092* and *plu4437–plu4436*, respectively, within *Photorhabdus luminescens* TT01. These "*Photorhabdus* insect-related proteins A and B" show similarity to both δ-endotoxins from *Bacillus thuringiensis*. In 2009, the oral toxicity of PirAB proteins of *Photorhabdus asymbiotica* expressed in *Escherichia coli* was assessed against the first-stage larvae of the dengue vectors *A. aegypti* and *A. albopictus*. PirAB proteins were highly toxic at a concentration of 0.13×10^6 cells ml^{-1} causing 100% larval mortality on day 1. When PirA or PirB proteins are cloned separately, they exhibit low larvicidal effects against mosquitoes (Ahantarig *et al.*, 2009) and *Galleria mellonella* (Yang *et al.*, 2017). A bioassay of PirAB protein was also performed on *Mesocyclops thermocyclopoides*, a crustacean biological control agent of mosquito larvae, and there was no *Mesocyclops thermocyclopoides* mortality. This result indicates the specificity of PirAB proteins against the dengue vectors. Thus, Ahantarig *et al.* (2009) argued that these binary toxins have high oral toxicity to mosquito larvae, making them ideal candidates for the control of mosquitoes that serve as vectors for malaria and other vector-borne disease-causing microorganisms.

Compared with *Photorhabdus* spp., significantly limited reports are available on identification of the mosquito larvicidal toxins in *Xenorhabdus* spp. A low-molecular-weight mosquitocidal lipopeptide toxin called Xlt (*Xenorhabdus* lipopeptide toxin) was isolated from *Xenorhabdus innexi* and described by Ensign *et al.* (2014) and patented. Xlt is both secreted from cells into the growth media and retained by the cells, especially on the cell surface. A concentration of 1–10 ng of Xlt in a 1 ml culture killed all mosquito larvae in 24 h, and it is equally active against *Anopheles*, *Aedes* and *Culex* larvae (Ensign *et al.*, 2014). Xlt is composed of amino acids such as histidine, glycine, asparagine/aspartate, diaminobutyric acid and serine. Data strongly indicated that Xlt belongs to the fabclavines family, which comprises hybrid secondary metabolites derived from NRPSs, polyunsaturated fatty acids and PKS (Fuchs *et al.*, 2014; Kim *et al.*, 2017a; Bode *et al.*, 2019; Kajla, 2019); fabclavines also have antibacterial, antifungal and antiprotozoal properties (Ozkan *et al.*, 2019; Wenski *et al.*, 2020; Cimen *et al.*, 2021). Kim *et al.* (2017b) demonstrated that Xlt toxin binds and creates pores in the apical membrane of anterior midgut cells, which disrupts midgut pH balance by disabling the H^+ vacuolar ATPase on the basal membrane, causing cell death in *A. aegypti* larvae, but pupae and adults were not affected in pathogenicity bioassays. The activity is stable with no loss of activity of crude or purified material after 2 years incubation at room temperature. The toxin is also not inactivated by heat, 0.1 M HCl or 0.1 M NaOH, ultraviolet light irradiation, freezing, drying or protease digestion (Ensign *et al.*, 2014; Cimen *et al.*, 2021).

Fabclavine and xenocoumacin, produced by *X. szentirmaii* and *X. nematophila*, respectively, have beeen determined to be the larvicidal compounds within the bacterial supernatant (Touray *et al.*, 2024a; Ulug *et al.*, 2024). Notably, fabclavine from *X. szentirmaii* exhibits deterrent properties, potentially disrupting mosquito oviposition in suitable habitats (Touray *et al.*, 2024b). This offers a promising avenue for mosquito population control by targeting their breeding cycle. The general structure of fabclavines is highly conserved and differs only in the specified moieties. Wenski *et al.* (2020) revealed 32 different types of fabclavine from *Xenorhabdus* strains; these derivatives might vary in their toxicity against mosquito larvae. As a result, fabclavines (or Xlt) are promising leads to investigate further as a new frontier in mosquitocides.

Besides mosquitoes, the effects of *Xenorhabdus* and *Photorhabdus* spp. have been assessed in other insect species of public health and agricultural importance (Waterfield *et al.*, 2001; Kim *et al.*, 2017b; El-Sadawy *et al.*, 2020). The crude and fractionated toxin complex of *X. nematophila* and *Photorhabdus luminescens laumondii* were tested on the second instar of *Phlebotomus papatasi* (Diptera: Psychodidae). The highest larval mortality ranged from 22% to 36% at the end of a 25-day incubation period. Thus, toxin complex proteins of *X. nematophila* and *Photorhabdus luminescens laumondii* added to larval food were deemed not to be effective against *Phlebotomus papatasi* larvae (El-Sadawy *et al.*, 2020).

In other arthropods, *Xenorhabdus* Xlt toxin was not toxic (or was minimally toxic) to *Galleria mellonella*, *Artemia* spp. larvae (brine shrimp), *Manduca sexta* (tobacco horn worm) larvae, *Leptinotarsa decemlineata* (Colorado potato beetle) larvae and *Drosophila* spp., (fruit fly) larvae upon ingestion (Ensign *et al.*, 2014; Kim *et al.*, 2017b). As far as we are aware, no study has been conducted to test bacterial metabolites of *Xenorhabdus* or *Photorhabdus* spp. against other potential important insect vectors such as thrips, aphids, house flies and cockroaches.

9.3 Herbivore Control

9.3.1 Insect control

Several researchers have assessed the potential use of *Xenorhabdus* and *Photorhabdus* spp. and their secondary metabolites in the control of herbivore insects. This line of research has been undertaken despite the fact that, thus far the symbiotic bacteria can only survive in association with EPNs and are only virulent when vectored by these nematodes or artificially injected into the insect hemoceol (Dunphy and Webster, 1988a,b; Mahar *et al.*, 2004). In the host hemocoel, the bacteria secrete virulence factors that can suppress insect immune responses by inhibiting eicosanoid synthesis, which mediates cellular and humoral immune responses against various microbial pathogens, induct hemolysis, degradation of antimicrobial peptides and inhibition of the activation of prophenoloxidase (PO) (Kim *et al.*, 2005; Sergeant *et al.*, 2006; Shrestha and Kim, 2007; Seo *et al.*, 2012; Ahmed and Kim, 2018; Kim *et al.*, 2018). These virulence factors are synthesized by NRPSs, PKS or both; for example, rhabdopeptide/xenortide peptides derived from NRPSs are structurally like protease inhibitors that can degrade various immune-related proteins. Another compound, rhabducin, is an isocyanide-containing compound that inhibits PO activity in *G. mellonella* (Crawford *et al.*, 2012). The symbiotic bacteria also produce indoles, oxindole, *cis*-proline-tyrosine and *p*-hydroxyphenyl propionic acid, which resulted in PO and melanization inactivation when injected in *Spodoptera exigua* and *Plutella xylostella* (Kim *et al.*, 2005; Seo *et al.*, 2012). Interestingly, a formulation of *Photorhabdus luminescens* cells with paraffin oil, Tween-20 and sucrose as a phagostimulant adjuvant sprayed on cabbage leaves infested with *Pieris brassicae* caused 100% larval mortality (Mohan *et al.*, 2003). Recently, Wu *et al.* (2022) applied a *Photorhabdus luminescens* and *Xenorhabdus bovienii* bacterial broth to pecan aphids (*Melanocallis caryaefoliae*) and the black-margined aphids (*Monellia caryella*), as well as their natural enemy, the Asian lady beetle (*Harmonia axyridis*). Treatments from both bacteria were as effective against these aphids as bifenthrin (a pyrethroid insecticide) and demonstrated less toxicity against *Harmonia axyridis*.

The use of *Xenorhabdus* and *Photorhabdus* spp. directly for the control of herbivore insects is generally not feasible under natural conditions. Instead, bioinsecticide formulations made of supernatants of *Xenorhabdus* and *Photorhabdus* spp. and other entomopathogens such as *Bacillus thuringiensis* spores could be used as an effective control agent against various pest species. Combining bacterial suspensions of *B. thuringiensis* and *X. nematophila* and incorporating the suspension into a *Plutella xylostella* and *Spodoptera exigua* artificial diet resulted in synergistic interactions whereby *B. thuringiensis* bound to and perforated the insect midgut epithelium through which *X. nematophila* can access and exert its effects in the insect hemocoel. With *Bacillus cereus* bacteria, only earlier stages of the insect host were affected (Jung and Kim, 2006, 2007). Similarly, a culture broth mixture of *Photorhabdus temperata temperata* and *B. thuringiensis tenebrionis* called Col-Kill showed high efficacy in control of the coleopteran, *Phaedon brassicae* (Kim *et al.*, 2015). Dual combination of *B. thuringiensis kurstaki* and *B. thuringiensis aizawai* with culture broth of *X. nematophila* was also effective against *Plutella xylostella* and *Spodoptera exigua* (Eom *et al.*, 2014).

In other cases, insecticidal toxic proteins secreted by *Xenorhabdus* and *Photorhabdus* spp. have been purified and their activity against various herbivore insects has been demonstrated (Bowen and Ensign, 1998; Bowen *et al.*, 1998; Waterfield *et al.*, 2001; Seo *et al.*, 2012; Mollah and Kim, 2020). It is also possible to transform the genes encoding these toxins into another bacterium or plants as a strategy for the control of insect pests (Bowen and Ensign, 1998). High oral toxicity was observed in *Manduca sexta* larvae when exposed to a combination of three toxin complex genes, tcdA, tcdB and tccC, of *Photorhabdus luminescens* W14 expressed in *E. coli* (Waterfield *et al.*, 2001). The supernatant of *X. nematophila* contains various soluble proteins associated with the outer membrane; these proteins with chitinase activity exhibited oral insecticidal activity against *Helicoverpa armigera* larvae (Khandelwal *et al.*, 2004). Seven toxins purified from *Xenorhabdus* sp. SY5 exhibited high insecticidal activity and/or growth inhibition against larvae of *Plutella xylostella*, *Spodoptera exigua*, *Ostrinia furnacalis*, *Mythimna separata* and *Tenebrio molitor* (Wang *et al.*, 2013).

9.3.2 Mite control

Like spiders and ticks, mites are arachnids that feed on fluids from live or dead plants or animals. Certain mite species cause severe damage to plants in agricultural areas. For example, the two-spotted spider mite, *Tetranychus urticae* (Acari: Tetranychidae), is a globally important mite pest species that attacks more than 1000 plant species, including crops such as strawberry, cotton, deciduous fruits, vegetables and ornamental plants (Migeon *et al.*, 2010). The rapid reproduction, high fecundity and short life cycle of *T. urticae* can result in resistance to many chemical pesticides in a very short period when they are used for mite suppression. The pesticides used against this pest also cause environmental pollution and adverse effects on non-target animals including humans. Therefore, biological control agents such as predatory mites (Phytoseiidae) are used in many countries as an alternative to chemical control of *T. urticae* (Cakmak *et al.*, 2006, 2009). Various secondary metabolites, such as from *Xenorhabdus* and *Photorhabdus* spp. may also be effective against mites (Seo *et al.*, 2012; Dhanasekaran and Thangaraj, 2014; Eroglu *et al.*, 2019). The first studies on the effects of secondary metabolites produced by *Xenorhabdus* and *Photorhabdus* spp. on mite species tested culture broth and cell-free supernatant of different bacterial species/strains against both female and male mushroom mites, *Luciaphorus* sp. and *Luciaphorus perniciosus* (Acari: Pygmephoridae). All tested *Xenorhabdus* and *Photorhabdus* spp. showed significant toxicity to the female mites within 48 h but were not toxic to the male mites, which have undeveloped mouth parts (Bussaman *et al.*, 2006, 2009, 2012; Sobanboa *et al.*, 2009; Namsena *et al.*, 2016). When the effects of bacterial culture, cell-free supernatant and crude cell masses of *Xenorhabdus* and *Photorhabdus* spp. were compared, the supernatants showed higher mortality than bacterial culture and crude cell mass (Bussaman *et al.*, 2006; Namsena *et al.*, 2016). Cell density in the bacterial culture and culture age also had significant effects on acaricidal activity (Bussaman *et al.*, 2009). Namsena *et al.* (2016) prepared wetted powder, liquid concentrate and cell-free supernatant formulations of *X. stockiae* bacterial metabolites and observed 90%, 86% and 92% mortality, respectively. Storage of these formulations at 4°C exhibited better activity than at room temperature.

In contrast to the above studies, a study reported the low toxicity of *Xenorhabdus* and *Photorhabdus* spp. metabolites against another mite species. Nermut *et al.* (2019) observed that none of the 20 *Xenorhabdus* and one *Photorhabdus* spp./strains tested against the bulb mite, *Rhizoglyphus robini* (Acari: Acaridae) caused more than 30% mortality. This difference in activity could be due to the following: (i) the researchers autoclaved the bacterial culture broth before obtaining the supernatants and this could have damaged the structure of bioactive acaricidal compounds; and/or (ii) the bacterial supernatants were applied to filter paper and there was no food in the experimental arena; therefore, the mites were unable to have oral contact with the metabolites. It is also possible that *R. robini* might be resistant to the tested bacterial metabolites.

Recently, detailed studies were carried out on various developmental stages of *T. urticae* using secondary metabolites produced by *Xenorhabdus szentirmaii*, *X. nematophila*, *X. bovienii*, *X. cabanillasii*, *Photorhabdus luminescens* and *Photorhabdus temperata*. The cell-free bacterial supernatants had little or no effect on the egg stage, as less than 4% mortality was recorded. Mortality in the other stages was 46–97% for larvae, 30–96% for protonymphs, 41–92% for deutonymphs, 92–100% for adult males and 46–93% for adult females depending on the bacterial supernatant. Among the supernatants tested, *X. szentirmaii* and *X. nematophila* were the most efficacious, resulting in mortality greater than 90% on the mobile stages of *T. urticae*. These bacteria were further tested in pot experiments. Both supernatants significantly reduced the *T. urticae* populations. However, a combined application of *X. szentirmaii* and *X. nematophila* supernatants did not increase *T. urticae* population reduction (Eroglu *et al.*, 2019). Subsequently, Cevizci *et al.* (2020) reported on the possible mode of entry of *X. szentirmaii* and *X. nematophila* supernatants into *T. urticae* adult females (Fig. 9.1). The bioactive acaricidal compound was most effective (86.5–89% mortality) when it was in contact with the entire integument of *T. urticae* compared with contact with the ventral side only (26.5–34%). The toxicity of the supernatants against the developmental stages of the predatory mites *Phytoseiulus persimilis* and *Neoseiulus californicus* (Acari: Phytoseiidae) was also assessed. The predatory mites were less affected by the bacterial metabolites than *T. urticae*. The mortality of the mobile stages (larva, protonymph, deutonymph, adult) ranged between 18.5% and 39.2%.

Most recently, the acaricidal compound in *X. nematophila* supernatant was identified as

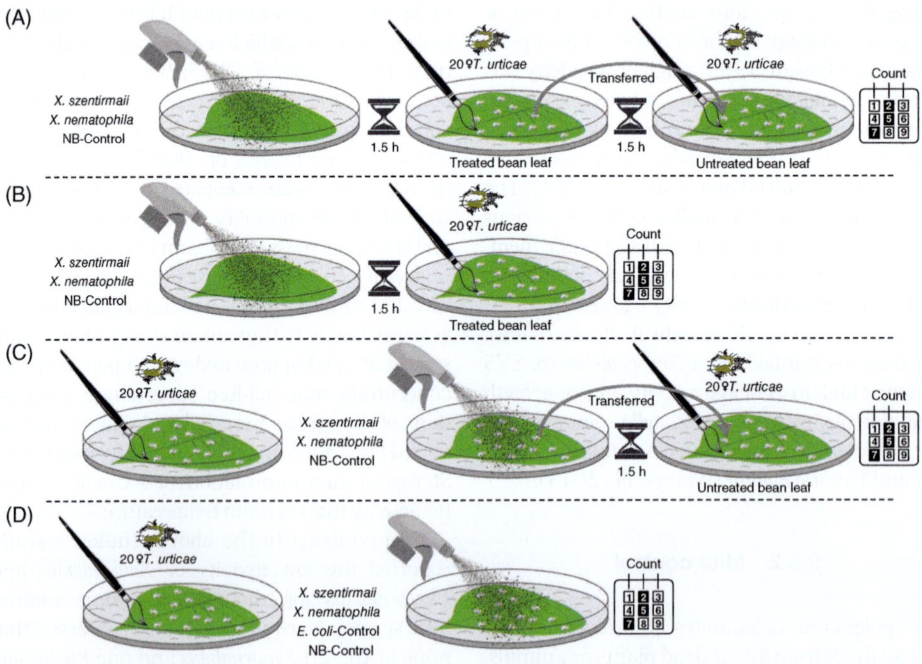

Fig. 9.1. Experimental set-up used to determine the mode of entry of secondary metabolites of *Xenorhabdus szentirmaii* or *Xenorhabdus nematophila* and nutrient broth (NB) on *Tetranychus urticae*. (A) Short-term exposure to ventral side. (B) Long-term exposure to ventral side. (C) Short-term exposure to dorsal and ventral sides. (D) Long-term exposure to dorsal and ventral sides. From Cevizci *et al.* (2020), used with permission from Elsevier.

xenocoumacin (Incedayi *et al.*, 2021) using the easyPACId approach (easy Promoter Activated Compound Identification) (Bode *et al.*, 2019). Cell-free supernatant of the induced promoter-exchange mutant of *X. nematophila* (pCEP_kan_XNC1_1711) and the extracted xenocoumacin were highly toxic with more than 96% mortality on the mobile stages of *T. urticae* at 7 days post-application in Petri dish and pot experiments. However, the same supernatant and extracted compound caused less than 36% mortality on the predatory mites *Phytoseiulus persimilis* and *Neoseiulus californicus*. A cytotoxicity assay conducted using the MRC-5 normal human fetal lung fibroblast cell line resulted in a 50% inhibitory concentration (IC_{50}) value for xenocoumacin extract of 17.71 $\mu g\ ml^{-1}$ after 48 h. These data indicate that xenocoumacin could potentially be used as bioacaricide in the control of *T. urticae* in the future.

9.3.3 Plant-parasitic nematode control

Plant-parasitic nematodes (PPNs) are serious pests of crops in agriculture. Over 4100 described species damage crops as endo- or ectoparasites causing annual crop losses of US$125 billion globally (Chitwood, 2003; Jones *et al.*, 2013). Usually, control methods such as crop rotation, cultural practices, use of chemical nematicides and combinations thereof are employed to control PPN populations. Use of several chemical nematicides such as methyl bromide is restricted or banned due to their negative impacts on the environment and public health, and there are reports of nematicide resistance development in PPN populations (Bi *et al.*, 2018; Abd-Elgawad and Askary, 2020; Touray *et al.*, 2021).

Soil, especially the rhizosphere, is a natural habitat for EPNs and other soil-living saprophytic nematodes. Although PPNs and EPNs do not share similar food sources, the presence of EPNs in soil has been observed to reduce PPN populations and associated damage (Bird and Bird, 1986; Ishibashi and Kondo, 1986). It was speculated that these nematodes compete for space as they migrate across a carbon dioxide gradient toward plant root systems (Gaugler *et al.*, 1980; Bird and Bird 1986; Robinson, 1995), and the metabolites produced by EPN–bacteria in infected insect cadavers exhibited nematicidal properties (Georgis and Kelly, 1997) or allelopathy (Grewal *et al.*, 1999). Several studies have assessed the effects of cell-free supernatants from different *Xenorhabdus* and *Photorhabdus* spp./strains against different PPN species such as *Aphelenchoides rhytium*, *Bursaphelenchus* spp., *Bursaphelenchus xylophilus*, *Meloidogyne incognita*, *Meloidogyne arenaria* and *Meloidogyne javanica* (Hu *et al.*, 1996, 1999; Samaliev *et al.*, 2000; Kepenekci *et al.*, 2016, 2018; Bi *et al.*, 2018; Abebew *et al.*, 2022). In pot and field experiments, *Steinernema carpocapsae*, *Steinernema feltiae*, *Steinernema glaseri* and *Heterorhabditis bacteriophora* nematodes, and supernatants from the bacteria *X. bovienii* and *Photorhabdus kayaii* significantly suppressed the root-knot nematodes *Meloidogyne incognita* and *Meloidogyne arenaria*, in tomato roots (Kepenekci *et al.*, 2016, 2018). In addition, supernatants from *X. nematophila*, *X. bovienii* and *Photorhabdus luminescens* have inhibitory effects on egg hatching of *Meloidogyne incognita* (Sayedain *et al.*, 2019).

Based on the effects of bacterial metabolites on PPNs, research was initiated to identify the bioactive nematicide compounds present in supernatants produced by *Xenorhabdus* and *Photorhabdus* spp. So far, the identified compounds in *Photorhabdus* spp. are 3,5-dihydroxy-4-isopropylstilbene (Hu *et al.*, 1996), *trans*-cinnamic acid (*t*-CA), (4E)-5-phenylpent-4-enoic acid (PPA), and indole (Kusakabe *et al.*, 2022), 4,4′-diisothiocyanatostilbene-2,2′-disulfonic acid, anthracene-9-carboxylic acid, 5-nitro-2-(3-phenylpropylamino) benzoic acid and indanyloxyacetic acid (Boina *et al.*, 2008). In *Xenorhabdus* spp., identified compunds are fabclavines, rhabdopeptides and xenocoumacins (Fig. 9.2) (Bi *et al.*, 2018; Abebew *et al.*, 2022). Fabclavines, rhabdopeptides and xenocoumacins caused nematode mortality of over 85% against *Meloidogyne javanica* but were also highly toxic to *Caenorhabditis elegans*, a non-target bacterivorous nematode (mortality range of 72–95%) (Abebew *et al.*, 2022). The bioactive compounds isolated from *Photorhabdus* and *Xenorhabdus* spp. have been found to be dose and target nematode species dependent; for example, 3,5-dihydroxy-4-isopropylstilbene was reportedly toxic against *Bursaphelenchus* spp., *Aphelenchoides* spp. and *Caenorhabditis elegans* but had no activity against *Meloidogyne incognita* (Hu *et al.*, 1996, 1999). *t*-CA and PPA, in contrast, exhibited nematicidal

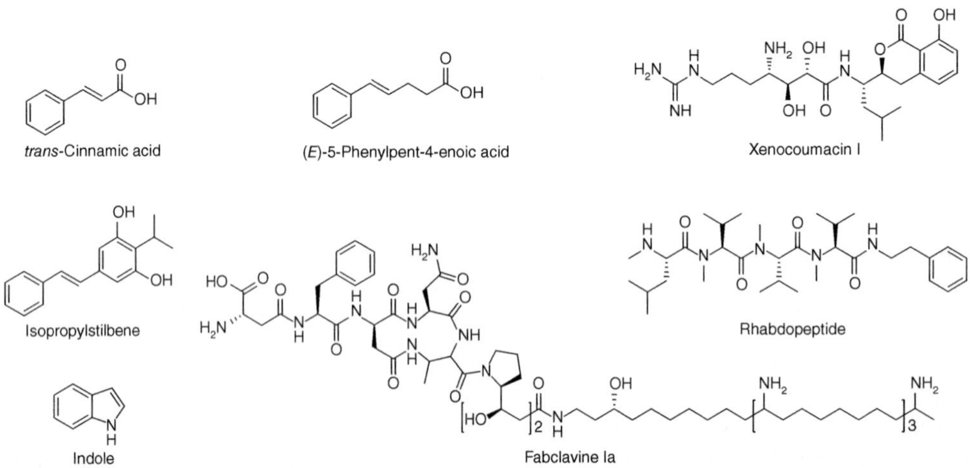

Fig. 9.2. Chemical structures of some of the natural products from *Xenorhabdus* and *Photorhabdus* spp.

activity toward *Meloidogyne incognita* and *Tylenchulus semipenetrans* but not on *Caenorhabditis elegans* or the EPN species *Steinernema carpocapsae* and *Heterorhabditis bacteriophora*, whereas indole had nematistatic activity on these nematodes. These metabolites from *Photorhabdus* spp. showed no cytotoxic effects on normal or neoplastic human cells. It was recommended that *t*-CA, PPA and indole, alone or in combination, could be used as new selective and environmentally safe compounds against PPNs (Kusakabe *et al.*, 2022).

9.4 Fungal Control

Pathogenic fungi impact human, animal and plant health. For instance, nearly 1.6 million people die and over 1 million are afflicted with fungal infections such as aspergillosis, candidiasis, cryptococcosis and dermatophytosis (Fisher *et al.*, 2018; Almeida *et al.*, 2019). In addition, damage caused by fungi to plants and crops leads to major losses (10–30% crop losses and over US$1 billion dollars of economic losses) in agricultural activities and food production (Peng *et al.*, 2021). Some fungal species are known to produce highly stable mycotoxins in food, making them unsafe for human/animal consumption (Savary *et al.*, 2012; Almeida *et al.*, 2019). A few antifungal classes are available for control of these organisms in humans including polyenes (e.g. amphotericin B), azoles (e.g. fluconazole), echinocandins (e.g. caspofungin) and pyrimidine analogs (e.g. flucytosine). For plants, more fungicides are available (about nine times as many) (Fisher *et al.*, 2018; Sun *et al.*, 2020) compared with those used against fungi affecting humans and other animals. Most fungicides used for the treatment of both clinical and crop maladies have harmful side effects because animals and plants share similar basic cell biology and physiological characteristics with fungi. Additionally, the emergence of resistant strains as a result of intensive fungicide use (Zubrod *et al.*, 2019) has necessitated the search for novel compounds to augment extant products from several promising sources, although this process is delayed by the long registration process.

EPNs must deal with other saprophytic fungi that try to invade nematode-infected host cadavers. To maintain monopoly over cadavers until the emergence of next generation of nematodes (which takes 1–2 weeks), *Xenorhabdus* and *Photorhabdus* spp. produce secondary metabolites with antifungal effects against most EPF or saprophytic fungi. Such metabolites have the potential to be used as alternative fungicidal substances against plant- or animal-pathogenic fungi. Thus, when cell cultures or cell-free supernatants of *Xenorhabdus* and *Photorhabdus* spp. are added to the growth medium for fungi, mycelial growth of important phytopathogens such as *Botrytis, Fusicladium, Fusarium, Monilinia, Moniliophthora* and *Rhizoctonia* spp. and *Neofusicoccum parvum* are adversely affected (Shapiro-Ilan *et al.*, 2009, 2014; San-Blas *et al.*, 2012,

2013; Bock *et al.*, 2014; Ullah *et al.*, 2015; Hazir *et al.*, 2016; Shi *et al.*, 2017; Chacón-Orozco *et al.*, 2020; Cimen *et al.*, 2021; Otoya-Martinez *et al.*, 2022). Similarly, oomycetes in the genus *Phytophthora* (Kaya *et al.*, 2006; Boszormenyi *et al.*, 2009; Fang *et al.*, 2011, 2014) are also negatively affected in medium augmented with cell cultures or cell-free supernatants of *Xenorhabdus* and *Photorhabdus* spp. A methanol extract or cell-free filtrate of *X. nematophila* TB also showed strong therapeutic and protective effects against *Botrytis cineria* and *Phytophthora capsici* in detached tomato fruits and plants, and in pepper plants (Fang *et al.*, 2014).

So far, the bioactive antifungal compound(s) identified from *Xenorhabdus* and *Photorhabdus* spp. with potential to be used as a substitute fungicide to control plant phytopathogens are *t*-CA from *Photorhabdus luminescens* (Bock *et al.*, 2014), stilbene derivative molecules (2-isopropyl-5-([E]-2-phenylethenyl), benzene-1,3-diol (syn. 3,5-dihydroxy-4-isopropyl stilbene), and 3-hydroxy-2-isopropyl-5-phenethyl phenyl carbamate) from *Photorhabdus temperata* SN259 (Shi *et al.*, 2017) and fabclavine from *X. szentirmaii* (Cimen *et al.*, 2021). *t*-CA, a cinnamic acid derivative previously only obtained from bark and leaves of cinnamon plants and now from *Photorhabdus luminescens*, is highly efficacious alone and synergizes with the antifungal efficacy of some commercial fungicides such as Elast® (dodine), Prophyt® (potassium phosphite), Regalia® (extract of *Reynoutria sachalinensis*), Serenade® (*Bacillus subtilis*) and PropiMax® (propiconazole) against fungal phytopathogens (Hazir *et al.*, 2017). Fabclavines, in contrast, are peptide–polyketide–polyamine hybrids encoded in highly conserved biosynthetic gene clusters identified in several *Xenorhabdus* spp. and *Photorhabdus asymbiotica*. These heat-stable peptides have broad-spectrum activity against bacteria, fungi and protozoa and are synthesized using NRPSs, PKS and polyunsaturated fatty acid enzyme systems (Wenski *et al.*, 2019).

9.5 Gene Regulation

Based on the variable environmental conditions, their extraordinary lifestyle, and interaction with nematode hosts, insects and food competitors, a complex and flexible regulation might be of great advantage for any aspect of *Xenorhabdus* and *Photorhabdus* spp. physiology (Herbert and Goodrich-Blair, 2007; Waterfield *et al.*, 2009). Therefore, it is not surprising that several regulatory mechanisms and regulators such as transcription factors have been described that also influence the generation of NPs in entomopathogenic bacteria (Joyce *et al.*, 2006; Goodrich-Blair and Clarke, 2007; Richards and Goodrich-Blair, 2009; Clarke, 2017; Eckstein and Heermann, 2019).

Despite these examples, the molecular details of such mechanisms are not well known with respect to the signals that are required to activate or inactivate such regulators, how the regulators themselves are in fact regulated, and how the regulatory network is coordinated and integrated to lead to the observed phenotypes of NP activation or inactivation.

However, recently, part of the regulatory network that is especially useful to identify and understand the function of most NPs in *Photorhabdus* and *Xenorhabdus* spp. was revealed: Hfq, HexA and *arcZ* (a *trans*-encoded small RNA). In this "trio," Hfq and *arcZ* regulate the activity of the repressor HexA.

Hfq was first described in *E. coli* as an essential factor for the replication of the RNA of bacteriophage Qβ. The RNA-binding protein is widespread among bacteria, archaea and eukaryotes. It interacts with all different types of RNA and is hence involved in post-transcriptional regulation (Chao and Vogel, 2010; Vogel and Luisi, 2011). Its regulatory influence can be either stabilizing, thereby accelerating mRNA translation, or initiating mRNA degradation, thereby preventing its translation. This differential activity is mediated by *trans*-encoded regulatory small RNAs that perform imperfect base pairing with their target mRNA, mediated by the Hfq protein.

Deletion of the *hfq* gene in *Photorhabdus* and *Xenorhabdus* spp. causes severe defects in gene expression level that lead to a different phenotype from the corresponding parental wild-type strain. A Δ*hfq* mutant shows almost no ability to produce NPs (Tobias *et al.*, 2017). Furthermore, and probably due to the NP loss, the development and growth of their respective symbiont feeding on bacterial lawns of Δ*hfq* mutant is no longer supported but can be restored by supplementing with an NP cocktail from the respective wild type, emphasizing the necessity of NPs in the developmental circle, as was shown for *Photorhabdus* spp. (Tobias *et al.*, 2017).

With respect to insect toxicity, the pathogenicity of *Photorhabdus luminescens* Δ*hfq* against insect larvae of *G. mellonella* was not affected. Infected larvae were killed during the same timeframe as was observed for the *Photorhabdus luminescens* wild type. Hence, the virulence seems also to be dominated by protein toxins or currently unknown factors, which underlie a different regulation mechanism independent of *hfq* in *Photorhabdus* spp.

The small RNA *arcZ* binds to the mRNA of *hexA* (Neubacher *et al.*, 2020). This interaction is mediated by the chaperone Hfq, leading to degradation of the *hexA* mRNA. HexA belongs to the family of LysR transcriptional regulators with an N-terminal DNA-binding domain and a C-terminal co-inducer binding domain of as-yet-unknown function. It was originally found in a transposon mutagenesis screen of non-bioluminescent and non-NP-producing secondary cells, restoring both phenotypes (Joyce and Clarke, 2003). Subsequently, it was shown to be highly abundant in secondary cells but not in primary cells and leading to phenotypic heterogeneity between both phenotypes (Langer *et al.*, 2017). With respect to NPs, it was shown previously that overexpression of HexA in the primary variant leads to a strong decrease in NP production (Engel *et al.*, 2017). Hence, Hfq and *arcZ* together regulate NP biosynthesis indirectly by preventing expression of *hexA* which acts as repressor on NP production with its detailed mode of action (either directly or indirectly via regulation of other transcriptional regulators) not yet elucidated.

Consequently, either deletion of *hfq* or *arcZ* or overexpression of *hexA* all result in a loss of NP production in *Photorhabdus* spp., while in a Δ*hfq*Δ*hexA* mutant the production of NPs is restored because the suppressor HexA is absent. Conclusively, Hfq is not acting directly on NP expression but via HexA in an indirect manner.

Although the full regulatory network, even regarding only these three players discussed above, is far from understood, it can be used effectively to produce and characterize NPs from *Photorhabdus* and *Xenorhabdus* spp. as follow: As a Δ*hfq* mutant shows hardly any NP production, activation of biosynthetic gene clusters via exchange of the NP against an inducible promoter results in so called easyPACId strains that exclusively produce the one NP class derived from the activated BGC (Bode *et al.*, 2019). Therefore, extracts from such easyPACId mutants can be used directly for bioactivity screening without the time-consuming NP isolation, which significantly speeds up NP identification and functional characterization beyond a classical promoter exchange approach in wild-type strains (Bode *et al.*, 2015). As *hfq* was also shown to regulate NP formation in *Pseudomonas* spp. and other proteobacteria (Matilla *et al.*, 2015, 2016; Bode *et al.*, 2019), the easyPACId approach also works here. Furthermore, it is worth mentioning that Δ*arcZ* mutants might also be suitable for easyPACId and that other general or specific regulatory mechanisms suitable for such approaches will be found in the future. Among such examples might be the global regulators Lrp (leucine-responsive regulator) and LeuO, which have already been investigated in detail in virulent bacteria from the genera *Salmonella* and *Vibrio*, and indicated a significant influence on NPs in *Photorhabdus luminescens* and *X. nematophila* (Cowles *et al.*, 2007; Engel *et al.*, 2017).

9.6 Conclusions

A recent pan-genomic analysis of 45 *Xenorhabdus* and *Photorhabdus* spp. genomes revealed the presence of 1000 biosynthetic gene clusters, of which 535 were not linked to any known NP (Shi *et al.*, 2022). This, and the fact that the complex life cycle of EPNs and their symbionts requires several different NPs with specific and complementary ecological functions, suggests that potent NPs still await identification. In combination with easyPACId, which is especially suitable in these bacteria, one can imagine the activation of all unknown biosynthetic gene clusters in a high-throughput manner. Recent examples such as darobactins (Imai *et al.*, 2019) and evybactins (Imai *et al.*, 2022), which are potent antibiotics from two different *Photorhabdus* strains, support this idea. Assuming we have only scratched the surface of *Photorhabdus* and *Xenorhabdus* spp. global diversity with many more strains to be isolated and analyzed, these bacteria might become as productive for their bioactive NPs as well-known NP producers such as myxobacteria or *Streptomyces* spp.

9.7 References

Abd-Elgawad, M.M. and Askary, T.H. (2020) Factors affecting success of biological agents used in controlling the plant-parasitic nematodes. *Egyptian Journal of Biological Pest Control* 30: 17.

Abebew, D., Sayedain, F.S., Bode, E. and Bode, H.B. (2022) Uncovering nematicidal natural products from *Xenorhabdus* bacteria. *Journal of Agricultural and Food Chemistry* 70, 498–506.

Adeolu, M., Alnajar, S., Naushad, S. and Gupta, R.S. (2016) Genome-based phylogeny and taxonomy of the 'Enterobacteriales': proposal for Enterobacterales ord. nov. divided into the families Enterobacteriaceae, Erwiniaceae fam. nov., Pectobacteriaceae fam. nov., Yersiniaceae fam. nov., Hafniaceae fam. nov., Morganellaceae fam. nov., and Budviciaceae fam. nov. *International Journal of Systematic and Evolutionary Microbiology* 66, 5575–5599.

Ahantarig, A., Chantawat, N., Waterfield, N.R., ffrench-Constant, R. and Pattamaporn Kittayapong, P. (2009) PirAB toxin from *Photorhabdus asymbiotica* as a larvicide against dengue vectors. *Applied and Environmental Microbiology* 75, 4627–4629.

Ahmed, S. and Kim, Y. (2018) Differential immunosuppression by inhibiting PLA_2 affects virulence of *Xenorhabdus hominickii* and *Photorhabdus temperata temperata*. *Journal of Invertebrate Pathology* 157, 136–146.

Ahn, J.Y., Lee, J.Y., Yang, E.J., Lee, Y.J., Koo, K.B. *et al.* (2013) Mosquitocidal activity of anthraquinones isolated from symbiotic bacteria *Photorhabdus* of entomopathogenic nematode. *Journal of Asia-Pacific Entomology* 16, 317–320.

Almeida, F., Rodrigues, M.L. and Coelho, C. (2019) The still underestimated problem of fungal diseases worldwide. *Frontiers in Microbiology* 10: 214.

Benelli, G., Jeffries, C.L. and Walker, T. (2016) Biological control of mosquito vectors: past, present, and future. *Insects* 7: 52.

Bi, Y., Gao, C. and Yu, Z. (2018) Rhabdopeptides from *Xenorhabdus budapestensis* SN84 and their nematicidal activities against *Meloidogyne incognita*. *Journal of Agricultural and Food Chemistry* 66, 3833–3839.

Bird, A.F. and Bird, J. (1986) Observations on the use of insect parasitic nematodes as a means of biological control of root-knot nematodes. *International Journal of Parasitology* 16, 511–516.

Bock, C.H., Shapiro-Ilan, D.I., Wedge, D. and Cantrell, C.H. (2014) Identification of the antifungal compound, *trans*-cinnamic acid, produced by *Photorhabdus luminescens*, a potential biopesticide. *Journal of Pest Science* 87, 155–162.

Bode, E., Brachmann, A.O., Kegler, C., Simsek, R., Dauth, C. *et al.* (2015) Simple "on-demand" production of bioactive natural products. *Chembiochem* 16, 1115–1119.

Bode, E., Heinrich, A.K., Hirschmann, M., Abebew, D., Shi, Y.N. *et al.* (2019) Promoter activation in Δhfq mutants as an efficient tool for specialized metabolite production enabling direct bioactivity testing. *Angewandte Chemie* 58, 18957–18963.

Boina, D.R., Lewis, E.E. and Bloomquist, J.R. (2008) Nematicidal activity of anion transport blockers against *Meloidogyne incognita*, *Ceanorhabditis elegans*, and *Heterorhabditis bacteriophora*. *Pest Management Science* 64, 646–653.

Boszormenyi, E., Ersek, T., Fodor, A., Fodor, A.M., Foldes, L.S. *et al.* (2009) Isolation and activity of *Xenorhabdus* antimicrobial compounds against the plant pathogens *Erwinia amylovora* and *Phytophthora nicotianae*. *Journal of Applied Microbiology* 107, 764–759.

Bowen, D.J. and Ensign, J.C. (1998) Purification and characterization of a high-molecular-weight insecticidal protein complex produced by the entomopathogenic bacterium *Photorhabdus luminescens*. *Applied and Environmental Microbiology* 64, 3029–3035.

Bowen, D., Rocheleau, T.A., Blackburn, M., Andreev, O., Golubeva, E. *et al.* (1998) Insecticidal toxins from the bacterium *Photorhabdus luminescens*. *Science* 280, 2129–2132.

Brachmann, A.O. and Bode, H.B. (2013) Identification and bioanalysis of natural products from insect symbionts and pathogens. *Advances in Biochemical Engineering/Biotechnology* 135, 123–155.

Bussaman, P., Sermswan, R.W. and Grewal, P.S. (2006) Toxicity of the entomopathogenic bacteria *Photorhabdus* and *Xenorhabdus* to the mushroom mite (*Luciaphorus* sp.; Acari: Pygmephoridae). *Biocontrol Science and Technology* 16, 245–256.

Bussaman, P., Sobanboa, S., Grewal, P.S. and Chandrapatya, A. (2009) Pathogenicity of additional strains of *Photorhabdus* and *Xenorhabdus* (Enterobacteriaceae) to the mushroom mite *Luciaphorus perniciosus* (Acari: Pygmephoridae). *Applied Entomology and Zoology* 44, 293–299.

Bussaman, P., Sa-Uth, C., Rattanasena, P. and Chandrapatya, A. (2012) Acaricidal activities of whole cell suspension, cell-free supernatant, and crude cell extract of *Xenorhabdus stokiae* against mushroom mite (*Luciaphorus* sp.). *Journal of Zhejiang University Science B* 13, 261–266.

Cakmak, I., Janssen, A. and Sabelis, M.W. (2006). Intraguild interactions between the predatory mites *Neoseiulus californicus* and *Phytoseiulus persimilis*. *Experimental and Applied Acarology* 38, 33–46.

Cakmak, I., Janssen, A., Sabelis, M.W. and Baspinar, H. (2009) Biological control of an acarine pest by single and multiple natural enemies. *Biological Control* 50, 60–65.

Cevizci, D., Ulug, D., Cimen, H., Touray, M., Hazir, S and Cakmak, I. (2020) Mode of entry of secondary metabolites of the bacteria *Xenorhabdus szentirmaii* and *X. nematophila* into *Tetranychus urticae*, and their toxicity to the predatory mites *Phytoseiulus persimilis* and *Neoseiulus californicus. Journal of Invertebrate Pathology* 174: 107418.

Chacón-Orozco, J.G., Bueno, C. Jr., Shapiro-Ilan, D.I., Hazir, S., Leite, L.G. and Harakava, R. (2020) Antifungal activity of the secondary metabolites and volatile compounds of *Xenorhabdus* spp. and *Photorhabdus* spp., against the soybean pathogenic *Sclerotinia scleortiorum. Scientific Reports* 10: 20649.

Challinor, V.L. and Bode, H.B. (2015) Bioactive natural products from novel microbial sources. *Annals of the New York Academy of Sciences,* 1354, 82–97.

Chao, Y. and Vogel, J. (2010) The role of Hfq in bacterial pathogens. *Current Opinion in Microbiology* 13, 24–33.

Chaston, J.M., Suen, G., Tucker, S.L., Andersen, A.W., Bhasin, A. *et al.* (2011). The entomopathogenic bacterial endosymbionts *Xenorhabdus* and *Photorhabdus*: convergent lifestyles from divergent genomes. *PLOS ONE* 6: e27909.

Chitwood, D.J. (2003) Research on plant-parasitic nematode biology conducted by the United States Department of Agriculture-Agricultural Research Service. *Pest Management Science* 59, 748–753.

Cimen, H., Touray, M., Gulsen, S.H., Erincik, O., Wenski, S.L. *et al.* (2021) Antifungal activity of different *Xenorhabdus* and *Photorhabdus* species against various fungal phytopathogens and identification of the antifungal compounds from *X. szentirmaii*. *Applied Microbiology and Biotechnology* 105, 5517–5528.

Clarke, D.J. (2017) The regulation of secondary metabolism in *Photorhabdus*. *Current Topics in Microbiology and Immunology* 402, 81–102.

Cowles, K.N., Cowles, C.E., Richards, G.R., Martens, E.C. and Goodrich-Blair, H. (2007) The global regulator Lrp contributes to mutualism, pathogenesis and phenotypic variation in the bacterium *Xenorhabdus nematophila*. *Cellular Microbiology* 9, 1311–1323.

Crawford, J.M., Portmann, C., Zhang, X., Roeffaers, M.B. and Clardy, J. (2012) Small molecule perimeter defense in entomopathogenic bacteria. *Proceedings of the National Academy of Sciences USA* 109, 10821–10826.

Dhanasekaran, D. and Thangaraj, R. (2014) Microbial secondary metabolites are an alternative approach against insect vector to prevent zoonotic diseases. *Asian Pacific Journal of Tropical Disease* 4, 253–261.

Dunphy, G.B. and Webster, J.M. (1988a) Interaction of *Xenorhabdus nematophilus* subsp. *nematophilus* with the haemolymph of *Galleria mellonella*. *Journal of Insect Physiology* 30, 883–889.

Dunphy, G.B. and Webster, J.M. (1988b) Virulence mechanisms of *Heterorhabditis heliothidi* and its bacterial associate, *Xenorhabdus luminescens*, in non-immune larvae of the dauer wax moth, *Galleria mellonella*. *International Journal of Parasitology* 18, 729–737.

Eckstein, S. and Heermann, R. (2019) Regulation of phenotypic switching and heterogeneity in *Photorhabdus luminescens* cell populations. *Journal of Molecular Biology* 431, 4559–4568.

El-Sadawy, H.A., Ramadan, M.Y., Abdel Megeed, K.N., Ali, H.H., El Sattar, S.A. and Elakabawy, L.M. (2020) Biological control of *Phlebotomus papatasi* larvae by using entomopathogenic nematodes and its symbiotic bacterial toxins. *Tropical Biomedicine* 37, 288–302.

Engel, Y., Windhorst, C., Lu, X., Goodrich-Blair, H. and Bode, H.B. (2017) The global regulators Lrp, LeuO, and HexA control secondary metabolism in entomopathogenic bacteria. *Frontiers in Microbiology* 8: 209.

Ensign, J.C., Lan, Q. and Dyer, D. (2014) Mosquitocidal *Xenorhabdus*, lipopeptide and methods. US Patent US20140274880A1. US Patent and Trademark Office, Washington, DC.

Eom, S., Park, Y., Kim, H. and Kim, Y. (2014) Development of a high efficient "Dual Bt-Plus" insecticide using a primary form of an entomopathogenic bacterium, *Xenorhabdus nematophila*. *Journal of Microbiology and Biotechnology* 24, 507–521.

Eroglu, C., Cimen, H., Ulug, D., Karagoz, M., Hazir, S. and Cakmak, I. (2019) Acaricidal effect of cell-free supernatants from *Xenorhabdus* and *Photorhabdus* bacteria against *Tetranychus urticae* (Acari: Tetranychidae). *Journal of Invertebrate Pathology* 1, 61–66.

Fang, X.L., Li, Z.Z., Wang, Y.H. and Zhang, X. (2011) *In vitro* and *in vivo* antimicrobial activity of *Xenorhabdus bovienii* YL002 against *Phytophthora capsici* and *Botrytis cinerea*. *Journal of Applied Microbiology* 111, 145–154.

Fang, X., Zhang, M., Tang, Q., Wang, Y. and Zhang, X. (2014) Inhibitory effect of *Xenorhabdus nematophila* TB on plant pathogens *Phytophthora capsici* and *Botrytis cinerea in vitro* and *in planta*. *Scientific Reports* 4: 4300.

Fisher, M.C., Hawkins, N.J., Sanglard, D. and Gurr, S.J. (2018) Worldwide emergence of resistance to antifungal drugs challenges human health and food security. *Science* 360, 739–742.

Fuchs, S.W., Grundmann, F., Kurz, M., Kaiser, M. and Bode, H.B. (2014) Fabclavines: bioactive peptide-polyketide-polyamino hybrids from *Xenorhabdus*. *ChemBioChem* 15, 512–516.

Fukruksa, C., Yimthin, T., Suwannaroj, M., Muangpat, P., Tandhavanant, S. *et al.* (2017). Isolation and identification of *Xenorhabdus* and *Photorhabdus* bacteria associated with entomopathogenic nematodes and their larvicidal activity against *Aedes aegypti*. *Parasites & Vectors* 10: 440.

Gaugler, R., Lebeck, L., Nakagaki, B. and Boush, G.M. (1980) Orientation of the entomogenous nematode *Neoaplectana carpocapsae* to carbon dioxide. *Environmental Entomology* 9, 649–652.

Georgis, R. and Kelly, J. (1997) Novel pesticidal substances from the entomopathogenic nematode-bacterium complex. In: Hedin. P.A. (ed.) *Phytochemicals for Pest Control*. American Chemical Society, Washington, DC, pp. 135–143.

Goodrich-Blair, H. and Clarke, D.J. (2007) Mutualism and pathogenesis in *Xenorhabdus* and *Photorhabdus*: two roads to the same destination. *Molecular Microbiology* 64, 260–268.

Grewal, P.S., Lewis, E.E. and Venkatachari, S. (1999) Allelopathy: a possible mechanism of suppression of plant-parasitic nematodes by entomopathogenic nematodes. *Nematology* 1, 735–743.

Gulcu, B., Cimen, H., Raja, R.K. and Hazir, S. (2017) Entomopathogenic nematodes and their mutualistic bacteria: their ecology and application as microbial control agents. *Biopesticide International* 13, 79–112.

Gulcu, B., Hazir, S., Lewis, E. and Kaya, H.K. (2018) Evaluation of responses of different ant species (Formicidae) to the scavenger deterrent factor associated with the entomopathogenic nematode–bacterium complex. *European Journal of Entomolology* 115, 312–317.

Hazir, S., Kaya, H.K., Stock, S.P. and Keskin, N. (2003) Entomopathogenic nematodes (Steinernematidae and Heterorhabditidae) for biological control of soil pests. *Turkish Journal of Biolology* 27, 181–202.

Hazir, S., Shapiro-Ilan, D.I., Bock, C.H., Hazir, C., Leite, L.G. and Hotchkiss, M.W. (2016) Relative potency of culture supernatants of *Xenorhabdus* and *Photorhabdus* spp. on growth of some fungal phytopathogens. *European Journal of Plant Pathology* 146, 369–381.

Hazir, S., Shapiro-Ilan, D., Bock, C.H. and Leite, L.G. (2017) *Trans*-cinnamic acid and *Xenorhabdus szentirmaii* metabolites synergize the potency of some commercial fungicides. *Journal of Invertebrate Pathology* 144, 22–31.

Hazir, S., Kaya, H.K., Touray, M., Cimen, H. and Shapiro-Ilan, D. (2022) Basic laboratory and field manual for conducting research with the entomopathogenic nematodes, *Steinernema* and *Heterorhabditis*, and their bacterial symbionts. *Turkish Journal of Zoology* 46, 305–350.

Heinz, F.X. and Stiasny, K. (2012) Flaviviruses and flavivirus vaccines. *Vaccine* 30, 4301–4306.

Herbert, E.E. and Goodrich-Blair, H. (2007) Friend and foe: the two faces of *Xenorhabdus nematophila*. *Nature Reviews Microbiology* 5, 634–646.

Hu, K., Li, J. and Webster, J.M. (1996) 3,5-Dihydroxy-4-isopropylstilbene: a selective nematicidal compound from the culture filtrate of *Photorhabdus luminescens*. *Canadian Journal of Plant Pathology* 18, 104–105.

Hu, K., Li, J. and Webster, J.M. (1999) Nematicidal metabolites produced by *Photorhabdus luminescens* (Enterobacteriaceae), bacterial symbiont of entomopathogenic nematodes. *Nematology* 1, 457–469.

Imai, Y., Meyer, K.J., Iinishi, A., Favre-Godal, Q., Green, R. *et al.* (2019). A new antibiotic selectively kills Gram-negative pathogens. *Nature* 576, 459–464.

Imai, Y., Hauk, G., Quigley, J., Liang, L., Son, S. *et al.* (2022) Evybactin is a DNA gyrase inhibitor that selectively kills *Mycobacterium tuberculosis*. *Nature Chemical Biology* 18, 1236–1244.

Incedayi, G., Cimen, H., Ulug, D., Touray, M., Bode, E. *et al.* (2021) Relative potency of a novel acaricidal compound from *Xenorhabdus*, a bacterial genus mutualistically associated with entomopathogenic nematodes. *Scientific Reports* 11: 11253.

Ishibashi, N. and Kondo, E. (1986) *Steinernema feltiae* (DD-136) and *S. glaseri*: persistence in soil and bark compost and their influence on native nematodes. *Journal of Nematology* 18, 310–316.

Jones, J.T., Haegeman, A., Danchin, E.G., Gaur, H.S., Helder, J. *et al.* (2013) Top 10 plant-parasitic nematodes in molecular plant pathology. *Molecular Plant Pathology* 14, 946–961.

Joyce, S.A. and Clarke, D.J. (2003) A *hexA* homologue from *Photorhabdus* regulates pathogenicity, symbiosis, and phenotypic variation. *Molecular Microbiology* 47, 1445–1457.

Joyce, S.A., Watson, R.J. and Clarke, D.J. (2006) The regulation of pathogenicity and mutualism in *Photorhabdus*. *Current Opinion in Microbiology* 9, 127–132.

Jung, S. and Kim, Y. (2006) Synergistic effect of *Xenorhabdus nematophila* K1 and *Bacillus thuringiensis* subsp. *aizawai* against *Spodoptera exigua* (Lepidoptera: Noctuidae). *Biological Control* 39, 201–209.

Jung, S. and Kim, Y. (2007) Potentiating effect of *Bacillus thuringiensis* subsp. *kurstaki* on pathogenicity of entomopathogenic bacterium *Xenorhabdus nematophila* K1 against diamondback moth (Lepidoptera: Plutellidae). *Journal of Economic Entomology* 100, 246–250.

Kajla, M.K. (2019) Symbiotic bacteria as potential agents for mosquito control. *Trends in Parasitology* 36, 4–7.

Kaya, H.K. (2002) Natural enemies and other antagonists. In: Gaugler, R. (ed.) *Entomopathogenic Nematology*. CAB International, Wallingford, UK, pp. 189–203.

Kaya, H.K., Aguillera, M.M., Alumai, A., Choo, H.Y., de la Torre, M. *et al.* (2006) Status of entomopathogenic nematodes and their symbiotic bacteria from selected countries or regions of the world. *Biological Control* 38, 134–155.

Kepenekci, I., Hazir, S. and Lewis, E.E. (2016) Evaluation of entomopathogenic nematodes and the supernatants of the *in vitro* culture medium of their mutualistic bacteria for the control of the root-knot nematodes *Meloidogyne incognita* and *M. arenaria*. *Journal of Pest Management Science* 72, 327–334.

Kepenekci, I., Hazir, S., Oksal, E. and Lewis, E.E. (2018) Application methods of *Steinernema feltiae*, *Xenorhabdus bovienii* and *Purpureocillium lilacinum* to control root-knot nematodes in greenhouse tomato systems. *Crop Protection* 108, 31–38.

Khandelwal, P., Bhatnagar, R., Choudhury, D. and Banerjee, N. (2004) Characterization of a cytotoxic pilin subunit of *Xenorhabdus nematophila*. *Biochemical and Biophysical Research Communications* 314, 943–949.

Kim, E., Jeoung, S., Park, Y., Kim, K. and Kim, Y. (2015) A novel formulation of *Bacillus thuringiensis* for the control of brassica leaf beetle, *Phaedon brassicae* (Coleoptera: Chrysomelidae). *Journal of Economic Entomology* 108, 2556–2565.

Kim, I.H., Aryal, S.K., Aghai, D.T., Casanova-Torres, Á.M., Hillman, K. *et al.* (2017a) The insect pathogenic bacterium *Xenorhabdus innexi* has attenuated virulence in multiple insect model hosts yet encodes a potent mosquitocidal toxin. *BMC Genomics* 18: 927.

Kim, I.H., Ensign, J., Kim, D.Y., Jung, H.Y., Kim, N.R. *et al.* (2017b) Specificity and putative mode of action of a mosquito larvicidal toxin from the bacterium *Xenorhabdus innexi*. *Journal of Invertebrate Pathology* 149, 21–28.

Kim, Y., Cho, J.D. and Park, Y. (2005) Two groups of entomopathogenic bacteria, *Photorhabdus* and *Xenorhabdus*, share an inhibitory action against phospholipase A2 to induce host immunodepression. *Journal of Invertebrate Pathology* 89, 258–264.

Kim, Y., Ahmed, S., Stanley, D. and An, C. (2018) Eicosanoid-mediated immunity in insects. *Developmental & Comparative Immunology* 83, 130–143.

Kusakabe, A., Wang, C., Xu, Y.M., Molnár, I. and Stock, S.P. (2022) Selective toxicity of secondary metabolites from the entomopathogenic bacterium *Photorhabdus luminescens sonorensis* against selected plant parasitic nematodes of the Tylenchina suborder. *Microbiology Spectrum* 10: e02577-21.

Langer, A., Moldovan, A., Harmath, C., Joyce, S.A., Clarke, D.J. and Heermann, R. (2017) HexA is a versatile regulator involved in the control of phenotypic heterogeneity of *Photorhabdus luminescens*. *PLOS ONE* 12: e0176535.

Mahar, A.N., Munir, M. and Mahar, A.Q. (2004) Studies of different application methods of *Xenorhabdus* and *Photorhabdus* cells and their toxin in broth solution to control locust (*Schistocerca gregaria*). *Asian Journal of Plant Sciences* 3, 690–695.

Matilla, M.A., Leeper, F.J. and Salmond, G.P. (2015) Biosynthesis of the antifungal haterumalide, oocydin A, in *Serratia*, and its regulation by quorum sensing, RpoS and Hfq. *Environmental Microbiology* 17, 2993–3008.

Matilla, M.A., Nogellova, V., Morel, B., Krell, T. and Salmond, G.P. (2016) Biosynthesis of the acetyl-CoA carboxylase-inhibiting antibiotic, andrimid in *Serratia* is regulated by Hfq and the LysR-type transcriptional regulator, AdmX. *Environmental Microbiology* 18, 3635–3650.

Migeon, A., Nouguier, E. and Dorkeld, F. (2010) Spider Mites Web: a comprehensive database for the Tetranychidae. In: Sabelis, M. and Bruin, J. (eds.) *Trends in Acarology*. Springer, Dordrecht, Netherlands, pp. 557–560.

Mohan, S., Raman, R. and Gaur, H.S. (2003) Foliar application of *Photorhabdus luminescens*, symbiotic bacteria from entomopathogenic nematode *Heterorhabditis indica*, to kill cabbage butterfly *Pieris brassicae*. *Current Science* 84, 1397–1397.

Mollah, M.I. and Kim, Y. (2020) Virulent secondary metabolites of entomopathogenic bacteria genera, *Xenorhabdus* and *Photorhabdus*, inhibit phospholipase A2 to suppress host insect immunity. *BMC Microbiology* 20: 359.
Namsena, P., Bussaman, P. and Rattanasena, P. (2016) Bioformulation of *Xenorhabdus stockiae* PB09 for controlling mushroom mite, *Luciaphorus perniciosus* Rack. *Bioresources and Bioprocessing* 3: 19.
Nermut J., Zemek, R., Mracek, Z., Palevsky, E. and Puza, V. (2019) Entomopathogenic nematodes as natural enemies for control of *Rhizoglyphus robini* (Acari: Acaridae)? *Biological Control* 128, 102–110.
Neubacher, N., Tobias, N.J., Huber, M., Cai, X., Glatter, T. *et al.* (2020) Symbiosis, virulence and natural-product biosynthesis in entomopathogenic bacteria are regulated by a small RNA. *Nature Microbiology* 5, 1481–1489.
Ozkan, H.D., Cimen, H., Ulug, D., Wenski, S., Yigit Ozer, S. *et al.* (2019) Nematode-associated bacteria: production of antimicrobial agent as a presumptive nominee for curing endodontic infections caused by *Enterococcus faecalis*. *Frontiers in Microbiology* 10: 2672.
Otoya-Martinez, N., Leite, L.G., Harakava, R., Touray, M., Hazir, S. *et al.* (2022). Disease caused by *Neofusicoccum parvum* in pruning wounds of grapevine shoots and its control by *Trichoderma* spp. and *Xenorhabdus szentirmaii*. *Fungal Biology* 127, 865–871.
Park, Y. (2015) Entomopathogenic bacterium, *Xenorhabdus nematophila* and *Photorhabdus luminescens*, enhances *Bacillus thuringiensis* Cry4Ba toxicity against yellow fever mosquito, *Aedes aegypti* (Diptera: Culicidae). *Journal of Asia-Pacific Entomology* 18, 459–463.
Park, Y., Jung, J.K. and Kim, Y. (2016) A mixture of *Bacillus thuringiensis* subsp. *israelensis* with *Xenorhabdus nematophila*-cultured broth enhances toxicity against mosquitoes *Aedes albopictus* and *Culex pipienspallens* (Diptera: Culicidae). *Journal of Economic Entomology* 109, 1086–1096.
Peng, Y., Li, S.J., Yan, J., Tang, Y., Tang, Y. *et al.* (2021) Research progress on phytopathogenic fungi and their role as biocontrol agents. *Frontiers in Microbiology* 12: 670135.
Richards, G.R. and Goodrich-Blair, H. (2009) Masters of conquest and pillage: *Xenorhabdus nematophila* global regulators control transitions from virulence to nutrient acquisition. *Cell Microbiology* 11, 1025–1033.
Robinson, A.F. (1995) Optimal release rates for attracting *Meloidogyne incognita, Rotylenchulus reniformis*, and other nematodes to carbon dioxide in sand. *Journal of Nematology* 27, 42–50.
Samaliev, H., Andreoglou, F., Elawad, S., Hague, N. and Gowen, S. (2000) The nematicidal effects of the bacteria *Pseudomonas oryzihabitans* and *Xenorhabdus nematophilus* on the root-knot nematode *Meloidogyne javanica*. *Nematology* 2, 507–514.
San-Blas, E., Carrillo, Z. and Parra, Y. (2012) Effect of *Xenorhabdus* and *Photorhabdus* bacteria and their exudates on *Moniliophthora roreri*. *Archives of Phytopathology and Plant Protection* 45, 1950–1967.
San-Blas, E., Parra, Y. and Carrillo, Z. (2013) Effect of *Xenorhabdus* and *Photorhabdus* bacteria (Enterobacteriales: Enterobacteriaceae) and their exudates on the apical rotten fruit disease caused by *Dothiorella* sp. in guava (*Psidium guajava*). *Archives of Phytopathology and Plant Protection* 46, 2294–2303.
Savary, S., Ficke, A., Aubertot, J.N. and Hollier, C. (2012) Crop losses due to diseases and their implications for global food production losses and food security. *Food Security* 4, 519–537.
Sayedain, F.S., Ahmadzadeh, M., Talaei-Hassanloui, R., Olia, M. and Bode, H.B. (2019) Nematicidal effect of cell-free culture filtrates of EPN-symbiotic bacteria on *Meloidogyne javanica*. *Biological Control of Pests & Plant Diseases* 8, 17–26.
Seo, S., Lee, S., Hong, Y. and Kim, Y. (2012) Phospholipase A_2 inhibitors synthesized by two entomopathogenic bacteria, *Xenorhabdus nematophila* and *Photorhabdus temperata* subsp. *temperata*. *Applied and Environmental Microbiology* 78, 3816–23.
Sergeant, M., Baxter, L., Jarrett, P., Shaw, E., Ousley, M. *et al.* (2006) Identification, typing, and insecticidal activity of *Xenorhabdus* isolates from entomopathogenic nematodes in United Kingdom soil and characterization of the *xpt* toxin loci. *Applied and Environmental Microbiology* 72, 5895–907.
Shah, F.A., Abdoarrahem, M.M., Berry, C., Touray, M., Hazir, S. and Butt, T.M. (2021) Indiscriminate ingestion of entomopathogenic nematodes and their symbiotic bacteria by *Aedes aegypti* larvae: a novel strategy to control the vector of chikungunya, dengue and yellow fever. *Turkish Journal of Zoology* 45: 372.
Shapiro-Ilan, D.I., Reilly, C.C. and Hotchkiss, M.W. (2009) Suppressive effects of metabolites from *Photorhabdus* and *Xenorhabdus* spp. on phytopathogens of peach and pecan. *Archives of Phytopathology and Plant Protection* 42, 715–728.
Shapiro-Ilan, D., Bock, C.H. and Hotchkiss, M.W. (2014) Suppression of pecan and peach pathogens on different substrates using *Xenorhabdus bovienii* and *Photorhabdus luminescens*. *Biological Control* 77, 1–6.

Shapiro-Ilan, D., Hazir, S. and Glazer, I. (2017) Basic and applied research: entomopathogenic nematodes. In: Lacey, L.A. (ed.) *Microbial Control of Insect and Mite Pests: From Theory to Practice*. Academic Press, London, pp. 91–105.

Shapiro-Ilan, D., Hazir, S. and Glaser, I. (2020) Advances in use of entomopathogenic nematodes in integrated pest management. In: Kogan, M. and Heinrichs, E.A. (eds.) *Integrated Management of Insect Pests: Current and Future Developments*. Burleigh Dodds Science Publishing, Cambridge, UK, pp. 649–678.

Shi, D., An, R., Zhang, W., Zhang, G. and Yu, Z. (2017) Stilbene derivatives from *Photorhabdus temperata* SN259 and their antifungal activities against phytopathogenic fungi. *Journal of Agricultural and Food Chemistry* 65, 60–65.

Shi, Y.M. and Bode, H.B. (2018) Chemical language and warfare of bacterial natural products in bacteria–nematode–insect interactions. *Natural Product Reports* 35, 309–335.

Shi, Y.M., Hirschmann, M., Shi, Y.N., Ahmed, S., Abebew, D. *et al.* (2022) Global analysis of biosynthetic gene clusters reveals conserved and unique natural products in entomopathogenic nematode-symbiotic bacteria. *Nature Chemistry* 14, 701–712.

Shrestha, S. and Kim, Y. (2007) An entomopathogenic bacterium, *Xenorhabdus nematophila*, inhibits hemocyte phagocytosis of *Spodoptera exigua* by inhibiting phospholipase A_2. *Journal of Invertebrate Pathology* 96, 64–70.

Shrestha, Y.K., Jang, E.K., Yu, Y.S., Kwon, M., Shin, J.H. and Lee, K.Y. (2011) Oral toxicity of symbiotic bacteria *Photorhabdus* spp. against immature stages of insects. *Journal of Asia-Pacific Entomology* 14, 127–130.

Silva, L.R.J., Undurraga, S.F., Eugênio, S.C., da Costa, M., Heermann, R. and Santos da Silva, O. (2017) Larvicidal and growth-inhibitory activity of entomopathogenic bacteria culture fluids against *Aedes aegypti* (Diptera: Culicidae). *Journal of Economic Entomology* 110, 378–385.

Silva, O.S., Prado, G.R., da Silva, J.L.R., Carlos Eugenio Silva, C.E., da Costa, M. and Heermann, R. (2013) Oral toxicity of *Photorhabdus luminescens* and *Xenorhabdus nematophila* (Enterobacteriaceae) against *Aedes aegypti* (Diptera: Culicidae). *Parasitology Research* 112, 2891–2896.

Sobanboa, S., Bussaman, P. and Chandrapatya, A. (2009) Efficacy of *Xenorhabdus* sp. (X1) as biocontrol against for controlling mushroom mites (*Luciaphorus* sp.). *Asian Journal of Food and Agro-industry* 2, 145–154.

Sun, S., Hoy, M.J. and Heitman, J. (2020) Fungal pathogens. *Current Biology* 30, R1163–R1169.

Tobias, N.J., Heinrich, A.K., Eresmann, H., Wright, P.R., Neubacher, N. *et al.* (2017) *Photorhabdus*–nematode symbiosis is dependent on *hfq*-mediated regulation of secondary metabolites. *Environmental Microbiology* 19, 119–129.

Touray, M., Cimen, H., Gulsen, S.H., Ulug, D., Erdogus, D. *et al.* (2021) The impact of chemical nematicides on entomopathogenic nematode survival and infectivity. *Journal of Nematology* 49, 1–17.

Touray, M., Ulug, D., Gulsen, S.H., Cimen, H., Hazir, C. *et al*. (2024a) Natural products from *Xenorhabdus* and *Photorhabdus* show promise as biolarvicides against *Aedes albopictus*. *Pest Management Science*. DOI: 10.1002/ps.8127

Touray, M., Cimen, H., Bode, E., Bode, H.B. and Hazir, S. (2024b) Effects of *Xenorhabdus* and *Photorhabdus* bacterial metabolites on the ovipositional activity of *Aedes albopictus*. *Journal of Pest Science*, 1–13.

Ullah, I., Khan, A.L., Ali, L., Khan, A.R.,Waqas, M. *et al.* (2015) Benzaldehyde as an insecticidal, antimicrobial, and antioxidant compound produced by *Photorhabdus temperata* M1021. *Journal of Microbiology* 53, 127–133.

Ulug, D., Touray, M., Gulsen, S.H., Cimen, H., Hazir, C., Bode, H.B. and Hazir, S. (2024) A taste of a toxin paradise: *Xenorhabdus* and *Photorhabdus* bacterial secondary metabolites against *Aedes aegypti* larvae and eggs. *Journal of Invertebrate Pathology* 205, 108126.

Vitta, A., Thimpoo, P., Meesil, W., Yimthin, T., Fukruksa, C. *et al.* (2018) Larvicidal activity of *Xenorhabdus* and *Photorhabdus* bacteria against *Aedes aegypti* and *Aedes albopictus*. *Asian Pacific Journal of Tropical Biomedicine* 8, 31–36.

Vogel, J. and Luisi, B.F. (2011) Hfq and its constellation of RNA. *Nature Reviews Microbiology* 19, 578–589.

Wagutu, G.K., Mwangi, W. and Waturu, C.N. (2017) Entomopathogenic bacteria: *Xenorhabdus* spp and *Photorhabdus* spp from *Steinernema karii* and *Heterorhabditis indica* for the control of mosquito larvae. *Journal of Agriculture, Science and Technology* 18, 21–38.

Wang, H., Dong, H., Qian, H., Xia, R. and Cong, B. (2013) Isolation, bioassay and characterisation of *Xenorhabdus* sp. SY5, a highly virulent symbiotic bacterium of an entomopathogenic nematode isolated from China. *Nematology* 15, 153–163.

Waterfield, N., Dowling, A., Sharma, S., Daborn, P.J., Potter, U. and ffrench-Constant, R.H. (2001) Oral toxicity of *Photorhabdus luminescens* W14 toxin complexes in *Escherichia coli*. *Applied and Environmental Microbiology* 67, 5017–5024.

Waterfield, N.R., Ciche, T. and Clarke, D. (2009) *Photorhabdus* and a host of hosts. *Annual Review of Microbiology* 63, 557–574.

Wenski, S.L., Kolbert, D., Grammbitter, G.L. and Bode, H.B. (2019) Fabclavine biosynthesis in *X. szentirmaii*: shortened derivatives and characterization of the thioester reductase FclG and the condensation domain-like protein FclL. *Journal of Industrial Microbiology and Biotechnology* 46, 565–572.

Wenski, S.L., Cimen, H., Berghaus, N., Fuchs, S.W., Hazir, S. and Bode, H.B. (2020) Fabclavine diversity in *Xenorhabdus* bacteria. *Beilstein Journal of Organic Chemistry* 16, 956–965.

World Health Organization (WHO) (2018) Global Malaria Programme, Malaria Policy Advisory Committee (MPAC) meeting report, April 2018. WHO, Geneva.

Wu, S., Toews, M.D., Cottrell, T.C., Schmidt, J.M. and Shapiro-Ilan, D.I. (2022). Toxicity of *Photorhabdus luminescens* and *Xenorhabdus bovienii* bacterial metabolites to pecan aphids (Hemiptera: Aphididae) and the lady beetle *Harmonia axyridis* (Coleoptera: Coccinellidae). *Journal of Invertebrate Pathology* 194: 107806.

Yang, Q., Zhang, J., Li, T., Liu, S., Song, P., Nangong, Z. and Wang, Q. (2017). PirAB protein from *Xenorhabdus nematophila* HB310 exhibits a binary toxin with insecticidal activity and cytotoxicity in *Galleria mellonella*. *Journal of Invertebrate Pathology* 148, 43–50.

Yooyangket, T., Muangpat, P., Polseela, R., Tandhavanant, S., Thanwisai, A. and Vitta, A. (2018) Identification of entomopathogenic nematodes and symbiotic bacteria from Nam Nao National Park in Thailand and larvicidal activity of symbiotic bacteria against *Aedes aegypti* and *Aedes albopictus*. *PLOS ONE* 13: e0195681.

Zubrod, J.P., Bundschuh, M., Arts, G., Brühl, C.A., Imfeld, G. *et al.* (2019) Fungicides: an overlooked pesticide class? *Environmental Science and Technology* 53, 3347–3365.

10 Mass Production

Richou Han[1]*, Luis G. Leite[2], Xun Yan[3], Julie G. Chacon-Orozco[2] and David I. Shapiro-Ilan[4]

[1]Institute of Zoology, Guangdong Academy of Sciences, Guangzhou, PR China; [2]Laboratório de Controle Biológico, Instituto Biológico, Campinas, São Paulo, Brazil; [3]Zhongkai University of Agriculture and Engineering, Guangzhou, PR China; [4]USDA-ARS, Southeastern Fruit and Tree Nut Research Station, Byron, Georgia, USA

Abstract

Entomopathogenic nematodes (EPNs) are parasites of insects that are used as biological control agents for suppression of various economically important pests. EPNs can be mass produced using *in vivo* and *in vitro* solid and liquid methods. Each approach has its advantages and disadvantages. *In vivo* culture uses insects as hosts and lacks economy of scale due to the costs of labor and insects; however, automation or grower-based techniques may alleviate these disadvantages to some extent. *In vitro* culture requires the purified symbiotic bacteria and a monoxenic system. *In vitro* liquid culture uses the largest capital inputs but provides the greatest economic efficiency. *In vitro* solid culture is intermediate between the other two methods, requiring less capital investment but more employees. This chapter summarizes the basic methods and the factors affecting efficiency of the state-of-the-art production technology. Each section will include a discussion on recent developments and advancements of the *in vivo* and *in vitro* solid and liquid production methods.

10.1 Introduction

Entomopathogenic nematodes (EPNs) are currently produced using *in vivo* or *in vitro* (solid and liquid culture) methods (Friedman, 1990; Ehlers and Shapiro-Ilan, 2005; Shapiro-Ilan *et al.*, 2012, 2023) as biological control agents on a commercial level by more than ten companies in Asia, Europe and North America (Kaya *et al.*, 2006; Peters, 2013; Shapiro-Ilan *et al.*, 2017, 2023). Each approach has its advantages and disadvantages in terms of EPN species, production cost, technical expertise, economies of scale and quality of the end product. *In vivo* production is an appropriate method for laboratory use and small-scale field testing, as well as for niche markets and cottage industry applications. *In vitro* production is economically reasonable to supply EPN products at high quality for the world market. We provide a summary of EPN production methods, crucial factors that affect efficiency and methods for improvement, as well as the future research and development directions needed.

10.2 *In vivo* Production

In vivo production of EPNs is based on a two-dimensional system that relies on suitable insect hosts for production in trays and shelves (Kaya

*Email: hanrc@giz.gd.cn

DOI: 10.1079/9781800620322.0010

and Stock, 1997; Shapiro-Ilan *et al.*, 2002, 2014, 2016, 2023). This method usually revolves around the concept of a White trap (White, 1927), which is a device used for harvesting infective juveniles (IJs) by taking advantage of the natural migration capacity of IJs from the host cadaver on emergence (Fig. 10.1) (Ehlers and Shapiro-Ilan, 2005; Hazir *et al.*, 2022).

The basic *in vivo* approach consists of host selection (usually the greater wax moth, *Galleria mellonella*), inoculation, harvest (e.g. using a White trap), concentration and (if necessary) decontamination. When the host is available, the IJs are inoculated on a dish or tray containing the selected insects and lined with absorbent paper or another material conducive to EPN infection such as soil, cheesecloth or plaster of Paris. After approximately 2–5 days post-inoculation at room temperature, infected insects are transferred to the White traps. It is important not to allow the infections to progress too long before transfer, otherwise the chance of the cadavers rupturing, or harm to the reproductive EPN stages, will increase (Shapiro-Ilan *et al.*, 2001). The IJs are harvested after they emerge and migrate into the surrounding water trap. The scale of the White trap in size and number can be expanded to commercial levels. For instance, trays as large as 1 m^2 are conceivable as long as mechanisms for the IJs to enter the water trap are available. Following harvest, concentration of EPNs can be accomplished by gravity settling (e.g. in large cone-shaped vessels) or vacuum filtration (Lindegren *et al.*, 1993). Gravity settling is simple and straightforward, but one potential drawback is that prolonged lack of aeration can be detrimental to the EPNs. Centrifugation is also feasible (Kaya and Stock, 1997), but, for commercial *in vivo* operations, the capital outlay for a centrifuge of sufficient capacity may be unwarranted.

Fig. 10.1. A modified White trap. Insect larvae (*Galleria mellonella*) infected with entomopathogenic nematodes (*Heterorhabditis bacteriophora*) are placed on moist filter paper in an inverted Petri dish lid (60 mm). The area around the smaller Petri dish is filled with water below the level of dish. The infective juveniles emerge from the insect cadavers and crawl over the small dish lid into the water trap.

Nematode yield in *in vivo* production is affected by nematode and host species, inoculation parameters and environmental conditions. Variation in yield among EPN species is generally inversely proportional to IJ size (Blinova and Ivanova, 1987; Grewal *et al.*, 1994; Flanders *et al.*, 1996; Hominick *et al.*, 1997), although some species simply have innately high reproductive capacities (e.g. *Heterorhabditis indica* and *Steinernema riobrave*). For example, yields of *S. riobrave* (average IJ body length 622 μm) may exceed 3 × 10^5 IJs per insect in *G. mellonella*, whereas for *Steinernema glaseri* (average IJ body length 1133 μm), yields do not exceed 5 × 10^4 IJs per insect in the same host (Grewal *et al.*, 1994; Stock and Hunt, 2005). The choice of insect host is also important to maximize yields and efficiency in production. Due to its high susceptibility to most EPN species, ease of rearing, wide availability and ability to produce high yields, the last instar of *G. mellonella* is the most common insect host used for *in vivo* laboratory and commercial EPN production (Woodring and Kaya, 1988; Hazir *et al.*, 2022). Only a few EPN species exhibit relatively poor reproduction in *G. mellonella* (due to extremes in host specificity); for example, *Steinernema kushidai* and *Steinernema scarabaei* appear to be especially adapted to hosts in the family Scarabaeidae (order Coleoptera) and *Steinernema scapterisci* appears to be especially adapted to the order Orthoptera (Mamiya, 1989; Nguyen and Smart, 1990; Kaya and Stock, 1997; Grewal *et al.*, 1999; Koppenhöfer and Fuzy, 2003). Significant research and commercial application have also been achieved for production of EPNs in the yellow mealworm, *Tenebrio molitor* (Blinova and Ivanova, 1987; Shapiro-Ilan *et al.*, 2002). Various other hosts have been studied for *in vivo* EPN production including the navel orangeworm, *Amyelois transitella*, tobacco budworm, *Heliothis virescens*, cabbage looper, *Trichoplusia ni*, pink bollworm, *Pectinophora gossypiella*, beet armyworm, *Spodoptera exigua*, corn earworm, *Helicoverpa zea*, spongy moth, *Lymantria dispar*, house cricket, *Acheta domesticus*, black soldier fly, *Hermetia illucens*, and various beetles (Coleoptera) (Lindegren *et al.*, 1979; Blinova and Ivanova, 1987; Cabanillas and Raulston, 1994; Grewal *et al.*, 1999; Elawad *et al.*, 2001; Tourtois *et al.*, 2017).

Inoculation parameters including dosage, host density and inoculation method can have a profound effect on the efficiency of production. For each nematode/host species combination, optimization of host density and inoculation rate is recommended to achieve maximum yields (Shapiro-Ilan *et al.*, 2002). In general, the number of hosts exhibiting patent signs of EPN infection increases with EPN concentration and decreases with host density per unit area (Shapiro-Ilan *et al.*, 2002). Intermediate dosages should be used to maximize yield (Boff *et al.*, 2000), because a dosage that is too low results in low host mortality, while a dosage that is too high has been suggested to potentially result in failed infections due to competition with secondary invaders (Woodring and Kaya, 1988). Rates of approximately 25–200 IJs per insect are usually sufficient (depending on EPN species and method of inoculation) for infecting *G. mellonella*, whereas higher rates are generally needed to infect *T. molitor* (e.g. 100–600 IJs per insect). Crowding of hosts can lead to oxygen deprivation or a buildup of ammonia, which suppresses EPN yield (Shapiro *et al.*, 2000; Shapiro-Ilan *et al.*, 2002). *In vivo* inoculation can be accomplished by pipetting or spraying EPNs on to a substrate, immersion of insects in a nematode suspension or applying the EPNs to the insect's food. Immersion of hosts is generally more time efficient but requires more nematodes than other procedures. However, some insects are more sensitive to being immersed than others. Shapiro-Ilan *et al.* (2002) observed immersion of *G. mellonella* to be approximately four times more efficient in time than pipetting inoculum on to the hosts.

Environmental factors including temperature, adequate aeration and moisture can affect yield (Burman and Pye, 1980; Woodring and Kaya, 1988; Friedman, 1990; Grewal *et al.*, 1994; Shapiro-Ilan *et al.*, 2002; Dolinski *et al.*, 2007). Temperature is critical during the rearing process as it affects both yield and life-cycle duration (time to emergence) (Grewal *et al.*, 1994). Grewal *et al.* (1994) determined the optimum rearing temperature in *G. mellonella* for 12 species and strains of EPNs; optimum temperatures varied from 18°C to 28°C. Usually, optimum rearing temperatures for species or strains isolated from cold regions were low, while those for EPNs isolated from warm regions were relatively high. In addition to appropriate temperatures, adequate aeration (Burman and Pye, 1980; Friedman, 1990) and high levels of humidity are important environmental factors that must be maintained throughout the production cycle (Woodring and Kaya, 1988). An optimum balance between aeration (to avoid build up ammonia or other harmful gases) and humidity should be reached, for example using HEPA-filtered air and a humidifying system.

The efficiency of *in vivo* production is limited by the cost of the insects and labor. Therefore, the economics of *in vivo* production can be improved substantially by mechanizing the process for nematode and insect production (thereby reducing labor) and by producing the insect hosts "in house." Various steps in the *in vivo* production can be mechanized (e.g. inoculation, harvest, concentration). Improved inoculation devices have been developed to allow mass inoculation on multiple shelves (Carr and Kolodny-Hirsch, 1993; Shapiro-Ilan *et al.*, 2009). A mechanized harvest device, LOTEK, allows automated collection of IJs from stacked trays; the IJs are pumped to a central collection tank, and unlike the White trap approach, harvesting does not require nematode migration to a water reservoir (Gaugler *et al.*, 2002). Shapiro-Ilan *et al.* (2011, 2016) introduced a single-tray design where infected host cadavers are placed on a substrate intended as the formulation carrier (e.g. a gel) within the final package; once IJs emerge the cadavers are removed and the final EPN product is ready for shipment or storage.

Advances have also been made in improving insect host production and nutrition for *in vivo* EPN production (Shapiro-Ilan *et al.*, 2016, 2023). The host's reproductive capacity can be enhanced by optimizing host density and life-cycle aspects of the rearing process (Morales-Ramos *et al.*, 2012). Optimization of insect diets can also lead to improved efficiency in insect host production (Morales-Ramos *et al.*, 2011). Different diets affected the costs and effectiveness of development and production of *G. mellonella* larvae and the EPN produced *in vivo* (Kotchofa and Baimey, 2019). Optimization of insect diets (e.g. lipids, salts) can also be directed to improving the quality and fitness of the resulting EPNs (as the improved nutrition is passed on from insect to nematode) (Shapiro-Ilan *et al.*, 2008, 2016).

Production and application of EPNs in infected hosts may be more efficient than other *in vivo* production methods because several labor-intensive steps are removed from the process. EPNs that emerge from the cadaver directly at the target site have been found to exhibit superior dispersal, infectivity and environmental tolerance relative to IJs applied in aqueous suspension (Shapiro and Glazer, 1996; Shapiro and Lewis, 1999; Gulzar *et al.*, 2020). Pest-control efficacy when using the cadaver application approach was reported to be superior to standard application in aqueous suspension, when the insect larvae are not actively moving in the soil (Shapiro-Ilan *et al.*, 2003).

Another novel approach to *in vivo* production is the development of a "grower-based" system where farmers carry out some or all of the production steps and thereby produce their own EPNs. In one example, adding IJs to plastic containers filled three-quarters full with sawdust and *G. mellonella* larvae has provided a "low technology" and "low labor" method for mass rearing of EPNs, in which emerged IJs remain within a non-aqueous, high-humidity environment with improved oxygen access without extra steps or equipment (Testa and Shields, 2017). However, in this sawdust approach, fresh inoculum must be supplied to the end-user (growers) in each production cycle and thus the process is not continuous. In a more recent approach, a continuous sustainable grower-based production method was developed by Oliveira-Hofman *et al.* (2023) that utilizes infected hosts and storage of IJs in gel to facilitate inoculation in subsequent rounds of production (Fig. 10.2).

10.3 *In vitro* Production

In vitro EPN production was first attempted on a solid medium axenically but was not successful because the nematodes' symbiotic bacteria were not known at that time (Glaser, 1931). Effective *in vitro* mass production requires the nematode's bacterial symbionts to be part of the process (Poinar and Thomas, 1966; Akhurst, 1980) and monoxenic culture conditions must be maintained (Bedding, 1981; Wouts, 1981; Friedman, 1990; Han *et al.*, 1990).

10.3.1 Recovery of IJs

The only stage of EPNs that can be used commercially is the IJ, a developmentally arrested juvenile, formed as a response to depleted food sources and adverse environmental conditions. IJs carry their symbiotic bacteria while searching for an insect host. They are activated to recover (i.e. to start to develop from the dauer stage to feeding stages) when they sense specific compounds in the insect haemocoel (Strauch and Ehlers, 1998). The symbiotic bacteria also produce signal components triggering IJ recovery and development (Strauch and Ehlers, 1998; Han and Ehlers, 2000; Aumann and Ehlers, 2001; Ciche and Ensign, 2003; Moshayov *et al.*, 2013). *Steinernema carpocapsae* nematodes can be induced to recover by some of the symbiotic *Photorhabdus* spp. bacteria; however, no *Heterorhabditis* spp. are induced to recover from the IJ stage by exposure to *Xenorhabdus* spp. bacteria (Han and Ehlers, 1998). No common signals from the bacteria induce IJ recovery of all *Steinernema* and *Heterorhabditis* spp., indicating the presence of different signals for IJ recovery among the bacterial species (Strauch and Ehlers, 1998; Han *et al.*, 2000). Cyclic guanosine monophosphate (cGMP) signaling can regulate IJ recovery in *Heterorhabditis* spp., and a membrane-permeable analog of cGMP, 8-bromo-cGMP, and the muscarinic agonists oxotremorine and arecoline stimulate exsheathment and regurgitation of *Heterorhabditis* spp. IJs (Hallem *et al.*, 2007). Stilbene antibiotics are secondary metabolites produced by *Photorhabdus* spp., which act as a signal to stimulate the recovery of the IJ to the adult hermaphrodite (Lango-Scholey *et al.*, 2013). A compound called isopropylstilbene from *Photorhabdus* spp. is also regarded as a recovery signal for *Heterorhabditis* spp. IJs (Shi and Bode, 2018; Hapeshi *et al.*, 2019). Synthesized ascarosides, the universal nematode pheromones, are found to improve IJ recovery in the supernatants of symbiotic bacterial cultures (Wang *et al.*, 2022).

The symbiotic bacteria not only trigger IJ recovery but also provide the essential food for EPN production. *Steinernema* and *Heterorhabditis* spp.

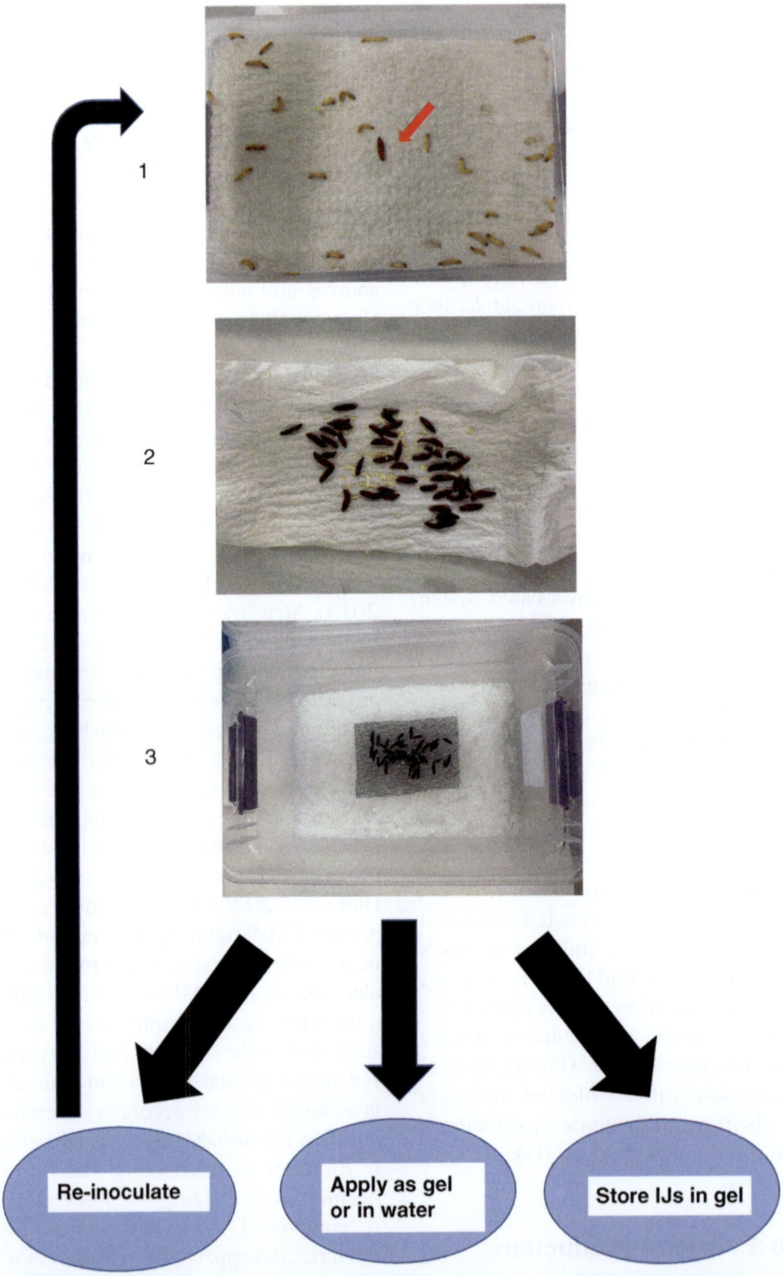

Fig. 10.2. Diagram of the grower-based method. Inoculation (1) is followed by mass incubation (2), and transfer to a polyacrylamide gel with a screen (3); the third step is then followed by re-inoculation, application for biocontrol or storage of the infective juveniles (IJs) in gel. From Oliveira-Hofman *et al.* (2023).

feed on their symbiotic bacteria (*Xenorhabdus* and *Photorhabdus* spp., respectively) and grow poorly when fed on other symbiotic or non-symbiotic bacteria genera, even if closely related (Akhurst, 1980; You *et al.*, 2006). It is therefore crucial to establish and maintain a monoxenic culture when growing EPNs in artificial media (Lunau *et al.*, 1993).

10.3.2 Monoxenic culture

A prerequisite to *in vitro* production (using solid or liquid media) is the establishment of a monoxenic bacterial culture. Key steps toward creating monoxenic cultures include isolation and purification of the symbiotic bacteria and establishment of microbe-free nematodes. The symbiotic bacteria can easily be isolated from EPN-infected insect larvae or surface-sterilized IJs. The bacteria can occur in two phase variants, phase I or primary form, and phase II or secondary form, which differ in several characteristics including dye absorption on indicator media such as nutrient bromothymol blue agar and MacConkey agar, response to biochemical tests, inclusion proteins, bioluminescence (for *Photorhabdus* spp.) and antibiotic production (Akhurst, 1980; Akhurst and Boemare, 1990). Although EPNs have been reported to grow on secondary variant symbionts, the primary variant is most conducive to growth and IJs tend only to retain the primary variant (Akhurst and Boemare, 1990; Han and Ehlers, 2001). The secondary variant is considered to possibly grow in the roots of plants outside the insects, and to act as an inhibitor for plant pathogens (Eckstein *et al.*, 2019; Regaiolo *et al.*, 2020). Stock cultures of bacteria are mixed with glycerol at 15% (v/v), and aliquots are frozen at –80°C.

Axenic EPNs need to be established to facilitate the association with the pure symbiotic bacteria and initiation of EPN production. In earlier work to create monoxenic cultures, surface-sterilized IJs or gravid females or hermaphrodites (for *Heterorhabditis* spp.) were added to a lawn of the bacterial symbionts (Akhurst, 1980; Bedding, 1981; Wouts, 1981; Han *et al.*, 1990). However, Lunau *et al.* (1993) suggested that surface sterilization of IJs is insufficient to establish monoxenicity because contaminating bacteria survive beneath the IJ's cuticle. Thus, an improved method was developed where axenic nematode eggs were obtained by rupturing gravid nematode females in an alkaline solution (Lunau *et al.*, 1993; Han and Ehlers, 1998). Specifically, in this method, fertile nematode eggs from gravid females (*Steinernema* spp.) or hermaphrodites (*Heterorhabditis* spp.) are collected in Ringer's solution and surface sterilized in a solution of 10% 4 M NaOH, 2% NaClO (12% solution) in distilled water. After rinsing twice in sterile Ringer's solution, the eggs are transferred to sterile liquid medium in which the eggs hatch. After 2 days, as long as no microbe contaminants are present in the medium, the hatched juveniles are combined with their symbiotic bacteria growing on fortified lipid agar (1.6% nutrient broth, 1.2% agar and 1% corn oil) (Wouts, 1981). Once the monoxenic cultures are established, the EPN culture can be scaled up in solid phase or liquid phase, as well as by a biphasic process.

10.3.3 Solid culture

The *in vitro* solid culture technique involves the following four key steps: (i) preparation of solid medium; (ii) inoculation with symbiotic bacteria; (iii) inoculation with monoxenic (or axenic) EPNs; and (iv) harvest. Solid culture was first accomplished in two-dimensional arenas such as Petri dishes, using various agar media (House *et al.*, 1965; Hara *et al.*, 1981; Wouts, 1981). Subsequently, *in vitro* solid culture was advanced considerably by the invention of a three-dimensional rearing system involving EPN culture on crumbled polyether polyurethane foam (Bedding, 1981). This major breakthrough opened the door for economic feasibility in commercial mass production. In the three-dimensional solid media method, a liquid medium is mixed with foam, autoclaved and then inoculated with symbiotic bacteria followed by the nematode inoculum. Nematodes are then harvested within 2–5 weeks (Bedding, 1981, 1984; Yan *et al.*, 2018) by placing the foam on to sieves immersed in water. The approach was later expanded to autoclavable bags with filtered air being pumped in (Bedding, 1984) and culture vessels comprising a tray with side walls and overlapping lids that allow gas exchange through a layer of foam. Large-scale production was advanced further through various mechanisms including using bags with gas-permeable Tyvac® strips for ventilation, automated mixing and autoclaving, simultaneous inoculation of nematodes and bacteria, sterile room technology and automated harvest through centrifugation (Gaugler and Han, 2002). The solid culture method is used to produce *Steinernema pakistanense* (94-1 strain), *H. indica* (212 strain) and *Steinernema ceratophorum* (64-2 strain) in the laboratory, with highest yields of 3.52×10^5, 7.08×10^5 and 2.09×10^6 IJs g^{-1} of medium, respectively, whereas *S. carpocapsae*, *S. feltiae*,

S. longicaudum, *S. scapterisci*, *Heterorhabditis bacteriophora* and *H. indica* are produced for commercial applications (Yan *et al.*, 2018, 2021).

Factors impacting EPN yield in *in vitro* solid production include media composition, nematode inoculum size, temperature and culture time. Media composition is a major factor that affects yield in solid culture. In two-dimensional culture system, various agar media such as those based on dog food, pork kidney, cattle blood and other animal products are used (House *et al.*, 1965; Hara *et al.*, 1981; Wouts, 1981). Wouts (1981) developed an improved medium (less expensive and more consistent from batch to batch) that included yeast extract, nutrient broth, vegetable oil and soy flour for nematode production in agar plates. In three-dimensional culture systems, media were initially based on animal products such as pork kidney or chicken offal (House *et al.*, 1965; Bedding, 1981; Hara *et al.*, 1981), but were later improved and contained various ingredients including peptone, yeast extract, eggs, soy flour and lard (Bedding, 1984; Han *et al.*, 1992, 1993). Increasing the lipid quantity and quality led to increases in EPN yield (Dunphy and Webster, 1989; Han *et al.*, 1992). The most suitable array of lipid components reflects the composition of the EPNs' natural host (Abu Hatab *et al.*, 1998; Abu Hatab and Gaugler, 2001). EPN yields were significantly affected by the addition of water, yeast extract and egg yolk for *S. carpocapsae* and by the addition of water and lard for *H. bacteriophora*, indicating the importance of selecting suitable medium components and combinations for each nematode species (Han *et al.*, 1995). Soy flour is one of the medium components both in solid and liquid cultures (Ehlers and Shapiro-Ilan, 2005). Media containing soy flour yielded the highest population densities compared with media without soy flour in *in vitro* solid production of eight Pakistani EPN strains (Salma and Shahina, 2012). Other medium ingredients that may affect EPN yield include salts and proteins (Dunphy and Webster, 1989). Addition of insect powder (*G. mellonella*, *T. molitor* or *Lucilia sericata*) to solid media led to high production yields of *S. feltiae* (SN strain) and enhanced the fitness of the harvested IJs (Zhen *et al.*, 2018).

The effect of EPN inoculum size (IJs per unit of medium) on yield varies among nematode species/strains (Han *et al.*, 1992, 1993; Wang and Bedding, 1998). For example, *S. carpocapsae* (Agriotos strain) produced optimum yields at an intermediate inoculum size (2×10^3 IJs g^{-1} of medium) (Han *et al.*, 1993), whereas *Steinernema* sp. (CB2B strain) and *H. bacteriophora* (H06 strain) were not significantly affected by inoculum size (Han *et al.*, 1992). Bacterial inoculum size does not appear to be important in yield determination in solid cultures (Han *et al.*, 1992, 1993).

Temperature adaptation of EPNs for maximum production is also related to the specific species or strains. For example, optimal production for *Steinernema* sp. (CB2B) and *H. bacteriophora* (H06) was 27°C and 25°C, respectively (Han *et al.*, 1992). Maximum yields of Pakistani strains of *S. pakistanense*, *S. asiaticum*, *S. abbasi*, *S. siamkayai*, *S. carpocapsae*, *H. indica* and *H. bacteriophora* were obtained at 32 ± 2°C, whereas *S. feltiae* produced the highest yields at 20 ± 2°C (Salma and Shahina, 2012). *S. pakistanense* (94-1) and *H. indica* (212-2) isolated from tropical zones with warm weather all year in China had higher yields when cultured at 30°C than at 25°C (Yan *et al.*, 2018).

Usually, the culture time of EPNs for optimum yield is inversely proportional to IJ inoculum size (Han *et al.*, 1992). It is worth noting that longer culture times can provide higher yields, but EPN mortality may also increase with time (Han *et al.*, 1992, 1993). Thus, culture time should be weighed against the cost of space and diminishing returns.

10.3.4 Liquid culture

The first attempt at EPN *in vitro* liquid culture was carried out by Stoll (1952) using raw liver extract in shaken flasks, and was further developed using different media ingredients in fermenters (Pace *et al.*, 1986; Buecher and Popiel, 1989; Friedman *et al.*, 1989; Friedman, 1990; Ehlers, 2001a; Shapiro-Ilan *et al.*, 2012), based on an improved understanding of nematode biology and the ideal development of bioreactors. Nowadays, modern bioreactors with controllers for various physical parameters (e.g. temperature, pH, dissolved oxygen, agitation speed, antifoam agent) are used for mass production of EPN in liquid media (Peters *et al.*, 2017).

The equipment and type of bioreactors used impact IJ yield. For example, although flat-blade impellers, bubble columns, and air-lift and internal loop bioreactors have all been tested successfully (Pace *et al.*, 1986; Surrey and Davies, 1996; Ehlers *et al.*, 1998), a direct comparison with flat-blade impeller-stirred tanks or air-lift bioreactors, internal loop bioreactors always yielded higher IJ concentrations (Ehlers and Shapiro-Ilan, 2005).

The selection of appropriate liquid media for EPN production largely follows economic aspects (Ehlers and Shapiro-Ilan, 2005). A variety of ingredients have been employed to formulate liquid culture media including soy flour, yeast extract, canola oil, corn oil, thistle oil, egg yolk, casein peptone, milk powder, liver extract and cholesterol (Surrey and Davies, 1996; Ehlers *et al.*, 2000; Yoo *et al.*, 2000; Cho *et al.*, 2011). Medium components and their quality greatly influence the culture process and times for harvest, which may be as long as 2–3 weeks (Friedman, 1990; Surrey and Davies, 1996; Ehlers *et al.*, 2000; Strauch and Ehlers, 2000; Yoo *et al.*, 2000; Chavarría-Hernández and de la Torre, 2001; Neves *et al.*, 2001). The osmotic strength of the medium must not surpass 600 mmol kg^{-1}. Improvements to the medium according to the requirements of different species are feasible approaches to increase yields (Ehlers, 2001b). Essential amino acid requirements have only been defined for *S. glaseri* (Jackson, 1973). Nematodes have nutritional demands for sterols, but they can metabolize the necessary sterols from a variety of steroid sources (Ritter, 1988), provided through the addition of lipids of animal or plant origin. According to Leite *et al.* (2016c), peanut oil provided numerically the highest growth of *S. feltiae* among six other plant-based oils, while pork fat was detrimental to EPN growth due to its solidification in the liquid medium at room temperature and consequently low solubility. For nitrogen source experiments, yeast extract provided the lowest yields, while egg yolk plus egg white increased EPN yield, especially when these two nitrogen sources were combined. The addition of 2.5% glucose also increased EPN yields (Leite *et al.*, 2016c). In general, *S. carpocapsae* requires proteins of animal origin (Yang *et al.*, 1997) and is unable to reproduce without the addition of lipid sources to the medium, whereas *H. bacteriophora* produces offspring in a liquid medium without the addition of lipids (Han and Ehlers, 2001). *Photorhabdus luminescens* provides or metabolizes essential lipids; however, lipids should always be added to increase the total IJ fat content. The lipid composition of the medium influences the fatty acid composition of the bacteria and IJs (Abu Hatab *et al.*, 1998); a low fatty content of IJs can reduce efficacy (Patel *et al.*, 1997a,b). Recently, it was discovered that IJ yields significantly increase (35% higher than the control at 15 days) in liquid medium with DMSO and ascr#11 (a universal nematode pheromone) at proper concentrations for *S. carpocapsae* (All strain) and for *H. bacteriophora* (H06), and the improved IJ yields in DMSO-supplemented liquid medium are correlated to the higher recovery rate, hermaphrodite numbers and eggs in the hermaphrodites in the critical development process of *H. bacteriophora* (H06) (Kong *et al.*, 2022).

Another important issue for EPN production in liquid culture is the viscosity of the medium. Medium viscosity increases naturally as the EPN production process progresses (Chavarría-Hernández *et al.*, 2010). Leite *et al.* (2016b) found higher recovery rates of *S. feltiae* IJs by increasing the viscosity of the medium with the addition of 0.2% agar. The addition of 0.2% agar to the liquid medium and increasing the aeration rate by using larger flasks with higher agitation speed (280 rpm) increased bacteria growth, EPN recovery and final yield. Media containing agar (0.4% and 0.6%) increased EPN yields only when cultures were grown at low agitation speed (180 rpm). These three conditions (agar, larger flask and higher agitation speed) favored the growth of bacteria and nematodes by improving oil distribution and dissolved oxygen in the medium.

Liquid culture starts with the inoculation of a pure culture of the symbiotic bacteria in their exponential growth phase into a suitable sterilized liquid medium at a concentration of approximately 0.1–1% (v/v) of the culture volume. Good-quality symbiotic bacteria should be used to trigger nematode development, to provide a suitable food source and to synchronize nematode development to obtain a pure IJ suspension at the end of the process (Peters *et al.*, 2017). As described above, the symbiotic *Xenorhabdus* and *Photorhabdus* spp. may switch morphologically and metabolically between phase variants (Akhurst, 1980; Boemare *et al.*, 1997), which can produce inferior yields in liquid culture processes of *Steinernema* spp. (Hirao and Ehlers, 2009) and *Heterorhabditis* spp. (Han and Ehlers, 2001). Secondary-phase bacteria result in a lower proportion of IJs developing than the primary form of the bacteria (Hirao and Ehlers, 2009). Phase variation may be induced by osmotic stress, heat treatment or by culturing bacteria at low oxygen levels (Krasomil-Osterfeld, 1994, 1995). To avoid a shift to the secondary phase, bacterial growth should use the same medium and avoid oxygen stress in all scale-up steps. The bacteria should then be cultured to the late stationary phase before EPNs are added (Johnigk *et al.*, 2004). High bacterial concentrations increased recovery and final yield in *S. carpocapsae* and *S. feltiae* (Hirao and Ehlers, 2009).

After the inoculated specific symbiont bacteria have been grown in liquid culture for 24–36 h, the IJ inoculum is introduced. The IJ inoculum should be maintained monoxenically and properly stored on shakers at 20 rpm and low temperature (generally 4°C, but for some EPN species it should be a higher temperature) for several months until they are introduced into the bioreactor. Strain collections of EPNs can be kept in liquid nitrogen (Popiel and Vasquez, 1991; Bai *et al.*, 2004). The IJ inoculum sizes vary depending on nematode species and medium composition (Han *et al.*, 1993; Han, 1996). An optimal number of adults per milliliter can be calculated, which is defined by the percentage of IJs likely to recover. The recovery of *S. feltiae* and *S. carpocapsae* decreased at lower bacterial densities (Hirao and Ehlers, 2009). If the concentration of adult EPNs feeding on the bacteria is too low, a larger proportion of the F_1 generation will develop into adults and start a second generation (Peters *et al.*, 2017). If the proportion is too high, there will not be enough bacteria left to support the formation of progeny. Generally, the EPN inoculum is between 5% and 10% of the culture volume.

Recovery rate is a limited factor in EPN *in vitro* liquid production, which determines the final yield of the IJs. *Steinernema* spp. (except *S. hermaphroditum*) occur only as males and females and are capable of mating in liquid culture (Strauch *et al.*, 1994). Maximization of mating is paramount and can be achieved through bioreactor design and regulation of aeration (Neves *et al.*, 2001). However, maximization of mating is not applicable for heterorhabditid production in liquid culture because the first generation is exclusively hermaphrodites and, although subsequent generations contain amphimictic forms, they cannot mate in liquid culture due to the effect of agitation on their mating behavior (Strauch *et al.*, 1994). Levels of *Heterorhabditis* spp. IJ recovery in liquid culture can range from 0% to 85% (Ehlers *et al.*, 2000; Jessen *et al.*, 2000; Yoo *et al.*, 2000). Recovery may be affected by various factors including nutrition, aeration, carbon dioxide, lipid content and temperature (Ehlers *et al.*, 2000; Jessen *et al.*, 2000; Strauch and Ehlers, 2000; Yoo *et al.*, 2000). Recovery can be influenced during the bacterial preculture phase, prior to EPN introduction. A higher bacterial density leads to a higher food signal concentration, which in turn produces a higher recovery rate (Hirao and Ehlers, 2009). Nematodes should therefore be inoculated during the late stationary growth phase of the bacteria (Johnigk *et al.*, 2004). The period starts with an increase in pH after its distinct minimum at pH <8.0 (Johnigk *et al.*, 2004). Glucose-fed batch reactors can increase bacterial density (Jeffke *et al.*, 2000) and enhance food signal production, and thus may be used to increase IJ recovery (Ehlers and Shapiro-Ilan, 2005). Jessen *et al.* (2000) reported that increasing the carbon dioxide concentration in the medium enhanced IJ recovery. However, the influence of decreasing pH caused by the carbon dioxide concentration was excluded. A pH below 6.5 significantly reduced IJ recovery. A protein source such as soybean can improve symbiotic bacterial growth and nematode recovery, which thereby promotes EPN production (Cho *et al.*, 2011). Recovery in secondary-phase bacteria is lower than in primary-phase bacteria (Han and Ehlers, 2001; Hirao and Ehlers, 2009). No significant influence of inoculum density on IJ recovery of *S. carpocapsae* and *S. feltiae* has been detected, but a significant correlation between recovery and the age of IJ inoculum was detected for *S. carpocapsae* (Hirao and Ehlers, 2010). The addition of 0.2% agar to liquid medium and increasing the aeration rate by using larger flasks with a higher agitation speed may increase EPN recovery and final yield (Leite *et al.*, 2016b).

Temperature has an influence on EPN development. For example, Ehlers *et al.* (2000) noted that recovery is influenced by culture temperature. Optimal culture temperature varies from species to species; for *S. carpocapsae* and *S. feltiae*, for example, the optimal culture temperature is 25°C.

The pH level appears to be regulated by the organisms themselves in the process. The culture medium should have a pH of between 5.5 and 7.0 when the culture is started. The pH is not controlled but is monitored to identify switches in the metabolic activity of the symbiotic bacteria (Peters *et al.*, 2017).

The oxygen supply must be maintained at approximately 30% saturation, which may help prevent the bacteria from shifting to the secondary phase. The aeration rate (l min^{-1} of sterile air supplied to the liquid) and the stirrer speed are used to control dissolved oxygen. Increasing the aeration and stirring/agitation rates often increases foaming, which can be reduced by changes in the bioreactor design (Strauch and Ehlers, 2000) and use of antifoam or defoaming agents (Gaugler and Han, 2002). The addition of agents such as silicon oil usually prevents foaming. However, it should be used carefully, because higher concentrations can be detrimental to the nematodes. Compared with the pure bacteria culture, the maximum stirrer speed is lowered after EPNs are added to avoid shear stress, which may kill the fragile hermaphrodites (Peters *et al.*, 2017). The oxygen uptake rate must therefore be balanced properly with stirrer speed and dissolved oxygen. In contrast, for production of inoculum inside Erlenmeyer flasks, the nematode is not affected by increasing the speed of rotation, and in fact favors its increase up to 280 rpm due to the increase of aeration in the liquid medium (Leite *et al.*, 2016a,b,c).

Nematodes can be harvested in 2–3 weeks depending on the medium and EPN species (Friedman, 1990; Surrey and Davies, 1996; Ehlers *et al.*, 2000; Strauch and Ehlers, 2000; Yoo *et al.*, 2000; Chavarría-Hernández and de la Torre, 2001; Neves *et al.*, 2001). The yield from liquid culture may be affected by nematode species, culture temperature and culture time (Han, 1996; Ehlers *et al.*, 2000). Similar to the other production approaches, EPN yield in liquid culture is generally inversely proportional to the size of the species (Ehlers *et al.*, 2000). Some of the maximum average yields reported include 3.0×10^5 and 3.2×10^5 IJs ml^{-1} for *H. bacteriophora* and *S. carpocapsae*, respectively (Han, 1996), 1.38×10^5 IJs ml^{-1} for *H. megidis* (Strauch and Ehlers, 2000), 7.147×10^5 IJs ml^{-1} for *S. feltiae* (Chavarría-Hernández and de la Torre, 2001) and 4.5×10^5 IJs ml^{-1} for *H. indica* (Ehlers *et al.*, 2000).

10.3.5 Biphasic process: liquid to solid

As outlined above, decades of research and development involving EPN mass production has resulted in two *in vitro* approaches: the sponge (polyurethane) process, soaked in a culture of the bacteria, and the fermentation process, submerged in the bacterial liquid culture (Ehlers, 2001a). Another approach to producing EPNs is the biphasic process, starting with a liquid culture and ending with a solid culture (Leite *et al.*, 2016a, 2023). For this process, a liquid culture suitable for mass production of EPNs using bioreactors is combined with the sponge process inside plastic bags.

The biphasic process was tested for the production of *Steinernema rarum*, and has some advantages because it requires smaller fermenters for the first stage of inoculum production in liquid culture, which also reduces the chances of contamination in the second stage, after the inoculation of the nematode in several bags containing the soaked sponge.

In this approach, the synthetic polyurethane sponge commonly used for solid production of EPNs was replaced by phenolic foam, which consists of a sterile substrate made of phenolic resin; the sponge is used as a substrate for rooting seedlings because it offers good water retention and aeration, which is essential for healthy root development. Phenolic foam is widely used to grow seedlings in agriculture and can be purchased in different sizes and divided in cells, with each cell being used to hold a seedling. The many advantages of using phenolic foam to grow EPNs are: (i) it is practical, as it is easy to handle and transport; (ii) it is hygienic, i.e. does not leave any residue; (iii) it is economical, with a cost equivalent to the synthetic (polyurethane) sponge commonly used in EPN solid culture; (iv) it is sterile; (v) it is versatile; (vi) it is inert; (vii) it eliminates the need to extract/harvest and formulate the nematode as it serves as a support for nematode growth as well as the final formulation; (viii) it enables storage of the nematode with high stability on the shelf without contaminants; (ix) it facilitates extraction of nematodes by crushing the sponge after the end of the production process; and (x) it allows direct application of the nematodes together with the foam debris/particles, just after its extraction, without clogging the nozzles of the application system, or alternatively can be formulated as wettable powder using the sponge debris/particle as a solid inert substrate (Leite *et al.*, 2023).

The production of *S. rarum* by the biphasic process using phenolic foam is carried out according to the following operational steps (Fig. 10.3) (Leite *et al.*, 2023):

1. The EPNs are initially grown together with their symbiotic bacteria in liquid culture, inside flasks kept on a shaker adjusted to 200 rpm. After its initial growth, the nematode culture is scaled up in bioreactors to increase production of the nematode inoculum.

Fig. 10.3 Stages of *Steinernema rarum* production by the biphasic process. (A) Liquid culture in flasks kept on a shaker. (B) Liquid culture of the nematode symbiotic bacteria inside a 6 l bioreactor, 2 days after bacterial inoculation. (C) Liquid culture of the nematode in the bacterial culture, 2 weeks after inoculation. (D) Final production of the nematode in phenolic sponge soaked in its symbiotic bacterial culture inside bags.

2. After 2–3 weeks in the bioreactor, the EPNs are inoculated into plastic bags with filters containing flaked phenolic foam soaked with 500 ml of the liquid culture medium, previously inoculated with the bacteria *Xenorhabdus szentirmaii*.
3. The plastic bags with the inoculated substrate are then closed and incubated at 20°C for about 4–7 weeks until the EPNs reach yields of around 200,000 IJs ml^{-1} of culture. After this time, the bags are ready to be supplied to farmers or stored at a cooler temperature (10–15°C) for a period of 2 months or longer. Under higher temperatures of 25°C and 35°C, the IJs viability will quickly drop.
4. Once supplied to the farmer, the product should be removed from inside the bags and crushed with the aid of a blender, mixer (e.g. a cake mixer), other tools or by hand. The formulation can then be added to the application tank and applied to target areas for pest suppression. A great advantage of using phenolic sponge for EPN production is that, after being crushed, the debris and nematode combination can be applied via most conventional spray or drench equipment.

In Brazil, *S. rarum* has been assessed against sugarcane pests, mainly against the sugarcane billbug, *Sphenophorus levis* (Coleoptera: Curculionidae), which infests more than 1 million ha of cane (Casteliani *et al.*, 2020). *S. rarum*, produced using the biphasic method and tested on sugarcane fields at rates of 1–3 × 10^8 IJs ha^{-1} resulted in up to 74% control of *S. levis* (Silva *et al.*, 2021). Similarly, following biphasic production, *S. rarum* has been used in Brazil to control *Sphenophorus levis* on sugarcane fields just after crop harvesting at rates of 1 × 10^8 IJs ha^{-1}. The EPNs are applied with implements that cut the straw and inject the EPN directly into the soil, or together with the application of vinasse, which is the liquid fraction generated from alcohol production and discarded on to sugarcane fields for fertigation.

10.4 Conclusions

With increasing application of EPNs for the biocontrol of many economically important pests, the market demand for these nematodes has increased markedly. Nematode production methods have been improved over the years. All four production approaches continue to be utilized for EPN production to meet the demands of the global market. The advantages and disadvantages of each EPN production approach are summarized in Table 10.1.

A major disadvantage of *in vivo* production is the labor costs. Based on the versatility for new/different EPN species, the low levels of technology and expertise required, and the lower investment, *in vivo* production may be most appropriate for small startup companies, niche markets, or in less developed countries or regions where the market is not too big. Due to the high labor costs, space requirements and cost of insects, economy of scale (cost efficiency) tends to be lowest for *in vivo* production. Despite the reduced economy of scale relative to other production methods, *in vivo* production has managed to sustain itself as a cottage industry. In some markets, *in vivo*-produced EPNs have remained competitive with *in vitro* products. Steps toward mechanization have enhanced the efficiency of *in vivo* production (Shapiro-Ilan *et al.*, 2016, 2023). Possibly, *in vivo* production will continue to expand based on additional advancements in mechanization and automation. Additionally, grower-based

Table 10.1. Comparison of production approaches for entomopathogenic nematodes.

	Production approach			
Issue	*In vivo*	*In vitro*: solid	*In vitro*: liquid	*In vitro*: biphasic
Capital outlay	Low	Intermediate	High	Intermediate
Required expertise	Low	Intermediate	Extensive	Extensive
Ease of achieving quality	Easy	Difficult	Difficult	Difficult
Labor required	High	Intermediate	Low	Intermediate
Economy of scale	Low	Intermediate	High	Intermediate
Ease of adaptation to new nematode species	Easy	Difficult	Difficult	Difficult

approaches may increase in popularity as sustainable production systems become more efficient and available.

In contrast, *in vitro* techniques require less labor and are ideal for developing economies of scale. Labor costs are lower for *in vitro* solid production and lowest for *in vitro* liquid production. Thus, on a global scale, the vast majority of EPN production is currently accomplished using *in vitro* liquid production. However, *in vitro* techniques may require substantial modification when adapting to new EPN species based on medium requirements, optimization of culture parameters and downstream processing.

In vitro solid production has several advantages. It does not require high technology inputs or large investments, and the effect of phase variation on yield is less than in liquid culture (Han and Ehlers, 2001). When it comes to large-scale production, the disadvantages can be overwhelming. *In vitro* solid production is labor intensive, vulnerable to contamination during up- and downstream processing, and difficult to monitor online. The uneven distribution of the EPNs in the medium prevents systematic sampling (Ehlers and Shapiro-Ilan, 2005), and thus improvement of the technique is required. However, due to its economy of scale, in developing countries, the *in vitro* solid culturing system is still superior to liquid culture technology (Ehlers *et al.*, 2000; Yan and Han, 2016). This technique has been transferred from China to Indonesia, Pakistan, the Democratic People's Republic of Korea, Rwanda, the Republic of Korea and some other countries (Yan and Han, 2016). Similar to *in vivo* production, efficiency can be greatly increased through mechanization or automation (labor reduction) and medium enhancement (Gaugler and Han, 2002; Shapiro-Ilan and Gaugler, 2002; Yan and Han, 2016).

The main advantage of liquid culture is the high economies of scale. The production units may be increased from a few milliliters to several cubic meters with hardly any increase in labor cost. EPN cultures in liquid are particularly vulnerable to contamination due to the even distribution of the fluids and organisms obtained through the mixing of liquid in the bioreactors and based on the long process time (Ehlers and Shapiro-Ilan, 2005).

The *in vitro* biphasic process requires bioreactors for production of the EPN inoculum, however, the investment is lower compared with that required for liquid culture production, which requires larger fermenters. By using phenolic sponge, the biphasic process may provide some advantages in comparison with both solid and liquid cultures. Specifically, the harvesting and formulation steps after growth in the phenolic sponge are eliminated because the sponge serves as a substrate for storage and application.

An issue that has been in contention is the relative quality of EPNs produced using the various methods. Several studies indicate that the quality of nematodes produced using solid culture is similar to that produced by an *in vivo* method (Gaugler and Georgis, 1991; Abu Hatab *et al.*, 1998; Abu Hatab and Gaugler, 1999). Yang *et al.* (1997) reported reduced fitness in *S. carpocapsae* produced in solid culture compared with *in vivo* culture. Furthermore, a number of reports also indicated reduced quality and field efficacy in *in vitro* liquid-produced EPNs relative to those produced *in vivo* (Gaugler and Georgis, 1991; Cottrell *et al.*, 2011), whereas others did not detect any differences (Gaugler and Georgis, 1991; Shapiro and McCoy, 2000). In contrast, to our knowledge there have no reports in which *in vitro*-produced EPNs caused superior field efficacy compared with *in vivo*-produced EPNs. In fact, the quality of *in vitro* production can vary from batch to batch. For example, in one field trial, Cottrell *et al.* (2011) observed reduced field efficacy in liquid-produced *S. carpocapsae* relative to *in vivo*-produced EPNs, yet there was no difference in another trial (the EPNs were from the same commercial source, but the trials were conducted using different batches). The quality of *in vivo*-produced EPNs may also vary based on production sources (Gaugler *et al.*, 2000). Therefore, it is conceivable that, at least in some (if not many) cases, differences observed in EPN quality stem from batch variation rather than innate differences in production technique. Indeed, despite a few reports otherwise, there are vast numbers of examples where liquid-produced EPNs have produced high levels of efficacy in a consistent manner. Additionally, with improvements in bioreactor design, media and other parameters, the quality of *in vitro* liquid-produced EPNs continues to advance (Chavarría-Hernández *et al.*, 2010; Hirao and Ehlers, 2010).

Despite years of commercial success and improved methodology for producing and utilizing

EPNs for biocontrol, additional research for further advancement is needed. *In vivo* production may become more competitive if large industrial-scale mechanization takes place in three-dimensional systems. The increased interest in mass production of insects for food and feed may also lead to reduced costs of insect media and automation, which could benefit *in vivo* approaches. Additional optimization of media and fermentation parameters for species that are already being produced and for additional EPN species that are not yet commercialized will expand markets and reduce costs. Grower-oriented technologies such as *in vivo* and biphasic systems are also likely to expand.

10.5 References

Abu Hatab, M. and Gaugler, R. (1999) Lipids of *in vivo* and *in vitro* cultured *Heterorhabditis bacteriophora. Biological Control* 15, 113–118.

Abu Hatab, M. and Gaugler, R. (2001) Diet composition and lipids of *in vitro*-produced *Heterorhabditis bacteriophora*. *Biological Control* 20, 1–7.

Abu Hatab, M., Gaugler, R. and Ehlers, R.-U. (1998) Influence of culture method on *Steinernema glaseri* lipids. *Journal of Parasitology* 84, 215–221.

Akhurst, R.J. (1980) Morphological and functional dimorphism in *Xenorhabdus* spp, bacteria symbiotically associated with insect pathogenic nematodes *Neoaplectana* and *Heterorhabditis*. *Journal of General Microbiology* 121, 303–309.

Akhurst, R.J. and Boemare, N.E. (1990) Biology and taxonomy of *Xenorhabdus*. In: Gaugler, R. and Kaya, H.K. (eds) (1990) *Entomopathogenic Nematodes in Biological Control.* CRC Press, Boca Raton, Florida, pp. 75–90.

Aumann, J. and Ehlers, R.-U. (2001) Physico-chemical properties and mode of action of a signal from the symbiotic bacterium *Photorhabdus luminescens* inducing dauer juvenile recovery in the entomopathogenic nematode *Heterorhabditis bacteriophora*. *Nematology* 3, 849–853.

Bai, C., Shapiro-Ilan, D.I., Gaugler, R. and Yi, S. (2004) Effect of entomopathogenic nematode concentration on survival during cryopreservation in liquid nitrogen. *Journal of Nematology* 36, 281–284.

Bedding, R.A. (1981) Low cost *in vitro* mass production of *Neoaplectana* and *Heterorhabditis* species (Nematoda) for field control of insect pests. *Nematologica* 27, 109–114.

Bedding, R.A. (1984) Large scale production, storage and transport of the insect-parasitic nematode *Neoaplectana* spp. and *Heterorhabditis* spp. *Annals of Applied Biology* 104, 117–120.

Blinova, S.L. and Ivanova, E.S. (1987) Culturing the nematode–bacterial complex of *Neoaplectana carpocapsae* in insects. In: Sonin, M.D. (ed.) *Helminths of Insects*. Brill, Leiden, Netherlands, pp. 22–26.

Boemare, N., Givaudan, A., Brehelin, M. and Laumond, C. (1997) Symbiosis and pathogenicity of nematode–bacterium complexes. *Symbiosis* 22, 21–45.

Boff, M., Wiegers, G.L., Gerritsen, L.J.M. and Smits, P.H. (2000) Development of the entomopathogenic nematode *Heterorhabditis megidis* strain NLH-E 87.3 in *Galleria mellonella. Nematology* 2, 303–308.

Buecher, E.J. and Popiel, I. (1989) Liquid culture of the entomogenous nematode *Steinernema feltiae* with its bacterial symbiont. *Journal of Nematology* 21, 500–504.

Burman, M. and Pye, A.E. (1980) *Neoaplectana carpocapsae*: respiration of infective juveniles. *Nematologica* 26, 214–219.

Cabanillas, H.E. and Raulston, J.R. (1994) Pathogenicity of *Steinernema riobravis* against corn earworm, *Helicoverpa zea* (Boddie). *Fundamental and Applied Nematology* 17, 219–223.

Carr, C.W. and Kolodny-Hirsch, D.M. (1993) Method and apparatus for mass producing insects, entomopathogens and entomoparasites. US Patent US5178094. US Patent and Trademark Office, Washington, DC.

Casteliani, A., de Fatima Martins, L., Cardoso, J.F.M., Silva, M.S.O., da Silva, R.S.A. *et al.* (2020) Behavioral aspects of *Sphenophorus levis* (Coleoptera: Curculionidae), damage to sugarcane and its natural infection by *Steinernema carpocapsae* (Nematoda: Rhabditidae). *Crop Protection* 137: 105262.

Chavarría-Hernández, N. and de la Torre, M. (2001) Population growth kinetics of the nematode, *Steinernema feltiae*, in submerged monoxenic culture. *Biotechnology Letters* 23, 311–315.

Chavarría-Hernández, N., Ortega-Morales, E., Vargas-Torres, A., Chavarría-Hernández, J. and Rodríguez-Hernández, A. (2010) Submerged monoxenic culture of the entomopathogenic nematode, *Steinernema carpocapsae* CABA01, in a mechanically agitated bioreactor: evolution of the hydrodynamic and mass transfer conditions. *Biotechnology and Bioprocess Engineering* 15, 580–589.

Cho, C.H., Whang, K.S., Gaugler, R. and Yoo, S.K. (2011) Submerged monoxenic culture medium development for *Heterorhabditis bacteriophora* and its symbiotic bacterium *Photorhabdus luminescens*: protein sources. *Journal of Microbiology and Biotechnology* 21, 869–873.

Ciche, T.A. and Ensign, J.C. (2003) For the insect pathogen *Photorhabdus luminescens*, which end of a nematode is out? *Applied and Environmental Microbiology* 69, 1890–1897.

Cottrell, T.E., Shapiro-Ilan, D.I., Horton, D.L. and Mizell, R.F. III. (2011) Laboratory virulence and orchard efficacy of entomopathogenic nematodes toward the lesser peachtree borer (Lepidoptera: Sesiidae). *Environmental Entomology* 104, 47–53.

Dolinski, C., Del Valle, E.E., Burla, R.S. and Machado, I.R. (2007) Biological traits of two native Brazilian entomopathogenic nematodes (Heterorhabditidae: Rhabditida). *Nematologia Brasileira* 31, 180–185.

Dunphy, G.B. and Webster, J.M. (1989) The monoxenic culture of *Neoaplectana carpocapsae* DD-136 and *Heterorhabditis heliothidis*. *Revue de Nematologie* 2, 113–123.

Eckstein, S., Dominelli, N., Brachmann, A. and Heermann, R. (2019) Phenotypic heterogeneity of the insect pathogen *Photorhabdus luminescens*: insights into the fate of secondary cells. *Applied and Environmental Microbiology* 85, e01910-19.

Ehlers, R.-U. (2001a) Achievements in research of EPN mass production. In Griffin, C.T., Burnell, A.M., Downes, M.J. and Mulder, R. (eds) *Developments in Entomopathogenic Nematode/Bacterial Research*. Publications Office of the European Union, Luxembourg, pp. 68–77.

Ehlers, R.-U. (2001b) Mass production of entomopathogenic nematodes for plant protection. *Applied Microbiology and Biotechnology* 56, 623–633.

Ehlers, R.-U. and Shapiro-Ilan, D.I. (2005) Mass production. In: Grewal, P.S., Ehlers, R.U. and Shapiro-Ilan, D.I. (eds) *Nematodes as Biocontrol Agents*. CAB International, Wallingford, UK, pp. 65–78.

Ehlers, R.-U., Lunau, S., Krasomil-Osterfeld, K.C. and Osterfeld, K.H. (1998) Liquid culture of the entomopathogenic nematode-bacterium complex *Heterorhabditis megidis/Photorhabdus luminescens*. *BioControl* 43, 77–86.

Ehlers, R.-U., Niemann, I., Hollmer, S., Strauch, O., Jende, D. *et al.* (2000) Mass production potential of the bacto-helminthic biocontrol complex *Heterorhabditis indica–Photorhabdus luminescens*. *Biocontrol Science and Technology* 10, 607–616.

Elawad, S.A., Gowen, S.R. and Hague, N.G.M. (2001) Progeny production of *Steinernema abbasi* in lepidopterous larvae. *International Journal of Pest Management* 47, 17–21.

Flanders, K.L., Miller, J.M. and Shields, E.J. (1996) *In vivo* production of *Heterorhabditis bacteriophora* 'Oswego' (Rhabditida: Heterorhabditidae), a potential biological control agent for soil inhabiting insects in temperate regions. *Journal of Economic Entomology* 89, 373–380.

Friedman, M.J. (1990) Commercial production and development. In: Gaugler, R. and Kaya, H.K. (eds) *Entomopathogenic Nematodes in Biological Control*. CRC Press, Boca Raton, Florida, pp. 153–172.

Friedman, M.J., Langston, S.L. and Pollit, S. (1989) Mass production in liquid culture of insect-killing nematodes. International Patent Application WO89/04602.

Gaugler, R. and Georgis, R. (1991) Culture method and efficacy of entomopathogenic nematodes (Rhabditida: Steinernematidae and Heterorhabditidae). *Biological Control* 1, 269–274.

Gaugler, R. and Han, R. (2002) Production technology. In: Gaugler, R. (ed.) *Entomopathogenic Nematology*. CAB International, Wallingford, UK, pp. 289–310.

Gaugler, R., Grewal, P., Kaya, H.K. and Smith-Fiola, D. (2000) Quality assessment of commercially produced entomopathogenic nematodes. *Biological Control* 17, 100–109.

Gaugler, R., Brown, I., Shapiro-Ilan, D.I. and Atwa, A. (2002) Automated technology for *in vivo* mass production of entomopathogenic nematodes. *Biological Control* 24, 199–206.

Glaser, R.W. (1931) The cultivation of a nematode parasite of an insect. *Science* 73, 614–615.

Grewal, P.S., Selvan, S. and Gaugler, R. (1994) Thermal adaptation of entomopathogenic nematodes – niche breadth for infection, establishment and reproduction. *Journal of Thermal Biology* 19, 245–253.

Grewal, P.S., Converse, V. and Georgis, R. (1999) Influence of production and bioassay methods on infectivity of two ambush foragers (Nematoda: Steinernematidae). *Journal of Invertebrate Pathology* 73, 40–44.

Gulzar, S., Usman, M., Wakil, W., Gulcu, B., Hazir, C. *et al.* (2020) Environmental tolerance of entomopathogenic nematodes differs among nematodes arising from host cadavers versus aqueous suspension. *Journal of Invertebrate Pathology* 175: 107452.

Hallem, E.A., Rengarajan, M., Ciche, T. and Sternberg, P.W. (2007) Nematodes, bacteria, and flies, a tripartite model for nematode parasitism. *Current Biology* 17, 898–904.

Han, R. (1996) The effects of inoculum size on yield of *Steinernema carpocapsae* and *Heterorhabditis bacteriophora* in liquid culture. *Nematologica* 42, 546–553.

Han, R. and Ehlers, R.-U. (1998) Cultivation of axenic *Heterorhabditis* spp. dauer juveniles and their response to non-specific *Photorhabdus luminescens* food signals. *Nematologica* 44, 425–435.

Han, R. and Ehlers, R.-U. (2000) Pathogenicity, development, and reproduction of *Heterorhabditis bacteriophora* and *Steinernema carpocapsae* under axenic *in vivo* conditions. *Journal of Invertebrate Pathology* 75, 55–58.

Han, R. and Ehlers, R.-U. (2001) Effect of *Photorhabdus luminescens* phase variants on the *in vivo* and *in vitro* development and reproduction of the entomopathogenic nematodes *Heterorhabditis bacteriophora* and *Steinernema carpocapsae*. *FEMS Microbiology Ecology* 35, 239–247.

Han, R., Wouts, W.M. and Li, L.Y. (1990) Development of *Heterorhabditis* spp. strains as characteristics of possible *Xenorhabdus luminescens* subspecies. *Revue de Nematologie* 13, 411–415.

Han, R., Cao, L. and Liu, X. (1992) Relationship between medium composition, inoculum size, temperature and culture time in the yields of *Steinernema* and *Heterorhabditis* nematodes. *Fundamental and Applied Nematology* 15, 223–229.

Han, R., Cao, L. and Liu, X. (1993) Effects of inoculum size, temperature, and time on *in vitro* production of *Steinernema carpocapsae* Agriotos. *Nematologica* 39, 366–375.

Han, R., Pang, X. and Li, L. (1995) Optimization of the medium components for the solid culture of entomopathogenic *Steinernema* and *Heterorhabditis* nematodes. *Natural Enemies of Insects* 17, 153–164.

Han, R.C., Cao, L., He, X.Y., Li, Q., Liu, X. *et al.* (2000) Recovery response of *Heterorhabditis bacteriophora* and *Steinernema carpocapsae* to different non-symbiotic microorganisms. *Entomologia Sinica* 7, 271–277.

Hapeshi, A., Benarroch, J.M., Clarke, D.J. and Waterfield, N.R. (2019) Iso-propyl stilbene, a life cycle signal? *Microbiology* 165, 516–526.

Hara, A.H., Lindegren, J.E. and Kaya, H.K. (1981) *Monoxenic Mass Production of the Entomogenous Nematode* Neoaplectana carpocapsae *Weiser on Dog Food/Agar Medium*. USDA Advances in Agricultural Technology, Vol. 16. US Department of Agriculture, Washington, DC.

Hazir, S., Kaya, H., Touray, M., Cimen, H. and Shapiro-Ilan, D. (2022) Basic laboratory and field manual for conducting research with the entomopathogenic nematodes, *Steinernema* and *Heterorhabditis*, and their bacterial symbionts. *Turkish Journaol of Zooogy* 46, 305–350.

Hirao, A. and Ehlers, R.U. (2009) Influence of cell density and phase variants of bacterial symbionts (*Xenorhabdus* spp.) on dauer juvenile recovery and development of biocontrol nematodes *Steinernema carpocapsae* and *S. feltiae* (Nematoda: Rhabditida). *Applied Microbiology and Biotechnology* 84, 77–85.

Hirao, A. and Ehlers, R-U. (2010) Influence of inoculum density on population dynamics and dauer juvenile yields in liquid culture of biocontrol nematodes *Steinernema carpocapsae* and *S. feltiae* (Nematoda: Rhabditida). *Applied Microbiology and Biotechnology* 85, 507–515.

Hominick, W.M., Briscoe, B.R., del Pino, F.G., Heng, J., Hunt, D.J. *et al.* (1997) Biosystematics of entomopathogenic nematodes: current status, protocols and definitions. *Journal of Helminthology* 71, 271–298.

House, H.L., Welch, H.E. and Cleugh, T.R. (1965) Food medium of prepared dog biscuit for the mass-production of the nematode DD-136 (Nematoda; Steinernematidae). *Nature* 206: 847.

Jackson, G.J. (1973) *Neoaplectana glaseri*: essential amino acids. *Experimental Parasitology* 34, 111–114.

Jeffke, T., Jende, D., Mätje, C., Ehlers, R.U. and Berthe-Corti, L. (2000) Growth of *Photorhabdus luminescens* in batch and glucose fed-batch culture. *Applied Microbiology and Biotechnology* 54, 326–330.

Jessen, P., Strauch, O., Wyss, U., Luttmann, R. and Ehlers, R.U. (2000) Carbon dioxide triggers recovery from dauer juvenile stage in entomopathogenic nematodes (*Heterorhabditis* spp.). *Nematology* 2, 319–324.

Johnigk, S.A., Ecke, F., Poehling, M. and Ehlers, R.U. (2004) Liquid culture mass production of biocontrol nematodes, *Heterorhabditis bacteriophora* (Nematoda: Rhabditida): improved timing of dauer juvenile inoculation. *Applied Microbiology and Biotechnology* 64, 651–658.

Kaya, H.K. and Stock, S.P. (1997) Techniques in insect nematology. In: Lacey, L. (ed.) *Manual of Techniques in Invertebrate Pathology*, 1st edn. Academic Press, San Diego, pp. 281–324.

Kaya, H.K., Aguillera, M.M., Alumai, A., Choo, H.Y., de la Toree, M. *et al.* (2006) Status of entomopathogenic nematodes and their symbiotic bacteria from selected countries or regions of the world. *Biological Control* 38, 134–155.

Kong, X., Huang, Z., Gu, X., Cui, Y., Li, J. *et al.* (2022) Dimethyl sulfoxide and ascarosides improve the growth and yields of entomopathogenic nematodes in liquid cultures. *Journal of Invertebrate Pathology* 193: 107800.

Koppenhöfer, A.M. and Fuzy, E.M. (2003) Ecological characterization of *Steinernema scarabaei*, a scarab-adapted entomopathogenic nematode from New Jersey. *Journal of Invertebrate Pathology* 83, 139–148.

Kotchofa, R. and Baimey, H. (2019) *In vivo* production of entomopathogenic nematodes using *Galleria mellonella*: costs and effect of diets on nematode pathogenicity. *Journal of Nematology* 51: e2019-66.

Krasomil-Osterfeld, K. (1994) Phase variation in *Photorhabdus*, *Xenorhabdus* and other bacteria: a review. In: Burnell, A.M., Ehlers, R.-U. and Masson, J.P. (eds.) *Genetics of Entomopathogenic Nematode-Bacterium Complexes*. European Commission Publications, Luxembourg, pp. 70–79.

Krasomil-Osterfeld, K.C. (1995) Influence of osmolarity on phase shift in *Photorhabdus luminescens*. *Applied and Environmental Microbiology* 61, 3748–3749.

Lango-Scholey, L., Brachmann, A.O., Bode, H.B. and Clarke, D.J. (2013) The expression of *stlA* in *Photorhabdus luminescens* is controlled by nutrient limitation. *PLOS ONE* 8: e82152.

Leite, L.G., Shapiro-Ilan, D.I., Hazir, S. and Jackson, M.A. (2016a) Effect of inoculum age and physical parameters on *in vitro* culture of the entomopathogenic nematode *Steinernema feltiae*. *Journal of Helminthology* 91, 686–695.

Leite, L.G., Shapiro-Ilan, D.I., Hazir, S. and Jackson, M.A. (2016b) The effects of nutrient concentration, addition of thickeners and agitation speed on liquid fermentation of *Steinernema feltiae*. *Journal of Nematology* 48, 126–133.

Leite, L.G., Shapiro-Ilan, D.I., Hazir, S. and Jackson, M. (2016c) A new medium for liquid fermentation of *Steinernema feltiae*: selection of lipid and protein sources. *Nematropica* 46, 147–153.

Leite, L.G., Chacon-Orozcoa, J.G., Shapiro-Ilan, D.I., Baldo, F.B. and Cardoso, J.M. (2023) Effects of temperature for optimizing production and storage of *Steinernema rarum* in a novel biphasic process, and efficacy of the nematode against *Sphenophorus levis*. *Biological Control* 187: 105381.

Lindegren, J.E., Hoffman, D.F., Collier, S.S. and Fries R.D. (1979) *Propagation and Storage of* Neoaplectana carpocapsae *Weiser using* Amyelois transitella *(Walker) Adults.* USDA Advances in Agricultural Technology. US Department of Agriculture, Berkeley, California.

Lindegren, J.E., Valero, K.A. and Mackey, B.E. (1993) Simple *in vivo* production and storage methods for *Steinernema carpocapsae* infective juveniles. *Journal of Nematology* 25, 193–197.

Lunau, S., Stoessel, S., Schmidt-Peisker, A.J. and Ehlers, R.-U. (1993) Establishment of monoxenic inocula for scaling up *in vitro* cultures of the entomopathogenic nematodes *Steinernema* spp. and *Heterorhabditis* spp. *Nematologica* 39, 385–399.

Mamiya, Y. (1989) Comparison of the infectivity of *Steinernema kushidai* (Nematoda: Steinernematidae) and other steinernematid and heterorhabditid nematodes from three different insects. *Applied Entomology and Zoology* 24, 302–308.

Morales-Ramos, J.A., Rojas, M.G., Shapiro-Ilan, D.I. and Tedders, W.L. (2011) Nutrient regulation in *Tenebrio molitor* (Coleoptera: Tenebrionidae): self-selection of two diet components by larvae and impact on fitness. *Environmental Entomology* 40, 1285–1294.

Morales-Ramos, J.A., Rojas, M.G., Shapiro-Ilan, D.I., Kay, S. and Tedders, W.L. (2012) Impact of adult weight, density, and age on reproduction of *Tenebrio molitor* (Coleoptera: Tenebrionidae). *Journal of Entomological Science* 47, 208–220.

Moshayov, A., Koltai, H. and Glazer, I. (2013) Molecular characterisation of the recovery process in the entomopathogenic nematode *Heterorhabditis bacteriophora*. *International Journal of Parasitology* 43, 843–852.

Neves, J.M., Teixeira, J.A., Simoes, N. and Mota, M. (2001) Effect of airflow rate on yield of *Steinernema carpocapsae* Az 20 in liquid culture in an external-loop airlift bioreactor. *Biotechnology and Bioengineering* 72, 369–373.

Nguyen, K.B. and Smart, G.C. Jr. (1990) *Steinernema scapterisci* n. sp. (Rhabditida: Steinernematidae). *Journal of Nematology* 22, 187–199.

Oliveira-Hofman, C., Steffan, S. and Shapiro-Ilan, D. (2023) A sustainable grower-based method for entomopathogenic nematodes production. *Journal of Insect Science* 23: 4.

Pace, W.G., Grote, W., Pitt, D.E. and Pitt, J.M. (1986) Liquid culture of nematodes. International Patent Application WO1986001074A1.

Patel, M.N., Perry, R.N. and Wright, D.J. (1997a) Desiccation survival and water contents of entomopathogenic nematodes, *Steinernema* spp. (Rhabditida: Steinernematidae). *International Journal of Parasitology* 27, 61–70.

Patel, M.N., Stolinski, M. and Wright, D.J. (1997b) Neutral lipids and the assessment of infectivity in entomopathogenic nematodes: observations on four *Steinernema* species. *Parasitology* 114, 489–496.

Peters, A. (2013) Application and commercialization of nematodes. *Applied Microbiology and Biotechnology* 97, 6181–6188.

Peters, A., Han, R., Yan, X. and Leite, L.G. (2017) Production of entomopathogenic nematodes. In: Lacey, L.A. (ed.) *Microbial Control of Insect and Mite Pests: From Theory to Practice*. Academic Press, London, pp. 157–167.

Poinar, GO. Jr. and Thomas, G.M. (1966) Significance of *Achromobacter nematophilus* Poinar and Thomas (Achromobacteriaceae: Eubacteriales) in the development of the nematode, DD-136 (*Neoaplectana* sp., Steinernematidae). *Parasitology* 56, 385–390.

Popiel, I. and Vasquez, E.M. (1991) Cryopreservation of *Steinernema carpocapsae* and *Heterorhabditis bacteriophora*. *Journal of Nematology* 23, 432–437.

Regaiolo, A., Dominelli, N., Andresen, K. and Heermann, R. (2020) The biocontrol agent and insect pathogen *Photorhabdus luminescens* interacts with plant roots. *Applied and Environmental Microbiology* 86: e00891-20.

Ritter, K.S. (1988) *Steinernema feltiae* (= *Neoaplectana carpocapsae*): effect of sterols and hypolipidemic agents on development. *Experimental Parasitology* 67, 257–267.

Salma, J. and Shahina, F. (2012) Mass production of eight Pakistani strains of entomopathogenic nematodes (Steinernematidae and Heterorhabditidae). *Pakistan Journal of Nematology* 30, 1–20.

Shapiro, D.I. and Glazer, I. (1996) Comparison of entomopathogenic nematode dispersal from infected hosts versus aqueous suspension. *Environmental Entomology* 25, 1455–1461.

Shapiro, D.I. and Lewis, E.E. (1999) Comparison of entomopathogenic nematode infectivity from infected hosts versus aqueous suspension. *Environmental Entomology* 28, 907–911.

Shapiro, D.I. and McCoy, C.W. (2000) Effect of culture method and formulation on the virulence of *Steinernema riobrave* (Rhabditida: Steinernematidae) to *Diaprepes abbreviatus* (Curculionidae). *Journal of Nematology* 32, 281–288.

Shapiro, D.I., Lewis, E.E., Paramasivam, S. and McCoy, C.W. (2000) Nitrogen partitioning in *Heterorhabditis bacteriophora*-infected hosts and the effects of nitrogen on attraction/repulsion. *Journal of Invertebrate Pathology* 76, 43–48.

Shapiro-Ilan, D.I. and Gaugler, R. (2002) Production technology for entomopathogenic nematodes and their bacterial symbionts. *Journal of Industrial Microbiology and Biotechnology* 28, 137–146.

Shapiro-Ilan, D.I., Lewis, E.E., Behle, R.W. and McGuire, M.R. (2001) Formulation of entomopathogenic nematode-infected-cadavers. *Journal of Invertebrate Pathology* 78, 17–23.

Shapiro-Ilan, D.I., Gaugler, R., Tedders, W.L., Brown, I. and Lewis, E.E. (2002) Optimization of inoculation for *in vivo* production of entomopathogenic nematodes. *Journal of Nematology* 34, 343–350.

Shapiro-Ilan, D.I., Lewis, E.E., Tedders, W.L. and Son, Y. (2003) Superior efficacy observed in entomopathogenic nematodes applied in infected-host cadavers compared with application in aqueous suspension. *Journal of Invertebrate Pathology* 83, 270–272.

Shapiro-Ilan, D.I., Guadalupe Rojas, M., Morales-Ramos, J.A., Lewis, E.E. and Tedders, W.L. (2008) Effects of host nutrition on virulence and fitness of entomopathogenic nematodes: lipid and protein based supplements in *Tenebrio molitor* diets. *Journal of Nematology* 40, 13–19.

Shapiro-Ilan, D.I., Tedders, W.L., Morales-Ramos, J.A. and Rojas, M.G. (2009) Insect inoculation system and method. US Patent Application No. 12636245, DN 0107.04. US Patent and Trademark Office, Washington, DC.

Shapiro-Ilan, D.I., Tedders, W.L., Morales-Ramos, J.A. and Rojas, M.G. (2011) System and method for producing beneficial parasites. US Patent US8677935B1. US Patent and Trademark Office, Washington, DC.

Shapiro-Ilan, D.I., Han, R. and Dolinski, C. (2012) Entomopathogenic nematode production and application technology. *Journal of Nematology* 44, 206–217.

Shapiro-Ilan, D.I., Han, R. and Qiu, X. (2014) Production of entomopathogenic nematodes. In: Morales-Ramos, J., Rojas, G. and Shapiro-Ilan, D.I. (eds) *Mass Production of Beneficial Organisms: Invertebrates and Entomopathogens*. Academic Press, San Diego, California, pp. 321–355.

Shapiro-Ilan, D.I., Morales-Ramos, J.A. and Rojas, M.G. (2016) *In vivo* production of entomopathogenic nematodes. In: Glare, T.R. and Moran-Diez, M.E. (eds) *Microbial-Based Biopesticides: Methods and Protocols*. Methods in Molecular Biology, Vol. 1477. Humana Press, New York, pp. 137–158.

Shapiro-Ilan, D.I., Hazir, S. and Glazer, I. (2017) Basic and applied research: entomopathogenic nematodes. In: Lacey, L.A. (ed.) *Microbial Control of Insect and Mite Pests: From Theory to Practice*. Academic Press, London, pp. 91–105.

Shapiro-Ilan, D.I., Garrigos Leite, L. and Han, R. (2023) Production of entomopathogenic nematodes. In: Morales-Ramos, J., Rojas, G. and Shapiro-Ilan, D.I. (eds) *Mass Production of Beneficial Organisms: Invertebrates and Entomopathogens*. Academic Press, San Diego, California, pp. 293–316.

Shi, Y.M. and Bode, H.B. (2018) Chemical language and warfare of bacterial natural products in bacteria–nematode–insect interactions. *Natural Product Reports* 35, 309–335.

Silva, M.S.O., Cardoso, J.F.M., Ferreira, M.E.P., Baldo, F.B., Silva, R.S.A. *et al.* (2021) An assessment of *Steinernema rarum* as a biocontrol agent in sugarcane with focus on *Sphenophorus levis*, host-finding ability, compatibility with vinasse and field efficacy. *Agriculture* 11: 500.

Stock, S.P. and Hunt, D.J. (2005) Morphology and systematics of nematodes used in biocontrol. In: Grewal, P.S., Ehlers, R.U. and Shapiro-Ilan, D.I. (eds) *Nematodes as Biocontrol Agents*. CAB International, Wallingford, UK, pp. 3–43.
Stoll, N.R. (1952) Axenic cultivation of the parasitic nematode, *Neoaplectana glaseri*, in a fluid medium containing raw liver extract. *Journal of Parasitology* 39, 422–444.
Strauch, O. and Ehlers, R.-U. (1998) Food signal production of *Photorhabdus luminescens* inducing the recovery of entomopathogenic nematodes *Heterorhabditis* spp. in liquid culture. *Applied Microbiology and Biotechnology* 54, 9–13.
Strauch, O. and Ehlers, R.U. (2000) Influence of the aeration rate on yields of the biocontrol nematodes *Heterorhabditis megidis* in monoxenic liquid cultures. *Applied Microbiology and Biotechnology* 54, 9–13.
Strauch, O., Stoessel, S. and Ehlers, R.-U. (1994) Culture conditions define automictic or amphimictic reproduction in entomopathogenic rhabditid nematodes of the genus *Heterorhabditis*. *Fundamental and Applied Nematology* 17, 575–582.
Surrey, M.R. and Davies, R.J. (1996) Pilot scale liquid culture and harvesting of an entomopathogenic nematode, *Heterorhabditis*. *Fundamental and Applied Nematology* 17, 575–582.
Testa, A.M. and Shields, E.J. (2017) Low labor "in vivo" mass rearing method for entomopathogenic nematodes. *Biological Control* 106, 77–82.
Tourtois, J., Ali, J.G. and Grieshop, M.J. (2015) Susceptibility of wounded and intact black soldier fly *Hermetia illucens* L. (Diptera: Stratiomyidae) to entomopathogenic nematodes. *Journal of Invertebrate Pathology* 150, 121–129.
Wang, J.X. and Bedding, R.A. (1998) Population dynamics of *Heterorhabditis bacteriophora* and *Steinernema carpocapsae* in *in vitro* solid culture. *Fundamental and Applied Nematology* 21, 165–171.
Wang, J., Cao, L., Huang, Z., Gu, X., Cui, Y. *et al.* (2022) Influence of the ascarosides on the recovery, yield and dispersal of entomopathogenic nematodes. *Journal of Invertebrate Pathology* 188: 107717.
White, G.F. (1927) A method for obtaining infective nematode larvae from cultures. *Science* 66, 302–303.
Woodring, J.L. and Kaya, H.K. (1988) *Steinernematid and Heterorhabditid Nematodes: A Handbook of Biology and Techniques*. Southern Cooperative Series Bulletin 331, Arkansas Agricultural Experiment Station, Fayetteville, Arkansas.
Wouts, W.M. (1981) Mass production of the entomogenous nematode *Heterorhabditis heliothidis* (Nematoda: Heterorhabditidae) on artificial media. *Journal of Nematology* 13, 467–469.
Yan, X. and Han, R. (2016) Production technology of the bio-insecticides: entomopathogenic nematodes. *Journal of Environmental Entomology* 38, 1044–1051.
Yan, X., Lin, Y., Huang, Z. and Han, R. (2018) Characterization of biological and biocontrol traits of entomopathogenic nematodes promising for control of striped flea beetle (*Phyllotreta striolata*). *Nematology* 20, 503–518.
Yan, X., Chen, G., Chen, Y., Sun, B., Gu, X. *et al.* (2021) Virulence of *Steinernema ceratophorum* against different pest insects and their potential for *in vivo* and *in vitro* culture. *Journal of Nematology* 53: e2021-46.
Yang, H., Jian, H., Zhang, S. and Zhang, G. (1997) Quality of the entomopathogenic nematode *Steinernema carpocapsae* produced on different media. *Biological Control* 10, 193–198.
Yoo, S.K., Brown, I. and Gaugler, R. (2000) Liquid media development for *Heterorhabditis bacteriophora*: lipid source and concentration. *Applied Microbiology and Biotechnology* 54, 759–763.
You, J., Liang, S., Cao, L., Liu, X. and Han, R. (2006) Nutritive significance of crystalline inclusion proteins of *Photorhabdus luminescens* in *Steinernema* nematodes. *FEMS Microbiology Ecology* 55, 178–185.
Zhen, S., Li, Y., Hou, Y., Gu, X., Zhang, L. *et al.* (2018) Enhanced entomopathogenic nematode yield and fitness via addition of pulverized insect powder to solid media. *Journal of Nematology* 50, 495–506.

11 Formulation and Application Technology for Entomopathogenic Nematodes

David I. Shapiro-Ilan[1]* and Edwin E. Lewis
[1]USDA-ARS, Southeastern Fruit and Tree Nut Research Station, Byron, Georgia, USA; [2]Department of Entomology, Plant Pathology and Nematology, University of Idaho, Moscow, Idaho, USA

Abstract

Entomopathogenic nematodes (EPNs) in the genera *Heterorhabditis* and *Steinernema* are potent biological control agents that are used to control a variety of economically important insect pests. The nematodes are usually applied in aqueous suspension and are amenable to distribution using most standard agricultural sprayers and irrigation systems. Biocontrol efficacy provided by EPNs is dependent on various factors including the quality of the formulation and application parameters. Abiotic and biotic factors play critical roles in the ability of EPNs to control insect pests. For example, among the most important abiotic factors affecting efficacy are the environmental conditions surrounding the application; EPN success can be severely limited or completely incapacitated under conditions of extreme temperature, desiccation or ultraviolet radiation. Additionally, the nature of the spray equipment used can be important. Key biotic factors include the choice of nematode and rate of nematodes applied; it is of the utmost importance to match the nematode species (or strain) to the target pest. A variety of approaches can be used to enhance EPN application technology such as protective gel adjuvants (that facilitate aboveground application) and pheromone boosters. In this chapter, we review and discuss EPN formulations and application technologies, and avenues for improvement.

11.1 Formulation Technology

Formulation of entomopathogenic nematodes (EPNs) serves two purposes: (i) to increase stability during storage; and (ii) to facilitate ease of handling during application. To enable levels of survival and shelf life that are acceptable for commercialization, the nematodes' metabolism can be slowed down by placing them in cool temperatures (e.g. 4–15°C, depending on the EPN species), physically constraining movement, and/or using formulation ingredients that reduce movement and thus reduce the nematode's metabolic rate (Grewal and Peters, 2005). The formulation must provide a suitable environment for nematode survival including adequate levels of air exchange and moisture (to avoid desiccation). A wide variety of formulations have been developed for EPNs including charcoal, alginate and polyacrylamide gels, clay, diatomaceous earth, paste, peat, polyurethane sponge, vermiculite and water-dispersible granules (Grewal, 2002; Shapiro-Ilan *et al.*, 2012a; Shapiro-Ilan and Dolinski, 2015).

*Email: David.Shapiro@usda.gov

DOI: 10.1079/9781800620322.0011

Formulations entail various levels of complexity and ease of handling. A simple approach is to concentrate the infective juvenile (IJ) nematodes into a paste using an air pump and filter (e.g. an aquarium air stone), and then spread the paste on to sponge or another substrate (Shapiro-Ilan *et al.*, 2016a). This approach is used by some researchers and small commercial producers. The type of sponge used can influence EPN survival (Touray *et al.*, 2020). The nematode paste can also be applied to various gels or other substrates. A drawback to the sponge formulation is that the nematodes must be separated from the substrate prior to application. Separation is done by rinsing or washing/squeezing the nematodes from the substrate. The separation step is labor intensive and thus reduces attractiveness to the end user. Some other formulations may also require separation prior to application, such as certain gels. Formulations that can be directly tank mixed are more user friendly. One example of a tank-mix-ready formulation is water-dispersible granules that reduce EPN activity via partial anhydrobiosis; this formulation can provide up to 6 months of storage capacity at room temperature (Grewal and Peters, 2005). Other formulations that can be tank mixed include those based on fine vermiculite or diatomaceous earth, which are currently the most widely distributed formulations on the market.

Survivability of EPNs in various substrates varies (Leite *et al.*, 2018; Kagimu and Malan, 2019; Askary and Ahmad, 2021). For example, using *Steinernema feltiae* (Filipjev) as the model nematode, Leite *et al.* (2018), compared seven different formulation substrates and combinations thereof at three temperatures. Peat and vermiculite provided the highest levels of survival at 35°C and 25°C, respectively, and polyacrylamide gel was superior at 15°C. Among the combinations, a mixture of vermiculite and polyacrylamide gel provided superior preservation with more than 80% EPN survival for at least 30 days at 35°C, 233 days at 25°C and 241 days at 15°C (Leite *et al.*, 2018).

In the past decade, several novel formulations have been developed. For example, ultraviolet (UV)-protecting nanoparticle-based formulations have been developed; the individual IJs are coated with the nanoparticle formulation (e.g. using titanium dioxide) like a coat of armor (Kotliarevski *et al.*, 2022; Wu *et al.*, 2022). Building on earlier encapsulation methods using calcium alginate (Kaya and Nelsen, 1985), several advancements have been made in encapsulated gel or pellet formulations, which may act as slow-release capsules (Matadamas-Ortiz *et al.*, 2014; Ruiz-Vega *et al.*, 2018; Kim *et al.*, 2021). An efficient labor-saving method of formulating EPNs produced *in vivo* captures IJs directly as they emerge from infected hosts into a gel formulation, which is already in the final package ready for storage and shipping (an all-in-one package) (Shapiro-Ilan *et al.*, 2016a). A significant breakthrough in expanding the utility of EPNs in aboveground application has been the development of various sprayable gels that protect against UV radiation and/or desiccation (Shapiro-Ilan *et al.*, 2017). The sprayable gel approach will be addressed later under improvements to application technology as the gels are more of an adjuvant (added to the mixture to enhance efficacy) than a formulation (the mixture intended for storage and transport).

11.2 Application Technology

EPNs can be applied as liquid-based products using most standard agricultural sprayers and irrigation systems such as microjet irrigation, mist blowers, pressurized sprayers, herbicide rigs, electrostatic sprayers, trunk sprayers, aerial sprays and handguns (Wright *et al.*, 2005; Shapiro-Ilan *et al.*, 2006a, 2012a, 2017). For small-scale applications, backpack sprayers or watering cans may be used. Other effective methods of application include baits or subsurface incorporation into soil (Wright *et al.*, 2005; Shapiro-Ilan and Dolinski, 2015). The most appropriate equipment for a particular application will depend on the target pest and cropping system involved. Factors to consider include volume, nozzle type, agitation, pressure, environmental conditions, the nature of the target site (e.g. soil, leaf, trunk) and the spray distribution pattern required (Fife *et al.*, 2005; Wright *et al.*, 2005; Shapiro-Ilan *et al.*, 2012a, 2017). The application equipment used must allow for adequate aeration and agitation to facilitate EPN survival in the spray tank and effective distribution. Application methods can mitigate various abiotic and biotic factors affecting the efficacy of EPN applications but not all.

11.2.1 Abiotic factors affecting efficacy

11.2.1.1 Environmental factors

Successful EPN application is dependent on suitable environmental conditions and choosing an EPN species that can tolerate these conditions. Foremost, biocontrol efficacy relies on the application site having appropriate moisture and relative humidity, temperatures that are conducive to infection and avoiding UV radiation (Kaya, 1990; Shapiro-Ilan *et al.*, 2006a, 2012a). Thus, successful EPN applications have primarily been targeted to soil or other protected habitats to avoid UV radiation and desiccation (as irrigation can be applied to overcome moisture deficiencies). Moreover, due to the antagonistic effects of UV radiation (Gaugler and Boush, 1978; Shapiro-Ilan *et al.*, 2015), nematode applications should be made in the evening or early morning when feasible. Aboveground applications outside protected environments have, for the most part, received less attention or resulted in failure (Arthurs *et al.*, 2004; Shapiro-Ilan *et al.*, 2006a, 2012a, 2017).

EPNs can desiccate, become immobile and die when the relative humidity is too low. Therefore, irrigation is recommended when EPNs are applied to soil (Downing, 1994; Shapiro-Ilan *et al.*, 2006a; Hazir *et al.*, 2022). Shapiro-Ilan *et al.* (2013) reported an exception in which irrigation in field plots did not improve EPN efficacy relative to an application without irrigation, but this situation is an outlier and presumably was the result of adequate rainfall and a soil with high water-holding capacity (Shapiro-Ilan *et al.*, 2013). Although irrigation is generally required, excess moisture in the soil can stress or kill EPNs due to a lack of aeration leading to anoxic conditions, and dispersal can be hindered (Kaya, 1990; Womersley, 1993; Koppenhöfer *et al.*, 1995). Optimum moisture levels and tolerance to desiccation vary by soil type and among nematode species (Koppenhöfer *et al.*, 1995; Shapiro-Ilan *et al.*, 2014a; Ramakrishnan *et al.*, 2022).

Optimal temperatures for EPN infection and reproduction vary among strains and species (Grewal *et al.*, 1994; Shapiro-Ilan *et al.*, 2014b). Some nematodes such as *Heterorhabditis megidis* Poinar, Jackson & Klein, *S. feltiae* and *Heterorhabditis marelata* Liu & Berry are relatively cold tolerant, whereas *Heterorhabditis floridensis* Nguyen, Gözel, Koppenhöfer and Adams, *Heterorhabditis indica* Poinar, Karunakar and David, *Steinernema glaseri* (Steiner) and *Steinernema riobrave* Cabanillas, Poinar and Raulston are relatively heat tolerant (Grewal *et al.*, 1994; Berry *et al.*, 1997; Shapiro-Ilan *et al.*, 2011, 2014b). Nematode species and strains also vary substantially in freezing tolerance; for example, *S. feltiae* and *Heterorhabditis georgiana* Nguyen, Shapiro-Ilan and Mbata are relatively tolerant, whereas *H. indica* and *Steinernema rarum* (de Doucet) are more sensitive (Shapiro-Ilan *et al.*, 2014a). Freeze tolerance may be a factor in determining the overwintering and recycling potential of EPNs.

Several other soil properties affect EPN application efficacy. Soil texture is an important factor that can impact nematode movement and survival (Kaya, 1990; Barbercheck, 1992; Kaspi *et al.*, 2010). Generally, lighter soils with higher proportions of sand (and lower clay content) tend to be more conducive to EPN dispersal and allow greater aeration compared with heavier soils (Georgis and Poinar, 1983; Molyneux and Bedding, 1984; Dolinski *et al.*, 2010). However, exceptions have been reported where heavier soils have been highly conducive for EPN efficacy and sometimes more conducive than lighter soils (Georgis and Gaugler, 1991; Shapiro *et al.*, 2000; Miklasiewicz *et al.*, 2002; Lankin *et al.*, 2020).

Soil pH can also impact EPN survival and efficacy although generally pH in the range of range of 4–8 is not likely to have any significant effect (Kung *et al.*, 1990). Khathwayo *et al.* (2021) characterized the pH tolerance of a broad array of EPNs (17 species and strains) when exposed to pH levels ranging from 2 to 11. Steinernematids generally had higher survival rates compared with heterorhabditids at extreme pH levels. Among all the EPNs tested, *Steinernema carpocapsae* and *S. riobrave* exhibited higher tolerance to both alkaline and acidic conditions.

Agricultural inputs such as fertilizers and chemical pesticides can also affect the efficacy of EPN applications. Interactions can be antagonistic, additive (neutral) or synergistic (Shapiro-Ilan *et al.*, 2017). Only excessive concentrations of organic or inorganic fertilizers tend to be harmful to EPNs, and regular recommended concentrations are generally compatible (Shapiro *et al.*, 1996a, 1999a; Bednarek and Gaugler, 1997). However, high concentrations of chemical fertilizers (e.g. urea at 560 kg N ha^{-1}) or fresh manure or can be detrimental to EPN persistence and efficacy (Shapiro *et al.*, 1996a, 1999a; Bednarek and Gaugler, 1997). Certain chemical pesticides are detrimental to EPNs (e.g. abamectin, acephate, aldicarb, dodine, fenamiphos, methomyl, parathion and teflubenzuron), whereas others are compatible or may cause synergistic levels of insect mortality when applied with EPNs (e.g. carbaryl, chlorpyrifos, dimethoate, endosulfan, fonofos, tefluthrin and imidacloprid) (Alumai and Grewal, 2004; Koppenhöfer and Grewal, 2005; Koppenhöfer and Fuzy, 2008). The nature of an interaction between a particular chemical and EPNs can vary based on the EPN species and chemical agent, and the timing and concentration of application (Koppenhöfer and Grewal, 2005; Amizadeh *et al.*, 2019; Mbata *et al.*, 2022), and therefore potential combination treatments should be considered specifically for each target pest and cropping system. Chapter 27 (this volume) provides a more comprehensive treatment of interactions between EPNs and agrochemicals.

11.2.1.2 Equipment

The choice of application equipment can impact EPN efficacy in biocontrol (Curran, 1992; Hayes *et al.*, 1999; Shields *et al.*, 1999; Brusselman *et al.*, 2010; Shapiro-Ilan *et al.*, 2012a, 2016b, 2017). For example, in one study, sprinkler irrigation was inferior to boom sprayer application in terms of IJ infectivity after application (Hayes *et al.*, 1999). Working on control of peachtree borer, *Synanthedon exitiosa* (Say) (Lepidoptera: Sesiidae), Shapiro-Ilan *et al.* (2016b) observed advantages in terms of biocontrol efficacy when using a handgun sprayer relative to a trunk sprayer or boom sprayer. The trunk sprayer, however, also has advantages for orchard applications because it senses the presence of the trunk and does not spray EPNs in between trees (thus avoiding waste) (Shapiro-Ilan *et al.*, 2016b). Injection of EPNs into irrigation systems is also an efficient method of application, but the potential settling of IJs within the irrigation lines should be monitored and, if needed, additional emitters can be added at the end of lines to increase flow (Duncan *et al.*, 1999). In contrast to some of the examples listed above, some studies did not detect efficacy differences when comparing EPN application equipment (Klein and Georgis, 1994; Lacey *et al.*, 2006a). Similar to other parameters, the suitability of application equipment will vary depending on the target pest and cropping system.

A number of characteristics associated with application equipment can impact EPN efficacy including pressure differentials and the types of nozzles and pumps used (Georgis, 1990; Shapiro-Ilan *et al.*, 2006a). The ability to withstand spray pressure varies among EPN species; for example, based on viability tests, Fife *et al.* (2003) recommended a maximum pressure of 1380 kPa for *H. megidis* and 2000 kPa for *S. carpocapsae* and *Heterorhabditis bacteriophora* Poinar. In contrast, Anifantis *et al.* (2020) did not detect significant effects of pressures up to 4000 kPa on the viability and infectivity of *S. feltiae* and *H. bacteriophora*. Regarding nozzle and pump options, greater reductions in EPN viability were observed when using a flat fan nozzle compared with a hollow cone type, and diaphragm and roller pumps (low-capacity pumps) were found to be better compared with a centrifugal pump (due to heat buildup in the latter) (Fife *et al.*, 2005, 2006, 2007). In another example, an ISO 04 standard flat fan nozzle was superior to the larger ISO 08, and the ISO 02 flat fan nozzle was found to clog for spraying *S. carpocapsae* on cauliflower (Beck *et al.*, 2013).

11.2.2 Biotic factors affecting efficacy

Perhaps the most critical biological factor in EPN efficacy is the choice of nematode species or strain to be applied. The appropriate nematode must be matched to the specific target pest and cropping system. A high level of virulence to the target pest is mandatory for success (Shapiro-Ilan *et al.*, 2002, 2017). An initial screening of virulence against the target pest can be achieved via laboratory bioassays. However, laboratory virulence may not necessarily translate into field efficacy; the suitability of the EPNs to environmental conditions at the target site is also critical for successful biocontrol (Shapiro-Ilan *et al.*, 2004a, 2012b). None the less, screening assays have been successful in identifying efficacious EPNs for biocontrol against numerous pests such as the plum curculio, *Conotrachelus nenuphar* (Herbst) (Coleoptera: Curculionidae) (Shapiro-Ilan *et al.*, 2011a, 2013; Pinero *et al.*, 2020), and the western corn rootworm, *Diabrotica virgifera virgifera* LeConte (Coleoptera: Chrysomelidae) (Toepfer *et al.*, 2005, 2008).

Another aspect of choosing the appropriate nematode for a biocontrol program that may be considered is the EPN's foraging strategy. EPN foraging strategies vary along a continuum between ambushers and cruisers; ambushers generally sit and wait for a passing host, whereas cruisers actively search for hosts (Lewis and Clarke, 2012). Some nematodes primarily exhibit characteristics of ambusher types (e.g. *S. carpocapsae*), while others display strong cruiser behavior (e.g. *S. glaseri*), but many EPNs also exhibit intermediate foraging strategies that have characteristics of both (Lewis, 2002; Lewis and Clarke, 2012). Moreover, in some cases, soil type or substrate has been reported to modulate foraging strategy (Kruitbos *et al.*, 2010). Nematodes that are ambushers may be more adapted to infecting mobile insects on or near the soil surface, whereas cruisers may be more adapted to infect sessile insects below the soil surface (Lewis *et al.*, 1992; Lewis, 2002). However, it should be noted that there are many exceptions. For example, the most efficacious nematode for controlling peachtree borer, a belowground root-boring pest, is *S. carpocapsae* (Shapiro-Ilan *et al.*, 2009; Wong *et al.*, 2022).

The rate of application is also a critical factor in achieving biocontrol efficacy when using EPNs. A general rule of thumb is that 2.5×10^9 IJs ha^{-1} (or 25 IJs cm^{-2}) is the minimum rate required to achieve acceptable levels of efficacy (Georgis *et al.*, 1995; Shapiro-Ilan *et al.*, 2006a, 2012a, 2017). Some insects that have relatively low susceptibility to infection, or that are found deep within the soil profile, can require higher rates to reach sufficient efficacy. For example, the root weevil *Diaprepes abbreviatus* (L.) (Coleoptera: Curculionidae) tends to require rates higher than 25 IJs cm^{-2} (McCoy *et al.*, 2000; Shapiro-Ilan *et al.*, 2002). However, if the pest is particularly susceptible, lower application doses may be effective. For foliar or other aboveground applications, the rate of application will vary depending on the target pest and system (Grewal *et al.*, 2005), and these rates are difficult to describe in terms of numbers of IJs per unit surface area, as is common for soil applications. For example, *S. carpocapsae* was applied at 50,000 IJs ml^{-1} sprayed to run off on peach limbs infested with lesser peachtree borer, *Synanthedon pictipes* (Grote and Robinson) (Lepidoptera: Sesiidae) (Shapiro-Ilan *et al.*, 2010a). Another example of foliar application was *H. baujardi* Phan *et al.* LPP7 and *S. carpocapsae* (All strain) tested at a concentration of 500 IJs ml^{-1} in a greenhouse against larvae of *Diatraea saccharalis* (Fabricius) (Lepidoptera: Crambidae) (Bellini and Dolinski, 2012); thus, one can see how aboveground application rates vary across target pests and arenas.

The potential for EPNs to recycle can also be considered. Generally, one can expect insect pest suppression due to EPN infection to last about 2–8 weeks post-application (assuming environmental conditions are conducive for survival and infection) (Kaya, 1990; Shapiro-Ilan *et al.*, 2002, 2017). Thus, for the most part, EPNs are applied as inundative biological control agents with the expectation that reapplication will be required each year or season. However, there have been exceptions reported where EPNs have provided multi-season or multi-year control, or where longer-term persistence has been documented (Klein and Georgis, 1992; Parkman *et al.*, 1994; Shields *et al.*, 1999, 2018). A variety of factors affect EPN persistence; the potential for recycling is impacted by soil type, ground cover, host and host density, and the nematode species or strain (Kaya, 1990; Klein and Georgis, 1992; Shapiro *et al.*, 1999b; Shapiro-Ilan *et al.*, 2002, 2006b).

Various biotic agents that co-occur with EPNs can have a positive, negative or neutral impact on EPN applications (Kaya, 2002; Koppenhöfer and Grewal, 2005; see Chapters 8 and 27, this volume, for a full treatment of this subject). Antagonists may include various nematode pathogens or predators such as phages, bacteria, nematophagous fungi, predacious mites and predatory nematodes (Kaya, 2002; Ulug *et al.*, 2014). EPNs also form phoretic relationships with some organisms such as mites, earthworms and isopods (Epsky *et al.*, 1988; Shapiro *et al.*, 1995; Eng *et al.*, 2005; Shapiro-Ilan and Brown, 2013). Interactions between other entomopathogens and EPNs can be synergistic, as observed with *Paenibacillus popilliae* (Dutky) (Thurston *et al.*, 1994), *Bacillus thuringiensis* Berliner (Koppenhöfer and Kaya, 1997), *Beauveria bassiana* (Balsamo) Vuillemin (Wakil *et al.*, 2017) and *Metarhizium anisopliae* (Metschnikoff) Sorokin (Ansari *et al.*, 2004, 2006; Acevedo *et al.*, 2007; Dlamini *et al.*, 2020), but other studies report antagonism, for example with *Beauveria bassiana* (Brinkman and Gardner, 2001) or *Cordyceps fumosorosea* (Wize) (Shapiro-Ilan *et al.*, 2004b). The relationship between EPNs and other entomopathogens (antagonism, additivity and synergism) can vary depending on the host or nematode species, or the rate and timing of application (Barbercheck and Kaya, 1990; Koppenhöfer and Kaya, 1997; Shapiro-Ilan *et al.*, 2004b; Acevedo *et al.*, 2007). Additionally, EPN species may compete or coexist with each other within the soil, depending on the specific combination and environment (Millar and Barbercheck, 2001; Duncan *et al.*, 2003).

11.3 Approaches to Improving Efficacy

The efficacy of EPN applications can be improved through various approaches. One approach that is highly valuable across ecosystem targets is to use superior nematode strains. There are several methods to improve EPNs including discovery, selection, hybridization and transgenics. Arguably, the simplest approach to obtaining superior EPN strains or species is to discover new strains in nature. This can be accomplished by conducting surveys for novel EPNs that may offer superior efficacy to a given target pest (Shapiro-Ilan *et al.*, 2003a, Bruck, 2004; Campos-Herrera *et al.*, 2008, Malan *et al.*, 2011). In some cases, indigenous strains may be better suited to the local environment (Dolinski and Moino, 2006; Khathwayo *et al.*, 2021). Therefore, it may also be feasible to obtain strains with specific desired traits from targeted ecosystems; for example, if a desiccation-tolerant strain is needed, surveys can be conducted in areas of moisture stress.

If the strains already in culture or obtained through survey cannot provide sufficient biocontrol efficacy, then genetic methods of improvement can be implemented (i.e. selection, hybridization or transgenics). To select for a desired character, a diverse population is exposed to selection pressure through multiple generations until the desired effect is achieved. Some examples of directed selection include improved EPN host finding (Gaugler *et al.*, 1989), nematicide resistance (Glazer *et al.*, 1997), dispersal (Bal *et al.*, 2014) and other biocontrol traits (Shapiro-Ilan *et al.*, 2023). Potential drawbacks to the selection approach include inadvertently selecting for lower biocontrol potential in one trait while obtaining superior levels of another (a trade-off), and reversion of a targeted trait to its original state once the selection pressure is removed (Gaugler, 1987; Gaugler *et al.*, 1990). Thus, these potential shortcomings should be monitored.

Hybridization is another approach to EPN strain improvement. The method entails transferring beneficial traits from one strain to another through controlled breeding. When crossing heterorhabditids, a difficulty arises due to the presence of both amphimictic forms and hermaphrodites. One must ensure that the progeny have arisen through mating rather than from self-replication; this can be accomplished through marker mutations (Shapiro *et al.*, 1997). Given that steinernematids only have amphimictic forms (except for *Steinernema hermaphroditum* Stock, Griffin and Chaerani), hybridization is more straightforward for species in this genus (Shapiro-Ilan *et al.*, 2005). It is also possible to combine selection and hybridization in a dual strain improvement program such as the approach taken to achieve superior environmental tolerance in *H. bacteriophora* by Mukuka *et al.* (2010).

Limited studies have addressed transgenic approaches to EPN strain improvement so far. Progress has also been made toward molecular approaches for improving EPNs. For example, an *H. bacteriophora* strain was improved for heat tolerance through transformation using a heat-shock protein that originated from *Caenorhabditis elegans* (Maupas) (Gaugler *et al.*, 1997).

Genomic sequencing of EPNs and their bacterial symbionts (e.g. Duchaud *et al.*, 2003; Bai and Grewal, 2007; Ciche, 2007; Bai *et al.*, 2009, 2013; Schwartz *et al.*, 2011) is expected to continue expanding and will serve to enhance the potential for strain improvement programs using molecular or non-molecular approaches (see Chapter 4, this volume, for a discussion of genomic studies on EPNs).

In addition to choosing the most suitable strain for a biocontrol program (or improving the strain using the methods described above), it is also important to stabilize the strain to deter attenuation. Trait deterioration can result from genetic processes such as inadvertent selection, inbreeding, genetic drift or cheaters, or from non-genetic factors (e.g. disease or nutrition) (Hopper *et al.*, 1993; Tanada and Kaya, 1993; Chaston *et al.*, 2011; Shapiro-Ilan and Raymond, 2016). Deterioration under laboratory conditions has been observed in EPN virulence, environmental tolerance, reproductive capacity and host finding (Shapiro *et al.*, 1996b; Wang and Grewal, 2002; Bai *et al.*, 2005; Bilgrami *et al.*, 2006; Sharifi-Far *et al.*, 2018). Both the nematodes and their bacterial symbionts may experience trait loss (Bilgrami *et al.*, 2006; Wang *et al.*, 2007). The cause of trait deterioration in EPNs (*H. bacteriophora*) was reported to be genetically based, with inbreeding depression being the prominent issue (Bai *et al.*, 2005; Adhikari *et al.*, 2009; Chaston *et al.*, 2011).

One approach to preserve various traits in EPN strains and reduce the impact of trait deterioration is cryopreservation (Bai *et al.*, 2004). Cryopreservation, however, can be problematic because it is expensive and may be prone to mechanical failure or human error, which can result in complete loss of genetic material (Wang and Grewal, 2002; Shapiro-Ilan *et al.*, 2023). Another approach to deter EPN trait deterioration is the creation of homozygous inbred lines (Bai *et al.*, 2005; Shapiro-Ilan *et al.*, 2023). The approach entails creating an array of inbred lines and selecting the superior lines for biocontrol application (Bai *et al.*, 2005; Shapiro-Ilan *et al.*, 2023). Bai *et al.* (2005) first demonstrated the approach using selected *H. bacteriophora* inbred lines that remained stable through serial culturing while their wild-type parent strains deteriorated in environmental tolerance, virulence and host finding. Thus, Bai *et al.* (2005) proposed that EPN producers develop numerous inbred lines from promising candidate strains and then select the lines that exhibit high levels of desirable traits for commercialization. In the method of Bai *et al.* (2005), individual inbred lines are created based on self-selection (e.g. starting a line with a single hermaphrodite for heterorhabditids and *S. hermaphroditum*) or sibling mating (for steinernematids) on sterile media plates. Multiple inbred heterorhabditid lines can also be created in liquid culture because the nematodes are not able to mate and therefore only reproduce by selfing (Anbesse *et al.*, 2012); the method is not applicable to steinernematids.

Improved formulation can also lead to enhanced biocontrol efficacy. Substantial progress in EPN formulation has been made in the last couple of decades, particularly in terms of aboveground applications. For example, a surfactant/polymer mixture led to improved aboveground control of diamondback moth larvae, *Plutella xylostella* (L.) (Lepidoptera: Plutellidae) (Schroer and Ehlers, 2005). Similarly, using Barricade® fire gel, Shapiro-Ilan *et al.* (2010a, 2016c) observed high levels of *S. pictipes* control with a dual or single spray (Fig. 11.1). The fire gel can also be used in soil applications to prevent EPN desiccation in lieu of irrigation, as demonstrated in the control of *Synanthedon exitiosa* (Shapiro-Ilan *et al.*, 2016b). Moreover, Barricade® fire gel can be improved further by addition of other UV protectants such as *p*-amino benzoic acid, octyl methoxycinnamate or titanium dioxide (Dito *et al.*, 2016; Acar and Sipes, 2022). A variety of other protective formulations or approaches have been developed for aboveground applications such as a superabsorbent polymer formulation, Zeba® (de Waal *et al.*, 2013), a wood flour foam (Lacey *et al.*, 2010) and leaf flooding with the addition of surfactants (Williams and Walters, 2000; Head *et al.*, 2004). Recently, a gel formulation based on carboxymethyl cellulose provided high levels of efficacy aboveground for control of the fall armyworm, *Spodoptera frugiperda* Smith (Lepidoptera: Noctuidae); the level of control from carboxymethyl cellulose plus nematodes was superior to EPN treatments with surfactant/polymer formulation (Fallet *et al.*, 2022).

Bait formulations have the potential to increase persistence and reduce the number of

Fig. 11.1. Barricade® fire gel sprayed on to peach tree limbs to protect entomopathogenic nematode applications from ultraviolet radiation and desiccation. The application was targeting the lesser peachtree borer, *Synanthedon pictipes*. The gel can be applied at full strength as shown, or it can be effective at a reduced concentration in a tank mix that can be used with standard agricultural sprayers. Photograph courtesy of USDA Agricultural Research Service.

EPNs required per unit area (Georgis 1990; Grewal, 2002). Thus, slow-release capsules have been developed from materials such as calcium alginate (Kaya and Nelsen, 1985). Subsequently, pellets have also been developed with other materials such as diatomaceous earth and attapulgite clay or combinations thereof (Matadamas-Ortiz *et al.*, 2014). Baits may be most appropriate for small arenas or cryptic habitats. However, baits may also be potentially useful in larger arenas, particularly when a host attractant is added (Hiltpold *et al.*, 2012).

In a novel approach, the infected host can be seen as "capsule" that serves as the vehicle for application. Application in EPN-infected hosts has been termed the "cadaver approach" (Shapiro and Glazer, 1996; Shapiro-Ilan *et al.*, 2003b; Dolinski *et al.*, 2015). The infected host cadavers are applied to the target site and pest suppression is subsequently achieved by the emerging IJs. EPNs applied in the cadaver approach have been observed to be superior to standard aqueous application in dispersal (Shapiro and Glazer, 1996), infectivity (Shapiro and Lewis, 1999), survival (Perez *et al.*, 2003), efficacy (Shapiro-Ilan *et al.*, 2003b) and environmental tolerance (Gulzar *et al.*, 2020). However, some other studies did not observe any benefits from cadaver application (Bruck *et al.*, 2005; Ramalingam *et al.*, 2015). To prevent cadavers from rupturing or sticking together, protective formulations have been developed (Shapiro-Ilan *et al.*, 2001, 2010b; Del Valle *et al.*, 2009). Moreover, equipment has been developed for field distribution of infected hosts (Zhu *et al.*, 2011). In a related approach, pre-infected hosts (but still living and mobile) can be applied to cryptic target sites (e.g. wood galleries), thereby facilitating enhanced control in locations that are hard to reach; pest suppression is achieved by the emerging IJs after the released insect host dies after entering the target location (Gumus *et al.*, 2015).

EPN efficacy can also be improved by manipulating the environment. For example, in nursery, greenhouse or field applications, the substrate can be optimized for optimum EPN performance (Ansari and Butt, 2011; Nielsen and Lewis, 2012; Duncan *et al.*, 2013). Environmental manipulations in field applications can also significantly enhance performance, such as increased persistence of *S. carpocapsae* via crop residues left in place to protect the EPNs from environmental extremes (Shapiro *et al.*, 1999b). Similarly, mulches have been used to increase EPN persistence in field applications (Lacey *et al.*, 2006b; de Waal *et al.*, 2011).

As indicated above, interactions between EPNs and other control agents can be synergistic. Thus, these synergistic combinations can be optimized to achieve improved control as a mechanism to improve EPN efficacy. The level of control and synergy will depend on the EPN and host species used, as well as timing and rate of application. Another combination that can result in enhanced biocontrol is the application of EPNs with phoretic hosts. For example, co-application of earthworms with EPNs (*S. carpocapsae*) resulted in improved control of pecan weevil, *Curculio caryae* (Horn) (Coleoptera: Curculionidae) (Shapiro-Ilan and Brown, 2013). The potential of combining EPNs with phoretic hosts (e.g. earthworms) in dual applications or enhancing phoretic host populations in the field through environmental manipulation deserves further exploration.

A relatively new and promising technology is to utilize compounds that direct nematode behavior to enhance biocontrol potential. When plants are attacked by insects, certain chemicals may be emitted that elicit a call-for-help signal; EPNs are attracted to these signals and thus "come to the rescue" by migrating to the plants and killing the pests (Hiltpold and Turlings, 2012). Conceivably, these call-for-help signals could be used to draw EPNs to target sites. The EPNs have been reported to learn to respond to specific chemicals following exposure, which may enhance the approach of using these signals for increasing biocontrol (Willett *et al.*, 2018).

Another novel utilization of EPN signaling compounds is to employ nematode pheromones for improved biocontrol. Ascaroside pheromones have been found to influence EPN dispersal, infectivity and reproductive capacity (Oliveira-Hofman *et al.*, 2019; Shapiro-Ilan *et al.*, 2019; Wang *et al.*, 2022). In laboratory and greenhouse experiments, exposure of IJs to EPN pheromone extracts prior to application led to enhanced control of *Curculio caryae* and black soldier fly, *Hermetia illucens* L. (Diptera: Stratiomyidae) (Oliveira-Hofman *et al.*, 2019). In 2021 and 2022, field trials also confirmed the ability of pheromone "boosters" to improve *Curculio caryae* control in a pecan orchard (D.I. Shapiro-Ilan, unpublished data).

11.4 Conclusions

Although biocontrol success using EPNs on a commercial scale has been achieved in many arenas, EPNs still represent a very small component of the pesticide industry (Grewal *et al.*, 2005; Shapiro-Ilan *et al.*, 2023). Fundamentally, successful implementation of EPNs in biocontrol must be competitive with more conventional approaches in two aspects: cost and pest control efficacy. Alternatively, the EPN product must be able to do something that chemical pesticides cannot. For example, the ability of EPNs to locate and infect insect hosts deep in the soil avoids direct competition with chemical pesticides for this use. EPN product costs and quality can be ameliorated through improved mass production (see Chapter 10, this volume). Moreover, as indicated above, biocontrol efficacy can be enhanced via strain improvement as well as improved formulation and application techniques, which can also lead to reduced costs.

To expand the utility of EPNs in biocontrol, all the areas discussed above can be developed further with novel technology. Strain development can be enhanced further through more comprehensive breeding/improvement programs that target multiple traits. Formulations that protect EPNs from environmental extremes, for aboveground use and otherwise, will benefit from incorporating novel materials (e.g. nanoparticles), especially those that are less expensive and compatible with organic cropping systems. Unique combinations with other tactics and novel distribution techniques are needed. Elucidation of chemical/signaling biology will allow us to leverage these interactions toward improved biocontrol, for example in the use of call-for-help signals (Hiltpold and Turlings, 2012), systemic acquired resistance (Thompson *et al.*, 2022) and "boosters" that enhance biocontrol traits such as dispersal and infectivity (Mbata *et al.*, 2019; Oliveira-Hofman *et al.*, 2019; Kaplan *et al.*, 2020).

EPN application has largely utilized existing technology developed for chemical insecticides, and this has worked well because growers do not need to adopt new methodology to their existing practices. However, given that EPNs are live organisms, they are subject to various biological constraints that are not applicable to chemical insecticides. Thus, novel technology is needed to effectively apply EPNs in an efficacious, low-cost and user-friendly manner.

11.5 References

Acar, I., Sipes, B. (2022) Enhancing the biological control potential of *Steinernema feltiae* with protection from desiccation and UV radiation. *Biological Control* 169: 104874.

Acevedo, J.P.M., Samuels, R.I., Machado, I.R. and Dolinski, C. (2007) Interactions between isolates of the entomopathogenic fungus *Metarhizium anisopliae* and the entomopathogenic nematode *Heterorhabditis bacteriophora* JPM4 during infection of the sugar cane borer *Diatraea saccharalis* (Lepidoptera: Pyralidae). *Journal of Invertebrate Pathology* 96, 187–192.

Adhikari, B.N., Lin, C.Y., Bai, X., Ciche, T.A., Grewal, P.S. *et al.* (2009) Transcriptional profiling of trait deterioration in the insect pathogenic nematode *Heterorhabditis bacteriophora*. *BMC Genomics* 10: 609.

Alumai, A. and Grewal, P.S. (2004) Tank-mix compatibility of the entomopathogenic nematodes, *Heterorhabditis bacteriophora* and *Steinernema carpocapsae*, with selected chemical pesticides used in turfgrass. *Biocontrol Science and Technology* 14, 725–730.

Amizadeh, M., Hejazi, M.J., Niknam, G. and Askari-Saryazdi, G. (2019) Interaction between the entomopathogenic nematode, *Steinernema feltiae* and selected chemical insecticides for management of the tomato leafminer, *Tuta absoluta*. *BioControl* 64, 709–721.

Anbesse, S., Sumaya, N.H., Dörfler, A.V., Strauch, S. and Ehlers, R.-U. (2012) Selective breeding for desiccation tolerance in liquid culture provides genetically stable inbred lines of the entomopathogenic nematode *Heterorhabditis bacteriophora. Applied Microbiology and Biotechnology* 97, 731–739.

Anifantis, A., Marziale, R., Pascuzzi, S., Ravlic, J., Nikola, G. and Eustachio, T. (2020) Effect of hydrostatic pressure on the viability of EPN native Italian strains *Steinernema feltiae* and *Heterorhabditis bacteriophora*. *Redia* 103, 115–120.

Ansari, M.A. and Butt, T.M. (2011) Effect of potting media on the efficacy and dispersal of entomopathogenic nematodes for the control of black vine weevil, *Otiorhynchus sulcatus* (Coleoptera: Curculionidae). *Biological Control* 58, 310–318.

Ansari, M.A., Tirry, L. and Moens, M. (2004) Interaction between *Metarhizium anisopliae* CLO 53 and entomopathogenic nematodes for the control of *Hoplia philanthus*. *Biological Control* 31, 172–180.

Ansari, M.A., Shah, F.A. and Moens, T.M. (2006) Field trials against *Hoplia philanthus* (Coleoptera: Scarabaeidae) with a combination of an entomopathogenic nematode and the fungus *Metarhizium anisopliae* CLO 53. *Biological Control* 39, 453–459.

Arthurs, S., Heinz, K.M. and Prasifka, J.R. (2004) An analysis of using entomopathogenic nematodes against above-ground pests. *Bulletin of Entomological Research* 94, 297–306.

Askary, T.H. and Ahmad, M.J. (2021) Survival and virulence capacity of native strain of entomopathogenic nematode, *Steinernema cholashanense* in different formulations. *Agricultural Research* 10, 390–397.

Bai, X. and Grewal, P.S. (2007) Identification of two down-regulated genes in entomopathogenic nematode *Heterorhabditis bacteriophora* infective juveniles upon contact with insect hemolymph. *Molecular and Biochemical Parasitology* 156, 162–166.

Bai, C., Shapiro-Ilan, D.I., Gaugler, R. and Yi, S. (2004) Effect of entomopathogenic nematode concentration on survival during cryopreservation in liquid nitrogen. *Journal of Nematology* 36, 281–284.

Bai, C., Shapiro-Ilan, D.I., Gaugler, R. and Hopper, K.R. (2005) Stabilization of beneficial traits in *Heterorhabditis bacteriophora* through creation of inbred lines. *Biological Control* 32, 220–227.

Bai, X., Adams, B.J., Ciche, T.A., Clifton, S., Gaugler, R. *et al.* (2009) Transcriptomic analysis of the entomopathogenic nematode *Heterorhabditis bacteriophora* TTO1. *BMC Genomics* 10: 205.

Bai, X., Adams, B.J., Ciche, T.A., Clifton, S., Gaugler, R. *et al.* (2013) A lover and a fighter: the genome sequence of an entomopathogenic nematode *Heterorhabditis bacteriophora*. *PLOS ONE* 8: 69618.

Bal, K.H., Michel, P.A. and Grewal, P.S. (2014) Genetic selection of the ambush foraging entomopathogenic nematode, *Steinernema carpocapsae* for enhanced dispersal and its associated trade-offs. *Evolutionary Ecology* 28, 923–939.

Barbercheck, M.E. (1992) Effect of soil physical factors on biological control agents of soil insect pests. *Florida Entomologist* 75, 539–548.

Barbercheck, M.E. and Kaya, H.K. (1990) Interactions between *Beauveria bassiana* and the entomogenous nematodes *Steinernema feltiae* and *Heterorhabditis heliothidis*. *Journal of Invertebrate Pathology* 55, 225–234.

Beck, B., Brusselman, E., Nuyttens, D., Moens, M., Pollet, S. *et al.* (2013) Improving foliar applications of entomopathogenic nematodes by selecting adjuvants and spray nozzles, *Biocontrol Science and Technology* 23, 507–520.

Bednarek, A. and Gaugler, R. (1997) Compatibility of soil amendments with entomopathogenic nematodes. *Journal of Nematology* 29, 220–227.

Bellini, L. and Dolinski, C. (2012) Foliar application of entomopathogenic nematodes (Rhabditida: Steinernematidae and Heterorhabditidae) for the control of *Diatraea saccharalis* in greenhouse. *Semina: Ciências Agrárias* 33, 997–1004.

Berry, R.E., Liu, J. and Groth, E. (1997) Efficacy and persistence of *Heterorhabditis marelatus* (Rhabditida: Heterorhabditidae) against root weevils (Coleoptera: Curculionidae) in strawberry. *Environmental Entomology* 26, 465–470.

Bilgrami, A.L., Gaugler, R., Shapiro-Ilan, D.I. and Adams, B.J. (2006) Source of trait deterioration in entomopathogenic nematodes *Heterorhabditis bacteriophora* and *Steinernema carpocapsae* during *in vivo* culture. *Nematology* 8, 397–409.

Brinkman, M.A. and Gardner, W.A. (2001) Use of diatomaceous earth and entomopathogen combinations against the red imported fire ant (Hymenoptera: Formicidae). *Florida Entomologist* 84, 740–741.

Bruck, D.J. (2004) Natural occurrence of entomopathogens in Pacific Northwest nursery soils and their virulence to the black vine weevil, *Otiorhynchus sulcatus* (F.) (Coleoptera: Curculionidae). *Environmental Entomology* 33, 1335–1343.

Bruck, D.J., Shapiro-Ilan, D.I. and Lewis, E.E. (2005) Evaluation of application technologies of entomopathogenic nematodes for control of the black vine weevil, *Otiorhynchus sulcatus*. *Journal of Economic Entomology* 98, 1884–1889.

Brusselman, E., Beck, B., Temmerman, F., Pollet, S., Steurbaut, W. *et al.* (2010) The spray pattern of entomopathogenic nematodes. *American Society of Agricultural and Biological Engineers Annual International Meeting* 6, 5054–5067.

Campos-Herrera, R., Gómez-Ros, J.M., Escuer, M., Cuadra, L., Barrios, L. and Gutiérrez, C. (2008) Diversity, occurrence, and life characteristics of natural entomopathogenic nematode populations from La Rioja (Northern Spain) under different agricultural management and their relationships with soil factors. *Soil Biology and Biochemistry* 40, 1474–1484.

Chaston, J.M., Dillman, A.R., Shapiro-Ilan, D.I., Bilgrami, A.L., Gaugler, R. *et al.* (2011) Outcrossing and crossbreeding recovers deteriorated traits in laboratory cultured *Steinernema carpocapsae* nematodes. *International Journal of Parasitology* 41, 801–809.

Ciche, T.A. (2007) The biology and genome of *Heterorhabditis bacteriophora*. In: *WormBook: The Online Review of C.* elegans *Biology*. Wormbook, Pasadena, California. Available at: https://www.ncbi.nlm.nih.gov/books/NBK19785/ (accessed 18 March 2024).

Curran, J. (1992) Influence of application method and pest population size on field efficacy of entomopathogenic nematodes. *Journal of Nematology* 24, 631–636.

de Waal, J.Y., Malan, A.P. and Addison, M.F. (2011) Evaluating mulches together with *Heterorhabditis zealandica* (Rhabditida: Heterorhabditidae) for the control of diapausing codling moth larvae, *Cydia pomonella* (L.) (Lepidoptera: Tortricidae). *Biocontrol Science and Technology* 21, 255–270.

de Waal, J.Y., Malan, A.P. and Addison, M.F. (2013) Effect of humidity and a superabsorbent polymer formulation on the efficacy of *Heterorhabditis zealandica* (Rhabditida: Heterorhabditidae) to control codling moth, *Cydia pomonella* (L.) (Lepidoptera: Tortricidae), *Biocontrol Science and Technology* 23, 62–78.

Del Valle, E.E., Dolinski, C., Barreto, E.L.S. and Souza, R.M. (2009) Effect of cadaver coatings on emergence and infectivity of the entomopathogenic nematode *Heterorhabditis baujardi* LPP7 (Rhabditida: Heterorhabditidae) and the removal of cadavers by ants. *Biological Control* 50, 21–24.

Dito, D.F., Shapiro-Ilan, D.I., Dunlap, C.A., Behle, R.W. and Lewis, E.E. (2016) Enhanced biological control potential of the entomopathogenic nematode, *Steinernema carpocapsae*, applied with a protective gel formulation. *Biocontrol Science and Technology* 26, 835–848.

Dlamini, B.E., Malan, A.P. and Addison, P. (2020) Combined effect of entomopathogenic fungi and *Steinernema yirgalemense* against the banded fruit weevil, *Phlyctinus callosus* (Coleoptera: Curculionidae), *Biocontrol Science and Technology* 30, 1169–1179.

Dolinski, C. and Moino, A. Jr. (2006) Use of native and exotic entomopathogenic nematodes: the risk of introductions. *Nematologia Brasileira* 30, 139–149.

Dolinski, C., Pinto, C.C.S., Robaina, R.R. and Bellini, L.L. (2010) Effect of soil texture on the mobility of *Heterorhabditis baujardi* 'LPP7' (Rhabditida: Heterorhabditidae). *Nematologia Brasileira* 34, 123–128.

Dolinski, C., Shapiro-Ilan, D.I. and Lewis, E.E. (2015) Insect cadaver applications: pros and cons. In: Campos-Herrera, R. (ed.) *Nematode Pathogenesis of Insects and Other Pests: Ecology and Applied Technologies for Sustainable Plant and Crop Protection*. Springer, Cham, Switzerland, pp. 207–230.

Downing, A.S. (1994) Effect of irrigation and spray volume on efficacy of entomopathogenic nematodes (Rhabditida: Heterorhabditidae) against white grubs (Coleoptera: Scarabaeidae). *Journal of Economic Entomology* 87, 643–646.

Duchaud, E., Rusniok, C., Frangeul, L., Buchrieser, C., Givaudan, A. *et al.* (2003) The genome sequence of the entomopathogenic bacterium *Photorhabdus luminescens*. *Nature Biotechnology* 21, 1307–1313.

Duncan, L.W., Shapiro, D.I., McCoy, C.W. and Graham, J.H. (1999) Entomopathogenic nematodes as a component of citrus root weevil IPM. In: Polavarapu, S. (ed.) *Optimal use of Insecticidal Nematodes in Pest Management*. Rutgers University, New Brunswick, New Jersey, pp. 69–78.

Duncan, L.W, Dunn, D.C., Bague, G. and Nguyen, K. (2003) Competition between entomopathogenic and free-living bactivorous nematodes in larvae of the weevil *Diaprepes abbreviatus*. *Journal of Nematology* 35, 187–193.

Duncan, L.W., Stuart, R.J., El-Borai, F.E., Campos-Herrera, R., Pathak, E. *et al.* (2013) Modifying orchard planting sites conserves entomopathogenic nematodes, reduces weevil herbivory and increases citrus tree growth, survival and fruit yield. *Biological Control* 64, 26–36.

Eng, M.S., Preisser, E.L. and Strong, D.R. (2005) Phoresy of the entomopathogenic nematode *Heterorhabditis marelatus* by a non-host organism, the isopod *Porcellio scaber*. *Journal of Invertebrate Pathology* 88, 173–176.

Epsky, N.D., Walter, D.E. and Capinera, J.L. (1988) Potential role of nematophagous microarthropods as biotic mortality factors of entomogenous nematodes (Rhabditida: Steinernematidae, Heterorhabditidae). *Journal of Economic Entomology* 81, 821–825.

Fallet, P., Bazagwira, D., Guenat, J.M., Bustos-Seguraa, C., Karangwa, P. *et al.* (2022) Laboratory and field trials reveal the potential of a gel formulation of entomopathogenic nematodes for the biological control of fall armyworm caterpillars (*Spodoptera frugiperda*). *Biological Control* 176: 105086.

Fife, J.P., Derksen, R.C., Ozkan, H.E. and Grewal, P.S. (2003) Effects of pressure differentials on the viability and infectivity of entomopathogenic nematodes. *Biological Control* 27, 65–72.

Fife, J.P., Ozkan, H.E., Derksen, R.C., Grewal, P.S. and Krause, C.R. (2005) Viability of a biological pest control agent through hydraulic nozzles. *Transactions of the ASAE* 48, 45–54.

Fife, J.P., Ozkan, H.E., Derksen, R.C. and Grewal, P.S. (2006) Using computational fluid dynamics to predict damage of a biological pesticide during passage through a hydraulic nozzle. *Biosystems Engineering* 94, 387–396.

Fife, J.P., Ozkan, H.E., Derksen, R.C. and Grewal, P.S. (2007) Effects of pumping on entomopathogenic nematodes and temperature increase within a spray system. *Applied Engineering in Agriculture* 23, 405–412.

Gaugler, R. (1987). Entomogenous nematodes and their prospects for genetic improvement. In: Maramorosch, K. (ed.) *Biotechnology in Invertebrate Pathology and Cell Culture*. Academic Press, San Diego, pp. 457–484.

Gaugler, R. and Boush, G.M. (1978) Effects of ultraviolet radiation and sunlight on the entomogenous nematode *Neoaplectana carpocapsae*. *Journal of Invertebrate Pathology* 32, 291–296.

Gaugler, R., Campbell, J.F. and McGuire, T.R. (1989) Selection for host-finding in *Steinernema feltiae*. *Journal of Invertebrate Pathology* 54, 363–372.

Gaugler, R., Campbell, J.F. and McGuire, T.R. (1990) Fitness of a genetically improved entomopathogenic nematode. *Journal of Invertebrate Pathology* 56, 106–116.

Gaugler, R., Wilson, M. and Shearer, P. (1997) Field release and environmental fate of a transgenic entomopathogenic nematode. *Biological Control* 9, 75–80.

Georgis, R. (1990) Formulation and application technology. In: Gaugler, R. and Kaya, H.K. (eds.) *Entomopathogenic Nematodes in Biological Control.* CRC Press, Boca Raton, Florida, pp. 173–194.

Georgis, R. and Gaugler, R. (1991) Predictability in biological control using entomopathogenic nematodes. *Journal of Economic Entomology* 84, 713–720.

Georgis, R. and Poinar, G.O. Jr. (1983) Effect of soil texture on the distribution and infectivity of *Neoaplectana carpocapsae* (Nematode: Steinernematidae). *Journal of Nematology* 15, 308–311.

Georgis, R., Dunlop, D.B. and Grewal, P.S. (1995) Formulation of entomopathogenic nematodes. In: Hall, F.R. and Barry, J.W. (eds.) *Biorational Pest Control Agents: Formulation and Delivery*. American Chemical Society, Washington, DC, pp. 197–205.

Glazer, I., Salame, L. and Segal, D. (1997) Genetic enhancement of nematicide resistance in entomopathogenic nematodes. *Biocontrol Science and Technology* 7, 499–512.

Grewal, P.S. (2002) Formulation and application technology. In: Gaugler, R. (ed.) *Entomopathogenic Nematology*. CAB International, Wallingford, UK, pp. 265–287.

Grewal, P.S. and Peters, A. (2005) Formulation and quality. In: Grewal, P.S., Ehlers, R.-U. and Shapiro-Ilan, D.I. (eds.) *Nematodes as Biological Control Agents*. CAB International, Wallingford, UK, pp. 79–90.

Grewal, P.S., Selvan, S. and Gaugler, R. (1994) Thermal adaptation of entomopathogenic nematodes: niche breadth for infection, establishment and reproduction. *Journal of Thermal Biology* 19, 245–253.

Grewal, P.S., Ehlers, R.-U. and Shapiro-Ilan, D.I. (eds.) (2005) *Nematodes as Biological Control Agents*. CAB International, Wallingford, UK.

Gulzar, S., Usman, M., Wakil, W., Gulcu, B., Hazir, C. *et al.* (2020) Environmental tolerance of entomopathogenic nematodes differs among nematodes arising from host cadavers versus aqueous suspension. *Journal of Invertebrate Pathology* 175: 107452.

Gumus, A., Karagoz, M., Shapiro-Ilan, D.I. and Hazir, S. (2015) A novel approach to biocontrol: release of live insect hosts pre-infected with entomopathogenic nematodes. *Journal of Invertebrate Pathology* 130, 56–60.

Hayes, A.E., Fitzpatrick, S.M. and Webster, J.M. (1999). Infectivity, distribution, and persistence of the entomopathogenic nematode *Steinernema carpocapsae* All strain (Rhabditida: Steinernematidae) applied by sprinklers or boom sprayer to dry-pick cranberries. *Journal of Economic Entomology* 92, 539–546.

Hazir, S., Kaya, H., Touray, M., Cimen, H. and Shapiro-Ilan, D. (2022) Basic laboratory and field manual for conducting research with the entomopathogenic nematodes, *Steinernema* and *Heterorhabditis*, and their bacterial symbionts. *Turkish Journal of Zoology* 46: 1.

Head, J., Lawrence, A.J. and Walters, K.F.A. (2004) Efficacy of the entomopathogenic nematode, *Steinernema feltiae*, against *Bemisia tabaci* in relation to plant species. *Journal of Applied Entomology* 128, 543–547.

Hiltpold, I. and Turlings, T.C.J. (2012) Manipulation of chemically mediated interactions in agricultural soils to enhance the control of crop pests and to improve crop yield. *Journal of Chemical Ecology* 38, 641–650.

Hiltpold, I., Hibbard, B.E., French, B.W. and Turlings, T.C.J. (2012) Capsules containing entomopathogenic nematodes as a Trojan horse approach to control the western corn rootworm. *Plant and Soil* 358, 11–25.

Hopper, K.R., Roush, R.T. and Powell, W. (1993) Management of genetics of biological-control introductions. *Annual Review of Entomology* 38, 27–51.

Kagimu, N. and Malan, A.P. (2019) Formulation of South African entomopathogenic nematodes using alginate beads and diatomaceous earth. *BioControl* 64, 413–422.

Kaplan, F., Perret-Gentil, A., Giurintano, J., Stevens, G., Erdogan, H. *et al.* (2020) Conspecific and heterospecific pheromones stimulate dispersal of entomopathogenic nematodes during quiescence. *Scientific Reports* 10: 5738.

Kaspi, R., Ross, A., Hodson, A.K., Stevens, G., Kaya, H.K. and Lewis, E. (2010) Foraging efficacy of the entomopathogenic nematode *Steinernema riobrave* in different soil types from California citrus groves. *Applied Soil Ecology* 45, 243–253.

Kaya, H.K. (1990) Soil ecology. In: Gaugler, R. and Kaya, H.K. (eds.) *Entomopathogenic Nematodes in Biological Control*. CRC Press, Boca Raton, Florida, pp. 93–116.

Kaya, H.K. (2002) Natural enemies and other antagonists. In Gaugler, R. (ed.) *Entomopathogenic Nematology*. CAB International, Wallingford, UK, pp. 189–204.

Kaya, H.K. and Nelsen, C.E. (1985) Encapsulation of steinernematid and heterorhabditid nematodes with calcium alginate: a new approach for insect control and other applications. *Environmental Entomology* 14, 572–574.

Khathwayo, Z., Ramakuwela, T., Hatting, J., Shapiro-Ilan, D.I. and Cochrane, N. (2021) Quantification of pH tolerance levels among entomopathogenic nematodes. *Journal of Nematology* 53: e2021-62.

Kim, J., Hiltpold, I., Jaffuel, G., Sbaiti, I., Hibbard, B.E. and Turlings, T.C.J. (2021) Calcium-alginate beads as a formulation for the application of entomopathogenic nematodes to control rootworms. *Journal of Pest Science* 94, 1197–1208.

Klein, M.G. and Georgis, R. (1992) Persistence of control of Japanese beetle (Coleoptera: Scarabaeidae) larvae with steinernematid and heterorhabditid nematodes. *Journal of Economic Entomology* 85, 727–730.

Klein, M.G. and Georgis, R. (1994) Application techniques for entomopathogenic nematodes. In: *Proceedings of the VI International Colloquium on Invertebrate Pathology and Microbial Control, Montpellier, France*, pp. 483–484.

Koppenhöfer, A.M. and Fuzy, E.M. (2008) Early timing and new combinations to increase the efficacy of neonicotinoid–entomopathogenic nematode (Rhabditida: Heterorhabditidae) combinations against white grubs (Coleoptera: Scarabaeidae). *Pest Management Science* 64, 725–735.

Koppenhöfer, A.M. and Grewal, P.S. (2005) Compatibility and interactions with agrochemicals and other biocontrol agents. In: Grewal, P.S., Ehlers, R.U. and Shapiro-Ilan, D.I. (eds.) *Nematodes as Biocontrol Agents*. CAB International, Wallingford, UK, pp. 363–381.

Koppenhöfer, A.M. and Kaya, H.K. (1997) Additive and synergistic interactions between entomopathogenic nematodes and *Bacillus thuringiensis* for scarab grub control. *Biological Control* 8, 131–137.

Koppenhöfer, A.M., Kaya, H.K. and Taormino, S.P. (1995) Infectivity of entomopathogenic nematodes (Rhabditida: Steinernematidae) at different soil depths and moistures. *Journal of Invertebrate Pathology* 65, 193–199.

Kotliarevski, L., Cohen, R., Ramakrishnan, J., Wu, S., Mani, K.A. *et al.* (2022) Individual coating of entomopathogenic nematodes with titania (TiO_2) nanoparticles based on oil-in-water pickering emulsion: a new formulation for biopesticides. *Journal of Agricultural and Food Chemistry* 70, 13518–13527.

Kruitbos, L.M., Heritage, S., Hapca, S. and Wilson, M.J. (2010) The influence of habitat quality on the foraging strategies of the entomopathogenic nematodes *Steinernema carpocapsae* and *Heterorhabditis megidis*. *Parasitology* 137, 303–309.

Kung, S., Gaugler, R. and Kaya, H.K. (1990) Influence of soil pH and oxygen on persistence of *Steinernema* spp. *Journal of Nematology* 22, 440–445.

Lacey, L.A., Arthurs, S.P., Unruh, T.R., Headrick, H. and Fritts, R. Jr. (2006a) Entomopathogenic nematodes for control of codling moth (Lepidoptera: Tortricidae) in apple and pear orchards: effect of nematode species and seasonal temperatures, adjuvants, application equipment and post-application irrigation. *Biological Control* 37, 214–223.

Lacey, L.A., Arthurs, S.P., Granatstein, D., Headrick, H. and Fritts, R. Jr. (2006b) Use of entomopathogenic nematodes (Steinernematidae) in conjunction with mulches for control of codling moth (Lepidoptera: Torticidae). *Journal of Entomological Science* 41, 107–119.

Lacey, L.A., Shapiro-Ilan, D.I. and Glenn, G.M. (2010) Post-application of anti-desiccant agents improves efficacy of entomopathogenic nematodes in formulated host cadavers or aqueous suspension against diapausing codling moth larvae (Lepidoptera: Tortricidae). *Biocontrol Science and Technology* 20, 909–921.

Lankin, G., Vidal-Retes, G., Allende, G., Castaneda-Alvarez, C., San-Blas, E. and Aballay, E. (2020) Soil texture, infective juvenile concentration, and soil organic matter influence the efficacy of *Steinernema feltiae* isolate Lican Ray. *Journal of Nematology* 52: e2020-07.

Leite, L.G., Shapiro-Ilan, D.I. and Hazir, S. (2018) Survival of *Steinernema feltiae* in different formulation substrates: improved longevity in a mixture of gel and vermiculite, *Biological Control* 126, 192–197.

Lewis, E.E. (2002) Behavioral ecology. In Gaugler, R. (ed.) *Entomopathogenic Nematology*. CAB International, Wallingford, UK, pp. 205–224.

Lewis, E.E. and Clarke, D.J. (2012) Nematode parasites and entomopathogens. In: Vega F.E. and Kaya, H.K. (eds.) *Insect Pathology*, 2nd edn. Elsevier, Amsterdam, pp. 395–424.

Lewis, E.E., Gaugler, R. and Harrison, R. (1992) Entomopathogenic nematode host finding: response to host contact cues by cruise and ambush foragers. *Parasitology* 105, 309–319.

Malan, A.P., Knoetze, R. and Moore, S.D. (2011) Isolation and identification of entomopathogenic nematodes from citrus orchards in South Africa and their biocontrol potential against false codling moth. *Journal of Invertebrate Pathology* 108, 115–125.

Matadamas-Ortiz, P.T., Ruiz-Vega, J., Vazquez-Feijoo, J.A., Cruz-Martínez, H. and Cortés-Martínez, C.I. (2014) Mechanical production of pellets for the application of entomopathogenic nematodes: factors that determine survival time of *Steinernema glaseri*. *Biocontrol Science and Technology* 24, 145–157.

Mbata, G.N., Li, Y. and Shapiro-Ilan, D.I. (2022) Evaluation of chemical and microbial control options for *Pangaeus bilineatus* (Say) (Hemiptera: Cydnidae) infesting peanut crop. *Pest Management Science* 78, 4719–4727.

Mbata, G.N., Shapiro-Ilan, D., Alborn, H.T. and Strand, M.R. (2019) Preferential infectivity of entomopathogenic nematodes in an envenomed host. *International Journal of Parasitology* 49, 737–745.

McCoy, C.W., Shapiro, D.I., Duncan, L.W. and Nguyen, K. (2000) Entomopathogenic nematodes and other natural enemies as mortality factors for larvae of *Diaprepes abbreviatus* (Coleoptera: Curculionidae). *Biological Control* 19, 182–190.

Miklasiewicz, T.J., Grewal, P.S., Hoy, C.W. and Malik, V.S. (2002) Evaluation of entomopathogenic nematodes for suppression of carrot weevil. *BioControl* 47, 545–561.
Millar, L.C. and Barbercheck, M.E. (2001) Interaction between endemic and introduced entomopathogenic nematodes in conventional-till and no-till corn. *Biological Control* 22, 235–245.
Molyneux, A.S. and Bedding, R.A. (1984) Influence of soil texture and moisture on the infectivity of *Heterorhabditis* sp. D1 and *Steinernema glaseri* for larvae of the sheep blowfly *Lucilia cuprina*. *Nematologica* 30, 358–365.
Mukuka, J., Strauch, O., Hoppe, C. and Ehlers, R.-U. (2010) Improvement of heat and desiccation tolerance in *Heterorhabditis bacteriophora* through cross-breeding of tolerant strains and successive genetic selection. *BioControl* 55, 511–521.
Nielsen, A.L. and Lewis, E.E. (2012) Designing the ideal habitat for entomopathogen use in nursery production. *Pest Management Science* 68, 1053–1061.
Oliveira-Hofman, C., Kaplan, F., Stevens, G., Lewis, E.E., Wu, S. *et al.* (2019) Pheromone extracts act as boosters for entomopathogenic nematodes efficacy. *Journal of Invertebrate Pathology* 164, 38–42.
Parkman, J.P., Frank, J.H., Nguyen, K.B. and Smart, G.C. Jr. (1994) Inoculative release of *Steinernema scapterisci* (Rhabditida: Steinernematidae) to suppress pest mole crickets (Orthoptera: Gryllotapidae) on golf courses. *Environmental Entomology* 23, 1331–1337.
Perez, E.E., Lewis, E.E. and Shapiro-Ilan, D.I. (2003) Impact of host cadaver on survival and infectivity of entomopathogenic nematodes (Rhabditida: Steinernematidae and Heterorhabditidae) under desiccating conditions. *Journal of Invertebrate Pathology* 82, 111–118.
Pinero, J.C., Shapiro-Ilan, D., Cooley, D.R., Tuttle, A.F., Eaton, E. *et al.* (2020) Toward the integration of an attract-and-kill approach with biological control involving entomopathogenic nematodes to control multiple life stages of plum curculio (Coleoptera: Curculionidae) in eastern North America. *Insects* 11: 375.
Ramakrishnan, J., Salame, L., Nasser, A., Glazer, I. and Ment, D. (2022) Survival and efficacy of entomopathogenic nematodes on exposed surfaces. *Scientific Reports* 12: 4629.
Ramalingam, K.R., Hazir, C., Gumus, A., Asan, C., Karagoz, M. and Hazir, S. (2015) Efficacy of the entomopathogenic nematode, *Heterorhabditis bacteriophora*, using different application methods in the presence or absence of a natural enemy. *Turkish Journal of Agriculture and Forestry* 39, 277–285.
Ruiz-Vega, J., Cortés-Martínez, C.I. and García-Gutiérrez, C. (2018) Survival and infectivity of entomopathogenic nematodes formulated in sodium alginate beads. *Journal of Nematology* 50, 273–280.
Schroer, S. and Ehlers, R.-U. (2005) Foliar application of the entomopathogenic nematode *Steinernema carpocapsae* for biological control of diamondback moth larvae (*Plutella xylostella*). *Biological Control* 33, 81–86.
Schwartz, H.T., Antoshechkin, I. and Sternberg, P.W. (2011) Applications of high-throughput sequencing to symbiotic nematodes of the genus *Heterorhabditis*. *Symbiosis* 55, 111–118.
Shapiro, D.I. and Glazer, I. (1996) Comparison of entomopathogenic nematode dispersal from infected hosts versus aqueous suspension. *Environmental Entomology* 25, 1455–1461.
Shapiro, D.I. and Lewis, E.E. (1999) Comparison of entomopathogenic nematode infectivity from infected hosts versus aqueous suspension. *Environmental Entomology* 28, 907–911.
Shapiro-Ilan, D.I. and Raymond, B. (2016) Limiting opportunities for cheating stabilizes virulence in insect parasitic nematodes. *Evolutionary Applications* 9, 462–470.
Shapiro, D.I., Tylka, G.L., Berry, E.C. and Lewis, L.C. (1995) Effects of earthworms on the dispersal of *Steinernema* spp. *Journal of Nematology* 27, 21–28.
Shapiro, D.I., Tylka, G.L. and Lewis, L.C. (1996a) Effects of fertilizers on virulence of *Steinernema carpocapsae*. *Applied Soil Ecology* 3, 27–34.
Shapiro, D.I., Glazer, I. and Segal, D. (1996b) Trait stability in and fitness of the heat tolerant entomopathogenic nematode *Heterorhabditis bacteriophora* IS5 strain. *Biological Control* 6, 238–244.
Shapiro, D.I., Glazer, I. and Segal, D. (1997) Genetic improvement of heat tolerance in *Heterorhabditis bacteriophora* through hybridization. *Biological Control* 8, 153–159.
Shapiro, D.I., Obrycki, J.J., Lewis, L.C. and Abbas, M. (1999a) The effects of fertilizers on black cutworm, *Agrotis ipsilon* (Lepidoptera: Noctuidae) suppression by *Steinernema carpocapsae*. *Journal of Nematology* 31, 690–693.
Shapiro, D.I., Obrycki, J.J., Lewis, L.C. and Jackson, J.J. (1999b) Effects of crop residue on the persistence of *Steinernema carpocapsae*. *Journal of Nematology* 31, 517–519.
Shapiro, D.I., McCoy, C.W., Fares, A., Obreza, T. and Dou, H. (2000) Effects of soil type on virulence and persistence of entomopathogenic nematodes in relation to control of *Diaprepes abbreviatus*. *Environmental Entomology* 29, 1083–1087.

Shapiro-Ilan, D.I. and Brown, I. (2013) Earthworms as phoretic hosts for *Steinernema carpocapsae* and *Beauveria bassiana*: implications for enhanced biological control. *Biological Control* 66, 41–48.

Shapiro-Ilan, D.I. and Dolinski, C. (2015) Entomopathogenic nematode application technology. In: Campos-Herrera, R. (ed.) *Nematode Pathogenesis of Insects and Other Pests: Ecology and Applied Technologies for Sustainable Plant and Crop Protection*. Springer, Cham, Switzerland, pp. 231–534.

Shapiro-Ilan, D.I., Lewis, E.E., Behle, R.W. and McGuire, M.R. (2001) Formulation of entomopathogenic nematode-infected cadavers. *Journal of Invertebrate Pathology* 78, 17–23.

Shapiro-Ilan, D.I., Gouge, D.H. and Koppenhöfer, A.M. (2002) Factors affecting commercial success: case studies in cotton, turf and citrus. In: Gaugler, R. (ed.) *Entomopathogenic Nematology*. CAB International, Wallingford, UK, pp. 333–355.

Shapiro-Ilan, D.I., Gardner, W.A., Fuxa, J.R., Wood, B.W., Nguyen, K.B. *et al.* (2003a) Survey of entomopathogenic nematodes and fungi endemic to pecan orchards of the Southeastern United States and their virulence to the pecan weevil (Coleoptera: Curculionidae). *Environmental Entomology* 32, 187–195.

Shapiro-Ilan, D.I., Lewis, E.E., Tedders, W.L. and Son, Y. (2003b) Superior efficacy observed in entomopathogenic nematodes applied in infected-host cadavers compared with application in aqueous suspension. *Journal of Invertebrate Pathology* 83, 270–272.

Shapiro-Ilan, D.I., Mizell, R.F. III, Cottrell, T.E. and Horton, D.L. (2004a) Measuring field efficacy of *Steinernema feltiae* and *Steinernema riobrave* for suppression of plum curculio, *Conotrachelus nenuphar*, larvae. *Biological Control* 30, 496–503

Shapiro-Ilan, D.I., Jackson, M., Reilly, C.C. and Hotchkiss, M.W. (2004b) Effects of combining an entomopathogenic fungi or bacterium with entomopathogenic nematodes on mortality of *Curculio caryae* (Coleoptera: Curculionidae). *Biological Control* 30, 119–126.

Shapiro-Ilan, D.I., Stuart, R.J. and McCoy, C.W. (2005) Targeted improvement of *Steinernema carpocapsae* for control of the pecan weevil, *Curculio caryae* (Horn) (Coleoptera: Curculionidae) through hybridization and bacterial transfer. *Biological Control* 34, 215–221.

Shapiro-Ilan, D.I., Gouge, G.H., Piggott, S.J. and Patterson Fife, J. (2006a) Application technology and environmental considerations for use of entomopathogenic nematodes in biological control. *Biological Control* 38, 124–133.

Shapiro-Ilan, D.I., Stuart, R.J. and McCoy, C.W. (2006b) A comparison of entomopathogenic nematode longevity in soil under laboratory conditions. *Journal of Nematology* 38, 119–129.

Shapiro-Ilan, D.I., Cottrell, T.E., Mizell, R.F. III, Horton, D.L. and Davis J. (2009) A novel approach to biological control with entomopathogenic nematodes: prophylactic control of the peachtree borer, *Synanthedon exitiosa*. *Biological Control* 48, 259–263.

Shapiro-Ilan, D.I., Cottrell, T.E., Mizell, R.F., Horton, D.L, Behle, R.W. and Dunlap, C.A. (2010a) Efficacy of *Steinernema carpocapsae* for control of the lesser peach tree borer, *Synanthedon pictipes*: improved aboveground suppression with a novel gel application. *Biological Control* 54, 23–28.

Shapiro-Ilan, D.I., Morales-Ramos, J.A., Rojas, M.G. and Tedders, W.L. (2010b) Effects of a novel entomopathogenic nematode-infected host formulation on cadaver integrity, nematode yield, and suppression of *Diaprepes abbreviatus* and *Aethina tumida* under controlled conditions. *Journal of Invertebrate Pathology* 103, 103–108.

Shapiro-Ilan, D.I., Leskey, T.C. and Wright, S.E. (2011a) Virulence of entomopathogenic nematodes to plum curculio, *Conotrachelus nenuphar*: effects of strain, temperature, and soil type. *Journal of Nematology* 43, 187–195.

Shapiro-Ilan, D.I., Cottrell, T.E. and Wood, B.W. (2011b) Effects of combining microbial and chemical insecticides on mortality of the pecan weevil (Coleoptera: Curculionidae). *Journal of Economic Entomology* 104, 14–20.

Shapiro-Ilan, D.I., Han, R. and Dolinski, C. (2012a) Entomopathogenic nematode production and application technology. *Journal of Nematology* 44, 206–217.

Shapiro-Ilan, D.I., Bruck, D.J. and Lacey, L.A. (2012b) Principles of epizootiology and microbial control. In: Vega, F.E. and Kaya, H.K. (eds.) *Insect Pathology*, 2nd edn. Elsevier, Amsterdam, pp. 29–72.

Shapiro-Ilan, D.I., Wright, S.E., Tuttle, A.F., Cooley, D.R. and Leskey, T.C. (2013) Using entomopathogenic nematodes for biological control of plum curculio, *Conotrachelus nenuphar*: effects of irrigation and species in apple orchards. *Biological Control* 67, 123–129.

Shapiro-Ilan, D.I., Brown, I. and Lewis, E.E. (2014a) Freezing and desiccation tolerance in entomopathogenic nematodes: diversity and correlation of traits. *Journal of Nematology* 46, 27–34.

Shapiro-Ilan, D.I., Blackburn, D., Duncan, L., El-Borai, F.E., Koppenhöfer, H. *et al.* (2014b) Characterization of biocontrol traits in *Heterorhabditis floridensis*: a species with broad temperature tolerance. *Journal of Nematology* 46, 336–345.

Shapiro-Ilan, D.I., Hazir, S. and Leite, L. (2015) Viability and virulence of entomopathogenic nematodes exposed to ultraviolet radiation. *Journal of Nematology* 47, 184–189.

Shapiro-Ilan, D.I., Morales-Ramos, J.A. and Rojas, M.G. (2016a) *In vivo* production of entomopathogenic nematodes. In: Glare, T.R. and Moran-Diez, M.E. (eds.) *Microbial-Based Biopesticides: Methods and Protocols*. Methods in Molecular Biology, Vol. 1477. Humana Press, New York, pp. 137–158.

Shapiro-Ilan, D.I., Cottrell, T.E., Mizell, R.F. III and Horton, D.L. (2016b) Curative control of the peachtree borer using entomopathogenic nematodes. *Journal of Nematology* 48, 170–176.

Shapiro-Ilan, D.I., Cottrell, T.E., Mizell, R.F. III and Horton, D.L. (2016c) Efficacy of *Steinernema carpocapsae* plus fire gel applied as a single spray for control of the lesser peachtree borer, *Synanthedon pictipes*. *Biological Control* 94, 33–36.

Shapiro-Ilan, D.I., Hazir, S. and Glazer, I. (2017) Basic and applied research: entomopathogenic nematodes. In: Lacey, L.A. *Microbial Agents for Control of Insect Pests: From Discovery to Commercial Development and Use*. Academic Press, Amsterdam, pp. 91–105.

Shapiro-Ilan, D.I., Kaplan, F., Oliveira-Hofman, C., Schliekelman, P., Alborn, H.T. and Lewis, E.E. (2019) Conspecific pheromone extracts enhance entomopathogenic infectivity. *Journal of Nematology* 51: e2019-82.

Shapiro-Ilan, D.I., Garrigos Leite, L. and Han, R. (2023) Production of entomopathogenic nematodes. In: Morales-Ramos, J., Rojas, G. and Shapiro- Ilan, D.I. (eds.) *Mass Production of Beneficial Organisms: Invertebrates and Entomopathogens*. Academic Press, San Diego, California, pp. 293–316.

Sharifi-Far, S., Shapiro-Ilan, D.I., Brownbridge, M. and Hallett, R.H. (2018) The combined approach of strain discovery and the inbred line technique for improving control of *Delia radicum* with *Heterorhabditis bacteriophora*. *Biological Control* 118, 37–43.

Shields, E.J., Testa, A., Miller, J.M. and Flanders, K.L. (1999) Field efficacy and persistence of the entomopathogenic nematodes *Heterorhabditis bacteriophora* 'Oswego' and *H. bacteriophora* 'NC' on Alfalfa snout beetle larvae (Coleoptera: Curculionidae). *Environmental Entomology* 28, 128–136.

Shields E.J., Testa, A.M. and O'Neil, W.J. (2018) Long-term persistence of native New York entomopathogenic nematode isolates across crop rotation. *Journal of Economic Entomology* 111, 2592– 2598.

Tanada, Y. and Kaya, H.K. (1993) *Insect Pathology*. Academic Press, San Diego, California.

Thompson, M.N., Grunseich, J.M., Marmolejo, L.O., Aguirre, N.M., Bradicich, P.A. *et al.* (2022) Undercover operation: belowground insect herbivory modifies systemic plant defense and repels aboveground foraging insect herbivores. *Frontiers in Ecology and Evolution* 10: 1033730.

Thurston, G.S., Kaya, H.K. and Gaugler, R. (1994) Characterizing the enhanced susceptibility of milky disease-infected scarabaeid grubs to entomopathogenic nematodes. *Biological Control* 4, 67–73.

Toepfer, S., Gueldenzoph, C., Ehlers, R.U. and Kuhlmann, U. (2005) Screening of entomopathogenic nematodes for virulence against the invasive western corn rootworm, *Diabrotica virgifera virgifera* (Coleoptera: Chrysomelidae) in Europe. *Bulletin of Entomological Research* 95, 473–482.

Toepfer, S., Peters, A., Ehlers, R.-U. and Kuhlmann, U. (2008) Comparative assessment of the efficacy of entomopathogenic nematode species at reducing western corn rootworm larvae and root damage in maize. *Journal of Applied Entomology* 132, 337–348.

Touray, M., Gulcu, B., Ulug, D., Gulsena, S.H., Cimen, H. *et al.* (2020) Evaluation of different sponge types on the survival and infectivity of stored entomopathogenic nematodes, *Journal of Invertebrate Pathology* 171: 107332.

Ulug, D., Hazir, S., Kaya, H.K. and Lewis, E.E. (2014) Natural enemies of natural enemies: the potential top-down impact of predators on entomopathogenic nematode populations. *Ecological Entomology* 39, 462–469.

Wakil, W., Yasin, M. and Shapiro-Ilan, D. (2017) Effects of single and combined applications of entomopathogenic fungi and nematodes against *Rhynchophorus ferrugineus* (Olivier). *Scientific Reports* 7: 5971.

Wang, J., Cao, L., Huang, Z., Gu, X., Cui, Y. *et al.* (2022) Influence of the ascarosides on the recovery, yield and dispersal of entomopathogenic nematodes. *Journal of Invertebrate Pathology* 188: 107717.

Wang, X. and Grewal, P.S. (2002) Rapid genetic deterioration of environmental tolerance and reproductive potential of an entomopathogenic nematode during laboratory maintenance. *Biological Control* 23, 71–78.

Wang, Y., Bilgrami, A.L., Shapiro-Ilan, D. and Gaugler, R. (2007) Stability of entomopathogenic bacteria, *Xenorhabdus nematophila* and *Photorhabdus luminescens*, during *in vitro* culture. *Journal of Industrial Microbiology and Biotechnology* 34, 73–81.

Willett, D.S., Alborn, H.T., Stelinski, L.L. and Shapiro-Ilan, D.I. (2018) Risk taking of educated nematodes. *PLOS ONE* 13: e0205804.

Williams, E.C. and Walters, K.F.A. (2000) Foliar application of the entomopathogenic nematode *Steinernema feltiae* against leafminers on vegetables. *Biocontrol Science and Technology* 10, 61–70.

Womersley, C.Z. (1993) Factors affecting physiological fitness and modes of survival employed by dauer juveniles and their relationship to pathogenicity. In Bedding, R.A., Akhurst, R. and Kaya, H.K. (eds.) *Nematodes and the Biological Control of Insect Pests*. CSIRO Publications, East Melbourne, Australia, pp. 79–88.

Wong, C., Oliveira-Hofman, C., Blaauw, B., Chavez, D.J., Jagdale, G. *et al.* (2022) Control of peachtree borer (*Synanthedon exitiosa*) using the nematode *Steinernema carpocapsae*: optimization of application rates and secondary benefits in control of root-feeding weevils. *Agronomy* 12: 2689.

Wright, D.J., Peters, A., Schroer, S. and Fife, J.P. (2015) Application technology. In: Grewal, P.S., Ehlers, R.U. and Shapiro-Ilan, D.I. (eds.) *Nematodes as Biocontrol Agents*. CAB International, Wallingford, UK, pp. 91–106.

Wu, S., Mechrez, G., Ment, M., Toews, M.D., Ananth, K. *et al.* (2022) Tolerance of *Steinernema carpocapsae* infective juveniles in novel nanoparticle formulations to ultraviolet radiation. *Journal of Invertebrate Pathology* 196: 107851.

Zhu, H., Grewal, P.S. and Reding, M.E. (2011) Development of a desiccated cadaver delivery system to apply entomopathogenic nematodes for control of soil pests. *Applied Engineering in Agriculture* 27, 317–324.

12 Safety and Regulation of Entomopathogenic Nematodes

Diana K. Londoño[1]* and Patrick Frettinger[2]

[1]BASF Corporation, Research Triangle Park, North Carolina, USA; [2]BASF France SAS, Ecully, France

Abstract

Biological insecticides are gaining momentum in multiple countries and are applied as solo tactics and within integrated pest-management programs. Entomopathogenic nematodes (EPNs) are one of the biological insecticides that are used widely. They are safer to the environment than chemical pesticides and can easily be combined with other control tactics. Many studies have been conducted to gauge the effects of these beneficial nematodes on non-target organisms; multiple studies using *Steinernema* and *Heterorhabditis* spp. have shown limited or no effects on non-target insects. For registration, no harmonized regulation systems exist across nations or regions around the world. Timelines to regulatory approval are very different from country to country, and it can take years to obtain approval. Many countries have implemented restrictions on the release of exotic species, which can affect the acceptability of EPN species produced in one country for sale in another. Most regulations allow the release of EPNs to provide farmers with biological solutions to control pests, but there are exceptions.

12.1 Introduction

Interest in using safer pesticides is ever increasing. Thus, biological insecticides are gaining momentum in multiple countries and are applied both as solo tactics and as components within integrated pest-management programs. Entomopathogenic nematodes (EPNs) comprise a group of diverse and specialized parasites that are used to control a wide variety of economically important insect pests and are one of the biological insecticides that are used widely. EPNs are safer to the environment than chemical pesticides and can easily be combined with other control tactics. Many studies have been conducted to gauge the effects of EPNs on non-target organisms; multiple studies using *Steinernema* and *Heterorhabditis bacteriophora* have shown limited or no effects on non-target insects (Georgis *et al.*, 1991). Under laboratory conditions, usually confined-area bioassays, EPNs have been shown to give different levels of control to a wide range of insect species. However, the number of species that can be killed under field conditions is reduced due to insect behavior and crop and environmental factors, as well as spatial and temporal differences between the applied EPN and the insects (application error). Smits (1996) underlined that EPNs are unlikely to produce adverse effects on non-targets due to the short period of persistence and their need to recycle frequently in the host to survive.

*Email: diana.londono@basf.com

 Entomopathogenic Nematodes as Biological Control Agents (D.I. Shapiro-Ilan and E.E. Lewis eds)
DOI: 10.1079/9781800620322.0012

Through the host insect, the EPNs can persist for years; however, when the host is not present, their survival declines by approximately 40% in the first few hours after application due to ultraviolet radiation (UV) and desiccation, after which other factors affect survival such as predation and depletion of energy (Smits, 2010).

Studies have been done to determine the effect of EPN applications on soil microbial biomass. De Nardo *et al.* (2006) found no negative effects in soil microbial biomass but an increase in ammonium in soil samples after the application of *Steinernema carpocapsae*, showing an indirect effect on soil food webs. The potential environmental impacts of EPNs were reviewed by Barbercheck and Millar (2000). They concluded that the non-target species most likely to be affected are those with soil-dwelling stages that are active when the nematodes are applied. However, only small reductions in field populations of non-target organisms have been documented (reviewed by Akhurst and Smith, 2002). Despite widespread application, no significant acute or chronic toxicity to humans or other vertebrates has been reported, and no significant long-term impact on non-target invertebrate populations has been established. Therefore, Akhurst and Smith (2002) concluded that there were no safety concerns that prevent continued use of EPNs in their current state as biological control agents.

12.1.1 Key elements influencing EPN efficacy

Nematode survival and infectivity are influenced by different factors such as soil composition, moisture retention, oxygen supply and texture (Toepfer *et al.*, 2010). Other factors need to be considered such as host range, application rate (affected by the high price of the nematodes) and time of application considering compatibility with other pesticides applied as a mixture or applied separately. The distance that the nematodes can travel depends on soil type and behavior, and therefore dispersal can play a role in the potential for encounters with non-target organisms. Persistence must be determined as part of the risk assessment. Gaugler *et al.* (1992) reported that *Steinernema glaseri* was still present in the soil 50 years after it was applied in 1939–1942 in high numbers to control Japanese beetle in New Jersey, USA. However, the trend of EPNs in an area is to decline after application until they are reduced to "background levels" (Smits, 1996). Georgis (1992) and Smits (1996) reported that populations of EPNs are undetectable in the soil 4 weeks after application.

12.1.2 Effects on non-target invertebrates

Numerous studies have been conducted on the effects of EPNs on non-target insects and plant-parasitic nematodes. Overall, the risk to non-target populations from EPNs was deemed to be substantially lower than chemical insecticide application. Under laboratory conditions, EPNs have been shown to give different levels of control to a wide range of insect species. *S. carpocapsae*, for example, has been shown to be able to kill over 200 species under laboratory conditions (Poinar, 1986). The effect of EPNs depends on many factors such as the interactions between insect and nematode and symbiotic bacteria (host range), application rate, soil type and time of application. EPNs are soil organisms, so they are usually in contact with soil insects of different stages. In 1996, experts in the field of EPNs for biological control met to discuss the safety of EPNs. They concluded that due to specific biological and ecological features, EPNs are "exceptionally safe" (Ehlers and Hokkanen, 1996). All the scientific evidence available supported the conclusion that EPNs are safe to the environment, as well as to production and application personnel, the general public and the consumers of agricultural products treated with them. Therefore, the conclusion that indigenous EPN species should not be regulated was supported by Ehlers and Hokkanen (1996).

For general studies including various *Steinernema* spp., Anes and Ganguly (2015) investigated the non-target effects of *Steinernema thermophilum*, *Steinernema glaseri* and *Heterorhabditis indica* on ants, spiders and earthworms. While earthworms were not susceptible, even at higher doses of EPNs, spiders and ants had a high mortality. Spiders could escape infection by EPNs, but ants could not. Capinera *et al.* (1982) demonstrated in laboratory assays that the earthworm *Aporrectodea* sp. was not susceptible to *S. carpocapsae*.

Georgis *et al.* (1991) studied the effect of steinernematids and heterorhabditids on non-target arthropods and found that they did not suppress non-targets compared with controls, while chemical insecticides significantly reduced non-target populations. Considering diverse non-target insects, Sandhi and Reddy (2019) concluded that non-target effects only occurred in high concentrations under laboratory conditions. In a field test assessing the effects of *S. carpocapsae* on the carabid beetles *Bembidion properans* and *Pterostichus cupreus*, no mortality was found (Ropek and Jaworska, 1994). Harvey *et al.* (2016) concluded that non-target effects on coleopteran communities were largely unaffected by EPNs. Exotic EPN species exhibited a lower risk than native EPN species due to the shorter persistence of the exotics. Powell and Webster (2004) reported the effects of *S. carpocapsae*, *Heterorhabditis bacteriophora* and *Steinernema feltiae* on *Aphidoletes aphidimyza*, an aphid predator, in both laboratory and greenhouse experiments. The authors concluded that application of *S. carpocapsae* or *H. bacteriophora* caused *A. aphidimyza* mortality, although only *S. carpocapsae* displayed the potential to affect *A. aphidimyza* efficacy. The timing of application may temper any negative effects.

The effects of EPNs on pollinators such as bees has been the focus of attention over many years. Adult bees might be infected by EPNs while foraging on flowers for nectar or pollen (Morse and Flottum, 1997); however, EPNs are unlikely to survive on flowers, making bee infection on flowers a minimal risk. Honey bees were found to be susceptible to EPNs, but the EPNs did not adversely affect the whole colonies. Taha and Abdelmegeed (2016) observed no mortality in *Apis mellifera*, when hives were treated with *H. bacteriophora* to control greater wax moth larvae, *Galleria mellonella*. Data from Kaya *et al.* (1982) suggest that the brood frame located in the center of the hive is not suitable for infection due to microclimatic conditions; the internal temperature ranges between 33.3°C and 35.2°C and has low humidity (40–78%).

Akhurst and Smith (2002) reviewed the effects of EPNs on non-target invertebrates and described how non-target invertebrate studies have been carried out under laboratory conditions with high nematode doses. They concluded that research on the extent and impact of EPNs on non-target invertebrates in the field with commercially used doses would be a more proper evaluation of their environmental impact. A review article by Peters (1996) concluded that EPNs generally rely on high host densities for long-term survival, and species that occur in geographically distinct regions are more specific than more widespread species. Based on these indications, introductions of non-indigenous EPN species can be expected to have a minor impact on native invertebrate fauna. Sandhi and Reddy (2019) indicated that few studies have shown effects on non-target arthropods, and those effects were minimal or none. Some of this information is included in the summary of efficacy of EPNs on beneficial and environmentally important insects shown in Table 12.1.

Not all non-target effects are negative. Somasekhar *et al.* (2002), and Lewis and Grewal (2004) demonstrated in a turfgrass system that the abundance of plant-parasitic nematodes was reduced when EPNs were applied, while the abundance of free-living nematodes was not affected. This can be interpreted as a beneficial non-target effect.

12.1.3 Safety for humans and other vertebrates

There is no record of pathogenicity of EPNs and their associated bacterial symbionts to warm-blooded vertebrates, including humans (Bera *et al.*, 2014). Boemare *et al.* (2010) found that there was "no risk to warm-blooded animals or plants in relation to the use of EPNs in biological control." It is further emphasized that symbiotic bacteria and EPNs cannot be considered harmful to humans. The lack of impact on warm-blooded organisms has been partly explained by the fact that EPNs cannot survive at temperatures greater than 37°C (Schmiege, 1963), and that the immune system of warm-blooded animals seems to be able to eliminate EPNs by encapsulation in macrophages. A summary of the studies is illustrated in Table 12.2.

Some studies were conducted in different amphibians, mainly under laboratory conditions. Overall, Kermarrec and Mauléon (1985) and Poinar and Thomas (1988) noted that tadpole susceptibility markedly decreased with age.

Table 12.1. Effects of entomopathogenic nematodes (EPNs) on beneficial and environmentally important insects.

Insect	Application/test	Outcome	Reference
Parasitoid *Apanteles militaris* (Hymenoptera: Braconidae)	Laboratory tests; dose = 100, 1000, 5000 and 10,000 EPNs per Petri dish	Indirect mortality in laboratory tests.	Kaya (1978)
Compsilura concinnata (Diptera: Tachinidae)	Laboratory tests; dose = 50 EPNs cm^{-2} of soil	Indirect mortality in laboratory tests.	Kaya (1984)
Mastrus ridibundus (Hymenoptera: Ichneumonidae)	Exposure to IJs in cocooned codling moth	70% mortality of *M. ridibundus* larvae in the laboratory (dose 10 IJs cm^{-2}; ~$LC_{80\text{–}90}$ IJs per codling moth larva). However, *M. ridibundus* females avoided ovipositing in IJ-infected codling moth larvae in laboratory choice trials (dose = 25 IJs cm^{-2}).	Lacey *et al.* (2003)
Liotryphon caudatus (Hymenoptera: Ichneumonidae)	Exposure to IJs in cocooned codling moth	85% mortality of *L. caudatus* larvae in the laboratory (dose 10 IJs cm^{-2}; ~$LC_{80\text{–}90}$ IJs per codling moth larva). However, *L. caudatus* females avoided ovipositing in IJ-infected codling moth larvae in laboratory choice trials (dose = 25 IJs cm^{-2}).	Lacey *et al.* (2003)
Two-spotted ladybird, *Adalia bipunctata* (Coleoptera: coccinellidae)	Laboratory bioassays	IJs could infect and kill ladybird larva at 15°C, 20°C and 25°C; concentration = 50, 250 and 500 IJs per larva, respectively.	Rojht *et al.* (2009)
Harlequin ladybird, *Harmonia axyridis* (Coleoptera: Coccinellidae)	Petri dish assays	Some beetles were temporarily paralysed and others were killed.	Lemire *et al.* (1996)
Lacewing, *Chrysoperla carnea* (Neuroptera: Chrysopidae)	Laboratory bioassays	IJs could infect and kill lacewing larva at 15°C, 20°C and 25°C; concentration = 500, 2500 and 5000 IJs ml^{-1}, respectively.	Rojht *et al.* (2009)
Two ground beetles, *Harpalus* sp. and *Pterostichus* sp., (Coleoptera: Carabidae); two tiger beetles, *Cicindela* sp. and *Tetracha* sp. (Coleoptera: Cicindelidae); a rove beetle, *Philonthus* sp. (Coleoptera: Staphylinidae) and the tawny earwig, *Labidura riparia* (Dermaptera: Labiduridae)	Laboratory bioassays	Immature stages but not adults killed. No effect on non-target populations in the field.	Georgis *et al.* (1991)
Ground beetles *Bembidion properans* and *Pterostichus cupreus* (Coleoptera: Carabidae)	Field bioassays	No effect on adults.	Ropek and Jaworska (1994)

IJ, infective juvenile; $LC_{80\text{–}90}$, 80–90% lethal concentration.

Table 12.2. Effects of *Steinernema carpocapsae* on mammals.

Mammal	Application/test	Outcome	Reference(s)
Rat, *Rattus norvegicus*	50,000 IJs applied orally and intraperitoneally	No pathology or effect on weight gain in laboratory bioassays.	Gaugler and Boush (1979)
Mouse, *Mus musculus domesticus*	Applied orally, subcutaneously and intraperitoneally (dose = 2 × 10^4 IJs) (Kobayashi *et al.*, 1987). Doses of 1000 and 10,000 (Poinar *et al.*, 1982). *Xenorhabdus nematophila* application to abraded skin (Kermarec *et al.*, 1991). Applications of *X. nematophila* by oral, intradermal, subcutaneous and intraperitoneal routes, and by inhalation and skin contact (Obendorf *et al.*, 1983). Fed a dose of 200,000 IJs over a 5-day period (Schmiege,1963)	Skin ulcers when administered subcutaneously. IJs administered into the stomach using a probe saw no disease symptoms develop. No associated pathologies were observed.	Schmiege (1963); Poinar *et al.* (1982); Obendorf *et al.* (1983); Kobayashi *et al.* (1987); Kermarrec *et al.* (1991)
Guinea pig, *Cavia porcellus*	Applications of *X. nematophila* by oral, intradermal, subcutaneous and intraperitoneal routes, and by inhalation and skin contact	No evidence of infectivity, pathogenicity or toxicity in laboratory bioassays.	Obendorf *et al.* (1983)
Rabbit, *Oryctolagus cuniculus*	Applications of *X. nematophila* by conjunctival inoculation	No evidence of infectivity, pathogenicity or toxicity.	Obendorf *et al.* (1983)

IJ, infective juvenile.

Kermarrec and Mauléon (1985) did not find effects on fish. Some of these studies are summarized in Table 12.3.

12.2 Registration: Import and Release

No harmonized regulation system exists across nations or regions around the world. Timelines are very different from country to country and it can take years to obtain regulatory approval. However, many countries have developed and use comparable local rules to protect the environment from invasive species. The regulations allow the release of EPNs to provide farmers with biological solutions to control pests. Overregulation can negatively impact the development of EPNs.

Registration of EPNs varies from country to country. One of the most important points that countries want to determine for requirements is the environmental impact of introduced species, even when they are natural enemies such EPNs. Unlike other biological control agents, EPNs are regulated in reference to the country of origin of the nematode. Importantly, the release of native or indigenous EPNs is generally not regulated, while exotic or non-indigenous EPN releases are authorized depending on the framework of the country's laws. The common target of regulations is to ensure the safety of humans and the environment following the release of EPNs. The potential risk of import and release of exotic Invertebrate Biological Control Agents (IBCAs) is

Table 12.3. Effects of *Steinernema carpocapsae* on amphibians and reptiles.

Invertebrate	Application/test	Outcome	Reference
Cane toad, *Bufo marinus*	Dose = 100 IJs ml^{-1} in 1000 ml of water	Killed young tadpoles after 5 days in laboratory bioassays	Kermarrec and Mauléon (1985)
Western tree frog, *Hyla regilla*	Laboratory bioassays on doses of 100, 200 and 400 IJs ml^{-1}	Killed young tadpoles	Poinar and Thomas (1988)
African clawed frog, *Xenopus laevis*	Laboratory bioassays on doses of 100, 200 and 400 IJs ml^{-1}	Killed young tadpoles	Poinar and Thomas (1988)
Eleutherodactylus sp. adults	Ingestion of 8×10^4 IJs in the laboratory	Remained unaffected	Kermarrec *et al.* (1991)
Anolis marmoratus	Oral application (dose = 80,000 IJs) in the laboratory	Death of adults and live nematodes were recovered from the body cavity	Kermarrec *et al.* (1991)
Anolis marmoratus	5000 IJs every 3 days for 30 days in laboratory bioassays	No mortality was observed	Kermarrec and Mauléon (1985)
Common snapping turtle, *Chelydra serpentina*	In laboratory, hatchlings challenged with 2000 IJs	No mortality or signs of disease were observed	G.O. Poinar and L.A. Miller (unpublished data, 1988); see Poinar (1989) for comment

associated with the potential establishment of an exotic species and possible negative impacts to the environment. The regulation of EPNs is under the control of similar authorities as for other pesticides (chemical, microbial and natural compounds), and the outcome varies by each country's specific requirements. For example, in Europe, there is no harmonized regulation as EPNs, similar to other IBCAs, are not covered by the EU 1107/2009 regulation. The main regulatory environments in some European countries are as follows, and a summary table of country policies is provided in Table 12.4:

- Austria: The release of EPNs is regulated by the Austrian Agency for Health and Food Safety (AGES) under the admission requirements according to the Plant Protection Products Ordinance 2011. Under this regulation, not only is the release of the nematode considered, but the complete product using the nematode will be also evaluated (formulation, efficacy and safety).
- Belgium: The federal government grants each region significant autonomy. In Wallonia, there are no limitations at the present. In Flanders, the EPN species must either be native or included in the European and Mediterranean Plant Protection Organization (EPPO)-positive list.
- Denmark; There is no approval process for macroorganisms if the species is native to the Danish environment or located on the EPPO list. No guidance for exotic nematodes is provided.
- France: ANSES (Agence Nationale de Sécurité Sanitaire de l'Alimentation, de l'Environnement et du Travail) evaluates EPNs at the strain level. This means that the EPN is considered native only if the strain used was isolated in the French territory. A non-native strain found in another country (even a neighboring country) will be considered exotic.
- Germany: The federal Nature Conservation Act stipulates the interdiction of release of exotic species, but the introduction and the use of native or exotic EPNs (IBCAs in general) do not need an approval and EPNs can be imported and sold. The users must obtain permission from a federal state agency.
- Netherlands: The National Reference Centre for Plant Protection of the Netherlands Food and Consumer Product Safety

Table 12.4. General overview of countries and associated authorities involved in entomopathogenic nematode (EPN) registration.

Country	Relevant authority or legal framework	Comments
Australia	Australian Pesticides and Veterinary Medicines Authority	Adequate genetic information to demonstrate that an exotic nematode is identical or extremely similar to other Australian species.
Austria	Austrian Agency for Health and Food Safety (AGES)	The authority will evaluate the safety (toxicity/ ecotoxicity) of the nematode, and the formulation containing the nematodes. The product should demonstrate an acceptable efficacy to be registered.
Belgium	Agency for Nature and Forests	The species must be found in the Belgian territory.
Brazil	ANVISA (Brazilian Health Regulatory Agency), IBAMA (Brazilian Institute of the Environment and Renewable Natural Resources) and Ministry of Agriculture, Livestock and Food Supply (MAPA)	Each relevant authority will evaluate the safety (toxicity/ecotoxicity) and consider the value for the farmer via evaluation of efficacy trials.
Canada	Canadian Food Inspection Agency	An import permit can be obtained for continental USA or all other countries after a risk assessment mainly driven by environmental impact of release.
Denmark	–	No approval process for macroorganisms, if the species is native to the Danish environment or located on the EPPO list.
France	French Agency for Food, Environmental and Occupational Health and Safety (ANSES)	EPN is native if isolated in the French territory; will assess both nematode and symbiotic bacteria safety.
Germany	Federal Nature Conservation Act	The species must be found in the German territory.
Netherlands	National Reference Centre for Plant Protection	With respect to the EPPO guidance, if the nematode occurs in more than five European countries but not in the Netherlands, the local authority will provide an authorization.
Norway	Norwegian Scientific Committee for Food and Environment	Will assess both nematode and symbiotic bacteria safety.
Sweden	Swedish Environmental Protection Agency	Has the principle not to approve species that have not yet been confirmed as established in Sweden, in particular if it can survive in winter.
USA	US Department of Agriculture-Animal and Plant Health Inspection Service (USDA-APHIS)	Will approve the use of a beneficial nematode if the species is geographically broadly present across the USA.

EPPO, European and Mediterranean Plant Protection Organization.

Authority will assess submission of the dossier on the release of a new EPN.

- UK: England, North Ireland, Scotland and Wales will independently evaluate and issue licenses for the release of a new exotic EPN.
- Sweden: The Swedish Environmental Protection Agency performs an evaluation based on EPPO-listed organisms.

General and common aspects are considered in the approval of the release of a new EPN in Europe. However, each authority will evaluate differently the risk of use and release of EPNs by considering the local environmental aspects (e.g. cold winters killing temperature-sensitive species) or the human safety (i.e. amateur or professional uses). However, most of the requirements follow the EPPO Standard PM 6/3 (List of biological control agents widely used in the EPPO region). As mentioned earlier, some countries strictly refer to the EPPO-positive list. Some key requirements to be listed are:

- the identity and classification of the organism;
- its geographical distribution;
- its main target pests and their host plants; and
- the establishment potential and non-target effects.

The EPPO-positive listing takes certain criteria of precedence into consideration; the proposed nematodes will be those species relevant for biocontrol of plant pests that are:

- indigenous or established and widespread in the EPPO region; or
- used for more than 5 years in at least five EPPO member countries without known adverse effect; or
- successful classical biological control agents.

Outside of Europe, for similar reasons of safety for humans and the environment, many countries have developed regulations or rules for imports and release of EPNs. In Australia, the release of EPNs is approved by the Australian Pesticides and Veterinary Medicines Authority (APVMA). Adequate genetic information to demonstrate that an exotic nematode is identical or extremely similar to other Australian species would diminish environmental concerns relating its release. In Brazil, each relevant authority (ANVISA, the Brazilian Health Regulatory Agency; IBAMA, the Brazilian Institute of Environment and Renewable Natural Resources, and MAPA, Ministry of Agriculture, Livestock and Food Supply) will evaluate the safety (toxicity/ecotoxicity) and consider the value for the farmer via evaluation of efficacy trials. In Canada, all non-indigenous EPNs require approval from the Canadian Food Inspection Agency (CFIA) before their first release into the Canadian environment. In the USA, the US Department of Agriculture regulates importation but movement between states is not regulated. The Environmental Protection Agency (EPA) approves the use of an EPN if the species is geographically broadly present across the USA. The EPA does not require EPNs to be registered under the Federal Insecticide, Fungicide, and Rodenticide Act (FIFRA).

Some elements must be considered regarding exotic and native species. EPNs do not respect borderlines (nor do the pests that EPNs attack). Thus, at a regional level (e.g. Europe, North America, Latin America), best regulatory practices and exchanges with EPN producers can support the development of the use of these biological control agents. "Mutual recognition" between European or South American countries, and EPPO processes in countries developing a regulation, appear to be a pragmatic way to approve the use/release of EPNs for the benefit of the farmers.

12.3 References

Akhurst, R. and Smith, K. (2002) Regulation and safety. In: Gaugler, R. (ed.) *Entomopathogenic Nematology*. CAB International, Wallingford, UK, pp. 311–332.

Anes, K.M. and Ganguly, S. (2015) Effect of entomopathogenic nematodes (Nematoda: Rhabditida) on earthworms, spiders and ants. *Research Journal of Agricultural and Forestry Sciences* 3, 19–22.

Barbercheck, M.E. and Millar, L.C. (2000) Environmental impacts of entomopathogenic nematodes used in biological control in soil. In: Follett, P.A. and Duan, J.J. (eds.) *Nontarget Effects of Biological Control*. Springer, New York, pp. 287–308.

Bera, T., Pandey, A.K., Maurya, R.P. and Mall, P. (2014) Role of entomopathogenic nematodes in the management of insect pests. *Journal of Entomological Research* 38, 301–310.

Boemare, N., Laumond, C. and Mauleon, H. (2010) The entomopathogenic nematode-bacterium complex: biology, life cycle and vertebrate safety. *Biocontrol Science and Technology* 6, 333–346.

Capinera, J.L., Blue, S.L. and Wheeler, G.S. (1982) Survival of earthworms exposed to *Neoaplectana carpocapsae* nematodes. *Journal of Invertebrate Pathology* 39, 419–421.

de Nardo, E., Grewal, P., McCartney, D. and Stinner, B. (2006) Nontarget effects of entomopathogenic nematodes on soil microbial community and nutrient cycling processes: a microcosm study. *Applied Soil Ecology* 34, 250–257.

Ehlers, R.-U. and Hokkanen, H.M.T. (1996) Insect biocontrol with non-endemic entomopathogenic nematodes (*Steinernema* and *Heterorhabditis* spp.): conclusions and recommendations of a combined

OECD and COST workshop on scientific and regulatory policy issues. *Biocontrol Science and Technology* 6, 295–302.
Gaugler, R. and Boush, G.M. (1979) Nonsusceptibility of rats to the entomogenous nematode, *Neoaplectana carpocapsae*. *Environmental Entomology* 8, 658–660.
Gaugler, R., Campbell, J.F., Selvan, S. and Lewis, E.E. (1992) Large-scale inoculative releases of the entomopathogenic nematode *Steinernema glaseri:* assessment 50 years later. *Biological Control* 2, 181–187.
Georgis, R. (1992) Present and future prospects for entomopathogenic nematode products. *Biocontrol Science and Technology* 2, 83–99.
Georgis, R., Kaya, H.K. and Gaugler, R. (1991) Effect of *Stenernematid* and *Heterorhabditid* nematodes (Rhabditida: *Steinernematidae* and *Heterorhabditidae*) on nontarget arthropods. *Environmental Entomology* 20, 815–822.
Harvey, C.D., Williams, C.D., Dillon, A.B. and Griffin, C.T. (2016) Inundative pest control: how risky is it? A case study using entomopathogenic nematodes in a forest ecosystem. *Forest Ecology and Management* 350, 242–251.
Kaya, H.K. (1978) Infectivity of *Neoaplectana carpocapsae* and *Heterorhabditis heliothidis* to pupae of the parasite of *Apanteles militaris. Journal of Nematology* 10, 241–244.
Kaya, H.K. (1984) Effect of the entomogenous nematode *Neoplectana carpocapsae* on the tachinid parasite *Compsilura concinnata* (Diptera: Tachinidae). *Journal of Nematology* 16, 9–13
Kaya, H.K., Marston, J.M., Lindegren, J.E. and Peng, Y. (1982) Low susceptibility of the honey bee, *Apis mellifera* L. (Hymenoptera: Apidae), to the entomogenous nematode, *Neoaplectana carpocapsae* Weiser. *Environmental Entomology* 11, 920–924.
Kermarrec, A. and Mauléon, H. (1985) Potential noxiousness of the entomogenous nematode *Neoaplectana carpocapsae* Weiser to the Antillan toad *Bufo marinus* L. *Mededelingen van de Faculteit Landbouwwetenschappen, Rijksuniversiteit (Gent)* 50, 831–838
Kermarrec, A., Mauléon, H., Sirjusingh, C. and Baud, L. (1991) Experimental studies on the sensitivity of tropical vertebrates (toads, frogs and lizards) to different species of entomoparasitic nematodes of the genera *Heterorhabditis* and *Steinernema*. In: Pavis, C. and Kermarrec, A. (eds.) *Recontres Caraïbes en Lutte Biologique*. Institut National de la Recherche Agronomique, Paris, pp. 193–204.
Kobayashi, M., Okano, H. and Kirihara, S. (1987) The toxicity of steinernematid and heterorhabditid nematodes to male mice. In: Ishibashi, N. (ed.) *Recent Advances in Biological Control of Insect Pests by Entomogenous Nematodes in Japan*. Saga University, Saga, Japan, pp. 153–157.
Lacey, A.L., Unruh, T.R. and Headrick, H.L. (2003) Interactions of two parasitoids (Hymenoptera: Ichneumonidae) of codling moth (Lepidoptera: Tortricidae) with the entomopathogenic nematode *Steinernema carpocapsae* (Rhabditida: Steinernematidae). *Journal of Invertebrate Pathology* 83, 230–239.
Lemire, S., Coderre, D., Vincent, C. and Bélair, G. (1996) Lethal and sublethal effects of the entomogenous nematode, *Steinernema carpocapsae*, on the coccinellid *Harmonia axyridis*. *Nematropica* 26, 284–285.
Lewis, E.E. and Grewal, P.S. (2004) Interactions with plant parasitic nematodes. In: Grewal, P.S., Ehlers, R.U. and Shapiro-Ilan, D.I. (eds.) *Nematodes as Biocontrol Agents*. CAB International, Wallingford, UK, pp. 349–362.
Morse, R.A. and Flottum, K. (1997) *Honeybee Pests, Predators, and Diseases*. A.I. Root Company, Ohio.
Obendorf, D.L., Peel, B., Akhurst, R.J. and Miller, L.A. (1983) Non-susceptibility of mammals to the entomopathogenic bacterium *Xenorhabdus nemtophilus*. *Environmental Entomology* 12, 368–370.
Peters, A. (1996) The natural host range of *Steinernema* and *Heterorhabditis* spp. and their impact on insect populations. *Biocontrol Science and Technology* 6, 398–402.
Poinar, G.O. Jr. (1986) Entomophagous nematodes. In: Franz, J.M. (ed.) *Biological Plant and Health Protection*. Gustav Fischer, Stuttgart, Germany.
Poinar, G.O. Jr. (1989) Non-insect hosts for the entomogenous rhabditoid nematodes *Neoaplectana* (Steinernematidae) and *Heterorhabditis* (Heterorhabditidae). *Revue de Nématologie* 12, 423–428.
Poinar, G.O. Jr. and Thomas, G.M. (1988) Infection of frog tadpoles (Amphibia) by insect parasitic nematodes (Rhabditida). *Experientia* 44, 528–553.
Poinar G.O. Jr., Thomas, G.M., Presser, S.B. and Hardy, J.L. (1982) Inoculation of entomogenous nematodes, *Neoaplectana* and *Heterorhabditis* and, their associated bacteria, *Xenorhabdus* spp., into chicks and mice. *Environmental Entomology* 11, 137–138.
Powell, J.R. and Webster, J.M. (2004) Intraguild antagonism between biological controls: impact of entomopathogenic nematode application on an aphid predator, *Aphidoletes aphidimyza* (Diptera: Cecidomyiidae). *Biological Control* 30, 110–118.

Rojht, H., Kač, M. and Trdan, S. (2009) Nontarget effect of entomopathogenic nematodes on larvae of twospotted lady beetle (Coleoptera: Coccinellidae) and green lacewing (Neuroptera: Chrysopidae) under laboratory condition. *Journal of Economic Entomology* 102, 1440–1443.

Ropek, D. and Jaworska, M. (1994) Effect of an entomopathogenic nematode, *Steinernema carpocapsae* Weiser (Nematoda, Steinernematidae), on carabid beetles in field trials with annual legumes. *Anzeiger für Schädlingskunde, Pflanzenschutz und Umweltschutz* 67, 97–100.

Sandhi, R.K. and Reddy, G.V.P. (2019) Effects of entomopathogenic nematodes and symbiotic bacteria on non-target arthropods. In: Khan, A.M. and Ahmad, W. (eds.) *Microbes for Sustainable Insect Pest Management*, Vol. 1. Springer, Cham, Switzerland, pp. 247–273.

Schmiege, D.C. (1963). The feasibility of using a neoaplectanid nematode for control of some forest insect pests. *Journal of Economic Entomology* 8, 133–134.

Smits, P.H. (1996) Post-application persistence of entomopathogenic nematodes. *Biocontrol Science and Technology* 6, 379–387.

Smits P. (2010) Post-application persistence of entomopathogenic nematodes. *Biocontrol Science and Technology* 6, 379–388.

Somasekhar, N., Grewal, P.S., de Nardo, E.A.B. and Stinner, B.R. (2002) Non-target effect of entomopathogenic nematodes on the soil nematode community. *Journal of Applied Ecology* 39, 735–744.

Taha, E.H. and Abdelmegeed, S.M. (2016) Effect of entomopathogenic nematodes, *Heterorhabditis bacteriophora*, on *Galleria mellonella* in beehives of *Apis mellifera*. *Journal of Biological Sciences* 16, 197–201.

Toepfer, S., Kurtz, B. and Kuhlmann, U. (2010) Influence of soil on the efficacy entomopathogenic nematodes in reducing *Diabrotica virgifera virgifera* in maize. *Journal of Pest Management Science* 83, 257–264.

13 Role of Marketing and Outreach for the Success of Entomopathogenic Nematodes

Surendra K. Dara[1,2]*

[1]*University of California Division of Agriculture and Natural Resources, San Luis Obispo, California, USA;* [2]*Current address: Department of Horticulture, Oregon State University, Corvallis, Oregon, USA*

Abstract

Entomopathogenic nematodes (EPNs) have been known for their efficacy to manage a variety of pests in multiple cropping systems. A number of commercial formulations of EPNs can be an important part of integrated pest management (IPM) strategies to maximize control efficacy. Successful use of EPNs depends on research studies that contribute to the development of effective use strategies and outreach efforts that increase the use of EPNs in IPM. Collaborative efforts involving academic institutions, private industry and the farming community will help address the knowledge gaps, popularize the potential of EPNs and increase their use for sustainable agriculture. This chapter provides an overview of EPN marketing and outreach, with some examples of successful pest suppression with EPNs, deficiencies in EPN awareness, practical challenges, and various research and outreach strategies for promoting their use.

13.1 Introduction

Several families of nematodes are entomopathogenic, and members of two families are commercially available as microbial control agents for controlling various pests (Koppenhöffer, 2007; Lacey, 2017). Species of the entomopathogenic genera *Heterorhabditis* and *Steinernema*, in association with their symbiotic bacteria *Photorhabdus* and *Xenorhabdus* spp., respectively, suppress various insects pests (primarily coleopteran, dipteran, lepidopteran and thysanopteran) that attack field crops, greenhouse crops, orchards, small fruits, vegetables, turfgrass and mushrooms; therefore, these beneficial nematodes have been successfully used for sustainable food production (Koppenhöffer *et al.*, 2020a; Devi, 2021). Depending on the pest, its targeted life stage and habitat, and the crop, entomopathogenic nematodes (EPNs) can be applied alone or used with other control options as part of an integrated pest management (IPM) program. Although EPNs are very attractive for use in organic production systems, they are also used widely in various conventional systems. This chapter provides an overview of the knowledge gaps and practical challenges in EPN use, various strategies for promoting research and outreach efforts, and the role of public and private institutions in addressing these needs.

*Email: surendra_d@hotmail.com

 Entomopathogenic Nematodes as Biological Control Agents (D.I. Shapiro-Ilan and E.E. Lewis eds)
DOI: 10.1079/9781800620322.0013

As EPNs are soil-borne organisms, they are well suited for controlling soil-inhabiting pests and pests such as borers and leaf miners that are in protected areas within the plant. Although there are a limited number of EPN formulations compared with other biological or microbial pesticides, the dispersing and foraging behavior of infective juveniles, high levels of efficacy against certain pests, safety to non-target organisms and the environment, and compatibility with other control options make EPNs an ideal choice for pest management. Several recent studies have demonstrated the efficacy of EPNs in pest suppression around the world (Lacey, 2017; Koppenhöffer *et al.*, 2020a; Shakeel *et al.*, 2022; Shapiro-Ilan *et al.*, 2022). For example, a study in Pakistan demonstrated the potential of EPNs, especially *Heterorhabditis bacteriophora*, as an IPM component against the soil-inhabiting stages of the tephritid fruit flies *Bactrocera dorsalis* and *Bactrocera zonata*, with 31–37% mortality under field conditions (Usman *et al.*, 2021). A study in Spain showed that 65% of the mature spotted-wing drosophila (*Drosophila suzukii*) adults and 89% of the teneral adults emerging from soil treated with *Steinernema carpocapsae* had a nematode infection (Garriga *et al.*, 2020). A study in the USA demonstrated reduced root damage from the western corn rootworm, *Diabrotica virgifera virgifera*, and increased corn yields in some fields when *H. bacteriophora* and *Steinernema feltiae* were used in combination with arbuscular mycorrhizal fungi and the bacteria *Pseudomonas chlororaphis* and *Pseudomonas protegens* (Jaffuel *et al.*, 2019). Similarly, applying *H. bacteriophora* in calcium alginate beads effectively delivered EPNs and helped reduce western corn rootworm damage in a field study in the USA (Kim *et al.*, 2021). Applying Egyptian isolates of *S. carpocapsae* in a calcium alginate-based formulation in greenhouse onions expanded the shelf life of EPNs and reduced onion thrips, *Thrips tabaci*, under heavy infestations (Hussein and El-Mahdi, 2019). Multiple *Heterorhabditis* and *Steinernema* spp. were also evaluated in laboratory, greenhouse and field studies against onion thrips in the USA (Gulzar *et al.*, 2021). Under field conditions, *H. bacteriophora* provided the best control of larvae and adults. Another US study also identified *S. carpocapsae* and *Steinernema riobrave* as potential control agents against the sugarbeet wireworm, *Limonius californicus*, in spring wheat (Sandhi *et al.*, 2020). While these examples demonstrate the potential of EPNs in experimental studies, there are also several examples of EPN use in commercial farming. A single application of native New York EPNs not only provided multi-year control of alfalfa snout beetle, *Otiorhynchus ligustici*, on thousands of acres, but also controlled corn rootworms when corn was grown after alfalfa (Shields, 2021). EPNs also provided successful control of several coleopteran, dipteran and lepidopteran pests, mole crickets and thrips on commercial production systems of berries, fruits and nuts, mushrooms, ornamentals and vegetables, and on turfgrass through inundative application (Shapiro-Ilan *et al.*, 2022). Similarly, the potential of EPNs in managing various pests has also been demonstrated in several such field studies (Lacey, 2017).

For effective use of EPNs, it is important to understand their interactions with other control options and agricultural inputs. This topic is covered extensively in Chapter 27 (this volume). Overall, EPNs are compatible with most agricultural inputs, which enhances the potential for incorporation into IPM programs. However, some interactions can be complex. For example, based on studies conducted in Turkey and the UK, Hummadi *et al.* (2021) reported that high concentrations of certain volatile organic compounds from the entomopathogenic fungus (EPF) *Metarhizium brunneum* are highly toxic to EPNs, while low concentrations attract them. Özdemir *et al.* (2020) evaluated the compatibility of certain insecticides and fungicides with *S. feltiae* and *H. bacteriophora* and found that the fungicides fluxapyroxad + difenoconazole and ametoctradin + dimethomorph, and the insecticides spinosad and spiromesifen negatively impacted the EPNs, while the insecticides cyflumetofen and imidacloprid had no adverse effect on the survival, reproduction and infectivity of EPNs in greater wax moth, *Galleria mellonella*, bioassays conducted in Turkey. Koppenhöffer *et al.* (2020b) found that combining imidacloprid and the EPNs *H. bacteriophora*, *S. carpocapsae* and *S. feltiae* had an additive effect against the annual bluegrass weevil, *Listronotus maculicollis*, in a US study. Under field conditions, *S. carpocapsae* provided 59–74% control when used alone and 75–84% when used in combination with imidacloprid. It appeared that splitting *S. carpocapsae* into two half-rate applications along with imidacloprid provided 88–95% control. Usman *et al.* (2020) also observed additive effects of EPN–EPF combinations from studies using *S. carpocapsae*, *S. riobravae* and the EPFs *Beauveria bassiana*, *Cordyceps javanica*, and *M. brunneum* against the pupae of the apple maggot, *Rhagoletis pomonella*, in another US study. These studies have helped to improve our understanding of EPN potential, strategies for using them as IPM tools, and the dissemination of the information to farmers and pest-management professionals.

13.2 Knowledge Gaps

In the past few decades, there have been significant strides in fundamental and applied research with EPNs and published research information in scientific journals. However, there are relatively few nematologists around the world who are working on EPNs and carrying the tremendous burden of generating applied research data. Very few of these nematologists are engaged in outreach activities, which further creates a wide knowledge gap for farmers and pest-management professionals. These circumstances have led to a lack of comprehensive IPM strategies where EPNs are key components and thus limited opportunities to educate and train farmers on EPN use. During multiple conversations with farmers and pest control professionals of nut crops, small fruits and vegetables in California, it appeared that many are unaware of EPNs, the use of EPNs as a control option, a commercial source of EPNs or a strategy to use them to address their pest issues (S.K. Dara, personal observations). Although the recent popularity of organic agriculture has helped promote the availability, knowledge and use of biopesticides, the potential of EPNs is largely underexplored. Limited expertise and funding to conduct applied research with EPNs, lack of outreach material and extension professionals to educate the farming community, limited availability of EPN-based biopesticides and their higher cost compared with conventional pesticides, and the general tendency to trust and choose synthetic pesticides have been seen as major challenges for optimal use of EPNs. However, increased funding from public and private entities and improved collaboration among researchers, extension educators, biopesticide manufacturers, pest-control professionals and farmers can help conduct experimental and demonstration field trials, provide highly implementable practical solutions, and enhance our understanding and use of EPNs.

13.3 Practical Challenges

In addition to the knowledge gaps that limit EPN use, some practical challenges also hinder the widespread acceptance of EPNs as an effective IPM tool. Although farmers generally prefer to minimize crop protection costs by using highly effective, economical and easy-to-use control options, many are increasingly open to exploring non-chemical alternatives due to resistance problems with conventional practices based on chemical pesticides and growing demand for sustainable production practices. As EPN-based pesticides contain live organisms, they have a shorter shelf life than some biopesticides and require special storage (i.e. refrigeration), handling, transportation and application requirements because of their sensitivity to environmental conditions. Although EPNs can be formulated in a user-friendly fashion (see Chapter 11, this volume), and applied using most standard agricultural equipment, EPN-based pesticides have several limitations compared with chemical insecticides. Storing and transporting in cooler conditions, avoiding exposure to warm temperatures and ultraviolet radiation, applying to moist soils or with large amounts of water and followed by irrigation, special application requirements related to the spray pressure, nozzle type and strainers, potential incompatibility with other agricultural inputs in the tank mixes, timing and periodicity of applications, variations in virulence among various EPN species, specificity of certain EPN species to certain pests and other factors are some of the practical challenges associated with using EPNs. If manufacturers can improve EPN-based formulations to increase the product stability, efficacy and adaptability to various field conditions, and optimize the cost, many of the practical challenges can be addressed. Additionally, farmers and pest-management professionals need to be aware of the limitations, requirements and various strategies for successfully using EPN-based pesticides. Manufacturers and extension educators can help fill these knowledge gaps. Some biological product distributors and manufacturers offer to store the products under refrigerated conditions until the farmers are ready to apply or bring them to the field at the time of application and further assist with application procedures. Such services can help address storage and handling issues and increase the willingness of farmers to use EPN-based pesticides and other biologicals.

13.4 Strategies for Promoting EPN Use

13.4.1 Marketing strategies

In general, most manufacturers prepare their educational or marketing materials based on studies where their products perform very well compared with industry standards. In many cases, the focus is to promote the use of a particular product rather than recommend an IPM strategy that includes multiple control options. However, some manufacturers have started marketing combinations of products or crop protection strategies that include multiple biological and synthetic inputs from various sources. Such efforts will encourage a collaborative environment among various manufacturers, contribute to the development of comprehensive strategies from evidence-based field data, improve pest management efficacy, optimize costs, and build farmers' confidence in EPNs and other biological solutions. It is also important for the manufacturers to invest in research to evaluate EPN efficacy against a variety of pests in various cropping systems and compatibility of EPNs with other products and farm practices, develop IPM strategies with EPNs, and support research and outreach efforts by universities and public institutions.

13.4.2 Academic strategies

There is a dearth of professionals who are trained in applied aspects of EPN use, invertebrate pathology and microbial control. Agricultural colleges and universities must develop academic programs to train students in these areas to maintain a continuous flow of pest-management professionals, researchers and educators experienced in EPNs and other biologicals. Even in programs where subjects related to sustainable agriculture and non-chemical control options are taught, expertise in EPNs and invertebrate pathology is vastly deficient around the world. Private organizations that are involved in sustainable agriculture, commodity groups, manufacturers of biopesticides and other entities could support and advocate for academic programs that help build expertise in EPNs and entomopathology. While food security and affordability are continuing problems, the urbanization of world populations is distancing the general public from food production practices. It will be very useful to introduce sustainable food production topics to children in their primary and secondary education levels to bring awareness and generate interest in agriculture and related topics. Early awareness of such topics helps them consider higher education and careers in those areas.

13.4.3 Research strategies

The efficacy of EPNs or any pesticide, in general, depends on numerous biotic and abiotic factors. As a result, multiple studies under varying crop, soil, environmental and pest conditions are required for researchers to have a good understanding of the product and its potential to develop appropriate field-use strategies. Although several scientific and extension publications report the efficacy of EPNs and other products, there are some challenges in promoting their use in commercial production systems. As evaluating efficacy in the laboratory, greenhouse or small-plot studies may or may not represent the performance of the product in a commercial field, data from large-scale field studies are also necessary. Similarly, evaluating biologicals along with highly effective synthetic pesticides, as seen in some studies, may not offer a true comparison of those products and could potentially give a wrong impression of the efficacy of biologicals. Although several biologicals can be as effective as certain synthetic pesticides, not all efficacy studies help build farmers' confidence in those products. As limited scientific data are available on using EPNs and other biologicals as a part of a comprehensive IPM strategy, it will be beneficial to increase the number of research studies that develop IPM strategies with support from the manufacturers of EPN-based products and funding from grant agencies. Research that addresses field stability, seasonal fluctuations, compatibility with other agricultural inputs or practices, application technologies, strategies for improving efficacy, field persistence or conservation, of EPNs, environmental and financial benefits in both the short and long term, and other such areas will help enhance EPN use in sustainable crop production.

Access to healthy and affordable food is a challenge in many parts of the world. Urbanization that reduces the availability of farmlands for food production, and migration of rural communities to urban areas, among other factors, have increased interest in urban farming in small areas, shipping containers or warehouses. While greenhouse production of fruits, vegetables and ornamentals has been around for a long time, vertical farming, container farming and other such systems are gaining momentum. EPNs can be an effective IPM tool in such systems where growing conditions are more ideal for EPNs and non-chemical alternatives are generally preferred. Research efforts to explore the potential of EPNs in urban agriculture will contribute to their increased use in these systems.

13.4.4 Outreach strategies

As a limited number of nematologists are involved in outreach activities, exploring opportunities to enhance the knowledge of extension researchers, government and university employees who train or advise farmers, and crop-care professionals who work with farmers about EPNs and their potential is one of the first steps. Although EPNs can be effective against a wide range of pests, not all crop management guidelines recommend EPNs as an option, mainly due to a lack of awareness or research experience. When more researchers consider EPN-based pesticides in their applied research studies, more information can be generated to educate farmers on their effective use. The following suggestions can support EPN outreach: (i) include EPN-based products as an IPM option for appropriate pests and life stages based on either locally developed data or research from other sources on the same or a similar pest; (ii) develop and disseminate outreach material (e.g. extension articles, factsheets, guides, videos) that increases the awareness of EPNs and promotes their use; (iii) organize field days, workshops or other events to train farmers and pest-management professionals in effectively using EPNs and other biologicals; (iv) develop web-based or smartphone decision-making tools for farmers and crop-care professionals; and (v) improve public relations by engaging the general public, news media, students and others in sustainable research and outreach efforts that include EPNs and other biologicals.

13.5 Conclusions

EPN-based pesticides can play a critical role in IPM practices in crop production, landscape management and other areas around the world. Limited expertise to promote their use, inherent challenges with EPN formulations, and a lack of knowledge about their potential and use strategies have limited their widespread use. To realize the full potential of EPNs for sustainable food production and maintaining environmental health, collaborative efforts from manufacturers, regulators, researchers, extension educators, farmers and pest management professionals are necessary. Academic training to build expertise in EPNs, applied research to develop practical solutions with EPN-based pesticides, improved formulations that enhance efficacy, training the farming community on EPN use, and developing and disseminating relevant crop-management literature are key areas to focus on for increased use of EPN-based pesticides and further advancing sustainable production practices (Fig. 13.1).

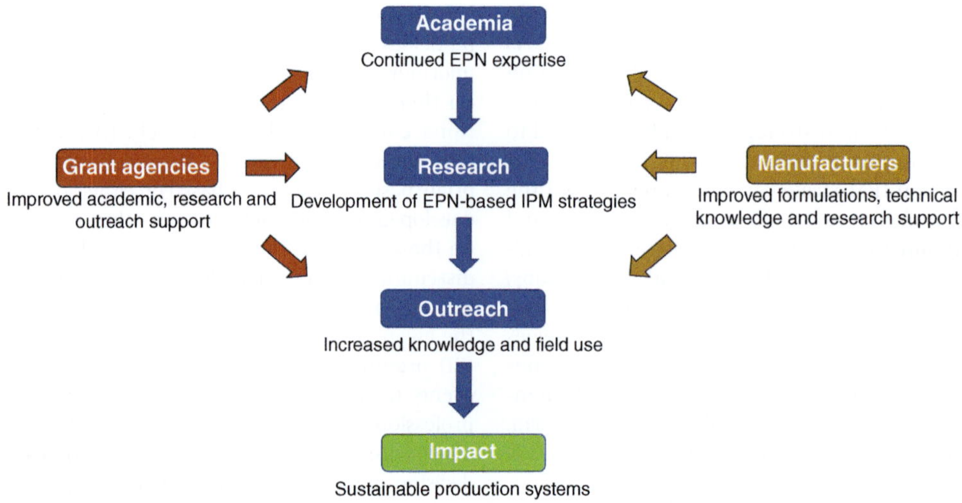

Fig. 13.1. Key players and strategies for improving entomopathogenic nematode (EPN) research and outreach for sustainable production practices. IPM, integrated pest management.

13.6 References

Devi, G. (2021) Management of coleopteran pests through entomopathogenic nematodes. *The Pharma Innovation* 10, 1021–1037.

Garriga, A., Morton, A., Ribes, A. and Garcia-del-Pino, F. (2020) Soil emergence of *Drosophila suzukii* adults: a susceptible period for entomopathogenic nematodes infection. *Journal of Pest Science* 93, 639–646.

Gulzar, S., Wakil, W. and Shapiro-Ilan, D.I. (2021) Potential use of entomopathogenic nematodes against the soil dwelling stages of onion thrips, *Thrips tabaci* Lindeman: laboratory, greenhouse and field trials. *Biological Control* 161: 104677.

Hummadi, E.H., Dearden, A., Generalovic, T., Clunie, B., Harrott, A. *et al.* (2021) Volatile organic compounds of *Metarhizium brunneum* influence the efficacy of entomopathogenic nematodes in insect control. *Biological Control* 155: 104527.

Hussein, M.A. and El-Mahdi, I.F.S. (2019) Efficiency of three formulated entomopathogenic nematodes against onion thrips, *Thrips tabaci* under aquaculture system. *Journal of Biopesticides* 12, 134–138.

Jaffuel, G., Imperiali, N., Shelby, K., Campos-Herrera, R., Geisert, R. *et al.* (2019) Protecting maize from rootworm damage with the combined application of arbuscular mycorrhizal fungi, *Pseudomonas* bacteria and entomopathogenic nematodes. *Scientific Reports* 9: 3127.

Kim, J., Hiltpold, I., Jaffuel, G., Sbaiti, I., Hibbard, B.E. and Turling, T.C.J. (2021) Calcium-alginate beads as a formulation for the application of entomopathogenic nematodes to control rootworms. *Journal of Pest Science* 94, 1197–1208.

Koppenhöffer, A.M. (2007) Nematodes. In: Lacey, L.A. and H.K. Kaya (eds.) *Field Manual of Techniques in Insect Pathology*. Springer, Dordrecht, Netherlands, pp. 249–264.

Koppenhöffer, A.M., Shapiro-Ilan, D.I. and Hiltpold, I. (2020a) Entomopathogenic nematodes in sustainable food production. *Frontiers in Sustainable Food Systems* 4: 125.

Koppenhöffer, A.M., Kostromytska, O.S. and Wu, S. (2020b) Optimizing the use of entomopathogenic nematodes for the management of *Listronotus maculicollis* (Coleoptera: Curculionidae): split applications and combinations with imidacloprid. *Crop Protection* 137: 105229.

Lacey, L.A. (ed.) (2017) *Microbial Control of Insect and Mite Pests: From Theory to Practice*. Academic Press, London.

Özdemir, E., İnak, E., Evlice, E. and Laznik, Z. (2020) Compatibility of entomopathogenic nematodes with pesticides registered in vegetable crops under laboratory conditions. *Journal of Plant Diseases and Protection* 127, 529–535.

Sandhi, R.K., Shapiro-Ilan, D., Sharma, A. and Reddy, G.V.P. (2020) Efficacy of entomopathogenic nematodes against the sugarbeet wireworm, *Limonius californicus* (Mannerheim) (Coleoptera: Elateridae). *Biological Control* 143: 104190.

Shakeel, Q., Shakeel, M., Raheel, M., Ali, S., Ashraf, W. *et al.* (2022) Entomopathogenic nematodes (EPNs): a green strategy for management of insect-pests of crops. In: de Mandal, S., Ramkumar, G., Karthi, S. and Jin, F. (eds.) *New and Future Development in Biopesticide Research: Biotechnological Exploration*. Springer, Singapore, pp. 115–135.

Shapiro-Ilan, D.I., Hazir, S. and Glazer, I. (2022) Entomopathogenic nematodes as models for inundative biological control. In: Glazer, I., Shapiro-Ilan, D.I. and Sternberg, P.W. (eds.) *Nematodes as Model Organisms*. CAB International, Wallingford, UK, pp. 293–308.

Shields, E. (2021) Biological control of corn rootworm with anative NY entomopathogenic nematodes. *The Manager* 2021, 1–4. Available at: https://hdl.handle.net/1813/103540 (accessed 18 March 2024).

Usman, M., Gulzar, S., Wakil, W., Wu, S., Piñero, J.C. *et al.* (2020) Virulence of entomopathogenic fungi to *Rhagoletis pomonella* (Diptera: Tephritidae) and interactions with entomopathogenic nematodes. *Journal of Economic Entomology* 113, 2627–2633.

Usman, M., Wakil, W. and Shapiro-Ilan, D.I. (2021) Entomopathogenic nematodes as biological control agent against *Batocera zonata* and *Batocera dorsalis* (Diptera: Tephritidae). *Biological Control* 163: 104706.

14 Applications of Entomopathogenic Nematodes for Insect Pest Control in Corn

Stefan Toepfer[1]*, Xun Yan[2] and Bart Vandenbossche[3]
[1]CAB International, Delemont, Switzerland; [2]Zhongkai University of Agriculture and Engineering, Guangzhou, PR China; [3]e-nema Ltd., Schwentinental, Germany

Abstract

Corn (*Zea mays*) is a major carbohydrate provider and therefore crucial for food security. Numerous coleopteran and lepidopteran key pests, as well as some minor pests attack corn in most agricultural regions. This leads to an influx of insecticides into this agroecosystem. Nevertheless, the use of entomopathogenic nematodes (EPNs) has been limited against those insect pests. This is likely because corn is, despite its importance, usually not a cash crop. Nevertheless, some EPN products have been registered for use against corn pests in several countries. Herewith we summarize field studies on the use of EPNs against the most important soil and aboveground insect pests in corn. It appears that soil and aboveground application techniques for EPNs have largely improved for field crops. Moreover, recent bans of several soil insecticides and insecticidal seed coatings have increased the demand for alternative solutions, and may lead to a wider use of EPNs in corn.

14.1 Introduction

Corn (*Zea mays*, maize) is, next to wheat and rice, one of the three major carbohydrate providers to humans. Corn accounts for an estimated 40–50% of the food calories consumed globally (FAO, 2021). The area of corn production is estimated to be 197 million ha producing more than 1 million t of corn every year (Shiferaw *et al.*, 2011; FAO, 2021). Therefore, corn production is essential to global and local food security; and this is through both large-scale production and small-holder farming. In most countries, corn is grown as grain corn for feeding livestock as well as for human consumption (Shiferaw *et al.*, 2011; Erenstein *et al.*, 2002). In some countries, corn is grown for silage or for biogas production. Corn is also grown as sweet corn and for seed production but to a much smaller extent. Despite the importance of corn as a field crop, it is, in most areas, not a cash crop based on profit per cropping area. Thus, cash that may be available for agricultural inputs is limited. Therefore, corn is often not treated with insecticides or planted with transgenic corn hybrids in many agricultural areas of the world. Other reasons for not using these options can be: (i) an acceptability of some level of pest damage, such as to roots or leaves, which may have limited effect on yield; (ii) the phasing out of many insecticides, particularly soil insecticides; (iii) the unavailability of transgenic corn with insecticidal toxins; or (iv) the lack of high clearance or other insecticide application equipment. However, in other agricultural

*Email: s.toepfer@cabi.org

DOI: 10.1079/9781800620322.0014

areas, corn is treated with insecticides because of: (i) extreme damage is caused by some corn pests; (ii) damage to corn cobs, which directly reduces yield; (iii) secondary infections by aflatoxin-producing fungi that hinder selling of corn, such as for animal feed; or (iv) because corn is grown in a high-value situation such as for sweet corn or seed production. In these cases, influx of insecticides into the corn agroecosystem can be significant, particularly when many fields are treated. With increasing environmental concerns, reports of threats to bees and declining insect diversity, insecticides are increasingly restricted, which limits the growers' toolbox for pest management. Consequently, less disruptive crop-protection products have been developed for some corn pests, such as egg parasitoids (e.g. *Trichogramma* spp., Hymenoptera; Trichogrammatidae) or nucleopolyhedroviruses against some lepidopteran pests. Another option is the use of entomopathogenic nematodes (EPNs). This is because some life stages of nearly every corn pest appear to be susceptible to certain EPN species and strains, and because some of these EPN species are cosmopolitan and are therefore often relatively easy to register or do not need registration at all. However, their adoption for the protection of field crops has been limited (Cabanillas, *et al.*, 2005).

This chapter reviews the research outputs and usage of EPNs against insect pests in corn from various data sources (Clarivate, 2022; CNKI, 2022; NAIS, 2022; CABI, 2024a,b). It is an update of the review by Cabanillas *et al.* (2005). We focus primarily on field studies on EPN usage against corn pests and less on laboratory research. We also focus on the most important corn pests only. These include soil pests such as rootworms (*Diabrotica* spp., Chrysomelidae), grubs (Scarabeidae, Melolonthidae) and wireworms (Elateridae) among the Coleoptera, and cutworms (e.g. *Agrotis* spp., Noctuidae), among the Lepidoptera. These also include aboveground pests such as corn borers (*Ostrinia* spp., Crambidae), armyworms (e.g. *Spodoptera* spp., Noctuidae), stalk or stem borers (e.g. *Chilo* spp., *Busseola* spp., Noctuidae), and boll or budworms (*Helicoverpa* spp., *Heliothis* spp., Noctuidae) among the Lepidoptera. We hope that our findings will encourage biological control agent manufacturers, agri-input suppliers and growers to adapt the usage of EPNs for the control of pests in large-scale crops such as corn.

14.2 EPNs to Control Corn Pests

There are by far fewer EPN products available and used against pests in field crops such as corn than in cash crops such as vegetables, berries or tree nurseries. Although there has been plenty of research on the use of EPNs against corn pests, only a few products have reached the market for pest management in corn. About one-quarter of countries seem to have EPN-based plant-protection products registered for use in all crops (10 out of 43 countries from all continents assessed, Table 14.1) (CABI, 2024a). Therefore, these products are, in principle, also allowed for pest management in corn. They include six EPN species, namely *Heterorhabditis bacteriophora*, *H. downesi* and *H. megidis* (Rhabditida:

Table 14.1. Species of entomopathogenic nematodes (EPNs) in plant-protection products registered for use in all crops and therefore also in corn. This is in ten out of 43 assessed countries by 2024, registered by 21 registrants, including seven manufacturers (CABI, 2024a). Note that some countries do not require specific registration of native EPNs.

EPN species	Countries
Heterorhabditis bacteriophora	Brazil, Canada, France, Germany, Morocco, Portugal, Spain, UK
Heterorhabditis downesi	Germany, UK
Heterorhabditis megidis	Spain, UK
Heterorhabditis spp.	Canada
Steinernema carpocapsae	Canada, France, Germany, Morocco, Portugal, Spain, Uganda, UK
Steinernema feltiae	Canada, Chile, France, Germany, Morocco, Portugal, Spain, Uganda, UK
Steinernema kraussei	Canada, France, Germany, Morocco, Portugal, Spain, UK
Steinernema spp.	Canada, France

Heterorhabditidae), *Steinernema carpocapsae*, *S. kraussei* and *S. feltiae* (Rhabditida: Steinernematidae).

Around one in ten countries has EPN-based plant-protection products registered specifically for use in corn (Table 14.2) (CABI, 2022a). These products include only three EPN species, namely *Heterorhabditis bacteriophora*, *Steinernema carpocapsae* and *S. feltiae*. They are registered for several pests, but only few of these are key pests in corn.

It should be noted that a number of countries do not require any specific registration of native macrobial biological control agents, such as EPNs (Ehlers, 1996; Ehlers and Hokkanen, 1996). This eases the commercialization of EPNs. In this case, commercial EPN products such as *Steinernema longicaudum* in China or *Steinernema riobrave* and *Heterorhabditis indica* in the USA that are used in a certain crop may also be used without restriction in corn.

14.3 EPNs against Coleopteran Corn Pests

14.3.1 Corn rootworms

Corn rootworms are chrysomelid corn pests in the genus *Diabrotica* (Coleoptera) and include the western corn rootworm, *Diabrotica virgifera virgifera*, Mexican corn rootworm *Diabrotica virgifera zeae*, northern corn rootworm, *Diabrotica barberi*, southern corn rootworm, *Diabrotica undecimpunctata howardi*, and a few others (Krysan and Smith, 1987). Most significant is the damage caused by their larvae to the roots of corn leading to lodging of plants. *Diabrotica* spp. are of American origin and most that are corn pests are found in North America. *D. v. virgifera* has invaded European corn-producing regions.

As those rootworms share the soil habitat with EPNs, there have been extensive studies on their use. Application technologies have been developed and a few commercial EPN products have reached the market.

EPNs are mainly virulent to the larval stages of corn rootworms and to a much less extent to pupae or adults (Jackson and Brooks, 1995; Kurtz *et al.*, 2009; Santos *et al.*, 2011). There are several different species and strains of EPN that are highly virulent to larvae of *D. v. virgifera*, such as *H. bacteriophora*, *H. megidis*, *H. indica*, *H. atacamensis*, *S. feltiae*, *S. arenarium*, *S. kraussei*, *S. diaprepesi* and a few others (Jackson and Brooks, 1989, 1995; Toepfer *et al.*, 2005; Geisert *et al.*, 2018). Some strains of *S. carpocapsae* and *S. riobrave* appeared to be virulent to *D. v. virgifera* but others were less so (Jackson and Brooks, 1989; Journey and Ostlie, 2000; Toepfer *et al.*, 2005; Geisert *et al.*, 2018; Bruno *et al.*, 2020), indicating that strains within EPN species differ in their virulence to rootworms (Jackson and Brooks, 1995). *Steinernema abbasi* and *Heterorhabditis*

Table 14.2. Species of entomopathogenic nematodes (EPNs) in plant-protection products registered for use specifically in corn. This is in five out of 43 assessed countries by 2024 (CABI, 2024a). These had been registered by seven registrants including five manufacturers. Note that some countries do not require specific registration of native EPNs.

EPN species	Corn pests
Heterorhabditis bacteriophora	*Autographa gamma* (ES), *Chrysodeixis chalcites* (ES), *Diabrotica virgifera virgifera* (DE, HU), *Helicoverpa armigera* (ES), *Heliothis peltigera* (ES), *Heliothis* spp. (ES), *Melolontha melolontha* (ES), white grubs (ES, CR), *Spodoptera exigua* (ES), *Spodoptera littoralis* (ES), *Spodoptera* spp.(ES)
Steinernema carpocapsae	*Agrotis segetum* (ES), *Agrotis* spp. (ES), cutworms (ES), *Autographa gamma* (ES), *Chrysodeixis chalcites* (ES), *Diabrotica virgifera virgifera* (ES), *Gryllotalpa gryllotalpa* (ES), *Helicoverpa armigera* (ES), *Heliothis peltigera* (ES), *Heliothis* spp. (ES), *Ostrinia furnacalis* (CN), *Sphenophorus parvulus* (ES), *Spodoptera exigua* (ES), *Spodoptera littoralis* (ES), *Spodoptera* spp. (ES)
Steinernema feltiae	*Agrotis* spp. (ES), *Autographa gamma* (ES), *Chrysodeixis chalcites* (ES), *Helicoverpa armigera* (ES), *Heliothis peltigera* (ES), *Heliothis* spp. (ES), *Spodoptera exigua* (ES), *Spodoptera littoralis* (ES), *Spodoptera* spp.(ES)

CN, China; CR, Costa Rica; DE, Germany; ES, Spain; HU, Hungary.

mexicana were found to possess medium virulence, while *Steinernema glaseri* and *S. rarum* had low virulence (Toepfer *et al.*, 2005; Geisert *et al.*, 2018; Bruno *et al.*, 2020). Larvae of *D. undecimpunctata* are also susceptible to some strains of *H. bacteriophora* and *S. carpocapsae* (Kaya *et al.*, 1989; Barbercheck *et al.*, 1995), *Diabrotica balteata* to some strains of *H. bacteriophora*, *H. mexicana*, *H. indica*, *H. atacamensis*, *S. riobrave* and *S. carpocapsae*, (Creighton *et al.*, 1968; Bruno *et al.*, 2020), and *D. barberi* to some strains of *S. feltiae* and *S. carpocapsae*, but less to *Steinernema bibionis* (Thurston and Yule, 1990; Wright *et al.*, 1993). Larvae of the South American *Diabrotica speciosa* are susceptible to some strains of *S. glaseri*, *Heterorhabditis amazonensis* and a few others (Santos *et al.*, 2011). Some mermithid species infect *Diabrotica* spp. (Poinar and Welch, 1968; Creighton and Fassuliotis, 1982; Heineck-Leonel and Salles, 1997), but are difficult to mass culture for biological control purposes. Some rootworms such as *D. v. virgifera* have defense mechanisms against EPNs, for example through benzoxazinoids that are sequestered from corn roots (Bruno *et al.*, 2020). Some EPNs can overcome these defenses or are little affected, allowing the biological control of rootworms. In addition, exchanges of bacterial symbionts with benzoxazinoid-resistant strains can further improve efficacy against rootworms (Machado *et al.*, 2020).

Under field conditions, few EPN species or strains have been compared in their efficacy to control corn rootworms. *H. bacteriophora*, *H. megidis* and *S. carpocapsae* are highly effective (Wright *et al.*, 1993; Jackson and Hesler, 1995; Toepfer *et al.*, 2008), and *S. feltiae* is moderately effective (Toepfer *et al.*, 2008).

There has been extensive research on application techniques for EPNs against rootworms in corn. Fluid and granular applications are technically possible with available farmer machinery and can achieve control of rootworms (Journey *et al.*, 1992; Toepfer *et al.*, 2010a). Currently, the most used technique is the fluid stream spray of EPNs into the furrow at the time of sowing (Fig. 14.1A) (Knuth and Toepfer, 2012; Toepfer *et al.*, 2010a,b, 2014a; Modic *et al.*, 2018). In this way, EPNs are applied a few weeks before hatching of the larvae in the soil but they persist long enough to control the pest (Pilz *et al.*, 2012). Another advantage is that soil applications require little water (Vörös *et al.*, 2022). Moreover, EPNs such as *H. bacteriophora* can control *D. v. virgifera* larvae in a wide range of soil conditions (Grabenweger *et al.*, 2010; Toepfer *et al.*, 2010c). However, persistence over several corn-growing seasons is low due to high disturbance of soils and limited alternative hosts in soils of many corn production systems (Millar and Barbercheck, 2002; Kurtz *et al.*, 2007; Pilz *et al.*, 2012). The trend toward conservation tillage and the increased use of intermediate cover crops may improve the persistence of EPNs (Shapiro *et al.*, 1999).

Other effective but less common application techniques are fluid stream sprays into the soil next to young corn plants during mechanical weed control or using injectors (Toepfer *et al.*, 2010a,b; Maurath, 2022) (Fig. 14.1C, D). Methods to apply granules or capsules need further research (Fig. 14.1B) (Toepfer *et al.*, 2010a; Pilz *et al.*, 2011; Hiltpold *et al.*, 2012). Application through irrigation systems can reduce *D. v. virgifera*, *D. barberi* and other rootworms but needs large numbers of EPNs (Wright *et al.*, 1993; Ellsbury *et al.*, 1996). Aboveground field applications against adults have not been studied.

H. bacteriophora has been commercialized for use against *D. v. virgifera* larvae in corn in several European countries, and also, in a few countries, *S. carpocapsae* (Table 14.2). Commercial *H. megidis* or other EPNs registered for all crops could also be considered in corn (Toepfer *et al.*, 2010c). These EPNs can reach control efficacies of *D. v. virgifera* to the same extent and at similar variability as soil insecticides and insecticide seed coatings (Toepfer *et al.*, 2014a; Modic *et al.*, 2020; Toth *et al.*, 2022).

In conclusion, EPN-based biological control of *D. v. virgifera* is well developed, effective and practical. It might be extended to other countries and/or rootworm species.

14.3.2 Wireworms

Wireworms are elaterid pests primarily from the genus *Agriotes* and a few from the genera *Melanotus*, *Limonius*, *Selatosomus* and *Gonocephalum* (Coleoptera) (Toepfer *et al.*, 2014b; CABI, 2022). The larvae of many of these species are polyphagous soil pests, with some also attacking the roots of corn. Although larger corn plants can often compensate for wireworm damage, early

Fig. 14.1. Commercial application of entomopathogenic nematodes (EPNs) in corn against soil insect pests (A–E) and against above-ground insect pests (F–H). (A) Application of EPNs as a fluid during sowing. (B) Application of EPNs as granules during sowing. (C) Application of EPNs as a fluid through injectors after corn germination. (D) Application of EPNs as a fluid during mechanical weed control. (E) Application of EPNs as a fluid stream spray near plant bases along rows. (F) Application of EPNs as a fluid, narrow flat spray across leaf whorls. (G) Application of EPNs as a fluid spot spray into leaf whorls. (H) Application of EPNs as a gel into leaf whorls. Photographs by S. Toepfer and R. Maurath.

attack can kill young plants. Wireworms can be pests of corn in most world regions.

Many wireworms stay in the soil for several years and have therefore developed defense mechanisms against antagonists, including EPNs (Eidt and Thurston, 1995; Rahatkhah *et al.*, 2015). Therefore, studies on the use of EPNs against wireworms in corn are limited. Wireworm species differ in their susceptibility to certain EPN species and strains. For example, *Agriotes lineatus* and *A. obscurus* are more susceptible to *S. carpocapsae* or *H. bacteriophora*, than *S. feltiae*, *H. indica* or other EPNs, while *Agriotes sputator* seems most susceptible to *H. bacteriophora* (Ansari *et al.*, 2009; Morton and Garcia-del-Pino, 2017; Öğretmen *et al.*, 2020). *Limonius* spp. such as *L. californicus* can be controlled by *H. bacteriophora* and *S. rarum* (Kaur Sandhi *et al.*, 2020), and to some extent by *S. feltiae* (Toba and Turner, 1983). *Selatosomus* spp. seem even less susceptible to EPNs than other wireworms.

Most field studies have assessed applications of EPNs into or on to the soil against wireworms, while a few have started to test whether attract-and-kill approaches such as those based on attractive potato extracts or yeasts and EPNs are feasible (La Forgia *et al.*, 2020). However, the studies showed that high dosages of EPNs are needed to reach sufficient control of wireworms under field conditions.

In conclusion, the medium susceptibility of some wireworms to EPNs and therefore the need for large numbers to reach acceptable control levels, combined with a high damage threshold for wireworms to corn, may challenge the development of commercially viable biocontrol solutions against this group of pests (Lehmhus, 2020).

14.3.3 White grubs

White grubs include some coleopteran key pests of corn from the families Scarabeidae and Melolonthidae (Toepfer *et al.*, 2014b; CABI, 2022). Grubs can be pests in most corn-growing regions of the world, representing different genera and species. These include species of *Anomala*, *Holotrichia*, *Popillia*, *Melolontha*, *Maladera*, *Phyllophaga* and others (Toepfer *et al.*, 2014b).

Many grubs are polyphagous soil pests, with some also attacking corn. Older and therefore larger larvae in particular can heavily damage the roots of corn and can be deadly to seedlings. Many grub species stay for several years in the soil. Sometimes, their adults may also damage corn.

Studies on the use of EPNs against grubs have a long history but are limited in the corn agro-ecosystem (van Lenteren, 2020). Although grubs, particularly the older instars, are generally less susceptible to EPNs than other soil insect pests, several effective EPN species and strains exist against certain grubs and are reviewed in Chapter 21 (this volume). They include, depending on the target, certain strains of *H. bacteriophora*, *S. glaseri*, *S. longicaudum* and in a few cases *S. carpocapsae*, as well as some others.

Some EPN applications against grubs in other crops may be adapted to corn protection. For example, *S. carpocapsae* (and to some extent *H. bacteriophora* and *S. feltiae*) work against *Phyllophaga vetula* under field conditions in Mexico (Ruiz Vega *et al.*, 2012). Moreover, *H. bacteriophora* products have been registered against white grubs in corn in some countries such as Spain and Costa Rica (Table 14.2).

However, as with many other pests of corn, the wider use of EPNs against grubs is currently prevented by the scale of growing corn and its relatively small value compared with cash crops.

14.4 EPNs against Lepidopteran Corn Pests

14.4.1 Armyworms

Armyworms include some lepidopteran key pests of corn from the noctuid genera of *Spodoptera* and *Mythimna*, as well as from a few other locally important genera (CABI, 2022). Most armyworms are polyphagous. Several of these prefer corn, such as the fall armyworm, *Spodoptera frugiperda*, or the African armyworm, *Spodoptera exempta*. Others do not have corn as a main host but may still cause damage under certain cropping conditions, such as *Spodoptera littoralis*, *Spodoptera exigua* or other *Spodoptera* spp. Most armyworms are aboveground pests with larval stages that attack corn leaves, vegetation points, tassels or corn ears (Fuxa, *et al.*, 1988; CABI, 2022). Some may also cut seedlings.

Armyworms of different species can be pests in most corn-growing regions of the world, except Europe. The most prominent species is the American-origin *Spodoptera frugiperda*, which has invaded most of Africa, as well as large parts of Asia and Australia (Guo *et al.*, 2020; Babendreier *et al.*, 2022; Kenis *et al.*, 2022).

Studies on the use of EPNs against spodopteran armyworms, particularly *Spodoptera frugiperda*, are numerous. This is due to the disastrous damage they cause and the high sensitivity of their larvae to EPNs (Kaya and Gaugler, 1993; Fallet *et al.*, 2022b).

For example, many EPN species and strains are virulent to *Spodoptera frugiperda* whether or not they are from the area of origin of *Spodoptera frugiperda* or the area of invasion, or are commercial EPNs. Highly virulent are certain strains of *S. carpocapsae*, *S. riobrave*, *S. arenarium*, *S. longicaudum*, *S. siamkayai*, *S. diaprepesi*, *S. sangi*, *S. surkhetense H. amazonensis*, *H. mexicana*, *H. zacatecana*, *H. bacteriophora*, *H. indica*, *H. ruandica*, *H. baujardi* and others (Fuxa, *et al.*, 1988; Epsky and Capinera, 1993; Raulston, *et al.*, 1997; Garcia *et al.*, 2008; Andaló *et al.*, 2010; Caccia *et al.*, 2014; Leyva-Hernandez *et al.*, 2018; Lalramnghaki *et al.*, 2021; Wattanachaiyingcharoen *et al.*, 2021; Fallet *et al.*, 2022b). However, there is a considerable variability among strains of EPN species in their virulence to *Spodoptera frugiperda* (Fallet *et al.*, 2022b). Although studies on other spodopteran corn pests are less numerous, they seem also susceptible to several EPNs (Shamseldean *et al.*, 1995; Cabanillas *et al.*, 2005; Maketon *et al.*, 2011; Rufai and Wahab, 2021). Certain strains of *S. riobrave*, *S. longicaudum* and *H. indica* seem also virulent to the pupae of *Spodoptera frugiperda* (Raulston, *et al.*, 1997; Leyva-Hernandez *et al.*, 2018; Acharya *et al.*, 2020; Lalramnghaki *et al.*, 2021; Patil *et al.*, 2022) and strains of *S. carpocapsae* to pupae of *Spodoptera exigua* or *Mythimna unipuncta* (Kaya and Hara, 1980).

Several studies have assessed the field efficacy of EPNs against spodopteran corn pests and different approaches to their application. EPNs have been applied on to or into the soil to control either: (i) the soil-surface crawling armyworm larvae or those moving into the soil for pupation (Andaló *et al.*, 2010); or (ii) the pre- or postpupal stages. However, aerial applications of EPNs on to soil require relatively large amounts of water and high numbers of EPNs to assure the target pests are reached. Moreover, proper timing of such applications remains a challenge (Yan *et al.*, 2019).

EPNs have also been applied above ground. Armyworms often feed hidden in whorls of leaves or cobs that also protect applied EPNs from unfavorable abiotic conditions compared with an open surface. Nevertheless, studies have revealed moderate efficacies of EPNs applied against armyworms under field conditions. For example, narrow strip sprays of 2.5×10^8 *S. carpocapsae* Mexican or All strain (old synonym *S. feltiae* in this study) ha^{-1} above corn plants reduced 30–40% of larvae in some but not all study fields (Richter and Fuxa, 1990). When $4–8 \times 10^4$ of the EPNs were applied as single sprays on to the cobs, some but not all cobs were protected from damage. Spot applications of 3000 *S. carpocapsae* RW14-G-R3a-2 or *H. ruandica* unformulated in 2 ml of water per plant did not lead to major control effects (Fallet *et al.*, 2022a). Naturally, conventional flat sprays led to limited larval mortality, as shown for *H. indica* or *S. carpocapsae* (Negrisoli *et al.*, 2010a). This was even the case when such EPN sprays were followed by irrigation, or when EPNs were applied through overhead irrigation (Garcia *et al.*, 2008).

Despite armyworms often feeding in leaf whorls, unfavorable abiotic conditions remain a challenge for aboveground EPN applications (Lacey and Georgis, 2012), as are fast re-infestations of corn by the pests. Several protective formulations of EPNs have been studied for use against armyworms. Surfactant-polymer formulations can improve the aboveground efficacy of EPNs to some extent (Garcia *et al.*, 2008; Fallet *et al.*, 2022a). However, spraying $1.5–3 \times 10^9$ *S. carpocapsae* mixed with 0.1% Tween-80 and 0.1% corn oil h^{-1} did not improve control efficacies (X. Yan, 2023, unpublished data). Therefore, more protective formulations seem to be the key to EPN applications against armyworm larvae (Fallet *et al.*, 2022b; Kotliarevski *et al.*, 2022). Promising approaches are individual coating of EPNs with titanium dioxide nanoparticles for better ultraviolet (UV) protection (Kotliarevski *et al.*, 2022) or gel formulations to help keep EPNs on the plant and protect them from desiccation (Fallet *et al.*, 2022a). In any case, further developments toward narrow flat sprays of well-formulated

EPNs over plants along corn rows or spot applications into the corn whorl appear most promising (Fallet *et al.*, 2022a,b).

In regions with temporally overlapping armyworm populations, corn fields are quickly re-infested and therefore may require repeat treatments. Patil *et al.* (2022) used two treatments with stream sprays of 2.5×10^{8} *H. indica* 1NBAIIH38 or *S. carpocapsae* NBAIRS59 ha^{-1} corn rows with a 20-day interval leading to slightly improved efficacies. Fallet *et al.* (2024) reported that up to four treatments of *S. carpocapsae* RW14G-R3a-2 were needed to reach high efficacies and constant damage prevention, which is a similar situation for insecticides. A number of studies have shown that a combination of EPNs with insecticides can improve the control of armyworms (Negrisoli *et al.*, 2010a,b), but the economic viability of such approaches may be questioned.

EPN products such as *S. carpocapsae*, *S. feltiae* and *H. bacteriophora* have been registered in a few countries against *Spodoptera* spp. including *Spodoptera exigua* and *Spodoptera littoralis* (Table 14.2). However, their wider use remains limited.

In conclusion, the high virulence of EPNs to most armyworms as well as recently developed protective formulations indicate that EPN applications could be further developed toward larger-scale use in corn.

14.4.2 Corn borers

Corn borers are key lepidopteran pests of corn in the genera of *Ostrinia* and *Diatraea* (Crambidae) (Zhou *et al.*, 2020; CABI, 2022). Most corn borers have larvae that are oligophagous with preference to corn (Wang *et al.*, 2000). Larvae that attack corn above ground affect the leaves, vegetation points, stems and tassels. Their late larval instars usually bore into stems or corn cobs. Corn borers can be pests in most corn-growing regions of the world. Prominent species are the Asian corn borer, *Ostrinia furnacalis* (Wang *et al.*, 2000), in Asia and Australia, and the European corn borer, *Ostrinia nubilalis*, which has also invaded North American corn-growing regions (Lassance, 2010).

There have been multiple studies on the use of EPNs against *Ostrinia* spp. larvae, but field studies in corn are limited. Larvae of *Ostrinia* spp. are susceptible to several EPN species and strains. The most virulent strains are often found among *S. feltiae* or *S. carpocapsae* (He *et al.*, 1991; Yang *et al.*, 1991; Ben-Yakir *et al.*, 1998; Cheng *et al.*, 1998, 1999).

Some studies have assessed the use of EPNs against *O. furnacalis* in corn fields in China. For example, *S. carpocapsae* A24 strain reduced 70–80% of damage when sprayed at 3000–4000 EPNs per leaf whorl (Sun and Li, 2004). *S. carpocapsae* Beijing, Agriotos (old synonym *S. feltiae* Agriotos), and DD-136 (old synonym *S. feltiae* DD-136) reduced 60–70% of third- to fourth-instar larvae of high population densities of *O. furnacalis* when applied at 2000 EPNs per leaf whorl in aqueous or starch-paste formulations (He *et al.*, 1991; Pan *et al.*, 1995; Cheng *et al.*, 1999). Cheng *et al.* (1999) reached similarly promising results using formulations of *S. carpocapsae* in pastes of water-retentive polymers, or emulsions with soybean or cottonseed oil, and spraying them on to corn leaves infested with *O. furnacalis*. However, only a few commercial products reached the Chinese market, such as those based on *S. carpocapsae*. These Chinese products have been used in, for example, Henan Province in central eastern China at rates of up to 6×10^{9} EPNs ha^{-1} against first- to second-instar larvae at the vegetative corn stage (Zhang *et al.*, 2020). Nevertheless, the commercial use of EPNs against *O. furnacalis* in China and elsewhere remains limited.

S. carpocapsae has also been studied under field conditions to control larvae of *O. nubilalis*. For example, the Mexican strain was applied in 6% Folicote and 0.05% Tween-80 suspensions to seed corn in Israel (Ben-Yakir *et al.*, 1998). Thus, 50,000 EPNs were sprayed on to the corn ears of each plant against young larvae about 4 days after the pest's egg laying period. This reduced economic damage by 5–10%. However, commercial use of EPNs against *O. nubilalis* has not yet been achieved.

14.4.3 Stalk and stem borers

Stalk and stem borers are a diverse group of lepidopteran corn pests. Key pests are found in the noctuid genera *Chilo*, *Busseola*, *Sesamia*, *Diatraea* and a few others (Zhou *et al.*, 2020; CABI, 2022). Most stalk and stem borers have larvae that are oligophagous but perform particularly well on corn. They are usually aboveground pests with larvae that attack leaves, vegetation points, stems or tassels (CABI, 2022). Their late larval instars usually bore into stems or corn cobs. Different stalk and stem borer species can be pests in most corn-growing regions of the world. Prominent species are the African corn stalk borer, *Busseola fusca*, as well as the spotted stemborer, *Chilo partellus*, in Asia, which has also invaded Africa.

Studies on the use of EPNs against stalk and stem borers in corn are scarce. Screenings revealed that different strains of *S. carpocapsae*, *S. feltiae* and *H. bacteriophora* are virulent to different stalk or stem borers (Saleh *et al.*, 2000; Halawa *et al.*, 2007; El-Wakeil and Hussein, 2009; Gozel and Gunes, 2013). El-Wakeil and Hussein (2009) reported high efficacies of 5000 *S. carpocapsae* BA2 or *H. bacteriophora* BA1 sprayed on to corn plants against the corn stalk borer, *Sesamia cretica*, in Egypt, but Halawa *et al.* (2007) reported a generally higher efficacy by certain strains of *S. carpocapsae* than of *H. bacteriophora*. Saleh *et al.* (2000) found comparable results using a local *Heterorhabditis taysearae* and a commercial *H. bacteriophora*. In contrast, field studies against *Chilo* and *Busseola* spp. are largely lacking.

In general, the development of commercial EPN solutions against stem or stalk borers in corn has been challenged by difficulties in effectively reaching the hidden pest stages through aboveground EPN applications, by the relatively low cash value of corn, and by limited research and developments on EPN solutions in the regions where these pests are most problematic.

14.4.4 Bollworms and budworms

There are several lepidopteran corn pests in the genera *Helicoverpa* and *Heliothis* (Noctuidae) (CABI, 2024b). Most have polyphagous larvae and are key pests in several crops while also attacking corn. Although they feed on corn leaves, whorls, tassels and silk, their most significant damage is usually on the corn ears. They can be pests in many corn-growing regions of the world but may represent different species. Most prominent are the corn ear worm, *Helicoverpa zea*, across the Americas, the cotton bollworm, *Helicoverpa armigera*, in Africa, Asia and Europe, and a few *Heliothis* spp. in the Americas, Europe and Asia.

Studies on the use of EPNs against *Helicoverpa* and *Heliothis* spp. in corn are relatively limited (Cabanillas *et al.*, 2005). In principle, larvae, prepupae and to some extent the pupae of *Helicoverpa* and *Heliothis* spp. are susceptible to several EPN species such as *S. riobrave*, *S. feltiae* and *S. carpocapsae* (Raulston *et al.*, 1992; Cabanillas and Raulston, 1994; Cabanillas *et al.*, 2005). *Helicoverpa armigera* is susceptible to *Heterorhabditis amazonensis* (Andaló *et al.*, 2021).

Moderate levels of suppression may be achieved when these EPNs are applied as fluids to the foliage or silks of the corn plants (Tanada and Reiner, 1962; Bong and Sikorowski, 1983; Bong, 1986; Richter and Fuxa, 1990; Purcell *et al.*, 1992).

After feeding above ground, the mature larvae drop to the ground and pupate in the soil. The prepupal stage is a promising target for EPNs. EPNs have been highly effective when applied via in-furrow irrigation or subsurface injection, and slightly less effective when applied to the soil surface before or after irrigation (Cabanillas and Raulston, 1995; Feaster and Steinkraus, 1996). *S. riobrave* has been used more successfully than *S. carpocapsae*, likely due to its greater tolerance to high temperatures (Cabanillas and Raulston, 1994, 1995). However, the prepupal stage in the soil remains difficult to monitor or predict (Cabanillas *et al.*, 2005).

EPN products such as *S. carpocapsae*, *S. feltiae* or *H. bacteriophora* are registered against *Heliothis* and *Helicoverpa* spp. (e.g. *H. armigera*) in a few countries (Table 14.2). However, their wider use has been challenged by difficulties in effectively reaching the aboveground larvae, by EPN applications not preventing cob damage in the current season when EPNs are applied against prepupae in the soil, and by the costs of application in corn.

14.4.5 Cutworms

Cutworms include some lepidopteran key pests of corn from the genus *Agrotis*, as well as a few less or locally important pests from the genera *Xestia*, *Mythimna*, *Euxoa*, *Feltia*, *Hadula*, *Richia* and *Striacosta* (Noctuidae), among others (CABI, 2022). Many cutworms are polyphagous and occasional pests of corn plants depending on the crop history of a field. Cutworms are largely below- and on-ground pests that attack corn roots, the base of stems and young leaves of seedlings (CABI, 2022). Most damage is usually caused through the cutting of seedlings at or near the soil surface. A few species among the cutworms may also attack corn ears. Different species of cutworms can be pests in most corn-growing regions of the world.

Numerous laboratory studies on EPNs against cutworms are available (Devi, 2020), but there have only been a few in the field. In principle, cutworm larvae are more susceptible to EPNs than other soil insect pests. *Agrotis* spp., for example, are susceptible to certain strains of *S. carpocapsae*, *S. feltiae*, *S. weiseri*, *S. kraussei*, *S. thesami*, *H. bacteriophora* and others; but *S. carpocapsae* seems the most studied (Epsky and Capinera, 1993; Ebssa and Koppenhöfer, 2012; Lortkipanidze *et al.*, 2017; Devi, 2020).

Field applications as low as 1×10^9 but usually of more than 2.5×10^9 and up to 5.4×10^9 *S. carpocapsae* ha^{-1} have been shown to reduce damage caused by *Agrotis* cutworms by 50–80% (Capinera *et al.*, 1988; Levine and Oloumi-Sadeghi, 1992; Yokomizo and Kashio, 1996). EPNs have been applied using different approaches, such as soil-surface fluid sprays followed by incorporation into the soil or injections of fluids, capsules or wheat bran with EPNs into the soil (Capinera *et al.*, 1988). However, simple surface sprays need too large a volume of water, even in case of stream sprays, to be practical (Levine and Oloumi-Sadeghi, 1992). EPNs are usually applied at the time of planting or at corn emergence, and can reach efficacies similar to those of insecticides (Capinera *et al.*, 1988; Levine and Oloumi-Sadeghi, 1992).

EPN products such as *S. carpocapsae* or *S. feltiae* are registered against *Agrotis* spp. cutworms in a few countries (Table 14.2), but several products are commercialized in countries that do not require EPN registration.

In conclusion, the biological control of some cutworms by EPNs is possible. Their wider use may be achieved through further developing commercial applicators for EPNs at planting or emergence of corn.

14.5 Application Technologies of EPNs in Corn

Applications of EPNs against corn pests are technically possible in several ways depending on whether the soil or aboveground stages of an insect pest are targeted (Fig. 14.1). Several carriers have been developed for EPNs, based on powders, granules, sponges and others. These are then usually diluted in water to apply the EPNs to the corn (Cabanillas and Raulston, 1995; Cabanillas *et al.*, 2005; Garcia *et al.*, 2008; Toepfer *et al.*, 2010a; Fallet *et al.*, 2022a). Application formulations such as gels, granules, capsules and thick surfactants are under development for larger-scale use in corn (Journey *et al.*, 1992; Toepfer *et al.*, 2010a; Pilz *et al.*, 2011; Hiltpold *et al.*, 2012; Fallet *et al.*, 2024). Coating seeds with EPNs is challenging, as EPNs need moisture to survive and seeds need to remain dry prior to sowing. On-farm seed coating just before sowing might be an option.

14.5.1 Belowground corn pests

EPNs can be applied in different ways for belowground or surface insect pests in corn, depending on the life cycle of the considered pest.

EPNs can be applied as fluids or potentially as powders, granules or with other carriers into the soil at the time of sowing (Fig. 14.1A, B) (Capinera *et al.*, 1988; Cabanillas and Raulston, 1995; Feaster and Steinkraus, 1996; Toepfer *et al.*, 2010a,b). Alternatively, they can be injected into the soil along rows of plants, for example using commercial injectors (Fig. 14.1C) (Capinera *et al.*, 1988; Toepfer and Zellner, 2017; Maurath, 2022) or during mechanical weeding (Fig. 14.1D). An advantage of these applications into the soil is that they require little water. They are usually applied at 200–300 l of water ha^{-1}, but this can likely be reduced (Vörös *et al.*, 2022). Against *Diabrotica* spp. rootworms, for example, currently the most used EPN application methods are fluid stream sprays into the sowing furrow, using sowing machines with applicators that apply EPNs behind the sowing wheel and prior to the soil-closing wheels, or into the soil along rows of corn seedlings using injectors (Fig. 14.1C) (Toepfer *et al.*, 2010a, 2010b, 2014a; Knuth and Toepfer, 2012; Modic *et al.*, 2018; Maurath, 2022).

Another way of EPN application is to apply them along corn rows as fluid stream sprays on to the soil near the stem bases of the corn (Fig. 14.1E)

(Capinera *et al.*, 1988; Levine and Oloumi-Sadeghi, 1992; Toepfer *et al.*, 2010a,b). However, these methods require larger amounts of water to allow EPNs to pass through the usually dry soil surface into the ground (≥1000 l water ha^{-1}) (Levine and Oloumi-Sadeghi, 1992; Toepfer *et al.*, 2014a).

Finally, EPNs could be applied through irrigation systems, with the disadvantage that large quantities of EPNs may be needed and sedimentation of EPNs in the irrigation system needs to be prevented (Wright *et al.*, 1993; Ellsbury *et al.*, 1996; Garcia *et al.*, 2008).

14.5.2 Aboveground corn pests

EPN usage against aboveground insect pests is challenging in corn. One challenge is the unfavorable abiotic conditions for EPNs that are encountered above ground (i.e. desiccation, UV radiation, heat) (Kaya and Gaugler, 1993; Lacey and Georgis, 2012; Kagimu *et al.*, 2017). Moreover, some of these pests may be hidden deep inside corn stubble, stems or cobs, such as some of the stem and stalk borers (e.g. *Busseola* and *Chilo* spp.) or corn borers (e.g. *Ostrinia* spp.). These pests may only be controlled in their young larval stage before entering corn stems, and proper timing and the low efficacy of surface sprays of EPNs become a challenge. Other aboveground pests such as spodopteran armyworms are accessible for longer periods as they remain reachable to applied EPNs when hidden in corn whorls. In such cases, EPNs may be applied in liquid formulations as spot or narrow flat sprays (Fig. 14.1F, G) or even through irrigation, although this requires large numbers of EPNs (Richter and Fuxa, 1990). To reduce the effects of abiotic factors on EPNs, they have been incorporated into protective formulations (Fallet *et al.*, 2022a; Kotliarevski *et al.*, 2022). For example, individual coating of EPNs with titanium dioxide nanoparticles and mineral oils in emulsions protect EPNs against UV radiation (Kotliarevski *et al.*, 2022). Several other UV protectants are available. Surfactant-polymer formulations can improve the aboveground efficacy of EPNs to some extent (Garcia *et al.*, 2008; Fallet *et al.*, 2022a). Another approach is using gels, such as those based on carboxymethyl cellulose, which help EPNs remain on the plant, particularly in the corn whorl, and protect them from desiccation (Fig. 14.1H) (Fallet *et al.*, 2022a, 2024). Such spot applications of gels can be implemented manually by small-plot farmers. For larger-scale farming, precision-agriculture approaches for a practical way of spot application of gel-based or otherwise formulated EPNs need to be further developed for top treatments into corn whorls.

Many aboveground lepidopteran pests of corn have dormant stages in the soil, such as *Helicoverpa* and *Heliothis* spp., armyworms, corn borers, and some stem and stalk borers. Therefore, EPNs could be applied on to or into the soil as described above for the control of soil insect pests (Capinera *et al.*, 1988; Cabanillas and Raulston, 1995; Feaster and Steinkraus, 1996; Andaló *et al.*, 2010; Lacey and Georgis, 2012). However, the pupae are often difficult to control by EPNs due to their limited susceptibility. In contrast, pre- or postpupal stages are more susceptible and are better targets for EPNs (Cabanillas and Raulston, 1995; Feaster and Steinkraus, 1996). However, monitoring for such specific life stages concealed in the soil and proper timing of treatments are challenging (Cabanillas *et al.*, 2005).

14.5.3 Economics

Despite several technical possibilities for applying EPNs against different target pests in corn below as well as above ground, economic viability needs to be considered (Journey *et al.*, 1992; Toepfer *et al.*, 2010a).

For small-plot farmers who manually manage their corn field, EPNs can be manually applied directly to the area where a target insect needs to be controlled, for example using a backpack sprayer to apply EPNs as a fluid stream or spot spray on to the seeding or at the stem base of corn against soil pests, as well as into leaf whorls against some lepidopteran larvae. Other formulations such as granules or gels can also be manually applied. Therefore, the EPN products themselves are the major cost factor and limitation in such farming situations.

For medium- to large-scale commercial growers, the application process of EPNs is a major cost factor. Although several technical possibilities are commercially available, a number of cost factors may hinder the use of EPNs. One factor is the cost of driving and applying EPNs across a field. For EPN applications during corn sowing or during mechanical weed control, no extra driving would be needed. Other types of EPN application usually require extra drives and therefore raise costs. In some cases, specific machinery such as high-wheel spraying machinery with narrow strip spray possibilities, or into-soil applicators may be a cost or availability factor. Finally, EPN products may, although already close to the prices of some synthetic insecticides, still be slightly more expensive. In some countries, express-shipment costs of EPNs to the grower and a relatively short shelf life may hinder the use of EPNs. However, many synthetic insecticides have been or are being phased out from use in corn production, increasing the need for alternative products. In some countries, subsidy programs for biological control products exist, facilitating their use (Wei *et al.*, 2019).

In conclusion, advances in the development of more practical and cheaper application techniques for EPNs in field crops offer possibilities for successful EPN application depending on the local situation and the pest species and development stage to be targeted. Moreover, decreasing product prices may allow a broader use of EPNs in the future.

14.6 Conclusions

The use of EPNs in field crops such as corn remains limited compared with their use in cash crops. Key factors behind the success of a plant-protection product are efficacy, safety, practicability, availability and economic viability.

Most corn pests are highly susceptible to numerous EPN species or strains. This is particularly true for the aboveground larvae of many lepidopteran corn pests, perhaps because they have not coevolved with the usually soil-inhabiting EPNs. However, it also holds true for several soil pests, such as corn rootworms, cutworms and some grub species. Many of the commercially available EPNs seem to be effective against a multitude of corn pests (Cabanillas *et al.*, 2005; Bruno *et al.*, 2020; Fallet *et al.*, 2022b). We suggest that the numerous researchers who conduct EPN screening also include commercial EPNs from their region as a comparison. Under field conditions, EPN usage has had medium or variable successes for pest control in corn. However, recent advances in application technologies have largely improved the efficacy and reliability of EPNs so that they have similar or even better or longer control effects than chemical insecticides. This is particularly true for EPNs injected into the soil, but also for aboveground applications if EPNs are properly formulated. Compared with synthetic and microbial insecticides, EPNs also have the advantage of being able to actively seek for the larvae that may feed hidden in the whorl (Richter and Fuxa, 1990) or in the roots of corn (Hiltpold and Turlings, 2008).

EPNs are considered safe to humans, livestock, crops and the environment. Moreover, they do not leave residues behind, and no maximum residue level consideration is needed. Considering the large scale of corn production and therefore the large scale of regular insecticide influx into these agroecosystems, the use of EPNs is a much safer and less disruptive approach. Therefore, many countries do not require registration processes for products based on EPNs. EPN usage has become more convenient with regard to large-scale use in field crops such as corn. For soil pest control, EPN products can be diluted in water, and no specific carriers or protectants are needed. Into-soil applicators for the use of EPNs during corn sowing or later at young plant stages are commercially available (Toepfer and Kuhlman, 2013; Maurath, 2022). Aboveground applications have become more feasible because protective formulations have been developed that keep the EPNs longer on the plant, such as in the leaf whorls of corn. In addition, strip sprays above the plants along corn rows are technically possible and effective. Further developments of precision agriculture toward into-whorl spot applications could further ease practicability. A disadvantage with regard to practicability is the relatively low shelf life of EPNs compared with chemical insecticides, and the need for cool storage and transport.

Although the number of countries with registered EPN products for corn, or for all crops including corn, has increased (Table 14.2), accessibility of products by growers remains a challenge in some regions. On the one hand, there is the problem of the relatively short shelf life of EPNs compared with insecticides. On the other hand, there are difficulties in express-shipping EPNs to growers in some regions. Although at least ten larger-scale commercial manufacturers of EPNs exist globally (CABI, 2024a), more may be needed in certain world regions. It could also be argued that more regional or local EPN manufacturers may be needed that rely on low- or medium-tech semi-solid approaches or small fermenters for mass producing EPNs (Holmes *et al.*, 2015), rather than on high-tech, large-volume fermenters.

With regard to economic viability, EPNs, as with any other plant-protection product, depend on the profitability of a crop and therefore cash availability by the grower. Unfortunately, corn is, despite its importance for food security, not a cash crop in many regions, hindering the use of EPNs. More positively, the price of EPN products has largely decreased over the last decade. Due to improved application technologies, the EPN dosage required has also been reduced from around 5 billion EPNs ha^{-1} for a cash crop down to 3 billion or even 2 billion ha^{-1} for aboveground control of corn pests, and to 2 billion or even 1 billion EPNs for into-soil applications. As a result, EPN usage often reaches costs that are nearly as low as the use of insecticides. Moreover, the recent and ongoing phase-out of a number of soil insecticides or insecticidal seed coatings in many world regions has increased the need for alternative solutions, including EPNs.

In conclusion, we believe that the importance of corn for food security and the improvements of application technologies for EPNs in field crops, as well as an increasing availability of EPNs, combined with a decreasing availability of insecticides may increase the use of EPNs in field crops in the near future.

14.7 References

Acharya, R., Hwang, H.S., Mostafiz, M.M., Yu, Y.S. and Lee, K.Y. (2020) Susceptibility of various developmental stages of the fall armyworm, *Spodoptera frugiperda*, to entomopathogenic nematodes. *Insects* 11: 868.

Andaló, V., Santos, V., Moreira, G.F., Moreira, C.C., Junior, A.M. *et al.* (2010) Evaluation of entomopathogenic nematodes under laboratory and greenhouses conditions for the control of *Spodoptera frugiperda*. *Ciencia Rural* 40, 1860–1866.

Andaló, V., de Faria, L.S., Carvalho, F.J., de ssis, G.A., Santos, V. *et al.* (2021) Entomopathogenic nematodes for the control of *Helicoverpa armigera* (Hübner) (Lepidoptera: Noctuidae) pupae. *Arquivos do Instituto Biológico (São Paulo)* 88: e00742019.

Ansari, M.A., Evans, M. and Butt, T.M. (2009) Identification of pathogenic strains of entomopathogenic nematodes and fungi for wireworm control. *Crop Protection* 28, 269–272.

Babendreier, D., Toepfer, S., Bateman, M. and Kenis, M. (2022) Management options for the invasive moth *Spodoptera frugiperda* in Europe. *Journal of Economic Entomlogy* 115, 1772–1782.

Barbercheck, M.E., Wang, J. and Hirsh, I.S. (1995) Host-plant effects on entomopathogenic nematodes. *Journal of Invertebrate Pathology* 66, 169–177.

Ben-Yakir, D., Efron, D., Chen, M. and Glazer, I. (1998) Evaluation of entomopathogenic nematodes for biocontrol of the European corn borer, *Ostrinia nubilalis*, on sweet corn in Israel. *Phytoparasitica* 26, 101–108.

Bong, C.F.J. (1986) Field control of *Heliothis zea* (Boddie) (Lepidoptera: Noctuidae) using parasitic nematodes. *Insect Science and Its Application* 7, 23–25.

Bong, C.F.J. and Sikorowski, P.P. (1983) Use of the DD 136 strain of *Neoaplectana carpocapsae* Weiser (Rhabditida: Steinernematidae) for control of corn earworm (Lepidoptera: Noctuidae). *Journal of Economic Entomology* 76, 590–593.

Bruno, P., Machado, R.A.R., Glauser, G., Köhler, A., Campos-Herrera, R. *et al.* (2020) Entomopathogenic nematodes from Mexico that can overcome the resistance mechanisms of the western corn rootworm. *Scientific Reports* 10: 8257.

Cabanillas, H.E. and Raulston, J.R. (1994) Pathogenicity of *Steinernema riobravis* against corn earworm, *Helicoverpa zea* (Boddie). *Fundamental and Applied Nematology* 17, 219–223.

Cabanillas, H.E. and Raulston, J.R. (1995) Impact of *Steinernema riobravis* (Rhabditida: Steinernematidae) on the control of *Helicoverpa zea* (Lepidoptera: Noctuidae) in corn. *Journal of Economic Entomology* 88, 58–64.

Cabanillas, H.E., Wright, R.J. and Vyas, R.V. (2005) Cereal, fibre, oilseed and medicinal crop applications. In: Grewal, P.S., Ehlers, R.U. and Shapiro-Ilan, D.I. (eds.) *Nematodes as Biocontrol Agents*. CAB International, Wallingford, UK, pp. 265–280.
CABI (2022) *Crop Protection Compendium*, 2nd edn. CAB International, Wallingford, UK. Available at: www.cabi.org/cpc (accessed 19 March 2024).
CABI (2024a) Bioprotection Portal. CAB International, Wallingford, UK. Available at: https://bioprotectionportal.com/ (accessed 19 March 2024).
CABI (2024b) CAB Abstracts. CAB International, Wallingford, UK. Available at: www.cabdirect.org/cabdirect (accessed 19 March 2024).
Caccia, M.G., Del Valle, E., Doucet, M.E. and Lax, P. (2014) Susceptibility of *Spodoptera frugiperda* and *Helicoverpa gelotopoeon* (Lepidoptera: Noctuidae) to the entomopathogenic nematode *Steinernema diaprepesi* (Rhabditida: Steinernematidae) under laboratory conditions. *Chilean Journal of Agricultural Research* 74, 123–126.
Capinera, J.L., Epsky, N.D., Menout, G.S. and Pelissier, D. (1988) Control of black cutworm, *Agrotis ipsilon* (Lepidoptera: Noctuidae), with entomogenous nematodes (Nematoda: Steinernematidae, Heterorhabditidae). *Journal of Invertebrate Pathology* 52, 427–435.
Cheng, C., Tang, L., Hou, R.F., Cheng, C.C. and Tang, L.C. (1998) Efficacy of the entomopathogenic nematode, *Steinernema carpocapsae* (Rhabditida: Steinernematidae), against the Asian corn borer, *Ostrinia furnacalis* (Lepidoptera: Pyralidae). *Chinese Journal of Entomology* 18, 51–60.
Cheng, C.C., Tang, L.C., Hou, R.F., Cheng, C.C., Tang, L.C. *et al.* (1999) Characteristics of a paste formulation containing entomopathogenic nematode, *Steinernema carpocapsae* (Nematoda: Steinermatidae) and its application for control of the Asian corn borer, *Ostrinia furnacalis* (Lepidoptera: Pyralidae). *Chinese Journal of Entomology* 19, 265–277.
Clarivate (2022) Web of Science. Available at: www.webofscience.com/wos/woscc/basic-search (accessed 19 March 2024).
CNKI (2022) China National Knowledge Infrastructure Database. Available at: www.cnki.net/ (accessed 19 March 2024).
Creighton, C.S. and Fassuliotis, G. (1982) Mass rearing a mermithid nematode, *Filipjevimermis leipsandra* (Mermithida: Mermithidae) on the banded cucumber beetle (*Diabrotica balteata*) (Coleoptera: Chrysomelidae). *Journal of Economic Entomology* 76, 615–618.
Creighton, C.S., Cuthbert F.P. Jr. and Reid W.J. Jr. (1968) Susceptibility of certain coleopterous larvae to the DD-136 nematode. *Journal of Invertebrate Pathology* 10, 368–373.
Devi, G. (2020) Management of cutworm by entomopathogenic nematodes – a review. *International Journal of Current Microbiology and Applied Sciences* 9, 2520–2526.
Ebssa, L. and Koppenhöfer, A.M. (2012) Entomopathogenic nematodes for the management of *Agrotis ipsilon*: effect of instar, nematode species and nematode production method. *Pest Management Science*. 68, 947–957.
Ehlers, R.-U. (1996) Current and future use of nematodes in biocontrol: practice and commercial aspects with regard to regulatory policy issues. *Biocontrol Science and Technology* 6, 303–316.
Ehlers, R.-U. and Hokkanen, H.M.T. (1996) Insect biocontrol with non-endemic entomopathogenic nematodes (*Steinernema* and *Heterorhabditis* spp.): conclusions and recommendations of a combined OECD and COST workshop on scientific and regulatory policy issues. *Biocontrol Science and Technology* 6, 295–302.
Eidt, D.C. and Thurston, G.S. (1995) Physical deterrents to infection by entomopathogenic nematodes in wireworm (Coleoptera: Elateridae) and other soil pests. *Canadian Entomologist* 127, 423–429.
Ellsbury, M.M., Jackson, J.J., Woodson, W.D., Beck, D.L. and Stange, K.A. (1996) Efficacy, application distribution, and concentration by stemflow of *Steinernema carpocapsae* (Rhabditia:Steinernematide) suspensions applied with a lateral-move irreigation system for corn rootworm (Coleoptera:Chrysomelidae) control in maize. *Journal of Economic Entomology* 85, 2425–2432.
El-Wakeil, N.E. and Hussein, M.A. (2009) Field performance of entomopathogenic nematodes and an egg parasitoid for suppression of corn borers in Egypt. *Archives of Phytopathology and Plant Protection* 42, 228–237.
Epsky, N.D. and Capinera, J.L. (1993) Quantification of invasion of two strains of *Steinernema carpocapsae* (Weiser) into three lepidopteran larvae. *Journal of Nematology* 25, 173–180.
Erenstein, O., Jaleta, M., Sonder, K., Mottaleb, K. and Prasanna, B.M. (2002) Global maize production, consumption and trade: trends and R&D implications. *Food Security* 14, 1295–1319.
Fallet, P., Bazagwira, D., Morgane Guenat, J., Bustos-Segura, C., Karangwa, P. *et al.* (2022a) Laboratory and field trials reveal the potential of a gel formulation of entomopathogenic nematodes for the biological control of fall armyworm caterpillars (*Spodoptera frugiperda*). *Biological Control* 176: 105086.

Fallet, P., de Gianni, L., Machado, R.A.R., Bruno, P., Bernal, J.S. *et al.* (2022b) Comparative screening of Mexican, Rwandan and commercial entomopathogenic nematodes to be used against invasive fall armyworm, *Spodoptera frugiperda*. *Insects* 13: 205.

Fallet, P., Bazagwira, D., Ruzzante, L., Ingabire, G., Levivier, S. *et al.* (2024) Entomopathogenic nematodes as an effective and sustainable alternative to control the fall armyworm in Africa. *PNAS Nexus* 3, 122–130.

FAO (Food and Agriculture Organization of the United Nations) (2021) FAO Statistics. FAO, Rome. Available at: www.fao.org/food-agriculture-statistics/en/ (accessed 19 March 2024).

Feaster, M.A.A. and Steinkraus, D.C. (1996) Inundative biological control of *Helicoverpa zea* (Lepidoptera: Noctuidae) with the entomopathogenic EPN *Steinernema riobravis* (Rhabditida: Steinernematidae). *Biological Control* 7, 38–43.

Fuxa, J.R., Richter, A.R. and Acudelo-Silva, F. (1988) Effect of host age and nematode strain on susceptibility of *Spodoptera frugiperda* to *Steinernema feltiae*. *Journal of Nematology* 20, 91–95.

Garcia, L.C., Raetano, C.G. and Leite, L.G. (2008) Application technology for the entomopathogenic nematodes *Heterorhabditis indica* and *Steinernema* sp. (Rhabditida : Heterorhabditidae and Steinernematidae) to control *Spodoptera frugiperda* (Smith) (Lepidoptera : Noctuidae) in corn. *Neotropical Entomology* 37, 305–311.

Geisert, R.W., Cheruiyot, D.J., Hibbard, B.E., Shapiro-Ilan, D.I., Shelby, K.S. and Coudron, T.A. (2018) Comparative assessment of four Steinernematidae and three Heterorhabditidae species for infectivity of larval *Diabrotica virgifera virgifera*. *Journal of Economic Entomology* 111, 542–548.

Gozel, U. and Gunes, C. (2013) Virulence of three entomopathogenic nematode species on corn stalk borer (*Sesamia cretica* Led. Lep: Noctuidae) at different temperatures. *African Journal of Agricultural Research* 8, 2178–2182.

Grabenweger, G., Pilz, C. and Heimbach, U. (2010) Influence of soil texture and density on the rate of survival of larvae of the western corn rootworm (*Diabrotica virgifera virgifera*). In: *Julius-Kühn-Archiv* 428, 188.

Guo, J., Wu, S., Zhang, F., Huang, C., He, K. (2020) Prospects for microbial control of the fall armyworm *Spodoptera frugiperda*: a review. *BioControl* 65, 647–662.

Halawa, S.M., El-Kholy, M.Y. and El-Zaher, T.R.A. (2007) Evaluation of entomopathogenic nematodes, a commercial bacterial bio-insecticide and the peppermint oil for the control the pink stem borer, *Sesamia cretica* Led. (Lepidoptera: Noctuidae). *Egyptian Journal of Biological Pest Control* 17, 1–11.

He, K.L., Zhou, D.R. and Yang, H.W. (1991) Biological control of Asian corn borer with the entomopathogenic nematode, *Steinernema feltiae* Agriotos. *Chinese Journal of Biological Control* 7, 1–6.

Heineck-Leonel, M.A. and Salles, L.A.B. (1997) Incidence of parasitoids and pathogens in adults of *Diabrotica speciosa* (Germ.) (Coleoptera: Chysomelidae) in Pelotas, RS. *Anais da Sociedade Entomologica do Brasil* 26, 81–85.

Hiltpold, I. and Turlings, T.C.J. (2008) Belowground chemical signaling in maize: when simplicity rhymes with efficiency. *Journal of Chemical Ecology* 34, 628–635.

Hiltpold, I., Hibbard, B.E., French, B.W. and Turlings, T.C.J. (2012) Capsules containing entomopathogenic nematodes as a Trojan horse approach to control the western corn rootworm. *Plant and Soil* 358, 11–25.

Holmes, K.A., Chen, J., Bollhalder, F., Ri, U., Waweru, B. *et al.* (2015) Designing factories for nematode-based biological control products for an alternative, environmentally friendly management of soil insect pests. *African Journal of Agricultural Research* 10, 4432–4448.

Jackson, J.J. and Brooks, M.A. (1989) Susceptibility and immune response of western corn rootworm larvae (Coleoptera: Chrysomelidae) to the entomogenous nematode, *Steinernema feltiae* (Rhabditida: Steinernematidae). *Journal of Economic Entomology* 82, 1073–1077.

Jackson, J.J. and Brooks, M.A. (1995) Parasitism of western corn rootworm larvae and pupae by *Steinernema carpocapsae*. *Journal of Nematology* 27, 15–20.

Jackson, J.J. and Hesler, L.S. (1995) Placement and application rate of the nematode *Steinernema carpocapsae* (Rhabditida: Steinernematidae) for suppression of the western corn rootworm (Coleoptera: Chrysomelidae). *Journal of the Kansas Entomological Society* 68, 461–467.

Journey, A.M. and Ostlie, K.R. (2000) Biological control of the western corn rootworm (Coleoptera: Chrysomelidae) using the entomopathogenic nematode, *Steinernema carpocapsae*. *Environmenal Entomology* 29, 822–831.

Journey, A.M., Ostlie, K.R. and Luedeman, L.J. (1992) Liquid options for corn rootworm control, 1991. *Insecticide and Acaricide Tests* 17, 198.

Kagimu, N., Ferreira, T. and Malan, A.P. (2017) The attributes of survival in the formulation of entomopathogenic nematodes utilised as insect biocontrol agents. *African Entomology* 25, 275–291.

Kaur Sandhi, R., Shapiro-Ilan, D. and Reddy, G.V.P. (2020) Montana native entomopathogenic nematode species against *Limonius californicus* (Coleoptera: Elateridae). *Journal of Economic Entomology* 113, 2104–2111.

Kaya, H.K. and Gaugler, R. (1993) Entomopathogenic nematodes. *Annual Review of Entomology* 38, 181–206.

Kaya, H.K. and Hara, A.H. (1980) Differential susceptibility of lepidopterous pupae to infection by the nematode *Neoaplectana carpocapsae*. *Journal of Invertebrate Pathology* 36, 389–393.

Kaya, H.K., Leong, K.L.H., Burlando, T.M., Smith, K. and Yoshimura, M.A. (1989) Entomogenous nematodes for biological control of the Western spotted cucumber beetle, *Diabrotica undecimpunctata*. *Journal of Nematology* 21, 568–569.

Kenis, M., Benelli, G., Biondi, A., Calatayud, P.-A., Day, R. *et al.* (2022) Invasiveness, biology, ecology, and management of the fall armyworm, *Spodoptera frugiperda*. *Entomologia Generalis* 43, 187–241.

Knuth, P. and Toepfer, S. (2012) Application, persistence and efficacy of entomopathogenic nematodes for controlling larvae of the western corn rootworm. *Julius-Kühn-Archiv* 438, 164.

Kotliarevski, L., Cohen, R., Ramakrishnan, J., Wu, S., Mani, K.A. *et al.* (2022) Individual coating of entomopathogenic nematodes with titania (TiO_2) nanoparticles based on oil-in-water pickering emulsion: a new formulation for biopesticides. *Journal of Agricultural and Food Chemistry* 70, 13518–13527.

Krysan, J.L. and Smith, R.F. (1987) Systematics of the virgifera species group of *Diabrotica* (Coleoptera: Chrysomelidae: Galerucinae). *Entomography* 5, 375–484.

Kurtz, B., Toepfer, S., Ehlers, R.U. and Kuhlmann, U. (2007) Assessment of establishment and persistence of entomopathogenic nematodes for biological control of western corn rootworm. *Journal of Applied Entomology* 131, 420–425.

Kurtz, B., Hiltpold, I., Turlings, T.C.J., Kuhlmann, U. and Toepfer, S. (2009) Comparative susceptibility of larval instars and pupae of the western corn rootworm to infection by three entomopathogenic nematodes. *BioControl* 54, 255–262.

La Forgia, D., Jaffuel, G., Campos-Herrera, R., Verheggen, F.J. and Turlings, T.C.J. (2020) Efficiency of an attract-and-kill system with entomopathogenic nematodes against wireworms (Coleoptera: Elateridae). *IOBC-WPRS Bulletin* 150, 91–95.

Lacey, L. and Georgis, R. (2012) Entomopathogenic nematodes for control of insect pests above and below ground with comments on commercial production. *Journal of Nematology* 44, 218–225.

Lalramnghaki, H.C., Lalramliana, Lalremsanga, H.T., Vanlalhlimpuia, Lalramchuani, M. and Vanramliana (2021) Susceptibility of the fall armyworm, *Spodoptera frugiperda* (J.E. Smith, 1797) (Lepidoptera: Noctuidae), to four species of entomopathogenic nematodes (Steinernematidae and Heterorhabditidae) from Mizoram, North-Eastern India. *Egyptian Journal of Biological Pest Control* 31: 110.

Lassance, J.M. (2010) Journey in the Ostrinia world: from pest to model in chemical ecology. *Journal of Chemical Ecology* 36, 1155–1169.

Lehmhus, J. (2020) Wireworm biology in Middle Europe – what are we facing? In: *Microbial and Nematode Control of Invertebrate Pests. IOBC-WPRS Bulletin* 150, 96–99.

Levine, E. and Oloumi-Sadeghi, H. (1992) Field evaluation of *Steinernema carpocapsae* (Rhabditida: Steinernematidae) against black cutworm (Lepidoptera: Noctuidae) larvae in field corn. *Journal of Entomological Science* 27, 427–435.

Leyva-Hernandez, H.A., Garcia-Gutierrez, C., Ruiz-Vega, J., Calderon-Vazquez, C.L. *et al.* (2018) Evaluation of the virulence of *Steinernema riobrave* and *Rhabditis blumi* against third instar larvae of *Spodoptera frugiperda*. *South Western Entomologist* 43, 189–197.

Lortkipanidze, M., Gorgadze, O., Babayev, V., Nwolisa, N. and Gabroshvili, N. (2017) Susceptibility of the cutworm, *Agrotis segetum* (Lepidoptera: Noctuidae) to entomopathogenic nematodes. *IOBC/WPRS Bulletin* 129, 120–124.

Machado, R.A.R., Thönen, L., Arce, C.C.M., Theepan, V., Prada, F. *et al.* (2020) Engineering bacterial symbionts of nematodes improves their biocontrol potential to counter the western corn rootworm. *Nature Biotechnology* 38, 600–608.

Maketon, M., Somsook, V. and Rattanakorn, P. (2011) Pathogenicity and culture of a *Heterorhabditis indica* isolate from Thailand. *Nematropica* 41, 52–61.

Maurath, R. (2022) Nematoden wirken gegen Maiswurzelbohrer. *Badische Bauernzeitung Pflanzenbau* 2022, 2.

Millar, L.C. and Barbercheck, M.E. (2002) Effects of tillage practices on entomopathogenic nematodes in a corn agroecosystem. *Biological Control* 25, 1–11.

Modic, Š., Žigon, P., Kolmanič, A., Godeša, T. and Razinger, J. (2018) Effectiveness of different control measures against western corn rootworm larvae *Diabrotica virgifera virgifera* LeConte, 1868. *Acta Agriculturae Slovenica* 111, 161–167.

Modic, Š., Žigon, P., Kolmanič, A., Trdan, S. and Razinger, J. (2020) Evaluation of the field efficacy of *Heterorhabditis bacteriophora* Poinar (Rhabditida: Heterorhabditidae) and synthetic insecticides for the control of western corn rootworm larvae. *Insects* 11: 202.

Morton, A. and Garcia-del-Pino, F. (2017) Laboratory and field evaluation of entomopathogenic nematodes for control of *Agriotes obscurus* (L.) (Coleoptera: Elateridae). *Journal of Applied Entomology* 141, 241–246.

NAIS (2022) National Agricultural Information System of China. Available at: www.nais.net.cn/ (accessed 19 March 2024).

Negrisoli, A.S., Garcia, M.S. and Barbosa Negrisoli, C.R.C. (2010a) Compatibility of entomopathogenic nematodes (Nematoda: Rhabditida) with registered insecticides for *Spodoptera frugiperda* (Smith, 1797) (Lepidoptera: Noctuidae) under laboratory conditions. *Crop Protection* 29, 545–549.

Negrisoli, A.S., Garcia, M.S., Barbosa Negrisoli, C.R.C., Bernardi, D. and da Silva, A. (2010b) Efficacy of entomopathogenic nematodes (Nematoda: Rhabditida) and insecticide mixtures to control *Spodoptera frugiperda* (Smith, 1797) (Lepidoptera: Noctuidae) in corn crops. *Crop Protection* 29, 677–683.

Öğretmen, A., Yüksel, E. and Canhilal, R. (2020) Susceptibility of larvae of wireworms (*Agriotes* spp.) (Coleoptera: Elateridae) to some Turkish isolates of entomopathogenic nematodes under laboratory and field conditions. *Biological Control* 149, 104320.

Pan, H.Y., Ding, L., Hu, H. and Zhang, X. (1995) Study on field control of Asian corn borer with *Steinernema carpocapsae* Agriotes. *Journal of Jilin Agricultural University* 17, 100–103.

Patil, J., Linga, V., Vijayakumar, R., Subaharan, K., Navik, O. *et al.* (2022) Biocontrol potential of entomopathogenic nematodes for the sustainable management of *Spodoptera frugiperda* (Lepidoptera: Noctuidae) in maize. *Pest Management Science* 78, 2883–2895.

Pilz, C., Heimbach, U. and Grabenweger, G. (2011) Entomoparasitic nematodes for control of the Western corn rootworm, *Diabrotica virgifera virgifera*. *IOBC-WPRS Bulletin* 66, 275–279.

Pilz, C., Toepfer, S., Knuth, P., Strimitzer, T., Heimbach, U. and Grabenweger, G. (2012) Persistence of the entomoparasitic nematode *Heterorhabditis bacteriophora* in maize fields. *Journal of Applied Entomology* 138, 202–212.

Poinar, J. and Welch, H.E. (1968) A new nematode, *Filipjevimermis leipsandra* sp. n. (Mermithidae), parasitic in chrysomelid larvae (Coleoptera). *Journal of Invertebrate Pathology* 12, 259–262.

Purcell, M., Johnson, M.W., Lebeck, L.M. and Hara, A.H. (1992) Biological control of *Helicoverpa zea* (Lepidoptera: Noctuidae) with *Steinernema carpocapsae* (Rhabditida: Steinernematidae) in corn used as a trap crop. *Environmental Entomology* 21, 1441–1447.

Rahatkhah, Z., Karimi, J., Ghadamyari, M. and Brivio, M.F. (2015) Immune defenses of *Agriotes lineatus* larvae against entomopathogenic nematodes. *BioControl* 60, 641–653.

Raulston, J.R., Pair, S.D., Loera, J. and Cabanillas, H.E. (1992) Prepupal and pupal parasitism of *Helicoverpa zea* and *Spodoptera frugiperda* (Lepidoptera: Noctuidae) by *Steinernema* sp. in cornfields in the lower Rio Grande valley. *Journal of Economic Entomology* 85, 1666–1670.

Raulston, J.R., Pair, S.D. and Cabanillas, E. (1997) *Steinernema* sp. nematode for suppression of *Helicoverpa zea* and *Spodoptera frugiperda*. US Patent 5932237. US Patent and Trademark Office, Washington, DC.

Richter, A.R. and Fuxa, J.R. (1990) Effect of *Steinernema feltiae* on *Spodoptera frugiperda* and *Heliothis zea* (Lepidoptera: Noctuidae) in corn. *Journal of Economic Entomology* 83, 1286–1291.

Rufai, M. and Wahab, A. (2021) Biocidal efficacy of *Heterorhabditis bacteriophora* against the African armyworm (*Spodoptera exempta*) towards pest control. *Pakistan Journal of Nematology* 39, 41–45.

Ruiz Vega, J., Aquino Bolaños, T., Silva Rivera, M.E. and Girón Pablo, S. (2012) Integrated control of white grubs *Phyllophaga vetula* Horn (Coleoptera: Melolonthidae) with entomopathogenic agents in Oaxaca, Mexico. *Revista Cientifica UDO Agricola* 12, 609–616.

Saleh, M.M.E., Matter, M.M. and Hussein, M.A. (2000) Efficiency of entomopathogenic nematodes in controlling *Sesamia cretica* (Lepidoptera, Noctuidae) in Egypt. *Bulletin of the National Research Centre (Cairo)* 25, 181–188.

Santos, V., Moino, A. Jr., Andaló, V., Moreira, C.C. and Olinda, R.A. de (2011) Virulence of entomopathogenic nematodes (Rhabditida: Steinernematidae and Heterorhabditidae) for the control of *Diabrotica speciosa* Germar (Coleoptera: Chrysomelidae). *Ciencia e Agrotecnologia* 35, 1149–1156.

Shamseldean, M.M., Abd-Elgawad, M.M. and Abd Elgawad, M.M. (1995) Survival and infectivity of entomopathogenic nematodes under environmental stress. *Anzeiger fur Schadlingskunde, Pflanzen- and Umweltschutz* 68, 31–33.

Shapiro, D.I., Obrycki, J.J., Lewis, L.C. and Jackson, J.J. (1999) Effects of crop residue on the persistence of *Steinernema carpocapsae*. *Journal of Nematology* 31, 517–519.

Shiferaw, B., Prasanna, B.M., Hellin, J. and Bänziger, M. (2011) Crops that feed the world 6. Past successes and future challenges to the role played by maize in global food security. *Food Security* 3, 307–327.

Sun, R. and Li, A. (2004) Test on entomopathogenic nematodes against corn borer. *Shandong Agricultural Sciences* 2004, 60–61.
Tanada, Y. and Reiner, C. (1962) The use of pathogens in the control of the corn earworm, *Heliothis zea* (Boddie). *Journal of Insect Pathology* 4, 139–154.
Thurston, G.S. and Yule, W.N. (1990) Control of larval northern corn rootworm (*Diabrotica barberi*) with two steinernematid nematode species. *Journal of Nematology* 22, 127–131.
Toba, H.H. and Turner, J.E. (1983) Evaluation of baiting techniques for sampling wireworms (Coleoptera: Elateridae) infesting wheat in Washington. *Journal of Economic Entomology* 76, 850–855.
Toepfer, S. and Kuhlman, U. (2013) Research and development for a nematode-based biological control solution for western corn rootworm in maize. *IOBC/WPRS Bulletin* 90, 277–282.
Toepfer, S. and Zellner, M. (2017) Entomopathogenic nematodes can replace soil insecticides in western corn rootworm control. *IOBC/WPRS Bulletin* 129, 144–156.
Toepfer, S., Gueldenzoph, C., Ehlers, R.U. and Kuhlmann, U. (2005) Screening of entomopathogenic nematodes for virulence against the invasive western corn rootworm, *Diabrotica virgifera virgifera* (Coleoptera: Chrysomelidae) in Europe. *Bulletin of Entomological Research* 95, 473–482.
Toepfer, S., Peters, A., Ehlers, R.U. and Kuhlmann, U. (2008) Comparative assessment of the efficacy of entomopathogenic nematode species at reducing western corn rootworm larvae and root damage in maize. *Journal of Applied Entomology* 132, 337–348.
Toepfer, S., Burger, R., Ehlers, R.U., Peters, A. and Kuhlmann, U. (2010a) Controlling western corn rootworm larvae with entomopathogenic nematodes: effect of application techniques on plant-scale efficacy. *Journal of Aplied Entomology* 134, 467–480.
Toepfer, S., Hatala-Zseller, I., Ehlers, R.U., Peters, A. and Kuhlmann, U. (2010b) The effect of application techniques on field-scale efficacy: can the use of entomopathogenic nematodes reduce damage by western corn rootworm larvae? *Agricultural and Forest Entomology* 12, 389–402.
Toepfer, S., Kurtz, B. and Kuhlmann, U. (2010c) Influence of soil on the efficacy of entomopathogenic nematodes in reducing *Diabrotica virgifera virgifera* in maize. *Journal of Pest Science* 83, 257–264.
Toepfer, S., Knuth, P., Glas, M. and Kuhlmann, U. (2014a) Successful application of entomopathogenic nematodes for the biological control of western corn rootworm larvae in Europe – a mini review. *Julius-Kühn-Archiv* 444, 59–66.
Toepfer, S., Li, H., Pak, S.G., Son, K.M., Ryang, Y.S. *et al.* (2014b) Soil insect pests of cold temperate zones of East Asia, including DPR Korea: a review. *Journal of Pest Sciences* 87, 567–595.
Toth, S., Toepfer, S., Szalai, M. and Kiss, J. (2022) Limited influence of abiotic and biotic factors on the efficacy of soil insecticides and entomopathogenic nematodes when managing the maize pest *Diabrotica v. virgifera* (Coleoptera: Chrysomelidae). *MDPI Agronomy*. 12.
van Lenteren, J.C. (2020) Biological control in Guatemala. In: van Lenteren, J.C., Bueno, V.H.P., Luna, M.G. and Colmenarez, Y.C. (eds.) *Biological Control in Latin America and the Caribbean: Its Rich History and Bright Future*. CAB International, Wallingford, UK, pp. 261–265.
Vörös, L., Ábrahám, R., Nagy, K., Tóth, S. and Toepfer, S. (2022) Can *Heterorhabditis bacteriophora* nematode still control western corn rootworm larvae when applied with low amounts of water? *Növényvédelem* 58, 192–200.
Wang, Z., Lu, X., He, K. and Zhou, D. (2000) Review of history, present situation and prospect of the Asian maize borer research in China. *Journal of Shenyang Agricultural University* 31, 402–412.
Wattanachaiyingcharoen, W., Lepcha, O., Vitta, A. and Wattanachaiyingcharoen, D. (2021) Efficacy of Thai indigenous entomopathogenic nematodes for controlling fall armyworm (*Spodoptera frugiperda*) (J.E. Smith) (Lepidoptera; Noctuidae). *Egyptian Journal of Biological Pest Control* 31: 149.
Wei, X., Zhao, L., Qiao, Y., Wang, B., Wan, M. and Toepfer, S. (2019) *Implementing Agri-policies on Pesticide Reduction through Subsidies and Plant Clinics in China*. CABI Working Paper 13. Available at: www.cabi.org/wp-content/uploads/Working-Paper-13-Full-English.pdf (accessed 19 March 2024).
Wright, R.J., Witkowski, J.F., Echtenkamp, G. and Georgis, R. (1993) Efficacy and persistence of *Steinernema carpocapsae* (Rhabditida: Steinernematidae) applied through a center-pivot irrigation system against larval corn rootworms (Coleoptera: Chrysomelidae). *Journal of Economic Entomology* 86, 1348–1354.
Yan, X., Gu, X.H. and Han, R.C. (2019) Research advances in using entomopathogenic nematodes to control the fall armyworm, *Spodoptera frugiperda*. *Journal of Environmental Entomology* 41, 695–700.
Yang, H., Zhang, G. and Zhang, S. (1991) Relationship between host exposure time to *Steinernema feltiae* (Nematoda: Steinernematidae) and the mortality and killing speed of *Ostrinia furnacalis* (Lep.: Pyralidae). *Chinese Journal of Biological Control* 7, 154–156.

Yokomizo, K. and Kashio, T. (1996) Application of an entomogenous nematode, *Steinernema carpocapsae*, for control of the cutworm, *Agrostis segetum*, in carrot fields. *Proceedings of the Association for Plant Protection of Kyushu* 42, 89–92.
Zhang, J., Zhu, X., Lian, S. and Li, M. (2020) Efficacy of *Steinernema carpocapsae* against Asian corn borer in corn field. *Anhui Agricultural Science Bulletin* 26, 91–92.
Zhou, N., Dong, Y., Qiao, P. and Yang, Z. (2020) Complete mitogenomic structure and phylogenetic implications of the genus *Ostrinia* (Lepidoptera: Crambidae). *Insects* 11: 232.

15 Entomopathogenic Nematodes as Biological Control Agents in Grain Crops

Ramandeep Kaur Sandhi[1] and Gadi V.P. Reddy[2*]
[1]FMC Corporation, Stine Research Center, Newark, Delaware, USA; [2]USDA-ARS, Southern Insect Management Research Unit, Stoneville, Mississippi, USA

Abstract

This chapter focuses on the use and potential of entomopathogenic nematodes (EPNs) for the management of insect pests of grains. We summarize progress made in the research and application of EPNs in insect pest management in important grain crops, including wheat, oats, barley, rice and sorghum. Significant factors affecting the success of EPNs in these crops are also reviewed. We also discuss the potential of EPNs for protecting postharvest grains from storage pests. Finally, the efficiency of EPNs in integrated pest-management programs, together with pesticides, entomopathogenic fungi, pheromones and/or predators/parasitoids for effective insect pest management, are discussed.

15.1 Introduction

Important biological control agents include entomopathogenic fungi, nematodes, bacteria and viruses. They can be significant, ecofriendly substitutes for hazardous chemical pesticides for managing crop pests. The control potential of entomopathogenic nematodes (EPNs) has been studied worldwide against a wide range of insect pests on different crop systems (Jess *et al.*, 2005; Dembilio *et al.*, 2010; Toepfer *et al.*, 2014; Malan and Hatting, 2015; Shields, 2015; Shapiro-Ilan *et al.*, 2016; Wraight *et al.*, 2017; Koppenhöfer *et al.*, 2020; and references therein). Recently, Koppenhöfer *et al.* (2020) reviewed the range of insect species and groups in different food production systems (orchards, small fruits, mushroom, corn, vegetable and tuber crops, and greenhouse production) in which EPNs have shown good control potential in field studies. They also discussed cost–benefit analyses, new developments for improving EPN efficacy and different factors influencing the commercial application of EPNs.

This chapter aims to summarize the progress made on the potential of EPNs as biological control agents against various insect pests attacking wheat and other grain crops. Several species of pest insects are polyphagous and can feed on several different cereal crops such as wheat, corn, rice and barley. Some of these insect species include the fall armyworm, *Spodoptera frugiperda* (Smith) (Lepidoptera: Noctuidae), and oriental armyworm, *Mythimna separata* Walker (Lepidoptera: Noctuidae). The virulence of EPNs against these insect pest species, irrespective of the major cereal host crops, is also reviewed. We also discuss the potential efficacy of EPNs against stored grain pests.

*Email: Gadi.Reddy@usda.gov

DOI: 10.1079/9781800620322.0015

15.2 Wheat, Barley and Oats

Wheat, *Triticum aestivum* L. (Poales: Poaceae), is a crucial staple food crop worldwide, providing approximately one-fifth of the calories and protein for human diets (Tilman *et al.*, 2011). Other important cereals that are often grown in the same cropping systems as wheat include oats (*Avena sativa* L.) and barley (*Hordeum vulgare* L.) (both Poales: Poaceae). Many factors can damage these cereal crops. For wheat, insects and diseases can significantly reduce yield and quality, while for barley and oats, such losses are less common. The planting of resistant varieties has usually controlled plant diseases that affect these grains. In contrast, insect pests have been controlled with chemical pesticides. However, some key insect pests have developed resistance to various insecticides, stimulating the search for alternative controls. EPNs are one such alternative, and many studies have examined their potential for controlling wheat pests (see Table 15.1 for details), as discussed below.

15.2.1 Wheat stem sawfly

Wheat stem sawfly, *Cephus cinctus* Norton (Hymenoptera: Cephidae), is a serious economic pest of grains such as wheat and barley in the northern Great Plains of the USA and the Canadian prairie provinces (Bekkerman, 2013). The adults emerge from wheat stubble in spring and after 4–6 weeks infest host plants by ovipositing in the stems of new growth. The larvae feed in the vascular tissue of the stems and, to overwinter, create a hibernaculum near the stem's base, where they chew a notch that causes the stem to lodge before harvest (Achhami *et al.*, 2021). Use of solid-stem, resistant cultivars and insecticides is less effective than in the past, which might be due to the adaptation to solid stems and insecticide resistance (Knodel *et al.*, 2009; Beres *et al.*, 2017). Foliar applications of the nematodes *Steinernema carpocapsae* (Weiser), *S. feltiae* Filipjev, *S. kraussei* Steiner, and *Heterorhabditis bacteriophora* Poinar reduced lodging caused by *C. cinctus* in Montana, and there were fewer sawfly larvae in EPN-treated plots (Tangtrakulwanich *et al.*, 2014). Chemical adjuvants added to EPN solutions (Penterra, Silwet L-77, Sunspray 11N or Syl-Tac) caused some mortality to *C. cinctus* in both the laboratory and field (Portman *et al.*, 2016), suggesting that the added adjuvants enhanced the survival and virulence of EPNs on *C. cinctus*.

15.2.2 Wireworms

Wireworms, the larval stage of click beetles (Coleoptera: Elateridae), are economically important pests in cereal crops with few management options throughout the world because they live in the soil and have long life cycles (Knodel and Shrestha, 2018; Sandhi *et al.*, 2020a). Wireworm damage can cause seedling emergence failure, delayed plant growth and plant death. Some neonicotinoids are registered as seed treatments in cereals but often fail to provide adequate protection because of resistance to most effective insecticides (Vernon *et al.*, 2009). Additionally, systemic neonicotinoids are not environmentally friendly and have been tentatively linked to harmful effects on pollinators in flowering crops (Labrie *et al.*, 2020).

Some entomopathogenic fungi (*Beauveria* and *Metarhizium* spp.), EPNs and bacteria have been evaluated against wireworms, with varying levels of efficacy in different crops (Ensafi *et al.*, 2018; Sharma *et al.*, 2020; Sandhi *et al.*, 2020a,bc, 2021). Several EPNs have been tested against *Limonius* (Table 15.1) and *Agriotes* spp. wireworms in cereal crops worldwide. Milosavljević *et al.* (2020) observed that the presence of different entomopathogens (fungi and nematodes) complemented each other, causing increased wireworm mortality, and was associated with higher wheat yield in the field. However, *Metarhizium brunneum* Petch (Hypocreales: Clavicipitaceae) alone had more potent effects than the other species investigated (including *H. bacteriophora* and *S. carpocapsae*) on *Limonius californicus* abundance (Milosavljević *et al.*, 2020) (Table 15.1).

Similar findings were reported by Sandhi *et al.* (2020a), where both *S. carpocapsae* and *Steinernema riobrave* (Cabanillas, Poinar & Raulston) killed *L. californicus* under laboratory and greenhouse conditions. However, these EPN species were not as effective in wheat and barley fields in Montana when applied in the form of EPN-infected cadavers (Sandhi *et al.*, 2021). Sandhi *et al.* (2020b) also found two native Montana

Table 15.1. Entomopathogenic nematodes (EPNs) evaluated against insect pests of wheat and other grains.

Crop	Pest species (stage)	EPN species tested	L/G/F studies	Dose	Efficacy	Reference
Spring wheat and barley	*Coptognathus curtipennis* (larva)	*Heterorhabditis bacteriophora* and *Steinernema yirgalemense*	L	0, 100, 250, 500 and 1000 IJs ml^{-1}	89% *C. curtipennis* mortality by *H. bacteriophora* at 1000 IJs per grub.	Anbesse *et al.* (2008)
Spring wheat, barley and oats	*Oulema melanopus* (larva)	Nemasys® C (*S. feltiae*) and a Slovenian strain of *Steinernema carpocapsae* (C101)	F	75,000 and 150,000 IJs m^{-2}	Nemasys® C treatment was associated with 6–7.4 t ha^{-1} yield compared with 6.8–8.2 t ha^{-1} yield found with Slovenian strain treatment.	Laznik *et al.* (2012)
Spring wheat	*Oscinella frit* (larva)	*S. carpocapsae*, *S. feltiae* and *H. bacteriophora*	L and F	0, 250, 500 and 1000 IJs ml^{-1} in L, 2.5×10^9 IJs ha^{-1} in F	*H. bacteriophora* was better and caused 100% *O. frit* mortality in the laboratory, but *S. carpocapsae* was more efficient in the field.	El-Wakeil and Volkmar (2013)
Winter wheat	*Cephus cinctus* (larva)	*S. carpocapsae* (Millenium®), *S. kraussei* (Steiner) (Nemasys® L), *Steinernema feltiae* (Nemasys®) and *H. bacteriophora* (Nemasys® G)	F	2.3×10^7 IJs ha^{-1}	EPNs reduced the plant damage caused by *C. cinctus* larvae and increased the yield.	Tangtrakulwanich *et al.* (2014)
Winter wheat	*C. cinctus* (larva)	*H. bacteriophora*, *S. carpocapsae*, *S. feltiae*, *S. glaseri*, *S. kraussei* and *S. riobrave*	L and F	50, 100, 200 and 500 IJs per larva for L and 1000 IJs ml^{-1} in 100 ml carrier solution in F	60–100% *C. cinctus* mortality observed in L, and adding Penterra, Silwet L-77, Sunspray 11N or Syl-Tacto enhanced EPN efficacy. *S. feltiae* added to 0.1% Penterra was able to cause 29% mortality in F.	Portman *et al.* (2016)

Continued

Table 15.1. Continued.

Crop	Pest species (stage)	EPN species tested	L/G/F studies	Dose	Efficacy	Reference
Spring wheat	*Limonius californicus* (larva)	*S. carpocapsae*	G	52,258 IJs per pot	EPNs caused only 14.3% wireworm mortality. They provided better seedling protection in sand than in peat.	Ensafi *et al.* (2018)
Spring wheat	*L. californicus* (larva)	*S. carpocapsae* (All and Cxrd) and *S. riobrave* (355 and 7-12)	L/G	>200 IJs cm^{-2}	*S. carpocapsae* (All and Cxrd strains) and *S. riobrave* (355 and 7-12 strains) killed >50% of *L. californicus* larvae.	Sandhi *et al.* (2020a)
Spring wheat	*L. californicus* (larva)	*S. feltiae* and *H. bacteriophora*	L/G	30, 60, 120 and 240 IJs cm^{-2} in L; 50 and 400 IJs cm^{-2} in G	No more than 25% mortality was observed.	Sandhi *et al.* (2020b,c)
Spring wheat	*Sitodiplosis mosellana* (Géhin) (larva)	*S. feltiae*	F	3×10^9 IJs ha^{-1}	*S. feltiae*, with Barricade® polymer gel 1%, reduced larval populations and produced a higher spring wheat yield.	Shrestha and Reddy (2019)
Rice	*Cnaphalocrocis medinalis* (larva)	*Heterorhabditis indica* (DRR-1) and *Steinernema asiaticum* (DRR-3)	L	25, 50, 75 and 100 IJs per larva	60% and 40% mortality was observed due to *H. indica* and *S. asiaticum*, respectively when applied at a rate of 75 IJs per larva.	Sankar *et al.* (2009)
Rice	*Maliarpha separatella* (Ragonot) (larva/pupa)	*H. indica*	F	6.7×10^4 IJs	Low *M. separatella* damage observed with *H. indica*, when applied 3 weeks after rice transplanting.	Kega *et al.* (2013)

Rice	*M. separatella* (larva)	*S. carpocapsae*, *S. karii*, *H. indica* and two Kenya native EPNsisolates, EXMOMBASA and EXNAKURU	L	50, 100 and 200 IJs per larva	*H. indica* followed by EX NAKURU, *S. carpocapsae*, EX MOMBASA and *S. karii* significantly reduced *M. separatella* larvae after 48 h at a rate of 200 IJs per larva.	Kega *et al.* (2020)
Rice	*Mythimna separata* (larva/pupa)	*Steinernema abbasi* (NBAIISa0), *S. carpocapsae* (NBAIISc05), *S. siamkayai* (NBAIRS92), and *H. indica* (NBAIIH38)	L	50, 100, 200, 300 or 400 IJs for larvae and 200, 400 or 600 IJs for pupae	*S. carpocapsae* caused 100% mortality at 400 IJs per larva and 75% mortality at 600 IJs pupa. *H. indica* was also able to cause >50% mortality.	Patil *et al.* (2020)
Soybean	*Anticarsia gemmatalis* Hübner (adult)	*Steinernema rarum* (OLI)	L and F	50 and 500 IJs per insect	*S. rarum* caused 33% and 80% mortality at 50 and 500 IJs per insect, respectively.	Cagnolo *et al.* (2011)

L, laboratory; G, greenhouse; F, field; IJs, infective juveniles.

EPN species (*S. feltiae* and *H. bacteriophora*) in local wheat fields, but these were not very effective against wireworms (Sandhi *et al.*, 2020c).

Recently, Williams *et al.* (2022) also found different *Steinernema* and *Heterorhabditis* spp. to be ineffective against *Melanotus communis* (Gyllenhal) (Coleoptera: Elateridae) in the laboratory, with smaller wireworms being more susceptible than larger wireworms.

15.2.3 Cereal leaf beetle

The cereal leaf beetle, *Oulema melanopus* (L.) (Coleoptera: Chrysomelidae), is an important insect pest of wheat, oat and barley. The larvae feed on the leaf surface, skeletonizing leaves and consuming the parenchyma tissue. Traditionally, EPNs are more effective against soil insect pests. Nevertheless, recently there have been reports of successful use of EPNs against different foliar insect pests and against those with cryptic habitats such as borers (Ebssa *et al.*, 2004; Shapiro-Ilan *et al.*, 2016; Wraight *et al.*, 2017; Sáenz-Aponte *et al.*, 2020). To the best of our knowledge, there has only been one study testing the efficacy of EPNs against this pest. Laznik *et al.* (2012) found that a commercial EPN product, Nemasys® C (comprising infective juveniles (IJs) of *S. carpocapsae*) and a Slovenian strain of *Steinernema carpocapsae* (C101) were effective against the larvae of *O. melanopus* in a winter wheat field when applied as foliar treatments. However, they emphasized that the activity of EPNs was influenced more by temperature than by the number of nematodes applied, although it depended on the EPN species (Table 15.1).

15.2.4 Frit fly

Frit fly, *Oscinella frit* (L.) (Diptera: Chloropidae), is an essential pest of cereals worldwide (El-Wakeil *et al.*, 2009). It overwinters as a larva within plant stems and pupates in spring. Damaged plants produce small panicles, leading to delayed maturity and yield loss (El-Wakeil *et al.*, 2009). *S. carpocapsae*, *S. feltiae*, *H. bacteriophora* and λ-cyhalothrin were evaluated against this pest in the laboratory and the field on two wheat varieties (Triso and Sakha 93) (Table 15.1) (El-Wakeil and Volkmar, 2013). In the laboratory, all EPN species caused 100% mortality of *O. frit* larvae within 7 days. Frit fly larval mortality was lower in Triso than Sakha 93 because, in Sakha 93, the frit fly infestation was high, providing more larvae for EPNs to find and parasitize. However, in the field, λ-cyhalothrin and *S. carpocapsae* were associated with higher spring wheat yield (5339 and 4776 kg ha^{-1}, respectively) compared with only 3823 kg ha^{-1} yield observed in control plots for Triso. The same trend was seen for Sakha 93 regarding crop yield (El-Wakeil and Volkmar, 2013).

15.2.5 Cutworms

Cutworms are soil-inhabiting insects that can attack wheat and other cereals. In the northern Great Plains of the USA, two important pest cutworm species are the army cutworm, *Euxoa auxiliaris* (Grote), and the pale western cutworm, *Agrotis orthogonia* Morrison (both Lepidoptera: Noctuidae). Both species spend part of their time in the soil, making them potential targets for EPNs. Army cutworms feed on foliage during the day but hide in the soil at night. However, the larvae of the pale western cutworm are subterranean and feed on plant stems below ground. The ability of EPNs to control cutworms in cereal crops has not been studied, but EPNs have been examined as controls of other cutworm species, such as the black cutworm, *Agrotis ipsilon* (Hufnagel) (Lepidoptera: Noctuidae). Different isolates of *S. carpocapsae* and *H. bacteriophora* have been found to cause over 60% mortality of *A. ipsilon* in laboratory bioassays (Ebssa and Koppenhöfer, 2012; Bélair *et al.*, 2013; Mahmoud *et al.*, 2016; Yüksel and Canhilal, 2018). There is a need to test available EPNs against the species of cutworms of importance in cereal crops.

15.2.6 Other minor pests

EPNs have also been considered for control of several less important pests of cereal crops, such as the sunn pest, *Eurygaster integriceps* Puton (Hemiptera: Scutelleridae). This bug can be destructive to wheat and barley in Western, Central Asia and South-east Europe (El Bouhssini *et al.*, 2009). Canhilal *et al.* (2007) found that *H. bacteriophora* isolate Tabeh-gazira, *H. bacterio-*

phora isolate El Ratla-1, *H. bacteriophora* isolate El Ratla-2, *H. bacteriophora* 8-14 strain, *S. carpocapsae* C3B strain and *S. riobrave* strains from Texas were effective against this pest at various concentrations in the laboratory.

The barley chafer grub, *Coptognathus curtipennis* Faimaire (Coleoptera: Dynastidae), is another important pest of barley and wheat in Ethiopia. Anbesse *et al.* (2008) tested *H. bacteriophora* and *Steinernema yirgalemense yirgalemense* Nguyen, Tesfamariam, Gozel, Gaugler & Adams (ANEPN21 and ANEPN22 strains) and three fungal isolates of *Metarhizium anisopliae* against *C. curtipennis* in the laboratory. Of these EPN species, *H. bacteriophora* and the *M. anisopliae* MM isolate were the most virulent to *C. curtipennis* when mortality was measured after 3–4 weeks of exposure. In addition, combined application of both *H. bacteriophora* and *M. anisopliae* MM isolate increased larval mortality of the pest even further (Table 15.1) (Anbesse *et al.*, 2008).

Subterranean termites are significant pests of wood structures and wood products globally. Although less well known, some termite species can also cause severe damage to several agricultural crops, including cereal crops, especially wheat and corn (Shrivastava *et al.*, 2014). Within the subterranean termites, *Odontotermes obesus* (Rambur) and *Microtermes obesi* Holmgren (both Isoptera: Termitidae) are the most damaging species (Dhadwal *et al.*, 2014). Control of termites almost solely relies on chemical insecticides. Due to subterranean termite residence in soil, EPNs can provide adequate control. A single application of Pusa Nemagel (a hydrogel formulation of *Steinernema thermophilum* Ganguly and Singh) reduced the subterranean levels of termites in wheat and pearl millet (*Pennisetum glaucum* L.; Poales: Poaceae) by 48–78% in India (Rathour *et al.*, 2014). There are no other studies on using EPN species to manage termite species in these crops (Wilson-Rich *et al.*, 2007; Baimey *et al.*, 2015).

White grubs (Coleoptera: Scarabaeidae) are soil-inhabiting pests that feed on turfgrass, pasture grasses and the roots of horticultural plants in many parts of the world. Damage to wheat, corn and rice due to white grubs has been reported (Chander *et al.*, 2003). Augmentative applications of EPNs have been investigated for control of various white grubs in high-value turf (Wu *et al.*, 2014; Guo *et al.*, 2017), but their use in low-value cereal crops has not been examined. Because of the relatively low value of cereal crops, the ability of EPNs as an affordable control solution for white grubs in cereals should be considered.

15.3 Rice

Rice, *Oryza sativa* L. (Poales: Poaceae) is a staple food for about half of the world's population (Babendreier *et al.*, 2020). Rice is grown in more than 100 countries, with a total harvested area of approximately 158 million ha, producing more than 700 million t annually. Insect pests attack all parts of the rice plant at all growth stages, and some transfer viruses that damage the plants. Rice is attacked by many species, including stem borers, delphacid planthoppers and the rice leaf folder (Babendreier *et al.*, 2020; Devi, 2020). In some regions, the water weevil, *Lissorhoptrus oryzophilus* Kuschel (Coleoptera: Curculionidae), and the gall midge, *Orseoia oryzae* (Wood-Mason) (Diptera: Cecidomyiidae), also cause severe damage to rice.

15.3.1 Rice leaf folder

Cnaphalocrocis medinalis (Goen.) is a destructive widespread pest of rice in South and South-east Asia (Padmavathi *et al.*, 2013; Karthikeyan and Razak, 2015), damaging rice by folding and feeding on leaves. Under laboratory conditions in India, *Heterorhabditis indica* Poinar, Karunakar & David killed 100% of *C. medinalis* within 18–20 h (Prasad *et al.*, 2006). Sankar *et al.* (2009) also reported 60% and 40% mortality of rice leaf folder by *H. indica* and *Steinernema asiaticum* Anis, Shahina, Reid & Rowe, respectively, while combined application of *H. indica* and *S. asiaticum* at 100 or 75 IJs per larva killed 100% of *C. medinalis* larvae within 25 or 28 h, respectively.

15.3.2 Rice stem borers

The striped rice stem borer, *Chilo suppressalis* (Walker) (Lepidoptera: Crambidae), is one of the most serious pests of rice worldwide. It can reduce crop yield significantly by affecting the tiller number and vigor (Devi, 2020). Females lay egg masses near the base of rice leaves or leaf sheaths, and the larvae bore into the tillers, feeding on the inner surfaces. Larval feeding interrupts the flow of water and nutrients within the plant. Damaged tillers turn brown and have "dead hearts." The panicles can turn white (whiteheads) and fail to produce grain if damage occurs after the formation of spikelets. When infested plants die, the larvae move to another stem and continue feeding. In addition to being a major pest of rice, *C. suppressalis* also feeds on many other wild grasses, including *Gigantochloa verticillata* (Willd.) Munro, *Echinochloa crus-galli crus-pavonis* (L.) Beauv, *Echinochloa stagnina* (Retz.) Beauv. and *Eleusine indica* (L.) Gaertn. Jung *et al.* (2018) studied the efficacy of the Korean EPNs *S. carpocapsae* Pocheon strain, *S. carpocapsae* Iksan strain, *S. monticolum* Stock, Choo & Kaya Hwasun strain and *Heterorhabditis megidis* Poinar, Jackson & Klein Hwasun strain against *C. suppressalis* in the laboratory and the artificially made Amur silver grass (*Miscanthus sacchariflorus* (Maxim.) Franch.; Poales: Poaceae) plantings in different provinces of Korea. In the laboratory, *S. carpocapsae* Pocheon strain caused higher mortality (49.6%) than *H. megidis* Hwasun strain (24.8%) or *S. carpocapsae* Iksan strain (20.4%). However, in the field, *S. carpocapsae* Iksan strain was the most effective (80.6% control), followed by the *H. megidis* Hwasun strain (58% control) (Jung *et al.*, 2018).

In West Africa, the white rice stem borer, *Maliarpha separatella* Ragonot (Lepidoptera: Pyralidae), is a key pest of rice. The larvae tunnel into tillers and stems and feed on the internal tissues. However, the symptoms associated with this borer's damage do not include dead hearts or whiteheads (Akinsola, 1979). Kega *et al.* (2013) evaluated the efficacy of *H. indica* against *M. separatella* under field conditions in resistant and susceptible rice cultivars (Table 15.1). Damage from *M. separatella* was low in plots treated with *H. indica* at 3 weeks after transplanting the rice. The authors recommended combining resistant varieties and EPNs in flooded rice crops. Kega *et al.* (2020) found that *H. indica* was the most effective EPN of the five tested against *M. separatella* in the laboratory. In Petri dish assays, EPNs reduced the test larvae by 100% within 48 h when applied at a rate of 200 IJs per larva (Table 15.1).

15.3.3 Oriental armyworm

Oriental armyworm, *Mythimna separata* (Walker) (Lepidoptera: Noctuidae), is a principal pest of rice but also damages sorghum (*Sorghum bicolor* Moench (L.)), pearl millet, corn (*Zea mays* L.), wheat and sugarcane (*Saccharum officinarum* L.). In Asia and Australia, it causes significant economic losses annually (Jiang *et al.*, 2011). This insect can cause severe damage to the vegetative and reproductive stages of its host plants. Patil *et al.* (2020) conducted pathogenicity tests with *H. indica*, *Steinernema abbasi* Elawad, Ahmad & Reid, *S. carpocapsae* and *S. siamkayai* Stock *et al* against second- and fourth-instar larvae and pupae of *M. separata* in the laboratory. *S. carpocapsae* killed 100% of the larvae and 75% of the pupae at the highest concentration, while *S. siamkayai* was the least virulent species (Patil *et al.*, 2020). Additionally, the most significant penetration rate was observed for *S. carpocapsae* (15%) and the lowest (4%) for *S. siamkayai* (Patil *et al.*, 2020).

15.3.4 Fall armyworm

The fall armyworm, *Spodoptera frugiperda*, is a polyphagous pest that causes significant economic losses in many crops, including corn, rice, wheat, sorghum, beans, potato and cotton, with corn being the pest's most crucial host (Acharya *et al.*, 2020). In rice, *Spodoptera frugiperda* larvae feed on the leaf tissue of young plants, and this feeding can cause severe stand losses. Fallet *et al.* (2020) found that Rwanda-native EPNs (*H. bacteriophora*) could infect *Spodoptera frugiperda* larvae in the laboratory and suggested using carriers with EPNs to protect the IJs from desiccation and ultraviolet (UV) light when applied to corn whorls against *Spodoptera frugiperda* larvae. Acharya *et al.* (2020) compared the virulence of seven EPN strains in Korea against larvae and pupae of *Spodoptera frugiperda* in a Petri dish assay and found that both *H. indica* and *S. carpocapsae* were found highly virulent against younger larvae, whereas *Steinernema arenarium* (Artyukhovsky) and *Steinernema longicaudum* Shen & Wang were highly virulent against older larvae. Similarly, Andaló *et al.* (2010) in Brazil found that *S. arenarium* and *Heterorhabditis* spp.

strain RSC02, each at a rate of 200 IJs per larva, caused 100% and 97.6% mortality in the laboratory and 77.5% and 87.5% mortality in the greenhouse compared with only 7.5% *Spodoptera frugiperda* mortality in controls. Meka *et al.* (2020) reported 100% mortality of *Spodoptera frugiperda* due to *Steinernema glaseri* (Steiner) when applied at different doses under laboratory conditions in India. Additionally, the toxicity to hosts of the symbiotic bacteria of different EPNs has been assessed for different insect pests. For example, *Photorhabdus luminescens* Boemare, Akhurst & Mourant SL0708 strain, a symbiont of *H. indica* SL0708 strain, when applied as 1×10^3–1×10^4 colony-forming units per larva, caused 100% mortality after 48 h (Salazar-Gutiérrez *et al.*, 2017).

15.3.5 Other insect pests

Steinernema masoodi Ali, Shaheen, Pervez & Hussain, *S. seemae* Ali, Shaheen, Pervez & Hussain, *S. carpocapsae*, *S. glaseri* and *S. thermophilum* were investigated against rice moth, *Corcyra cephalonica* Stainton (Lepidoptera: Pyralidae), in a potted plant study in India (Ali *et al.*, 2008). Of the tested EPN species, *S. masoodi*, *S. seemae* and *S. carpocapsae* were the most virulent to *Corcyra cephalonica* but did not cause more than 40% mortality (Ali *et al.*, 2008).

Rice blue beetle, *Leptispa pygmaea* Baly (Coleoptera: Chrysomelidae), another pest of rice, especially in India, is a surface feeder and mainly scrapes the leaves, making them membranous and dry. Under laboratory conditions, *H. indica* caused 67–92% mortality of *Leptispa pygmaea* at concentrations of 5–9 IJs ml^{-1} (Karthikeyan and Jacob, 2009). Karthikeyan and Jacob (2010) found a reduction in leaf damage caused by *Leptispa pygmaea* when *H. indica* was applied in the rice field. They also reported reduced rice damage due to several other pests, *Orseoia oryzae* and *Cnaphalocrocis medinalis* because of *H. indica* application at concentrations of 1500, 2000, 2500 and 3000 IJs ml^{-1} (Karthikeyan and Jacob 2010).

The rice water weevil, *Lissorhoptrus oryzophilus* Kuschel (Coleoptera: Curculionidae), is another destructive rice pest in the USA, especially in Louisiana (Stout *et al.*, 2001). Both larvae and adults feed on rice, but the larvae cause more economic losses. Adult weevils feed on the rice foliage, scarring the leaves, while the larvae feed on the roots, leading to stunted plants and yield loss (Stout *et al.*, 2001). In laboratory tests, *S. feltiae* and *H. bacteriophora* caused 82% and 97% mortality of *Lissorhoptrus oryzophilus* adults in China after 10 days of treatment (Li *et al.*, 2007).

15.4 Sorghum

The fall armyworm, *Spodoptera frugiperda*, is also a pest of sorghum (Day *et al.*, 2017). As above in the discussion of rice pests (see also Table 15.1), various studies have examined the pathogenicity of several EPN species against this pest, with some studies showing high efficacy of EPNs (Andaló *et al.*, 2010; Acharya *et al.*, 2020). However, foliar application of *H. indica* and *Steinernema* sp. at a rate of 88 million IJs ha^{-1} with 0.01% ethoxylate tensioactive agent to corn plants under controlled environmental conditions (24 ± 2°C, relative humidity 65 ± 10 %) failed to control *Spodoptera frugiperda* (Garcia *et al.*, 2008).

The cotton bollworm, *Helicoverpa armigera* (Hübner) (Lepidoptera: Noctuidae), although a major pest of cotton, is polyphagous and also attacks soybean, corn and sorghum (Fitt, 1989). EPNs have been assessed for their infectivity toward *Helicoverpa armigera* in the laboratory. Ebrahimi *et al.* (2018) tested an Iranian isolate of *S. feltiae* against late prepupae of *H. armigera* and found the 20% lethal concentration (LC_{20}), LC_{50} and LC_{80} values to be 4.5, 19 and 76 IJs per insect. Caoili *et al.* (2015) assessed Philippine populations of *H. indica* and *S. abbasi* against *H. armigera* and *Spodoptera litura* (F.) and found them to be infective under laboratory conditions. Whether EPNs can be effective against sorghum pests under field conditions has not yet been assessed.

15.5 Stored Grains

Many beetle and moth species damage stored grains or processed food commodities and can severely damage the amount and quality of many stored grains and starch-based products. Different management tactics for controlling these pests have been investigated, with insecticides being the most used method. However, the health and environmental risks, resistance to key pests and the need for residue-free foods drive the search for reduced-risk control tactics. Entomopathogens or microbials, such as bacteria, fungi, protozoa, viruses and EPNs, are promising alternatives to traditional insecticides for stored-product pest control (Moore *et al.*, 2000). EPNs have received very little attention in this area because of their need for moisture (Moore *et al.*, 2000). Research on the efficacy of EPNs against stored-product pests has been summarized previously by Rumbos and Athanassiou (2017). Table 15.2 shows the work that has been done on the efficacy of steinernematid and heterorhabditid nematodes against different stored-product insect pests.

In actual storage facilities, the level of virulence toward pests, as seen in laboratory trials, can be reduced by factors such as temperature, relative humidity and available host life stage. As stored-product insects are mostly thermophilic, they often show reduced activity at lower temperatures, which in turn can reduce the production of the stimuli used by EPNs for host location, lowering their efficacy (Trdan *et al.*, 2006). Trdan *et al.* (2006) showed that *S. feltiae* was more effective against the granary weevil, *Sitophilus granarius* (L.) (Coleoptera: Curculionidae), and the saw-toothed grain beetle, *Oryzaephilus surinamensis* (L.) (Coleoptera: Silvanidae), at 20°C or 25°C than at 15°C. The same results were reported by Javed *et al.* (2020) as EPN efficacy increased with higher temperatures (Table 15.2). However, this positive response to higher temperatures has limits. Athanassiou *et al.* (2010) and Karanastasi *et al.* (2020) observed reduced EPN efficacy at 30°C against different stored insect pests (Table 15.2).

An important reason why there has been much less exploration of EPNs targeting storage insect pests is the requirement for dry conditions in grain- or food-storage facilities. Ramos-Rodríguez *et al.* (2007) found that changes in temperature levels (25°C versus 30°C) or relative humidity levels (43%, 57%, 75% and 100%) did not affect the efficacy of *S. riobrave* against *Tribolium castaneum* larvae, pupae or adults under laboratory conditions. Additionally, they observed that *S. riobrave* caused 80% and 99% larval mortality of *T. castaneum* and *Plodia interpunctella*, respectively, in field trials simulating empty grain bins (Ramos-Rodríguez *et al.*, 2007).

Similarly, the insect life stage and EPN dose can also affect EPN efficacy, as shown in some of the studies in Table 15.2. The larval stages of postharvest insects are generally considered more susceptible to EPNs than the adult stages, at least for coleopterans (Athanassiou *et al.*, 2010; Rumbos and Athanassiou, 2012). In adults, the sclerotized cuticle is thought to act as a barrier to EPNs, and the greater mobility of adults is believed to allow them to move to avoid contact with IJs. However, this generality is not always the case, and differences in results have been observed in different studies (Table 15.2) (Trdan *et al.*, 2006; Rumbos and Athanassiou, 2012; Karanastasi *et al.*, 2020).

15.6 Entomopathogenic Nematodes in Integrated Pest Management

EPNs can be well suited for integrated pest management (IPM) as they are safer for non-target organisms, can be applied using conventional spray equipment and are compatible with many agrochemicals (Grewal, 2012). Stevens and Lewis (2017) highlighted the ways that EPNs can be used in IPM in different agroecosystems in the USA. We address a few recent examples where EPNs were used in integration with other pest-management practices and showed mixed outcomes for pest management.

Some insecticides such as abamectin, acephate, aldicarb, dodine, fenamiphos, methomyl, parathion and teflubenzuron are toxic to EPNs. In contrast, other pesticides (carbaryl, chlorpyrifos, dimethoate, endosulfan, fonofos, tefluthrin and imidacloprid) were found to be compatible with EPNs (Shapiro-Ilan and Dolinski, 2015). Recently, in a pot experiment with wheat seedlings, Sandhi *et al.* (2021) reported synergism between imidacloprid and *S. carpocapsae* or *S. riobrave* for killing *Limonius californicus* larvae. Exploring the compatibility of different pesticides with EPNs can be an important step in dealing with insecticide resistance problems in cereal crops. Similarly, entomopathogenic fungi can also have an additive effect on EPN

Table 15.2. Assessment of entomopathogenic nematodes (EPNs) against stored-product pests.

Pest species	Stage	EPN species tested	L/G/F studies	Dose	Efficacy	Reference
Sitophilus oryzae (L.), *Rhyzopertha dominica* (F.), *Tribolium castaneum* (Herbst), *Oryzaephilus surinamensis* (L.), *Trogoderma variabile* Ballion, *Tenebrio molitor* L., *Plodia interpunctella* Hübner and *Ephestia kuehniella* Zeller	Larva/pupa/adult	*Steinernema carpocapsae* (Weiser), *S. feltiae* (Filipjev) and *S. riobrave* Cabanillas, Poinar and Raulston	L	0, 1, 5, 10, 20, 50, 100 or 200 IJs per individual	*S. riobrave* at 10 IJs caused ≥80% mortality of larvae of *P. interpunctella*, *E. kuehniella*, *T. castaneum* and *O. surinamensis*, pupae of *T. castaneum* and *T. molitor*, and adults of *T. molitor* and the two moth species.	Ramos-Rodríguez *et al.* (2006)
Tribolium castaneum and *P. interpunctella*	Larva/pupa/adult	*S. riobrave*	L/F	300 IJs cm^{-2}	*S. riobrave* reduced survival of larvae, pupae and adults of *T. castaneum*. Field trials simulating empty grain bin treatments showed 27% and 42% survival of both insect species.	Ramos-Rodríguez *et al.* (2007)
Tribolium confusum du Val, *E. kuehniella*, *R. dominica* and *S. oryzae*	Larva/adult	*Heterorhabditis bacteriophora* Poinar, *S. carpocapsae* and *S. feltiae*	L	10, 50, 100, 150, 500, 1000 and 2000 IJs per insect	*H. bacteriophora* was the least virulent, with *S. feltiae* causing 36–78% larval mortality of *E. kuehniella* with no mortality caused by *S. carpocapsae* at 20°C or 30°C. However, *S. feltiae* and *S. carpocapsae* were not effective against *R. dominica* adults at 20°C.	Athanassiou *et al.* (2010)
S. oryzae	Adult	*S. feltiae* (B30, B49 and 3162)	L	125, 250, 500, 1000 and 2000 IJs/individual	All *S. feltiae* strains were most effective at 25°C but did not cause more than 30% mortality.	Laznik *et al.* (2010)

Continued

Table 15.2. Continued.

Pest species	Stage	EPN species tested	L/G/F studies	Dose	Efficacy	Reference
Lasioderma serricorne (F.) and *T. confusum*	Larva	*H. bacteriophora*, *H. megidis* Poinar, Jackson and Klein, *S. carpocapsae* and *S. feltiae*	L	0,10,100, 500, 1000 and 2000 IJs per insect	*S. carpocapsae* was the most effective species and caused almost 60% mortality of *L. serricorne* adults at ~2000 IJs per insect in treated semolina but only 19% larval mortality was found.	Rumbos and Athanassiou (2012)
S. oryzae, *Acanthoscelides obtectus* (Say), *T. molitor*, *Anagasta kuehniella*, *Sitophilus zeamais* Motschulsk	Larva/pupa/adult	*H. bacteriophora* (RS33, RS56, RS57, RS72 and RS88), *S. riobrave* (RS59), *S. rarum* (Doucet) (RS106) and *S. carpocapsae*	L	10, 20, 50, 100 and 200 IJs per insect	Most of the EPN strains were highly virulent to *A. kuehniella* and *T. molitor* larvae and *A. obtectus* adults at the lower doses. However, *H. bacteriophora* strains needed a higher dose (200 IJs) to kill >55% of *S. zeamais* and *S. oryzae*.	Negrisoli *et al.* (2013)
R. dominica	Adult	*S. feltiae* (isolate 09-31), *S. carpocapsae* (Black Sea isolate) and *H. bacteriophora* (isolate 09-43)	L	500, 1 000 and 2000 IJs per insect	Mortality rates were significantly higher for all the nematode treatments at 20°C and 25°C. *S. carpocapsae* caused the highest mortality of 54% at 2000 IJs per insect.	Tülek *et al.* (2015)
Sitophilus granarius (L.), *R. dominica* and *E. kuehniella*	Larva/adult	*S. feltiae* UTP-5, *H. bacteriophora* UMK-7, *S. feltiae* DDKB-17 and *H. bacteriophora* AVB-15	L	50, 500 and 1000 IJs per adult for *S. granarius* and *R. dominica*; 10, 50 and 100 IJs per larva for *E. kuehniella*	*S. feltiae* DDKB-17 caused >80% *S. granarius* adult mortality at all temperatures with mortality not exceeding 44% for *R. dominica* adults. In contrast, UMK-7 and AVB-15 were more virulent toward *E. kuehniella* larvae (100% mortality). Overall, the mortality rates were higher at 25°C than at 15° or 20°C.	Yüksel *et al.* (2019)

Trogoderma granarium Everts	Larva	*H. bacteriophora*, *S. carpocapsae* and *S. feltiae*	L	10, 50, 100, 500, 1000 and 5000 IJs per larva	*S. feltiae* killed 99% of small and 88% of large larvae, followed by *S. carpocapsae*, with 91% small and 63% large larval mortality. *H. bacteriophora* did not cause > 77% mortality.	Karanastasi *et al*. (2020)
Tribolium confusum and *R. dominica*	Adult	*Steinernema pakistanense* (Shahina, Anis, Reid, Rowe and Maqbool) (LM-07), *S. bifurcatum* (Shahina, Xun, Lihong, Richou, Mehreen, Tabassum and Salma) (LM-30), *S. affinae* (Bovien) (GB-14), and *S. cholashanense* Nguyen, Půža, and Mráček (GB-22)	L	50, 100 and 150 IJs per beetle	*S. affine* and *S. cholashanense* killed 70–75% of *T. confusum* and *R. dominica* at 150 IJs per beetle at 20°C, and *S. pakistanense* was able to cause 100% mortality at 30°C at 100 IJs per beetle.	Javed *et al*. (2020)
S. oryzae	Adult	*S. feltiae* (UKK-1, DTK-6 and UIP-19)	L	50, 100, 250, 500, 1000 and 2000 IJs per adult	The highest mortality rates were 80% and 92% within 4 and 8 days, with *S. feltiae* causing 92% mortality at 2000 IJs per adult.	Canhilal and Yüksel (2020)

L, laboratory; G, greenhouse; F, field; IJs, infective juveniles.

efficacy, as reported by Ensafi *et al.* (2018) and Milosavljević *et al.* (2020), where different entomopathogens (fungi and nematodes) complemented each other to some extent in killing wireworms in greenhouse and field experiments. Similarly, the combination of entomopathogenic fungi and EPNs can also be effective against foliage or stem insect pests such as cereal leaf beetle, frit fly and wheat stem sawfly.

Integration of pheromones with EPNs can be an effective strategy for the biological control of different insect pests and has been studied by some researchers. Oliveira-Hofman *et al.* (2019) also observed that pheromone extracts increased the dispersal and efficacy of *S. carpocapsae* and *S. feltiae* toward *Tenebrio molitor.* Pheromone and bait trapping methods can be implemented against certain insect pests of cereal crops (e.g. wireworms), which can be an effective component of an IPM program against this pest.

Sometimes, several management strategies for a given pest are employed in a crop, and intraguild interactions between these natural enemies may occur, potentially affecting their efficacy. The chances of intraguild predation between predators/parasitoids of foliage insect pests (e.g. aphids, wheat midges and leaf folders) and EPNs are low in cases where the predators/parasitoids do not have a soil-dwelling stage in their life cycle. Moreover, Mbata and Shapiro-Ilan (2010) reported some possible benefits of the combined application of *H. indica* and *Habrobracon hebetor* (Hymenoptera: Braconidae) for controlling *Plodia interpunctella*, provided that EPNs were applied at times that minimized intraguild predation by nematodes on parasitoid adults.

15.7 Concluding Remarks and Future Trends

EPNs have shown efficacy against several grain crop insect pests, especially the soil-inhabiting species and those with cryptic habitats that shelter inside plant tissues or other structures. This impact makes EPNs an alternative to chemical insecticides in certain cases. The compatibility of EPNs with other biological control agents and pesticides facilitates their inclusion in IPM programs for cereal crops and should be explored. One advantage of EPNs is their exemption (or nearly so) from registration requirements compared with chemical insecticides, due to the recognized safety of EPNs. However, successful use of EPNs against aboveground pests is limited, and EPNs have little potential for control of some insect pests (e.g. cereal leaf beetle, frit fly) because the available nematode species or strains are ineffective, or the pest resides in niches with unfavorable environmental conditions. New EPN formulations and field application technologies may overcome some of these limitations.

More research is needed to find more virulent EPN species or strains that are resistant to the environments where they are intended to be applied. Different cereal crops have different temperature and humidity requirements for their growth and development. Understanding the interactions between the requirements of EPNs, the target pests and the crop will enable selection of suitable EPN species in an IPM program. Rice is grown mainly in relatively warm environments with high humidity, which can be a conducive environment for the survival, dispersal, establishment and infection ability of EPNs against rice insect pests. However, the sensitivity of EPNs to desiccation under high temperatures, as well as chances of death due to UV light, make them difficult to use for aboveground insect pests. However, some crops, such as wheat, are grown in relatively cooler areas, which may be an obstacle for some EPN species or strains that might not be able to perform well at very low temperatures. These limitations indicate a need for field-based EPN efficacy trials to find cold-tolerant EPN species that might work best against wheat pests. A need also exists to find EPN species that are resistant and effective at high and low humidity and temperatures, such as *S. riobrave*. Some potential also exists for genetic manipulation to improve the virulence, survival and persistence of EPNs (Shapiro-Ilan and Dolinski, 2015).

The requirement for dry conditions for post-harvest storage of crop products is the biggest obstacle blocking the use of EPNs against stored-product insects. Most EPNs cannot survive or persist for long periods under such conditions. Moreover, the direct application EPNs on to edible grains and other stored food commodities may be objectionable to consumers. EPNs are slow acting compared with chemical insecticides, and reliance on EPNs to kill the storage pests may provide opportunities for the pest to mate and lay eggs before dying, leading to damage by further pest generations (Moore *et al.*, 2000). Future research is warranted to find alternative innovative methods to formulate EPNs and use them as killing agents in lure-and-kill systems in storage areas. The combined use of insect growth regulators with EPNs might also be useful for insect pests of stored produce. For example, hydroprene causes high molt failure in *Tribolium castaneum* (Arthur and Hoernemann, 2004), and the additive effects of these growth regulators in combination with nematodes should be studied.

Above all, the use of EPNs in IPM programs, including the use of pesticides, entomopathogenic fungi, pheromones and/or predators/parasitoids, may be more suitable for effective insect pest management of cereal crops as well as stored-product insect pests and warrants further studies.

15.8 Acknowledgements

We are thankful to the USDA-ARS Research Project #6066-22000-090-00D-Insect Control and Resistance Management in Corn, Cotton, Sorghum, Soybean, and Sweet Potato, and Alternative Approaches to Tarnished Plant Bug Control in the Southern United States.

15.9 References

Acharya, R., Hwang, H.S., Mostafiz, M.M., Yu, Y.S. and Lee, K.Y. (2020) Susceptibility of various developmental stages of the fall armyworm, *Spodoptera frugiperda*, to entomopathogenic nematodes. *Insects* 11: 868.

Achhami, B.B., Reddy, G.V.P., Hofland, M.L., Sherman, J.D., Peterson, R.K. and Weaver, D.K. (2021) Plant volatiles and oviposition behavior in the selection of barley cultivars by wheat stem sawfly (Hymenoptera: Cephidae). *Environmental Entomology* 50, 940–947.

Akinsola, E.A. (1979) The biology and ecology of rice stem borers in Nigeria. PhD thesis, University of Ibadan, Ibadan, Nigeria.

Ali, S.S., Pervez, R., Hussain, M.A. and Ahmad, R. (2008) Susceptibility of three lepidopteran pests to five entomopathogenic nematodes and *in vivo* mass production of these nematodes. *Archives of Phytopathology and Plant Protection* 41, 300–304.

Anbesse, S.A., Adge, B.J. and Gebru, W.M. (2008) Laboratory screening for virulent entomopathogenic nematodes (*Heterorhabditis bacteriophora* and *Steinernema yirgalemense*) and fungi (*Metarhizium anisopliae* and *Beauveria bassiana*) and assessment of possible synergistic effects of combined use against grubs of the barley chafer *Coptognathus curtipennis*. *Nematology* 10, 701–709.

Andaló, V., Santos, V., Moreira, G.F., Moreira, C.C. and Moino, A. Jr (2010) Evaluation of entomopathogenic nematodes under laboratory and greenhouses conditions for the control of *Spodoptera frugiperda*. *Ciência Rural* 40, 1860–1866.

Arthur, F.H. and Hoernemann, C.K. (2004) Impact of physical and biological factors on susceptibility of *Tribolium castaneum* and *Tribolium confusum* (Coleoptera: Tenebrionidae) to new formulations of hydroprene. *Journal of Stored Products Research* 40, 251–268.

Athanassiou, C.G., Kavallieratos, N.G., Menti, H. and Karanastasi, E. (2010) Mortality of four stored product pests in stored wheat when exposed to doses of three entomopathogenic nematodes. *Journal of Economic Entomology* 103, 977–984.

Babendreier, D., Hou, M., Tang, R., Zhang, F., Vongsabouth, T. *et al.* (2020) Biological control of lepidopteran pests in rice: a multi-nation case study from Asia. *Journal of Integrated Pest Management* 11: 5.

Baimey, H., Zadji, L., Afouda, L., Moens, M. and Decraemer, W. (2015) Influence of pesticides, soil temperature and moisture on entomopathogenic nematodes from southern Benin and control of underground termite nest populations. *Nematology* 17, 1057–1069.

Bekkerman, A. (2013) The changing landscape of Northern Great Plains wheat markets. *Choices* 28, 1–6.

Bélair, G., Simard, L. and Dionne, J. (2013) Canadian entomopathogenic nematode isolates: virulence against black cutworm (Lepidoptera: Noctuidae). *Phytoprotection* 93, 43–46.

Beres, B.L., Hill, B.D., Cárcamo, H.A., Knodel, J.J., Weaver, D.K. and Cuthbert, R.D. (2017) An artificial neural network model to predict wheat stem sawfly cutting in solid-stemmed wheat cultivars. *Canadian Journal of Plant Science* 97, 329–336.

Cagnolo, S.R., Peschiutta, M.L. and Bertolotti, M.A. (2011) Susceptibility of adults of *Anticarsia gemmatalis* Hübner, 1818 (Lepidoptera: Noctuidae) to the entomopathogenic nematode *Steinernema rarum* (Doucet, 1986) Mamiya, 1988 (Rhabditida: Steinernematidae) under laboratory conditions 13, 373–376.

Canhilal, R. and Yüksel, E. (2020) Laboratory evaluation of some native isolates of *Steinernema feltiae* (Filipjev) against the rice weevil, *Sitophilus oryzae* (L.) (Coleoptera: Curculionidae). *Türkiye Biyolojik Mücadele Dergisi* 11, 65–69.

Canhilal, R., Reid, W., Kutuk, H. and El-Bouhssini, M. (2007) Susceptibility of sunn pest, *Eurygaster integriceps* Puton (Hemiptera: Scutelleridae), to various entomopathogenic nematodes (Rhabditida: Steinernematidae and Heterorhabditidae). *Journal of Agricultural and Urban Entomology* 24, 19–26.

Caoili, B.L., Mercado, C.D.M., Sandoval, R.F.C. and Latina, R.A. (2015) Pathogenicity of Philippine populations of *Steinernema abbasi* and *Heterorhabditis indica* to major lepidopterous pests of corn and sweet sorghum. *Philippine Entomologist* 29, 225–226.

Chander, S., Aggarwal, P.K., Kalra, N. and Swaruparani, D.N. (2003) Changes in pest profiles in rice-wheat cropping system in Indo-gangetic plains. *Annals of Plant Protection Sciences* 11, 258–263.

Day, R., Abrahams, P., Bateman, M., Beale, T., Clottey, V. *et al.* (2017) Fall armyworm: impacts and implications for Africa. *Outlooks on Pest Management* 28, 196–201.

Dembilio, O., Llácer, E., Martínez de Altube, M.D.M. and Jacas, J.A. (2010) Field efficacy of imidacloprid and *Steinernema carpocapsae* in a chitosan formulation against the red palm weevil *Rhynchophorus ferrugineus* (Coleoptera: Curculionidae) in *Phoenix canariensis*. *Pest Management Science* 66, 365–370.

Devi, G. (2020) Entomopathogenic nematodes against insect pests of rice. *International Journal of Environment, Agriculture and Biotechnology* 5, 1143–1150.

Dhadwal, R., Sharma, P.K., Vashisth, S., Kumar, S. and Verma, K.S. (2014) Insect pest complex of wheat (*Triticum aestivum* L.) in Himachal Pradesh. *Journal of Entomological Research* 38, 147–152.

Ebrahimi, L., Shiri, M. and Dunphy, G.B. (2018) Effect of entomopathogenic nematode, *Steinernema feltiae*, on survival and plasma phenoloxidase activity of *Helicoverpa armigera* (Hb) (Lepidoptera: Noctuidae) in laboratory conditions. *Egyptian Journal of Biological Pest Control* 28: 12.

Ebssa, L. and Koppenhöfer, A.M. (2012) Entomopathogenic nematodes for the management of *Agrotis ipsilon*: effect of instar, nematode species and nematode production method. *Pest Management Science* 68, 947–957.

Ebssa, L., Borgemeister, C. and Poehling, H.M. (2004) Effectiveness of different species/strains of entomopathogenic nematodes for control of western flower thrips (*Frankliniella occidentalis*) at various concentrations, host densities, and temperatures. *Biological Control* 29, 145–154.

El Bouhssini, M., Street, K., Joubi, A., Ibrahim, Z. and Rihawi, F. (2009) Sources of wheat resistance to Sunn pest, *Eurygaster integriceps* Puton, in Syria. *Genetic Resources and Crop Evolution* 56, 1065–1069.

El-Wakeil, N.E. and Volkmar, C. (2013) Efficacy of entomopathogenic nematodes against the frit fly *Oscinella frit* (L.) (Diptera: Chloropidae) on spring wheat. *Journal fur Kulturpflanzen* 65, 9–18.

El-Wakeil, N.E., Sallam, A.A. and Volkmar, C. (2009) Ecological studies on frit fly *Oscinella frit* (L.) and its control in summer wheat in Central Germany. *Mitteilungen der Deutschen Gesellschaft für Allgemeine und Angewandte Entomologie* 17, 215–220.

Ensafi, P., Crowder, D.W., Esser, A.D., Zhao, Z., Marshall, J.M. and Rashed, A. (2018) Soil type mediates the effectiveness of biological control against *Limonius californicus* (Coleoptera: Elateridae). *Journal of Economic Entomology* 111, 2053–2058.

Fallet, P., Machado, R.A., Toepfer, S., Ye, W., Kajuga, J. *et al.* (2020) A Rwandan survey of entomopathogenic nematodes that can potentially be used to control the fall armyworm. *IOBC-WPRS Bulletin* 150, 87–90.

Fitt, G.P. (1989) The ecology of *Heliothis* species in relation to agroecosystems. *Annual Review of Entomology* 34, 17–52.

Garcia, L.C., Raetano, C.G. and Leite, L.G. (2008) Application technology for the entomopathogenic nematodes *Heterorhabditis indica* and *Steinernema* sp. (Rhabditida: Heterorhabditidae and Steinernematidae) to control *Spodoptera frugiperda* (Smith) (Lepidoptera: Noctuidae) in corn. *Neotropical Entomology* 37, 305–311.

Grewal, P.S. (2012) Entomopathogenic nematodes as tools in integrated pest management. In: Abrol, D.P. and Shankar, U. (eds) *Integrated Pest Management: Principles and Practice*. CABI Publishing, Wallingford, UK, pp. 162–236.

Guo, W., Yan, X., Zhao, G. and Han, R. (2017) Increased efficacy of entomopathogenic nematode–insecticide combinations against *Holotrichia oblita* (Coleoptera: Scarabaeidae). *Journal of Economic Entomology* 110, 41–51.

Javed, S., Khanum, T.A. and Khan, S. (2020) Biocontrol potential of entomopathogenic nematode species against *Tribolium confusum* (Jac.) (Coleoptera: Tenebrionidae) and *Rhyzopertha dominica*

(Fab.) (Coleoptera: Bostrichidae) under laboratory conditions. *Egyptian Journal of Biological Pest Control* 30: 5.

Jess, S., Schweizer, H., Kilpatrick, M., Grewal, P.S., Ehlers, R.U. and Shapiro-Ilan, D.I. (2005) Mushroom applications. In: Grewal, P.S., Ehlers, R.U. and Shapiro-Ilan, D.I. (eds) *Nematodes as Biocontrol Agents*. CAB International, Wallingford, UK, pp. 191–213

Jiang, X., Luo, L., Zhang, L., Sappington, T.W. and Hu, Y. (2011) Regulation of migration in *Mythimna separata* (Walker) in China: a review integrating environmental, physiological, hormonal, genetic, and molecular factors. *Environmental Entomology* 40, 516–533.

Jung, Y.H., Choo, H.Y., Choi, J.Y. and Choo, Y.M. (2018) Occurrence and environmentally friendly control of rice stem borer, *Chilo suppressalis*, in Amur Silver Grass (*Mischanthus sacchariflorus*) plantation. *Korean Journal of Pesticide Science* 22, 36–44.

Karanastasi, E., Kavallieratos, N.G., Boukouvala, M.C., Christodoulopoulou, A.D. and Papadopoulou, A.A. (2020) Effect of three entomopathogenic nematode species to *Trogoderma granarium* Everts (Coleoptera: Dermestidae) larvae on stored wheat. *Journal of Stored Products Research* 88: e101641.

Karthikeyan, K. and Jacob, S. (2009) Bioefficacy of white muscardine fungus, *Beauveria bassiana* (Balsamo) Vuillemin and entomopathogenic nematode, *Heterorhabditis indica* (Poinar) against rice blue beetle, *Leptispa pygmaea* Baly. *Journal of Biological Control* 23, 79–81.

Karthikeyan, K. and Jacob, S. (2010) *Pseudomonas flourescens* and *Heterorhabditis indica* Poinar for the management of major insect pests of rice. *Journal of Biopesticides* 3, 96–99.

Karthikeyan, S. and Razak, T.A. (2015) Evaluation of efficacy of *Steinernema feltiae* and isolated native strain *S. kushidai* against *Cnaphalocrosis medinalis* Guenee. *Current Nematology* 26, 39–44.

Kega, V.M., Kasina, M., Olubayo, F. and Nderitu, J.H. (2013) Management of *Maliarpha separatella* Rag using effective entomopathogenic nematodes and resistant rice cultivars. *Journal of Entomology* 10, 103–109.

Kega, V.M., Kasina, M., Olubayo, F. and Nderitu, J.H. (2020) Effectiveness of entomopathogenic nematodes against the African white rice stem borer *Maliarpha separatella* Rag. *Universal Journal of Agricultural Research* 8, 41–45.

Knodel, J.J. and Shrestha, G. (2018) Pulse crops: pest management of wireworms and cutworms in the Northern Great Plains of United States and Canada. *Annals of the Entomological Society of America* 111, 195–204.

Knodel, J.J., Beauzay, P.B., Eriksmoen, E.D. and Pederson, J.D. (2009) Pest management of wheat stem maggot (Diptera: Chloropidae) and wheat stem sawfly (Hymenoptera: Cephidae) using insecticides in spring wheat. *Journal of Agricultural and Urban Entomology* 26, 183–197.

Koppenhöfer, A.M., Shapiro-Ilan, D.I. and Hiltpold, I. (2020) Entomopathogenic nematodes in sustainable food production. *Frontiers in Sustainable Food Systems* 4: 125.

Labrie, G., Gagnon, A.È., Vanasse, A., Latraverse, A. and Tremblay, G. (2020) Impacts of neonicotinoid seed treatments on soil-dwelling pest populations and agronomic parameters in corn and soybean in Quebec (Canada). *PLOS ONE* 15: e0229136.

Laznik, Ž., Tóth, T., Lakatos, T., Vidrih, M. and Trdan, S. (2010) The activity of three new strains of *Steinernema feltiae* against adults of *Sitophilus oryzae* under laboratory conditions. *Journal of Food Agriculture and Environment* 88, 150–154.

Laznik, Ž., Vidrih, M., Vučajnk, F. and Trdan, S. (2012) Is foliar application of entomopathogenic nematodes (Rhabditida) an effective alternative to thiametoxam in controlling cereal leaf beetle (*Oulema melanopus* L.) on winter wheat. *Journal of Food Agriculture and Environment* 10, 716–719.

Li, J.J., Qian, H.T., Dong, H. and Cong, B. (2007) Laboratory test on the infectivity of entomopathogenic nematode to *Lissorhoptrus oryzophilus*. *Chinese Journal of Biological Control* 23: 188.

Mahmoud, M.F., Mahfouz, H.M. and Mohamed, K.M. (2016) Compatibility of entomopathogenic nematodes with neonicotinoids and azadirachtin insecticides for controlling the black cutworm, *Agrotis ipsilon* (Hufnagel) in canola plants. *International Journal of Research in Environmental Science* 2, 11–18.

Malan, A.P. and Hatting, J.L. (2015) Entomopathogenic nematode exploitation: case studies in laboratory and field applications from South Africa. In: Campos-Herrera, R. (ed.) *Nematode Pathogenesis of Insects and Other Pests: Ecology and Applied Technologies for Sustainable Plant and Crop Protection*. Springer, Cham, Switzerland, pp. 477–508.

Mbata, G.N. and Shapiro-Ilan, D.I. (2010) Compatibility of *Heterorhabditis indica* (Rhabditida: Heterorhabditidae) and *Habrobracon hebetor* (Hymenoptera: Braconidae) for biological control of *Plodia interpunctella* (Lepidoptera: Pyralidae). *Biological Control* 54, 75–82.

Meka, M., Anita, B., Vetrivelkalai, P. and Muthukrishnan, N. (2020) Infectivity of entomopathogenic nematode, *Steinernema glaseri* on fall army worm (FAW), *Spodoptera frugiperda* (Smith, 1797) in maize (*Zea mays*). *Journal of Entomology and Zoology Studies* 8, 1023–1028.

Milosavljević, I., Esser, A.D., Rashed, A. and Crowder, D.W. (2020) The composition of soil-dwelling pathogen communities mediates effects on wireworm herbivores and wheat productivity. *Biological Control* 149: 104317.

Moore, D., Lord, J.C. and Smith, S.M. (2000) Pathogens. In: Subramanyam, B. amd Hagstrum, D.W. (eds) *Alternatives to Pesticides in Stored-product IPM*. Springer, Boston, Massachusetts, pp. 193–227.

Negrisoli, C.R.D.C.B., Júnior, A.S.N., Bernardi, D. and Garcia, M.S. (2013) Activity of eight strains of entomopathogenic nematodes (Rhabditida: Steinernematidae, Heterorhabditidae) against five stored product pests. *Experimental Parasitology* 134, 384–388.

Oliveira-Hofman, C., Kaplan, F., Stevens, G., Lewis, E., Wu, S. *et al.* (2019) Pheromone extracts act as boosters for entomopathogenic nematodes efficacy. *Journal of Invertebrate Pathology* 164, 38–42.

Padmavathi, C., Katti, G., Padmakumari, A.P., Voleti, S.R. and Subba Rao, L.V. (2013) The effect of leaffolder *Cnaphalocrocis medinalis* (Guenee) [Lepidoptera: Pyralidae] injury on the plant physiology and yield loss in rice. *Journal of Applied Entomology* 137, 249–256.

Patil, J., Vijayakumar, R., Linga, V. and Sivakumar, G. (2020) Susceptibility of oriental armyworm, *Mythimna separata* (Lepidoptera: Noctuidae) larvae and pupae to native entomopathogenic nematodes. *Journal of Applied Entomology* 144, 647–654.

Portman, S.L., Krishnankutty, S.M. and Reddy, G.V.P. (2016) Entomopathogenic nematodes combined with adjuvants presents a new potential biological control method for managing the wheat stem sawfly, *Cephus cinctus* (Hymenoptera: Cephidae). *PLOS ONE* 11: e0169022.

Prasad, J.S., Katti, G., Padmakumari, A.P. and Shankar, M. (2006) Bioefficacy of indigenous entomopathogenic nematodes against leaf folder, *Cnaphalocrocis medinalis* Guenee in rice. In: *Proceedings of the 2nd International Congress, 9–13 October*, *National Academy of Agricultural Sciences*, *New Delhi*, p. 418.

Ramos-Rodríguez, O., Campbell, J.F. and Ramaswamy, S.B. (2006) Pathogenicity of three species of entomopathogenic nematodes to some major stored-product insect pests. *Journal of Stored Products Research* 42, 241–252.

Ramos-Rodríguez, O., Campbell, J.F. and Ramaswamy, S.B. (2007) Efficacy of the entomopathogenic nematode *Steinernema riobrave* against the stored-product insect pests *Tribolium castaneum* and *Plodia interpunctella*. *Biological Control* 40, 15–21.

Rathour, K.S., Ganguly, S., Das, T.K., Singh, P., Kumar, A. and Somvanshi, V.S. (2014). Biological management of subterranean termites (*Odontotermes obesus*) infesting wheat and pearl millet crops by entomopathogenic nematodes. *Indian Journal of Nematology* 44, 97–100.

Rumbos, C.I. and Athanassiou, C.G. (2012) Insecticidal effect of six entomopathogenic nematode strains against *Lasioderma serricorne* (F.) (Coleoptera: Anobiidae) and *Tribolium confusum* Jacquelin du Val (Coleoptera: Tenebrionidae). *Journal of Stored Products Research* 50, 21–26.

Rumbos, C.I. and Athanassiou, C.G. (2017) The use of entomopathogenic nematodes in the control of stored-product insects. *Journal of Pest Science* 90, 39–49.

Sáenz-Aponte, A., Correa-Cuadros, J.P. and Rodríguez-Bocanegra, M.X. (2020) Foliar application of entomopathogenic nematodes and fungi for the management of the diamond back moth in greenhouse and field. *Biological Control* 142: 104163.

Salazar-Gutiérrez, J.D., Castelblanco, A., Rodríguez-Bocanegra, M.X., Teran, W. and Sáenz-Aponte, A. (2017) *Photorhabdus luminescens* subsp. akhurstii SL0708 pathogenicity in *Spodoptera frugiperda* (Lepidoptera: Noctuidae) and *Galleria mellonella* (Lepidoptera: Pyralidae). *Journal of Asia-Pacific Entomology* 20, 1112–1121.

Sandhi, R.K., Shapiro-Ilan, D., Sharma, A. and Reddy, G.V.P. (2020a) Efficacy of entomopathogenic nematodes against the sugarbeet wireworm, *Limonius californicus* (Mannerheim) (Coleoptera: Elateridae). *Biological Control* 143: 104190.

Sandhi, R.K., Pothula, R., Pothula, S.K., Adams, B.J. and Reddy, G.V.P. (2020b) First record of native entomopathogenic nematodes from Montana agroecosystems. *Journal of Nematology* 52: e2020-60.

Sandhi, R.K., Shapiro-Ilan, D. and Reddy, G.V.P. (2020c) Montana native entomopathogenic nematode species against *Limonius californicus* (Coleoptera: Elateridae). *Journal of Economic Entomology* 113, 2104–2111.

Sandhi, R.K., Shapiro-Ilan, D., Ivie, M. and Reddy, G.V.P. (2021) Biocontrol of wireworms (Coleoptera: Elateridae) using entomopathogenic nematodes: the impact of infected host cadaver application and soil characteristics. *Environmental Entomology* 50, 868–877.

Sankar, M., Prasad, J.S., Padmakumari, A.P., Katti, G. and Divya, K. (2009) Combined application of two entomopathogenic nematodes, *Heterorhabditis indica* and *Steinernema asiaticum* to control the rice leaf folder, *Cnaphalocrosis medinalis* (Goen.). *Journal of Biopesticides* 2, 135–140.

Shapiro-Ilan, D. and Dolinski, C. (2015) Entomopathogenic nematode application technology. In: Campos-Herrera, R. (ed.) *Nematode Pathogenesis of Insects and Other Pests: Ecology and Applied Technologies for Sustainable Plant and Crop Protection*. Springer, Cham, Switzerland, pp. 231–254.

Shapiro-Ilan, D.I., Cottrell, T.E., Mizell, R.F. III, and Horton, D.L. (2016) Efficacy of *Steinernema carpocapsae* plus fire gel applied as a single spray for control of the lesser peachtree borer, *Synanthedon pictipes*. *Biological Control* 94, 33–36.

Sharma, A., Jaronski, S. and Reddy, G.V.P. (2020) Impact of granular carriers to improve the efficacy of entomopathogenic fungi against wireworms in spring wheat. *Journal of Pest Science* 93, 275–290.

Shields, E.J. (2015) Utilizing persistent entomopathogenic nematodes in a conservation or a more classical biological control approach. In: Campos-Herrera, R. (ed.) *Nematode Pathogenesis of Insects and Other Pests: Ecology and Applied Technologies for Sustainable Plant and Crop Protection*. Springer, Cham, Switzerland, pp. 165–184.

Shrestha, G. and Reddy, G.V.P. (2019) Field efficacy of insect pathogen, botanical, and jasmonic acid for the management of wheat midge *Sitodiplosis mosellana* and the impact on adult parasitoid *Macroglenes penetrans* populations in spring wheat. *Insect Science* 26, 523–535.

Shrivastava, S.K., Verma, R.K. and Singh, B. (2014) Integrated pest management in wheat. In: Shukla, R., Mishra, P., Chatrath, R., Gupta, R., Tomar, S. and Sharma, I. (eds) *Wheat: Recent Trends on Production Strategies of Wheat in India*. Jawaharlal Nehru Krishi Vishwa Vidyalaya (JNKVV) and ICAR – Indian Institute of Wheat and Barley Research, Jabalpur, India, pp. 197–209.

Stevens, G. and Lewis, E. (2017) Status of entomopathogenic nematodes in integrated pest management strategies in the USA. In: Abd-Elgawad, M.M., Askary, T.H. and Coupland, J. (eds) *Biocontrol Agents: Entomopathogenic and Slug Parasitic Nematodes*. CAB International, Wallingford, UK, pp. 289–311.

Stout, M.J., Rice, W.C., Linscombe, S.D. and Bollich, P.K. (2001) Identification of rice cultivars resistant to *Lissorhoptrus oryzophilus* (Coleoptera: Curculionidae), and their use in an integrated management program. *Journal of Economic Entomology* 94, 963–970.

Tangtrakulwanich, K., Reddy, G.V.P., Wu, S., Miller, J.H., Ophus, V.L. and Prewett, J. (2014) Efficacy of entomopathogenic fungi and nematodes, and low risk insecticides against wheat stem sawfly, *Cephus cinctus* (Hymenoptera: Cephidae). *Journal of Agricultural Science* 6, 1–9.

Tilman, D., Balzer, C., Hill, J. and Befort, B.L. (2011) Global food demand and the sustainable intensification of agriculture. *Proceedings of the National Academy of Sciences USA* 108, 20260–20264.

Toepfer, S., Knuth, P., Glas, M. and Kuhlmann, U. (2014) Successful application of entomopathogenic nematodes for the biological control of western corn rootworm larvae in Europe – a mini review. *Julius-Kühn-Archiv* 444, 59–66.

Trdan, S., Vidrih, M. and Valic, N. (2006) Activity of four entomopathogenic nematode species against young adults of *Sitophilus granarius* (Coleoptera: Curculionidae) and *Oryzaephilus surinamensis* (Coleoptera: Silvanidae) under laboratory conditions. *Journal of Plant Diseases and Protection* 113, 168–173.

Tülek, A., Erturk, S., Kepenekci, I. and Atay, T. (2015) Efficacy of native entomopathogenic nematodes against the stored grain insect pest, *Rhyzopertha dominica* (F.) (Coleoptera: Bostrichidae) adults. *Egyptian Journal of Biological Pest Control* 25, 251–254.

Vernon, R.S., van Herk, W.G., Clodius, M. and Harding, C. (2009) Wireworm management I: stand protection versus wireworm mortality with wheat seed treatments. *Journal of Economic Entomology* 102, 2126–2136.

Williams, L., Cherry, R. and Shapiro-Ilan, D. (2022) Effect of host size on susceptibility of *Melanotus communis* (Coleoptera: Elateridae) wireworms to entomopathogens. *Journal of Nematology* 54: 20220033.

Wilson-Rich, N., Stuart, R.J. and Rosengaus, R.B. (2007) Susceptibility and behavioral responses of the dampwood termite *Zootermopsis angusticollis* to the entomopathogenic nematode *Steinernema carpocapsae*. *Journal of Invertebrate Pathology* 95, 17–25.

Wraight, S.P., Lopes, R.B. and Faria, M. (2017) Microbial control of mite and insect pests of greenhouse crops. In: Lacey, L.A. (ed.) *Microbial Control of Insect and Mite Pests: From Theory to Practice*. Academic Press, London, pp. 237–252.

Wu, S., Youngman, R.R., Kok, L.T., Laub, C.A. and Pfeiffer, D.G. (2014) Interaction between entomopathogenic nematodes and entomopathogenic fungi applied to third instar southern masked chafer white grubs, *Cyclocephala lurida* (Coleoptera: Scarabaeidae), under laboratory and greenhouse conditions. *Biological Control* 76, 65–73.

Yüksel, E. and Canhilal, R. (2018) Evaluation of local isolates of entomopathogenic nematodes for the management of black cutworm, *Agrotis ipsilon* Hufnagel (Lepidoptera: Noctuidae). *Egyptian Journal of Biological Pest Control* 28: 82.

Yüksel, E., Canhilal, R. and Imren, M. (2019) Potential of four Turkish isolates of entomopathogenic nematodes against three major stored products insect pests. *Journal of Stored Products Research* 83, 317–321.

16 Entomopathogenic Nematodes in Citrus Integrated Pest Management in South Africa and Florida

Larry Duncan[1]* and Antoinette P. Malan[2]

[1]*University of Florida, IFAS, Citrus Research and Education Center, Lake Alfred, Florida, USA;* [2]*Department of Conservation Ecology and Entomology, Faculty of AgriSciences, Stellenbosch University, Matieland, Stellenbosch, South Africa*

Abstract

In South Africa, two decades of research on key citrus pests characterized the susceptibility of false codling moth, fruit fly and citrus mealybug to local isolates of entomopathogenic nematodes (EPNs). Possible non-target displacement of native species by exotic species, an important concern in South Africa, has driven extensive laboratory and field studies, demonstrating strong potential for control of important citrus pests by local EPNs, one of which is now registered for use against false codling moth. By contrast, commercially available, exotic EPN species have been used by Florida growers to help manage citrus root weevils for more than 30 years. Factors affecting their availability and use have included uneven product quality, non-persistent activity against a resilient target pest, and grower cost constraints due to the introduction of new pests and diseases, most notably citrus greening disease (also known as huanglongbing). Citrus orchards in both Florida and South Africa have diverse, often abundant, native EPN species with suites of traits that remain to be fully exploited through commercialization and conservation.

16.1 Introduction

Citrus orchards present extensive opportunities for biological control due to their economic importance, above- and belowground habitat stability, and the potential to manage the diversity of alternate host plant species in borders and between rows. Indeed, the first successful effort at classical biological control targeted the cottony cushion scale in California citrus orchards (Caltagirone and Doutt, 1989). *Steinernema scapterisci*, a rare example of an entomopathogenic nematode (EPN) imported for classical biocontrol of mole cricket in turf (Dolinski *et al.*, 2012), is now also well established in Florida citrus orchards (Campos-Herrera *et al.*, 2013a). Citrus orchards are similarly suited to conservation biocontrol, another tactic infrequently studied with reference to EPNs (Stuart *et al.*, 2008; Dritsoulas *et al.*, 2022). By contrast, research on EPNs for augmentation biocontrol of citrus pests is widely reported. Here, we focus, with some exceptions, on work done in the citrus industries of South Africa and Florida because sustained research programs in both regions have characterized the wide diversity of EPN communities in these

*Email: lwduncan@ufl.edu

DOI: 10.1079/9781800620322.0016

orchards and have fostered substantial interest in the use of EPNs in citrus integrated pest management (IPM) programs.

16.2 Citrus Industries in South Africa and Florida

Worldwide citrus (oranges, grapefruit, lemons, limes and specialty varieties) production in 2020/21 was 98 million t, of which about half were oranges. The USA, primarily California and Florida, produced about 10% of the world's oranges (behind the leading producers Brazil, China and Europe). South Africa produced 3.5% of the world's oranges (USDA, 2024).

The production of citrus in South Africa dates to the 1600s, when the first citrus fruit trees were planted in the Cape of Good Hope. In South Africa, citrus is produced with the main aim of fresh fruit export. The country is the 14th largest citrus producer in the world and the second largest citrus exporter (CGA, 2021a,b). During the 2020 citrus season, approximately 76% of all production was exported to foreign markets. The remaining citrus was either sold to local markets (6%) or processed into other citrus-related products (18%). The majority of South African citrus is exported to Europe (40%), South-east Asia (17%) and the Middle East (17%) (CGA, 2021b).

The citrus industry in South Africa currently hosts over 60 major and minor species of insect pests, of which nine species are classified as economically important (Grout and Moore, 2015). These pests are citrus psylla, *Trioza erytreae* (Del Guercio) (Hemiptera: Triozidae); citrus thrips, *Scirtothrips aurantii* (Faure) (Thysanoptera: Thripidae); false codling moth (FCM), *Thaumatotibia leucotreta* (Meyrick); fruit fly (Diptera: Tephritidae), including the Mediterranean fruit fly, *Ceratitis capitata* (Wiedemann) and the Natal fruit fly, *Ceratitis rosa* (Karsch); citrus mealybug, *Planococcus citri* (Risso), oleander mealybug, *Paracoccus burnerae* (Brain) and long-tailed mealybug, *Pseudococcus longispinus* (Targioni Tozzetti) (all Hemiptera: Pseudococcidae); and red scale, *Aonidiella aurantii* (Maskell) (Hemiptera: Diaspididae) (Grout and Moore, 2015; Manrakhan *et al.*, 2018). The most important citrus pests that are of quarantine importance are fruit flies and FCM because they are internal pests of fruits, which may be exported without detection (Grout and Moore, 2015). FCM is regarded as a key pest of citrus in South Africa, due to its phytosanitary status (Venette *et al.*, 2003; Moore, 2021). Navel oranges and mandarin varieties are the most susceptible citrus cultivars for FCM infestation, due to the chemical compounds that are present within such varieties (Love *et al.*, 2014). Grapefruit and soft citrus have been shown to be the least susceptible cultivars to FCM infestation, whereas lemons are regarded as a non-host (Moore *et al.*, 2015).

In contrast to the high-value (blemish-free) fresh fruit that are marketed by South Africa and other industries with Mediterranean climates, the majority of citrus in Florida and Brazil is processed for juice, due to the higher cost in subtropical regions of maintaining the appearance of the fruit. The Florida citrus industry has contended since 2005 with citrus greening disease (also known as huanglongbing (HLB)), with a resulting production decline of 80% (Singerman *et al.*, 2018). Nevertheless, the state remains the second largest producer of orange juice, following Brazil. The humid, subtropical climate has a rainy season during summer to early fall and a dry season requiring low-volume irrigation, primarily using microjets, during fall to late spring. Citrus in the state is grown in two ecoregions, defined primarily by the depth to groundwater (Fig. 16.1). A 'central ridge' of deep sandy soils extends from north to south in the center of the peninsula. Prior to adoption of irrigation, trees on the central ridge required a deep root system to encounter groundwater during the dry season. Coastal and central 'flatwood' regions surrounding the central ridge have much shallower groundwater, usually requiring the construction of elevated beds to provide adequate root growth. Soils in the two regions are primarily sandy, with up to 97% sand content common on the central ridge, and somewhat finer and more variable textures in the flatwoods.

Besides HLB, Florida growers contend with introduced foliar/fruit diseases such as citrus black spot and citrus canker, as well as long-endemic greasy spot, citrus scab and melanose (Diepenbrock *et al.*, 2022). Several soil-borne nematodes and *Phytophthora* spp. are economically important, as is citrus blight, a vascular disease of unknown origin. The more important insect pests include the Asian citrus psyllid vector

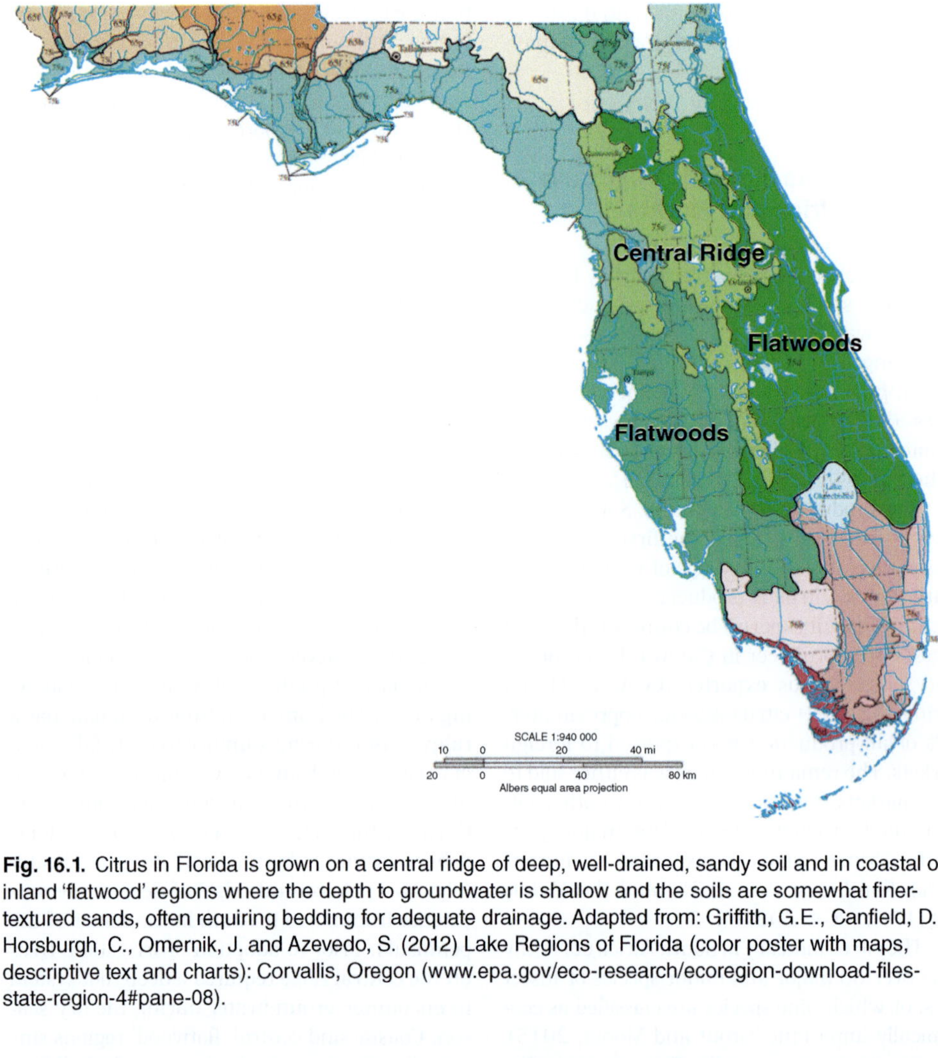

Fig. 16.1. Citrus in Florida is grown on a central ridge of deep, well-drained, sandy soil and in coastal or inland 'flatwood' regions where the depth to groundwater is shallow and the soils are somewhat finer-textured sands, often requiring bedding for adequate drainage. Adapted from: Griffith, G.E., Canfield, D., Horsburgh, C., Omernik, J. and Azevedo, S. (2012) Lake Regions of Florida (color poster with maps, descriptive text and charts): Corvallis, Oregon (www.epa.gov/eco-research/ecoregion-download-files-state-region-4#pane-08).

of HLB, rust mite, leafminer and various secondary pests when pesticides are overused to manage psyllids. Several curculionid root herbivores are serious economic pests and are considered below.

16.3 Pest Targets of Entomopathogenic Nematodes

In South Africa, research conducted from 2006 onward focused on the susceptibility of key pests of citrus, grapevine and deciduous fruits to local EPN isolates (Malan and Hatting, 2015; Hatting and Malan, 2017; Malan and Ferreira, 2017). The mostly fundamental research was key to determining the occurrence and distribution of local isolates of EPNs and choosing suitable species and isolates that would serve against key pests for purposes of mass culture and formulation. In the case of citrus, most of the research was targeted at FCM, fruit fly and the citrus mealybug, using the local EPN isolates available at the time of the research.

16.3.1 False codling moth

Thaumatotibia leucotreta is endemic and indigenous mostly to the tropical and subtropical regions of sub-Saharan Africa, being widespread in Senegal, the Ivory Coast, Togo, the Upper Volta, South Africa, and islands in the Atlantic and Indian oceans, such as Mauritius and Madagascar (Carstens and Moore, 2021). The moth is known to be a relatively poor invader, as it has only successfully established itself in two regions where it is not indigenous: the Western Cape province of South Africa (Giliomee and Riedl, 1998; Hofmeyr *et al.*, 2015) and Israel (Wyoski, 1986). FCM causes economic damage to several citrus cultivars, as well as deciduous and vegetable crops throughout South Africa (Grout and Moore, 2015; Malan *et al.*, 2018). FCM larvae are known to cause fruit and feeding lesions, adversely affecting the marketability of the fruit. The moth threatens both large-scale commercial and subsistence production (Stotter, 2009; Grout and Moore, 2015; Malan *et al.*, 2018).

A lengthy soil stage distinguishes the FCM life cycle from that of codling moths (de Waal *et al.*, 2011). The last-instar larvae of FCM leave the fruit on a silken thread to reach the soil. Preferring dry soil, the larvae burrow down a few millimeters, spinning themselves within a soil/silk pupal cocoon (Love *et al.*, 2014). Thus, the last-instar larvae, prepupae, pupae and emerging moth are all in contact with the soil for as long as 14 days, depending on temperature (Daiber, 1979; 1980), offering many opportunities for infection by EPNs. In contrast to codling moths, no diapause exists for FCM (Terblanche *et al.*, 2014).

A wide range of biological control options are currently available for the control of FCM (Malan *et al.*, 2018; Moore, 2021). However, no IPM treatment is currently directed toward the soil stages of the moth, leaving EPNs with the potential to fill this important niche. Inoculative EPN release in citrus orchards is of critical importance as an ecosystem service to prevent the eruption of insect populations under optimum environmental conditions.

16.3.2 Fruit fly

Ceratitis capitata is a highly invasive species indigenous to most of the sub-Saharan region, perhaps originating from East Africa. The species occurs in the fruit-growing regions of southern Africa, the Mediterranean region, the Middle East, South America and Australia (de Meyer, 2021; Weldon 2021). The distribution of *Ceratitis rosa* is limited primarily to eastern Africa (from Kenya southwards to South Africa) and the continent's adjoining islands. In South Africa, the species is seldom found in the hotter, drier Western Cape (Grout and Moore, 2015).

The life cycle of both fruit fly species is very similar, with the entire life cycle lasting approximately 3 weeks at an optimum temperature of 26°C. Mediterranean and Natal fruit flies provide only a small window of opportunity for infection with EPNs. Fruit fly larvae drop to the ground to form a puparium in the soil. However, in the soil the fruit fly larvae pupate in less than 24 h (Aluja *et al.*, 2005). The puparium remains for a relatively long period in contact with the soil, with the flies tending to emerge 10–20 days after pupation, depending on the climatic conditions. The fruit flies are also susceptible to EPN infection during their emergence from the puparium, while still soft and in contact with the soil.

16.3.3 Citrus mealybug

The cosmopolitan pest *Planococcus citri* (citrus mealybug) is the most important mealybug in citrus production throughout southern Africa. Its origin remains speculative due to widespread international trade; however, since 1990, mealybugs have existed at economically damaging population densities (Hattingh and Tate, 1996). From a phytosanitary perspective, *P. citri* is a major constraining factor for exporting citrus fruit to foreign markets (Hattingh *et al.*, 1998; Wakgari and Giliomee, 2005).

No soil connection exists in the life cycle of the citrus mealybug, which occurs throughout the citrus tree canopy. The adult female mealybugs secrete thin, waxy threads along the dorsal length of the body, which develop into a cottony mass, in which 300–600 eggs are laid (Annecke and Moran, 1982; Smith *et al.*, 1997). Although the females are primarily found on the leaves and twigs, some migrate to the main branches or trunk of their host tree to lay their eggs (Martinez-Ferrer *et al.*, 2006). Mealybugs tend to overwinter in cracks and crevices or leaf axils throughout the tree canopy (Hattingh, 1993; Smith *et al.*, 1997). During spring and early summer in South Africa, shortly after fruit set, the first-generation nymphs tend to move to the foliage to colonize the young growth and small fruit (Martinez-Ferrer *et al.*, 2006), settling in such cryptic habitats as the fruit calyx (Hattingh and Moore, 2003). The highest population density of *P. citri* on grapefruit and orange varieties occurs between mid- and late summer (Franco *et al.*, 2004). In South Africa, the life cycle may vary from 4 weeks in midsummer to several months during winter. Each nymphal stage lasts for slightly over 2 weeks, with oviposition starting about 2 weeks after the final molt. EPN application should, accordingly, be targeted at the aboveground life stages.

16.3.4 Citrus root weevil

The root weevils *Diaprepes abbreviatus* and *Pachnaeus litus* are the only pests targeted for EPN use in Florida citrus. *D. abbreviatus* was introduced into Florida from the Caribbean in the mid-1960s, becoming the most serious economic pest of citrus until the arrival four decades later of the Asian citrus psyllid, which vectors HLB (Woodruff, 1964; Dolinski *et al.*, 2012). The adults of both weevil species feed on the new foliage of citrus and many of the native plants in row middles and surrounding the orchards. The eggs are laid on leaves, and the neonate larvae fall to the soil where they feed on progressively larger roots as they develop over several months to about 2 cm in length (Wolcott, 1936). The larvae feed on the cortex of larger, woody structural roots, creating distinctive channels (Fig. 16.2) that can girdle the root, disrupt nutrient flow and create a pest-disease complex by providing entry to *Phytophthora* spp. Extensive damage to the crown area of the root system can kill the tree. Although less widespread than *D. abbreviatus*, *P. litus* biology and damage to trees is much the same.

These insects are not obvious candidates for biological control by EPNs because they do not aggregate at a predictable time and place where EPN applications will be most effective and affordable (e.g. Gouge *et al.*, 1996; Shapiro-Ilan

Fig. 16.2. (A) Adult *Diaprepes abbreviatus* aggregating and mating in a citrus tree canopy. (B) Typical root channeling and girdling caused by larval feeding.

et al., 2016). Rather, they are multivoltine, with life stages developing simultaneously in the rhizospheres and the canopies of a wide range of host plants during all but the coldest winter months (Wolcott, 1936; Futch *et al.*, 2005). In the absence of persistent pesticides, adults and eggs killed in the canopy are soon replaced by adults emerging from soil. Newly hatched larvae falling from leaves replace larvae killed following use of non-persistent EPNs in the rhizosphere (Duncan *et al.*, 2007). Nevertheless, growers have used EPNs for four decades to help manage the weevils because they are devastating pests with no registered, effective soil-applied chemical pesticides. EPNs are used as components of IPM systems that employ cultural, chemical and biological control tactics (Duncan and Mannion, 2022).

16.4 Naturally Occurring Entomopathogenic Nematode Species in Citrus Orchards

Citrus orchards in South Africa, North America and the Mediterranean have been surveyed exclusively, or as part of multi-habitat surveys for indigenous and exotic EPN species. Orchards on moist, sandy soils were the only habitats found to support EPNs, exclusively heterorhabditids, in Israel's Negev desert (Glazer *et al.*, 1991). Populations in one Negev orchard tended to be nearly undetectable until the peak seasonal abundance of endemic scarab beetle larvae induced rapid EPN population growth throughout the site (Efron *et al.*, 2001). In California, the EPNs *Heterorhabditis bacteriophora* and *Heterorhabditis taysearae* (= *H. sonorensis*) were detected in just five out of 36 orchards throughout the state. The occurrence of these endemic EPNs was unrelated to soil properties; however, the ability of commercial *Steinernema riobrave* to forage for hosts in laboratory studies was inversely related to the clay content of the California soils (Kaspi *et al.*, 2010). Citrus had the highest mean EPN richness and diversity among four perennial plant habitats in a survey of 50 sites in the southern Algarve region of Portugal. At least one of four species (*Steinernema feltiae*, *S. arenarium*, *H. bacteriophora* and *H. indica*) was found in half the orchards using species-specific quantitative polymerase chain reaction (qPCR) tools (Campos-Herrera *et al.*, 2019b). Validation of these results by metabarcoding DNA from the Portugal samples revealed strong correlations between data from both molecular methods, but compared with qPCR, metabarcoding detected EPNs in more sites including two additional species (Dritsoulas *et al.*, 2020). A survey of 58 sites in 13 Egyptian citrus orchards using metabarcoding detected three EPN species of each genus (*Steinernema* sp., *S. scapterisci*, *S. glaseri*, *H. bacteriophora*, *H. indica* and *H. taysearae*) in 29% of samples from fine-textured soils of the Nile Delta, but just the three heterorhabditid species, mainly *H. indica*, in 60% of the samples from sandier soils in the reclaimed desert orchards (Dritsoulas *et al.*, 2021). A similar survey of a north–south transect of 62 Greek citrus orchards detected *Steinernema carpocapsae* and *S. feltiae* in 60% of sites in Crete, *S. carpocapsae*, *S. feltiae* and *H. bacteriophora* in 40% of sites in Epirus, and *S. carpocapsae* in 10% of sites in north-eastern Peloponnese (A. Dritsoulas, unpublished data). The occurrence of steinernematids was unrelated to soil texture, but in Epirus *H. bacteriophora* was strongly associated with sand content that ranged between 15–71%.

Malan *et al.* (2011) surveyed citrus orchards in the Western Cape, the Eastern Cape and the Mpumalanga provinces of South Africa and estimated EPN diversity and frequency of occurrence. Soil trapping was done with FCM, the greater wax moth, *Galleria mellonella*, and the yellow mealworm beetle, *Tenebrio molitor*. The main aim of the survey was to obtain nematodes as biological control agents against FCM. From a total of 202 samples, 35 (17%) tested positive for EPNs. Of these, four isolates (11%) were steinernematids and 31 isolates (89%) were heterorhabditids. Morphometrics and morphologically different life stages were used to support molecular identification and characterization. The identified steinernematids were *Steinernema khoisanae* (Nguyen *et al.*, 2006), *S. yirgalemense* (Malan *et al.*, 2011) and *S. citrae* (Stokwe *et al.*, 2011). This survey was the first to report *S. yirgalemense* in South Africa, and *S. citrae* was the second new steinernematid identified in South Africa. The identified heterorhabditids were *H. bacteriophora*, *H. zealandica* and an unknown species later described as *H. noenieputensis* (Malan *et al.*, 2014).

An early study of natural control of *D. abbreviatus* larvae in Florida citrus orchards found as much as 70% infection of buried sentinel larvae by nematodes (EPNs and free living) during the summer months and a 27% incidence of "*S. carpocapsae*" and 44% of *Heterorhabditis* sp. in 55 surveyed orchards (Beavers *et al.*, 1983). Although *S. carpocapsae* does not occur in Florida citrus, a survey that employed species-specific primer/probe sets detected EPNs in each of 52 orchards studied (Campos-Herrera *et al.*, 2013a). *Steinernema diaprepesi* occurred in nearly all orchards on the central ridge and sporadically, as did *Steinernema khuongi*, in flatwoods sites. *H. indica* was detected in all but six orchards, equally distributed among ecoregions, and *H. zealandica* was found sporadically in the sites on the central ridge. Three other commercially available species (*S. scapterisci*, *S. riobrave* and *S. feltiae*) were detected in several sites. A second survey of 91 natural sites adjacent to citrus orchards found the same seven species plus *Heterorhabditis floridensis* and *S. glaseri*; however, in contrast to the orchards, EPNs were detected in just 60% of the natural sites, supporting several reports of generally high EPN incidence in orchards compared with other habitats (Campos-Herrera *et al.*, 2008, 2016). Several variables related to soil moisture (groundwater depth, water-holding capacity, clay and organic matter content) appeared to be significant drivers of the EPN community structure in both the orchards and natural areas.

Comparison of EPN diversity and prevalence in the citrus surveys regionally is complicated by important differences in methodology; nevertheless, some conclusions appear warranted. To date, the subtropical Florida orchards are unique for their combination of species richness and prevalence. The only other subtropical survey described here, in Mpumalanga province in South Africa, had twice the incidence of positive detections, as did orchards in the two Mediterranean regions of South Africa. Florida orchard soils are also uniquely coarse textured, generally having over 90% sand content. Indeed, whereas higher sand content favored EPN occurrence or activity in some of the studies, in Florida it was the clay content that best predicted EPN community structure, likely because of its importance in maintaining soil moisture in such porous sands. These kinds of interactions between the edaphic properties driving community structure require more information to be fully understood. The adoption of next-generation sequencing methods to characterize soil food webs is facilitating this effort, which is critical to understand the properties that support EPNs and where EPNs are sufficiently abundant and active to merit conservation as a management strategy (see section 16.8).

16.5 Commercial Application of EPNs and Challenges to Product Development

The only commercial application of EPNs in citrus orchards to test their efficacy against the soil stages of FCM was undertaken between 2011 and 2013. Formulated, imported *H. bacteriophora* and *S. feltiae* were applied, using either a spray machine or a microsprinkler irrigation system, in three provinces of South Africa. To evaluate EPN efficacy, sentinel FCM larvae were buried in cages and revealed an infection rate of up to 80% (Moore *et al.*, 2023). In a few trials, the insect mortality was increased by the occurrence of naturally occurring EPNs in the orchards surveyed. The most important environmental factor, in terms of efficacy, was identified as the amount of moisture that was present, due to the thorough irrigation that was supplied both before and after the application. The persistence of the two nematodes concerned was recorded over a period of up to 4 months.

Laboratory screening of FCM indicated outperformance by locally isolated species in most cases compared with the performance of imported *H. bacteriophora*. Especially good results were obtained using imported *S. feltiae*. However, this species is exotic to South Africa, and should preferably be avoided for commercial use due to biological pollution and potential masking of a naturally occurring South African isolate. Nevertheless, according to Ehlers (2003), the risk from the use of exotic species is extremely low, and where *S. feltiae* has been applied in citrus orchards, no establishment of the species has yet been reported (S. Moore, personal communication).

In South Africa, a great deal of industry-funded research has been done in academia to select appropriate EPN candidates against the key citrus pests under laboratory conditions and to establish protocols for their mass production in solid or liquid media (Ramakuwela *et al.*, 2016; Dunn *et al.*, 2020). Studies are currently focused on formulations to improve shelf life and on field trials needed for registration of the formulated active ingredient. The current challenge in South Africa is the high capital input required from small- to medium-sized enterprises to mass culture and formulate nematodes as commercial products. The criteria for registration of EPN products in South Africa are based on those for the use of chemicals. However, full-scale field evaluation is complicated by the need for a comparable control and a containment method for measuring the immediate effect of the EPN on the target pest after application. There is also a need to have a mass cultured formulated product available for field trials.

In Florida, following reports that *S. carpocapsae* and *H. bacteriophora* reduced *D. abbreviatus* by up to 70% in field trials (Schroeder, 1987, 1992), the nematodes quickly became commercially available and widely used. However, the failure of a commercial product in a 4-year trial (Adair, 1994) and subsequent laboratory (Schroeder, 1994) and field (Duncan *et al.*, 1996; Duncan and McCoy, 1996; Bullock *et al.*, 1999; Shapiro *et al.*, 1999) research resulted in changing the commercial species for weevil management from *S. carpocapsae* and *H. bacteriophora* to *S. riobrave* and *H. indica*. Thereafter, an eventual change in the company licensed to produce *S. riobrave* resulted in quality-control problems, which again reduced the use of EPNs substantially from their initial, near-universal adoption for weevil IPM. In 2001, Becker Underwood Ltd. released a high-quality formulation of *S. riobrave* but eventually discontinued its manufacture when the arrival of HLB caused growers to reduce management of other pests and diseases (Dolinski *et al.*, 2012). Nevertheless, over time, the severe tree debilitation by HLB has convinced many that weevil IPM is critically important. BASF, having acquired Becker Underwood Ltd., resumed the sale of *S. riobrave* under the name Nemasys® R in 2016. The product is produced in the UK and shipped to Florida. Despite a refrigerated shelf life of up to 2 months, the logistics of long-distance transport of Nemasys® R has frequently disrupted product availability. Ongoing commercial research is also focused on strain development to improve the production capacity and efficacy of the nematodes (Stuart *et al.*, 2004).

16.6 Research Using Commercial Entomopathogenic Nematodes

16.6.1 False codling moth

Interest in EPNs started with the aim of controlling FCM in citrus (Malan *et al.*, 2006, 2011; Malan and Hatting, 2015) and codling moth in deciduous fruit (de Waal *et al.*, 2011). Locally isolated, known and newly described EPN species, were demonstrated in laboratory bioassays to be highly virulent toward FCM larvae at low nematode concentrations (Fig. 16.3) (Malan *et al.*, 2011; W.P. Steyn *et al.*, 2017a,b, 2019; V.M. Steyn *et al.*, 2021). The species screened against FCM included *H. bacteriophora* (Steyn *et al.*, 2017a), *H. baujardi*, *H. safricana*, *H. zealandica*, *Steinernema jeffreyense* (Malan *et al.*, 2016), *S. khoisanae*, *S. litchi* (Steyn *et al.*, 2017a) and *S. yirgalemense* (Malan *et al.*, 2011). Laboratory bioassays using 24-well bioassay plates showed that the local isolates were all highly virulent against the last instar of FCM. Pilot trials showed that a rate of 100 infective juveniles (IJs) per larva was highly effective against FCM and they were eventually reduced by half. *S. yirgalemense* at 50 IJs per FCM larva caused up to 100% mortality and showed significant differences between the different isolates. FCM pupae were less susceptible to EPN infection than the last instars (Malan *et al.*, 2011), and similar results were obtained by V.M. Steyn *et al.* (2019), using fully formed older pupae. In a sand bioassay, *S. yirgalemense* at 20 IJs cm^{-2} infected 93% of cocooned pupae and emerging moths. The emerging moths were shown to be infected with nematodes, which might have aided in the control and extended the dispersal of the EPNs (Malan *et al.*, 2011). Laboratory bioassays, using sieved orchard soil and *H. bacteriophora*, showed the 50% lethal dose (LD_{50}) to be approximately 3 IJs per FCM larva and the LD_{90} to be around 58 IJs per larva.

The strong potential of local EPNs to infect FCM in the laboratory prompted a semi-field trial

Fig. 16.3. Different life stages of the false codling moth, *Thaumatotibia leucotreta*, infested with entomopathogenic nematodes.

in a citrus orchard at Welgevallen, Stellenbosch (Malan and Moore, 2016). Application of 20, 40 and 80 IJs cm^{-2} of *H. bacteriophora* resulted in FCM larval infection rates of up to 90%, with significant differences between the lower rates but not the highest. No further infections occurred 3 weeks after application (Malan and Moore, 2016). Six days after application of *H. zealandica*, no significant differences were found between three nematode rates (5, 10 and 20 IJs cm^{-2}) that produced FCM mortality of over 80%. After 21 and 35 days, non-significant differences in mortality were apparent between the three treatments. In a field trial using *H. bacteriophora*, *H. zealandica* and S. *khoisanae*, *H. zealandica* provided significant control at each evaluation time for 49 days (Malan and Moore, 2016). The trials demonstrated that local EPN species have great potential to manage the soil stages of FCM with good persistence. The results were supported by subsequent semi-field trials in vineyards, avocado, lychee and macadamia, using the same methods to evaluate *S. yirgalemense* and *S. jeffreyense* against FCM (V.M. Steyn *et al.*, 2019).

Large commercial-scale field trials using *H. bacteriophora* and *S. feltiae* imported from Germany were conducted from 2011 to 2013 in citrus orchards in the Eastern Cape, the Western Cape and the Mpumalanga provinces of South Africa. The trials aimed to test the effect of full orchard application in terms of different environmental conditions and application and irrigation methods (Moore *et al.*, 2023). *H. bacteriophora* and *S. feltiae* were applied underneath tree canopies in large orchards using tractor-drawn spray machinery or by means of the irrigation system. A successful FCM infection rate sometimes reaching 80% was recorded in several trials, where larval infestation of fruit was reduced as much as 88%, with persistence of up to 5 months (Moore *et al.*, 2023). However, the results of many of the trials were disappointing, with low soil moisture seen as the main environmental factor influencing the low infection rate. South Africa is a water-scarce country and the use of drippers in citrus orchards is the preferred irrigation method to conserve water. The first 4 mm of soil tends to dry quickly and last-instar FCM larvae prefer to pupate in dry soil, thus lessening the chance of infection. Thus, applying nematodes by means of microsprinklers currently seems to be the most effective application method (Moore *et al.*, 2023).

16.6.2 Fruit fly

Malan and Manrakhan (2009) studied the effect of locally isolated EPNs on the larvae, pupae and adult *C. capitata* and *C. rosa* flies in 24-well bioassay plates. The insect larvae, pupae and adults were exposed to 200 IJs of *S. khoisanae*, *H. zealandica* SF41 strain (blue variant) and three isolates of *H. bacteriophora*. *C. capitata* was generally more susceptible to EPN infection than *C. rosa*. A reduced susceptibility of the adult flies was ascribed to the grooming behavior of the insect. The *Heterorhabditis* spp. isolates were generally more virulent than *S. khoisanae*, perhaps due to IJ size differences between the species. Infection of *C. capitata* larvae ranged between 80% and 95% for *Heterorhabditis* spp. isolates and between 30% and 65% for *C. rosa*. The lowest infection rate of 60% was found for *C. capitata*, with 15% being found for *S. khoisanae*. No infection of fully formed pupae occurred for either fruit fly species, whereas *Heterorhabditis* spp. isolates infected between 35% and 60% of adult *C. capitata*, and between 15% and 60% of *C. rosa* adults. A second study measured the virulence of five local South African EPN species (50 IJs per larva) against third-instar *C. capitata* larvae (James *et al.*, 2018). The most virulent EPN species, *H. noenieputensis* and *S. yirgalemense*, were further tested using 2000 IJs in 100 ml of sterilized sand. After 24 h, ten third-instar *C. capitata* larvae were added to the sand. A high mortality of the larvae after 48 h of exposure to *H. noenieputensis* revealed good potential for use of local EPN species as biological control agents against *C. capitata*.

EPN infection of the fruit fly puparium is critical because the last-instar larvae are in the soil only a few hours, whereas pupae remain in the soil for up to 20 days. Malan and Manrakhan (2009) reported that late larval infection can result in malformed infected puparia (Fig. 16.4) but that fully formed puparia are impregnable. Subsequent studies found EPN infection of the fruit fly puparium (Torrini *et al.*, 2017; Godjo *et al.*, 2018), albeit at lower rates than for younger pupae. These findings agree with Gazit *et al.* (2010), who reported that *S. riobrave* caused up to 20% mortality in the young pupae of *C. capitata*. In addition, Minas *et al.* (2016) showed up to 100% mortality of *C. capitata* 1-day-old pupae at high concentrations. However, Langford *et al.* (2014) and Yee and Lacey (2003) observed no pupal susceptibility of either *Bactrocera tryoni* or *Rhagoletis indifferens* to EPNs. The body of a pupa is hardened by sclerotization and possesses no opening through which the nematode can penetrate. Further research with different EPNs is needed, especially using older pupae, to better assess fruit fly susceptibility to nematodes.

16.6.3 Citrus mealybug

The citrus mealybug is the most important mealybug on citrus in South Africa. In the first study to evaluate the potential of EPNs to control citrus mealybug, females were found to be susceptible to six indigenous nematode species, with *S. yirgalemense* and *H. zealandica* being the most virulent, causing 97% and 91% infection respectively, at 200 IJs per insect (van Niekerk and Malan, 2012). Both EPNs were able to complete their life cycles in the mealybug and produce new cohorts of IJs (Fig. 16.5). Additional bioassays detected a linear relationship between mealybug mortality and IJ concentration, with the highest control at 80 IJs per insect (van Niekerk and Malan, 2012). The study showed that the first 2–4 h post-application are critical for establishing successful infection of mealybugs, during which the humidity and temperature must be optimum (van Niekerk and Malan, 2012). IJs were able to infect their target after exposure that was as short as 30 min. *S. yirgalemense* infected *Planococcus citri* faster than *H. zealandica* and was twice as tolerant to relatively low levels of free water. However, no significant difference was found in infection at 15°C, 20°C or 25°C.

Fig. 16.4. Different life stages of the Mediterranean fruit fly, *Ceratitis capitata*, infected with entomopathogenic nematodes.

Fig. 16.5. Change in appearance of female *Planococcus citri*. (A) Before infection. (B–D) At 48 h after infection with *Heterorhabditis zealandica* (B) or *Steinernema yirgalemense* (C, D). Copyright: S. van Niekerk.

The compatibility of *H. zealandica* and *S. yirgalemense* with biological control agents and agrochemicals that are likely to be used in an IPM program for citrus in South Africa was investigated. *Cryptolaemus montrouzieri* (Fig. 16.6) is a predatory coccinellid beetle that is produced commercially and used against mealybugs. The beetle larvae were found to be susceptible to EPN at a rate of 80 IJs per insect, with 80% infection occurring for *H. zealandica* and 92% for *S. yirgalemense*. Adults were less susceptible, with 64% and 30% infection by *S. yirgalemense* and *H. zealandica*, respectively. Application of EPNs prior to release of *C. montrouzieri* in citrus orchards is indicated (van Niekerk and Malan, 2014a). *H zealandica* proved to be highly compatible with aqueous solutions of the chemical and biological insecticides Cyperfos 500 EC®, Cryptogran™ and Helicovir™, and the adjuvants Nu-Film-P® and Zeba®. In addition, 24 h exposure to the formulations did not affect the ability of *H. zealandica* to infect *Tenebrio molitor*. However, *S. yirgalemense* mortality increased after 12 h exposure to Nu-Film-P® and Cyperfos 500 EC®, and after 6 h exposure to Nu-Film-P®. Nevertheless, *S. yirgalemense* infectivity of insects was unaffected by the products (van Niekerk and Malan, 2014b).

Adequate aboveground control of *P. citri* by EPNs requires the use of adjuvants to maintain humidity (Fig. 16.7). In a growth chamber simulating glasshouse conditions of 75% relative humidity and 22°C, the addition of 0.3% Zeba® and 0.6% Nu-Film-P® to EPN suspensions of *H. zealandica* and *S. yirgalemense* did not significantly increase the levels of infection achieved, although they prolonged the IJ infectivity and survival by up to 2–3 h (van Niekerk

Fig. 16.6. (A) Adult and (B) larval stages of the mealybug predatory coccinellid beetle, *Cryptolaemus montrouzieri*.

Fig. 16.7. Insect containment methods used to evaluate the aboveground application of *Heterorhabditis zealandica* and *Steinernema yirgalemense*. (A) Petri dish with leaves pretreated with a nematode suspension. (B) Plastic container with citrus fruit. (C) Pockets made of fine-mesh netting, containing a citrus leaf with female mealybugs. Copyright: S. van Niekerk.

and Malan, 2015). In a semi-field trial, higher control of *P. citri* (up to 53%) was achieved when *S. yirgalemense* was used with 0.3% Zeba® to retard desiccation. The addition of Nu-Film-P® and Zeba® to *H. zealandica* suspensions did not retard runoff on citrus leaves but did retard sedimentation, thereby increasing the number of nematodes deposited on leaf disks. Xanthan gum (0.2%) was highly effective at retarding sedimentation, with 72% of the initial nematode number remaining in suspension after 1 h (van Niekerk and Malan, 2015).

16.6.4 Citrus root weevil

Almost immediately following its description by Cabanillas *et al.* (1993), *S. riobrave* was shown in numerous laboratory assays to be superior to other EPN species against *D. abbreviatus*, *P. litus* (Schroeder, 1994; Shapiro and McCoy, 2000a,b) and similar weevil pests of citrus such as the Fuller rose beetle, *Naupactus godmanni*, in California (Gulcu *et al.*, 2019). The effectiveness of *S. riobrave* varied widely in field trials. When used at rates of 20–108 IJs cm^{-2}, the nematode reduced larvae in soil and adult emergence by 50–90% in trials in Florida and California (Duncan and McCoy, 1996; Duncan *et al.*, 1996, 2003b, 2007, 2010; Bullock *et al.*, 1999; McCoy *et al.*, 2002; Bender *et al.*, 2014). In contrast, commercial products with *S. riobrave* and *H. indica* failed to reduce sentinel weevils in an orchard on sandy clay loam soil, perhaps due to the finer texture compared with most trial sites (McCoy *et al.*, 2002). Although *S. riobrave* performed well in fine-textured soil in some studies (Shapiro and McCoy, 2000b; Jenkins *et al.*, 2008), others have demonstrated that these nematodes generally perform better in the coarse sands typical of the central ridge compared with finer-textured soils in some flatwood sites (Kaspi *et al.*, 2010; El-Borai *et al.*, 2012).

16.7 Grower Adoption and Nature of Use in Integrated Pest Management Programs

EPNs are currently available for use in IPM programs against FCM in South Africa (Moore and Hattingh, 2004; Moore, 2021). The "systems approach" varies but essentially consists of a combination of risk-management measures that cumulatively increase quarantine security. Various biocontrol technologies, including the commercial use of EPNs, are available and continue to be studied for the control of FCM (Hatting *et al.*, 2009; Moore, 2021). The South African government has restricted the use of several insecticides since the late 1970s, encouraging the development and use of biological control agents to support the South African National Bio-Economy Strategy (DST, 2013).

In Florida, weevil management recommendations are guided by the seasonal dynamics of weevil populations as modulated by ecoregions. All life stages of *D. abbreviatus* other than eggs are present in soil throughout the year, and adult weevils emerge from the soil during all but the coldest winter months (Duncan *et al.*, 2001; McCoy *et al.*, 2003; Nigg *et al.*, 2003). In both the flatwoods and central ridge, a distinct peak adult emergence occurs in spring and a second, smaller peak is occasionally seen in fall. However, the amplitude of the peaks and abundance of the weevils in flatwoods can exceed those on the ridge by an order of magnitude (Futch *et al.*, 2005). Consequently, weevil IPM is employed more intensively in flatwoods than on the central ridge. Growers decide from experience whether the tree condition and population density warrant the cost of management. Weevils are sometimes so few on the central ridge that it takes many years before damage to roots causes trees to decline noticeably. By contrast, weevils are sometimes so numerous in certain flatwood regions that they prevent the establishment of newly planted trees. Weevil IPM includes attention to proper soil drainage (Lapointe and Shapiro, 1999), rootstock selection for tolerance of regionally prevalent *Phytophthora* spp. (Graham *et al.*, 2003), population monitoring (Duncan *et al.*, 2001; Futch *et al.*, 2005), and judicious use of adulticides and ovicides only at peak adult emergence to avoid disrupting endemic biological control agents (Duncan and Mannion, 2022). EPN application via under-canopy microjet irrigation systems when weevil soil populations are high in early fall or in spring can provide immediate, non-persistent reduction of larvae. EPN applications are best timed well prior to anticipated peaks of adult emergence and egg laying to avoid a brief reduction in natural control by native EPNs that sometimes occurs following inundation with commercial nematodes (Duncan *et al.*, 2003b, 2007; Dritsoulas *et al.*, 2022).

Although expensive, growers in Texas, California and Florida are increasingly installing landscape fabric as barriers to weevil entry or exit from the soil (McKenzie *et al.*, 2001; Duncan *et al.*, 2009; Bender *et al.*, 2014; Phillips, 2017). EPNs applied via microsprinklers and other spray equipment in the field were able to penetrate the fabric and kill *D. abbreviatus* sentinels but at a lower rate than in uncovered soil (L.W. Duncan *et al.*, unpublished data).

16.8 Conservation and Tactics Adopted by Growers

A very low incidence of EPNs, ranging from 5% to 7% of samples, occurred in many of the relevant South African surveys; however, a survey of citrus orchards, which used three highly susceptible hosts for trapping, produced a recovery rate of 17% (Malan *et al.*, 2011). These abundance estimates are in stark contrast to those obtained in citrus surveys in Florida, where most of the citrus orchards were populated with native EPNs, often with up to three sympatric species (Campos-Herrera *et al.*, 2012b, 2013a; Pathak *et al.*, 2017). The low recovery rate of EPNs from orchards in South Africa may be ascribed in part to the application of nematicides to control the citrus nematode, *Tylenchulus semipenetrans*, or to the warm, dry summers that tend to prevail in the Mediterranean climate of the Western Cape province. For example, Manrakhan *et al.* (2014) reported that naturally occurring *H. zealandica* (SF41 strain) and its symbiotic "blue variant" *Photorhabdus heterorhabditis* subsp. *heterorhabditis* (Booyzen *et al.*, 2022) reduced citrus fruit infestation by FCM within an orchard by 59%. The extent of control achieved by the endemic nematode was reduced by more than half when the soil was treated with the nematicide cadusafos. Further research is warranted to determine whether non-target effects of nematicides materially affect the biological control of soil-borne pests.

The commercialization and study of EPN products in Florida citrus orchards revealed that formulated EPNs were competing with several known and undescribed endemic species. Adair (1994) found a high incidence of "*H. bacteriophora*" (later found to be *H. indica*) in trials using commercial *S. carpocapsae* for weevil management and suggested the potential use of conservation methods to enhance the services of endemic EPNs. Caged, buried sentinel larvae of *D. abbreviatus* used to measure the efficacy of commercial *S. riobrave* on the central ridge (Fig. 16.8) were infected by several endemic EPN species in

Fig. 16.8. (A) Stainless steel in-line filter modified with caps on each end to hold moist soil and a weevil larva. The cages are buried prior to application of entomopathogenic nematodes on the soil surface and recovered after several days. The mesh screen provides access to the nematodes but excludes larger predators such as ants. (B) A ground trap for recovery of weevil adults emerging from the soil. (C) A Tedders trap used to recover weevils moving aboveground, which mistake the base for a tree trunk and are caught in the boll weevil trap.

untreated plots at rates as high as 55% (Duncan *et al.*, 2003b, 2007). Moreover, overall sentinel insect mortality of over 70% in some months was likely due mainly to EPNs, based on the emergence of free-living nematode competitors of EPNs (Duncan *et al.*, 2003a, 2007) from some cadavers. During the same survey in a flatwood orchard, only *H. indica* was detected in untreated plots, infecting the larvae at rates no higher than 8% (Duncan *et al.*, 2003b).

These studies provided evidence that large regional differences in abundance of citrus root weevils is due in part to regional variations in levels of biocontrol by EPNs. Support for this hypothesis came from pot experiments showing that *S. diaprepesi* (a ubiquitous Florida "ridge" species) and *S. khuongi* (sporadically found in flatwoods) are much more effective than *H. indica* (the primary flatwood species) in reducing *D. abbreviatus*, and that both steinernematids reduced weevil damage to roots more effectively in the coarse sandy central ridge soils than in soils from flatwood orchards (El-Borai *et al.*, 2012). Surveys (Campos-Herrera *et al.*, 2013a, 2016, 2019a) and experiments (Duncan and McCoy, 2001; Campos-Herrera *et al.*, 2013b, 2014; El-Borai *et al.*, 2016) to discover soil properties that might be managed to conserve or increase activities of native EPNs revealed pH and several variables related to soil moisture (groundwater depth, water-holding capacity, clay and organic matter content) to be likely drivers of the EPN community structure (see section 16.4) in Florida orchards.

Several tactics based on the above findings have been proposed for EPN conservation, some of which are practiced by growers. Tree planting holes in a *D. abbreviatus*-infested flatwood orchard with poor drainage and a shallow water table were filled with either the local soil or sandy soil from the central ridge. *H. indica*, the only species found naturally in the orchard, was introduced to the planting holes along with three other "Florida" species (see section 16.4). Over 6 years, EPNs killed fewer sentinel weevil larvae, more weevil adults were trapped and tree mortality was fivefold higher in native compared with sandy soil (Duncan *et al.*, 2013). Predictably, the two "flatwood" species eventually displaced the species introduced from the central ridge (Fig. 16.9). The introduction of sand in

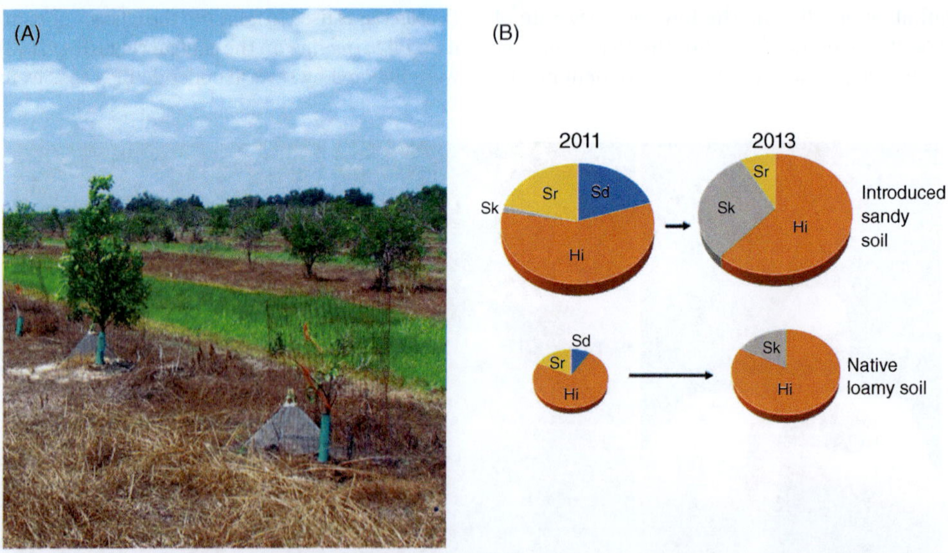

Fig. 16.9. (A) Growth response of trees planted into native loamy soil (proximal tree) or introduced sandy soil (distal tree) in a flatwood orchard heavily infested with *Diaprepes abbreviatus* and inoculated with four species of entomopathogenic nematode (EPN). (B) Note how eventually the EPN communities in both soils are dominated by the two species best adapted to flatwood habitats (*Steinernema khuongi* and *Heterorhabditis indica*) rather than one adapted to drier conditions on the central ridge (*Steinernema diaprepesi*). *Steinernema riobrave* is exotic to Florida. Hi, *Heterorhabditis indica*; Sd, *Steinernema diaprepesi*; Sk, *Steinernema khuongi*; Sr, *Steinernema riobrave*.

planting holes is employed occasionally in the flatwoods, primarily for drainage. Similarly, most growers now maintain soil pH at or below 6.5 to reduce stress to HLB-affected trees. A low soil pH also helps reduce adherence by spores of an ectoparasitic *Paenibacillus* sp. to the cuticle of *S. diaprepesi* (Campos-Herrera *et al.*, 2019a). The spores impair EPN movement, which has been associated with fewer *S. diaprepesi* and more root weevils in laboratory assays, field experiments, and spatial and temporal orchard surveys (El-Borai *et al.*, 2005; Campos-Herrera *et al.*, 2011, 2012a, 2013a,b, 2014, 2019a).

The recommended application timing of commercial EPNs avoids use immediately prior to peak weevil egg laying because use of commercial *S. riobrave* (20 and 30 IJs cm^{-2}) was shown to increase larval mortality by 46% during the treatment months but to significantly decrease weevil mortality during the intervening months, perhaps by increasing the abundance of nematode natural enemies (Duncan *et al.*, 2003b, 2007; El-Borai and Duncan, 2007; El-Borai *et al.* 2009, 2011). Rootstocks that produce the volatile sesquiterpene pregeijerene in response to herbivory by weevil larvae (i.e. Swingle citrumelo), rather than constitutively (e.g. Cleopatra mandarin), may be preferable because proximity to pregeijerene increases larval infection by EPNs (Ali *et al.*, 2012). However, the discovery that foliar feeding by psyllids and infection by CLas (*Candidatus* Liberibacter asiaticus) bacteria can induce pregeijerene in the rhizosphere of seedlings shows that additional research is needed (Rivera *et al.*, 2017).

The introduction of EPNs into depauperate orchards merits additional research. *S. khuongi* is better adapted to flatwood habitats than *S. diaprepesi* (El-Borai *et al.*, 2016), but is only occasionally present in flatwood sites (Campos-Herrera *et al.*, 2013a, 2016). Both species were superior to *H. indica* for control of *D. abbreviatus* in greenhouse trials (El-Borai *et al.*, 2012). If introduced, *S. khuongi* may be capable of persisting and providing some regulation of weevils (Fig. 16.9) (Duncan *et al.*, 2013).

16.9 Synthesis and Prospects

Currently in South Africa, no local production of EPNs occurs. However, the prevailing situation is likely to change soon, as most species isolated from South African soil are virulent to economically important pests. With positive results having been obtained using local EPNs, both in the laboratory and in semi-field trials, the *H. bacteriophora*-based product Cryptonem (River Bioscience, South Africa) has been registered for use against FCM in citrus (Hatting *et al.*, 2019; Moore, 2021).

Relatively few ecological studies have focused on South African EPNs and the results reported recently concerning natural regulation of FCM by *H. zealandica* reflect the potential importance of conservation management in orchards conducive to some of the many native species detected to date. Similarly, the possible non-target displacement of native species by the introduction of effective products with exotic species is an important concern in South Africa. Despite this, there is much interest in the exotic *S. feltiae*, a cold-weather species, because EPN management of both FCM and codling moth is most feasible during winter when the soil temperatures are generally below 18°C. In Africa, endemic *S. feltiae* has been reported only in Algeria (Tarasco and Triggiani, 2010). With relatively few surveys on the occurrence and distribution of EPNs in South Africa, the introduction of *S. feltiae* could mask the natural occurrence of the nematode. Perhaps the history of EPN use in Florida citrus is instructive here, because widespread use of *S. riobrave* for decades has had no documented long-term impact on EPN diversity in the orchards. Indeed, while other species remain abundant, *S. riobrave* is rarely recovered in research samples within weeks of its application. Similarly, despite large-scale applications in the different provinces of South Africa, *S. feltiae* appears not to have established (S. Moore, personal communication).

Profitable management of weevil larvae with EPNs in Florida citrus today is constrained by HLB occurrence and EPN cost. HLB complicates IPM, especially for pests and diseases of the root system, because HLB reduces the tree fibrous roots by at least 40% before symptoms are seen in the canopy (Graham *et al.*, 2013). Although there are no studies of the economic outcome of weevil IPM in orchards with HLB, the disease clearly impairs the capacity of trees to respond favorably to practices that reduce root herbivory. Additionally, the high cost of treating the understorey of mature trees restricts the use of non-persistent EPN products to once or twice annually, providing long intervals during which larvae are controlled only by naturally occurring enemies. Nevertheless, the widespread, continual replanting of groves due to HLB may provide new opportunities to adapt weevil IPM tactics.

Individual protective tree covers have recently been shown to effectively block psyllids from feeding and transmitting HLB (Alferez *et al.*, 2020). Current cover designs can be used for the first 3 years of tree growth. When used in orchards heavily infested by *D. abbreviatus*, the covers protect trees from both HLB and weevils by denying insects access to the tree canopy. These covers provide an opportunity for a new approach to weevil IPM. When the covers are removed, the young healthy trees still require a very restricted irrigation pattern compared with the patterns required by large, mature trees. EPNs can be applied via microjets or by hand multiple times annually to these trees for less cost than a single application in mature trees. If growers apply *S. riobrave* frequently (i.e. 4–6-week intervals) to just the area around the crown of the root system, they may succeed in preventing serious damage to this critical juncture of all the lateral roots and may maintain trees in a healthier condition for a much longer time than currently. Indeed, in the absence of HLB, focusing EPNs at just the crown may be profitable, even as trees reach maturity.

16.10 References

Adair, R. Jr. (1994) A four-year field trial of entomopathogenic nematodes for control of *Diaprepes abbreviatus* in a flatwoods citrus grove. *Proceedings of the Florida State Horticultural Society* 107, 63–68.

Alferez, F., Gaire, S., Albrecht, U., Batuman, O., Qureshi, J. and Zekri, M. (2020) Research Update on Individual Protective Covers. *Citrus Industry* November, 18–20. Available at: https://crec.ifas.ufl.edu/media/crecifasufledu/extension/extension-publications/2020/2020_nov_research.pdf (accessed 17 April 2024).

Ali, J.G., Alborn, H.T., Campos-Herrera, R., Kaplan, F., Duncan, L.W. *et al.* (2012) Subterranean, herbivore-induced plant volatile increases biological control activity of multiple beneficial nematode species in distinct habitats. *PLOS ONE* 7: e38146.

Aluja, M., Sivinsk, J., Rull, J. and Hodgson, P.J. (2005) Behavior and predation of fruit fly larvae (*Anastrepha* spp.) (Diptera: Tephritidae) after exiting fruit in four types of habitats in tropical Veracruz, Mexico. *Environmental Entomology* 34, 1507–1516.

Annecke, D.P. and Moran, V.C. (1982) *Insects and Mites of Cultivated Plants in South Africa*. Butterworths, Durban, South Africa.

Beavers, J.B., McCoy, C.W. and Kaplan, D.T. (1983) Natural enemies of subterranean *Diaprepes abbreviatus* (Coleoptera: Curculionidae) larvae in Florida. *Environmental Entomology* 12, 840–843.

Bender, G.S., Bates, L.M., Bethke, J.A., Lewis, E., Tanizaki, G. *et al.* (2014) Evaluation of insecticides, entomopathogenic nematodes, and physical soil barriers for control of *Diaprepes abbreviatus* (Coleoptera: Curculionidae) in citrus. *Journal of Economic Entomology* 107, 2137–2146.

Booyzen, E., Malan, A.P. and Dicks, L.T.D. (2022) Colour of *Heterorhabditis zealandica*-infected-*Galleria mellonella* dependent on the *Photorhabdus* symbiont, with two new nematode-symbiotic associations reported. *Journal of Invertebrate Pathology* 189: 107729.

Bullock, R.C., Pelosi, R.R. and Killer, E.E. (1999) Management of citrus root weevils (Coleoptera: Curculionidae) on Florida citrus with soil-applied entomopathogenic nematodes (Nematoda: Rhabditida). *Florida Entomologist* 82, 1–7.

Cabanillas, H.E., Poinar, G.O. and Raulston, J.R. (1993) *Steinernema riobravis* n. sp. (Rhabditida: Steinernematidae) from Texas. *Fundamental and Applied Nematology* 17, 123–131.

Caltagirone, L.E. and Doutt, R.L. (1989) The history of the vedalia beetle importation to California and its impact on the development of biological control. *Annual Review of Entomology* 34, 1–16.

Campos-Herrera, R., Gomez-Ros, J.M., Escuer, M., Cuadra, L., Barrios, L. and Gutierrez, C. (2008) Diversity, occurrence, and life characteristics of natural entomopathogenic nematode populations from La Rioja (Northern Spain) under different agricultural management and their relationships with soil factors. *Soil Biology and Biochemistry* 40, 1474–1484.

Campos-Herrera, R., El-Borai, F.E., Stuart, R.J., Graham, J.H. and Duncan, L.W. (2011) Entomopathogenic nematodes, phoretic *Paenibacillus* spp., and the use of real time quantitative PCR to explore soil food webs in Florida citrus groves. *Journal of Invertebrate Pathology* 108, 30–39.

Campos-Herrera, R., El-Borai, F.E. and Duncan, L.W. (2012a) Real-time PCR as an effective technique to assess the impact of phoresy by *Paenibacillus* sp. bacteria on *Steinernema diaprepesi* nematodes in nature. *Molecular Ecology Resources* 12, 885–893.
Campos-Herrera, R., El-Borai, F.E. and Duncan, L.W. (2012b) Wide interguild relationships among nematodes and fungi in soil as measured by real time qPCR. *Journal of Invertebrate Pathology* 111, 126–135.
Campos-Herrera, R., Pathak, E., El-Borai, F.E., Schumann, A., Abd-Elgawad, M.M. and Duncan, L.W. (2013a) New citriculture system suppresses native and augmented entomopathogenic nematodes. *Biological Control* 66, 183–194.
Campos-Herrera, R., Pathak E., El-Borai, F.E., Stuart, R.J., Gutiérrez, C. *et al.* (2013b) Geospatial patterns of soil properties and the biological control potential of entomopathogenic nematodes in Florida citrus groves. *Soil Biology and Biochemistry* 66, 163–174.
Campos-Herrera, R., El-Borai, F.E., Ebert, T.E., Schumann, A. and Duncan, L.W. (2014) Management to control citrus greening alters the soil food web and severity of a pest–disease complex. *Biological Control* 76, 41–51.
Campos-Herrera, R., El-Borai, F.E., Rodriguez Martin, J.A. and Duncan, L.W. (2016) Entomopathogenic nematode food web assemblages in Florida natural areas. *Soil Biology and Biochemistry* 93, 105–114.
Campos-Herrera, R., Stuart, R.J., Pathak, E., El-Borai, F.E. and Duncan, L.W. (2019a) Temporal patterns of entomopathogenic nematodes in Florida citrus orchards: evidence of natural regulation by microorganisms and nematode competitors. *Soil Biology and Biochemistry* 128, 193–204.
Campos-Herrera, R., Blanco-Perez, R., Bueno-Pallero, F.A., Duarte, A., Nolasco, G. *et al.* (2019b) Vegetation drives assemblages of entomopathogenic nematodes and other soil organisms: evidence from the Algarve, Portugal. *Soil Biology and Biochemistry* 128, 150–163.
Carstens, E. and Moore, S. (2021) *Thaumatotibia leucotreta* (false codling moth (FCM)). In: *CABI Compendium*. CAB International, Wallingford, UK. Available at: www.cabi.org/isc/datasheet/6904 (accessed 20 March 2024).
CGA (Citrus Growers Association) (2021a) *Annual Report 2021*. CGA, Durban, South Africa. Available at: www.citrusresourcewarehouse.org.za/home/document-home/internal-citrus-industry-organisation-publications/annual-reports/cga-annual-reports/7460-cga-annual-report-2021/file (accessed 20 March 2024).
CGA (Citrus Growers Association) (2021b) *Industry Statistics, 2020 Export Season*. CGA, Durban, South Africa. Available at: www.citrusresourcewarehouse.org.za/home/document-home/information/cga-key-industry-statistics/7253-cga-key-industry-statistics-2021/file (accessed 20 March 2024).
Daiber, C.C. (1979) A study of the biology of the false codling moth [*Cryptophlebia leucotreta* (Meyr.)]: the cocoon. *Phytophylactica* 11, 151–157.
Daiber, C.C. (1980) A study of the biology of the false codling moth *Cryptophlebia leucotreta* (Meyr.): the adult and generations during the year. *Phytophylactica* 12, 187–193.
de Meyer, M. (2021) *Ceratitis rosa* (Natal fruit fly). In: *CABI Compendium*. CAB International, Wallingford, UK. Available at: www.cabi.org/isc/datasheet/12378 (accessed 20 March 2024).
de Waal, J.Y., Malan, A.P. and Addison, M.F. (2011) Efficacy of entomopathogenic nematodes (Rhabditida: Heterorhabditidae and Steinernematidae) against codling moth, *Cydia pomonella* (Lepidoptera: Tortricidae) in temperate regions. *Biocontrol Science and Technology* 20, 489–502.
Diepenbrock, L.M., Dewdney, M.M. and Vashisth, T. (eds.) (2022) *2022–2023 Florida Citrus Production Guide*. University of Florida Institute of Food and Agricultural Sciences, Gainesville, Florida. Available at: https://edis.ifas.ufl.edu/publication/CG101 (accessed 20 March 2024).
Dolinski, C., Choo, H.Y. and Duncan, L.W. (2012) Grower acceptance of entomopathogenic nematodes: case studies on three continents. *Journal of Nematology* 44, 226–235.
Dritsoulas, A., Campos-Herrera, R., Blanco-Perez, R. and Duncan, L.W. (2020) Comparing high-throughput sequencing and real time PCR for characterizing entomopathogenic nematode biogeography. *Soil Biology and Biochemistry* 145: 107793.
Dritsoulas, A., El-Borai, F.E., Shehata, I.E., Hammam, M.M., El-Ashry, R.M. *et al.* (2021) Reclaimed desert habitats favor entomopathogenic nematode and microarthropod abundance compared to ancient farmlands in the Nile Basin. *Journal of Nematology* 53: e2021-47.
Dritsoulas, A., Wu, S.-Y., Regmi, H. and Duncan, L.W. (2022) Arthropod community responses reveal potential predators and prey of entomopathogenic nematodes in a citrus orchard. *Agronomy* 12: 2502.
DST (Department of Science and Technology) (2013) *The Bio-Economy Strategy*. DST, Pretoria. Available at: www.innovus.co.za/media/Bioeconomy_Strategy.pdf (accessed 20 March 2024).

Duncan, L.W. and Mannion, C. (2022) Citrus root weevils. In: Diepenbrock, L.M., Dewdney, M.M. and Vashisth, T. (eds.) (2022) *2022–2023 Florida Citrus Production Guide*. University of Florida Institute of Food and Agricultural Sciences, Gainesville, Florida. Available at: https://edis.ifas.ufl.edu/publication/cg006 (accessed 20 March 2024).

Duncan, L.W. and McCoy, C.W. (1996) Vertical distribution in soil, persistence, and efficacy against citrus root weevil (Coleoptera: Curculionidae) of two species of entomogenous nematodes (Rhabditida, Steinernematidae, Heterorhabditidae). *Environmental Entomology* 25, 174–178.

Duncan, L.W. and McCoy, C.W. (2001) Hydraulic lift increases herbivory by *Diaprepes abbreviatus* larvae and persistence of *Steinernema riobrave* in dry soil. *Journal of Nematology* 33, 142–146.

Duncan, L.W., McCoy, C.W. and Terranova, A.C. (1996) Estimating sample size and persistence of entomogenous nematodes in sandy soils and their efficacy against the larvae of *Diaprepes abbreviatus* in Florida. *Journal of Nematology* 28, 56–67.

Duncan, L.W., McCoy, C.W., Stansly, P.A., Graham, J.H. and Mizell, R.F. (2001) Estimating the relative abundance of citrus root weevils with modified Tedders traps. *Environmental Entomology* 30, 939–946.

Duncan, L.W., Dunn, D.C., Bague, G. and Nguyen, K. (2003a) Competition between entomopathogenic and free-living bactivorous nematodes in larvae of *Diaprepes abbreviatus. Journal of Nematology* 35, 187–193.

Duncan, L.W., Graham, J.H., Dunn, D.C., Zellers, J., McCoy, C.W. and Nguyen, K. (2003b) Incidence of endemic entomopathogenic nematodes following application of *Steinernema riobrave* for control of *Diaprepes abbreviatus*. *Journal of Nematology* 35, 178–186.

Duncan, L.W., Graham, J.H., Zellers, J., Bright, D., Dunn, D.C. *et al.* (2007) Food web responses to augmenting the entomopathogenic nematodes in bare and animal manure-mulched soil. *Journal of Nematology* 39, 176–189.

Duncan, L.W., Stuart, R.J. and Lapointe, S.L. (2009) Use of landscape fabric to manage *Diaprepes* root weevil in citrus groves. *Florida Entomologist* 92, 74–79.

Duncan, L.W., Dewdney, M. and Graham, J.H. (2010) Remember *Diaprepes*? It's still a problem. *Citrus Industry Magazine*, July, 9–13.

Duncan, L.W., Stuart, R.J., El-Borai, F.E., Campos-Herrera, R., Pathak, E. and Graham, J.H. (2013) Modifying orchard planting sites conserves entomopathogenic nematodes, reduces weevil herbivory and increases citrus tree growth, survival and fruit yield. *Biological Control* 64, 26–36.

Dunn, M.D., Belur, P.D. and Malan, A.P. (2020) *In vitro* liquid culture and optimization of *Steinernema jeffreyense*, using shake flasks. *BioControl* 65, 223–233.

Efron, D., Nestel, D. and Glazer, I. (2001) Spatial analysis of entomopathogenic nematodes and insect hosts in a citrus grove in a semi-arid region in Israel. *Environmental Entomology* 30, 254–261.

Ehlers, R.U. (2003) Biocontrol nematodes. In: Hokkanen, H.M.T. and Hajek, A. (eds.) *Environmental Impacts of Microbial Insecticides*. Kluwer Scientific, Dordrecht, Netherlands, pp. 177–220.

El-Borai, F.E. and Duncan, L.W. (2007) Suppression of *Diaprepes abbreviatus* in potted citrus by combinations of entomopathogenic nematodes with different lifespans. *Nematropica* 37, 33–41.

El-Borai, F.E., Duncan, L.W. and Preston, J.F. (2005) Bionomics of a phoretic association between *Paenibacillus* sp. and the entomopathogenic nematode *Steinernema diaprepesi*. *Journal of Nematology* 37, 18–25.

El-Borai, F.E., Bright, D.B., Graham, J.H., Stuart, R.J., Cubero, J. and Duncan, L.W. (2009) Differential susceptibility of entomopathogenic nematodes to nematophagous fungi from Florida citrus orchards. *Nematology* 11, 233–243.

El-Borai, F.E., Campos-Herrera, R., Stuart, R.J., and Duncan, L.W. (2011) Substrate modulation, group effects and the behavioral responses of entomopathogenic nematodes to nematophagous fungi. *Journal of Invertebrate Pathology* 106, 347–356.

El-Borai, F.E., Stuart, R.J., Campos-Herrera, R., Pathak, E. and Duncan, L.W. (2012) Entomopathogenic nematodes, root weevil larvae, and dynamic interactions among soil texture, plant growth, herbivory, and predation. *Journal of Invertebrate Pathology* 109, 134–142.

El-Borai, F.E., Killiny, N. and Duncan, L.W. (2016) Concilience in entomopathogenic nematode responses to water potential and their geospatial patterns in Florida. *Frontiers in Microbiology* 7: 356.

Franco, J., Suma, P., da Silva, E., Blumberg, D. and Mendel, Z. (2004) Management strategies of mealybug pests of citrus in Mediterranean countries. *Phytoparasitica* 32, 507–522.

Futch, S.H., Duncan, L.W. and Zekri, M. (2005) Validation of an area-wide extension program to estimate the seasonal abundance of adult citrus root weevils with un-baited pyramidal traps. *Proceedings of the Florida State Horticultural Society* 117, 143–147.

Gazit, Y., Rossler, Y. and Glazer, I. (2010) Evaluation of entomopathogenic nematodes for the control of Mediterranean fruit fly (Diptera: Tephritidae). *Biocontrol Science and Technology* 10, 157–164.

Giliomee, J.H. and Riedl, H. (1998) A century of codling moth control in South Africa 1. Historical perspective. *Journal of the South African Society for Horticultural Sciences* 8, 27–31.

Glazer, I., Liran, N. and Steinberger, Y. A. (1991) A survey of entomopathogenic nematodes (Rhabditida) in the Negev desert. *Phytoparasitica* 19, 291–300.

Godjo, A., Zadji, L., Decraemer, W., Williams, A. and Afouda, L. (2018) Pathogenicity of indigenous entomopathogenic nematodes from Benin against mango fruit fly (*Bactrocera dorsalis*) under laboratory conditions. *Biological Control* 117, 68–77.

Gouge, D.H., Reaves, L.L., Stoltman, M.M., van Berkum, J.R., Burke, R.A. *et al.* (1996) Control of pink bollworm *Pectinophora gossypiella* (Saunders) (Lepidoptera: Gelechiidae) larvae in Arizona and Texas cotton fields using the entomopathogenic nematode *Steinernema riobrave* (Cabanillas, Poinar, and Raulston) (Rhabditida: Steinernematidae). In: *Proceedings of the Beltwide Cotton Conference, Vol.* 2. National Cotton Council, Memphis, Tennessee, 1078–1082.

Graham, J.H., Bright, D.B. and McCoy, C.W. (2003) *Phytophthora–Diaprepes* weevil complex: *Phytophthora* spp. relationship with citrus rootstocks. *Plant Disease* 87, 85–90.

Graham, J.H., Johnson, E.G., Gottwald, T.R. and Irey, M.R. (2013) Presymptomatic fibrous root decline in citrus trees caused by huanglongbing and potential interaction with *Phytophthora* spp. *Plant Disease* 97, 1195–1199.

Grout, T.G. and Moore, S.D. (2015) Citrus. In: Prinsloo, G.L. and Uys, V.M. (eds.) *Insects of Cultivated Plants and Natural Pastures in Southern Africa*. Entomological Society of Southern Africa, Pretoria, pp. 448–499.

Gulcu, B., Hodson, A., Omaleki, V., Ross, A.B. and Lewis, E.E. (2019) A biological control approach to reducing *Naupactus godmani* (Curculionidae) populations in citrus using entomopathogenic nematodes. *Crop Protection* 115, 99–103.

Hatting, J.L. and Malan, A.P. (2017) Status of entomopathogenic nematodes in integrated pest management strategies in South Africa. In: Abd-Elgawad, M.M., Askary, T.H. and Coupland, J. (eds.) *Biocontrol Agents: Entomopathogenic and Slug Parasitic Nematodes*. CAB International, Wallingford, UK, pp. 409–428.

Hatting, J., Stock, S.P. and Hazir, S. (2009) Diversity and distribution of entomopathogenic nematodes (Steinernematidae, Heterorhabditidae) in South Africa. *Journal of Invertebrate Pathology* 102, 120–128.

Hatting, J.L., Moore, S.D. and Malan, A.P. (2019) Microbial control of phytophagous invertebrate pests in South Africa: current status and future. *Journal of Invertebrate Pathology* 165, 54–66.

Hattingh, V. (1993) Mealybugs and cottony cushion scale on citrus in Southern Africa. *Citrus Journal* 3, 20–22.

Hattingh, V. and Moore, S.D. (2003) Mealybugs. In: Grout, T.G. (ed.) *Integrated Production Guidelines for Export Citrus, Vol. 3: Integrated Pest and Disease Management*. Citrus Research International, Mbombela, South Africa, pp. 65–69.

Hattingh, V. and Tate, B.A. (1996) The pest status of mealybugs on citrus in Southern Africa. *Proceedings of the International Society of Citriculture* 1, 560–563.

Hattingh, V., Cilliers, C.J. and Bedford, E.C.G. (1998) Citrus mealybugs. In: Bedford, E.C.G., van den Berg, M.A. and de Villiers, E.A. (eds.) *Citrus Pests in the Republic of South Africa*. ARC-Institute for Tropical and Subtropical Crops, Mbombela, South Africa, pp. 112–120.

Hofmeyr, J.H., Carpenter, J.E., Bloem, S., Slabbert, J.P., Hofmeyr, M. and Groenewald, S.S. (2015) Development of the sterile insect technique to suppress false codling moth *Thaumatotibia leucotreta* (Lepidoptera: Tortricidae) in citrus fruit: research to implementation (Part 1). *African Entomology* 23, 180–186.

James, M., Addison, P. and Malan, A.P. (2018) Surveying and screening South African entomopathogenic nematodes for the control of the Mediterranean fruit fly, *Ceratitis capitata* (Wiedemann). *Crop Protection* 105, 41–28.

Jenkins, D.A., Shapiro-Ilan, D.I. and Goenaga, R. (2008) Efficacy of entomopathogenic nematodes versus *Diaprepes abbreviatus* (Coleoptera: Curculionidae) larvae in a high clay-content oxisol soil: greenhouse trials with potted *Litchi chinensis*. *Florida Entomologist* 91, 75–78.

Kaspi, R., Ross, A., Hodson, A.K., Stevens, G.N., Kaya, H.K. and Lewis, E.E. (2010) Foraging efficacy of the entomopathogenic nematode *Steinernema riobrave* in different soil types from California citrus groves. *Applied Soil Ecology* 45, 243–253.

Langford, E.A., Nielsen, U.N., Johnson, S.N. and Riegler, M. (2014) Susceptibility of Queensland fruit fly, *Bactrocera tryoni* (Froggatt) (Diptera: Tephritidae), to entomopathogenic nematodes. *Biological Control* 69, 34–39.

Lapointe, S.L. and Shapiro, J.P. (1999) Effect of soil moisture on development of *Diaprepes abbreviatus* (Coleoptera: Curculionidae). *Florida Entomologist* 82, 291–299.

Love, C.N., Hill, M.P. and Moore, S.D. (2014) *Thaumatotibia leucotreta* and the Navel orange: ovipositional preferences and host susceptibility. *Journal of Applied Entomology* 138, 600–611.

Malan, A.P. and Ferreira, T. (2017) Entomopathogenic nematodes. In: Fourie, H., Spaull, V.W., Jones, R.K., Daneel, M.S. and de Waele, D. (eds.) *Nematology in South Africa: A View from the 21st Century*. Springer International, Cham, Switzerland, pp. 459–480.

Malan, A.P. and Hatting, J. (2015) Entomopathogenic nematode exploitation: case studies in laboratory and field applications from South Africa. In: Campos-Herrera, R. (ed.) *Sustainability in Plant and Crop Protection: Ecology and Applied Technologies for Sustainable Plant and Crop Protection*. Springer, Cham, Switzerland, pp. 475–506.

Malan, A.P. and Manrakhan, A. (2009) Susceptibility of the Mediterranean fruit fly (*Ceratitis capitata*) and the Natal fruit fly (*Ceratitis rosa*) to entomopathogenic nematodes. *Journal of Invertebrate Pathology* 100, 47–49.

Malan, A.P. and Moore, S.D. (2016) Evaluation of local entomopathogenic nematodes for the control of false codling moth, *Thaumatotibia leucotreta* (Meyrick, 1913), in a citrus orchard in South Africa. *African Entomology* 24, 489–501.

Malan, A.P., Nguyen, K.B. and Addison, M.F. (2006) Entomopathogenic nematodes (Steinernematidae and Heterorhabditidae) from the southwestern parts of South Africa. *African Plant Protection* 12, 65–69.

Malan, A.P., Knoetze, R. and Moore, S.D. (2011) Isolation and identification of entomopathogenic nematodes from citrus orchards in South Africa and their biocontrol potential against false codling moth. *Journal of Invertebrate Pathology* 108, 115–125.

Malan, A.P., Knoetze, R. and Tiedt, L. (2014) *Heterorhabditis noenieputensis* n. sp. (Rhabditida: Heterorhabditidae), a new entomopathogenic nematode from South Africa. *Journal of Helminthology* 88, 139–151.

Malan, A.P., Knoetze, R. and Tiedt, T.R. (2016) *Steinernema jeffreyense* n. sp. (Rhabditida: Steinernematidae), a new entomopathogenic nematode from South Africa. *Journal of Helminthology* 90, 262–278.

Malan, A.P., von Diest, J.I., Moore, S.M. and Addison, P. (2018) Control options for false codling moth, *Thaumatotibia leucotreta* (Lepidoptera: Tortricidae), in South Africa, with emphasis on the potential use of entomopathogenic nematodes and fungi. *African Entomology* 26, 14–29.

Manrakhan, A., Daneel, J.-H. and Moore, S.D. (2014) The impact of naturally occurring entomopathogenic nematodes on false codling moth, *Thaumatotibia leucotreta* (Lepidoptera: Tortricidae), in citrus orchards. *Biocontrol Science and Technology* 24, 241–245.

Manrakhan, A., Daneel, J.-H., Beck, R., Theron, C.D., Weldon, C.W. *et al.* (2018) Non-host status of commercial export grade lemon fruit (*Citrus limon* (L.) Burman f. cv. Eureka) for *Ceratitis capitata*, *Ceratitis rosa*, *Ceratitis quilicii* and *Bactrocera dorsalis* (Diptera: Tephritidae) in South Africa. *African Entomology* 26, 202–214.

Martinez-Ferrer, M.T., Ripolles, J.L. and Garcia-Mari, F. (2006) Enumerative and binomial sampling plans for citrus mealybug (Homoptera: Pseudococcidae) in citrus groves. *Journal of Economic Entomology* 99, 993–1001.

McCoy, C.W., Shapiro, D.I., Duncan, L.W. and Nguyen, K. (2000) Entomopathogenic nematodes and other natural enemies as mortality factors for larvae of *Diaprepes abbreviatus* (Coleoptera: Curculionidae). *Biological Control* 19, 182–190.

McCoy, C.W., Stuart, R.J., Duncan, L.W. and Nguyen, K. (2002) Field efficacy of two commercial preparations of entomopathogenic nematodes against larvae of *Diaprepes abbreviatus* (Coleoptera: Curculionidae) in alfisol type soil. *Florida Entomologist* 85, 537–544.

McCoy, C.W., Stuart, R.J. and Nigg, H.N. (2003) Seasonal life stage abundance of *Diaprepes abbreviatus* in irrigated and non-irrigated citrus plantings in central Florida. *Florida Entomologist* 86, 34–42.

McKenzie, C.L., Lapointe, S.L. and Duncan, L.W. (2001) Landscape fabric as a physical barrier to neonate *Diaprepes abbreviatus* (Coleoptera: Curculionidae). *Florida Entomologist* 84, 721–722.

Minas, R.S., Souza, R.M., Dolinski, C., Carvalho, R.S. and Burla, R.S. (2016) Potential of entomopathogenic nematodes (Rhabditida: Heterorhabditidae) to control Mediterranean fruit fly (Diptera: Tephritidae) soil stages. *Nematoda* 3: e02016.

Moore, S.D. (2021) Biological control of a phytosanitary pest (*Thaumatotibia leucotreta*): a case study. *International Journal of Environmental Research Public Health* 18: 1198.
Moore, S.D. and Hattingh, V. (2004) Augmentation of natural enemies for control of citrus pests in South Africa: a guide for growers. *South African Fruit Journal* 3, 45–47.
Moore, S.D., Kirkman, W. and Hattingh, V. (2015) The host status of lemons for the false codling moth, *Thaumatotibia leucotreta* (Meyrick) (Lepidoptera: Tortricidae) with particular references to export protocols. *African Entomology* 23, 519–525.
Moore, S.D., Ehlers, R.-U., Manrakhan, A., Gilbert, M., Kirkman, W. *et al.* (2023) Field-scale efficacy of entomopathogenic nematodes to control false codling moth, *Thaumatotibia leucotreta* (Lepidoptera: Tortricidae), in citrus orchards in South Africa. *Crop Protection* 179: 106610.
Nguyen, K.B., Malan, A.P. and Gozel, U. (2006) *Steinernema khoisanae* sp. n. (Rhabditida: Steinernematidae), a new entomopathogenic nematode from South Africa. *Nematology* 8, 157–175.
Nigg, H.N., Simpson, S.E., Stuart, R.J., Duncan, L.W., McCoy, C.W. and Gmitter, F.G. Jr. (2003) Abundance of *Diaprepes abbreviatus* (L.) (Coleoptera, Curculionidae) neonates falling to the soil under tree canopies in Florida citrus groves. *Journal of Economic Entomology* 96, 835–843.
Pathak, E., Campos-Herrera, R., El-Borai, F.E. and Duncan, L.W. (2017) Spatial relationships between entomopathogenic nematodes and nematophagous fungi in Florida citrus orchards. *Journal of Invertebrate Pathology* 144, 37–46.
Phillips, K. (2017) New citrus planting method stops bugs, yields additional benefits. *Farm and Ranch*, September 30. Available at: https://agrilifetoday.tamu.edu/2017/09/25/new-citrus-planting-method-stops-bugs-yields-additional-benefits/ (accessed 20 March 2024).
Ramakuwela, T., Hatting, H., Laing, M.D., Hazir, S. and Thiebaut, N. (2016) *In vitro* solid-state production of *Steinernema innovationi* with cost analysis. *Biocontrol Science and Technology* 26, 792–808.
Rivera, M.J., Pelz-Stelinski, K.S., Martini, X. and Stelinski, L.L. (2017) Bacterial phytopathogen infection disrupts belowground plant indirect defense mediated by tritrophic cascade. *Ecological Evolution* 7, 4844–4854.
Schroeder, W.J. (1987) Laboratory bioassays and field trials of entomogenous nematodes for control of *Diaprepes abbreviatus*. *Environmental Entomology* 16, 987–989.
Schroeder, W.J. (1992) Entomopathogenic nematodes for control of root weevils of citrus. *Florida Entomologist* 75, 563–567.
Schroeder, W.J. (1994) Comparison of two steinernematid species for control of the root weevil *Diaprepes abbreviatus*. *Journal of Nematology* 26, 360–362.
Shapiro, D.I. and McCoy, C.W. (2000a) Virulence of entomopathogenic nematodes to *Diaprepes abbreviatus* (Coleoptera: Curculionidae) in the laboratory. *Journal of Economic Entomology* 93, 1090–1095.
Shapiro, D.I. and McCoy, C.W. (2000b) Susceptibility of *Diaprepes abbreviatus* (Coleoptera: Curculionidae) larvae to different rates of entomopathogenic nematodes in the greenhouse. *Florida Entomologist* 83, 1–9.
Shapiro, D.I., Cate, J.R., Pena, J., Hunsberger, A. and McCoy, C.W. (1999) Effects of temperature and host age on suppression of *Diaprepes abbreviatus* (Coleoptera: Curculionidae) by entomopathogenic nematodes. *Journal of Economic Entomology* 92, 1086–1092.
Shapiro-Ilan, D.I., Cottrell, T.E., Mizell, R.F. and Horton, D.L. (2016) Curative control of the peachtree borer using entomopathogenic nematodes. *Journal of Nematology* 48, 170–176.
Singerman, A., Burani-Arouca, M. and Futch, S. (2018) The profitability of new citrus plantings in Florida in the era of huanglongbing. *Hortscience* 53, 1655–1663.
Smith, D., Beattie, G.A.C. and Broadley, R. (1997) *Citrus Pests and Their Natural Enemies: Integrated Pest Management in Australia*. Queensland Department of Primary Industries, Queensland, Australia.
Steyn, V.M., Malan, A.P. and Addison, A. (2019) Control of false codling moth, *Thaumatotibia leucotreta* (Lepidoptera: Tortricidae), using *in vitro*-cultured *Steinernema jeffreyense* and *S. yirgalemense*. *Biological Control* 138: 104052.
Steyn, V.M., Malan, A.P. and Addison, A. (2021) Efficacy of entomopathogens against *Thaumatotibia leucotreta* (Lepidoptera: Tortricidae) under laboratory conditions. *Entomologia Experimentalis et Applicata* 169, 449–461.
Steyn, W.P., Knoetze, R., Tiedt, T.R. and Malan, A.P. (2017a) *Steinernema litchii* n. sp. (Rhabditida: Steinernematidae), a new entomopathogenic nematode from South Africa. *Nematology* 19: 1157–1177.
Steyn, W.P., Malan, A.P., Daneel, M.S., Retha, M. and Slabbert, R.M. (2017b) Entomopathogenic nematodes from north-eastern South Africa and their virulence against false codling moth, *Thaumatotibia leucotreta* (Lepidoptera: Tortricidae). *Biocontrol Science and Technology* 11, 1265–1278.

Steyn, W.P., Daniel, M.S. and Malan, A.P. (2019) Field application of entomopathogenic nematodes against *Thaumatotibia leucotreta* in South African avocado, litchi and macadamia orchards. *BioControl* 64, 401–411.

Stokwe, N.F., Malan, A.P., Nguyen, B.K., Knoetze, R. and Tiedt, L. (2011) *Steinernema citrae* n. sp. (Rhabditida: Steinernematidae), a new entomopathogenic nematode from South Africa. *Nematology* 13, 567–587.

Stotter, R.L. (2009) Spatial and temporal distribution of false codling moth across landscapes in the Citrusdal area. MSc thesis. University of Stellenbosch, Stellenbosch, South Africa.

Stuart, R.J., Shapiro-Ilan, D.I., James, R.I., Nguyen, K.B. and McCoy, C.W. (2004) Virulence of new and mixed strains of the entomopathogenic nematode *Steinernema riobrave* to larvae of the citrus root weevil *Diaprepes abbreviatus*. *Biological Control* 30, 439–445.

Stuart, R.J., El-Borai, F.E. and Duncan L.W. (2008) From augmentation to conservation of entomopathogenic nematodes. Trophic cascades, habitat manipulation and enhanced biological control of *Diaprepes abbreviatus* in Florida citrus groves. *Journal of Nematology* 40, 73–84.

Tarasco, T.E. and Triggiani, O. (2010) Desiccation tolerance of Algerian strains of entomopathogenic nematodes. *Redia* 93, 75–77.

Terblanche, J.S., de Jager, Z., Boardman, L. and Addison, P. (2014) Physiological traits suggest limited diapause response in false codling moth, *Thaumatotibia leucotreta* (Lepidoptera: Tortricidae). *Journal of Applied Entomology* 138, 683–691.

Torrini, G., Mazza, G., Benvenuti, C. and Roversi, P.F. (2017) Susceptibility of olive fruit fly, *Bactrocera oleae* (Diptera: Tephritidae) pupae to entomopathogenic nematodes. *Journal of Plant Protection Research* 57, 318–320.

USDA (US Department of Agriculture) (2024) *Citrus: World Markets and Trade*. USDA Foreign Agricultural Service, Washington, DC. Available at: http://apps.fas.usda.gov/psdonline/circulars/citrus.pdf (accessed 20 March 2024).

van Niekerk, S. and Malan, A.P. (2012) Potential of South African entomopathogenic nematodes (Heterorhabditidae and Steinernematidae) for control of the citrus mealybug, *Planococcus citri* (Pseudococcidae). *Journal of Invertebrate Pathology* 111, 166–174.

van Niekerk, S. and Malan, A.P. (2014a) Compatibility of *Heterorhabditis zealandica* and *Steinernema yirgalemense* with agrochemicals and biological control agents. *African Entomology* 22, 49–56.

van Niekerk, S. and Malan, A.P. (2014b) Evaluating the efficacy of a polymer-surfactant formulation to improve control of *Planococcus citri* (Risso) (Hemiptera: Pseudococcidae) under simulated natural conditions. *African Plant Protection* 17, 1–9.

van Niekerk, S. and Malan, A.P. (2015) Adjuvants to improve aerial control of the citrus mealybug *Planococcus citri* (Hemiptera: Pseudococcidae) using entomopathogenic. *Journal of Helminthology* 98, 189–195.

Venette, R.C., Davis, E.E., DaCosta, M., Heisler, H. and Larson, M. (2003) Mini risk assessment. False codling moth, *Thaumatotibia* (=*Cryptophlebia*) *leucotreta* (Meyrick) [Lepodoptera:Tortricidae]. University of Minnesota, St Paul, Minnesota. Available at: www.cdfa.ca.gov/plant/fcm/pdfs/publications/Venette_et_al_2003-FCM_PRA.pdf (accessed 20 March 2024).

Wakgari, W.M. and Giliomee, J.H. (2005) Description of adult and immature females of six mealybug species (Hemiptera: Pseudococcidae) found on citrus in South Africa. *African Entomology* 13, 281–332.

Weldon, C. (2021) *Ceratitis capitata* (Mediterranean fruit fly). In: *CABI Compendium*. CAB International, Wallingford, UK. Available at: https://www.cabi.org/isc/datasheet/12367 (accessed 20 March 2024).

Wolcott, G.N. (1936) The life history of *Diaprepes abbreviatus* at Rio Piedras, Puerto Rico. *Journal of Agriculture of the University of Puerto Rico* 20, 883–914.

Woodruff, R.E. (1964) A Puerto Rican weevil new to the United States (Coleoptera: Curculionidae). Florida Department of Agriculture. *Plant Industry Entomology Circular* 77, 1–4.

Wyoski, M. (1986) New records of lepidopterous pest of macadamia in Israel. *Phytoparasitica* 14: 147.

Yee, W.L. and Lacey, L.A. (2003) Stage-specific mortality of *Rhagoletis indifferens* (Diptera: Tephritidae) exposed to three species of *Steinernema* nematodes. *Biological Control* 27, 349–356.

17 Other Orchard Systems

David I. Shapiro-Ilan[1]* and Steven P. Arthurs[2]
[1]*USDA-ARS, Southeastern Fruit and Tree Nut Research Station, Byron, Georgia, USA;* [2]*BioBee USA, Salisbury, Maryland, USA*

Abstract

Temperate orchards have several environmental features that make them conducive to biocontrol applications involving entomopathogenic nematodes (EPNs). Favorable abiotic conditions are provided by adequate soil moisture and canopy shading (protection from harmful ultraviolet light) while the perennial crop provides stability. This chapter reviews and analyzes the use of EPNs (*Heterorhabditis* and *Steinernema* spp.) in temperate orchards, including pome fruit, stone fruit and nut crops. Emphasis is placed on pest complexes that have received the most attention as targets for EPNs in North America. Examples of pests that have successfully been controlled with EPNs include codling moth (*Cydia pomonella*), pecan weevil (*Curculio caryae*), peachtree borer (*Synanthedon exitiosa*), lesser peachtree borer (*Synanthedon pictipes*) and plum curculio (*Conotrachelus nenuphar*). In some cases, improvements in efficacy have resulted from careful selection of nematode strains or species, as well as recent advances in formulation (e.g. the use of nanoparticles or gel substrates) or application techniques (e.g. the use of pheromones to enhance nematode activity). While commercially produced EPNs are applied to an increasing number of insect pests in temperate orchard systems, based on current trends outlined in this chapter, there is great potential for expansion.

17.1 Introduction

Orchards consist of perennial tree plantings producing fruit and nut products. A number of economically important arthropod pests occur in most types of orchards. Many orchards contain attributes that are amenable to insect suppression using entomopathogenic nematodes (EPNs), including hosts available through much of the year, favorable soil conditions (moist, sandy) and shade (Shapiro-Ilan *et al.*, 2017a). Additionally, orchard crops are often relatively high value, favoring the economic feasibility of nematode applications. As a result, several major orchard pests have been studied extensively for their potential as targets for EPNs, with some commercial success stories resulting from this research. In this chapter, we review significant research on EPN control of orchard pests, highlighting successes and failures, recent advances, research priorities and future potential.

17.2 Pome and Stone Fruits

17.2.1 Codling moth

Codling moth, *Cydia pomonella*, remains a key worldwide tortricid pest of apple, pear, walnut

*Email: David.Shapiro@usda.gov

DOI: 10.1079/9781800620322.0017

and other fruit. The most vulnerable stages of *C. pomonella* to EPNs are the last instars after they exit fruit in the fall but before they pupate the following spring. *C. pomonella* overwinter as full-grown cocooned larvae in cryptic habitats (under bark, within prop/refuse piles, fruit bins and leaf litter). Control of diapausing *C. pomonella* larvae can reduce or prevent fruit damage the following growing season. Earlier research has indicated that there is good potential for *Steinernema carpocapsae* (which was originally isolated from *C. pomonella* (Weiser, 1955)) and a number of other nematode species to be used against diapausing *C. pomonella* larvae (Kaya *et al.*, 1984; Sledzevskaya, 1987; Nachtigall and Dickler, 1992; Lacey and Unruh, 1998; Lacey and Chauvin, 1999; Unruh and Lacey, 2001).

Protocols for the field evaluation of EPNs against cocooned *C. pomonella* larvae were presented by Lacey *et al.* (2000, 2007). In general, the major obstacles for successful *C. pomonella* control are low temperatures and desiccation of infective juveniles (IJs) after application but prior to infection (Unruh and Lacey, 2001; Lacey *et al.*, 2006a,b). Several studies have investigated environmental conditions favoring different EPN species. In a series of orchard trials in the US Pacific North-west, *Steinernema feltiae* was generally more effective than *S. carpocapsae* under cooler conditions (10–15°C post-application), whereas *S. carpocapsae* was more effective at temperatures above 15°C (>90% mortality) (Lacey *et al.*, 2006a). Significantly improved control can be achieved when target sites are kept moist for 8 h after application, for example through post-application wetting with spray or irrigation equipment (Lacey and Unruh, 1998; Unruh and Lacey, 2001). Larval control has been improved by using adjuvants or mulches that protect EPNs from desiccation (Lacey *et al.*, 2006b, 2010; Nevaneethan *et al.*, 2010; de Waal *et al.*, 2011). Although temperatures below 10°C are considered limiting, in a study of commercial organic orchards in Argentina, *S. feltiae* applied at 1.5 billion IJs ha^{-1} under cold temperatures (<5°C) was reported to provide over 85% mortality of diapausing *C. pomonella* larvae in two of the four test sites (Eliceche *et al.*, 2023).

When applying EPNs against diapausing *C. pomonella*, application rates in the range of 1–2 billion IJs ha^{-1} (or over 1 million IJs per tree) can provide effective control of cocooned larvae under permissive environmental conditions (Lacey *et al.*, 2006a, 2007), although this can depend on the time of day that the EPNs are applied (de Waal *et al.*, 2018). As a result of successful field trials, there has been increased commercial adoption of EPNs against *C. pomonella* in the Pacific North-west (S.P. Arthurs, personal observations). Typically, applications are made in orchard "hotspots" using *S. carpocapsae* (or *S. feltiae* in cooler conditions in early spring or late fall) using traditional air blast sprayers configured to apply EPNs. Promising results for control of diapausing *C. pomonella* larvae were reported from South Africa, Italy and Germany using *Heterorhabditis zealandica* and *Steinernema* spp. (de Waal *et al.*, 2010; Nevaneethan *et al.*, 2010; de Luca *et al.*, 2015). In more recent studies, *H. zealandica* provided high levels of efficacy in semi-field conditions (85–100% control) (de Waal *et al.*, 2018).

Wooden fruit bins infested with cocooned *C. pomonella* can be a significant source of invading moths in mid- to late summer when they are placed in orchards for harvest. EPNs offer potential for decontaminating fruit bins when they are submerged in drop tanks (Lacey and Chauvin, 1999; Odendaal *et al.*, 2016) or sprayed with drenchers used for treating or cooling fruit (Cossentine *et al.*, 2002). De Waal *et al.* (2010) reported that *C. pomonella* control in infested bins using *H. zealandica* was enhanced when bins were pre-wetted for at least 1 min and subsequently maintained at over 95% relative humidity post-treatment for 3 days. Studies indicate that EPNs are compatible with other biological control agents (e.g. ichneumonid parasitoids) for *C. pomonella* control (Lacey *et al.*, 2003) and mating disruption technologies (Odendaal *et al.*, 2015).

17.2.2 Plum curculio

Plum curculio, *Conotrachelus nenuphar*, is a key native pest of pome and stone fruit in eastern North America (Horton and Johnson, 2005). Adult weevils emerge from overwintering sites in the spring to feed, mate and oviposit in fruit. Infested fruits may abort or become non-saleable, with up to 85% damage reported at harvest (Lampasona *et al.*, 2020). Larvae continue their development in fallen fruit, exiting as fourth instars, and burrow into the soil to pupate at depths of approximately 1–8 cm. Subsequently, the adults emerge and feed on fruit prior to migrating to overwintering sites in the orchard or surrounding area (Olthof and Hagley, 1993). In the southern USA, a second generation may occur prior to overwintering (Horton and Johnson, 2005).

Laboratory research indicates that *C. nenuphar* is susceptible to various EPN strains and species (Olthof and Hagley, 1993; Shapiro-Ilan *et al.*, 2002, 2011; Alston *et al.*, 2005). Overall, *Steinernema riobrave* has proven to be the most effective EPN for control of *C. nenuphar* following fruit drop. In multiple field trials conducted in peach and apple orchards across the eastern USA, soil applications of *S. riobrave* consistently averaged over 92% control (Shapiro-Ilan *et al.*, 2004, 2008, 2013). Additionally, Pereault *et al.* (2009) observed 80–89% reduction of *C. nenuphar* by *S. riobrave* in Michigan apple and cherry orchards, whereas *Heterorhabditis bacteriophora* was ineffective. In contrast to *S. riobrave*, equivalent tests with *S. feltiae* applications failed (Shapiro-Ilan *et al.*, 2004), or produced low (22–39%) (Alston *et al.*, 2005) or intermediate (58%) (Shapiro-Ilan *et al.*, 2013) levels of control.

A substantial drawback to EPN applications targeting *C. nenuphar* in the soil (belowground stages) is that this approach does not control the adults that immigrate into the orchard canopy. Soil applications can only prevent later damage resulting from subsequent generations of the pest (Shapiro-Ilan *et al.*, 2005a). Therefore, an integrated approach combining EPNs in combination with other tactics targeting adult stages of *C. nenuphar* is often needed.

Although not yet widely adopted in commercial orchards, recent research suggests that the use of EPNs as a component of an integrated pest-management program for *C. nenuphar* has potential for success. In one approach that targets multiple *C. nenuphar* stages, volatile lures attract adult weevils to selected sentinel "trap" trees on the orchard perimeter. The trap trees are then sprayed with insecticides that kill adult *C. nenuphar* (in the canopy) while the other (non-sentinel) trees remain pesticide free. Limiting pesticide treatments to trap trees can reduce insecticide treatment by more than 90% (Leskey *et al.*, 2008), and this approach showed similar effectiveness (in terms of overall *C. nenuphar* injury to fruit) to conventionally managed plots (Piñero *et al.*, 2011). As a result of the lures concentrating *C. nenuphar* to perimeter trees, some fruit damage in the trap trees will occur in these areas. As a way to further mitigate this damage, once the infested fruit in the trap tree drops, EPNs can be applied to suppress the ground-dwelling stages (Shapiro-Ilan *et al.*, 2013). Piñero *et al.* (2020) demonstrated the efficacy of the trap-tree approach in a 6-year study conducted in seven commercial apple orchards in New England, USA; in these tests, both *S. riobrave* and *S. carpocapsae* consistently reduced *C. nenuphar* populations (both species performed at a similar level of efficacy).

17.2.3 Borers

17.2.3.1 Peachtree borer

Various borer pests of pome and stone fruits have shown susceptibility to entomopathogens (Deseo and Miller, 1985; Cossentine *et al.*, 1990; Kahounova and Mracek, 1991; Kain and Agnello, 1999; Grewal *et al.*, 2005; Shapiro-Ilan *et al.*, 2017a). Larvae of clearwing moths (Lepidoptera: Sesiidae), particularly *Synanthedon* spp., are generally very susceptible to EPNs (Grewal *et al.*, 2005; Shapiro-Ilan *et al.*, 2017a). One example of a highly susceptible sesiid borer is the peachtree borer, *Synanthedon exitiosa*, which is a major pest of peach and other *Prunus* spp. (Horton and Johnson, 2005). Adult *S. exitiosa* moths emerge and mate during late summer and early fall. The mated females oviposit on or near the base of host plants. Damage occurs from larvae, which bore into the trunk of stone fruit trees near the soil surface, tunnel toward the roots, and feed at the crown and on other major roots.

Multiple field trials have consistently shown that applications of *S. carpocapsae* are as effective as the standard recommended chemical insecticides (Shapiro-Ilan *et al.*, 2009, 2015, 2016a; Wong *et al.*, 2022). EPNs can be applied prophylactically for *S. exitiosa* control in the same manner as chemical standards (i.e. during the oviposition period). For example, applying *S. carpocapsae* preventatively to soil surrounding the tree base (1.5 million IJs per tree) during the late summer or early fall reduced subsequent

S. exitiosa infestations by 78–100% (Shapiro-Ilan *et al.*, 2009). In a more recent study, Wong *et al.* (2022) reported that rates of 0.5 million IJs per tree provided similar control levels as a rate of 1.5 million IJs per tree. EPNs can be applied effectively for *S. exitiosa* control using standard orchard equipment that growers already use, including boom sprayers, trunk sprayers, handguns or watering cans (Shapiro-Ilan *et al.*, 2015, 2016a). Generally, for the EPNs to be effective against aboveground borers, supplemental irrigation is needed. However, in cases where this is not feasible, such as with non-irrigated orchards, the use of sprayable fire-retardant gels (e.g. Barricade®) (or conceivably other gels or protective formulations) can be used to protect nematodes from desiccation in lieu of irrigation (Shapiro-Ilan *et al.*, 2015; Wong *et al.*, 2022).

In addition to applying EPNs for *S. exitiosa* control in a preventative manner, EPNs can also be used curatively against existing infestations in the spring. The first report of successful curative *S. exitiosa* control using EPNs was from a study in Canada using *H. bacteriophora*, which reported an 80% reduction in moth emergence using this approach (Cossentine *et al.*, 1990). Subsequently, an 88% reduction in moth emergence was observed following a late-spring application of *S. carpocapsae* using a lower rate of 3 × 10^5 IJs per tree (Cottrell and Shapiro-Ilan, 2006). Shapiro-Ilan *et al.* (2016a) reported over 80% curative control when *S. carpocapsae* was applied at 1 million IJs per tree, whereas chlorpyrifos applications were ineffective (presumably because the chemical was unable to penetrate into the root system where the larvae were located. Wong *et al.* (2022) observed highly effective results (~88% control of *S. exitiosa* in peaches in Georgia) when applying *S. carpocapsae* in consecutive prophylactic and curative sprays with rates as low as 0.5 million IJs per tree. Moreover, these latter EPN applications also reduced root-feeding weevil populations; thus, secondary benefits were obtained from the applications targeting *S. exitiosa* (Wong *et al.*, 2022).

In conclusion, for both preventative and curative control applications, the high levels of efficacy when using *S. carpocapsae*, combined with potentially low application rates (e.g. >300 million IJs ha^{-1}), strongly indicate the economic feasibility of incorporating EPNs into integrated pest-management programs for *S. exitiosa* and similar targets. For these reasons, *S. carpocapsae* is now recommended for *S. exitiosa* control in peaches in Georgia (Blaauw *et al.*, 2020).

17.2.3.2 Lesser peachtree borer

The lesser peachtree borer, *Synanthedon pictipes*, is another important pest of *Prunus* spp. in the eastern USA (Horton and Johnson, 2005). Adult emergence typically begins in March and peaks in April and May, with a second emergence between July and September. Adults oviposit in bark cracks in tree trunks and scaffold limbs, or near injured areas (Horton and Johnson, 2005). The larvae tunnel into the cambium and inner bark where they feed and develop, causing severe damage, and overwinter in the galleries.

Similar to *S. exitiosa*, while traditional control recommendations for *S. pictipes* focus on intensive chemical insecticide use, EPNs may be a viable alternative. Laboratory studies screened the virulence of ten EPN strains and species to *S. pictipes* larvae, and overall, *S. carpocapsae* was the most effective (Shapiro-Ilan and Cottrell, 2006; Cottrell *et al.*, 2011). The challenge is that aboveground applications of EPNs leave the nematode vulnerable to degradation due to ultraviolet (UV) radiation and/or desiccation (Shapiro-Ilan *et al.*, 2017a). Indeed, when EPNs were initially applied in aqueous suspension to *S. pictipes*-infested wounds on peach trees, the treatments failed to control the pest (Shapiro-Ilan *et al.*, 2010). Therefore, improved formulations and/or protectants are needed to facilitate EPN survival and efficacy for aboveground applications for *S. pictipes* control.

Among several protective formulations or covering agents tested, the most effective treatment was Barricade®, a fire gel commonly used to protect houses and other structures from wildfire. In field trials over 2 years, treatment with Barricade® plus *S. carpocapsae* (1 million IJs per infested wound) resulted in 100% control of *S. pictipes* in the first year and 70% control in the second year (Shapiro-Ilan *et al.*, 2010). Initially, due to the high viscosity of Barricade®, the material was applied as a cover following the nematode application (the standard 4% v/v Barricade® was too viscous to be mixed directly with the nematodes). Subsequently, it was discovered that Barricade® diluted to 2% v/v and tank mixed with *S. carpocapsae* was equally effective as the full Barricade® concentration and controlled *S. pictipes* at the same level as chlorpyrifos (Shapiro-Ilan *et al.*, 2016b). The use of Barricade® or potentially other sprayable gels has broad potential to improve the efficacy of aboveground applications of EPNs in various temperate orchard systems.

17.2.4 Fruit flies

Fruit flies are among the most harmful pests of stone and pome fruit. Substantial research has demonstrated that several economically important species including *Rhagoletis* spp., *Bactrocera* spp. and *Drosophila suzukii* are susceptible to EPNs (Beavers and Calkins, 1984; Lindegren and Vail, 1986; Lindegren *et al.*, 1990; Gazit *et al.*, 2000; Usman *et al.*, 2020; Evans, and Renkema, 2021).

However, to date, most investigations have been limited to laboratory research. For example, in a broad laboratory screening, among seven EPN species tested, *S. riobrave*, *S. carpocapsae* and *S. feltiae* indicated the most promise for control of the apple maggot, *Rhagoletis pomonella* (Usman *et al.*, 2020). Following promising laboratory screening, field testing to target *R. pomonella* under the orchard canopy prior to adult emergence should ensue. In subsequent tests against *Bactrocera zonata* and *Bactrocera dorsalis*, *H. bacteriophora* was identified to be the most virulent among ten EPN species tested in the laboratory, and subsequently was reported to reduce emergence of *B. zonata* and *B. dorsalis* from infested fruit by up to 75% under field conditions (Usman *et al.*, 2021). Interactions between *H. bacteriophora* and *S. carpocapsae* and entomopathogenic fungi (e.g. *Beauveria bassiana*) were also explored in laboratory, greenhouse and field conditions. In this case, the potential for additive or synergistic levels of control when combining both EPNs and fungi against *B. zonata* and *B. dorsalis* were demonstrated (Wakil *et al.*, 2022).

The western cherry fruit fly, *Rhagoletis indifferens*, a serious pest of sweet cherries in the western USA, has also been investigated as a target for EPNs under laboratory and field conditions (Patterson Stark and Lacey, 1999; Yee and Lacey, 2003). Yee and Lacey (2003) evaluated *S. carpocapsae*, *S. feltiae* and *S. intermedium* in soil against *R. indifferens* larvae, pupae and adults in the laboratory. Among the life stages tested, the larvae were the most susceptible, with *S. carpocapsae* and *S. feltiae* producing mortality rates of 62–100% at rates of 50 and 100 IJs cm^{-2}. The authors noted that the pupae of *R. indifferens* were not infected, but adult flies were infected by all three nematode species in the laboratory at a rate of 100 IJs cm^{-2}. In field trials, *S. carpocapsae* and *S. feltiae* were equally effective against larvae (59–85% mortality) when applied to soil under cherry trees at 50–100 IJs cm^{-2}. Further studies are needed to demonstrate the feasibility of this approach under operational conditions where fruit flies are controlled using insecticides. However, Yee and Lacey (2003) also proposed the use of EPNs for control of *R. indifferens* in abandoned orchards and trees in yards of homeowners, as these situations represent a threat to commercial cherry orchards by providing significant sources of invading flies.

17.2.5 Other pests of pome and stone fruit

Beyond the *Synanthedon* spp. indicated above, a variety of other lepidopterans are important regional pests of temperate apple, pear and stone fruits. EPNs offer a specific window of opportunity to control the soil stages of defoliating lepidopteran species. For example, the mature larvae of several noctuid species are vulnerable to EPNs as prepupae or when they search for a soil site in which to pupate. Laboratory studies have demonstrated fair to good activity of several nematode species against leaf rollers and other torticids that construct retreats in rolled leaves or shoots. For example, several *Steinernema* spp. showed activity against the oblique-banded leaf roller, *Choristoneura rosaceana* (Poinar, 1991; Belair *et al.*, 1999), with foliar application of *S. carpocapsae* All strain providing up to 37% larval control under field conditions. The oriental fruit moth, *Grapholita molesta*, a major pest of pome and stone fruits, has been controlled in fruit bins using *S. feltiae* (Riga *et al.*, 2006) and under field conditions in peach orchards using *Steinernema rarum* or *H. bacteriophora* applied to larvae on peach tree trunks (causing 94–97% mortality) (Negrisoli *et al.*, 2013).

There is also a wide variety of non-lepidopteran pests of apple, pears and peaches that are potential targets for EPNs. Among the few examples, Vincent and Belair (1992) and Belair *et al.* (1998) evaluated EPNs against the apple sawfly, *Hoplocampa testudinea*, with mixed results. In field tests over 3 years, repeated applications of *S. carpocapsae* every 2–3 days from early May until mid-June reduced primary damage caused by larvae of *Hoplocampa testudinea* by 98–100% in two seasons (1992–1993), but treatments were ineffective in the following year (Belair *et al.*, 1998). Western flower thrips, *Frankliniella occidentalis*, are pests of several temperate fruits, including apple, pear and cherry. Potential for control of *F. occidentalis* with EPNs, especially the non-feeding life stages in the soil, has also been demonstrated (Helyer *et al.*, 1995; Ebssa *et al.*, 2001a,b; Dlamini *et al.*, 2020).

The Mediterranean flat-headed rootborer, *Capnodis tenebrionis*, is an economically important pest of stone fruit trees, particularly in Mediterranean regions (Morton and Garcia-del-Pino, 2008). High levels of *C. tenebrionis* control (88–97%) were observed in a cherry orchard following applications of *S. feltiae* (Morton and Garcia-del-Pino, 2008). In later tests, Şahin and Gözel (2021) also reported high efficacy of several commercial and native EPN species. In this case, EPNs applied to the soil in peach orchards at 50 IJs cm^{-2} resulted in 100% mortality of *C. tenebrionis* within 5 days of application.

17.3 Nut Crops

17.3.1 Navel orangeworm

The navel orangeworm, *Amyelois transitella*, is a primary pest of almonds, walnuts and pistachio (Rice, 1978a,b) and the most important pest of almonds in the western USA. The larval stage invades nuts during hull split and feeds on the nutmeats. The larvae infest mature nuts on the tree and nut mummies on the tree and ground. Conventional control of *A. transitella* during the growing season is through the application of insecticides, postharvest removal and destruction of nuts, and, more recently, mating disruption with pheromones and releases of sterile moths (Reger *et al.*, 2020; Gordon *et al.*, 2023).

Orchard sanitation remains an important aspect of navel orangeworm control. In this process, nut mummies are removed from the trees by shaking, polling and pruning, and blown into furrows for disking (in pistachios) or flail mowing (in almonds), rendering most of the nuts unsuitable for development of larvae. However, larvae that survive this treatment pose a significant threat to nuts in the following season.

The use of EPNs offers an alternative means of *A. transitella* control that will help reduce the need for traditional insecticides and improve air quality. However, initial investigations in this area were not especially promising. Summer-time field application of *S. carpocapsae* to open-hulled almonds resulted in over 65% mortality in baited *A. transitella* (Lindegren *et al.*, 1987), whereas dormant-season (winter) EPN applications to trees resulted in substantially lower control (Agudelo-Silva *et al.*, 1995). Siegel *et al.* (2004) studied the efficacy of *S. carpocapsae* and *S. feltiae* applied to *A. transitella* larvae infesting almond and pistachio nut mummies on the ground. The larvae were almost completely controlled with *S. carpocapsae* at 10^5 IJs m^{-2} and to a lesser extent by *S. feltiae* at the same dosage. The low rate of applications used to achieve these high levels of control indicates that ground application of EPNs as a sanitation tool for *A. transitella* warrants further attention. Another potential advantage is that EPNs can persist in almond and pistachio orchards, offering the potential of recycling within the *A. transitella* population (Agudelo-Silva *et al.*, 1987; Siegel *et al.*, 2004, 2006).

17.3.2 Pecan weevil

The pecan weevil, *Curculio caryae*, is a key pest of pecans throughout the south-eastern USA, as well as portions of Kansas, Oklahoma and Texas (Shapiro-Ilan, 2003). Adults emerge from the soil in late July to August, and feed and oviposit on developing nuts. Larvae develop in the nuts, drop to the soil, burrowing to a depth of 8–25 cm, and form a soil cell where they spend 1 or 2 years before pupating and molting to adulthood. Adult *C. caryae* spend approximately an additional 9 months in the soil before emerging

(Harris, 1985). Control recommendations for the pecan weevil currently consist solely of aboveground applications of chemical insecticides (mainly carbaryl) to suppress adults (Barman *et al.*, 2023).

Initial work with EPNs showed that the susceptibility of *C. caryae* varies based on developmental stage and nematode species. In laboratory bioassays, *C. caryae* larvae had low to moderate susceptibility to nine EPN species, whereas adults were highly susceptible, especially to *S. carpocapsae* (Shapiro-Ilan 2001a,b; Shapiro-Ilan *et al.*, 2003). Field tests initially targeted adult weevils during emergence; *S. carpocapsae* (applied to the soil under the canopy) caused approximately 60% mortality, which was insufficient for grower adoption (Shapiro-Ilan *et al.*, 2006). To improve efficacy, subsequent studies focused on multiple applications of EPNs across seasons/years. In this way, multiple stages are targeted (e.g. larvae in the first year and adults in the second year). Field experiments using multiple applications of *S. carpocapsae* resulted in high levels of *C. caryae* control; for example, in a 2-year trial, only 1% of the weevils survived in nematode-treated plots (Shapiro-Ilan and Gardner, 2012; Shapiro-Ilan *et al.*, 2017b).

In addition to implementing a multi-stage, multi-season approach, other methods to improve EPN efficacy in *C. caryae* control have been studied. Strain improvement via hybridization of *S. carpocapsae* strains was accomplished for the first time by transferring the bacterial symbiont (*Xenorhabdus nematophila*) from the Italian strain to the DD-136 strain, using *C. caryae* as a target pest (Shapiro-Ilan *et al.*, 2005b). Given that some *S. carpocapsae* strains exhibited high virulence but low environmental tolerance (e.g. to heat and desiccation), and other strains possessed low virulence but high levels of environmental tolerance the strains were subjected to controlled crosses and resulting hybrid strains exhibited high levels of all traits (Shapiro-Ilan *et al.*, 2005b).

Another novel approach to enhancing EPN efficacy against *C. caryae* is based on using nematode pheromones to direct EPN behavior. Nematode pheromones (ascarosides) were discovered to increase EPN dispersal and infection behavior (Oliveira-Hofman *et al.*, 2019; Shapiro-Ilan *et al.*, 2019). It was demonstrated that under laboratory and greenhouse conditions, exposure of EPNs to pheromone extracts improved *C. caryae* control compared with equivalent treatments without pheromone exposure (Oliveira-Hofman *et al.*, 2019). Additionally, these results (boosted efficacy from pheromone exposure) were also observed in consecutive trials in a pecan orchard (D.I. Shapiro-Ilan, unpublished data).

17.4 Banana

The banana stem borer, *Odoiporus longicollis*, and banana weevil borer, *Cosmopolites sordidus*, are important weevil pests of banana in many tropical and subtropical countries. Nematodes have been used to control these pests in Australia and China with encouraging results (Shapiro-Ilan *et al.*, 2005a). Both pest species usually occur all year around, where larvae and adults feed on the plant and bore into the stem, weakening or killing the plant (Azad and Singh, 2019). In southern China, *O. longicollis* has six generations per year with two population peaks in March–June and November–December. Usually, the corms of the trees are cut at the base in winter after harvest. Approximately 90% of the overwintering populations are larvae in the residual stems, which attack the banana stems the following year.

Nematodes have been used to control *O. longicollis* and *C. sordidus* in Australia and China with encouraging results. EPNs are capable of migrating through living stem tissue to kill the borers; field trial results indicated that the majority of *O. longicollis* (76–90% of overwintering larvae, 68–92% of the pupae and 25–80% of adults) were controlled by spraying 3–6 million IJs of *S. carpocapsae* (A24 strain) into each residual stem base (Xu *et al.*, 1991).

The larvae of *C. sordidus* burrow into corms, producing severe damage, which can be exacerbated by subsequent fungal or bacterial infection. Laumond *et al.* (1979) demonstrated the pathogenicity of *S. carpocapsae* to adult *C. sordidus* in laboratory trials. Figueroa (1990) further demonstrated the pathogenicity of additional EPN species (*S. carpocapsae*, *S. feltiae* and *S. glaseri*) to *C. sordidus* larvae, and observed 100% mortality in greenhouse tests using 4000 IJs per plant. Field applications of *S. carpocapsae* in a water-thickening gel (used to maintain IJs near the target site) to cuts or holes made in the residual banana rhizomes also provided control of larvae and adults that were attracted to the wounds (Treverrow *et al.*, 1991). Kermarrec and Mauléon (1989) reported that the effects of *S. carpocapsae* on *C. sordidus* can be enhanced through synergistic interactions with chemical insecticides (e.g. chlordecone).

17.5 Lychee

Lychee (*Litchi chinensis*) is a high-value crop in several subtropical and tropical regions. Key lychee pests include the lychee stem borer *Indarbela dea* (Lepidoptera: Cossidae), and the lychee longhorn beetle, *Aristobia testudo* (Srivastava *et al.*, 2016). However, other pests with broader plant host ranges can also be serious pests of lychee. For example, in Puerto Rico, the root weevil *Diaprepes abbreviatus* can cause substantial damage to seedlings (Jenkins *et al.*, 2008).

17.5.1 Lychee stem borer

I. dea has one generation per year, with the most damaging larval stage lasting up to 9 months, beginning in June. First-instar *I. dea* damage the lower bark of the lychee tree and then bore into the trunk as they mature. The resulting damage can weaken the trees or cause death depending on the lychee strain, age and location.

Larvae of *I. dea* are susceptible to *S. carpocapsae* (A24). The nematodes are applied by conventional sprayers around the borer holes. The *I. dea* larvae are usually active just outside the borer holes at night, providing an ideal place for contact between the nematodes and the insects. Thus, the nematodes do not need to be applied directly into the borer holes (e.g. by injection), which would increase labor. Over 86% *I. dea* mortality was obtained by spraying 1000 IJs around each borer hole (Xu and Yang, 1992). In later trunk applications, Schulte *et al.* (2009) demonstrated that two applications of *S. carpocapsae* provided 95–100% mortality, whereas 100% mortality was observed following a single application of *S. carpocapsae* plus the fungus *Beauveria bassiana*.

17.5.2 Lychee longhorn beetle

The most important lychee pest is *Aristobia testudo*, which causes great economic losses to the crop. Similar to *I. dea*, *A. testudo* has one generation per year. Adult beetles lay eggs between the crotches of lychee trees. The hatched larvae, whose distribution is aggregative, bore into the stem and develop in the holes for up to 9 months (Han *et al.*, 1994). Without control, the infested branches wither and die, resulting in no fruit yield.

Most tactics to control *A. testudo* are ineffective due to the inaccessibility of the larvae in tunnels. Mobile nematodes, however, actively search for the larvae of the beetle in their feeding galleries and have been shown to produce over 73% mortality following injection of 3000 IJs of *S. carpocapsae* (A24) into the freshly bored holes of the beetle (Xu *et al.*, 1995; Han *et al.*, 1996). Using a similar approach, successful control of *A. testudo* by EPNs was achieved in 1700 ha of lychee orchards in Guangdong, China (Shapiro-Ilan *et al.*, 2005a). As a result, farmers' interest in utilization of EPNs as a safe and effective control of these pests has been generated. However, competition from chemical insecticides and the relatively limited hectarage occupied by lychee mean limited commercial production and distribution of EPNs for this market. Nonetheless, the successful field demonstrations indicate the potential of EPNs to control these pests. Further research will focus on strain selection and formulation development to enhance migration ability and desiccation tolerance.

17.5.3 Other lychee pests

EPNs have been evaluated against a small number of lychee pests. In potted seedling trials, *S. riobrave*, *Heterorhabditis megidis* and *Steinernema diaprepesi* all provided significant control of *Diaprepes abbreviatus*; the commercialized 355 strain of *S. riobrave* resulted in 100% control (Jenkins *et al.*, 2008). Another pest that affects various fruit crops including lychee is the false codling moth, *Thaumatotibia leucotreta* (Steyn *et al.*, 2019). Semi-field trials in avocado, lychee and macadamia orchards demonstrated that four species of EPNs applied at 30 IJs cm^{-2}

reduced *T. leucotreta* populations, with *Steinernema yirgalemense* being most effective (86% *T. leucotreta* mortality) (Steyn *et al.*, 2019).

17.6 Summary and Conclusions

EPNs are being applied commercially for control of several important insect pests of temperate orchards, but clearly there is significant room for expansion. Existing barriers to commercial adoption of EPNs in temperate orchards likely include competition with pesticides, price/supply issues and technology transfer to improve awareness. Advances in research and manufacturing are required, particularly in reducing costs of production and application, and methodology to improve the survival and persistence of nematodes in soil or in the canopy. In terms of improving production, novel approaches such as grower-based production or biphasic solid culture may increase efficiency, lower costs and increase quality (Oliveira-Hofman *et al.*, 2023; see Chapter 10, this volume).

The survival and persistence of EPNs in orchards can be enhanced through novel formulations. As indicated above, the use of Barricade® and other gels can enhance EPN efficacy (Fallet *et al.*, 2022; Wong *et al.*, 2022), particularly against aboveground targets. Moreover, additional improvements may be possible with other inexpensive formulations to further protect EPNs from UV radiation or desiccation, and that are compatible with organic farming systems (Wu *et al.*, 2023).

EPN efficacy can also be improved via strain improvement or boosting agents. As indicated above, pheromone-based (e.g. ascaroside) boosters can increase biocontrol potential, but other boosters remain unexplored (Mbata *et al.*, 2019). Biocontrol potential may also be increased through combination with other biocontrol or chemical insecticides (synergists) (see Chapter 27, this volume). EPN strains that persist longer in the environment may improve efficacy or reduce costs (reducing application rates or frequency of application; Lauriault *et al.*, 2020). Moreover, strains can be improved via various genetic mechanisms such as hybridization, selection and transgenics (Glazer and Shapiro-Ilan, 2022).

Additionally, inoculative or conservation approaches to biocontrol with nematodes must be explored. Various conditions associated with orchards as agroecosystems may facilitate these approaches, such as plant species and structural diversity, soil conditions and stability (Kaya, 1990; Barbosa, 1998; Lewis *et al.*, 1998; see Chapter 28, this volume). Various characteristics and management practices, including soil types, fertilization, irrigation and cover crops should be investigated for specific orchard systems to determine their effects on EPN ecology and their potential to improve long-term efficacy.

17.7 References

Agudelo-Silva, F., Lindegren, J.E. and Valero, K.A. (1987) Persistence of *Neoaplectana carpocapsae* (Kapow Selection) infectives in almonds under field conditions. *Florida Entomologist* 70, 288–291.

Agudelo-Silva, F., Zalom, F.G., Hom, A. and Hendricks, L. (1995) Dormant season application of *Steinernema carpocapsae* (Rhabditida: Steinernematidae) and *Heterorhabditis* sp. (Rhabditida: Heterorhabditidae) on almond for control of overwintering *Amyelois transitella* and *Anarsia lineatella* (Lepidoptera: Gelchidiidae). *Florida Entomologist* 78, 516–523.

Alston, D.G., Rangel, D.E.N., Lacey, L.A., Golez, H.G., Kim, J.J. and Roberts, D.W. (2005) Evaluation of novel fungus and nematode isolates for control of *Conotrachelus nenuphar* (Coleoptera: Curculionidae) larvae. *Biological Control* 35, 163–171.

Azad, N.D. and Singh, V.K. (2019) Impact of infestation of banana pest *Odoiporous longicollis* and *Cosmopolites sordidus* on four common varieties of banana plant namely, Malbhog, Alpan, Champa and Harichhal in Koshi and Vaishali District, Bihar, India. *Journal of Pharmacognosy and Phytochemistry* 8 (Special Issue 5), 134–138.

Barbosa, P. (1998) Agroecosystems and conservation biological control. In: Barbosa, P. (ed.) *Conservation Biological Control*. Academic Press, San Diego, California, pp. 39–54.

Barman, A., Hudson, W. and Sawyer, A. (2023) Commercial pecan insect control (bearing trees). In: Wells, L. (ed.) *2023 Commercial Pecan Spray Guide*. Bulletin 841. University of Georgia Extension, Athens, Georgia, pp. 2–6.

Beavers, J.B. and Calkins, C.O. (1984) Susceptibility of *Anastrepha suspensa* (Diptera: Tephritidae) to steinernematid and heterorhabditid nematodes in laboratory studies. *Environmental Entomology* 13, 137–139.

Belair, G., Vincent, C. and Chouinard, G. (1998) Foliar sprays with *Steinernema carpocapsae* against early season apple pests. *Journal of Nematology* 30, 599–606.

Belair, G., Vincent, C., Lemire, S. and Coderre, D. (1999) Laboratory and field assays of entomopathogenic nematodes for the management of the obliquebanded leafroller, *Choristoneura rosaceana* (Harris) (Tortricidae). *Journal of Nematology* 31, 684–689.

Blaauw, B., Brannen, P., Lockwood, D., Schhnabel, G. and Ritchie, D. (2020) *Southeastern Peach, Nectarine, and Plum Pest Management and Culture Guide*. Bulletin 1171. University of Georgia Extension, Athens, Georgia.

Cossentine, J.E., Banham, F.L. and Jensen, L.B. (1990) Efficacy of the nematode, *Heterorhabditis heliothidis* (Rhaditida: Heterorhabditidae) against the peachtree borer, *Synanthedon exitiosa* (Lepidoptera: Sesiidae) in peach trees. *Journal of the Entomological Society of British Columbia* 87, 82–84.

Cossentine, J.E., Jensen, L.B. and Moyls, L. (2002) Fruit bins washed with *Steinernema carpocapsae* (Rhabditida: Steinernematidae) to control *Cydia pomonella* (Lepidoptera: Tortricidae). *Biocontrol Science and Technology* 12, 251–258.

Cottrell, T.E. and Shapiro-Ilan, D.I. (2006) Susceptibility of the peachtree borer, *Synanthedon exitiosa*, to *Steinernema carpocapsae* and *Steinernema riobrave* in laboratory and field trials. *Journal of Invertebrate Pathology* 92, 85–88.

Cottrell, T.E., Shapiro-Ilan, D.I., Horton, D.L. and Mizell, R.F. III. (2011) Laboratory virulence and orchard efficacy of entomopathogenic nematodes toward the lesser peachtree borer (Lepidoptera: Sesiidae). *Environmental Entomology* 104, 47–53.

de Luca, F., Clausi, M., Troccoli, A., Curto, G., Giancarlo Rappazzo, G. and Tarasco, E. (2015) Entomopathogenic nematodes in Italy: occurrence and use in microbial control strategies. In: Campos-Herrera, R. (ed.) *Nematode Pathogenesis of Insects and Other Pests: Ecology and Applied Technologies for Sustainable Plant and Crop Protection*. Springer, Cham, Switzerland, pp. 431–449.

de Waal, J.Y., Malan, A.P., Levings, J. and Addison, M.F. (2010) Key elements in the successful control of diapausing codling moth, *Cydia pomonella* (Lepidoptera: Tortricidae) in wooden fruit bins with a South African isolate of *Heterorhabditis zealandica* (Rhabditida: Heterorhabditidae). *Biocontrol Science and Technology* 20, 489–502.

de Waal, J.Y., Malan, A.P. and Addison, M.F. (2011) Efficacy of entomopathogenic nematodes (Rhabditida: Heterorhabditidae and Steinernematidae) against codling moth, *Cydia pomonella* (Lepidoptera: Tortricidae) in temperate regions. *Biocontrol Science and Technology* 20, 489–502.

de Waal, J.Y., Addison, M.F. and Malan, A.P. (2018) Potential of *Heterorhabditis zealandica* (Rhabditida: Heterorhabditidae) for the control of codling moth, *Cydia pomonella* (Lepidoptera: Tortricidae) in semi-field trials under South African conditions. *International Journal of Pest Management* 64, 102–109.

Deseo, K.V. and Miller, L.A. (1985) Efficacy of entomogenous nematodes, *Steinernema* spp., against clearwing moths, *Synanthedon* spp., in North Italian apple orchards. *Nematologica* 31, 100–108.

Dlamini, T.M., Allsopp, E. and Malan, A.P. (2020) Application of *Steinernema yirgalemense* to control *Frankliniella occidentalis* (Thysanoptera: Thripidae) on blueberries. *Crop Protection* 128: 105016.

Ebssa, L., Borgemeister, C., Berndt, O. and Poehling, H.M. (2001a) Efficacy of entomopathogenic nematodes against soil-dwelling life stages of western flower thrips, *Frankliniella occidentalis* (Thysanoptera: Thripidae). *Journal of Invertebrate Pathology* 78, 119–127.

Ebssa, L., Borgemeister, C., Berndt, O. and Poehling, H.M. (2001b) Impact of entomopathogenic nematodes on different soil-dwelling stages of western flower thrips, *Frankliniella occidentalis* (Thysanoptera: Thripidae), in the laboratory and under semi-field conditions. *Biocontrol Science and Technology* 11, 515–525.

Eliceche, D., Rusconi, M., Rosales, M., Rossi, J., Salas, A. *et al.* (2023) The efficacy of *Steinernema feltiae* in the control of *Cydia pomonella* in organic apple and pear orchards of Patagonia Argentina and its impact on local populations of entomopathogenic nematodes. *BioControl* 68, 459–470.

Evans, B.G. and Renkema, J.M. (2021) Initial tests of *Steinernema riobrave* against spotted wing drosophila, *Drosophila suzukii* Matsumura (Diptera: Drosophilidae) and comparison to other entomopathogenic nematode species in the laboratory. *Biocontrol Science and Technology* 31, 336–342.

Fallet, P., Bazagwira, D., Guenat, J.M., Segura, C.B., Karangwa, P. *et al.* (2022) Laboratory and field trials reveal the potential of a gel formulation of entomopathogenic nematodes for the biological control of fall armyworm caterpillars (*Spodoptera frugiperda*). *Biological Control* 176: 105086.

Figueroa, W. (1990) Biocontrol of the banana root borer weevil, *Cosmopolites sordidus* (Gemar), with steinernematid nematodes. *Journal of Agriculture of the University of Puerto Rico* 74, 15–19.

Gazit, Y., Rossler, Y. and Glazer, I. (2000) Evaluation of entomopathogenic nematodes for the control of Mediterranean fruit fly (Diptera: Tephritidae). *Biocontrol Science and Technology* 10, 157–164.

Glazer, I. and Shapiro-Ilan, D.I. (2022) Genetic improvement of beneficial organisms. In: Glazer, I., Shapiro-Ilan, D.I. and Sternberg, P.W. (eds) *Nematodes as Model Organisms*. CAB International, Wallingford, UK, pp. 346–364.

Gordon, P.E., Goodrich, B.K. and Wilson, H. (2023) Adoption of *Amyelois transitella* (navel orangeworm) monitoring and management practices across California tree nut crops. *Journal of Integrated Pest Management* 14: 16.

Grewal, P.S., Ehlers, R.U. and Shapiro-Ilan, D.I. (eds) (2005) *Nematodes as Biocontrol Agents*. CAB International, Wallingford, UK.

Han, R., Xu, J., Cao, L. and Liu, X. (1994) Spatial distribution pattern and sampling technique of *Aristobia testudo* (Coleoptera: Cerambycidae) in litchi orchards. *Natural Enemies of Insects* 16, 175–182.

Han, R., Liu, X., Xie, R. and Xu, J. (1996) Effects of entomopathogenic nematodes on the spatial distribution pattern of the larvae of litchi beetle *Aristobia testudo* (Coleoptera: Cerambycidae) in litchi orchards and the economic threshold. *Natural Enemies of Insects* 18, 99–105.

Harris, M.K. (1985) Pecan phenology and pecan weevil biology and management. In: Neel, W.W. (ed.) *Pecan Weevil: Research Perspective*. Quail Ridge Press, Brandon, Mississippi, pp. 51–58.

Helyer, N.L., Brobyn, P.J., Richardson, P.N. and Edmondson, R.N. (1995) Control of Western flower thrips (*Frankliniella occidentalis* Pergande) pupae in compost. *Annals of Applied Biology* 127, 405–412.

Horton, D. and Johnson, D. (eds) (2005) *Southeastern Peach Growers' Handbook*. GES Handbook Vol. 1. University of Georgia Extension, Athens, Georgia.

Jenkins, D.A., Shapiro-Ilan, D. and Goenaga, R. (2008) Efficacy of entomopathogenic nematodes versus *Diaprepes abbreviatus* (Coleoptera: Curculionidae) larvae in a high clay-content oxisol soil: greenhouse trials with potted *Litchi chinensis*. *Florida Entomologist* 91, 75–78.

Kahounova, L. and Mracek, Z. (1991) Larval mortality of *Synanthedon myopaeformis* (Lepidoptera, Sesiidae) in apple trees sprayed with *Steinernema* sp. strain *Hylobius* (*Steinernema*, Steinernematidae). *Acta Entomologica Bohemoslovaca* 88, 205–210.

Kain, D.F. and Agnello, A.M. (1999) Pest status of American plum borer (Lepidoptera: Pyralidae) and fruit tree borer control with synthetic insecticides and entomopathogenic nematodes in New York State. *Journal of Economic Entomology* 92, 193–200.

Kaya, H.K. (1990) Soil ecology. In: Gaugler, R. and Kaya, H.K. (eds) *Entomopathogenic Nematodes in Biological Control.* CRC Press, Boca Raton, Florida, pp. 93–116.

Kaya, H.K., Joos, J.L., Falcon, L.A. and Berlowitz, A. (1984) Suppression of the codling moth (Lepidoptera: Olethreutidae) with the entomogenous nematode, *Steinernema feltiae* (Rhabditida: Steinernematidae). *Journal of Economic Entomology* 77, 1240–1244.

Kermarrec, A. and Mauléon, H. (1989) Synergy between chlordecone and *Neoaplectana carpocapsae* Weiser (Nematoda; Steinernematidae) in the control of *Cosmopolites sordidus* (Coleoptera: Curculionidae). *Revue de Nematologie* 12, 324–325.

Lacey, L.A. and Chauvin, R.L. (1999) Entomopathogenic nematodes for control of codling moth in fruit bins. *Journal of Economic Entomology* 92, 104–109.

Lacey, L.A. and Unruh, T.R. (1998) Entomopathogenic nematodes for control of codling moth: effect of nematode species, dosage, temperature and humidity under laboratory and simulated field conditions. *Biological Control* 13, 190–197.

Lacey, L.A., Knight, A. and Huber, J. (2000) Microbial control of lepidopterous pests of apple orchards. In: Lacey, L. and Kaya, H.K (eds.) *Field Manual of Techniques in Invertebrate Pathology*. Springer, Dordrecht, Netherlands, pp. 557–576.

Lacey, L.A., Unruh, T.R. and Headrick, H.L. (2003) Interactions of two idiobiont parasitoids (Hymenoptera: Ichneumonidae) of codling moth (Lepidoptera: Tortricidae) with the entomopathogenic nematode *Steinernema carpocapsae* (Rhabditida: Steinernematidae). *Journal of Invertebrate Pathology* 83, 230–239.

Lacey, L.A., Arthurs, S.P., Unruh, T.R., Headrick, H. and Fritts, R. Jr (2006a) Entomopathogenic nematodes for control of codling moth (Lepidoptera: Tortricidae) in apple and pear orchards: effect of nematode species and seasonal temperatures, adjuvants, application equipment and post-application irrigation. *Biological Control* 37, 214–223.

Lacey, L.A., Granatstein, D., Arthurs, S.P., Headrick, H. and Fritts, R. Jr (2006b) Use of entomopathogenic nematodes (Steinernematidae) in conjunction with mulches for control of overwintering codling moth (Lepidoptera: Tortricidae). *Journal of Entomological Science* 41, 107–119.

Lacey, L.A., Arthurs, S.P., Knight, A.L. and Huber, J. (2007) Microbial control of lepidopteran pests of apple orchards. In: Lacey, L.A. and Kaya, H.K. (eds), *Field Manual of Techniques in Invertebrate Pathology*, 2nd edn. Springer, Dordrecht, Netherlands, pp. 527–546.

Lacey, L.A., Shapiro-Ilan, D.I. and Glenn, G.M. (2010) The effect of post-application anti-desiccant agents and formulated host-cadavers on entomopathogenic nematode efficacy for control of diapausing codling moth larvae (Lepidoptera: Tortricidae). *Biocontrol Control Science and Technology* 20, 909–921.

Lampasona, T.P., Rodriguez-Saona, C., Leskey, T.C. and Nielsen, A.L. (2020) A review of the biology, ecology, and management of plum curculio (Coleoptera: Curculionidae). *Journal of Integrated Pest Management* 11: 22.

Laumond, C., Mauléon, H. and Kermarrec, A. (1979) Données sur le spectre d'hôtes et le parasitisme du nematode entomophage *Neoaplectana carpocapsae*. *Entomophaga* 24, 13–27.

Lauriault, L.M., Shields, E.J., Testa, A.M. and Porter, R.P. (2020) Persistence of select introduced entomopathogenic nematodes in the US Southwest as potential biological control for whitefringed beetle in alfalfa. *Southwestern Entomologist* 45, 41–49.

Leskey, T.C., Piñero, J.C. and Prokopy, R.J. (2008) Oder-baited trap trees: a novel management tool for plum curculio (Coleoptera: Curculionidae). *Journal of Economic Entomology* 101, 1302–1309.

Lewis, E.E., Campbell, J.F. and Gaugler, R. (1998) A conservation approach to using entomopathogenic nematodes in turf and landscapes. In: Barbosa, P. (ed.) *Conservation Biological Control*. Academic Press, San Diego, California, pp. 235–254.

Lindegren, J.E. and Vail, P.V. (1986) Susceptibility of Mediterranean fruit fly, melon fly, and oriental fruit fly (Diptera: Tephritidae) to the entomogenous nematode *Steinernema feltiae* in laboratory tests. *Environmental Entomology* 15, 465–468.

Lindegren, J.E., Agudelo-Silva, F., Valero, K.A. and Curtis, C.E. (1987) Comparative small-scale field application of *Steinernema feltiae* for navel orangeworm control. *Journal of Nematology* 19, 503–504.

Lindegren, J.E., Wong, T.T. and McInnis, D.O. (1990) Response of Mediterranean fruit fly (Diptera: Tephritidae) to the entomogenous nematode *Steinernema feltiae* in field tests in Hawaii. *Environmental Entomology* 19, 383–386.

Mbata, G.N., Shapiro-Ilan, D., Alborn, H.T. and Strand, M.R. (2019) Preferential infectivity of entomopathogenic nematodes in an envenomed host. *International Journal of Parasitology* 49, 737–745.

Morton, A. and Garcia-del-Pino, F. (2008) Field efficacy of the entomopathogenic nematode *Steinernema feltiae* against the Mediterranean flat-headed rootborer *Capnodis tenebrionis*. *Journal of Applied Entomology* 132, 632–637.

Nachtigall, G. and Dickler, E. (1992) Experiences with field applications of entomoparasitic nematodes for biological control of cryptic living insects in orchards. *Acta Phytopathologica et Entomologica Hungarica* 27, 485–490.

Negrisoli, D.B., Negrisoli, A.S., Garcia, M.S., Dolinski, C. and Bernardi, D. (2013) Control of *Grapholita molesta* (Busck, 1916) (Lepidoptera: Tortricidae) with entomopathogenic nematodes (Rhabditida: Heterorhabditidae, Steinernematidae) in peach orchards. *Experimental Parasitology* 135, 466–470.

Nevaneethan, T., Strauch, O., Besse, S., Bonhomme, A. and Ehlers, R.U. (2010) Influence of humidity and a surfactant-polymer-formulation on the control potential of the entomopathogenic nematode *Steinernema feltiae* against diapausing codling moth larvae (*Cydia pomonella* L.) (Lepidoptera: Tortricidae). *BioControl* 55, 777–788.

Odendaal, D., Addison, M.F. and Malan, A.P. (2015) Control of codling moth (*Cydia pomonella*) (Lepidoptera: Tortricidae) in South Africa with special emphasis on using entomopathogenic nematodes. *African Entomology* 23, 259–274.

Odendaal, D., Addison, M.F. and Malan, A.P. (2016) Control of diapausing codling moth, *Cydia pomonella* (Lepidoptera: Tortricidae) in wooden fruit bins, using entomopathogenic nematodes (Heterorhabditidae and Steinernematidae). *Biocontrol Science and Technology* 26, 1504–1515.

Oliveira-Hofman, C., Kaplan, F., Stevens, G., Lewis, E.E., Wu, S. *et al.* (2019) Pheromone extracts act as boosters for entomopathogenic nematodes efficacy. *Journal of Invertebrate Pathology* 164, 38–42.

Oliveira-Hofman, C., Steffan, S. and Shapiro-Ilan, D. (2023) A sustainable grower-based method for entomopathogenic nematodes production. *Journal of Insect Science* 23: 4.

Olthof, T.H. and Hagley, E. (1993) Laboratory studies of the efficacy of steinernematid nematodes against the plum curculio (Coleoptera: Curculionidae). *Journal of Economic Entomology* 86, 1078–82.

Patterson Stark, J.E. and Lacey, L.A. (1999) Susceptibility of western cherry fruit fly (Diptera: Tephritidae) to steinernematid and heterorhabditid nematodes in laboratory studies. *Journal of Invertebrate Pathology* 74, 206–208.

Pereault, R.J., Whalon, M.E. and Alston, D.G. (2009) Field efficacy of entomopathogenic fungi and nematodes targeting caged last-instar plum curculio (Coleoptera: Curculionidae) in Michigan cherry and apple orchards. *Environmental Entomology* 38, 1126–1134.

Piñero J.C., Shapiro-Ilan, D., Cooley, D.R., Tuttle, A.F., Eaton, A. *et al.* (2020) Toward the integration of an attract-and-kill approach with biological control involving entomopathogenic nematodes to control multiple life stages of plum curculio (Coleoptera: Curculionidae) in eastern North America. *Insects* 11: 375.

Piñero, J.C., Agnello, A.M., Tuttle, A., Leskey, T.C., Faubert, H. *et al.* (2011) Effectiveness of odor-baited trap trees for plum curculio (Coleoptera: Curculionidae) monitoring in commercial apple orchards in the Northeast. *Journal of Economic Entomology* 104, 1613–1621.

Poinar, G.O. Jr (1991) Nematode parasites. In: van der Geest, L.P.S. and Evenhius, H.H (eds) *Tortricid Pests, Their Biology, Natural Enemies and Control*. Elsevier Science Publishers, Amsterdam, pp. 273–281.

Reger, J., Wenger, J., Brar, G., Burks, C. and Wilson, H. (2020) Evaluating response of mass-reared and irradiated navel orangeworm, *Amyelois transitella* (Lepidoptera: Pyralidae), to crude female pheromone extract. *Insects* 11: 703.

Rice, R.E. (1978a) Insect and mite pests of almonds. In: Micke, W. and Keaster, D. (eds) *Almond Orchard Management*. Division of Agriculture and Natural Resources Publication No. 4092. University of California, Berkeley, California, pp 117–124.

Rice, R.E. (1978b) Navel orangeworm, a pest of pistachio nuts in California. *Journal of Economic Entomology* 71, 822–824.

Riga, E., Lacey, L.A., Guerra, N. and Headrick, H.L. (2006) Control of the oriental fruit moth, *Grapholita molesta*, using entomopathogenic nematodes in laboratory and fruit bin assays *Journal of Nematology* 38, 168–171.

Şahin, Ç. and Gözel, U. (2021) Efficacy and persistence of native entomopathogenic nematodes against *Capnodis tenebrionis* in peach (*Prunus persica*) orchard in Turkey. *Phytoparasitica* 49, 613–621.

Schulte, M.J., Martin, K., Büchse, A. and Sauerborn, J. (2009) Entomopathogens (*Beauveria bassiana* and *Steinernema carpocapsae*) for biological control of bark-feeding moth *Indarbela dea* on field-infested litchi trees. *Pest Management Science* 65, 105–112.

Shapiro-Ilan, D.I. (2001a) Virulence of entomopathogenic nematodes to pecan weevil larvae *Curculio caryae* (Coleoptera: Curculionidae) in the laboratory. *Journal of Economic Entomology* 94, 7–13.

Shapiro-Ilan, D.I. (2001b) Virulence of entomopathogenic nematodes to pecan weevil adults (Coleoptera: Curculionidae). *Journal of Entomological Science* 36, 325–328.

Shapiro-Ilan, D.I. (2003) Microbial control of the pecan weevil, *Curculio caryae* (Coleoptera: Curculionidae). *Southwestern Entomologist* Suppl. 27, 101–114.

Shapiro-Ilan, D.I. and Cottrell, T.E. (2006) Susceptibility of the lesser peachtree borer (Lepidoptera: Sesiidae) to entomopathogenic nematodes under laboratory conditions. *Environmental Entomology* 35, 358–365.

Shapiro-Ilan, D.I., Mizell, R.F. and Campbell, J.F. (2002) Susceptibility of plum curculio, *Conotrachelus nenuphar*, to entomopathogenic nematodes. *Journal of Nematology* 24, 246–249.

Shapiro-Ilan, D.I., Stuart, R. and McCoy, C.W. (2003) Comparison of beneficial traits among strains of the entomopathogenic nematode, *Steinernema carpocapsae*, for control of *Curculio caryae* (Coleoptera: Curculionidae). *Biological Control* 28, 129–136.

Shapiro-Ilan, D.I, Mizell, R.F., Cottrell, T.E. and Horton, D.L. (2004) Measuring field efficacy of *Steinernema feltiae* and *Steinernema riobrave* for suppression of plum curculio, *Conotrachelus nenuphar*, larvae. *Biological Control* 30, 496–503.

Shapiro-Ilan, D.I., Duncan, L.W., Lacey, L.A. and Han, R. (2005a) Orchard crops. In: Grewal, P.S., Ehlers, R.U. and Shapiro-Ilan, D.I. (eds) *Nematodes as Biocontrol Agents*. CAB International, Wallingford, UK, pp. 215–230.

Shapiro-Ilan, D.I., Stuart, R.J. and McCoy, C.W. (2005b) Targeted improvement of *Steinernema carpocapsae* for control of the pecan weevil, *Curculio caryae* (Horn) (Coleoptera: Curculionidae) through hybridization and bacterial transfer. *Biological Control* 34, 215–221.

Shapiro-Ilan, D.I., Cottrell, T.E., Brown, I., Gardner, W.A., Hubbard, R.K. and Wood, B.W. (2006) Effect of soil moisture and a surfactant on entomopathogenic nematode suppression of the pecan weevil, *Curculio caryae*. *Journal of Nematology* 38, 474–482.

Shapiro-Ilan, D.I., Mizell, R.F. III, Cottrell, T.E. and Horton, D.L. (2008) Control of plum curculio, *Conotrachelus nenuphar* with entomopathogenic nematodes: effects of application timing, alternate host plant, and nematode strain. *Biological Control* 44, 207–215.

Shapiro-Ilan, D.I., Cottrell, T.E., Mizell, R.F. III, Horton, D.L. and Davis, J. (2009) A novel approach to biological control with entomopathogenic nematodes: prophylactic control of the peachtree borer, *Synanthedon exitiosa*. *Biological Control* 48, 259–263.

Shapiro-Ilan, D.I., Cottrell, T.E., Mizell, R.F., Horton, D.L, Behle, R.W. and Dunlap, C.A. (2010) Efficacy of *Steinernema carpocapsae* for control of the lesser peachtree borer, *Synanthedon pictipes*: improved aboveground suppression with a novel gel application. *Biological Control* 54, 23–28.

Shapiro-Ilan, D.I., Leskey, T.C. and Wright, S.E. (2011) Virulence of entomopathogenic nematodes to plum curculio, *Conotrachelus nenuphar*: effects of strain, temperature, and soil type. *Journal of Nematology* 43, 187–195.

Shapiro-Ilan, D.I. and Gardner, W.A. (2012) Improved control of *Curculio caryae* (Coleoptera: Curculionidae) through multi-stage pre-emergence applications of *Steinernema carpocapsae*. *Journal of Entomological Science* 47, 27–34.

Shapiro-Ilan, D.I., Wright, S.E., Tuttle, A.F., Cooley, D.R. and Leskey, T.C. (2013) Using entomopathogenic nematodes for biological control of plum curculio, *Conotrachelus nenuphar*: Effects of irrigation and species in apple orchards. *Biological Control* 67, 123–129.

Shapiro-Ilan, D.I., Cottrell, T.E., Mizell, R.F. III., Horton, D.L. and Abdo, Z. (2015) Field suppression of the peachtree borer, *Synanthedon exitiosa*, using *Steinernema carpocapsae*: effects of irrigation, a sprayable gel and application method. *Biological Control* 82, 7–12.

Shapiro-Ilan, D.I., Cottrell, T.E., Mizell, R.F. III. and Horton, D.L. (2016a) Curative control of the peachtree borer using entomopathogenic nematodes. *Journal of Nematology* 48, 170–176.

Shapiro-Ilan, D.I., Cottrell, T.E., Mizell, R.F. III. and Horton, D.L. (2016b) Efficacy of *Steinernema carpocapsae* plus fire gel applied as a single spray for control of the lesser peachtree borer, *Synanthedon pictipes*. *Biological Control* 94, 33–36.

Shapiro-Ilan, D.I., Arthurs, S.P. and Lacey, L.A. (2017a) Microbial control of arthropod pests of orchards in temperate climates. In: Lacey, L.A. *Microbial Agents for Control of Insect Pests: From Discovery to Commercial Development and Use*. Academic Press, Amsterdam, UK, pp. 253–267.

Shapiro-Ilan, D.I., Cottrell, T.E., Bock, C., Mai, K., Boykin, D. *et al.* (2017b) Control of pecan weevil with microbial biopesticides. *Environmental Entomology* 46, 1299–1304.

Shapiro-Ilan, D.I., Kaplan, F., Oliveira-Hofman, C., Schliekelman, P., Alborn, H.T. and Lewis, E.E. (2019) Conspecific pheromone extracts enhance entomopathogenic infectivity. *Journal of Nematology* 51: e2019-82.

Siegel, J., Lacey, L.A., Fritts, R. Jr., Higbee, B.S. and Noble, P. (2004) Use of steinernematid nematodes for post harvest control of navel orangeworm (Lepidoptera: Pyralidae, *Amyelois transitella*) in fallen pistachios. *Biological Control* 30, 410–417.

Siegel, J., Lacey, L.A., Higbee, B.S., Noble, P. and Fritts, R. Jr. (2006) Effect of application rates and abiotic factors on *Steinernema carpocapsae* for control of overwintering navel orangeworm (Lepidoptera: Pyralidae, *Amyelois transitella*) in fallen pistachios. *Biological Control* 36, 324–330.

Sledzevskaya, E.R. (1987) Study of the factors determining the activity of the nematode *Neoaplectana carpocapsae* and its efficacy against orchard insect pests. In: Sonin, M.D. (ed.) *Helminths of Insects.* Amerind Publishing Co., New Delhi, pp. 152–155.

Srivastava, K., Pandey, S.D., Patel, R.K., Sharma, D. and Nath, V. (2016) Insect pest management practices in litchi. In: Pandey, A.K. and Mall, P. (eds) *Insect Pests Management of Fruit Crops*. BioTech Books Ltd, New Delhi, pp. 127–149.

Steyn, W.P., Daneel, M.S. and Malan, A.P. (2019) Field application of entomopathogenic nematodes against *Thaumatotibia leucotreta* in South African avocado, litchi and macadamia orchards. *BioControl* 64, 401–411.

Treverrow, N., Bedding, R.A., Dettmann, E.B. and Maddox, C. (1991) Evaluation of entomopathogenic nematodes for control of *Cosmopolites sordidus* Gemar (Coleoptera: Curculionidae), a pest of bananas in Australia. *Annals of Applied Biology* 119, 139–145.

Unruh, T.R. and Lacey, L.A. (2001) Control of codling moth, *Cydia pomonella* (Lepidoptera: Tortricidae) with *Steinernema carpocapsae*: effects of supplemental wetting and pupation site on infection rate. *Biological Control* 20, 48–56.

Usman, M., Gulzar, S., Wakil, W., Piñero, J.C., Leskey, T.C. *et al.* (2020) Potential of entomopathogenic nematodes against the pupal stage of the apple maggot *Rhagoletis pomonella* (Walsh) (Diptera: Tephritidae). *Journal of Nematology* 52: e2020-79.

Usman, M., Wakil, W., Sufyan, M. and Shapiro-Ilan, D. (2021) Entomopathogenic nematodes as biological control agent against *Bactrocera zonata* and *Bactrocera dorsalis* (Diptera: Tephritidae). *Biological Control* 163: 104706.

Vincent, C. and Belair, G. (1992) Biocontrol of the apple sawfly, *Hoplocampa testudinea*, with entomogenous nematodes. *Entomophaga* 37, 575–582.

Wakil, W., Usman, M., Piñero, J.C., Wu, S., Toews, M.D. and Shapiro-Ilan, D.I. (2022) Combined application of entomopathogenic nematodes and fungi against fruit flies, *Bactrocera zonata* and *B. dorsalis* (Diptera: Tephritidae): laboratory cups to field study. *Pest Management Science* 78, 2779–2791.

Weiser, J. (1955) *Neoaplectana carpocapsae* n. sp. (Anguillaluata: Steinernematinae), novy cizopasník housenek obalece jablecného, *Carpocapsa pomonnella* L. *Bestniik Ceskoslovenske Zoologické Spolecnosti* 19, 44–51.

Wong, C., Oliveira-Hofman, C., Blaauw, B., Chavez, D.J., Jagdale, G. *et al.* (2022) Control of peachtree borer (*Synanthedon exitiosa*) using the nematode *Steinernema carpocapsae*: optimization of application rates and secondary benefits in control of root-feeding weevils. *Agronomy* 12: 2689.

Wu, S., Li., Y., Toews, M.D., Mbata, G. and Shapiro-Ilan, D.I. (2023) Novel formulations improve the environmental tolerance of the entomopathogenic nematodes. *Biological Control* 186: 105329.

Xu, J. and Yang, P. (1992) The application of *Steinernema carpocapsae* nematodes for the control of the litchi stem borer, *Arbela dea*. *Acta Phytophylacica Sinica* 19, 217–222.

Xu, J., Yang, P. and Wu, S. (1991) The control of banana borer with *Steinernema carpocapsae* nematodes. *Chinese Journal of Tropical Crops* 12, 77–82.

Xu, J., Han, R., Liu, X., Cao, L. and Yang, P. (1995) The application of *Steinernema carpocapsae* nematodes for the control of the litchi longhorn beetle, *Aristobia testudo*. *Acta Phytophylacica Sinica* 22, 12–16.

Yee, W.L. and Lacey, L.A. (2003) Stage-specific mortality of *Rhagoletis indifferens* (Diptera: Tephritidae) exposed to three species of *Steinernema* nematodes. *Biological Control* 27, 349–356.

18 Entomopathogenic Nematodes in Berry Crops

Cesar Rodriguez-Saona[1]* and Surendra K. Dara[2,3]
[1]*PE Marucci Center for Blueberry and Cranberry Research and Extension, Rutgers University, Chatsworth, New Jersey, USA;* [2]*University of California Division of Agriculture and Natural Resources, San Luis Obispo, California, USA;* [3]*Current address: Department of Horticulture, Oregon State University, Oregon, USA*

Abstract
This chapter reviews the potential of using entomopathogenic nematodes (EPNs) as biological agents to control foliar- and root-feeding insect pests of berry crops including strawberries, caneberries (raspberries and blackberries), blueberries and cranberries. These are perennial crops that, in general, provide a suitable environment of high humidity and low disturbance for the persistence of EPNs applied to the soil. The chapter particularly highlights the progress made in the research of EPNs in insect pest management in these crops, as well as the factors affecting their success and the potential barriers for farmer adoption. An increasing interest in alternatives to synthetic insecticides due to legislative restrictions in their use and in organic and pick-your-own farming is opening up opportunities for EPNs. However, there is the need to further improve the efficacy of both commercially produced and naturally occurring EPN species/strains under different soil conditions and to optimize formulations, application methods and costs.

18.1 Introduction

Berry crops, such as strawberries, caneberries (raspberries and blackberries), blueberries and cranberries, are woody and herbaceous perennials grown commercially under a wide variety of field (i.e. soil and moisture) and climatic conditions. These crops are grown in many countries around the world with the exception of cranberries, which are restricted mainly to certain parts of the USA and Canada due to more specific cultivation needs.

Berry crops are often threatened by soil-inhabiting insect pests such as root weevils and white grubs, which can injure the crop and reduce yields, and entomopathogenic nematodes (EPNs) offer an environmentally safe alternative to chemical control for managing these and other pests. For example, different EPN species have recently been explored for controlling the spotted-wing drosophila (*Drosophila suzukii* (Matsumura) (Diptera: Drosophilidae)), which has become the most economically important insect pest of several berry crops, including strawberries, caneberries and blueberries, worldwide.

In this chapter, we discuss the opportunities for using EPNs to manage key insect pests of the most important berry crops produced in the

*Email: crodriguez@njaes.rutgers.edu

DOI: 10.1079/9781800620322.0018

world, namely strawberries, raspberries, blackberries, blueberries and cranberries.

18.2 Strawberries

The demand for fresh market and processed strawberries makes it a high-value commercial crop around the world. Strawberry, *Fragaria*× *ananassa* Duch. (Rosales: Rosaceae), is vulnerable to seven insect pests that are good candidates for using EPNs (Table 18.1).

Strawberry is a temperate crop that grows under mild environmental conditions, has a canopy that is close to the ground and protects the soil around the base, and grows in beds that are covered with plastic or straw mulch. These characteristics provide protective soil and environmental conditions within the crop microclimate that are ideal for EPNs. Currently, several synthetic pesticide options address strawberry pests, but their preharvest and restricted entry intervals complicate application timing. Additionally, because strawberries are harvested multiple times each week, farm workers operate in the field for several months, which poses a safety concern if synthetic pesticides are applied. These issues create an opportunity to explore non-chemical control options such as EPNs. In general, pesticide resistance is also a widespread problem in many crops including strawberries, and a biocontrol option such as EPNs helps maintain control efficacy without the risk of resistance development.

18.2.1 Root weevils

Several studies have explored the potential of using EPN for controlling root weevils in strawberries (Table 18.1). Root weevil adults feed on foliage, and larvae feed on roots and overwinter in the soil. In one of the older studies, Bedding and Miller (1981) found 87% infection in the black vine weevil (*Otiorhynchus sulcatus* (F.)) (Coleoptera: Curculionidae) in potted strawberry of an isolate (T327) of *Heterorhabditis bacteriophora* Poinar and suggested EPNs as an effective and economical option in greenhouses and nurseries. In an American field study, *Heterorhabditis marelata* Liu and Berry provided up to 77% control of the larvae and pupae of *O. sulcatus* and the strawberry root weevil (*Otiorhynchus ovatus* (L.)) (Coleoptera: Curculionidae), and *H. bacteriophora* provided up to 74% control at 20 days after application (Berry *et al.*, 1997). In a German study, Susurluk and Ehlers (2008) observed 90–96% mortality of *O. sulcatus* larvae in potted strawberry with the application of *H. bacteriophora*.

Many factors influence the efficacy of EPNs against the target pest; some of these factors include the suitability of an isolate or species for the target pest, EPN virulence, soil conditions, EPN environmental tolerance at both low and high temperatures, type of host plant, pest life stage and where it is present, and ability of the EPN to locate its host. For example, Georgis and Poinar (1984) reported that efficacy was influenced by the nematode species, placement of the application, stage of the nematode larvae and nematode depth in the soil in their greenhouse study against early- and late-instar larvae of *O. sulcatus* using *H. bacteriophora*, *Steinernema carpocapsae* (Weiser) and *Steinernema glaseri* (Steiner). When EPNs were applied to the soil surface, *H. bacteriophora* performed better than *Steinernema* spp. against early-instar larvae, but all three species were equally effective against late instars. When the late-instar larvae were 20 cm deep in the soil, only *S. glaseri* provided effective control from surface application. However, all species were equally effective against the late-instar larvae at 5, 10 and 20 cm depths when nematodes were injected 5 cm deep into the soil. Long *et al.* (2000) evaluated two isolates (L017 and L137) of *Steinernema kraussei* (Steiner) from the UK and commercial products based on *Steinernema feltiae* (Filipjev) and *Heterorhabditis megidis* Poinar, Jackson and Klein against early- and late-instar larvae and pupae of *O. sulcatus* at 6°C, 10°C and 18°C. *S. kraussei* isolate L137 was the most virulent with 79% and 83% infection in early-instar larvae at 6°C and 10°C, respectively. A higher efficacy of *S. kraussei* against second- and third-instar larvae than that of fourth- to seventh-instar larvae of *O. sulcatus* was also reported by Guy *et al.* (2017). In a follow-up study to Long *et al.* (2000), Willmott *et al.* (2002) found 81% control of *O. sulcatus* overwintering larvae in potted strawberry with the L137 isolate and found that nematodes survived prolonged winter field conditions (2.7°C average).

Table 18.1. A list of entomopathogenic nematodes tested against insect pests of strawberries.

Target pest	Order, family	Target stage	Nematode species	Target site	Reference(s)
Black vine weevil (*Otiorhynchus sulcatus* (F.))	Coleoptera, Curculionidae	Larva	*Heterorhabditis bacteriophora* Poinar	Soil	Bedding and Miller (1981); Georgis and Poinar (1984)
		Larva, pupa	*Heterorhabditis marelata* Liu and Berry, *H. bacteriophora*	Soil	Berry *et al*. (1997)
		Larva	*H. bacteriophora*	Soil	Susurluk and Ehlers (2008)
		Larva, pupa	*Heterorhabditis megidis* Poinar, Jackson and Klein	Soil	Long *et al*. (2000)
		Larva	*H. megidis*	Soil	Haukeland and Lola-Luz (2010)
		Larva	*Steinernema carpocapsae* (Weiser)	Soil	Georgis and Poinar (1984)
		Larva	*Steinernema glaseri* Steiner	Soil	Georgis and Poinar (1984)
		Larva, pupa	*Steinernema kraussei* (Steiner)	Soil	Long *et al*. (2000)
		Larva	*S. kraussei*	Soil	Willmott *et al*. (2002); Ansari *et al*. (2010); Haukeland and Lola-Luz (2010); Ansari and Butt (2013); Guy *et al*. (2017)
		Larva, pupa	*Steinernema feltiae* (Filipjev)	Soil	Long *et al*. (2000)
Strawberry root weevil (*Otiorhynchus ovatus* (L.))	Coleoptera, Curculionidae	Larva, pupa	*H. marelata*, *H. bacteriophora*	Soil	Berry *et al*. (1997)
Strawberry blossom weevil (*Anthonomus rubi* Herbst.)	Coleoptera, Curculionidae	Larva	*Steinernema* sp., *S. feltiae*	NA	Sigareva and Kharchenko (2018)
Diaprepes root weevil (*Diaprepes abbreviatus* (L.))	Coleoptera, Curculionidae	Larva	*Steinernema riobrave* Cabanillas, Poinar and Raulston	Soil	Renkema *et al*. (2021)
Strawberry root grub (*Temnorhynchus baal* (Reiche and Saulcy))	Coleoptera, Scarabaeidae	Larva, adult	*H. bacteriophora*	NA	Shamseldean and Atwa 2004
		Larva	*H. bacteriophora*	NA	Atwa and Hassan (2014)
		Larva	*H. bacteriophora*	NA and soil	Shehata *et al*. (2020)
		Larva	*Heterorhabditis indica* Poinar, Karunakar and David	Soil	Shehata *et al*. (2019)
		Larva, adult	*Steinernema* sp., *S. carpocapsae*, *Steinernema kushidai* Mamiya	NA	Shamseldean and Atwa (2004)
		Larva	*S. glaseri*	NA	Atwa and Hassan (2014)
Spotted-wing drosophila (*Drosophila suzukii* (Matsumura))	Diptera, Drosophilidae	Larva, pupa, adult	*S. carpocapsae*, *S. kraussei*, *S. feltiae*	NA	Cuthbertson *et al*. (2014)

		Larva, pupa, adult	*H. bacteriophora*, *Steinernema* sp. (*glaseri* group), *S. carpocapsae*, *S. feltiae*	Fruit	Garriga *et al*. (2018)
		Larva, adult	*H. bacteriophora*, *S. carpocapsae*, *S. feltiae*	Fruit	Woltz *et al*. (2015)
		Larva, pupa	*H. bacteriophora*, *S. carpocapsae*, *S. kraussei*, *S. feltiae*	Soil	Cuthbertson and Audsley (2016)
		Adult	*H. bacteriophora*, *S. carpocapsae*, *S. feltiae*	Soil	Garriga *et al*. (2020)
Western flower thrips (*Frankliniella occidentalis*)	Thysanoptera, Thripidae	Larva, adult	*S. feltiae*	Foliar	Renkema *et al*. (2018)
		Larva, prepupa, pupa	*H. bacteriophora*	Soil	Ebssa *et al*. (2001, 2004)
		Larva, prepupa, pupa	*H. bacteriophora*, *H. indica*	Soil	Ebssa *et al*. (2006)
		Larva, prepupa, pupa	*S. carpocapsae*, *S. feltiae*	Soil	Ebssa *et al*. (2001, 2004)

NA, not applicable; laboratory study done in Petri dishes or in cages.

However, in field-grown strawberry studies conducted in Ireland and Norway, the performance of *H. megidis* and *S. kraussei* varied depending on the temperature and soil type (Haukeland and Lola-Luz, 2010). Spring application of *H. megidis* was effective against *O. sulcatus* larvae, as soil temperatures above 10°C were ideal for this nematode species, but the cold-active *S. kraussei* was not as effective, regardless of the application rates and season.

In addition to choosing the correct species and application rates, other strategies can also help improve pest control efficacy. For example, a preconditioning step consisting of cold storage at 9°C for 3–6 weeks improved the efficacy of *H. megidis* and *S. carpocapsae* but did not have any effect on the cold-active *S. kraussei* against *O. sulcatus* in glasshouse studies (Guy *et al.*, 2017). Similarly, a combination of the entomopathogenic fungus *Metarhizium anisopliae* (Metschn.) Sorokin (strain V275) and the cold-active *S. kraussei* at a reduced rate had an additive or synergistic effect with 100% control of third-instar overwintering larvae of *O. sulcatus* in an unheated glasshouse in the UK (Ansari *et al.*, 2010). Application method and concentration influenced the efficacy of an EPN–entomopathogenic fungus combination in a multi-year study in which a low dose of *S. kraussei* and a commercial formulation of *Metarhizium brunneum* (Petch) provided 100% control of *O. sulcatus* larvae in field-grown strawberry (Ansari and Butt, 2013).

The root weevil *Diaprepes abbreviatus* (L.) is an occasional pest of strawberry in Florida. The adults feed on foliage and the larvae feed on the roots. A recent small-scale field study explored the potential of *Steinernema riobrave* Cabanillas, Poinar and Raulston as a control tool (Renkema *et al.*, 2021). EPNs were as effective as imidacloprid against the late-instar larvae in reducing plant wilting. Because pesticides like imidacloprid cannot be applied close to the fruit harvest, EPNs can be a good option with the flexibility to apply at any time during the crop season.

18.2.2 Strawberry blossom weevil

The strawberry blossom weevil, *Anthonomus rubi* Herbst. (Coleoptera: Curculionidae), is a major pest of strawberry, blackberry and raspberry in Europe, and controlling this pest with synthetic insecticides is challenging (Innocenzi *et al.*, 2001; Tonina *et al.*, 2021). By cutting the flower buds, adult females can cause severe yield losses. A Ukrainian laboratory study found that an isolate of *Steinernema* sp. and a commercial formulation of *S. feltiae* cause 100% mortality in *A. rubi* larvae 3 days after infection (Sigareva and Kharchenko, 2018). Although information on the field efficacy of EPNs against *A. rubi* is not currently available, EPNs could be a potential control option for this pest (Table 18.1).

18.2.3 Strawberry root grub

The strawberry root grub, *Temnorhynchus baal* (Reiche and Saulcy) (Coleoptera: Scarabaeidae), is a serious root-feeding pest of strawberry in Egypt (Atwa and Hassan, 2014). Multiple Egyptian studies have explored the potential of native strains of *Heterorhabditis indica* Poinar, Karunakar and David, *H. bacteriophora, S. carpocapsae, S. glaseri, Steinernema kushidai* Mamiya and *Steinernema* spp. against *T. baal* in laboratory or field studies evaluating the efficacy, application rate, type of soil and field persistence (Shamseldean and Atwa, 2004; Atwa and Hassan, 2014; Shehata *et al.*, 2019, 2020). Some of the findings of these studies (Table 18.1) include the following: nearly 100% mortality from *H. indica* populations in sandy and loamy soils; 26% infectivity from *H. bacteriophora* (Hb-EG strain) 5 months after application under field conditions; 52–93% mortality from *S. glaseri* depending on the application rate in sandy soils and 95–100% mortality depending on the larval stage in laboratory assays; and 28–45% mortality from *S. carpocapsae* (EGBX strain) and *H. bacteriophora* (EKB20 strain), 52–65% mortality from *S. kushidai* (EBN32 strain), 45–55% mortality from *Steinernema* sp. (EBNE strain) and 55–70% mortality from *Steinernema* sp. (EGG4 strain) in laboratory assays. They also found longer soil persistence and infectivity in soil up to 30 cm deep from *S. glaseri*. Overall, these studies provided valuable information on the selection of species and strains and their use.

18.2.4 Spotted-wing drosophila

The spotted-wing drosophila, *D. suzukii*, is an invasive pest native to south-eastern Asia that attacks soft-skinned fruits including raspberries, blueberries, blackberries, strawberries and cherries (Walsh *et al.*, 2011). Since 2008, this pest has spread rapidly through North and South America and Europe (Hauser, 2011; Cini *et al.*, 2012; Deprá *et al.*, 2014; Asplen *et al.*, 2015). Female flies possess a serrated, sclerotized ovipositor that allows them to oviposit in ripening and healthy fruit. Although oviposition can render fruit susceptible to secondary pathogen infection (Ioriatti *et al.*, 2018), damage is attributed mainly to the larvae feeding inside, which can result in unmarketable fruits due to the zero tolerance for larval-infested fruit in the fresh fruit market (Bruck *et al.*, 2011). Larvae feed inside the fruit and last instars come out, fall to the ground and pupate in the soil. The invasion of this pest has resulted in an increased use of insecticide sprays during the ripening period (van Timmeren and Isaacs, 2013). These issues have prompted the need for alternative methods of control, such as biological control (Lee *et al.*, 2019).

Earlier laboratory studies in the UK and USA in blueberries could not demonstrate the potential of *H. bacteriophora*, *S. carpocapsae*, *S. feltiae* and *S. kraussei* as good control options for *D. suzukii* compared with other options (Cuthbertson *et al.*, 2014; Woltz *et al.*, 2015). However, in a later assay with higher application rates, *H. bacteriophora* caused 95% larval mortality (Cuthbertson and Audsley, 2016). Another laboratory study in Spain showed 84% and 94% larval mortality on filter paper from *S. carpocapsae* and *S. feltiae*, respectively, and *S. feltiae* reduced the adult emergence from strawberry fruit by 35% (Garriga *et al.*, 2018). A follow-up laboratory study showed that soil application of *S. carpocapsae* caused 65–89% infections in emerging adults depending on their maturity stage (Garriga *et al.*, 2020).

18.2.5 Western flower thrips

The western flower thrips, *Frankliniella occidentalis* (Pergande) (Thysanoptera: Thripidae), can be a significant strawberry pest due to its direct damage and the role it plays in transmitting the virus decline disease of strawberry (Dara, 2018). Larvae and adults scrape the leaf and fruit surfaces and feed on the plant juices, and pupation occurs on the substrate. A study conducted in Florida strawberries with *S. feltiae* did not find EPN treatments effective against *F. occidentalis* compared with synthetic pesticides, possibly due to hot and dry weather conditions (Renkema *et al.*, 2018). However, depending on the EPN species, application rates, host densities and temperature, up to 60% mortality in the soil-dwelling stages of *F. occidentalis* was observed in laboratory studies conducted in Germany (Ebssa *et al.*, 2001, 2004). When EPNs were used in combination with predatory mites, control efficacy improved significantly against the foliage-feeding and soil-dwelling life stages of *F. occidentalis* compared with the individual application of EPNs and predatory mites (Ebssa *et al.*, 2006).

18.3 Caneberries

Caneberries (blackberries and raspberries) include a multitude of species in the genus *Rubus* (Rosales: Rosaceae) that are perennial crops grown for their high-value, soft-bodied fruits. Well-drained loam or sandy loam soils are ideal for caneberries and, depending on the local environmental conditions, they are grown in open fields or high tunnels, and supported by trellises. As for strawberries, harvesting of blackberry and raspberry crops continues for a few weeks, with harvest and crop care crews regularly present in the field. Therefore, application of pesticides poses a risk to these workers. As pollinators are typically used for ensuring optimal fruit yields, their presence also restricts the application of certain pesticides during the fruit production season. EPNs can be a good biological option for controlling four insect pests of caneberries in an integrated pest management (IPM) program in caneberries (Table 18.2).

18.3.1 Root weevils

In an Australian study, Bedding and Miller (1981) reported that *Heterorhabditis heliothidis* Khan, Brooks and Kirschmann (T327 isolate) caused 100% infection in *O. sulcatus* larvae in potted raspberry in a nursery. However, studies

Table 18.2. A list of entomopathogenic nematodes tested against insect pests of caneberries.

Target pest	Order, family	Target stage	Nematode species	Target site	Reference
Black vine weevil (*Otiorhynchus sulcatus* (F.))	Coleoptera, Curculionidae	Larva	*Heterorhabditis bacteriophora* Poinar	Soil	Bedding and Miller (1981)
			Steinernema carpocapsae (Weiser)	Soil	Booth *et al.* (2002)
Armadillo weevil (*Otiorhynchus armadillo* (Rossi) subsp. *obsitus*)	Coleoptera, Curculionidae	Larva	*Heterorhabditis megidis* Poinar, Jackson and Klein, *Steinernema feltiae* (Filipjev)	Soil	Grassi *et al.* (2003)
Raspberry crown borer (*Pennisettia marginata* (Harris))	Lepidoptera, Seiidae	Larva	*H. bacteriophora*, *S. carpocapsae*, *S. feltiae*	Soil	McKern *et al.* (2007)
Ground pearl (*Eurhizococcus colombianus* Jakubski)	Hemiptera, Margarodidae	Immature, adult female	*H. bacteriophora*, *Steinernema colombiense* López-Nuñez, Plichta, Gongora-Botero and Stock	Soil	Aristizábal *et al.* (2015)

conducted in red raspberry fields in Washington State with *H. bacteriophora* and *S. carpocapsae* did not cause a significant reduction in *O. sulcatus* populations (Booth *et al.*, 2002). Cooler soil temperature was thought to be the reason, among others, for the low EPN efficacy during this study. In an Italian study in a red raspberry field, a commercial formulation of *H. megidis* provided 60% control against a related species, *Otiorhynchus armadillo* (Rossi) subsp. *obsitus*, whereas *S. feltiae* provided only 17% control under mean soil temperatures of 14–15°C (Grassi *et al.*, 2003).

18.3.2 Raspberry crown borer

A native of North America, the raspberry crown borer, *Pennisetia marginata* (Harris) (Lepidoptera: Sesiidae), is a serious pest of caneberries (McKern *et al.*, 2007). The larvae bore into the crown or cane, and their feeding damage results in yield losses or the death of canes or the entire plant. A study conducted in Arkansas blackberry explored the potential of using several synthetic pesticides and EPNs against *P. marginata* (McKern *et al.*, 2007). Some synthetic pesticide applications resulted in more than 89% control of the caterpillars, while *H. bacteriophora*, *S. feltiae* and *S. carpocapsae* reduced larval numbers by 33%, 46% and 53%, respectively. While McKern *et al.* (2007) believed EPNs to be "less successful than insecticides," their results show that EPNs can be an important part of an IPM program.

18.3.3 Ground pearl

The ground pearl, *Eurhizococcus colombianus* Jakubski (Hemiptera: Margarodidae), is a pearly-looking root-feeding insect related to scales. It is a pest of various fruit crops and is a particularly significant pest of caneberries, where feeding damage reduces the plant's vigor (Aristizábal *et al.*, 2015). In a Colombian study in blackberry, a local strain of *Steinernema colombiense* López-Nuñez, Plichta, Gongora-Botero & Stock (SNIO198 strain) at a high concentration caused a significant reduction in adult female *E. colombianus* populations, but *H. bacteriophora* (Fresno strain) was not as effective (Aristizábal *et al.*, 2015). The synthetic pesticide carbofuran provided moderate control.

18.4 Blueberries

Blueberries are native to North America and belong to the genus *Vaccinium* (Ericales: Ericaceae). Several species of blueberries are grown commercially worldwide; they include lowbush blueberry (*Vaccinium angustifolium* Aiton, *V. boreale* Hall and Alders, *V. myrtilloides* Michaux., *V. pallidum* Aiton, and *V. tenellum* Aiton), southern (*V. darrowii* Camp) and northern (*V. corymbosum* L.) highbush blueberry, and rabbiteye blueberry (*V. ashei* Reade) (Vorsa, 1997). Blueberries are perennial crops that thrive on well-drained, acidic (pH 4.0–5.0), sandy to sandy loam soils with high organic matter content (Kender and Brightwell, 1966). This information is important because soil characteristics such as pH, soil type (e.g. sandy and sandy loam) and amount of organic matter can affect the infectivity and persistence of EPNs (Koppenhöfer and Fuzy, 2006). In fact, *S. glaseri* was the most abundant EPN in soils of commercial highbush blueberry farms in New Jersey; however, soils associated with wild blueberries contained higher EPN diversity than those associated with cultivated blueberries (Rivera *et al.*, 2016). Soils from highbush blueberry fields had higher soil moisture and organic matter and lower pH than soils associated with wild blueberries (Rivera *et al.*, 2016), which could explain the lower EPN diversity.

EPNs have been tested for their ability to control at least nine insect pests of blueberries (Table 18.3). Most of these pests belong to the order Coleoptera, which spend part of their life cycle as immatures in the soil. The most tested and effective EPNs for the control of these pests are *S. riobrave*, *S. carpocapsae* and *H. bacteriophora* (Table 18.3).

18.4.1 Spotted-wing drosophila

Several EPNs have been tested against *D. suzukii* adults, larvae and pupae in blueberries, including *S. riobrave*, *S. carpocapsae*, *S. feltiae*, *H. bacteriophora*, *H. amazonensis* (Andaló, Nguyen and Moino), *H. indica* and *Oscheius onirici* Torini (Table 18.3). These studies have targeted the fruit, both on the plant and on the ground, to control the larvae feeding inside, as well as the soil to control the larvae, pupae and emerging adults because third-instar larvae can leave the fruit to pupate in the soil (Woltz and Lee, 2017). Although these studies show the potential of EPNs to kill different *D. suzukii* stages, the results have, in general, been mixed. For example, *S. feltiae* and *S. carpocapsae* were more efficient at killing larvae in the soil, at application rates of 25–400 infective juveniles (IJs) cm^{-2}, than *H. bacteriophora* (Hübner *et al.*, 2017). In addition, 65% of *D. suzukii* adults emerging from the soil were infected with *S. carpocapsae*, whereas infection with *H. bacteriophora* and *S. feltiae* was less than 4% (Garriga *et al.*, 2020). In laboratory studies, although *S. carpocapsae* (IBCBn 02 isolate) could kill *D. suzukii* pupae, *H. indica* (IBCBn 05 isolate) and *H. amazonensis* (IBCBn 24 isolate) were more effective (de Brida *et al.*, 2019). However, another study showed little effect of *H. bacteriophora*, *S. feltiae* and *S. carpocapsae* on *D. suzukii* survival in the laboratory (at 100–200 IJs cm^{-2}) and in the field (at 100 IJs cm^{-2}) (Woltz *et al.*, 2015). These discrepancies could be due to differences in the application method, the EPN concentration and the target stage because *D. suzukii* pupae seem to be less susceptible to EPN attack than the larvae (Hübner *et al.*, 2017; Garriga *et al.*, 2020).

Recently, a rare nematode, *O. onirici* (Fig. 18.1A), discovered from Wisconsin wild marshlands was found to be highly virulent against *D. suzukii* larvae, reducing their survival by 78% with 10,000 IJs applied to infested berries (Foye and Steffan, 2020). Another promising EPN is *S. riobrave*, which was effective at killing 80–100% of *D. suzukii* larvae (Evans and Renkema, 2021). Nevertheless, additional large-scale field studies are needed to determine the potential of EPNs to control *D. suzukii*. It is also possible that the efficacy of these EPNs could be enhanced by combining them with other complementary biological control agents (Renkema and Cuthbertson, 2018).

18.4.2 Oriental beetle and other white grubs

The oriental beetle, *Anomala orientalis* (Waterhouse) (Coleoptera: Scarabaeidae), is part of a complex of white grub species, which includes the Japanese beetle (*Popillia japonica* Newman), the Asiatic garden beetle (*Maladera castanea*

Table 18.3. A list of entomopathogenic nematodes tested against insect pests of blueberries.

Crop	Target pest	Order, family	Target stage	Nematode species	Target site	Reference(s)
Blueberry (all)	Spotted-wing drosophila (*Drosophila suzukii* (Matsumura))	Diptera, Drosophilidae	Larva, pupa, adult	*Steinernema riobrave* Cabanillas, Poinar and Raulston	Fruit	Evans and Renkema (2021)
				Oscheius onirici Torini	Fruit	Foye and Steffan (2020)
				Steinernema carpocapsae (Weiser)	Soil	Garriga *et al.* (2020) Hübner *et al.* (2017)
				Heterorhabditis bacteriophora (Poinar)	Fruit	Renkema and Cuthbertson (2018)
				Steinernema feltiae (Filipjev)	Soil	Hübner *et al.* (2017)
				Heterorhabditis amazonesis (Andaló, Nguyen and Moino), *Heterorhabditis indica* (Poinar, Karunakar and David), *S. carpocapsae*, *S. feltiae*	NA	de Brida *et al.* (2019)
	Western flower thrips (*Frankliniella occidentalis* (Pergande))	Thysanoptera, Thripidae	Nymph	*Steinernema yirgalemense* Nguyen, Tesfamariam, Gozel, Gaugler and Adams	Soil, foliar	Dlamini *et al.* (2020)
Blueberry (northern highbush)	Oriental beetle (*Anomala orientalis* (Waterhouse))	Coleoptera, Scarabaeidae	Larva	*Steinernema scarabaei* Stock and Koppenhöfer	Soil	Polavarapu *et al.* (2007)
	Plum curculio (*Conotrachelus nenuphar* (Herbst))	Coleoptera, Curculionidae	Larva	*S. riobrave*	Soil	Sousa *et al.* (2022)
	Blueberry mealybug (*Dysmicoccus vaccinii* (Miller and Polavarapu))	Hemiptera, Pseudococcidae	Nymph	*H. bacteriophora*, *Heterorhabditis hawaiiensis* Gardner, Stock and Kaya, *H. indica*	Soil	Polavarapu *et al.* (1996); Stuart *et al.* (1997)
Blueberry (southern highbush)	Diaprepes root weevil (*Diaprepes abbreviatus* (L.))	Coleoptera, Curculionidae	Larva	*S. riobrave*	Soil	Phillips *et al.* (2019)
	White grubs (*Cyclocephala longula* (LeConte))	Coleoptera, Scarabaeidae	Larva	*H. bacteriophora*	Soil	Haviland and Hernandez (2012)
Blueberry (lowbush)	Blueberry flea beetle (*Altica sylvia* Malloch)	Coleoptera, Chrysomelidae	Larva	*S. riobrave*, *S. carpocapsae*	Foliar	Drummond and Groden (2000)
	Blueberry spanworm (*Itame argillacearia* (Packard))	Lepidoptera, Geometridae	Larva	*S. carpocapsae*	Foliar	Drummond and Groden (2000)

NA, not applicable; laboratory study done in Petri dishes.

(Arrow)) and the white grub *Cyclocephala longula* (LeConte) that attack blueberries in the USA. In the north-eastern USA, *A. orientalis* has become the most important white grub pest of highbush blueberries (Polavarapu, 1996), although it can cause losses to other berry crops such as cranberries (Averill and Sylvia, 1998). Like other white grub species of economic importance, *A. orientalis* has a 1-year life cycle. Adults live about 2 weeks and do not cause any crop damage. In New Jersey, adult flight occurs from early June to early August and peaks in late June to early July (Polavarapu, 1996). Females lay eggs around the roots of plants, and the larvae feed on the roots and become third instars by September; the larvae overwinter as third instars and pupate in late spring.

Currently, the only chemical control option to manage root-feeding white grubs in blueberries is the neonicotinoid insecticide imidacloprid (Polavarapu *et al.*, 1998). However, imidacloprid is effective only when applied to early instars (Potter, 1998), and concerns exist on their potential negative effects on pollinators (Blacquiere *et al.*, 2012). Therefore, effective environmentally safe and IPM-compatible alternatives are needed to manage white grubs in blueberries, particularly those that can be effective against late instars such as curative spring or fall applications of EPNs. In one study, two EPNs, *Steinernema scarabaei* Stock and Koppenhöfer (Fig. 18.1B) and *H. bacteriophora*, were tested in greenhouse and field studies against third-instar *A. orientalis* (Polavarapu *et al.*, 2007). The researchers showed that *S. scarabaei* at 200–750 × 10^3 IJs m^{-2} provided 50–95% larval control, whereas *H. bacteriophora* was ineffective, confirming previous findings by Polavarapu

Fig. 18.1. (A) Close-up view of the entomopathogenic nematode *Oscheius onirici*. Photograph courtesy of Shawn Steffan, University of Wisconsin-Madison. (B) Oriental beetle (*Anomala orientalis*) larva infected by *Steinernema scarabaei*. Photograph courtesy of Albrecht Koppenhöfer, Rutgers University. (C) Red-banded flea beetle (*Systena frontalis* F.) adult with many *O. onirici* infective juveniles emerging from the cadaver. Photograph courtesy of Shawn Steffan and Shane Foye, University of Wisconsin-Madison.

et al. (1998) that *H. bacteriophora* does not provide *A. orientalis* larval control. In addition, previous studies showed that combined applications of EPNs and imidacloprid can work synergistically to increase the mortality of third-instar *A. orientalis* (Koppenhöfer and Kaya, 1998; Koppenhöfer *et al.*, 2020). However, there was no effect of combining imidacloprid with *S. scarabaei* or *H. bacteriophora* at 5 billion IJs ha^{-1} compared with imidacloprid or nematode treatments alone (Polavarapu *et al.*, 2007). For various reasons that included the development of a practical and economic method for mass rearing, a product containing *S. scarabaei* (Nemagard™; Purely Organic Products LLC, Portsmouth, New Hampshire, USA) has only recently become commercially available. In contrast, *H. bacteriophora* (Terranem™; Koppert Biological Systems, Inc., Romulus, Michigan, USA) applied at 1 billion IJs ha^{-1} was shown to effectively control *C. longula* larvae, the predominant white grub species damaging southern highbush blueberries in California (Haviland and Hernandez, 2012). Although nematode applications in April killed only 8.3% of *C. longula* third-instar larvae, they led to an epizootic within the grub population that effectively controlled the following generation. This study suggests that long-term field-level evaluations of EPN persistence in the soil are needed to better understand their potential as biological control agents of white grubs in blueberries.

18.4.3 Plum curculio

Native to North America, the plum curculio, *Conotrachelus nenuphar* (Herbst) (Coleoptera: Curculionidae), is considered a major pest of pome and stone fruits including apples, peaches, plum, pears, cherries and apricots in the eastern USA (Racette *et al.*, 1992; Lampasona *et al.*, 2020). Although *C. nenuphar* is not considered a pest in most blueberry-growing regions, populations can reach levels that cause economic damage in regions such as New Jersey (Polavarapu *et al.*, 2004). Adults overwinter in leaf litter and become active during spring, moving from the overwintering sites into blueberry fields during bloom (in April–May) in New Jersey (Tewari *et al.*, 2014). Females lay eggs on young, green fruit, causing characteristic crescent-shaped oviposition scars (Mampe and Neunzig, 1967). After hatching, larvae develop inside the fruit. The infested fruit turn blue prematurely and usually drop to the ground before harvest (Antonelli *et al.*, 1992). Mature larvae exit the fruit to pupate in the soil, and the next generation of adults emerges in July and August (Tewari *et al.*, 2014). Newly emerged adults eventually move to their overwintering sites for hibernation.

Recently, Piñero *et al.* (2020) described a multiple life-stage approach to manage *C. nenuphar* in apples. This approach consists of combining low-risk strategies to control the adults, such as the use of an attract-and-kill approach that combines odor-baited trap trees with insecticide sprays (Leskey *et al.*, 2008), with the use of EPNs to control the soil-inhabiting larvae and pupae. They found a low emergence of pest adults from soils treated with *S. riobrave* (355 strain) applied at 100–200 IJs cm^{-2} (Piñero *et al.*, 2020). A similar strategy to manage multiple life stages of *C. nenuphar* is being explored in blueberries. Early studies showed the feasibility of using odor-baited trap bushes to concentrate *C. nenuphar* activity and subsequent injury in blueberry fields (Rodriguez-Saona *et al.*, 2019). In blueberries, *S. riobrave* also reduced adult emergence from the soil (Sousa *et al.*, 2022) (Table 18.3). Overall, studies from apples and blueberries show a high virulence of *S. riobrave* against *C. nenuphar* larvae (Shapiro-Ilan *et al.*, 2004, 2011, 2013). Future studies in blueberries need to evaluate the efficacy of this EPN in reducing *C. nenuphar* populations in commercial farms.

18.4.4 Other pests of blueberries

Other insect pests of blueberries have been targeted for control by EPNs (Table 18.3); three of these species, namely the western flower thrips (*F. occidentalis*), the blueberry mealybug (*Dysmicoccus vaccinii* (Miller and Polavarapu)), and the root weevil *D. abbreviatus*, attack highbush blueberries and spend part of their life cycle in the soil; thus, they can be targeted by soil applications of EPNs. In contrast, the blueberry flea beetle (*Altica sylvia* Malloch; Coleoptera: Chrysomelidae) and the blueberry spanworm (*Itame argillacearia* (Packard); Lepidoptera: Geometridae) are pests of lowbush blueberries and are the only pests targeted for foliar EPN applications (Table 18.3), likely due to differences in crop characteristics. Compared with highbush blueberries, lowbush blueberries are smaller plants that grow in dense canopies closer to the ground, which could allow more suitable environments for EPNs to persist in the canopy.

18.5 Cranberries

The American cranberry, *Vaccinium macrocarpon* Aiton (family Ericaceae), is a perennial crop native to North America that is commercially grown in acidic, sandy and wet to moist soils with high organic matter (Eck, 1990). The cranberry plants are woody, low-growing vines that form a thick, continuous mat over the surface of cultivated beds also known as bogs or marshes. Cranberry beds can thus provide suitable environmental conditions for EPNs because the soil around the roots is often moist and humid and is protected from harmful ultraviolet light and temperatures.

Studies on the use of EPNs for insect control in cranberries were initiated about three decades ago. These EPNs have been tested for control of at least eight insect pests of cranberries belonging to the orders Coleoptera and Lepidoptera (Table 18.4). The focus of these studies has been the control of life stages inhabiting the soil (Table 18.4). Various species of EPNs from the genus *Steinernema* and *Heterorhabditis* have been tested and shown to be effective for controlling these pests. Undoubtedly, the most successful and widely adopted case in cranberries is the use *S. carpocapsae* for control of the cranberry girdler, *Chrysoteuchia topiaria* (Zeller) (Lepidoptera: Crambidae). Notably, an EPN species discovered recently in Wisconsin cranberry marshland, *O. onirici* (Fig. 18.1A), has been shown to have broad virulence against several cranberry pests (Table 18.4) (Ye *et al.*, 2018; Foye and Steffan, 2019).

18.5.1 Root weevils

Two species of root-feeding weevils (Coleoptera: Curculionidae), the black vine weevil (*O. sulcatus*) and the strawberry root weevil (*O. ovatus*), are important pests of cranberries (Warner and Negley, 1976; Averill and Sylvia, 1998). The adults feed on the foliage of a wide variety of plants; however, in cranberries, the damage caused by adults is unimportant (Averill and Sylvia, 1998). The larvae of these weevils feed on the roots, which can result in the death of the vines and yield reduction. They have one generation per year. Females are nocturnal and active from May to August. The eggs are laid from late June to August, and the larvae feed on the cranberry roots until the plants become dormant. The larvae overwinter and pupate in the spring.

Due to the polyphagous habits of *Otiorhynchus* spp., many early studies tested the potential of EPNs against these pests (Bedding and Miller, 1981; Georgis and Poinar, 1984; Klingler, 1988). Shanks and Agudelo-Silva (1990) were the first to test the potential of EPNs to manage *Otiorhynchus* spp. in cranberries. They found that *H. heliothidis* applied to cranberries at 160 IJs cm^{-2} reduced *O. sulcatus* larvae and pupae by more than 70%, and persisted in the soil for at least 10 months and were as effective as *H. bacteriophora* (HP88 isolate) and *Neoaplectana carpocapsae* Weiser (= *S. feltiae* (Filipjev)) (All strain). Later, Simser and Roberts (1994) reported *O. ovatus* larval mortality as high as 69% and 74% with applications of *H. bacteriophora* (NC-1 strain) and *S. carpocapsae* (A11 strain), respectively. Booth *et al.* (2002) also showed a high (96–100%) reduction of *O. sulcatus* populations after applications of *S. carpocapsae* and *S. glaseri* at 30 IJ cm^{-2}; however, *S. carpocapsae* persisted at higher levels and for a longer period than *S. glaseri*. These studies suggest that several heterorhabditid and steinernematid nematode species and strains have the potential to control root weevil pests of cranberries (Table 18.4).

18.5.2 Cranberry white grubs and other scarabs

Several species of scarabs (Coleoptera: Scarabaeidae) feed on the root systems of cranberries including the oriental beetle (*A. orientalis*), the cranberry root grub (*Lichnanthe vulpina* Hentz), the cranberry white grubs (*Phyllophaga georgiana* (Horn) and *Phyllophaga anxia* (LeConte)), the striped colaspis (*Colaspis costipennis* Crotch),

Table 18.4. A list of entomopathogenic nematodes tested against insect pests of cranberries.

Target pest	Order, family	Target stage	Nematode species	Target site	Reference
Black vine weevil (*Otiorhynchus sulcatus* (F.))	Coleoptera, Curculionidae	Larva, pupa	*Steinernema carpocapsae* (Weiser)	Soil	Booth *et al.* (2002)
		Larva, pupa	*Steinernema glaseri* Steiner, *Heterorhabditis bacteriophora* Poinar, *Heterorhabditis* sp. (HP88 isolate), *Neoaplectana carpocapsae* Weiser (= *Steinernema feltiae* (Filipjev))	Soil	Shanks and Agudelo-Silva (1990)
Strawberry root weevil (*Otiorhynchus ovatus* (L.))	Coleoptera, Curculionidae	Larva, pupa	*S. carpocapsae*, *H. bacteriophora*	Soil	Simser and Roberts (1994)
Cranberry white grub (*Phyllophaga georgiana* Horn)	Coleoptera, Scarabaeidae	Larva	*Steinernema scarabaei* Stock and Koppenhöfer	Soil	Koppenhöfer *et al.* (2008)
Cranberry rootworm (*Rhabdopterus picipes* (Olivier))	Coleoptera, Chrysomelidae	Larva	*H. bacteriophora*	Soil	Stuart and Polavarapu (1997)
		Larva	*S. glaseri*	Soil	Polavarapu *et al.* (2000)
Red-banded flea beetle (*Systena frontalis* F.)	Coleoptera, Chrysomelidae	Larva, pupa	*Oscheius onirici* Torini, *Heterorhabditis georgiana* Nguyen	Soil	Foye and Steffan (2019)
Cranberry girdler (*Chrysoteuchia topiaria* (Zeller))	Lepidoptera, Crambidae	Larva	*Heterorhabditis megidis* Poinar, Jackson and Klein	Soil	Simard *et al.* (2002)
					Henderson and Singhai (1999)
		Larva	*H. bacteriophora*	Soil	
		Larva	*Heterorhabditis marelatus* Liu and Berry	Soil	Berry and Liu (1998)
		Larva	*S. feltiae*	Soil	Smith *et al.* (1993)
		Larva	*S. carpocapsae*, *S. glaseri*	Soil	Dapsis (1993)
Sparganothis fruitworm (*Sparganothis sulfureana* Clemens)	Lepidoptera, Tortricidae	Larva	*O. onirici*, *H. georgiana*	Soil	Foye and Steffan (2019)

and two *Hoplia* spp. (*Hoplia modesta* (Haldeman) and *Hoplia equina* (LeConte)) (Averill and Sylvia 1998). Larval feeding causes stunting and, eventually, death of plants, resulting in bare patches that are often colonized by weeds.

Early studies by Dapsis (1991) reported that *S. feltiae* was ineffective in suppressing *L. vulpina* populations. Later, Weber and Henderson (1998) reported about 20% mortality of *L. vulpina* larvae with *H. bacteriophora*, although as many as 60% of the recovered grubs had nematode infections, suggesting a strong immune response of grubs against the nematode. Low and high concentrations of *H. bacteriophora* provided 30% and 60% larval mortality, respectively, of *Hoplia* spp. (Weber and Henderson, 1998).

Phyllophaga spp. require 2 or 3 years to complete development (Vittum *et al.*, 1999; Koppenhöfer *et al.*, 2008). Adults are active in June–July (Robbins *et al.*, 2009), and young larvae occur during July–August. They seem to overwinter as second instars, which resume feeding the following year. In the following year, third instars enter the prepupal stage in the fall, overwinter, and pupate in the spring (Koppenhöfer *et al.*, 2008). The cranberry white grub, *P. anxia*, is the most common among the scarab grubs infesting cranberries in the USA. However, in New Jersey, *P. georgiana* seems to be the most abundant white grub in cranberries (Robbins *et al.*, 2009). Early evaluations showed no or limited effects of *S. carpocapsae*, *S. glaseri*, *S. feltiae*, or *H. bacteriophora* on white grub populations (Dapsis, 1991, 1993; Dittl, 1996; Weber and Henderson, 1998). More recent laboratory and greenhouse studies compared the virulence of *H. bacteriophora* (GPS11 strain), *Heterorhabditis zealandica* Poinar (X1 strain), and *S. scarabaei* (AMK001 strain) against second- and third-instar *P. georgiana* (Koppenhöfer *et al.*, 2008). Among the three EPN species, *S. scarabaei* was the most effective with 76–100% larval control at a rate of 2.5×10^9 IJs ha^{-1} in the greenhouse. In the laboratory, *S. scarabaei* was more effective against third instars than against second instars but not in the greenhouse. These results suggest that *S. scarabaei* may have the potential for management of *Phyllophaga* spp. in cranberry. A product (Nemagard™) that contains *S. scarabaei* as the active ingredient is now sold commercially. Additional research is needed to determine whether *S. scarabaei*, and other promising EPN species or strains, can control the full complex of white grub species feeding on cranberries under field conditions.

18.5.3 Cranberry girdler

The cranberry girdler (also known as the subterranean sod webworm), *C. topiaria*, causes vine death in cranberries by feeding on subterranean vines and girdling and sometimes severing them (Franklin, 1950; Roberts and Mahr, 1986). This insect is found throughout the USA and Canada and is also considered a pest of turfgrass, grass seed production and coniferous seedling plants (Potter, 1998; Hodgson and Roe, 2007). It has one generation per year (Averill and Sylvia, 1998). The adults emerge in mid-June and are active for 6–8 weeks. The larvae, the damaging life stage, soon after emergence burrow into the thatch layer or upper soil layer to feed on the stem and runners and to a lesser extent on the roots; they then continue to feed for about 2 months (July–mid-September). In the fall, the larvae spin a cocoon or hibernacula and overwinter in the soil (Hodgson and Roe, 2007). Pupation takes place in the spring (usually May–early June) (Averill and Sylvia, 1998).

Limited chemical control options exist against *C. topiaria* larvae and most are ineffective because the insecticide needs to penetrate the thatch layer. Thus, EPNs offer an effective alternative to chemical control (Table 18.4). Initial work on *C. topiaria* was conducted by Dapsis (1993) who demonstrated 60% *S. carpocapsae* infection of newly emerged larvae in laboratory assays and 44–88% larval mortality in field trials. Additional field studies by Smith *et al.* (1993) showed that applications of *S. carpocapsae* reduced *C. topiaria* larval populations by 92%. Subsequently, Berry and Liu (1998) reported a reduction of larval populations by *H. marelata* and *H. bacteriophora*. Henderson and Singhai (1999) compared the efficacy of *H. bacteriophora*, *S. carpocapsae* and *S. kraussei* against *C. topiaria* and found significantly lower larval populations in plots treated with *H. bacteriophora* and *S. carpocapsae*, but *S. kraussei* had no effect. In Petri dish experiments, although all four EPN species tested were good at killing *C. topiaria* larvae, *H. megidis* and *S. glaseri* were more virulent (significantly lower 50% lethal concentration values) than *S. carpocapsae* and *S. feltiae* (Simard *et al.*, 2002). Products containing *S. carpocapsae* are currently sold commercially for control of *C. topiaria* at recommended rates of 7.5 billion IJs ha^{-1} (e.g. TerraLink Horticulture Inc., Abbotsford, British Columbia, Canada).

18.5.4 Cranberry rootworm

The cranberry rootworm, *Rhabdopterus picipes* (Olivier) (Coleoptera: Chrysomelidae), has a 1-year life cycle. The adults feed on the cranberry foliage in late spring to early summer (early June–July), causing holes in the leaves and damaging new growth (Olivier and Chapin, 1980; Averill and Sylvia, 1998). However, larval feeding damage to the roots is most important in cranberries due to potential yield reduction and economic losses. The larvae move down in the soil in the fall to overwinter and pupate in the spring (late May). This pest can also be a serious pest of ornamentals (Olivier and Chapin, 1980).

Stuart and Polavarapu (1997) evaluated several species of steinernematids and heterorhabditids against *R. picipes* larvae in laboratory (Petri dish) assays and found that *H. bacteriophora* and *S. glaseri* at 50–500 IJs per dish infected and killed them successfully. However, heterorhabditids were generally more effective than steinernematids (Polavarapu, 1999). In field trials, *H. bacteriophora* at 5×10^9 IJs ha^{-1} reduced *R. picipes* larval populations by 50–80% compared with the untreated control and provided similar control to an imidacloprid treatment (Polavarapu *et al.*, 2000). Overall, these laboratory and field results indicate that *H. bacteriophora* is a suitable biological control agent to manage this pest.

18.5.5 Other pests of cranberries

Other cranberry pests that have been targeted by the use of EPNs, albeit to a much lesser extent, are the red-banded flea beetle (*Systena frontalis* F.; Coleoptera: Chrysomelidae) (Fig. 18.1C), the Sparganothis fruitworm (*Sparganothis sulfureana* Clemens; Lepidoptera: Tortricidae) and the cranberry fruitworm (*Acrobasis vaccinii* Riley; Lepidoptera: Pyralidae) (Table 18.4) (Foye and Steffan, 2019). Two EPNs, namely *O. onirici* and *Heterorhabditis georgiana* Nguyen, resulted in 75% and 65% *S. sulfureana* larval mortality, respectively, in laboratory experiments (Foye and Steffan, 2019). Both EPNs also reduced *S. frontalis* populations under field conditions (Foye and Steffan, 2019).

18.6 Conclusions

Many opportunities exist for the use of EPNs in berry crops; however, a number of challenges are also present. Soil conditions, such as water content, structure, cover (shading and insulation) and temperature, can vary widely across berry farms, both locally and regionally, which can influence EPN efficacy. For example, highbush blueberry farmers in New Jersey maintain bare soils, which could reduce the efficacy of EPNs by increased exposure to ultraviolet light from the sun. Organic farmers are more likely to practice cultural control tactics such as middle-row cover crops that can protect EPNs from direct sunlight exposure. This might also be true for lowbush blueberry and cranberry farmers because the thicker plant canopy can protect EPNs from the sun. EPNs can also die if the soil is too dry; this information is important because soil moisture in berry farms can vary, largely based on the type of irrigation (i.e. drip or overhead). In fact, sandy soils typical of blueberry farms need to be irrigated regularly after an EPN application to maintain the water-filled pores to allow EPN movement. Soil moisture might be less of a problem in crops such as cranberries where water content in beds is usually high. This suggests that the use of EPNs need to be tailored to specific soil conditions depending on the region and farming practices. The efficacy of EPNs in berry crops can also be affected by the method of application such as via sprinklers or boom sprayers (Hayes *et al.*, 1999). However, this equipment is usually readily available in most berry farms for irrigation and pest management, which could facilitate farmer adoption.

For berry farmers to adopt EPNs for insect pest control, the cost of products needs to be comparable to other methods of control and the product needs to be easy to apply (i.e. it uses already-existing farm equipment). Most berry farmers are likely to be unfamiliar with the use of EPNs because they are rarely used for insect pest control; thus, extension personnel need to inform farmers on their efficacy and use (see Chapter 13, this volume). Because insect pests do not always overlap in their phenology, particularly at the life stages vulnerable to EPN infection, and because EPN species or strains do not always provide the same level of control against a complex of pest species, farmers will likely need to tailor specific EPN species/strains for specific insect pests and apply them at different times during the growing season. In general, berry farmers are most likely to adopt the use of EPNs if there are limited alternative options for the control of insect pests. This is mostly true for root-feeding pests because there are restrictions on the use of chemical controls applied to the soil due to contamination concerns. Further studies are needed to optimize the use of EPNs including strain enhancement, formulations and application technology. Despite these challenges, berry farmers are likely to invest in effective alternatives to synthetic pesticides, such as the use of EPNs, in high-value commodity crops such as strawberries, caneberries, blueberries and cranberries.

18.7 References

Ansari, M.A. and Butt, T.M. (2013) Influence of the application methods and doses on the susceptibility of black vine weevil larvae *Otiorhynchus sulcatus* to *Metarhizium anisopliae* in field-grown strawberry. *BioControl* 58, 257–267.

Ansari, M.A., Shah, F.A. and Butt, T.M. (2010) The entomopathogenic nematode *Steinernema kraussei* and *Metarhizium anisopliae* work synergistically in controlling overwintering larvae of the black vine weevil, *Otiorhynchus sulcatus*, in strawberry growbags. *Biocontrol Science and Technology* 20, 99–105.

Antonelli, A., Elsner, E. and Shanks, C. (1992) Arthropod management. In: Pritts, M.P. and Hancock, J.F. (eds) *Highbush Blueberry Production Guide*. Northeast Regional Agricultural Engineering Service-55. Cooperative Extension, Ithaca, New York, pp. 55–75.

Aristizábal, L.F., Ortiz, A.L., Quintero, J.C., López-Nuñez, J.C., Toro, H. and Arthurs, S.P. (2015) Effect of Colombian strains of *Steinernema colombiense* (Rhabditida: Steinernematidae) and *Heterorhabditis bacteriophora* (Rphabditida: Heterorhabditidae) against *Eurhizococcus colombianus* (Hemiptera: Margarodidae) and *Aeneolamia* sp. (Hemiptera: Cercopidae). *Florida Entomologist* 98, 981–983.

Asplen, M.K., Anfora, G., Biondi, A., Choi, D.-S., Chu, D. *et al.* (2015) Invasion biology of spotted wing Drosophila (*Drosophila suzukii*): a global perspective and future priorities. *Journal of Pest Science* 88, 469–494.

Atwa, A.A. and Hassan, S.H. (2014) Bioefficacy of two entomopathogenic nematodes against *Spodoptera littoralis* Boisduval (Lepidoptera) and *Temnorhynchus baal* Reiche (Coleoptera) larvae. *Journal of Biopesticides* 7, 104–109.

Averill, A.L. and Sylvia, M.M. (1998) *Cranberry Insects of the Northeast*. UMass Cranberry Experimental Station, East Wareham, Massachusetts.

Bedding, R.A. and Miller, L.A. (1981) Use of a nematode, *Heterorhabditis heliothidis*, to control black vine weevil, *Otiorhynchus sulcatus*, in potted plants. *Annals of Applied Biology* 99, 211–216.

Berry, R.E. and Liu, J. (1998) *Insect Control in Cranberry with Indigenous Entomopathogenic Nematodes*. Cranberry Institute, Wareham, Massachusetts.

Berry, R.E., Liu, J. and Growth, E. (1997) Efficacy and persistence of *Heterorhabditis marelatus* (Rhabditida: Heterorhabditidae) against root weevils (Coleoptera: Curculionidae) in strawberry. *Biological Control* 26, 465–470.

Blacquiere, T., Smagghe, G., van Gestel, C.A.M. and Mommaerts, V. (2012) Neonicotinoids in bees: a review on concentrations, side-effects and risk assessment. *Ecotoxicology* 21, 973–992.

Booth, S.R., Tanigoshi, L.K. and Shanks, C.H. (2002) Evaluation of entomopathogenic nematodes to manage root weevil larvae in Washington State cranberry, strawberry, and red raspberry. *Environmental Entomology* 31, 895–902.

Bruck, D.J., Bolda, M., Tanigoshi, L., Klick, J., Kleiber, J. *et al.* (2011) Laboratory and field comparisons of insecticides to reduce infestation of *Drosophila suzukii* in berry crops. *Pest Management Science* 67, 1375–1385.

Cini, A., Ioriatti, C. and Anfora, G. (2012) A review of the invasion of *Drosophila suzukii* in Europe and a draft research agenda for integrated pest management. *Bulletin of Insectology* 65, 149–160.

Cuthbertson, A.G.S. and Audsley, N. (2016) Further screening of entomopathogenic fungi and nematodes as control agents for *Drosophila suzukii*. *Insects* 7: 24.

Cuthbertson, A.G.S., Collins, D.A., Blackburn, L.F., Audsley, N. and Bell, H.A. (2014) Preliminary screening of potential control products against *Drosophila suzukii*. *Insects* 5, 488–498.

Dapsis, L. (1991) *White Grub and Cranberry Root Grub Trials*. Cranberry Agricultural Research, Ocean Spray Cranberries, Lakeville-Middleborough, Massachusetts.

Dapsis, L. (1993) *Management of Cranberry Girdler and Cranberry White Grub with Entomopathogenic Nematodes*. Cranberry Agricultural Research, Ocean Spray Cranberries, Lakeville-Middleborough, Massachusetts.

Dara, S.K. (2018) Virus decline of strawberry in California and the role of insect vectors and associated viruses. *Plant Health Progress* 16, 211–215.

de Brida, A.L., Wilcken, S.R.S., Leite, L.G. and Garcia, F.R.M. (2019) Virulence of entomopathogenic nematode to pupae and adults of *Drosophila suzukii* in laboratory. *Revista Cientifica Rural* 21, 126–138.

Deprá, M., Poppe, J.L., Schmitz, H.J., de Toni, D.C. and Valente, V.L.S. (2014) The first records of the invasive pest *Drosophila suzukii* in the South American continent. *Journal of Pest Science* 87, 379–383.

Dittl, T. (1996) *Management of Cranberry White Grub in Wisconsin with Entomopathogenic Nematodes*. Cranberry Agricultural Research, Ocean Spray Cranberries, Lakeville-Middleborough, Massachusetts.

Dlamini, T.M., Allsopp, E. and Malan, A.P. (2020) Application of *Steinernema yirgalemense* to control *Frankliniella occidentalis* (Thysanoptera: Thripidae) on blueberries. *Crop Protection* 128: 105016.

Drummond, F.A. and Groden, E. (2000) *Evaluation of Entomopathogens for Biological Control of Insect Pests of Lowbush (Wild) Blueberry*. Technical Bulletin 172. Maine Agricultural and Forest Experiment Station, University of Maine, Orono, Maine.

Ebssa, L., Borgemeister, C., Berndt, O. and Poehling, H.-M. (2001) Efficacy of entomopathogenic nematodes against soil-dwelling life stages of western flower thrips, *Frankliniella occidentalis* (Thysanoptera: Thripidae). *Journal of Invertebrate Pathology* 78, 119–127.

Ebssa, L., Borgemeister, C. and Poehling, H.-M. (2004) Effectiveness of different species/strains of entomopathogenic nematodes for control of western flower thrips (*Frankliniella occidentalis*) at various concentrations, host densities, and temperatures. *Biological Control* 29, 145–154.

Ebssa, L., Borgemeister, C. and Poehling, H.-M. (2006) Simultaneous application of entomopathogenic nematodes and predatory mites to control western flower thrips *Frankliniella occidentalis*. *Biological Control* 39, 66–74.

Eck, P. (1990) *The American Cranberry*. Rutgers University Press, New Brunswick, New Jersey.

Evans, B.G. and Renkema, J.M. (2021) Initial tests of *Steinernema riobrave* against spotted wing drosophila, *Drosophila suzukii* Matsumura (Diptera: Drosophilidae) and comparison to other entomopathogenic nematode species in the laboratory. *Biocontrol Science and Technology* 31, 336–342.

Foye, S. and Steffan, S. (2019) Two native Wisconsin nematodes represent virulent biocontrol agents in cranberries. *Biological Control* 138: 104042.

Foye, S. and Steffan, S.A. (2020) A rare, recently discovered nematode, *Oscheius onirici* (Rhabditida: Rhabditidae), kills *Drosophila suzukii* (Diptera: Drosophilidae) within fruit. *Journal of Economic Entomology* 113, 1047–1051.

Franklin, H.J. (1950) *Cranberry Insects in Massachusetts, Parts II–VII*. Bulletin 445. Massachusetts Agricultural Experiment Station, Cape Cod, Massachusetts.

Garriga, A., Morton, A. and Garcia-del-Pino, F. (2018) Is *Drosophila suzukii* as susceptible to entomopathogenic nematodes as *Drosophila melanogaster*? *Journal of Pest Science* 91, 789–798.

Garriga, A., Morton, A., Ribes, A. and Garcia-del-Pino, F. (2020) Soil emergence of *Drosophila suzukii* adults: a susceptible period for entomopathogenic nematodes infection. *Journal of Pest Science* 93, 639–646.

Georgis, R. and Poinar, G.O. Jr (1984) Greenhouse control of the black vine weevil *Otiorhynchus sulcatus* (Coleoptera: Curculionidae) by heterorhabditid and steinernematid nematodes. *Environmental Entomology* 13, 1138–1140.

Grassi, A., Maines, R. and Zini, M. (2003) Field application and effectiveness of commercial entomopathogenic nematode formulations against *Otiorhynchus armadillo* subsp. *obsitus* Gyllenhal (Coleoptera: Curculionidae) larvae on raspberry. *IOBC-WPRS Bulletin* 26, 61–66.

Guy, A., Gaffney, M., Kapranas, A. and Griffin, C.T. (2017) Conditioning the entomopathogenic nematodes *Steinernema carpocapsae* and *Heterorhabditis megidis* by pre-application storage improves efficacy against black vine weevil, *Otiorhynchus sulcatus* (Coleoptera: Curculionidae) at low and moderate temperatures. *Biological Control* 108, 40–46.

Haukeland, S. and Lola-Luz, T. (2010) Efficacy of the entomopathogenic nematodes *Steinernema kraussei* and *Heterorhabditis megidis* against the black vine weevil *Otiorhynchus sulcatus* in open field-grown strawberry plants. *Agricultural and Forest Entomology* 12, 363–369.
Hauser, M. (2011) A historic account of the invasion of *Drosophila suzukii* (Matsumura) (Diptera: Drosophilidae) in the continental United States, with remarks on their identification. *Pest Management Science* 67, 1352–1357.
Haviland, D.R. and Hernandez, N.M. (2012) Development of management programs for white grubs in California blueberries. *International Journal of Fruit Science* 12, 114–123.
Hayes, A.E., Fitzpatrick, S.M. and Webster, J.M. (1999) Infectivity, distribution, and persistence of the entomopathogenic nematode, *Steinernema carpocapsae* All strain (Rhabditida: Steinernematidae) applied by sprinklers or boom sprayer to dry-pick cranberries. *Journal of Economic Entomology* 92, 539–546.
Henderson, D. and Singhai, R. (1999) Efficacy study of three species of parasitic nematodes for control of cranberry girdler, *Chrysoteuchia topiaria*, in British Columbia, Canada. In: Deziel, G. and Hogan, M. (eds) *Cranberry Research Compilation. Progress and Final Reports on Cranberry Research Conducted in 1998*. Cranberry Institute, Wareham, Massachusetts, pp. 167–176.
Hodgson, E.W. and Roe, A.H. (2007) *Cranberry Girdler*. Publication ENT-42-07. Utah State University Extension and Utah Plant Pest Diagnostic Laboratory, Logan, Utah.
Hübner, A., Englert, C. and Herz, A. (2017) Effect of entomopathogenic nematodes on different developmental stages of *Drosophila suzukii* in and outside fruits. *BioControl* 62, 669–680.
Innocenzi, P.J., Hall, D.R. and Cross, J.V. (2001) Components of male aggregation pheromone of strawberry blossom weevil, *Anthonomus rubi* Herbst. (Coleoptera: Curculionidae). *Journal of Chemical Ecology* 27, 1203–1218.
Ioriatti, C., Guzzon, R., Anfora, G., Ghidoni, F., Mazzoni, V. *et al.* (2018) *Drosophila suzukii* (Diptera: Drosophilidae) contributes to the development of sour rot in grape. *Journal of Economic Entomology* 111, 283–292.
Kender, W.J. and Brightwell, W.T. (1966) Environmental relationships. In: Eck, P. and Childers, N.F. (eds) *Blueberry Culture*. Rutgers University Press, New Brunswick, New Jersey, pp. 75–93.
Klingler J. (1988) Investigations on the parasitism of *Otiorhynchus salicicola* and *O. sulcatus* (Col.: Curculionidae) by Heterorhabditis sp. (Nematoda). *Entomophaga* 33, 325–331.
Koppenhöfer, A.M. and Fuzy, E.M. (2006) Effect of soil type on infectivity and persistence of the entomopathogenic nematodes *Steinernema scarabaei*, *Stenernema glaseri*, *Heterorhabditis zealandica*, and *Heterorhabditis bacteriophora*. *Journal of Invertebrate Pathology* 92, 11–22.
Koppenhöfer, A.M. and Kaya, H.K. (1998) Synergism of imidacloprid and an entomopathogenic nematode: a novel approach to white grub control in turfgrass. *Journal of Economic Entomology* 91, 618–623.
Koppenhöfer, A.M., Rodriguez-Saona, C.R., Polavarapu, S. and Holdcraft, R.J. (2008) Entomopathogenic nematodes for control of *Phyllophaga georgiana* (Coleoptera: Scarabaeidae) in cranberries. *Biocontrol Science and Technology* 18, 21–31.
Koppenhöfer, A.M., Shapiro-Ilan, D.I. and Hiltpold, I. (2020) Entomopathogenic nematodes in sustainable food production. *Frontiers in Sustainable Food Systems* 4: 125.
Lampasona, T.P., Rodriguez-Saona, C., Leskey, T.C. and Nielsen, A.L. (2020). A review of the biology, ecology, and management of plum curculio (Coleoptera: Curculionidae). *Journal of Integrated Pest Management* 11: 22.
Lee, J.C., Wang, X., Daane, K.M., Hoelmer, K.A., Isaacs, R. *et al.* (2019) Biological control of spotted-wing drosophila (Diptera: Drosophilidae) – current and pending tactics. *Journal of Integrated Pest Management* 10: 13.
Leskey, T.C., Pinero, J.C. and Prokopy, R.J. (2008) Odor-baited trap trees: a novel management tool for the plum curculio, *Conotrachelus nenuphar* (Herbst) (Coleoptera: Curculionidae). *Journal of Economic Entomology* 101, 1302–1309.
Long, S.J., Richardson, P.N. and Fenlon, J.S. (2000) Influence of temperature on the infectivity of entomopathogenic nematodes (*Steinernema* and *Heterorhabditis* spp.) to larvae and pupae of the vine weevil *Otiorhynchus sulcatus* (Coleoptera: Curculionidae). *Nematology* 3, 309–317.
Mampe, C.D. and Neunzig, H.H. (1967) The biology, parasitism and population sampling of the plum curculio on blueberry in North Carolina. *Journal of Economic Entomology* 60, 807–812.
McKern, J.A., Johnson, D.T. and Lewis, B.A. (2007) Biology and control of the raspberry crown borer (Lepidoptera: Sesiidae). *Journal of Economic Entomology* 100, 398–404.
Olivier, A.D. and Chapin, J.B. (1980) The cranberry rootworm: adult seasonal history, and factors affecting its status as a pest of woody ornamentals in Louisiana. *Journal of Economic Entomology* 73, 96–100.

Phillips, D.A., Liburd, O.E. and Duncan, L.W. (2019) *Diaprepes Root Weevil on Southern Highbush Blueberry in Florida*. Publication ENY999. UF/IFAS Extension Service, University of Florida, Gainesville, Florida.

Piñero, J.C., Shapiro-Ilan, D., Cooley, D.R., Tuttle, A.F., Eaton, A. *et al.* (2020) Toward the integration of an attract-and-kill approach with entomopathogenic nematodes to control multiple life stages of plum curculio (Coleoptera: Curculionidae). *Insects* 11: 375.

Polavarapu, S. (1996) Species composition of scarab grubs and seasonal life-history of oriental beetle in blueberries. *Horticultural News* 76, 8–11.

Polavarapu, S. (1999) *Insect Parasitic Nematodes for Cranberry Pest Management*. Publication FS029. Rutgers Cooperative Extension, New Jersey Agricultural Experiment Station, New Brunswick, New Jersey.

Polavarapu, S., Stuart, R.J. and Lewis, E.E. (1996) Laboratory efficacy of selected species of entomopathogenic nematodes against blueberry mealybug, 1995. *Arthropod Management Tests* 21: 414.

Polavarapu, S., Peng, H. and Stuart, R.J. (1998) Evaluation of Admire 2F and an entomopathogenic nematode to manage oriental beetle populations on blueberries, 1997. *Arthropod Management Tests* 22: 51.

Polavarapu, S., Stuart, R.J. and Schiffhauer, D.F. (2000) Evaluation of insecticides against cranberry rootworm. *Arthropod Management Tests* 25, 66–67.

Polavarapu, S., Kyryczenko-Roth, V. and Barry, J.D. (2004) Phenology and infestation patterns of plum curculio (Coleoptera: Curculionidae) on four highbush blueberry cultivars. *Journal of Economic Entomology* 97, 1899–1905.

Polavarapu, S., Koppenhöfer, A.M., Barry, J.D., Holdcraft, R.J. and Fuzy, E.M. (2007) Entomopathogenic nematodes and neonicotinoids for remedial control of oriental beetle, *Anomala orientalis* (Coleoptera: Scarabaeidae), in highbush blueberry. *Crop Protection* 26, 1266–1271.

Potter, D.A. (1998) *Destructive Turfgrass Insects: Biology, Diagnosis, and Control*. Ann Arbor Press, Chelsea, Michigan.

Racette, G., Chouinard, G., Vincent, C. and Hill, S.B. (1992) Ecology and management of the plum curculio, *Conotrachelus nenuphar* (Herbst) (Coleoptera: Curculionidae). *Phytoprotection* 73, 85–100.

Renkema, J.M. and Cuthbertson, A.G.S. (2018) Impact of multiple natural enemies on immature *Drosophila suzukii* in strawberries and blueberries. *BioControl* 63, 719–728.

Renkema, J.M., Evans, B. and Devkota, S. (2018) Management of flower thrips in Florida strawberries with *Steinernema feltiae* (Rhabditida: Steinernematidae) and the insecticide sulfoxaflor. *Florida Entomologist* 1010, 102–108.

Renkema, J.M., Krey, K.L. and Lahiri, S. (2021) Control of *Diaprepes abbreviates* (Coleoptera: Curculionidae) with *Steinernema riobrave* (Rhabditida: Steinernematidae) in plasticulture Florida strawberry. *Florida Entomologist* 104, 124–130.

Rivera, M.J., Rodriguez-Saona, C., Egizi, A., Fonseca, D.M., Jennings, D.E. and Koppenhöfer, A.M. (2016) Cultivation and domestication of highbush blueberry (*Vaccinium corumbosum*) alters abundance, diversity and virulence of entomopathogenic nematodes. *Agriculture, Ecosystems & Environment* 222, 148–155.

Robbins, P.S., Nojima, S., Polavarapu, S., Koppenhöfer, A.M., Rodriguez-Saona, C. *et al.* (2009) Pheromone of the scarab beetle *Phyllophaga* (*Phytalus*) *georgiana* (Horn). *Journal of Chemical Ecology* 35, 336–341.

Roberts, S.L. and Mahr, D.L. (1986) Development of cranberry girdler, *Chrysoteuchia topiaria* (Lepidoptera: Pyralidae) in relation to temperature. *The Great Lakes Entomologist* 19, 85–90.

Rodriguez-Saona, C., Nielsen, A., Shapiro-Ilan, D., Tewari, S., Kyryczenko-Roth, V. *et al.* (2019) Exploring an odor-baited "trap bush" approach to aggregate plum curculio (Coleoptera: Curculionidae) injury in blueberries. *Insects* 10: 113.

Shamseldean, M.M. and Atwa, A.A. (2004) Laboratory and field tests of entomopathogenic nematodes against the scarab beetle *Temnorhynchus baal* (Reiche) a novel insect pest of strawberry in Egypt. *Egyptian Journal of Biological Pest Control* 14, 127–133.

Shanks, C.H. and Agudelo-Silva, F. (1990) Field pathogenicity and persistence of heterorhabditid and steinernematid nematodes (Nematoda) infecting black vine weevil larvae (Coleoptera: Curculionidae) in cranberry bogs. *Journal of Economic Entomology* 83, 107–110.

Shapiro-Ilan, D.I., Mizell, R.F., Cottrell, T.E. and Horton, D.I. (2004) Measuring field efficacy of *Steinernema feltiae* and *Steinernema riobrave* for suppression of plum curculio, *Conotrachelus nenuphar*, larvae. *Biological Control* 30, 496–503.

Shapiro-Ilan, D.I., Leskey, T.C. and Wright, S.E. (2011) Virulence of entomopathogenic nematodes to plum curculio, *Conotrachelus nenuphar*: effects of strain, temperature, and soil type. *Journal of Nematology* 43, 187–197.

Shapiro-Ilan, D.I., Wright, S.E., Tuttle, A.F., Cooley, D.R. and Leskey, T.C. (2013) Using entomopathogenic nematodes for biological control of plum curculio, *Conotrachelus nenuphar*: effects of irrigation and species in apple orchard. *Biological Control* 67, 123–129.
Shehata, I.E., Hammam, M.M.A., El-Borai, F.E., Duncan, L.W. and Abd-Elgawad, M.M.M. (2019) Comparison of virulence, reproductive potential, and persistence among local *Heterorhabditis indica* populations for the control of *Temnorhynchus baal* (Reiche & Saulcy) (Coleoptera: Scarabaeidae) in Egypt. *Egyptian Journal of Biological Pest Control* 29: 32.
Shehata, I.E., Hammam, M.M.A., El-Borai, F.E., Duncan, L.W. and Abd-Elgawad, M.M.M. (2020) Traits of the entomopathogenic nematode, *Heterorhabditis bacteriophora* (Hb-EG strain), for potential biocontrol in strawberry fields. *Egyptian Journal of Biological Pest Control* 30: 40.
Sigareva, D. and Kharchenko, V. (2018) Reproduction of entomopathogenic nematodes of the genus *Steinernema* Travassos, 1927 (Rhabditida: Steinernematidae) within strawberry blossom weevil larvae *Anthonomus rubi* Herbst, 1795 (Coleoptera: Curculionidae). *Interdepartmental Thematic Scientific Collection of Plant Protection and Quarantine* 64, 184–190.
Simard, L., Bélair, G. and Brodeur, J. (2002) Susceptibility of cranberry girdler to entomopathogenic nematodes. *Canadian Entomologist* 134, 329–330.
Simser, D. and Roberts, S. (1994) Suppression of strawberry root weevil, *Ottorhynchis ovatus*, in cranberries by entomopathogenic nematodes (Nematoda: Steinernematidae and Heterorhabditidae). *Nematologica* 40, 456–462.
Smith, K., Fisher, G.C., Poole, A. and Jackson, D. (1993) *Effectiveness of Biosafe-N*, Steinernema carpocapsae *for Control of Cranberry Girdler, 1992*. Cranberry Agricultural Research, Ocean Spray Cranberries, Lakeville-Middleborough, Massachusetts.
Sousa, A.L., Rodriguez-Saona, C., Holdcraft, R., Kyryczenko-Roth, V. and Koppenhöfer, A.M. (2022) Entomopathogenic nematodes for the management of plum curculio in highbush blueberry. *Biology* 11: 45.
Stuart, R.J. and Polavarapu, S. (1997) Susceptibility of the cranberry rootworm to entomopathogenic nematodes. *Horticultural News* 78: 15.
Stuart, R.J., Polavarapu, S., Lewis, E.E. and Gaugler, R. (1997) Differential susceptibility of *Dysmicoccus vaccinii* (Homoptera: Pseudococcidae) to entomopathogenic nematodes (Rhabditida: Heterorhabditidae and Steinernematidae). *Journal of Economic Entomology* 90, 925–932.
Susurluk, A. and Ehlers, R.-U. (2008) Sustainable control of black vine weevil larvae, *Otiorhynchus sulcatus* (Coleoptera: Curculionidae) with *Heterorhabditis bacteriophora* in strawberry. *Biocontrol Science and Technology* 18, 627–632.
Tewari, S., Rodriguez-Saona, C. and Polk, D. (2014) *Plum Curculio: A Key Pest of Blueberries in New Jersey*. Publication FS1229. New Jersey Agricultural Experimental Station, Rutgers University, New Brunswick, New Jersey.
Tonina, L., Zanettin, G., Miorelli, P., Puppato, S., Cuthbertson, A.G.S. and Grassi, A. (2021) *Antonomus rubi* on strawberry fruit: its biology, ecology, damage, and control from and IPM perspective. *Insects* 12: 701.
van Timmeren, S. and Isaacs, R. (2013) Control of spotted wing drosophila, *Drosophila suzukii*, by specific insecticides and by conventional and organic crop protection programs. *Crop Protection* 54, 126–133.
Vittum, P.J., Villani, M.G. and Tashiro, H. (1999) *Turfgrass Insects of the United States and Canada*, 2nd edn. Cornell University Press, Ithaca, New York.
Vorsa, N. (1997). On a wing: the genetics and taxonomy of *Vaccinium* species from a pollination perspective. *Acta Horticulturae* 446, 59–66.
Walsh, D.B., Bolda, M.P., Goodhue, R.E., Dreves, A.J., Lee, J. *et al.* (2011) *Drosophila suzukii* (Diptera: Drosophilidae): invasive pest of ripening soft fruit expanding its geographic range and damage potential. *Journal of Integrated Pest Management* 2, G1–G7.
Warner, R.E. and Negley, F.B. (1976) The genus *Otiorhynchus* in America north of Mexico (Coleoptera: Curculionidae). *Proceedings of the Entomological Society of Washington* 78, 240–262.
Weber, D. and Henderson, D. (1998) Beneficial nematodes in cranberry pest management. Paper presented at the *Integrated Pest Management in Cranberry Production Workshop, 19–20 April, Montreal, Quebec, Canada.*
Willmott, D.M., Hart, A.J., Long, S.J., Edmondson, R.N. and Richardson, P.N. (2002) Use of a cold-active entomopathogenic nematode *Steinernema kraussei* to control overwintering larvae of the black vine weevil *Otiorhynchus sulcatus* (Coleoptera: Curculionidae) in outdoor strawberry plants. *Nematology* 4, 925–932.
Woltz, J.M. and Lee, J. (2017) Pupation behavior and larval and pupal biocontrol of *Drosophila suzukii* in the field. *Biological Control* 110, 62–69.

Woltz, J.M., Donahue, K.M., Bruck, D.J. and Lee, J.C. (2015) Efficacy of commercially available predators, nematodes and fungal entomopathogens for augmentative control of *Drosophila suzukii*. *Journal of Applied Entomology* 139, 759–770.

Ye, W., Foye, S., MacGuidwin, A.E. and Steffan, S. (2018) Incidence of *Oscheius onirici* (Nematoda: Rhabditidae), a potential entomopathogenic nematode from the marshlands of Wisconsin, USA. *Journal of Nematology* 50, 9–26.

19 Role of Entomopathogenic Nematodes in an Integrated Pest Management Strategy for Grapevine: South Africa as a Model System

Antoinette P. Malan[1]* and Rinus Knoetze[1,2]

[1]*Department of Conservation Ecology and Entomology, Faculty of AgriSciences, Stellenbosch University, Stellenbosch, South Africa;* [2]*Plant Protection Division, Agricultural Research Council (ARC), Infruitec-Nietvoorbij, Stellenbosch, South Africa*

Abstract

The application of entomopathogenic nematodes (EPNs) as biological control agents against aboveground and soil insect pests in viticulture in South Africa is especially suitable in the case of table grape production. Table grapes are export driven and during harvesting, when chemicals cannot be used, application of nematodes to the bunches in the shaded canopy of the trellis system is highly recommended. Although ten of the most important grapevine pests have been tested against locally isolated EPNs, large-scale field trials are lacking because of the unavailability of locally produced mass-cultured and formulated EPN products. Some of the pests, such as ground pearls, Trimen's tiger moth, Cape leaf miner and katydids, tested against EPNs are sporadic, occurring only in specific viticulture areas in South Africa. In such cases, hot-spot application using EPNs will not disrupt the current insect control management plan. The South African vine industry also focuses on the production of organic wines, which makes EPNs especially attractive for insect pest control.

19.1 Introduction

Viticulture, as an agricultural industry in South Africa, embraces the production of wine, spirits and table grapes, as well as the manufacture of dried grape products. The first grapevines were planted by Dutch settlers in Cape Town in 1655, with the first wines being produced there in February 1659. However, the arrival of the French Huguenots in 1688 is regarded as having been the major influence on viticulture at the Cape of Good Hope (James, 2013).

Grapevines are grown both for wine and for table grape production in South Africa, with a small percentage of the latter being produced as raisins. Although most of South Africa's viticulture occurs in the Western Cape province, a limited number of wine grapes are produced in the KwaZulu-Natal province. Apart from for the Western Cape province, table grapes are also grown in the inland regions of the Northern Cape, both along the Orange River and in isolated pockets in the North-west, Limpopo and Mpumalanga provinces. During the 2020/21

*Email: apm@sun.ac.za

 Entomopathogenic Nematodes as Biological Control Agents (D.I. Shapiro-Ilan and E.E. Lewis eds)
DOI: 10.1079/9781800620322.0019

season, 117,041 ha were under grape production, with 92,005 ha for wine grapes, 18,028 ha for table grapes and 6939 ha for raisins, totaling to 1,342,112 t of grapes (SAWIS, 2020).

19.1.1 Wine grapes

Most of the wine grape production areas of South Africa are located near the coastal influences of the Atlantic and Indian Oceans. The five main regions are: the Coastal region, encompassing areas such as Stellenbosch, Paarl and the West Coast; the Breede River Valley; the Olifants River region; the Cape South Coast; and the Klein Karoo. These areas mainly have a Mediterranean climate, with a wide range of macroclimates and vineyard soil types present in and on the numerous mountain slopes and valleys. Generally, the soils tend to retain moisture and to drain well, having a significant proportion of clay, with low pH levels of around 4. The pH levels of the soils are often adjusted with lime and calcium treatment. Summers are marked by intense sunlight and dry heat, while winters tend to be cold and wet with potential snowfall at higher elevations. As South Africa is a water-scarce country, irrigation is essential for the maintenance of viticulture in many of these regions (Stevenson, 2005).

19.1.2 Table grapes

While the cooler coastal regions specialize in providing premium wine grapes, table grapes are grown for export in the hotter, inland regions. The five major growing regions in South Africa are the Northern province, Orange River, Olifants River, Berg River and Hex River regions. The difference in the soil and climate enables the growers to supply the international market from November to May. In the Northern province region, the early summer and warm climate combine to produce early-maturing varieties, while in the Hex River region, mid-season to late-season grapes are produced. The 2020/21 table grape harvest amounted to 74.9 million cartons inspected and passed for export (SATI, 2021).

19.2 Entomopathogenic Nematodes: South African Perspective

As most entomopathogenic nematode (EPN)-related research has been undertaken into grapevine insect pest management in South Africa, this chapter will use South Africa as a model but with reference to research from other countries. The South African wine industry is mostly focused on the production of organically produced wine, with the use of approved materials for organic production. In the case of table grapes, EPNs are a perfect solution for managing some insect pests, specifically in terms of overcoming the problems with maximum residue limits, and with the withholding period before and during the harvesting of grapes for export. Research into the use of EPNs to control grapevine pests in South Africa includes the study of fruit flies, grapevine mealybug, weevils, false codling moth, ground pearls, Trimen's tiger moth, Cape leaf miner, Western flower thrips, lobesia moth and katydids, in order of economic importance (Table 19.1).

Another potential important grapevine pest is *Vitacea polistiformis* (Harris) (Lepidoptera: Sesiidae), the grape root borer, which has not been reported from South Africa, and which is regarded as a pest of commercial vineyards in the eastern USA (Bergh, 2012). The larvae feed initially on the roots of grapevine, leading to the decline and eventual death of the vines. The larvae bore into the root cortex, where they spend 1–3 years, offering an opportunity for control using EPNs (Rijal and Bergh, 2017). Williams *et al.* (2002) found, in laboratory bioassays, that infection rates with *H. bacteriophora* were 92% and with *H. zealandica* were 86%, while greenhouse tests showed lower infection rates of 16% and 53%, respectively. However, the success of EPNs in terms of locating, infecting and proliferating in grape borer larvae in the roots of grapevine under glasshouse conditions is regarded as promising.

19.3 Research into the Use of Entomopathogenic Nematodes to Control Grapevine Pests

No surveys to date specifically targeting grapevine in South Africa have been undertaken. Reports of the occurrence of EPNs in vineyards

Table 19.1. Common grapevine pests of South African vineyards. The most important pests, discussed in this chapter, are shown in bold. Adapted from Allsopp *et al.* (2015) and du Preez (2019).

Order	Family	Common name	Species
Coleoptera	Curculionidae	**Banded fruit weevil**	***Phlyctinus callosus* (Schönherr)** and ***Phlyctinus xerophilus* (Haran)**
		Black snout beetle	*Eremnus atratus* (Sparrman)
		Gray weevil	*Eremnus setulosus* Boheman
		Speckled beetle	*Eremnus cerealis* Marshall
		Vine weevil	*Eremnus chevrolati* Oberprieler
Diptera	Tephritidae	**Mediterranean fruit fly**	*Ceratitis capitata* and *Ceratitis rosa*
Hemiptera	Cicadellidae	Grapevine leafhoppers	*Acia lineatifrons* (Naudé) and *Mgenia fuscovaria* (Stål)
	Phylloxeridae	Grapevine phylloxera	*Daktulosphaira vitifoliae* (Fitch)
	Margarodidae	**Ground pearls**	***Margarodes prieskaensis* (Jakubski)**, *M. capensis* Giard, *M. greeni* Brian, *M. trimeni* Giard and *M. vredendalensis* De Klerk
	Pseudococcidae	**Grapevine mealybug**	***Planococcus ficus* (Signoret)**
		Bud nibbler	*Tanyrhynchus carinatus* Boheman
Lepidoptera	Cossidae	Apple trunk borer	*Coryphodema tristis* (Drury)
	Heliozelidae	**Cape leaf miner**	*Holocacista capensis*
	Tortricidae	**False codling moth**	***Thaumatotibia leucotreta* (Meyrick)**
		Pear leaf roller	*Epichoristodes acerbella* (Walker)
	Agaristidae	**Trimen's false tiger**	***Agoma trimenii* (Felder)**
	Noctuidae	African bollworm	*Helicoverpa armigera* (Hübner)
		Lobesia moths	***Lobesia vanillana***
Orthoptera	Tettigoniidae	**Katydid**	***Plangia graminea* (Serville)**
Thysanoptera	Thripidae	**Western flower thrips**	***Frankliniella occidentalis* (Pergande)**
		Guava thrips	*Heliothrips sylvanus* Faure

have, so far, mostly come from non-targeted surveys (Table 19.2). Evidently, the most common species of EPNs is *Heterorhabditis bacteriophora*, followed by *Heterorhabditis zealandica* (with red and green bacteria) and *Heterorhabditis indica* (with red and yellow bacteria) (V.M. Steyn *et al.*, 2021; Booysen *et al.*, 2022).

19.3.1 Fruit flies

The Mediterranean fruit fly, *Ceratitis capitata* (also known as medfly) is widely distributed across 80 countries throughout the world, while the natal fruit fly, *Ceratitis rosa*, is endemic to Africa and the Indian Ocean Islands. The two species are known to infest over 260 species of plants, including all deciduous fruit and grapevines in the Western Cape (Blomefield *et al.*, 2015). *C. capitata* is one of the world's most destructive fruit pests because of its wide distribution, host range and ability to adapt to cooler climates.

Only the larval stages of both species cause damage. Eggs are deposited under the skin of ripening host fruits, with the resulting larvae feeding on the fruit tissue, tunneling toward the fruit center. Subsequent infection by fungi and bacteria can reduce the fruit to a pulpy mass. When the larvae reach maturity, they leave the fruit, fall to the ground and pupate just below the surface of the soil (Blomefield *et al.*, 2015). After the adult flies emerge, they crawl up to the soil surface. The management of fruit flies includes chemical control, sanitation, application of bait mixed with insecticide, augmentative release of parasitoids and use of the sterile insect technique (Venter *et al.*, 2021). Chemical control restrictions and regulations have led to the withdrawal of many chemicals from the market, with growers searching for effective alternative control options such as biological control.

The potential of local EPN isolates of *H. bacteriophora*, *H. zealandica* and *Steinernema khoisanae* to infect the pupariating larvae, pupae and adults of *C. capitata* and *C. rosa* has been investigated

Table 19.2. *Heterorhabditis*, *Steinernema* and *Oscheius* spp. and their symbionts reported from grapevine in South Africa.

Species	Isolate no.	Color	ITS[a]	Symbiont	Province	Nearest town	Reference(s)
Heterorhabditis bacteriophora	SF351	Red	FJ455843	*Photorhabdus laumondii*	Western Cape	Wellington	Malan *et al*. (2006); Geldenhuys *et al*. (2016)
H. bacteriophora	301	Red	MH333236	–	Western Cape	Grabouw	B.E. Dlamini *et al*. (2020a)
H. bacteriophora	INF 16	–	–	–	Western Cape	Stellenbosch	Hatting *et al*. (2009)
H. bacteriophora	INF 29	–	–	–	Western Cape	Stellenbosch	Hatting *et al*. (2009)
H. bacteriophora	INF 61	–	–	–	Western Cape	Hex River Valley	Hatting *et al*. (2009)
H. bacteriophora	INF 79	–	–	–	Western Cape	Stellenbosch	Hatting *et al*. (2009)
H. bacteriophora	UP2A1	Red	MF372924	–	Northern Cape	Upington	James *et al*. (2018)
H. bacteriophora	UP2A2	Red	MF033536	–	Northern Cape	Upington	James *et al*. (2018)
H. bacteriophora	UP3A1	Red	MF372925	–	Northern Cape	Upington	James *et al*. (2018)
H. bacteriophora	UP3A2	Red	MF372926	–	Northern Cape	Upington	James *et al*. (2018)
H. bacteriophora	UP3B1	Red	MF372928	–	Northern Cape	Upington	James *et al*. (2018)
H. bacteriophora	UP3B1	Red	MF372927	–	Northern Cape	Upington	James *et al*. (2018)
H. bacteriophora	VS7	Red	MN932135	–	Western Cape	Robertson	Steyn *et al*. (2021)
H. bacteriophora	VS9	Red	MN932136	–	Western Cape	Halfmanshof	Steyn *et al*. (2019)
H. bacteriophora	VS10	Red	MN932137	–	Western Cape	Riebeeck Kasteel	Steyn *et al*. (2019)
H. bacteriophora	VS11	Red	MN932138	–	Western Cape	Riebeeck Kasteel	Steyn *et al*. (2019)
H. bacteriophora	VS12	Red	MN932139	–	Western Cape	Halfmanshof	Steyn *et al*. (2019)
H. bacteriophora	VS13	Red	MN932140	–	Western Cape	Halfmanshof	Steyn *et al*. (2019)
H. bacteriophora	VS14	Red	MN932141	–	Western Cape	Riebeeck Kasteel	Steyn *et al*. (2019)
H. bacteriophora	VS15	Red	MN932142	–	Western Cape	Riebeeck Kasteel	Steyn *et al*. (2019)
H. bacteriophora	VS18	Red	MN932143	–	Western Cape	Halfmanshof	Steyn *et al*. (2019)
Heterorhabditis indica	VS8	Red	MN922779	–	Western Cape	Halfmanshof	Steyn *et al*. (2019)
H. indica	VS5	Yellow	MN922778	–	Western Cape	Halfmanshof	Steyn *et al*. (2019)
H. indica	SGS	Red	GQ377411	–	Western Cape	Bonnievale	Steyn *et al*. (2019)
Heterorhabditis zealandica	MJ1C	Green	MF167295	*Photorhabdus thracensis*	Northern Cape	Hex River Valley	James *et al*. (2018); Booysen *et al*. (2022)
H. zealandica	MJ2B2	Red	MF185663	*P. laumondii* subsp. *laumondii*	Northern Cape	Hex River Valley	James *et al*. (2018); Booysen *et al*. (2022)

H. zealandica	VS1	Green	MN932149	*P. thracensis*	Western Cape	De Doorns	Steyn *et al.* (2021); Booysen *et al.* (2022)
H. zealandica	VS2	Green	MN932150	*P. thracensis*	Western Cape	De Doorns	Steyn *et al.* (2021); Booysen *et al.* (2022)
Oscheius myriophilus	UP2B	Blue	–	–	Northern Cape	Upington	James *et al.* (2018)
O. myriophilus	VS17	Red	MN932153	–	Western Cape	Paarl	Steyn *et al.* (2019); Booysen *et al.* (2022)
Steinernema innovationi	–	Black	–	Unknown sp.	Western Cape	Clanwilliam	Hatting *et al.* (2009)
Steinernema khoisanae	–	Black	–	*Xenorhabdus khoisanae*	Western Cape	De Berg	Hatting *et al.* (2009); Ferreira *et al.* (2013)
Steinernema tophus	ROOI 352 (sp. 2)	Yellow	KJ701241	Unknown sp.	Western Cape	Clanwilliam	Çimen *et al.* (2014)
S. tophus	INF 23	–	–	Unknown sp.	Western Cape	Stellenbosch	Hatting *et al.* (2009)
Steinernema sp.	IFN 73	–	–	–	Western Cape	De Doorns	Hatting *et al.* (2009)

–, Unidentified.
[a]GenBank accession number of internal transcribed spacer region.

using 24-well plates in laboratory bioassays (Fig. 19.1). The pupariating larvae and adult flies were found to be susceptible to nematode infection, with no infection being recorded for the fully formed pupae. The nematodes develop in pupariating larvae infected before pupation, with, in some cases, the pupae being formed during different stages of transition leading to malformed pupae (Fig. 19.2). The pupariating larvae of *C. capitata* were generally found to be more susceptible to infection than those of *C. rosa*. Significantly more larvae of *C. capitata* were infected by *H. bacteriophora* (>80%) than those of *C. rosa* (65%). For *C. rosa*, the highest infectivity of larvae was obtained with *H. zealandica* (90%). In the case of *H. zealandica*, the mortality of *C. capitata* larvae was 90%. In contrast, the adults of both species were highly infected by *S. khoisanae*, with infection rates of 75% and 80% for *C. capitata* and *C. rosa*, respectively (Malan and Manrakhan, 2009).

In a study conducted by James *et al.* (2018), the main aim was to identify the South African EPN species that would be highly pathogenic against *C. capitata* larvae after they entered the soil to pupate. The virulence of five local EPN species was screened, using 24-well bioassay plates, by inoculating single third-instar *C. capitata* larvae with 100 infective juveniles (IJs) in 50 μl of water. After 48 h, high levels of *C. capitata* larval infection were obtained, with no significant difference being found among the different nematode species. The experiment was repeated at a lower rate of 50 IJs per insect, with both *Heterorhabditis noenieputensis* and *H. indica* found to cause a mean mortality of over 80%. The virulence of *H. noenieputensis*, along with *Steinernema yirgalemense*, was further tested by inoculating 100 ml of sterilized sand with 2000 IJs. After 24 h, ten third-instar *C. capitata* larvae were added to the sand. A high infection rate (>90%) of *C. capitata* larvae was obtained using *H. noenieputensis*, with the adults in the soil being infected during eclosion (James *et al.*, 2018).

Soil application of EPNs will not provide adequate control of *C. capitata*. The last-instar larvae and the adult fly are both highly susceptible

Fig. 19.1. Last-instar *Ceratitis rosa* larva infected with *Heterorhabditis bacteriophora*. (A) First generation. (B) Infective juveniles leaving the depleted cadaver.

Fig. 19.2. Malformed medfly (*Ceratitis capitata*) pupae infected with *Heterorhabditis bacteriophora*. (A) Prepupae. (B) Fully formed infected pupae.

to EPNs, but only a very small window of opportunity of a few hours is available for nematode infection. The last-instar larvae infected with EPNs can sometimes develop into pupae as well as emerging in the soil. Maintaining high residential EPN populations in vineyards is likely to affect the control of fruit fly indirectly. The high infection rate of *C. capitata* larvae, using local EPN isolates, despite the short window of opportunity for infection, provides assurance that EPNs can be used as a biological control agent against *C. capitata* as one component of an integrated pest management (IPM) system. Although field trials with EPN applications against fruit flies have not yet been attempted in South Africa, a strong likelihood exists that the control of fruit flies will have a secondary effect in terms of EPN soil applications against other insect pest soil stages, such as false codling moth.

19.3.2 Vine mealybug

The vine mealybug, *Planococcus ficus*, is regarded as a key pest insect of the South African table grape and wine industries. They are phytophagous, soft-bodied insects with piercing, sap-sucking mouth parts. Mealybugs are severe agricultural pests, as their presence and feeding cause direct damage to plants by means of lowering production and rendering the fruit concerned unmarketable, while also transmitting various plant viruses (Holm, 2008). Vine mealybugs are sexually dimorphic. Adult females are covered with a fine, white, powdery wax layer and the fringe of the body has waxy hair-like extensions. All life stages can be present throughout the year. Although they shelter beneath loose bark during winter, they can also hibernate on grapevine roots and weeds. As soon as the temperatures rise in spring, crawlers emerge, in search of new growth. Male *P. ficus* adults are tiny, delicate insects, with a pair of inconspicuous, transparent wings.

Mealybugs are controlled with a combination of chemical and biological control. The most prevalent enemies of *P. ficus* in South Africa include hymenopteran parasites such as *Coccidoxenoides perminutus* Girault (Hymenoptera: Encyrtidae), which is commercially produced and available for augmentative release. The mutualistic relationship between mealybugs and honeydew-seeking ants has been shown to reduce the efficacy of biological control significantly. Ants can exacerbate mealybug pest problems by disrupting the processes of augmentative and natural biological control, and by aiding in their dispersal (Addison, 2002; Le Vieux and Malan, 2013b).

The life cycle of *P. ficus* is mostly spent above ground, while the colonies tend to hide in cracks on the grapevine stem. In 2000, the possibility that *P. ficus* might survive on grapevine root remnants was postulated and confirmed by Walton and Pringle (2004). This agrees with reports of *Pseudococcus calceolariae* (Maskell) from New Zealand, which has frequently been found to occur on the roots of grapevine and on numerous weed hosts (Bell *et al.*, 2009).

Laboratory bioassays have been conducted to establish the potential of EPNs as biological control agents of *P. ficus*. Six indigenous and two commercially available nematode species were screened using 24-well bioassay plates for their efficacy in infecting female *P. ficus*. The two indigenous species with the most promising results were *H. zealandica* and *S. yirgalemense*, which caused 96% and 65% infection, respectively (Fig. 19.3). Tests were conducted to compare the efficacy of *H. bacteriophora* and an exotic *Steinernema feltiae* produced both *in vivo* and *in vitro*. *H. bacteriophora* showed no significant difference in efficacy between the two production methods, except that *in vivo*-cultured *S. feltiae* produced significantly higher mean mortality (40%), in contrast to the 19% mean infection rate obtained with the *in vitro*-produced IJs (Le Vieux and Malan, 2013a). On the screening of EPN species against *P. ficus*, Platt *et al*. (2018) found high (90%) infectivity for *H. noenieputensis*. Further investigation into the effect of temperature showed that *S. yirgalemense* yielded the highest mortality at 25°C of 72%, followed by 45% mortality at 30°C and by only 9% mortality at 15°C. *S. yirgalemense* performed best at 100% relative humidity, resulting in 70% mortality.

The ability of both *H. zealandica* and *S. yirgalemense* to complete their life cycles in the host and to produce a new cohort of IJs has been demonstrated. The bioassays indicated a concentration-dependent susceptibility of *P. ficus* to *H. zealandica*, *S. yirgalemense* and commercially produced *H. bacteriophora*, with 50% lethal concentrations (LC_{50}) and LC_{90} values of 19 and 82, 13 and 80, and 36 and 555 IJs per mealybug, respectively. In the laboratory, both *H. zealandica* and *S. yirgalemense* were able to move 15 cm downwards and to infect *P. ficus* buried in pierced Eppendorf tubes, with respective mortalities of 82% and 95%. This study showed *P. ficus* to be a suitable host for both *H. zealandica* and *S. yirgalemense*, with both nematode species showing considerable potential for future use in the field control of *P. ficus* (Le Vieux and Malan, 2013a).

Adjuvants are generally defined as additives to an insecticide solution that are intended to increase, or modify their effects (Krogh *et al*., 2003). Mixing nematode suspensions with adjuvants, or a combination of adjuvants, should facilitate the use of biological control agents in aboveground areas that have previously been considered inaccessible for nematode application (Platt *et al*., 2020). As the use of EPNs against the vine mealybug would be an aboveground application, Platt *et al*. (2019a) assessed the ability of adjuvants to increase the survival rate of *S. yirgalemense* on grapevine leaves. The combination of both Nu-Film-P® (poly-1-*p*-menthene, spreader, sticker; Hydrotech, Pretoria, South Africa) and Zeba® (starch-g-poly (2-propenamideco-2-propenoic acid) potassium salt; Tongaat Hulett Starch, Germiston, South Africa) resulted in significantly more IJs being deposited over a 4 cm^2 of grapevine leaf disk than before. The application of *S. yirgalemense* together with both Zeba® and Nu-Film-P® to *P. ficus* on leaf disks in a growth chamber resulted in 84% mortality, with similar results being observed in a glasshouse trial. A combination of *S. yirgalemense* with adjuvants

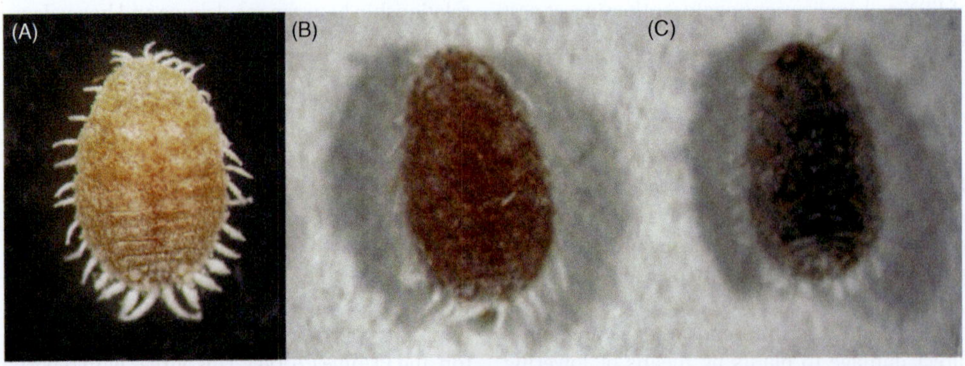

Fig. 19.3. Common color differences of the female vine mealybug (*Planococcus ficus*) 48 h after infection. (A) Control (uninfected). (B) Infected with *Steinernema yirgalemense*. (C) Infected with *Heterorhabditis zealandica*. (P. Le Vieux, Stellenbosch University, South Africa).

gave significantly more control of *P. ficus* on the grapevine foliage than when the EPNs were used alone. The next step for testing the additions of both adjuvants in semi-field trial conditions was then taken (Platt *et al.*, 2019b). A trial was performed to assess *S. yirgalemense* survival on grapevine foliage when applied in the morning at high humidity and low temperatures compared with in the afternoon at a time of high temperatures and low humidity. The application of the combination of the two adjuvants Nu-Film-P® and Zeba® to the nematode suspension resulted in 66% control of *P. ficus*. A significantly higher EPN survival rate was recorded at each time interval in the morning compared with in the afternoon. This study demonstrated the ability of *S. yirgalemense*, when applied with adjuvants in a morning application, to control the presence of *P. ficus* on grapevine under semi-field conditions (Platt *et al.*, 2019b).

To test whether the vine mealybug migrated to the grapevine roots during the winter months, *S. yirgalemense* was applied to the soil of two vineyards, with the aim of controlling the soil stages. The efficacy of EPN applications was evaluated by the administration of *P. ficus*, contained in pierced Eppendorf tubes, which were buried at a depth of 15 cm in the soil. Mortality of up to 50% infection was obtained after 48 h exposure in the soil environment before retrieval. However, when the persistence of *S. yirgalemense* in the two EPN-treated vineyards was measured using codling moth (*Cydia pomonella*) larvae in pierced Eppendorf tubes, the mortality was found to be zero in one vineyard, whereas in the other it was 70%, 12 weeks after application. The loss of IJs could be ascribed to plowing between the rows and to an ant attack of the codling moth larvae in the tubes, leading to the zero result in one vineyard (Le Vieux and Malan, 2015). The potential of *S. yirgalemense*, under laboratory conditions, however, was shown to last longer than 6 months (Le Vieux and Malan, 2015).

Olfactometer tests, with an option of IJs moving to contained mealybugs, damaged the grapevine roots or a control, indicating a significant difference concerning the higher number of *S. yirgalemense* IJs that were attracted to the damaged *Vitis vinifera* roots, compared with the case of *P. ficus* (Le Vieux and Malan, 2015). Allsopp and Fourie (2019) undertook additional surveys three times over the growing season in vineyards in the wine grape-growing districts to assess whether *P. ficus* commonly occurs on grapevine roots in the Western Cape. The results indicated that mealybugs were rarely found on grapevine roots in either summer or fall over three seasons, and that mealybugs were found mainly only below ground on the stem of a grapevine in each of the two vineyards in the McGregor district on one occasion (Allsopp and Fourie, 2019).

19.3.3 Weevils

The banded fruit weevil, *Phlyctinus callosus*, which is indigenous to South Africa, is polyphagous, feeding on a wide variety of weeds, grasses and woody plants. A recent study conducted by Haran *et al.* (2020), showed *P. callosus* to be a complex of closely related species. By finding morphological and molecular evidence to distinguish between the different species involved, a total of six species were identified from the *P. callosus* complex, with five being described as new species. The mitochondrial gene (cytochrome oxidase I) indicated a new species, *Phlyctinus xerophilus* (Haran *et al.*, 2020), which, together with *P. callosus*, has been found to be present as mixed populations in association with agriculture in specific areas. This indicates that, in previous studies, the correct species identification of weevil species might no longer be valid. The larvae tend to feed on the roots of various host plants, including grapevine.

The adult weevils feed above ground on the fruit, leaves, shoots and fruit stalks, whereas the larvae feed on the roots of grapevine. Although damage to the roots by the larvae is usually not evident on the aerial parts of the plant, feeding by the adults causes shot-holing of leaves, stem notching and berry scarring in grapevine (Fig. 19.4) (Allsopp *et al.*, 2015).

Weevils are adapted to hot, dry summers and wet winters, with the eggs being laid in weeds and leaf litter. The larvae are subterranean, with the pupae cocooning in a soil cell, from which they emerge as non-flying adults that crawl up the trunk of the vine to feed on the leaves, shoots and berries in the grapevine canopy. Overwintering takes place during the larval stage (Blomefield *et al.*, 2015). An integrated management approach using chemical, physical

Fig. 19.4. (A) The adult banded fruit weevil (*Phlyctinus callosus*). (B) Shot-hole damage on grapevine leaves, caused by feeding of *P. callosus*.

and cultural control is implemented when populations cause a significant amount of damage in vineyards. Natural control with biological agents is limited, making chemical control, together with some physical and cultural control practices, the best solution. Such practices include exclusion barriers, kaolin particle film technology and weed control (Blomefield *et al.*, 2015).

The soil stages that offer the best window of opportunity for EPN soil application are the life stages that are in contact with the soil, which include the larval stage, the pupa and the emerging soft sclerotized adult, which moves through the soil to the surface (B.E. Dlamini *et al.*, 2019b). For IJs to be in contact with the adult weevil, either an aboveground canopy application should be considered, or novel ideas such as trunk bands with embedded IJs over or under which the adults must crawl or hide. The soil stages of the weevil offer a 2–4-week window of opportunity of exposure after soil application of EPNs.

The first laboratory study conducted on the virulence of EPNs on *P. callosus sensu lato* larvae and adults to EPNs was conducted by Ferreira and Malan (2014). Weevil larvae were screened at a rate of 400 IJs per insect and evaluated for infection (visually confirmed by dissection) after 4 days in 24-well bioassay trays. The nematode species used were two isolates of *H. bacteriophora* and one of *H. zealandica*. The results indicated that both species were able to infect weevil larvae, with an infection rate of 73% for *H. zealandica* and 42–67% for the *H. bacteriophora* isolates. The LC_{50} and LC_{90} values after 4 days of incubation were 96 and 278 IJs per larva, respectively. In the case of the field-collected adult banded fruit weevil, *H. bacteriophora* caused 45% infection, followed by *H. zealandica* at 16% and *S. khoisanae* at 13%. The nematodes were found to be inactive against weevil larvae at a temperature lower than 10°C and to infect only 11% of weevil larvae at 15°C, with the highest mortality (54%) occurring at 25°C. *H. zealandica* was able to complete its life cycle successfully after a period of 22 days in sixth-instar banded fruit weevil larvae (Ferreira and Malan, 2014).

H. zealandica was used to investigate the effect of the vertical movement of nematodes in sand only and in sand versus sandy loam soil columns on their virulence against weevil larvae (Ferreira and Malan, 2014). Barnes (1989) indicated that weevil larvae can occur up to 30 cm deep on the roots of plants. Plastic centrifuge tubes (15 cm long, 1.5 cm wide) were used. A banded fruit weevil larva was placed in an Ependorf tube pierced with ten holes using a heated needle (Kehres *et al.*, 2001). The tubes were placed at depths of 0, 5, 10 and 15 cm in each of ten sand-filled centrifuge tubes. *H. zealandica* was added by pipetting the IJs on to the surface of the sand of each tube, with the control treatments receiving water only. After closing the tubes with the lid, they were left vertically in a growth chamber at 25°C for 7 days. The mortality of the weevil larvae caused by infection was confirmed by means of dissection. A high percentage infection rate was obtained for weevil larvae for the different depths, ranging between 71% and 80%, with no significant difference between them. A higher percentage mortality of

82% was obtained using sandy loam soil in place of sand (68%) (Ferreira and Malan, 2014).

In a later study, B.E. Dlamini *et al.* (2019b) screened different EPN species for virulence against the different life stages of *P. callosus*, also using 24-well bioassay plates. *S. yirgalemense*, *H. noenieputensis* and an imported isolate of *S. feltiae* caused infection rates of 43–66% of *P. callosus* larvae, using 100 IJs per insect after 48 h, with no significant difference between the treatments. In the case of the *P. callosus* pupae, *H. indica* caused 70% infection and *H. baujardi* 67% compared with *H. noenieputensis* (55%). *H. indica* (95%) and *S. yirgalemense* (94%) induced a higher infection of *P. callosus* adults compared with the infection rate attained with the three other nematode species tested. *H. bacteriophora* field-collected adults induced significantly higher ($P < 0.05$) infection (53%) of the adult banded fruit weevil than *Heterorhabditis safricana* (37%) (B.E. Dlamini *et al.*, 2020a). The results indicate that all of the EPNs tested were effective against the pupal, larval and adult stages of *P. callosus* under laboratory conditions, with differences in infection among the EPN species tested (Fig. 19.5).

In field trials using *S. yirgalemense* at rates of 20 and 40 IJs cm^{-2}, infection rates of 69% and 78% were obtained for *P. callosus* larvae after only 48 h of field exposure in the orchard soil environment. Accordingly, *S. yirgalemense*, at low concentration, was found to control *P. callosus* effectively under field conditions. However, large-scale field trials are recommended to demonstrate the potential use of this biological control agent within IPM programs (B.E. Dlamini *et al.*, 2019a).

Fig. 19.5. Banded fruit weevil (*Phlyctinus callosus*) larva infected with *Steinernema yirgalemense* 4 days after infection.

In another study, conducted by B.E. Dlamini *et al.* (2020b), control of banded fruit weevil was improved by combining entomopathogenic fungi (EPFs) with EPNs under laboratory conditions. The combination of commercial isolates of *Beauveria bassiana* (Bb1 and Bb2), *Metarhizium anisopliae* (Ma1) and a South African isolate *M. anisopliae* (MaEA2), together with *S. yirgalemense* against banded fruit weevil adults and larvae was evaluated in laboratory bioassays. Four treatment regimens were compared: the banded fruit weevils were exposed, separately, to the EPFs and EPNs, as well as to a combination of EPFs and *S. yirgalemense*, either simultaneously, or 1 or 2 weeks apart. The combined application of the different EPF isolates together with *S. yirgalemense* increased the mortality of the weevils. With application of *S. yirgalemense* 7 or 14 days after application of the Bb2 and Bb1 isolates, 100% mortality was obtained for both the larvae and the adults. Synergy was noted with Bb1, Bb2, Ma1 and MaEA2 when *S. yirgalemense* was applied 7 or 14 days after EPF application. The interactions showed that the EPFs and *S. yirgalemense* had an additive effect when they were applied simultaneously, with a synergistic effect being achieved when they were applied 7 or 14 days after application of the different EPFs. These results indicate that a combination of EPFs and EPNs shows potential as an economically viable control strategy that should be investigated under field conditions (B.E. Dlamini *et al.*, 2020a).

In the case of weevils, the application of EPNs should be targeted against the soil stages concerned, as such stages were shown to be much more susceptible than the aboveground adult stage. The sporadic occurrence and flare-ups of weevils in vineyards could be reduced by inoculation of vineyards lacking any residential EPNs of their own and yearly maintenance. The inoculation of hot spots in vineyards with EPNs should also have no effect either on the current IPM systems or on the natural enemies that are present in the vineyards.

19.3.4 False codling moth

False codling moth (FCM), *Thaumatotibia leucotreta*, which is endemic to Africa, is an important pest of several deciduous, subtropical fruit and vegetable crops in most production areas in

South Africa, including grapevine. FCM larvae burrow into the fruit, with feeding lesions diminishing the marketability of the fruit. Primary damage is caused by larval feeding within the fruit, which induces premature ripening, fruit drop and phytosanitary issues. Secondary damage is caused by the larval entry wounds into the fruit, which serve as pathways for the establishment of fungal and bacterial infections, resulting in postharvest decay. Most of the overwintering population consists of pupae, which are formed on the soil surface (Malan *et al.*, 2018). Current FCM control options include IPM methods such as chemical insecticides, orchard sanitation, sterile insect technique and mating disruption.

Most of the research that has been conducted over the last 15 years in South Africa has been targeted at the use of EPNs against FCM in citrus orchards, with it previously having been regarded as a pest on citrus only in South Africa. However, over the last few years, it has become clear that this pest has switched to other hosts (Malan *et al.*, 2018). For table grapes in particular, FCM has been of great importance to South Africa. Currently, FCM is an important pest in most South African industries, with the most important being citrus, grapevine, persimmon and avocado. No control treatment is currently directed toward the soil stages of FCM, with EPNs having the potential to fill this existing niche.

Laboratory bioassays using 24-well bioassay disks showed the six local EPN species found during the survey by Malan *et al.* (2011) to be highly virulent against the last instar of FCM larvae. *S. yirgalemense* at a rate of 50 IJs per larva caused 100% mortality, and caused 74% mortality at a rate of 200 IJs per pupa. Using a sand bioassay, *S. yirgalemense* gave 93% control of the cocooned pupae and emerging moths at a rate of 20 IJs cm^{-2}. This was the first report on the potential use of EPNs to control the soil-borne life stages of FCM, including larvae, pupae and emerging moths. Emerging FCM moths in the soil examined were found to be infected with nematodes, which might have aided their control and dispersal (Malan *et al.*, 2011). Laboratory bioassays using *H. bacteriophora* conducted on orchard soil showed the LC_{50} to be 3 IJs and the LC_{90} to be 58 IJs per FCM larva (Fig. 19.6) (Malan and Moore, 2016). In a field trial, after application of three rates (20, 40 and 80 IJs cm^{-2}) of *H. bacteriophora*, mortality of over 90% was obtained for the FCM larvae, with a significant difference being found among the lower, but not among the higher, rates. At 21 days after application, FCM control ceased. With the field application of *H. zealandica*, after 6 days no significant differences were found in the FCM mortality of

Fig. 19.6. Late-instar false codling moth (*Thaumatotibia leucotreta*) larvae infected with *Heterorhabditis bacteriophora*.

over 80% between three different nematode rates (5, 10 and 20 IJs cm^{-2}). After 21 and 35 days, respectively, no significantly different FCM mortality was found for any of the treatments. In a field trial using three nematode species, treatment with *H. zealandica* resulted in significant control for each day of evaluation, up to day 49. Thus, the results obtained from field trials conducted in citrus orchards showed that local EPN species have potential for the control of the soil stages of FCM, with the added possibility of good persistence (Malan and Moore, 2016).

Steyn *et al.* (2019) investigated the infection rate, quality and age of *in vitro* liquid-cultured *Steinernema jeffreyense* in the laboratory and in field trials. Additionally, pre- and post-application pathogenicity tests were conducted with FCM larvae to assess the EPN virulence of the *in vitro*-cultured nematodes. Field trials were conducted by applying IJs to the vineyard soil. Four different *S. jeffreyense* rates (0, 10, 20 and 30 IJs cm^{-2}) were applied to 40 × 1 m^2 experimental plots that were artificially infested with FCM larvae. To test the immediate effect of exposure to the nematodes in the soil, the insects were retrieved from the soil after 48 h of exposure. The FCM larvae-loaded cages were replaced over a period of 4 weeks to determine the persistence of the original application. In a second trial, following the same procedure, *S. jeffreyense* and *S. yirgalemense* were compared for their virulence and persistence over a 4-week period. In the laboratory, the *in vitro*-cultured EPNs proved to be of similar quality to the *in vivo*-cultured *S. jeffreyense*, with a high mortality of over 80%. The semi-field study showed promising results, with the immediate effect yielding up to 77% mortality of the larvae, which remained at more than 35% over the 4-week period following application. These results compare favorably with those from previous field studies using *in vivo* EPNs, proving that EPNs can form a valuable part of an IPM program aimed at the control of FCM (Steyn *et al.*, 2019).

In a study conducted by Steyn *et al.* (2021), EPNs were tested against FCM larvae and fully formed pupae, using both *in vivo*- and *in vitro*-cultured (Dunn *et al.*, 2020) EPN IJs. The eggs were shown to be susceptible (30–65%) to several different EPN species at 200 IJs per 50 µl. However, only when the fully formed larva chewed a hole through the eggshell were the IJs able to access the egg and penetrate the larva before it left the shell. *Xenorhabdus indica*, the symbiotic bacterium of *S. yirgalemense*, was also tested for mortality of FCM eggs, with significantly more mortality being obtained with the test sample than with the control. The EPNs were found to be highly virulent against the FCM larvae, with *S. yirgalemense* causing 100% mortality in all laboratory trials conducted. The fully formed pupae were the most resistant of the immature stages, with the EPNs causing low mortality (6–33%) at 100 IJs per insect. The results obtained differed in virulence from those of Malan *et al.* (2011), who experienced a much higher infection rate (27–74%). However, in the latter study, the incompletely formed pupa was used, which could explain the higher rate of infection.

The FCM larvae were highly susceptible to both *S. yirgalemense* and *Metarhizium robertsii* under laboratory conditions (Steyn *et al.*, 2021). The use of these two biologicals in combination should be further investigated, both separately and in combination, under field conditions, to illustrate the benefits of including them in an FCM IPM program.

19.3.5 Western flower thrips

Western flower thrips (WFT), *Frankliniella occidentalis*, are a polyphagous species with an extremely wide host range, including table grapes. Feeding by the larvae on developing grape berries results in russet-like scarring, whereas feeding by the larvae on young leaves and growth tips causes severe scarring and stunted growth. Egg laying by females causes halo spot scarring around the oviposition site. Shoot scarring and stunted growth are serious problems in those table grape areas where grapes are harvested early in the growing season (T.M. Dlamini *et al.*, 2020). WFT is known to rapidly develop resistance to pesticides (Allsopp *et al.*, 2015). Although thrips are not currently regarded as a major pest of grapevine, problems are experienced with this pest in several different regions and are carefully monitored. EPNs can be effective in protected conditions. In South Africa, shade netting is used in the cultivation of table grapes, which offers an opportunity for aboveground application of EPNs.

The results from laboratory pre-screening of local and one exotic EPNs (five *Heterorhabditis* and seven *Steinernema* spp.) showed varying mortality, ranging between 11% and 67%, at a rate of 100 IJs per insect against the second-instar larva of WFT (T.M. Dlamini *et al.*, 2019b), with *Heterorhabditis* spp. (57%) being found to outperform *Steinernema* spp. (32%). WFT was found by Griffin *et al.* (2005) and Belay *et al.* (2005) to be more susceptible to *Heterorhabditis* spp. than to *Steinernema* spp. *Heterorhabditis baujardi* outperformed the others, although it did not differ significantly from *S. yirgalemense* (Fig. 19.7). The difference involved might relate to various factors, including nematode body size, as *H. baujardi* is smaller than *S. yirgalemense*. Low levels of mortality were recorded with *Steinernema litchi*, *S. khoisanae*, *S. innovationi* and *S. jeffreyense*, which was expected, as they belong to the *Khoisanae* clade, with IJs greater than 1 mm long. Their size might explain their poor infectivity to WFT, whose second-instar larva is approximately 1.49–1.79 mm in length. With the size of the nematode depending on the insect to be controlled, the impact on virulence was found to be significant; similar results were also obtained in research on *Bradysia impatiens* (Bastidas *et al.*, 2014; Katumanyane *et al.*, 2018).

Fig. 19.7. Second-instar larva of thrips (*Frankliniella occidentalis*) infected *Steinernema yirgalemense*.

Currently, EPNs are not considered for full-scale aboveground application against WFT. However, should mealybugs be targeted for the undercover production of table grapes, the control of thrips will most probably be a secondary effect.

19.3.6 Ground pearls

Ground pearls, *Margarodes* spp., are indigenous to South Africa, with *Margarodes prieskaensis* being found in the summer rainfall regions of the Orange River Valley in the Northern Cape province, as well as in the table grape-growing areas in the Limpopo and Mpumalanga provinces of South Africa (Allsopp *et al.*, 2015).

Feeding by the nymphs on the roots results in reduced vigor of the grapevines, eventually leading to death. *M. prieskaensis* causes major losses in the lower Orange River area, where it is regarded as a serious pest of table grapes. Once the nymph has moved on to a host plant root, where it stays, it starts to feed on the available plant sap. The second-instar nymph secretes layers of wax that form a characteristic cyst, which is resistant to adverse conditions and which can remain viable for many years in the soil. Females emerge from the cyst, move to the soil surface and secrete pheromones to attract males. After mating, the females burrow back into the soil, where they lay their eggs near the roots (Allsopp *et al.*, 2015). All life stages of *Margarodes* spp., except for the adult males, remain in contact with the soil. This type of life cycle indicates a good opportunity for the application of EPNs for purposes of control.

T.M. Dlamini *et al.* (2019a) investigated five locally isolated EPN species (*H. bacteriophora*, *H. baujardi*, *H. indica*, *H. noenieputensis* and *S. yirgalemense*), which they tested in the laboratory for virulence against the females of *M. prieskaensis*. The EPN species tested caused different mortality rates, following inoculation with a high concentration of nematodes. However, only in extremely rare cases were the nematodes able to develop inside the *M. prieskaensis* females (Fig. 19.8), although the life cycle of the nematodes involved was not completed. From this study, it can be concluded that the five local EPNs tested cannot be used as a biological control option for *M. prieskaensis* (T.M. Dlamini *et al.*, 2019a).

Fig. 19.8. (A) A female ground pearl (*Margarodes prieskaensis*). (B) Extremely rare case of infection of a female *M. prieskaensis* with *Heterorhabditis bacteriophora*.

However, other EPN species, especially those in the other provinces, should be tested for pathogenicity against *Margarodes* spp. as a potential future biological control agent.

19.3.7 Trimen's false tiger moth

Trimen's false tiger moth, *Agoma trimenii*, is an indigenous moth that is widespread throughout southern Africa. The larvae target the young shoots and leaves of wine, table and raisin grapes in the Northern Cape and Limpopo provinces of South Africa (Morris *et al.*, 2020a). Larval damage on grapevine leaves results in defoliation of vines. The larvae are often seen suspended from silken threads in vineyards, later pupating in the soil at the base of the grapevines (Allsopp *et al.*, 2015). The pupal stage, which is a soil cocoon that is open to the soil surface, forms near the trunk of the grapevine plant, offering an opportunity for the soil application of EPNs (Fig. 19.9). The rest of the life cycle, from eggs to the prepupal phase, should be open to aboveground applications (Morris *et al.*, 2020a).

The susceptibility of *A. trimenii* larvae and pupae to *S. yirgalemense* and *H. noenieputensis* was tested on site by mimicking laboratory conditions. The field-collected larvae of different larval instars were inoculated using a concentration of 100 IJs per 50 µl of water. The inoculated larvae were left in a plastic container. After 48 h, 100% mortality of the larval stage was found (Morris *et al.*, 2020b). Excellent results were obtained with the larvae of all stages, and these trials should be repeated with lower concentrations of EPNs. Movement of these insects for research purposes is currently restricted due to the ease of distribution to other grapevine-producing areas.

The same procedures were followed with the heavily armored pupae (Morris *et al.*, 2020b). To determine whether the pupae were dead or alive, their tail end was touched on to a hot surface, which results in movement if they are alive. Unfortunately, no infection was obtained for the pupae, which were subjected to infection with the two nematode species. This was ascribed to the heavily protected pupal case, which did not allow penetration of the nematodes. In future studies, the prepupal soil stage of *A. trimenii* should be screened for susceptibility to EPNs, as the fully formed pupae are resistant nematode infection.

The results of this study indicate the excellent potential of EPNs as an aboveground

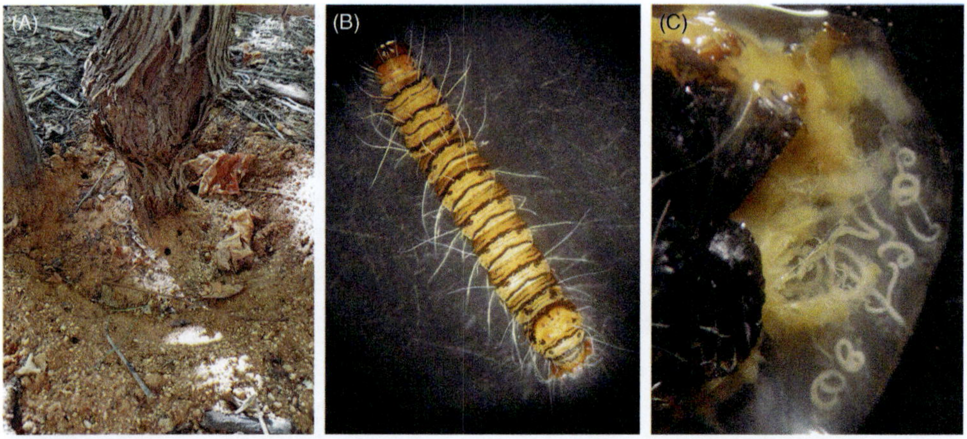

Fig. 19.9. (A) Pupal case of Trimen's false tiger moth (*Agoma trimenii*) open to the surface of the soil, close to the trunk of the vine. (B) *A. trimenii* larva. (C) Larva infected with *Steinernema yirgalemense*. Photographs courtesy of C. Morris, Stellenbosch University, South Africa.

biological control agent against the larvae of *A. trimenii*, because of their high susceptibility, especially in terms of application to small areas with high infestation, without disrupting an IPM program (Morris *et al.*, 2020a,b).

19.3.8 Cape leaf miner

The larval instars of the cape leaf miner, *Holocacista capensis*, are obligate leaf miners that are present throughout the grapevine-growing season. *H. capensis*, which is widely distributed throughout the Western Cape province, occurs at relatively high abundance in some of the major table grape-producing regions (Steyn *et al.*, 2020b).

The mining activities of the larvae cause physical damage to infested grapevine leaves. High-density infestations are usually found late in the growing season. Leaf miners overwinter in the pupal stage within a cocoon casing, usually either under the under the bark of grapevine stems or in the leaf litter.

The complete life cycle of *H. capensis* occurs above ground. Their eggs are laid singly in the leaves, with the hatched larvae having four instars, feeding only on the grapevine leaf tissue concerned. After first encasing itself in a cocoon, the final-instar larva then descends in its cocoon casing from the leaf by means of a silken thread that it attaches to objects in its immediate surroundings such as leaves, trellis posts or grape bunches, whereupon the cocooned larva fastens itself firmly to the object. The cocoon casings are problematic when they are present on fruit intended for consumption (van Nieukerken and Geertsema, 2015; Torrance, 2016).

Steyn *et al.* (2020a) conducted a study to determine the susceptibility of the leaf-mining *H. capensis* larvae to seven different EPN species. Occupied mines (each containing a single larva) were carefully excised from field-collected leaves. The leaf pieces containing leaf-mining galleries, occupied by the feeding larvae, were inoculated with 200 IJs per insect. A high level of infection of the leaf-mining larvae was obtained with *H. baujardi* (92%), *H. noenieputensis* (85%) and *H. indica* (83%) (Fig. 19.10). Concentration assays were conducted, with the lethal concentration being determined for the three most virulent species using rates of 0, 25, 50, 100, 200 and 400 IJs per leaf-mining larva. Almost double the number of *H. noenieputensis* (34 nematodes per insect) penetrated the insect larvae, compared with the other two EPNs. However, the relative potency of *H. baujardi* was found to be 3.56 times higher than that for *H. indica*, while that of *H. indica* was 2.57 times higher than for

Fig. 19.10. Cape leaf miner (*Holocacista capensis*) infected with first-generation *Heterorhabditis noenieputensis*. Photograph courtesy of Leigh Steyn, Stellenbosch University, South Africa).

H. noenieputensis. The results obtained in the laboratory were found to be encouraging, especially regarding the nematodes' ability to penetrate the leaf-mining galleries present and to infect the larvae successfully (Steyn *et al.*, 2020a).

In the case of table grapes, an aboveground application to the canopy of the grapevine could be effective, as it was shown that *H. capensis* larvae in a leaf gallery tend to be highly susceptible to nematode infection.

19.3.9 Lobesia moth

Lobesia moth, *Lobesia vanillana*, which is a sporadic pest of wine grapes in the Western Cape of South Africa, has been reported from the Breede River Valley. In some instances, feeding damage, if not properly controlled, quickly reaches economic thresholds. Signs of larval infestation are frass, webbing and bunch rot. Biological control agents for *L. vanillana* are unknown, and the growers concerned tend to have to depend on the use of synthetic pesticides (du Preez, 2019).

Observations of *L. vanillana* indicate that the species lacks a soil stage. They have a life cycle of 4–5 weeks, and overwinter as pupae, being regarded as a sporadic pest of table grapes. *L. vanillana* was successfully reared from field-collected larvae using an agar-based modified codling moth diet (du Preez, 2019). EPNs are promising as biological control agents of this pest, especially in terms of application to table grapes during harvesting when they are part of an IPM program.

Laboratory-cultured last-instar larvae of *L. vanillana* were evaluated against *S. jeffreyense* and *S. yirgalemense* with regard to pathogenicity under laboratory conditions (Fig. 19.11) (du Preez *et al.*, 2021). The two nematode species used in the study were cultured both *in vivo* (van Zyl and Malan, 2014a,b, 2015) and *in vitro* (Dunn *et al.*, 2020). The results from 24-well bioassay plates showed that *L. vanillana* larvae were susceptible to all nematode treatments, resulting in more than 72% infection. No significant difference in infection between the *in vivo*- and *in vitro*-cultured nematodes of the same EPN species was found. However, in the *in vitro*

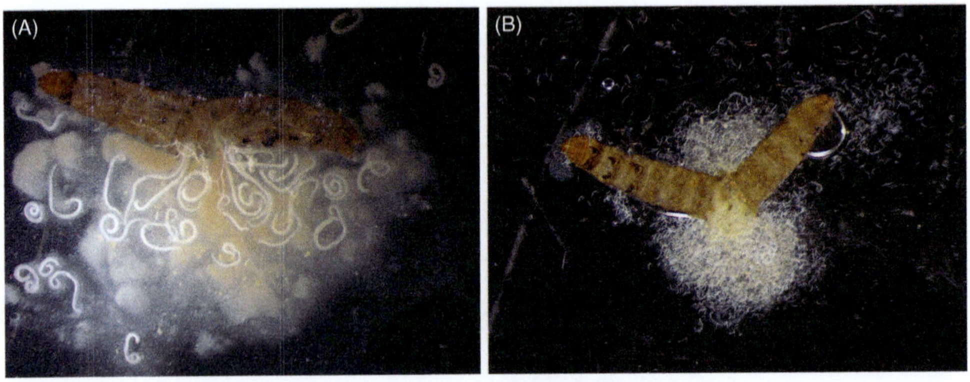

Fig. 19.11. *Steinernema yirgalemense*-infected last-instar larvae of lobesia moth (*Lobesia vanillana*). (A) First-generation larva. (B) Infective juveniles leaving the cadaver.

culture, *S. yirgalemense* (98%) performed significantly better than *S. jeffreyense* (73%).

Confirmation of the ability of IJs to complete their life cycle *in vivo* in *L. vanillana*, coupled with their ability to produce a new cohort of IJs, suggests that they might be able to provide control in optimum environments. An application of EPNs above ground in hot spots can be considered for the control of sporadic pest flares.

19.3.10 Katydid

Katydid, *Plangia graminea*, is a minor and sporadic pest of vineyards in South Africa. Locally known as "krompokkels" (Afrikaans), the genus is still taxonomically unclear, with little being known about the biology and ecology of its members. Although, under normal conditions, katydids are unobtrusive, with little agricultural importance, observations conducted since 2012 indicate an increase in katydid abundance and damage intensity. On grapevines, nymphs and adults tend to feed on foliage within the canopy (Fig. 19.12) (du Preez, 2019).

Observations on the biology and ecology of *P. graminea* on South African grapevine suggest that they lack a soil-dwelling life stage, having only a single generation per year (Hemp *et al.*, 2015; du Preez *et al.*, 2022). As the *P. graminea* life cycle is completely above ground, EPN application needs to take place in the canopy of the vines.

In a study conducted by du Preez *et al.* (2022), the nymphs of *P. graminea* were evaluated against 12 *in vivo*-cultured EPNs in laboratory bioassays. As the katydids did not perform well in captivity, field-collected nymphs had to be used in the laboratory bioassays. Therefore, the nymphs that were used in this study were hand collected and treated soon after collection. The pathogenicity and reproductive potential of the nematodes in the host were investigated. Seven locally occurring nematode species achieved a significant amount of infection, with *H. zealandica*, *H. indica*, *S. jeffreyense* and *S. yirgalemense* performing the best, with an infection rate of over 90%. *H. zealandica* (SF41 strain), *H. indica* (SGS strain), *Steinernema jeffreyense* (J192 strain), *S. yirgalemense* (157-C strain) and *H. baujardi* (MT19 strain), resulted in over 82% mortality of *P. graminea* nymphs after 48 h. By being able to complete their life cycle, their *in vivo* reproduction ability was also confirmed.

As katydids are currently regarded as a sporadic pest of grapevine in South Africa, their presence has been investigated in table grapes for the potential of hot-spot, sporadic application. EPN applications do not interrupt any management system in vineyards.

19.4 Groundcover Management in Vineyards

In South Africa, the use of annual cover crops is an established cultivation practice in vineyards, as it is generally regarded as being essential for all the added benefits involved, such as moisture

Fig. 19.12. (A) Adult male katydid (*Plangia graminea*). (B) Nymphs of *P. graminea* on a grapevine leaf, with visible feeding damage. Photographs courtesy of F. du Preez, University of Stellenbosch, South Africa.

retention and increased water infiltration (Fourie, 2022). During 2013, research into the use of cover crops, including species of Brassicaceae with biofumigation properties, with the aim of suppressing plant-parasitic nematodes when used as a green manure, were conducted (Kruger *et al.*, 2013, 2015a,b). Although biofumigation was relatively ineffective, due to the limitation in irrigation water, a steady decrease in the number of plant-parasitic nematodes was found over a period of 3 years. The effect of biofumigation on EPN occurrence has not, however, been investigated.

In Spain, the impact of cover-crop management on EPNs was investigated (Blanco-Péreza *et al.*, 2020), as well as the sowing of annual seed, flower-driven and spontaneous cover-crop management on EPNs. The results from the study showed increased EPN abundance and activity in spontaneous cover crops. To support conservation biological control, especially in terms of EPNs, the use of spontaneous cover was found to be the most promising strategy of all (Blanco-Péreza *et al.*, 2020).

19.5 Potential Use of Entomopathogenic Nematodes in an Integrated Pest Management System

In South Africa, the grapevine industries, especially in the case of table grapes, are willing to invest in research regarding the use of biological control agents in an IPM system. In table grapes in particular, the aboveground application of EPNs to the grapevine canopy bearing grapevine bunches is likely to prove advantageous regarding the maximum residue limits in terms of the export of fruit. Table grapes also offer ideal conditions for aboveground application of EPNs, as such applications can occur in the shade, with drip-off on to the soil.

19.6 Availability and Application of Commercial Entomopathogenic Nematode Products

Currently, in South Africa, registration is available for different imported products of both *H. bacteriophora* and *S. feltiae* (Hatting *et al.*, 2019; Moore, 2021). However, *S. feltiae* is not found in South Africa, nor on the rest of the African continent, except for Ethiopia, where it has been found to be one of the most common EPN species present. Applying inundative quantities of a formulated exotic product, with *S. feltiae* as the active ingredient, is regarded by some as biological pollution, with such pollution possibly impacting on the natural occurrence of *S. feltiae* and other EPN species in South Africa.

19.7 Advances in Mass Culture and Formulation

In a recent review, the conditions required for optimal EPN production in liquid culture and how these conditions might be optimized for South African EPN species were assessed (Dunn *et al.*, 2021a; see also Chapter 10, this volume). In this study, an initial attempt was made using *in vitro* liquid mass culture of a local South African isolate, *S. jeffreyense*, in 250 ml Erlenmeyer flasks, together with their mutualistic bacterial species, *Xenorhabdus khoisanae*. After successful *in vitro* production of the *S. jeffreyense* inoculum, different parameters for optimizing IJ recovery (during the development step, when the IJs molt to initiate the life cycle) and yield were investigated. With 30 ml of liquid medium, the mean percentage recovery of IJs after 6 days was 86%, with a yield of approximately 122,000 IJs ml^{-1} being achieved after 14 days (Dunn *et al.*, 2020, 2021b,c).

For EPNs to be commercially applied in the case of grapevine in South Africa, it is important to develop a formulation for mass-cultured nematodes, as the large-scale use of EPNs as biological control agents is impaired by their short shelf life (see Chapter 11, this volume). Three local South African EPN species, *S. yirgalemense*, *S. jeffreyense* and *H. bacteriophora*, were investigated in formulation. Encapsulation of the IJs in either alginate beads and diatomaceous earth was investigated. Survival of the IJs in the formulations was determined at 6°C, 14°C and 25°C for 4 weeks. Of the IJs concerned, 10–20% were observed to escape from the beads, depending on the prevailing temperature, and they readily survived the encapsulation process. The diatomaceous earth did not cause desiccation of the nematodes, with there still being a relatively low mortality rate by week 4 of the study. In both formulations, the survival and virulence rates differed significantly at 6°C compared with 14°C and 25°C, with a drastic decrease over time for *S. yirgalemense*. The EPN species revealed poor survival and loss of virulence at low temperatures in both formulations. The alginate beads, which successfully retained most of the IJs concerned, can be stored for a relatively long length of time (Kagimu and Malan, 2019).

19.8 Conclusions and Prospects

The South African viticulture industry is very much focused on the organic production of grapevine. In biodynamic wine production, using organic farming methods such as compost as a fertilizer, combined with the use of cover crops and the minimum use of pesticides, EPNs have the potential to play an important role in the integration of an IPM system. The same applies in the production of table grapes, as far as maximum residue limit problems are concerned regarding the export of grapes. The South African grapevine industry, which currently has a positive outlook regarding the funding of applied research in terms of making a local EPN available as a commercial product in South Africa, is actively involved in developing a commercial product that will become available for producers in the near future.

19.9 References

Addison, P. (2002) Chemical stem barriers for the control of ants (Hymenoptera: Formicidae) in vineyards. *South African Journal of Enology Viticulture* 23, 1–8.

Allsopp, E. and Fourie, J.C. (2019) Survival of vine mealybug, *Planococcus ficus* (Signoret) (Hemiptera: Pseudococcidae), on grapevine root remnants in soil in the Western Cape province, South Africa. *South African Journal of Enology and Viticulture* 40, 253–259.

Allsopp, E., Barnes, B.N., Blomefield, T.L. and Pringle, K.L. (2015) Grapevine. In: Prinsloo, G.L. and Uys, V.M. (eds) *Insects of Cultivated Plants and Natural Pastures in Southern Africa*. Entomological Society of Southern Africa, Pretoria, pp. 420–437.

Barnes, B.N. (1989) Embryonic and immature stages of *Phlyctinus callosus* Boh. (Coleoptera, Curculionidae): aspects of biology and behaviour with respect to control in deciduous fruit orchards. *Journal of the Entomological Society of Southern Africa* 52, 165–178.

Bastidas, B., Portillo, E. and San-Blas, E. (2014) Size does matter: the life cycle of *Steinernema* spp. in micro-insect hosts. *Journal of Invertebrate Pathology* 121, 46–55.

Belay, D., Ebssa, L. and Borgemeister, C. (2005) Time and frequency of applications of entomopathogenic nematodes and their persistence for control of western flower thrips *Frankliniella occidentalis*. *Nematology* 7, 611–622.

Bell, V.A., Bonfiglioli, R.G.E., Walker, J.T.S., Lo, P.L., Mackay, J.F. and McGregor, S.E. (2009) Grapevine leafroll-associated virus 3 persistence in *Vitis vinifera* remnant roots. *Journal of Plant Pathology* 91, 527–533.

Bergh, J.C. (2012) Grape root borer. In: Bostanian, N., Vincent, C. and Isaacs, R. (eds) *Arthropod Management in Vineyards*. Springer, Dordrecht, Netherlands.

Blanco-Péreza, R., Sáenz-Romob, M.G., Vicente-Díeza, I., Ibáñez-Pascuala, S., Martínez-Villarb, E. *et al.* (2020) Impact of vineyard ground cover management on the occurrence and activity of entomopathogenic nematodes and associated soil organisms. *Agriculture, Ecosystems & Environment* 303, 107028.

Blomefield, T.L., Barnes, B.N. and Gilliomee, J.H. (2015) Peach and nectarine. In: Prinsloo, G.L. and Uys, V.M. (eds) *Insects of Cultivated Plants and Natural Pastures in Southern Africa*. Entomological Society of Southern Africa, Pretoria, pp. 320–339.

Booysen, E., Malan, A.P. and Dicks, L.T.M. (2022) Colour of *Heterorhabditis zealandica*-infected-*Galleria mellonella* dependent on the *Photorhabdus* symbiont, with two new nematode–symbiotic associations reported. *Journal of Invertebrate Pathology* 189, 107729.

Çimen, H., Lee, M.-M., Hatting, J., Hazir, S. and Stock, S.P. (2014) *Steinernema tophus* sp. n. (Nematoda: Steinernematidae), a new entomopathogenic nematode from South Africa. *Zootaxa* 3821, 337–353.

Dlamini, B.E., Malan, A.P. and Addison, P. (2019a) Control of the banded fruit weevil *Phlyctinus callosus* (Coleoptera: Curculionidae) using entomopathogenic nematodes. *Austral Entomology* 58, 687–695.

Dlamini, B.E., Malan, A.P. and Addison, P. (2019b) A review of the biology and control of *Phlyctinus callosus* (Schönherr) (Coleoptera: Curculionidae), with special reference to biological control using entomopathogenic nematodes and fungi. *African Entomology* 27, 279–288.

Dlamini, B.E., Malan, A.P. and Addison, P. (2020a) Combined effect of entomopathogenic fungi and *Steinernema yirgalemense* against the banded fruit weevil, *Phlyctinus callosus* (Coleoptera: Curculionidae). *Biocontrol Science and Technology* 30, 1169–1179.

Dlamini, B.E., Malan, A.P. and Addison, P. (2020b) Entomopathogens from agricultural soil and their potential to control the banded fruit weevil, *Phlyctinus callosus* (Schönherr) (Coleoptera: Curculionidae). *African Entomology* 28, 374–384.

Dlamini, T.M., de Klerk, A. and Malan, A.P. (2019a) Controlling 'ground pearls' in vineyards. *SA Fruit Journal* 18, 45–47.

Dlamini, T.M., Allsopp, E. and Malan, A.P. (2019b) Efficacy of entomopathogenic nematodes against western flower thrips, *Frankliniella occidentalis* (Thysanoptera: Thripidae), under laboratory conditions. *African Entomology* 27, 322–335.

Dlamini, T.M., Allsopp, E. and Malan, A.P. (2020) Management of *Frankliniella occidentalis* (western flower thrips), and the potential use of entomopathogenic nematodes: a South African perspective. *African Entomology* 27, 265–278.

du Preez, F. (2019) Biological control of two sporadic grapevine pests, *Plangia graminea* and *Lobesia vanillana*, using entomopathogenic nematodes. MSc thesis, Stellenbosch University, Stellenbosch, South Africa.

du Preez, F., Malan, A.P. and Addison, P. (2021) Potential of *in vivo*- and *in vitro*-cultured entomopathogenic nematodes to infect *Lobesia vanillana* (Lepidoptera: Tortricidae) under laboratory conditions. *PLOS ONE* 16: e0242645.

du Preez, F., Malan, A.P. and Addison, P. (2022) Potential of entomopathogenic nematodes for the control of *Plangia graminea* (Orthoptera: Tettigoniidae) under laboratory conditions. *South African Journal of Enology and Viticulture* 43, 103–112.

Dunn, M.D., Belur, P. and Malan, A.P. (2020) *In vitro* liquid culture and optimization of *Steinernema jeffreyense*, using shake flasks. *BioControl* 65, 223–233.

Dunn, M., Belur, P.D. and Malan, A.P. (2021a) Effect of glucose, agar supplementation and bacterial cell density on the *in vitro* liquid culture of *Steinernema jeffreyense. African Entomology* 29, 423–434.

Dunn, M.D., Belur, P.D. and Malan, S.P. (2021b) Development of cost-effective media for the *in vitro* liquid culture of entomopathogenic nematodes. *Nematology* 24, 1–13.

Dunn, M.D., Belur, P.D. and Malan, A.P. (2021c) A review of the *in vitro* liquid mass culture of entomopathogenic nematodes. *Biocontrol Science and Technology* 31, 1–21.

Ferreira, T. and Malan, A.P. (2014) Potential of entomopathogenic nematodes for the control of the banded fruit weevil, *Phlyctinus callosus* (Schönherr) (Coleoptera: Curculionidae). *Journal of Helminthology* 88, 293–301.

Ferreira, T., van Reenen, C.A., Endo, A., Spröer, C., Malan, A.P. and Dicks, L.M.T. (2013) Description of *Xenorhabdus khoisanae* sp. nov., a symbiont of the entomopathogenic nematode *Steinernema khoisanae*. *International Journal of Systematic and Evolutionary Microbiology* 63, 3220–3224.

Fourie, J. (2022) *Cover Crops in South African Vineyards*. Wine Industry Network of Expertise & Technology, Suider Paarl, South Africa.

Geldenhuys, J., Malan, A.P. and Dicks, L.M.T. (2016) First report of the symbiotic bacterium *Photorhabdus luminescens* subsp. *laumondii* associated with *Heterorhabditis safricana* from South Africa. *Current Microbiology* 73, 790–795.

Griffin, C.T., Boemare, N.E. and Lewis, E.E. (2005) Biology and behaviour. In: Grewal, P.S., Ehlers, R.U. and Shapiro-Ilan, D.I. (eds) *Nematodes as Biocontrol Agents*. CAB International, Wallingford, UK, pp. 47–75.

Haran, J.M., Hansen, S., Benoit, L. and Addison, P. (2020) Description of five new species in the genus *Phlyctinus* Schoenherr (Coleoptera, Curculionidae): a first step in deciphering the *P. callosus* complex. *European Journal of Taxonomy* 669, 1–29.

Hatting, J.L., Stock, S.P. and Hazir, S. (2009) Diversity and distribution of entomopathogenic nematodes (Steinernematidae, Heterorhabditidae) in South Africa. *Journal of Invertebrate Pathology* 102, 120–128.

Hatting, J.L., Moore, S.D. and Malan, A.P. (2019) Microbial control of phytophagous invertebrate pests in South Africa: current status and future. *Journal of Invertebrate Pathology* 165, 54–66.

Hemp, C., Klaus-Gerhard Heller, K.-G., Warchałowska-Śliwa, E., Grzywacz, B. and Hemp, A. (2015) Review of the *Plangia graminea* (Serville) complex and the description of new *Plangia* species from East Africa (Orthoptera: Phaneropteridae, Phaneropterinae) with data on habitat, bioacoustics, and chromosomes. *Organisms Diversity and Evolution* 15, 471–488.

Holm, K. (2008) Construction of a cDNA library for the vine mealybug, *Planococcus ficus* (Signoret). MSc thesis, Stellenbosch University, Stellenbosch, South Africa.

James, M., Addison, P. and Malan, A.P. (2018) Surveying and screening South African entomopathogenic nematodes for the control of the Mediterranean fruit fly, *Ceratitis capitata* (Wiedemann). *Crop Protection* 105, 41–48.

James, T. (2013) *Wines of the New South Africa: Tradition and Revolution*. University of California Press, Oakland, California.

Kagimu, N. and Malan, A.P. (2019) Alginate and diatomaceous earth formulations of entomopathogenic nematodes. *BioControl* 64, 413–422.

Katumanyane, A., Ferreira, T. and Malan, A.P. (2018) Potential use of local entomopathogenic nematodes to control *Bradysia impatiens* (Diptera: Sciaridae) under laboratory conditions. *African Entomology* 26, 337–349.

Kehres, J., Denon, D. and Mauleon, H. (2001) A simple technique to estimate, *in situ*, population densities of an entomopathogenic nematode (*Heterorhabditis indica*) in sandy soils. *Nematology* 3, 285–287.

Krogh, K.A., Halling-Sørensen, B., Mogensen, B.B. and Vejrup, K.V. (2003) Environmental properties and effects of nonionic surfactant adjuvants in pesticides: a review. *Chemosphere* 50, 871–901.

Kruger, D.H.M., Fourie, J.C. and Malan, A.P. (2013) Cover crops with biofumigation properties for the suppression of plant-parasitic nematodes: a review. *South African Journal of Enology and Viticulture* 34, 287–295.

Kruger, D.H.M., Fourie, J.C. and Malan, A.P. (2015a) Control potential of Brassicaceae cover crops as green manure and their host status for *Meloidogyne javanica* and *Criconemoides xenoplax*. *South African Journal of Enology and Viticulture* 36, 165–174.

Kruger, D.H.M., Fourie, J.C. and Malan, A.P. (2015b) The effect of cover crops and the management thereof on plant-parasitic nematodes in vineyards. *South African Journal of Enology and Viticulture* 36, 195–209.

Le Vieux, P.D. and Malan, A.P. (2013a) The potential use of entomopathogenic nematodes to control *Planococcus ficus* (Signoret) (Hemiptera: Pseudococcidae). *South African Journal of Enology and Viticulture* 34, 296–306.

Le Vieux, P.D. and Malan, A.P. (2013b) An overview of the vine mealybug (*Planococcus ficus*) in South African vineyards and the use of entomopathogenic nematodes as potential biocontrol agents. *South African Journal of Enology and Viticulture* 34, 34–45.

Le Vieux, P.D. and Malan, A.P. (2015) Prospects for using entomopathogenic nematodes to control the vine mealybug, *Planococcus ficus*, in South African vineyards. *South African Journal of Enology and Viticulture* 36, 59–70.

Malan, A.P. and Manrakhan, A. (2009) Susceptibility of the Mediterranean fruit fly (*Ceratitis capitata*) and the Natal fruit fly (*Ceratitis rosa*) to entomopathogenic nematodes. *Journal of Invertebrate Pathology* 100, 47–49.

Malan, A.P. and Moore, S.D. (2016) Evaluation of local entomopathogenic nematodes for the control of false codling moth, *Thaumatotibia leucotreta* (Meyrick, 1913), in a citrus orchard in South Africa. *African Entomology* 24, 489–501.

Malan, A.P., Nguyen, K.B. and Addison, M.F. (2006) Entomopathogenic nematodes (Steinernematidae and Heterorhabditidae) from the southwestern parts of South Africa. *African Plant Protection* 12, 65–69.

Malan, A.P., Knoetze, R. and Moore, S.D. (2011) Isolation and identification of entomopathogenic nematodes from citrus orchards in South Africa and their biocontrol potential against false codling moth. *Journal of Invertebrate Pathology* 108, 115–125.

Malan, A.P., von Diest, J.I., Moore, S.M. and Addison, P. (2018) Control options for false codling moth, *Thaumatotibia leucotreta* (Lepidoptera: Tortricidae), in South Africa, with emphasis on the potential use of entomopathogenic nematodes and fungi. *African Entomology* 26, 14–29.

Moore, S.D. (2021) Biological control of a phytosanitary pest (*Thaumatotibia leucotreta*): a case study. *International Journal of Environmental Research and Public Health* 18: 1198.

Morris, C.A., Johnson, S.A., de Waal, J.Y. and Malan, A.P. (2020a) Review of Trimen's false tiger moth, *Agoma trimenii* (Lepidoptera: Agaristidae): seasonal biology, potential monitoring and control techniques. *South African Journal of Enology and Viticulture* 42, 128–132.

Morris, C.A., Malan, A.P., de Waal, J.Y. and Johnson, S. (2020b) Laboratory bioassays on the susceptibility of Trimen's false tiger moth, *Agoma trimenii* (Lepidoptera: Agaristidae), to entomopathogenic nematodes and fungi. *South African Journal of Enology and Viticulture* 41, 183–188.

Platt, T., Stokwe, N.F. and Malan, A.P. (2018) Potential of local entomopathogenic nematodes for control of the vine mealybug, *Planococcus ficus*. *South African Journal of Enology and Viticulture* 39, 208–215.

Platt, T., Stokwe, N.F. and Malan, A.P. (2019a) Foliar application of *Steinernema yirgalemense* to control *Planococcus ficus*: assessing adjuvants to improve efficacy. *South African Journal of Enology and Viticulture* 40, 13–19.

Platt, T., Stokwe, N.F. and Malan, A.P. (2019b) Grapevine leaf application of *Steinernema yirgalemense* to control *Planococcus ficus* in vineyard conditions. *South African Journal of Enology and Viticulture* 40, 75–83.

Platt, T., Stokwe, N.F. and Malan, A.P. (2020) A review of the potential use of entomopathogenic nematodes to control above-ground insect pests in South Africa. *South African Journal of Enology and Viticulture* 41, 1–16.

Rijal, J.P. and Bergh, J.C. (2017) Grape root borer (Lepidoptera: Sesiidae): an economic pest of commercial vineyards in the eastern United States. *Journal of Integrated Pest Management* 8: 30.

SATI (South African Table Grape Industry) (2021) Statistics of table grapes in South Africa. SATI, Paarl, South Africa. Available at www.satgi.co.za/statistics-booklet/ (accessed 21 March 2024).

SAWIS (South African Wine Industry Information and Systems NPC) (2020) South African wine industry 2020 statistics no. 45. SAWIS, Paarl, South Africa. Available at www.sawis.co.za (accessed 21 March 2024).

Stevenson, T. (2005) *The Sotheby's Wine Encyclopedia*. Dorling Kindersley Publishing, London.

Steyn, V.M., Malan, A.P. and Addison, P. (2019) Control of false codling moth, *Thaumatotibia leucotreta* (Lepidoptera: Tortricidae), using *in vitro*-cultured *Steinernema jeffreyense* and *S. yirgalemense*. *Biological Control* 138, 104052.

Steyn, L.A.I., Addison, A. and Malan, A.P. (2020a) Potential of South African entomopathogenic nematodes to control the leaf miner, *Holocacista capensis* (Lepidoptera: Heliozelidae). *South African Journal of Enology and Viticulture* 40, 301–309.

Steyn, L.A.I., Geertsema, H., Malan, A.P. and Addison, P. (2020b) A review of leaf-mining insects and control options for their management, with special reference to *Holocacista capensis* (Lepidoptera: Heliozelidae) in vineyards in South Africa. *South African Journal of Enology and Viticulture* 42, 318–332.

Steyn, V.M., Malan, A.P. and Addison, P. (2021) Efficacy of entomopathogens against *Thaumatotibia leucotreta* (Lepidoptera: Tortricidae) under laboratory conditions. *Entomologia Experimentalis et Applicata* 169, 449–461.

Torrance, L.A.I. (2016) The bio-ecology of the Cape grapevine leafminer, *Holocacista capensis* (Lepidoptera: Heliozelidae), in the Western Cape. MSc thesis, Stellenbosch University, Stellenbosch, South Africa.

van Nieukerken, E.J. and Geertsema, H. (2015) A new leafminer on grapevine and *Rhoicissus* (Vitacea) in South Africa within an expanded generic concept of *Holocacista* (Insecta, Lepidoptera, Heliozelidae). *ZooKeys* 507, 41–97.

van Zyl, C. and Malan, A.P. (2014a) Optimisation of inoculation techniques for *in vivo* mass culture of entomopathogenic nematodes through nematode and insect host manipulation. *African Entomology* 24, 405–416.

van Zyl, C. and Malan, A.P. (2014b) The role of nematodes as biological control agents, the history of nematode mass culture and *in vivo* nematode production. *African Entomology* 22, 236–249.

van Zyl, C. and Malan, A.P. (2015) Cost-effective culturing of *Galleria mellonella* and *Tenebrio molitor* and nematode production in various hosts. *African Entomology* 23, 361–375.

Venter, J.-H., Baard, C.W.L. and Barnes, B.N. (2021) Area-wide management of Mediterranean fruit fly with the sterile insect technique in South Africa: new production and management techniques pay dividends. In: Hendrichs, J., Pereira, R., Marc, J.B. and Vreysen, M.J.B. (eds) *Area-wide Integrated Pest Management*. CRC Press, Boca Raton, Florida, pp. 129–141.

Walton, V.M. and Pringle, K.L. (2004) A survey of mealybugs and associated natural enemies in vineyards in the Western Cape province, South Africa. *South African Journal of Enology and Viticulture* 25, 23–25.

Williams, R.N., Fickle, D.S., Grewal, P.S. and Meyer, J.R. (2002) Assessing the potential of entomopathogenic nematodes to control the grape root borer, *Vitacea polistiformis* (Lepidoptera: Sesiidae) through laboratory and greenhouse bioassays. *Biocontrol Science and Technology* 12, 35–42.

20 Vegetables

Weibin Ruan[1]* and Sheng-Yen Wu[2]

[1]College of Life Sciences, Nankai University, Tianjin, PR China; [2]State Key Laboratory of Ecological Pest Control for Fujian and Taiwan Crops, Ministerial and Provincial Joint Innovation Centre for Safety Production of Cross-Strait Crops, Fujian Agriculture and Forestry University, Fuzhou, PR China

Abstract

Vegetables contain vitamins, minerals, antioxidants and other plant secondary metabolites with health benefits. Some vegetables maintain a higher nutritional value when consumed raw. Insect pest infestations and damage can limit vegetable production globally. The major insect pests affecting various vegetables include beetles, leaf miners, aphids, whiteflies, thrips, cutworms, caterpillars, fungus gnats and mites. Pesticides are indispensable and widely used chemical compounds in agriculture used to reduce populations of pest insects and meet the increasing food needs associated with the rapid growth of the world population. However, if used incorrectly, they may harm consumers who ingest vegetables contaminated with pesticide residues. The increasing public concern over pesticide safety and potential environmental risks has led to growing attention toward environmentally friendly approaches for agricultural pest control. Entomopathogenic nematodes are lethal obligate parasites and can be powerful biological control agents for vegetable pests.

20.1 Introduction

The global vegetable farming industry has experienced intensive growth in recent decades. Fruits and vegetables, known for their concentrations of vitamins (especially vitamins C and A), minerals (particularly electrolytes) and phytochemicals (notably antioxidants) have historically been included in dietary guidance plans. Additionally, fruits and vegetables are recommended as a source of dietary fiber (Slavin and Lloyd, 2012). Including vegetables in the daily diet has been strongly associated with overall good health. Consumption of vegetables helps improve the immune system, gastrointestinal health and vision, and reduces the risks of various chronic diseases such as cancer, heart disease, stroke and diabetes (Peluso *et al.*, 2018). Increased vegetable production and consumption are expected to strengthen global vegetable exports and imports. The global production of vegetables reached 1.15 billion t in 2021 (FAO, 2022). Recent estimates suggest that China produces half of the world's vegetables (Wang *et al.*, 2021).

Insect pests and diseases pose serious threats to vegetable production, affecting crops from seedlings to harvest and potentially causing catastrophic losses for commercial vegetable producers. Crop contamination from insect excrement and other factors can also impact farm profits. Approximately half of the Earth's insect species (400,000–500,000) feed directly on the tissues

*Email: ruanweibin@nankai.edu.cn

 Entomopathogenic Nematodes as Biological Control Agents (D.I. Shapiro-Ilan and E.E. Lewis eds)
DOI: 10.1079/9781800620322.0020

of living plants, with every plant species serving as food for at least one insect species. Insects that feed on plants often act as vectors for plant pathogens that cause disease, spreading viruses, mycoplasmas, bacteria, protozoa, fungi and nematodes. Pesticides play a crucial role in preventing or controlling pests, diseases and other plant-related issues in vegetable production systems. However, despite their effectiveness, pesticides present serious environmental risks due to their adverse effects on non-target organisms in agroecological habitats (Schmidt-Jeffris *et al.*, 2021). Health risks have also increased due to pesticide residues on food and in drinking water for the general population (Guan, 2021).

Farmers must minimize the risk of chemical pesticides to consumers and workers by using the correct chemical agents at the proper dosage and applying them correctly to avoid overuse or inappropriate use. Therefore, there is an urgent need for effective and safe alternatives to conventional pesticides. One potential option is the use of bioinsecticides, including *Bacillus thuringiensis* and entomopathogenic nematodes (EPNs), which are effective biological control agents for many important insect pests of vegetable crops.

In this chapter, we describe how and under what circumstances EPNs can be employed for pest management in vegetable production. EPNs are often used as part of an integrated pest management (IPM) system, and are thus used in conjunction with other natural and chemical inputs to these systems.

20.2 Herbivores in Vegetables Controlled by Entomopathogenic Nematodes

20.2.1 Diamondback moth

Diamondback moth (DBM), *Plutella xylostella* (Lepidoptera: Plutellidae), is a major destructive pest of cruciferous crops such as cauliflower, cabbage and mustard, causing significant economic losses to farmers. The EPNs *Steinernema carpocapsae, Steinernema feltiae* and *Heterorhabditis bacteriophora* exhibited high levels of virulence against *P. xylostella*. The mortality varied with the insect host developmental stage (Eivazian Kary *et al.*, 2019), although only one generation of *S. carpocapsae* completed development in *P. xylostella* larvae (Ratnasinghe and Hague, 1998). Additionally, three EPN species in the genus *Heterorhabditis* from India showed sufficient virulence to induce 100% mortality in *P. xylostella* larvae under laboratory conditions, suggesting the potential of EPNs for integrated management practices (Vashisth *et al.*, 2017). In field studies, three Kenyan EPN isolates, *Steinernema* sp., *Steinernema weiseri* and *Heterorhabditis indica*, caused significant reductions in *P. xylostella* populations and plant damage (Nyasani *et al.*, 2008). Similarly, *Steinernema thermophilus* exhibited high virulence against *P. xylostella* infesting cabbage in a 2-year field experiment (Somvanshi *et al.*, 2006).

The application of EPNs to cabbage leaves against *P. xylostella* larvae requires the addition of formulation adjuvants to achieve satisfactory control. In field trials, a single application of 0.5 million *S. carpocapsae* infective juveniles (IJs) m^{-2}, combined with a surfactant polymer formulation containing 0.3% xanthan and 0.3% rimulgan, achieved a significant reduction in the number of insects per plant with more than 50% control after 7 days of exposure (Schroer *et al.*, 2005a). Regarding *S. carpocapsae* against *P. xylostella*, the addition of xanthan gum or potassium alginate for foliar application of *S. carpocapsae* resulted in a twofold increase in insect mortality at 80% relative humidity (RH) and a fivefold increase at 60% RH (Schroer *et al.*, 2005b). The best results were achieved with the surfactant rimulgan at 0.3% combined with 0.3% xanthan to retard sedimentation of EPNs in 500 ml of spraying mixture, resulting in *P. xylostella* mortality of over 90% at 80% RH and over 70% at 60% RH (Schroer *et al.*, 2005c). The addition of antidesiccants such as fire retardant gel (e.g. Barricade®) could improve the virulence of *Steinernema* spp. to control *P. xylostella* in greenhouse tests (Noosidum *et al.*, 2016).

The symbiotic bacteria of EPNs also have potential as pest-management materials in their own right. *P. xylostella* that fed on *Xenorhabdus nematophila* had higher mortality in the early instars than in the late instars (Jung and Kim, 2007). Nevertheless, the late instars of *P. xylostella* feeding on a mixture of *X. nematophila* and *B. thuringiensis* had significantly enhanced mortality because *X. nematophila* cells could be recovered from the hemocoel of *P. xylostella* in the feeding of the mixture, rather than with *X. nematophila* alone (Jung and Kim, 2007). A protein complex isolated from *X. nematophila* HB310 strain induced the death of *P. xylostella* by destroying midgut tissues and inhibiting midgut protease activities (Wang *et al.*, 2012).

20.2.2 Cutworms and armyworms

The black cutworm, *Agrotis ipsilon* (Lepidoptera: Noctuidae), is a widely distributed polyphagous insect pest of vegetables and field crops. In Petri dish trials, the nematode *H. bacteriophora* HP88 strain and four EPN isolates indigenous to Egypt (*H. bacteriophora* FLH-4-H, *H. indica* 216-H, *S. carpocapsae* E76-S and *S. feltiae* E76-S) were successfully used against the larvae of *A. ipsilon* (Yüksel and Canhilal, 2018).

In field trials, *S. carpocapsae* Mexican strain and *H. indica* LN2 strain reduced the damage caused by *A. ipsilon* and increased cabbage yield when compared with the control, showing better control effects than cyfluthrin and *Bacillus thuringiensis* (Yan *et al.*, 2014). However, in another study, the combination of both EPNs and *B. thuringiensis* did not result in significantly greater control than the nematodes used alone (Shamseldean and Ismail, 1997).

In Iran, larvae of the turnip moth, *Agrotis segetum* (Lepidoptera: Noctuidae), are one of the most serious pests that can attack nearly all vegetable crops. *S. carpocapsae* isolated from the Moghan region of Iran showed high efficacy against *A. segetum* under laboratory conditions (Ebrahimi *et al.*, 2019). It was demonstrated that two indigenous EPN species, *H. bacteriophora* and *S. carpocapsae*, isolated from Iran had good potential to control *A. segetum* under laboratory conditions, especially the final instar, which was the most susceptible stage followed by the prepupa and then the pupa (Goudarzi *et al.*, 2015). In a pot trial, the mortality level of the potato cutworm, *Agrotis deprivata*, applied as either an EPN aqueous suspension or EPN-infected cadaver was equivalent to that of treatment with the synthetic insecticides chlorpyrifos and cypermethrin (Lankin *et al.*, 2020).

In China, a native EPN species, *Steinernema ceratophorum* collected in Yunnan, showed great virulence against armyworms, including *Spodoptera litura*, *Spodoptera exigua*, *Spodoptera frugiperda* and *Helicoverpa armigera*, in Petri dish bioassays (Yan *et al.*, 2021). If the entomopathogenic fungus *Beauveria bassiana* was used to infect *Spodoptera littoralis* prior to the application of EPNs, more offspring of *S. riobrave* and *H. bacteriophora* were produced within the insect under laboratory conditions (Shaurub *et al.*, 2016). The mortality of *Spodoptera litura* larvae with foliar application of *S. carpocapsae* plus 0.25% or 0.5% Barricade® treatment was significantly higher (66.0% and 61.5%, respectively) than in the *S. carpocapsae* plus tap water treatment (29.5%), and this treatment also resulted in significantly less leaf damage (11.0–11.1%) compared with control treatments (33.8–61.6%) (Noosidum *et al.*, 2016). In addition, *H. bacteriophora* had an additive effect against *Spodoptera littoralis* larvae when all the tested additive compounds (ascorbic acid, citric acid, glucose, l-alanine, potassium chloride and sodium carbonate) were combined with it under laboratory conditions, except for glucose (0.1%), which was antagonistic to *H. bacteriophora* and decreased the mortality of the larvae (Amer *et al.*, 2014). Similarly, all compounds had an additive interaction with the EPN *Steinernema abbasi* against *Spodoptera littoralis* larvae, except for ascorbic acid (0.025%), glucose (0.1%) and sodium carbonate (0.025%), which had antagonistic interactions with *S. abbasi* (Amer *et al.*, 2014). Moreover, the symbiotic bacteria *Xenorhabdus* sp. and *Photorhabdus* sp., isolated from *Steinernema monticolum* and *Heterorhabditis megidis* respectively, showed a synergistic effect in killing *Spodoptera exigua* by oral treatment in a mixture with *B. thuringiensis* (Jung and Kim, 2006). EPNs showed ovicidal activity toward *Helicoverpa armigera* and *Spodoptera litura*, and both phase I and phase II of the symbiotic bacteria *Xenorhabdus indica* were virulent to these insects, indicating that phase variation did not affect the level of virulence (Kalia *et al.*, 2014). Chapter 27 (this volume) provides further information on interactions between EPNs and other inputs.

20.2.3 Cabbage butterfly

Cabbage butterfly, *Pieris brassica* (Lepidoptera: Pieridae), poses a significant threat to cabbage and other brassica crops, causing annual yield losses exceeding 50%. In laboratory conditions, seven EPN species including *H. bacteriophora* and *Steinernema glaseri* demonstrated 100% mortality of *P. brassicae* larvae at a concentration of 1500 IJs ml^{-1} after 48 h of exposure (Abbas *et al.*, 2021). A regional survey in Iran identified native EPN isolates, including *H. bacteriophora* and *S. feltiae*, causing high mortality among fourth- and fifth-instar *P. brassicae* larvae (Abdolmaleki *et al.*, 2015). Similarly, *Heterorhabditis pakistanensis* exhibited high virulence against *P. brassicae* under both laboratory and field conditions (Ahmad *et al.*, 2023).

A mixture of *H. bacteriophora* and *S. glas*eri at a dose of 1500 IJs ml^{-1} was found to be the best treatment tested, compared with either species alone at the same concentration or bacterial metabolites from the nematodes' symbiotic bacteria. Mortality reached a maximum of about 80% of *P. brassicae* larvae in comparison with the other treatments in a field trial (Abbas *et al.*, 2022).

20.2.4 Thrips

Onion thrips, *Thrips tabaci* Lindeman (Thysanoptera: Thripidae), stands out as one of the most destructive insect pests of onion (*Allium cepa* L.), a globally significant agricultural crop. In a field study by Azazy *et al.* (2018), it was demonstrated that the tested EPN isolates were efficient against both adult and nymph stages of the thrips. Notably, *H. indica* (EGAZ3 isoalte) exhibited higher efficacy, causing a significant reduction in *T. tabaci* populations (adults and nymphs) after 24 and 48 h at a concentration of 10^4 IJs ml^{-1} via foliar application. *S. carpocapsae* isolates were less effective in controlling *T. tabaci* populations compared with isolates from the genus *Heterorhabditis*.

In a two consecutive-year field experiment, *H. bacteriophora* demonstrated the highest efficacy against *T. tabaci*, leading to a reduction of more than 75% in the number of nymphs and adults per plant in *H. bacteriophora*-treated plots compared with control plots (Gulzar *et al.*, 2021b). In general, combined application of EPNs and fungi resulted in greater mortality than single applications of each pathogen when tested in both a potted soil bioassay and a field trial, suggesting their potential use in IPM of *T. tabaci* in onion production systems. Notably, the prepupal stage was identified as the more susceptible stage for entomopathogen applications (Gulzar *et al.*, 2021a).

Additionally, Belay *et al.* (2005) proposed that early application (10 days after adult *Frankliniella occidentalis* release) of *H. bacteriophora* at 400 IJs cm^{-2} controlled *F. occidentalis* better than late applications, indicating that the time of EPN application is crucial in thrips control. This highlights the importance of strategic nematode application for effective and sustainable onion thrips management.

20.2.5 Whitefly

Whitefly, *Bemisia tabaci* (Gennadius) (Hemiptera: Aleyrodidae), boasts an extensive host range and is recognized as one of the most destructive insect pests affecting vegetables and ornamental crops globally. The B biotype is particularly economically concerning as it serves as an effective vector for over 111 viruses from various families, with a notable affinity for geminiviruses (Cuthbertson *et al.*, 2011). The escalating record of increased insecticide resistance makes *B. tabaci* increasingly challenging to control.

The oral toxicity of cell-free culture media from the symbiotic bacteria *Photorhabdus temperata* and *Photorhabdus luminescens*, associated with the EPNs *H. megidis* and *H. bacteriophora*, respectively, proved lethal to *B. tabaci* adults 60 h after treatment (Shrestha and Lee, 2012). Following treatment with *S. feltiae* in a small-scale glasshouse test, the mortality of *B. tabaci* on tomato, cucumber, verbena and chrysanthemum reached 32%, 28%, 22% and 22%, respectively (Head *et al.*, 2004). The combined application of EPNs and the chemical insecticide imidacloprid resulted in significantly higher *B. tabaci* second-instar mortality under controlled laboratory and glasshouse conditions (>90% and >80%, respectively) compared with either treatment alone on tomato and verbena plants (Cuthbertson *et al.*, 2003). In a greenhouse setting, a mixture of *S. feltiae* (10^4 IJs ml^{-1}) and adjuvant (1% horticultural spraying oil or 0.1% Triton X-100) increased the mortality of second-instar *B. tabaci* nymphs by 83% and 89% on collard and hibiscus, respectively. This suggests that the use of adjuvants significantly enhances infection of *B. tabaci* by *S. feltiae* (Qiu *et al.*, 2008).

Beyond foliar application, soil application of *S. carpocapsae*-infected waxworm cadavers serves as a complementary method, inducing broad-spectrum systemic resistance in tomato plants and impairing egg hatch of sweet potato whitefly *B. tabaci* (An *et al.*, 2016). Two native strains of *H. bacteriophora* and *H. indica* from Sinaloa, Mexico, along with a commercial product *S. feltiae* at concentrations of 250–1000 IJs ml^{-1}, killed 70–95% of *B. tabaci* B-biotype nymphs on tomato leaves under laboratory conditions (Meza-Garcia *et al.*, 2014).

20.2.6 Leaf miners

Leaf miner larvae, encompassing various species from the orders Lepidoptera, Diptera and Hymenoptera, are notorious for residing within plant leaves, causing damage and squiggly patterns. These insects pose significant agricultural threats worldwide. *Tuta absoluta* (Lepidoptera: Gelechiidae), commonly known as the tomato leaf miner moth, has emerged as one of the most destructive pests impacting tomato plants and fruits.

In laboratory experiments, the EPNs *S. carpocapsae*, *S. feltiae* and *H. bacteriophora* exhibited high larval mortality (78.6–100%) and low pupal mortality (<10%). In a leaf bioassay, these nematodes demonstrated a remarkable capacity to kill larvae within leaf galleries, with a high level of larval parasitization (77.1–91.7%). Pot experiments further revealed a substantial decrease (87–95%) in insect infection of tomato plants with nematode treatment (Batalla-Carrera *et al.*, 2010). Additionally, foliar applications of *S. carpocapsae* and *H. bacteriophora* targeting larvae at a rate of 50 IJs cm^{-2} led to a 50% reduction in the emergence of adult *T. absoluta*, even under high initial pest density, demonstrating the potential for effective control (Kamali *et al.*, 2018).

The susceptibility of *T. absoluta* larvae to cell-free supernatants from the symbiotic bacteria *Xenorhabdus* and *Photorhabdus* spp. of EPNs was notable. The first- and second-instar larvae were more susceptible, achieving the highest mortality rates in contact treatments with the supernatants (Yüksel, 2022). Moreover, *S. feltiae* application to lettuce under greenhouse conditions achieved 82% mortality of another leaf miner, *Liriomyza huidobrensis* (Diptera: Agromyzidae), surpassing the efficacy of treatment with the chemical heptenophos (Williams and Walters, 2000).

In a 12-well tissue-culture plate bioassay, *H. bacteriophora* demonstrated a mortality rate of 65.1% at 24 h and 73.8% 48 h after inoculation for late-instar larvae of *Liriomyza trifolii*, using an inoculation rate of 100 IJs per larva (Gayatri and Duraimurugan, 2019). This highlights the potential of EPNs in managing various leaf miner species, showcasing their versatility in pest-control strategies.

20.2.7 Flea beetles and wireworms

Flea beetles, particularly *Phyllotreta* spp. (Coleoptera: Chrysomelidae), pose economic challenges to crucifer vegetables globally. Notably, the crucifer flea beetle, *Phyllotreta cruciferae*, is a dominant pest of canola (*Brassica napus* L.) in the Northern Great Plains of the USA, while the striped flea beetle, *Phyllotreta striolata*, can cause significant harm to cruciferous crops in both larval and adult stages. In a field experiment conducted in Slovenia, *S. feltiae*, *S. carpocapsae* and *H. bacteriophora* proved effective for controlling adult *Phyllotreta* spp. during the warm summer months (Trdan *et al.*, 2008).

In laboratory bioassays, 11 out of 20 tested EPN isolates caused over 80% mortality of *P. striolata*, with local EPN isolates such as *H. indica* 212-2 strain and *S. pakistanense* 94-1 strain in South China showing greater potential than commercial isolates of *S. feltiae* and *S. carpocapsae* (Xu *et al.*, 2010). Field studies demonstrated that *H. indica* 212-2, with higher pathogenicity, heat tolerance and reproductive potential, effectively controlled *P. striolata*, suppressing larvae at a dose of 1.5×10^9 IJs ha^{-1} or higher (Yan *et al.*, 2018). Similarly, *S. carpocapsae* All strain and *H. indica* LN2 strain reduced *P. striolata* larvae, thereby mitigating damage to host leaves caused by the adult-stage beetles (Yan *et al.*, 2013). As an alternative to conventional chemical seed treatment, *S. feltiae* combined with a sprayable polymer gel (1%) effectively managed flea beetles in field conditions (Antwi and Reddy, 2016). Foliar treatment with *S. feltiae* and Barricade® in the field reduced feeding injuries by *P. cruciferae*, demonstrating comparable efficacy to chemical seed treatments (Briar *et al.*, 2018).

Wireworms, the larval stage of several species of click beetles, including *Agriotes lineatus* (Coleoptera: Elateridae), are economically important soil-dwelling pests affecting various field crops globally. *H. bacteriophora* Poinar UWS1 strain caused significantly greater mortality (67%) to *A. lineatus* than other EPN species 3 weeks after inoculation in laboratory assays (Ansari *et al.*, 2009). In laboratory studies, *S. feltiae* SSK strain demonstrated potential in controlling sugarbeet wireworm, *Limonius californicus* (Nikoukar *et al.*, 2021), and *S. carpocapsae* B14 strain showed great potential against dusky wireworm *Agriotes obscurus* (Morton and Garcia-del-Pino, 2017). In a field study, *S. carpocapsae* E-76 caused 55% mortality in the larval stage of *Agriotes* spp., outperforming *H. indica* 216-H (Ogretmen *et al.*, 2020).

20.2.8 Weevils

The weevil *Phyrdenus muriceus* Germar (Coleoptera: Curculionidae) has emerged as a pest in eggplant (*Solanum melongena* L.) crops in the humid Pampa of Argentina. Both its larval and adult stages were found to be susceptible to *H. bacteriophora* isolated from Argentina. Application of these nematodes resulted in a slight reduction in *P. muriceus* populations and decreased foliar damage in eggplant crops (Eliceche *et al.*, 2020). In Peru, a natural EPN strain of *Heterorhabditis* sp. demonstrated strong virulence against Andean potato weevils, *Premnotrypes suturicallus* (Coleoptera: Curculionidae), with a 50% lethal concentration of 5.9 IJs per larva. At this concentration, the mortality of *P. suturicallus* increased over 7 days, resulting in 66% of prepupae, 65% of pupae and 52% of adults being killed in laboratory bioassays (Parsa *et al.*, 2006).

In laboratory bioassays, various EPNs, including *S. carpocapsae*, *S. riobrave*, *S. feltiae*, *H. megidis* and *H. bacteriophora*, induced high levels of larval mortality in the carrot weevil, *Listronotus oregonensis* (Coleoptera: Curculionidae). In the field, application of *H. bacteriophora* not only produced higher-quality carrots but also resulted in higher mortality of carrot weevil 1 week after treatment compared with the control (Miklasiewicz *et al.*, 2002). The sweet potato weevil, *Cylas formicarius* (Coleoptera: Apionidae), a major quarantine pest causing severe damage to crops, showed the lowest numbers on plants treated with *S. carpocapsae* All strain, followed by those treated with chemical insecticides, *H. bacteriophora* HP88 strain and non-treated plants (Jansson and Lecrone, 1997). Similarly, *S. carpocapsae* induced the highest mortality (>90%) in both sexes of the adult sweet potato weevil and successfully reduced root damage compared with the control in field assays (Yamaguchi *et al.*, 2006). Recently, it was demonstrated that *H. indica* isolated from the Hawaiian Islands killed 88%, 96% and 4% of *C. formicarius* in the larval, pupal and adult stages, respectively, after 48 h of exposure in a laboratory bioassay (Myers *et al.*, 2020).

20.2.9 Fungus gnats

Fungus gnats, *Bradysia* spp. (Diptera: Sciaridae), are a source of economic losses for horticultural producers in greenhouses due to their damage to young root systems, which leads to the spread of soil-borne diseases because the damaged roots are more susceptible to infection by many pathogens. This reduces crop marketability. In a cucumber greenhouse, significant mortality of the fungus gnat *Bradysia impatiens* was observed after 21 days of exposure to *S. yirgalemense* (Katumanyane *et al.*, 2018a). Four EPN species, namely *S. yirgalemense*, *H. noenieputensis*, *H. indica* and *H. zealandica*, demonstrated high efficacy, killing more than 80% of *B. impatiens* at 25°C or 30°C (Katumanyane *et al.*, 2018b).

Out of 145 EPN isolates screened, the best nine were further tested and displayed a strong performance against the chive gnat *Bradysia odoriphaga*. *S. feltiae* JY-90 strain exhibited the most robust suppression against *B. odoriphaga*, suggesting the potential of EPNs in the integrated management of *B. odoriphaga* 14 days after application in the field (Ma *et al.*, 2013b). However, the virulence of different EPN isolates against *B. odoriphaga* varied (Ma *et al.*, 2013a). For instance, *H. bacteriophora* caused higher mortality of *B. odoriphaga* than *S. carpocapsae* (Bai *et al.*, 2016). Notably, waterlogging pretreatment combined with later exposure to EPNs showed synergistic effects on the mortality of *B. odoriphaga* compared with using either waterlogging or EPNs alone (Chen *et al.*, 2020). Additionally, the application of EPN-infected cadavers resulted in higher mortality compared with treatment with EPNs applied in water (Bai *et al.*, 2016).

20.2.10 Root-knot nematodes

Infections caused by plant-parasitic nematodes continue to be a major hindrance in agricultural systems, with root-knot nematodes (RKNs) from the genus *Meloidogyne* being particularly problematic due to their wide host range and challenges in control. EPNs and their symbiotic bacteria exhibit antagonistic effects against various RKN species such as *Meloidogyne incognita*, *M. arenaria*, *M. javanica*, *M. enterolobii* and *M. mayaguensis*.

Applications of *S. feltiae* have shown efficacy in reducing *M. incognita* galling, egg production and hatching (Lewis *et al.*, 2001). The use of insect cadavers infected by *H. bacteriophora* and *Steinernema diaprepesi* reduced damage caused by *M. incognita* in pepper and summer squash plants in greenhouse experiments, although the effectiveness varied among EPN species (Del Valle *et al.*, 2013). EPN-colonized cadavers were found to be effective carriers of biological control agents, leading to the highest reduction in RKN pathogenicity indices for *M. javanica* in growth chamber and greenhouse conditions compared with other application methods (Sayedain *et al.*, 2021).

Several EPN species, including *Steinernema brazilense*, *S. rarum*, *S. amazonensis*, *S. feltiae* and *H. bacteriophora*, have demonstrated the ability to reduce galls, egg masses, egg hatching and reproduction of *M. enterolobii* in a pot experiment (Damascena *et al.*, 2019). Low application rates of *S. glaseri* suppressed *M. incognita* penetration and reproduction in tomato roots in growth chamber experiments (Perez and Lewis, 2004). However, in a potted plant experiment, pretreatment of *H. bacteriophora* with certain metal salts increased *M. incognita* penetration into cucumber roots compared with the untreated control (Sun *et al.*, 2016).

EPN symbiotic bacteria, such as *Xenorhabdus bovienii* associated with *S. feltiae*, have been shown to suppress the negative impact of RKNs on tomatoes in large, in-ground tomato production areas and significantly reduced the gall index compared with the control plants (Kepenekci *et al.*, 2018). Additionally, cell-free extracts from symbiotic bacteria (isolated from *H. bacteriophora*, *Steinernema* sp. and *S. rarum*) were found to reduce the reproduction of *Meloidogyne hapla* in tomato plants in the greenhouse (Caccia *et al.*, 2018).

The mechanisms behind the antagonistic effects of EPNs on RKNs are not fully understood, but potential mechanisms include competition for space, the influence of exudates from EPNs and their bacteria (Grewal *et al.*, 1999) and the induction of plant defense systems (Fig. 20.1). Exudates from EPNs, including nematode pheromones (Manohar *et al.*, 2020), may prime host plant defenses and directly impact RKN behavior, leading to reduced penetration, egg hatching and mortality. The complex interactions between EPNs, their symbiotic bacteria and RKNs necessitate further research to unravel the precise mechanisms involved in the suppression process.

20.3 Current Status and Analysis

EPNs have great potential as biological control agents for insect pests and for supporting IPM programs. To advance the discipline, we make the following suggestions:

1. Establishment of a permanent storage facility for EPN strains. EPNs are used in the biological control of pest insects, but their potential may be limited by strain availability from different bioregions and effectiveness against specific pests. Establishing an EPN strain collection will facilitate finding suitable strains for controlling corresponding insect hosts and matching particular environmental conditions. New species of nematodes have been described in Brazil, such as *Heterorhabditis amazonensis* Andaló, Nguyen & Moino (Andaló *et al.*, 2006) and *Steinernema brazilense* Nguyen, Ginarte & Leite (Nguyen *et al.*, 2010). The use of these species is closely related to the adaptation of the nematode to the environment and their targeting of specific insects (Kaya and Gaugler, 1993; Andaló *et al.*, 2014). Surveys have been conducted in almost all parts of the world to isolate locally adapted EPN species or

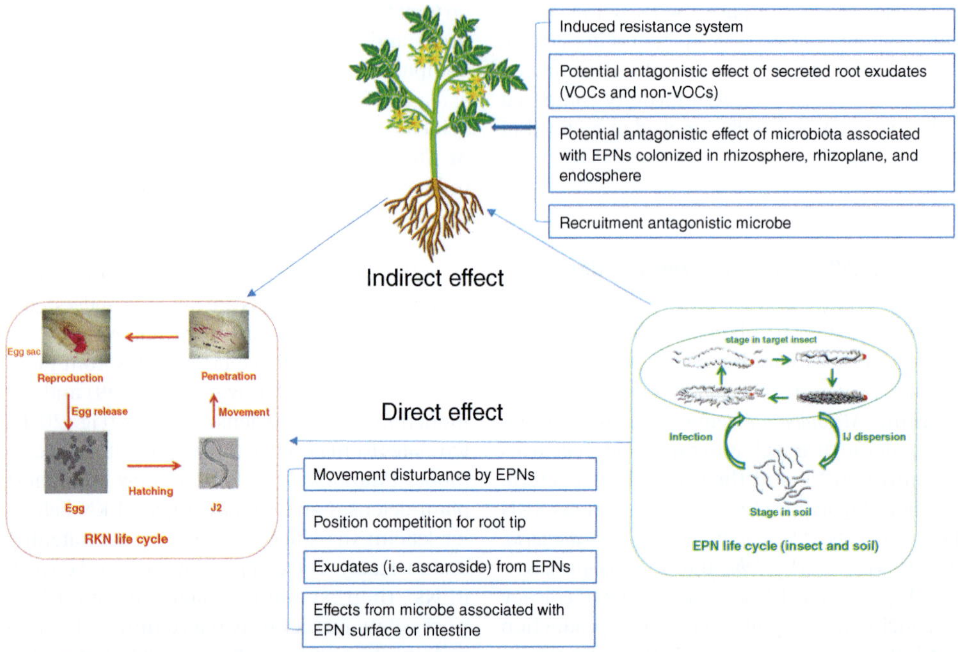

Fig. 20.1. A potential pathway for entomopathogenic nematodes (EPNs) leading to direct and indirect impact on root-knot nematodes. IJ: infective juvenile of EPN; J2, second-stage juvenile of RKN; RKN, root-knot nematode; VOC, volatile organic compound.

isolates for formulation and commercialization. Although surveys for indigenous strains have been done in most parts of the world, the commercialization of these strains is substantially limited. In practice, most of the time, only the very few available commercialized strains are used for the target pest insect. Collecting and preserving diverse EPN strains to exploit their various characteristics are the prerequisites for commercialization.

2. Development of new application technologies. Specific techniques are required for EPN application due to the sensitivity of EPNs to ultraviolet (UV) radiation and desiccation. To control aboveground pest insects, adjuvants (UV protectants, antidesiccants) are key to enhancing EPN survival and efficiency. In general, the efficiency of up-to-date reported adjuvants is disappointing. Extending the viability of EPNs on the leaf surface using formulations is crucial for EPN field application on a large scale. Intensive studies are needed to exploit low-cost and efficient UV protectants and antidesiccants in the future. In general, low oxygen might harm the vitality of EPNs during the delivery and transportation of EPN suspensions. Instead, we suggest that using agents that retard sedimentation might be employed for EPN aqueous solutions during the process of transport and spraying.

Regarding belowground applications, infected-insect cadavers might be a good option as the cadaver may be a refuge for IJs of EPNs in the complex soil context. IJs can emerge from the cadaver once the surrounding conditions (i.e. moisture) are optimal for release. In addition, IJs released from the cadaver might have a number of advantages, including long persistence, high virulence and more prolonged survival than EPNs applied in aqueous solution. However, commercial mass production of infected cadavers has not yet been successfully achieved and we should pay more attention to developing the relevant application technologies to accelerate the application of EPN-infected cadavers under field conditions. In addition, technologies need to be developed to apply EPN aqueous suspensions directly into the soil to avoid potential desiccation and UV radiation stress.

20.4 References

Abbas, W., Javed, N., Ul Haq, I. and Ahmed, S. (2021) Pathogenicity of entomopathogenic nematodes against cabbage butterfly (*Pieris brassicae*) Linnaeus (Lepidoptera: Pieridae) in laboratory conditions. *International Journal of Tropical Insect Science* 41, 525–531.

Abbas, W., Javed, N., Haq, I.U. and Ahmed, S. (2022) Virulence potential of two entomopathogenic nematodes, their associated bacteria, and its metabolites to larvae of *Pieris brassicae* L. (Lepidoptera, Pieridae) in cabbage under greenhouse and field bioassays. *International Journal of Tropical Insect Science* 42, 557–563.

Abdolmaleki, A., Maafi, Z.T., Dastjerdi, H.R. and Lewis, E. (2015) Potential efficacy of Iranian isolates of *Heterorhabditis bacteriophora* and *Steinernema feltiae* on *Pieris brassicae* (Lepidoptera: Pieridae). *Russian Journal of Nematology* 23, 91–97.

Ahmad, M.J., Mohiudin S. and Manzar, A. (2023) Field efficacy of entomopathogenic nematode against two Lepidopteran pests of *Brassica oleracea* var. Khanyari in Hilly areas of Kashmir valley, Inida. *Journal of Experimental Zoology India* 26, 545–553.

Amer, R.A.M., Nouh, G.M., Yacoub, S.S. and Elhosary, R.A. (2014) Role of some additives for enhancing entomopathogenic biocontrol agents against larvae of the cotton leaf worm, *Spodoptera littoralis* (Boisd.) (Lepidoptera: Noctuidae). *Egyptian Journal of Biological Pest Control* 24, 399–405.

An, R.S., Orellana, D., Phelan, L.P., Canas, L. and Grewal, P.S. (2016) Entomopathogenic nematodes induce systemic resistance in tomato against *Spodoptera exigua*, *Bemisia tabaci* and *Pseudomonas syringae*. *Biological Control* 93, 24–29.

Andaló, V., Nguyen, K.B. and Moino, A. Jr. (2006) *Heterorhabditis amazonensis* n. sp. (Rhabditida: Heterorhabditidae) from Amazonas, Brazil. *Nematology* 8, 853–867.

Andaló, V., Moreira, G.F. and Moino, A. (2014) *Heterorhabditis amazonensis* RSC5 (Rhabditida: Heterorhabditidae) movement and host recognition. *Revista Colombiana de Entomología* 40, 91–97.

Ansari, M.A., Evans, M. and Butt, T.M. (2009) Identification of pathogenic strains of entomopathogenic nematodes and fungi for wireworm control. *Crop Protection* 28, 269–272.

Antwi, F.B. and Reddy, G.V.P. (2016) Efficacy of entomopathogenic nematodes and sprayable polymer gel against crucifer flea beetle (Coleoptera: Chrysomelidae) on canola. *Journal of Economic Entomology* 109, 1706–1712.

Azazy, A.M., Abdelall, M.F.M., El-Sappagh, I.A. and Khalil, A.E.H. (2018) Biological control of the onion thrips, *Thrips tabaci* Lindeman (Thysanoptera: Thripidae), in open fields using Egyptian entomopathogenic nematode isolates. *Egyptian Journal of Biological Pest Control* 28: 27.

Bai, G.Y., Xu, H., Fu, Y.Q., Wang, X.Y., Shen, G.S. *et al.* (2016) A comparison of novel entomopathogenic nematode application methods for control of the chive gnat, *Bradysia odoriphaga* (Diptera: Sciaridae). *Journal of Economic Entomology* 109, 2006–2013.

Batalla-Carrera, L., Morton, A. and Garcia-del-Pino, F. (2010) Efficacy of entomopathogenic nematodes against the tomato leafminer *Tuta absoluta* in laboratory and greenhouse conditions. *Biocontrol* 55, 523–530.

Belay, D., Ebssa, L. and Borgemeister, C. (2005) Time and frequency of applications of entomopathogenic nematodes and their persistence for control of western flower thrips *Frankliniella occidentalis*. *Nematology* 7, 611–622.

Briar, S.S., Antwi, F., Shrestha, G., Sharma, A. and Reddy, G.V.P. (2018) Potential biopesticides for crucifer flea beetle, *Phyllotreta cruciferae* (Coleoptera: Chrysomelidae) management under dryland canola production in Montana. *Phytoparasitica* 46, 247–254.

Caccia, M., Marro, N., Duenas, J.R., Doucet, M.E. and Lax, P. (2018) Effect of the entomopathogenic nematode–bacterial symbiont complex on *Meloidogyne hapla* and *Nacobbus aberrans* in short-term greenhouse trials. *Crop Protection* 114, 162–166.

Chen, C.Y., Ma, H.K., Ma, M.Y., Li, J.J., Zheng, S.Y. *et al.* (2020) An innovative strategy for control of fungus gnats using entomopathogenic nematodes alone or in combination with waterlogging. *Journal of Nematology* 52: e2020-57.

Cuthbertson, A.G.S., Head, J., Walters, K.F.A. and Murray, A.W.A. (2003) The integrated use of chemical insecticides and the entomopathogenic nematode, *Steinernema feltiae*, for the control of sweetpotato whitefly, *Bemisia tabaci*. *Nematology* 5, 713–720.

Cuthbertson, A.G.S., Blackburn, L.F., Eyre, D.P., Cannon, R.J.C., Miller, J. and Northing, P. (2011) *Bemisia tabaci*: the current situation in the UK and the prospect of developing strategies for eradication using entomopathogens. *Insect Science* 18, 1–10.

Damascena, A.P., Ferreira, J.C.A., Costa, M.G.S., de Araujo, L.M. and Wilcken, S.R.S. (2019) Hatching and mortality of *Meloidogyne enterolobii* under the interference of entomopathogenic nematodes *in vitro*. *Journal of Nematology* 51: e2019-58.

Del Valle, E.E., Lax, P., Duenas, J.R. and Doucet, M.E. (2013) Effects of insect cadavers infected by *Heterorhabditis bacteriophora* and *Steinernema diaprepesi* on *Meloidogyne incognita* parasitism in pepper and summer squash plants. *Ciencia E Investigacion Agraria* 40, 109–118.

Ebrahimi, L., TanhaMaafi, Z. and Sharifi, P. (2019) First report of the entomopathogenic nematode, *Steinernema carpocapsae*, from Moghan region of Iran and its efficacy against the turnip moth, *Agrotis segetum* Denis and Schiffermuller (Lepidoptera: Noctuidae), larvae. *Egyptian Journal of Biological Pest Control* 29: 66.

Eivazian Kary, N., Chahardoli, S., Mohammadi, D. and Dillon, A.B. (2019) Virulence of entomopathogenic nematodes against developmental stages of *Plutella xylostella* (Lepidoptera: Plutellidae) – effect of exposure time. *Nematology* 21, 293–300.

Eliceche, D., Rusconi, M., Rosales, M., Salas, A. and Achinelly, F. (2020) Field assay using a native entomopathogenic nematode for biological control of the weevil *Phyrdenus muriceus* in organic eggplant crops in Argentina. *Biocontrol* 65, 613–621.

FAO (Food and Agriculture Organization of the United Nations) (2022) *Agricultural Production Statistics 2000–2021*. FAOSTAT Analytical Brief Series No. 60. FAO, Rome. Available at: https://doi.org/10.4060/cc3751en (accessed 22 March 2024).

Gayatri, B. and Duraimurugan, P. (2019) Bio-efficacy of *Heterorhabditis bacteriophora* (Rhabditida: Heterorhabditidae) against serpentine leafminer, *Liriomyza trifolii* Burgess (Diptera: Agromyzidae) in oilseed crops. *Journal of Biological Control* 33, 169–172.

Goudarzi, M., Moosavi, M.R. and Asadi, R. (2015) Effects of entomopathogenic nematodes, *Heterorhabditis bacteriophora* (Poinar) and *Steinernema carpocapsae* (Weiser), in biological control of *Agrotis segetum* (Denis & Schiffermuller) (Lepidoptera: Noctuidae). *Turkish Journal of Entomology* 39, 239–250.

Grewal, P.S., Lewis, E.E. and Venkatachari, S. (1999) Allelopathy: a possible mechanism of suppression of plant-parasitic nematodes by entomopathogenic nematodes. *Nematology* 1, 735–743.

Guan, H. (2021) Dispersive solid-phase extraction of multiresidue pesticides in food and water samples. *Methods in Molecular Biology* 2326, 289–299.

Gulzar, S., Wakil, W. and Shapiro-Ilan, D.I. (2021a) Combined effect of entomopathogens against *Thrips tabaci* Lindeman (Thysanoptera: Thripidae): laboratory, greenhouse and field trials. *Insects* 12: 456.

Gulzar, S., Wakil, W. and Shapiro-Ilan, D.I. (2021b) Potential use of entomopathogenic nematodes against the soil dwelling stages of onion thrips, *Thrips tabaci* Lindeman: laboratory, greenhouse and field trials. *Biological Control* 161: 104677.

Head, J., Lawrence, A.J., and Walters, K.F.A. (2004) Efficacy of the entomopathogenic nematode, *Steinernema feltiae*, against *Bemisia tabaci* in relation to plant species. *Journal of Applied Entomology* 128, 543–547.

Jansson, R.K. and Lecrone, S.H. (1997) Effect of sweet potato cultivar on efficacy and persistence of entomopathogenic nematodes for control of *Cylas formicarius*. Nematropica 27, 41–52.

Jung, S.C. and Kim, Y. (2006) Synergistic effect of entomopathogenic bacteria (*Xenorhabdus* sp. and *Photorhabdus temperata* ssp. *temperata*) on the pathogenicity of *Bacillus thuringiensis* ssp. *aizawai* against *Spodoptera exigua* (Lepidoptera: Noctuidae). *Environmental Entomology* 35, 1584–1589.

Jung, S.C. and Kim, Y.G. (2007) Potentiating effect of *Bacillus thuringiensis* subsp. *kurstaki* on pathogenicity of entomopathogenic bacterium *Xenorhabdus nemtophila* K1 against diamondback moth (Lepidoptera: Plutellidae). *Journal of Economic Entomology* 100, 246–250.

Kalia, V., Sharma, G., Shapiro-Ilan, D.I. and Ganguly, S. (2014) Biocontrol potential of *Steinernema thermophilum* and its symbiont *Xenorhabdus indica* against Lepidopteran pests: virulence to egg and larval stages. *Journal of Nematology* 46, 18–26.

Kamali, S., Karimi, J. and Koppenhofer, A.M. (2018) New insight into the management of the tomato leaf miner, *Tuta absoluta* (Lepidoptera: Gelechiidae) with entomopathogenic nematodes. *Journal of Economic Entomology* 111, 112–119.

Katumanyane, A., Ferreira, T. and Malan, A.P. (2018a) Greenhouse application of *Steinernema yirgalemense* to control fungus gnats, *Bradysia impatiens*. *Biocontrol* 63, 729–738.

Katumanyane, A., Ferreira, T. and Malan, A.P. (2018b) Potential use of local entomopathogenic nematodes to control *Bradysia impatiens* (Diptera: Sciaridae) under laboratory conditions. *African Entomology* 26, 337–349.

Kaya, H.K. and Gaugler, R. (1993) Entomopathogenic nematodes. *Annual Review of Entomology* 38, 181–206.

Kepenekci, I., Hazir, S., Oksal, E. and Lewis, E.E. (2018) Application methods of *Steinernema feltiae*, *Xenorhabdus bovienii* and *Purpureocillium lilacinum* to control root-knot nematodes in greenhouse tomato systems. *Crop Protection* 108, 31–38.

Lankin, G., Castaneda-Alvarez, C., Vidal-Retes, G. and Aballay, E. (2020) Biological control of the potato cutworm *Agrotis deprivata* (Lepidoptera: Noctuidae) with *Steinernema feltiae* LR (Nematoda: Steinernematidae): influence of the temperature, host developmental stage, and application mode on its survival and infectivity. *Biological Control* 144: 104219.

Lewis, E.E., Grewal, P.S. and Sardanelli, S. (2001) Interactions between the *Steinernema feltiae Xenorhabdus bovienii* insect pathogen complex and the root-knot nematode *Meloidogyne incognita*. *Biological Control* 21, 55–62.

Ma, J., Chen, S.L., Moens, M., de Clercq, P., Li, X.H. and Han, R.C. (2013a) Characterization in biological traits of entomopathogenic nematodes isolated from North China. *Journal of Invertebrate Pathology* 114, 268–276.

Ma, J., Chen, S.L., Moens, M., Han, R.C. and de Clercq, P. (2013b) Efficacy of entomopathogenic nematodes (Rhabditida: Steinernematidae and Heterorhabditidae) against the chive gnat, *Bradysia odoriphaga*. *Journal of Pest Science* 86, 551–561.

Manohar, M., Tenjo-Castano, F., Chen, S.Y., Zhang, Y.K., Kumari, A. *et al.* (2020) Plant metabolism of nematode pheromones mediates plant-nematode interactions. *Nature Communications* 11: 208.

Meza-Garcia, J.L., Elias-Santos, M., Cortez-Mondaca, E., Guerrero-Olazaran, M., Viader-Salvado, J.M. *et al.* (2014) Evaluation of *Heterorhabditis indica* (Rhabditida: Heterorhabditidae) nematode strain from Sinaloa, Mexico, against *Bemisia tabaci* immatures under laboratory conditions. *Southwestern Entomologist* 39, 727–738.

Miklasiewicz, T.J., Grewal, P.S., Hoy, C.W. and Malik, V.S. (2002) Evaluation of entomopathogenic nematodes for suppression of carrot weevil. *Biocontrol* 47, 545–561.

Morton, A. and Garcia-del-Pino, F. (2017) Laboratory and field evaluation of entomopathogenic nematodes for control of *Agriotes obscurus* (L.) (Coleoptera: Elateridae). *Journal of Applied Entomology* 141, 241–246.

Myers, R.Y., Sylva, C.D., Mello, C.L. and Snook, K.A. (2020) Reduced emergence of *Cylas formicarius elegantulus* (Coleoptera: Curculionidae) from sweet potato roots by *Heterorhabditis indica*. *Journal of Economic Entomology* 113, 1129–1133.

Nguyen, K.B., Ginarte, C.M., Leite, L.G., Santos, J.M. and Harakava, R. (2010) *Steinernema brazilense* n. sp. (Rhabditida: Steinernematidae), a new entomopathogenic nematode from Mato Grosso, Brazil. *Journal of Invertebrate Pathology* 103, 8–20.

Nikoukar, A., Ensafi, P., Lewis, E.E., Crowder, D.W. and Rashed, A. (2021) Efficacy of naturally occurring and commercial entomopathogenic nematodes against sugar beet wireworm (Coleoptera: Elateridae). *Journal of Economic Entomology* 114, 2241–2244.

Noosidum, A., Satwong, P., Chandrapatya, A. and Lewis, E.E. (2016) Efficacy of *Steinernema* spp. plus anti-desiccants to control two serious foliage pests of vegetable crops, *Spodoptera litura* F. and *Plutella xylostella* L. *Biological Control* 97, 48–56.

Nyasani, J.O., Kimenju, J.W., Olubayo, F.M. and Wilson, M.J. (2008) Laboratory and field investigations using indigenous entomopathogenic nematodes for biological control of *Plutella xylostella* in Kenya. *International Journal of Pest Management* 54, 355–361.

Ogretmen, A., Yuksel, E. and Canhilal, R. (2020) Susceptibility of larvae of wireworms (*Agriotes* spp.) (Coleoptera: Elateridae) to some Turkish isolates of entomopathogenic nematodes under laboratory and field conditions. *Biological Control* 149: 14320.

Parsa, S., Alcazar, J., Salazar, J. and Kaya, H.K. (2006) An indigenous Peruvian entomopathogenic nematode for suppression of the Andean potato weevil. *Biological Control* 39, 171–178.

Peluso, I., Raguzzini, A., Catasta, G., Cammisotto, V., Perrone, A. *et al.* (2018) Effects of high consumption of vegetables on clinical, immunological, and antioxidant markers in subjects at risk of cardiovascular diseases. *Oxidative Medicine and Cellular Longevity* 2018: 5417165.

Perez, E.E. and Lewis, E.E. (2004) Suppression of *Meloidogyne incognita* and *Meloidogyne hapla* with entomopathogenic nematodes on greenhouse peanuts and tomatoes. *Biological Control* 30, 336–341.

Qiu, B.L., Mandour, N.S., Xu, C.X. and Ren, S.X. (2008) Evaluation of the entomopathogenic nematode *Steinernema feltiae* as a biological control agent of the whitefly, *Bemisia tabaci*. *International Journal of Pest Management* 54, 247–253.

Ratnasinghe, G. and Hague, N.G.M. (1998) The invasion, development and reproduction of *Steinernema carpocapsae* (Rhabditida: Steinernematidae) in the diamondback moth, *Plutella xylostella* (Lepidoptera: Yponomeutidae). *Nematropica* 28, 1–6.

Sayedain, F.S., Ahmadzadeh, M., Fattah-Hosseini, S. and Bode, H.B. (2021) Soil application of entomopathogenic nematodes suppresses the root-knot nematode *Meloidogyne javanica* in cucumber. *Journal of Plant Diseases and Protection* 128, 215–223.

Schmidt-Jeffris, R.A., Beers, E.H. and Sater, C. (2021) Meta-analysis and review of pesticide non-target effects on phytoseiids, key biological control agents. *Pest Management Science* 77, 4848–4862.

Schroer, S., Sulistyanto, D. and Ehlers, R.U. (2005a) Control of *Plutella xylostella* using polymer-formulated *Steinernema carpocapsae* and *Bacillus thuringiensis* in cabbage fields. *Journal of Applied Entomology* 129, 198–204.

Schroer, S., Yi, X.L. and Ehlers, R.U. (2005b) Evaluation of adjuvants for foliar application of *Steinernema carpocapsae* against larvae of the diamondback moth (*Plutella xylostella*). *Nematology* 7, 37–44.

Schroer, S., Ziermann, D. and Ehlers, R.U. (2005c) Mode of action of a surfactant-polymer formulation to support performance of the entomopathogenic nematode *Steinernema carpocapsae* for control of diamondback moth larvae (*Plutella xylostella*). *Biocontrol Science and Technology* 15, 601–613.

Shamseldean, M.M.M. and Ismail, A.A. (1997) Effect of the nematode *Heterorhabditis bacteriophora* and the bacterium *Bacillus thuringiensis* as integrated biocontrol agents of the black cutworm. *Anzeiger für Schaedlingskunde Pflanzenschutz Umweltschutz* 70, 77–79.

Shaurub, E.S.H., Reyad, N.F., Abdel-Wahab, H.A. and Ahmed, S.H. (2016) Mortality and nematode production in *Spodoptera littoralis* larvae in relation to dual infection with *Steinernema riobrave*, *Heterorhabditis bacteriophora*, and *Beauveria bassiana*, and the host plant. *Biological Control* 103, 86–94.

Shrestha, Y.K. and Lee, K.Y. (2012) Oral toxicity of *Photorhabdus* culture media on gene expression of the adult sweetpotato whitefly, *Bemisia tabaci*. *Journal of Invertebrate Pathology* 109, 91–96.

Slavin, J.L. and Lloyd, B. (2012) Health benefits of fruits and vegetables. *Advances in Nutrition* 3, 506–516.

Somvanshi, V.S., Ganguly, S. and Paul, A.V.N. (2006) Field efficacy of the entomopathogenic nematode *Steinernema thermophilum* Ganguly and Singh (Rhabditida: Steinernematidae) against diamondback moth (*Plutella xylostella* L.) infesting cabbage. *Biological Control* 37, 9–15.

Sun, Y., Bai, G.Y., Wang, Y.X., Zhang, Y.Y., Pan, J. *et al.* (2016) The impact of Cu, Zn and Cr salts on the relationship between insect and plant parasitic nematodes: a reduction in biocontrol efficacy. *Applied Soil Ecology* 107, 108–115.

Trdan, S., Vidrih, M., Valic, N. and Laznik, Z. (2008) Impact of entomopathogenic nematodes on adults of *Phyllotreta* spp. (*Coleoptera: Chrysomelidae*) under laboratory conditions. *Acta Agriculturae Scandinavica Section B: Soil and Plant Science* 58, 169–175.

Vashisth, S., Chandel, Y.S., Chandel, R.S. and Sharma, P.K. (2017) Pathogenicity of *Heterorhabditis* nematodes isolated from north-western Himalaya against the larvae of *Plutella xylostella* (L.) (Lepidoptera: Plutellidae). *Annales de la Societe Entomologique de France* 53, 204–210.

Wang, Q.Y., Nangong, Z.Y., Yang, J., Song, P., Wang, Y. *et al.* (2012) Toxic activity of a protein complex purified from *Xenorhabdus nematophila* HB310 to *Plutella xylostella* larvae. *Insect Science* 19, 329–336.

Wang, X., Dou, Z., Shi, X., Zou, C., Liu, D. *et al.* (2021) Innovative management programme reduces environmental impacts in Chinese vegetable production. *Nature Food* 2, 47–53.

Williams, E.C. and Walters, K.F.A. (2000) Foliar application of the entomopathogenic nematode *Steinernema feltiae* against leafminers on vegetables. *Biocontrol Science and Technology* 10, 61–70.

Xu, C.X., de Clercq, P., Moens, M., Chen, S.L. and Han, R.C. (2010) Efficacy of entomopathogenic nematodes (Rhabditida: Steinernematidae and Heterorhabditidae) against the striped flea beetle, *Phyllotreta striolata*. *Biocontrol* 55, 789–797.

Yamaguchi, T., Kawasoe, K. and Izumi, S. (2006) Pathogenicity of steinernematid nematodes on the sweet potato weevil, *Cylas formicarius* (Fabricius) (Coleoptera: Brentidae) and their feasibility for suppression of the weevil population. *Japanese Journal of Applied Entomology and Zoology* 50, 217–224.

Yan, X., Han, R.C., Moens, M., Chen, S.L. and de Clercq, P. (2013) Field evaluation of entomopathogenic nematodes for biological control of striped flea beetle, *Phyllotreta striolata* (Coleoptera: Chrysomelidae). *Biocontrol* 58, 247–256.

Yan, X., Wang, X.D., Han, R.C. and Qiu, X.H. (2014) Utilisation of entomopathogenic nematodes, *Heterorhabditis* spp. and *Steinernema* spp., for the control of *Agrotis ipsilon* (Lepidoptera, Noctuidae) in China. *Nematology* 16, 31–40.

Yan, X., Lin, Y.Y., Huang, Z.M. and Han, R.C. (2018) Characterisation of biological and biocontrol traits of entomopathogenic nematodes promising for control of striped flea beetle (*Phyllotreta striolata*). *Nematology* 20, 503–518.

Yan, X., Chen, G.M., Chen, Y.Q., Sun, B.J., Gu, X.H. *et al.* (2021) Virulence of *Steinernema ceratophorum* against different pest insects and their potential for *in vivo* and *in vitro* culture. *Journal of Nematology* 53: e2021-46.

Yüksel, E. (2022) Biocontrol potential of endosymbiotic bacteria of entomopathogenic nematodes against the tomato leaf miner, *Tuta absoluta* (Meyrick) (Lepidoptera: Gelechiidae). *Egyptian Journal of Biological Pest Control* 32: 135.

Yüksel, E. and Canhilal, R. (2018) Evaluation of local isolates of entomopathogenic nematodes for the management of black cutworm, *Agrotis ipsilon* Hufnagel (Lepidoptera: Noctuidae). *Egyptian Journal of Biological Pest Control* 28: 82.

21 Turfgrass and Pasture Applications

Albrecht M. Koppenhöfer[1,2]* **and Ana Luiza Sousa**[1,2]
[1]*Department of Entomology, Rutgers University, New Brunswick, New Jersey, USA;*
[2]*Current address: Syngenta Seeds Development—R&D Innovation Center, Malta, Illinois, USA*

Abstract
Use of entomopathogenic nematodes (EPNs) in turfgrass is very limited due to high turf quality standards and competition from synthetic insecticides. White grubs are the most studied group of turfgrass pest with regard to EPN use. White grub species differ in their susceptibility to EPNs, and the relative virulence of different EPN species varies among white grub species. In general, *Steinernema scarabaei*, *S. kushidai*, *S. glaseri*, *Heterorhabditis bacteriophora*, *H. zealandica* and *H. megidis* are more virulent against white grubs than *H. indica*, *H. marelata*, *S. arenarium*, *S. carpocapsae*, *S. feltiae* and *S. riobrave*. White grub instars may differ in their susceptibility, but the effect varies with white grub and EPN species. *Neoscapteriscus* spp. mole crickets can be controlled with *S. scapterisci* and *S. riobrave*. Due to its close association with these mole crickets, *S. scapterisci* can effectively recycle and spread in them, providing long-term suppression after inoculative releases. Other turfgrass insect groups that have been effectively controlled with EPNs are billbugs, annual bluegrass weevil, and various species of cutworms, armyworms and sod webworms. However, poor shelf life and poor ease of use impinge on the wider use of EPN products. Use of EPNs could be increased by more systematically exploiting their ability to recycle and provide long-term pest suppression, either through the use of more host-specific and better-adapted species of major insect pests such as seen with *S. scapterisci* and *S. scarabaei*, or through the use of native EPN strains, maintained to preserve their ability to persist in the environment. Significant increases in biological control agent use will likely happen only through education and legislation.

21.1 Introduction

Grasses are the dominant vegetation in many environments that vary in size and composition from the great prairies to manicured golf courses and home lawns. Grasslands, improved by sowing and managing desirable species, support livestock industries around the world. Wear-tolerant grass species are used to create recreational spaces in the urban environment. The turfgrass market in the USA is by far the largest in the world, with amenity turfgrasses occupying about 200,000 km^2 and the size of the industry estimated at US$40 billion year^{-1} (Breuninger *et al.*, 2013), consisting of about 66% home lawns, with the rest being parks, athletic fields, golf courses, cemeteries, roadsides and sod farms. In most countries, turf hectarages consist primarily of golf courses, followed by parks and athletic fields, with residential turf being insignificant, except for Canada where home lawns constitute a significant area (Breuninger *et al.*, 2013). Grasses provide

*Email: a.koppenhofer@rutgers.edu

 Entomopathogenic Nematodes as Biological Control Agents (D.I. Shapiro-Ilan and E.E. Lewis eds)
DOI: 10.1079/9781800620322.0021

many benefits in addition to recreational and livestock uses, including controlling soil erosion; capturing and cleaning runoff water from urban areas; providing soil improvement and restoration; sequestering carbon; moderating temperature; reducing glare, noise, pests, pollen and human disease exposure; creating good wildlife habitats; and improving the physical and mental health of urban populations (Beard and Green, 1994).

The purpose of this chapter is to review the pest complexes in turf where EPNs are used, and to assess the performance of EPNs in different turf conditions.

21.2 Major Pests

Grasses provide habitat for many invertebrate species, less than 1% of which acquire pest status. In natural grasslands, insect outbreaks occur rarely. However, in managed grasslands, sowing of palatable grass species and increasing fertility can enable some herbivorous species to become key pests. In amenity turf, high cosmetic value and low damage thresholds have led to many arthropod species being regarded as pests. Their behavior and feeding location and distribution vary considerably (Table 21.1). Pests of grasslands and turf have been reviewed by Delfosse (1993), Potter (1998), and Vittum (2020).

21.3 White Grubs

Root-feeding larvae of scarabaeid beetles are among the most damaging pests of turfgrass and pastures in different parts of the world (Jackson, 1992). The adults of some species can cause extensive damage to foliage or flowers of ornamentals and fruit trees. Important endemic scarab pests include *Cyclocephala* spp. and *Phyllophaga* spp. in the Americas, *Holotrichia* and *Heteronychus* spp. throughout Asia and Africa, *Melolontha*, *Amphimallon* and *Phyllopertha* spp. in Europe, and *Anomala* spp. in Japan and Korea. Exotic invasive species include the Japanese beetle, *Popillia japonica*, the oriental beetle, *Anomala orientalis*, the European chafer, *Rhizotrogus majalis*, and the Asiatic garden beetle, *Maladera castanea* in North America, and the South African beetle, *Heteronychus arator*, in New Zealand and Australian pastures.

Most important grub species have annual life cycles and three instars (Potter, 1998; Vittum, 2020). Adults emerge in late spring or summer, and females lay eggs in the soil among

Table 21.1. Major turfgrass and pasture pests, part of the plant attacked and geographic problem areas. Modified from Grewal *et al.* (2005).

Plant part attacked	Pests	Pest life stage	Geographic location
Roots	White grubs	Larva	Worldwide
	Mole crickets	Nymph and adult	Southern USA
Stem/crown	Annual bluegrass weevil	Larva	Eastern North America
	Billbugs	Larva and adult	USA, Japan, New Zealand, Australia
	Crane flies	Larva	Europe, north-western and north-eastern USA, south-western and south-eastern Canada
Leaf/stem	Armyworms	Larva	Worldwide
	Cutworms	Larva	Worldwide
	Sod webworms	Larva	USA
	Chinch bugs	Nymph and adult	Central and eastern USA, south-eastern Canada, Japan
	Greenbug aphids	Nymph and adult	USA
	Mites	Larva, nymphs and adult	USA
	Spittlebugs	Nymph and adult	Eastern USA, Brazil
	Scales	Nymph and adult	Southern USA, Japan
	Mealybugs	Nymph and adult	Southern USA, New Zealand

grass roots. The grubs feed on the roots near the soil surface. At high larval densities and under warm dry conditions, this can lead to wilting and gradual thinning of the turf and death of large turf areas. Mammals and birds foraging for the larvae can further disrupt the turf surface. In annual white grub species, most larvae reach the third-instar stage in late summer to early fall. The larvae overwinter deeper in the soil to resume feeding in the spring before pupating and emerging as adults. Some species, such as those in the genera *Melolontha* and *Amphimallon* and some *Phyllophaga* spp., have multi-year life cycles, with damage dependent on the larval stage and species present.

21.3.1 Nematodes for white grub control

Entomopathogenic nematodes (EPNs) are the most extensively studied parasites of white grubs, and many species use white grubs as natural hosts (Poinar, 1992; Peters, 1996; Stock and Koppenhöfer, 2003). Much of the research has focused on the potential use of EPNs for inundative applications. Several species (e.g. *Heterorhabditis bacteriophora*, *H. zealandica*, *H. megidis*, *H. marelata*, *Steinernema glaseri* and *S. scarabaei*) are currently available commercially for grub control. However, because of their coevolution with soil pathogens, white grubs have developed defense mechanisms including infrequent carbon dioxide output, sieve plates over their spiracles, frequent defecation, defensive and evasive behaviors, a dense peritrophic membrane and a strong immune response, which make them relatively resistant to infection by EPNs (Grewal *et al.*, 2005).

21.3.2 Nematode field efficacy

S. glaseri and *Heterorhabditis* spp. are generally more effective than *Steinernema carpocapsae* and *S. feltiae* against white grubs (Klein, 1993). Georgis and Gaugler (1991) related most control failures against *P. japonica* to unsuitable environmental conditions or EPN species/strain. Later research on discovery of new species and strains and factors affecting efficacy has shown that white grub species differ in their susceptibility to EPNs and that the relative virulence of different EPN species varies among white grub species (e.g. Grewal *et al.*, 2002; Koppenhöfer *et al.*, 2004, 2006). In general, *S. scarabaei*, *S. kushidai*, *S. glaseri*, *H. bacteriophora*, *H. zealandica* and *H. megidis* are more virulent against white grubs than *H. indica*, *H. marelata*, *S. arenarium*, *S. carpocapsae*, *S. feltiae* and *S. riobrave* (Tables 21.2 and 21.3). Moreover, strains of the same EPN species may vary considerably in their virulence to different white grub species (Grewal *et al.*, 2002). However, differences among application rates, evaluation timing, irrigation regimes and EPN quality may have affected outcomes. Below, we summarize results from more recent field trials concentrating primarily on similar rates (~2.5×10^9 infective juveniles (IJs) ha^{-1}) when available. Most studies have concentrated on short-term efficacy of EPN applications against white grubs. Long-term effects are discussed in section 21.9.

Against *P. japonica*, *S. scarabaei* AMK001 strain, *H. zealandica* X1 strain and *H. bacteriophora* GPS11 and TF strains were more effective than all the other EPN species tested (Table 21.2). Against *A. orientalis*, *S. scarabaei* AMK001 was the most effective species; other EPNs that provided some control included *Heterorhabditis* sp. Gyeongsan, *Steinernema longicaudum*, *H. bacteriophora* GPS11, *S. glaseri*, and *S. carpocapsae* (Table 21.2). In a greenhouse trial, *S. kushidai* provided 88–94% control of *A. orientalis* (Table 21.3). Against *Cyclocephala borealis*, *H. zealandica* X1, *S. scarabaei* and *H. bacteriophora* GPS11 were the most effective (Table 21.2). Against other *Cyclocephala* spp., none of the EPNs tested provided effective control.

For other scarab species, limited field and/or greenhouse data are available (Tables 21.2 and 21.3). Against the garden chafer, *Phyllopertha horticola*, *H. bacteriophora* has provided better control than *H. megidis*. Against a mixture of *Phyllophaga anxia*, *Phyllophaga fusca* and *Polyphylla comes*, *H. bacteriophora* and *S. carpocapsae* strains provided variable results and showed no dosage effects, but overall *S. carpocapsae* All strain provided the best control. Against *R. majalis*, only *S. scarabaei* has provided good control, whereas *S. glaseri* and *H. bacteriophora* TF are ineffective. *M. castanea* follows the same pattern as *R. majalis*, but appears to be somewhat less susceptible to all EPNs tested. For *Hoplia philanthus*, only greenhouse data are published that suggest

Table 21.2. Field efficacy of entomopathogenic nematodes (*Heterorhabditis* and *Steinernema* spp.) against white grubs in turfgrass (late summer/early fall applications only).[a] Modified from Grewal *et al.* (2005).

Grub species	Nematode species[a]	Strain	Rate ($\times 10^9$ IJs ha^{-1})	Mean control (%)	Duration (days)	References
Anomala orientalis	*H. bacteriophora*	TF	1.0	11–40	21–39	Koppenhöfer and Fuzy (2003a)
			1.25	40	21	Koppenhöfer *et al.* (2002); Koppenhöfer and Fuzy (2003a)
			2.5	0–52	21–39	Koppenhöfer *et al.* (2002); Koppenhöfer and Fuzy (2003b); Koppenhöfer and Fuzy (2003c)
		GPS11	0.31	5	21–23	Koppenhöfer and Fuzy (2008a)
			0.63	2–20	21–23	Koppenhöfer and Fuzy (2008a); Koppenhöfer and Fuzy (2008b)
			1.25	3–42	21–23	Koppenhöfer and Fuzy (2008a); Koppenhöfer and Fuzy (2008b)
			2.5	20–63	21	Koppenhöfer and Fuzy (2008b)
	Heterorhabditis sp.	Gyeongsan	0.5	54	28	Lee *et al.* (2002)
			1.0	67	28	Lee *et al.* (2002)
	S. carpocapsae	Pocheon	1.0	56	28	Lee *et al.* (2002)
	S. glaseri	Dongrae	1.0	50	28	Lee *et al.* (2002)
			2.5	49	21	Koppenhöfer *et al.* (1999)
		Mungyeong	1.0	50	28	Lee *et al.* (2002)
		Biosys #326	1.24	0	28–44	Yeh and Alm (1995)
			2.47	21–70	28–44	Yeh and Alm (1995)
			4.9	54–68	28–44	Yeh and Alm (1995)
	S. longicaudum	Gongju	1.0	56	28	Lee *et al.* (2002)
		Nonsan	1.0	41–55	28	Lee *et al.* (2002)
	S. scarabaei	AMK001	0.06	10	34	Koppenhöfer and Fuzy (2009)
			0.12	50–82	31–34	Koppenhöfer and Fuzy (2009)
			0.25	72–77	31–34	Koppenhöfer and Fuzy (2009)
			0.4	43	21	Koppenhöfer and Fuzy (2003a)
			0.4	63–100	31–39	Koppenhöfer and Fuzy (2009)
			0.63	87–93	31–34	Koppenhöfer and Fuzy (2009)
			1.0	60–89	21	Koppenhöfer and Fuzy (2003a)
			1.0	100	31–39	Koppenhöfer and Fuzy (2009)
			2.5	87–96	21	Koppenhöfer and Fuzy (2003a)
			2.5	100	31–39	Koppenhöfer and Fuzy (2009)

Continued

Table 21.2. Continued.

Grub species	Nematode species[a]	Strain	Rate (× 10^9 IJs ha^{-1})	Mean control (%)	Duration (days)	References
Aphodius contaminatus	*H. bacteriophora*	EN0043	5.0	55	29	Sulistyanto and Ehlers (1996)
	H. megidis	HSH2	5.0	40	42	Sulistyanto and Ehlers (1996)
Ataenius spretulus	*S. carpocapsae*	All	4.9	46	15	Alm *et al*. (1992)
	S. glaseri	Biosys #2	4.9	14	15	Alm *et al*. (1992)
Cyclocephala borealis	*H. bacteriophora*	GPS11	2.5	47–83	28–35	Grewal *et al*. (2004)
		TF	1.0	6	21	Koppenhöfer and Fuzy (2003a)
			2.5	20	21	Koppenhöfer and Fuzy (2003a)
	H. zealandica	X1	2.5	72–96	28–35	Grewal *et al*. (2004)
	S. glaseri	MB	2.5	0	28	Grewal *et al*. (2004)
	S. kraussei	UK	2.5	50	21	Grewal *et al*. (2004)
	S. scarabaei	AMK001	1.0	58	21	Koppenhöfer and Fuzy (2003a)
			2.5	84	21	Koppenhöfer and Fuzy (2003a)
Cyclocephala hirta	*H. bacteriophora*	NC1	2.5	16	18	Koppenhöfer *et al*. (1999)
			2.5	34–48	26	Koppenhöfer *et al*. (2000a)
			5.0	13	18	Koppenhöfer *et al*. (2000b)
	S. glaseri	NC	2.5	9	20	Koppenhöfer *et al*. (2000a)
	S. kushidai	–	5.0	33	18	Koppenhöfer *et al*. (2000b)
Cyclocephala pasadenae	*H. bacteriophora*	NC1	2.5	8	18	Koppenhöfer *et al*. (1999)
Maladera castanea	*H. bacteriophora*	TF	2.5	12–33	14–21	Koppenhöfer and Fuzy (2003b)
	S. scarabaei	AMK001	1.0	51–60	14–21	Koppenhöfer and Fuzy (2003b)
			2.5	71–86	14–21	Koppenhöfer and Fuzy (2003b)
Phyllopertha horticola	*H. bacteriophora*	EN0043	5.0	55–74	21–42	Sulistyanto and Ehlers (1996); Ehlers and Peters (1998)
	H. megidis	HSH2	5.0	40	42	Sulistyanto and Ehlers (1996); Ehlers and Peters (1998)
Phyllophaga anxia, *Phyllophaga fusca*, *Polyphylla comes*	*H. bacteriophora* (= *H. heliothidis*)	?	1.35	61	35–42	Kard *et al*. (1988)
			2.69	0	35–42	Kard *et al*. (1988)
			5.38	44	35–42	Kard *et al*. (1988)
	S. carpocapsae (= *S. feltiae*)	DD-136	1.35	67	35–42	Kard *et al*. (1988)
			2.69	48	35–42	Kard *et al*. (1988)
			5.38	22	35–42	Kard *et al*. (1988)
		Mexican	1.08	50	35–42	Kard *et al*. (1988)
			2.69	68	35–42	Kard *et al*. (1988)

			5.38	40	35–42	Kard *et al*. (1988)
		All	1.08	87	35–42	Kard *et al*. (1988)
			2.69	77	35–42	Kard *et al*. (1988)
			5.38	61	35–42	Kard *et al*. (1988)
Popillia japonica	*H. bacteriophora*	GPS11	0.31	8	23	Koppenhöfer and Fuzy (2008a)
			0.63	5–21	21–23	Alm *et al*. (1992); Koppenhöfer and Fuzy (2008a)
			1.25	25–63	21–23	Koppenhöfer and Fuzy (2008a,b)
			2.5	34–97	21–35	Grewal *et al*. (2004); Koppenhöfer and Fuzy (2008b)
		HP88	2.5	52–74	22–35	Georgis and Gaugler (1991); Grewal *et al*. (2004)
			5.0	51	21	Selvan *et al*. (1993)
			7.5	67	28–35	Georgis and Gaugler (1991)
		NC	2.5	57	28–35	Georgis and Gaugler (1991)
			7.5	62	28–35	Georgis and Gaugler (1991)
		NC1	1.0	40	25	Koppenhöfer *et al*. (2000a)
			2.0	85	25	Koppenhöfer *et al*. (2000a)
		NJ2	5.0	70	21	Selvan *et al*. (1993)
		TF	1.25	58	22	Koppenhöfer *et al*. (2002)
			2.5	65–92	21–22	Koppenhöfer *et al*. (2000a, 2002); Koppenhöfer and Fuzy (2003a,c)
			5.0	51–63	18	Koppenhöfer *et al*. (2000b)
		ItH-LU1	10.0	93	33	Marianelli *et al*. (2018)
		Koppert	2.5	46	28	Marianelli *et al*. (2018)
	H. indica	man 16	2.5	9	21	Koppenhöfer *et al*. (2000a)
	H. zealandica	X1	2.5	73–98	22–35	Grewal *et al*. (2004)
	S. arenarium	Ryazan	5.0	40	21	Selvan *et al*. (1994)
	S. carpocapsae	All	2.5	38	28–35	Georgis and Gaugler (1991)
			4.9	66	20	Alm *et al*. (1992)
			7.5	45	28–35	Georgis and Gaugler (1991)
		ItS-CAO1	10.0	37	33	Marianelli *et al*. (2018)
		ItS-Ole1	10.0	48	33	Marianelli *et al*. (2018)
	S. feltiae	Biosys #27	4.9	10–15	28–42	Alm *et al*. (1992)
		Biosys #980	4.9	13–19	42	Alm *et al*. (1992)
	S. glaseri	Biosys #2	4.9	39	20	Alm *et al*. (1992)
		Biosys #326	1.24	0	21	Yeh and Alm (1995)
			2.47	0–47	21–25	Yeh and Alm (1995)
			4.9	55–82	21–25	Yeh and Alm (1995)

Continued

Table 21.2. Continued.

Grub species	Nematode species[a]	Strain	Rate ($\times 10^9$ IJs ha^{-1})	Mean control (%)	Duration (days)	References
		NC	2.5	62	21	Koppenhöfer *et al.* (2000a)
			5.0	49–52	21	Selvan *et al.* (1993, 1994)
		NJ	2.5	20	28	Grewal *et al.* (2004)
		NJ43	5.0	70–72	21	Selvan *et al.* (1993, 1994)
		MB	2.5	41–58	28	Grewal *et al.* (2004)
		SI-12	5.0	72	21	Selvan *et al.* (1994)
	S. kraussei	UK	2.5	30	2121	Grewal *et al.* (2004)
	S. kushidai	–	5.0	37–73	18	Koppenhöfer *et al.* (2000b)
	S. riobrave	RGV	5.0	32	21	Selvan *et al.* (1994)
	S. scarabaei	AMK001	1.0	100	14	Koppenhöfer and Fuzy (2003a)
			2.5	100	14	Koppenhöfer and Fuzy (2003a)
Rhizotrogus majalis	*H. bacteriophora*	TF	2.5	38	21	Cappaert and Koppenhöfer (2003)
	S. scarabaei	–	1.0	75	21	Cappaert and Koppenhöfer (2003)
			2.5	89	21	Cappaert and Koppenhöfer (2003)

IJ, infective juvenile.

[a]Data are only shown from tests that were conducted under conducive conditions (e.g. sufficiently high soil temperature, post-treatment irrigation), where data are separated by white grub species if more than one species was present, and only at rates of 7.5×10^9 IJs ha^{-1} or less.

Table 21.3. Efficacy of entomopathogenic nematodes (*Heterorhabditis* and *Steinernema* spp.) against white grubs in turfgrass greenhouse/pot experiments.[a] Modified from Grewal *et al.* (2005).

Grub species	Nematode species	Strain	Rate (× 10^9 IJs ha^{-1})	Mean control (%)	References
Anomala orientalis	*H. bacteriophora*	NC1	0.5	6	Koppenhöfer *et al.* (2000a)
		TF	1.25	24–40	Koppenhöfer *et al.* (2002); Koppenhöfer and Fuzy (2003a,c)
		CT	1.25	35	Koppenhöfer and Fuzy (2003a)
		O	1.25	44	Koppenhöfer and Fuzy (2003a)
		GPS11	0.16	11	Koppenhöfer *et al.* (2006)
			0.31	20	Koppenhöfer *et al.* (2006)
			0.5	23–30	Koppenhöfer and Fuzy (2008a); Ebssa *et al.* (2012)
			0.63	39	Koppenhöfer *et al.* (2006)
			1.0	36–48	Koppenhöfer and Fuzy (2008a)
			1.25	40–53	Koppenhöfer *et al.* (2006); Ebssa *et al.* (2012)
			2.5	56	Koppenhöfer *et al.* (2006)
	H. megidis	IN	1.25	11	Koppenhöfer *et al.* (2002)
		UK211	1.25	28	Koppenhöfer *et al.* (2002)
	H. zealandica	X1	0.16	8	Koppenhöfer *et al.* (2006)
			0.31	20	Koppenhöfer *et al.* (2006)
			0.63	40	Koppenhöfer *et al.* (2006)
			1.25	53	Koppenhöfer *et al.* (2006)
			2.5	60	Koppenhöfer *et al.* (2006)
	S. feltiae	SN	1.25	0	Koppenhöfer *et al.* (2002)
	S. glaseri	NC	0.63	5–55	Koppenhöfer *et al.* (2000a,b)
			1.25	30–72	Koppenhöfer *et al.* (2002); Koppenhöfer and Fuzy (2003a)
		38	1.25	36	Koppenhöfer and Fuzy (2003a)
	S. kushidai	–	0.63	88	Koppenhöfer *et al.* (2000b)
			1.25	94	Koppenhöfer *et al.* (2000b)
	S. scarabaei	AMK001	0.16	58–63	Koppenhöfer and Fuzy (2003a); Koppenhöfer *et al.* (2006)
			0.31	68–73	Koppenhöfer and Fuzy (2003a); Koppenhöfer *et al.* (2006)
			0.63	85–91	Koppenhöfer and Fuzy (2003a); Koppenhöfer *et al.* (2006)
			1.25	95–96	Koppenhöfer and Fuzy (2003a); Koppenhöfer *et al.* (2006)
			2.5	97	Koppenhöfer *et al.* (2006)

Continued

Table 21.3. Continued.

Grub species	Nematode species	Strain	Rate (× 10^9 IJs ha^{-1})	Mean control (%)	References
Cyclocephala borealis	*H. bacteriophora*	NC	0.5	0	Koppenhöfer *et al.* (2000a)
		TF	1.25	38–48	Koppenhöfer and Fuzy (2003a,c)
			2.5	46	Koppenhöfer and Fuzy (2003c)
		O	1.25	59	Koppenhöfer and Fuzy (2003a)
		GPS11	0.31	16	Koppenhöfer *et al.* (2006)
			0.5	5–10	Ebssa *et al.* (2012)
			0.63	31	Koppenhöfer *et al.* (2006)
			1.0	15	Koppenhöfer and Fuzy (2008a)
			1.25	15–37	Koppenhöfer *et al.* (2006); Ebssa *et al.* (2012)
			2.5	48	Koppenhöfer *et al.* (2006)
			0.31	16	Koppenhöfer *et al.* (2006)
			0.63	35	Koppenhöfer *et al.* (2006)
			1.25	39	Koppenhöfer *et al.* (2006)
			2.5	50	Koppenhöfer *et al.* (2006)
	S. glaseri	NC	0.6	0	Koppenhöfer *et al.* (2000a)
			1.25	13	Koppenhöfer and Fuzy (2003a)
		38	1.25	12	Koppenhöfer and Fuzy (2003a)
	S. scarabaei	AMK001	0.31	30–42	Koppenhöfer and Fuzy (2003a); Koppenhöfer *et al.* (2006)
			0.63	35–55	Koppenhöfer and Fuzy (2003a); Koppenhöfer *et al.* (2006)
			1.25	38–68	Koppenhöfer and Fuzy (2003a); Koppenhöfer *et al.* (2006)
			2.5	43–76	Koppenhöfer and Fuzy (2003a); Koppenhöfer *et al.* (2006)
Cyclocephala hirta	*H. bacteriophora*	NC1	0.4	13–53	Koppenhöfer and Kaya (1998); Koppenhöfer *et al.* (1999, 2000a)
			0.6	20	Koppenhöfer *et al.* (2000b)
			1.2	22–29	Koppenhöfer *et al.* (2000a,b)
	S. glaseri	NC	0.4	13	Koppenhöfer *et al.* (2000a)
			1.2	37–39	Koppenhöfer *et al.* (2000a)
	S. kushidai	–	0.4	48	Koppenhöfer *et al.* (2000b)
			0.6	40–63	Koppenhöfer *et al.* (2000b)
			1.2	52–55	Koppenhöfer *et al.* (2000a,b)
Cyclocephala pasadenae	*H. bacteriophora*	NC1	0.5	20	Koppenhöfer *et al.* (2000a)
			1.0	14–24	Koppenhöfer and Kaya (1998); Koppenhöfer *et al.* (1999)
			2.0	12	Koppenhöfer *et al.* (1999)
	S. glaseri	NC	0.5	22	Koppenhöfer *et al.* (2000a)

	S. kushidai	–	0.5	39	Koppenhöfer *et al.* (2000a)
Hoplia philanthus	*H. megidis*	VBM30	2.5	12	Ansari *et al.* (2003)
			5.0	32	Ansari *et al.* (2003)
			7.5	37	Ansari *et al.* (2003)
	S. feltiae	MA40	2.5	12	Ansari *et al.* (2003)
			5.0	6	Ansari *et al.* (2003)
			7.5	14	Ansari *et al.* (2003)
	S. glaseri	NC	2.5	35	Ansari *et al.* (2003)
			5.0	49	Ansari *et al.* (2003)
			7.5	54	Ansari *et al.* (2003)
Maladera castanea	*H. bacteriophora*	TF	1.25	13	Koppenhöfer and Fuzy (2003b)
	S. glaseri	NC	1.25	17	Koppenhöfer and Fuzy (2003b)
	S. scarabaei	AMK001	1.25	71	Koppenhöfer and Fuzy (2003b)
			2.5	94	Koppenhöfer and Fuzy (2003b)
Popillia japonica	*H. bacteriophora*	TF	0.31	77	Koppenhöfer and Fuzy (2003a)
			0.5	20–25	Koppenhöfer *et al.* (2000a)
			1.25	81	Koppenhöfer and Fuzy (2003a)
		GPS11	0.16	31	Koppenhöfer *et al.* (2006)
			0.2	27	Koppenhöfer and Fuzy (2008a)
			0.31	50	Koppenhöfer *et al.* (2006)
			0.63	53	Koppenhöfer *et al.* (2006)
			1.25	65	Koppenhöfer *et al.* (2006)
			2.5	73	Koppenhöfer *et al.* (2006)
	H. zealandica	X1	0.16	28	Koppenhöfer *et al.* (2006)
			0.31	38	Koppenhöfer *et al.* (2006)
			0.63	62	Koppenhöfer *et al.* (2006)
			1.25	74	Koppenhöfer *et al.* (2006)
			2.5	76	Koppenhöfer *et al.* (2006)
	S. glaseri	NC	0.2	28–38	Ebssa *et al.* (2012)
			0.5	3–48	Koppenhöfer *et al.* (2000a); Ebssa *et al.* (2012)
			1.25	81	Koppenhöfer and Fuzy (2003a)
	S. kushidai	–	0.5	36	Koppenhöfer *et al.* (2000a)
	S. scarabaei	AMK001	0.07	20–25	Ebssa *et al.* (2012)
			0.16	59–67	Koppenhöfer and Fuzy (2003a); Koppenhöfer *et al.* (2006)
			0.21	40–45	Ebssa *et al.* (2012)

Continued

Table 21.3. Continued.

Grub species	Nematode species	Strain	Rate (× 10^9 IJs ha^{-1})	Mean control (%)	References
			0.32	77–88	Koppenhöfer and Fuzy (2003a); Koppenhöfer *et al*. (2006)
			0.63	85–90	Koppenhöfer and Fuzy (2003a); Koppenhöfer *et al*. (2006)
			1.25	92–96	Koppenhöfer and Fuzy (2003a); Koppenhöfer *et al*. (2006)
			2.5	98	Koppenhöfer *et al*. (2006)
Rhizotrogus majalis	*H. bacteriophora*	TF	1.25	27	Cappaert and Koppenhöfer (2003)
	S. glaseri	NC	1.25	38	Cappaert and Koppenhöfer (2003)
	S. scarabaei	AMK001	0.16	87	Cappaert and Koppenhöfer (2003)
			0.31	91	Cappaert and Koppenhöfer (2003)
			0.63	91	Cappaert and Koppenhöfer (2003)
			1.25	98	Cappaert and Koppenhöfer (2003)

[a]All data shown are control rates observed at 14 days after treatment targeting third instars.

that *S. glaseri* and *H. megidis* may provide acceptable control whereas *S. feltiae* is ineffective.

21.3.3 Biotic factors affecting nematode efficacy

Differences in the virulence of EPN species have been attributed to differences in foraging behavior (Kaya and Gaugler, 1993), infectivity (Grewal *et al.*, 2002), ability to escape from the host immune response (Wang *et al.*, 1995; Grewal *et al.*, 2002), virulence of the symbiotic bacteria (Yeh and Alm, 1992; Ansari *et al.*, 2003) and the number of symbiont cells carried (Selvan *et al.*, 1993).

To understand differences in field efficacy toward white grubs among EPN species and in EPN susceptibility among white grub species, several parallel studies examined EPN virulence (Koppenhöfer *et al.*, 2006), EPN infectivity and infection routes (Koppenhöfer *et al.*, 2007), EPN attraction to hosts (Koppenhöfer and Fuzy, 2008c; Koppenhöfer *et al.*, 2013), white grub aggressive and evasive behaviors in response to EPN attack (P.S. Grewal, unpublished data) and persistence of IJs in different substrates (Koppenhöfer and Fuzy, 2006, 2007). These studies were conducted against combinations of *Popillia japonica*, *A. orientalis*, *C. borealis* and *R. majalis* with the EPNs *S. glaseri* NC, *S. scarabaei* AMK001, *H. zealandica* X1 and *H. bacteriophora* GPS11. *S. scarabaei*'s high efficacy is based on its limited elicitation of host defensive behaviors, high infectivity, outstanding virulence and long persistence; these factors were correlated with the species' degree of efficacy against different white grub species (*P. japonica* > *A. orientalis* = *R. majalis* >> *C. borealis*). *S. scarabaei*'s low dispersal rate was not limiting in turfgrass where white grub density can be very high (>100 m^{-2} concentrated in the top 5 cm of soil). It was less effective against *C. borealis* due to its lower virulence and infectivity to this host. Good efficacy of *H. bacteriophora* and *H. zealandica* against *P. japonica* can be ascribed to high virulence (*H. bacteriophora*), high infectivity (*H. zealandica*) and high dispersal rates (*H. bacteriophora*, *H. zealandica*) but limited persistence. The low efficacy of *S. glaseri* and the *Heterorhabditis* spp. against *A. orientalis* and especially *R. majalis* is due to a combination of low virulence (especially to *R. majalis*) or low infectivity (*H. bacteriophora*, *S. glaseri*) and limited persistence (*Heterorhabditis* spp.). As strains of the same EPN species can differ to some extent in factors affecting virulence and efficacy (e.g. Grewal *et al.*, 2002; Koppenhöfer and Fuzy, 2003a), it is possible that using different EPN populations would have affected the outcome of the above studies.

White grub instars may differ in their susceptibility, but the effect varies with white grub and EPN species. *Melolontha melolontha* first and early second instars were the most susceptible stages to *S. glaseri* and *Heterorhabditis* sp. (Deseö *et al.*, 1990). *Maladera matrida* third instars were more susceptible than first and second instars to *H. bacteriophora* (Glazer and Gol'berg, 1989, 1993). *Maladera castanea* third instars were more susceptible than second instars to *S. scarabaei* (Koppenhöfer and Fuzy, 2004). *Phyllopertha horticola* susceptibility increased significantly in older instars for *S. glaseri*, *H. bacteriophora* and *H. downesi* but not for *S. arenarium* (= *anomali*) and *S. carpocapsae* (Smits *et al.*, 1994). *Anomala cuprea* third instars were more susceptible to *S. kushidai* than second and first instars (Fujiie *et al.*, 1993). *A. orientalis* susceptibility decreases from second to third instars for *H. bacteriophora* (Koppenhöfer and Fuzy, 2004, 2008b) and for *Heterorhabditis* sp. Gyeongsan, *S. carpocapsae*, *S. glaseri* and *S. longicaudum* (Lee *et al.*, 2002), and from young third instars to older third instars for *H. bacteriophora* (Koppenhöfer and Fuzy, 2004), but there was no difference in susceptibility between second and third instars for *S. scarabaei* or *S. glaseri* (Koppenhöfer and Fuzy, 2004). *P. japonica* susceptibility decreased from second to third instars for *H. bacteriophora* (Grewal *et al.*, 2004; Koppenhöfer and Fuzy, 2004; Power *et al.*, 2009) but not for *S. scarabaei* (Koppenhöfer and Fuzy, 2004) or *H. zealandica* (Grewal *et al.*, 2004). Hence, EPN applications should target the most susceptible instars provided other conditions are also conducive at that time. At least with annual white grub species, applications made against earlier grub instars may allow more time for EPN recycling in the grub populations, leading to more secondary infections and enhanced grub control during the season.

Effects of host density on EPN efficacy have only been studied in greenhouse pot experiments. In combinations of different EPN species with different white grub species representing somewhat resistant to highly susceptible hosts, no significant host density effects were detected (Ebssa *et al.*, 2012). The expected exhaustion of available IJ populations to less lethal levels by high host numbers may have been counteracted by factors such as increased chances for IJ–host contact and increased host susceptibility due to stress via reduced food resources and increased aggression among the larvae.

EPN application rates of $0.5–12.5 \times 10^9$ IJs ha^{-1} have been tested in various studies. Georgis and Gaugler (1991) recommended a rate of 2.5×10^9 IJs ha^{-1} for *P. japonica*. Any increases in control from $2.5–5 \times 10^9$ IJs ha^{-1} were generally not significant (e.g. Grewal *et al.*, 2004). Rates exceeding 2.5×10^9 IJs ha^{-1} are also generally not economically acceptable (Grewal and Georgis, 1998), at least not for larger-scale commercial applications. Lower rates may be effective against more EPN-susceptible white grub species under optimal environmental conditions and for highly effective EPN species such as *S. scarabaei* (Koppenhöfer and Fuzy, 2003a, 2009).

21.3.4 Abiotic factors affecting nematode efficacy

The influence of abiotic factors on EPN efficacy is discussed here for white grubs, as this system has seen the greatest amount of research. However, the observations are more or less transferrable to other insect species in turfgrass and pastures.

Due to the thick canopy of turfgrass, after spray applications many IJs may get stuck on leaves and stems where they may be rapidly inactivated by heat and ultraviolet radiation. To minimize this IJ loss, spray volumes of 750–1890 l ha^{-1} are recommended (Georgis *et al.*, 1995). Depending on the spray volume used and the environmental conditions, spraying needs to be followed as soon as possible by irrigation (Selvan *et al.*, 1993), with a minimum of 7.4 mm of post-application irrigation required for establishment of the IJs in turfgrass (Shetlar *et al.*, 1988). As soil moisture plays a crucial role in IJ movement and survival, irrigation volume and frequency and soil moisture are positively related to EPN efficacy against white grubs (Georgis and Gaugler, 1991; Grewal *et al.*, 2004). An irrigation frequency of 1–4-day intervals was optimal for grub control with EPNs to maintain moderate soil moisture for at least 2 weeks after application (Georgis and Gaugler, 1991). Because soil types vary in their ability to absorb and retain water, irrigation regimes need to be adjusted accordingly. *H. bacteriophora* was more effective against *P. japonica* in fine-textured soils because finer soils retain moisture better and restrict IJ movement to the upper soil layers where most of the white grubs are found (Georgis and Gaugler, 1991).

IJ downward movement can be restricted by excessive thatch, an accumulation of organic matter between the soil and turfgrass foliage (Georgis and Gaugler, 1991; Zimmerman and Cranshaw, 1991). Anionic and non-ionic wetting agents can enhance water permeability through the thatch and thereby help transfer IJs into the soil when thick thatch layers are present (Schroeder and Sieburth, 1997).

Soil temperature affects EPN activity. Applications of *H. bacteriophora* against *P. japonica* were more effective in late summer/early fall than in spring (Georgis and Gaugler, 1991; Marianelli *et al.*, 2018), and Georgis and Gaugler (1991) suggested that soil temperature should be over 20°C for optimal activity of *H. bacteriophora*. Other EPN species such as *S. scarabaei* may be more tolerant of lower temperatures and can provide effective control in late spring (e.g. Koppenhöfer and Fuzy, 2003b, 2009).

21.4 Mole Crickets

After accidental introduction from South America around 1900, three *Neoscapteriscus* spp. mole crickets have become important turfgrass and pasture pests throughout the coastal plain region of the USA from eastern Texas to North Carolina. Nymphs and adults cause damage by feeding on grass roots and shoots and through their extensive tunneling activity. In broadcast applications (2.5×10^9 IJs ha^{-1}), *Steinernema scapterisci* and *S. riobrave* were more effective (75% control) in controlling mole crickets than *S. carpocapsae* (58%) (Georgis and Poinar, 1994). Only the mole cricket-specific *S. scapterisci* effectively reproduces in *Neoscapteriscus* spp. and provides long-term suppression of mole cricket populations. It is more effective against the southern mole cricket (*Neoscapteriscus borellii*) than against the tawny mole cricket (*Neoscapteriscus vicinus*) and least effective against the short-winged mole cricket (*Neoscapteriscus abbreviatus*); it is most effective against adult mole crickets and is ineffective against small nymphs (Parkman and Frank, 1992). *S. scapterisci* was isolated from mole crickets in Argentina and Uruguay and was successfully established by using electronic mating callers to attract mole crickets to the site of application and after inoculative application of *S. scapterisci*-infected cricket cadavers (Parkman *et al.*, 1993). Its introduction into the USA was the first successful use of an EPN in classical biological control (Parkman and Smart, 1996; Frank and Walker, 2006).

In more damage-sensitive areas, *S. scapterisci* use has been limited by competition from insecticides and nematicide applications required for the control of plant-parasitic nematodes. Success has also been hampered by the fact that *S. scapterisci* has to be applied in spring or fall when adults are present, whereas control measures are typically necessary in summer against nymphs. In pastures, *S. scapterisci* has been applied using slit injectors in strips covering 12.5% of the area, from which it then spreads over a period of several years (Frank and Walker, 2006), making it more cost-effective than chemical insecticides. Nematac® S, a *S. scapterisci*-based product, appeared to become a success story, especially for uses in pastures and low-maintenance turf, where some damage is tolerated, but it was discontinued in 2012 due to insufficient demand. Because of the nematode's slow spread from inoculation sites, widespread use of a commercial product or a large-scale inoculation program would be necessary to accelerate its spread. In areas of Florida where both *S. scapterisci* and a parasitoid wasp, *Larra bicolor* (Hymenoptera: Crabonidae), also introduced from South America, have become established, mole cricket populations are significantly suppressed causing much fewer problems (Frank and Walker, 2006; Leppla *et al.*, 2007).

21.5 Weevils

Billbugs, *Sphenophorus* spp., cause damage to turfgrass primarily in the larval stage when the young larvae tunnel into the stems and crown and older larvae feed externally at the base of the stems, crown and other belowground parts of the plant. Seasonal life cycles vary depending on species and latitude. No detailed studies on billbug–EPN interaction have been published. In the USA, applications at 2.5×10^9 IJs ha^{-1} of *S. carpocapsae* (average 78% control) and *H. bacteriophora* (74%) provided control of bluegrass billbug (*Sphenophorus parvulus*) larvae in the soil, which is similar to levels of control provided by standard insecticides (Georgis and Poinar, 1994; Smith, 1994). Before registration of the neonicotinoid insecticide imidacloprid, *S. carpocapsae* was widely used for control of the hunting billbug, *Sphenophorus venatus vestitus*, on golf courses in Japan, providing around 84% control at application rates of 2.5×10^9 IJs ha^{-1} due to favorable temperature and rainfall and the adoption of EPN-friendly application protocols (Smith, 1994; Kinoshita and Yamanaka, 1998).

Annual bluegrass weevil (*Listronotus maculicollis*) larvae are an important pest of golf course turfgrass in eastern North America. *S. carpocapsae*, *S. feltiae*, *H. bacteriophora* and combinations of these species gave variable control (0–94% control at 2.5×10^9 IJs ha^{-1}) in part due to limited IJ persistence, but *S. carpocapsae* gave the most consistent control (McGraw *et al.*, 2010). Splitting *S. carpocapsae* applications into two applications at half rate, applied 1 week apart, improved the control from 62–80% for the single application to 75–88% for the split applications (Koppenhöfer *et al.*, 2020). Although the destructive nature of *L. maculicollis* necessitates high levels of control, the dearth of effective insecticides for insecticide-resistant populations may increase opportunities for EPN use.

21.6 Cutworms, Webworms and Armyworms

Larvae of various moth species cause turf or pasture damage by feeding on the foliage and stems; a few species also feed on the roots. Many of these are amenable to management by EPNs, but detailed studies on EPN use have been published only for a few species.

Black cutworm (*Agrotis ipsilon*) larvae are a cosmopolitan pest of golf course greens and tees.

Commercial strains of several EPN species were highly virulent to the larvae in the laboratory (Ebssa and Koppenhöfer, 2012; Bélair *et al.*, 2013). Under field conditions targeting fourth-instar larvae (2.5×10^9 IJs ha^{-1}), *S. carpocapsae* was the best-performing species due to a combination of high (average 83%) and most consistent (70–90%) control rates (at 7 days after treatment) and fastest kill rate (average 68% at 4 days after treatment); *S. feltiae* (average 68%) and *H. bacteriophora* (average 62%) were less consistent, whereas *S. riobrave* was the least effective (average 32%) (Ebssa and Koppenhöfer, 2011). Combinations of *S. carpocapsae* with *S. feltiae* or *H. bacteriophora* at half rate each did not provide significantly better control than either of the two species alone at full rate (Koppenhöfer *et al.*, 2022). Syringing (i.e. a twice daily small amount of irrigation) under warm, sunny conditions provided some limited improvement of *S. carpocapsae* efficacy, and split application (two applications at half rate 3 days apart) improved *S. carpocapsae* efficacy by around 20% (Koppenhöfer *et al.*, 2022). Despite this high efficacy, EPNs are not widely used for black cutworm control on golf courses due to extremely low tolerance for damage.

In laboratory sod arenas, Turkish isolates of *S. carpocapsae* and *H. bacteriophora* (2.5×10^9 IJs ha^{-1}) provided 77% and 29% control of fifth and sixth instars of the armyworm *Spodoptera cilium* (Gulcu *et al.*, 2014). A commercial *S. carpocapsae* product (2.5×10^9 IJs ha^{-1}) applied to greenhouse pots with grass infested with larvae of the tropical sod webworm, *Herpetogramma phaeopteralis*, provided 80%, 86% and 90% control of first and second, third and fourth, and fifth and sixth instars, respectively (Tofangsazi *et al.*, 2014). In pasture plots infested with sixth-instar common armyworm, *Mythimna* (*Pseudaletia*) *unipuncta*, larvae, local Azorean EPN strains (10^{10} IJs ha^{-1}) provided only limited control (*H. bacteriophora* 43%, *S. carpocapsae* 32%); however, efficacy might have been higher against earlier instars and with an exposure longer than 3 days (Rosa and Simões, 2004). In a field study with fifth instars of the armyworm *Mythimna* (*Pseudaletia*) *separata*, a low rate (10^9 IJs ha^{-1}) of Korean strains of *S. carpocapsae* (GSN1 strain), *S. longicaudum* (Nosan strain), *Heterorhabditis* sp. (HG strain) and *S. monticolum* (Jiri strain), provided 65%, 60%, 50%, and 45% control, respectively (Jung *et al.*, 2013).

21.7 Crane Flies

The larvae of European cranefly, *Tipula paludosa*, are susceptible to several *Heterorhabditis* spp. and particularly to *S. feltiae* (Ehlers and Gerwien, 1993; Simard *et al.*, 2006). Susceptibility to *S. feltiae* decreased with larval development (Peters and Ehlers, 1994). In field trials, *S. carpocapsae* (All strain) (5×10^9 IJs ha^{-1}) provided 75-82% control when applied against first and second instars but was ineffective applied against second to third instars, likely due to decreasing soil temperatures and lower susceptibility of older larvae; *S. feltiae* (OBSIII strain) was ineffective (0-16%), irrespective of application timing (Oestergaard *et al.*, 2006).

21.8 Nematode Ecology and Long-term Effects on Pests

Turfgrass and pastures are perennial systems with very limited mechanical disruption. They contain a complex of soil and aboveground arthropod pests and a multitude of non-pest arthropods, provided these are not excessively curtailed by pesticide applications. Natural EPN populations are very common in turfgrass areas and in the temperate north-eastern USA usually consist primarily of *S. carpocapsae* and *H. bacteriophora*, with *S. feltiae*, *S. glaseri* and *S. scarabaei* being less common (Campbell *et al.*, 1995, 1996, 1998; Alumai *et al.*, 2006; Koppenhöfer and Fuzy, 2009; McGraw and Koppenhöfer, 2009).

Natural populations of *H. bacteriophora* and *S. carpocapsae* can be recovered throughout the year (Campbell *et al.*, 1995; Koppenhöfer and Fuzy, 2009; McGraw and Koppenhöfer, 2009; Elmowitz *et al.*, 2013). In lower-maintenance turfgrass (mown once or twice a week at ≥38 mm height, with limited pesticide input), EPN populations tend to be relatively stable with no clear relationship between potential host species and *S. carpocapsae* population distribution but an inverse relationship between populations of *H. bacteriophora* and *P. japonica* larvae (Campbell *et al.*, 1995, 1998). On golf course fairways with limited insecticide use, *S. carpocapsae* and *H. bacteriophora* exhibited a distinct seasonality, with high densities in the weeks immediately following high densities of *L. maculicollis* larvae and pupae, and distribution of both EPN species cycling dynamically between aggregated and uniform throughout the season (McGraw and Koppenhöfer, 2009).

Turfgrass is a fairly uniform habitat. None the less, native EPN populations are spatially very patchy (Campbell *et al.*, 1996, 1998). Even after inundative applications, IJ populations decline quickly and return to the aggregated pattern typical of natural populations within weeks, whether they were applied uniformly to the surface (Campbell *et al.*, 1998; Ebssa and Koppenhöfer, 2011) or in patches (Wilson *et al.*, 2003).

It is likely that turf system type and maintenance intensity affect EPN populations. On golf courses in Ohio, Alumai *et al.* (2006) found natural EPNs in 57% of roughs (lower maintenance) and 43% of fairways (mown three to four times weekly at around 12 mm height, with intermediate level of pesticide input) but 0% of the putting greens (mown daily at around 3 mm, sand based or top dressed with sand, with highest pesticide input). However, after applications of *S. carpocapsae*, *S. feltiae*, *S. riobrave* or *H. bacteriophora* to greens (soil and sand based), fairways and roughs, there was no consistent effect of turf type. Persistence was often longer on the greens, but generally a rapid decline of all species was observed (Ebssa and Koppenhöfer, 2011).

The patchy distribution of EPN populations limits reliable long-term effects on pest populations by natural or applied EPN populations except in some scenarios with very mobile key hosts such as with *S. scapterisci* and mole crickets. The patchiness is likely the result of the patchy distribution of EPN hosts, EPN natural enemies, the aggregative behavior of some EPN species and the limited lifespan of IJs of many EPN species. EPN strains that have been cultured in the laboratory for many years, and especially commercial strains selected for high virulence and ability to be mass produced, may have lost some of their persistence potential (Wang and Grewal, 2002). Inoculative applications of native EPN strains, maintained to preserve their ability to persist in the environment, may under the right conditions (system, hosts) suppress pest populations in the long term after one initial application (Shields, 2015; Shields and Testa, 2015). After inoculative application of such native persistent isolate mixes of *S. carpocapsae*, *H. bacteriophora* and their combination on golf course fairways and adjacent roughs, EPN populations increased initially in plots treated with the respective species and stayed high for most of the 3-year experiment for *S. carpocapsae* but stabilized at lower levels for *H. bacteriophora* (Koppenhöfer and Sousa, 2024). Averaged across the 3 years, the combination treatment reduced *L. maculicollis* larvae (59% reduction) and adults (74%) in the fairway; *H. bacteriophora* (67%) and the combination (63%) reduced white grubs in the rough; all treatments reduced black turfgrass ataenius (*Ataenius spretulus*) adults in rough and fairway (43–62%); *S. carpocapsae* reduced sod webworm larvae in the fairway (75%) and the rough (100%); and *S. carpocapsae* (88%) and the combination (75%) reduced cutworm larvae in the fairway. Inoculative applications of fresh field isolate mixes of endemic EPNs may be a feasible approach to long-term suppression of insect pests in turfgrass but may require periodic reapplications.

Long-term effects of EPN applications may be more likely through the use of more host-specific EPNs, provided that the target pest is a key pest that is present in susceptible stages for extended periods of the year. The success with *S. scapterisci* (see section 21.4) is in part due to the mobility of the target, adult mole crickets, which can survive for several days after infection and, by walking on the surface and especially through flight, can initiate new *S. scapterisci* population foci over significant distances. *S. scarabaei*, in three 3–4-year experiments targeting *A. orientalis*, provided not only high control within 1 month of application (77–100% control at $0.25–2.5 \times 10^9$ IJs ha^{-1}), but also additional control the following spring (86–100% control at $0.1–2.5 \times 10^9$ IJs ha^{-1}), and persisted and suppressed *A. orientalis* for up to 4 years after application (Koppenhöfer and Fuzy, 2009). The success of *S. scarabaei* is based on its excellent adaptation to white grubs (high virulence and recycling) and outstanding persistence of individual IJs (Koppenhöfer and Fuzy, 2006, 2007). However, *S. scarabaei* is not likely to spread quickly over longer distances due to the limited dispersal of its IJs and *S. scarabaei*-infected white grubs.

Any long-term effects of inundative or inoculative EPN applications would also be aided by measures that conserve their populations (see Chapter 28, this volume). Unfortunately, two of the major factors that contribute to limiting EPN populations are not very "negotiable" with most turfgrass managers as they are seen as essential to maintain acceptable turfgrass quality: the use of insecticides that limit the ability of EPNs to recycle, and the use of nematicides in warmer climates where plant-parasitic nematodes are major turfgrass pests. These hindrances can only be overcome if a more holistic approach to turfgrass management can be developed and implemented that makes the system more tolerant of pests, for example more pest-tolerant or -resistant turfgrass cultivars and improved cultural methods.

21.9 Conclusions and Recommendations

Despite considerable efforts in research and development over the last decades, the use of EPNs in the management of turfgrass insects remains very limited. In the largest turfgrass market, the USA, EPNs are mostly used by homeowners who value reduced exposure to synthetic insecticides more than they value product efficacy, ease of use and price. As a result of concerns about health risks and environmental hazards of pesticides, in the USA and Canada pesticide legislation has resulted in the loss of many insecticides, especially organophosphates, for turfgrass uses (Bélair *et al.*, 2010). In Canada, the use of pesticides for cosmetic purposes is banned in two provinces (Québec and Ontario), and more than 152 municipalities have adopted by-laws restricting or banning the use of landscape pesticides (Bélair *et al.*, 2010). Golf courses in Canada may still use pesticides but are required to comply with new stricter regulations. However, in the USA, due to the larger turfgrass market and more uniform pesticide regulations, combined with less cumbersome and faster regulatory process, the old chemistries have largely been replaced by new active ingredients that are considered low-risk insecticides (e.g. neonicotinoids, oxadiazines, diamides). However, public concerns about insecticide safety persist, recently driven by findings that the widely used group of neonicotinoid insecticides may contribute to the decline of, or have detrimental effects on, endemic pollinator populations and commercial honeybees (Charles *et al.*, 2014). As a result, many of the non-agricultural neonicotinoid uses including turfgrass may be lost or at least restricted, leaving turf managers with fewer and more expensive chemical alternatives that could make the use of EPNs economically more competitive.

Taking advantage of the growing interest in alternatives to synthetic insecticides and organic land care will require improvements with respect to EPN efficacy (more virulent strains from field populations or through biotechnology), improved application technologies (see Chapter 11, this volume), lower costs (better production technologies; see Chapter 10, this volume) and especially ease of use (extended shelf life of formulations and tolerance to temperature extremes; see Chapter 12, this volume). Advances in these areas will probably be gradual but may suffice to significantly increase EPN use at least where pesticide regulations, local ordinances and public opinion already impinge on synthetic insecticide use. Better exploitation of the ability of EPNs to recycle in hosts and provide long-term pest suppression could also increase EPN use in many situations. Most likely, significant increases in EPN use, whether as biopesticides or inoculative agents, will happen through education and legislations that will encourage changes in synthetic insecticide use through legislative incentives, regulations and restrictions.

21.10 References

Alm, S.R., Yeh, T., Hanula, J.L. and Georgis, R. (1992) Biological control of Japanese, oriental, and black turfgrass ataenius beetle (Coleoptera: Scarabaeidae) larvae with entomopathogenic nematodes (Nematoda: Steinernematidae, Heterorhabditidae). *Journal of Economic Entomology* 85, 1660–1665.
Alumai, A., Grewal, P.S., Hoy, C.W. and Willoughby, D.A. (2006) Factors affecting the natural occurrence of entomopathogenic nematodes in turfgrass. *Biological Control* 36, 368–374.

Ansari, M.A., Tirry, L. and Moens, M. (2003) Entomopathogenic nematodes and their symbiotic bacteria for the biological control of *Hoplia philanthus* (Coleoptera: Scarabaeidae). *Biological Control* 28, 111–117.

Beard, J.B. and Green, R.L. (1994) The role of turfgrass in environmental protection and their benefits to humans. *Journal of Environmental Quality* 23, 452–460.

Bélair, G., Koppenhöfer, A.M., Dionne, J. and Simard, L. (2010) Current and potential use of pathogens in the management of turfgrass insects as affected by new pesticide regulations in North America. *International Journal of Pest Management* 56, 51–60.

Bélair, G., Simard, L. and Dionne, J. (2013) Canadian entomopathogenic nematode isolates: virulence against black cutworm (Lepidoptera: Noctuidae). *Phytoprotection* 93, 43–46.

Breuninger, J.M., Welterlen, M.S., Augustin, B.J., Cline, V. and Morris, K. (2013) The turfgrass industry. In: Stier, J.C., Horgan, B.P. and Bonos, S.A. (eds.) *Turfgrass: Biology, Use, and Management*. Agronomy Monograph 56. American Society of Agronomy, Crop Science Society of America, Soil Science Society of America, Madison, Wisconsin. pp. 933–1006.

Cappaert, D.C. and Koppenhöfer, A.M. (2003) *Steinernema scarabaei*, an entomopathogenic nematode for control of the European chafer. *Biological Control* 28, 379–386.

Campbell, J.F., Lewis, E., Yoder, F. and Gaugler, R. (1995) Entomopathogenic nematode (Heterorhabditidae and Steinernematidae) seasonal population dynamics and impact on insect populations in turfgrass. *Biological Control* 5, 598–606.

Campbell, J.F., Lewis, E., Yoder, F. and Gaugler, R. (1996) Entomopathogenic nematode (Heterorhabditidae and Steinernematidae) spatial distribution in turfgrass. *Parasitology* 113, 473–482.

Campbell, J.F., Orza, G., Yoder, F., Lewis, E. and Gaugler, R. (1998) Spatial and temporal distribution of endemic and released entomopathogenic nematode populations in turfgrass. *Entomologia Experimentalis et Applicata* 86, 1–11.

Charles, H., Godfray, J., Blacquière, T., Field, L.M., Hails, R.S. *et al.* (2014) A restatement of the natural science evidence base concerning neonicotinoid insecticides and insect pollinators. *Proceedings of the Royal Society B* 281: 20140558.

Delfosse, E.S. (1993) *Pests of Pastures: Weed, Invertebrate and Disease Pests of Australian Sheep Pastures.* CSIRO Publishing, Canberra, Australia.

Deseö, K.V., Bartocci, R., Tartaglia, A. and Rovesti, L. (1990) Entomopathogenic nematodes for control of scarab larvae. *IOBC/WPRS Bulletin* 14, 57–58.

Ebssa, L. and Koppenhöfer, A.M. (2011) Efficacy and persistence of entomopathogenic nematodes for black cutworm control in turfgrass. *Biocontrol Science and Technology* 21, 779–796.

Ebssa, L. and Koppenhöfer, A.M. (2012) Entomopathogenic nematodes for black cutworm management: effect of instar, nematode species, and nematode production method. *Pest Management Science* 68, 947–957.

Ebssa, L., Fuzy, E.M., Bickerton, M.W. and Koppenhöfer, A.M. (2012) Host density effects on efficacy of entomopathogenic nematodes against white grub (Coleoptera: Scarabaeidae) species. *Biocontrol Science and Technology* 22, 117–123.

Ehlers, R.-U. and Gerwien, A. (1993) Selection of entomopathogenic nematodes (Steinernematidae and Heterorhabditidae, Nematoda) for the biological control of cranefly larvae *Tipula paludosa* (Tipulidae, Diptera). *Zeitschrift für Pflanzenkrankheiten und Pflanzenschutz* 100, 343–353.

Ehlers, R.-U. and Peters, A. (1998) Bekämpfung von Engerlingen auf Sportsrasen. *Rasen Turf Gazon* 1998, 1–8.

Elmowitz, D.E., Ebssa, L. and Koppenhöfer, A.M. (2013) Overwintering behavior of the entomopathogenic nematodes *Steinernema scarabaei* and *Heterorhabditis bacteriophora* (Rhabditida: Steinernematidae and Heterorhabditidae) and their white grub hosts (Coleoptera: Scarabaeidae). *Entomologia Experimentalis et Applicata* 148, 246–258.

Frank, J.H. and Walker, T.J. (2006) Permanent control of pest mole crickets (Orthoptera: Gryllotalpidae) in Florida. *American Entomologist* 52, 139–144.

Fujiie, A., Yokohama, T., Fujikata, M., Sawada, M. and Hasegawa, M. (1993) Pathogenicity of an entomogenous nematode, *Steinernema kushidai* Mamiya (Nematoda: Steinernematidae), on *Anomala cuprea* (Coleoptera: Scarabaeidae). *Japanese Journal of Applied Entomology and Zoology* 37, 53–60.

Georgis, R. and Gaugler, R. (1991) Predictability in biological control using entomopathogenic nematodes. *Journal of Economic Entomology* 84, 713–720.

Georgis, R. and Poinar, G.O. Jr. (1994) Nematodes as bioinsecticides in turf and ornamentals. In: Leslie, A. (ed.) *Integrated Pest Management for Turf and Ornamentals*. CRC Press, Boca Raton, Florida, pp. 477–489.

Georgis, R., Dunlop, D.B. and Grewal, P.S. (1995) Formulation of entomopathogenic nematodes. In: Hall, F.R. and Barry, J.W. (eds.) *Biorational Pest Control Agents: Formulation and Delivery*. ACS Symposium Series No. 595. American Chemical Society, Washington, DC, pp. 197–205.

Glazer, I. and Gol'berg, A. (1989) Laboratory evaluation of steinernematid and heterorhabditid nematodes for control of the beetle *Maladera matrida* Argaman (Coleoptera: Scarabaeidae). *Phytoparasitica* 17, 3–11.

Glazer, I. and Gol'berg, A. (1993) Field efficacy of entomopathogenic nematodes against the beetle *Maladera matrida* (Coleoptera: Scarabaeidae). *Biocontrol Science and Technology* 3, 367–376.

Grewal, P.S. and Georgis, R. (1998) Entomopathogenic nematodes. In: Hall, F.R. and Menn, J.J. (eds) *Biopesticides: Use and Delivery*. Humana Press, Totowa, New Jersey, pp. 271–299.

Grewal, P.S., Grewal, S.K., Malik, V.S. and Klein, M.G. (2002) Differences in the susceptibility of introduced and native white grub species to entomopathogenic nematodes from various geographic localities. *Biological Control* 24, 230–237.

Grewal, P.S., Power, K.T., Grewal, S.K., Suggars, A. and Haupricht, S. (2004) Enhanced consistency in biological control of white grubs (Coleoptera: Scarabaeidae) with new strains of entomopathogenic nematodes. *Biological Control* 30, 73–82.

Grewal, P.S., Koppenhöfer, A.M. and Choo, H.Y. (2005) Lawn, turfgrass, and pasture applications. In: Grewal, P.S., Ehlers, R.U. and Shapiro-Ilan, D.I. (eds.) *Nematodes as Biocontrol Agents*. CAB International, Wallingford, UK, pp. 115–146.

Gulcu, B., Ulug, D., Hazir, C., Karagoz, M. and Hazir, S. (2014) Biological control potential of native entomopathogenic nematodes (Steinernematidae and Heterorhabditidae) against *Spodoptera cilium* (Lepidoptera: Noctuidae) in turfgrass. *Biocontrol Science and Technology* 24, 965–970.

Jackson, T.A. (1992) Scarabs – pests of the past or future? In: Jackson, T.A. and Glare, T.R. (eds.) *Use of Pathogens in Scarab Pest Management*. AgResearch, Lincoln, New Zealand, pp. 1–6.

Jung, Y.H., Kim, J.J., You, E.J., Lee, C.M., Choo, H.Y. and Lee, D.W. (2013) Evaluation of entomopathogenic nematodes against armyworm, *Pseudaletia separata* on tall fescue, *Festuca arundinacea*. *Weed & Turfgrass Science* 2, 312–317.

Kard, B.M.R., Hain, F.P. and Brooks, W.M. (1988) Field suppression of three white grub species (Coleoptera: Scarabaeidae) by the entomogenous nematodes *Steinernema feltiae* and *Heterorhabditis heliothidis*. *Journal of Economic Entomology* 81, 1033–1039.

Kaya, H.K. and Gaugler, R. (1993) Entomopathogenic nematodes. *Annual Review of Entomology* 38, 181–206.

Kinoshita, M. and Yamanaka, S. (1998) Development and prevalence of entomopathogenic nematodes in Japan. *Japanese Journal of Nematology* 28, 42–45.

Klein, M.G. (1993) Biological control of scarabs with entomopathogenic nematodes. In: Bedding, R.A., Akhurst, R. and Kaya, H.K. (eds.) *Nematodes and the Biological Control of Insect Pests*. CSIRO Publications, East Melbourne, Australia, pp. 49–58.

Koppenhöfer, A.M. and Fuzy, E.M. (2003a) *Steinernema scarabaei* for the control of white grubs. *Biological Control* 28, 47–59.

Koppenhöfer, A.M. and Fuzy, E.M. (2003b) Biological and chemical control of the Asiatic garden beetle, *Maladera castanea* (Coleoptera: Scarabaeidae). *Journal of Economic Entomology* 96, 1076–1082.

Koppenhöfer, A.M. and Fuzy, E.M. (2003c) Effects of turfgrass endophytes (Clavicipitaceae: Ascomycetes) on white grub (Coleoptera: Scarabaeidae) control by the entomopathogenic nematode *Heterorhabditis bacteriophora* (Rhabditida: Heterorhabditidae). *Environmental Entomology* 32, 392–396.

Koppenhöfer, A.M. and Fuzy, E.M. (2004) Effect of white grub developmental stage on susceptibility to entomopathogenic nematodes. *Journal of Economic Entomology* 97, 1842–1849.

Koppenhöfer, A.M. and Fuzy, E.M. (2006) Effect of soil type on infectivity and persistence of the entomopathogenic nematodes *Steinernema scarabaei*, *Steinernema glaseri*, *Heterorhabditis zealandica*, and *Heterorhabditis bacteriophora*. *Journal of Invertebrate Pathology* 92, 11–22.

Koppenhöfer, A.M. and Fuzy, E.M. (2007) Soil moisture effects on infectivity and persistence of the entomopathogenic nematodes *Steinernema scarabaei*, *Steinernema glaseri*, *Heterorhabditis zealandica*, and *Heterorhabditis bacteriophora*. *Applied Soil Ecology* 35, 128–139.

Koppenhöfer, A.M. and Fuzy, E.M. (2008a) Effects of the anthranilic diamide insecticide, chlorantraniliprole, on *Heterorhabditis bacteriophora* (Rhabditida: Heterorhabditidae) efficacy against and reproduction in white grubs (Coleoptera: Scarabaeidae). *Biological Control* 45, 93–102.

Koppenhöfer, A.M. and Fuzy, E.M. (2008b) Early timing and new combinations to increase the efficacy of neonicotinoid-entomopathogenic nematodes (Rhabditida: Heterorhabditidae) combinations against white grubs (Coleoptera: Scarabaeidae). *Pest Management Science* 64, 725–735.

Koppenhöfer, A.M. and Fuzy, E.M. (2008c) Attraction of four entomopathogenic nematode species to four white grub species. *Journal of Invertebrate Pathology* 99, 227–234.

Koppenhöfer, A.M. and Fuzy, E.M. (2009) Long-term effects and persistence of *Steinernema scarabaei* applied for suppression of *Anomala orientalis* (Coleoptera: Scarabaeidae). *Biological Control* 48, 63–72.

Koppenhöfer, A.M. and Kaya, H.K. (1998) Synergism of imidacloprid and an entomopathogenic nematode: a novel approach to white grub (Coleoptera: Scarabaeidae) control in turfgrass. *Journal of Economic Entomology* 91, 618–623.

Koppenhöfer, A.M. and Sousa, A.L. (2024) Long-term suppression of turfgrass insect pests with native persistent entomopathogenic nematodes. *Journal of Invertebrate Pathology* 204: 108123.

Koppenhöfer, A.M., Choo, H.Y., Kaya, H.K., Lee, D.W. and Gelernter, W.D. (1999) Increased field and greenhouse efficacy against scarab grubs with a combination of an entomopathogenic nematode and *Bacillus thuringiensis*. *Biological Control* 14, 37–44.

Koppenhöfer, A.M., Brown, I.M., Gaugler, R., Grewal, P.S., Kaya, H.K. and Klein, M.G. (2000a) Synergism of entomopathogenic nematodes and imidacloprid against white grubs: greenhouse and field evaluation. *Biological Control* 19, 245–251.

Koppenhöfer, A.M., Wilson, M.G., Brown, I., Kaya, H.K. and Gaugler, R. (2000b) Biological control agents for white grubs (Coleoptera: Scarabaeidae) in anticipation of the establishment of the Japanese beetle in California. *Journal of Economic Entomology* 93, 71–80.

Koppenhöfer, A.M., Cowles, R.S., Cowles, E.A., Fuzy, E.M. and Baumgartner, L. (2002) Comparison of neonicotinoid insecticides as synergists for entomopathogenic nematodes. *Biological Control* 24, 90–97.

Koppenhöfer, A.M., Fuzy, E.M., Crocker, R., Gelernter, W. and Polavarapu, S. (2004) Pathogenicity of *Steinernema scarabaei*, *Heterorhabditis bacteriophora* and *S. glaseri* to twelve white grub species. *Biocontrol Science and Technology* 14, 87–92.

Koppenhöfer, A.M., Grewal, P.S. and Fuzy, E.M. (2006) Virulence of the entomopathogenic nematodes *Heterorhabditis bacteriophora*, *H. zealandica*, and *Steinernema scarabaei* against five white grub species (Coleoptera: Scarabaeidae) of economic importance in turfgrass in North America. *Biological Control* 38, 397–404.

Koppenhöfer, A.M., Grewal, P.S. and Fuzy, E.M. (2007) Differences in penetration routes and establishment rates of four entomopathogenic nematode species into four white grub species. *Journal of Invertebrate Pathology* 94, 184–195.

Koppenhöfer, A.M., Ebssa, L. and Fuzy, E.M. (2013) Storage temperature and duration affect *Steinernema scarabaei* dispersal and attraction, virulence, and infectivity to a white grub host. *Journal of Invertebrate Pathology* 112, 129–137.

Koppenhöfer, A.M., Kostromytska, O.S. and Wu, S. (2020) Optimizing the use of entomopathogenic nematodes for the management of *Listronotus maculicollis* (Coleoptera: Curculionidae): split applications and combinations with imidacloprid. *Crop Protection* 137: 105229.

Koppenhöfer, A.M., Kostromytska, O.S. and Ebssa, L. (2022) Species combinations, split applications, and syringing to optimize the efficacy of entomopathogenic nematodes against *Agrotis ipsilon* (Lepidoptera: Noctuidae) larvae in turfgrass. *Crop Protection* 155: 104927.

Lee, W., Choo, H.Y., Kaya, H.K., Lee, S.M., Smitley, D.R. *et al.* (2002) Laboratory and field evaluation of Korean entomopathogenic nematode isolates against the oriental beetle *Exomala orientalis* (Coleoptera: Scarabaeidae). *Journal of Economic Entomology* 95, 918–926.

Leppla, N.C., Frank, J.H., Adjei, M.B. and Vicente, N.E. (2007) Management of pest mole crickets in Florida and Puerto Rico with a nematode and parasitic wasp. *Florida Entomologist* 90, 229–233.

Marianelli, L., Paoli, F., Torrini, G., Mazza, G., Benvenuti, C. *et al.* (2018) Biological control of *Popillia japonica* (Coleoptera, Scarabaeidae): managing the Italian outbreak. *Journal of Applied Entomology* 142, 311–318.

McGraw, B.A. and Koppenhöfer, A.M. (2009) Population dynamics and interactions between endemic entomopathogenic nematodes and annual bluegrass weevil populations in golf course turfgrass. *Applied Soil Ecology* 41, 77–89.

McGraw, B.A., Cowles, R., Vittum, P.J. and Koppenhöfer, A.M. (2010) Field evaluation of entomopathogenic nematodes for the biological control of the annual bluegrass weevil, *Listronotus maculicollis* (Coleoptera: Curculionidae) in golf course turfgrass. *Biocontrol Science and Technology* 20, 149–163.

Oestergaard, J., Belau, C., Strauch, O., Ester, A., van Rozen, K. and Ehlers, R.-U. (2006) Biological control of *Tipula paludosa* (Diptera: Nematocera) using entomopathogenic nematodes (*Steinernema* spp.) and *Bacillus thuringiensis* subsp. *israelensis*. *Biological Control* 39, 525–531.

Parkman, J.P. and Frank, J.H (1992) Infection of sound-trapped mole crickets, *Scapteriscus* spp., by *Steinernema scapterisci*. *Florida Entomologist* 75, 163–165.

Parkman, J.P. and Smart, G.C. Jr. (1996) Entomopathogenic nematodes, a case study: introduction of *Steinernema scapterisci* in Florida. *Biocontrol Science and Technology* 6, 413–419.

Parkman, J.P., Hudson, W.G., Frank, J.H., Nguyen, K.B. and Smart, G.C. Jr. (1993) Establishment and persistence of *Steinernema scapterisci* (Rhabditida: Steinernematidae) in field populations of *Scapteriscus* spp. mole crickets (Orthoptera: Gryllotalpidae). *Journal of Entomological Science* 28, 182–190.

Peters, A. (1996) The natural host range of *Steinernema* and *Heterorhabditis* spp. and their impact on insect populations. *Biocontrol Science and Technology* 6, 389–402.

Peters, A. and Ehlers, R.-U. (1994) Susceptibility of leatherjackets (*Tipula paludosa* and *T. oleracea*; Tipulidae: Nematocera) to the entomopathogenic nematode *Steinernema feltiae*. *Journal of Invertebrate Pathology* 63, 163–171.

Poinar, G.O. Jr. (1992) Nematodes associated with Scarabaeidae. In: Jackson, T.A. and Glare, T.R. (eds.) *Use of Pathogens in Scarab Pest Management*. AgResearch, Lincoln, New Zealand, pp. 93–109.

Potter, D.A. (1998) *Destructive Turfgrass Insects: Biology, Diagnosis, and Control*. Ann Arbor Press, Chelsea, Michigan.

Power, K.T., An, R. and Grewal, P.S. (2009) Effectiveness of *Heterorhabditis bacteriophora* strain GPS11 applications targeted against different instars of the Japanese beetle, *Popillia japonica*. *Biological Control* 48, 232–236.

Rosa, J.S. and Simões, N. (2004) Evaluation of twenty-eight strains of *Heterorhabditis bacteriophora* isolated in Azores for biological control of the armyworm, *Pseudaletia unipunctata* (Lepidoptera: Noctuidae). *Biological Control* 29, 409–417.

Schroeder, W.J. and Sieburth, P.J. (1997) Impact of surfactants on control of the root weevil *Diaprepes abbreviatus* larvae with *Steinernema riobravis*. *Journal of Nematology* 29, 216–219.

Selvan, S., Gaugler, R. and Campbell, J.F. (1993) Efficacy of entomopathogenic nematode strains against *Popillia japonica* (Coleoptera: Scarabaeidae) larvae. *Journal of Economic Entomology* 86, 353–359.

Selvan, S., Grewal, P.S., Gaugler, R. and Tomalak, M. (1994) Evaluation of steinernematid nematodes against *Popillia japonica* (Coleoptera: Scarabaeidae) larvae: species, strains, and rinse after application. *Journal of Economic Entomology* 87, 605–609.

Shetlar, D.J., Suleman, P.E. and Georgis, R. (1988) Irrigation and use of entomogenous nematodes, *Neoaplectana* spp. and *Heterorhabditis heliothidis* (Rhabditida: Steinernematidae and Heterorhabditidae), for control of Japanese beetle (Coleoptera: Scarabaeidae) grubs in turfgrass. *Journal of Economic Entomology* 81, 1318–1322.

Shields, E.J. (2015) Utilizing persistent entomopathogenic nematodes in a conservation or a more classical biological control approach. In: Campos-Herrera, R. (ed.) *Nematode Pathogenesis of Insects and Other Pests: Ecology and Applied Technologies for Sustainable Plant and Crop Protection*. Springer, Cham, Switzerland, pp. 165–184.

Shields, E.J. and Testa, A.M. (2015) New York case study: biological control of *Otiorhynchus ligustici* with native persistent entomopathogenic nematodes using a more classical approach. In: Campos-Herrera, R. (ed.) *Nematode Pathogenesis of Insects and Other Pests: Ecology and Applied Technologies for Sustainable Plant and Crop Protection*. Springer, Cham, Switzerland, pp. 285–307.

Simard, L., Bélair, G., Gosselin, M.E. and Dionne, J. (2006) Virulence of entomopathogenic nematodes (Rhabditida: Steinernematidae, Heterorhabditidae) against *Tipula paludosa* (Diptera: Tipulidae), a turfgrass pest on golf courses. *Biocontrol Science and Technology* 16, 789–801.

Smith, K.A. (1994) Control of weevils with entomopathogenic nematodes. In: Smith, K.A. and Hatsukade, M. (eds.) *Control of Insect Pests with Entomopathogenic Nematodes*. Food and Fertilizer Technology Center, Taipei, pp. 1–13.

Smits, P.H., Wiegers, G.L. and Vlug, H.J. (1994) Selection of insect parasitic nematodes for biological control of the garden chafer, *Phyllopertha horticola*. *Entomologia Experimentalis et Applicata* 70, 77–82.

Stock, S.P. and Koppenhöfer, A.M. (2003) *Steinernema scarabaei* n. sp. (Rhabditida: Steinernematidae), a natural pathogen of scarab beetle larvae (Coleoptera: Scarabaeidae) from New Jersey, USA. *Nematology* 5, 191–204.

Sulistyanto, D. and Ehlers, R.-U. (1996) Efficacy of the entomopathogenic nematodes *Heterorhabditis megidis* and *Heterorhabditis bacteriophora* for control of grubs (*Phyllopertha horticola* and *Aphodius contaminatus*) in golf course turf. *Biocontrol Science and Technology* 6, 247–250.

Tofangsazi, N., Cherry, R.H. and Arthurs, S.P. (2014) Efficacy of commercial formulations of entomopathogenic nematodes against tropical sod webworm, *Herpetogramma phaeopteralis* (Lepidoptera: Crambidae). *Journal of Applied Entomology* 138, 656–661.

Vittum, P.J. (2020) *Turfgrass Insects of the United States and Canada*, 3rd edn. Cornell University Press, Ithaca, New York.

Wang, X. and Grewal, P.S. (2002) Rapid deterioration of environmental tolerance and reproductive potential of an entomopathogenic nematode during laboratory maintenance. *Biological Control* 23, 71–78.
Wang, Y., Campbell, J.F. and Gaugler, R. (1995) Infection of entomopathogenic nematodes *Steinernema glaseri* and *Heterorhabditis bacteriophora* against *Popillia japonica* (Coleoptera: Scarabaeidae) larvae. *Journal of Invertebrate Pathology* 66, 178–184.
Wilson, M.J., Lewis, E.E., Yoder, F. and Gaugler, R. (2003) Application pattern and persistence of the entomopathogenic nematode *Heterorhabditis bacteriophora*. *Biological Control* 26, 180–188.
Yeh, T. and Alm, S.R. (1992) Effects of entomopathogenic nematode species, rate, soil moisture, and bacteria on control of Japanese beetle (Coleoptera; Scarabaeidae) larvae in the laboratory. *Journal of Economic Entomology* 85, 2144–2148.
Yeh, T. and Alm, S.R. (1995) Evaluation of *Steinernema glaseri* (Nematoda: Steinernematidae) for biological control of Japanese and oriental beetles (Coleoptera: Scarabaeidae). *Journal of Economic Entomology* 88, 1251–1255.
Zimmerman, R.J. and Cranshaw, W.S. (1991) Short-term movement of *Neoaplectana* spp. (Rhabditida: Steinernematidae) and *Heterorhabditis* "HP88" strain (Rhabditida: Heterorhabditidae) through turfgrass thatch. *Journal of Economic Entomology* 84, 875–878.

22 Home and Garden Pests

Dawn H. Gouge*
University of Arizona, Maricopa Agricultural Center, Maricopa, Arizona, USA

Abstract

Most entomopathogenic nematode (EPN) products are developed for agricultural and horticultural pest control. They are ideal but underutilized tools when it comes to pest control around the home, garden and landscapes. EPNs are safe to handle and apply in areas where children and pets play. They can be used on vegetables and on fruiting trees and bushes. EPNs have some specificity to host targets and compatibility with other control tools. Lower levels of efficacy are acceptable in most home, garden and landscape situations, and products are often cost-effective options. There are ample opportunities to increase the use of EPNs in the built environment. This includes established markets for home pests as well as those with susceptibilities established in agricultural production systems. Perhaps the most exciting prospect is the expansion of use to address previously untested pests. With growing urban populations and an increasing appreciation of the benefits of fresh fruits and vegetables, consumption of home-grown and community garden produce has expanded over time. Spurred by a post-pandemic citizenry with improved awareness of their own personal responsibility toward sustainability and resilience, this trend will likely continue to gain momentum. Arguably, the most significant increases in EPN use could be in our towns and cities.

22.1 Introduction

Buildings and landscapes are complex ecosystems that often provide conducive conditions for synanthropic pests to thrive. Building users and residents have a variety of tolerance levels and make decisions about pest management based on very different criteria compared with profit-motivated commercial interests. Residents can also decide to undertake pest-management actions themselves or contract with a professional pest management or landscape company. Unlike crop or livestock producers, managing pests in and around your home rarely has to meet specific cost-effective criteria. Arthropods that pose a health risk to building inhabitants or pets are often the highest priority when it comes to remediation. This includes disease-causing pathogen vectors, those with a venomous sting or bite, allergen producers, and those that contaminate food and food preparation areas. Additionally, pests that cause damage to structures, landscape and garden plants, and those that cause distress due to population explosions are also priority pests (Baldwin *et al.*, 2008).

While most people have insufficient knowledge to identify insect pests or necessarily appreciate the many methods used to effectively control

*Email: dhgouge@arizona.edu

 Entomopathogenic Nematodes as Biological Control Agents (D.I. Shapiro-Ilan and E.E. Lewis eds)
DOI: 10.1079/9781800620322.0022

populations, surveys indicate that more residents undertake pest management themselves than contract with a service (Ralph and Kelvin, 2013). Irrespective of who undertakes the responsibility, family, personal and pet safety is a top concern.

Relative to many over-the-counter chemical pesticides, EPNs (Rhabditida: Steinernematidae and Heterorhabditidae) pose relatively low levels of risk to applicators, building occupants, landscape and grounds users, vertebrate pets and non-target organisms in general. This is especially important considering that residents and property owners seeking to control pests have little or no pesticide safety training and little in the way of specialized application equipment or personal protective equipment. Additionally, EPNs are safe to use on and around edible plants and when used in the natural nematode–bacterial complex cause no phytotoxicity (Vasseghian *et al.*, 2022).

Methods to improve EPN efficacy in non-soil environments have progressed since their initial introduction into home pest-management markets. The inclusion of guidance on product labels covering weather conditions (de Luca *et al.*, 2015), time-of-day application (de Waal *et al.*, 2017), addition of adjuvants (Shapiro-Ilan *et al.*, 2010; de Waal *et al.*, 2013) and humidity levels (Odendaal *et al.*, 2016a) improve product performance and add utility to EPN products.

Another significant benefit of using EPN products in homes and gardens is their ease of use within integrated pest-management approaches. EPNs can be used in conjunction with chemical pesticides and other biopesticides (see Chapter 27, this volume) or may replace chemical insecticides. Decreasing residential and landscape insecticide use would reduce resistance pressure on products used against pathogen-vectoring arthropods and pests of agricultural and horticultural crops (Peng *et al.*, 2021). Some biocontrol combinations are not complementary; van Niekerk and Malan (2014) documented that larva of the predatory ladybird beetle *Cryptolaemus montrouzieri* (Mulsant) (Coleoptera: Coccinellidae) were highly susceptible to multiple EPN species. Inclusion of product incompatibilities on product labels is advisable.

This chapter does not attempt to provide a comprehensive account of all home and landscape pest species tested for EPN susceptibility but rather highlights a number of well-established groups of common pests that commercially available EPNs have proven effective against. Gozel and Gozel (2016) list numerous crop pests for which EPNs have generated useful levels of efficacy, many of which are pests in home gardens and landscapes. Pest species listed include fruit tree borers, grape and berry fruit, vegetable, turf and ornamental plant pests. Nevertheless, it is likely that EPN products have a broader utility than is currently appreciated by consumers.

EPN products are increasingly available throughout the Americas, Europe, Australia, New Zealand, and parts of Asia and Africa. Sales for home garden and landscape use are likely greatest in North America and Europe (van Lenteren, 2012; Abate *et al.*, 2017).

22.2 Landscape and Garden Pests

In 2018, the United Nations estimated that 55% of the world population lived in urban areas, with a projected increase to 68% by 2050 (Department of Economic and Social Affairs, 2020). As human populations become increasingly separated from nature, the benefits associated with natural or agrarian landscapes, urban farms, residential gardens and community grounds are increasingly important. EPNs are of great value in home gardens and community farms when used on or around edible fruit and vegetable plants as no chemical residue results. In fact, EPN products used for common pests in the built environment often adhere to the principles of sustainable food production (Koppenhöfer *et al.*, 2020) and eco-landscaping.

22.2.1 Tree, fruit and ornamental plants

While foliar application of a naturally soil-dwelling group of nematodes presents inherent challenges, they have proven useful biological control agents of insects in protective niches. Control of lepidopterans with cryptic larvae has been particularly successful. Examples include insects that shelter under bark (Odendaal *et al.*, 2016b,c), construct shelters such as bagworms (Gill and Raupp, 1994), leaf miner species (van Damme *et al.*, 2016) and pests that sequester themselves between fruits and branches (Shapiro-Ilan *et al.*, 2016).

Tree fruit and nut pests are covered in Chapter 17 (this volume) in detail, but those most pertinent to home garden and community landscapes with well-established EPN biocontrol solutions include codling moth, *Cydia pomonella* (L.) (Lepidoptera: Tortricidae), oriental fruit moth, *Grapholita molesta* (Busck) (Lepidoptera: Tortricidae), navel orangeworm, *Amyelois transitella* (Walker) (Lepidoptera: Pyralidae), and multiple clearwing moth species (Lepidoptera: Sesiidae) such as the peachtree borer, *Synanthedon exitiosa* (Say), and the lesser peachtree borer, *Synanthedon pictipes* (Grote & Robinson).

There are at least 22 significant pests of landscape trees and shrubs in North America. These include but are not limited to the American hornet moth, *Sesia apiformis* (Clerck), ash borer, *Podosesia syringae* Harris, Douglas-fir pitch moth, *Synanthedon novaroensis* (Edwards), peachtree borer, red-belted clearwing, *Synanthedon myopaeformis* (Borkhausen), sequoia pitch moth, *Synanthedon sequoiae* (Edwards), sycamore borer, *Synanthedon resplendens* (Edwards), and western poplar clearwing moth, *Paranthrene robiniae* (Edwards). The feeding of pitch moth larvae causes injury to conifers that results in globular resin masses forming on the outside of tree bark.

The use of EPNs as biological control agents of clearwing landscape pests has not been evaluated fully. However, Davidson *et al.* (1992) documented significant reductions in the dogwood borer, *Synanthedon scitula* (Harris), a serious landscape pest of flowering dogwood *Cornus florida* L. under field conditions using *Steinernema carpocapsae*.

Banana moth, *Opogona sacchari* Bojer (Lepidoptera: Tineidae), is native to the humid tropical and subtropical regions of sub-Saharan Africa. However, the moth has been introduced into South and Central America and parts of Europe and regions in the USA. The moth is attracted to stressed, injured and pruned plants. In landscapes, it can infest a wide variety of palms, yucca, *Strelitzia*, cacti and dragon tree (*Dracaena*) species. The larvae tunnel through stems and can cause the entire plant to lodge. There is scant information in the literature, but Pena *et al.* (1990) reported good efficacy using *Steinernema feltiae* (Filipjev) and *Heterorhabditis bacteriophora* Poinar (Rhabditida: Heterorhabditidae) in laboratory and greenhouse trials, and several EPN products list the moth as a target.

Van der Linden *et al.* (2022) presented data on the effective use of EPNs for the control of the boxtree moth, *Cydalima perspectalis* Walker (Lepidoptera: Crambidae). The moth is an invasive pest of boxtree/boxwood (*Buxus* spp.) plants, which are ornamental shrubs widely used in public and private landscapes in Europe and the USA. The authors concluded that the characteristics of *Buxus* spp. foliage provide supportive conditions for control of moths using EPNs, with *S. carpocapsae* Weiser being particularly effective (Göttig and Herz, 2018).

The false codling moth, *Thaumatotibia leucotreta* (Meyrick) (Lepidoptera: Tortricidae), and plum curculio, *Conotrachelus nenuphar* (Herbst) (Coleoptera: Curculionidae) have immature stages that feed in fruit, causing fruit fall, and pupate in the soil. *T. leucotreta* is native to sub-Saharan Africa, and in residential gardens it is most likely to infest citrus, stone fruit, avocado and pomegranate. Within the native range, a commercially available indigenous *H. bacteriophora*-based product has been used (Cryptonem™; River Bioscience, South Africa) with success (Ndlela *et al.*, 2022). *C. nenuphar* is native to eastern regions of the USA and Canada, and attacks pome fruits including apples, crab apples, pears and quince, as well as stone fruits including cherry, peach, apricot, plum, wild plum and others. Adults overwinter in leaf litter, and females puncture fruit and lay eggs in the puncture holes. Due to the feeding of grubs, infested fruit falls to the ground and late instars leave the fruit to pupate in the soil. Shapiro-Ilan *et al.* (2013) established *Steinernema riobrave* Cabanillas, Poinar & Raulston as a highly effective option for ground-dwelling stages of *C. nenuphar*.

Many armyworm (Lepidoptera: Noctuidae) species impact ornamental and vegetable plants. The beet armyworm, *Spodoptera exigua* Hübner, is one of the most notorious and is found worldwide defoliating many crops. In gardens, they can impact beans and peas, beets, eggplant, lettuce and tomato plants (Robinson *et al.*, 2023). Together with the cotton cutworm, *Spodoptera litura* F., and the American bollworm, *Helicoverpa armigera* Hübner, the three species are significant pests of cut flowers (Burana *et al.*, 2016). In gardens and landscapes, there are a vast number of ornamental plants impacted including chrysanthemum, gerbera, roses, marigolds and potted plants. Good levels of control of *Helicoverpa armigera* have been demonstrated by Navon *et al.* (2002) and many others. The susceptibility of *Spodoptera litura* and *Spodoptera exigua* have been recently reconfirmed using *S. carpocapsae* and *Heterorhabditis indica* (Poinar, Karunaka & David) by Yan *et al.* (2020) in marigolds.

A number of root weevils (Coleoptera: Curculionidae) are common garden pests. These include *Diaprepes* root weevil, *Diaprepes abbreviatus* (L.), strawberry root weevil, *Otiorhynchus ovatus* (L.), and black vine weevil, *Otiorhynchus sulcatus* (F.). All are covered in other chapters, but in gardens, *D. abbreviatus* will feed on citrus, hibiscus, palm, birch, roses, guava, loquat, holly and other ornamentals. The immature stages of the weevil feed on roots, which can kill delicate plants, and the adults chew semi-circular notches in leaf margins. *O. ovatus* larvae feed on roots of strawberry plants, as well as a wide variety of conifers including spruce (*Picea* spp.), pine (*Pinus* spp.) and yew (*Taxus* spp.) trees and other plants. The weevils are considered harmless to established trees but can damage stressed trees and stunt the growth of strawberry plants. The larger *O. sulcatus* can cause greater devastation to host plants and is a polyphagous pest of garden strawberries and many ornamental plants. Black vine weevils are a serious pest of azalea, euonymus and rhododendron in the USA and southern Canada.

Mealybugs (Hemiptera: Pseudococcidae) are among one of the most important pest groups in many agricultural and horticultural crops and are frequently the target of residential insecticide applications on fruit-bearing and ornamental plants. Several species of mealybugs appear to be susceptible to nematodes with good efficacy, established by Le Vieux and Malan (2013), using commercially available *H. bacteriophora* and *S. feltiae* for the vine mealybug *Planococcus ficus* (Signoret). This is a species that migrates up and down plants depending on environmental conditions but spends a lot of time under bark or underground in root systems. *P. ficus* lives on a large range of host plants damaging fruit trees and ornamental plants as well as grapevines. Preliminary laboratory studies on the coffee root mealybug, *Dysmicoccus texensis* Tinsley, a pest of many ornamental plants in western USA, were carried out by Andaló *et al.* (2004) and resulted in convincing levels of efficacy, but field studies have shown only limited success (Alves *et al.*, 2009).

Leaf miner species (Diptera: Agromyzidae) are worldwide pests of a wide variety of plants. The larvae feed on numerous field crops, ornamentals and vegetables. The damage is caused by larval mining within leaf surfaces, which significantly reduces the photosynthetic capacity of plants. If severe, leaf fall can result (Johnson *et al.*, 1983). Research on *Liriomyza trifolii* (Burgess) by LeBeck *et al.* (1993) first described infective juvenile (IJ) *S. carpocapsae* entering a leaf mine via the oviposition puncture created by the female fly during egg laying. Others have documented control of other *Liriomyza* spp., expanded the nematode species useful for control and targeted leaf miner species in other insect orders. For example, van Damme *et al.* (2016) conducted tests using *S. carpocapsae* targeting larval *Tuta absoluta* Meyrick (Lepidoptera: Gelechiidae) in tomato leaves.

Beyond the pest species listed in this section, there is ample opportunity to expand the use of EPNs within residential and landscape markets, not only for the specific pests listed but also for closely related species such as Curculionidae root weevils: the cribrate weevil, *Otiorhynchus cribricollis* Gyllenhal, lilac or privet weevil, *Otiorhynchus meridionalis* Gyllenhal, and the rough strawberry root weevil, *Otiorhynchus rugosostriatus* Goeze. Additionally, research opportunities exist for a plethora of species that inhabit similar microniches. In some cases, positive results are dated, or field tests are lacking. For example, little is known about EPN efficacy on the fuller rose beetle, *Asynonychus godmoni* Crotch (Morse and Lindegren, 1996). Efficacy toward other species may simply not be widely known, for example for the large pine weevil, *Hylobius abietis* (McNamara *et al.*, 2018), or may be completely lacking, as in the case of the Douglas-fir twig weevil, *Cylindrocopturus furnissi* Buchanan.

22.2.2 Turf and lawn

A number of turf pests have been managed successfully using EPNs. Information published is covered in detail in Chapter 21 (this volume) and in many instances is taken from research addressing sod farm and golf course turf in which very high levels of control are necessary. There are a multitude of opportunities to manage pests in schools, parks, cemeteries, zoos, botanical gardens, theme parks, common area landscapes and athletic fields. The EPN utility for pest management in residential lawns alone is significant. Costs are not prohibitive and are often comparable with retail products. For example, Shoeb *et al.* (2016) established good results against the African mole cricket, *Gryllotalpa africana* Palisot de Beauvois, but did so using a nematode that is no longer commercially available. African mole crickets are pests of both turf and vegetables in Asia and parts of Europe, and commercially available EPNs may prove effective.

Low water use management practices are ever more important in drought-affected regions of the world. Control solutions for pests that negatively impact heat- and drought-tolerant turf and alternative groundcover plants must be found that can be used effectively under these conditions. To date, relatively little work has been undertaken using EPNs within these systems.

22.2.3 Vegetables

The diamondback moth, *Plutella xylostella* L. (Lepidoptera: Plutellidae), is a pest of brassicas including broccoli, Brussels sprouts, cabbage, cauliflower, collards, kale and others grown at cool temperatures. It is an important pest of Brassicaceae vegetables globally, and economic losses due to the pest are estimated in the billions of US dollars (Zalucki *et al.*, 2012). For many years, control of this pest has been achieved through the constant use of organophosphate, carbamate and pyrethroid insecticides, which ultimately led to resistance issues for growers (Furlong *et al.*, 2013). Vashisth *et al.* (2017) compared *P. xylostella* larval efficacy due to commercially available *H. indica* and a locally isolated *H. bacteriophora* and found that *H. indica* was more pathogenic than the other species and strains of EPNs tested. A year earlier, Correa-Cuadros *et al.* (2016) published their success using *H. bacteriophora* in combination with the fungal pathogens *Beauveria bassiana* (Bals.-Criv.) Vuill. (Hypocreales: Cordycipitaceae) and *Metarhizium anisopliae* (now *Metarhizium robertsii* (Metchnikoff) Sorokin) (Hypocreales: Clavicipitaceae) and described an early antagonistic interaction followed by a synergistic result on day 6. The group concluded that the combination of *H. bacteriophora* with *M. anisopliae* and *B. bassiana* could be considered an innovative alternative for the control of *P. xylostella* in broccoli crops. For home-grown brassicas, all three biocontrol options are more attractive than chemical insecticide alternatives.

A vegetable pest that bores into onions, leeks,and other *Allium* spp. is the leek moth, *Acrolepiopsis assectella* Zeller (Lepidoptera: Acrolepiidae). The larvae mine and perforate leaves and bore into bulbs (Nyrop *et al.*, 1989). Work published by García del Pino and Morton (2008) showed the larvae to be highly susceptible to *S. feltiae* in laboratory and pot tests. The researchers also documented survival of nematodes within overlapping leaves of leek plants causing 97% mortality of leek moth larvae inside leaf galleries.

While there are many lepidopteran families that include vegetable pest species, the family Noctuidae contains several genera considered critically important pests of agricultural, vegetable and forage crops globally. Several polyphagous noctuids are considered among the most devastating pests on the planet. In many cases, chemical insecticides have failed due to the development of resistance (Le Goff and Nauen, 2021). Unsurprisingly, many of these insects impact garden vegetables. All those listed in Table 22.1 have been documented as susceptible to EPNs in research tests.

More specifically, late-instar larvae and pupae of *Spodoptera frugiperda* J.E. Smith were found to be highly susceptible to *H. indica*, *S. carpocapsae* and other species not currently commercially available in soil column and pot assays by Acharya *et al.* (2020). Following this, Fallet *et al.* (2022) conducted field trials and found that *S. carpocapsae* formulated in carboxymethyl cellulose gel reduced the number of larvae per corn plant by 40%, which was equivalent to the control achieved using a standard dose of a pyrethroid insecticide with cypermethrin as the active ingredient.

Shaurub *et al.* (2016) combined nematode species and the fungal pathogen *B. bassiana* in applications to control *Spodoptera littoralis* on cabbage, cowpea and three wild host plants. *Spodoptera littoralis* reared on cabbage as opposed to other host plants was highly resistant to infection by *S. riobrave* and *H. bacteriophora*, indicating that the host plant that the target pest feeds on could be critically significant in some cases.

Table 22.1. Published susceptibility data for noctuid pests susceptible to entomopathogenic nematodes.

Common name	Species name	Reference
Beet armyworm	*Spodoptera exigua*	Canhilal (2013)
Fall armyworm	*Spodoptera frugiperda* (J.E. Smith)	Acharya *et al.* (2020)
African/Egyptian cotton leafworm	*Spodoptera littoralis* (Boisduval)	Shaurub *et al.* (2016)
African armyworm	*Spodoptera exempta* Walker	Rufai *et al.* (2021)
Gram pod borer/cotton bollworm	*Helicoverpa armigera* (Hübner)	Kary *et al.* (2012)
Corn earworm	*Helicoverpa zea* (Boddie)	Feaster and Steinkraus (1996)
Black cutworm	*Agrotis ipsilon* (Hufnagel)	Yan *et al.* (2014)
Turnip cutworm	*Agrotis segetum* Denis & Schiffermüller	Vashisth *et al.* (2018)
Variegated cutworm	*Peridroma saucia* Hübner	Yuksel *et al.* (2018)
Cotton bollworm	*Spodoptera litura*	Noosidum *et al.* (2016)

Interactions between *S. riobrave* and *H. bacteriophora* and *B. bassiana* were found to be synergistic, and the researchers concluded that efficient control of *Spodoptera littoralis* can be achieved when multiple biological control agents are simultaneously or sequentially applied.

Researchers in India combined chemical insecticides with EPNs to determine the compatibility of *S. carpocapsae* and *H. indica* used for *Helicoverpa armigera* control in pigeon pea. Their studies revealed that some but not all chemicals showed a strong sublethal effect on nematode reproductive potential, limiting their in-field longevity (Devindrappa *et al.*, 2017). One of their conclusions was that generalizations regarding EPN tolerance to insecticides are likely to be inaccurate. While this may be a significant constraint in agricultural crops, in the home vegetable garden there are likely to be fewer concerns.

Spodoptera litura and *Spodoptera exigua* have both been mentioned previously as ornamental pests. In vegetables, there is additional research documented. Noosidum *et al.* (2016) treated kale plants infested with *Spodoptera litura* with *S. carpocapsae* plus a protective additive, Barricade®, and established better nematode survival and reduced leaf damage. The susceptibility of the black cutworm (*Agrotis ipsilon*), first mentioned as a turf pest, has been studied by researchers around the world. Notable work by Ebssa and Koppenhöfer (2012) established that a number of commercially available EPN species are capable of infecting larvae including *S. carpocapsae*, *S. riobrave*, *H. bacteriophora* and *H. megidis*. Fourth- and/or fifth-instar stages were the most susceptible to most EPN species, and pupae the least susceptible. Furthermore, *H. bacteriophora*, *H. megidis* and *S. carpocapsae* successfully reproduced in fifth and sixth instars and in pupae.

Rootworms (Coleoptera: Chrysomelidae) in the genus *Diabrotica* contain many destructive species. The western corn rootworm, *Diabrotica virgifera virgifera* LeConte, is an important insect pest of corn in North America and Central and Eastern Europe. As with many previous pests, insecticide resistance has hampered control of these insects (Pereira *et al.*, 2017). Early work by Toepfer *et al.* (2008) showed that the larvae were susceptible to a range of commercially available EPN species. Since then, researchers have undertaken field trials in various parts of the world, and work has expanded to include several *Diabrotica* spp. Successful tests have also been published covering EPN susceptibility to the cucurbit beetle, *Diabrotica speciosa* Germar (Santos *et al.*, 2011).

Other root-feeding pests of vegetables established as susceptible to EPNs include the root weevil *Diaprepes abbreviatus*, which is covered earlier under tree, fruit and ornamental plant pests, the sweet potato weevil, *Cylas formicarius elegantulus* (Summers) (Myers *et al.*, 2020), and the sugarbeet weevil *Temnorhinus mendicus* Gyllenhal (Boselli *et al.*, 1997).

There are many more vegetable pests of significance that have only been tested minimally. For example, the South American bollworm, *Helicoverpa gelotopoeon* (Dyar), is a noctuid that damages pods and seeds of soybean and other legumes. The insect has been found to be susceptible to an EPN species that is not commercially available (Caccia *et al.*, 2014). Unfortunately, no additional work has been undertaken. Additionally, there are many more pests that have no readily available findings published in the literature, such as the western yellow-striped armyworm, *Spodoptera praefica* Grote, and the vegetable weevil, *Listroderes costirostris* Schönherr.

22.3 Social Insects: Hymenoptera

Social behavior in the class Insecta covers a broad range of activities and interactions. Subsocial behavior is widespread in many insect orders including cockroaches, earwigs, webspinners, true bugs, aphids and various beetles. However, here the focus is on insect groups classed as eusocial in nature. Eusocial insects are characterized by parental care of young, overlapping generations and the reproductive division of labor. Although the organisms in this section provide essential ecosystem services, each can be problematic in and around the built environment.

Bees, wasps and ants of the Aculeate Hymenoptera have the ability to sting and as such can be hazardous to people. Even wild honey bee as a human-mediated hybrid of the East African lowland bee, *Apis mellifera scutellata* Lepeletier (Hymenoptera: Apidae), and various European and Middle Eastern honeybee subspecies are considered potentially deadly to people and animals (Feás *et al.*, 2022). Similarly, while we are dependent on termites and many ant species to provide the beneficial breakdown and recycling of plant materials, some species can devastate plants or structures.

Social insects demonstrate the most spectacular evolutionary transitions: competition to cooperation and individualism to altruism between adults. Certainly, the greatest barrier to social pest management is limited understanding of social insect behavioral ecology, as well as insect immunity and physiology. Additionally, species have complex behaviors linked with their social structure and environmental conditions and may nest in cryptic or diffuse nest sites. Another important consideration is that relatively large worker mortality may fail to cause colony death. Despite these challenges, social insects have some inherent vulnerabilities. There is relatively low genetic variability among individuals in a social insect colony, which could mean that the efficacy of a biological control agent is either compromised or enhanced (Sherman *et al.*, 1988). The highly organized colony in which individuals participate in self- or allogrooming, along with nest hygiene, could eliminate or facilitate the spread of pathogens (Oi and Pereira, 1993). Hygiene adaptations are particularly important for social insects, which are at an especially high risk of pathogen transfer due to high-density living and high site fidelity (Cole *et al.*, 2021). The highly volatile chemical nature of social insect communication could also influence biological control organisms (Schildknecht and Koob, 1971; Story *et al.*, 1991). Once within a social nest, however, a pathogen is likely to be in a highly regulated and protected environment, with a stable temperature and humidity, high host density and little to no ultraviolet light exposure. Perhaps as a reflection of these challenges, Peters (1996), in a comprehensive study on the natural host range of *Steinernema* and *Heterorhabditis* spp. known at the time, found only two references to the natural infection of social insects. More associations have come to light in recent years as EPNs have been used more widely for pest management.

22.3.1 Apidae: honey bee and bumble bees

The original reports by Dutky and Hough (1955) and Cantwell *et al.* (1972) indicated that adult honey bee were resistant to *S. carpocapsae*. However, Hackett and Poinar (1973) showed that adult worker western honey bee, *Apis mellifera* L., could be parasitized and killed by *S. carpocapsae* when IJs were offered to the bees in sugar, honey or fruit solutions. Kaya *et al.* (1982) attributed the low susceptibility of the honeybee brood stages to high temperatures in the hives (33.3–35.2°C) and low humidity (40–78%), and noted that after nematode application, there was a notable increase in adult worker mortality, but the overall colony activity was not affected. Direct spraying of nematodes on to caged workers resulted in 4–10% mortality. It was concluded that *S. carpocapsae* can safely be used against insect pests in areas where honey bee occur by following normal spray precautions, and while close-proximity spraying might kill some bees, this would not be detrimental to the colony overall. Similarly, Stock (1996) tested the pathogenicity of *Heterorhabditis argentinensis* Stock on insects associated with alfalfa crops. Stock concluded that *H. argentinensis* could be used against damaging insects of lucerne crops without causing serious damage to the beneficial insects surveyed including *A. mellifera*. More recent work by Taha and Abdelmegeed (2016) investigated the practicality of controlling the larvae of the wax moth, *Galleria mellonella* L. (Lepidopera: Pyralidae), in *A. mellifera* hives and found that even high doses of *H. bacteriophora* could be applied causing 100% mortality of *G. mellonella* with minimal impact on honey bee and no effect on colony health.

Baur *et al.* (1995) established *A. mellifera* to be EPN resistant by exposing workers and brood to four EPN species sprayed directly on to combs. The mortality of adult bees exposed to any of the nematode species was less than 10%, and there was no evidence of nematode infection when dead adults were dissected. The authors decided to assess the impact of nematodes on brood by looking at smaller-size honeycombs placed in the second story (super) of a hive and large brood combs placed in the main section of the hive. The results were inconsistent between these two experiments and did not support the hypothesis that high-temperature-tolerant species of nematodes (*S. riobrave*) are more infective to honey bee. All things considered, EPNs are not considered a threat to natural or commercial honeybee colonies and would likely not be a good control option for treating problem wild colonies.

Conversely, Dutka *et al.* (2015) documented the high susceptibility of the ground-nesting buff-tailed bumble bee, *Bombus terrestris* L., when two commercial EPN products were applied to soil in a laboratory exposure test. Although not explicitly stated, it can be deduced that adult workers were exposed to commercial EPN products, one containing *Steinernema kraussei* Steiner and the other a mix of unspecified *Heterorhabditis* and *Steinernema* spp. Both products caused high levels of bee mortality. The group confirmed that *S. kraussei* was not able to reproduce in the bumble bees, but an unidentified nematode from the mixed product did. Although the susceptibility was surprising and concerning, field studies are needed to assess the actual risk to wild *B. terrestris* and other bumble bees. Bumble bees are rarely considered a pest species, but occasionally nests establish in inconvenient locations and need to be eliminated.

22.3.2 Vespidae: yellowjackets

Vespula spp. can pose a serious hazard to humans and domesticated animals due to their attraction to food items and their ability to sting repeatedly. Human deaths are rare but similar to honeybee stings in that a proportion of society is hypersensitive to the venom and can develop life-threatening reactions. Akhurst (1980) was the first to report a naturally infected *Vespula* sp. in Tasmania. Later, Bedding (1984) discovered a naturally infected queen German yellowjacket, *Vespula germanica* F. Bedding found the wasp hibernating beneath *Eucalyptus* bark and identified *S. carpocapsae* from the insect.

Laumond *et al.* (1979), Wassink and Poinar (1984) and Wojcik and Georgis (1987) all included Vespidae species in EPN susceptibility tests, finding some susceptible species. Poinar and Ennik (1972) reported that workers of the western yellowjacket, *Vespula pensylvanica* Saussure, the insectivorous prairie yellowjacket, *Vespula rufa atropilosa* (now *Vespula atropilosa* Miller), and a *Vespula* species from the Netherlands were all susceptible to infection by *S. carpocapsae*. Infection occurred after the adult yellowjackets ingested IJs placed in fruit concentrate or on sugar cubes. Similarly, Wojcik and Georgis (1988) observed fatalities after they mixed IJs of *S. feltiae* in a sucrose solution and allowed adult *V. pensylvanica* to drink the solution.

Gambino (1984) tested the susceptibility of *V. pensylvanica* to three species of EPNs in the laboratory, noting 100% mortality in those exposed to *S. carpocapsae* and an unspecified *Steinernema* sp., and 98% mortality in those exposed to *H. bacteriophora*. All three nematodes tested were able to reproduce inside the wasp hosts, and preliminary tests indicated that the same three nematodes could kill and reproduce in the common yellowjacket, *Vespula vulgaris* L.

Guzman (1984) evaluated *S. feltiae* against *V. germanica* larvae in the laboratory noting 100% mortality. When adult workers captured in the field were kept in wire cages and fed IJs in a sucrose bait, mortality ranged from 76% to 100%. When workers were dissected, Guzman found *S. feltiae* reproducing.

In the 1990s, research focused less on nematode host ranges under laboratory conditions and to a greater extent on natural host ranges and considered the impact of augmentation of nematode populations on natural insect populations (Peters, 1996). Field tests using *S. feltiae* drenches on yellowjacket colonies were undertaken by Gambino *et al.* (1992). They found that in all treated colonies, worker activity was reduced by at least 50% after 1 week. Some treated colonies partially recovered, while some were destroyed. Interestingly *Vespula* workers were observed removing nematode-infected wasps from the colony. The results suggest that, under certain conditions, nematodes could be used as biological control agents of yellowjackets; however, prompt removal of infected individuals by workers would eliminate recycling of the nematodes and reduce ongoing infection. Therefore, efficacy was obviously limited due to the social cooperation of the workers.

Despite the limited control effect, both Drlik (1994) and Lind (1998) are proponents of pouring *S. feltiae* into the holes of ground-nesting yellowjackets as the least toxic method of management. Rose *et al.* (1999) agreed and concluded from their pathogen screening tests that EPNs may be useful as inundative control agents but do not have potential to establish and provide long-term or permanent suppression. As yellowjackets in many circumstances may be considered beneficial (predators of plant-feeding pests), the impermanent effect of EPNs can be considered an advantage from an integrated pest management point of view.

The early work described offers two application strategies to manage yellowjacket pests using EPNs. One is the application of nematodes directly to the nest and the second is to attract workers to nematode-treated baits. Wojcik and Georgis (1987) used desiccated nematodes successfully in their studies with *V. pennsylvanica*. Although the use of baits to deliver IJs may reduce worker numbers, it might not be sufficient to affect the colony as a whole. The use of baits to reduce (but not eliminate) populations is a sound approach, and repeated drench applications to nests is a useful management option for areas not requiring immediate or total elimination. A third deterrent strategy was investigated by Gulcu *et al.* (2012) who studied the in-field response of scavenging insect species to *Galleria mellonella* killed with *H. bacteriophora*, and meat substrates treated with the *Heterorhabditis* bacterial symbiont *Photorhabdus luminescens* (Thomas and Poinar). The researchers observed that *Vespa orientalis* L. and *Paravespula* spp. consumed or collected 1-day-old nematode-killed *G. mellonella* cadavers but not 2-day-old nematode-killed cadavers. Additionally, the wasps were deterred from feeding on meat or liver substrates treated with supernatant from *Photorhabdus luminescens* culture.

22.3.3 Formicidae: ants

In the USA, ants are often quoted as one of the most frequently occurring home pests reported by clients to pest-management professionals. Globally, the ant species that are considered pests vary greatly between different regions. Many can be considered simply as nuisance infestations, but more serious impacts do occur. For example, in the USA, Pharaoh ants, *Monomorium pharaonis* L., are considered one of the top forensic pests as they can phoretically transmit many food-borne pathogens. Carpenter ants, *Camponotus* spp., can be damaging pests of buildings, and fire ants (*Solenopsis* spp.) especially the red imported fire ant, *Solenopsis invicta* Buren, injure people, domesticated animals and livestock.

The individual ant itself appears as a veritable fortress against invading pathogens, as adult ant anatomy has several characteristics that limit EPN infection. Via the oral route, debris in food such as dirt and other particulate matter is filtered before it enters the prepharynx and is collected in the infrabuccal pocket. When full, the ant empties the pocket into an area within or external to the nest designated for waste. Many ant species have an infrabuccal filter but usually in the adult stage only. Spinules (spines) in the spiracular chambers limit entrance through the spiracles, and pilosity (setae) around the anus also limits entry. Many of the adaptations are primarily water-conservation structures but certainly double as defensive structures against relatively large invading parasites. Adding to the structural defenses is the fact that most ant species have an efficient colony protection system. Most fundamentally, the loss of many foraging workers has a minimal effect on the colony in general. Management using toxic baits is achieved when foraging ants collect the bait and return it to the colony for dissemination between nest mates and ultimately the egg-laying queen.

Products for ant management in and around buildings are often deployed in bait stations in which the bait is contained, protected from the environment but accessible to ant targets. Other landscape products are applied in granular formulations that are watered into the ground. Successful delivery of a slow-acting biological control agent using either method would have great utility around homes and in landscapes.

Early studies established that *S. carpocapsae* was able to infect adult tropical fire ants *Solenopsis geminata* F.; the larvae and adults of a carpenter ant *Camponotus* sp.; the larvae, pupae and adults of parasol leaf-cutter ants, *Acromyrmex octospinosus* Reich; and workers of *Myrmica* spp. (Kermarrec, 1975; Laumond *et al.*, 1979; Bedding, 1984). Poole (1976) conducted studies using *S. carpocapsae* on the black imported fire ant, *Solenopsis richteri* Buren, and *S. invicta*. Both reproductives and brood were susceptible under laboratory conditions, but workers were less susceptible. In field tests, Poole (1976) recorded greater mortality occurring after fall applications compared with spring and concluded that *S carpocapsae* could be an effective biological control agent if timing of application, mound size (smaller mounds were more easily controlled) and climatic conditions were favorable.

Quattlebaum (1980) evaluated fungal and EPN effects on fire ants by applying *S. carpocapsae* and *H. bacteriophora* using a compressed air sprayer directly into *Solenopsis invicta* and *Solenopsis richteri* mounds. Both nematode species reduced field populations of *Solenopsis invicta*, and *S. carpocapsae* generated the greatest control. Efficacy was influenced by soil texture, and synergistic effects were found when *S. carpocapsae* was applied with various carbamate and organophosphate insecticides. Drenching outdoor ant nests with EPNs has produced mixed results. During developmental phases, fire ant larvae may be fed additional food relative to those destined to become workers, and food-supplemented larvae will grow larger than workers and become reproductive members of the colony. Drees *et al.* (1992) applied *Steinernema* and *Heterorhabditis* spp. EPNs to the reproductive stages of *Solenopsis invicta* under laboratory conditions. The mortality of reproductive larvae, pupae and alates ranged from 28% to 100%. *S. carpocapsae* generated the most consistent results of the species tested causing 82–94%, 64–96% and 38–99% mortality of larvae, pupae and alates, respectively. Workers were resistant to nematode infection and were observed preening nematodes from brood, alates and themselves. In field tests, the researchers developed an ant activity assessment method to reduce colony movement due to physical disturbance. Researchers lightly tapped mounds and if 50 or more ants exited the mound, the nest was considered active. In addition, satellite mound activity was evaluated by examining the immediate 9 m radius around each mound. In field tests, *S. carpocapsae* was applied in water suspension to active *Solenopsis invicta* mounds. Six weeks after treatment, activity in mounds treated with nematodes ranged from 52% to 80%. Satellite mound activity accounted for 0–24% of overall activity. Soil drench applications of *S. carpocapsae* show potential as a control option for *Solenopsis invicta*, but colony relocation after nematode treatment could limit overall efficacy unless application techniques are developed to overcome or take advantage of colony movement. Jouvenaz *et al.* (1990) and Jouvenaz and Martin (1992) also evaluated commercially available *S. carpocapsae* and *S. feltiae* to control *Solenopsis invicta*. The earlier study indicated moderate (40–58%) queen mortalities, and in separate tests they observed that colonies of ants vacated soil areas treated with high concentrations of nematodes. The later test involved the application of nematodes to *Solenopsis invicta* nesting in nursery pots containing *Pittosporum* shrubs. Despite being contained, *Solenopsis invicta* could not be eliminated from 1-gallon pots even using high doses of nematodes (10,000 nematodes per square inch of soil surface). Conflicting results were reported by Morris *et al.* (1990), who found that treatment of *Solenopsis invicta* with *S. carpocapsae* gave control comparable to that obtained with amidinohydrazone over a 6-week period. Castellanos *et al.* (1997) tested the susceptibility of the tropical fire ant, *Solenopsis geminata*, to *H. bacteriophora* and found that the IJs were only able to infect adult males.

Studies by Zhou *et al.* (2002) found that *Xenorhabdus nematophila* can deter ants under some circumstances. They studied the response of ants to deterrent factors produced by

X. nematophila and *Photorhabdus luminescens* and showed that an active ingredient present in the supernatants of bacterial cultures is filterable, heat stable, acid sensitive, and passes through a 10 kDa pore-size membrane. Thus, the ant-deterring factor appears to be comprised of a small, extracellular and possibly non-proteinaceous compound(s). They also established that the amount of repellency depended on the ant species, the sucrose concentration used in *in vitro* assays, and the strain, form and age of the bacteria. They concluded that the symbiotic bacteria of some species of EPNs produce a compound or number of compounds that deters ants and could potentially protect nematodes from being eaten during reproduction within insect cadavers. Dudney (1997), later updated by Gouge *et al.* (2001), filed patents for use based on the insecticidal effect of *X. nematophila* on *Solenopsis invicta*. Given that in subsequent research, the deterrence behavior was found to extend to other insects including some flies, crickets and wasps (see preceding section), the deterrent factors were renamed from ant-deterrent factors to scavenger-deterrent factors (Gulcu *et al.*, 2012; see Chapter 8, this volume). However, further research is needed to isolate, synthesize and evaluate the potential of ant-deterrent factors for the management of ants.

As the ecology and biology of different ant species is so diverse and inherently sensitive to environmental conditions, it is difficult to anticipate how EPNs would affect different populations of even the same species. Repeated drench applications of large doses would perhaps be more effective on the species that are less mobile as a colony. There may be opportunities to exclude certain ant species using applications of deterrent factors from *Photorhabdus* and possibly *Xenorhabdus* spp. symbionts should they be reliably refined, and there remains much to be investigated using bait stations for nematode delivery.

22.4 Termites: Blattodea (Rhinotermitidae) and Others

Termites are an ecologically significant insect group comprising at least 3000 described species (Krishna *et al.*, 2013), and provide vital ecosystem services due to their ability to digest lignocellulose, thus recycling plant material into soil nutrients (Ebeling, 1978). Accordingly, they are abundant in nature worldwide. Pest species are concentrated in a few clades of termites, with approximately 97 species considered to be major pests, and, of these, 39 species are found in the subterranean termite genera *Reticulitermes*, *Heterotermes* and *Coptotermes* (Krishna *et al.*, 2013). Native species in all genera are found around the world (Bourguignon *et al.*, 2016), and due to the movement of goods as well as many natural phenomena, 28 species of invasive termites also belong to these genera (Evans, 2011; Evans *et al.*, 2013).

Subterranean termites (Rhinotermitidae) are of economic importance as pests in agriculture, forestry and the built environment. The global economic impact of termites is estimated to be in the region of US$40 billion annually, with subterranean termites responsible for an estimated 80% of costs (Oi, 2022). Similar to the other social insect groups, termite control is challenging due to the inherent complexity of termite biology and ecology, and historically, many approaches to prophylactic and mitigation measures against termites have failed (Oi, 2022).

With regard to the practicality of using EPNs for control purposes, it is important to consider that termites either create nests in or generate for themselves moist, dark and protected colonies. These conditions are ideal for the survival of EPNs (Lewis *et al.*, 1992; Kaya and Gaugler, 1993). However, like ants, many termite species have refined sanitary behaviors, including allogrooming, removing, entombing or feeding on cadavers, and produce antibiotics that limit bacterial disease outbreaks. Generally, only when colony vigor is weakened by age or by chemical or environmental stress do epizootics naturally develop and colonies succumb to disease. Some termiticides act synergistically with soil microorganisms to cause a more rapid decline in termite populations (Lenz, 2004). Chouvenc *et al.* (2011) examined 50 years of research on the microbial control of termites to understand why commercial products have failed to be widely adopted while research findings maintain that there is great potential. Among other things, the researchers found a disconnect between laboratory bioassays testing pathogens and real-world assessment of efficacy.

22.4.1 *Coptotermes* spp.

The genus *Coptotermes* is thought to have originated in South-east Asia but has been distributed worldwide. There are a number of destructive species in the genus, but the most notorious is the Formosan termite, *Coptotermes formosanus* Shiraki. Early investigations beginning in the 1960s were undertaken by University of Hawaii researchers who established the EPN susceptibility of *C. formosanus* in laboratory-based assays, and additionally documented limited field test success (Reese, 1971; Fujii, 1975; Tamashiro, 1976). Reese (1971) infected *C. formosanus* and returned them to their original colonies. Individuals were seen collecting and isolating the parasitized termites, walling them behind earthen barriers. The overall lack of success was attributed to nest cleaning and the large number of termites in a single colony. Fujii (1975) also described *S. carpocapsae*-infected *C. formosanus* being walled off with this strategy limiting colony infection. Nevertheless, around the same time, Poinar (1979) established the susceptibility of members of the genera *Coptotermes*, *Nasutitermes* and *Termes* to *S. carpocapsae* under laboratory conditions.

The susceptibility of constrained *C. formosanus* and the Japanese termite, *Reticulitermes speratus* Kolbe, to *S. feltiae* was investigated by Wu *et al.* (1991). They found nematodes invading termites via the mouth, and all termite castes were susceptible to nematodes except for the eggs. The susceptibility of the termites to *S. feltiae* was evaluated under an imitative artificial nest environment, and the mean nematode-associated host mortality 1–3 weeks after exposure reached 98% for the highest dose applied. Again, the researchers observed that termites were able to detect the nematodes and presumably sought ways to avoid them by extending their earthen tunnels to the outside of the nest; however, a considerable number of the termites had been parasitized before escaping.

Wang *et al.* (2002) ran laboratory tests involving multiple EPN species against the eastern subterranean termite, *Reticulitermes flavipes* (Kollar), and *C. formosanus*, documenting infection of *C. formosanus* by all EPNs tested. Interestingly, *S. riobrave* failed to infect *R. flavipes*, even at the highest application rate. Most notably, *H. indica* repelled termites at high concentrations in a sand and vermiculite medium, with the longevity of repellency influenced by nematode concentration.

Mankowski *et al.* (2005) looked at EPN attachment and infectivity of *S. carpocapsae* and *H. indica* on soldiers and workers of *Coptotermes vastator* Light (now synonymous with *Coptotermes gestroi* (Wasmann)) and *C. formosanus*. The researchers found differences in nematode attachment to termite casts and cast susceptibility to infection. *S. carpocapsae* caused the highest mortality of soldiers of either termite species compared with *H. indica*, but the mortality of workers was higher after exposure to *H. indica*. The researchers concluded that in a mixed group of soldiers and workers, or workers alone, the grooming behavior of the workers serves as a mechanism to reduce nematode infection. Soldiers are adapted for colony defense, with large heads and mandibles to ward off predators, but their defensive adaptations prevent them from grooming. However, the proportion of a colony that is made up of soldiers varies, and 2–18% was reported for *C. formosanus* by Wells and Henderson (1993).

To obfuscate any general conclusions entirely, Yu *et al.* (2010) documented that even different strains of the same EPN can cause very different levels of mortality. In prior studies, *S. riobrave* 355 strain exhibited a high level of efficacy applied to the desert subterranean termite, *Heterotermes aureus* (Snyder), but a poor or moderate level of efficacy when applied to *R. flavipes* and *C. formosanus*. However, in laboratory assays, the group compared the performance of three novel strains of *S. riobrave* (3-8b, 7-12 and TP) with the 355 strain on worker termites of all three species. While *Heterotermes aureus* was found to be very susceptible to all the *S. riobrave* strains, the TP strain of *S. riobrave* caused far greater mortality in *R. flavipes* and *C. formosanus* compared with other strains. The variance of strain virulence has been documented repeatedly for many target insects.

New nematode species are periodically tested such as *Steinernema karii* Waturu, Hunt & Reid, investigated by Wagutu *et al.* (2017), but again researchers have concluded that while laboratory assays show some susceptibility, field performance needs to be improved.

22.4.2 *Reticulitermes* and *Heterotermes* spp.

Reticulitermes spp. are found in most temperate regions including much of Asia and the Middle East, Western Europe and North America, although they are uncommon in much of Canada and Alaska, and are most common in the warm and humid regions. *Heterotermes* spp. are found in many tropical and subtropical regions, although *Heterotermes aureus* is distributed through the desert south-western USA and the lower deserts of north-western Mexico.

Mix and Beal (1985) covered areas of a heavily *Reticulitermes*-infested pine forest floor with pine boards attractive to the termites. Relatively low levels of nematodes were applied, and it was noted that certain nematode species and strains were more effective against termites than others; overall, however, the researchers concluded that EPNs did not look very promising. Differential susceptibility to nematodes species was also noted by Poinar and Georgis (1989) who detailed the comparison of *S. carpocapsae* and *H. bacteriophora* efficacy against the western subterranean termite, *Reticulitermes hesperus* Banks, in laboratory susceptibility experiments. They concluded that while both species caused the same level of mortality, *S. carpocapsae* killed the termites twice as quickly.

A study testing the efficacy of *S. carpocapsae* against foraging workers of the arid-land subterranean termite, *Reticulitermes tibialis* Banks, in pastureland (Epsky and Capinera, 1988) established 50% lethal dose values for specific nematode and termite combinations, showing that large numbers of *S. carpocapsae* were required per termite in filter-paper assays. When IJs were applied directly beneath baited traps, a significant decrease in the number of termites per trap resulted. They found that the pattern lasted for 2–3 weeks and noted that termites entering traps with nematodes entered via a corner or from the top, suggesting that termites may be avoiding contact with the nematodes. The transient nature of the avoidance documented over time in this study suggests that either nematode applications would need to be applied frequently to maintain a termite-free area or that the entire colony must be remediated rather than relying on deterring the foraging few. Where EPNs were not applied in a continuous barrier, or in places where they had been removed by ground substrate removal, termites could be seen exploiting the gaps allowing access to the food source in traps. The exploitation of gaps in chemical termiticide barriers has since also been established (Forschler, 1994).

Mauldin and Beal (1989) determined the efficacy of nematodes in preventing or eliminating *R. flavipes* in the laboratory and infestations of an unspecified species of *Reticulitermes* in field tests. In the laboratory study, termites quickly moved from a nest container through a tube containing a mixture of sand, vermiculite and water to reach a chamber in which one of two strains of *S. carpocapsae*, *S. feltiae* or *H. bacteriophora* had been released. After 9.5 weeks, termite survival rates in treatments and in an untreated control did not differ significantly. In field studies, EPNs failed to eliminate or control termites in soil treatments under concrete slabs or in logs naturally infested with termites. Conversely, Poinar and Georgis (1989) found that the *Reticulitermes* spp. were susceptible (80–87% mortality) when applying high doses of nematodes and carefully maintaining moisture levels under laboratory conditions. There seems no doubt that many termite species can be killed under controlled conditions ideal for nematode infection, but success under field conditions is highly variable.

In laboratory bioassays, Yu *et al.* (2008) compared the developmental characteristics of *S. riobrave*, *S. carpocapsae*, *S. feltiae* and *H. bacteriophora* in *Heterotermes aureus* workers. *S. carpocapsae*, *S. riobrave* and *H. bacteriophora* successfully reproduced and IJs exited the termite cadavers. However, no progeny were produced by *S. feltiae*, and dissection of the termites showed that *S. feltiae* had developed inside the termite but formed smaller-sized female adults. IJs of *S. carpocapsae* formed two distinct size groups in which the average total body length for the smaller group was 299.5 μm, and the average total body length for the larger group was 545.6 μm. The researchers found that small *S. carpocapsae* IJs infected, reproduced and formed normal-sized IJs after subsequent development in *Galleria mellonella* larvae. In termite mortality tests, IJs from stock cultures (reared in *G. mellonella*) caused higher termite mortality compared with IJs cycled through termites, then *G. mellonella* and then applied to termites. Although they had recovered in size, their fitness had been impacted. Thus, the successful generation of offspring in *Heterotermes aureus* does not necessarily translate to a self-sustaining epizootic developing within a termite colony.

Yu (2009) described *Heterotermes aureus* repelled from EPN-treated areas for up to 10 days. The repellency threshold varied among the nematode species tested and it was hypothesized that it could be the physical movement of the nematodes that repels the termites. Yu (2009) also investigated temperature effects and noted that after a gradual heat adaptation process, *S. riobrave* and *H. bacteriophora* caused significantly higher *Heterotermes aureus* mortality at 32°C compared with original laboratory strains cultured at lower temperatures, indicating that temperature conditioning could be used to boost efficacy in laboratory assays. Yu (2009) established preliminary field study results and confirmed short-term EPN protection of a structure from *Heterotermes aureus*, but termites began to re-infest after 4 weeks.

As in the case of many other target pests, some investigators have combined EPNs with chemical insecticides in laboratory and field studies. Manzoor (2012) compared applications of *S. carpocapsae* and *H. bacteriophora* alone and in combination with imidacloprid against workers and nymphs of *R. flavipes*. The results showed that neither nematode applied alone caused significant mortality, but a synergistic effect was found when using imidacloprid and nematodes together. Even so, only a little over 50% mortality was achieved in the three field colonies observed.

22.4.3 Other termites

Bedding and Stanfield (1981) reported that large numbers of the primitive Australian termite *Mastotermes darwiniensis* Froggatt (Mastotermitidae) could be killed using *Heterorhabditis* spp. nematodes but also noted that termites tended to be repelled from treated areas. Nematode-infected termites were recovered from neighboring untreated trees within a week of treatment. The cadavers yielded approximately 10,000 IJs, and the dead termites were not walled off sufficiently to prevent exit and spread of nematode progeny. After injections of a *Heterorhabditis* sp. isolate from tropical Australia into eucalypt trunks in which *M. darwiniensis* foragers were active, masses of dead termites were found. However, due to the complex biology of *M. darwiniensis*, including its diffuse nest system, the presence of multiple sets of reproductives, large territory size and simultaneous use of many feeding sites by termites, it remained uncertain what the impact of the treatment on the colonies as a whole was (R.A. Bedding, Canberra, Australia, 2004, personal communication).

The termite family Kalotermitidae are relatively primitive morphologically, with simple colony behavior and organization. Termites in this family have no need to contact soil and can live within excavated wood. This family includes species classed as drywood termites. A number of species have been established as susceptible to EPNs including *Neotermes* spp. in the South Pacific. EPNs were injected into the main trunks of plants to eliminate infestations in coconut palms, *Citrus*, cocoa and American mahogany trees (*Swietenia macrophylla*). *Neotermes* spp. were successfully eliminated from unbranched coconut palms, but their efficacy was variable in trees with branches (Lenz and Runko, 1992; Lenz *et al.*, 2000). The researchers noted that the branch structures allowed a proportion of the termite population to retreat and block off connections to the main trunk, thus preventing the spread of the nematodes to all areas occupied by a colony.

The termite family Archotermopsidae includes dampwood termite species, one of which was investigated by Georgis *et al.* (1982). Only Petri dish assays were undertaken, but the authors reported high mortality of *Zootermopsis* and *Reticulitermes* spp. exposed to *S. carpocapsae* and *H. bacteriophora*. Subsequent laboratory-based work by Samarasinghe (1996) again demonstrated the susceptibility of the dampwood termite *Zootermopsis* spp. and *Reticulitermes santonensis* (now synonymous with the eastern subterranean termite *Reticulitermes flavipes*) to EPNs in the genera *Heterorhabditis* and *Steinernema* as well as the fungal pathogen *Metarhizium anisopliae* (now *Metarhizium robertsii* (Metchnikoff) Sorokin (1883)).

Later work undertaken by Wilson-Rich *et al.* (2007) on the dampwood termite *Zootermopsis angusticollis* (Hagen) included accounts of the termites altering their behavior in response to *S. carpocapsae* IJs. Relative to their baseline behavior, termites increased the frequency and duration of allogrooming and vibratory displays, as well as abdominal tip-raising and self-scratching. While allogrooming and vibratory displays are documented as responses to other pathogens and parasites in other termites, certain behaviors had not been documented for *Z. angusticollis*. They conclude that behavioral responses in termites are important in the mitigation of pathogens and parasites. This work may also have documented newly established behaviors.

The termite family Termitidae is the largest family of termites and are commonly referred to as the higher termites. They are the most specialized group of the infraorder Isoptera. Fungus-growing termites in the subfamily Macrotermitinae destroy crops in the tropics of Africa, the Middle East, and southern and south-eastern Asia. Rouland *et al.* (1996) studied the sensitivity of *Pseudacanthotermes spiniger* Sjostedt and *Ancistrotermes guineensis* Silvestri reproductives and found the alates of both species to be highly susceptible to *H. bacteriophora*, *S. carpocapsae* and *Steinernema kushidai* Mamiya. Susceptibility to *Steinernema* spp. was higher. Benmoussa-Haichour *et al.* (1998) further measured the sensitivity of the different castes of *P. spiniger* infected by *Heterorhabditis* and *Steinernema* spp. The researchers found that the nematodes were incapable of completing their life cycle in workers and soldiers but could complete their life cycle in alates. They theorized that this could be due to the biochemical composition of different castes. Because of this, they deemed it unlikely that nematodes could produce an epizootic in large colonies.

Yu *et al.* (2006) studied the long-jawed desert termite, *Gnathamitermes perplexus* Banks, the dark southern subterranean termite, *Reticulitermes virginicus* (Banks), *Heterotermes aureus* and *R. flavipes*, conducting sand bioassays with a range of EPN species. *S. riobrave* and *S. feltiae* caused relatively low levels of *R. virginicus* mortality, but significant mortality (80%) of *Heterotermes aureus* and *G. perplexus* workers was caused by *S. riobrave*.

It is often not clear whether applications of EPNs causing termite population reductions of 50% or lower have utility. As early as 1990, it was concluded that although it may be possible to protect some trees and crops from termites using nematodes, the protection of homes and buildings is not very practical (Klein, 1990). Klein (1990) listed that the low economic threshold for termite damage to homes, the large number of nematodes required to kill termites, the presence of secondary reproductives and the behavior of socially oriented termites (utilizing barrier gaps and removing infected individuals) precluded nematodes as a viable termite management option.

22.5 Pests Invading and Infesting Homes

The cat flea, *Ctenocephalides felis* (Bouché) (Siphonaptera: Pulicidae), has a global distribution and is an extremely common ectoparasite of domestic cats, dogs and rodents. Flea control options include insecticidal products applied to the environment, oral medication, collars and spot-on treatments for animals, and insecticidal shampoos. While pesticide resistance to older chemistries including carbamate and organophosphate classes is widespread (Coles and Dryden, 2014), there are many flea control products with active ingredients in the neonicotinoid, phenylpyrazole, semicarbazone, spinosyn and isoxazoline classes that are effective. However, many pet owners have concerns regarding the safety of pesticides applied to pets from a pet health perspective as well as human health exposure concerns, especially when children and other chemically vulnerable people are in close contact with pets.

Additionally, *C. felis* is an ectoparasite of livestock including sheep, goats, calves and horses, all of which can be severely affected. *C. felis* can even cause the death of animals (Yeruham *et al.*, 1989; Dryden *et al.*, 1992). The ectoparasite causes stress, dermatitis and loss of blood, and the flea vectors various disease-causing pathogens to animals. They are competent vectors of dog tapeworm, *Dipylidium caninum* L., and flea-borne typhus, *Rickettsia typhi* (Wolbach and Todd, 1920), to humans (Bitam *et al.*, 2010).

The efficacy of EPNs on off-host larval and silk-cocooned pupae is well established. Multiple groups including Henderson *et al.* (1995) recorded success using *S. carpocapsae* applied to different substrates including soil, sand and gravel in controlled environment and outdoor tests, recording reductions in adult flea emergence of 70–100%. This is not surprising, as flea larvae prefer to develop in areas protected from rainfall, irrigation and direct sunlight, with 75% relative humidity or higher and temperatures of 21–32°C. In an established infestation, adult fleas comprise less than 10% of the population and remain on host animals for the remainder of their lives. Eggs laid on animals fall off anywhere the animal wanders, inside and outside. Because of this, simply treating animals directly often fails to control fleas sustainably. Animal treatments must be undertaken along with indoor vacuuming, the laundering of bedding, and outside garden and patio area treatment.

Recently, Samish *et al.* (2020) evaluated the various life stages of *C. felis* exposed to EPNs and fungal pathogens at different ambient temperatures and levels of relative humidity. Interestingly, they conducted tests on filter paper, sand and carpet. The researchers found that on all substrates, *S. feltiae* was the most effective against larvae, pupae in cocoons and adults, while *H. bacteriophora* was highly effective against cocooned pupae. Overall, EPNs proved most effective at 28°C and 95% relative humidity. The fungal pathogen *Metarhizium robertsii* also proved highly virulent against adult fleas. Flea eggs were resistant to both nematodes and fungal infection. The researchers suggested that a combined approach of *M. robertsii* animal treatments along with EPN treatment of environmental media could provide an entire biological control approach to flea management. The practicalities of treating interior operational carpets with EPNs has yet to be tested, but treatment of exterior areas is very logical.

Hard ticks (Ixodida: Ixodidae) transmit pathogens responsible for the majority of vector-borne diseases of humans in temperate North America, Europe and Asia. Hard ticks as a group are cosmopolitan in distribution, and are found in natural habitats and suburban and urban environments. Pet animals and humans can acquire them in all of these locations, although some regions have far higher populations than others.

Significant work has been undertaken evaluating EPN utility against ticks that impact cattle, and a brief review and the latest work has been published by Monteiro *et al.* (2020). There is, however, relatively little covering tick species found in and/or around homes, despite the human disease burden.

Only referencing US diseases of greatest concern here, the blacklegged tick (*Ixodes scapularis* (Say)) is important as a vector of *Borrelia burgdorferi* and *Borrelia mayonii* (Lyme disease), *Anaplasma phagocytophilum* (anaplasmosis), *Borrelia miyamotoi* (a form of relapsing fever), *Ehrlichia muris eauclairensis* (ehrlichiosis), *Babesia microti* (babesiosis) and Powassan virus (Powassan virus disease). *I. scapularis* has an extensive geographic distribution in highly diverse environmental conditions in eastern North America, from Mexico through the USA and into Canada. Much of the work on ticks is covered in Chapter 26 (this volume).

Zhioua *et al.* (1995) reported the high susceptibility of engorged adult *I. scapularis* females to *S. carpocapsae* and *S. glaseri*, but not males or other tick stages irrespective of sex and feeding status. Nematodes were observed infecting via genital pores, and nematode development and reproduction was confirmed by dissecting tick cadavers. Hill (1998) documented similar results, additionally finding that *S. riobrave* and *H. megidis* killed ticks the most rapidly, but all EPN species tested caused tick mortality. However, no emergence of IJs was observed from the tick cadavers. Most significantly, only four of 702 nematode-infected, engorged females produced egg masses, and only one mass hatched. The geographic range and human population impacted by *I. scapularis* continues to increase. In high-density tick regions, residential lawns, school playing fields and parks next to forested areas are treated with acaricides extensively, but little information is available on the deployment of EPNs in these situations.

Kocan *et al.* (1998) reported that *S. riobrave* could infect and kill replete females of the American dog tick, *Dermacentor variabilis* (Say), the brown dog tick *Rhipicephalus sanguineus* (Latreille) and others less commonly found around homes. *S. feltiae* was able to infect and kill replete females of *D. variabilis*, and *Rhipicephalus sanguineus*. Ticks that survived produced lower mean egg mass weight. EPNs again failed to develop inside tick cadavers. The tick species studied in this study are all critically important pathogen vectors. *D. variabilis* is found in the continental USA, Mexico and southern Canada, and is widely distributed east of the Rocky Mountains and in limited areas on the Pacific coast. Again only referencing US diseases of greatest concern,

D. variabilis is a vector of *Francisella tularensis* (tularemia) and *Rickettsia rickettsii* (Rocky Mountain spotted fever). *Rhipicephalus sanguineus* is the most widespread tick in the world but is found more commonly in the warmer climates of south-western Europe, parts of northern Africa and the Americas. This tick also vectors *Rickettsia rickettsii*. Environmental applications of bifenthrin pesticides are commonly used in regions with elevated risk of Rocky Mountain spotted fever, but efficacy is relatively poor, even when environmental pesticides are used in tandem with animal treatments. *Rhipicephalus sanguineus* can persist in abundance around and sometimes inside homes.

Laboratory studies undertaken by Samish *et al.* (1999) found no difference in *H. bacteriophora*-induced mortality of adult *Rhipicephalus sanguineus* irrespective of sex or feeding status (all >80%) indicating differences in IJ infection route. More recently, Samish *et al.* (2000) undertook an overview of EPN work involving ticks and concluded that, in most cases, *Heterorhabditis* spp. were the most virulent nematodes tested and they could infect unfed adult ticks. They also stated that, as most tick species spend most of their life cycle in the upper humid layer of the ground, essentially sharing the same ecological niche as EPNs, ticks as targets hold good potential.

Several cockroach pest species transmit disease-causing pathogens to food and food contact surfaces. The German cockroach, *Blattella germanica* (L.) (Blattodea: Ectobiidae), lives in indoor environments and has a worldwide distribution. It is an allergen-producing cockroach impacting allergy-sensitive asthmatic people. The brown-banded cockroach, *Supella longipalpa* (F.) (Blattodea: Ectobiidae), is also an indoor species found mainly in the continental USA but has become established in areas in Europe. The following cockroach species live primarily outdoors but invade or infest structures under some conditions. The American cockroach, *Periplaneta americana* (L.) (Blattodea: Blattidae), has a virtually cosmopolitan distribution. The smoky brown cockroach, *Periplaneta fuliginosa* (Serville) (Blattodea: Blattidae), is found widely in the warmer regions of the world. The oriental cockroach, *Blatta orientalis* L. (Blattodea: Blattidae), is common worldwide.

Early work demonstrated *Blattella germanica* infection by *S. feltiae* and *S. carpocapsae* in laboratory tests (Skierska *et al.*, 1976; Zukowski, 1984; Locatelli and Parleaz, 1987). All key pest species were investigated by Koehler *et al.* (1992), who exposed cockroaches to *S. carpocapsae* IJs on filter paper in Petri dishes and using feeding stations. All cockroach species died rapidly (<24 h) in Petri dish assays with the exception of *P. americana* (up to 5 days). All species were observed actively grooming nematodes from their legs and antennae. Mortality due to nematodes delivered in bait stations took far longer or was completely lacking in the case of *P. americana*.

Additional work exposing *Blattella germanica* to *S. carpocapsae* was undertaken by Appel *et al.* (1993), and observations again noted increased antennal preening but also the general avoidance of areas with IJs present. Following this, Appel and Benson (1994) documented that, although female cockroaches could be infected and killed by *S. carpocapsae*, fully formed oothecae were not penetrated by the nematodes. However, if exposed females were carrying partially formed oothecae, the egg cases were 100% infested with nematodes. Work presented by Manweiler *et al.* (1993) showed preliminary success delivering *S. carpocapsae* in a control station like those sold containing insecticidal baits today.

Baker *et al.* (2012) compared *S. carpocapsae* and *H. bacteriophora* on *Blattella germanica* establishing that *S. carpocapsae* was the more pathogenic EPN, while Půža and Mráček (2010) tested two sympatric EPNs, *Steinernema affine* (Bovien) Wouts, Mráček, Gerdin & Bedding and *S. kraussei*, on *Blattella germanica* confirming that both EPNs could reproduce and generate IJs. Morton and García-del-Pino (2013) looked at *S. carpocapsae* infection of *P. americana*, noting that males were more susceptible than females, and more nematodes were found inside the genitalia of males than that of females, concluding that the genital aperture of males was an easier entry point.

Shaban *et al.* (2010) investigated the cellular and humoral immune responses of adult *P. americana* by injecting *Steinernema* spp. IJs into cockroaches. The group results suggested that the *Xenorhabdus* spp. bacteria possess efficient immune-suppressive mechanisms that resulted in septicemia and rapid death of cockroaches. Thus, IJ infection of *P. americana* is likely to be the limiting pathogenic factor. More information can be found in Chapter 3 (this volume).

A relatively new invasive in the continental USA is the Turkestan cockroach, *Blatta lateralis* Walker (Blattodea: Blattidae). This is primarily an outdoor species that occasionally enters buildings and sometimes infests under certain conditions. Özdemir and Evlice (2020) applied two Turkish strains of *S. feltiae* and *H. bacteriophora* to nymphs in Petri dish assays. Both EPNs killed the cockroaches in under a week. Unfortunately, little else is known about the EPN susceptibility of this cockroach.

Considering how often populations of *P. americana*, *P. fuliginosa*, *Blatta orientalis* and *Blatta lateralis* build up in subterranean, damp and protected environments such as irrigation boxes and under home crawl spaces, it would appear that field studies in these environments are lacking and warrant investigation.

Fungus gnats, *Bradysia* and *Lycoriella* spp. (Diptera: Sciaridae), are small fly pests associated with indoor potted plants. Control of sciarids is covered in Chapters 23 and 25 (this volume) in detail. However, integrating the application of EPNs during routine watering as part of in-home plant maintenance brings relief from indoor fly populations, which can build up to intolerable levels. EPNs also have utility in atrium and interior plantscape control, although they are currently not routinely used for this purpose.

There are numerous opportunities to expand the use of EPNs to control home-invading and -infesting pests, extrapolating from experiences with agricultural and horticultural pests and also expanding research within the groups touched on here. For example, the dog flea, *Ctenocephalides canis* Curtis, also occurs globally but at lower rates than the cat flea on domestic canines, and the oriental rat flea, *Xenopsylla cheopis* Rothschild, a vector of *Yersinia pestis*, the bacterium that causes plague outbreaks in prairie dogs, may be a legitimate mitigation tool when plague is detected in rural areas.

While *I. scapularis* is the primary vector of the spirochete *Borrelia burgdorferi* to humans in the eastern USA, in the western USA, the western blacklegged tick, *Ixodes pacificus* Cooley & Kohls, is the primary vector to humans. A closely related species to *D. variabilis* is *Dermacentor similis* n. sp., a newly described species found west of the Rocky Mountains (Lado *et al.*, 2021). The lone star tick, *Amblyomma americanum* (L.), is distributed widely across the eastern USA and Mexico and is another tick vector posing a risk to human health. The preliminary work involving *Rhipicephalus sanguineus* might be expanded on now that lineage differences between northern and southern types are better understood. What is clear is that as bioclimatic variables have changed, human populations have expanded and the regional movement of animals has increased, so has the range of important tick species increased (Clow *et al.*, 2017). Research teams are applying increasingly sophisticated models to predict future vector and disease ranges with variable results. Generally, models show a tendency for northern expansions, and for some species southern retractions (Alkishe *et al.*, 2021). Tick vectors are increasing public health burdens worldwide and this unfortunate trend is likely to continue in the years ahead. There appears to be a lot more work that could be done to understand the full potential and limitations of EPNs in this field.

22.6 Summary and Conclusions

Most research studies on EPNs have naturally focused on agricultural and horticultural crop pest control needs, but as a general rule, garden and landscape pest suppression will not necessitate the same level of efficacy or cost return incentives to be considered successful. It is likely that EPNs, as environmentally friendly pest control alternatives to chemical insecticides, have great utility as biological control agents in home gardens and built environment landscaping. Pest pressure is often lower in landscape and home gardens, which are heterogeneous and dense mono-plantings are not the norm. The exception to this is turf to some extent. Nevertheless, there is growing concern regarding pesticide use and ever-increasing interest in developing biological control products to replace chemical products in sustainable systems (van Lenteren *et al.*, 2018). Reducing pesticide use has become a goal shared by several European countries due to the negative impacts of pesticides on the environment and on human health (Jacquet *et al.*, 2022).

Chouvenc *et al.* (2011) encouraged scientists engaged in the development of biological control options to use a cross-disciplinary team to develop a deeper understanding of the complex biology of the target pest and their defense mechanisms. While funding mechanisms for unbiased research can sometimes limit cross-disciplinary collaborations, only in this way are we likely to find mechanisms to optimize research supportive of successful biological control adoption and implementation.

Finally, Gould *et al.* (2018) warned that pest resistance to insecticides and herbicides is a sociobiological dilemma and that, while breakthroughs in chemistry and molecular biology may continue to provide new pesticides and novel methods for pest control, there is also an increasing possibility that the evolution of pesticide resistance will outpace innovation. Enormous reforms in approach to agricultural pest management may someday become a necessity, but in the meantime, residents responsible for managing homes and landscapes, together with managers and administrators of public cityscapes, parks and neighborhoods, can work toward policy and practices keeping the preservation of ecosystems and the protection of biodiversity as cornerstones of sustainability. This author concludes that EPN products for garden and landscape pest control are underutilized, and the unrealized possibilities are vast.

22.7 References

Abate, B.A., Wingfield, M.J., Slippers, B. and Hurley, B.P. (2017) Commercialisation of entomopathogenic nematodes: should import regulations be revised? *Biocontrol Science and Technology* 27, 149–168.

Acharya, R., Hwang, H.-S., Mostafiz, M.M., Yu, Y.-S. and Lee, K.-Y. (2020) Susceptibility of various developmental stages of the fall armyworm, *Spodoptera frugiperda*, to entomopathogenic nematodes. *Insects* 11: 868.

Akhurst, R.J. (1980) Morphological and functional dimorphism in *Xenorhabdus* spp., bacteria symbiotically associated with insect pathogenic nematodes *Neoaplectana* and *Heterorhabditis*. *Journal of General Microbiology* 121, 303–309.

Alkishe, A., Raghavan, R.K. and Peterson, A.T. (2021) Likely geographic distributional shifts among medically important tick species and tick-associated diseases under climate change in North America: a review. *Insects* 12: 225.

Alves, V.S., Moino, A. Jr., Santa-Cecilia, L.V.C., Rohde, C. and Silva, M.A.T. (2009) Tests for the control of coffee root mealybug *Dysmicoccus texensis* (Tinsley) (Hemiptera, Pseudococcidae) with *Heterorhabditis* (Rhabditida, Heterorhabditidae). *Revista Brasileira de Entomologia* 53, 139–143.

Andaló, V., Moino, A. Jr., Santa-Cecilia, L.V.C. and Souza, G.C. (2004) Seleção de isolados de fungos e nematóides entomopatogênicos para a cochonilha-da-raiz-do-cafeeiro *Dysmicoccus texensis* (Tinsley). *Arquivos do Instituto Biologico* 71, 181–187.

Appel, A.G. and Benson, E.P. (1994) Pathogenicity and limited transoothecal transmission of *Steinernema carpocapsae* (Nematoda: Steinernematidae) in adult female German cockroaches (Dictyoptera: Blattellidae). *Journal of Medical Entomology* 31, 127–131.

Appel, A.G., Benson, E.P., Ellenberger, J.M. and Manweiler, S.A. (1993) Laboratory and field evaluations of an entomogenous nematode (Nematoda: Steinernematidae) for German cockroach (Dictyoptera: Blattellidae) control. *Journal of Economic Entomology* 86, 777–784.

Baker, N.R., Ali, H.B. and Gowen, S. (2012) Reproduction of entomopathogenic nematodes *Steinernema carpocapsae* and *Heterorhabditis bacteriophora* on the German cockroach *Blattella germanica* at different temperatures. *Iraqi Journal of Science* 53, 505–512.

Baldwin, R.W., Koehler, P.G., Pereira, R.M. and Oi, F.M. (2008) Public perceptions of pest problems. *American Entomologist* 54, 73–79.

Baur, M.E., Kaya, H.K., Peng, Y.S. and Jiang, J. (1995) Nonsusceptibility of the honey bee, *Apis mellifera* (Hymenoptera: Apidae) to steinernematid and heterorhabditid nematodes. *Journal of Nematology* 27, 378–381.

Bedding, R.A. (1984) Nematode parasites of Hymenoptera. In: Nickle, W.R. (ed.) *Plant and Insect Nematodes.* Marcel Dekker, New York, pp. 755–795.

Bedding, R.A. and Stanfield, M. (1981) Nematodes for insect control. *Mastotermes*. In: *Annual Report*. CSIRO Division of Entomology, pp. 79–80.

Benmoussa-Haichour, D., Reversat, G. and Rouland, C. (1998) Sensitivity of the different castes of higher termites to an infestation with entomopathogenic nematodes – role of the lipids composition. *Actes des Colloques Insectes Sociaux* 11, 45–52.

Bitam, I., Dittmar, K., Parola, P., Whiting, M.F. and Didier, R. (2010) Fleas and flea-borne diseases. *International Journal of Infectious Disease* 14, e667–e676.
Boselli, M., Curto, G.M. and Tacconi, R. (1997) Field efficacy of entomopathogenic nematodes against the sugar-beet weevil *Temnorhinus* (= *Conorrhynchus*) *mendicus* Gyll. (Coleoptera: Curculionidae). *Biocontrol Science and Technology* 7, 231–238.
Bourguignon, T., Lo, N., Šobotník, J., Sillam-Dussès, D., Roisin, Y. and Evans, T.A. (2016) Oceanic dispersal, vicariance and human introduction shaped the modern distribution of the termites *Reticulitermes, Heterotermes* and *Coptotermes*. *Proceedings of the Royal Society B: Biological Sciences* 283: 20160179.
Burana, K., Jamjanya, T. and Nimkingrat, P. (2016) Monitoring of the marigold flower's insect pest during flowering stage in Mahasarakham province. *Songklanakarin Journal Plant Science* 3, 1–7.
Caccia, M.G., Del Valle, E., Doucet, M.E. and Lax, P. (2014) Susceptibility of *Spodoptera frugiperda* and *Helicoverpa gelotopoeon* (Lepidoptera: Noctuidae) to the entomopathogenic nematode *Steinernema diaprepesi* (Rhabditida: Steinernematidae) under laboratory conditions. *Chilean Journal of Agricultural Research* 74, 123–126.
Canhilal, R. (2013) Virulence of new South Carolinian heterorhabditid isolates (Rhabditida: Heterorhabditidae) to the beet armyworm, *Spodoptera exigua*. *African Journal of Microbiology Research* 7, 5669–5676.
Cantwell, G.E., Lehnert, T. and Fowler, J. (1972) Are biological insecticides harmful to the honey bee? *American Bee Journal* 112, 255–258.
Castellanos, L., Rodriguez, I. and Martinez, M. (1997) Susceptibility of the ant *Solenopsis geminata* (Hymenoptera: Formicidae) to the parasitism by *Heterorhabditis bacteriophora* (Rhabditida: Heterorhabditidae). *Revista de Proteccion Vegetal* 12, 159–162.
Chouvenc, T., Su, N.Y. and Grace, J.K. (2011) Fifty years of attempted biological control of termites – analysis of a failure. *Biological Control* 59, 69–82.
Clow, K.M., Leighton, P.A., Ogden, N.H., Lindsay, L.R., Michel, P. *et al.* (2017) Northward range expansion of *Ixodes scapularis* evident over a short timescale in Ontario, Canada. *PLOS ONE* 12: e0189393.
Cole, M., Ceja-Navarro, J. and Mikaelyan, A. (2021). The power of poop: defecation behaviors and social hygiene in insects. *PLOS Pathogens* 17: e1009964.
Coles, T.B. and Dryden, M.W. (2014) Insecticide/acaricide resistance in fleas and ticks infesting dogs and cats. *Parasites & Vectors* 7: 8.
Correa-Cuadros, J.P., Sáenz-Aponte, A. and Rodríguez-Bocanegra, M.X. (2016) *In vitro* interaction of *Metarhizium anisopliae* Ma9236 and *Beauveria bassiana* Bb9205 with *Heterorhabditis bacteriophora* HNI0100 for the control of *Plutella xylostella*. *SpringerPlus* 5: 2068.
Davidson, J.A., Gill, S.A. and Raupp, M.J. (1992) Controlling clearwing moths with entomopathogenic nematodes: the dogwood borer case study. *Journal of Arboriculture* 18, 81–84.
de Luca, F., Clausi, M., Troccoli, A., Curto, G., Rappazzo, G. and Tarasco, E. (2015) Entomopathogenic nematodes in Italy: Occurrence and use in microbial control strategies. In: Campos-Herrera, R. (ed). *Sustainability in Plant and Crop Protection: Ecology and Applied Technologies for Sustainable Plant and Crop Protection*. Springer, Cham, Switzerland, 431–449.
de Waal, J.Y., Malan, A.P. and Addison, M.F. (2013) Effect of humidity and a superabsorbent polymer formulation on the efficacy of *Heterorhabditis zealandica* (Rhabditida: Heterorhabditidae) to control codling moth, *Cydia pomonella* (L.) (Lepidoptera: Tortricidae). *Biocontrol Science and Technology* 23, 62–78.
de Waal, J.Y., Addison, M.F. and Malan, A.P. (2017) Evaluation of a South African isolate of *Heterorhabditis zealandica* (Rhabditida: Heterorhabditidae) for the control of codling moth, *Cydia pomonella* (L.) (Lepidoptera: Tortricidae) in field applications. *International Journal of Pest* Management 64, 102–109.
Department of Economic and Social Affairs (2020) 68% of the world population projected to live in urban areas by 2050, says UN 2018. United Nations, New York. Available at: www.un.org/development/desa/en/news/population/2018-revision-of-world-urbanization-prospects.html (accessed 24 March 2024).
Devindrappa, J.P., Gowda, M.T. and Vijayakumar, R. (2017) Compatibility of *Steinernema carpocapsae* and *Heterorhabditis indica* with insecticides registered against *Helicoverpa armigera* (Lepidoptera: Noctuidae). *Journal of Biological Control* 31, 95–101.
Drees, B.M., Miller, R.W., Vinson, S.B. and Georgis, R. (1992) Susceptibility and behavioral response of red imported fire ant (Hymenoptera: Formicidae) to selected entomogenous nematodes (Rhabditida: Steinernematidae & Heterorhabditidae). *Journal of Economic Entomology* 85, 365–370.
Drlik, T. (1994) Solving yellowjacket problems. *IPM Practitioner* 16: 19.
Dryden, M.W., Moore, W.E. and Broke, A.B. (1992) Severe flea infestation in dairy calves. In: *National Conference on Urban Entomology, 23–26 February, College Park, Maryland*, p. 153.

Dudney, R.A. (1997) Use of *Xenorhabdus nematophilus* Im/1 and 19061/1 for fire ant control. US Patent US5616318 970401. US Patent and Trademark Office, Washington, DC.
Dutka, A., McNulty, A. and Williamson, S.M. (2015) A new threat to bees? Entomopathogenic nematodes used in biological pest control cause rapid mortality in *Bombus terrestris*. *PeerJ* 3: e1413.
Dutky, S.R. and Hough, W.S. (1955) Note on a parasitic nematode from codling moth larvae, *Carpocapsa pamonetta*. *Proceedings of the Entomological Society of Washington* 57, 244.
Ebeling, W. (1978) *Urban Entomology.* Division of Agricultural Sciences, University of California, California.
Ebssa, L. and Koppenhöfer, A.M. (2012) Entomopathogenic nematodes for the management of *Agrotis ipsilon*: effect of instar, nematode species and nematode production method. *Pest Management Science* 68, 947–957.
Epsky, N. and Capinera, J.L. (1988). Efficacy of the entomogenous nematode *Steinernema feltiae* against a subterranean termite, *Reticulitermes tibialis* (Isoptera: Rhinotermitidae). *Journal of Economic Entomology* 81, 1313–1317.
Evans, T.A. (2011) Invasive termites. In: Bignell, D.E., Roisin, Y. and Lo, N. (eds.) *Biology of Termites: A Modern Synthesis.* Springer, Dordrecht, Netherlands, pp. 519–562.
Evans, T.A., Forschler, B.T. and Grace, J.K. (2013) Biology of invasive termites: a worldwide review. *Annual Revue of Entomology* 58, 455–474.
Fallet, P., Bazagwira, D., Guenat, J.M., Bustos-Segura, C., Karangwa, P. *et al.* (2022) Laboratory and field trials reveal the potential of a gel formulation of entomopathogenic nematodes for the biological control of fall armyworm caterpillars (*Spodoptera frugiperda*). *Biological Control* 176: 105086.
Feás, X., Vidal, C. and Remesar, S. (2022) What we know about sting-related deaths? Human fatalities caused by hornet, wasp and bee stings in Europe (1994–2016). *Biology* 11: 282.
Feaster, M.A. and Steinkraus, D.C. (1996) Inundative biological control of *Helicoverpa zea* (Lepidoptera: Noctuidae) with the entomopathogenic nematode *Steinernema riobravis* (Rhabditida: Steinernematidae). *Biological Control* 7, 38–43.
Forschler, B.T. (1994) Survivorship and tunneling activity of *Reticulitermes flavipes* (Kollar) (Isoptera: Rhinotermitidae) in response to termiticide toil barriers with and without gaps of untreated soil. *Journal of Entomological Science* 29, 43–54.
Fujii, J.K. (1975) Effects of an entomogenous nematode, *Neoaplectana carpocapsae* Weiser, on the Formosan subterranean termite, *Coptotermes formosanus* Shiraki, with ecological and biological studies on *C. formosanus*. PhD thesis, University of Hawaii, Honolulu.
Furlong, M.J., Wright, D.J. and Dosdall, L.M. (2013) Diamondback moth ecology and management: problems, progress, and prospects. *Annual Review of Entomology* 58, 517–541.
Gambino, P. (1984) Susceptibility of the western yellowjacket *Vespula pensylvanica* to three species of entomogenous nematodes. *International Research Communications System Medical Science: Microbiology, Parasitology and Infectious Diseases* 12: 264.
Gambino, P., Pierluisi, G.J. and Poinar, G.O. Jr. (1992) Field test of the nematode *Steinernema feltiae* (Nematoda: Steinernematidae) against yellowjacket colonies (Hym.: Vespidae). *Entomophaga* 37, 107–114.
García del Pino, F. and Morton, A. (2008) Efficacy of *Steinernema feltiae* against the leek moth *Acrolepiopsis assectella* in laboratory and field conditions. *BioControl* 53, 643–650.
Georgis, R., Poinar, G.O. Jr. and Wilson, A.P. (1982) Susceptibility of damp-wood termites and soil and wood-dwelling termites to the entomogenous nematode *Neoaplectana carpocapsae*. *International Research Communications System Medical Science: Microbiology, Parasitology and Infectious Diseases* 10: 563.
Gill, S.A. and Raupp, M.J. (1994) Using entomopathogenic nematodes and conventional and biorational pesticides for controlling bagworm. *Journal of Arboriculture & Urban Forestry* 20, 318–318.
Göttig, S. and Herz, A. (2018) Susceptibility of the box tree pyralid *Cydalima perspectalis* Walker (Lepidoptera: Crambidae) to potential biological control agents Neem (NeemAzal®-T/S) and entomopathogenic nematodes (Nemastar®) assessed in laboratory bioassays and field trials. *Journal of Plant Diseases and Protection* 125, 365–375.
Gouge, D.H., Dudney, R.A. and Smith, K.A. (2001) Bacteria for insect control. US Patent (expired).
Gould, F., Brown, Z.S. and Kuzma, J. (2018) Wicked evolution: can we address the sociobiological dilemma of pesticide resistance? *Science* 360, 728–732.
Gozel, U. and Gozel, C. (2016) Entomopathogenic nematodes in pest management. In: Harsimran Kaur, G. and Gaurav, G. (eds.) *Integrated Pest Management (IPM): Environmentally Sound Pest Management*. IntechOpen. Available at: www.intechopen.com/books/5252 (accessed 25 March 2024).

Gulcu, B., Hazir, S. and Kaya, H. (2012) Scavenger deterrent factor (SDF) from symbiotic bacteria of entomopathogenic nematodes. *Journal of Invertebrate Pathology* 110, 326–333.

Guzman, R.F. (1984) Preliminary evaluation of the potential of *Steinernema feltiae* for controlling *Vespula germanica*. *New Zealand Journal of Zoology* 11: 100.

Hackett, K.J. and Poinar, G.O. Jr (1973) The ability of *Neoaplectana carpocapsae* Weiser (Steinernematidae: Rhabditoidea) to infect adult honey bees (*Apis mellifera*, Apidae: Hymenoptera). *American Bee Journal* 113: 100.

Henderson, G., Manweiler, S.A., Lawrence, W.J., Tempelman, R.J. and Foil, L.D. (1995) The effects of *Steinernema carpocapsae* (Weiser) application to different life stages on adult emergence of the cat flea *Ctenocephalides felis* (Bouché). *Veterinary Dermatology* 6, 159–163.

Hill, D.E. (1998) Entomopathogenic nematodes as control agents of developmental stages of the black-legged tick, *Ixodes scapularis*. *Journal of Parasitology* 84, 1124–1127.

Jacquet, F., Jeuffroy, M.H., Jouan, J., Le Cadre, E., Litrico, I. *et al.* (2022) Pesticide-free agriculture as a new paradigm for research. *Agronomy for Sustainable Development* 42: 8.

Johnson, M.W., Welter, S.C., Toscano, N.C., Ting, P., Trumble, J.T. (1983) Reduction of tomato leaflet photosynthesis rates by mining activity of *Liriomyza sativae* (Diptera: Agromyzidae). *Journal of Economic Entomology* 76, 1061–1063.

Jouvenaz, D.P. and Martin, W.R. (1992) Evaluation of the nematode *Steinernema carpocapsae* to control fire ants in nursery stock. *Florida Entomologist* 75, 148–151.

Jouvenaz, D.P., Lofgren, C.S. and Miller, R.W. (1990) Steinernematid nematode drenches for control of fire ants, *Solenopsis invicta*, in Florida. *Florida Entomologist* 73, 190–193.

Kary, N.E., Golizadeh, A., Dastjerdi, H.R., Mohammadi, D., Afghahi, S. *et al.* (2012) A laboratory study of susceptibility of *Helicoverpa armigera* (Hübner) to three species of entomopathogenic nematodes. *Munis Entomology & Zoology* 7, 372–379.

Kaya, H.K. and Gaugler, R. (1993) Entomopathogenic nematodes. *Annual Review of Entomology* 38, 181–206.

Kaya, H.K., Marston, J.M., Lindegren, J.E. and Peng, Y.S. (1982) Low susceptibility of the honey bee, *Apis mellifera* L. (Hymenoptera: Apidae), to the entomogenous nematode, *Neoaplectana carpocapsae* Weiser. *Environmental Entomology* 11, 920–924.

Kermarrec, A. (1975) Etude des relations synécologiques entres les nématodes et la fourmi manioc: *Acromyrmex octospinosus* Reich. *Annales de Zoologie Ecologie Animale* 7, 27–44.

Klein, M.G. (1990). Efficacy against soil-inhabiting insect pests. In: Gaugler, R. and Kaya, H.K. (eds.) *Entomopathogenic Nematodes in Biological Control.* CRC Press, Boca Raton, Florida, pp. 195–214.

Kocan, K.M., Pidherney, M.S., Blouin, E.F., Claypool, P.L., Samish, M. and Glazer, I. (1998) Interaction of entomopathogenic nematodes (Steinernematidae) with selected species of ixodid ticks (Acari: Ixodidae). *Journal of Medical Entomology* 35, 514–520.

Koehler, P.G., Patterson, R.S. and Martin, W.R. (1992). Susceptibility of cockroaches (Dictyoptera: Blattellidae, Blattidae) to infection by *Steinernema carpocapsae*. *Journal of Economic Entomology* 85, 1184–1187.

Koppenhöfer, A.M., Shapiro-Ilan, D.I. and Hiltpold, I. (2020) Entomopathogenic nematodes in sustainable food production. *Frontiers in Sustainable Food Systems* 4: 125.

Krishna, K., Grimaldi, D.A. and Engel, M.S. (2013) Treatise on the Isoptera of the world. *Bulletin of the American Museum of Natural History* 377, 2435–2704.

Lado, P., Glon, M.G. and Klompen, H. (2021) Integrative taxonomy of *Dermacentor variabilis* (Ixodida: Ixodidae) with description of a new species, *Dermacentor similis* n. sp. *Journal of Medical Entomology* 58, 2216–2227.

Laumond, C., Mauléon, H. and Kermarrec, A. (1979) New data on the host spectrum and parasitism of the entomophagous nematode, *Neoaplectana carpocapsae*. *Entomophaga* 24, 13–27.

Le Goff, G. and Nauen, R. (2021) Recent advances in the understanding of molecular mechanisms of resistance in noctuid pests. *Insects* 12: 674.

Le Vieux, A.P. and Malan, P.D. (2013) An overview of the vine mealybug (*Planococcus ficus*) in South African vineyards and the use of entomopathogenic nematodes as potential biocontrol agent. *South African Journal of Enology and Viticulture* 34, 108–118.

LeBeck, L.M., Gaugler, R., Kaya, H.K., Hara, A.H. and Johnson, M.W. (1993) Host stage suitability of the leafminer *Liriomyza trifolii* (Diptera: Agromyzidae) to the entomopathogenic nematode *Steinernema carpocapsae* (Rhabditida: Steinernematidae). *Journal of Invertebrate Pathology* 62, 58–63.

Lenz, M. (2004) Biological control in termite management – the potential of nematodes and fungal pathogens. In: *Proceedings of the 35th Annual Meeting of IRG*. IRG/WP 04-10521. International Research Group on Wood Preservation, Stockholm, Sweden.

Lenz, M. and Runko, S. (1992) Use of microorganisms to control colonies of the coconut termite *Neotermes rainbowi* (Hill) on Vaitupu, Tuvalu. *Commonwealth Scientific and Industrial Research Organisation, Division of Entomology, Termite Group Report* 92, 16–47.

Lenz, M., Kamath, M.K., Lal, S. and Senivasa, E. (2000) *Status of the Tree-damaging Neotermes sp. in Fiji's American Mahogany Plantation and Preliminary Evaluation of the Use of Entomopathogens for their Control.* Project Report. ACIAR Small Project No. FST/96/205. ACIAR, Canberra.

Lewis, E., Gaugler, R. and Harrison, R. (1992) Entomopathogenic nematode host finding: response to host contact cues by cruise and ambush foragers. *Parasitology* 105, 309–315.

Lind, P. (1998) Solving yellowjacket problems. *Journal of Pesticide Reform* 18, 22–23.

Locatelli, D.P. and Parleaz, E. (1987) Valutazione in Laboratorio dell'attivita di *Steinernema* spp. e *Heterorhabditis* spp. su *Blattella germanica* (L.). *La Difesa delle Piante* 10, 339–348.

Mankowski, M.E, Kaya, H.K., Grace, K. and Sipes, B. (2005) Differential susceptibility of subterranean termite castes to entomopathogenic nematodes. *Biocontrol Science and Technology* 15, 367–77.

Manweiler, S., Appel, A. and Weber, T. (1993) Nematode-based biological control of German cockroaches. In: Wildey, K.B. and Robinson, W.H. (eds.) *Proceedings of the 1st International Conference on Insect Pests in the Urban Environment*. Organising Committee, pp. 173–180.

Manzoor, F. (2012) Synergism of imidacloprid and entomopathogenic nematodes for the control of eastern subterranean termite, *Reticulitermes flavipes* (Isoptera: Rhinotermitidae). *Pakistan Journal of Zoology* 44, 1397–1403.

Mauldin, J.K. and Beal, R.H. (1989) Entomogenous nematodes for control of subterranean termites, *Reticulitermes* spp. (Isoptera: Rhinotermitidae). *Journal of Economic Entomology* 82, 1638–1642.

McNamara, L., Kapranas, A., Williams, C.D., O'Tuama, P., Kavanaugh, K. and Griffin, C.T. (2018) Efficacy of entomopathogenic fungi against large pine weevil, *Hylobius abietis*, and their additive effects when combined with entomopathogenic nematodes. *Journal of Pest Science* 91, 1407–1419.

Mix, J. and Beal, R.H. (1985) Can nematodes halt subterranean termites? *Pest Control* 53, 22–23.

Monteiro, C., Vale, F.L., Vieira, M.S., Perinotto, W.M.D.S., Auad, A.M. *et al.* (2020) Efficacy of *Heterorhabditis baujardi* (Rhabditida: Heterorhabditidae) against *Rhipicephalus microplus* (Acari: Ixodidae) in presence of susceptible and alternate insect hosts. *Biocontrol Science and Technology* 30, 1316–1329.

Morris, J.R., Stewart, K.W. and Hassage, R.L. (1990) Use of the nematode *Steinernema carpocapsae* for control of the red imported fire ant (Hymenoptera: Formicidae). *Florida Entomologist* 73, 675–677.

Morse, J.G. and Lindegren, J.E. (1996). Suppression of fuller rose beetle (Coleoptera: Curculionidae) on citrus with *Steinernema carpocapsae* (Rhabditida: Steinernematidae). *Florida Entomologist* 79, 373–384.

Morton, A. and García-del-Pino, F. (2013) Sex-related differences in the susceptibility of *Periplaneta americana* and *Capnodis tenebrionis* to the entomopathogenic nematode *Steinernema carpocapsae*. *Journal of Invertebrate Pathology* 112, 203–207.

Myers, R.Y., Sylva, C.D., Mello, C.L. and Snook, K.A. (2020) Reduced emergence of *Cylas formicarius elegantulus* (Coleoptera: Curculionidae) from sweet potato roots by *Heterorhabditis indica*. *Journal of Economic Entomology* 113, 1129–1133.

Navon, A., Nagalakshmi, V.K., Levski, S., Salame, L. and Glazer, I. (2002) Effectiveness of entomopathogenic nematodes in an alginate gel formulation against lepidopterous pests. *Biocontrol Science and Technology* 12, 737–746.

Ndlela, S., Niassy, S. and Mohamed, S.A. (2022) Important alien and potential native invasive insect pests of key fruit trees in Sub-Saharan Africa: advances in sustainable pre- and post-harvest management approaches. *CABI Agriculture and Bioscience* 3, 1–46.

Noosidum, A., Satwong, P., Chandrapatya, A. and Lewis, E. (2016) Efficacy of *Steinernema* spp. plus anti-desiccants to control two serious foliage pests of vegetable crops, *Spodoptera litura* F. and *Plutella xylostella* L. *Biological Control* 97, 48–56.

Nyrop, J., Shelton, A. and Theunissen, J. (1989) Value of a control decision rule for leek moth infestation in leek. *Entomologia Experimentalis et Applicata* 53, 167–176.

Odendaal, D., Addison, M.F. and Malan, A.P. (2016a) Evaluation of aboveground application of entomopathogenic nematodes for the control of diapausing codling moth (*Cydia pomonella* L.) under natural conditions. *African Entomology* 24, 61–74.

Odendaal, D., Addison, M.F. and Malan, A.P. (2016b) Control of diapausing codling moth, *Cydia pomonella* (Lepidoptera: Tortricidae) in wooden fruit bins, using entomopathogenic nematodes (Heterorhabditidae and Steinernematidae). *Biocontrol Science and Technology* 26, 1504–1515.

Odendaal, D., Addison, M.F. and Malan, A.P. (2016c) Entomopathogenic nematodes for the control of the codling moth (*Cydia pomonella* L.) in field and laboratory trials. *Journal of Helminthology* 90, 615–623.

Oi, D.H. and Pereira, R.M. (1993) Ant behavior and microbial pathogens. *Florida Entomologist* 76, 63–74.
Oi, F. (2022) A review of the evolution of termite control: a continuum of alternatives to termiticides in the United States with emphasis on efficacy testing requirements for product registration. *Insects* 13: 50.
Özdemir, E. and Evlice, E. (2020) Assessment of the susceptibility of the Turkestan cockroach, *Blatta lateralis* to Turkish isolates of entomopathogenic nematodes. *Türkiye Biyolojik Mücadele Dergisi* 11, 129–137.
Pena, J.E., Schroeder, W.J. and Osborne, L.S. (1990) Use of entomogenous nematodes of the families Heterorhabditidae and Steinernematidae to control banana moth (*Opogona sachari*). *Nematropica* 20, 51–55.
Peng, Y., Li, S.J., Yan, J., Tang, Y., Cheng, J.P. *et al.* (2021) Research progress on phytopathogenic fungi and their role as biocontrol agents. *Frontiers in Microbiology* 12: 670135.
Pereira, A.E., Souza, D., Zukoff, S.N., Meinke, L.J. and Siegfried, B.D. (2017) Cross-resistance and synergism bioassays suggest multiple mechanisms of pyrethroid resistance in western corn rootworm populations. *PLOS ONE* 12: e0179311.
Peters, A. (1996) The natural host range of *Steinernema* and *Heterorhabditis* spp. and their impact on insect populations. *Biocontrol Science and Technology* 6, 389–402.
Poinar, G.O. Jr. (1979) *Nematodes for Biological Control of Insects*. CRC Press, Boca Raton, Florida.
Poinar, G.O. Jr. and Ennik, F. (1972) The use of *Neoaplectana carpocapsae* (Steinernematidae: Rhabditoidea) against adult yellowjackets (*Vespula* spp., Vespidae: Hymenoptera). *Journal of Invertebrate Pathology* 19, 331–334.
Poinar, G.O. Jr. and Georgis, R. (1989) Biological control of social insects with nematodes. In: Leslie, A.R. and Metcalf, R.L. (eds) *Integrated Pest Management for Turfgrass and Ornamentals*. US Environmental Protection Agency, Washington, DC.
Poole, M.A. (1976) Survey and control efficacy of endoparasites of *Solenopsis richteri* Forel and *Solenopsis invicta* Buren in Mississippi. PhD thesis, Mississippi State University, Starkville, Mississippi.
Půža, V. and Mráček, Z. (2010) Does scavenging extend the host range of entomopathogenic nematodes (Nematoda: Steinernematidae)? *Journal of Invertebrate Pathology* 104, 1–3.
Quattlebaum, E.D. (1980) Evaluation of fungal and nematode pathogens to control the red imported fire ant, *Solenopsis invicta* Buren. PhD thesis, Clemson University, Clemson, South Carolina.
Ralph, N.L. and Kelvin, E.A. (2013) Extermination strategies reported in the NYC Health and Nutrition Survey, 2004: prevalence and predictors. *Journal of Exposure Science and Environmental Epidemiology* 23, 392–398.
Reese, K.M. (1971) Navy fights Formosan termite in Hawaii. *Chemical and Engineering News* 49: 52.
Robinson, G.S., Ackery, P.R., Kitching, I., Beccaloni, G.W. and Hernández, L.M. (2023) HOSTS – a database of the world's lepidopteran hostplants. Natural History Museum. Available at: https://data.nhm.ac.uk/dataset/hosts/resource/877f387a-36a3-486c-a0c1-b8d5fb69f85a (accessed 25 March 2024).
Rose, E.A.F., Harris, R.J. and Glare, T.R. (1999) Possible pathogens of social wasps (Hymenoptera: Vespidae) and their potential as biological control agents. *New Zealand Journal of Zoology* 26, 179–190.
Rouland, C., Benmoussa, D., Reversat, G. and Laumond, C. (1996) Study of the sensitivity of sexual termites Macrotermitinae to an infestation with entomopathogenic nematodes of genera *Heterorhabditis* and *Steinernema*. *Comptes Rendus de l Academie des Sciences Series III: Sciences de la Vie* 319, 997–1001.
Rufai, M.A., Wahab, A.A., Fasasi, K.A., Adeshina, Q.O. and Awotidebe, M.T. (2021) Biocidal efficacy of *Heterorhabditis bacteriophora* against the African armyworm (*Spodoptera exempta*) towards pest control. *Pakistan Journal of Nematology* 39, 41–45.
Samarasinghe, S.M. (1996) Pathogenicity of entomopathogenic agents against termites employing insecticidal bioassay with borax-toxicity method. *Sri Lanka Journal of Tea Science* 63, 78–93.
Samish, M., Alekseev, E. and Glazer, I. (1999) Interaction between ticks (Acari: Ixodidae) and pathogenic nematodes (Nematoda): susceptibility of tick species at various developmental stages. *Journal of Medical Entomology* 36, 733–740.
Samish, M., Alekseev, E. and Glazer, I. (2000) Biocontrol of ticks by entomopathogenic nematodes: research update. *Annals of the New York Academy of Sciences* 916, 589–594.
Samish, M., Rot, A., Gindin, G., Ment, D., Behar, A. and Glazer, I. (2020) Biocontrol of the cat flea, *Ctenocephalides felis*, by entomopathogenic nematodes and fungi. *Biological Control* 149: 104301.
Santos, V., Junior, A.M., Andaló, V., Costa Moreira, C. and Alves de Olinda, R. (2011) Virulence of entomopathogenic nematodes (Rhabditida: Steinernematidae and Heterorhabditidae) for the control of *Diabrotica speciosa* Germar (Coleoptera: Chrysomelidae). *Ciência e Agrotecnologia* 35, 1149–1156.
Schildknecht, H. and Koob, K. (1971) Myrmicacin, the first insect herbicide. *Angewandte Chemie* 10, 124–125.

Shaban, H.E., Helal, I.B., Shamseldean, M.M. and Seif, A.I. (2010) Virulence of *Steinernema* and *Heterorhabditis* entomopathogenic nematodes (Rhabditida) from Egypt against the greater wax moth *Galleria mellonella* (Lepidoptera: Pyralidae) and American cockroach, *Periplaneta americana* (Blattodea: Blattidae). *Egyptian Journal of Experimental Biology (Zoology)* 6, 239–245.
Shapiro-Ilan, D.I., Cottrell, T.E., Mizell, R.F. III, Horton, D.L., Behleand, R.W. and Dunlap, C.A. (2010) Efficacy of *Steinernema carpocapsae* for control of the lesser peachtree borer, *Synanthedon pictipes*: improved aboveground suppression with a novel gel application. *Biological Control* 54, 23–28.
Shapiro-Ilan, D.I., Wright, S.E., Tuttle, A.F., Cooley, D.R. and Leskey, T.C. (2013) Using entomopathogenic nematodes for biological control of plum curculio, *Conotrachelus nenuphar*: effects of irrigation and species in apple orchards. *Biological Control* 67, 123–129.
Shapiro-Ilan, D., Cottrell, T., Mizell, R. and Horton, D. (2016) Efficacy of *Steinernema carpocapsae* plus fire gel applied as a single spray for control of the lesser peachtree borer, *Synanthedon pictipes. Biological Control* 94, 33–36.
Shaurub, E.S.H., Reyad, N.F., Abdel-Wahab, H.A. and Ahmed, S.H. (2016) Mortality and nematode production in *Spodoptera littoralis* larvae in relation to dual infection with *Steinernema riobrave*, *Heterorhabditis bacteriophora*, and *Beauveria bassiana*, and the host plant. *Biological Control* 103, 86–94.
Sherman, P.W., Seeley, T.D. and Reeve, H.K. (1988) Parasites, pathogens, and polyandry in social Hymenoptera. *American Naturalist* 131, 602–610.
Shoeb, M.A., El-Mahdi, I.F. and Shairra, S.A. (2016) Virulence of *Steinernema scapterisci* and *Beauveria bassiana* and their effects to the biochemistry of the mole cricket *Gryllotalpa africana* (Orthoptera: Gryllotalpidae). *Egyptian Journal of Zoology* 66, 205–216.
Skierska, B., Szadziewska, M. and Stanuszek, S. (1976) Laboratory tests for the usability of the entomophilic nematodes Steinernematidae Chitwood and Chitwood 1937 in biological control of some noxious arthropods. *Bulletin of the Institute of Maritime and Tropical Medicine in Gdynia* 27, 207–227.
Stock, S.P. (1996) Pathogenicity of the entomopathogenic nematode *Heterorhabditis argentinensis* (Nemata: Heterorhabditidae) on insects associated with alfalfa crops. *Revista de la Sociedad Entomologica Argentina* 55, 175–177.
Story, G.K., Vander Meer, R.K., Boucias, D.G. and McCoy, C.W. (1991) Effect of fire ant (*Solenopsis invicta*) venom alkaloids on the *in vitro* germination and development of selected entomopathogenic fungi. *Journal of Invertebrate Pathology* 58, 88–95.
Taha, E.H. and Abdelmegeed, S.M. (2016) Effect of entomopathogenic nematodes, *Heterorhabditis bacteriophora* on *Galleria mellonella* in bee hives of *Apis mellifera. Journal of Biological Sciences* 16, 197–201.
Tamashiro, M. (1976) *Susceptibility of Termites to Microbes*. Final Technical Report. Office of Naval Research Department of Entomology, University of Hawaii, Honolulu.
Toepfer, S., Peters, A., Ehlers, R.-U. and Kuhlmann, U. (2008) Comparative assessment of the efficacy of entomopathogenic nematode species at reducing western corn rootworm larvae and root damage in maize. *Journal of Applied Entomology* 132, 337–348.
van Damme, V., Beck, B., Berckmoes, E., Moerkens, R., Wittemans, L. *et al.* (2016) Efficacy of entomopathogenic nematodes against larvae of *Tuta absoluta* in the laboratory. *Pest Management Science* 72, 1702–1709.
van der Linden, C.F.H., Fatouros, N.E. and Kammenga, J.E. (2022) The potential of entomopathogenic nematodes to control moth pests of ornamental plantings. *Biological Control* 165: 104815.
van Lenteren, J.C. (2012) The state of commercial augmentative biological control: plenty of natural enemies, but a frustrating lack of uptake. *BioControl* 57, 1–20.
van Lenteren, J.C., Bolckmans, K., Köhl, J., Ravensberg, W.J. and Urbaneja, A. (2018) Biological control using invertebrates and microorganisms: plenty of new opportunities. *BioControl* 63, 39–59.
van Niekerk, S. and Malan, A.P. (2014) Compatibility of biological control agents and agrochemicals to entomopathogenic nematodes, *Steinernema yirgalemense and Heterorhabditis zealandica*. *African Entomology* 22, 49–56.
Vashisth, S., Chandel, Y.S., Chandel, R.S. and Sharma, P.K. (2017) Pathogenicity of *Heterorhabditis* nematodes isolated from north-western Himalaya against the larvae of *Plutella xylostella* (L.) (Lepidoptera: Plutellidae). *Annales de la Société Entomologique de France* 53, 204–210.
Vashisth, S., Chandel, Y.S. and Chandel, R.S. (2018) Biological control potential of North West Himalayan strains of heterorhabditid nematodes against the turnip moth, *Agrotis segetum* (Denis & Schiffermuller) (Lepidoptera: Noctuidae). *Egyptian Journal of Biological Pest Control* 28: 37.
Vasseghian, Y., Arunkumar, P., Joo, S.W., Gnanasekaran, L., Kamyab, H. *et al.* (2022) Metal-organic frame work enabled pesticides are an emerging tool for sustainable cleaner production and environmental hazard reduction. *Journal of Cleaner Production* 373, 133966.

Wagutu, G.K., Kang'ethe, L.N. and Waturu, C.N. (2017) Efficacy of entomopathogenic nematode (*Steinernema karii*) in control of termites (*Coptotermes formosanus*). *Journal of Agriculture, Science and Technology* 18, 55–64.

Wang, C., Powell, J.E. and Nguyen, K. (2002) Laboratory evaluations of four entomopathogenic nematodes for control of subterranean termites (Isoptera: Rhinotermitidae). *Environmental Entomology* 31, 381–387.

Wassink, H. and Poinar, G.O. Jr. (1984) Use of the entomogenous nematode, *Neoaplectana carpocapsae* Weiser (Steinernematidae: Rhabditida), in Latin America. *Nematropica* 14, 97–109.

Wells, J.D. and Henderson, G. (1993) Fire ant predation on native and introduced subterranean termites in the laboratory: effect of high soldier number in *Coptotermes formosanus*. *Ecological Entomology* 18, 270–274.

Wilson-Rich, N., Stuart, R.J. and Rosengaus, R.B. (2007) Susceptibility and behavioral responses of the dampwood termite *Zootermopsis angusticollis* to the entomopathogenic nematode *Steinernema carpocapsae*. *Journal of Invertebrate Pathology* 95, 17–25.

Wojcik, W.F. and Georgis, R. (1987) Infection of selected insect species with desiccated *Steinernema feltiae* (Nematoda). *Proceedings of the Society of Invertebrate Pathology* 20, 80.

Wojcik, W.F. and Georgis, R. (1988) Infection of adult western yellowjackets with desiccated *Steinernema feltiae* (Nematoda). *Journal of Invertebrate Pathology* 52, 183–184.

Wu, H.J., Wang, Z.N., Ou, C.F., Tsai, R.S. and Chow, Y.S. (1991) Susceptibility of two Formosan termites to the entomogenous nematode *Steinernema feltiae* Filipjev. *Bulletin of the Institute of Zoology, Academia Sinica* 30, 31–39.

Yan, X., Wang, X., Han, R. and Qiu, X. (2014) Utilization of entomopathogenic nematodes, *Heterorhabditis* spp. and *Steinernema* spp., for the control of *Agrotis ipsilon* (Lepidoptera, Noctuidae) in China. *Nematology* 16, 31–40.

Yan, X., Shahid Arain, M., Lin, Y., Gu, X., Zhang, L., Li, J. and Han, R. (2020) Efficacy of entomopathogenic nematodes against the tobacco cutworm, *Spodoptera litura* (Lepidoptera: Noctuidae). *Journal of Economic Entomology* 113, 64–72.

Yeruham, I., Rosen, S. and Hadani, A. (1989) Mortality in calves, lambs and kids caused by severe infestation with the cat flea *Ctenocephalides felis felis* (Bouché, 1835) in Israel. *Veterinary Parasitology* 30, 351–356.

Yu, H. (2009) Parasitism of subterranean termites (Isoptera: Rhinotermitidae: Termitidae) by entomopathogenic nematodes (Nematoda: Steinernematidae: Heterorhabditidae). PhD thesis, University of Arizona, Tucson, Arizona.

Yu, H., Gouge, D.H. and Baker, P. (2006) Parasitism of subterranean termites (Isoptera: Rhinotermitidae: Termitidae) by entomopathogenic nematodes (Rhabditida: Steinernematidae; Heterorhabditidae). *Journal of Economic Entomology* 99, 1112–119.

Yu, H., Gouge, D.H., Stock, S.P. and Baker, P.B. (2008) Development of entomopathogenic nematodes (Rhabditida: Steinernematidae; Heterorhabditidae) in the desert subterranean termite *Heterotermes aureus* (Isoptera: Rhinotermitidae). *Journal of Nematology* 40, 311–317.

Yu, H., Gouge, D.H. and Shapiro-Ilan, D.I. (2010) A novel strain of *Steinernema riobrave* (Rhabditida: Steinernematidae) possesses superior virulence to subterranean termites (Isoptera: Rhinotermitidae). *Journal of Nematology* 42, 91–95.

Yuksel, E., Taskesen, Y.E., Erarslan, D. and Canhilal, R. (2018) Effectiveness of different entomopathogenic nematode species against the variegated cutworm, *Peridroma saucia* (Hubner) (Lepidoptera: Noctuidae). *Egyptian Journal of Biological Pest Control* 28: 8.

Zalucki, M.P., Rehan Silva, A.S., Adamson, D., Shu-Sheng, L. and Furlong, M.J. (2012) Estimating the economic cost of one of the world's major insect pests, *Plutella xylostella* (Lepidoptera: Plutellidae): just how long is a piece of string? *Journal of Economic Entomology* 105, 1115–1129.

Zhioua, E., Lebrun, R.A., Ginsberg, H.S. and Aeschlimann, A. (1995) Pathogenicity of *Steinernema carpocapsae* and *S. glaseri* (Nematoda: Steinernematidae) to *Ixodes scapularis* (Acari: Ixodidae). *Journal of Medical Entomology* 32, 900–905.

Zhou, X., Kaya, H.K., Heungens, K. and Goodrich-Blair, H. (2002) Response of ants to a deterrent factor(s) produced by the symbiotic bacteria of entomopathogenic nematodes. *Applied and Environmental Microbiology* 68, 6202–6209.

Zukowski, K. (1984) Badania laboratoryjne roli nicieni *Neoaplectana carpocapsai* Weiser w redukcji prusakow (*Blattella germanica*). *Roczniki Panstwowego Zaklad Higieny* 35 451–457.

23 Applications for Nursery Crop Production

T. Stefanovska[1]*, Edwin E. Lewis[2] and Regina K. Cruzado-Gutiérrez[3]

[1]Department of Entomology Integrated Pest Management and Plant Quarantine, National University of Life and Environmental Science, Kiev, Ukraine; [2]Department of Entomology, Plant Pathology and Nematology, University of Idaho, Moscow, Idaho, USA; [3]Institute for Health in the Human Ecosystem, University of Idaho, Moscow, Idaho, USA

Abstract

The concept of "nursery crops" is a challenging one. What is included? For this chapter, we have included crops grown under glass including floriculture and other ornamental plants, and outdoor nursery stock production. Biological control methods are used in these crops with great success. One of the main reasons for this success is that the environmental conditions in which the plants are produced are usually under more control of the user than typical crops. For example, greenhouse temperatures can be controlled and ultraviolet light can be limited. Furthermore for entomopathogenic nematode (EPN) use, the medium in which the plants are grown (and to which the nematode would be applied) can be manipulated. We cannot provide an exhaustive treatment of all important nursery crop pests in this chapter, as there are so many. For example, a large commercial nursery might produce over 400 plant species, each with a different pest complex. However, there are some stand-out nursery pests on which we concentrate. We have emphasized pests where evidence, either from industry or laboratory/greenhouse experiments, shows that EPNs are effective or may have promise of efficacy for insect pest management. EPN use in nursery crops will continue to increase and improve, especially with the ability to have control over at least some environmental conditions where plants are produced.

23.1 Introduction

23.1.1 Nursery markets

Nursery crop production and the maintenance of wholesale and retail nursery stock for ornamental landscape use is an important segment of US and European agriculture, as is vegetable production under glass. Nursery production is difficult to quantify accurately because there are outdoor and greenhouse facilities, which can include hydroponic production, container plants and in-ground production. There are ornamental plants, which can be further divided into landscape plants and floriculture, and production agricultural plants, which include vegetables and, more recently, cannabis. We restrict "nursery production" here to in-ground and container plant production of ornamental (including landscape), vegetable and floriculture plants. Within the USA, nursery production revenues in 2009 exceeded many key agronomic crops; nursery

*Email: steftat@hotmail.com

 Entomopathogenic Nematodes as Biological Control Agents (D.I. Shapiro-Ilan and E.E. Lewis eds)
DOI: 10.1079/9781800620322.0023

and greenhouse crops ranked first in ten of the 12 states in the northeastern USA. Floriculture and nursery comprise almost 30% of the specialty crops grown in the USA, and they are grown in every state. Fifty-five counties in California produce nursery and floriculture crops, generating 170,000 jobs worth US$3.78 billion to the state's economy (Carman, 2011; Parrella and Lewis, 2017).

Nursery plant production is challenging due to numerous plant species and cultivars, each with unique pathogens, nematode and insect pests, factors affecting plant growth, potting media/soil type and methods of culturing (Johnson and Lyon, 1991; Dreistadt *et al.*, 2004; Alford, 2012). Individual commercial operations may grow up to 400 plant genera (Yeager *et al.*, 2010). Another challenge is that feeding injury on ornamental plants directly affects marketability or aesthetic value, and the threshold for damage is low. These conditions cause potentially high initial product cost inputs and intensive labor demands during production. For example, nursery liners (transplants) range in initial cost from US$1 to US$25 each, and after planting require significant management and manual labor (Jeffers *et al.*, 2009, 2010). Despite efforts to protect high-value nursery crops, the green industry experiences extensive annual losses attributed to insects and plant diseases.

23.1.2 Special considerations for nursery pests

Pest management for ornamental plants relies largely on pesticide application, often on a calendar basis (Grewal, 2012; LeBude *et al.*, 2012) and efficacy is unpredictable. The chemical paradigm remains strong within this industry, with neonicotinoids the most common insecticides used. Overreliance on insecticides has led to insecticide resistance, resurgence of pests due to elimination of beneficial species and removal of pesticides from the marketplace. Nursery growers commonly use integrated pest management (IPM) practices, but some components are not widely adopted due to the lack of demonstrated success (LeBude *et al.*, 2012). US growers often perceive the risk of biological control as uncomfortably high; chemical programs may be more expensive, but the risk of value loss due to insect damage overrides cost.

Nevertheless, entomopathogenic nematodes (EPNs) have been used effectively to reduce damage caused by insects in soil and other cryptic habitats, and have seen some success against foliar insects. The range of EPN targets has expanded over the last 20 years due to new EPN isolates with different host ranges, formulations that are amenable to new application strategies and application methods that favor EPN efficacy where it was low in the past. We will address some of the successes and failures seen in the management of insect pests of nursery crops. We will attempt to provide some reasoning behind the outcomes we discuss and finally to predict where nursery and greenhouse IPM will move forward.

23.2 Root Weevils and Factors Affecting Entomopathogenic Nematode Efficacy

23.2.1 Black vine weevil

Black vine weevil (BVW), *Otiorhynchus sulcatus*, is one of the most important pests of cranberries, strawberries and nursery ornamentals in the USA, Canada and Western Europe. Crop damage and crop rejections due to infestation with BVW larvae can result in 100% losses. Insecticide-based BVW control is made more difficult by the withdrawal of persistent products for prophylactic use in growing media, current European Commission restriction of neonicotinoid insecticides and growing demands for insecticide-free products. The EC Sustainable Use Directive states that all growers must use IPM where practical, and BVW populations have been increasing recently in some localities due to these restrictions. There is growing interest in adopting biological control methods within the IPM framework. A survey asked 29 key UK hardy nursery stock industry representatives the question "What kind of management programs, including biological and chemical control of BVW, are used currently?" The answers indicated that all but one grower relied on IPM programs for management of BVW and other pests (Daughtrey and Benson, 2005).

BVW adults eclose in late spring. All individuals are female and begin to lay eggs about 2 weeks later. The larvae develop through six instars, and late instars overwinter in soil. EPNs are applied to target overwintered larvae and pupae in the spring or late-instar larvae in the fall (Sampson, 1994; Georgis *et al.*, 2006). EPNs were more effective at targeting third and fourth instars than first and second instars (Georgis and Poinar, 1984; Kakouli-Duarte *et al.*, 1997; Lola-Luz *et al.*, 2005). In fields with natural infestations, at the end of the summer, a wide range of instars makes efficacy less predictable than spring applications where all larvae are later instars. Results obtained in spring at temperatures over 19°C demonstrated 60–80% control by *Steinernema feltiae*, *Steinernema carpocapsae*, two *Heterorhabditis bacteriophora* strains and *Heterorhabditis megidis* (Stimmann *et al.*, 1985; Hanula, 1993). Efficacy against late instars in spring is tied to soil temperature at the time of application. Kakouli-Duarte *et al.* (1997) conducted field trials against BVW larvae in strawberries using *S. carpocapsae* and *H. megidis*. EPNs were applied via drip irrigation along and across raised beds. *S. carpocapsae* treatment in late summer produced 49.5% reduction of young-instar BVW larvae, whereas late spring application provided 65% control of later instars. In the same study, spring application of *H. megidis* caused 26% mortality of late instars. This topic has been researched extensively and the details below pertain to many different systems.

23.2.1.1 Nematode species and application temperatures

When applied during periods of favorable soil temperatures, EPNs provide excellent BVW control. Commercially available EPNs have provided over 90% efficacy but are limited by cool soil temperatures. Sensitivity to soil temperature is a critical factor limiting EPN control of BWV (van Tol and Raupp, 2005; Georgis *et al.*, 2006; Lacey *et al.*, 2015), and tolerance of cool temperatures is a key aspect of choosing an EPN species that will be effective. *H. bacteriophora*, *H. megidis*, *H. marelata* (OH10 strain), *H. downesi*, *S. carpocapsae*, *S. feltiae* and *S. glaseri* are effective at controlling BVW. Application of *H. marelata* and *H. bacteriophora* outdoors (soil temperature 11.5°C) and in the glasshouse (soil temperature 20°C) to control BWV on impatiens provided almost 100% mortality (Bruck *et al.*, 2005). Van Toll and Raupp (2005) summarized the efficacy of different EPN species/strains for BWV control in fall and spring when soil temperature was warmer than 12°C. *H. megidis* outperformed *S. kraussei* at warmer temperatures, while *H. megidis* at 50% the recommended application rate was superior to *S. kraussei* applied at full rate in the greenhouse and in field-grown strawberries when the spring soil temperature was warm (Haukeland and Lola-Luz, 2010). However, *S. kraussei* was the best EPN species for BVW control at cooler temperatures. In a winter application with temperatures varying from 1.5°C to 7.3°C, Willmott *et al.* (2002) found that *S. kraussei* provided better control of BVW larvae than *S. carpocapsae*. *S. kraussei* (L137 strain) was consistently the most virulent species at cooler temperatures. Thus, *S. kraussei* can be applied earlier in the spring than most *Steinernema* and *Heterorhabditis* spp.

Van Tol and Raupp (2005) presented results of long-term studies regarding the impact of temperature on EPN efficacy in natural settings in fall. Studies were conducted by inoculating pots or field plots with BWV eggs followed by EPN application at rates of 0.5×10^6 per pot and 1×10^6 m^{-2} in open fields (van Tol, 1993a,b, 1994, 1996a,b; van Tol *et al.*, 1998; Fitters *et al.*, 2001). Results indicated that most EPN species tested provided good control of BWV at 12°C, but the time it took to infect BVW larvae varied with the type of test. In pots, sufficient control obtained in a few days, whereas field trials took much longer (van Tol, 1993c; Fitters *et al.*, 2001). Because of this temperature- and situation-dependent efficacy, growers should be advised that control can be achieved only if the soil is projected to be at least 12°C for at least 1 week (van Toll and Raupp, 2005).

23.2.1.2 Special application methods

EPNs can be conditioned by appropriate storage regimes to optimize control across a range of conditions, including but not restricted to cool temperatures. Fitters *et al.* (2001) showed that "conditioning" *H. megidis* improved control of BVW when applied to potted plants at 9°C. Guy *et al.* (2017) demonstrated similar results for *S. carpocapsae*; EPNs conditioned for 3–6 weeks at 9°C before application at 9°C provided better efficacy against BVW in strawberries at 0–12°C application temperature than after conditioning at 15°C. A commercial product based on the cold-active species *S. kraussei* was used as a positive control in this study. At 9°C, *S. kraussei* outperformed both conditioned and non-conditioned *H. megidis* and *S. carpocapsae*, whereas at 15°C, both conditioned and non-conditioned *H. megidis* applied at half the recommended rate outperformed *S. kraussei* applied at the full rate.

EPN infective juveniles (IJs) applied in aqueous solutions by sprays or irrigation are available in several commercial formulations (Grewal *et al.*, 2002; Cruz-Martínez *et al.*, 2017). However, EPNs can also be applied via infected cadavers to achieve reliable BWV control (Shapiro-Ilan *et al.*, 2003; Lacey and Shapiro-Ilan, 2008; Dolinski *et al.*, 2015). Several studies demonstrated that IJs emerging directly from their host cadaver have superior dispersal (Shapiro and Glazer, 1996), infectivity (Shapiro and Lewis, 1999), survival (Pérez, 2003) and environmental tolerance (Gulzar *et al.*, 2020) compared with nematodes applied in aqueous suspension. Greenhouse studies comparing virulence of EPNs applied in water versus as cadavers against the *Diaprepes* root weevil, *Diaprepes abbreviatus* (Coleoptera: Curculionidae), and BVW showed cadaver applications to provide superior efficacy (Shapiro-Ilan *et al.*, 2003). The increased efficacy observed in the cadaver applications may be due to additional physiological stress in the aqueous application during storage in water or on application. However, using cadavers for application would be a costly and time-consuming application (Bruck *et al.*, 2005).

23.2.1.3 Application methods and rate

Using a sprayer to apply EPNs is effective. Under standard nursery conditions, 90–100 % control of BVW larvae was achieved using *H. bacteriophora* at 66,000 IJs per pot on *Bergenia cordifolia* (soil temperature 23°C) when applied to a mixture of peat, composted leaves, sand and pine bark (Gill *et al.*, 2001). However, Godliman (1991) found that nematodes sprayed on cyclamen plants did not penetrate the leaf canopy in sufficient quantities, and that a drench application was more effective. Dipping the roots of strawberry plants into a nematode suspension before planting was proposed by Susurluk and Ehlers (2008). A suspension of 1500 *H. bacteriophora* IJs ml^{-1} was mixed with carboxymethylcellulose to promote EPN adhesion to the roots and reduce sedimentation. Each plant submerged in the suspension retained 6000–7500 IJs. BWV eggs were added monthly for 3 months in pot trials, and larval mortality exceeded 90% in two trials (18°C and 26°C).

Based on field studies, a rate of 0.5×10^6 IJs per 1-litre pot or 1.0×10^6 IJs m^{-2} surface area provides excellent BVW control (Stimmann *et al.*, 1985; van Tol, 1998; van Tol *et al.*, 1998). EPN applications below the recommended rates provided reduced levels of control, but doubling the recommended rate did not improve efficacy (Hanula, 1993). Similar results were obtained in container testing; application rates ranging from 10,000 to 15,000 IJs per container in container-grown azalea demonstrated that applications of *S. carpocapsae* and *H. bacteriophora* IJs at 10,000 IJs per pot provided acceptable (68–83%) control (Cowles, 1997). Application of *S. feltiae* at 15,000 IJs per plant can provide 100% control (Mráček *et al.*, 1993).

23.2.1.4 Impact of soil/media type

Soil factors such as texture, particle size, moisture content and chemistry influence EPN dispersal and host-finding behavior, and hence their success (Stuart *et al.*, 2015). An experiment on field-grown strawberries in Norway found that in soils with greater clay content, such as silty loam sand (5–10%) and silty clay loam (10–25%), EPNs had the lowest efficacy (Haukeland and Lola-Luz, 2010). Different potting media components (e.g. peat, bark, coir, and peat blended with compost green waste or sand) can also impact EPN efficacy against BVW (Ansari and But, 2011). *Heterorhabditis* spp. caused 100% mortality across all media, whereas *Steinernema* spp. caused 100% larval mortality only in peat blended with 20% compost green waste. Dispersal was also impacted; *H. bacteriophora* dispersed more widely in all media than *H. megidis*, *H. downesi* or *S. kraussei*, whereas *S. feltiae* and *S. carpocapsae* scattered substantially less widely and were best in peat blended with 20% compost green waste. Kakouli-Duarte *et al.* (1997) suggested that *H. megidis* and *S. carpocapsae* (Biosys 252) worked better in bark growbags compared with peat-based compost in pots at similar temperature and moisture. Nielsen and Lewis (2011) suggested that potting media could be customized for EPN use, showing that some components, such as redwood sawdust, were detrimental to EPNs and should be avoided.

23.2.2 *Diaprepes* root weevil

Diaprepes root weevil (DRW), *D. abbreviatus*, attacks over 270 plant species belonging to 59 plant families, including citrus, ornamental and agronomic crops (Simpson *et al.*, 1996). DRW was first found in Florida in 1964 and quickly spread throughout the state (Woodruff, 1968). DRW control is essential for protecting plants and meeting regulatory requirements for shipping. Citrus growers rely on chemical insecticides and insect growth regulators, but they disrupt natural enemy complexes that reduce other pest species. Insecticides without long residual effects leave large windows of opportunity between applications for weevils to lay eggs (Duncan *et al.*, 1999). DRW has been a major commercial target for EPNs for more than 20 years (Dolinski *et al.*, 2012), but the focus of most studies is on DRW on citrus (see Chapter 16, this volume).

The first commercially produced EPNs used to control DRW were *S. carpocapsae*, *H. bacteriophora* and later *H. indica* (Adair, 1994; Shapiro *et al.*, 1999). In orchard soils and in greenhouse conditions, *S. riobrave* (Cabanillas *et al.*, 1994; Duncan and McCoy, 1996; Shapiro and McCoy, 2000; Shapiro-Ilan *et al.*, 2002) provided greater DRW reduction than other EPNs (Bullock *et al.*, 1999). Field efficacy of *S. riobrave* varied from 46% at 25 IJs cm^{-2} soil surface to 90% at higher rates (Duncan *et al.*, 2003), and adult emergence from the soil was decreased by 50% when *S. riobrave* was used at the recommended rates (Duncan *et al.*, 2007; El-Borai *et al.*, 2007). However, estimates of *S. riobrave*'s DRW control efficacy depends to a large extent on the soil type and texture (McCoy *et al.*, 2000, 2002; Jenkins *et al.*, 2007, 2008).

Shapiro *et al.* (2010b) suggested applying EPN-infected *Tenebrio molitor* cadavers enclosed in paper packets. At 15 days post-application, the infected host treatments caused up 91% control in DRW in the laboratory and 75% in the greenhouse. Due to the variations in weevil parasitism rates, it was suggested that EPNs may play a role in regional spatial patterns of weevils, which are more numerous and destructive to citrus in some eco-regions than others (Futch *et al.*, 2005). The prevalence and diversity of EPNs in the Florida peninsula and the possibility that nematodes modulate DRW abundance more in some habitats than others suggests the need to study how soil physical properties affect EPNs in order to implement cultural practices that conserve these nematodes.

23.2.3 Scarab grubs

There is significant variability in the levels of control provided by EPNs toward different species of scarabs (see also Chapter 21, this volume, for control of scarabs in turf); these species are among the more difficult insects to control with EPNs because they have evolved different physical and behavioral anti-infection defenses (Klein *et al.*, 2007; Lacey and Georgis, 2012). Japanese beetle, *Popillia japonica*, northern masked chafer, *Cyclocephala borealis*, southern masked chafer, *Cyclocephala lurida*, western masked chafer, *Cyclocephala hirta*, oriental beetles *Anomala orientalis* and *Exomala orientalis*, and European chafer, *Rhizotrogus majalis*, cause major problems in nurseries. Root feeding negatively impacts plant growth while foliar feeding causes aesthetic damage, and plant quarantine compliance requires a high level of control.

Currently, there are 12 steinernematid and six heterorhabditid nematodes that have been reported to be pathogenic to scarab grubs (Wu, 2013). Several pot and field trials have shown that EPNs have potential against root-feeding white grubs in nurseries. Villani and Wright (1988) studied the efficacy of *S. feltiae*, *S. glaseri*, *H. bacteriophora* and *Heterorhabditis* sp. Holand strain against *P. japonica* and *R. majalis* in container-grown Japanese yew (*Taxus cuspidata*). *H. bacteriophora* and *Heterorhabditis* sp. (Holand) provided 60–90% *P. japonica* control, whereas *S. feltiae* and *S. kraussei* reduced grub populations ranging from 0% to 86%. Nielson and Cowles (1998) performed a field experiment to show that EPN applications were unsuccessful controlling *P. japonica*, *R. majalis* and *E. orientalis*.

Several studies have shown that increasing application rates improves scarab control. Doubling of the *H. marelata* application rate from 2.5 billion to 5 billion IJs ha^{-1} doubled *P. japonica* mortality (Mannion *et al.*, 2001). Wright *et al.* (1988) observed greater *P. japonica* grub mortality due to increased rates of several EPN species from 46 to 385 IJs cm^{-2}. Although four nematode strains provided poor or moderate control of *R. majalis*, mortality was marginally higher when the EPN application rate was increased from 46 to 92 IJs cm^{-2}. Converse and Grewal (1998) evaluated four strains of *S. glaseri* in laboratory trials and found that all strains caused 100% larval mortality of western masked chafer, *Cyclocephala hirta*, 6 days after treatment, but the same strains of *S. glaseri* and *S. feltiae* (Argentina strain) reduced the larval population by only 16%, and *S. scapterisci* (Colon strain), *S. carpocapsae* (All and Mexican strains) did not cause any mortality. *S. riobrave* provided 45–71% larval reduction at 30°C, whereas at 25°C it was ineffective.

Progress in white grub control in the last 20 years has resulted in the discovery of new, highly virulent strains of *Steinernema* and *Heterorhabditis* spp. Identification of new species and thorough examination of nematode–host specificity have explained the differences in efficacy (Grewal *et al.*, 2002, 2005; Lee *et al.*, 2002; Koppenhöfer and Fuzy, 2003a,b, 2007; Koppenhöfer *et al.*, 2006, 2007; Lacey and Georgis, 2012). The promising *Steinernema scarabaei* shows high levels of virulence and also persists well in soil (Koppenhöfer and Fuzy, 2003a,b; Stock and Koppenhöfer, 2003; Koppenhöfer and Fuzy, 2009, 2013). Several of these species/strains have been tested in turfgrass (see Chapter 21, this volume). *S. scarabaei* was most virulent to *R. majalis* and *P. japonica*, and least virulent to *C. borealis* (Koppenhöfer *et al.*, 2006), whereas *Heterorhabditis zealandica* and *H. bacteriophora* were slightly virulent against *R. majalis*, and had marginal virulence to *P. japonica*, *A. orientalis*, *C. borealis*, and *Maladera castanea*. The susceptibility of white grubs to EPNs differed depending on the nematode/host type and strain. *P. japonica* had a greater susceptibility to *H. bacteriophora* and *S. glaseri* compared with *A. orientalis*, *Cyclocephala* spp., *M. castanea*, *R. majalis* and *Phyllophaga* spp. (Koppenhöfer *et al.*, 2000, 2004, 2006; Simard *et al.*, 2001; Grewal *et al.*, 2002, 2005; Shapiro-Ilan *et al.*, 2002; Cappaert and Koppenhöfer, 2003; Koppenhöfer and Fuzy, 2003a,b).

S. scarabaei appears to be more virulent against grubs of *A. orientalis*, *R. majalis* and *P. japonica* (Cappaert and Koppenhöfer, 2003; Koppenhöfer *et al.*, 2004) but less pathogenic to *C. borealis* (Koppenhöfer *et al.*, 2006). *H. bacteriophora* GPS11 strain was more virulent to *A. orientalis*, *C. borealis* and *P. japonica* than to other strains tested but had limited infectivity against *R. majalis* (Grewal *et al.*, 2002). Field trials of EPNs often do not reflect laboratory results because of soil conditions and other environmental constraints.

23.2.4 Fungus gnats

Bradysia spp., also commonly known as fungus gnats, are important insect pests that can cause significant economic losses in crops in protected environments such as greenhouses, nursery beds, houseplants and mushroom compost. The adults can also be nuisance pests in retail operations when customers are repelled by gnats in potted plants for sale. They have been referred to as generalist opportunistic herbivores. Most of the fungus gnat's life is spent as a larva and pupa in organic matter or soil, so most control methods target these immature stages rather than the short-lived adults (Cloyd, 2008; Katumanyane, *et al.*, 2018a,b).

Bradysia spp. control with chemical pesticides leaves harmful residues on plants, which potentially threaten human health. In some crops, such as Chinese chive, *Allium tuberosum*, where pesticides are applied until shortly before harvest, chemical control can be a risk to human health (Rajendra, *et al.*, 2019; Chen *et al.*, 2020). Likewise, the overuse of pesticides has also led to the development of pesticide resistance in *Bradysia* spp. Thus, environmentally friendly approaches such as EPNs have been used to control these pests.

EPNs of the genera *Steinernema* and *Heterorhabditis* have been used to control various *Bradysia* spp. pests (Bai *et al.*, 2016; Katumanyane *et al.*, 2018a,b; Rajendra, *et al.*, 2019; Chen *et al.*, 2020). The effect of EPNs on gnat larvae might be dose dependent, and it may be affected by characteristics of the potting medium (Katumanyane *et al.*, 2018a,b). Although EPNs reduce the larval populations, the efficacy of EPN control is highly enhanced when they are combined with other organisms or management measures. Chen *et al.* (2020) found that the mortality of chive gnat (*Bradysia odoriphaga*) larvae increased when the strategy of control included the combination of waterlogging and EPNs compared with waterlogging or EPNs alone. This synergistic effect is also observed when EPNs are combined with other biological agents. For instance, control of *Bradysia impatiens* was greatly enhanced by using a combination of two species of predatory mites (*Androlaelaps casalis* and *Stratiolaelaps scimitus*), along with two species of EPNs (*H. indica* and *S. carpocapsae*) (Rajendra *et al.*, 2019). The efficacy of the control measure was enhanced because EPNs and mite species targeted different instars of the gnat larvae. The EPNs most likely infected the older larvae, while the predatory mites showed the best control of the younger larvae of *B. impatiens*.

EPNs are currently used for controlling fungus gnats, but the results remain unpredictable. Thus, further investigation is needed into the integrated management of *Bradysia* spp. EPNs have been shown to control fungus gnats by themselves, but using them in combination with other measures will likely be an optimal solution.

23.3 Wood-boring Insects

23.3.1 Coleoptera

The sheltered habitat of wood-boring insects provides some opportunities for EPN use. Several species of bark beetles (Coleoptera: Curculionidae) and longhorn beetles (Coleoptera: Cerambycidae) have been targeted for control. *S. carpocapsae* isolates infect a range of coleopteran wood-boring species, including European elm bark beetle, *Scolytus scolytus* (F.) (Finney, 1977; Finney and Walker, 1979) in laboratory trials. However, applying a suspension of *S. carpocapsae* to heavily infested logs in the field in summer to control this insect was ineffective (Finney and Walker, 1977). Infected larvae and adults were found beneath the bark, but there was no difference in the number of emergence holes found compared with controls. Finney and Walker (1979) found that application of *S. carpocapsae* (DD-136 strain) to the bark of elm logs infested with *S. scolytus* reduced emergence by 15–40%. Six species of EPNs (*S. feltiae*, *S. carpocapsae* and *H. bacteriophora*) isolated from Turkey (*S. feltiae* Aydin isolate, *S. carpocapsae* Black Sea isolate, *H. bacteriophora* Aydin isolate and *H. bacteriophora* Çanakkale isolate) and from Kyrgyzstan (*S. feltiae* KG3 isolate and *H. bacteriophora* KG81 isolate) were tested against black timber bark beetle, *Xylosandrus germanus* (Blandford), under laboratory conditions (Kepenekci *et al.*, 2018). Adult mortality was higher in all EPN treatments than in controls. Turkish EPN isolates *S. carpocapsae* (Black Sea) and *S. feltiae* (KG3) caused the highest mortality (98.66%). Kanzaki and Kosaka (2009) recorded seven parasitic and phoretic nematode species naturally associated with Scolytinae and Platypodinae beetles, but no practical implementation was found.

Qin *et al.* (1988) demonstrated moderate levels of mortality (62%) of Asian long-horned beetle (*Anoplophora glabripennis*) with *S. feltiae* applied into galleries. Fallon *et al.* (2004) screened *S. feltiae* SN strain, *S. glaseri* NJ strain, *S. riobrave* TX strain, *S. carpocapsae* Sal strain, *S. carpocapsae* All strain and *H. marelata* IN strain for efficacy against laboratory colonies of *A. glabripennis*. The most effective were *S. feltiae* and *S. carpocapsae* Sal, which caused 100% mortality. Against third, sixth and seventh instars, *S. feltiae* was more virulent than *S. carpocapsae* Sal. Later instars were unaffected by *S. riobrave*, *S. glaseri* and *H. marelata*, whereas *S. feltiae* and *S. carpocapsae* Sal infected 71–100% of mid- to late instars.

Two strains of *S. carpocapsae* were tested against *Melanophila decastigma* Le Conte (Coleoptera: Buprestidae) in Chinese poplars (Liu *et al.*, 1998). EPNs were injected into active gallery openings and efficacy varied from 66.7% to 89.5%. However, this application method would not be practical for large areas.

The potential to use EPNs to control wood-boring beetles was studied most extensively and successfully with the large pine weevil, *Hylobius abietis* L., which is a serious pest of temperate forestry throughout Scandinavia, the UK, the Republic of Ireland and Eastern Europe. In the absence of control measures, 100% mortality of planted saplings frequently occurs. Due to the successful realization of several projects, the most effective nematodes for weevil control were selected and methods for applying nematodes on an operational scale were developed further. This work is covered extensively in Chapter 24 (this volume).

23.3.2 Lepidoptera

The peachtree borer, *Synanthedon exitiosa*, lesser peachtree borer, *Synanthedon pictipes*, clearwing borers *Synanthedon culiciformis* and *Paranthrene robiniae*, and banded ash borer, *Podoesia aureocincta* (Lepidoptera: Sesiidae) attack ornamental trees and shrubs in nurseries. *S. carpocapsae* and *S. feltiae* have been used successfully to control *Synanthedon exitiosa*, *Synanthedon pictipes* and *Podoesia aureocincta* (Miller and Bedding, 1982; Deseö and Miller, 1985; Kaya and Brown, 1986; Begley, 1990; Nachtigall and Dickler, 1992; Williams *et al.*, 2002; Shapiro-Ilan *et al.*, 2010a). Gill *et al.* (1994) demonstrated greater levels of population reduction of *Podoesia aureocincta* using *S. carpocapsae* with a lower concentration of 77.5 IJs cm^{-2} compared with 145 IJs cm^{-2}.

Shapiro-Ilan and Cottrell (2006) showed that S. *feltiae* and *S. carpocapsae* perform better than *H. bacteriophora* against *Synanthedon pictipes*. Application of *S. carpocapsae* in orchards has significantly reduced *Synanthedon exitiosa* infestations compared with *S. riobrave* in field and laboratory experiments; *S. carpocapsae* caused 88% reduction of *Synanthedon exitiosa* when applied in spring to an existing infestation (Cottrell and Shapiro-Ilan, 2006), and when nematodes were applied prophylactically in the fall, during the oviposition period of the target, infestations were reduced by 77–100% (Shapiro-Ilan *et al.*, 2009). In these studies, applications were made by pouring IJs on to the ground at the base of the tree, covering the application site with a thin layer of soil and watering after application. This method can protect IJs from ultraviolet (UV) radiation and desiccation but requires individual watering for 2 weeks post-application. This manual method may work for small or backyard producers but is probably not feasible for larger commercial growers.

The use of syringes to inject *S. feltiae* directly into borer galleries has proven to be efficient in eliminating carpenter worms, *Holcocerus insularis*, in urban trees (Qin *et al.*, 1988). *S. feltiae* and *S. carpocapsae* injected directly into the galleries of the clearwing borers *Synanthedon culiciformis* and *Paranthrene robiniae* in alder and birch caused 86% and 93% larval mortality, respectively (Kaya and Brown, 1986). In these investigations, EPNs applied directly to the bark of trees produced less consistent and effective control.

Most methods of using EPNs to control wood-boring Lepidoptera have a number of disadvantages. Because of the sensitivity of EPN to UV light and desiccation, they have limited time for host location. Thus, successful control of clearwing and carpenter moth larvae requires high-volume sprays that deliver the IJs close to the feeding galleries. Direct injection of EPN solutions into larval feeding galleries with a syringe or fine-spray stream are effective, but this is more labor intensive.

To improve the efficacy of aboveground applications, IJs have been applied in combination with various protective materials to preserve moisture and shield them from the sun. One of the first such efforts was to combine *S. carpocapsae* IJs with a sprayable fire-retardant gel (Barricade®) to control *Synanthedon pictipes* (Shapiro-Ilan *et al.*, 2010a). Applying *S. carpocapsae* combined with Barricade® around the tree base for preventive control of *Synanthedon exitiosa* has also proven effective (Shapiro-Ilan *et al.*, 2015). The Barricade® plus nematode treatments provided equal suppression when compared with nematode treatments with irrigation and pesticide application, while nematode application without irrigation failed. Furthermore, because only the tree's base must be treated, the cost of EPNs for *Synanthedon exitiosa* treatment (curative or preventive) may be acceptable. The findings of Shapiro-Ilan *et al.* (2016) and Wong *et al.* (2022) have corroborated and expanded on previous investigations showing that *S. carpocapsae* provided significant curative control of *Synanthedon exitiosa*.

23.4 Foliage-feeding Insects

23.4.1 Sawflies

EPN application to foliage is less effective compared with more cryptic habitats (Arthurs *et al.*, 2014). Anti-desiccants have been found to improve efficacy against specific pest species, but desiccation of IJs limits the effectiveness of application of EPNs to leaf surfaces (Begley, 1990; Lacey and Georgis, 2012). There are few studies on the susceptibility of sawflies to EPNs. Application of *S. feltiae* to foliage or pupation sites provided moderate control of the larch sawfly (*Cephalaria lariciphila* (Wacht)) larvae with a range of 2–61%. Soil application of 200 nematodes cm^{-2} caused 61% mortality of sawfly prepupae and 17.3% of pupae, but foliar sprays of 5000–20,000 IJs per 100 cm^2 of branch provided only 3.4–29% control (Georgis and Hague, 1988). Later instars were more susceptible to nematode infection. This study clearly indicated that EPN soil applications have the potential to control sawfly prepupae.

Tomalak (2006) reported that late instars of the black plum sawfly, *Hoplocampa minuta*, were susceptible to *H. megidis* and *S. feltiae*, but larvae inside the pupal cocoon were not infected once the cocoon was constructed. This was confirmed by Ulu *et al.* (2015). In several studies, EPNs caused high mortality to apple and pear sawflies (Vincent and Bélair, 1992; Zijp and Blommers, 1993; Curto *et al.*, 2007). *H. bacteriophora*, *S. feltiae* and *S. carpocapsae* caused 100% mortality of apple sawfly (*Hoplocampa testudinea*) larvae after 72 h of exposure (Vincent and Bélair, 1992). *S. carpocapsae* treatments at 40 and 80 IJs cm^{-2} in the next year resulted in 72% and 68% decreases in captured adults, respectively. Nježić and Ehlers (2020) demonstrated that in foliar treatment plum sawfly (*Hoplocampa minuta* and *H. flava*) larvae were not infected by *S. feltiae*, *S. carpocapsae* or *H. bacteriophora*. However, EPN treatment of sawfly larvae in the field prior to their migrating into soil for overwintering caused a significant reduction of emerging adults ranging from 62% to 92%. Using the EPNs against adults just before they emerged provided 100 % reduction in fruit infection, but results varied with weather conditions.

23.4.2 Lepidoptera

Several studies suggest that leaf-eating Lepidoptera can be susceptible to EPNs at high humidity; levels of control vary by EPN species/strain, application concentration and host developmental stage. The winter moths *Operophtera brumata* and *Operophtera fagata* were successfully controlled (95%) by *H. megidis*, whereas S. *feltiae* provided only 27% control, but only after dropping to the soil to pupate (Tomalak, 2004). *S. carpocapsae* and *S. feltiae* as a foliar application to containerized evergreen provided 41% and 100% control of bagworm, *Thyridopteryx ephemeraeformis* (Gill and Raupp, 1994). The lowest mortality (41–58%) was found when EPNs were applied to last instars, whereas earlier instars had higher mortalities (70–94%). Yamanaka *et al.* (1986) applied *S. feltiae* and obtained 95–100% mortality at maximum rates against the fall webworm, *Hyphantria cunea*, in landscape conditions; however, Kaya and Reardon (1982) found that a foliar application to spruce caused lower mortality ranging from 0% to 52%.

23.4.3 Coleoptera

Sharghi *et al.* (2019) used *H. bacteriophora* against the elm leaf beetle, *Xanthogaleruca luteola*, and suggested it as a potential candidate for further studies as a biological control agent of important pests in urban green spaces. Fard *et al.* (2020) evaluated the susceptibility of third-instar larvae and adults of *X. luteola* to *S. feltiae*, *S. carpocapsae* and *H. bacteriophora* in laboratory assays. *S. feltiae* was the most effective species in controlling larvae and adults of *X. luteola*. Tomalak (2004) tested the biocontrol potential of *H. megidis* and *S. feltiae* on the alder leaf beetle, *Agelastica alni*, under laboratory and semi-field conditions and reported that *H. megidis* caused significant mortality against last instars pupating in the soil. Bayramoglu *et al.* (2018) tested *H. bacteriophora* (ZET35 strain), *S. feltiae* (ZET31 strain) and *S. websteri* (AS-1 strain) against prepupae and adults of *A. alni* under laboratory conditions and ranked *S. websteri* as the most effective on both prepupae and adults of *A. alni* at a concentration of 1000 IJs ml^{-1}.

23.4.4 Thrips and leaf miners

Several species of thrips are a significant problem to nursery production, with western flower thrips, *Frankliniella occidentalis*, being the most widespread. Although some EPN species show high virulence against thrips (Chyzik *et al.*, 1996; Ebssa *et al.*, 2001; Campos-Herrera *et al.*, 2021), most studies suggest that applying any EPN species against thrips is most effective when targeting the soil stages (pupa and prepupa) (Arthurs and Heinz, 2006). Promising results have been obtained when EPNs are used as one of several different tools for thrips control. Arthurs and Heinz (2006) suggested combining EPNs with the thrips-specialist nematode *Thripinema nicklewoodi* for more complete control, as these nematodes are associated with different life stages. The drawback is that *T. nicklewoodi* is not available commercially. Ebssa *et al.* (2006) suggested that application of EPNs combined with releases of predatory mites to attack thrips as larvae is promising for thrips biological control.

EPNs are naturally adapted to subterranean environments, but they also can affect aboveground pests such as the tomato leaf miner, *Tuta absoluta* (Meyrick) (Lepidoptera: Gelechiidae), and *F. occidentalis* (Pergande) (Thysanoptera: Thripidae) (Ebssa *et al.*, 2001; Laznik and Trdan, 2008; Batalla-Carrera *et al.*, 2010). One of the most studied leaf miners for which EPN control measures were developed is *T. absoluta*, an important pest of tomato (Batalla-Carrera *et al.*, 2010; Biondi *et al.*, 2018; Taha *et al.*, 2021). Adults of *T. absoluta* emerge from pupae and, after mating, oviposition occurs on foliage. After hatching, the larvae enter the stems, leaves or fruits to feed and develop. While the larva is feeding, it produces mines and galleries through all areas invaded, which affects the crop's yield (Biondi *et al.*, 2018). Under laboratory conditions, *T. absoluta* larvae and pupae have been shown to be susceptible to EPN species such as *S. carpocapsae*, *S. feltiae* and *H. bacteriophora* but also to subtropical EPN species such as *S. yirgalemense* and *S. jeffreyense* (Batalla-Carrera *et al.*, 2010; Bonginkhosi *et al.*, 2020). Nematodes can go through the galleries and reach the larvae; this strategy can result in a high level of larval parasitism (77–91.7%) in leaf galleries (Batalla-Carrera *et al.*, 2010).

The efficacy of EPNs as a potential biological control agent of aboveground pests depends on several factors such as relative humidity, temperature, the time required by EPNs to enter a leaf and the number of applications. Taha *et al.* (2021) showed that relative humidity is a crucial factor because it had a negative impact on the efficacy and survival of *S. feltiae*, *S. carpocapsae* and *H. bacteriophora* when they were used to control the first- and second-instar larvae of *T. absoluta*. Optimal conditions to maximize EPN efficacy can be achieved by increasing relative humidity by adding adjuvants or protective gel (Barricade®) to nematode suspensions (Shapiro-Ilan *et al.*, 2010a; Dito *et al.*, 2016). The potential use of EPNs as biological control agents of aboveground pests has been shown; therefore, they should be tested in large-scale field trials to demonstrate their use within integrated pest-management programs.

23.5 Innovations Improving Success of Entomopathogenic Nematode Applications

Nursery settings allow more manipulations of environmental conditions in ways that favor EPN efficacy than most application sites. Thus, approaches employing conservation biocontrol strategies can lead to expanded use of EPNs and other microbial control agents. Taking advantage of the malleable nature of the nursery habitat should underpin other methods that can be employed. Temperature and moisture levels, both essential to EPN success, can be optimized relatively easily in these locations. In addition, the impact of UV light can be mitigated using shades or by coating the greenhouse with materials that limit light penetration. Finally, when growers use soil-less potting mixes, these can be customized to exclude materials that are harmful to EPNs, such as redwood chips (Nielsen and Lewis, 2011).

One such strategy is to combine EPNs with other materials. Compatibility and synergy are treated in detail in Chapter 27 (this volume). Several studies suggest that combining entomopathogenic fungi and EPNs can interact in a synergistic (Ansari *et al.*, 2008; Correa-Cuadros *et al.*, 2016; Wakil *et al.*, 2017) or additive (Barbercheck and Kaya, 1991; Shapiro-Ilan *et al.*, 2004; Acevedo *et al.*, 2007; Ansari *et al.*, 2008; Wu *et al.*, 2014) manner. True synergies can significantly reduce the cost of pest control (Ansari *et al.*, 2008, 2010; Hussein *et al.*, 2016). The underlying mechanisms of these synergies are unclear; however, one agent may stress or alter the feeding or movement of the target insect, making it more susceptible to the other agent.

Another strategy is specialized formulations or application methods. Again, these are made practical by the contained nature of greenhouses and nurseries. Applying EPNs as infected cadavers is most likely to be possible and effective in these controlled environments. Using materials such as Barricade® fire retardant is another tactic that may be most effective in controlled environments. Moving forward, taking advantage of controlled environmental conditions and the ability to make changes in temperature and humidity and in the substrate used for planting may take a more central role in using EPNs in nursery and greenhouse conditions.

23.6 References

Acevedo, J.P.M., Samuels, R.I., Machado, I.R. and Dolinski, C. (2007) Interactions between isolates of the entomopathogenic fungus *Metarhizium anisopliae* and the entomopathogenic nematode *Heterorhabditis bacteriophora* JPM4 during infection of the sugar cane borer *Diatraea saccharalis* (Lepidoptera: Pyralidae). *Journal of Invertebrate Pathology* 96, 187–192.

Adair, R.C. (1994) A four-year field trial of entomopathogenic nematodes for control of *Diaprepes abbreviatus* in a Flatwoods citrus grove. *Proceedings of the Florida State Horticultural Society* 107, 63–68.

Alford, D.V. (2012) *Pests of Ornamental Trees, Shrubs and Flowers: A Colour Handbook*, 2nd edn. CRC Press. Boca Raton, Florida.

Ansari, M.A., and Butt, T.M. (2011) Effects of successive subculturing on stability, virulence, conidial yield, germination and shelf-life of entomopathogenic fungi. *Journal of Applied Microbiology* 110, 460–1469.

Ansari, M.A., Shah, F.A. and Butt, T.M. (2008) Combined use of entomopathogenic nematodes and *Metarhizium anisopliae* as a new approach for black vine weevil, *Otiorhynchus sulcatus*, control. *Entomologica Experimentalis et Applicata* 129, 340–347.

Ansari, M.A., Shah, F.A. and Butt, T.M. (2010) The entomopathogenic nematode *Steinernema kraussei* and *Metarhizium anisopliae* work synergistically in controlling overwintering larvae of the black vine weevil, *Otiorhynchus sulcatus*, in strawberries growbags. *Biocontrol Science and Technology* 20, 99–105.

Arthurs, S. and Heinz, K.M. (2006) Evaluation of the nematodes *Steinernema feltiae* and *Thripinema nicklewoodi* as biological control agents of western flower thrips *Frankliniella occidentalis* infesting chrysanthemum. *Biocontrol Science and Technology* 16, 141–155.

Arthurs, S., Heinz, K.M. and Prasifka, J.R. (2004) An analysis of using entomopathogenic nematodes against above-ground pests. *Bulletin of Entomological Research* 94, 297–306.

Bai, G., Xu, H., Fu, Y.Q., Wang, X.Y., Shen, G.S. *et al.* (2016) A comparison of novel entomopathogenic nematodes application methods for control of the chive gnat, *Bradysia odoriphaga* (Diptera: Sciaridae). *Journal of Economic Entomology* 109, 2006–2013.

Barbercheck, M. and Kaya, H.K. (1991) Competitive interactions between entomopathogenic nematodes and *Beauveria bassiana* (Deuteromycotina: Hyphomycetes) in soilborne larvae of *Spodoptera exigua* (Lepidoptera: Noctuidae). *Environmental Entomology* 20, 707–712.

Batalla-Carrera, L., Morton, A. and Garcıa-del-Pino, F. (2010) Efficacy of entomopathogenic nematodes against the tomato leafminer *Tuta absoluta* in laboratory and greenhouse conditions. *BioControl* 55, 523–530.

Bayramoglu, Z., Demir, I., Inan, C. and Demirbag, Z. (2018) Efficacy of native entomopathogenic nematodes from Turkey against the alder leaf beetle, *Agelastica alni* L. (Coleoptera: Chrysomelidae), under laboratory conditions. *Egyptian Journal of Biological Pest Control* 28: 17.

Begley, J.W. (1990) Efficacy against insects in habitats other than soil. In: Gaugler, R. and Kaya, H.K. (eds.) *Entomopathogenic Nematodes in Biological Control.* CRC Press, Boca Raton, Florida, pp. 215–220.

Biondi, A., Guedes, R.N.C., Wan, F.-H. and Desneux, N. (2018) Ecology, worldwide spread, and management of the invasive South American tomato pinworm, *Tuta absoluta*: past, present, and future. *Annual Review Entomology* 63, 239–258.

Bonginkhosi, E., Dlamini, N.D., Masarirambi, M.T. and Kwanele, N.A. (2020) Control of the tomato leaf miner, *Tuta absoluta* (Meyrick) (Lepidoptera: Gelechiidae) larvae in laboratory using entomopathogenic nematodes from subtropical environment. *Journal of Nematology* 52, 1–8.

Bruck, D.J., Shapiro-Ilan, D.I. and Lewis, E.E. (2005) Evaluation of application technologies of entomopathogenic nematodes for control of the black vine weevil. *Journal of Economic Entomology* 98, 1884–1889.

Bullock, R.C., Pelosi, R.R. and Killer, E.E. (1999) Management of citrus root weevils (Coleoptera: Curculionidae) on Florida citrus with soil-applied entomopathogenic nematodes (Nematoda: Rhabditida). *Florida Entomologist* 82, 1–7.

Cabanillas, H.E., Poinar, G.O. and Raulston, J.R. (1994) *Steinernema riobravis* sp. nov. (Rhabditida: Steinernematidae) from Texas. *Fundamental and Applied Nematology* 17, 123–131.

Campos-Herrera, R., Vicente-Díez, I., Galeano, M., Chelkha, M., Del Mar González-Trujillo, M. *et al.* (2021) Intraspecific virulence of entomopathogenic nematodes against the pests *Frankliniella occidentalis*

(Thysanoptera: Thripidae) and *Tuta absoluta* (Lepidoptera: Gelechiidae). *Journal of Nematology* 53: e2021-102.

Cappaert, D.L. and Koppenhöfer, A.M. (2003) *Steinernema scarabaei*, an entomopathogenic nematode for control of the European chafer. *Biological Control* 28, 379–386.

Carman, H.F. (2011) *Economic Aspects of the California Nursery and Floral Industry, 2001–2009*. Giannini Foundation Information Series Report 11-1. Division of Agriculture and Natural Resources, University of California, Davis, California.

Chen, C., Ma, H., Li, J., Zheng, S., Song, Q. *et al.* (2020) An innovative strategy for control of fungus gnats using entomopathogenic nematodes alone or in combination with waterlogging. *Journal of Nematology* 52: e2020-57.

Chyzik, R., Glazer, I. and Klein, M. (1996) Virulence and efficacy of different entomopathogenic nematode species against western flower thrips (*Frankliniella occidentalis*). *Phytoparasitica* 24, 103–110.

Cloyd, R.A. (2008) Management of fungus gnats (*Bradysia spp.*) in greenhouses and nurseries. *Floriculture Ornamental Biotechnology* 2, 84–89.

Converse, V. and Grewal, P.S. (1998) Virulence of entomopathogenic nematodes to the western masked chafer *Cyclocephala hirta* (Coleoptera: Scarabaeidae). *Journal of Economic Entomology* 91, 428–432.

Correa-Cuadros, J.P., Sáenz-Aponte, A. and Rodríguez-Bocanegra, M.X. (2016) *In vitro* interaction of *Metarhizium anisopliae* Ma9236 and *Beauveria bassiana* Bb9205 with *Heterorhabditis bacteriophora* HNI0100 for the control of *Plutella xylostella*. *SpringerPlus* 5, 1–8.

Cottrell, T.E. and Shapiro-Ilan, D.I. (2006) Susceptibility of the peachtree borer, *Synanthedon exitiosa*, to *Steinernema carpocapsae* and *Steinernema riobrave* in laboratory and field trials. *Journal of Invertebrate Pathology* 92, 85–88.

Cowles, R.S. (1997) Several methods reduce insecticide use in control of black vine weevils. *Frontiers of Plant Science* 49, 2–4.

Cruz-Martínez, H., Ruiz-Vega, J., Matadamas-Ortíz, P.T., Cortés-Martínez, C.I. and Rosas-Diaz, J. (2017) Formulation of entomopathogenic nematodes for crop pest control – a review. *Plant Protection Science* 53, 15–24.

Curto, G., Boselli, M., Vergnani, S. and Reggiani, A. (2007) Effectiveness of entomopathogenic nematodes in the control of sawflies (*Hoplocampa brevis*) in pear orchards. *IOBC/WPRS Bulletin* 30, 13–17.

Daughtrey, M.L. and Benson, D.M. (2005) Principles of plant health management for ornamental plants. *Annual Review of Phytopathology* 43, 141–169.

Deseö, K.V. and Miller, L.A. (1985). Efficacy of entomogenous nematodes *Steinernema* spp., against clearwing moths, *Synanthedon* spp., in north Italian apple orchards. *Nematologica* 31, 100–108.

Dito, D.F., Shapiro-Ilan, D.I., Dunlap, C.A., Behle, R.W. and Lewis, E.E. (2016) Enhanced biological control potential of the entomopathogenic nematode, *Steinernema carpocapsae*, applied with a protective gel formulation. *Biocontrol Science and Technology* 26, 835–848.

Dolinski, C., Choo, H.Y. and Duncan, L.W. (2012) Grower acceptance of entomopathogenic nematodes: case studies on three continents. *Journal of Nematology* 44, 226–235.

Dolinski, C., Shapiro-Ilan, D. and Lewis, E.E. (2015) Insect cadaver applications: pros and cons. In: Campos-Herrera, R. (ed.) *Nematode Pathogenesis of Insects and Other Pests: Ecology and Applied Technologies for Sustainable Plant and Crop Protection*. Springer, Cham, Switzerland, pp. 207–229.

Dreistadt, S.H., Clark, J.K. and Flint, M.L. (2004) *Pests of Landscape Trees and Shrubs: An Integrated Pest Management Guide.* Publication 3359. University of California Division of Agriculture and Natural Resources, Oakland, California.

Duncan, L.W. and McCoy, C.W. (1996) Vertical distribution in soil, persistence, and efficacy against citrus root weevil (Coleoptera: Curculionidae) of two species of entomogenous nematodes (Rhabditida: Steinernematidae; Heterorhabditidae). *Environmental Entomology* 25, 174–178.

Duncan, L.W., Shapiro, D.I., McCoy, C.W. and Graham, J.H. (1999) Entomopathogenic nematodes as a component of citrus root weevil IPM. In: Polavarapu, S. (ed.) *Proceedings of the Workshop on the Optimal Use of Insecticidal Nematodes in Pest Management*. Rutgers University, New Brunswick, New Jersey, pp. 28–30.

Duncan, L.W., Dunn, D.C., Bague, G. and Nguyen, K. (2003) Competition between entomopathogenic and free-living bactivorous nematodes in larvae of the weevil *Diaprepes abbreviatus*. *Journal of Nematology* 35, 187–193.

Duncan, L.W., Graham, J.H., Zellers, J., Bright, D., Dunn, D.C. *et al.* (2007) Food web responses to augmenting the entomopathogenic nematodes in bare and animal manure-mulched soil. *Journal of Nematology* 39, 176–189.

Ebssa, L., Borgemeister, C., Berndt, O. and Poehling, H.M. (2001) Efficacy of entomopathogenic nematodes against soil-dwelling life stages of western flower thrips, *Frankliniella occidentalis* (Thysanoptera: Thripidae). *Journal of Invertebrate Pathology* 78, 119–127.

Ebssa, L., Borgemeister, C., Berndt, O. and Poehling, H.M. (2006) Simultaneous application of entomopathogenic nematodes and predatory mites to control western flower thrips *Frankliniella occidentalis*. *Biological Control* 39, 66–74.

El-Borai, F.E., Zellers, J.D. and Duncan, L.W. (2007) Suppression of *Diaprepes abbreviatus* in potted citrus by combinations of entomopathogenic nematodes with different life spans. *Nematropica* 37, 33–42.

Fallon, D.J., Solter, L.F., Keena, M., McManus, M., Cate, J.R. and Hanks, L.M. (2004) Susceptibility of Asian longhorned beetle, *Anoplophora glabripennis* (Motchulsky) (Coleoptera: Cerambycidae) to entomopathogenic nematodes. *Biological Control* 30, 430–438.

Fard, N.Z., Abbasipour, H., Saeedizadeh, A. and Karimi, J. (2020) Laboratory assay of entomopathogenic nematodes against the elm leaf beetle, *Xanthogaleruca luteola* Müller (Col.: Chrysomelidae). *Journal of Forest Science* 66, 524–531.

Finney, J. (1977) The DD-136 strain of *Neoaplectana* sp. as a potential biological control agent for the European elm bark beetle, *Scolytus scolytus*. *Journal of Invertebrate Pathology* 29, 7–9.

Finney, J.R. and Walker, C. (1977) The DD-136 strain of *Neoaplectana* sp. as a potential biological control agent for the European elm bark beetle, *Scolytus scolytus*. *Journal of Invertebrate Pathology* 29, 7–9.

Finney, J.R. and Walker, C. (1979) Assessment of a field trial using the DD-136 strain of *Neoaplectana* sp. for the control of *Scolytus scolytus*. *Journal of Invertebrate Pathology* 33, 239–241.

Fitters, P.F.L., Dunne, R. and Griffin, C. (2001) Vine weevil control in Ireland with entomopathogenic nematodes: optimal time of application. *Irish Journal of Agricultural and Food Research* 40, 199–213.

Futch, S.H., Duncan, L. and Zekri, M. (2005) Validation of an area-wide extension program to estimate the seasonal abundance of adult citrus root weevils with unbaited pyramidal traps. *Proceedings of the Florida State Horticultural Society* 117, 143–147.

Georgis, R. and Hague, N.G. (1988) Field evaluation of *Steinernema feltiae* against the web-spinning larch sawfly *Cephalcia lariciphila*. *Journal of Nematology* 20, 317–320.

Georgis, R. and Poinar, G.O. (1984) Greenhouse control of the black vine weevil *Otiorhynchus sulcatus* (Coleoptera: Curculionidae) by heterorhabditid and steinernematid nematodes. *Environmental Entomology* 13, 1138–1140.

Georgis, R., Koppenhöfer, A.M., Lacey, L.A., Bélair, G., Duncan, L.W. *et al.* (2006) Successes and failures in the use of parasitic nematodes for pest control. *Biological Control* 38, 103–123.

Gill, S. and Raupp, M.J. (1994) Using entomopathogenic nematodes and conventional and biorational pesticides for controlling bagworm. *Journal of Arboriculture* 20, 318–322.

Gill, S., Davidson, J.A., MacLachlan, W. and Potts, W. (1994) Controlling banded ash clearwing moth borer using entomopathogenic nematodes. *Journal of Arboriculture* 20, 146–149.

Gill, S., Lutz, J., Shrewsbury, P. and Raupp, M. (2001) Evaluation of biological and chemical control methods for black vine weevil, *Otiorhynchus sulcatus* (Fabricius) (Coleoptera: Curculionidae), in container grown perennials. *Journal of Environmental Horticulture* 19, 166–170.

Godliman, J. (1991) Control of vine weevil larvae on cyclamen with insect-parasitic nematodes. *IOBC/WPRS Bulletin* 14, 54–54.

Grewal, P.S. (2012) Entomopathogenic nematodes as tools in integrated pest management. In: Abrol, D.P. and Shankar, U. (eds.) *Integrated Pest Management: Principles and Practice*. CABI Publishing, Wallingford, UK, pp. 162–236.

Grewal, P.S., Wang, X. and Taylor, R.A.J. (2002) Dauer juvenile longevity and stress tolerance in natural populations of entomopathogenic nematodes: is there a relationship. *International Journal for Parasitology* 32, 717–725.

Grewal, P.S., Ehlers, R.-U. and Shapiro-Ilan, D.I. (eds) (2005) *Nematodes as a Biological Control Agents.* CAB International, Wallingford, UK.

Gulzar, S., Usman, M., Wakil, W., Gulcu, B., Hazir, C., Karagoz, M. *et al.* (2020). Environmental tolerance of entomopathogenic nematodes differs among nematodes arising from host cadavers versus aqueous suspension. *Journal of Invertebrate Pathology* 175: 107452.

Guy, A., Gaffney, M., Apostolos, K. and Griffin, C.T. (2017) Conditioning the entomopathogenic nematodes *Steinernema carpocapsae* and *Heterorhabditis megidis* by pre-application storage improves efficacy against black vine weevil, *Otiorhynchus sulcatus* (Coleoptera: Curculionidae) at low and moderate temperatures. *Biological Control* 108, 40–46.

Hanula, J.L. (1993) Vertical distribution of black vine weevil (Coleoptera: Curculionidae) immatures and infection by entomogenous nematodes in soil columns and field soil. *Journal of Econonic Entomology* 86, 340–347.

Haukeland, S. and Lola-Luz, T. (2010) Efficacy of the entomopathogenic nematodes *Steinernema kraussei* and *Heterorhabditis megidis* against the black vine weevil *Otiorhynchus sulcatus* in open field-grown strawberry plants. *Agricultural and Forest Entomology* 12, 363–369.

Husin, T.O.B. and Port, G.R. (2021) Efficacy of entomopathogenic nematodes against *Tuta absoluta*. *Biological Control* 160: 104699.

Hussein, H.M., Skoková Habuštová, O., Půža, V. and Zemek, R. (2016) Laboratory evaluation of *Isaria fumosorosea* CCM 8367 and *Steinernema feltiae* Ustinov against immature stages of the Colorado potato beetle. *PLOS ONE* 11: e0152399.

Jeffers, A.H., Palma, M., Klingeman, W.E., Hall, C.R., Buckley, D.S. and Kopsell, D.A. (2009) Assessments of bare-root liner quality and purchasing decisions made by green industry professionals. *HortScience* 44, 717–724.

Jeffers, A.H., Klingeman, W.E., Hall, C.R., Palma, M.A., Buckley, D.S. and Kopsell, D.A. (2010) Estimated nursery liner production costs for woody ornamental plant stock. *HortTechnology* 20, 804–811.

Jenkins, D.A., Shapiro-Ilan, D. and Goenaga, R. (2007) Virulence of entomopathogenic nematodes against *Diaprepes abbreviatus* in an oxisol. *Florida Entomologist* 90, 401–403.

Jenkins, D.A., Shapiro-Ilan, D. and Goenaga, R. (2008) Efficacy of entomopathogenic nematodes versus *Diaprepes abbreviatus* (Coleoptera: Curculionidae) larvae in a high clay-content oxisol soil: greenhouse trials with potted *Litchi chinensis*. *Florida Entomologist* 91, 75–78.

Johnson, W.T. and Lyon, H.H. (1991) *Insects that Feed on Trees and Shrubs: An Illustrated Practical Guide*, 2nd edn. Cornell University Press, Ithaca, New York.

Kakouli-Duarte, T., Labuschagne, L. and Hague, N.M. (1997) Biological control of the black vine weevil, *Otiorhynchus sulcatus* (Coleoptera: Curculionidae) with entomopathogenic nematodes (Nematoda: Rhabditida). *Annals of Applied Biology* 131, 11–27.

Kanzaki, N. and Kosaka, H. (2009) Relationship between the nematodes and bark and the ambrosia beetles. *Journal of the Japanese Forest Society* 91, 446–460.

Katumanyane, A., Ferreira, T. and Malan, A.P. (2018a) A review of *Bradysia* spp. (Diptera: Sciaridae) as pests in nursery and glasshouse crops, with special reference to biological control using entomopathogenic nematodes. *African Entomology* 26, 1–13.

Katumanyane, A., Ferreira, T. and Malan, A.P. (2018b) Greenhouse application of *Steinernema yirgalemense* to control fungus gnats, *Bradysia impatiens*. *BioControl* 63, 729–738.

Kaya, H.K. and Brown, L.R. (1986) Field application of entomogenous nematodes for biological control of clear-wing moth borers in alder and sycamore trees. *Journal of Arboriculture* 12, 150–154.

Kaya, H.K. and Reardon, R. (1982) Evaluation of *Neoaplectana carpocapsae* for biological control of the western spruce budworm, *Choristoneura occidentalis*: ineffectiveness and persistence of tank mixes. *Journal of Nematology* 14, 595–597.

Kepenekci, I., Toksöz, S., Atay, T. and Saruhan, I. (2018) Efficacy of of entomopathogenic nematode isolates from Turkey and Kyrgyzstan against black timber bark beetle, *Xylosandrus germanus* (Blandford) (Coleoptera: Curculionidae: Scolytinae) adults. *Pakistan Journal of Nematology* 36, 65–70.

Klein, M.G., Grewal, P.S., Jackson, T.A. and Koppenhöfer, A.M. (2007) Lawn, turf and grassland pests. In Lacey, L.A. and Kaya, H.K. (eds.), *Field Manual of Techniques in Invertebrate Pathology*, 2nd edn. Springer, Dordrecht, Netherlands, pp. 655–675.

Koppenhöfer, A.M. and Fuzy, E.M. (2003a) *Steinernema scarabaei* for the control of white grubs. *Biological Control* 28, 47–59.

Koppenhöfer, A.M. and Fuzy, E.M. (2003b) Ecological characterization of *Steinernema scarabaei*: a natural pathogen of scarab larvae. *Journal of Invertebrate Pathology* 83, 139–148.

Koppenhöfer, A.M. and Fuzy, E.M. (2007) Soil moisture effects on infectivity and persistence of the entomopathogenic nematodes *Steinernema scarabaei*, *S. glaseri*, *Heterorhabditis zealandica*, and *H. bacteriophora*. *Applied Soil Ecology* 35, 128–139.

Koppenhöfer, A.M. and Fuzy, E.M. (2009) Long-term effects and persistence of *Steinernema scarabaei* applied for suppression of *Anomala orientalis* (Coleoptera: Scarabaeidae). *Biological Control* 48, 63–72.

Koppenhöfer, A.M., Brown, I.M., Gaugler, R., Grewal, P.S., Kaya, H.K. and Klein, M.G. (2000) Synergism of entomopathogenic nematodes and imidacloprid against white grubs: greenhouse and field evaluation. *Biological Control* 19, 245–251.

Koppenhöfer, A.M., Fuzy, E.M., Crocker, R.L., Gelernter, W.D. and Polavarapu, S. (2004) Pathogenicity of *Heterorhabditis bacteriophora*, *Steinernema glaseri*, and *S. scarabaei* (Rhabditida: Heterorhabditidae, Steinernematidae) against 12 white grub species (Coleoptera: Scarabaeidae). *Biocontrol Science and Technology* 14, 87–92.

Koppenhöfer, A.M., Grewal, P.S. and Fuzy, E.M. (2006) Virulence of the entomopathogenic nematodes *Heterorhabditis bacteriophora*, *Heterorhabditis zealandica*, and *Steinernema scarabaei* against five white grub species (Coleoptera: Scarabaeidae) of economic importance in turfgrass in North America. *Biological Control* 38, 397–404.

Koppenhöfer, A.M., Grewal, P.S. and Fuzy, E.M. (2007) Differences in penetration routes and establishment rates of four entomopathogenic nematode species into four white grub species. *Journal of Invertebrate Pathology* 94, 184–195.

Koppenhöfer, A.M., Ebssa, L. and Fuzy, E.M. (2013) Storage temperature and duration affect *Steinernema scarabaei* dispersal and attraction, virulence, and infectivity to a white grub host. *Journal of Invertebrate Pathology* 112, 129–137.

Lacey, L.A. and Georgis, R. (2012) Entomopathogenic nematodes for control of insect pests above and below ground with comments on commercial production. *Journal of Nematology* 44, 218–225.

Lacey, L.A. and Shapiro-Ilan, D.I. (2008) Microbial control of insect pests in temperate orchard systems: potential for incorporation into IPM. *Annual Review of Entomology* 53, 121–144.

Lacey, L.A., Grywacz, D., Shapiro-Ilan, D.I., Frutos, R., Brownbridge, M. and Goettel, M.S. (2015) Insect pathogens as biological control agents: back to the future. *Journal of Invertebrate Pathology* 132, 1–41.

Laznik, Z. and Trdan, S. (2008) Entomopathogenic and entomoparasitic nematodes as biological control agents of thrips. *Acta Phytopathologica et Entomologica Hungarica* 43, 317–322.

LeBude, A.V., White, S.A., Fulcher, A.F., Frank, S., Klingeman, W.E., III *et al.* (2012) Assessing the integrated pest management practices of southeastern US ornamental nursery operations. *Pest Management Science* 68, 1278–1288.

Lee, D.W., Choo, H.Y., Kaya, H.K., Lee, S.M., Smitley, D.R. *et al.* (2002) Laboratory and field evaluation of Korean entomopathogenic nematode isolates against the oriental beetle *Exomala orientalis* (Coleoptera: Scarabaeidae). *Journal of Economic Entomology* 95, 918–926.

Liu, H.X., Chang, Q., Ma, F. and Liu, H.X. (1998) Field experiment of *Anoplophora glabripennis* and *Melanophila decastigma* control with pathogenic nematode *Steinernema carpocapsae*. *Ningxia Journal of Agriculture and Forestry Science and Technology* 5, 15–17.

Lola-Luz, T., Downes, M. and Dunne, R. (2005) Control of black vine weevil *Otiorhynchus sulcatus* (Fabricus) (Coleoptera: Curculionidae) in grow bags outdoors with nematodes. *Agricultural Forest Entomology* 7, 121–126.

Mannion, C.M., McLane, W., Klein, M.G., Moyseenko, J., Oliver, J.B. and Cowan, D. (2001) Management of early-instar Japanese beetle (Coleoptera: Scarabaeidae) in field-grown nursery crops. *Journal of Economic Entomology* 94, 1151–1161.

McCoy, C.W., Shapiro, D.I., Duncan, L.W. and Nguyen, K. (2000) Entomopathogenic nematodes and other natural enemies as mortality factors for larvae of *Diaprepes abbreviatus* (Coleoptera: Curculionidae) *Biological Control* 19, 182–190.

McCoy, C.W., Stuart, R.J., Duncan, L.W. and Nguyen, K. (2002) Field efficacy of two commercial preparations of entomopathogenic nematodes against larvae of *Diaprepes abbreviatus* (Coleoptera: Curculionidae) in alfisol type soil. *Florida Entomologist* 85, 537–544.

Miller, L.A. and Bedding, R.A. (1982) Field testing of the insect parasitic nematode, *Neoaplectana bibionis* [Nematoda: Steinernematidae] against currant borer moth, *Synanthedon tipuliformis* [Lep.: Sesiidae] in black currants. *Entomophaga* 27, 109–114.

Mráček, Z., Jiskra, K. and Kahounova, L. (1993) Efficiency of steinernematid nematodes (Nematoda: Steinernematidae) in controlling larvae of the black vine weevil, *Otiorrhynchus sulcatus* (Coleoptera: Curculionidae) in laboratory and field experiments. *European Journal of Entomology* 90, 71–76.

Nachtigall, G. and Dickler, E. (1992) Experiences with field applications of entomopathogenic nematodes for biological control of cryptic living insects in orchards. *Acta Phytopathologica et Entomologica Hungarica* 27, 485–490.

Nielsen, D.G. and Cowles, R.S. (1998) Preventing white grub infestation in container-grown nursery stock. *Journal of Environmental Horticulture* 16, 202–207.

Nielsen, A.L. and Lewis, E.E. (2011) Designing the ideal habitat for entomopathogen use in nursery production. *Pest Management Science* 68, 1053–1061.

Nježić, B. and Ehlers, R.-U. (2020) Entomopathogenic nematodes control plum sawflies (*Hoplocampa minuta* and *H. flava*). *Journal of Applied Entomology* 144, 491–499.

Parrella, M.P. and Lewis, E.E. (2017) Biological control in greenhouse and nursery production: present status and future directions. *American Entomologist* 63, 237–250.

Pérez, E.E., Lewis, E.E. and Shapiro-Ilan, D.I. (2003) Impact of host cadaver on survival and infectivity of entomopathogenic nematodes (Rhabditida: Steinernematidae and Heterorhabditidae). *Journal of Invertebrate Pathology* 82, 111–118.

Qin, X.X., Kao, R.T., Yang, H.W. and Zhang, G.Y. (1988) Study on application of entomopathogenic nematodes *Steinernema bibionis* and *S. feltiae* to control *Anoplophora glabripennis* and *Holocercus insularis*. *Forestry Research* 1, 179–185.

Rajendra, A., Hwang, H.-S., Shim, J.-K., Yu, Y.-S. and Lee, K.-Y. (2019) Control efficacy of fungus gnat, *Bradysia impatiens*, enhanced by a combination of entomopathogenic nematodes and predatory mites. *Biological Control* 138: 104071.

Sampson, A.C. (1994) Control of *Otiorhynchus sulcatus* in soft fruit using drench treatments of *Steinernema carpocapsae*. *Brighton Crop Protection Conference – Pests and Diseases* 6A, 601–608.

Shapiro, D.I. and Glazer, I. (1996) Comparison of entomopathogenic nematode dispersal from infected hosts versus aqueous suspension. *Environmental Entomology* 25, 1455–1461.

Shapiro, D.I. and Lewis, E.E. (1999) Comparison of entomopathogenic nematode infectivity from infected hosts versus aqueous suspension. *Environmental Entomology* 28, 907–911.

Shapiro, D.I. and McCoy, C.W. (2000) Susceptibility of *Diaprepes abbreviatus* (Coleoptera: Curculionidae) larvae to different rates of entomopathogenic nematodes in the greenhouse. *Florida Entomologist* 83, 1–9.

Shapiro, D.I., Cate, J.R., Pena, J., Hunsberger, A. and McCoy, C.W. (1999) Effects of temperature and host age on suppression of *Diaprepes abbreviatus* (Coleoptera: Curculionidae) by entomopathogenic nematodes. *Journal of Economic Entomology* 92, 1088–1092.

Shapiro-Ilan, D.I. and Cottrell, T.E. (2006) Susceptibility of the lesser peachtree borer (Lepidoptera: Sesiidae) to entomopathogenic nematodes under laboratory conditions. *Environmental Entomology* 35, 358–365.

Shapiro-Ilan, D.I., Gouge, D.H. and Koppenhöfer, A.M. (2002) Factors affecting commercial success: case studies in cotton, turf and citrus. In: Gaugler, R. (ed.) *Entomopathogenic Nematology*. CAB International, Wallingford, UK, pp. 333–355.

Shapiro-Ilan, D.I., Lewis, E.E., Son, Y. and Tedders, W.L. (2003) Superior efficacy observed in entomopathogenic nematodes applied in infected-host cadavers compared with application in aqueous suspension. *Journal of Invertebrate Pathology* 83, 270–272.

Shapiro-Ilan, D.I., Jackson, M., Reilly, C.C. and Hotchkiss, M.W. (2004) Effects of combining an entomopathogenic fungi or bacterium with entomopathogenic nematodes on mortality of *Curculio caryae* (Coleoptera: Curculionidae). *Biological Control* 30, 119–126.

Shapiro-Ilan, D.I., Cottrell, T.E., Mizell, R.F. III, Horton, D.L. and Davis, J. (2009) A novel approach to biological control with entomopathogenic nematodes: prophylactic control of the peachtree borer, *Synanthedon exitiosa*. *Biological Control* 48, 259–263.

Shapiro-Ilan, D.I., Cottrell, T.E., Mizell, R.F., Horton, D.L., Behle, R.W. and Dunlap, C.A. (2010a) Efficacy of *Steinernema carpocapsae* for control of the lesser peachtree borer, *Synanthedon pictipes*: improved aboveground suppression with a novel gel application. *Biological Control* 54, 23–28.

Shapiro-Ilan, D.I., Morales-Ramos, J.A., Rojas, M.G. and Tedders, W.L. (2010b) Effects of a novel entomopathogenic nematode-infected host formulation on cadaver integrity, nematode yield, and suppression of *Diaprepes abbreviatus* and *Aethina tumida*. *Journal of Invertebrate Pathology* 103, 103–108.

Shapiro-Ilan., D.I, Cottrell, T.E., Mizell, R.F., Horton, D.L. and Abdo, Z. (2015) Field suppression of the peachtree borer, *Synanthedon exitiosa*, using *Steinernema carpocapsae*: effects of irrigation, a sprayable gel and application method. *Biological Control* 82, 7–12.

Shapiro-Ilan, D.I., Cottrell, T.E., Mizell, R.F. and Horton, D.L. (2016) Curative control of the peachtree borer using entomopathogenic nematodes. *Journal of Nematology* 48, 170–176.

Sharghi, H., Eivazian, K.N. and Mohammadi, D. (2019) Susceptibility of *Polyphylla olivieri* and *Xanthogaleruca luteola* to an entomopathogenic nematode, *Heterorhabditis bacteriophora*. In: *Proceedings of 3rd Iranian International Congress of Entomology, 17–19 August, University of Tabriz, Tabriz, Iran*, p. 180.

Simard, L., Bélair, G. and Brodeur, J. (2001) Susceptibility of the European chafer (Coleoptera: Scarabaeidae) to entomopathogenic nematodes (Rhabditida: Steinernematidae, Heterorhabditidae). *Journal of Nematology* 33, 297–301.

Simpson, S.E., Nigg, H.N., Coile, N.C. and Adair, R.A. (1996) *Diaprepes abbreviatus* (Coleoptera: Curculionidae): host plant associations. *Environmental Entomology* 25, 333–349.

Stimmann, M.W., Kaya, H.K., Burlando, T.M. and Studdert, J.P. (1985) Black vine weevil management in nursery plants. *California Agriculture* 39, 25–26.

Stock, S.P. and Koppenhöfer, A. (2003). *Steinernema scarabaei* n. sp. (Rhabditida: Steinernematidae), a natural pathogen of scarab beetle larvae (Coleoptera: Scarabaeidae) from New Jersey, USA. *Nematology* 5, 191–204.

Stuart, R.J., Barbercheck, M.E. and Grewal, P.S. (2015) Entomopathogenic nematodes in the soil environment: distributions, interactions and the influence of biotic and abiotic factors. In: Campos-Herrera, R. (ed.) *Nematode Pathogenesis of Insects and Other Pests: Ecology and Applied Technologies for Sustainable Plant and Crop Protection*. Springer, Cham, Switzerland, pp. 97–137.

Susurluk, A. and Ehlers, R.-U. (2008) Field persistence of the entomopathogenic nematode *Heterorhabditis bacteriophora* in different crops. *BioControl* 53, 627–641.

Tomalak, M. (2004) Infectivity of entomopathogenic nematodes to soil-dwelling developmental stages of the tree leaf beetles, *Altica quercetorum* and *Agelastica alni*. *Entomologia Experimentalis et Applicata* 110, 125–133.

Tomalak, M. (2006) Potencjał nicieni owadobójczych w biologicznym zwalczaniu szkodliwych błonkówek atakujących drzewa owocowe i parkowe. *Progress in Plant Protection* 46, 249–255.

Ulu, T.C., Sadic, B., Susurluk, I.A. and Aksit, T. (2015) Virulence of four entomopathogenic nematode species for plum sawfly, *Hoplocampa flava L.* (Hymenoptera: Tenthredinidae). *Invertebrate Survival Journal* 12, 274–277.

van Tol, R.W.H.M. (1993a) Control of the black vine weevil (*Otiorhynchus sulcatus*) with different isolates of *Heterorhabditis* sp. and *Metarhizium anisopliae* in nursery stock. *Proceedings of the Section Experimental and Applied Entomology of the Netherlands Entomological Society, Amsterdam* 4, 181–186.

van Tol, R.W.H.M. (1993b) Efficacy of control of the black vine weevil *Otiorhynchus sulcatus* with strains of *Heterorhabditis* sp, *Steinernema* sp. and the fungus *Metarhizium anisopliae* in nursery stock. *Mededelingen van de Faculteit der Landbouwwetenschappen, Rijksuniversiteit Gent* 58, 461–467.

van Tol, R.W.H.M. (1993c) Efficacy of control of the black vine weevil (*Otiorhynchus sulcatus*) with strains of *Heterorhabditis* sp., *Steinernema* sp. and the fungus *Metarhizium anisopliae* in nursery stock. *Mededelingen Faculteit Landbouwwetenschappen Universiteit Gent* 58, 461–467.

van Tol, R.W.H.M. (1994) Influence of temperature on the control of the black vine weevil with strains of some insect-parasitic nematodes. *IOBC/WPRS Bulletin* 17, 116–119.

van Tol, R.W.H.M. (1996a) A strategy for control of the black vine weevil *Otiorhynchus sulcatus* in an integrated pest management programme in nursery stock. In: Alford, D.V. and Backhaus, G.F. (eds.) *Second International Workshop on Vine Weevil Otiorynchus sulcatus Fabr (Coleoptera: Curculionidae)*. Blackwell, Berlin, pp. 76–80.

van Tol, R.W.H.M. (1996b) Prospects for biological control of black vine weevil (*Otiorhynchus sulcatus*) in nursery stock. *Mitteilungen aus der Biologischen Bundesanstalt für Land- und Forstwirtschaft* 316, 69–75.

van Tol, R.W.H.M. (1998) Biological and chemical control of the vine weevil (*Otiorhynchus sulcatus*) in nursery stock. *BPO Report Vine Weevil* 1991–1997.

van Tol, R.W.H.M. and Raupp, M.J. (2005) Nursery and tree applications. In: Grewal, P.S., Ehlers, R.U. and Shapiro-Ilan, D.I. (eds.) *Nematodes as Biocontrol Agents*. CAB International, Wallingford, UK, pp. 167–190.

van Tol, R.W.H.M., van Bezooijen, J. and Ketelaars, T.A.C.M. (1998) Searching behaviour of entomopathogenic nematodes: roots and soil temperature determine success of black vine weevil (*Otiorhynchus sulcatus*) control. *International Organisation of Biological Control Bulletin* 21, 187–191.

Villani, M.G. and Wright, R.J. (1988) Entomogenous nematodes as biological control agents of European chafer and Japanese beetle (Coleoptera: Scarabaeidae) larvae infesting turfgrass. *Journal of Economic Entomology* 81, 484–487.

Vincent, C. and Bélair, G. (1992) Biocontrol of the apple sawfly, *Hoplocampa testudinea*, with entomogenous nematodes. *Entomophaga* 37, 575–582.

Wakil, W., Yasin, M. and Shapiro-Ilan, D. (2017). Effects of single and combined applications of entomopathogenic fungi and nematodes against *Rhynchophorus ferrugineus* (Olivier). *Scientific Reports* 7: 5971.

Williams, R.N., Fickle, D.S., Grewal, P.S. and Meyer, J.R. (2002) Assessing the potential of entomopathogenic nematodes to control the grape root borer *Vitacea polistiformis* (Lepidoptera: Sesiidae) through laboratory and greenhouse bioassays. *Biocontrol Science and Technology* 12, 35–40.

Willmott, D.M., Hart, A.J., Long, S.J., Edmondson, R.N. and Richardson, P.N. (2002) Use of a cold-active entomopathogenic nematode *Steinernema kraussei* to control overwintering larvae of the black vine weevil *Otiorhynchus sulcatus* (Coleoptera: Curculionidae) in outdoor strawberry plants. *Nematology* 4, 925–932.

Wong, C., Oliveira-Hofman, C., Blaauw, B., Chavez, D.J., Jagdale, G. *et al.* (2022) Control of peachtree borer (*Synanthedon exitiosa*) using the nematode *Steinernema carpocapsae*: optimization of application rates and secondary benefits in control of root-feeding weevils. *Agronomy* 12: 2689.

Woodruff, R.E. (1968) The present status of a West Indian weevil (*Diaprepes abbreviata* (L.)) in Florida (Coleoptera: Curculionidae). *Florida Department of Agriculture, Division of Plant Industry, Entomology Circular* 77, 1–4.

Wright, R.J., Villani, M.G. and Agudelo-Silva, F. (1988) Steinernematid and heterorhabditid nematodes for control of larval European chafers and Japanese beetles (Coleoptera: Scarabaeidae) in potted yew. *Journal of Economic Entomology* 81, 152–157.

Wu, S. (2013) Efficacy of entomopathogenic nematodes and entomopathogenic fungi against masked chafer white grubs, *Cyclocephala* spp. (Coleoptera: Scarabaeidae). PhD thesis, Virginia Polytechnic Institute and State University, Blacksburg, Virginia.

Wu, S., Youngman, R.R., Kok, L.T., Laub, C.A. and Pfeiffer, D.G. (2014) Interaction between entomopathogenic nematodes and entomopathogenic fungi applied to third instar southern masked chafer white grubs, *Cyclocephala lurida* (Coleoptera: Scarabaeidae), under laboratory and greenhouse conditions. *Biological Control* 76, 65–73.

Yamanaka, K., Seta, K. and Yasuda, M. (1986) Evaluation of the use of entomogenous nematode, *Steinernema feltiae* (Str. Mexican) for the biological control of the fall webworm, *Hyphantria cunea* (Lepidoptera: Arctiidae). *Japanese Journal of Nematology* 16, 26–32.

Yeager, T., Million, J., Larsen, C. and Stamps, B. (2010) Florida nursery best management practices: past, present and future. *HortTechnology* 20, 82–88.

Zijp, J.P. and Blommers, L. (1993) *Lathrolestes ensator*, a parasitoid of the apple sawfly. *Proceedings of the Section Experimental and Applied Entomology of the Netherlands Entomological Society* 4, 237–242.

24 Forest Applications of Entomopathogenic Nematodes

Christopher D. Williams*
School of Biological and Environmental Sciences, Liverpool John Moores University, Liverpool, UK

Abstract
The application of entomopathogenic nematodes (EPNs) for the control of insect pests of trees in the broadest sense is reviewed. First considered are pests of trees outside of woodlands and forests, typically in urban environments. Next, pests of deciduous forests and their control are dealt with before moving on to pests of coniferous forests. It should be noted that of the 99 serious pests of forestry identified by Day and Leather (1997), only 16 (16.2%) have had any research on applications of EPNs for their control. These examples and others are presented in this chapter, as well as more broadly addressing the challenges faced in controlling these pests. Finally, a case study of the control of the most serious pest of forestry in Europe, *Hylobius abietis*, is presented, which takes account of its life cycle, current control methods, factors affecting the efficacy of EPNs, environmental safety of EPNs and interaction with other biological control agents.

24.1 Introduction

There has been much basic and applied research on entomopathogenic nematodes (EPNs) since the publication of the first major edited volume on all aspects of their biology (Gaugler and Kaya, 1990). Forest applications are no exception to this trend, and it is the aim of this chapter to consider forestry in its widest sense and to show how continuous research on the application of EPNs to the control of pests of trees and forests is making advances toward more sustainable forest-management systems throughout the world.

In this chapter, I will first consider EPN applications to tree pests outside of woodlands and forests. I will then consider pests in deciduous forests and pests of coniferous forests. Finally, I will consider, as a case study, the application of EPNs in the control of the large pine weevil, *Hylobius abietis*, the most serious insect pest of clear-fell forest restocking in Europe. Throughout the chapter, I will highlight not only where there has been successful application of EPNs to forest pests but also where there is potential to apply EPNs to pests that have not currently been explored in any detail. There will inevitably be some crossover between other chapters in the present volume, most notably with Chapters 16 and 17 on orchard applications of EPNs, Chapter 6 on behavioral ecology of EPNs and Chapter 7 on population dynamics. It is hoped, however, that such crossover will enhance the integration of the chapter in the volume as a whole and not be an unwanted distraction to the reader.

*Email: C.D.Williams@ljmu.ac.uk

 Entomopathogenic Nematodes as Biological Control Agents (D.I. Shapiro-Ilan and E.E. Lewis eds)

DOI: 10.1079/9781800620322.0024

Biological control can take one of four forms. The first, and least intrusive, is so-called "conservation biocontrol." In this method, habitats and parts of habitats are modified to promote natural enemies of the pest that are already resident in the ecosystem. The populations of such natural ecosystem service providers may be boosted by appropriate management. One of the most successful examples of this method has been the establishment of "beetle banks" in tillage field systems. The fields are split with strips of perennial grasses, which are sown to provide overwintering habitats for natural insect enemies of agricultural pests. Another type of biocontrol is "augmentative biological control." In this system, native populations of natural enemies are supplemented by release of additional individuals of the same species. Release of predatory mites and parasitoids under glasshouse conditions have been a successful application of this method. Classical biological control involves the inoculative release of non-native natural enemies of pest species into a geographical area where the agent would not naturally occur. Although there have been examples of major environmental impacts (e.g. the displacement and local extinction of native ladybirds following the escape of *Harmonia axyridis* from glasshouses), when adequate screening and impacts are assessed, the method can be very effective; for example, the control of a non-native psyllid pest of eucalyptus plantations in south-west Ireland through the release of an exotic parasitoid has been both effective and environmentally benign. The release of the predator *Rhizophagus grandis* of the bark beetle pest *Dendroctonus micans* has been a major success in the UK. The final type of biological control, and the method dealt with in the research described in the present chapter, is that of "inundative biological control" or "biopesticides." For this method, the aim is not to establish a population of control agents in the ecosystem but to rather apply vast numbers of agents that will knock down the pest at the critical time and then no longer persist in the environment. EPNs and entomopathogenic fungi are prime examples of organisms used in inundative biological control. With that said, EPNs have been used as classical (inoculative) biological control agents, for example in the control of mole crickets (Gryllotalpidae) by *Steinernema scapterisci* (Parkman *et al.*, 1996).

It is instructive to consider EPN applications to the 99 major pests of forestry in Europe identified by Day and Leather (1997) for two reasons. First, Europe has one of the best-known insect faunas in the world, and knowledge of the pest status of insects there is probably higher than in any other part of the world. Second, Europe is a major producer of EPN products and their likelihood of uptake is high. Nevertheless, of the 99 serious pests of forestry identified by Day and Leather (1997), only 16 (16.2%) have had any research on applications of EPNs for their control, and in some cases, this has been restricted to identification of natural nematode infections (Table 24.1). We may, therefore, tentatively conclude that there remains a huge opportunity to apply EPNs to these and other serious forest pests.

24.2 Entomopathogenic Nematode Applications to Tree Pests Outside of Woodlands and Forests

Trees do not just occur in woodlands and forests. A large proportion of trees are found in urban environments such as parks and gardens and as street trees. These trees have great amenity value and provide many ecosystem services for people who live and share their urban environment with them (Livesley *et al.*, 2016). These urban trees are just as susceptible to pests and pathogens as trees that are found in woodlands and forestry. It has been suggested that urban trees could act as sentinels for emerging pests and pathogens in a country (H. Evans, personal communication). This, combined with an active citizen scientist program of surveillance (Pocock *et al.*, 2020), could be important in early reporting of exotic, non-native, serious pests and pathogens of woody plants in a country. As an example of the increasing pests and pathogens on a national scale, O'Hanlon *et al.* (2021) document the establishment of non-native pests of trees on the island of Ireland showing that, even in a country with a modest forestry estate, protected zone status for many pests and a relatively isolated biogeographical area, invasive species are a serious problem, which is only getting worse.

There has been some uptake of EPN applications for urban tree protection, particularly in China. Wang and Li (2017) reviewed the status

Table 24.1. Major pests of forestry in Europe. Modified from Day and Leather (1997).

Major forest pests in Europe	Common name	Use of entomopathogenic nematodes
Hemiptera		
Adelges abietis (L.)	Pineapple gall woolly aphid	No
Adelges laricis Vallot	Larch woolly aphid	No
Aradus cinnamomeus Panzer	Pine bark bug	No
Cinara piceae (Panzer)	Great black spruce bark aphid	No
Cinara pilicornis (Hartig)	Brown spruce shoot aphid	No
Cinara pinea (Mordvilko)	Large pine aphid	No
Cryptococcus fagisuga Lindinger	Felted beech coccus	No
Drepanosiphum platanoidis (Schrank)	Sycamore aphid	No
Dreyfusia piceae (Ratzeburg)	Balsam woolly aphid	No
Dreyfusia nordmannianae (Eckstein)	Silver fir woolly aphid	No
Gilletteela cooleyi (Gillette)	Douglas fir woolly aphid	No
Elatobium abietinum (Walker)	Green spruce aphid	No
Eucallipterus tiliae (L.)	Lime aphid	No
Eulachnus agilis (Kaltenbach)	Spotted pine aphid	No
Kermes quercus (L.)	Oak scale or pox	No
Matsucoccus josephi (Bodenheimer & Harpaz)	Pine bast scale	No
Mindarus abietinus (Koch)	Balsam twig aphid	No
Myzus cerasi (F.)	Cherry blackfly	*Steinernema feltiae*, *S. carpocapsae* and *Heterorhabditis bacteriophora*
Pachypappella vesicalis (Koch)	Spruce root aphid	No
Pachypappella lactea (Tullgren)	Spruce root aphid	No
Phyllaphis fagi (L.)	Beech woolly aphid	No
Pineus pini (Macquart)	Pine woolly aphid	No
Schizolachnus pineti (F.)	Gray pine-needle aphid	No
Stagona pini (Burmeister)	Pine root aphid	No
Lepidoptera		
Bupalus piniaria (L.)	Pine looper moth	No
Choristoneura murinana (Hübner)	Fir bud moth	No
Coleophora laricella (Hübner)	Larch case bearer	No
Cydia conicolana (Heylaerts)	Pine cone moth	No
Cydia fagiglandana (Zeller)	Beech seed moth	*S. feltiae*, *S. ichnusae*, *S. carpocapsae*, *S. vulcanicum*, *S. kraussei*, *Heterorhabditis bacteriophora* and *H. megidis*
Cydia strobilella (L.)	Spruce cone tortrix	No
Dendrolimus pini (L.)	Pine lappet moth	*S. feltiae* and *H. megidis*
Dioryctria abietella (Denis & Schiffermüller)	Pine knot-horn moth	No
Epinotia nanana (Treitschke)	Dwarf spruce bell moth	No
Epinotia tedella (Clerck)	Spruce bell moth	No
Epirrita autumnata (Borkhausen)	Autumnal moth	No
Euproctis chrysorrhoea L.	Brown-tail moth	*S. carpocapsae* and *H. bacteriophora*

Continued

Table 24.1. Continued.

Major forest pests in Europe	Common name	Use of entomopathogenic nematodes
Lymantria dispar L.	Gypsy moth	*Steinernema* sp., *S. borjomiense*, *S. thesami*, *Parasitorhabditis* sp., *Phasmarhabditis* sp.
Lymantria monacha L.	Nun moth	No
Malacosoma neustria (L.)	Lackey moth	Bacteria isolated from septicemic *Galleria mellonella*
Operophtera brumata (L.)	Winter moth	No
Orgyia antiqua (L.)	Vapourer moth	No
Panolis flammea (Denis & Schiffermüller)	Pine beauty moth	No
Prays fraxinella Bjerkander	Ash bud moth	No
Retinia resinella (L.)	Pine resin gall moth	No
Rhyacionia buoliana (Denis & Schiffermüller)	Pine shoot moth	No
Thaumetopoea pityocampa (Denis & Schiffermüller)	Pine processionary moth	*S. carpocapsae*, *S. feltiae* and *H. bacteriophora*
Tortrix viridana (L.)	Green oak tortrix, oak leaf roller moth	No
Yponomeuta evonymella (L.)	Bird cherry ermine moth	No
Zeiraphera diniana (Guenée)	Gray larch tortrix, larch bud moth	No
Diptera		
Dasyneura laricis (Löw)	Larch bud midge	No
Phytoagromya populicolia Haliday	Poplar leaf miner	No
Resseliella picea Seitner	Fir seed gall midge	No
Thecodiplosis brachyntera (Schwägrichen)	Pine-needle shortening gall midge	No
Hymenoptera		
Andricus quercuscalicis (Burgsdorff)	Knopper gall wasp	No
Cephalcia abietis (L.)	Web-spinning fir sawfly	No
Cephalcia lariciphila (Wachtl)	Web-spinning larch sawfly	No
Diprion pini (L.)	Large pine sawfly	No
Gilpinia hercyniae (Hartig)	European spruce sawfly	No
Megastigmus pinus Parfitt	Silver fir seed wasp	No
Megastigmus spermotrophus (Wachtl)	Douglas fir seed wasp	No
Nematus melanaspis Hartig	Gregarious poplar sawfly	No
Nematus pavidus Lepeletier	Lesser willow sawfly	No
Nematus salicis (L.)	Large willow sawfly	No
Neodiprion lecontei (Fitch)	Red-headed pine sawfly	Undescribed *Neoaplectana* sp.
Neodiprion sertifer (Geoffroy)	European pine sawfly, fox-colored sawfly	No
Pristiphora abietina Christ	Gregarious spruce sawfly	No
Pristiphora erichsonii (Hartig)	Large larch sawfly	Undescribed *Neoaplectana* (= *Steinernema*) sp.
Pristiphora testacea (Jurine)	Birch sawfly	No
Pristiphora wesmaeli (Tischbein)	Large sawfly	No
Sirex noctilio F.	Steely blue wood wasp	No
Urocerus gigas (L.)	Giant wood wasp	No

Continued

Table 24.1. Continued.

Major forest pests in Europe	Common name	Use of entomopathogenic nematodes
Coleoptera		
Brachyonyx pineti (Paykull & G. de)	Pine-needle feeding weevil	No
Chrysomela populi L.	Large red poplar leaf beetle	No
Dendroctonus micans (Kugelann)	Great spruce bark beetle	*S. carpocapsae*, *S. feltiae* and *H. bacteriophora*
Galerucella lineola (F.)	Brown willow beetle	No
Hylastes ater Paykull	Black pine beetle	No
Hylastes brunneus Erichson	Black pine beetle	No
Hylastes cunicularius Erichson	Black spruce beetle	No
Hylobius abietis (L.)	Large pine weevil	*S. carpocapsae*, *S. kraussei*, *S. feltiae*, *Heterorhabditis downesi* and *H. megidis*
Hylobius pinastri (Gyllenhal)	Small pine weevil	No
Hylurgopinus rufipes (Eichoff)	Canadian elm bark beetle	No
Ips acuminatus Gyllenhal	Engraver beetle	Natural nematodes isolated
Ips cembrae Heer	Large larch bark beetle	Natural nematodes isolated
Ips typographus (L.)	Eight-toothed spruce bark beetle	Natural nematodes isolated
Otiorhynchus singularis (L.)	Clay-colored weevil	*S. kraussei*
Phyllobius pyri (L.)	Common leaf weevil	No
Phratora vitellinae (L.) (previously *Phyllodecta vitellinae*)	Brassy willow beetle	No
Phratora vulgatissima (L.) (previously *Phyllodecta vulgatissima*)	Blue willow beetle	No
Pissodes pini (L.)	Banded pine weevil	No
Pissodes validirostris Sahlberg	Pine cone weevil	No
Pityogenes bidentatus (Herbst)	Two-toothed pine beetle	No
Plagiodera versicolora (Laicharting)	Broader willow leaf beetle	No
Rhynchaenus fagi (L.)	Beech leaf miner	No
Saperda populnea (L.)	Small poplar long-horned beetle	No
Scolytus multistriatus (Marsham)	Smell elm beetle	No
Scolytus scolytus F.	Large elm beetle	No
Tomicus minor (Hartig)	Lesser pine shoot beetle	No
Tomicus piniperda (L.)	Pine shoot beetle	*H. bacteriophora*
Xyloterus lineatus (Olivier)	Conifer ambrosia beetle	No

of EPNs in integrated pest management in China. In a subsection of their chapter, they considered the operational application of EPNs in street trees. They noted that carpenter worm, *Holcocerus insularis*, a serious pest that attacks ash, *Fraxinus pennsylvanica*, the Chinese scholar tree, *Sophora japonica*, and willows (*Salix* spp.) mostly in Northern China, has been treated at an operational level by the injection of a *Steinernema feltiae* suspension into boring tunnels left by the pest. Yang and Zhang (1990) report 97% larval mortality following such treatment. Wang and Li (2017) noted that several other cities in the region used *Steinernema carpocapsae* with similar success. Citrus long-horned beetle, *Anoplophora chinensis*, and Asian long-horned beetle, *Anoplophora glabripennis*, are serious boring pests that pose a threat to European forestry; an example is the report of a local outbreak in Kent (Straw *et al.*, 2015). These pests are also economically important in China. Zhang *et al.* (2007) used strains of *Heterorhabditis bacteriophora* and *S. carpocapsae* against these pests and reported over 80% control. Such control methods could be brought into Europe and North America in eradication efforts targeting these pests as and when they emerge, although it should be noted that these potential programmes may not be approved.

Another pest that threatens street trees is the red palm weevil, *Rhynochophorus ferrugineus*, which attacks palm trees in cities throughout the Mediterranean. Although not currently treated operationally, there has been a detailed laboratory and field trial. Atwa and Hegazi (2014) compared the susceptibility of different life stages of *R. ferrugineus* to EPNs. Twelve EPN species were tested under no-choice and five-stage choice experiments. The five-stage choice experiment was used to test the hypothesis that nematode infestation was related to the life stage of *R. ferrugineus*. Each box was provided with five weevils: three larval stages comprising one young second- to third-instar larva, one medium fifth- to sixth-instar larva and one full-grown larva, a newly formed cocoon with a live pupa, and an adult (Atwa and Hegazi, 2014). These EPNs were *H. bacteriophora* (HP88 strain), *Steinernema abbasi*, *S. anomali*, *S. carpocapsae* (All strain), *S. feltiae*, *S. glaseri*, *S. riobrave*, *Steinernema* sp. (EGG4 strain), *S. ritterai* (EGBS strain), *S. egyptens* (EBNGE strain), *S. kushidai* and *Heterorhabditis* sp. (EKB20 strain). The highest mean total mortality was in those treated with *Steinernema* sp. (EGG4) and *S. egyptens*. These species along with *Heterorhabditis* sp. (EKB20) were used in curative assays on infected palms with the percentage palm recovery highest (88.1%) on those treated with *Steinernema* sp. (EGG4). There is obviously huge potential in using EPNs against this serious pest of palms.

Another pest of street trees that poses a public health hazard because of their urticating hairs is the oak processionary moth, *Thaumetopoea processionea*. This pest is currently present in the Greater London area of the UK, and current control strategies include spraying with pesticide. There is potential to use EPNs against this pest, although studies on this have not yet (to the best of the author's knowledge) been undertaken. However, given the success of EPN usage on other street trees and trees in urban situations, there is huge potential to extend the use of EPNs to other emerging pests.

24.3 Entomopathogenic Nematode Applications to Pests of Trees in Deciduous Forests

EPNs have been identified in deciduous forest soils throughout the world. In the Czech Republic, Mráček *et al.* (2005) identified 88 strains of EPN from fruit tree soils and 105 strains from "other" deciduous forest tree soils. These consisted of nine species of EPN, all Steinernematidae. In Italy, Tarasco *et al.* (2015) identified 38 strains of EPN from deciduous forest soils. These consisted of six species of EPN, again all Steinernematidae. In Nepal, Khatri-Chhetri *et al.* (2010) identified strains of EPN from a wide variety of habitats: 9.09% of their strains were identified from deciduous forests and 5.26% from orchards. Of the species identified from deciduous forests and orchards, three were as-yet-undescribed species of Steinernematidae (*Steinernema* sp. A, *Steinernema* sp. B and *Steinernema* sp. E). Strains of *Heterorhabditis indica* were also isolated, although the authors noted that this species was not a habitat specialist, a point made previously by Griffin *et al.* (2000).

One of the major pests of forestry identified by Day and Leather (1997) is the cherry blackfly, *Myzus cerasi*. In Turkey, Kepenekci *et al.* (2014) used local isolated EPNs of *S. feltiae*, *S. carpocapsae* and *H. bacteriophora* against the adult blackfly. They applied these nematodes at concentrations of 25, 50 and 100 infective juveniles (IJs) in 0.2 ml per adult insect at 15°C and 25°C. After 96 h, mortality remained below 50%, with *S. feltiae* being the most efficacious. At 25°C, mortality increased to over 50% for *H. bacteriophora* but stayed low for *S. feltiae* and lower still for *S. carpocapsae*. Further laboratory screening and field trials are recommended to test for other strains of EPN against this pest.

In applications against the beech seed moth, *Cydia fagiglandana*, Clausi *et al.* (2014) tested the pathogenicity of 17 strains of the following species: *S. feltiae*, *S. ichusae*, *S. carpocapsae*, *Steinernema vulcanicum*, *S. kraussei*, *H. bacteriophora* and *H. megidis*. The most efficacious strains were *S. feltiae* ESA and *S. feltiae* EPP.

The serious foliar pest of forests known as the brown-tail moth, *Euproctis chrysorrhoea*, was investigated by Nikdel *et al.* (2010) in laboratory bioassays using *S. carpocapsae* and *H. bacteriophora*. Third-, fourth- and fifth-instar *E. chrysorrhoea* were exposed to the EPNs at concentrations of 0, 500, 1000, 1500, 3000 and 5000 IJs ml^{-1}. *S. carpocapsae* caused significantly higher mortality of fourth and fifth instars compared with *H. bacteriophora*.

Gorgadze *et al.* (2020) reported mortality of 98.4% when an unidentified *Steinernema* sp. was used against the gypsy moth, *Lymantria dispar*. Yu *et al.* (2017) reported the use of *S. kraussei* against the clay-colored weevil, *Otiorhynchus singularis*. However, this application was in blueberry crops, rather than forests, where this weevil is also a serious pest.

Despite the wide prevalence of EPNs (especially steinernematids) in deciduous forest soils, there has been limited application of EPNs as biological control agents of boring or foliar pests of deciduous forests where the primary commercial interest is wood. There has, however, been some application of EPN control in fruit-bearing deciduous trees in orchards. Chapter 18 (this volume) deals, in greater detail than given here, with applications of EPNs to berry pests. Lakatos and Tóth (2006) investigated the use of *S. glaseri* and *Heterorhabditis downesi* to control the European cockchafer, *Melolontha melolontha*, a serious pest of the roots of fruit trees. Kuske *et al.* (2005) investigated the application of EPNs in the control of serious pests of nut trees and orchards. They applied *S. carpocapsae*, *S. feltiae*, *H. bacteriophora*, *H. indica* and *H. megidis* to the European cherry fruit fly, *Rhagoletis cerasi*, western cherry fruit fly, *Rhagoletis indifferens*, the hazelnut weevil, *Balaninus nucum*, and the chestnut pest, *Curculio elephas*. Şahin and Gözel (2021) investigated EPN efficacy against the flat-headed woodborer, *Capnodis tenebrionis*, a serious pest of peach orchards in Turkey. They found *H. bacteriophora* to be more effective than *S. feltiae*, *S. affine* and *S. carpocapsae*. They did not speculate as to whether this was due to different foraging strategies of the applied EPNs. Lacey *et al.* (2006) investigated the efficacy of *S. feltiae* and *S. carpocapsae* in controlling the codling moth, *Cydia pomonella* (Lepidoptera: Tortricidae). Despite this active program of research, Belien (2018) noted low uptake among growers and attributed this to reduced control activity and persistence due to low temperature regimes and desiccation. Orchard applications of EPNs have been reviewed by Shapiro-Ilan *et al.* (2005) and elsewhere in the present volume and so will not be considered further in this chapter.

24.4 Entomopathogenic Nematode Applications to Pests of Trees in Coniferous Forests

There is a long history of application of EPNs for the control of pests of conifer forests. In 1963, Schmiege investigated the feasibility of using an as-then-undescribed species of *Neoaplectana* (= *Steinernema*) nematode in the control of some forestry pests. Schmiege (1963) investigated, as part of this study, the susceptibility of the red-headed pine sawfly, *Neodiprion lecontei*, and the large larch sawfly, *Pristiphora erichsonii*. The study incorporated an analysis of the temperature and moisture requirements of the nematode. However, since this study, there has been no follow-up work on *N. lecontei* or *P. erichsonii*.

EPNs have been identified from coniferous forest soils from around the world. In a comprehensive study (270 soil samples) from 30 different habitats in ten geographic regions of California, Stock *et al.* (1999) recovered seven species of rhabditid nematode, which were isolated from 26.3% of their samples. In this study of the distribution of EPNs in natural habitats in California, they noted that most of the *Steinernema* spp. were recovered from coniferous and oak forests. *S. kraussei* was isolated above 700 feet from coniferous forests and a grassland in the Sierra Nevada mountains. Mráček *et al.* (1992) and Steiner (1996) also found this species in high-elevation coniferous forests, and additionally found *Steinernema longicaudum* in lodgepole pine (*Pinus contorta*) forests in the Sierra Nevada range.

In Jordan, Stock *et al.* (2008) recovered four EPN species (three *Steinernema* spp. and one *Heterorhabditis* sp.). In loamy soil of a coniferous forest plot, the authors recovered *Steinernema anatoliense* (EU 200356), associated with the symbiotic bacterium *Xenorhabdus nematophila*.

Tumialis *et al.* (2019) carried out laboratory bioassays to infect pine lappet moth, *Dendrolimus pini*, with isolates of *S. feltiae* and *H. megidis*.

In general, they found that *S. feltiae* was more efficacious than *H. megidis* in terms of percentage parasitism (86.7–100% for *S. feltiae* compared with 20–100% for *H. megidis*). There were no temperature- or concentration-dependent effects on parasitism for *S. feltiae*, while there was concentration dependence in *H. megidis*. However, it should be noted that the authors used high concentrations of IJs, which is not ideal to determine concentration effects. The authors suggested further studies including field trials to help apply EPNs to this serious pest of pine.

Another serious pest of pine is the pine processionary moth, *Thaumetopoea pityocampa*. As well as defoliating trees, the urticating hairs associated with this pest also make it a considerable public health hazard. Gözel and Gözel (2019) conducted temperature-controlled laboratory bioassays on the susceptibility of this pest to various species of EPN. The EPN species tested were *S. carpocapsae*, *S. feltiae* and *H. bacteriophora*. At 10°C, larvae treated with 200 IJs of *S. carpocapsae* and *S. feltiae* reached 80% mortality by day 6. In contrast, the two strains of *H. bacteriophora* reached a maximum mortality of 70% by day 7. At 25°C, all species and strains reached 100% mortality by day 3. These promising early results could pave the way for field trials of these nematodes against this serious pest.

The great spruce bark beetle, *Dendroctonus micans*, is a serious boring pest of spruce trees (*Picea* spp.). Current control methods for this pest rely on sanitation felling, restricted transport of infected trees and inoculative classical biological control using the predatory beetle *Rhizophagus grandis* (Evans and Fielding, 1994). Despite this rather successful approach in the UK, in Turkey there has been some investigation of the possible application of EPNs for the control of this pest. Kepenekci and Atay (2014) tested, under temperature-controlled laboratory conditions, the efficacy of strains of *S. carpocapsae*, *S. feltiae* and *H. bacteriophora* against larval *D. micans* isolated from Turkey. At a concentration of 1000 IJs per cup and 25°C, they reported 98.04% and 94.04% mortality for *S. feltiae* and *H. bacteriophora*, respectively. *S. carpocapsae* did not attain mortality higher than 40% under any combination of temperature and IJ concentration.

Grucmanová and Holuša (2013) described phoretic relationships between nematodes and various *Ips* spp., the most frequent phoretic nematodes being *Bursaphelenchus eidmanni* and *Micoletzkya buetschlii*. Endoparasitic nematodes were also identified including *Contortylenchus*, *Cryptaphelenchus*, *Ektaphelenchus*, *Parasitaphelenchus*, *Parasitorhabditis* and *Parasitylenchus* spp. The most frequent endoparasite was *Parasitylenchus dispar* and *Contortylenchus diplogaster*. It is doubtful whether any of these associated nematodes could act as an effective biological suggest the possibility of using *Ips* as a "living insect bomb" as suggested by Gumus *et al.* (2015), especially given the cryptic nature of the life history of *Ips* spp.

Triggiani (1983) conducted field trials in Italy using *H. bacteriophora* against the pine shoot beetle, *Tomicus piniperda*, and showed that adults excavating their nuptial chambers in pine trunks could be infected by nematodes if a suspension was injected into the tunnels.

The large pine weevil, *Hylobius abietis*, is one of the most serious pests of clear-fell forests in Europe. As such, there has been much research on the use of the following EPN species: *S. carpocapsae*, *S. kraussei*, *S. feltiae*, *H. dowensi* and *H. megidis* against this pest. There have also been many studies of efficacy, a meta-analysis on factors driving success, much work on interactions with other biological control agents, and investigations of environmental safety and non-target effects. Given this body of research and given the lessons that can be learned in application to other pests, the following section deals with *Hylobius abietis* as a case study.

24.5 Case Study: Application of Entomopathogenic Nematodes in Control of the Large Pine Weevil, *Hylobius abietis*

24.5.1 Life cycle of *H. abietis* and its pest status

The large pine weevil, *H. abietis*, is the most serious pest of clear-fell forests in Europe, but from the outset it is important to note that the problems associated with this pest are essentially and fundamentally man-made problems. Silvicultural practices that promote the use of clear-fell (clear-cut) forestry and large areas of breeding sites (stumps) with even-aged plantings provide an optimum ecological niche for *Hylobius* spp.

Since the establishment of plantation forestry in Europe, *H. abietis* L. has been a major pest (Munro, 1927). In newly clear-felled stands, adult female weevils are attracted by the volatiles and oviposit just under the bark of the stumps. Weevil larvae subsequently develop in the protected environment under the bark for 1–3 years depending on temperature (Leather *et al.*, 1999; Inward *et al.*, 2012). Following emergence, the adults feed on the bark of young trees, and replanted sites can suffer up to 100% mortality of newly planted trees if no control measures are taken. *H. abietis* is estimated to cost the UK economy £2 million year^{-1} (Weslien, 1998; Leather *et al.*, 1999). The set of environmental conditions provided by clear-fell forestry—large amounts of volatile cues that attract gravid females, copious breeding habitats in the form of stumps and fresh food for newly emerging adults—provide the ideal niche for pine weevils and so it is important to note that *H. abietis* is in fact a man-made problem (Evans *et al.*, 2004): there is very little impact of pine weevil in mixed-aged stands subject to selective felling. However, as long as plantation forestry, subject to clear-felling, remains an important part of the forestry estate, control of *H. abietis* will be necessary. Damage to saplings occurs on weevil emergence and when the weevil undergoes a period of maturation feeding (Fig. 24.1).

24.5.2 Control of *H. abietis* with EPNs

Current control measures include the synthetic chemicals α-cypermethrin and cypermethrin, which are administered in nursery pretreatment either via Electrodyn application or dipping of young trees prior to planting, and/or through on-site postplanting spray. However, with concerns over potential environmental impacts, cypermethrin is being phased out across Europe (European Commission, 2012). Also, under Forest Stewardship Council (FSC) guidelines, α-cypermethrin and cypermethrin are considered "highly hazardous chemicals" applied only under derogation, so there is an obligation on FSC-certified companies to find alternatives to chemical control. Furthermore, current pesticides have a repellent effect on the pine weevil and, while this protects young plants, it does little to impact the local populations of the pest (Leather *et al.*, 1999; Torr *et al.*, 2005).

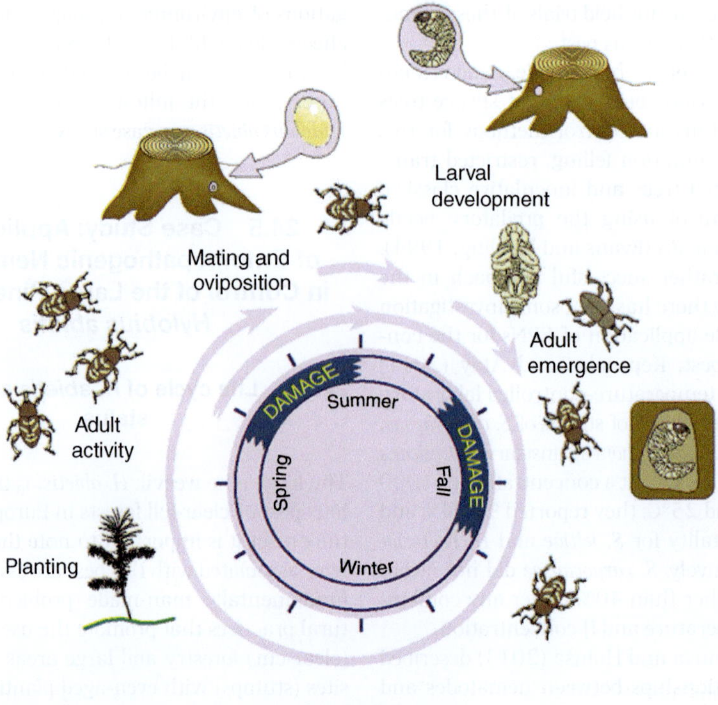

Fig. 24.1. The life cycle of the large pine weevil, *Hylobius abietis*.

Alternatives to chemical control of pine weevil include changes in silviculture practices, including mounding, planting later in the season and leaving sites fallow for a number of years (von Sydow, 1997; Örlander and Nilsson, 1999; Örlander and Nordlander; 2003). Another cultural tactic is application of the fungus *Phlebiopsis gigantea* (Fr.: Fr.) Jülich for biocontrol of the fungus *Heterobasidion annosum* (Fr.) Bref., which has the additional benefit of making stumps unsuitable for weevil oviposition and development (Skrzecz, 1996, 2001). Another approach to weevil control is to manage forest blocks in a landscape context with regard to pine weevil meta-population dynamics; to this end, a *Hylobius*-integrated management system using geographic information systems (GIS) technology has been developed (Wainhouse *et al.*, 2001; Evans *et al.*, 2004).

EPNs are widely used in the UK to control pine weevils and, until recently, were also widely used in Ireland. At an operational level, *S. carpocapsae* is applied by pressure hose from a tank mixer mounted on a modified forwarder, at an average rate of 3.5 million IJs per stump (Torr *et al.*, 2005). These nematodes are targeted to kill the weevil larvae, pupae and callow adults that are developing in the stump under the bark. For experiments, however, EPNs are usually applied by hand as a subsurface drench at the same rate. For experiments, weevil control is measured either as percentage parasitism by destructively sampling stumps (Fig. 24.2), or through monitoring the percentage reduction in emergence relative to a control, using emergence traps (Fig. 24.3). There have been many studies assessing the efficacy of EPNs in the control of pine weevil (Brixey *et al.*, 2006; Dillon *et al.*, 2006, 2007, 2008; Torr *et al.*, 2007; Williams *et al.*, 2013a).

Brixey *et al.* (2006), reporting on experiments carried out in 1997, were the first to report on EPN efficacy in clear-fell forestry. They used a subsurface drench and a surface spray to apply *S. carpocapsae* IJs to Sitka spruce stumps. They report a higher susceptibility of callow adults than either larvae or older adults, and this is consistent with the results of Williams *et al.* (2015) who showed that pupae that are infected with nematodes will metamorphose into callow adults before being killed by IJs. This is the case even if the pupae are washed in Tween after infection, suggesting that IJs are waiting in spiracles or some other orifice. Brixey *et al.* (2006) report relatively poor rates of control (5.2–18.3%) and suggested that the timing of applications is important.

Dillon *et al.* (2006), in field trials, tested the efficacy of the three commercially available EPNs (*H. megidis*, *S. feltiae* and *S. carpocapsae*) against pine weevil and also *H. downesi*. All EPNs parasitized larvae 40–49 cm from the bole of the stump and to a depth of 30–39 cm below the soil

Fig. 24.2. A destructively sampled stump after monitoring percentage weevil parasitism under the bark, which has been removed.

Fig. 24.3. An emergence trap surrounding a stump to monitor the percentage reduction of weevil emergence relative to the control.

surface. *H. downesi* was the most efficacious EPN used, parasitizing 55–63% of developing weevils. Dillon *et al.* (2007) showed that even half doses of EPN applied to pine stumps still maintained efficacies of 75–79% reduction relative to the control for *H. downesi*. In contrast to Brixey *et al.* (2006), Dillon *et al.* (2007) reported that a higher percentage of larvae were parasitized (60%) compared with pupae (36%) or callow adults (18%).

Torr *et al.* (2007) compared *S. carpocapsae* with *S. kraussei*, as the latter has been isolated from soils in northern England, is efficacious at a lower temperature and adopts a "cruiser" rather than "ambusher" foraging strategy. Laboratory and field trials showed no difference in efficacy, although *S. kraussei* did persist for longer in the field. Overall, the authors noted very little advantage to swapping control strategies from *S. carpocapsae* to *S. kraussei*.

24.5.3 Soil, tree stump species and pest density effects on efficacy

Williams *et al.* (2013b) analyzed existing data together with previously unpublished data from the COFORD (Council for Forest Research and Development)-funded project ABATE and the INTERREG (European Territorial Cooperation) IVA project IMPACT, to determine which factors are most important in determining the efficacy of EPNs in the control of *Hylobius abietis*. Altogether, there were 28 field trials in the data set. They had data for the efficacy of two nematode species: *S. carpocapsae*, a so-called "ambush forager," which is typically thought to wait and stand on its tail to latch on to passing insects (this is also the species used at an operational level), and *H. downesi*, a so-called "cruise forager," which is typically thought to seek out insect prey (this species was reported by Dillon *et al.* 2006, 2007) to be the most efficacious of the nematodes that they tested). Efficacy was measured by two methods in the datasets: (i) by measuring the reduction in weevil emergence from treated compared with untreated control stumps by placing tent-like emergence traps, which collect emerging weevils, over the stumps (Fig. 24.3); and (ii) by comparing levels of parasitism between treated and untreated stumps by removing the bark and counting the number of infected pine weevils in the stumps (Fig. 24.2). General linear models were constructed with percentage parasitism and percentage reduction as responses, and soil type (peat versus mineral), tree species (spruce versus pine) and weevil population as explanatory variables. The most parsimonious models for both nematode species, both measures of efficacy and at both a site and stump level, consistently showed that the most significant factor was soil type, with EPNs being more efficacious on sites with peaty soils than on sites with mineral soil. Neither tree species nor weevil population significantly affected EPN efficacy. They also found that *H. downesi* was significantly more efficacious than *S. carpocapsae*, but the latter did surprisingly well. A number of practical recommendations can be made from this meta-analysis. First, it may be worthwhile biological control companies investigating the feasibility of mass-producing *H. downesi*. Second, EPN applications should be focused on peaty sites where they are likely to be more effective. However, peaty sites have their own difficulties in terms of ease of access. As efficacy was independent of both the tree species of the stumps and the weevil population in the stumps, it would be beneficial to focus on spruce, which typically has lower populations of weevils so that equivalent percentage reductions are more likely to reduce the weevil population below the absolute economic threshold of about five weevils per stump. The results of this analysis are presented in detail in Williams *et al.* (2013b).

In a follow-up study, Kapranas *et al.* (2017) tested for soil type and weevil density effects in field trials using *S. carpocapsae* and *H. downesi*. Whereas Williams *et al.* (2013b) used general linear models, Kapranas *et al.* (2017) used generalized linear models (Crawley, 1997). They assumed quasi-binomial error variance for parasitism (proportional) data, and the significance of effects was assessed by the change in deviance when a variable was removed from the full model. Williams *et al.* (2013b) also used a mixed-effect logistic regression analysis to explore parasitism rates in relation to depth below the soil surface and the horizontal distance from the bole of the stump (Kapranas *et al.*, 2017). Kapranas *et al.* (2017) reported a significant negative density dependence of parasitism rate when the soil type was taken into account.

24.5.4 EPN use in combination with other control agents

The first study to combine EPNs with other control strategies in *Hylobius abietis* control efforts was that of Dillon *et al.* (2008). They used both "top-down" (*S. carpocapsae* and *H. downesi*) and "bottom-up" methods of population reduction (the wood-colonizing fungus *Trichoderma koningii*). They also monitored the levels of natural control by the braconid *Bracon hylobii*. Experiments were conducted in Sitka spruce stumps. *H. downesi* reduced the amount of emergence relative to controls and EPNs had no effect on parasitism rates by the braconid. *T. koningii* did not suppress weevils but did advance weevil development. Whereas the fungus had a negative effect on EPN parasitism, it positively affected parasitism by the braconid, suggesting that it interfered with competition between EPNs and *B. hylobii*.

Other potential biological control agents of pine weevil are entomopathogenic fungi (EPFs) of the genera *Beauveria* and *Metarhizium*. Ansari and Butt (2012) showed with laboratory assays that larvae, pupae and adults of pine weevils are all susceptible to EPFs. The efficacy of EPFs, EPNs, and mixtures of EPNs and EPFs was investigated in five field trials over 3 years (Williams *et al.*, 2013a). These trials focused on pine stumps on peaty substrates, but similar trials were undertaken on spruce stumps in Wales (H. Evans, personal communication). In general, treatments that included EPNs were more efficacious than those that did not, and in only one trial did an EPF-only treatment provide a reduction in weevil populations, relative to untreated controls, approaching significance ($P = 0.058$). Half-dose treatments allowed the assessment of synergy between agents. When agents are more efficacious together than when each is applied alone and the effect is significantly greater than an additive effect, they are said to act synergistically. In all of the trials that allowed this analysis, the effects were additive. Interestingly, half-dose treatments were often seen to be as effective, if not more effective, than full doses of EPNs. This suggests that wider operational-level field trials of reduced-dose EPN applications may be an avenue for future study. All these results were qualitatively similar to those of our Welsh colleagues (F. McAllister, personal communication). When a biological control agent fails to reduce pest populations to an economically acceptable level, a number of processes may be occurring: (i) it may be that the agent is getting to the target but not killing it; (ii) it may be that the agent is not getting to the target and so has no opportunity to kill it; and (iii) it may be that although some of the agent is getting to some of the target and killing it, not enough is getting to enough targets to reduce them below the level required. Given the results of Ansari and Butt (2012), it is unlikely that EPFs are getting to the weevils and not killing them, so either all or none of the EPF is reaching the target. Williams *et al.* (2013a) developed a molecular protocol using restriction fragment length polymorphism to distinguish between the native species of *Beauveria* commonly found on pine weevil under bark in untreated forests, *Beauveria caledonica*, and *Beauveria bassiana*, the species that was applied. They found that there was a high association between *B. bassiana* infection and application of this agent. As they measured the depth in the soil and distance from the bole of the stump of recovered weevils, they also showed that weevils could be infected by applied fungus up to a depth of 18 cm below the soil surface and up to 28 cm distant from the bole of the stump, under the bark. Interestingly, native *B. caledonica* infected weevils up to 30 cm below the soil and up to 33 cm distant from the stump. Future research should therefore, perhaps, focus on native EPFs and agent delivery technology to improve the efficacy of EPFs in weevil control. Soil from around stumps (EPF-treated and untreated control) at three of the field sites were monitored for EPF presence. Soil samples were baited with wax moth larvae, and EPF mycelia from killed moths were grown on agar that was selective for EPFs. DNA was then extracted and the internal transcribed spacer region was sequenced to identify the EPF species. It was apparent that applied EPF persisted for at least 3 years in soil around stumps, but there was no apparent spread of inoculum from fungus-treated stumps to untreated control stumps. A number of interesting EPFs were also identified, including *Pochonia bulbillosa* (a species previously isolated from forest soils in New Zealand; Ali, 2012) and *Beauveria pseudobassiana* (a species previously isolated from South America). All of these native isolates have potential for future study as biocontrols of pine weevil (C.D. Williams, unpublished data).

McNamara *et al.* (2018) also reported additive results when using commercially available *Metarhizium brunneum* and *B. bassiana* and a strain of *B. caledonica* isolated from the pest's habitat, along with *S. carpocapsae* and *H. downesi*. They also note that applied EPFs can survive in the soil for up to 2 years (especially *M. brunneum*).

24.5.5 Environmental safety of EPNs in forest ecosystems

S. carpocapsae is native to Britain but to date has not been isolated from natural populations in Ireland. Obviously, many environmentalists have legitimate concerns over the release of a non-native species at a rate of 3.5 million IJs per stump over vast tracts of the forestry estate in Ireland. A number of studies have been undertaken at the National University of Ireland to assess the significance of the risk posed by EPNs. These studies considered the following risks: (i) what is the risk to important ecosystem service provider; (ii) what are the risks to non-target, stump-inhabiting beetles; and (iii) what are the effects on native EPNs (i.e. do applied EPNs displace the native EPN community)? Ecosystem services is currently a "hot-button topic" in ecology with the realization that services provided by nature, and traditionally regarded as economic externalities of zero value, are actually worth US\$33 trillion year^{-1} to the world economy (Costanza *et al.*, 1997; note that at this time, world total gross national product was estimated at US\$18 trillion). *Rhagium bifasciatum* is a native long-horned beetle of plantation forests in Ireland. The species typically inhabits logs and branches of brash rather than stumps, where it develops and breaks up the interior of the wood, allowing clearing of the brash and recycling of nutrients into the forest soil. Harvey *et al.* (2011) showed in the laboratory that realistic doses of nematodes can easily penetrate logs of brash and will subsequently infect and kill *R. bifasciatum* inhabiting logs. Field collections of brash logs, however, showed a relatively low infection rate (<10% of logs infected and <4% of *R. bifasciatum* individuals) from sites that had been operationally treated with EPNs and that levels of infection quickly dropped off as distance to the nearest stump increased. Thus, whereas there is potential impact here, this is largely mitigated through targeted application of control agents.

Another ecosystem service provider is *Bracon hylobii*, a native parasitoid of pine weevil. Any effect of nematode infection on wasp populations is likely to be detrimental to the provision of this service. In a series of elegant experiments, reported by Harvey and Griffin (2012), it was shown that *B. hylobii* uses vibration cues to select host larvae under bark, and wasps will parasitize nematode-infected hosts as long as they move. The non-specificity of host choice was further demonstrated by eliciting oviposition even on abnormal hosts (i.e. larvae of insects not usually found under the bark). Hence, there is potential here for competition between *B. hylobii* and inundatively applied EPNs. It was suggested that EPNs be applied just after most adult wasps have eclosed in late spring but when parasitism of weevils is still quite low, to mitigate competition between the two species.

One might predict that applying millions of insect-killing nematodes on to a stump is likely to have a major effect on non-target insect species. Indeed, EPNs have a wide host range, and laboratory experiments suggest potential non-target susceptibility. Surprisingly, however, Dillon *et al.* (2012) showed that this is actually not the case. Non-target beetles were collected from a number of sites, 1 and 2 years after application of EPNs. All beetles that emerged from both nematode-treated and untreated control stumps were identified to the species level. There was no significant effect of nematode treatment on the total numbers of beetles, the number of beetle species, their diversity or the beetle community composition. Nor was there an effect on these metrics for the subset of stump-inhabiting (saproxylic) species. There was, however, not surprisingly, a large effect of both site and tree species on these metrics. It is therefore important to note that "potential" effects identified under controlled conditions in the laboratory need not necessarily become "realized" effects in the field.

Another set of organisms potentially disrupted by the application of EPNs is the native EPN species that would ordinarily inhabit the site. Harvey and Griffin (2016) showed that applied EPNs only persist at very low levels 3 years after application and that persistence decreased with increasing distance from the stump. They also showed a very limited amount of off-site spread of EPNs. Five years after application, sampling at the same sites showed no persistence of the applied EPNs and there was recolonization of sites by the native species, *S. feltiae*, probably as a result of ecological succession of sites with native vegetation and the associated insect fauna suitable to the native EPN species (Harvey and Griffin, 2016). We may therefore conclude that there are many potential negative impacts in releasing the non-native *S. carpocapsae* to control pine weevils in Ireland, but in fact most of these risks have not been realized in the field. Harvey *et al.* (2016) provided a rigorous, quantitative evaluation of five risk categories in the use of EPNs to control *Hylobius abietis*. These categories were: establishment, dispersal, host range, direct non-target effects and indirect non-target effects. Overall, EPNs were given a lower score (35–51 out of 125) compared with conventional control with pyrethroid or neonicotinoid insecticides.

24.6 Conclusions

EPNs are a classic example of inundative (biopesticide) biological control. Although there has been some work on trees outside of forests, deciduous forests and coniferous forests, of the 99 serious pests of European Forests, only 16.2% have had any investigation of the applicability of EPNs to sustainably control these pest populations. EPNs (especially Steinernematidae) have been isolated from forestry soils throughout the world, so the dangers posed by introducing EPNs en masse appears to be negligible, but it is important to consider environmental safety and non-target effects. One area that has received a lot of attention and has been scaled up to operational levels is the biological control of the large pine weevil, *Hylobius abietis*. This is the most serious insect pest of coniferous forests in Europe. I have reviewed the various aspects of EPN research in relation to its control, effects of stump species, soil type and weevil populations on EPN efficacy, the use of EPNs with other control agents and the environmental safety of EPNs in forestry. It is hoped that in the future such studies will become common on the other serious pests of forestry throughout the world.

24.7 References

Ali, F. (2012) Cold tolerance mechanisms of entomopathogenic nematodes. PhD thesis, University of Otago, Otago, new Zealand.

Ansari, M.A. and Butt, T. (2012) Susceptibility of different developmental stages of large pine weevil *Hylobius abietis* (Coleoptera: Curculionidae) to entomopathogenic fungi and effect of fungal infection to adult weevils by formulation and application methods. *Journal of Invertebrate Pathology* 111, 33–40.

Atwa, A.A. and Hegazi, E.M. (2014) Comparative susceptibilities of different life stages of the red palm weevil (Coleoptera: Curculionidae) treated by entomopathogenic nematodes. *Journal of Economic Entomology* 107, 1339–1347.

Belien, T. (2018) Entomopathogenic nematodes as biocontrol agents of insect pests in orchards. *CAB Reviews* 2018, 1–11.

Brixey, J.M., Moore, R. and Milner, A.D. (2006) Effect of entomopathogenic nematode (*Steinernema carpocapsae* Weiser) application technique on the efficacy and distribution of infection of the large pine weevil (*Hylobius abietis* L.) in stumps of Sitka spruce (*Picea sitchensis* Carr.) created at different times. *Forest Ecology and Management* 226, 161–172.

Clausi, M., Leone, D., Vinciguerra, M.T., Rappazzo, G. and Tarasco, E. (2014) Laboratory tests on the biocontrol of Chestnut insect pests on Etna (Sicily, Italy) by means of entomopathogenic nematodes. *Redia* 96, 133–136.

Costanza, R., d'Arge, R., de Groot, R., Farber, S., Grasso, M. *et al.* (1997) The value of the world's ecosystem services and natural capital. *Nature* 387, 253–260.

Crawley, M.J. (1997) *GLIM for Ecologists*. Blackwell Science, Oxford, UK.

Day, K.R. and Leather, S.R. (1997) Threats to forestry by insect pests in Europe. In: Watt, A.D., Stork, N.E. and Hunter, M.D. (eds.) *Forests and Insects*. Chapman & Hall, London, pp. 177–205.

Dillon, A.B., Ward, D., Downes, M.J. and Griffin, C.T. (2006) Suppression of the large pine weevil *Hylobius abietis* (L.) (Coleoptera: Curculionidae) in pine stumps by entomopathogenic nematodes with different foraging strategies. *Biological Control* 38, 217–226.

Dillon, A.B., Downes, M.J., Ward, D. and Griffin, C.T. (2007) Optimizing application of entomopathogenic nematodes to manage large pine weevil, *Hylobius abietis* L. (Coleoptera: Curculionidae) populations developing in pine stumps, *Pinus sylvestris*. *Biological Control* 40, 253–263.

Dillon, A.B., Rolston, A.N., Meade, C.V., Downes, M.J. and Griffin, C.T. (2008) Establishment, persistence, and introgression of entomopathogenic nematodes in a forest ecosystem. *Ecological Applications* 18, 735–747.

Dillon, A., Foster, A., Williams, C.D. and Griffin, C. (2012) Environmental safety of entomopathogenic nematodes – effects on abundance, diversity and community structure of non-target beetles in a forest ecosystem. *Biological Control* 63, 107–114.

European Commission (2012) Proposal for a revised directive of the European Parliament and of the Council on Priority Substances in the field of water quality. MEMO 12/59. European Commission, Brussels.

Evans, H.F. and Fielding, N.J. (1994) Integrated management of *Dendroctonus micans* in the UK. *Forest Ecology and Management* 65, 17–30.

Evans, H., Moore, R., Heritage, S. and Wainhouse, D. (2004) *Developments in the Integrated Management of Pine Weevil, a Pest of Restocking in Conifer Plantations*. Forest Research Annual Reports and Accounts 2003–2004, Edinburgh, UK.

Gaugler, R. and Kaya, H.K. (eds.) (1990) *Entomopathogenic Nematodes in Biological Control*. CRC Press, Boca Raton, Florida.

Gorgadze, O., Kereselidze, M., Beruashvili, M. and Bagaturia, N. (2020) Fauna of nematodes of Georgian hazelnut soils and its entomopathogenic forms. *Bulletin of the Georgian National Academy of Sciences* 14, 95–101.

Gözel, Ç. and Gözel, U. (2019) The efficacy of native entomopathogenic nematodes against the pine processionary moth, *Thaumetopoea pityocampa* Den. & Schiff. (Lepidoptera: Thaumetopoeidae). *Türkiye Biyolojik Mücadele Dergisi* 10, 118–126.

Griffin, C.T., Chaerani, R., Fallon, D., Reid, A.P. and Downes, M.J. (2000) Occurrence and distribution of entomopathogenic nematodes *Steinernema* spp. and *Heterorhabditis indica* in Indonesia. *Journal of Helminthology* 74, 143–150.

Grucmanová, Š. and Holuša, J. (2013) Nematodes associated with bark beetles, with focus on the genus *Ips* (Coleoptera: Scolytinae) in Central Europe. *Acta Zoologica Bulgarica* 65, 547–556.

Gumus, A., Karagoz, M., Shapiro-Ilan, D. and Hazir, S. (2015) A novel approach to biocontrol: release of live insect hosts pre-infected with entomopathogenic nematodes. *Journal of Invertebrate Pathology* 130, 56–60.

Harvey, C. and Griffin, C. (2016) Local host-dependent persistence of the entomopathogenic nematode *Steinernema carpocapsae* used to control the large pine weevil *Hylobius abietis*. *BioControl* 65, 185–193.
Harvey, C.D. and Griffin, C.T. (2012) Host activity and wasp experience affect parasitoid wasp foraging behaviour and oviposition on nematode-infected larvae of the forestry pest *Hylobius abietis*. *Ecological Entomology* 37, 269–282.
Harvey, C.D., Alameen, K.M. and Griffin, C.T. (2011) The impact of entomopathogenic nematodes on a non-target, service-providing longhorn beetle is limited by targeted application when controlling forestry pest *Hylobius abietis*. *Biological Control* 62, 173–182.
Harvey, C.D., Williams, C.D., Dillon, A.B. and Griffin, C.T. (2016) Inundative pest control: how risky is it? A case study using entomopathogenic nematodes in a forest ecosystem. *Forest Ecology and Management* 380, 242–251.
Inward, D.J.G., Wainhouse, D. and Peace, A. (2012) The effect of temperature on the development and life cycle regulation of the pine weevil *Hylobius abietis* and the potential impacts of climate change. *Agricultural and Forest Entomology* 14, 348–357.
Kapranas, A., Malone, B., Quinn, S., McNamara, L., Williams, C.D. *et al.* (2017) Efficacy of entomopathogenic nematodes for control of large pine weevil, *Hylobius abietis*: effects of soil type and pest density and spatial distribution. *Journal of Pest Science* 90, 495–505.
Kepenekci, I. and Atay, T. (2014) Evaluation of aqueous suspension and entomopathogenic nematodes infected cadaver applications against the great spruce bark beetle *Dendroctonus micans* (Kugelann), (Coleoptera: Scolytidae). *Egyptian Journal of Biological Pest Control* 24, 335–339.
Kepenekci, I., Inal, B., Atay, T. and Yeşilayer, A. (2014) Effect of entomopathogenic nematodes, on the black cherry aphid, *Myzus cerasi* (F.) (Hemiptera: Aphididae). *Egyptian Journal of Biological Pest Control* 24, 157–161.
Khatri-Chhetri, H.B., Waeyenberge, L., Manandhar, H.K. and Moens, M. (2010) Natural occurrence and distribution of entomopathogenic nematodes (Steinernematidae and Heterorhabditidae) in Nepal. *Journal of Invertebrate Pathology* 103, 74–78.
Kuske, S., Daniel, C., Wyss, E., Sarraquigne, J., Jermini, M. *et al.* (2005) Biocontrol potential of entomopathogenic nematodes against nut and orchard pests. *IOBC/WPRS Bulletin* 28, 163–167.
Lacey, L.A., Arthurs, S.P., Unruh, T.R., Headrick, H. and Fritts, R. Jr. (2006) Entomopathogenic nematodes for control of codling moth (Lepidoptera: Tortricidae) in apple and pear orchards: effect of nematode species and seasonal temperatures, adjuvants, application equipment, and post-application irrigation. *Biological Control* 37, 214–223.
Lakatos, T. and Tóth, T. (2006). Biological control of European cockchafer larvae (*Melolontha melolontha* L.) – preliminary results. *Journal of Fruit and Ornamental Plant Research* 14, 73–77.
Leather, S.R., Day, K.R. and Salisbury, A.N.J. (1999) The biology and ecology of the large pine weevil, *Hylobius abietis* (Coleoptera: Curculionidae): a problem of dispersal? *Bulletin of Entomological Research* 89, 3–16.
Livesley, S.J., McPherson, E.G. and Calfapietra, C. (2016) The urban forest and ecosystem services: impacts on urban water, heat, and pollution cycles at the tree, street, and city scale. *Journal of Environmental Quality* 45, 119–124.
McNamara, L., Kapranas, A., Williams, C.D., O'Tuama, P., Kavanagh, K. and Griffin, C.T. (2018) Efficacy of entomopathogenic fungi against large pine weevil, *Hylobius abietis*, and their additive effects when combined with entomopathogenic nematodes. *Journal of Pest Science* 91, 1407–1419.
Mráček, Z., Weiser, J., Bures, M. and Kahounova, L. (1992) *Steinernema kraussei* (Steiner, 1923) (Nematoda: Steinernematidae): rediscovery of its type locality in Germany. *Folia Parasitologica* 39, 181–182.
Mráček, Z., Bečvář, S., Kindlmann, P. and Jersáková, J. (2005) Habitat preference for entomopathogenic nematodes, their insect hosts and new faunistic records for the Czech Republic. *Biological Control* 34, 27–37.
Munro, J.W. (1927) The present position of forest entomology in Great Britain. *Forestry* 1, 44–46.
Nikdel, M., Niknam, G. and Dordaei, A.A. (2010) Evaluation of susceptibility of the brown tail moth, *Euproctis chrysorrhoea*, to entomopathogenic nematodes under laboratory conditions. *Nematologia Mediterranea* 38, 3–6.
O'Hanlon, R., Ryan, C., Choiseul, J., Murchie, A. and Williams, C.D. (2021) Catalogue of pests and pathogens of trees on the island of Ireland. *Biology and Environment: Proceedings of the Royal Irish Academy Series B* 121B, 21–45.
Örlander, G.R. and Nilsson, U. (1999) Effect of reforestation methods on pine weevil (*Hylobius abietis*) damage and seedling survival. *Scandinavian Journal of Forest Research* 14, 341–354.

Örlander, G.R. and Nordlander, G.R. (2003) Effects of field vegetation control on pine weevil (*Hylobius abietis*) damage to newly planted Norway spruce seedlings. *Annals of Forestry Science* 60, 667–671.

Parkman, J.P., Frank, J.H., Walker, T.J. and Schuster, D.J. (1996) Classical biological control of *Scapteriscus* spp. (Orthoptera: Gryllotalpidae) in Florida. *Environmental Entomology* 25, 1415–1420.

Pocock, M.J.O, Marzano, M., Bullas-Appleton, E., Dyke, A., de Groot, M. *et al.* (2020) Ethical dilemmas when using citizen science for early detection of invasive tree pests and diseases. *Management of Biological Invasions* 11, 720–732.

Şahin, Ç. and Gözel, U. (2021) Efficacy and persistence of native entomopathogenic nematodes against *Capnodis tenebrionis* in peach (*Prunus persica*) orchard in Turkey. *Phytoparasitica* 49, 613–621.

Schmiege, D.C. (1963) The feasibility of using a neoaplectanid nematode for control of some forest insect pests. *Journal of Economic Entomology* 56, 427–431.

Shapiro-Ilan, D.I., Duncan, L.W., Lacey, L.A. and Han, R. (2005) Orchard applications. In: Grewal, P.S., Ehlers, R.U. and Shapiro-Ilan, D.I. (eds.) *Nematodes as Biocontrol Agents*. CAB International, Wallingford, UK, pp. 215–229.

Skrzecz, I. (1996) Impact of *Phlebia gigantea* (Fr.: Fr) Donk on the colonization of Scots pine (*Pinus sylvestris* L.) stumps by the large pine weevil (*Hylobius abietis* L.). *Folia Forestalia Polonica* 38, 89–101.

Skrzecz, I. (2001) Large pine weevil (*Hylobius abietis*) abundance and the extent of damage in plantations established on clearcuts with pine stumps treated with the fungus *Phlebiopsis gigantea* (Fr.: Fr.) Julich. *Folia Forestalia Polonica* 43, 127–142.

Steiner, W. (1996) Distribution of entomopathogenic nematodes in the Swiss Alps. *Suisse Journal de Zoologie* 103, 439–452.

Stock, S.P., Pryor, B.M. and Kaya, H.K. (1999) Distribution of entomopathogenic nematodes (Steinernematidae and Heterorhabditidae) in natural habitats in California, USA. *Biodiversity & Conservation* 8, 535–549.

Stock, S.P., Al Banna, L., Darwish, R. and Katbeh, A. (2008) Diversity and distribution of entomopathogenic nematodes (Nematoda: Steinernematidae, Heterorhabditidae) and their bacterial symbionts (γ-Proteobacteria: Enterobacteriaceae) in Jordan. *Journal of Invertebrate Pathology* 98, 228–234.

Straw, N.A., Fielding, N.J., Tilbury, C., Williams, D.T. and Inward, D. (2015) Host plant selection and resource utilisation by Asian longhorn beetle *Anoplophora glabripennis* (Coleoptera: Cerambycidae) in southern England. *Forestry* 88, 84–95.

Tarasco, E., Clausi, M., Rappazzo, G., Panzavolta, T., Curto, G. *et al.* (2015) Biodiversity of entomopathogenic nematodes in Italy. *Journal of Helminthology* 89, 359–366.

Torr, P.S., Wilson, M.J. and Heritage, S. (2005) Forestry applications. In: Grewal, P.S., Ehlers, R.U. and Shapiro-Ilan, D.I. (eds.) *Nematodes as Biocontrol Agents*. CAB International, Wallingford, UK, pp. 47–64.

Torr, P., Heritage, S. and Wilson, M.J. (2007) *Steinernema kraussei*, an indigenous nematode found in coniferous forests: efficacy and field persistence against *Hylobius abietis*. *Agricultural and Forest Entomology* 9, 181–188.

Triggiani, O. (1983) Sensibilità del *Tomicus (Blastophagus) piniperda* L. (Col. Scolytidae) ai Nematodi della famiglia Steinernematidae e Heterorhabditidae. *Entomologica* 18, 215–223.

Tumialis, D., Skrzecz, I., Mazurkiewicz, A. and Pezowicz, E. (2019) Sensitivity of caterpillars of the pine tree lappet moth *Dendrolimus pini* to native isolates of entomopathogenic nematodes. *International Journal of Pest Management* 65, 332–337.

von Sydow, F. (1997) Abundance of pine weevils (*Hylobius abietis*) and damage to conifer seedlings in relation to silvicultural practices. *Scandinavian Journal of Forest Research* 12, 157–167.

Wainhouse, D., Evans, H., Moore, R., Webber, J., Thorpe, K. and Stanly, J. (2001) *The Integrated Forest Management Programme*. Forest Research Annual Report and Accounts 2000–2001, Edinburgh, U.K.

Wang, C. and Li, C. (2017) Status of entomopathogenic nematodes in integrated pest management strategies in China. In: Abd-Elgawad, M.M., Askary, T.H. and Coupland, J. (eds.) *Biocontrol Agents: Entomopathogenic and Slug Parasitic Nematodes*. CAB International, Wallingford, UK, pp. 457–472.

Weslien, J. (1998) Vad kostar snytbaggeskadorna? [How much does the pine weevil damage cost?] *Kunglige Skogs-och Lantbruksakademins Tidskrift* 137, 19–22.

Williams, C.D., Dillon, A.B., Hennessy, R., McNamara, L., Harvey, C.D. and Griffin, C.T. (2013a) Control of a major pest of forestry, *Hylobius abietis*, with entomopathogenic nematodes and fungi using eradicant and prophylactic strategies. *Forest Ecology and Management* 305, 212–222.

Williams, C.D., Dillon, A.B., Girling, R. and Griffin, C.T. (2013b) Organic soils promote the efficacy of entomopathogenic nematodes, with different foraging strategies, in the control of a major forest pest: a meta-analysis of studies to date. *Biological Control* 65, 357–364.

Williams, C.D., Dillon, A.B., Ennis, D., Hennessy, R. and Griffin, C.T. (2015) Differential susceptibility of pine weevil, *Hylobius abietis* (Coleoptera: Curculionidae), larvae and pupae to entomopathogenic nematodes and death of adults infected as pupae. *BioControl* 60, 537–546.

Yang, H. and Zhang, G. (1990) Biological control of *Holcocerus insularis* with *Steinernema feltiae*. *Journal of Chinese Landscape Architecture* 1: 55.

Yu, Q., Sun, F. and Coupland, J. (2017) Status of entomopathogenic nematodes in integrated pest management strategies in Canada. In: Abd-Elgawad, M.M., Askary, T.H. and Coupland, J. (eds.) *Biocontrol Agents: Entomopathogenic and Slug Parasitic Nematodes*. CAB International, Wallingford, UK, pp. 312–326.

Zhang, G., Cheng, G., Zhang, S. and Yan, L. (2007) Preliminary study of the control of *Anoplophora glabripennis* with insect pathogenic nematodes in Harbin. *Forest By-product and Specialty in China* 89, 34–36.

25 Glasshouse Production

Ganpati B. Jagdale*
Extension Nematology Laboratory, Department of Plant Pathology, University of Georgia, Athens, Georgia, USA

Abstract
In North America, the glasshouse industry, which was valued at US$16.9 billion in 2006, is the fastest-growing agricultural segment. It encompasses 300,000 ha of land, with 65% dedicated to vegetables and 35% to ornamentals, yielding billions in annual sales. Glasshouses allow control over abiotic conditions crucial for plant growth but also foster pests such as insects, mites, nematodes, slugs and snails. Integrated pest management faces challenges due to diverse plant species and various media. Traditionally, growers relied on cultural practices and chemical pesticides, but restrictions arise from resistance, environmental concerns and regulations. Entomopathogenic nematodes (EPNs) have emerged as effective biological control agents in glasshouses. Notably, *Steinernema feltiae* demonstrates high efficacy under glasshouse conditions, achieving 80–100%, 96–100%, 76% and 95% control of fungus gnats, shore flies, Western flower thrips and leaf miners, respectively. Continuous research is crucial for developing new EPN isolates, refining EPN application methods and meeting modern glasshouse production demands while ensuring successful and environmentally safe pest management.

25.1 Introduction

The worldwide glasshouse industry, generally referred to as the greenhouse industry in North America, is the most rapidly growing segment of agriculture, worth US$16.9 billion in 2006, with more than 300,000 ha of land under cultivation of vegetables (65% area) and ornamentals (35% area) worth billions of dollars in annual sales (Albajes *et al.*, 1999; Parrella *et al.*, 1999; Jerardo, 2004, 2007). Of the total land under glasshouse cultivation, 250,000 ha are under plastic cover and 50,000 ha under glass cover. Glasshouses are regularly used for propagation, overwintering and the full production cycle for many plant species, which are generally grown on a variety of organic and mineral substrates. Covered houses allow a degree of control over many abiotic environmental conditions required for survival and proper growth of plants. Unfortunately, these conditions favor rapid growth and multiplication of many economically important pests and diseases. Arrays of pests that threaten the glasshouse industry include insects, mites, nematodes, slugs and snails. The presence of even a few of these pests, dead or alive, and their damage can cause millions of dollars in losses to the glasshouse industry because of rejected and returned shipments by retailers and wholesalers. For example, nurseries in the Pacific North-west

*Email: gbjagdal@uga.edu

 Entomopathogenic Nematodes as Biological Control Agents
(D.I. Shapiro-Ilan and E.E. Lewis eds)
DOI: 10.1079/9781800620322.0025

of the USA spend more than US$1 million year^{-1} scouting for black vine weevils and relinquish over US$500,000 year^{-1} on shipments of plant stocks returned due to their presence.

In glasshouses, although integrated pest management is a common practice increasingly adopted by many growers, its development and implementation is made difficult by the complexity of plant species being grown on a variety of media, the presence of a wide range of insect pests and the incompatibility of many biological control agents with pesticides. Most growers have relied on cultural practices (plant hygiene and light/sticky traps) and/or chemical pesticides to manage insect pests of glasshouse crops (Lindquist *et al.*, 1985), but use of pesticides has been increasingly restricted because of the development of resistance, environmental pollution, human health concerns (van Lenteren, 2003) and statutory reductions in the availability of effective pesticides. In addition, pesticides pose a serious threat to beneficial organisms that are frequently used in pollination and plant-protection programs in glasshouses. Currently, over 30 biological control agents including parasitoids, predatory insects and mites, and pathogens including bacteria, fungi and entomopathogenic nematodes (EPNs) are commercially available to control several glasshouse pests (van Lenteren, 2003). So far, the efficacy of only a few EPN species has been evaluated; some of those tested have proven effective against key target pests in glasshouses (Table 25.1). This chapter focuses on the application of EPNs in glasshouses to manage root and foliage feeding insect pests.

25.2 The Glasshouse Environment

The controlled glasshouse environment favors excellent growth and production of many plant species, development and rapid multiplication of

Table 25.1. List of species/strains of entomopathogenic nematodes (EPNs) used for control of insect pests in glasshouses.

Insect pest	Target stages	Plant parts infested	EPN species	References
Fungus gnats, *Bradysia* spp.	Larvae	Roots and stems	*S. feltiae* SN strain, *S. carpocapsae* All strain, *S. anomaly*, *S. glaseri*, *S. riobrave*, *H. bacteriophora*, *H. indica*, *H. zealandica*	Lindquist *et al.* (1994); Harris *et al.* (1995); Jagdale *et al.* (2004, 2007); Kim *et al.* (2004); G.P. Jagdale and P.S. Grewal (unpublished data); M. Tomalak (unpublished data)
Greenhouse whiteflies, *Trialeurodes vaporariorum*	Nymphs, pupae, adults	Leaves	*S. feltiae*, *H. bacteriophora*	Rezaei *et al.* (2015); Laznik *et al.* (2011)
Leaf miners, *Liriomyza* spp.	Larvae	Leaves	*S. feltiae*, *Heterorhabditis* sp.	Williams and MacDonald (1995); Head and Walters (2003)
Shore fly, *Scatella stagnalis*	Larvae	Tender plant parts	*S. feltiae*, *S. carpocapsae*, *S. arenarium*, *H. megidis*	Gouge (1994); Morton and García del Pino (2003)
Western flower thrip, *Frankliniella occidentalis*[a]	Adults and nymphs	Stems and leaves	*S. feltiae*, *H. bacteriophora*, *S. abassi*, *S. arenarium*, *S. bicornutum*, *S. carpocapsae*, *H. indica*, *H. marelatus*, *Heterorhabditis* sp.	Ebssa *et al.* (2001a,b, 2004)

[a]Vectors of viral diseases.

several pests and diseases, and survival and effectiveness of beneficial organisms including pollinators and biological control agents (Hussey and Scopes, 1985; Albajes *et al.*, 1999; Parrella *et al.*, 1999; van Lenteren, 2003). Year-round warm temperatures help maintain high-density pest populations, a main food source for biological control agents, but temperature fluctuations can affect the activity and efficacy of biological control agents including EPNs (Grewal *et al.*, 1994).

Glasshouse crops are grown in a broad range of soils, soil mixtures and soil-less media with different chemical and physical properties. The most commonly used rooting and plant-growing media consist of soil mixtures prepared by mixing peat, vermiculite, perlite, composted bark, composted wastes or sewage sludge with soil to modify its texture and structure to that required for proper growth and development of plants (Adams and Fonteno, 2003). However, certain media serve as excellent substrates for the development and reproduction of insect pests (Lindquist *et al.*, 1985; Olson *et al.*, 2002; Jagdale *et al.*, 2004). For example, hardwood bark was the most conducive medium for fungus gnat colonization when compared with Ball-Mix, Metro-Mix, Pro-Mix and pine bark (Lindquist *et al.*, 1985; Jagdale *et al.*, 2004). Peat-based mixes have also provided conditions for survival and development of both fungus gnats (Olson *et al.*, 2002) and black vine weevils (Moorhouse *et al.*, 1992). In contrast, the rock wool-based non-soil media, which have become very popular in cultivation of vegetables, slow development and reduce reproduction of both thrips and leaf miners when compared with mineral soil. Specific conditions (pH or moisture) and ingredients (sandy soil, soils with high organic content, peat moss, composted bark, rock wool or mixtures of all of these with perlite, vermiculite, etc.) prevailing in these media can also affect the survival, recycling, persistence and efficacy of biological control agents including EPNs (Oetting and Latimer, 1991; Gouge and Hague, 1994, 1995a,b; Jagdale *et al.*, 2004).

In the integrated pest management approach, broad-spectrum chemical insecticides are important components used for suppressing insect pests of various crops, and compatibility with these chemicals is essential for the survival of biological control agents during and after their applications. As many studies have shown that EPNs are relatively compatible with many chemical pesticides used in plant protection (see Chapter 27, this volume), no special limitations are imposed on their use in the routine integrated pest control programs. However, as a safe practice, it is generally recommended that EPNs should be applied separately because of the potential deleterious effects of osmotic pressure on infective juveniles (IJs) if mixed with chemical pesticides or fertilizers.

25.3 Soil Application

25.3.1 Fungus gnats

Fungus gnats, *Bradysia* spp. (Diptera: Sciaridae), are relatively small (3–4 mm) flies commonly associated with compost and soils with high organic contents and are one of the most common pests of production nurseries and glasshouse crops (Harris *et al.*, 1995; Vaughan *et al.*, 2011; Cloyd, 2015). Although several species of fungus gnats are present in the glasshouse environment, *Bradysia paupera* and *Bradysia coprophila* are economically the most important species reported on many crops in Europe and North America, respectively (Harris *et al.*, 1995). Adult females often lay about 200 eggs in small batches in a lifetime on the media or soil surface (Meers and Cloyd, 2005). Eggs hatch within 4–6 days; larvae develop through four instars within 12–14 days, pupate in the soil for 3–4 days and then emerge as adults. Thus, the egg-to-egg life cycle can be completed within 20–25 days at 20–25°C (Wilkinson and Daugherty, 1970).

Fungus gnat larvae feed primarily on fungi and organic matter (Freeman, 1983), but they can also cause serious damage to the roots of many ornamentals including African violets, carnations, chrysanthemums, cyclamen, lilies, geraniums, impatiens and poinsettias. Commercially, these plants are propagated using stem cuttings, and feeding by larvae on these fresh cuttings can prevent callus development and root formation. In already-rooted plants, larvae often feed on the roots and stems by chewing/stripping and tunneling, respectively (Binns, 1973). Severely injured plants generally lose their healthy appearance, turn off-color and eventually dry. In addition, direct injuries caused by larvae to the roots can become the major route of entry for many soil-borne pathogens such as *Fusarium*, *Phoma*, *Pythium* and *Verticillium* spp., which are generally responsible for root and stem rots (Scarlett *et al.*, 2014). Larvae can transmit fungal pathogens during feeding (Ludwig and Oetting, 2001; Braun *et al.*, 2012; Scarlett *et al.*, 2014), whereas adult flies are a nuisance to people and are known to disseminate fungal spores from plant to plant when they migrate through the glasshouse (Gillespie and Menzies, 1993). As seedlings, rooted stocks and young plants (shortly after transplantation) are most sensitive to fungus gnat damage, the greatest economic losses are observed in nurseries. It has been estimated that fungus gnats and shore flies cause about 5% of the annual economic losses in direct crop yields, aesthetic marketing values, and contamination of ornamental and herbal plants (Chandler *et al.*, 2010).

Continuous overlapping generations of fungus gnats in glasshouses have made most control strategies ineffective. Chemical insecticides such as diazinon and oxamyl are not very effective for controlling fungus gnats and can also be phytotoxic to seedlings and young plants. The application of pyrethroids against adult flies is ineffective because of continuous immigration and the emergence of new-generation adults. The first attempts to use EPNs for biological control of fungus gnats in glasshouses (Nedstam and Burman, 1990) included several species of nematodes: *Steinernema feltiae*, *S. carpocapsae*, *S. arenarium* (= *S. anomali*), *S. riobrave* and *Heterorhabditis bacteriophora*; however, only *S. feltiae* proved to be as effective as chemical insecticides (Harris *et al.*, 1995). According to Gouge and Hague (1994), *S. feltiae* usually enters the fungus gnat larva through both the anus and the mouth, and once inside, kills the larva within 20 h. These authors and Tomalak (1994a) noted that, due to the small size of fungus gnat larvae, nematodes completed only one generation inside the cadaver and produced about 1000 IJs per cadaver within 6–7 days of infection. Currently, the use of commercially produced EPNs, especially *S. feltiae* in Europe has become common practice to control fungus gnats in glasshouse production, but in the USA, commercial success has been limited.

25.3.1.1 EPN application rate

Determining an appropriate concentration of EPNs is a crucial step toward cost-effective control of fungus gnats in glasshouses. Both Gouge and Hague (1995b) and Lindquist and Piatkowski (1993) used relatively high rates of 7.8×10^5 and 8.86×10^5 IJs of *S. feltiae* m^{-2}, respectively, and obtained up to 92% control of *B. paupera*. In contrast, Jagdale *et al.* (2004) applied only 2.5×10^5 IJs of *S. feltiae* m^{-2} and obtained up to 100% control of *B. coprophila*, while Harris *et al.* (1995) using the same rate of *S. feltiae* obtained about 80% control. Katumanyane *et al.* (2018) made two applications, one of 5×10^5 IJs m^{-2} and the second of 2.5×10^5 IJs m^{-2} of *Steinernema yirgalemense* and obtained 77% population reduction of *Bradysia impatiens* 14 days after application. San-Blas *et al.* (2017) applied *Heterorhabditis amazonensis* at a rate of 3.8×10^5 IJs m^{-2} and obtained 85% control of fungus gnats. Jagdale *et al.* (2007) reported that four applications of even a double rate of six different EPN species did not improve control of fungus gnats compared with a single application.

25.3.1.2 Method of application

EPNs are applied in water suspensions to pots and top spraying is convenient. This is a common practice to control fungus gnats infesting small seedlings and compost-filled trays before transplanting. However, some EPN IJs can stick to the surface of leaves, so washing the leaves with water spray or flood irrigation for larger plants ensures that EPNs make it to the medium surface. Applying EPNs via drip irrigation is less feasible because of aggregations of IJs caused by slow flow of the suspension inside the tubes when they are too long. The short distance of nematode movement in the pots makes precise spraying and even distribution of IJs a general requirement for good control of insects, although adult fungus gnats infected with *S. feltiae* can occasionally help to disperse nematodes to untreated pots (Gouge and Hague, 1995b).

25.3.1.3 Life stage

Synchronization of EPN application with the most susceptible development stage of the target pest is important, especially when persistence of the EPNs is low. The second and fourth larval instars of the fungus gnat *B. coprophila* were significantly more susceptible to *S. feltiae* than the pupae (Harris *et al.*, 1995). The third and fourth instars of another fungus gnat, *Bradysia agrestis*, were highly susceptible to *S. carpocapsae* Pocheon strain (Kim *et al.*, 2004). Similarly, the third and fourth instars of *B. impatiens* were more susceptible to *S. carpocapsae* and *Heterorhabditis indica* compared with second instars (Acharya *et al.*, 2019). Thus, targeting cohorts of second, third or fourth instars is a vital step for suppressing fungus gnat populations. After inoculation of mature *B. coprophila* adults in a growth chamber, Jagdale *et al.* (2004) reported that EPNs applied 16 days after transplanting when fourth instars were expected significantly suppressed fungus gnat populations, but not when they were applied after 0, 4 and 8 days. Acharya *et al.* (2019) demonstrated that both *H. indica* and *S. carpocapsae* caused the highest mortality of fourth-instar larvae (70% and 53%, respectively) compared with the mortality of second-instar (10% and 3%, respectively) and third-instar (40% and 37%, respectively) larvae.

25.3.1.4 Potting media

In glasshouse production, the potting medium can be an important factor for the survival and infectivity of EPNs (Oetting and Latimer, 1991). IJs of *S. feltiae* actively searched for fungus gnat larvae and persisted in the media over 60 days after application to the soil or compost surface (Gouge and Hague, 1994, 1995b). Application of *S. feltiae* in Ball-Mix containing pine wood bark mix (Ball Horticulture, Chicago, Illinois, USA), Metro-Mix containing 30–40% coconut coir pith, 20–30% vermiculite, 20–30% compost pine bark and 10–20% horticultural perlite (Scotts Miracle-Gro, Marysville, Ohio, USA), or Pro-Mix containing 75–85% Canadian sphagnum peat and 15–25% perlite, vermiculite and limestone (Premier Tech, Rivière-du-Loup, Quebec, Canada) reduced the overall population of *B. coprophila* compared with the control by 40%, 50% and 56%, respectively (Jagdale *et al.*, 2004). Similarly, *S. feltiae* provided 60% control of fungus gnats compared with the control when applied in an equal mixture of bark and peat, but control was comparatively low when applied in a mixture containing both peat and coir (Cloyd and Zaborski, 2004). However, EPN application to nursery mix containing 60% pine wood, 20% hardwood and 20% peat resulted in a 27% reduction in fungus gnats. Katumanyane *et al.* (2018) compared different types of potting media for the survival and efficacy of *S. yirgalemense* against *B. impatiens* and demonstrated that mortality of *B. impatiens* larvae was over 76% in all three EPN-treated potting media including a coir and vermiculite mixture, pine sawdust and coco coir. Acharya *et al.* (2019) demonstrated that application of *H. indica* and *S. carpocapsae* in nursery media containing 68% coco peat, 7% perlite, 15% peat moss 3% vermiculite and 7% zeolite resulted in 54% and 33% mortality of *B. impatiens* infesting bean seedlings, respectively. These authors confirmed the findings of Jagdale *et al.* (2007), who reported that *H. indica* caused significantly higher mortality of fungus gnats than the mortality caused by *S. carpocapsae* in Pro-Mix. Jagdale *et al.* (2007) also demonstrated that the efficacy of EPNs against fungus gnats differed with potting medium (nursery mix or Pro-Mix) and species. When EPNs were applied as a single application in nursery mix, only *H. bacteriophora* significantly reduced fungus gnat emergence relative to the control throughout the experimental period of 63 days, whereas *S. feltiae* reduced fungus gnats for 7 days after treatment. In Pro-Mix, only *H. indica* was effective against fungus gnats from 14 days after treatment. Furthermore, both *Heterorhabditis marelatus* and *H. zealandica* were effective against fungus gnats in the Pro-Mix and *S. carpocapsae* was only effective in the nursery mix for up to 3 days after treatment, suggesting that these species are incompatible with this potting medium. This agreed with the findings of Koppenhöfer and Fuzy (2006), who reported that the persistence of *H. zealandica* in potting mix (pine bark:peat moss:sand at 3:2:1 by volume) was poor (0–6%) and shorter than that of *H. bacteriophora* and *Steinernema glaseri*. According to Hoitink (1989), continued decomposition of hardwood bark during the growing season increases water-holding capacity and decreases air porosity, which in turn increases fungus gnat populations and suppresses plant-parasitic nematode populations. Therefore, the low efficacy of EPNs against fungus gnats in nursery mix obtained by Jagdale *et al.* (2004) may be due to the unfavorable environment for *S. feltiae*. Also, the addition of perlite in the growing substrates or use of perlite alone as a medium can adversely affect the efficacy of EPNs (A. Peters, personal communication, Germany, 2004). This may be due to the sharp edges of the perlite particles, which injure and kill moving IJs (M. Tomalak, unpublished data).

25.3.1.5 *Host plant*

Host plant species can influence fungus gnat colonization and the efficacy of *S. feltiae*. Jagdale *et al.* (2004) found that poinsettia supported significantly higher numbers of fungus gnats than impatiens. They also found that the EPN efficacy against fungus gnats was higher in impatiens than poinsettia. In addition, the efficacy of different EPN rates against fungus gnats in the glasshouse was host plant dependent. *S. feltiae* applied at 1.25×10^5 IJs m^{-2} caused a 55% reduction in fungus gnats in impatiens but only an 18% reduction in poinsettia 30 days after treatment. When applied at 2.5×10^5 IJs m^{-2}, the reduction was 41% in impatiens and only 20% in poinsettia 12 days after treatment. As many plant species can affect both chemical and physical properties of the potting medium (Argo, 1998), the efficacy of EPNs in poinsettia may be low due to unfavorable conditions in the rhizosphere. This area needs further investigation.

25.3.1.6 *Temperature*

In the glasshouse, the efficacy of EPNs against fungus gnats is temperature and species dependent. *S. feltiae* is a cold-adapted species with infection occurring between 8°C and 30°C and reproduction between 10°C and 25°C (Grewal *et al.*, 1994). Poor persistence and lack of reproduction of *S. feltiae* at warm temperatures poses a serious constraint for their use in glasshouses where temperatures often exceed 30°C during the summer. Gouge and Hague (1994) reported that the efficacy of cold-adapted *S. feltiae* against fungus gnats was reduced if soil temperatures in the glasshouse remained above 25°C for prolonged periods and suggested that *S. feltiae* should be used against fungus gnats at temperatures between 15°C and 26°C for satisfactory results. In a subsequent study, these researchers demonstrated that warm-adapted *Heterorhabditis* spp. and *S. anomali* and *S. riobrave* provided better control of fungus gnats than *S. feltiae* at 30°C (Gouge and Hauge, 1995a,b). Jagdale *et al.* (2004) demonstrated that *S. feltiae* was more effective at 22 ± 1°C) in the growth chamber (73–80%) than at 25 ± 5°C) in the glasshouse (34–41%). In an effort to find a suitable warm-adapted species that can be used to 30°C during the summer, G.B. Jagdale and P.S. Grewal (unpublished data, 2003) compared the efficacies of *H. bacteriophora* (GPS11 strain), *H. indica*, *H. zealandica* and *S. carpocapsae* against fungus gnats with *S. feltiae* in a growth chamber at fluctuating temperatures from 22°C to 29°C. Of the four warm-adapted species, *H. bacteriophora* (GPS11) and *H. indica* were significantly more effective than *S. feltiae* in controlling *B. coprophila* infesting poinsettia. These findings supported the findings of Jagdale *et al.* (2007), who also compared the effectiveness of the cold-adapted *S. feltiae* and *H. marelatus* with an intermediate *S. carpocapsae* and warm-adapted *H. bacteriophora*, *H. indica* and *H. zealandica* strains at 20/28 ± 1°C (night/day) against *Bradysia difformis* infesting poinsettia (*Euphorbia pulcherrima*) in the glasshouse. They reported that *H. bacteriophora* (52%) and *H. indica* (54%) performed better than *S. feltiae* (38%) in causing mortality of fungus gnats at temperatures above 25°C. When compared with the control treatment, both *H. bacteriophora* and *H. indica* significantly reduced the population of fungus gnats at 29 ± 1°C. Similarly, Katumanyane (2017) demonstrated that *H. indica*, *H. noenieputensis*, *H. zealandica* and *S. yirgalemense* caused over 80% mortality of *B. impatiens* at 25°C and 30°C compared with *S. feltiae*, which caused 82% mortality of *B. impatiens* at 13°C. In contrast, neither intermediate *S. carpocapsae* nor warm-adapted *H. zealandica* were effective against fungus gnats when average night-time and daytime temperatures were 20°C and 29°C, respectively (Jagdale *et al.*, 2007). These findings were similar to those of Lindquist and Piatkowski (1993), who also reported that when compared with *S. feltiae*, *S. carpocapsae* performed poorly against fungus gnats at 18°C (night) and 24°C (day) temperatures. However, Katumanyane *et al.* (2018) reported that *S. yirgalemense* caused 75% mortality of *B. impatiens* infesting cucumber plants grafted on pumpkin stalks in the glasshouse at 18°C (night) and 31°C (day) temperatures.

25.3.2 Western flower thrips

Western flower thrips (WFT), *Frankliniella occidentalis* (Thysanoptera: Thripidae), are considered a major pest of many field- and glasshouse-grown vegetables and ornamentals throughout the world (Yudin *et al.*, 1986; Mantel and van de Vrie, 1988; Tommasini and Maini, 1995). WFT can produce large populations within a few generations, and their life cycle has eggs, two larval instars, and prepupal, pupal and adult stages. Adults lay eggs in the parenchymal tissue of plants. After hatching, first-instar larvae feed by piercing the cell contents of aerial plant parts. Second-instar larvae feed voraciously until they move to the soil or to cryptic habitats for pupation, which lasts for 2–5 days. Newly emerged adults leave the soil, immediately disperse in the glasshouse, and begin feeding on leaves and flowers. In the glasshouse, WFT generally take about 10–20 days to complete the egg-to-egg life cycle at 25–30°C. The direct feeding of adults by piercing and scraping the tissues of leaves, stems, flowers and fruits leads to discoloration and drying of the wounded area, which give a flecked and malformed appearance to the infested plant parts. In addition to direct feeding, WFT are also capable of transmitting plant viruses (Ullman *et al.*, 1997), leading to substantial losses in the horticultural industry.

Resistance to chemical insecticides and their cryptic behavior make WFT control very difficult. Moreover, the use of many chemical compounds is limited due to concerns over the safety of beneficial organisms, which are concurrently released for pollination and control of glasshouse pests. Biological control agents including mites (*Amblyseius* and *Neoseiulus* spp.), entomopathogenic fungi (*Metarhizium* spp.) and bugs (*Orius* spp.) are currently available (Ebssa *et al.*, 2006; Ansari *et al.*, 2007, 2008; Pozzebon *et al.*, 2014; Saito and Brownbridge, 2018) and could provide an alternative to chemical pesticides, but they are only partially effective in controlling WFT. Several species of EPNs have been used as potential biological control agents and have shown great promise for controlling the soil-dwelling late second-instar nymphs, and prepupal and pupal stages of thrips (Tomalak, 1994b; Helyer *et al.*, 1995; Chyzik *et al.*, 1996; Ebssa *et al.*, 2001a,b, 2004, 2006; Premachandra *et al.*, 2003a,b; Dlamini *et al.*, 2019). It has been reported that the IJs of *S. feltiae* infected and killed both the prepupal and pupal stages of WFT within 2–4 h after infection. The actual death of WFT stages occurred because of the direct body damage resulting from the vigorously moving EPNs, and the small body size of the pupae prevented further development and reproduction of EPNs (Tomalak, 1994b).

25.3.2.1 EPN species/strain

In general, the efficacies of different EPN species and strains vary against soil-dwelling stages of WFT, and heterorhabditid nematodes tend to be more effective than steinernematids (Ebssa *et al.*, 2001a,b, 2004; Premachandra *et al.*, 2003a,b). Ebssa *et al.* (2004) screened six species of *Steinernema* (*S. abbasi*, *S. arenarium*, *S. bicornutum*, *S. carpocapsae*, *S. feltiae* and *Steinernema* sp.) and three species of *Heterorhabditis* (*H. bacteriophora*, *H. indica* and *H. marelatus*) against mixed stages of WFT in the soil and demonstrated that *Heterorhabditis* spp. nematodes caused overall higher WFT mortality (24–60%) than *Steinernema* spp. nematodes (2.6–54%). According to Ebssa *et al.* (2004), of the four strains of *S. carpocapsae* (A1 B5, S.N2, DD136 and S.S2) and *H. bacteriophora* (PS8 hybrid, PALH04, PALH05 and HK3), only DD136, S.S2, PALH04, PALH05 and HK3 caused significant mortality of mixed stages of WFT in the soil. In a laboratory bioassay, a similar trend was observed by Premachandra *et al.* (2003a) who demonstrated that two strains of *H. bacteriophora*, HD01 and HK3, caused significantly higher mortality of the prepupal and pupal stages of WFT than the HBN strain. In addition, the IS5 strain of *H. bacteriophora* was less effective in controlling WFT than the HP88 strain and two *Steinernema* spp. (Chyzik *et al.*, 1996). Recently, Gulzar *et al.* (2021a) tested the efficacy of eight species of EPNs including *H. bacteriophora*, *H. indica*, *H. georgiana*, *H. floridensis*, *S. carpocapsae*, *S. feltiae*, *S. rarum* and *S. riobrave* against tobacco thrips (*Frankliniella fusca*) using different bioassays and concluded that *H. bacteriophora*, *H. indica*, *S. feltiae* and *S. riobrave* have great potential for control of *F. fusca*. For example, results of both broad virulence screening and potting soil bioassays showed that treatment with *H. bacteriophora* resulted in the lowest (18.3% and 16.6%, respectively) emergence of thrips adults from pupae compared with *S. feltiae* (26.6% and 20%, respectively) and *S. riobrave* (35% and 22%, respectively). In a subsequent study, Gulzar *et al.* (2021b,c), also reported that the emergence of adult onion thrips (*Thrips tabaci*) was also significantly reduced in pots treated with *H. bacteriophora* (28.33 ± 1.66 adults) and *S. feltiae* (17.50 ± 0.00 adults) compared with the control after 1–7 days and 3–7 days of treatment, respectively.

Although all the available data suggest that EPNs can be effective against WFT in soil, relatively high application rates are needed to obtain satisfactory control. It is probable that further improvement of EPN performance and optimization of dosages against WFT could be obtained by finding new, more effective strains of EPNs. Another possible approach could be genetic selection/improvement of EPNs. Intraspecific hybridization and selection of *S. feltiae* resulted in development of a series of improved strains with much greater ability to control the pest populations in the glasshouse (Tomalak, 1994b). Application of the best strains at a rate of 1 million IJs of *S. feltiae* m^{-2} of soil surface resulted in 44–76% WFT mortality, and this increased efficacy was apparently related to the increased proportion of small IJs (<600 μm long) present in the EPN populations resulting from the intraspecific hybridization and selection. Unfortunately, despite strong selection pressure, a high proportion of individuals with the "small" phenotypes could not be maintained after relaxation of selection pressure (M. Tomalak, unpublished data). Therefore, further research is needed to produce stable populations of improved strains, perhaps through mutagenesis.

25.3.2.2 EPN application rate

To achieve economically feasible control of WFT, the appropriate application dose of EPNs needs to be determined. Application of *S. feltiae* at a rate of 1 million IJs m^{-2} of soil surface caused only 10–18% reduction in the emerging WFT adults. In contrast, application of *S. carpocapsae* at 25 × 10^4 IJs l^{-1} of compost produced 76.6% control for the prepupal and pupal stages of WFT (Helyer *et al.*, 1995), but this rate is very high and uneconomical for use in glasshouses. Tests with rates ranging from 100 to 1000 IJs of several different EPN species cm^{-2} of soil/medium concluded that a rate of 4 million IJs m^{-2} of the soil surface was the best for achieving more than 50% control of the soil-dwelling stages of WFT (Chyzik *et al.*, 1996; Ebssa *et al.*, 2001a,b, 2004; Premachandra *et al.*, 2003a,b). In a dose–response bioassay, Gulzar *et al.* (2021a) found that application rates of 25–100 IJs cm^{-2} of different EPNs tested against the pupae of *F. fusca* did not differ except for *H. bacteriophora* causing lower (60%) emergence of adult thrips from pupae than *S. carpocapsae* (70%) when applied at a rate of 250,000 IJ m^{-2}. Combined applications of ten predatory adult mites (*Amblyseius cucumeris*) and 2 million IJs of *H. indica* or *H. bacteriophora* m^{-2} caused a significantly higher reduction in the population of thrips (53%) than the untreated control (Ebssa *et al.*, 2006).

25.3.2.3 Life stage

Although EPNs are effective against mixed stages of WFT in the soil, EPN efficacies differ among the WFT stages. For example, the efficacy of *H. bacteriophora* HK3 strain was highest against both larvae and prepupae but lowest against pupae (Ebssa *et al.*, 2001a). Similarly, of three strains of *S. feltiae* (CR, OBSIII and Sylt), OBSIII and Sylt caused higher larval mortality, OBSIII caused the highest mortality of prepupae and Sylt caused the highest mortality of pupae. Between the species, both *H. bacteriophora* HK3 and *S. feltiae* Sylt caused overall highest mortality of all the three soil-dwelling stages including larvae, prepupae and pupae (Ebssa *et al.*, 2001a). According to Premachandra *et al.* (2003a), both *Steinernema* and *Heterorhabditis* spp. showed higher efficacy against the larval stages than the pupal stages, but according to Ebssa *et al.* (2001b) they were more effective against the prepupal and pupal stages than the late larval stages.

25.3.2.4 Temperature and moisture

The OBSIII strain of *S. feltiae* was more effective against late larval and prepupal stages with adequate soil moisture but less effective against pupal stages under comparatively dry conditions (Ebssa *et al.*, 2001a,b). It has been reported that the cold-adapted species *S. bicornutum* was more effective in reducing populations of WFT at 25°C or below than at higher temperatures. Similarly, the warm-adapted *H. indica* LN2 strain was more effective against WFT at higher (25–30°C) than lower (20°C) temperatures (Ebssa *et al.*, 2004). However, Ebssa *et al.* (2006) concluded that high temperature (47°C) and low relative humidity (15%) in the glasshouse may have helped build up the population of thrips and reduced the efficacies of predatory mites and of *H. indica* and *H. bacteriophora*.

25.3.3 Shore fly

The shore fly, *Scatella stagnalis*, is a frequent and occasionally numerous insect present in glasshouse cultures. Shore fly larvae feed on blue-green algae that develops on the surface of a variety of organic and mineral plant-growing substrates in moist habitats. Adult insects can reduce the aesthetic value of ornamentals by leaving fecal spots on leaves and flowers (Foote, 1977; Zack and Foote, 1978), whereas the larvae damage tender crop tissues causing them to dry, which affects the overall productivity and quality of the plants (Ciampolini and Suss, 1994). Moreover, the larvae are capable of ingesting spores of the fungal plant pathogens *Pythium* and *Fusarium* spp., which survive in their intestine and are transmitted to new hosts by adult flies (Goldberg and Stanghellini, 1990; Corbaz and Fischer, 1994).

In laboratory trials, *Scatella stagnalis* showed a very high susceptibility to *S. feltiae* (Gouge, 1994; Morton and García del Pino, 2003), *S. carpocapsae* (Gouge, 1994), *S. arenarium* and *H. megidis* (Morton and García del Pino, 2003). According to Morton and García del Pino (2003), all three EPN species caused 65–100% *S. stagnalis* larval mortality, which was dose dependent. EPN application doses as low as 3 IJs cm^{-2} of soil surface were sufficient to obtain 87% insect mortality. The most effective EPN species, *S. feltiae*, when applied at a dose of 500,000 IJs m^{-2}, caused 100% mortality within 2 days of treatment; *H. megidis* and *S. arenarium* applied at the same rate caused over 96% control. The rate of penetration of IJs into the insect hemocoel increased with increased rate of all the EPN species except *H. bacteriophora*, and the greatest penetration was observed in the case of IJs of *S. arenarium*. In a subsequent laboratory bioassay conducted in 24-well plates with the wells filled with algae, Morton and García del Pino (2007) reported that both *S. feltiae* and *S. arenarium* when applied at a rate of 500,000 IJs m^{-2} caused over 94% mortality of shore flies.

Surprisingly, glasshouse trials conducted by researchers in Europe and the USA provided poor results in terms of EPN efficacy against shore flies. Satisfactory control was not obtained on maize seedlings when *S. feltiae* was applied even at the higher dose of 2.5 × 10^{6} IJs m^{-2} on compost over 10 weeks (Lindquist *et al.*, 1994). Furthermore, applications of either *S. feltiae* (ScP) or *H. megidis* at a rate of 5 × 10^{6} IJs m^{-2} in commercial gerbera culture showed no visible effects on the population of shore flies in Poland (M. Tomalak, unpublished data, 1996–1997). Although these failed attempts have clearly prevented EPNs from being recommended against shore flies in commercial glasshouse cultures, more recent studies by Morton and Garcia del Pino (2003), who reported 65–100% larval mortality, and Vanninen and Koskula (2003), who reported 83% control of shore flies at a rate of 5 million *S. feltiae* IJs m^{-2}, suggest that more research is needed to develop alternative application methods to achieve a cost-effective control of shore flies with EPNs.

25.3.4 Glasshouse whitefly

The glasshouse whitefly, *Trialeurodes vaporariorum*, is one of the most economically important pests of over 250 plant species, especially glasshouse ornamentals and vegetables including poinsettia, gerbera, roses, beans, cucumber, tomatoes, peppers, lettuce, squash and pumpkin (Johnson *et al.*, 1992; Capinera, 2008). Worldwide, whiteflies cause millions of dollars of crop losses annually (Gelman *et al.*, 2005). Both adults and nymphs cause direct damage by sucking cell sap from leaves and twigs, which in turn reduces the plant vigor, and causes stunting, yellowing and premature loss of damaged leaves. Whiteflies also cause indirect damage by secreting honeydew, which encourages growth of sooty mold on leaf surfaces, hindering photosynthesis (Inbar and Gerling, 2008) and causing plant sickness (Gelman *et al.*, 2005). They also transmit plant viruses (Jones, 2003).

Chemical pesticides are used to control *T. vaporariorum*, but their use is avoided in controlled settings due to their detrimental effects on human health, non-target organisms and the environment, and the possibility of resistance development (Gorman *et al.*, 2007). Currently, biological control agents such as parasitoids (*Encarsia formosa* and *Eretmocerus eremicus*) and predatory mites (*Amblyseius swirskii*) have been successfully used as safe alternatives to control *T. vaporariorum* (van Lenteren *et al.*, 1996; de Vis and van Lenteren, 2008; Mortazavi *et al.*, 2019). In addition to these biological control agents, EPNs have shown potential for use as biological control agents against *T. vaporariorum* (Laznik *et al.*, 2011; Rezaei *et al.*, 2015). For example, both *S. feltiae* and *H. bacteriophora* were effective in killing both the second-instar nymphs and adults of *T. vaporariorum*, but *S. feltiae* provided a lower 50% lethal concentration (LC_{50}) for both nymphs (85 IJs cm^{-2}) and adults (143 IJs cm^{-2}) compared with that for *H. bacteriophora* for nymphs (94 IJs cm^{-2}) and adults (198 IJs cm^{-2}) under laboratory conditions (Rezaei *et al.*, 2015). These authors also reported that the mortality of both nymphs and adults of *T. vaporariorum* infesting cucumber and pepper significantly increased with increased rates of *S. feltiae* and *H. bacteriophora* from 25 to 250 IJs cm^{-2} under greenhouse conditions. It has been also demonstrated that the mortality of sweetpotato whiteflies (*Bemisia tabaci* B biotype) on cucumber increased by 20% when *S. feltiae* application rates were increased from 5000 to 15,000 IJs ml^{-1} (Qiu *et al.*, 2008).

25.4 Foliar Application

Several attempts have been made to use EPNs as biological control agents against pests located on foliage, but early results were not encouraging in glasshouses (Hara *et al.*, 1993) or fields (Kaya *et al.*, 1981; Hara *et al.*, 1993; Mason and Wright, 1997; Bélair *et al.*, 1998; Grewal and Georgis, 1998). However, recent development of more effective application techniques/tools for EPNs promises to improve their efficacy against foliar pests, particularly under glasshouse conditions.

25.4.1 Western flower thrips

The foliar feeding stages of WFT damage the rapidly growing regions of plants. In the early stages of flower development, a small amount of damage can lead to excessive aesthetic damage upon the opening of flower buds. This is also true for many other crops including food crops, where young shoot damage by WFT can often lead to the production of malformed and unmarketable products. Although most researchers have considered soil application of EPNs as a useful strategy to control WFT, foliar applications of EPNs may be considered more practical after reported successes in controlling leaf miners. One of the first positive results using EPNs against the foliar feeding stages of WFT was reported by Bennison *et al.* (1998), who achieved a significant reduction of nymphal stages on the leaves in both laboratory and glasshouse tests. This research led to the first successful use of commercially produced EPNs for controlling WFT by foliar applications (Wardlow *et al.*, 2001). Thereafter, several sites in the UK were treated with *S. feltiae* at rates of 125–250 × 10^6 IJs in 100 l per 1000 m^2, which in addition to successful control of the WFT populations also led in several cases to population reductions that were better than those achieved by the standard chemical treatments. This success was attainable by the prudent use of adjuvants to enable suitable targeting of the nematodes to the WFT on the foliage. Currently, throughout Europe and North America, use of commercially produced EPNs is being seen as a viable solution to control WFT when chemical compounds are withdrawn due to legislation. In contrast, Arthurs and Heinz (2006) reported that foliar application of *S. feltiae* at 1.25–2.5 × 10^3 IJs ml^{-1} and 1000–2000 l ha^{-1} at 3-day intervals did suppress populations of WFT. However, the thrips-parasitic nematode *Thripinema nicklewoodi*, when released through parasitized thrips, established and infected 83% of adult female WFT infesting flowers and buds of chrysanthemum (Arthurs and Heinz, 2006).

25.4.2 Leaf miners

Serpentine leaf miners (*Liriomyza* spp.) are economically important polyphagous pests in many countries. *Liriomyza* spp. commonly infest indoor vegetables including eggplant, beans, beet, carrots, celery, cucumbers, lettuce, melons, onions, peas, peppers, potatoes, squash and tomatoes. They also infest flowering plants including chrysanthemum, gerbera and gypsophila. Adult leaf miners measure less than 2 mm in length and live for about 13–18 days depending on temperature. Females lay eggs in punctures just beneath the epidermis on either the under or upper side of the leaf depending on the species. The eggs hatch within 4–8 days; the larvae immediately begin feeding on the mesophyll and undergo three molts within 7–13 days. The mature larvae eventually cut through the leaf epidermis and move to the soil for pupation, or may overwinter if they mature in autumn. Adults emerge within 3 weeks of pupation in the summer (Parrella, 1987).

Major damage is often in the form of punctures caused by females during feeding and oviposition and by larvae during mining, all of which result in the destruction of leaf mesophyll and a stippled appearance on foliage, especially at the leaf tip and along the leaf margins. Noticeable damage occurs 3–4 days after oviposition and increases with increasing size of larvae and mines. Both extensive mining and stippling on the leaves can greatly decrease the level of photosynthesis, which results in premature leaf drop (Parrella *et al.*, 1985). Excessive leaf drop can reduce the amount of shade, which in turn causes sun scalding of fruits. Wounding of the foliage also allows entry of bacterial and fungal disease-causing pathogens. Floriculture crops are generally less tolerant to leaf-miner damage than vegetable crops such as tomatoes, which are quite resilient and capable of withstanding considerable leaf damage.

In many countries, *Liriomyza* spp. are statutory pests and there is a general requirement for their eradication on imported crops and in outbreak situations. However, due to the protection of immature stages from conventional pesticide sprays and their ability to develop resistance to many insecticides, leaf miners are difficult pests to control (Mason *et al.*, 1987; Parrella *et al.*, 1999). Therefore, biological control agents including parasites and EPNs are deployed against leaf miners in glasshouses as an alternative to chemical pesticides (Harris *et al.*, 1990; Olthof and Broadbent, 1992; Williams and Walters, 1994; Parrella *et al.*, 1999).

25.4.2.1 EPN species and application rate

Head and Walters (2003) conducted glasshouse trials involving insecticide-resistant leaf-miner populations infesting lettuce and demonstrated that *S. feltiae* caused over 85% larval mortality, which exceeded the mortality achieved by chemical treatments. Similarly, foliar application of *S. feltiae* caused 82% mortality of the leaf miner *Liriomyza huidobrensis*, which was significantly higher than the chemical treatment, Heptenophos (Williams and Walters, 2000). These results agree with the findings of Harris *et al.* (1990), who demonstrated that *S. carpocapsae*, when applied at a rate of 5×10^8 IJs ha^{-1}, was equally as effective as abamectin (0.17 a.i. ha^{-1}) in causing mortality of *Liriomyza trifolii* infesting chrysanthemum foliage.

EPN efficacy against leaf miners is species/strain specific. Both *S. carpocapsae* and *S. feltiae*, when applied at same rate (9×10^9 IJs ha^{-1}), produced higher mortality (>63%) of *L. trifolii* on bean leaves than *Heterorhabditis* sp. (<33%) in a foghouse that was specially constructed with a fiberglass roof and polypropylene shade cloth sides and equipped with a fogging system to maintain the relative humidity above 80% (Hara *et al.*, 1993). However, when tested against another species of leaf miner (*L. huidobrensis*), *S. feltiae* and *Heterorhabditis* sp. were equally effective in reducing pupal formation (Williams and MacDonald, 1995). In addition, *S. feltiae* significantly and equally reduced the pupal production of three species of leaf miners, *L. huidobrensis*, *L. bryoniae* and *Chromatomyia syngenesiae* (Williams and Walters, 2000). In a laboratory test, strains of *S. feltiae* (SN, MG-24, MG-14 and MG25R3), *Heterorhabditis* sp. (H-3, MB-7, OK-3) and *S. carpocapsae* (S.20) were equally effective, causing 67–80% mortality of *L. trifolii* (Hara *et al.*, 1993). When *S. carpocapsae*, *S. feltiae* and *H. bacteriophora* were sprayed on to tomato leaves at a concentration of 1000 IJs ml^{-1} with 0.05% of the adjuvant Addit® (Koppert) against the tomato leaf miner, *Tuta absoluta*, both *S. feltiae* and *H. bacteriophora* caused 95% mortality of the leaf miners, whereas *S. carpocapsae* caused 87% mortality (Batalla-Carrera *et al.*, 2010; García-del-Pino *et al.*, 2018). According to Kamali *et al.* (2018), when applied at rates of 200,000 and 500,000 IJs m^{-2} in the soil, both *S. carpocapsae* and *H. bacteriophora* can cause 99% mortality of *T. absoluta* under growth chamber conditions but they can cause only 48–51% mortality of *T. absoluta* in greenhouse conditions.

25.4.2.2 *Life stage*

As the efficacy of EPNs varies among the developing stages of leaf miners, timing of application is important for targeting susceptible stages of this pest. Harris *et al.* (1990) reported that the prepupal and pupal stages of leaf miners were not susceptible to *S. carpocapsae* (Mexican strain). LeBeck *et al.* (1993) reported that all the larval and prepupal stages, except the pupal stage older than 1 h, were susceptible to *S. carpocapsae* All strain. Among the different larval stages of *L. trifolii*, second-stage larvae were most susceptible to *S. carpocapsae*, which caused 93% larval mortality (LeBeck *et al.*, 1993). Foliar application of *S. feltiae* also caused high mortality of all three larval stages of *L. huidobrensis*, the second instar being more susceptible than the first and third instars (Williams and MacDonald, 1995). It has been demonstrated that EPNs can infect and kill all four larval instars of *T. absoluta* (Batalla-Carrera *et al.*, 2010; van Damme *et al.*, 2016). For example, *H. bacteriophora*, *S. carpocapsae* and *S. feltiae* caused 33–86%, 41–95% and 59–91% mortality of all instars of *T. absoluta*, respectively (van Damme *et al.*, 2016). Batalla-Carrera *et al.* (2010) reported that *S. carpocapsae*, *S. feltiae* and *H. bacteriophora* caused higher mortality of late-instar larvae (78.6–100%) than of pupae (1.7–8.3%) for *T. absoluta* whether they were applied at a rate of 25 or 50 IJs cm^{-2} in a Petri dish bioassay. García-del-Pino *et al.* (2013) also reported that *S. carpocapsae*, *H. bacteriophora* and *S. feltiae* caused greater mortality to larvae (10%, 96.7% and 52.3%, respectively) than to pupae of *T. absoluta* in the soil under laboratory conditions. It has been demonstrated that the second and early third instars of leaf miners are susceptible to EPNs, so repeated EPN applications could target these instars (Williams and Walters, 2000). As some leaf miner species complete their life cycle on the underside of foliage, improved application technology is needed to target leaf-miner stages underneath the leaf.

25.4.2.3 *Temperature and relative humidity*

Temperature and relative humidity significantly affect the efficacy of EPNs in controlling leaf miners. Studies by Williams and MacDonald (1995) demonstrated that the foliar application of *S. feltiae* and *Heterorhabditis* sp. effectively killed second-stage larvae of *L. huidobrensis* at 20°C and over 80% relative humidity. *S. feltiae* remained effective against these larvae at temperatures between 10°C and 30°C, but relative humidity of over 90% was optimal for its efficacy.

S. feltiae also exhibited the highest efficacy against pupal production of *L. huidobrensis* at 10°C and over 90% relative humidity. Williams and Walters (2000) further confirmed the effectiveness of *S. feltiae* against three leaf-miner species including *L. huidobrensis*, *L. bryoniae* and *C. syngenesiae* at 20°C and over 90% relative humidity. *S. carpocapsae* All strain also showed promise, achieving 50–67% leaf-miner mortality at 22–23°C and 92–95% relative humidity (Olthof and Broadbent, 1992; Hara *et al.*, 1993). Maintaining 20°C temperature and over 90% relative humidity in glasshouses appears ideal for effective leaf-miner control. Adequate water usage is crucial for EPN mobility and successful leaf-miner infection, but excess water can lead to EPN runoff and reduced efficacy. Strategies such as use of anti-desiccants may enhance water retention on leaf surfaces and improve EPN host-seeking capabilities (Glazer *et al.*, 1992). Evening applications are favored, as higher evening humidity levels support EPN survival and targeting of leaf miners compared with sunny mornings. Hara *et al.* (1993) emphasized the variability in EPN strains and greenhouse humidity levels, highlighting the importance of precise water management during application (Olthof and Broadbent, 1992; Hara *et al.*, 1993; Williams and Walters, 1994).

25.5 Potential New Target Pests and Entomopathogenic Nematode Isolates

The progress made in recent years on EPN applications against glasshouse pests holds promise that other insect species and crops can be targeted. New *Steinernema* and *Heterorhabditis* spp. and strains are continuously being discovered that potentially carry new characteristics of practical value. Better sensitivity to host-specific cues, better desiccation tolerance and better adaptation to high moisture conditions in the soil are some of the characters that would be particularly welcome in new nematode products. Research on further improvements of application techniques and equipment, including better timing and improved formulations for foliar treatments, is still under way (see Chapter 11, this volume).

As the technology of glasshouse production develops further, new environmental conditions may lead to new problems for pest control. One such problem has recently been identified in tomato and cucumber cultures grown on rock wool. It was observed that many populations of sciarids (*Bradysia paupera*) have adapted to environmental conditions prevailing in the rock wool and occasionally become serious pests and plant pathogen vectors in the glasshouse. Although *S. feltiae* has been proven to be effective against this pest, direct application of EPNs to the surface of the growing substrate is difficult due to the polyethylene lining of the rock wool blocks. A series of laboratory and glasshouse experiments revealed that EPN infectivity to sciarid larvae exposed in the rock wool was not significantly different from that in horticultural compost. However, effective EPN application techniques had to be developed (M. Tomalak, unpublished data).

25.6 Conclusions

EPNs have become widely accepted as effective biological control agents for several soil-dwelling and foliage pests of many crops in glasshouses. They can be used safely with most chemical pesticides and all the other beneficial organisms routinely employed in this environment. Control of fungus gnats with EPNs has become a standard procedure in glasshouse crops. New target pests and new niches for EPN application are continuously being identified and positively verified by small- and large-scale glasshouse trials. Particularly promising is the recent commercial success of *S. feltiae*-based products in control of WFT and leaf miners on plant foliage. Results of research on the control of other pests, such as shore flies and soil-dwelling developmental stages of WFT, are also promising. Nevertheless, further research on the development of new EPN isolates, application methods and equipment are needed to meet the growing demands of modern glasshouse production.

25.7 References

Acharya, R., Hwang, H.-S., Shim, J.-K., Yu, Y.-S. and Lee, K.-Y. (2019) Control efficacy of fungus gnat, *Bradysia impatiens*, enhanced by a combination of entomopathogenic nematodes and predatory mites. *Biological Control* 138: 104071.

Adams, R. and Fonteno, R. (2003) Media. In: Hamrick, D. (ed.) *Ball Redbook Crop Production*. Ball Publishing, Batavia, Illinois, pp. 19–27.

Albajes, R., Lodovica Gullino, M., van Lenteren, J.C. and Elad, Y. (eds.) (1999) *Integrated Pest and Disease Management in Greenhouse Crops*. Kluver Academic Publishers, Dordrecht, Netherlands.

Ansari, M.A., Brownbridge, M., Shah, F.A. and Butt, T.M. (2008) Efficacy of entomopathogenic fungi against soil-dwelling stages of western flower thrips, *Frankliniella occidentalis*, in plant-growing media. *Entomologia Experimentalis Applicata* 27, 80–87.

Ansari, M.A., Shah, F.A., Whittaker, M., Prasaad, M. and Butt, T.M. (2007) Control of western flower thrips (*Frankliniella occidentalis*) pupae with *Metarhizium anisopliae* in peat and peat alternative growing media. *Biological Control* 40, 293–297.

Argo, W.R. (1998) Root medium chemical properties. *HortTechnology* 8, 481–485.

Arthurs, S. and Heinz, K.M. (2006) Evaluation of the nematodes *Steinernema feltiae* and *Thripinema nicklewoodi* as biological control agents of western flower thrips *Frankliniella occidentalis* infesting chrysanthemum. *Biocontrol Science and Technology* 16, 141–155.

Batalla-Carrera, L., Morton, A. and García-del-Pino, F. (2010) Efficacy of entomopathogenic nematodes against the tomato leafminer *Tuta absoluta* in laboratory and greenhouse conditions. *BioControl* 55, 523–530.

Bélair, G., Vincent, C. and Chouinard, G. (1998) Foliar sprays with *Steinernema carpocapsae* against early-season apple pests. *Journal of Nematology* 30, 599–606.

Bennison, J.A., Maulden, K.A., Wardlow, L.R., Pow, E.M. and Wadhams, L.J. (1998) Novel strategies for improving biological control of western flower thrips on protected ornamentals – potential new biological control agents. In: *Brighton Crop Protection Conference: Pests and Disease, 16–19 November, Brighton, UK, Vol.* 1. British Crop Protection Council, Alton, UK, pp. 193–198.

Binns, E.S. (1973) Fungus gnats (Diptera: Mycetophilidae, Sciaridae) and the role of mycophagy in soil: a review. *Revue d'Ecologie et de Biologie du Sol* 18, 77–90.

Braun, S.E., Sanderson, J.P. and Wraight, S.P. (2012) Larval *Bradysia impatiens* (Diptera: Sciaridae) potential for vectoring *Pythium* root rot pathogens. *Phytopathology* 102, 283–289.
Capinera, J.L. (2008) Greenhouse whitefly, *Trialeurodes vaporariorum* (Westwood) (Hemiptera: Aleyrodidae). In: Capinera, J.L. (ed.) *Encyclopedia of Entomology*. Springer, Dordrecht, Netherlands, pp. 1723–1726.
Chandler, D., Hri, W., Prince, G. and Bennison, J. (2010) *New Approaches to Microbial Control of Insect Pests in Protected Crops and their Interactions with Waste-based Growing Media*. Project No. PC283. University of Warwick, Warwick, UK.
Chyzik, R., Glazer, I. and Klein, M. (1996) Virulence and efficacy of different entomopathogenic nematode species against western flower thrips (*Frankliniella occidentalis*). *Phytoparasitica* 24, 103–110.
Ciampolini, M. and Suss, L. (1994) *Scatella stagnalis* Fallen (Diptera Ephydridae) dannosa allo stadio adulto a piante ortive e floreali in serra. *Bollettino di Zoologia Agraria e di Bachicoltura* 26, 115–126.
Cloyd, R.A. (2015) Ecology of fungus gnats (*Bradysia* spp.) in greenhouse production systems associated with disease-interactions and alternative management strategies. *Insects* 6, 325–332.
Cloyd, R.A. and Zaborski, E.R. (2004) Fungus gnats, *Bradysia* spp. (Diptera: Sciaridae), and other arthropods in commercial bagged soilless growing media and rooted plant plugs. *Journal of Economic Entomology* 97, 503–510.
Corbaz, R. and Fischer, S. (1994) La mouche *Scatella stagnalis* Fall. (*Diptera, Ephydridae*) responsable de la dispersion de *Fusarium oxysporum* f.sp. *lycopersici* dans les cultures de tomates hors sol. *Revue Suisse de Viticulture, d'Arboriculture et d'Horticulture* 26, 383–385.
de Vis, R.M.J. and van Lenteren, J.C. (2008) Biological control of *Trialeurodes vaporariorum* by *Encarsia formosa* on tomato in unheated greenhouses in the high-altitude tropics. *Bulletin of Insectology* 61, 43–57.
Dlamini, T., Allsopp, E. and Malan, A. (2019) Management of *Frankliniella occidentalis* (western flower thrips), and the potential use of entomopathogenic nematodes: a South African Perspective. *African Entomology* 27, 265–278.
Ebssa, L., Borgemeister, C., Berndt, O. and Poehling, H.M. (2001a) Efficacy of entomopathogenic nematodes against soil-dwelling life stages of western flower thrips, *Frankliniella occidentalis* (Thysanoptera: Thripidae). *Journal of Invertebrate Pathology* 78, 119–127.
Ebssa, L., Borgemeister, C., Berndt, O. and Poehling, H.M. (2001b) Impact of entomopathogenic nematodes on different soil-dwelling stages of western flower thrips, *Frankliniella occidentalis* (Thysanoptera: Thripidae), in the laboratory and under semi-field conditions. *Biocontrol Science and Technology* 11, 515–525.
Ebssa, L., Borgemeister, C. and Poehling, H.M. (2004) Effectiveness of different species/strains of entomopathogenic nematodes for control of western flower thrips (*Frankliniella occidentalis*) at various concentrations, host densities and temperatures. *Biological Control* 29, 145–154.
Ebssa, L., Borgemeister, C. and Poehling, H.M. (2006) Simultaneous application of entomopathogenic nematodes and predatory mites to control western flower thrips *Frankliniella occidentalis*. *Biological Control* 39, 66–74.
Foote, B.A. (1977) Utilization of blue-green algae by larvae of shore flies. *Environmental Entomology* 6, 812–814.
Freeman, P. (1983) Sciarid flies, Diptera; Sciaridae. In: *Handbooks for the Identification of British Insects, Vol.* 9, Part 6. Royal Entomological Society, London, p. 68.
García-del-Pino, F., Alabern, X. and Morton, A. (2013) Efficacy of soil treatments of entomopathogenic nematodes against the larvae, pupae and adults of *Tuta absoluta* and their interaction with the insecticides used against this insect. *BioControl* 58, 723–731.
García-del-Pino, F., Morton, A. and Shapiro-Ilan, D. (2018) Entomopathogenic nematodes as biological control agents of tomato pests. In: Wakil, W., Brust, G.E. and Perring, T.M. (eds.) *Sustainable Management of Arthropod Pests of Tomato*. Academic Press, London, pp. 269–282.
Gelman, D.B., Blackburn, M.B. and Hu, J.S. (2005) Identification of the molting hormone of the sweet potato (*Bemisia tabaci*) and greenhouse (*Trialeurodes vaporariorum*) whitefly. *Journal of Insect Physiology* 51, 47–53.
Gillespie, D.R. and Menzies, J.G. (1993) Fungus gnat vector *Fusarium oxysporum* f. sp. *radicislycopersici*. *Annals of Applied Biology* 123, 539–544.
Glazer, I., Klein, M., Navon, A. and Nakache, Y. (1992) Comparison of efficacy of entomopathogenic nematodes combined with antidesiccants applied by canopy sprays against three cotton pests. *Journal of Economic Entomology* 85, 1636–1640.
Goldberg, N.P. and Stanghellini, M.E. (1990) Ingestion–egestion and aerial transmission of *Pythium aphanidermatum* by shore flies (*Ephydrinae: Scatella stagnalis*). *Phytopathology* 80, 1244–1246.

Gorman, K., Devine, G., Bennison, J., Coussons, P., Punchard, N. and Denholm, I. (2007) Report of resistance to the neonicotinoid insecticide imidacloprid in *Trialeurodes vaporariorum* (Hemiptera: Aleyrodidae). *Pest Management Science* 63, 555–558.

Gouge, D.H. (1994) Biological control of sciarid and other flies with entomopathogenic nematodes. PhD thesis, University of Reading, Reading, UK.

Gouge, D.H. and Hague, N.G.M. (1994) Development of *Steinernema feltiae* (Steinernematidae: Nematoda) in *Bradysia paupera* (Sciaridae: Diptera). *IOBC/WPRS Bulletin* 17, 132–135.

Gouge, D.H. and Hague, N.G.M. (1995a) The susceptibility of different species of sciarid flies to entomopathogenic nematodes. *Journal of Helminthology* 69, 313–318.

Gouge, D.H. and Hague, N.G.M. (1995b) Glasshouse control of fungus gnats, *Bradysia paupera*, on fuchsias by *Steinernema feltiae*. *Fundamental and Applied Nematology* 18, 77–80.

Grewal, P.S. and Georgis, R. (1998) Entomopathogenic nematodes. In: Hall, F.R. and Menn, J.J. (eds.) *Biopesticides: Use and Delivery*. Humana Press, Totowa, New Jersey, pp. 271–299.

Grewal, P.S., Selvan, S. and Gaugler, R. (1994) Thermal adaptation of entomopathogenic nematodes: niche breadth for infection, establishment, and reproduction. *Journal of Thermal Biology* 19, 245–253.

Gulzar, S., Usman, M., Wakil, W., Wu, S., Oliveira-Hofman, C. *et al.* (2021a) Virulence of entomopathogenic nematodes to pupae of *Frankliniella fusca* (Thysanoptera: Thripidae). *Journal of Economic Entomology* 114, 2018–2023.

Gulzar, S., Wakil, W. and Shapiro-Ilan, D.I. (2021b) Combined effect of entomopathogens against *Thrips tabaci* Lindeman (Thysanoptera: Thripidae): laboratory, greenhouse and field trials. *Insects* 12*:* 456.

Gulzar, S., Wakil, W. and Shapiro-Ilan, D.I. (2021c) Potential use of entomopathogenic nematodes against the soil dwelling stages of onion thrips, *Thrips tabaci* Lindeman: laboratory, greenhouse and field trials. *Biological Control* 161: 104677.

Hara, A.H., Kaya, H.K., Gaugler, R., Lebeck, L.M. and Mello, C.L. (1993) Entomopathogenic nematodes for biological control of the leafminer, *Liriomyza trifolii* (Dipt.: Agromyzidae). *Entomophaga* 38, 359–369.

Harris, M.A., Begley, J.W. and Warkentin, D.L. (1990) *Liriomyza trifolii* (Diptera: Agromyzidae) suppression with foliar application *Steinernema carpocapsae* (Rhabditida: Steinernematidae) and abamectin. *Journal of Economic Entomology* 83, 2380–2384.

Harris, M.A., Oetting, R.D. and Gardner, W.A. (1995) Use of entomopathogenic nematodes and a new monitoring technique for control of fungus gnats, *Bradysia coprophila* (Diptera: Sciaridae), in floriculture. *Biological Control* 5, 412–418.

Head, J. and Walters, K.F.A. (2003) Augmentation biological control utilising the entomopathogenic nematode, *Steinernema feltiae*, against the South American leafminer, *Liriomyza huidobrensis*. In: *Proceedings of the 1st International Symposium on Biological Control, 13–18 January 2002, Honolulu, Hawaii*. USDA Forest Service FHTET-03-05. Forest Health Technology Enterprise Team, Washington, DC, pp. 136–140.

Helyer, N.L., Brobyn, P.J., Richardson, P.N. and Edmondson, R.N. (1995) Control of Western flower thrips (*Frankliniella occidentalis* Pergande) pupae in compost. *Annals of Applied Biology* 127, 405–412.

Hoitink, H.A.J. (1989) Integrated control of *Pythium*, *Rhizoctonia* and *Fusarium* diseases. In: Daughtrey, M. and Begley, J. (eds.) *Proceedings of the Fifth Conference on Insect and Disease Management on Ornamentals, 26–28 February, Orlando, Florida*. Society of American Florists, Alexandria, Virginia, pp. 116–121.

Hussey, N.W. and Scopes, N. (eds.) (1985) *Biological Pest Control: The Glasshouse Experience*. Blandford Press, Poole, UK.

Inbar, M., and Gerling, D. (2008) Plant-mediated interactions between whiteflies, herbivores, and natural enemies. *Annual Review of Entomology* 53, 431–448.

Jagdale, G.B., Casey, M.L., Grewal, P.S. and Lindquist, R.K. (2004) Application rate and timing, potting medium, and host plant on efficacy of *Steinernema feltiae* against fungus gnat, *Bradysia coprophila*, in floriculture. *Biological Control* 29, 296–305.

Jagdale, G.B., Casey, M.L., Grewal, P.S. and Cañas, L. (2007) Effect of entomopathogenic nematode species, split application and potting medium on the control of the fungus gnat, *Bradysia difformis* (Diptera: Sciaridae), in the greenhouse at alternating cold and warm temperatures. *Biological Control* 43, 23–30.

Jerardo, A. (2004) Floriculture and nursery crops outlook. Electronic Outlook Report from the Economic Research Service. Available at: https://usda.library.cornell.edu/concern/publications/bn999674d?locale=en (accessed 26 March 2024).

Jerardo, A. (2007) Floriculture and nursery crops yearbook. Economic Research Service, USDA, Washington, DC. Available at: https://downloads.usda.library.cornell.edu/usda-esmis/files/5m60qr88b/tt44pr36t/n009w541n/FLO-yearbook-10-04-2007.pdf (accessed 26 March 2024).

Johnson, M.W., Caprio, L.C., Coughlin, J.A., Tabashnik, B.E., Rosenheim, J.A. and Wel Ter, S.C. (1992) Effect of *Trialeurodes vaporariorum* (Homoptera: Aleyrodidae) on yield of fresh market tomatoes. *Journal of Economic Entomology* 85, 2370–2376.

Jones, D.R. (2003) Plant viruses transmitted by whiteflies. *European Journal of Plant Pathology* 109, 159–219.

Kamali, S., Karimi, J. and Koppenhöfer, A.M. (2018) New insight into the management of the tomato leaf miner, *Tuta absoluta* (Lepidoptera: Gelechiidae) with entomopathogenic nematodes. *Journal of Economic Entomology* 111, 112–119.

Katumanyane, A. (2017) Prospects for using entomopathogenic nematodes as a biocontrol agent against fungus gnats, *Bradysia* spp. (Diptera: Sciaridae) in nursery and glasshouse crops. MSc thesis, University of Stellenbosch, Stellenbosch, South Africa.

Katumanyane, A., Ferreira, T. and Malan, A.P. (2018). Greenhouse application of *Steinernema yirgalemense* to control fungus gnats, *Bradysia impatiens*. *BioControl* 63, 729–738.

Kaya, H.K., Hara, A.H. and Reardon, R.C. (1981) Laboratory and field evaluation of *Neoplectana carpocapsae* (Rhabditida: Steinernematidae) against the elm leaf beetle (Coleoptera: Chrysomelidae) and the western spruce budworm (Lepidoptera: Tortricidae). *Canadian Entomologist* 113: 787.

Kim, H.H., Choo, H.Y., Kaya, H.K., Lee, D.W., Lee, S.M. and Jeon, H.Y. (2004) *Steinernema carpocapsae* (Rhabditida: Steinernematidae) as a biological control agent against fungus gnat *Bradysia agrestis* (Diptera: Sciaridae) in propagation houses. *Biocontrol Science and Technology* 14, 171–183.

Koppenhöfer, A.M. and Fuzy, E.M. (2006) Effect of soil type on infectivity and persistence of the EPNs *Steinernema scarabaei*, *Steinernema glaseri*, *Heterorhabditis zealandica*, and *Heterorhabditis bacteriophora*. *Journal of Invertebrate Pathology* 92, 11–22.

Laznik, Z., Znidarcic, D. and Trdan, S. (2011) Control of *Trialeurodes vaporariorum* (Westwood) adults on glasshouse-grown cucumbers in four different growth substrates: an efficacy comparison of foliar application of *Steinernema feltiae* (Filipjev) and spraying with thiamethoxam. *Turkish Journal of Agriculture and Forestry* 35, 631–640.

LeBeck, L.M., Gaugler, R., Kaya, H.K., Hara, A.H. and Johnson, M.W. (1993) Host stage suitability of the leafminer *Liriomyza trifolii* (Diptera: Agromyzidae) to the entomopathogenic nematode *Steinernema carpocapsae* (Rhabditida: Steinernemattidae). *Journal of Invertebrate Pathology* 62, 58–83.

Lindquist, R.K. and Piatkowski, J. (1993) Evaluation of entomopathogenic nematodes control of fungus gnat larvae. *IOBC/WPRS Bulletin* 16, 97–100.

Lindquist, R.K., Faber, W.R. and Casey, M.L. (1985) Effect of various soilless root media and insecticides on fungus gnats. *HortScience* 20, 358–360.

Lindquist, R., Buxton, J. and Piatkowski, J. (1994) Biological control of sciarid flies and shore flies in glasshouses. *Crop Protection Conference: Pests and Diseases* 3, 1067–1072.

Ludwig, S.W. and Oetting, R.D. (2001) Evaluation of medium treatments for management of *Frankliniella occidentalis* (Thripidae: Thysanoptera) fungus gnats, *Bradysia coprophila* (Diptera: Sciaridae). *Pest Management Science* 57, 1114–1118.

Mantel, W.P. and van de Vrie, M. (1988) De Californische trios, *Frankliniella occidentalis*, een nieuwe Schadelijke tripssoort in de tuinbouw onder glas in Nederland. *Entomologische Berichten* 48, 140–144.

Mason, J.M. and Wright, D.J. (1997) Potential of the control of *Plutella xylostella* larvae with entomopathogenic nematodes. *Journal of Invertebrate Pathology* 70, 234–242.

Mason, G.A., Tabashnik, B.E. and Johnson, M.W. (1987) Susceptibility of *Liriomyza sativae* and *L. trifolii* (Diptera: Agromyzidae) to permethrin and fenvalerate. *Journal of Economic Entomology* 80, 1262–1266.

Meers, T.L. and Cloyd, R.A. (2005) Egg-laying preference of female fungus gnat *Bradysia* sp. nr. *coprophila* (Diptera: Sciaridae) on three different soilless substrates, *Journal of Economic Entomology* 98, 1937–1942.

Moorhouse, E.R., Charnley, A.K. and Gillespie, A.T. (1992) A review of the biology and control of the vine weevil, *Otiorhynchus sulcatus* (Coleoptera: Curculionidae). *Annals of Applied Biology* 121, 431–454.

Mortazavi, N., Fathipour, Y. and Talebi, A.A. (2019) The efficiency of *Amblyseius swirskii* in control of *Tetranychus urticae* and *Trialeurodes vaporariorum* is affected by various factors. *Bulletin of Entomological Research* 109, 365–375.

Morton, A. and García del Pino, F. (2003) Potential of entomopathogenic nematodes for the control of shore flies (*Scatella stagnalis*). In: *Growing Biocontrol Markets Challenge Research and Development, 9th European Meeting of the IOBC/WPRS Working Group "Insect Pathogens and Entomopathogenic Nematodes"*, Abstract 67.

Morton, A. and García del Pino, F. (2007) Susceptibility of shore fly *Scatella stagnalis* to five entomopathogenic nematode strains in bioassays. *BioControl* 52, 533–545.

Nedstam, B. and Burman, M. (1990) The use of nematodes against sciarids in Swedish greenhouses. *IOBC/WPRS Bulletin* 13, 147–148.
Oetting, R.D. and Latimer, J.G. (1991) An entomogenous nematode *Steinernema carpocapsae* is compatible with potting media environments created by horticultural practices. *Journal of Entomological Science* 26, 390–394.
Olson, D.L., Oetting, R.D. and van Iersel, M.W. (2002) Effect of soilless media and water management on development of fungus gnats (Diptera: Sciaridae) and plant growth. *HortScience* 37, 919–923.
Olthof, T.H.A. and Broadbent, A.B. (1992) Evaluation of steinernematid nematodes for control of a leafminer, *Liriomyza trifolii*, in greenhouse chrysanthemums. *Journal of Nematology* 24: 612.
Parrella, M.P. (1987) Biology of *Liriomyza*. *Annual Review of Entomology* 32, 201–224.
Parrella, M.P., Jones, V.P., Youngman, R.R. and LeBeck, L.M. (1985) Effect of leaf mining and leaf stippling of *Liriomyza* spp. on photosynthetic rates of chrysanthemum. *Annals of Entomological Society of America* 78, 90–93.
Parrella, M.P., Hansen, L.S. and van Lenteren, J. (1999) Glasshouse environments. In: Bellows, T.S. and Fisher, T.W. (eds.) *Handbook of Biological Control*. Academic Press, San Diego, California, pp. 819–839.
Pozzebon, A., Boaria, A. and Duso, C. (2014) Single and combined releases of biological control agents against canopy- and soil-dwelling stages of *Frankliniella occidentalis* in cyclamen. *BioControl* 60, 341–350.
Premachandra, W.T.S.D., Borgemeister, C., Berndt, O., Ehlers, R.-U. and Poehling, H.-M. (2003a) Laboratory bioassays of virulence of entomopathogenic nematodes against soil-inhabiting stages of *Frankliniella occidentalis* Pergand (Thysanoptera: Thripidae). *Nematology* 5, 539–547.
Premachandra, W.T.S.D., Borgemeister, C., Berndt, O., Ehlers, R.-U. and Poehling, H.-M. (2003b) Combined release of entomopathogenic nematodes and the predatory mite *Hypoaspis aculeifer* to control soil-dwelling stages of western flower thrips *Frankliniella occidentalis*. *BioControl* 48, 529–541.
Qiu, B.L., Mandour, N.S., Xu, C.X. and Ren, S.X. (2008) Evaluation of the entomopathogenic nematode *Steinernema feltiae* as a biological control agent of the whitefly, *Bemisia tabaci*. *International Journal of Pest Management* 54, 247–253.
Rezaei, N., Karimi, J., Hosseini, M., Goldani, M. and Campos-Herrera, R. (2015) Pathogenicity of two species of entomopathogenic nematodes against the greenhouse whitefly, *Trialeurodes vaporariorum* (Hemiptera: Aleyrodidae), in laboratory and greenhouse experiments. *Journal of Nematology* 47, 60–66.
Saito, T. and Brownbridge, M. (2018) Compatibility of soil-dwelling predators and microbial agents and their efficacy in controlling soil dwelling stages of western flower thrips, *Frankliniella occidentalis*. *Biological Control* 92, 92–100.
San-Blas, E., Luzardo, M., Portillo, E., Fuenmayor, Y. and Bastidas, B. (2017) *Heterorhabditis amazonensis* for biological control of fungus gnats *Bradysis difformis* in daisy gerberas. *BioControl* 62, 847–855.
Scarlett, K., Tesoriero, L. and Guest, D. (2014) Sciarid and shore flies as aerial vectors of *Fusarium oxysporum* f. sp. *cucumerinum* in greenhouse cucumbers. *Journal of Applied Entomology* 138, 368–377.
Tomalak, M. (1994a) Selective breeding of *Steinernema feltiae* (Filipjev) (Nematoda: Steinernematidae) for improved efficacy in control of a mushroom fly, *Lycoriella solani* Winnertz (Diptera: Sciaridae). *Biocontrol Science and Technology* 4, 187–198.
Tomalak, M. (1994b) Genetic improvement of *Steinernema feltiae* for integrated control of the Western flower thrips, *Frankliniella occidentalis*. *IOBC/WPRS Bulletin* 17, 17–20.
Tommasini, M. and Maini, S. (1995) *Frankliniella occidentalis* and other thrips harmful to vegetables and ornamental crops in Europe. In: *Biological Control of Thrip Pests*. Wageningen Agriculture University Papers No. 95. Wageningen Agriculture University, Wageningen, Netherlands.
Ullman, D.E., Sherwood, J.L. and German, T.L. (1997) Thrips as vectors of plant pathogens. In: Lewis, T. (ed.) *Thrips as Crop Pests*. CAB International, Wallingford, UK, pp. 539–565.
van Damme, V.M., Beck, B.K., Berckmoes, E., Moerkens, R., Wittemans, L. *et al.* (2016) Efficacy of entomopathogenic nematodes against larvae of *Tuta absoluta* in the laboratory. *Pest Management Science* 72, 1702–1709.
van Lenteren, J.C. (2003) Commercial availability of biological control agents. In: van Lenteren, J.C. (ed.) *Quality Control and Production of Biological Control Agents: Theory and Testing Procedures*. CAB International, Wallingford, UK, pp. 167–179.
van Lenteren, J.C., van Roermund, H.J.W. and Sütterlin, S. (1996) Biological control of greenhouse whitefly (*Trialeurodes vaporariorum*) with the parasitoid *Encarsia formosa*: how does it work? *Biological Control* 6, 1–10.
Vanninen, I. and Koskula, H. (2003) Biological control of the shore fly (*Scatella tenuicosta*) with steinernematid nematodes and *Bacillus thuringiensis* var. *thuringiensis* in peat and rockwool. *Biocontrol Science and Technology* 13, 47–63.

Vaughan, M., Tholl, D. and Tokuhisa, J. (2011) An aeroponic culture system for the study of root herbivory on *Arabidopsis thaliana*. *Plant Methods* 7, 5–14.

Wardlow, L.R., Piggott, S. and Goldsworthy, R. (2001) Foliar application of *Steinernema feltiae* for the control of flower thrips. *Medeelingen Faculteit Landbouwkundige en Toegepaste Biologische Wetenschappen Universiteit Gent* 66, 285–291.

Wilkinson, J.D. and Daugherty, D.M. (1970) Comparative development of *Bradysia impatiens* (Diptera: Sciaridae) under constant and variable temperatures. *Annals of the Entomological Society of America* 63, 1079–1083.

Williams, E.C. and MacDonald, O.C. (1995) Critical factors required by the nematode *Steinernema feltiae* for the control of the leafminers *Liriomyza huidobrensis*, *Liriomyza bryoniae* and *Chromatomyia syngenesiae*. *Annals of Applied Biology* 127, 329–341.

Williams, E.C. and Walters, K.F.A. (1994) Nematode control of leafminers: efficacy, temperature and timing. In: *Proceedings of the Brighton Crop Protection Conference: Pests and Diseases, 21–24 November, Brighton, UK, Vol.* 3. British Crop Protection Council, Alton, UK, pp. 1079–1084.

Williams, E.C. and Walters, K.F.A. (2000) Foliar application of the entomopathogenic nematode *Steinernema feltiae* against leafminers on vegetables. *Biocontrol Science and Technology* 10, 61–70.

Yudin, L.S., Cho, J.J. and Mitchell, W.C. (1986) Host range of western flower thrips, *Frankliniella occidentalis* (Thysanoptera: Thripidae) with special references to *Leucaena glauca*. *Environmental Entomology* 15, 1292–1295.

Zack, R.S. and Foote, B.A. (1978) Utilization of algal monoculture by larvae of *Scatella stagnalis*. *Environmental Entomology* 7, 509–511.

26 Applications for the Control of Pests of Humans and Animals

John A. Goolsby[1*], Itamar Glazer[2], Michael Samish[3] and Fernando Garcia del Pino[4]

[1]US Department of Agriculture, Agricultural Research Service, Cattle Fever Tick Research Unit, Edinburg, Texas, USA; [2]Department of Nematology, ARO, Volcani Center, Bet-Dagan, Israel; [3]Kimron Veterinary Institute, Bet-Dagan, Israel; [4]Department of Animal Biology, Vegetal Biology and Ecology, Faculty of Bioscience, Universitat Autònoma de Barcelona, Bellaterra, Barcelona, Spain

Abstract

Biological control of plant insect pests using entomopathogenic nematodes (EPNs) is developing rapidly, but relatively little attention is being paid as yet to their potential use against veterinary and human pests. Entomophilic mermithids were studied intensively for use against mosquitoes during the 1960s, but the initiation of use of steinernematids and heterorhabditids against veterinary and human pests has been explored only in the last two decades. Progress has been made using *Steinernema riobrave* nematodes applied directly to free-ranging, tick-infested wildlife. Due to increasing levels of resistance in cockroaches to conventional pesticides, EPNs may be a new strategy for their biocontrol. *Steinernema carpocapsae* applied directly to cockroach habitats or used in bait stations has been shown to be effective against this cosmopolitan pest. This chapter provides the current knowledge on this particular issue and is intended to encourage further exploration of EPN use against veterinary and human pests.

26.1 Introduction

Thousands of species of insects, mites and ticks worldwide have the ability to attack animals. Fortunately, only a few are significant within a given area as pests of pets, domestic livestock, poultry and wildlife (Lancaster and Meisch, 1986; Tsao *et al.*, 2021). For example, in the USA, only 45 species are reported to be pests of cattle and 75 of fowl. Economic damage caused by such pests is in the range of tens of billions of US dollars. The damage is related not only to the direct effect of the pests on animal mortality and the reduction in productivity and reproduction, but also to the ability of certain pests to transmit diseases. During the past 100 years, control of veterinary pests and the diseases they transmit has largely been through the application of insecticides and acaricides as sprays or dips. Short-interval applications of pesticides, in conjunction with control of animal movement, quarantine and slaughter, can prevent transmission of the parasites. However, the development of pesticide resistance has been a major problem. This has been compounded by the increasing cost of pesticides and poor management, and also by inadequate maintenance. Another complication associated with the use of pesticides in husbandry

*Email: goolsby1600@gmail.com

 Entomopathogenic Nematodes as Biological Control Agents (D.I. Shapiro-Ilan and E.E. Lewis eds)

DOI: 10.1079/9781800620322.0026

practice is their role as environmental pollutants, and they may also contaminate animal products such as milk and meat.

Insects and other arthropods also parasitize humans. The most important pests for humans are mosquitoes, black flies, fleas, ticks, lice, houseflies and cockroaches. As in the case of animals, the devastating effect of these pests is not only their direct damage and nuisance but also transmission of diseases. Control measures for human pests include chemicals (pesticides and repellents) and physical methods (nets and elimination of incubation sites). Control of animal pests is applied mainly to the animal directly or to the site it inhabits, and in the case of human pests, the application is also done at the breeding sites of the pests.

The biocontrol of plant insect pests is developing rapidly, but relatively little attention is being paid as yet to its potential use against veterinary and human pests (Samish and Rehacek, 1999). The only worldwide success story of a biological control agent against human pests is the use of the bacterium *Bacillus thuringiensis israelensis* for control of mosquito larvae (Margalit and Ben-Dov, 2000). Although the pathogenicity of entomopathogenic nematodes (EPNs) was tested against thousands of insect species, little attention was given by entomopathogenic nematologists to veterinary and human pests. Entomophilic mermithids have been studied intensively for use against mosquitoes during the 1960s (see Chapter 22, this volume), but the initiation of use of steinernematids and heterorhabditids against veterinary and human pests has been explored only in the last two decades. This chapter provides the current knowledge on this particular issue and is intended to encourage further exploration of EPN use against veterinary and human pests.

26.2 Ticks

Ticks (Ixodidae; approximately 850 species) inhabit highly varied ecological niches. Ticks are economically very important pests worldwide, mainly as vectors of animal and human diseases. Approximately 80% of the world's cattle population of 1281 million is at risk from ticks and tick-borne diseases (Sutherst *et al.*, 1982). Over four decades ago, McCosker (1979) estimated the global cost of their control and of productivity losses to be US\$7,000 million year^{-1} (= US\$7 per head year^{-1}). In Africa, with 186 million head of cattle, ticks and tick-borne diseases are the most serious constraints to increased production. Tick-borne diseases also potentially affect wildlife (Ginsberg, 1993; Sonenshine and Roe, 1993). Furthermore, in the USA, tick-borne diseases are by far the most commonly reported vector-borne illnesses affecting humans, especially Lyme disease and Rocky Mountain spotted fever (Centers for Disease Control, 1996).

Ticks are the sole or major vectors of many domestic animal pathogens, such as *Anaplasma*, *Babesia*, *Cowdria*, *Ehrlichia* and *Theileria* spp., and human diseases, such as African tick-borne relapsing fever, Q fever, tick-borne encephalitis, Lyme disease and Rocky Mountain spotted fever. In addition, ticks can also provoke anemia, toxicosis and sometimes also paralysis.

Ticks are obligatory blood-sucking arthropods. They have three blood-sucking stages, larvae, nymphs and adults. In some species, all three stages drop off the animal to the ground when fully engorged, while in others only two stages or only fully engorged adults drop off. During most of their life cycle, the ticks remain in the upper layer of the ground. Engorged female ticks require several pre-oviposition days after they drop off the host. The other tick stages that drop off their hosts to the ground also need several days before they molt, finish their prefeeding period and become active. During these resting periods, the ticks in the upper layer of the ground can serve as ideal targets for nematode attack (Fig. 26.1). Tick control is based worldwide almost solely on chemical acaricides, although vaccines are under development for *Rhipicephalus microplus*, *Rhipicephalus annulatus* and *Ixodes scapularis* (de la Fuente and Contreras, 2022). Application of acaricides often causes not only the development of resistance but also acaricide poisoning of domestic animals. Until recently, little attention has been paid to the potential use of nematodes against ticks (Samish and Rehacek, 1999). *Steinernema riobrave* has been developed for application to nilgai, *Boselaphus tragocamelus*, and white-tailed deer, *Odocoileus virginianus*, infested with *R. microplus*. Remotely operated sprayers have been developed to directly spray infested nilgai and deer as they transit fence crossings (Goolsby *et al.*, 2019). These remotely operated

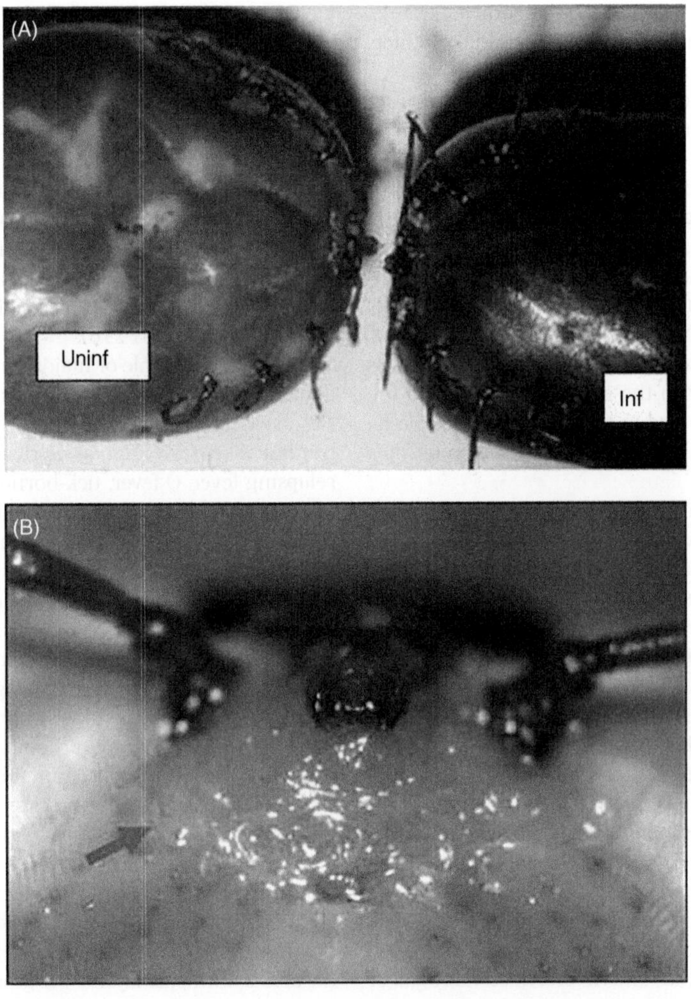

Fig. 26.1. Infectivity of entomopathogenic nematodes to ticks. (A) Infected (Inf) and uninfected (Uninf) engorged females of the cattle tick, *Rhipicephalus annulatus.* (B) Aggregation of infective juveniles of *Steinernema carpocapsae* near the mouth region of an engorged female of *Amblyomma americanum*. Photographs courtesy of K.M. Kocan and E.F. Blouin.

sprayers are programed to work every few hours, wetting the foliage and soil at fence crossings, allowing passive transfer of EPNs to the body of the infested animal (Goolsby et al., 2018; Singh *et al.*, 2019). Treatment of off-host *R. microplus* larvae on foliage has also shown promise as a way to treat infested pastures. The use of the adjuvant Barricade® may reduce desiccation of EPNs and increase their infectivity in this setting. This research against ticks on nilgai and deer shows promise for direct application to other wildlife species and other areas such as hiking trails and pastures where questing tick larvae or nymphs may attach to humans or wildlife (Shapiro-Ilan and Goolsby, 2021).

26.2.1 Infectivity

Samish and Glazer (1991a,b) were the first to report that EPNs are capable of killing engorged females of *R. annulatus*. Studies during the last decade have shown that EPNs are also pathogenic to other tick species (El-Sadawy and Habeeb,

1998; Samish *et al.*, 2001). Out of 16 ixodid tick species from six genera and three argasid species from two genera tested, only one species seemed not to be susceptible to nematodes (Table 26.1). However, engorged female ticks from several species in Petri dish trials demonstrated very large differences in susceptibility to nematodes (Table 26.1). Studies in soil-filled cups showed that engorged Ixodidae (*Hyalomma dromedarii*) females were more susceptible to nematodes than argasid ticks (Hassanain *et al.*, 1999; Kaaya *et al.*, 2000). Fully engorged argasid and ixodid female ticks were generally most sensitive to EPNs, with unfed adult ticks being less sensitive and pre-imaginal stages the least sensitive, while the eggs were fully resistant (Samish and Glazer, 1991a,b, 1992; Samish *et al.*, 1996, 2001). Ovipositing females seem to be more sensitive than their pre-oviposition stage (Mauleon *et al.*, 1993; Samish *et al.*, 2001). During their feeding stage on a host, ticks are resistant to nematodes except on very moist feeding sites (Fig. 26.2) (Kocan *et al.*, 1998a; M. Samish *et al.*, 2002, unpublished data). In Brazil, EPNs were applied to pasture grasses infested with spittlebugs, which allowed reproduction on this innocuous insect (Monteiro *et al.*, 2020). Subsequently, EPNs moved down the foliage killing engorged adult *R. microplus* females lodged at the crown of the grass clump.

The 42 nematode strains tested for their anti-tick activity showed varying degrees of virulence (Table 26.1). In laboratory tests, heterorhabditids were generally more virulent to ticks than steinernematid nematodes (Mauleon *et al.*, 1993; Hill, 1998; Hassanain *et al.*, 1999; Kaaya *et al.*, 2000; Glazer *et al.*, 2001; Singh *et al.*, 2018). Strains virulent to one tick species and one stage were found, in most cases, also to be highly virulent to other tick species and stages (Hassanain *et al.*, 1999; Kaaya *et al.*, 2000; Samish *et al.*, 2001). The concentration of nematodes and the time required to kill 50% of the ticks (LC_{50} and LT_{50}, respectively) were lowest in trials with engorged *Rhipicephalus annulatus* (Table 26.1). The LC_{50} for engorged females was mostly lower than for unfed adults. Also, unfed ixodid females were killed up to six times more quickly than engorged ticks. The LT_{50} for unfed *Rhipicephalus bursa* females is only 1 day compared with 6 days for engorged females (Samish *et al.*, 2000b). The rate of mortality post-infection may be affected by the antibacterial activity that was found in ticks after engorgement (Samish *et al.*, 2000c). At high nematode concentrations and optimum conditions, the nematodes killed the engorged females before they had time to lay eggs (Samish *et al.*, 2000a). However, this still has to be demonstrated under field conditions with ticks that have a shorter pre-oviposition period.

EPNs have also been tested against populations of *R. microplus* that are known to have high levels of resistance to conventional acaricides. Singh *et al.* (2018) did not find any significant differences in mortality rates of engorged female *R. microplus* between susceptible and resistant populations. Given the challenges worldwide with resistance to acaricides in tick species, incorporating the use EPNs could lead to more sustainable pest management (Perez de Leon *et al.*, 2009). EPNs can be tank mixed with many conventional pesticides, thus increasing the efficacy of conventional pesticides (see Chapter 27, this volume). This combination of EPNs and pesticides warrants further research, especially for use in dipping vats and other applications directed at tick pests on livestock and poultry.

26.2.2 Host–parasite interactions

Nematodes are known to enter the bodies of insects mainly via natural orifices. Nematodes virulent to engorged *Amblyomma americanum* ticks were attracted toward the natural apertures of the females (Zhioua *et al.*, 1995). However, their mode of invasion has not yet been proven. No obvious relationship was observed between the size of spiracles, genital openings or cuticle thickness and the relative susceptibility to two strains of nematodes (Mauleon *et al.*, 1993). Exposure of *Rhipicephalus annulatus* ticks to nematodes for only 1 h resulted in some mortality. The mortality increased linearly up to 100% when the exposure time was increased up to 32 h (Samish *et al.*, 1996; Kocan *et al.*, 1998b). Ticks can be killed by injection of a single nematode (Glazer and Samish, 1993), but axenic nematodes are unable to kill ticks even though they are pathogenic to insects (M. Samish *et al.*, 2002, unpublished data). This demonstrates the crucial role of the symbiotic bacteria (*Photorhabdus* spp. for *Heterorhabditis* spp. nematodes and *Xenorhabdus* spp. for *Steinernema* spp. nematodes) of nematodes in killing ticks (Kocan *et al.*, 1998a,b).

Table 26.1. Infectivity data for ticks (Petri dish tests).

		Most virulent nematodes		Susceptibility of developmental stages			
Species	No. of strains tested	Species	Strain	Engorged nymph	Unfed adult	Engorged female[a]	Reference(s)
Amblyomma americanum	5	*Steinernema riobrave*	TX	–	–	<35	Kocan *et al*. (1998b)
Amblyomma cajennense	1	*S. riobrave*	TX	–	–	>180	Kocan *et al*. (1998a)
Amblyomma gemma	4	*Steinernema carpocapsae*	DT	++	–	250	Kaaya *et al*. (2000)
Amblyomma maculatum	1	*S. riobrave*	TX	–	–	>180	Kocan *et al*. (1998a)
Amblyomma variegatum	21	*S. carpocapsae*	Mexican	+	+	20	Samish and Glazer (1992); Mauleon *et al*. (1993); Samish *et al*. (2000c)
Argas persicus	3	*Heterorhabditis bacteriophora*	HP88	+	+	70	Hassanain *et al*. (1999)
Dermacentor variabilis	2	*S. riobrave*	TX	–	–	>180	Kocan *et al*. (1998a)
Hyalomma dromedarii	5	*Steinernema* sp.	S1	+	–	50	El-Sadawy and Habeeb (1998)
Hyalomma excavatum	5	*Heterorhabditis* sp.	IS-5	+	+	–	Samish *et al*. (1996, 2000b)
Ixodes scapularis	15	*Heterorhabditis megidis*	M145	–	–	–	Zhioua *et al*. (1995); Hill (1998)
Ornithodoros moubata	3	*S. carpocapsae*	DT	+	–	–	M. Samish *et al*. (unpublished data)
Ornithodoros tholozani	4	*S. carpocapsae*	DT	+	–	–	M. Samish *et al*. (unpublished data)
Rhipicephalus annulatus	7	*Heterorhabditis* sp.	IS-5	++	++	<2.6	Samish and Glazer (1992); Mauleon *et al*. (1993); Samish *et al*. (2000a,b,c)
R. annulatus	9	*S. carpocapsae*	Mexican	–	–	25	Mauleon et al. (1993)
Rhipicephalus appendiculatus	5	*S. riobrave*	TX	+	+	<50	Kaaya *et al*. (2000)
Rhipicephalus bursa	5	*Heterorhabditis* sp.	IS-3	–	–		Samish *et al*. (1999, 2000a,b,c)
Rhipicephalus decoloratus	4	*S. riobrave*	TX	–	–	<50	Kaaya *et al*. (2000)
Rhipicephalus evertsi	5	*S. carpocapsae*	DT	+	+	<50	Kaaya *et al*. (2000)
Rhipicephalus microplus	14	None		–	–	–	Mauleon et al. (1993)
Rhipicephalus microplus	6	*S. riobrave*	355	–	–	+	Singh *et al*. (2018)
Rhipicephalus sanguineus	7	*S. riobrave*	TX	+	+	>180	Samish *et al*. (1999); Samish and Glazer (1992); Kocan *et al*. (1998a)

–, Not susceptible; +, low susceptibility; + +, moderate susceptibility.
[a]50% lethal concentration (LC^{50}), given as IJs cm^{-2}.

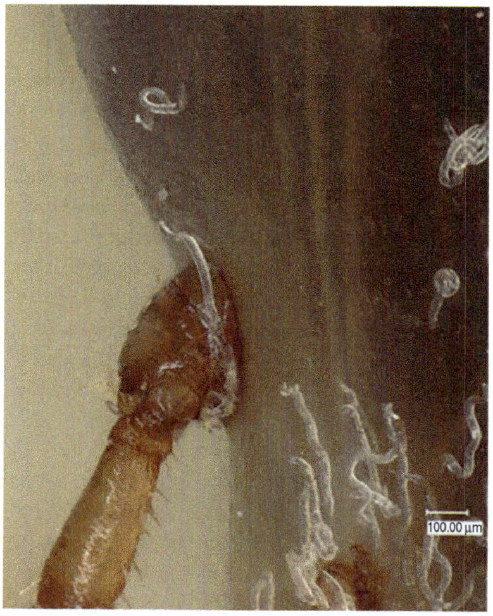

Fig. 26.2. *Steinernema riobrave* infective juvenile larvae searching a *Rhipicephalus microplus* tick for an opening to penetrate.

The number of nematodes that were recovered from infected ticks increased over the course of the first 3–4 days of exposure, but longer exposure did not increase their quantity per cadaver (Samish *et al.*, 1996; Kocan *et al.*, 1998a,b). Between 16 and 140 nematodes were found in each engorged *B. annulatus* female infected with various nematode strains. The LC_{50} of the various nematode strains to ticks was not related to the average number of nematodes recovered (Samish and Glazer, 1991b). A few days after infective juvenile (IJ) nematodes penetrated or were injected into ticks, all or most of them died inside their tick host. However, in rare cases, they survived as IJs or even started to develop within the tick but did not complete their life cycle (Mauleon *et al.*, 1993; El-Sadawy and Habeeb, 1998; Hill, 1998; Kaaya *et al.*, 2000). When the cuticle of ticks was slit artificially before their infection, the nematodes were able to complete their life cycle (Zhioua *et al.*, 1995; Samish *et al.*, 1998).

26.2.3 Application

Nematodes have been used successfully to control ticks in simulated field conditions but have only recently been tried in large-scale field trials. For example, ticks were killed in 10 l buckets with soil inoculated with any of five nematode strains (Mauleon *et al.*, 1993; Hassanain *et al.*, 1999; Kaaya *et al.*, 2000). A large-scale test of *S. riobrave* (Nemasys® R, BASF Agricultural Solutions, New Jersey, USA) using remotely operated sprayers on several thousand hectares of ranchland containing infested cattle, nilgai and deer was conducted in South Texas. Tick levels were lower on animals on the treated ranches compared with ranches without sprayers. In general, commercial use of EPNs for control of ticks has been slow to develop, most likely due to the dominance and overall effectiveness of conventional acaricides. Recent considerations for the use of EPNs for tick control include acaricide resistance and environmental safety, especially in ecologically sensitive habitats where tick infestations can occur on wildlife (Goolsby *et al.*, 2019).

There is a tenuous connection between virulence assays in Petri dishes containing filter paper and those conducted in soil environments such as buckets (Samish *et al.*, 1997, 2000b). For example, the steinernematids *Steinernema carpocapsae* DT strain and *S. carpocapsae* Mexican strain killed *B. annulatus* ticks on soil more rapidly than in Petri dishes. Heterorhabditid strains (HP88, IS-3 and IS-5), however, were more efficient in Petri dishes than on soil (Kocan *et al.*, 1998a). These differences could be due to different search strategies of the nematodes (Lewis, 2002) or to different abilities to stay virulent under various soil environment stress factors (Glazer, 2002). All nematode strains tested were most efficient against ticks at about 26°C. Some strains were far less efficient at 22°C or 30°C, while others had a wider range of efficiency (between 18°C and 34°C) (Samish *et al.*, 1996). Placing ticks on sandy soil 3 days after it was sprayed with nematodes resulted in 100% mortality at 10 days post-infection. However, the mortality of ticks on sandy soil with 25% (v/v) cattle manure or soil containing 40–50% silt was only 45% or 25%, respectively (Samish *et al.*, 1995). Some ticks prefer humid environments such as the upper soil layer, under stones or in leaf litter conditions, which also favor EPNs. Goolsby and Shapiro-Ilan (2020) found that *S. riobrave* could infect wax moth (*Galleria mellonella*) larvae from foliage and soil misted for up to 3 h post-application. New formulations of EPNs consisting of nanoparticles can provide additional extension of viability and activity to the nematodes under these conditions (Ramakrishnan *et al.*, 2023). Adjuvants that protect against desiccation and ultraviolet light also show promise in improving efficacy under field conditions (Wu *et al.*, 2023). Thus, nematodes could be used as part of an integrated tick-control strategy in target sites such as human and animal rest areas, animal paths, water sources and wildlife refuges.

26.3 Flies

The housefly, *Musca domestica*, is a well-known cosmopolitan pest of animals. This is the most common species found on pig and poultry farms, horse stables and ranches. Houseflies are always found in association with humans or activities of humans. Not only are they a nuisance, but they can also transport disease-causing organisms. Excessive fly populations are obnoxious to farm workers, and a public health hazard to nearby human habitations. Flies commonly develop in large numbers as a serious problem in poultry manure of caged hens. The control of *M. domestica* is vital to human health and comfort in many areas of the world. The most important damage related to this insect is the annoyance and indirect damage produced by the potential transmission of more than 100 pathogens associated with this fly, which may cause disease in humans and animals, including typhoid, cholera, bacillary dysentery, tuberculosis, anthrax, ophthalmia and infantile diarrhea, as well as parasitic worms (Scott *et al.*, 2014). Pathogenic organisms are picked up by flies from garbage, sewage and other sources of filth, and then transferred on their mouth parts and other body parts, through their vomitus and feces, to human and animal food. *M. domestica* has a complete metamorphosis with distinct egg, larva, pupal and adult stages. It overwinters either in the larval or pupal stage under manure piles or in other protected locations. Warm summer conditions are generally optimum for development of *M. domestica*; it can complete its life cycle in as little as 7–10 days, and have as many as 10–12 generations in one summer.

26.3.1 Infectivity

Geden *et al.* (1986) demonstrated the infectivity of EPNs to different developmental stages of *M. domestica*. Second- and third-instar larvae and adults of this muscid were highly susceptible to *S. carpocapsae* and *Heterorhabditis heliothidis* (= *H. bacteriophora*) when hosts were confined in Petri dishes containing nematode-treated filter paper. The larvae were not susceptible to *Steinernema glaseri*, and the fly pupae were refractory to infection by all three species. When second- and third-instar larvae were exposed to 5000 nematodes per host in rearing medium, *S. carpocapsae* caused higher mortality (55–61%) than *H. heliothidis* (11–26%). Both *S. carpocapsae* and *H. heliothidis* were more infective for third-instar larvae (21–29%) than for second-instar larvae (2–6%) at this dosage in poultry droppings. When adult flies were offered *S. carpocapsae* suspensions in 5% sucrose bait in cotton balls, mortality ranged from 53% to 67% at dose rates ranging from 1000 to 100,000 nematodes ml^{-1} of bait (Table 26.2).

Taylor *et al.* (1998) screened 40 strains representing eight *Heterorhabditis* spp. and five *Steinernema* spp. for virulence toward third-instar larvae of *M. domestica* in a filter paper assay. None of the 22 strains of *Heterorhabditis* infecting larvae caused significant levels of mortality in this assay. Of ten strains of *Steinernema*-infected larvae, seven strains (four *S. carpocapsae*, two *S. feltiae* and one *S. scapterisci*) caused significant mortality. Ten *Heterorhabditis* strains and ten *Steinernema* strains successfully reproduced for two generations in larvae. Taylor *et al.* (1998) also selected six strains of *Steinernema* for ten generations on larvae and then compared them with the unselected parental strains. No difference in pathogenicity between selected and unselected strains was observed (Table 26.2).

26.3.2 Host–parasite interactions

Renn (1998a) determined the routes of penetration of the *S. feltiae* IJs into larval and adult houseflies. IJs aggregate on the proboscis and anal aperture of male and female houseflies after 1 h. The nematodes penetrate female flies after 2 h by moving through the cloaca, then along the

Table 26.2. Infectivity data for insects.

Pest	Bioassay	Infectivity measure	Nematode concentration	No. of strains tested	Most virulent nematode		Mortality of developmental stage (%)				Reference
					Species	Strain	Young larva	Old larva	Pupa	Adult	
Housefly (*Musca domestica*)	PDA	Insect mortality	5000 IJs per host	3	*Steinernema feltiae*[a]	DD-136	100	92	–	–	Geden *et al.* (1986)
	PDA	Insect mortality	100	40	*S. feltiae*	SN	–	50.4	–	–	Taylor *et al.* (1998)
	SFC – poultry manure	Insect mortality	5000 IJs per host	3	*S. feltiae*[a]	DD-136	2.4	28.2	–	–	Geden *et al.* (1986)
	SFC – poultry manure	Insect mortality	1000 IJs per host	3	*S. feltiae*[a]	All	–	36.9	–	–	Georgis *et al.* (1987)
	SFC – poultry manure	Emerging adults	4 × 10^6 IJs m^{-2}	1	*Heterorhabditis heliothidis*	NC1	–	–	–	86	Belton *et al.* (1987)
Louse (*Pediculus humanus humanus*)	PDA	Insect mortality	1500 IJs per dish	4	*Steinernema glaseri*	–	75–90	80–95	–	92–97	Weiss *et al.* (1993)
Louse (*P. humanus capitis*)	Insect mortality	Insect mortality	100 IJs per host	4	*Heterorhabditis bacteriophora*	RN	100	100	–	84	Doucet *et al.* (1998)
Cat flea (*Ctenocephalides felis*)	PDA + soil	Insect	50 IJs cm^{-2}	1	*Steinernema carpocapsae*	All	87–100	87–100	95–100		Manweiler (1994)
	SFC	Insect mortality-Emerging adults	50 IJs cm^{-2}	1	*S. carpocapsae*	All	–	–	–	67	Manweiler (1994)

PDA, Petri dish assay; SFC, simulated field conditions.
[a]*Steinernema feltiae* = *Steinernema carpocapsae*.

oviduct and through the ovaries. Male houseflies are penetrated via the cloaca, and the nematodes then enter the hemocoel by penetrating the wall of the ejaculatory sac. All larval stages are penetrated via the anal aperture. Nematodes then move through the hindgut and penetrate the wall of the ileum, immediately posterior to the pylorus.

Histopathological analysis of the effect of different dosages (50, 100, 200, 500 and 1000 IJs per larvae) of *S. feltiae* on the larval tissues of *M. domestica* was conducted by Ghally *et al.* (1991). *S. feltiae* nematodes invaded the fat tissue, gut, cuticle and muscle tissue of the host. All of these tissues, along with the gut epithelium, showed signs of disintegration before the death of the host. The tissue of the gut and fat body were the most severely damaged by the presence of *S. feltiae*. The damage described in this investigation depended mainly on the time and intensity of infection.

26.3.3 Application

EPNs have been evaluated as a control measure for housefly populations in various animal farming environments.

26.3.3.1 Cattle

Two strains of *S. feltiae* (SN and UNK-36) and two of the best *Heterorhabditis* strains (*H. bacteriophora* OSWEGO and *H. megidis* HF-85) were tested in a fresh bovine manure substrate (Taylor *et al.*, 1998). All four strains produced significant fly mortality in the manure substrate; the *S. feltiae* strains had significantly lower LC_{50} values than the *Heterorhabditis* spp. The most promising strain, *S. feltiae* SN, gave LC_{50} and LC_{99} values of 4 and 82 IJs per larva, respectively. These doses were equivalent to 2.7 and 55 IJs g^{-1} of manure and 5.1 and 104 IJs cm^{-2} (Taylor *et al.*, 1998). Shapiro *et al.* (1996, 1999) reported fresh cow manure to be detrimental to *S. carpocapsae* virulence, whereas composted cow manure had no effect.

26.3.3.2 Pigs

The efficacy of *S. feltiae* and *H. megidis* after formulation into a housefly bait was compared with a commercial bait formulation of methomyl for the control of houseflies in a UK pig farm (Renn, 1998b). The housefly infestation was confined to the farrowing unit, which consisted of ten farrowing houses, where two adjacent houses were sequentially restocked with pregnant sows at weekly intervals. Shortly after restocking, one house was baited with one of the nematode species and the other with methomyl. Significantly fewer flies were counted in the houses baited with *S. feltiae* or *H. megidis* than in those baited with methomyl. The efficacy of *S. feltiae* sprayed on to the manure was also compared with methomyl bait. Counts of houseflies carried out in the farrowing cycle before this treatment were not significantly different; however, significantly fewer flies occurred after *S. feltiae* was sprayed.

26.3.3.3 Poultry

Belton *et al.* (1987) evaluated the use of *H. heliothidis* (= *H. bacteriophora*) (North Carolina strain) for control of *M. domestica* larvae in chicken manure. Laboratory tests showed that *H. heliothidis* applied at rates of 2×10^6 and 4×10^6 IJs m^{-2} significantly decreased fly emergence over 24 days compared with the control (310 and 227 versus 1570, respectively). Treatment of chicken manure in barns with the nematode after a 10-week period showed fly populations of about 1×10^7 in the control barns and 2×10^6 in the treated barn. No fly pupae were found parasitized by *H. heliothidis* at any time. Georgis *et al.* (1987) demonstrated that the nematodes were severely hampered by poultry manure. Exposure of IJs of *S. carpocapsae*, *S. feltiae* and *H. bacteriophora* to manure resulted in high nematode mortality (70–100%) within 18 h. Nematodes exposed to manure slurry for more than 6 h had drastically reduced efficacy against larvae of *M. domestica*. It was concluded that "poor survival and limited movement of nematodes in the poultry manure appear to make them unlikely candidates for biocontrol of filth flies in this habitat" (Georgis *et al.*, 1987). To overcome this problem, Renn (1995) tested the use of the nematodes *S. feltiae* and *H. megidis* encapsulated in calcium alginate against the larvae of the housefly in chicken manure. Aliquots of capsules (15 ml) containing either 1,000,000, 500,000, 250,000 or 125,000 nematodes were added to 70 ml portions of grass meal diet containing either eggs or first-, second- or third-instar larvae. Complete mortality of the larvae was achieved within 6 days.

Another experiment was carried out where immature houseflies were placed in chicken manure. The emergence of houseflies as adults was used to measure the effect of the encapsulated nematodes. The treatment with encapsulated *S. feltiae* resulted in 55–96% reduction in adult housefly emergence, whereas treatment with encapsulated *H. megidis* resulted in 35–98% reduction in emergence. Finally, when encapsulated nematodes were presented as bait to adult houseflies, little infectivity was observed. Renn and Wright (2000) further evaluated the effect of artificial substrates on the virulence of *S. feltiae* to adult houseflies, demonstrating that certain substrates with high porosity are most effective, giving LD_{50} values of 2812–8912 IJs in solution in the bait. In contrast, poor substrates with low porosity required 531,450–899,930 *S. feltiae* IJs to achieve LD50. The use of advanced formulation and baiting approaches may allow the use of EPNs against this important pest.

26.4 Fleas

Fleas are annoying and medically important pests of humans and animals. In addition to their bites and induced dermatitis, fleas are important vectors of dog tapeworm, murine typhus and plague (Strand, 1977). Fleas can be a significant problem outdoors near homes, as well as in indoor areas. Persistence of outdoor populations of fleas can be a continuous source of flea re-infestations for companion animals (Henderson *et al.*, 1995).

26.4.1 Infectivity

Silverman *et al.* (1982) were the first to demonstrate the susceptibility of the cat flea, *Ctenocephalides felis*, to *S. carpocapsae*. Henderson *et al.* (1995) evaluated the effect of *S. carpocapsae* on cat flea larvae and pupae in different substrates. Application of the nematodes to potting soil, sand or gravel substrates containing different developmental stages of *C. felis* reduced adult flea emergence by 70–100%. It was demonstrated in this study that the silk cover of the cocoon did not prevent infection by nematodes.

26.4.2 Application

In the mid-1990s, Biosys developed and launched a *S. carpocapsae*-based product for outdoor control of fleas (Manweiler, 1994; see also Table 26.2). It was suggested that nematode application would be part of a flea-control program including the use of chemicals (insect growth regulators) for indoor treatments against adults, shampoos or animal dips, and EPNs applied to the area where fleas were developing, to eliminate outdoor sources of re-infestations (Manweiler, 1994). The nematode-containing product that was sold in substantial quantities in southern USA was withdrawn from the market due to competition from new and effective chemicals and widespread fungal contamination of the newly developed water-dispersible granular formulation (Grewal and Georgis, 1999). Flea control using EPNs is currently continuing in the USA on a smaller scale (D.I. Shapiro-Ilan, 2003, personal communication).

26.5 Lice

Other than malaria-carrying mosquitoes, no other insect has caused more death to mankind than lice, particularly the body louse, *Pediculus humanus humanus*. This ectoparasite, which thrives in conditions of overcrowding, filth and famine, can have a devastating effect on humans due to its association with endemic typhus (caused by *Rickettsia prowazekii*), trench fever (caused by *Rochalimaea quintana*) and louse-borne relapsing fever (caused by *Borrelia recurrentis*), diseases that probably have changed the course of history of mankind.

26.5.1 Infectivity

The susceptibility of *P. humanus humanus* to infection by EPNs was first demonstrated by Weiss *et al.* (1993). Exposure of female lice to IJs of *S. carpocapsae* and *S. glaseri* in a Petri dish assay resulted in over 85% mortality within 24 h (Table 26.2). *H. bacteriophora* HP88 strain was found to be less effective, causing only 45% mortality after 42 h exposure. The mortality of female lice was directly related to an increased dosage of *S. glaseri* IJs from 100 to 800 IJs per dish. Complete mortality was achieved at a concentration of 400 IJs per dish. Exposure of lice to 800 *H. bacteriophora* HP88 IJs per dish caused only 27% mortality. Nematodes can be observed in the transparent body of the adult louse (Fig. 26.3).

Doucet *et al.* (1999) showed that the head louse, *Pediculus humanus capitis*, is also susceptible to EPN infection (Table 26.2). The authors showed substantial differences in infectivity of various nematode strains. The least infectious was *S. feltiae*, while *H. bacteriophora* strains readily killed adults and nymphs of the insect. The above studies showed that the small size of the insect hinders nematode penetration into the insect (Doucet *et al.*, 1998) and causes abnormal development of the nematode inside the body (Weiss *et al.*, 1993).

26.5.2 Application

No further development of EPNs for control of lice has been reported. It seems that the use of nematodes on the human head or body will be impractical unless an appropriate formulation is developed. Doucet *et al.* (1998) suggested using the toxins produced by the nematodes. It may also be possible to use toxins from the symbiotic bacteria as anti-louse agents.

26.6 Cockroaches

Cockroaches are one of the oldest and more important household and commercial establishment pests associated with humans around the world. Of the approximately 4000 living species of cockroaches in the world, only a few species inhabit human dwellings. The most common are the German, American and oriental cockroaches. The German cockroach, *Blattella germanica*, is the most important domiciliary pest worldwide, especially in regions that provide a favorable environment during cold or dry months. *Blattella germanica* infestations are associated with warm, humid environments such as commercial food preparation and storage facilities, or private kitchens (Cornwell, 1976). The American cockroach, *Periplaneta americana*, is one of the largest of the house-infesting cockroaches, with a worldwide distribution. It generally prefers warm and humid

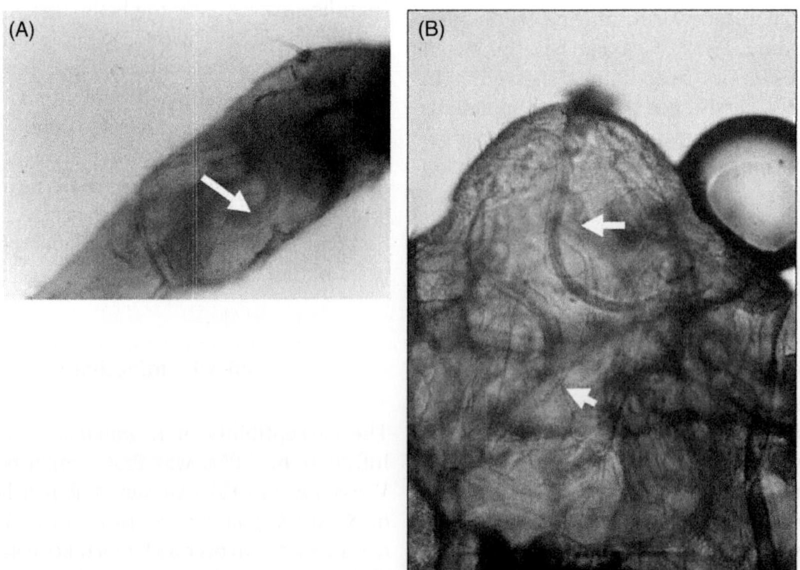

Fig. 26.3. Infective juveniles (IJs) (arrows) of *Steinernema carpocapsae* in the legs (A) and abdomen (B) of an body louse, *Pediculus humanus humanus*.

conditions and is active from 21°C to 33°C. American cockroaches are found in residences but are more common in larger commercial buildings such as restaurants, bakeries, food-processing plants and grocery stores. During the summer, American cockroaches can be found outdoors in yards and alleys. They are the most common species found in city sewer systems (Smith and Whitman, 1992). The oriental cockroach, *Blatta orientalis*, is adapted to areas that combine summer heat and moderately cold winter temperatures. It can successfully infest outdoor habitats year-round where it finds harborage and perhaps food in plant material or crevices in sidewalks and buildings, but it will move indoors or to protected harborages during cold weather (Robinson, 1996). Indoors, they are typically found in crawl spaces, cellars, basements and on the first floor but at times on higher floors, especially around water pipes on which they climb (Smith and Whitman, 1992).

The importance of cockroaches as a pest is traditionally linked to their medical importance. They are known to be capable of carrying many common disease-causing pathogens such as fungi, viruses, protozoa and about 40 species of bacteria that are pathogenic to vertebrates. The more common of these disease agents include bacteria such as *Salmonella* spp. and *Escherichia coli* (food poisoning), *Mycobacterium leprae* (leprosy), *Yersinia pestis* (bubonic plague), *Shigella* spp. (dysentery and diarrhea in children), *Pseudomonas* spp. (urinary tract infections), *Staphylococcus* spp., *Streptococcus* spp., *Bacillus* spp. and *Clostridium* spp. Cockroaches also cause diverse allergic responses in sensitive individuals. Fragments of dead domestic cockroaches and exuviae, and cockroach fecal material and saliva have been recognized as strong indoor allergenic agents (Pope *et al.*, 1993) and are as important as dust mites in the etiology of house dust allergies (Hullet and Dockhorn, 1979).

Currently available pesticides reduce cockroach populations by 85–100%, but resistance to organophosphates, carbamates and pyrethroids is already widespread within cockroach populations (Rahimian *et al.*, 2019). The continuous use of pesticides in human living spaces can also harm humans following exposure. Biological control has gradually become a promising strategy for cockroach control (Pan and Zhang, 2020), and EPNs could be a good tool if incorporated into an integrated pest management approach for cockroaches.

26.6.1 Infectivity

The susceptibility of cockroaches to infection by EPNs was first observed by Skierska *et al.* (1976). Exposure of adult *Blattella germanica* to 200–800 IJs of *S. carpocapsae* (DD 136 strain) in a Petri dish assay resulted in 95% mortality. However, Laumond *et al.* (1979), in an analogous experiment using 10,000 IJs per dish of the same nematode strain, reported that *Blattella germanica* was resistant to these nematodes. Subsequent studies have shown that EPNs are pathogenic to several cockroach species. *S. carpocapsae* is the most widely tested species in infectivity studies with household cockroaches (Table 26.3). Other species of steinernematids have been tested against cockroaches in laboratory studies. *S. feltiae* evaluation in Petri dish assays resulted in 50% mortality for adults and 30% mortality for nymphs of *Blattella germanica* (García del Pino and Morton, 2001), and 92–100% mortality against *Blatta orientalis*.

In similar experiments, *Steinernema arenarium* caused 45–63% mortality of *Blattella germanica* (García del Pino and Morton, 2001), and *S. scapterisci* killed 66% of *B. germanica* but was ineffective against *P. americana* (Grewal *et al.*, 1993). Maketon *et al.* (2010) observed no mortality of American or German cockroaches caused by *S. glaseri*, *H. bacteriophora* and *H. indica*, but 86.7% and 40% mortality caused by *S. carpocapsae*, respectively, in bait treatments. However, *Steinernema rarum* caused a mortality of 89% of *P. americana* adults in Petri dish assays (Aguirre Varela *et al.*, 2021).

In general, heterorhabditids have been used less in infectivity studies with cockroaches. *H. bacteriophora* caused 50–70% mortality of *Blattella germanica* (García del Pino and Morton, 2001) and 73–95% mortality of *Blatta orientalis* (Koehler *et al.*, 1992). *H. heliothidis* (= *H. bacteriophora*) has been tested against *P. americana*, with 88% mortality of nymphs but only 9% mortality of adults (Zervos and Webster, 1989). *Blattella germanica* is more susceptible to EPN infection. Appel *et al.* (1993) observed a high susceptibility (LT_{50} 2.1 days) of adults of this cockroach to

Table 26.3. Infectivity data for cockroaches.

Species	Bioassay	Nematode: No. of strains tested	Nematode: Species	Nematode: Strain	Nematode concentration	Infectivity measure	Mortality of developmental stages (%): Nymphs	Adults females	Adults males	Reference
Blattella germanica	PDA	7	*Steinernema feltiae*[a]	All	500 per dish	% mortality	–	100	100	Locatelli and Parleaz (1987)
	PDA	1	*Steinernema bibionis*	Bib	500 per dish	% mortality	–	100	100	Locatelli and Parleaz (1987)
	PDA	1	*Heterorhabditis heliothidis*	HH	1000 per dish	% mortality	–	40	80	Locatelli and Parleaz (1987)
	PDA	3	*Heterorhabditis* sp.	HW	500 per dish	% mortality	–	100	100	Locatelli and Parleaz (1987)
	PDA	1	*Steinernema carpocapsae*	All	500,000 per dish	LT_{50} (days)	<1	<1	<1	Koehler *et al.* (1992)
	Bait	1	*S. carpocapsae*	All	500,000 per bait	LT_{50} (days)	4.9	3.5	3.5	Koehler *et al.* (1992)
	PDA	1	*Steinernema scapterisci*	Uruguay	8000 per dish	% mortality	–	66	–	Grewal *et al.* (1993)
	PDA	1	*S. carpocapsae*	All	0.5×10^6 per dish	LT_{50} (days)	–	–	2.1	Appel *et al.* (1993)
	SFC (bait)	1	*S. carpocapsae*	All	2×10^6 per bait	% reduction	–	67.1	–	Appel *et al.* (1993)
	PDA	4	*Steinernema feltiae*	e-NEMA	40,000 per dish	% mortality	30	50	–	García del Pino and Morton (2001)
	PDA	1	*Heterorhabditis bacteriophora*	e-NEMA	40,000 per dish	% mortality	50	70	–	García del Pino and Morton (2001)
	PDA	1	*Steinernema arenarium*	S2	40,000 per dish	% mortality	45	62.5	–	García del Pino and Morton (2001)
	Terraria (bait)	1	*S. carpocapsae*	–	0.27×10^6	% mortality	–	–	80	Pye *et al.* (2001)
	Bait	1	*S. carpocapsae*	–	1.7×10^5 ml^{-1}	% mortality	86.7	86.7	86.7	Maketon *et al.* (2010)
	Bait	1	*Steinernema glaseri*	–	1.7×10^5 ml^{-1}	% mortality	0	0	0	Maketon *et al.* (2010)
	Bait	1	*H. bacteriophora*	–	1.7×10^5 ml^{-1}	% mortality	0	0	0	Maketon *et al.* (2010)
	Bait	1	*Heterorhabditis indica*	–	1.7×10^5 ml^{-1}	% mortality	0	0	0	Maketon *et al.* (2010)
	Glass cage (bait)	1	*S. carpocapsae*	–	1500 ml^{-1}	% mortality	85	90	90	El-Kady *et al.* (2015)

	Glass cage (spray)	1	*S. carpocapsae*	–	1500 per bait	% mortality	100	95	95	El-Kady *et al.* (2015)
Blatta orientalis	PDA	1	*S. carpocapsae*	All	500,000 per dish	LT_{50} (days)	1.4	1	<1	Koehler *et al.* (1992)
	SFC (bait)	1	*S. carpocapsae*	All	500,000 per bait	LT_{50} (days)	0	12	7	Koehler *et al.* (1992)
	PDA	1	*S. feltiae*	Low87	1000 per dish	% mortality	96.7	91.7	100	Kotlarska-Mordzinska *et al.* (2000)
	PDA	1	*H. bacteriophora*	PL	1000 per dish	% mortality	73.3	78.3	95	Kotlarska-Mordzinska *et al.* (2000)
Periplaneta americana	PDA	1	*H. heliothidis*	T327	1000 per dish	% mortality	88	9	–	Zervos and Webster (1989)
	PDA	1	*S. carpocapsae*	All	500,000 per dish	LT_{50} (days)	1	5	2.6	Koehler *et al.* (1992)
	SFC (bait)	1	*S. carpocapsae*	All	500,000 per bait	LT_{50} (days)	0	0	0	Koehler *et al.* (1992)
	PDA	1	*S. scapterisci*	Uruguay	8000 per dish	% mortality	–	0	0	Grewal *et al.* (1993)
	Bait	1	*S. carpocapsae*	–	1.7×10^5 ml^{-1}	% mortality	40	40	40	Maketon *et al.* (2010)
	Bait	1	*S. glaseri*	–	$1.7 \times 10^{5\ ml-1}$	% mortality	0	0	0	Maketon *et al.* (2010)
	Bait	1	*H. bacteriophora*	–	1.7×10^5 ml^{-1}	% mortality	0	0	0	Maketon *et al.* (2010)
	Bait	1	*H. indica*	–	1.7×10^5 ml^{-1}	% mortality	0	0	0	Maketon *et al.* (2010)
	PDA	2	*Steinernema rarum*	OLI	6400 per dish	% mortality	–	89	89	Aguirre Varela *et al.* (2021)
Periplaneta fuliginosa	PDA	1	*S. carpocapsae*	All	500,000 per dish	LT_{50} (days)	<1	<1	1.2	Koehler *et al.* (1992)
	Bait	1	*S. carpocapsae*	All	500,000 per bait	LT_{50} (days)	0	6.7	9.8	Koehler *et al.* (1992)
	SFC (spray)	1	*S. carpocapsae*	–	500,000 l^{-1}	% reduction	–	0	–	Smith *et al.* (1997)
Periplaneta brunnea	Glass jars (3.75 l)	1	*S. carpocapsae*	All	300,000 per jar	% mortality	–	–	92	Corpus and Sikorowski (1992)
Supella longipalpa	PDA	1	*S. carpocapsae*	All	500,000 per dish	LT_{50} (days)	<1	<1	<1	Koehler *et al.* (1992)
	SFC (bait)	1	*S. carpocapsae*	All	500,000 per bait	LT_{50} (days)	4	2.6	3.7	Koehler *et al.* (1992)

PDA, Petri dish assay; SFC, simulated field conditions.
[a]*Steinernema feltiae* = *Steinernema carpocapsae*.

0.5×10^6 IJs of *S. carpocapsae* in a Petri dish bioassay. Locatelli and Parleaz (1987) showed that 500 IJs per dish of *S. feltiae* (= *S. carpocapsae*), *S. bibionis* (= *S. feltiae*) and *Heterorhabitis* sp. caused 100% mortality of *Blattella germanica*. El-Kady *et al.* (2015) evaluated the efficacy of *S. carpocapsae* against *Blattella germanica* with bait and spray treatments (1500 IJs ml^{-1}) in a glass cage (60 × 30 × 40 cm) as a kitchen model. They showed that the highest mortality percentage for the adult stage was recorded with baits (90%), and the spray method gave the highest mortality for the nymph stage (100%). Koehler *et al.* (1992) determined the susceptibility of *P. americana*, the smoky-brown cockroach (*Periplaneta fuliginosa*), *Blatta orientalis*, *Blattella germanica* and the brown-banded cockroach (*Supella longipalpa*) to *S. carpocapsae* (All strain). Two different experiments were conducted: (i) a forced-exposure bioassay in Petri dishes with 500,000 IJs; and (ii) a non-forced-exposure bioassay with bait stations. In forced-exposure bioassays, the relative order of susceptibility was: *Blattella germanica* = *Supella longipalpa* > *Periplaneta fuliginosa* > *Blatta orientalis* > *P. americana*. In bait bioassays the susceptibility was: *Supella longipalpa* > *Blattella germanica* > *Blatta orientalis* > *P. fuliginosa* > *P. americana*. In both bioassays, *P. americana* was the least susceptible to *S. carpocapsae*, and in bait stations the nematodes were ineffective in killing late-stage nymphs, males or females.

Behavioral, morphological, physiological or immunological responses may be the reason for the lower susceptibility of *P. americana* to infection. Zervos and Webster (1989) observed encapsulation and melanization as immunological responses in adult *P. americana* to *H. bacteriophora*. These immunological responses caused nematode mortality but not before bacteria were released. Koehler *et al.* (1992) observed that in a forced bioassay, late-stage nymphs, males and females of *Blattella germanica* and *Supella longipalpa* had equal susceptibility to *S. carpocapsae*. Nevertheless, different instars of cockroaches usually have different susceptibilities to EPNs. In general, EPNs are less effective at killing nymphs than adults, and in adults, males are more susceptible than females (Fig. 26.4). In *Periplaneta fuliginosa*, the relative order of susceptibility was: females > males > nymphs; in *Blatta orientalis*, it was males > females > nymphs; and in *P. americana*, it was nymphs > males > females (Koehler *et al.*, 1992).

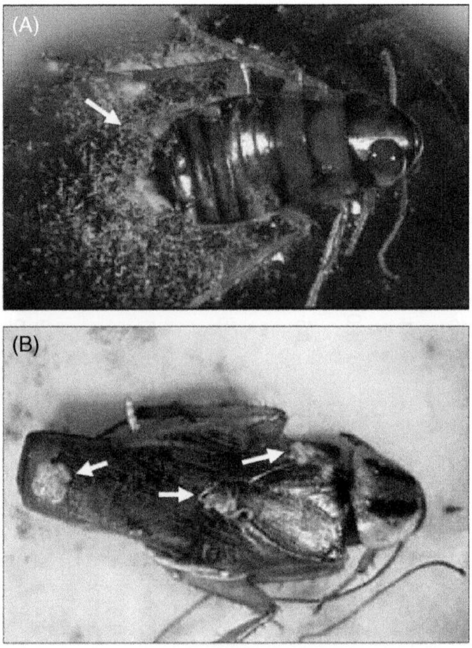

Fig. 26.4. Infection of nymphs of the American cockroach, *Periplaneta americana*, (A) and adults of the German cockroach, *Blattella germanica*, (B) by entomopathogenic nematodes (arrows).

Morton and García-del-Pino (2013) showed sex-related differences in the susceptibility of *P. americana* to *S. carpocapsae*, with males being more susceptible than females. These authors also detected a significantly higher number of nematodes in gonads and ducts of males than of females, demonstrating that genital apertures are an important route for EPNs to enter adult cockroaches.

Appel and Benson (1994) showed that mature oothecae of *Blattella germanica* are not affected by *S. carpocapsae* and there is no transoothecal transmission of this nematode in infected females.

26.6.2 Application

Although the pathogenicity of *Steinernema* and *Heterorhabditis* spp. to cockroaches has been proved in many laboratory bioassays with Petri dish or bait stations, only a few experiments have been carried out in field conditions. Smith *et al.* (1997) compared the efficacy of *S. carpocapsae* and chemical insecticides against *P. fuliginosa* in outdoor applications. They applied the nematodes as a spray solution (500,000 IJs l^{-1}) in the perimeter area with vegetation or with soil and permanent shade of single-family detached houses. Nematodes did not affect the abundance of *P. fuliginosa*, and no cockroaches were recovered that exhibited symptoms of nematode infection. The use of EPNs against *Blattella germanica* in indoor applications was tested by Appel *et al.* (1993). They treated apartments with baits containing 2×10^6 nematodes presented in moisture-retaining stations patented by Weber *et al.* (1992). This nematode concentration had low repellence and good formulation and survival characteristics. Nematode treatment of the apartments was as effective as the standard insecticide bait treatment, with 67% reduction of cockroaches at 8 weeks after treatment. They concluded that the optimization of the exposure station, the number of stations needed per apartment and the station location will likely increase field performance. Pye *et al.* (2001) evaluated a commercial bait station with *S. carpocapsae* against *Blattella germanica* for use in structures. The bait station contained 0.27×10^6 IJs in polysaccharide gel with food and pheromone attractant. It killed more than 80% of small populations ($n = 25$) in 4 days, and the killing activity continued for about 1 month.

26.7 Conclusions

Important veterinary and human pests are highly susceptible to EPNs. In some cases, laboratory bioassay results have led to commercial utilization (i.e. for fleas and houseflies). In general, most of the off-host stages of these pests hide for days and even for weeks in the dark, humid upper layer of the ground, and this particular environment is also the natural habitat of EPN IJs (Kaya, 1990). Most pests can be found in confined areas, where nematodes can be applied in high concentrations. In particularly, the off-host stages of most pests drop off mainly at the resting areas of the mammalian host. EPNs are especially well suited for an integrated control regime as they are tolerant to many insecticides (Glazer *et al.*, 1997; see Chapter 19, this volume). In addition, as certain compounds are applied mainly on the host and the nematodes are commonly applied on the ground, the two control methods could serve as complementary treatments.

Table 26.4. Comparative considerations for the control of animal and human pests with those for control of plant pests.

Consideration	Plant pests	Veterinary and human pests
Inhabiting site of the pest	In the soil Plant foliage	On host On soil surface In the soil
Hazardous and antagonistic factors affecting EPN efficacy	Desiccation Ultraviolet radiation Soil fauna	Body heat of host Animal manure
Application of EPNs	Large surface in the field or greenhouse Formulations for storage and/or ultraviolet radiation and desiccation protection	Small surface on hosts and also where hosts are confined Formulation for survival on and off host body
Economic	Relatively lower value Direct damage of pests to the crop has higher economic threshold	Higher value and ability to pay more per treatment Indirect damage as disease vectors lower economic threshold
End user	Open for testing new technologies and pest management approaches	Conservative and resistant toward new technologies and pest management approaches

EPN, entomopathogenic nematode.

Housefly (*Musca domestica*)

Life cycle 10–12 days[a]	Eggs	First larvae	Second larvae	Third larvae	Pupae	Adult
Location	In/on ground	In/on ground	In/on ground	In/on ground	In/on ground	On/off host
Susceptibility to EPNs	–	+ + +	+ + +	+ + +	+ + +	+ + (in baits)

Cat flea (*Ctenocephalides felis*)

Life cycle 14–25 days	Eggs	First larvae	Second larvae	Third larvae	Pupae	Adult
Location	In/on ground	In/on ground	In/on ground	In/on ground	In/on ground	On host
Susceptibility to EPNs	–	+ + +	+ + +	+ + +	+ + +	+ + +

Body louse (*Pediculus humanus humanus*)

Life cycle 21–27 days[a]	Eggs	First nymph	Second nymph	Third nymph	Adult
Location	On host	On host	On host	On host	On host
Susceptibility to EPNs	–	+ +	+ + +	+ + +	+ + +

German cockroach (*Blattella germanica*)

Life cycle 80–92 days[a]	Eggs	First nymph	Second nymph	Third nymph	Fourth nymph	Fifth nymph	Sixth nymph	Adult
Location	Encountering EPNs is possible only in bait							
Susceptibility to EPNs	–	+ + +	+ + +	+ + +	+ + +	+ + +	+ + +	+ + +

Ticks: one host (*Rhipicephalus microplus*)

Life cycle 60–120 days	Eggs	Larvae			Nymphs			Adults		
		Unfed	Fed	Engorged	Unfed	Fed	Engorged	Unfed	Fed	Engorged
Location	Off host	On host	On host	On host	On host	On host	On host	On host	On/off host	Off host
Susceptibility to EPNs	–	+	++	++	+	++	+++	++	++	+ + +

Ticks: two hosts (*Rhipicephalus bursa*)

Life cycle 141–223 days	Eggs	Larvae			Nymph			Adults		
		Unfed	Fed	Engorged	Unfed	Fed	Engorged	Unfed	Fed	Engorged
Location	Off host	On host	On host	Off host	Off host	On host	On host	On host	On host	Off host
Susceptibility to EPNs	–	+ + +	–	–	–	–	+	+ +	–	+ +

Ticks: three hosts (*Rhipicephalus sanguineus*)

Life cycle 71–128 days	Eggs	Larvae			Nymph			Adults		
		Unfed	Fed	Engorged	Unfed	Fed	Engorged	Unfed	Fed	Engorged
Location	Off host	Off host	On host	Off host	Off host	On host	On host	Off host	On host	Off host
Susceptibility to EPNs	–	+ + +	++	+ +	+ + +	++	+	+ + +	++	+ + +

Fig. 26.4. Comparison of susceptibility levels of various developmental stages in the course of the pest life cycle with their natural habitat. –, Not susceptible; +, low susceptibility; + +, moderate susceptibility; + + +, high susceptibility. [a]Under laboratory conditions.

The wide genetic diversity of nematode strains suggests that screening, selection and/or genetic manipulation could further improve their virulence. This, and the frequent similarity of the preferred ecological habitats of the nematodes and the off-host stages, makes it desirable to devote further efforts toward the development of a commercial nematode agent.

A comparison of the susceptibility level of various developmental stages in the course of the pest life cycle with their natural habitat (Fig. 26.5) indicates that in many cases the most susceptible stages inhabit an environment in which EPNs can be applied, mainly the soil. For certain pests, such as lice and cockroaches, particularly those inhabiting substrates such as manure, special formulations are required to ensure the effectiveness of the nematodes. When development of EPNs for the control of veterinary pests is considered, there is a natural tendency to use the knowledge and approaches used in plant protection. However, there are substantial differences between the considerations for plants versus veterinary and human pest management considerations (see Table 26.4).

26.8 References

Aguirre Varela, A.D.A., Bertolotti, M.A. and Cagnolo, S.R. (2021) Susceptibility of *Periplaneta americana* (Blattodea: Blattidae) to two isolates of *Steinernema rarum* (Rhabditida: Steinernematidae) of the province of Cordoba, Argentina. *Revista de la Sociedad Entomologica Argentina* 80, 40–47.

Appel, A.G. and Benson, E.P. (1994) Pathogenicity and limited transoothecal transmission of *Steinernema carpocapsae* (Nematoda: Steinernematidae) in adult female German cockroaches (Dictyoptera: Blattellidae). *Journal of Medical Entomology* 31, 127–131.

Appel, A.G., Benson, E.P., Ellenberger, J.M. and Manweiler, S.A. (1993) Laboratory and field evaluations of entomogenous nematode (Nematoda: Steinernematidae) for German cockroach (Dictyoptera: Blattellidae) control. *Journal of Economic Entomology* 86, 777–784.

Belton, P., Rutherford, T.A., Trotter, D.B. and Webster, J.M. (1987) *Heterorhabditis heliothidis*: a potential biological control agent of house flies in caged-layer poultry barns. *Journal of Nematology* 19, 263–266.

Centers for Disease Control and Prevention (CDC) (1996) Summary of notifiable diseases, United States 1996. *Morbidity and Mortality Report* 45, 1–87.

Cornwell, P.B. (1976) *The Cockroach, Vol.* 2. Associated Business Programmes, London.

de la Fuente, J. and Contreras, M. (2022) Additional considerations for anti-tick vaccine research. *Expert Review of Vaccines* 21, 1019–1021.

Doucet, M.M.A., Miranda, M.B. and Bertolotti, M.A. (1998) Infectivity of entomogenous nematodes (Steinernematidae and Heterorhabditidae) to *Pediculus humanus capitis* de Geer (Anoplura: Pediculidae). *Fundamental and Applied Nematology* 21, 13–16.

Doucet, M.M.A., Bertolotti, M.A., Giayetto, A.L. and Miranda, M.B. (1999) Host range, specificity, and virulence of *Steinernema feltiae*, *Steinernema rarum*, and *Heterorhabditis bacteriophora* (Steinernematidae and Heterorhabditidae) from Argentina. *Journal of Invertebrate Pathology* 73, 237–242.

El-Sadawy, H.A.E. and Habeeb, S.M. (1998) Testing some entomopathogenic nematodes for the biocontrol of *Hyalomma dromedrii* Koch (Acarina: Ixodidae). *Journal of Union Arab Biology, Cairo* 10A, 1–11.

El-Kady, G.A., El-Badry, Y.S., El-Bahrawy, A.F., El-Sharabasy, H.M. and Mahmoud, M.F. (2015) Evaluation of entomopathogenic nematode, *Steinernema carpocapsae* against the German cockroach, *Blattella germanica* (L.) under laboratory conditions. *Egyptian Journal of Biological Pest Control* 25, 355–358.

García del Pino, F. and Morton, A. (2001) Los nematodos entomopatogenos para el control biologico de la cucaracha alemana, *Blattella germanica* (Dictyoptera: Blattellidae). In: *Proceedings of the III Congreso Nacional de Entomologia Aplicada, Pamplona, Spain*, p. 122.

Geden, C.J., Axtell, R.C. and Brooks, W.M. (1986) Susceptibility of the house fly, *Musca domestica* (Diptera: Muscidae), to the entomogenous nematodes *Steinernema feltiae*, *S. glaseri* (Steinernematidae), and *Heterorhabditis heliothidis* (Heterorhabditidae). *Journal of Medical Entomology* 23, 326–332.

Georgis, R., Mullens, B.A. and Meyer, J.A. (1987) Survival and movement of insect parasitic nematodes in poultry manure and their infectivity against *Musca domestica. Journal of Nematology* 19, 292–295.

Ghally, S.E., Kamel, E.G. and Nasr, N.M. (1991) *Steinernema* parasitism of larvae of the house fly *Musca domestica* Linnaeus. *Journal of Egyptian Society of Parasitology* 21, 633–640.

Ginsberg, H.S. (1993) Transmission risk of Lyme disease and implications for tick management. *American Journal of Epidemiology* 138, 65–73.

Glazer, I. (2002) Survival biology. In: Gaugler, R. (ed.) *Entomopathogenic Nematology*. CAB International, Wallingford, UK, pp. 169–188.
Glazer, I. and Samish, M. (1993) Suitability of *Boophilus annulatus* repleted female ticks as hosts of the nematode *Steinernema carpocapsae. Journal of Invertebrate Pathology* 61, 220–222.
Glazer, I., Salame, L. and Segal, D. (1997) Genetic enhancement of nematicidal resistance of entomopathogenic nematodes. *Biocontrol Science and Technology* 7, 499–512.
Glazer, I., Alekseev, E. and Samish, M. (2001) Factors affecting the virulence of entomopathogenic nematodes to engorged female *Boophilus annulatus* ticks. *Journal of Parasitology* 87, 808–812.
Goolsby, J.A. and Shapiro-Ilan, D.I. (2020) Passive transfer of *Steinernema riobrave* entomopathogenic nematodes with potential implications for treatment of cattle fever tick-infested nilgai. *Biocontrol Science and Technology* 30, 1330–1339.
Goolsby, J.A., Singh, N.K., Shapiro-Ilan, D.I., Miller, R.J., Moran, P.J. and de Leon, A.A.P. (2018) Treatment of cattle with *Steinernema riobrave* and *Heterorhabditis floridensis* for control of the southern cattle fever tick, *Rhipicephalus* (=*Boophilus*) *microplus*. *Southwestern Entomologist* 3, 295–301.
Goolsby, J.A., Cantu, D., Vasquez, A. and Racelis, A. (2019) Development of a remotely activated field sprayer and evaluation of temperature and aeration on the longevity of *Steinernema riobrave* entomopathogenic nematodes for treatment of cattle fever tick-infested nilgai. *Subtropical Agriculture and Environments* 70, 1–5.
Grewal, P.S. and Georgis, R. (1999) Entomopathogenic nematodes. In: Hall, F.R. and Menn, J.J. (eds.) *Biopesticides: Use and Delivery*. Humana Press, Totowa, New Jersey, pp. 271–299.
Grewal, P.S., Gaugler, R., Kaya, H.K. and Wusaty, M. (1993) Infectivity of the entomopathogenic nematode *Steinernema scapterisci* (Nematoda: Steinernematidae). *Journal of Invertebrate Pathology* 62, 22–28.
Hassanain, M.A., Derbala, A.A., Abdel-Barry, N.A., El-Sherif, M.A. and El-Sadawe, H.A. (1999) Biological control of ticks (Argasidae) by entomopathogenic nematodes. *Egyptian Journal of Biological Pest Control* 7, 41–46.
Henderson, G., Manweiler, S.A., Lawrence, W.J., Tempelman, R.J. and Foil, L.D. (1995) The effect of *Steinernema carpocapsae* (Weiser) application to different life stages on adult emergence of the cat flea *Ctenocephalides felis* (Bouche). *Veterinary Dermatology* 6, 159–163.
Hill, D.E. (1998) Entompathogenic nematodes as control agents of developmental stages of the black-legged tick, *Ixodes scapularis. Journal of Parasitology* 84, 1124–1127.
Hullet, A.C. and Dockhorn, R.J. (1979) House dust mite (*D. Farinae*) and cockroach allergy in a midwestern population. *Annals of Allergy* 42, 160–165.
Kaaya, G.P., Glazer, I. and Samish, M. (2000) Laboratory evaluation of pathogenicity of entomopathogenic nematodes to African tick species. *Annals of the New York Academy of Science* 916, 303–308.
Kaya, H.K. (1990) Soil ecology. In: Gaugler, R. and Kaya, H.K. (eds.) *Entomopathogenic Nematodes in Biological Control.* CRC Press, Boca Raton, Florida, pp. 93–115.
Kocan, K.M., Pidherney, M.S., Bluin, E.F. and Claipool, P.L. (1998a) Interaction of entomopathogenic nematodes (Steinernematidae) with selected species of ixodid ticks (Acari: Ixodidae). *Journal of Medical Entomology* 35, 514–520.
Kocan, K.M., Bluin, E.F., Pidenherney, M.S., Claipool, P.L., Samish, M. and Glazer, I. (1998b) EPNs as a potential biological control method for ticks. *Annals of the New York Academy of Science* 849, 355–356.
Koehler, P.G., Patterson, R.S. and Martin, W.R. (1992) Susceptibility of cockroaches (Dictyoptera: Blattellidae, Blattidae) to infection by *Steinernema carpocapsae*. *Journal of Economic Entomology* 85, 1184–1187.
Kotlarska-Mordzinska, J., Goszczynski, W. and Bednarek, A. (2000) Susceptibility of the cockroach, *Blatta orientalis* L. to entomopathogenic nematodes. *IOBC/WPRS Bulletin* 23, 39–40.
Lancaster, J.L. Jr. and Meisch, M.V. (1986) *Arthropods in Livestock and Poultry Production*. Ellis Horwood, London.
Laumond, C., Mauleion, H. and Kermarrec, A. (1979) Données nouvelles sur le spectre d'hôtes et le parasitisme du nématode entomophage *Neoaplectana carpocapsae*. *Entomophaga* 24, 13–27.
Lewis, E.E. (2002) Behavioral ecology. In: Gaugler, R. (ed.) *Entomopathogenic Nematology*. CAB International, Wallingford, UK, pp. 205–224.
Locatelli, D.P. and Parleaz, E. (1987) Valutazione in laboratorio de ll'attivitai di *Steinernema* spp. e *Heterorhabditis* spp. su *Blattella germanica* (L.). *La Difensa Delle Piante* 10, 339–348.
Maketon, M., Hominchan, A. and Hotaka, D. (2010) Control of American cockroach (*Periplaneta americana*) and German cockroach (*Blattella germanica*) by entomopathogenic nematodes. *Revista Colombiana de Entomologia* 36, 249–253.

Manweiler, S.A. (1994) Development of the first cat flea biological control product employing the entomopathogenic nematode *Steinernema carpocapsae*. In: *Proceedings of the Brighton Crop Protection Conference: Pests and Diseases, 21–24 November, Brighton, UK, Vol.* 3. British Crop Protection Council, Alton, UK, pp. 1005–1012.

Margalit, Y. and Ben-Dov, E. (2000) Biological control by *Bacillus thuringiensis* subs. *israelensis*. In: Rechcigel, J.E. and Rechcigel, N.A. (eds.) *Insect Pest Management Techniques for Environmental Protection*. CRC Press, Boca Raton, Florida, pp. 243–301.

Mauleon, H., Barre, N. and Panoma, S. (1993) Pathogenicity of 17 isolates of entomophagous nematodes (Steinernematidae and Heterorhabditidae) for the ticks *Amblyomma variegatum* (Fabricius), *Boophilus microplus* (Canestrini) and *Boophilus annulatus* (Say). *Experimental and Applied Acarology* 17, 831–838.

McCosker, P.J. (1979) Global aspects of the management and control of ticks of veterinary importance. In: *Recent Advances in Acarology II*. Academic Press, New York, pp. 45–53.

Monteiro, C., Vale, F.L., Vieira, M.S., Marcelo de Souza Perinotto, W., Auad, A.M. *et al.* (2020) Efficacy of *Heterorhabditis baujardi* (Rhabditida: Heterorhabditidae) against *Rhipicephalus microplus* (Acari: Ixodidae) in presence of susceptible and alternate insect hosts. *Biocontrol Science and Technology* 30, 1316–1329.

Morton, A. and García-del-Pino, F. (2013) Sex-related differences in the susceptibility of *Periplaneta americana* and *Capnodis tenebrionis* to the entomopathogenic nematode *Steinernema carpocapsae*. *Journal of Invertebrate Pathology* 112: 203–207.

Pan, X.Y. and Zhang, F. (2020) Advances in biological control of the German cockroach, *Blattella germanica* (L.). *Biological Control* 142: 104104.

Perez de Leon, A.A., Pete, T.D., Auclair, A.N., Messenger, M.T., Guerrero, F. *et al.* (2012) Integrated strategy for sustainable cattle fever tick eradication in USA is required to mitigate the impact of global change. *Frontiers in Physiology* 3: 195.

Pope, A.M., Patterson, R. and Burge, H. (1993) *Indoor Allergens: Assessing and Controlling Adverse Health Effects*. National Academy Press, Washington, DC.

Pye, N., Holbrook, G. and Pye, A. (2001) New bait station with *Steinernema carpocapsae* nematodes kills cockroaches. In: *Proceedings of 34th Annual Meeting of the Society of Invertebrate Pathology, 25–30 August, Noordwijkerhout, Netherlands*. Society for Invertebrate Pathology, Wageningen, Netherlands, p. 62.

Rahimian, A.A., Hanafi-Bojd, A.A., Vatandoost, H. and Zaim, M. (2019) A review on the insecticide resistance of three species of cockroaches (Blattodea: Blattidae) in Iran. *Journal of Economic Entomology* 112, 1–10.

Ramakrishnan, J., Salame, L., Mani, A.M., Feldbaum, R., Karavani, E. *et al.* (2023) Increasing the survival and efficacy of entomopathogenic nematodes on exposed surfaces by Pickering emulsion formulations offers new venue for foliar pest management. *Journal of Invertebrate Pathology* 199: 107938.

Renn, N. (1995) Mortality of immature houseflies (*Musca domestica* L.) in artificial diet and chicken manure after exposure to encapsulated entomopathogenic nematodes (Rhabditida: Steinernematidae, Heterorhabditidae). *Biocontrol Science and Technology* 5, 349–359.

Renn, N. (1998a) Routes of penetration of the entomopathogenic nematode *Steinernema feltiae* attacking larval and adult houseflies (*Musca domestica*). *Journal of Invertebrate Pathology* 72, 281–287.

Renn, N. (1998b) The efficacy of EPNs for controlling housefly infestations of intensive pig units. *Medical Veterinary Entomology* 12, 46–51.

Renn, N. and Wright, E. (2000) The effect of artificial substrates on the pathogenicity of *Steinernema feltiae* (Rhabditida: Steinernematidae) to adult *Musca domestica* (Diptera: Muscidae). *Nematology* 2, 217–222.

Robinson, W.H. (1996) *Urban Entomology: Insect and Mite Pests in the Human Environment*. Chapman & Hall, London.

Samish, M. and Glazer, I. (1991a) Killing ticks with parasitic nematodes of insects. *Journal of Invertebrate Pathology* 58, 281–282.

Samish, M. and Glazer, I. (1991b) Pathogenicity of parasitic nematodes to ticks. In: Dusbabek, R. and Bukva, V. (eds.) *Tick-Borne Diseases and their Vectors*. SPB Academic Publishers, The Hague, Netherlands, pp. 629–632.

Samish, M. and Glazer, I. (1992) Infectivity of EPNs (Steinernematidae and Heterorhabditidae) to female ticks of *Boophilus annulatus* (Arachnida: Ixodidae). *Journal of Medical Entomology* 29, 614–618.

Samish, M. and Rehacek, J. (1999) Pathogens and predators of ticks and their potential in biological control. *Annual Review of Entomology* 44, 159–182.

Samish, M., Alekseev, E. and Glazer, I. (1995) The development of entomopathogenic nematodes in the tick *Boophilus annulatus*. In: Coons, L. and Rothschild, M. (eds.) *Proceedings of the Second International Conference on Tick Borne Pathogens at the Host Vector Interface: A Global Perspective, 28 August–1 September, Kruger National Park, South Africa*, pp. 2–4.

Samish, M., Glazer, I. and Alekseev, E.A. (1996) The susceptibility of the development stages of ticks (Ixodidae) to entomopathogenic nematodes. In: Rodger, M., Horn, D.J., Needham, G.R. and Welbourn, W.C. (eds) *Acarology IX*. Ohio Biological Survey, Columbus, Ohio, pp. 121–123.

Samish, M., Alekseev, E. and Glazer, I. (1997) The effect of substrate composition on the success of entomopathogenic nematodes to kill ticks. *Israel Journal of Medical Science* 33: 705.

Samish, M., Alekseev, E. and Glazer, I. (1998) The effect of the soil environment upon anti-tick activity of entomopathogenic nematodes. *Annals of the New York Academy of Sciences* 849, 402–404.

Samish, M., Alekseev, E. and Glazer, I. (1999) Interaction between ticks (Acari: Ixodidae) and pathogenic nematodes (Nematoda): susceptibility of tick species at various developmental stages. *Journal of Medical Entomology* 36, 733–740.

Samish, M., Alekseev, E. and Glazer, I. (2000a) Efficacy of entomopathogenic nematode strains against engorged *Boophilus annulatus* (Say) female ticks (Arachnida: Ixodida) under seminatural conditions. *Journal of Medical Entomology* 36, 727–732.

Samish, M., Alekseev, E. and Glazer, I. (2000b) Interaction between ticks and pathogenic nematodes: susceptibility of tick species at various developmental stages. *Journal of Medical Entomology* 36, 733–740.

Samish, M., Alekseev, E. and Glazer, I. (2000c) Mortality rate of adult ticks due to infection by entomopathogenic nematodes. *Journal of Parasitology* 86, 679–684.

Samish, M., Alekseev, E. and Glazer, I. (2001) EPNs for the biocontrol of ticks. *Trends in Parasitology* 17, 368–371.

Scott, J.G., Warren, W.C., Beukeboom, L.W., Bopp, D., Clark, A.G. *et al.* (2014) Genome of the house fly, *Musca domestica* L., a global vector of diseases with adaptations to a septic environment. *Genome Biology* 15: 466.

Shapiro, D.I., Tylka, G.L. and Lewis, L.C. (1996) Effects of fertilizers on virulence of *Steinernema carpocapsae*. *Applied Soil Ecology* 3, 27–34.

Shapiro, D.I., Lewis, L.C., Obrycki, J.J. and Abbas, M. (1999) Effects of fertilizers on suppression of black cutworm (*Agrotis ipsilon*) damage with *Steinernema carpocapsae*. *Journal of Nematology* 31 (Suppl. 4), 690–693.

Shapiro-Ilan, D.I., and Goolsby, J.A. (2021) Evaluation of Barricade® to enhance survival of entomopathogenic nematodes on cowhide. *Journal of Invertebrate Pathology* 184: 107592.

Silverman, J., Platzer, E.G. and Rust, M.K. (1982) Infection of cat flea, *Ctenocephalides felis* (Bouche) by *Neoaplectana carpocapsae* (Weiser). *Journal of Nematology* 14, 394–497.

Singh, N.K., Goolsby, J.A., Shapiro-Ilan, D.I., Miller, R.J., Setamou, M. and Perez de Leon, A.A. (2018) Effect of immersion time on efficacy of entomopathogenic nematodes against engorged females of cattle fever tick, *Rhipicephalus* (=*Boophilus*) *microplus*. *Southwestern Entomologist* 43, 19–28.

Singh, N.K., Goolsby, J.A., Jyoti, J., Shapiro-Ilan, D.I., Miller, R.J. and Perez de Leon, A.A. (2019) Comparative efficacy of entomopathogenic nematodes against a multi-acaricide resistant strain of southern cattle fever tick, *Rhipicephalus microplus*. *Southwestern Entomologist* 44, 143–153.

Skierska, B., Szadziewska, M. and Stanuszek, S. (1976) Laboratory test for the usability of the entomophilic nematodes Steinernematidae Chitwood and Chitwood 1937 in biological control of some noxious arthropods. *Bulletin of the Institute of Maritime Tropical Medicine, Gdynia* 27, 207–227.

Smith, E.H. and Whitman, R.C. (1992) *Field Guide to Structural Pests*. National Pest Control Association, Dunn Loring, Virginia.

Smith, L.M., Appel, A.G., Mack, T.P., Keever, G.J. and Benson, E.P. (1997) Evaluation of methods of insecticide application for control of smokybrown cockroaches (Dictyoptera: Blattidae). *Journal of Economic Entomology* 90, 1232–1242.

Sonenshine, D.E. and Roe, R.M. (eds.) (1993) *Biology of Ticks, Vol.* 2. Oxford University Press, New York.

Strand, M.A. (1977) Pathogens of *Siphonaptera*. *Bulletin of the World Health Organization* 55 (Suppl. 1), 279–283.

Sutherst, R.W., Jones, R.J. and Schnitzerling, H.J. (1982) Tropical legumes of the genus *Stylosanthes* immobilize and kill cattle ticks. *Nature* 295, 320–321.

Taylor, D.B., Szalanski, A.L., Adams, B.J. and Peterson, R.D. (1998) Susceptibility of house fly (Diptera: Muscidae) larvae to EPNs (Rhabditida: Heterorhabditidae, Steinernematidae). *Environmental Entomology* 7, 1514–1519.

Tsao, J.I., Hamer, S.A., Han, S., Sidge, J.L. and Hickling, G.J. (2021). The contribution of wildlife hosts to the rise of ticks and tick-borne diseases in North America. *Journal of Medical Entomology* 58, 1565–1587.

Weber, T., Georgis, R., Pruitt, P. and Wren, J. (1992) Insect trap. US Patent No. 5,172,514. US Patent and Trademark Office, Washington, DC.

Weiss, M., Glazer, I., Mumcuoglu, K.Y., Elking, Y. and Galun, R. (1993) Infectivity of steinernematid and heterorhabditid nematodes for the human body louse *Pediculus humanus humanus* (Anoplura: Pediculidae). *Fundamental and Applied Nematology* 16, 489–493.

Wu, S., Li, Y., Toews, M.D., Mbata, G. and Shapiro-Ilan, D.I. (2023). Novel formulations improve the environmental tolerance of the entomopathogenic nematodes. *Biological Control* 186: 105329.

Zervos, S. and Webster, J.M. (1989) Susceptibility of the cockroach *Periplaneta americana* to *Heterorhabditis heliothidis* (Nematoda: Rhabditoidea) in the laboratory. *Canadian Journal of Zoology* 67, 1609–1611.

Zhioua, E., Lebrum, R.A., Ginsberg, H.S. and Aeschlimann, A. (1995) Pathogenicity of *Steinernema carpocapsae* and *S. glaseri* (Nematoda: Steinernematidae) to *Ixodes scapularis* (Acari: Ixodidae). *Journal of Medical Entomology* 32, 900–905.

27 Interactions between Agrochemicals and Biological Control Agents

Albrecht M. Koppenhöfer[1*] and Shane Foye[1,2]

[1]*Department of Entomology, Rutgers University, New Brunswick, New Jersey, USA;*
[2]*Current address: Department of Entomology, University of Wisconsin, Madison, Wisconsin, USA*

Abstract

Entomopathogenic nematode (EPN) performance may be affected by many agrochemicals and control agents (CAs). Infective juveniles of EPNs can tolerate short-term exposure to many agrochemicals at concentrations high enough to allow them to be tank mixed. If EPNs and agrochemicals are not tank-mix compatible, CAs must be applied 1–2 weeks apart. When combining EPNs with another CA for improved pest control, they most commonly do not interact, resulting in additive target mortality. However, in some combinations, antagonist or synergistic interactions may result in significantly lower or higher target mortality, respectively. When synergistic EPN–insecticide interactions occurred, the insecticide affects either EPN behavior or, more commonly, the behavior of the pest due to neurotoxic insecticide activity. EPN–pathogen combinations mostly result in additive or weak synergistic target mortality, and, rarely, antagonism or strong synergism. Observations for most combinations are limited. Different trends may be observed with different EPN species, insect species, and developmental stages, pathogens or environmental conditions. Co-applications against the same target likely trigger competition for resources, although under field conditions coinfections are less likely. Currently, combinations of EPNs with imidacloprid and with commercially unavailable *Bacillus thuringiensis* ssp. *japonensis* in white grubs are the only systems with sufficient greenhouse and field trials to support informed recommendations. EPN–imidacloprid combinations could be an effective tool for curative white grub control. Because EPNs are expensive, the feasibility of the combination depends on reducing EPN rates while retaining acceptable pest control.

27.1 Introduction

Entomopathogenic nematode (EPN) performance may be affected by many agrochemicals and control agents (CAs). Interactions, occurring at different intensities, are least likely if the EPNs are applied simultaneously or shortly after other materials. Commonly, chances of interactions are much higher if EPNs are mixed with other agents in spray tanks for convenience, due to the other agent's higher concentration. Interactions may be intentional when EPNs are combined with other CAs to improve pest control through additive or, preferably, synergistic effects on pest mortality. Interactions in combined applications include effects on CA viability and/or pathogenicity, CA recycling in the pest, and increased pest susceptibility to the CA.

*Email: a.koppenhofer@rutgers.edu

 Entomopathogenic Nematodes as Biological Control Agents (D.I. Shapiro-Ilan and E.E. Lewis eds)

DOI: 10.1079/9781800620322.0027

27.2 Compatibility

27.2.1 Chemicals

Infective juveniles (IJs) of EPNs can tolerate short-term exposure (2–24 h) to many agrochemicals at concentrations high enough to allow them to be tank mixed (Tables 27.1–27.4). This facilitates EPN use in integrated pest management systems. The application system and spray volume will affect IJ exposure to agrochemicals (Alumai and Grewal, 2004). If EPNs and agrochemicals are not tank-mix compatible, CAs must be applied separately, according to the agrochemical's persistence (Grewal *et al.*, 1998). Without interval information, EPNs should be applied at least 1 week after synthetic insecticides and 2 weeks after nematicides. As EPN species differ in compatibility with synthetic or biological materials (Tables 27.1–27.4) and some inert ingredients or adjuvants can be toxic to IJs (Krishnayya and Grewal, 2002), EPN–agrochemical formulation compatibility should be evaluated.

27.2.2 Biological control agents

Most studies on compatibility of EPNs with parasitoids and predators have been conducted in the laboratory, without quantifying effects on the EPN.

27.2.2.1 Parasitoids

EPNs might not hinder egg parasitoids. For example, *Heterorhabditis bacteriophora* did not affect *Trichogramma chilonis* emergence from parasitized rice moth (*Corcyra cephalonica*) eggs (Lalitha *et al.*, 2012). However, larval parasitoids are generally hindered by EPN infection of the host or the parasitoid. For example, *Steinernema carpocapsae* killed *Ganaspidium utilis*, a eucoilid parasitoid of *Liriomyza trifolii* (Hara *et al.*, 1989). *Steinernema feltiae* infected *Liriomyza huidobrensis* larvae that were already parasitized in their mines by *Diglyphus isaea* or *Dancus sibirica* (Head *et al.*, 2003). Tachinid larvae (*Myxexoristops* sp.) did not complete development in parasitized sawfly (*Cephalcia abietis*) larvae treated with *Steinernema kraussei* (Mracek and Spitzer, 1983). Externally feeding and cocooned larvae of *Bracon hylobii* on pine weevil, *Hylobius abietis*, were readily infected by *Heterorhabditis downesi* (Everard *et al.*, 2009). *Steinernema* spp. soil applications targeting the sawfly *Cephalcia arvensis* reduced sawfly parasitoid (*Xenoschesis fulvipes*) emergence (Battisti, 1994). In adult *Scapteriscus vicinus* mole crickets, *Steinernema scapterisci* and tachinid larvae (*Ormia depleta*) combinations resulted in a higher percentage of hosts producing EPN progeny but a lower percentage with tachinid progeny versus single-agent exposure; 10–31% of hosts produced both EPN and tachinid progeny (Parkman and Frank, 2002).

EPNs damage parasitoids to various degrees depending on EPN application timing. *S. carpocapsae* host infection reduced braconid *Apanteles militaris* (Kaya, 1978) and ichneumonid *Hyposoter exiguae* (Kaya and Hotchkin, 1981) survival inside *Pseudaletia unipuncta* larvae and braconid *Meteorus rubens* inside *Agrotis ipsilon* larvae (Zaki *et al.*, 1997). Parasitoid survival increased with host larva age at EPN infection. EPNs reproduced inside the parasitized hosts. Larvae of the ichneumonids *Mastrus ridibundus* and *Liotryphon caudatus*, parasitoids of cocooned codling moth larvae, were susceptible to *S. carpocapsae* until their cocoons were formed (Lacey *et al.*, 2003). When *Spodoptera littoralis* larvae parasitized by the braconid *Microplitis rufiventris* were exposed to *H. bacteriophora* or *S. carpocapsae*, both EPNs reproduced. All parasitoid larvae were killed when exposed during the parasitoid's egg or young larval stages, but 50% of wasps emerged when exposed during late larval stages (Atwa *et al.*, 2013). *S. carpocapsae* reduced braconid *Cardiochiles diaphaniae* survival in *Diaphania* spp. hosts, but the parasitoid's pupae were not infected (Shannag and Capinera, 2000). *Heterorhabditis indica* infected only larval *Habrobracon hebetor* (Mbata and Shapiro-Ilan, 2010), and *H. indica* and *Steinernema glaseri* exhibited preferential infection of Indian meal moth (*Plodia interpunctella*) tachinid (*Compsilura concinnata*) larvae that were more than 3 days old inside armyworm hosts and eclosing adult flies but not pupae (Kaya, 1984); when tachinid larvae were more than 3 days old, both the EPN and parasitoid could develop.

Under field conditions, parasitoid avoidance may attenuate EPN damage. *Mastrus ridibundus* and *Liotryphon caudatus* avoided ovipositing on *S. carpocapsae*-infected codling moth larvae (Lacey *et al.*, 2003). *Diglyphus begini* did not oviposit on

Table 27.1. Compatibility of entomopathogenic nematodes (*Steinernema* and *Heterorhabditis* spp.) with acaricides, insecticides and nematicides.

Active ingredient	Trade name/ formulation	Nematode	Compatibility	Reference(s)
Abamectin	Dynamec	Sf	–	Head *et al.* (2000)
	Vertimec	Hb, Sc/Sf, Sk	+/#	Laznik and Trdan (2014)
	Gol Shimi 1.8 EC	Sc	#	Eizavian Kary *et al.* (2018)
Abamectin + pyrethrin	Triathlon CS	Hb, Sc/Sf, Sk	+/–	Bajc *et al.* (2017)
Acephate	Orthene 97 PE	Hb, Sf	–	Rovesti *et al.* (1988); de Nardo and Grewal (2003)
	Starthene	Hi	+	Chavan *et al.* (2018)
Acetamiprid	Acetamiprid 70 WP	Hb, Hm, Sc/Sf	+/#	Koppenhöfer *et al.* (2003)
Aldicarb	Temik	Hb, Sc, Sf	–	Rovesti *et al.* (1988); Rovesti and Deseö (1990)
Amitraz	Edrizar 20.5 EC	Hb/Sc, Sf	+/–	Rovesti *et al.* (1988); Rovesti and Deseö (1990)
Azadirachtin	Azatin 3 EC	Sf	+	Grewal *et al.* (1998)
	Margosan-O/0.25%	Sf	+	Grewal *et al.* (1998)
	NeemAzal-T/S	Hb, Sf, Sk/Hi/Sc	+/+/–	Laznik and Trdan (2014); Chavan *et al.* (2018)
	Nimbecidine EC	Sf/Hi	–/+	Krishnayya and Grewal (2002); Chavan *et al.* (2018)
	Multineem EC	Hi	+	Chavan *et al.* (2018)
Azocyclotin	Perporal WP	Sf	+	Radová (2010)
Bacillus thuringiensis israelensis	ABG-6193	Hb, Sc	+	Poinar *et al.* (1990)
	Gnatrol	Hb, Sc/Sf	+/#	Poinar *et al.* (1990); de Nardo and Grewal (2003)
B. thuringiensis kurstaki	Delfin WG	Sc, Sf/Hb, Sk	+/#	Barbarossa *et al.* (1996); Laznik and Trdan (2014)
	Dipel	Hb, Sc	+	Poinar *et al.* (1990)
B. thuringiensis san diego	M-One	Hb, Sc	+	Poinar *et al.* (1990)
Bendiocarb	Turcam 76W	Sf, Sc, Hb	–	Zimmerman and Cranshaw (1990)
Binapacryl	Morocode	Sc	–	Prakasa Rao *et al.* (1975)
Buprofezin	Kaal SC	Hi	+	Chavan *et al.* (2018)
Carbaryl	Sevin SL	Hb/Sc/Sf	–/+/#	Das and Divakumar (1987); Zimmerman and Cranshaw (1990); Alumai and Grewal (2004)
Carbofuran	Curater 5 G	Hb/Sc, Sf	–/+	Rovesti *et al.* (1988); Rovesti and Deseö (1990, 1991)
Carbosulfan	Marshal 5 G	Hb/Sc, Sf	–/+	Rovesti *et al.* (1988); Rovesti and Deseö (1990, 1991)
Cartap hydrochloride	Caldan SP	Hi	–	Chavan *et al.* (2018)
Chlorantraniliprole	Coragen 20SC	Hi, Sc	+	Khan *et al.* (2021)
	Acelepryn	Hb	+	Koppenhöfer and Fuzy (2008)
Chlorfenapyr	Pirate 36SC	Hi, Sc	+	Khan *et al.* (2021)
Chlorfenvinphos	Birlane	Sc	–	Prakasa Rao *et al.* (1975)
Chloropyriphos-ethyl	Terial 40 L/40.8 EC	Hb/Sc/Sf	+/+/–	Rovesti *et al.* (1988); Rovesti and Deseö (1990)
Chlorpyrifos	Dursban 4 E	Hb, Sc, Sf	–	Zimmerman and Cranshaw (1990)

Continued

Table 27.1. Continued.

Active ingredient	Trade name/ formulation	Nematode	Compatibility	Reference(s)
	Dursban Pro 23.5%	Hb, Sc	+	Alumai and Grewal (2004)
	Lorsban 480 BR	Hi, Sc, Sg	+	Negrisoli *et al.* (2010)
	Vexter	Hi, Sc, Sg	+	Negrisoli *et al.* (2010)
	Dursban EC	Hi	+	Chavan *et al.* (2018)
Clofentezin	Apollo SC	Sf, Sk	#	Radová (2010); Bajc *et al.* (2017)
		Hb, Sc	+	Bajc *et al.* (2017)
Cyfluthrin	Baythroid 50	Sc	–	Barbarossa *et al.* (1996)
	Turbo	Hi, Sc, Sg	–	Negrisoli *et al.* (2010)
Cyhexatin	Plictran L 51.8 Flo	Hb, Sc, Sf	+	Rovesti *et al.* (1988); Rovesti and Deseö (1990)
γ-Cyhalothrin	Stallion 150 CS	Hi, Sc, Sg	+	Negrisoli *et al.* (2010)
λ-Cyhalothrin	Karate Zeon 250 CS	Hi, Sc/Sg	–/#	Negrisoli *et al.* (2010)
	Karate	Sc	–	Barbarossa *et al.* (1996)
	Karate Zeon 5 CS	Hb, Sc/Sf/Sk	+/#/–	Laznik and Trdan (2014)
Cypermethrin	Galgotrin	Hi, Sc, Sg	+	Negrisoli *et al.* (2010)
α-Cypermethrin	Fastac	Hi, Sc, Sg	–	Negrisoli *et al.* (2010)
Cyproconazole	Alto 100 SL	Sc	+	Barbarossa *et al.* (1996)
Deltamethrin	Decis	Hb/Hi/Sc	+	Rovesti *et al.* (1988); Rovesti and Deseö (1990); Negrisoli *et al.* (2010)
		Sf	+/–	Rovesti and Deseö (1990); Head *et al.* (2000)
		Sg	–	Negrisoli *et al.* (2010)
Diafenthurion	Polo SC	Sf	+	Radová (2010)
Diazinon	Basudin 20 E	Hb	#	Rovesti *et al.* (1988); Rovesti and Deseö (1991)
	Diazinon AG 500	Hb/Sc, Sf	–/#	Rovesti and Deseö (1990, 1991); Zimmerman and Cranshaw (1990)
Dicofol	Kelthane MF 42 E	Hb/Sc, Sf	+	Rovesti *et al.* (1988); Rovesti and Deseö (1990)
Difenoconzole	Score 250 EC	Sc	–	Barbarossa *et al.* (1996)
Diflubenzuron	Adept IGR	Sc, Sf	+	Rovesti and Deseö (1990); de Nardo and Grewal (2003)
	Dimilin	Hb, Hi, Sc, Sf, Sg	+	Rovesti *et al.* (1988); Grewal and Richardson (1993); Negrisoli *et al.* (2010)
Dimethoate	Danadim	Sf	+	Head *et al.* (2000)
Dinotefuran	Token SG	Hi	+	Chavan *et al.* (2018)
DNOC	Sclinon 50 EC	Hb, Sc, Sf	+	Rovesti *et al.* (1988); Rovesti and Deseö (1990)
Emamectin benzoate	Proclaim 019EC	Hi, Sc	+	Khan *et al.* (2021)
Endosulfan	Thiodan 32.9 WP	Hb, Sc, Sf	+	Das and Divakumar (1987); Rovesti and Deseö (1990)
Ethiprole + imidacloprid	Glamore WG	Hi	+	Chavan *et al.* (2018)
Etoxazole	Borneo SC	Sf	+	Radová (2010)
Exitiazox	Matacar 10 WP	Hb, Sc, Sf	+	Rovesti *et al.* (1988); Rovesti and Deseö (1990)
Fenamiphos	Nemacur 4.8 G	Hb, Sc, Sf	–	Rovesti *et al.* (1988); Rovesti and Deseö (1990)
Fenbutatinoxide	Torque-L SC	Sf	+	Radová (2010)

Continued

Table 27.1. Continued.

Active ingredient	Trade name/ formulation	Nematode	Compatibility	Reference(s)
Fenitrothion	Sumithion	Sc	–	Prakasa Rao *et al.* (1975)
Fenoxycarb	Precision 25 WP	Sf	–	de Nardo and Grewal (2003)
Fenpyroximate	Naja SC	Sf	–	Radová (2010)
	Ortus 5 SC	Hb, Sc/Sf, Sk	+/–	Bajc *et al.* (2017)
Fenthion	Lebaycid	Sc	+	Prakasa Rao *et al.* (1975)
Fipronil	Regent SC	Hi	+	Chavan *et al.* (2018)
Flubendiamide	Belt 48SC	Hi, Sc	+	Khan *et al.* (2021)
Flubenzimine	Cropotex 50 WP	Hb, Sc, Sf	–	Rovesti *et al.* (1988); Rovesti and Deseö (1990, 1991)
Fluroxypyr	Starane 180	Sc	+	Barbarossa *et al.* (1996)
Fluvalinate	Klartan	Sc	+	Barbarossa *et al.* (1996)
Fonofos	Dyfonate 5 G	Hb, Sc, Sf	+	Rovesti *et al.* (1988); Rovesti and Deseö (1990)
Formothion	Anthio	Sc	–	Prakasa Rao *et al.* (1975)
Halofenozide	Mach2 1.5 G	Hb/Sc	–/+	Alumai and Grewal (2004)
Haloxyfop	Gallant super	Sc	+	Barbarossa *et al.* (1996)
Heptenophos	Hostacquick	Sc, Sf	–	Barbarossa *et al.* (1996); Head *et al.* (2000)
Imidacloprid	Merit 75 WP	Hb, Hm, Sc, Sf, Sg	+	Koppenhöfer and Kaya (1998); Koppenhöfer *et al.* (2003); Alumai and Grewal (2004)
	Confidor 200 SL	Hb/Sc, Sf, Sk	+/#	Laznik and Trdan (2014)
	Confidor SL	Hi	+	Chavan *et al.* (2018)
Indoxacarb	Steward 150EC	Hi, Sc	+	Khan *et al.* (2021)
Isofenphos + phoxim	Oftanol combination	Hb	+/–	Poinar *et al.* (1990)
Kinoprene	Enstar EC	Sf	+	Radová (2010)
Lecithin	Bio blatt	Sc	–	Barbarossa *et al.* (1996)
Lindane (γ-HCH)	Lintox 3G	Hb, Sc, Sf	+	Rovesti *et al.* (1988); Rovesti and Deseö (1990)
Lufenuron	Match	Hi, Sc, Sf, Sg	+	Negrisoli *et al.* (2010); Radová (2010); Khan *et al.* (2021)
	Match 050 SC	Hb, Sf/Sk/Sc	+/#/–	Laznik and Trdan (2014)
Malathion	Malathion 50 EC	Sc	–	Das and Divakumar (1987)
Metham sodium	Geort 32.7 WSL	Hb, Sc, Sf	–	Rovesti *et al.* (1988); Rovesti and Deseö (1990)
Methomyl	Lannate 25 WP	Hb, Sc, Sf	–	Rovesti *et al.* (1988); Rovesti and Deseö (1990, 1991)
	Lannate SL	Sf	+	Radová (2010)
Methoprene	Apex 5E	Sf	+	Grewal *et al.* (1998)
Methoxyfenozide	Intrepid 240 SC	Hi, Sg/Sc	#/–	Negrisoli *et al.* (2010)
	Runner 240SC	Hi, Sc, Sf	+	Radová (2010); Khan *et al.* (2021)
Metramitron	Goltix 70 WG	Sc	–	Barbarossa *et al.* (1996)
Methyl parathion	Folidol 600	Hi, Sc, Sg	–	Negrisoli *et al.* (2010)
Monocrotophos	Azodrin	Sc	–	Prakasa Rao *et al.* (1975)
	Monostar SL	Hi	–	Chavan *et al.* (2018)
Neem	Neem kernel	Hb/Sc, Sf, Sg	–/+	Prakasa Rao *et al.* (1975); Rovesti *et al.* (1988)
Neem oil	Neem oil	Sf	+	Krishnayya and Grewal (2002)
	Neem oil + soap	Sf	–	Krishnayya and Grewal (2002)
Novaluron	Corvus 10EC	Hi, Sc	+	Khan *et al.* (2021)
Oxamyl	Vydate EC	Sf	#	Radová (2010)

Continued

Table 27.1. Continued.

Active ingredient	Trade name/ formulation	Nematode	Compatibility	Reference(s)
Parathion	E 605 FT	Hb, Sc, Sf	–	Rovesti *et al.* (1988); Rovesti and Deseö (1990)
Penconazole	Omnex WP 10	Sc	–	Barbarossa *et al.* (1996)
Permethrin	Talcord 250 CE	Hi, Sg/Sc	+/#	Negrisoli *et al.* (2010)
	Velon 384 EC	Hi, Sc, Sg	–	Negrisoli *et al.* (2010)
Phenthioate	Elsan	Sc	+	Prakasa Rao *et al.* (1975)
Phorate	Thimet 5G	Hb/Sc/Sf	–/–/+	Rovesti *et al.* (1988); Rovesti and Deseö (1990); Gupta and Siddiqui (1999)
Phosalone	Zolone Flo	Sc	#	Prakasa Rao *et al.* (1975); Barbarossa *et al.* (1996)
Phosmet	Imidan	Sc	–	Barbarossa *et al.* (1996)
Phosphamidon	Dimecron 20 EC	Hb/Sc/Sf	#/+/–	Rovesti *et al.* (1988); Rovesti and Deseö (1990, 1991)
Piperonyl butoxide	Spruzit EC	Sf	+	Radová (2010)
Pirimicarb	Pirimor 50 WG	Hb, Sc, Sf/Sk	+/#	Laznik and Trdan (2014)
Propargite	Kelaran 57 EC	Hb/Sc/Sf	#/–/+	Rovesti *et al.* (1988); Rovesti and Deseö (1990, 1991)
Pymetrozine	Chess WG	Hb, Sc, Sf/Sk/Hi	+/#/+	Laznik and Trdan (2014); Chavan *et al.* (2018)
Pyridimine phosphate	PP-511	Sc	–	Prakasa Rao *et al.* (1975)
Pyridproxyfen	Admiral EC	Sf	+	Radová (2010)
Spinetoram	Radiant 120SC	Hi, Sc	+	Khan *et al.* (2021)
Spinosad	Conserve SC	Hb, Sc, Sf	+	de Nardo and Grewal (2003); Alumai and Grewal (2004)
	Tracer	Hi, Sc, Sg	+	Negrisoli *et al.* (2010); Khan *et al.* (2021)
Spriodiclofen	Envidor 240 SC	Hb, Sc/Sf/Sk	+/#/–	Bajc *et al.* (2017)
Tebuconazole	Folicur 250 EC	Sc	–	Barbarossa *et al.* (1996)
Tebufenozide	Mimic SC	Sf	+	Radová (2010)
Tebufenpyrad	Masai WG	Sf	–/#	Radová (2010); Bajc *et al.* (2017)
		Hb, Sc/Sk	+/–	Bajc *et al.* (2017)
Teflubenzuron	Nomolt	Sc	–	Barbarossa *et al.* (1996)
Terbufos	Gyanater 2G	Hb/Sc, Sf	#/+	Rovesti *et al.* (1988); Rovesti and Deseö (1990, 1991)
Thiamethoxam	Meridian 25%	Hb	–/+	Koppenhöfer *et al.* (2003); Alumai and Grewal (2004)
		Hm, Sf, Sg	+	Koppenhöfer *et al.* (2003)
		Sc	+	Koppenhöfer *et al.* (2003); Alumai and Grewal (2004)
	Actara WG	Hi	+	Chavan *et al.* (2018)
Triazophos	Hostathion 400 BR	Sg/Hi, Sc	+/–	Negrisoli *et al.* (2010)
Trichlorfon	Dipterex 80	Sf	+	Head *et al.* (2000)
Trichlorfon	Dylox 80	Hb, Sc	–	Alumai and Grewal (2004)
Triflumuron	Certero	Sg/Hi/Sc	+/#/–	Negrisoli *et al.* (2010)
Vamidothion	Kilval	Sc	+	Prakasa Rao *et al.* (1975)
Verticillium lecanii	Micro Germin	Sc	–	Barbarossa *et al.* (1996)
White oil	Biancolio 80 E	Hb, Sc, Sf	+	Rovesti *et al.* (1988); Rovesti and Deseö (1990)

Hb, *Heterorhabditis bacteriophora*; Hi, *H. indica*; Hm, *H. megidis*; Sc, *Steinernema carpocapsae*; Sf, *S. feltiae*; Sg, *S. glaseri*; Sk, *S. kraussei*; +, compatible; #, limited negative effects; –, non-compatible.

Table 27.2. Compatibility of entomopathogenic nematodes (*Steinernema* and *Heterorhabditis* spp.) with fungicides.

Active ingredient	Trade name/ formulation	Nematode	Compatibility	Reference(s)
Abamectine	Vertimec	Sc	–	Barbarossa *et al.* (1996)
Aluminum tris	Aliette 80 WDG	Hb, Sc	–	Alumai and Grewal (2004)
Azoxystrobin	Abound 22.9%	Sf	+	Krishnayya and Grewal (2002)
Azoxystrobin	Heritage 50%	Sf	+	de Nardo and Grewal (2003)
	Quadris	Sf/Hd, Sc	+/–	Laznik *et al.* (2012)
Benomyl	Benlate 76 W	Hb, Sc, Sf	+	Zimmerman and Cranshaw (1990)
Boscalid + pyraclostrobin	Bellis	Hd, Sc/Sf	–/+	Laznik *et al.* (2012)
Bupirimate	Nimrod 25.6 EC	Hb, Sc, Sf	+	Rovesti *et al.* (1988); Rovesti and Deseö (1990)
Calcium hypochlorite	Chlorine	Sf	–	Grewal *et al.* (1998)
Captan	Captan WG	Sf	+	Radová (2010)
Carbendazim	Bavistin 50 WP	Hb/Sc/Sf	–/#/+	Rovesti *et al.* (1988); Rovesti and Deseö (1990); Barbarossa *et al.* (1996); Gupta and Siddiqui (1999)
	Benfil SC	Hi	+	Chavan *et al.* (2018)
Carbendazim + mancozeb	Companion WP	Hi	#	Chavan *et al.* (2018)
Carboxin + thiram	Blekritt 31+30 WP	Hb, Sc, Sf	+	Rovesti *et al.* (1988); Rovesti and Deseö (1990)
Chloridazon	Pyramin	Sc	+	Barbarossa *et al.* (1996)
Chlorothalonil	Bravo 500 40.4%	Sf	+	Grewal *et al.* (1998)
Chlorothalonil	Clortosip 40 L Flo	Hb, Sc, Sf	+	Rovesti *et al.* (1988); Zimmerman and Cranshaw (1990)
Cinameldehyde	Cinamate 30%	Sf	–	Krishnayya and Grewal (2002)
Cinameldehyde	Vertigo 50%	Sf	–	Grewal *et al.* (1998)
Copper (II) hydroxide	Cuprablau-Z	Sd, Sf/Sc	#/+	Laznik *et al.* (2012)
Copper (II) hydroxide + metalaxyl-M	Ridomil Gold Plus 42.5 WP	Hd, Sc, Sf	–	Laznik *et al.* (2012)
Copper sulfate	Poltiglia bord.	Hb	+	Laznik *et al.* (2012)
Cycloxydim	Focus	Sc	+	Barbarossa *et al.* (1996)
Cymoxanil + Folpet	Folcarb comb. 8.4 + 64 WP	Hb/Sc, Sf	–/+	Rovesti *et al.* (1988); Rovesti and Deseö (1990)
Diafenthiuron	Polo	Sc	+	Barbarossa *et al.* (1996)
Dinocap	Karathane PB 19.5 WP	Hb, Sc, Sf	+	Rovesti *et al.* (1988); Rovesti and Deseö (1990)
	Sabithane	Hd/Sc, Sf	#/–	Laznik *et al.* (2012)
Dithianon	Delan 75 WP	Hb, Sc/Sf	+	Rovesti *et al.* (1988); Rovesti and Deseö (1990); Barbarossa *et al.* (1996)

Continued

Table 27.2. Continued.

Active ingredient	Trade name/ formulation	Nematode	Compatibility	Reference(s)
Dodine	Melprex 65 WP	Hb, Sc, Sf	–	Rovesti *et al.* (1988); Rovesti and Deseö (1990)
Etridiazole	Terrazole 35 WP	Sf	+	de Nardo and Grewal (2003)
Fenarimol	Rubigan 6 PB 6 WP	Hb, Sc, Sf	+	Rovesti *et al.* (1988); Rovesti and Deseö (1990)
Fenhexamid	Teldor WG	Sf	+	Radová (2010)
	Teldor SC 500	Sd, Sf/Sc	+/–	Laznik *et al.* (2012)
Fenpropimorph	Corbel 79.5 EC	Hb, Sc, Sf	+	Rovesti *et al.* (1988); Rovesti and Deseö (1990)
Fentin-acetate	Brestan	Hb/Sc/Sf	–/+/–	Rovesti *et al.* (1988); Rovesti and Deseö (1990)
Fluquinconazole	Clarinet	Hd, Sc/Sf	–/#	Laznik *et al.* (2012)
Flutdioxonil	Medallion 50%	Sf	+	de Nardo and Grewal (2003)
Flutolanil	Prostar 70 WP	Sf	+	de Nardo and Grewal (2003)
Folpet	Foltan 47.5 WP	Hb, Sc, Sf	+	Rovesti *et al.* (1988); Rovesti and Deseö (1990)
	Folpan 80 WDG	Hd, Sf/Sc	#/+	Laznik *et al.* (2012)
Formaldehyde	Formalin 37%	Sf	–	Grewal *et al.* (1998)
Fosethyl	Aliette	Sc/Sf/Hd	+/#/–	Barbarossa *et al.* (1996); Laznik *et al.* (2012)
Hexaconzole	Contaf EC	Hi	+	
Hydrogen peroxide and peroxyacetic acid mixture	ZeroTol	Sf	–	Krishnayya and Grewal (2002)
Iprodione	Chipco 26GT	Sf	+	de Nardo and Grewal (2003)
Kresoxim-methyl	Candit WG	Sf	+	Radová (2010)
Mancozeb	Indofil M-45 WP	Hi	+	Chavan *et al.* (2018)
Maneb	Dithane M-45	Hd, Sc, Sf	–	Laznik *et al.* (2012)
Maneb + propamocarb	Tattoo	Sf/Sc/Hd	+/#/–	Laznik *et al.* (2012)
Mannose	Dithane M 45 80 WP	Hb, Sc, Sf	+	Rovesti *et al.* (1988); Rovesti and Deseö (1990)
Mefenoxam	Subdue 2X 45 WSP	Hb, Sc	+	Alumai and Grewal (2004)
Mefenoxam	Subdue Maxx 21.3 ME	Sf	+	de Nardo and Grewal (2003)
Mepronil	?	Hb, Sg	+	Lee *et al.* (1999)
Mercoprop-p	Duplosan	Sc	+	Barbarossa *et al.* (1996)
Mercurious chloride (60%) + mercuric chloride (30%)	Calco Clor	Hb, Sc, Sf	–	Zimmerman and Cranshaw (1990)
Metalaxyl + Folpet	Eucritt F 10 + 40 WP	Hb/Sc, Sf	–/+	Rovesti *et al.* (1988); Rovesti and Deseö (1990)

Continued

Table 27.2. Continued.

Active ingredient	Trade name/ formulation	Nematode	Compatibility	Reference(s)
Metazachlor	Butisan	Sc	+	Barbarossa *et al.* (1996)
Methoprene	Apex 5E 65.7%	Sf	+	Grewal *et al.* (1998)
Metiram	Polyram DF	Sc/Sf/Hd	+/#/–	Laznik *et al.* (2012)
Nuarimol	Tridal SC	Sf	+	Radová (2010)
Pentachloronitrobenzene	Terraclor 75%	Hb, Sc, Sf	+	Zimmerman and Cranshaw (1990)
Prochloraz	Sportak 40 EC	Hb/Sc, Sf	–/+	Rovesti *et al.* (1988); Rovesti and Deseö (1990)
Prochloraz-manganese	Sporgon 50%	Sf	+	Grewal *et al.* (1998)
Procymidone	Sialex 50 WP	Hb/Sc, Sf	–/+	Rovesti *et al.* (1988); Rovesti and Deseö (1990)
Propamocarb	Previcur 66.5 WSL	Hb/Sc, Sf	–/+	Rovesti *et al.* (1988); Rovesti and Deseö (1990)
	Previcur 607 SL	Sf/Hd, Sc	+/–	Laznik *et al.* (2012)
Propiconazole	Tilt EC	Hb, Hi, Sc, Sf	+	Rovesti *et al.* (1988); Zimmerman and Cranshaw (1990); Chavan *et al.* (2018)
Pyrazophos	Afugan 15 EC	Hb, Sc, Sf	+	Rovesti *et al.* (1988); Rovesti and Deseö (1990)
Rape seed oil	Telmion	Sc	+	Barbarossa *et al.* (1996)
Streptomyces griseoviridis K61 strain	Mycostop 30%	Sf	+	de Nardo and Grewal (2003)
Sulfur	Kumulus	Sc	+	Barbarossa *et al.* (1996)
	Tiolvit 79.9 WP	Hb, Sc, Sf	+	Rovesti *et al.* (1988); Rovesti and Deseö (1990)
	Pipelin	Sc, Sf /Hd	#/–	Laznik *et al.* (2012)
Tebuconazole + spiroxamine + triadimenol	Falcon EC-460	Hd, Sc, Sf	–	Laznik *et al.* (2012)
Thiobendazole	Mertect 340-F 42.28%	Sf	–	Grewal *et al.* (1998)
Thiophanatemethyl	Fungo Flo 4.5 F	Sf	+	de Nardo and Grewal (2003)
Thiophanatmethyl	Topsin M	Sf	+	Barbarossa *et al.* (1996)
Thiram	Pomarsol F	Sc	+	Barbarossa *et al.* (1996)
Thiram	TMTD 49 WP	Hb, Sc, Sf	+	Rovesti *et al.* (1988); Rovesti and Deseö (1990)
Tolclofosmethyl	Rizolex 50 WP	Hb, Sc, Sf, Sg	+	Rovesti *et al.* (1988); Lee *et al.* (1999)
Tolylfluanide	Euparen M	Sc	+	Barbarossa *et al.* (1996)
Trichoderma harzianum	Rootshield 1.15%	Sf	+	de Nardo and Grewal (2003)
Tricyclazole	Baan WP	Hi	+	Chavan *et al.* (2018)
Triflumizole	Terraguard 50 W	Sf	+	de Nardo and Grewal (2003)
Triforine	Saprol 18 EC	Hb, Sc, Sf	+	Rovesti *et al.* (1988); Rovesti and Deseö (1990)

?, Not mentioned in study; Hb, *Heterorhabditis bacteriophora*; Hd, *H. downesi*; Hi, *H. indica*; Sc, *Steinernema carpocapsae*; Sf, *S. feltiae*; Sg, *S. glaseri*; +, compatible; #, limited negative effects; –, non-compatible.

Table 27.3. Compatibility of entomopathogenic nematodes (*Steinernema* and *Heterorhabditis* spp.) with herbicides.

Active ingredient	Trade name/ formulation	Nematode	Compatibility	Reference(s)
2,4-D	Dicofex	Sf, Sk/Hb, Sc	+/–	Laznik and Trdan (2016)
2,4-D sodium salt	2–4 D	Sc	–	Gupta and Siddiqui (1999)
	Superhit WP	Hi	+	Chavan *et al.* (2018)
Alachlor	Lasso 43.1 EC	Hb, Sc, Sf	–	Rovesti *et al.* (1988); Rovesti and Deseö (1990)
Atrazine	Gesaprim 44.4 Flo	Hb, Sc, Sf	+	Rovesti *et al.* (1988); Rovesti and Deseö (1990)
Bispyrac sodium	Nominee EC	Hi	+	Chavan *et al.* (2018)
Butachlor	Butachlor EC	Hi	+	Chavan *et al.* (2018)
Chloridazon	Pyramin 39 Flo	Hb/Sc, Sf	–/+	Rovesti *et al.* (1988); Rovesti and Deseö (1990)
Chlorsulfuron	Glean 75 Df	Hb, Sc, Sf	+	Rovesti *et al.* (1988); Rovesti and Deseö (1990)
Chlorthal-dimethyl	Dacthal 75 WP	Hb, Sc, Sf	+	Rovesti *et al.* (1988); Rovesti and Deseö (1990)
Chlortoluron	Dicuran L 45 Flo	Hb, Sc, Sf	+	Rovesti *et al.* (1988); Rovesti and Deseö (1990)
Clethodim	Envoy 12.6%	Sf	+	de Nardo and Grewal (2003)
Clopyralid	Cirloxin 10 WSL	Hb, Sc, Sf	+	Rovesti *et al.* (1988); Rovesti and Deseö (1990)
Cyhalofop butyl	Clincher EC	Hi	+	Chavan *et al.* (2018)
Dalapon	Dalascam 85 WP	Hb, Sc, Sf	+	Rovesti *et al.* (1988); Rovesti and Deseö (1990)
Dicamba	Banvel	Hb, Sc, Sf	+	Zimmerman and Cranshaw (1990)
Diphenamid	Enide 50 WP	Hb/Sc, Sf	+	Rovesti *et al.* (1988); Rovesti and Deseö (1990)
Diquat dibromide	Reglone 200 SL	Sk/Hb, Sc, Sf	#/–	Laznik and Trdan (2016)
Fluazifop butyl	Fusilade 26 EC	Hb, Sc, Sf	+	Rovesti *et al.* (1988); Rovesti and Deseö (1990)
Fomesafen	Flex 29 WSL	Hb, Sc, Sf	+	Rovesti *et al.* (1988); Rovesti and Deseö (1990)
Glufosinate-ammonium	Basta-15	Hb, Sc, Sf, Sk	–	Laznik and Trdan (2016)
Glyphosate	Roundup 30.4 WSL	Hb/Sc, Sf	+	Rovesti *et al.* (1988); Rovesti and Deseö (1990, 1991)
	Boom efect	Hb, Sk/Sf/Sc	+/#/–	Laznik and Trdan (2016)
	?	Hb, Sf	+	García-del-Pino and Morton (2010)
Glyphosate + MCPA	?	Hb, Sf	–	García-del-Pino and Morton (2010)
Ioxynil	Cipotril 24.8 E	Hb, Sc, Sf	+	Rovesti *et al.* (1988); Rovesti and Deseö (1990)
L-flampropisopropyl	Effix 20.9 EC	Hb, Sc, Sf	+	Rovesti *et al.* (1988); Rovesti and Deseö (1990)
Lenacil	Venazar 29 WSL	Hb, Sc, Sf	+	Rovesti *et al.* (1988); Rovesti and Deseö (1990)
Linuron	Afalon DS 37.6 Flo	Hb	+/–	Rovesti *et al.* (1988); Rovesti and Deseö (1991)
	Afalon	Hb, Sc, Sf, Sk	–	Laznik and Trdan (2016)

Continued

Table 27.3. Continued.

Active ingredient	Trade name/ formulation	Nematode	Compatibility	Reference(s)
MCPA	?	Sf, Hb	+	García-del-Pino and Morton (2010)
MCPP (mecoprop)	U 46 KV 51 WSL	Hb, Sc, Sf	+	Rovesti *et al.* (1988); Rovesti and Deseö (1990)
Metazachlor	Fuego	Sk/Sf/Hb, Sc	+/#/–	Laznik and Trdan (2016)
Metribuzin	Sencor 35 WP	Hb, Sc, Sf	+	Rovesti *et al.* (1988); Rovesti and Deseö (1990)
	Sencor SC 600	Hb, Sc, Sf, Sk	–	Laznik and Trdan (2016)
Oxyfluorfen	Goal 23.6 EC	Hb/Sc, Sf	+/–	Rovesti *et al.* (1988); Rovesti and Deseö (1990)
	Goal	Hb, Sc, Sf, Sk	–	Laznik and Trdan (2016)
Paraquat	Gramoxone 17.9 WSL	Hb, Sc, Sf	–	Rovesti *et al.* (1988); Rovesti and Deseö (1990, 1991)
Pendimethalin	Stomp 31.7 E	Hb/Sc, Sf	–/+	Rovesti *et al.* (1988); Rovesti and Deseö (1990)
	Swalpendi	Hi	+	Chavan *et al.* (2018)
Penoxsulam	Granite EC	Hi	+	Chavan *et al.* (2018)
Phenmedipham	Betanal 15.9 EC	Hb	+	Rovesti *et al.* (1988)
Propyzamide	Kerb 50 WP	Hb/Sc/Sf	–/+/–	Rovesti *et al.* (1988); Rovesti and Deseö (1990)
Trifluralin	Treflan 45.8 EC	Hb, Sc	–	Rovesti *et al.* (1988); Rovesti and Deseö (1990, 1991)

?, Not mentioned in study; Hb, *Heterorhabditis bacteriophora*; Hi, *H. indica*; Sc, *Steinernema carpocapsae*; Sf, *S. feltiae*; Sk, *S. kraussei*; +, compatible; #, limited negative effects; –, non-compatible.

Table 27.4. Compatibility of entomopathogenic nematodes (*Steinernema* and *Heterorhabditis* spp.) with fertilizers, growth regulators and surfactants.

Active ingredient	Trade name/ formulation	Nematode	Compatibility	Reference(s)
Fertilizers				
Ammonium nitrate	Ammonium nitrate	Sf, Hb	–	Sahin and Sursuluk (2018)
Ammonium sulfate	Ammonium sulfate	Sc, Sf, Hb	+	Prakasa Rao *et al.* (1975); Sahin and Sursuluk (2018)
Calcium nitrate	Calcium nitrate	Sf	+	Bednarek and Gaugler (1997)
Diammonium phosphate	Diammonium phosphate	Sf, Hb	–	Sahin and Sursuluk (2018)
Cow manure, fresh	Cow manure, fresh	Sc	–	Shapiro *et al.* (1999)
Cow manure, composted	Cow manure, composted	Sc	+	Shapiro *et al.* (1999)
Nitrogen + potassium oxide mix	Gromore	Sc	+	Prakasa Rao *et al.* (1975)
Nitrogen + potassium oxide mix	Sulphala	Sc	+	Prakasa Rao *et al.* (1975)

Continued

Table 27.4. Continued.

Active ingredient	Trade name/ formulation	Nematode	Compatibility	Reference(s)
NP fertilizer	NP fertilizer	Sf, Hb	–	Sahin and Sursuluk (2018)
NPK fertilizer	NPK fertilizer	Sf, Hb	–	Bednarek and Gaugler (1997); Sahin and Sursuluk (2018)
Potassium nitrate	Potassium nitrate	Sf	–	Bednarek and Gaugler (1997)
Potassium oxide	Muriate of Potash	Sc	+	Prakasa Rao *et al*. (1975)
Potassium oxide	Superphosphate	Sc	+	Prakasa Rao *et al*. (1975)
Potassium phosphate	Potassium phosphate	Sf	–	Bednarek and Gaugler (1997)
Urea	Urea	Sc, Sf, Hb	+	Prakasa Rao *et al*. (1975); Shapiro *et al*. (1999); Sahin and Sursuluk (2018)
Growth regulators				
Ancymidol	A-Rest 0.4%	Sf	+	de Nardo and Grewal (2003)
Paclobutrazol	Bonzi 0.4%	Sf	+	de Nardo and Grewal (2003)
Surfactants				
Antibacterial liquid soap	Ajax	Sf	–	Krishnayya and Grewal (2002)

Hb, *Heterorhabditis bacteriophora*; Sc, *Steinernema carpocapsae*; Sf, *S. feltiae*; +, compatible; –, non-compatible.

S. carpocapsae-infected *Liriomyza trifolii* larvae (Sher *et al*., 2000). *B. hylobii* did not oviposit on *H. downesi*-infected *Hylobius abietis* larvae (Everard *et al*., 2009).

27.2.2.2 Predators

Many predators are resistant to EPN infections. Larvae of two sawfly predators, the therevid *Thereva handlirschi* and the rhagionid *Rhagio notatus*, were not infected by *S. kraussei* (Mracek and Spitzer, 1983). Immature and adult earwigs, *Labidura riparia*, were resistant to infection by *H. bacteriophora* and *S. carpocapsae* (Georgis *et al*., 1991) and *S. scapterisci* did not infect the European earwig, *Forficula auricularia* (Grewal *et al*., 1993). The staphylinid *Dalotia coriaria* was not susceptible to *S. feltiae* (Jandricic *et al*., 2006). *H. bacteriophora* did not affect egg hatching or survival of larvae or adults of the lacewing *Chrysoperla zastrowi sillemi* (Lalitha *et al*., 2012). *S. feltiae* was compatible with the predatory mite *Stratiolaelaps scimitus* but not the predatory mite *Gaeolaelaps gillespiei* (Saito and Brownbridge, 2016).

Other predators were susceptible to EPNs under laboratory conditions. The staphylinid *Philonthus* sp. was infected by *H. bacteriophora* and *S. carpocapsae*, but the adult was less susceptible than third-instar larvae (Georgis and Hague, 1991). *H. bacteriophora* and *S. carpocapsae* but not *S. feltiae* reduced adult emergence from the soil of the cecidomyiid aphid predator *Aphidoletes aphidimyza* (Powell and Webster, 2004). Coccinellid *Adalia bipunctata* larvae were more susceptible to *S. feltiae* and *S. carpocapsae* than *H. bacteriophora*. Larvae of *Chrysoperla carnea* were more susceptible to *S. carpocapsae* than *S. feltiae* and *H. bacteriophora* (Rojht *et al*., 2009). *F. auricularia* was susceptible to *S. carpocapsae* (Hodson *et al*., 2011).

In the field, EPN applications had only temporally and spatially limited effects on natural populations of carabids, staphylinids, histerids and formicids (Georgis *et al*., 1991; Bathon, 1996; Wang *et al*., 2001).

27.3 Interactions with Other Control Agents

Combining two CAs may improve the control (i.e. mortality) and/or speed of pest control. If CAs do not interact, the control is additive (complementary). Antagonistic interaction between CAs results in significantly lower control than additive control. Synergism (or potentiation), the joint (or supplemental) action of two agents resulting in a greater effect than the sum of the activities of the agents acting alone, results in significantly higher levels of control. Unless the interaction mechanism is known, statistical tests determine the interaction type, based on whether the observed effect is significantly different from an additive effect. Tables 27.5 and 27.6 summarize interactions between EPNs and chemicals or other pathogens.

27.3.1 Chemicals

The majority of insecticides listed in Table 27.5 did not interact with the respective EPN species (i.e. additive control). Some insecticides were either synergistic or additive. Larger data sets indicate that variability within a specific chemical (e.g. imidacloprid, clothianidin) is affected by EPN species and/or target pest species. Interactions between EPNs and chemicals can change based on development of the pest. For example, weaker interactions occurred between EPNs and imidacloprid in second-instar compared with third-instar white grubs (Koppenhöfer and Fuzy, 2008b). Extreme antagonism occurred in *Phthorimaea operculella* pupae, whereas synergism occurred in prepupae exposed to *S. carpocapsae*–abamectin combinations (Eivazian Kary *et al.*, 2018).

The best-studied system is the synergistic interaction between EPNs and the neonicotinoid imidacloprid in white grubs. Imidacloprid reduces larval defensive behaviors, increasing EPN attachment and penetration (Koppenhöfer *et al.*, 2000b). Imidacloprid and *H. bacteriophora* synergized more consistently against older third-instar larvae but provided better control against younger larvae due to their greater susceptibility to both CAs, with rates as low as 25% and 50% of the full rates for *H. bacteriophora* and imidacloprid, respectively (Koppenhöfer and Fuzy, 2008b). Two rather scarab-specific and highly virulent species, *Steinernema kushidai* and *S. scarabaei*, generally did not interact with imidacloprid. Imidacloprid did not compromise EPN recycling (Koppenhöfer *et al.*, 2003). The neonicotinoids thiamethoxam and clothianidin interacted less consistently with EPNs (Koppenhöfer *et al.*, 2002; Koppenhöfer and Fuzy, 2008b). The diamide chlorantraniliprole synergized with *H. bacteriophora* to control scarab larvae without compromising nematode recycling (Koppenhöfer and Fuzy, 2008a).

27.3.2 Pathogens

27.3.2.1 EPN–EPN interactions

Two EPN species may coinfect the same host (Kondo, 1989; Alatorre-Rosas and Kaya, 1991; Koppenhöfer *et al.*, 1995). Hosts infected with other EPNs repel steinernematids if the first species establishes in the host's hemocoel within 6–9 h (Glazer, 1997; Grewal *et al.*, 1997). If only one host is available, this avoidance likely initiates competitive exclusion of one species (Koppenhöfer *et al.*, 1996). For immediate control of one pest, combinations generally result in additive mortality (Table 27.6). If several hosts are present that differ in susceptibility to the two EPNs due to differences in the EPN species' search strategies and/or virulence, combinations may provide effective control of the pests (Kaya *et al.*, 1993) and the EPN species may coexist (Koppenhöfer and Kaya, 1996).

27.3.2.2 EPN–bacterium interactions

EPN–bacterial combinations have resulted in antagonistic or additive interactions (Table 27.6). In these combinations, EPN reproduction tended to be reduced (Bednarek, 1986) or inhibited (Kaya and Burlando, 1989; Poinar *et al.*, 1990), unless the EPNs were inoculated before the bacterium. When the β-exotoxin of *Bacillus thuringiensis* ssp. *thuringiensis* was combined with economically unfeasible rates of *S. carpocapsae*, the combination caused synergistic European crane fly (*Tipula paludosa*) larval mortality (Lam and Webster, 1972). Third-instar mortality in various white grubs exposed to combinations of *B. thuringiensis* ssp. *japonensis* and *S. glaseri* or *H. bacteriophora* was mostly additive, or sometimes synergistic, particularly if younger larvae were exposed (Koppenhöfer *et al.*, 1999). The effect on EPN reproduction was not studied.

Third-instar masked chafer, *Cyclocephala hirta*, was more susceptible to *H. bacteriophora* and *S. glaseri* when infected with the milky disease bacterium *Paenibacillus popilliae*, due to enhanced midgut penetration by the EPNs (Thurston *et al.*, 1993, 1994). EPN reproduction was not affected

Table 27.5. Effect of combining entomopathogenic nematodes (*Steinernema* and *Heterorhabditis* spp.) with insecticides on insect control.

Active ingredient	Trade name/ formulation	Rate (active ingredient ha^{-1})	Nematode	Target pest	Arena	Interaction	Reference(s)
Abamectin	Abamectin-Gol Shimi 1.8 EC	50–90 ppm	Sc	*Phthorimaea operculella* (pupa)	L	–	Eivazian Kary *et al.* (2018)
				P. operculella (prepupa)	L	*	Eivazian Kary *et al.* (2018)
		50–300 ppm	Sc	*P. operculella* (prepupa)	G	*/+	Eivazian Kary *et al.* (2018)
Acephate	?	250 ppm[a]	Sc	*Pieris rapae crucivora*	F	+	Ishibashi (1993)
				Spodoptera litura	F	+	Ishibashi (1993)
Azadirachtin	NeemAzal-T 5%	0.15–2.5%	Sf	*Bactrocera zonata*	L	*/+	Mahmoud (2007)
	Neemix 4.5%	0.1–2.0%	Sf	*B. zonata*	L	*/+	Mahmoud (2007)
Chlorantraniliprole	Acelepryn	125–400 g	Hb	*Anomala orientalis*	G, F	*/+	Koppenhöfer and Fuzy (2008a)
				Popillia japonica	G, F	*/+	Koppenhöfer and Fuzy (2008a)
				Cyclocephala borealis	G	*/+	Koppenhöfer and Fuzy (2008a)
	Coragen 20SC	6 ppm	Hi, Sc	*S. litura*	L	+	Khan *et al.* (2021)
Chlorfenapyr	Pirate 36SC	17 ppm	Hi, Sc	*S. litura*	L	+	Khan *et al.* (2021)
Clothianidin	Arena 50 WDG	50–150 g	Hb, Hz	*A. orientalis*	L, F	*/+	Koppenhöfer and Fuzy (2008a)
				P. japonica	L, F	*/+	Koppenhöfer and Fuzy (2008a)
				C. borealis	L	*/+	Koppenhöfer and Fuzy (2008a)
				A. orientalis	L	*/+	Koppenhöfer and Fuzy (2008a)
				P. japonica	L	*/+	Koppenhöfer and Fuzy (2008a)
				C. borealis	L	*/+	Koppenhöfer and Fuzy (2008a)
		123–225 g	Hb	*Amphimallon majalis*	L, G, F	*/+	Morales and Peck (2009)
Chlorpyrifos-methyl	Dursban 2 EC	9.4 kg	HG, Sl	*A. orientalis*	F	*	Lee *et al.* (2002)
Cryomazine	Trigard 75WP	0.8–50 g kg^{-1} seed[c]	Hb	*Delia antiqua*	L	+	Yildrim and Hoy (2003)
Diazinon	Diazinon 4 EC	700 kg	Sk	*Cyclocephala hirta*	G	+	Koppenhöfer *et al.* (2000c)
				A. orientalis	G	+	Koppenhöfer *et al.* (2000c)
Dichlorvos	?	250 ppm[a]	Sc	*P. rapae crucivora*	F	+	Ishibashi (1993)
				S. litura	F	+	Ishibashi (1993)
Emamectin benzoate	Proclaim 019EC	3 ppm	Hi, Sc	*S. litura*	L	+	Khan *et al.* (2021)
Flubendiamide	Belt 48SC	7 ppm	Hi, Sc	*S. litura*	L	+	Khan *et al.* (2021)
Gonofos	Techn. grade	0.06–0.25 ppm[b]	Sc	*Diabrotica virgifera virgifera*	L	–/+	Nishimatsu and Jackson (1998)
Halofenozide	Mach2/60 WP	1.1–2.3 kg	Hm	*P. japonica*	L	+	Mannion *et al.* (2000)

Continued

Table 27.5. Continued.

Active ingredient	Trade name/ formulation	Rate (active ingredient ha^{-1})	Nematode	Target pest	Arena	Interaction	Reference(s)
Imidacloprid	Merit 75 WP	200 g	Hb	*C. borealis*	G, L	*/+	Koppenhöfer *et al*. (2000a); Koppenhöfer and Fuzy (2008b)
				C. hirta	G, F	*/+	Koppenhöfer and Kaya (1998); Koppenhöfer *et al*. (2000a)
				Cyclocephala pasadenae	G	*	Koppenhöfer and Kaya (1998); Koppenhöfer *et al*. (2000a)
				A. orientalis	L, G, F	*/+	Koppenhöfer *et al*. (2000a, 2002); Koppenhöfer and Fuzy (2008b)
				P. japonica	L, G, F	*/+	Koppenhöfer *et al*. (2000a, 2002); Koppenhöfer and Fuzy (2008b); Morales and Peck (2009)
		123–225 g	Hb	*A. majalis*	L, G, F	*/+	Morales and Peck (2009)
			Hz	*A. orientalis*	L	*/+	Koppenhöfer and Fuzy (2008b)
				P. japonica	L	*/+	Koppenhöfer and Fuzy (2008b)
				C. borealis	L	*/+	Koppenhöfer and Fuzy (2008b)
			Hm, He	*A. orientalis*	G	*	Koppenhöfer *et al*. (2002)
			Sg	*C. borealis*	G, F	*	Koppenhöfer and Kaya (1998); Koppenhöfer *et al*. (2000a)
				C. hirta	G, F	*	Koppenhöfer and Kaya (1998); Koppenhöfer *et al*. (2000a)
				C. pasadenae	G, F	*	Koppenhöfer and Kaya (1998); Koppenhöfer *et al*. (2000a)
				A. orientalis	G	*	Koppenhöfer *et al*. (2000a, 2002)
				P. japonica	G, F	*	Koppenhöfer *et al*. (2000a)
			Ss	*C. borealis*	G	+	Koppenhöfer and Fuzy (2003a)
				A. orientalis	G, F	*/+	Koppenhöfer and Fuzy (2003a)
				P. japonica	G	+	Koppenhöfer and Fuzy (2003a)
				Maladera castanea	G	+	Koppenhöfer and Fuzy (2003b)

				A. majalis	G	–	Cappaert and Koppenhöfer (2003)
		186–336 g	Hb, Sc, Sf	*Listronotus maculicollis*	G	+	Koppenhöfer *et al.* (2020)
		336 g	Sc	*L. maculicollis*	F	+	Koppenhöfer *et al.* (2020)
	Technical grade	5.6–9.6 ng per larva	Sc	*Arge ochropus*	L	*	Sheykhnejad *et al.* (2014)
			Hb	*Arge ochropus*	L	+	Sheykhnejad *et al.* (2014)
Indoxacarb	Steward 150SC	11 ppm	Hi, Sc	*S. litura*	L	+	Khan *et al.* (2021)
Lufenuron	Match 050 EC	8 ppm	Hi, Sc	*S. litura*	L	+	Khan *et al.* (2021)
Methoxyfenozide	Runner 240SC	21 ppm	Hi, Sc	*S. litura*	L	+	Khan *et al.* (2021)
Novaluron	Corvus 10EC	30 ppm	Hi, Sc	*S. litura*	L	+	Khan *et al.* (2021)
Oxamyl	?/1% G	3 kg	Sc	*Agrotis segetum*	F	+/*	Ishibashi (1993)
Spinetoram	Radiant 120SC	9 ppm	Hi, Sc	*S. litura*	L	+	Khan *et al.* (2021)
Spinosad	Tracer 24SC	15 ppm	Hi, Sc	*S. litura*	L	+	Khan *et al.* (2021)
Tefluthrin	Techn. grade	0.06–0.25 ppm[b]	Sc, Hb	*D. virgifera virgifera*	L	+/*	Nishimatsu and Jackson (1998)
Terbufos	Techn. grade	0.025–0.1 ppm[b]	Sc	*D. virgifera virgifera*	L	–/+	Nishimatsu and Jackson (1998)
Thiamethoxam	Meridian 25 WG	200 g	Sg	*A. orientalis*	G	*	Koppenhöfer *et al.* (2002)
			Hb	*A. orientalis*	G, F	+	Koppenhöfer *et al.* (2002)
			Hm, He	*A. orientalis*	G	+	Koppenhöfer *et al.* (2002)

?, Not mentioned in study; Hb, *H. bacteriophora*; Hm, *H. marelatus*; He, *H. megidis*; Hi = *H. indica*; Hz, *H. zealandica*; HG, *Heterorhabditis* sp. Gyeongsan; Sc, *S. carpocapsae*; Sg, *S. glaseri*; Sl, *S. longicaudum*; Ss, *S. scarabaei*; F, field; G, greenhouse/pot; L, laboratory; –, antagonistic; +, additive; *, synergistic; /, both interaction types observed.

[a]Foliar application.

[b]Mixed in soil.

[c]The active ingredient was film-coated on onion seed.

Table 27.6. Effect of combining entomopathogenic nematodes (*Steinernema* and *Heterorhabditis* spp.) with other pathogens on insect control.

Other pathogen	Nematode	Target pest	Arena	Interaction	Reference(s)
Nematodes					
H. bacteriophora	Sk, Sg	*Cyclocephala hirta*	G, F	+	Lam and Webster (1972)
Steinernema glaseri	Sk	*Anomala orientalis*	G	+	Lam and Webster (1972)
Heterorhabditis bacteriophora	Sc, Sr, SH	*Diabrotica undecimpunctata undecimpunctata*	L, G	+	Choo *et al.* (1996)
Steinernema sp. Hanrim	Hb, Sr	*D. undecimpunctata undecimpunctata*	L, G	+	Choo *et al.* (1996)
Steinernema carpocapsae	Hb, Sf	*Agrotis ipsilon*	G, F	+	Koppenhöfer *et al.* (2022)
H. bacteriophora	Sf	*Agrotis ipsilon*	G	+	Koppenhöfer *et al.* (2022)
Bacteria					
Bacillus thuringiensis kurstaki	Sc	*Plutella xylostella*	L	+/–[a]	Baur *et al.* (1998)
B. thuringiensis aizawai	Sc	*P. xylostella*	F	+	Baur *et al.* (1998)
B. thuringiensis kurstaki	Sc	*Platyptilia carduidactyla*	F	–	Bari and Kaya (1984)
B. thuringiensis kurstaki	Sc	*Lymantria dispar*	L	–	Bednarek (1986)
B. thuringiensis thuringiensis (β-exotoxin)	Sc	*Tipula paludosa*	L	*	Lam and Webster (1972)
B. thuringiensis japonensis	Hb	*C. hirta*	L, G, F	+/*	Koppenhöfer and Kaya (1997); Koppenhöfer *et al.* (1999)
		Cyclocephala pasadenae	L, G, F	+/*	Koppenhöfer and Kaya (1997); Koppenhöfer *et al.* (1999)
		Popillia japonica	F	+	Koppenhöfer *et al.* (2000c)
	Sg	*C. hirta*	L	+/*	Koppenhöfer and Kaya (1997)
		A. orientalis	L, F	*/+	Koppenhöfer *et al.* (1999)
	Sk	*C. hirta*	L	+	Koppenhöfer and Kaya (1997)
Paenibacillus popilliae[b]	Hb	*C. hirta*	G	+	Thurston *et al.* (1994)
	Sg	*C. hirta*	G	*	Thurston *et al.* (1994)
Serratia marcescens	Hi	*Curculio caryae*	L	–	Shapiro-Ilan *et al.* (2004)
	Sc	*C. caryae*	L	–/+	Shapiro-Ilan *et al.* (2004)
Fungi					
Beauveria bassiana	Sc	*Galleria mellonella*	L	Accel.	Kamionek *et al.* (1974a)
		Tribolium castaneum	L	Decel.	Kamionek *et al.* (1974a)
	Sc, Hb	*Spodoptera exigua*	L	+	Babercheck and Kaya (1991a)
	Hb	*P. xylostella*	L	+/*[c]	Correa-Cuadros *et al.* (2016)
	Hi	*C. caryae*	L	–	Shapiro-Ilan *et al.* (2004)

Continued

	Sc	*C. caryae*	L	–/+	Shapiro-Ilan *et al.* (2004)
	Hb	*Rhynchophorus ferrugineus*	L	*[d]	Wakil *et al.* (2017)
Metarhizium anisopliae	Hi	*C. caryae*	L	+	Shapiro-Ilan *et al.* (2004)
	Sc	*C. caryae*	L	–	Shapiro-Ilan *et al.* (2004)
	Hb	*P. xylostella*	L	+/*[c]	Correa-Cuadros *et al.* (2016)
	He, Sg	*Hoplia philanthus*	L, G	+/*[e]	Ansari *et al.* (2004)
	Hb	*H. philanthus*	F	+/*[f]	Ansari *et al.* (2006)
	Hb	*Otiorhynchus sulcatus*	G	*[g]	Ansari *et al.* (2008)
	Sf	*O. sulcatus*	G	+/*[g]	Ansari *et al.* (2008)
	Hb	*R. ferrugineus*	L	+/*[d]	Wakil *et al.* (2017)
Paecilomyces farinosus	Sc	*G. mellonella*	L	Accel.	Kamionek *et al.* (1974b)
		T. castaneum	L	+	Kamionek *et al.* (1974b)
		Trogoderma granarium	L	+	Kamionek *et al.* (1974b)
	Hi, Sc	*C. caryae*	L	–	Shapiro-Ilan *et al.* (2004)
Viruses					
Lymantria dispar NPV	Sc	*L. dispar*	L	–	Bednarek (1986)
Spodoptera exigua NPV	Sc	*S. exigua*	L, F	+	Gothama *et al.* (1995, 1996)

Hb, *H. bacteriophora*; He, *H. megidis*; Hi, *H. indica*; Sc, *S. carpocapsae*; Sg, *S. glaseri*; SH, *Steinernema* sp. Hanrim; Sk, *S. kushidai*; Sr, *S. riobrave*; L, laboratory; G, greenhouse/pot; F, field; –, antagonistic; +, additive; *, synergistic; /, both interaction types observed; accel./decel., speed of kill accelerated/decelerated; NPV, nucleopolyhedrovirus.
[a]Additive effect in *B. thuringiensis*-susceptible *P. xylostella* strain, antagonistic effect in *B. thuringiensis*-resistant strain.
[b]Third-instar larvae were exposed for 15 days to *P. popilliae* before exposure to nematodes.
[c]Antagonistic/synergistic when Hb was applied 0–4/6–days after fungus.
[d]Interaction tended to increase with interval between fungus and EPN application (0, 1 or 2 weeks) and decrease with larval instar (second, fourth or sixth).
[e]Additive if EPNs were applied within 2 weeks after fungus, synergistic if 3–4 weeks after fungus.
[f]Hb was applied 4 weeks after fungus.
[g]EPNs were applied 2 weeks after fungus.

by *P. popilliae* infection (Thurston *et al.*, 1993). The feasibility of this combination is limited by the slow establishment of milky disease in white grub populations and the lack of *in vitro* bacterial production methods.

27.3.2.3 EPN–fungus interactions

EPN–fungus combinations cause additive or weak synergistic pest mortality, with rare antagonistic effects (Table 27.6). Pest mortality tended to increase with delayed EPN application (Ansari *et al.*, 2004; Correa-Cuadros *et al.*, 2016; Wakil *et al.*, 2017). EPNs and fungi often did not co-reproduce in hosts (Barbercheck and Kaya, 1991a) because the EPNs' symbiotic bacteria and the fungi excluded each other. The agent infecting first often excluded the second. Antagonistic interactions may be avoided in the field because the EPNs may avoid fungus-infected hosts (Barbercheck and Kaya, 1991b). Occasionally, both pathogens reproduced, but EPN production declined with increasing fungal inoculum (e.g. Ansari *et al.*, 2004).

27.3.2.4 EPN–virus interactions

Combining *S. carpocapsae* with the nucleopolyhedrovirus (NPV) of the beet armyworm, *Spodoptera exigua*, caused additive *Spodoptera exigua* larval mortality without affecting *S. carpocapsae* reproduction (Gothama *et al.*, 1995, 1996). Combinations of *S. carpocapsae* and *Lymantria dispar* NPV showed antagonistic *L. dispar* larval mortality (Bednarek, 1986) despite EPNs developing in NPV-infected larvae (Kaya and Burlando, 1989). EPN infection had to occur early enough to allow the EPNs to finish development before they were exposed to desiccation when the insect's cuticle disintegrated due to the virus infection (Bednarek, 1986; Kaya and Burlando, 1989). In sixth-instar *Pseudaletia unipuncta* infected with a granulovirus (Oregonian strain), *S. carpocapsae* reproduced normally (Kaya and Brayton, 1978).

27.3.3 Parasitoids and predators

Few studies have quantified the effects of combinations of EPNs with parasitoids or predators on pests. Combinations of *S. carpocapsae* with *Ganaspidium utilis* or *Diglyphus begini* caused higher mortality of *Liriomyza trifolii* than either agent alone (Hara *et al.*, 1989; Sher *et al.*, 2000). The parasitoid *Habrobracon hebetor* and *H. indica* alone caused 71% and 62% mortality of *Plodia interpunctella* larvae, respectively; their combination resulted in 99% mortality (Mbata and Shapiro-Ilan, 2010). *H. bacteriophora* or *S. feltiae* combinations with the predatory laelapid mite *Hypoaspisa aculeifer* were more effective than the individual agents in controlling soil-dwelling stages of the thrips *Frankliniella occidentalis* in pots (Premachandra *et al.*, 2003).

27.4 Synthesis and Recommendations

Many agrochemicals can negatively affect EPN survival. There is no obvious trend regarding EPN sensitivity to different agrochemical classes, so new combinations need to be evaluated for compatibility. Otherwise, the agents should be applied 1–2 weeks apart.

Regarding improved pest control, when selecting combinations for study and development, one should consider whether the combination will be economical, practical and/or safer to users and the environment than the alternatives. Combined applications are only useful if the target mortality is synergistically increased. If a combination provides additive control, it is easier to use only the better agent. Additive combinations may have advantages. Thus, on golf course fairways, *S. carpocapsae* at half rate combined with imidacloprid provided *Listronotus maculicollis* larvae control equivalent to the full rate of *S. carpocapsae* alone, but the combination provided more consistent control, was less expensive due to the lower cost of imidacloprid, and imidacloprid simultaneously controlled white grubs, another key pest group (Koppenhöfer *et al.*, 2020).

When synergistic EPN–insecticide interactions occurred, the insecticide either affected EPN behavior or, more commonly, the behavior of the pest due to neurotoxic insecticide activity. Experimental arenas must therefore allow normal EPN and insect behaviors.

EPN–pathogen combinations mostly resulted in additive or weak synergistic target mortality, or rarely antagonism or strong synergism. Observations for most combinations are limited. Different trends may be observed with different insects, pathogens or environmental conditions. Co-applications against the same target likely trigger competition for resources, although under field conditions coinfections are less likely.

Currently, combinations of EPNs with imidacloprid and commercially unavailable *B. thuringiensis* ssp. *japonensis* in white grubs are the only systems with sufficient greenhouse and field trials to support informed recommendations. EPN–imidacloprid combinations could be an effective tool for curative white grub control. EPNs are expensive, so the feasibility of the combination depends on reducing EPN rates, while retaining acceptable pest control.

27.5 References

Alatorre-Rosas, R. and Kaya, H.K. (1991) Interaction between two entomopathogenic nematode species in the same host. *Journal of Invertebrate Pathology* 57, 1–6.

Alumai, A. and Grewal, P.S. (2004) Compatibility of the entomopathogenic nematodes, *Heterorhabditis bacteriophora* and *Steinernema carpocapsae*, with selected chemical pesticides used in turfgrass. *Biocontrol Science and Technology* 14, 613–618.

Ansari, M.A., Tirry, L. and Moens, M. (2004) Interaction between *Metarhizium anisopliae* CLO 53 and entomopathogenic nematodes for the control of *Hoplia philanthus*. *Biological Control* 31, 172–180.

Ansari, M.A., Shah, F.A., Tirry, L. and Moens, M. (2006) Field trials against *Hoplia philanthus* (Coleoptera: Scarabaeidae) with a combination of an entomopathogenic nematode and the fungus *Metarhizium anisopliae* CLO 53. *Biological Control* 39, 453–459.

Ansari, M.A., Shah, F.A. and Butt, T.M. (2008) Combined use of entomopathogenic nematodes and *Metarhizium anisopliae* as a new approach for black vine weevil, *Otiorhynchus sulcatus*, control. *Entomologia Experimentalis et Applicata* 129, 340–347.

Atwa, A.A., Hegazi, A.M., Khafagi, W.A. and Abd El-Aziz, G.M. (2013) Interaction of the koinobiont parasitoid *Microplitis rufiventris* of the cotton leafworm, *Spodoptera littoralis*, with two entomopathogenic rhabditids, *Heterorhabditis bacteriophora* and *Steinernema carpocapsae*. *Journal of Insect Science* 13, 84–91.

Bajc, N., Držaj, U., Trdan, S. and Laznik, Ž. (2017) Compatibility of acaricides with entomopathogenic nematodes (*Steinernema* and *Heterorhabditis*). *Nematology* 19, 891–898.

Barbarossa, B., Frulla, R., Grazzi, G. and Rovesti, L. (1996) Compatibility of *Steinernema carpocapsae* with pesticides. *International Organization for Biological and Integrated Control Bulletin* 19, 136–139.

Barbercheck, M.E. and Kaya, H.K. (1991a) Competitive interactions between entomopathogenic nematodes and *Beauveria bassiana* (Deuteromycotina: Hyphomycetes) in soilborne larvae of *Spodoptera exigua* (Lepidoptera: Noctuidae). *Environmental Entomology* 20, 707–712.

Barbercheck, M.E. and Kaya, H.K. (1991b) Effect of host condition and soil texture on host finding by the entomogenous nematodes *Heterorhabditis bacteriophora* (Rhabditida: Heterorhabditidae) and *Steinernema carpocapsae* (Rhabditida: Steinernematidae). *Environmental Entomology* 20, 582–589.

Bari, M.A. and Kaya, H.K. (1984) Evaluation of the entomogenous nematode *Neoaplectana carpocapsae* (=*Steinernema feltiae*) Weiser (Rhabditida: Steinernematidae) and the bacterium *Bacillus thuringiensis* Berliner var. *kurstaki* for suppression of the artichoke plume moth (Lepidoptera: Pterophoridae). *Journal of Economic Entomology* 77, 225–229.

Bathon, H. (1996) Impact of entomopathogenic nematodes on non-target hosts. *Biocontrol Science and Technology* 6, 421–434.

Battisti, A. (1994) Effects of entomopathogenic nematodes on the spruce web-spinning sawfly *Cephalcia arvensis* Panzer and its parasitoids in the field. *Biocontrol Science and Technology* 4, 95–102.

Baur, M.E., Kaya, H.K., Tabashnik, B.E. and Chilcutt, C.F. (1998) Suppression of diamondback moth (Lepidoptera: Plutellidae) with an entomopathogenic nematode (Rhabditida: Steinernematidae) and *Bacillus thuringiensis* Berliner. *Journal of Economic Entomology* 91, 1089–1095.

Bednarek, A. (1986) Development of the *Steinernema feltiae* (Filipjev) entomogenous nematode (Steinernematidae) in the conditions of occurrence in the insect's body cavity of other pathogens. *Annals of Warsaw Agricultural University: Animal Science* 20, 69–74.

Bednarek, A. and Gaugler, R. (1997) Compatibility of soil amendments with entomopathogenic nematodes. *Journal of Nematology* 29, 220–27.

Cappaert, D.L. and Koppenhöfer, A.M. (2003) *Steinernema scarabaei*, and entomopathogenic nematode for control of the European chafer. *Biological Control* 28, 379–386.

Chavan, S.N., Somasekhar, N. and Katti, G. (2018) Compatibility of entomopathogenic nematode *Heterorhabditis indica* (Nematoda: Heterorhabditidae) with agrochemicals used in the rice ecosystem. *Journal of Entomology and Zoology Studies* 6, 527–32.

Choo, H.Y., Koppenhöfer, A.M. and Kaya, H.K. (1996) Combination of two entomopathogenic nematode species for suppression of an insect pest. *Journal of Economic Entomology* 89, 97–103.

Correa-Cuadros, J.P., Sáenz-Aponte, A. and Rodríguez-Bocanegra, M.X. (2016) *In vitro* interaction of *Metarhizium anisopliae* Ma9236 and *Beauveria bassiana* Bb9205 with *Heterorhabditis bacteriophora* HNI0100 for the control of *Plutella xylostella*. *SpringerPlus* 5: 2068.

Das, J.N. and Divakumar, B.J. (1987) Compatibility of certain pesticides with DD-136 nematode. *Plant Protection Bulletin* 39, 20–22.

de Nardo, E.A.B. and Grewal, P.S. (2003) Compatibility of *Steinernema feltiae* (Nematoda: Steinernematidae) with microbial and chemical pesticides and plant growth regulators used in glasshouse plant production. *Biocontrol Science and Technology* 13, 441–448.

Eivazian Kary, N., Sanatipour, Z., Mohammadi, D. and Koppenhöfer, A.M. (2018) Developmental stage affects the interaction of *Steinernema carpocapsae* and abamectin for the control of *Phthorimaea operculella* (Lepidoptera, Gelechidae). *Biological Control* 122, 18–23.

Everard, A., Griffin, C.T. and Dillon, A.B. (2009) Competition and intraguild predation between the braconid parasitoid *Bracon hylobii* and the entomopathogenic nematode *Heterorhabditis downesi*, natural enemies of the large pine weevil, *Hylobius abietis*. *Bulletin of Entomological Research* 99, 151–161.

García-del-Pino, F. and Morton, A. (2010) Synergistic effect of the herbicides glyphosate and MCPA on survival of entomopathogenic nematodes. *Biocontrol Science and Technology* 20, 483–88.

Georgis, R. and Hague, N.G.M. (1991) Nematodes as biological insecticides. *Pesticide Outlook* 2, 29–32.

Georgis, R., Kaya, H.K. and Gaugler, R. (1991) Effect of steinernematid and heterorhabditid nematodes (Rhabditida: Steinernematidae and Heterorhabditidae) on nontarget arthropods. *Environmental Entomology* 20, 815–822.

Glazer, I. (1997) Effects of infected insect on secondary invasion of steinernematid entomopathogenic nematodes. *Parasitology* 114, 597–604.

Gothama, A.A.A., Sikorowski, P.P. and Lawrence, G.W. (1995) Interactive effects of *Steinernema carpocapsae* and *Spodoptera exigua* nuclear polyhedrosis virus on *Spodoptera exigua* larvae. *Journal of Invertebrate Pathology* 66, 270–276.

Gothama, A.A.A., Lawrence, G.W. and Sikorowski, P.P. (1996) Activity and persistence of *Steinernema carpocapsae* and *Spodoptera exigua* nuclear polyhedrosis virus against *S. exigua* larvae on soybean. *Journal of Nematology* 28, 68–74.

Grewal, P.S. and Richardson, P.N. (1993) Effects of application rates of *Steinernema feltiae* (Nematoda: Steinernematidae) on biological control of the mushroom fly *Lycoriella auripila* (Diptera: Sciaridae). *Biocontrol Science and Technology* 3, 29–40.

Grewal, P.S., Gaugler, R., Kaya, H.K. and Wustly, M. (1993) Infectivity of the entomopathogenic nematode, *Steinernema scapterisci* (Nematoda: Steinernematidae). *Journal of Invertebrate Pathology* 62, 22–28.

Grewal, P.S., Lewis, E.E. and Gaugler, R. (1997) Response of infective stage parasites (Nematoda: Steinernematidae) to volatile cues from infected hosts. *Journal of Chemical Ecology* 23, 503–515.

Grewal, P.S., Weber, T.A. and Betterley, D.A. (1998) Compatibility of *Steinernema feltiae* with chemicals used in mushroom production. *Mushroom News* 46, 6–10.

Gupta, P. and Siddiqi, M.R. (1999) Compatibility studies on *Steinernema carpocapsae* with some pesticidal chemicals. *Indian Journal of Entomology* 61, 220–225.

Hara, A.H., Mello, C.L., Tsutsumi, L.H.A. and Kaya, H.K. (1989) Laboratory susceptibility of some tropical pest species and a non-target organism to the entomopathogenic nematode, *Steinernema carpocapsae*. *Journal for Hawaiian and Pacific Agriculture* 2, 6–9.

Head, J., Walters, K.F.A. and Langton, S. (2000) The compatibility of the entomopathogenic nematode, *Steinernema feltiae*, and chemical insecticides for the control of the South American leafminer, *Liriomyza huidobrensis*. *BioControl* 45, 345–353.

Head, J., Palme, L.F. and Walters, K.F.A. (2003) The compatibility of control agents used for the control of the South American leafminer, *Liriomyza huidobrensis*. *Biocontrol Science and Technology*, 13, 77–86.

Hodson, A.K., Friedman, M.L., Wu, L.N. and Lewis, E.E. (2011) European earwig (*Forficula auricularia*) as a novel host for the entomopathogenic nematode *Steinernema carpocapsae*. *Journal of Invertebrate Pathology* 107, 60–64.

Ishibashi, N. (1993) Integrated control of insect pests by *Steinernema carpocapsae*. In: Bedding, R.A., Akhurst, R. and Kaya, H.K. (eds.) *Nematodes and the Biological Control of Insect Pests*. CSIRO Publications, East Melbourne, Australia, pp. 105–113.

Jandricic, S., Scott-Dupree, C.D., Broadbent, A.B., Harris, C.R. and Murphy, G. (2006) Compatibility of *Atheta coriaria* with other biological control agents and reduced-risk insecticides used in greenhouse floriculture integrated pest management programs for fungus gnats. *Canadian Entomologist* 138, 712–722.

Kamionek, M., Sandner, H. and Seryczynska, H. (1974a) The combined action of *Beauveria bassiana* (Bals/Vuill) (Fungi Imperfecti: Moniliales) and *Neoaplectana carpocapsae* Weiser (Nematoda: Steinernematidae). *Bulletin de l'Academie Polonaise des Sciences* 22, 253–257.

Kamionek, M., Sandner, H. and Seryczynska, H. (1974b) Combined action of *Paecilomyces farinosus* Dicks (Brown et Smith) (Fungi imp.: Moniliales) and *Neoaplectana carpocapsae* Weiser, 1955 (Nematoda: Steinernematidae). *Acta Parasitologica* 22, 357–363.

Kaya, H.K. (1978) Interaction between *Neoaplectana carpocapsae* (Nematoda: Steinernematidae) and *Apanteles militaris* (Hymenoptera: Braconidae), a parasitoid of the armyworm *Pseudaletia unipuncta*. *Journal of Invertebrate Pathology* 31, 358–364.

Kaya, H.K. (1984) Effect of the entomogenous nematode *Neoaplectana carpocapsae* on the tachinid parasite *Compsilura concinnata*. *Journal of Nematology* 16, 9–13.

Kaya, H.K. and Brayton, M.A. (1978) Interaction between *Neoaplectana carpocapsae* and a granulosis virus of the armyworm *Pseudaletia unipuncta*. *Journal of Nematology* 10, 350–354.

Kaya, H.K. and Burlando, T.M. (1989) Development of *Steinernema feltiae* (Rhabditida: Steinernematidae) in diseased insect hosts. *Journal of Invertebrate Pathology* 53, 164–168.

Kaya, H.K. and Hotchkin, P.G. (1981) The nematode *Neoaplectana carpocapsae* Weiser and its effect on selected ichneumonid and braconid parasites. *Environmental Entomology* 10, 474–478.

Kaya, H.K., Burlando, T.M. and Thurston, G.S. (1993) Two entomopathogenic nematode species with different search strategies for insect suppression. *Environmental Entomology* 22, 859–864.

Khan, R.R., Arshad, M., Aslam, A. and Arshad, M. (2021) Additive interactions of some reduced-risk biocides and two entomopathogenic nematodes suggest implications for integrated control of *Spodoptera litura* (Lepidoptera: Noctuidae). *Scientific Reports* 11: 1268.

Kondo, E. (1989) Studies on the infectivity and propagation of entomogenous nematodes, *Steinernema* spp. (Rhabditida: Steinernematidae), in the common cutworm, *Spodoptera litura* (Lepidoptera: Noctuidae). *Bulletin of the Faculty of Agriculture, Saga University* 67, 1–88.

Koppenhöfer, A.M. and Fuzy, E.M. (2003a) *Steinernema scarabaei* for the control of white grubs. *Biological Control* 28, 47–59.

Koppenhöfer, A.M. and Fuzy, E.M. (2003b) Biological and chemical control of the Asiatic garden beetle, *Maladera castanea* (Coleoptera: Scarabaeidae). *Journal of Economic Entomology* 96, 1076–1082.

Koppenhöfer, A.M. and Fuzy, E.M. (2008a) Effect of the anthranilic diamide insecticide, chlorantraniliprole, on *Heterorhabditis bacteriophora* (Rhabditida: Heterorhabditidae) efficacy against white grubs (Coleoptera: Scarabaeidae). *Biological Control* 45, 93–102.

Koppenhöfer, A.M. and Fuzy, E.M. (2008b) Early timing and new combinations to increase the efficacy of neonicotinoid-entomopathogenic nematode (Rhabditida: Heterorhabditidae) combinations against white grubs (Coleoptera: Scarabaeidae). *Pest Management Science* 64, 725–735.

Koppenhöfer, A.M. and Kaya, H.K. (1996) Coexistence of two steinernematid nematode species (Rhabditida: Steinernematidae) in the presence of two host species. *Applied Soil Ecology* 4, 221–230.

Koppenhöfer, A.M. and Kaya, H.K. (1997) Additive and synergistic interaction between entomopathogenic nematodes and *Bacillus thuringiensis* for scarab grub control. *Biological Control* 8, 131–137.

Koppenhöfer, A.M. and Kaya, H.K. (1998) Synergism of imidacloprid and an entomopathogenic nematode: a novel approach to white grub (Coleoptera: Scarabaeidae) control in turfgrass. *Journal of Economic Entomology* 91, 618–623.

Koppenhöfer, A.M., Kaya, H.K., Shanmugam, S. and Wood, G.L. (1995) Interspecific competition between steinernematid nematodes within an insect host. *Journal of Invertebrate Pathology* 66, 99–103.

Koppenhöfer, A.M., Baur, M.E. and Kaya, H.K. (1996) Competition between two steinernematid nematode species for an insect host at different soil depths. *Journal of Parasitology* 82, 34–40.

Koppenhöfer, A.M., Choo, H.Y., Kaya, H.K., Lee, D.W. and Gelernter, W.D. (1999) Increased field and greenhouse efficacy against scarab grubs with a combination of an entomopathogenic nematode and *Bacillus thuringiensis*. *Biological Control* 14, 37–44.

Koppenhöfer, A.M., Brown, I.M., Gaugler, R., Grewal, P.S., Kaya, H.K. and Klein, M.G. (2000a) Synergism of entomopathogenic nematodes and imidacloprid against white grubs: greenhouse and field evaluation. *Biological Control* 19, 245–251.

Koppenhöfer, A.M., Grewal, P.S. and Kaya, H.K. (2000b) Synergism of imidacloprid and entomopathogenic nematodes against white grubs: the mechanism. *Entomologia Experimentalis et Applicata* 94, 283–293.

Koppenhöfer, A.M., Wilson, M.G., Brown, I., Kaya, H.K. and Gaugler, R. (2000c) Biological control agents for white grubs (Coleoptera: Scarabaeidae) in anticipation of the establishment of the Japanese beetle in California. *Journal of Economic Entomology* 93, 71–80.

Koppenhöfer, A.M., Cowles, R.S., Cowles, E.A., Fuzy, E.M. and Baumgartner, L. (2002) Comparison of neonicotinoid insecticides as synergists for entomopathogenic nematodes. *Biological Control* 24, 90–97.

Koppenhöfer, A.M., Cowles, R.S., Cowles, E.A., Fuzy, E.M. and Baumgartner, L. (2003) Effect of neonicotinoid insecticide synergists on entomopathogenic nematode fitness. *Entomologia Experimentalis et Applicata* 106, 7–18.

Koppenhöfer, A.M., Kostromytska, O.S. and Wu, S. (2020) Optimizing the use of entomopathogenic nematodes for the management of *Listronotus maculicollis* (Coleoptera: Curculionidae): split applications and combinations with imidacloprid. *Crop Protection* 137: 105229.

Koppenhöfer, A.M., Kostromytska, O.S. and Ebssa, L. (2022) Species combinations, split applications, and syringing to optimize the efficacy of entomopathogenic nematodes against *Agrotis ipsilon* (Lepidoptera: Noctuidae) larvae in turfgrass. *Crop Protection* 155: 105927.

Krishnayya, P.V. and Grewal, P.S. (2002) Effect of neem and selected fungicides on viability and virulence of the entomopathogenic nematode *Steinernema feltiae*. *Biocontrol Science and Technology* 12, 259–266.

Lacey, L.A., Unruh, T.R. and Headrick, H.L. (2003) Interactions of two idiobiont parasitoids (Hymenoptera: Ichneumonidae) of codling moth (Lepidoptera: Tortricidae) with the entomopathogenic nematode *Steinernema carpocapsae* (Rhabditida: Steinernematidae). *Journal of Invertebrate Pathology* 83, 230–239.

Lalitha, Y., Nagesh, M. and Jalali, S.K. (2012) Intraguild predation and biosafety of entomopathogenic nematode, *Heterorhabditis bacteriophora* Poinar *et al.*, and its bacterial symbiont, *Photorhabdus luminescens*, to parasitoid, *Trichogramma chilonis* Ishii and predator *Chrysoperla zastrowi sillemi* (Esben, Petersen). *Journal of Biological Control* 26, 334–340.

Lam, A.B.Q. and Webster, J.M. (1972) Effect of the DD-136 nematode and of a β-exotoxin preparation of *Bacillus thuringiensis* var. *thuringiensis* on leatherjackets, *Tipula paludosa* larvae. *Journal of Invertebrate Pathology* 20, 141–149.

Laznik, Ž. and Trdan, S. (2014) The influence of insecticides on the viability of entomopathogenic nematodes (Rhabditidae: Steinernematidae and Heterorhabditidae) under laboratory conditions. *Pest Management Science* 70, 784–789.

Laznik, Ž. and Trdan, S. (2016) The influence of herbicides on the viability of entomopathogenic nematodes (Rhabditidae: Steinernematidae and Heterorhabditidae). *International Journal of Pest Management* 63, 105–111.

Laznik, Ž., Vidrih, M. and Trdan, S. (2012) The effects of different fungicides on the viability of entomopathogenic nematodes *Steinernema feltiae* (Filipjev), *S. carpocapsae* Weiser, and *Heterorhabditis downesi* Stock, Griffin & Burnell (Nematoda: Rhabditidae) under laboratory conditions. *Chilean Journal of Agricultural Research* 72, 62–67.

Lee, D.W., Choo, H.Y. and Choi, E.J. (1999) Effect of aqueous solution of pesticides on survival of entomopathogenic nematodes. *Korean Journal of Turfgrass Science* 13, 213–222.

Lee, D.W., Choo, H.Y., Kaya, H.K., Lee, S.M., Smitley, D.R. *et al.* (2002) Laboratory and field evaluation of Korean entomopathogenic nematode isolates against the oriental beetle *Exomala orientalis* (Coleoptera: Scarabaeidae). *Journal of Economic Entomology* 95, 918–926.

Mannion, C.M., Winkler, H.E., Shapiro, D.I. and Gibb, T. (2000) Interaction between halofenozide and entomopathogenic nematode *Heterorhabditis marelatus* for control of Japanese beetle (Coleoptera: Scarabaeidae) larvae. *Journal of Economic Entomology* 93, 48–53.

Mbata, G.N. and Shapiro-Ilan, D. (2010) Compatibility of *Heterorhabditis indica* (Rhabditida: Heterorhabditidae) and *Habrobracon hebetor* (Hymenoptera: Braconidae) for biological control of *Plodia interpunctella* (Lepidoptera: Pyralidae). *Biological Control* 54, 75–82.

Mbata, G.N., Shapiro-Ilan, D., Alborn, H.T. and Strand, M.R. (2019) Preferential infectivity of entomopathogenic nematodes in an envenomed host. *International Journal of Parasitology* 49, 737–745.

Morales-Rodriguez, A. and Peck, D. (2009) Synergies between biological and neonicotinoid insecticides for the curative control of the white grubs *Amphimallon majale* and *Popillia japonica*. *Biological Control* 51, 169–180.

Mracek, Z. and Spitzer, K. (1983) Interaction of predators and parasitoids of the sawfly, *Cephalcia abietis* (Pamphilidae: Hymenoptera) with its nematode *Steinernema kraussei*. *Journal of Invertebrate Pathology* 42, 397–399.

Negrisoli, A.S., Garcia, M.S. and Negrisoli, C.R.C.B. (2010) Compatibility of entomopathogenic nematodes (Nematoda: Rhabditida) with registered insecticides for *Spodoptera frugiperda* (Smith, 1797) (Lepidoptera: Noctuidae) under laboratory conditions. *Crop Protection* 29, 545–549.

Nishimatsu, T. and Jackson, J.J. (1998) Interaction of insecticides, entomopathogenic nematodes, and larvae of the western corn rootworm (Coleoptera: Chrysomelidae). *Journal of Economic Entomology* 91, 410–418.

Parkman, J.P. and Frank, J.H. (2002) Interaction between *Ormia depleta* (Diptera: Tachinidae) and *Steinernema scapterisci* (Nematoda: Steinernematidae), natural enemies of pest mole crickets (Orthoptera: Gryllotalpidae). *Environmental Entomology* 31, 1226–1230.

Poinar, G.O. Jr., Thomas, G.M. and Lighthart, B. (1990) Bioassay to determine the effect of commercial preparation of *Bacillus thuringiensis* on entomogenous rhabditoid nematodes. *Agriculture Ecosystems and Environment* 30, 195–202.

Powell, J.R. and Webster, J.M. (2004) Interguild antagonism between biological controls: impact of entomopathogenic nematode application on an aphid predator, *Aphidoletes aphidimyza*. *Biological Control* 30, 110–118.

Prakasa Rao, P.S., Das, P.K. and Pandhi, G. (1975) Note of compatibility of DD-136 (*Neoaplectana dutkyi*), an insect parasitic nematode with some insecticides and fertilizers. *Indian Journal of Agricultural Sciences* 45, 275–277.

Premachandra, W.T.S.D., Borgemeister, C., Berndt, O., Ehlers, R.-U. and Poehling, H.-M. (2003) Combined releases of entomopathogenic nematodes and the predatory mite *Hypoaspis aculeifer* to control soil-dwelling stages of western flower thrips *Frankliniella occidentalis*. *BioControl* 48, 529–541.

Radová, S. (2010) Effect of selected pesticides on the vitality and virulence of the entomopathogenic nematode *Steinernema feltiae* (Nematoda: Steinernematidae). *Plant Protection Science* 46, 83–88.

Rojht, H., Kac, M. and Trdan, S. (2009) Nontarget effect of entomopathogenic nematodes on larvae of twospotted lady beetle (Coleoptera: Coccinellidae) and green lacewing (Neuroptera: Chrysopidae) under laboratory conditions. *Journal of Economic Entomology* 102, 1440–1443.

Rovesti, L. and Deseö, K.V. (1990) Compatibility of chemical pesticides with entomopathogenic nematodes, *Steinernema carpocapsae* Weiser and *S. feltiae* Filipjev (Nematoda: Steinernematidae). *Nematologica* 36, 237–245.

Rovesti, L. and Deseö, K.V. (1991) Compatibility of pesticides with entomopathogenic nematodes, *Heterorhabditis heliothidis*. *Nematologica* 37, 113–122.

Rovesti, L., Heinzpeter, E.W., Tagliente, F. and Deseö, K.V. (1988) Compatibility of pesticides with the entomopathogenic nematode *Heterorhabditis bacteriophora* Poinar (Nematoda: Heterorhabditidae). *Nematology* 34, 462–476.

Sahin, Y.S. and Susurluk, A. (2018) Effects of some inorganic fertilizers on entomopathogenic nematodes, *Steinernema feltiae* (Tur-S3) and *Heterorhabditis bacteriophora* (Hb.H). *Türkiye Biyolojik Mücadele Dergisi* 9, 102–109.

Saito, T. and Brownbridge, M. (2016) Compatibility of soil-dwelling predators and microbial agents and their efficacy in controlling soil-dwelling stages of western flower thrips *Frankliniella occidentalis*. *Biological Control* 92, 92–100

Shannag, H.K. and Capinera, J.L. (2000) Interference of *Steinernema carpocapsae* (Nematoda: Steinernematidae) with *Cardiochiles diaphaniae* (Hymenoptera: Braconidae), a parasitoid of melonworm and pickleworm (Lepidoptera: Pyralidae). *Environmental Entomology* 29, 612–617.

Shapiro, D.I., Lewis, L.C., Obrycki, J.J. and Abbas, M. (1999) Effects of fertilizers on suppression of black cutworm (*Agrotis ipsilon*) damage with *Steinernema carpocapsae*. *Journal of Nematology* 31 (Suppl.), 690–693.

Shapiro-Ilan, D.I., Jackson, M., Reilly, C. and Hotchkiss, M. (2004) Effects of combining an entomopathogenic fungi or bacterium with entomopathogenic nematodes on mortality of *Curculio caryae* (Coleoptera: Curculionidae). *Biological Control* 30, 119–126.

Sher, R.B., Parrella, M.P. and Kaya, H.K. (2000) Biological control of the leafminer *Liriomyza trifolii* (Burgess): Implications for intraguild predation between *Diglyphus begini* Ashmead and *Steinernema carpocapsae* (Weiser). *Biological Control* 17, 155–163.

Sheykhnejad, H., Ghadamyari, M., Koppenhöfer, A.M. and Karimi, J. (2014) Interactions between entomopathogenic nematodes and imidacloprid for rose sawfly control. *Biocontrol Science Technology* 24, 1481–1486.

Thurston, G.S., Kaya, H.K., Burlando, T.M. and Harrison, R.E. (1993) Milky disease bacterium as a stressor to increase susceptibility of scarabaeid larvae to an entomopathogenic nematode. *Journal of Invertebrate Pathology* 61, 167–172.

Thurston, G., Kaya, H.K. and Gaugler, R. (1994) Characterizing the enhanced susceptibility of milky disease-infected scarabaeid grubs to entomopathogenic nematodes. *Biological Control* 4, 67–73.

Wakil, W., Yasin, M. and Shapiro-Ilan, D. (2017) Effects of single and combined applications of entomopathogenic fungi and nematodes against *Rhynchophorus ferrugineus* (Olivier). *Scientific Reports* 7: 5971.

Wang, Y., Crocker, R.L., Wilson, L.T., Smart, G., Wei, X. *et al.* (2001) Effect of nematodes and fungal treatments on nontarget turfgrass-inhabiting arthropod and nematode populations. *Environmental Entomology* 30, 196–203.

Yildrim, E. and Hoy, C.W. (2003) Interaction between cryomazine and the entomopathogenic nematode *Heterorhabditis bacteriophora* Poinar "GPS11" for control of onion maggot, *Delia antiqua* (Meigen). *Crop Protection* 22, 923–927.

Zaki, F.N., Awadallah, K.T. and Gesraha, M.A. (1997) Competitive interaction between the braconid parasitoid *Meteorus rubens* Nees and the entomogenous nematode, *Steinernema carpocapsae* (Weiser) on larvae of *Agrotis ipsilon* Hufn. (Lepidoptera, Noctuidae). *Journal of Applied Entomology* 121, 151–153.

Zimmerman, R.J. and Cranshaw, W.S. (1990) Compatibility of three entomogenous nematodes (Rhabditida) in aqueous solutions of pesticides used in turfgrass maintenance. *Journal of Economic Entomology* 83, 97–100.

28 Conservation of Entomopathogenic Nematodes in Agroecosystems

Mary Barbercheck*
Department of Entomology, The Pennsylvania State University, University Park, Pennsylvania, USA

Abstract

Conservation biological control (CBC) is a pest-management approach based on manipulation of the environment to promote pest suppression by naturally occurring natural enemies. This chapter reviews information relevant to factors that affect entomopathogenic nematodes (EPNs) as CBC agents within the framework of soil health and soil health management principles. Valuable insights for predicting where CBC with EPNs should be possible have been gained from laboratory, greenhouse and field experiments, and from biogeographical surveys that include environmental information. Even so, a remaining challenge is the general lack of reporting on the relationship between the effect of management and environmental factors on changes in EPN prevalence, biological control services, and the suppression of target pests or improvements in crop yield, quality or profitability. New and emerging technologies offer powerful tools to elucidate and manipulate EPN and plant characteristics. Research that builds on current knowledge and combines traditional and new techniques will likely provide a fuller understanding of the complexity of interactions in soil food webs that can be used to improve our ability to use EPNs in CBC programs.

28.1 Conservation Biological Control

Conservation biological control (CBC) is a pest-management approach based on manipulation of the environment to promote pest suppression by naturally occurring predators, parasitoids and pathogens by promoting their survival, reproduction and efficacy (Barbosa, 1998). In agroecosystems, CBC is based on the premise that improving environmental conditions, remediating habitat loss and reducing environmental disturbance associated with intensive crop production will conserve natural enemies that contribute to pest suppression (Tscharntke *et al.*, 2007; Begg *et al.*, 2017; Shields *et al.*, 2019). Habitat management, such as intercropping, providing in-field or field-edge refugia from physical or chemical disturbance, and providing vegetative and floral resources for parasitoids and predators, are examples of practices that have been used to promote CBC by arthropod natural enemies. When achieved, CBC can provide relatively stable, persistent, and cost-effective control of target pests, although this has not always been the case globally (e.g. Finke and Denno, 2006; Coll, 2009; Karp *et al.*, 2018).

Entomopathogenic nematodes (EPNs) are generalist predators of arthropods in soil and have been studied extensively as inundatively applied biological control agents. Historically, most of the research on CBC has focused on arthropod natural enemies of arthropods, and relatively little has intentionally focused on experimentally examining factors in the field

*Email: meb34@psu.edu

 Entomopathogenic Nematodes as Biological Control Agents (D.I. Shapiro-Ilan and E.E. Lewis eds)
DOI: 10.1079/9781800620322.0028

that result in CBC by EPNs. Earlier reviews on the potential for CBC by EPNs have drawn on an extensive scientific literature reporting results from manipulative experiments in the laboratory, greenhouse and field to determine the effects of abiotic and biotic factors on the survival, persistence and efficacy of EPNs; inundative releases to provide biological control of pests in agricultural and other managed systems; and surveys to detect the occurrence and distribution of endemic populations of EPNs, especially where environmental data are included (e.g. Lewis *et al.*, 1998; Koppenhöfer *et al.*, 2020).

28.1.1 Ecological concepts in CBC

The ecological theories that underpin CBC suggest that existing, colonizing or applied natural enemies should be able to persist in a conducive environment and increase the likelihood of maintaining pest populations below economic thresholds (Heimpel and Mills, 2017; Settele and Settle, 2018). Two foundational theories that explain the mechanisms of CBC of arthropod pests include the "enemies hypothesis" (Root, 1973; Russell, 1989) and the "resource concentration hypothesis" (Root, 1973). The enemies hypothesis predicts a positive correlation between plant species richness and natural enemy diversity and abundance, which results in the regulation of herbivores at lower population densities in diverse vegetation than in monocultures. The resource concentration hypothesis predicts that denser or larger stands of a single crop species will recruit more herbivores per unit plant than a more diverse stand, and this effect will be strongest for specialist herbivores (Chaplin-Kramer and Kremen, 2012).

Another key concept in biological control is that of a trophic cascade, which relates the effects of natural enemies on pest populations to the performance of plants (Pace *et al.*, 1999). Positive impacts of CBC on plant protection, quality and productivity should be achieved when a trophic cascade, in which predators alter the biomass of plants as an indirect consequence of their direct effects on herbivores, is established and functions effectively (Finke and Denno, 2006). EPNs can benefit plants through their effects on pest arthropods in trophic cascades (Denno *et al.*, 2008; Ram *et al.*, 2008; Stuart *et al.*, 2008). Effective CBC by EPNs in a trophic cascade has been described in the extensively studied example of a natural system involving *Heterorhabditis marelata*, which indirectly benefits the bush lupine, *Lupinus arboreus*, through suppression of the soil-dwelling larvae of the ghost moth, *Hepialus californicus*, which feed on the roots of the bush lupine (Strong, 2002; Preisser, 2003; Ram *et al.*, 2008). A trophic cascade in an agroecosystem involving *Steinernema diaprepesi*, *Heterorhabditis indica* and *Heterorhabditis zealandica*, and the *Diaprepes* root weevil, *Diaprepes abbreviatus*, has been studied extensively in citrus (Stuart *et al.*, 2008; see Chapter 16, this volume).

The complexity of the soil environment, with its numerous interacting biotic and abiotic factors that affect EPN occurrence, survival and persistence and an incomplete understanding of the structure and function of soil food webs, has challenged the wider consideration of EPNs as CBC agents (Labaude and Griffin, 2018). Recent advances in molecular and omics methods, as well as the chemical ecology of plant–EPN interactions, are producing knowledge that can elucidate interactions in soil and environmental factors that can inform management to support EPNs as CBC agents (Rasmann *et al.*, 2005, 2012; Hiltpold *et al.*, 2010a,b; Campos-Herrera *et al.*, 2011a,b, 2012a,b,c, 2013, 2015, 2019; Ali *et al.*, 2013; Jaffuel *et al.*, 2016, 2017, 2018).

28.2 Soil Health

This chapter reviews information relevant to factors that affect EPNs as CBC agents in agroecosystems within the framework of soil health and soil health management principles. Soil health is an increasingly widely used term to describe the general condition or quality of the soil resource (Romig *et al.*, 1995; Stott and Moebius-Clune, 2017; Lehmann *et al.*, 2020). Some essential functions of a healthy soil include: regulation of water, buffering of organic and inorganic pollutants, nutrient cycling, physical stability and support for plants and structures, and, of relevance here, sustenance of plant and animal life, which includes organisms involved in biological control. Among soil organisms that provide ecosystem services, EPNs function as natural mortality agents of arthropods and therefore can be considered an important component of a healthy soil (Lu *et al.*, 2020).

In agroecosystems, soil health has been defined as the continued capacity of the soil to function as a vital living ecosystem that sustains plants, animals and humans (Doran and Parkin, 1994; Stott and Moebius-Clune, 2017; USDA, 2024). Soil health integrates biological, chemical and physical properties, and the resulting level of soil function and productivity. The fundamental concept of soil health, which is over 100 years old, has evolved over time, changing from a primary focus on soil fertility to a holistic perspective that includes soil fertility, ecosystem functioning and the support of organismal health. Over the past decade, awareness of the critical importance of soil health has grown among diverse stakeholders in the environmental and agricultural sectors, so much so that the concept has become institutionalized in the US Department of Agriculture (USDA). To aid land managers, farmers and ranchers in improving the soil resource, the USDA Natural Resources Conservation Service (NRCS) created a Soil Health Division to increase the capacity of the NRCS to incentivize producers to broadly implement science-based, effective and economically viable soil health management systems (USDA, 2024). Management for soil health and soil function relies on maintaining habitat that supports the soil food web. The NRCS has developed and promotes four guiding principles for the management of agricultural systems for soil health: (i) minimize soil disturbance; (ii) maximize diversity of crops, animals and amendments; (iii) keep the soil covered as much of the time as possible through surface-managed plant residues and/or plant canopies; and (iv) maximize the amount of time during which living roots are actively growing (Stott and Moebius-Clune, 2017; USDA, 2024)

28.2.1 Soil health management, abiotic and biotic factors, and EPNs

Agricultural management practices can impact EPNs through their effects on the soil abiotic and biotic environment in which EPNs and their hosts reside. The effects of soil abiotic factors on the occurrence, movement and persistence of EPNs have been extensively studied and reviewed (reviewed by Stuart *et al.*, 2008, 2015). In agroecosystems, EPNs are affected by various abiotic soil properties such as soil texture, structure, moisture, temperature, chemistry and organic matter, some of which can be altered by agricultural management practices. Soil moisture, in particular, is a critical abiotic factor for EPN efficacy and survival, as soil nematodes require water films of adequate but not excessive thickness and continuity to allow movement (e.g. Wallace, 1968; Grant and Villani, 2003; Salame and Glazer, 2015). Soil texture, structure and chemistry are also critical factors influencing the occurrence, movement and survival of EPNs (e.g. Kung *et al.*, 1990, 1991; Kapranas *et al.*, 2017; Quist *et al.*, 2019). Although local abiotic conditions can influence EPN occurrence and survival, phenotypic and genotypic variability in tolerance to abiotic stresses (see Chapter 5, this volume) and metapopulation processes, such as colonization and extinction, and patchy refugia may override local limitations and may allow EPNs to persist despite local seasonal variability in biotic and abiotic conditions (see Chapters 6 and 7, this volume). It is important to note that most research on the effect of abiotic factors on EPNs has been conducted using infective juveniles (IJs). The role of the host cadaver on the interaction between EPNs and environmental stresses has been less commonly studied. Results from these studies indicate that the cadaver may provide protection of EPNs from both abiotic and biotic stresses (e.g. Shapiro-Ilan *et al.*, 2001; Perez *et al.*, 2003).

Numerous biotic factors, such as potential host arthropods (alive or EPN-infected cadavers), vegetation, plant species, and competitors and natural enemies (e.g. nematophagous fungi and bacteria, predatory and saprophagic nematodes, mites, collembolans and other microarthropods) can affect EPNs and their interactions with other organisms in the soil food web (Stuart *et al.*, 2015; see Chapter 8, this volume). In addition to biotic factors that affect the soil-inhabiting IJ stage, EPN-infected insect cadavers represent a resource with which soil organisms can interact. Both increased attractiveness and repellency of EPN-infected cadavers, depending on the EPN species and antagonist or natural enemy, have been demonstrated (Baur *et al.*, 1998; Zhou *et al.*, 2002; Greenwood *et al.*, 2011; Gulçu *et al.*, 2012; Ulug *et al.*, 2014; Stuart *et al.*, 2015; Zhang *et al.*, 2019).

The remainder of this chapter will focus on how the four NRCS management principles for soil health influence EPN occurrence and persistence, and therefore the potential for CBC.

28.3 Soil Health Management Principle 1: Minimize Soil Disturbance

The abundance and diversity of EPNs and other soil organisms and their interactions can be significantly altered by physical and chemical disturbance of the soil (reviewed by Kladivko, 2001; Stuart *et al.*, 2015; Rowen *et al.*, 2020). Mechanical or physical disturbance from tillage and other field operations can have profound impacts on the biological and functional properties of the soil. For example, soil faunal biomass often drops with increased agricultural usage, especially where inversion tillage is practiced (Rowen *et al.*, 2020). In managed systems, reduced physical disturbance and favorable soil conditions, such as adequate moisture, aeration and structure, have favored the successful use of inundative application of EPNs (Stuart *et al.*, 2008, 2015). Under a conventional (inversion) tillage regime, the soil surface experiences greater fluctuations in temperature and water content than under conservation tillage management. Compaction and removal of surface residue resulting from tillage also contribute to a reduction in soil moisture and pore space for soil organisms.

28.3.1 Physical disturbance: tillage

The effects of tillage on EPNs are variable and depend on the EPN species and their associated foraging strategies (see Chapter 6, this volume). Several authors have observed negative effects on the prevalence and persistence of EPNs related to disturbance from tillage (Hummel *et al.*, 2002; Susurluk and Ehlers, 2007; Hoy *et al.*, 2008; Campos-Herrera *et al.*, 2015; Jaffuel *et al.*, 2016). The greater complexity of the soil environment associated with relatively high levels of crop residue in conservation tillage regimes may influence the prevalence of EPNs through provision of a greater number and diversity of hosts. The seasonal detection of naturally occurring *Steinernema carpocapsae* and *Heterorhabditis bacteriophora* was significantly greater in vegetable production systems managed with conservation (chisel plow and disk) compared with conventional (inversion) tillage (Hummel *et al.*, 2002). The authors suggested that the more moderate temperatures and greater soil moisture levels measured in the conservation compared with the conventional tillage system contributed to the greater prevalence of EPNs.

Hoy *et al.* (2008) investigated nematode establishment and persistence of *H. bacteriophora* in different crops and found that the percentage of nematode-positive soil samples dropped considerably after tillage. The authors suggested that the influence of tillage on factors such as invertebrate abundance, organic matter and soil moisture retention can be influenced by soil-management practices and increases EPN abundance by increasing the nematode food web enrichment and structure indices, and leading to lower phosphorus (P), higher potassium (K) and a lower carbon:nitrogen (C:N) ratio.

In a field experiment comparing the effects of tillage in a no-till and conventional-till corn system, Millar and Barbercheck (2001, 2002) found no effect of tillage on the detection of endemic *H. bacteriophora* but a significant negative effect of tillage on the detection of endemic *S. carpocapsae* in corn. In the same study, tillage had a significantly positive effect on the detection of inundatively applied, non-endemic *Steinernema riobrave*, which was detectable for more than 1 year after application. The authors suggested that dissimilar sensitivities of EPN species to tillage could be partly explained by differences in environmental tolerances and differences in tendencies to disperse deeper in the soil profile to avoid disturbance from tillage (Millar and Barbercheck, 2001, 2002). Others (e.g. Campos-Herrera *et al.*, 2015; Bal *et al.*, 2017) have observed no effects of tillage on detection, dispersal or resulting distributions of either heterorhabditid or steinernematid nematodes in the field.

28.3.2 Chemical disturbance: pesticides and fertilizers

In addition to physical disturbance caused mainly by tillage, a major source of disturbance in agroecosystems is chemical, through the use of inputs such as synthetic and natural pesticides (i.e. insecticides, fungicides and other microbicides, herbicides) and fertilizers (Bitew and Alemayehu, 2017; Rowen *et al.*, 2019, 2020; see Chapter 27, this volume,). Natural populations of EPNs in most agroecosystems are likely to be routinely exposed to pesticide active ingredients, adjuvants, fertilizers and other soil amendments. However, it is challenging to generalize about the effects of agricultural inputs on soil-dwelling IJs due to the reported variability among EPN species and strains in response to exposure to agrochemicals, depending on formulations, concentrations, environmental conditions (e.g. temperature) and timing of exposure (see Chapter 27, this volume).

Much of the research on the effects of pesticides has focused on laboratory bioassays that directly expose the IJs of EPNs to pesticides or other chemicals to determine compatibility in tank mixes using commercially available EPN species and strains. Some species of EPN may be sensitive to a material, while exposure to others may result in no impact or even a positive effect in terms of infection and mortality of the arthropod species and life stage of arthropod used in the assay. For example, Özdemir *et al.* (2020) examined the mortality, infectivity and reproductive capacity of pesticide-treated EPNs. The authors found that the insecticides imidacloprid and cyflumetofen caused no adverse effect on the survival, infectivity and reproduction rates of *Steinernema feltiae* and *H. bacteriophora*. However, the fungicides fluxapyroxad + difenoconazole and ametoctradin + dimethomorph caused significant mortality to *S. feltiae*. Only fluxapyroxad + difenoconazole negatively affected the survival of *H. bacteriophora*. In addition to the active ingredients of pesticides, research on the combination of EPNs and adjuvants has shown that even short exposure to some adjuvants can decrease the viability of EPNs (Beck *et al.*, 2013).

Environmental conditions during the bioassay can also affect results of the exposure to EPNs. For example, Laznik and Trdan (2014) assessed the compatibility of the IJs of steinernematid and heterorhabditid EPNs with eight chemical and biological insecticides under laboratory conditions. The results varied according to the duration of exposure, temperature and EPN species. The IJs of *Steinernema kraussei* and *S. carpocapsae* were sensitive to all tested insecticides at exposure durations of 6 and 24 h. *S. feltiae* was compatible with azadirachtin, imidacloprid and the toxin of *Bacillus thuringiensis* subsp. *kurstaki*, while *H. bacteriophora* was sensitive to abamectin and lufenuron. The greatest percentage of IJs surviving after 6 h of exposure occurred at 15°C (82%) and 20°C (80%), while IJ survival was lowest at 25°C (76%). After 24 h of exposure, there were no differences in IJ survival at 15°C (55%) and 20°C (55%), but at 25°C, IJ survival was greatest (59%).

Management of soil fertility with inorganic fertilizers and manures or compost represents a chemical disturbance to the soil (Edmeades, 2003). Fertilizers have variable short- and long-term and species-specific impacts on EPN mortality, infectivity and persistence (Shapiro *et al.*, 1996, 1999a; Nádasy and Nádasy, 2006; Duncan *et al.*, 2007; Şahin and Susurluk, 2018; Shehata *et al.*, 2021). In laboratory assays, Karunakar *et al.* (2002) found differences in sensitivity to inorganic fertilizers depending on EPN species and fertilizer concentration. Survival of IJs of *Steinernema glaseri*, *S. feltiae* and *H. indica* were not affected by exposures to urea and muriate of potash at concentrations of 0–400 ppm for 72 h. *S. glaseri* and *S. feltiae* were compatible with superphosphate up to a concentration of 100 ppm, whereas *H. indica* was observed to be compatible up to 400 ppm.

In a long-term study comparing conventional, synthetic, chemical-based with organic cropping systems, Jaffuel *et al.* (2016) found that seven species of EPN were randomly distributed among the conventional and organic cropping systems. The authors concluded that long-term application of organic fertilizers in a manure-based cropping system had no impact on the abundance or the activity of EPNs, but that physical disturbance in annual cropping systems may mask the effects of chemical disturbance. In comparable studies comparing organic with conventional annual cropping systems, Campos-Herrera *et al.* (2008, 2019) did not detect any effect of cropping system type on EPNs, although EPNs were rarely detected at the experimental sites.

In contrast, Bednarek and Gaugler (1997) showed differential effects of inorganic and organic fertilizers on three EPN species, *Steinernema anomali*, *S. feltiae* and *H. bacteriophora*. Prolonged exposure (up to 20 days) of *S. feltiae* and *H. bacteriophora* to high concentrations of inorganic fertilizer in laboratory assays inhibited their infectivity and reproduction in greater wax moth (*Galleria mellonella*) larvae, whereas short (1 day) exposures increased the infectivity of *S. feltiae* (Bednarek and Gaugler, 1997). *H. bacteriophora* was more sensitive to adverse effects of exposure to inorganic fertilizer than *S. feltiae* and *S. anomali*. Şahin and Susurluk (2018) conducted laboratory assays to determine the effects of six inorganic fertilizers on the mortality of IJs of *S. feltiae* Tur-S3 strain and *H. bacteriophora* HBH strain. IJs were exposed to concentrations of 1–10 g l^{-1} for 10 days. *S. feltiae* (Tur-S3) was more resistant to inorganic fertilizers than *H. bacteriophora* HBH, and both EPN species were more adversely affected by diammonium phosphate, NPK and NP compared with other fertilizers tested.

In greenhouse studies to determine the effects of N fertilizers on the nitrate content of lettuce and on the mortality of cockchafer grubs due to *S. glaseri*, cockchafer grub mortality from *S. glaseri* was inversely related to the concentration of N fertilizer applied, ranging from 100 to 300 mg kg^{-1} of N (Nádasy and Nádasy, 2006). Shapiro *et al.* (1996) conducted laboratory and field experiments to determine the effects of inorganic fertilizers and raw and composted manure on the virulence of *S. carpocapsae* against *G. mellonella*. In laboratory experiments, the authors observed a loss of EPN virulence in urea and fresh manure treatments relative to unmodified soil with no amendment or amended with composted manure.

In field experiments, EPN virulence was significantly lower in soil amended with fresh manure than in all other treatments in one out of 2 years (Shapiro *et al.*, 1996). However, also in field studies, Bednarek and Gaugler (1997) found that the application of organic manure resulted in increased densities of a native population of *S. feltiae*, whereas NPK fertilizer suppressed EPN densities regardless of manure applications. The authors concluded that inorganic fertilizers are likely to be compatible with EPNs applied in tank mixes but may interfere with inoculative releases for long-term establishment. Similarly, Shehata *et al.* (2021) found that, although all inorganic fertilizers tested had a negative effect on *S. glaseri* relative to the no-fertilizer control, P had more adverse effects than K, NPK and nitrogenous fertilizers on the virulence of *S. glaseri*. The authors concluded that 1% fertilizer concentrations are compatible with *S. glaseri* in tank mixes for short-term (1–7 days) insect control, but longer-term effects on EPNs may be negative.

28.4 Soil Health Management Principle 2: Maximize Diversity

Soil is among the most biodiverse and densely inhabited environments on Earth (Delgado-Baquerizo *et al.*, 2020). Huge numbers of microbial, plant, and invertebrate and vertebrate animal taxa comprise soil biodiversity, and are dependent on climate and geography but can also be affected by agricultural management. Where plant and insect species diversity is high, one would expect that the diversity of natural enemies, including EPNs and other nematodes, would also be high (Finke and Snyder, 2008; Tamburini *et al.*, 2020). Both plant diversity and plant identity can influence the level of predation and predator abundance above ground, but how the level of predation in the soil and the abundance of predatory soil fauna are related to plant diversity and identity remains less well known.

Both local and landscape-level management can affect biodiversity conservation in intensively managed agricultural landscapes (Tscharntke *et al.*, 2007; Gonthier *et al.*, 2014; Karp *et al.*, 2018). The local species pool of natural enemies in cultivated crops is influenced by the composition of the surrounding landscape, and fields located in complex landscapes with a low proportion of annual crops and a high diversity of other vegetation types tend to have a higher diversity of natural enemies than fields in simpler landscapes dominated by annual agriculture (Chaplin-Kramer and Kremen, 2012; Delgado-Baquerizo *et al.*, 2020). Although it may be optimal to manage the environment at multiple scales to conserve biodiversity, this section will focus on farm-scale practices, as these are manageable by most growers and land managers (Thiele-Bruhn *et al.*, 2012). Aboveground and belowground communities can be tightly linked (Wardle *et al.*, 2003, 2004), and there are numerous agricultural practices that affect the reciprocal interactions between plants and soil organisms that can increase and conserve biodiversity (Chaplin-Kramer and Kremen, 2012; Tamburini *et al.*, 2020; Beillouin *et al.*, 2021).

Examples of local agricultural management practices that can increase biodiversity include crop rotation, staggered planting dates, planting crop mixtures and multiple crop cultivars, intercropping or strip cropping, cover cropping, integrating perennial phases or livestock grazing into a rotation, conservation tillage and maintaining surface residues, and incorporation of organic matter into the soil in the form of animal and green manures, cover crop and crop residues and organic amendments. Additionally, some weed species allowed to grow between growing seasons or within crops can contribute to biodiversity and provide resources to natural enemies (Baraibar *et al.*, 2021; Barbercheck and Wallace, 2021). These practices and conditions within and adjacent to fields can provide natural enemies with food resources, including host insects, refuge from disturbance, shelter and a favorable microclimate (Coll, 2009). The remainder of this section will focus on the effects of within-field plant diversification on the conservation of EPNs.

28.4.1 Cropping systems and crop species

Research on the effects of cropping systems that vary across multiple characteristics, such as stability (perennial versus annual), crop species and cultivars, rotations, cover crop use and species, and types of fertility and pest-management inputs, has reported variable effects on EPNs. In a survey to evaluate the impact of ecosystem type, habitat and insect host species on EPN prevalence, Mráček *et al.* (2005) detected EPNs more frequently in closed and semi-open perennial habitats with trees, light soils and abundant potential insect hosts than in open habitats, such as natural meadows or managed pastures. Campos-Herrera *et al.* (2019) surveyed four perennial habitats (citrus groves, palmetto, oaks and pines) in two eco-regions of Portugal (calcareous versus non-calcareous) for EPNs and associated organisms, and found that oaks and calcareous soils favored EPNs. These results agreed with those of Mráček *et al.* (2005) who recorded annual crops as the habitat with the lowest frequency of occurrence of EPNs compared with natural areas and perennial crops. Similarly, Campos-Herrera *et al.* (2008) detected the greatest frequency of occurrence of EPNs in perennial crops, followed by natural areas and rarely in annual crops.

Conversely, Jaffuel *et al.* (2018) compared the soil food web, including EPNs, in 40 sites in Switzerland categorized as arable agriculture, forest or grassland, and found no differences in EPN abundance, richness or diversity. However, the authors reported greatest EPN activity, measured as ability to kill insects, in grasslands. The prevalence and diversity of nematophagous fungi were greater in agricultural areas, and the authors suggested that abiotic conditions in the agricultural areas might have favored their presence. Six soil abiotic factors (pH, elevation, clay content, soil water content, temperature and rainfall) explained 54% of the variation in the EPN soil food web community among their sites (Jaffuel *et al.*, 2018). Similar to Campos-Herrera *et al.* (2019), the authors suggested that the intensive management in annual crops disrupts EPN persistence by creating unfavorable physical conditions, decreasing host abundance and increasing EPN natural enemies, or a combination of these.

The effects of plant diversity and identity within perennial systems (e.g. grasslands and tree and vine crops) on EPNs have also been studied (see Chapters 16–19 and 25, this volume). Kostenko *et al.* (2015) examined the effects of plant diversity and identity in an experimental grassland on infection rates of *G. mellonella* by *Heterorhabditis* and *Steinernema* spp. and found that infection by *Heterorhabditis* spp. was significantly greater in vegetated compared with bare plots but was not related to plant diversity or the proportion of legumes, grasses or forbs in vegetated plots. Infection rates by *Steinernema* spp. were higher in monocultures of *Festuca rubra* and *Trifolium pratense* than in monocultures of other forb species (*Lotus corniculatus*, *Trifolium repens*, *Achillea millefolium*, *Hypochaeris radicata*, *Leucanthemum vulgare*, *Plantago lanceolata* and *Tanacetum vulgare*). Structural equation modeling revealed that legumes positively affected infection rates by *Steinernema* spp., whereas plant diversity indirectly affected infection rates by *Heterorhabditis* spp. via effects on the abundance of soil insects, which was positively related to plant diversity.

In orchards, vineyards and other tree-based systems, ground-cover management affects plant diversity and soil-dwelling organisms. Blanco-Pérez *et al.* (2020) conducted experiments in a Spanish vineyard (*Vitis vinifera* var. Tempranillo, rootstock 110 Richter) to compare the effects of soil-management systems on naturally occurring EPNs and other components of the soil food web. Their treatments included management with conventional tillage practices and seeded grass (*Bromus catharticus*) or flowering cover crops, or with natural inter-row vegetation remaining. The authors recorded variability of treatment effects among the five naturally occurring EPN species. Overall greatest activity of EPNs and lowest numbers of antagonists of EPNs were observed in treatments with grass cover crops and natural vegetation. There was no difference between treatments with bare soil or flowering cover crops. Overall numbers of EPN IJs were greatest in treatments with grass-seeded covers, but EPN activity appeared to be suppressed by the high prevalence of endoparasitic nematophagous fungi in these treatments. The authors concluded that cover provided by spontaneous vegetation could be a promising strategy to support CBC provided by the naturally occurring EPN species in vineyards.

28.4.1.1 Annual cropping systems

Jaffuel *et al.* (2016) investigated how communities of natural EPN populations, their associated natural enemies and competitors in annual cropping systems are affected by crop type (wheat, corn and grass-clover ley) and farming system (conventional, organic and biodynamic). The authors did not detect differences in EPN abundance or diversity among the farming systems, but crop type and the crop history (crops in rotation) affected EPNs, which were more abundant in winter-wheat following potato than in corn and grass-clover ley. Overall, the authors detected very low numbers of EPNs in arable systems, similar to other studies, and suggested that disturbance or lack of host arthropods limits the abundance of EPNs in cultivated systems, and that the natural presence of EPNs would not be sufficient to suppress an outbreak of root-feeding pests or to protect yield. Conversely, Shields *et al.* (2018) examined the potential of local EPN isolates inoculated at a low rate to maintain population levels for multiple years in a corn–alfalfa crop rotation. The authors found that EPN populations established in prior years could persist in continuous and rotated crops and responded positively to an increase of susceptible insect hosts.

28.4.1.2 Cover crops and weeds

Cover crops planted in rotation with cash crops to conserve soil and water and suppress weeds during a time when soil would otherwise be bare add diversity to cropping systems. Cover crops can also provide resources to support arthropods and their natural enemies and can have short-term and species-specific impacts on EPN persistence (Araj *et al.*, 2009; Jaffuel *et al.*, 2017). For example, Susurluk and Ehlers (2008) compared the persistence of *H. bacteriophora* in beans (*Vicia faba*), wheat (*Triticum sativum*), pasture, oilseed rape (*Brassica napus*), lupine (*Lupinus angustifolius*), potato (*Solanum tuberosum*) and pea (*Pisum sativum*), and recorded the greatest persistence of EPNs in beans followed in rotation by wheat with red clover as a cover crop, and suggested that larvae of the pea weevil, *Sitona lineatus*, in the red clover cover crop sustained *H. bacteriophora*.

Increasing plant diversity through inter-cropping or planting species-rich habitat strips within monocultures or along field edges in annual cropping systems may increase the occurrence and movement of EPNs by providing habitat or refuge from abiotic extremes or by supporting potential host insects that feed on roots or diapause in relatively undisturbed areas (Jabbour and Barbercheck, 2008). However, Bal *et al.* (2017) found that the long-term dispersal of *H. bacteriophora* from grassy borders to adjacent cultivated fields was not influenced by the ground cover, grass or cultivated land. In an experiment comparing the effects of vegetable production systems on entomopathogens in soil, Hummel *et al.* (2002) found that ground cover from cereal rye mulch and a red clover intercrop significantly reduced soil temperature and increased soil moisture compared with black plastic mulch or bare ground, which was related to greater EPN activity in plots with ground cover and the intercrop. Conversely, Jaffuel *et al.*

(2017) did not find a strong effect of cover crops on the persistence of native or augmented EPN persistence. In only one sample period, the authors observed a significant positive effect of pea (*P. sativum*) and mustard (*Brassica juncea*) cover crops on the abundance of augmented *S. feltiae*, but not *H. bacteriophora*, compared with bare plots.

Weeds can also add diversity and habitat complexity to monocultures. Jabbour and Barbercheck (2008) compared the effects of simple (corn) and more complex (corn interplanted with a mixed annual forb refuge) habitats on the persistence and dispersal of *S. carpocapsae* applied to soil as nematode-killed insects. *S. carpocapsae* detection at the point of application was not affected by habitat complexity but was associated with soil bulk density, plant density and soil moisture. *S. carpocapsae* dispersed farther within complex habitats, where soil moisture was greater, than into the adjacent corn crop, where soil moisture was lower. Dispersal of *S. carpocapsae* was similar in simple and complex habitat treatments when plant density was similar due to weeds. The authors suggested that movement of *S. carpocapsae* is not necessarily dependent on habitat complexity but may respond to variations in factors associated with plant density such as soil moisture.

28.5 Soil Health Management Principle 3: Maintain Surface Residues

On average, soil organisms are more abundant and diverse in agricultural systems that maintain surface residues. Surface residues from crops, cover crops and mulches reduce fluctuations in soil moisture, temperature and organic matter, and can enhance the occurrence, movement, persistence and efficacy of EPNs through these environmental effects and protection from ultraviolet light (Shapiro *et al.*, 1999b). For example, Duncan *et al.* (2007) examined the effects of the application of composted manure mulch on naturally occurring *S. diaprepesi* and *H. zealandica* and augmented *S. riobrave* and their associated soil food web, and on the mortality of sentinel *D. abbreviatus* in citrus. The mulch treatments included a 10 cm layer of composted horse manure and stable straw mixture or a similar layer of aged poultry manure. Soil moisture was greater in the mulched than in the unmulched treatment plots in one year of the 2-year study, and the cumulative percentage mortality of sentinel *D. abbreviatus* from EPNs was greater in soil in mulched compared with non-mulched treatments. Additionally, the prevalence of the two endemic EPNs was consistently greater in mulched plots than in unmulched plots.

Lacey *et al.* (2006) applied S. *carpocapsae* or *S. feltiae* to codling moth (*Cydia pomonella*) cocoons in cardboard strips covered with one of four types of mulch (clover, shredded paper, grass hay or wood chips). The authors observed *C. pomonella* pupal mortalities from applied *S. carpocapsae* and *S. feltiae* of 97% and 98%, respectively, in mulched plots compared with 80% and 76% in bare plots. Overall, 25% and 14% of the sentinel cocoons in the soil in hay- and paper-mulched treatments, respectively, became infected by a naturally occurring *Heterorhabditis* sp. Similarly, de Waal *et al.* (2011) conducted a laboratory study to determine the ability of *H. zealandica* to control diapausing *C. pomonella* in different types of mulch, including pine chips, wheat straw, pine wood shavings, blackwood chips and apple wood chips. Higher levels of *C. pomonella* mortality were obtained in treatments with pine wood (*Pinus radiata*) shaving mulch (88%) compared with pine chip, wheat straw (*Triticum aestivum*), blackwood (*Acacia mearnsii*) and apple wood (*Malus domestica*) chip mulch (41–88%). The authors demonstrated that EPNs were able to move upward from soil into moist mulch and infect larvae located up to 10 cm above the soil. In two associated field trials, *C. pomonella* mortality was greater in the treatment plots using apple wood chip mulch compared with wheat straw mulch. The authors suggested that wood chips are better able to absorb moisture and maintain high levels of humidity, which are critical for EPN survival, in contrast to wheat straw, which tended to dry out rapidly.

In contrast, Leite *et al.* (2015) examined the effects of sugarcane residues on the persistence of applied *Steinernema brazilense* and found that mulch did not affect EPN persistence compared with bare soil treatments but did favor the movement of *S. brazilense*. It is possible that, in addition to creating an environment that can conserve EPNs, mulch creates conditions that favor the natural enemies and competitors of EPNs, such as nematophagous and entomopathogenic fungi and bacteria, bacteriophagous, predatory and omnivorous nematodes, and predatory and omnivorous microarthropods.

28.6 Soil Health Management Principle 4: Maximize the Amount of Time during which Living Roots Are Present

In CBC of arthropods, low availability of key resources for natural enemies, such as hosts, alternate food and overwintering sites, can limit biological control efficacy in many agroecosystems (Begg *et al.*, 2017), and it has been suggested that an unsupportive environment and lack of hosts can impact EPN occurrence and persistence. Plants play a key role in the ecology of EPN interactions because they affect abiotic and biotic characteristics of the soil environment. EPNs potentially interact with plant roots in several ways. For example, roots can impede EPN movement through the soil when they act as a physical barrier, or they can facilitate EPN movement through the soil when they serve as physical pathways; roots can provide host-finding cues or repellent compounds through the production of semiochemicals and exudates; and they provide food for arthropods that are hosts of EPNs or they can support natural enemies of EPNs with the potential for directly or indirectly affecting EPN population dynamics (see Chapters 6 and 7, this volume).

28.6.1 Root morphology and density and nematode movement

Roots can impede EPN movement through the soil by acting as a physical barrier, or can facilitate EPN movement when they serve as physical pathways. In general, higher root density or structural complexity has been negatively correlated with EPN movement through soil and host finding (Choo *et al.*, 1989; Choo and Kaya, 1991; Cutler and Webster, 2003). Demarta *et al.* (2014) designed three artificial model roots with different degrees of complexity and connectivity to investigate the impact of root architecture on the host-finding behavior of *Heterorhabditis megidis*. *H. megidis* performed better in the presence than in the absence of model roots. However, the host-finding performance of *H. megidis* declined with increasing model-root complexity. Significantly more *G. mellonella* larvae were infected by *H. megidis* in pots planted with straight-pattern model roots with one level of branching than in treatments with the most complex model root, where the number of infected bait larvae did not differ from the infection level measured in the controls. Similarly, in pot experiments using spruce (*Picea sitchensis)* twigs as surrogate roots, the mortality of pine weevil (*Hylobius abietis*) larvae from *S. carpocapsae* was lowest in the absence of twigs and increased significantly when twigs linked the nematodes and insects (Ennis *et al.*, 2010). The authors suggested that roots serving as physical pathways may be an important factor in the movement of EPNs toward subterranean hosts, especially for ambush predators such as *S. carpocapsae*.

28.6.2 Root-associated chemicals and nematode host finding

Roots can provide host-finding cues or repellent compounds through the production of semiochemicals. Host-finding behavior by EPNs is mediated by both long-range cues that facilitate finding the root zone and shorter-range cues that facilitate host location within the root zone (Zhang *et al.*, 2021). EPNs can use gradients of carbon dioxide and plant- and insect-derived cues to locate biological activity and to locate host insects from a distance (see Chapter 6, this volume). For example, Demarta *et al.* (2014) overcame reductions in host finding by *H. megidis* associated with root complexity by adding synthetic root volatiles to their assays.

Studies involving several plant species have shown that EPNs use herbivore-induced plant volatiles to aggregate in the vicinity of plant roots and preferentially orient toward herbivore-wounded roots compared with healthy roots, resulting in improved efficacy of biological control (Rasmann *et al.*, 2005, 2012; Hiltpold *et al.*, 2010a,b; Ali *et al.*, 2013). For example, corn roots infested with larvae of the western corn rootworm, *Diabrotica virgifera virgifera*, release terpenoids, typically (E)-β-caryophyllene, which attracts *H. megidis* (Degenhardt and Gershenzon, 2000). In field experiments, the infection rate of *Diabrotica v. virgifera* larvae by *H. megidis* on a corn variety that produced (E)-β-caryophyllene was fivefold higher than on a variety that did not produce the chemical, and spiking the soil near the non-caryophyllene-producing corn variety decreased the emergence of adult *Diabrotica v. virgifera* to less than half (Rasmann *et al.*, 2005; Turlings *et al.*, 2012). Citrus roots recruited significantly more *S. diaprepesi* when infested with *Diaprepes abbreviatus* larvae than non-infested or mechanically damaged roots, suggesting that EPNs can use herbivore-induced plant volatiles as a host-finding cue (Ali *et al.*, 2013).

EPNs can also be non-responsive to or repelled by plant-derived compounds, or insects can use plant compounds to avoid infection by EPNs, depending on the EPN and plant species. For example, *S. kraussei* was repelled by glucosinolate breakdown products from the roots of *Brassica nigra* under attack by cabbage root fly (*Delia radicum*) larvae in an *in vitro* assay (Jagodič *et al.*, 2017). In pot experiments, roots of broad bean, *Vicia faba*, did not affect the host-finding ability of a local *Steinernema* sp., but hindered the ability of *H. bacteriophora* to find and infect *G. mellonella* (Cutler and Webster, 2003). The effects of herbivore-induced plant volatiles and other plant-based cues produced in response to root herbivory on EPN host-finding behavior and efficacy may be offset by the simultaneous attraction of competitors or natural enemies of EPN to the same cues (see Chapters 6 and 8, this volume).

28.6.3 Impact of arthropod host plant on EPN biology

Arthropods in soil are the primary hosts of EPNs, and roots and other belowground plant structures are the primary source of food for many soil-dwelling or soil-associated arthropod species. As such, roots and other belowground plant structures can affect both EPN virulence and arthropod host susceptibility through tritrophic-level effects of the host plant on the host arthropod, EPNs and their symbiotic bacteria (Cory and Hoover, 2003). Phytochemicals produced in the host plant can directly affect the insect immune system and alter their susceptibility to infection from EPNs, either increasing or decreasing susceptibility (Klemola *et al.*, 2007; Gassmann *et al.*, 2010). The mortality of larval southern corn rootworm, *Diabrotica undecimpunctata howardi*, feeding on the roots of corn (*Zea mays*), peanut (*Arachis hypogea*) or squash (*Cucurbita maxima*) from EPNs varied according to the host plant on which the rootworms had fed (Barbercheck, 1993; Barbercheck *et al.*, 1995, 2003). Mortality from *S. carpocapsae* was lower for rootworms that had fed on peanut roots than for rootworms that had fed on squash or corn roots, whereas mortality from *H. bacteriophora* was lower for rootworms that had been reared on corn roots than for rootworms that had fed on peanut or squash roots. In experiments to test the infectivity and reproduction of *S. riobrave* in larval corn earworm, *Helicoverpa zea*, in tobacco (*Nicotiana tabacum*), eggplant (*Solanum melongena*) and tomato (*Solanum lycopersicum*), Hazir *et al.* (2016) found that *S. riobrave* virulence and reproduction in *Helicoverpa zea* reared on tobacco was lower relative to those reared on tomato and eggplant. Specific phytochemicals involved in plant defense can affect EPN–insect interactions. Robert *et al.* (2017) found that 6-methoxy-2-benzoxazolinone *N*-glucoside (MBOA-Glc), a plant benzoxazinoid detoxification product that is released from the plant and sequestered by *D. v. virgifera*, rendered the insect highly resistant to *H. bacteriophora*.

The plant on which an arthropod feeds can be a selective force on EPNs, although the ability of EPNs and their associated bacteria to adapt to particular phytochemicals can vary with EPN species and isolate and the particular plant species. Progeny production of multiple isolates of *S. carpocapsae* and *H. bacteriophora* was reduced from EPN-infected *D. u. howardi* that had fed on a variety of the squash *Cucurbita pepo* that contained the terpenoid cucurbitacin compared with those that had fed on a variety lacking cucurbitacin (Barbercheck, 1993; Barbercheck *et al.*, 1995; Barbercheck and Wang, 1996; Hirsh and Barbercheck, 1996). When three isolates of *S. carpocapsae* (Agriotos, Mexican and a hybrid) were continuously cultured in corn-fed or squash-fed *D. u. howardi* for 25 passages, there was no increase in the ability to kill rootworms in selected EPNs compared with unselected EPNs, although there were changes in virulence based on the host plant on which the rootworms had fed. The ability of EPNs selected on corn-fed rootworms to kill corn-fed rootworms compared with squash-fed rootworms improved in the Mexican and Agriotos isolates; however, the squash-selected Mexican isolate lost virulence against rootworms. Hiltpold *et al.* (2010a) selected a strain of *H. bacteriophora* for attraction to (E)-β-caryophyllene, a sesquiterpene compound produced by corn roots in some varieties. After six generations of selection, the selected strain moved twice as rapidly toward a (E)-β-caryophyllene source than the original strain. In subsequent field tests, the selected strain was significantly more effective than the original strain in reducing *D. v. virgifera* in treatment plots with a corn variety that released (E)-β-caryophyllene than in ones that did not. Hazir *et al.* (2016) found that *S. riobrave* subjected to successive selection for five passages overcame negative impacts of tobacco on virulence and reproduction in *Helicoverpa zea* and the authors concluded that negative effects of host plant may be overcome through directed selection, and that host plant species should be considered when designing biocontrol programs using EPNs.

28.7 Knowledge Gaps and Conclusions

CBC is a complex pest management strategy influenced by many ecological processes, operating at multiple scales and mediated by agricultural management practices. Given this complexity, the evidence-based understanding needed to reliably exploit EPNs for CBC seems daunting, but the ability to predictably rely on EPNs as CBC organisms to achieve reliable control of pests across many types of agroecosystems could provide many environmental and economic benefits through enhanced crop yield and quality, avoided costs associated with pesticide use and yield losses, lower production costs, reduced risks to water, soil and air quality, and insect-related ecosystem services such as pollination, and human and animal health (Losey and Vaughn, 2006; Cullen *et al.*, 2008; Begg *et al.*, 2017). Similarly, managing soil to promote soil health can help to conserve soil and water resources, and can reduce costs for fuel and labor associated with tillage, machinery use, repair and maintenance, and water treatment, and for risks associated with the use of fertilizers and pesticides (Stott and Moebius-Clune, 2017; Lehmann *et al.*, 2020).

Although there is a considerable and growing research- and evidence-based literature focused on how biotic and abiotic factors and management practices affect EPNs in specific contexts, further research is needed before the goal of predictable conservation biocontrol with EPNs can be realized beyond a case-by-case basis. From the literature, it appears that more intensively managed agricultural systems, such as those characterized by reliance on pesticides, simplified landscapes and those dominated by disturbed habitats, are not highly conducive to CBC with EPNs. More stable perennial systems, such as orchards, vineyards and meadows, appear to be a more likely near-term target for CBC with EPNs, and more research is needed to determine if and how these and more intensively managed or highly disturbed production systems can be ameliorated to improve the environment to be more conducive to CBC.

How habitat manipulation through implementing soil health practices relates to EPN populations, activity and interactions with other natural enemies of EPNs and target pests remains relatively unknown. In many cases, management guided by soil health principles of minimizing soil disturbance, maximizing diversity, keeping the soil covered through surface-managed plant residues or crop canopy, and maximizing the time during which living roots are actively growing should improve the environment for EPN occurrence, persistence and activity, but the literature indicates that this is not always the case. If soil health practices increase EPN and natural enemy diversity, biological control may be improved (e.g. Snyder *et al.*, 2006; Straub and Snyder, 2006, 2008; Jabbour *et al.*, 2011), but increasing natural enemy diversity can also have a negative effect on biological control through intraguild predation or interspecific competition or antagonism (Rosenheim *et al.*, 1995; Finke and Denno, 2006; Straub *et al.*, 2008; Araj *et al.*, 2009). Although there is evidence that some agricultural practices can decrease negative interactions among EPNs and their natural enemies (e.g. Jaffuel *et al.*, 2016, 2017, 2018), this has not been widely studied across diverse cropping systems. Therefore, even when practices to support soil health are associated with greater prevalence of EPNs, greater EPN numbers may fail to result in improved or more effective pest control, crop yield or quality. A remaining challenge to the development of effective CBC, not unique to EPNs, is the general lack of reporting on the relationship between changes in EPN prevalence, biological control services and the suppression of target pests (as opposed to indicator species such as *G. mellonella*), or improvements in crop yield, quality or profitability (Letourneau and Bothwell, 2008; Chaplin-Kramer *et al.*, 2011).

Laboratory and greenhouse experiments have provided, and doubtless will continue to provide, invaluable insights into possible mechanisms by which many abiotic and biotic factors and interactions can affect EPN populations and activity. However, more field-based research is needed to understand how these factors interact in the context of agricultural production systems. Field-based research focused on soil food webs has been very informative in increasing our understanding about how and why particular practices or production systems affect EPNs and indicate that both short- and long-term research in agroecosystems is needed to further evaluate the impact of farming systems on soil health, fauna and productivity.

It is clear from the literature that the use of molecular techniques can be applied to greatly increase our understanding of EPN ecology in food webs (e.g. Campos-Herrera *et al.*, 2012b,c, 2015). New and emerging technologies, such as gene editing, may also offer powerful tools to elucidate and manipulate nematode and plant characteristics to further increase our understanding of EPN ecology in agroecosystems (Gurr and You, 2016). Further advances in technology will likely provide a fuller understanding of the complexity of interactions in soil food webs that can be used to improve our ability to use EPNs in CBC programs.

The interplay between EPNs and plants, and the ecology of above- and belowground effects, are other potentially fruitful areas for research to improve CBC with EPNs. This is reflected in research in both perennial and annual systems that has sought to understand and exploit herbivore-induced plant volatiles to attract EPNs to attacked plants and increase the efficacy of CBC (Stelinski *et al.*, 2019). A greater development of general principles could possibly inform crop variety development to capitalize on these interactions. Research indicates that it is possible to increase EPN efficacy against pests of particular plant species or cultivars based on adaptation to plant-based cues (e.g. Hiltpold *et al.*, 2010a). A related area for research would be to determine if and to what extent this adaptation can occur naturally in the field. This area, as well as others, will benefit by a greater focus on naturally occurring host communities as CBC targets, rather than relying on indicators of EPN activity, such as *G. mellonella.*

In addressing these knowledge gaps, the specifics related to location, cropping system, and the dominant pest and natural enemy pest communities will create challenges to making general "rules" for implementing CBC with EPNs. Research that builds on current knowledge and combines traditional and new techniques to understand the complex interactions that affect natural populations of EPNs should enable us to achieve more predictable and robust use of EPNs for CBC.

28.8 References

Ali, J.G., Campos-Herrera, R., Alborn, H.T., Duncan, L.W. and Stelinski, L.L. (2013) Sending mixed messages: a trophic cascade produced by a belowground herbivore-induced cue. *Journal of Chemical Ecology* 39, 1140–1147.

Araj, S.A., Wratten, S.D., Lister, A.J. and Buckley, H.L. (2009) Adding floral resources to improve biological control: potential pitfalls of the fourth trophic level. *Basic and Applied Ecology* 10, 554–562.

Bal, H.K., Acosta, N., Cheng, Z., Grewal, P.S. and Hoy, C.W. (2017) Effect of habitat and soil management on dispersal and distribution patterns of entomopathogenic nematodes. *Applied Soil Ecology* 121, 48–59.

Baraibar, B., White, C.M., Hunter, M.C., Finney, D.M., Barbercheck, M.E. *et al.* (2021) Weeds in cover crops: context and management considerations. *Agriculture* 11: 193.

Barbercheck, M.E. (1993) Tritrophic level effects on entomopathogenic nematodes. *Environmental Entomology* 22, 1166–1171.

Barbercheck, M.E. and Wallace, J. (2021) Weed–insect interactions in annual cropping systems. *Annals of the Entomological Society of America* 114, 276–291.

Barbercheck, M.E. and Wang, J. (1996) Effect of cucurbitacin-D on *in vitro* growth of *Xenorhabdus* and *Photorhabdus* spp., symbiotic bacteria of entomopathogenic nematodes. *Journal of Invertebrate Pathology* 68, 141–145.

Barbercheck, M.E., Wang, J. and Hirsh, I.S. (1995) Host plant effects on entomopathogenic nematodes. *Journal of Invertebrate Pathology* 66, 169–177.

Barbercheck, M.E., Wang, J. and Brownie, C. (2003) Adaptation of the entomopathogenic nematode, *Steinernema carpocapsae*, to insect food plant. *Biological Control* 27, 81–94.

Barbosa, P. (ed.) (1998) *Conservation Biological Control*. Academic Press, San Diego, California.

Baur, M.E., Kaya, H.K. and Strong, D.R. (1998) Foraging ants as scavengers on entomopathogenic nematode-killed insects. *Biological Control* 12, 231–236.

Beck, B., Brusselman, E., Nuyttens, D., Moens, M., Pollet, C. *et al.* (2013) Improving foliar applications of entomopathogenic nematodes by selecting adjuvants and spray nozzles. *Biocontrol Science and Technology* 23, 507–520.

Bednarek, A. and Gaugler, R. (1997) Compatibility of soil amendments with entomopathogenic nematodes. *Journal of Nematology* 29, 220–227.

Begg, S.G., Cook, S.M., Dye, R., Ferrante, M., Franck, P. *et al.* (2017) A functional overview of conservation biological control. *Crop Protection* 97, 145–158.

Beillouin, D., Ben-Ari, T., Malézieux, E., Seufert, V. and Makowski, D. (2021). Positive but variable effects of crop diversification on biodiversity and ecosystem services. *Global Change Biology* 27, 4697–4710.

Bitew, Y. and Alemayehu, M. (2017) Impact of crop production inputs on soil health: a review. *Asian Journal of Plant Sciences* 16, 109–131.

Blanco-Pérez, R., Sáenz-Romo, M.G., Vicente-Díez, I., Ibáñez-Pascual, S., Martínez-Villar, E. *et al.* (2020) Impact of vineyard ground cover management on the occurrence and activity of entomopathogenic nematodes and associated soil organisms. *Agriculture, Ecosystems & Environment* 301: 107028.

Campos-Herrera, R., Gómez-Ros, J.M., Escuer, M., Cuadra, L., Barrios, L., and Gutiérrez, C. (2008) Diversity, occurrence, and life characteristics of natural entomopathogenic nematode populations from La Rioja (Northern Spain) under different agricultural management and their relationships with soil factors. *Soil Biology and Biochemistry* 40, 1474–1484.

Campos-Herrera, R., El-Borai, F.E., Stuart, R.J., Graham, J.H. and Duncan, L.W. (2011a) Entomopathogenic nematodes, phoretic *Paenibacillus* spp., and the use of real time quantitative PCR to explore soil food webs in Florida citrus groves. *Journal of Invertebrate Pathology* 108, 30e39.

Campos-Herrera, R., Johnson, E.G., El-Borai, F.E., Stuart, R.J., Graham, J.H., and Duncan, L.W. (2011b) Long-term stability of entomopathogenic nematode spatial patterns measured by sentinel insects and real-time PCR assays. *Annals of Applied Biology* 158, 55e68.

Campos-Herrera, R., Barbercheck, M., Hoy, C.W. and Stock, S.P. (2012a) Entomopathogenic nematodes as a model system for advancing the frontiers of ecology. *Journal of Nematology* 44, 162–176.

Campos-Herrera, R., El-Borai, F.E. and Duncan, L.W. (2012b) Wide interguild relationships among entomopathogenic and free-living nematodes in soil as measured by real time qPCR. *Journal of Invertebrate Pathology* 111, 126e135.

Campos-Herrera, R., El-Borai, F.E., Duncan, L.W. (2012c) Real-time PCR as an effective technique to assess the impact of phoresy by *Paenibacillus* sp. bacteria on *Steinernema diaprepesi* nematodes in nature. *Molecular Ecology Resources* 12, 885–893.

Campos-Herrera, R., Pathak, E., El-Borai, F.E., Stuart, R.J., Gutiérrez, C. *et al.* (2013) Geospatial patterns of soil properties and the biological control potential of entomopathogenic nematodes in Florida citrus groves. *Soil Biology and Biochemistry* 66, 163–174.

Campos-Herrera, R., Jaffuel, G., Chiriboga, X., Blanco-Pérez, R., Fesselet, M., Půža, V., Mascher, F. and Turlings, T. (2015) Traditional and molecular detection methods reveal intense interguild competition and other multitrophic interactions associated with native entomopathogenic nematodes in Swiss tillage soils. *Plant and Soil* 389, 237–255.

Campos-Herrera, R., Blanco-Pérez, R., Bueno Pallero, F.A., Duarte, A., Nolasco, G. *et al.* (2019) Vegetation drives assemblages of entomopathogenic nematodes and other soil organisms: Evidence from the Algarve, Portugal. *Soil Biology and Biochemistry* 128, 150–163.

Chaplin-Kramer, R. and Kremen, C. (2012) Pest control experiments show benefits of complexity at landscape and local scales. *Ecological Applications* 22, 1936–1948.
Chaplin-Kramer, R., O'Rourke, M.E., Blitzer, E.J. and Kremen, C. (2011) A meta-analysis of crop pest and natural enemy response to landscape complexity. *Ecology Letters* 14, 922–932.
Choo, H.Y. and Kaya, H.K. (1991) Influence of soil texture and presence of roots on host finding by *Heterorhabditis bacteriophora*. *Journal of Invertebrate Pathology* 58, 279–280.
Choo, H.Y., Kaya, H.K., Burlando, T.M. and Gaugler, R. (1989) Entomopathogenic nematodes host-finding ability in the presence of plant-roots. *Environmental Entomology* 18, 1136–1140.
Coll, M. (2009) Conservation biological control and the management of biological control services: are they the same? *Phytoparasitica* 37, 205–208.
Cory, J.S. and Hoover, K. (2003) Plant-mediated effects in insect–pathogen interactions. *Trends in Ecology and Evolution* 21, 278–286.
Cullen, R., Warner, K.D., Jonsson, M. and Wratten, S.D. (2008) Economics and adoption of conservation biological control. *Biological Control* 45, 272–280.
Cutler, G.C. and Webster, J.M. (2003) Host-finding ability of three entomopathogenic nematode isolates in the presence of plant roots. *Nematology* 5, 601–608.
de Waal, J.Y., Malan, A.P. and Addison, M.F. (2011) Evaluating mulches together with *Heterorhabditis zealandica* (Rhabditida: Heterorhabditidae) for the control of diapausing codling moth larvae, *Cydia pomonella* (L.) (Lepidoptera: Tortricidae). *Biocontrol Science and Technology* 21, 225–270.
Degenhardt, J. and Gershenzon, J. (2000) Demonstration and characterization of (*E*)-nerolidol synthase from maize: a herbivore-inducible terpene synthase participating in (3*E*)-4,8-dimethyl-1,3,7-nonatriene biosynthesis. *Planta* 210, 815–822.
Delgado-Baquerizo, M., Reich, P.B., Trivedi, C., Eldridge, D.J., Abades, S. *et al.* (2020) Multiple elements of soil biodiversity drive ecosystem functions across biomes. *Nature Ecology and Evolution* 4, 210–220.
Demarta, D., Hibbard, B.E., Bohn, M.O. and Hiltpold, I. (2014) The role of root architecture in foraging behavior of entomopathogenic nematodes. *Journal of Invertebrate Pathology* 122, 32–39.
Denno, R.F., Gruner, D.S., and Kaplan, I. (2008). Potential for entomopathogenic nematodes in biological control: A meta-analytical synthesis and insights from trophic cascade theory. *Journal of Nematology* 40, 61–72.
Doran, J.W. and Parkin, T.B. (1994) Defining and assessing soil quality. In: Doran, J.W., Coleman, D.C., Bezdicek, D.F. and Stewart, B.A. (Eds), *Defining Soil Quality for a Sustainable Environment*. Soil Science Society of America, Madison, pp. 1–21.
Duncan, L.W., Graham, J.H., Zellers, J., Bright, D., Dunn, D.C. *et al.* (2007) Food web responses to augmenting the entomopathogenic nematodes in bare and animal manure-mulched soil. *Journal of Nematology* 39, 176–89. Erratum: *Journal of Nematology* 40, 55.
Edmeades, D.C. (2003) The long-term effects of manures and fertilisers on soil productivity and quality: a review. *Nutrient Cycling in Agroecosystems* 66, 165–180.
Ennis, D.E., Dillon, A.B. and Griffin, C.T. (2010) Simulated roots and host feeding enhance infection of subterranean insects by the entomopathogenic nematode *Steinernema carpocapsae*. *Journal of Invertebrate Pathology* 103, 140–143.
Finke, D.L. and Denno, R.F. (2006) Spatial refuge from intraguild predation: implications for prey suppression and trophic cascades. *Oecologia* 149, 265–275.
Finke, D.L. and Snyder, W.E. (2008) Niche partitioning increases resource exploitation by diverse communities. *Science* 321, 1488–1490.
Gassmann, A.J., Stock, S.P., Tabashnik, B.E. and Singer, M.S. (2010) Tritrophic effects of host plants on an herbivore–pathogen interaction. *Annals of the Entomological Society of America* 103, 371–378.
Gonthier, D.J., Ennis, K.K., Farinas, S., Hsieh, H.-Y., Iverson, A.L. *et al.* (2014) Biodiversity conservation in agriculture requires a multi-scale approach. *Proceedings of the Royal Society B* 281: 20141358. doi:10.1098/rspb.2014.1358
Grant, J.A. and Villani, M.G. (2003) Soil moisture effects on entomopathogenic nematodes. *Environmental Entomology* 32, 80– 87.
Greenwood, C.M., Barbercheck, M.E. and Brownie, C. (2011) Short term response of soil microinvertebrates to application of entomopathogenic nematode–infected insects in two tillage systems. *Pedobiologia* 54, 177–186.
Gulçu, B., Hazir, S. and Kaya, H.K. (2012) Scavenger deterrent factor (SDF) from symbiotic bacteria of entomopathogenic nematodes. *Journal of Invertebrate Pathology* 110, 326–333.

Gurr, G.M. and You, M. (2016) Conservation biological control of pests in the molecular era: new opportunities to address old constraints. *Frontiers in Plant Science*. 6, 1255, DOI: 10.3389/fpls.2015.01255

Hazir, S., Shapiro-Ilan, D.I., Hazir, C., Leite, L.G., Cakmak, I. and Olson, D. (2016) Multifaceted effects of host plants on entomopathogenic nematodes. *Journal of Invertebrate Patholology* 135, 53–59.

Heimpel, G.E. and Mills, N.J. (2017) *Biological Control: Ecology and Applications*. Cambridge University Press, New York, 386 pp.

Hiltpold, I., Baroni, M., Toepfer, S., Kuhlmann, U., and Turlings, T.C.J. (2010a) Selection of entomopathogenic nematodes for enhanced responsiveness to a volatile root signal helps to control a major root pest. *Journal of Experimental Biology* 213, 2417–2423.

Hiltpold, I., Toepfer, S., Kuhlmann, U. and Turlings, T.C.J. (2010b) How maize root volatiles affect the efficacy of entomopathogenic nematodes in controlling the western corn rootworm. *Chemoecology* 20, 155–162.

Hirsh, I.S. and Barbercheck, M.E. (1996) Effects of host plant and cucurbitacin on growth of larval *Diabrotica undecimpunctata howardi*. *Entomologia Experimentalis et Applicata* 81, 47–51.

Hoy, C.W., Grewal, P.S., Lawrence, J.L., Jagdale, G. and Acosta, N. (2008). Canonical correspondence analysis demonstrates unique soil conditions for entomopathogenic nematode species compared with other free–living nematode species. *Biological Control* 46, 371–379.

Hummel, R., Walgenbach, J.F., Barbercheck, M.E., Kennedy, G.G., Hoyt, G.O. and Arellano, C. (2002) Effects of production practices on soil borne entomopathogens in western North Carolina vegetable systems. *Environmental Entomology* 31, 84–91.

Jabbour, R. and Barbercheck, M.E. (2008) Soil and habitat complexity effects on movement of the entomopathogenic nematode, *S. carpocapsae*, in maize. *Biological Control* 47, 235–243.

Jabbour, R., Crowder, D.W., Aultman, E.A. and Snyder, W.E. (2011) Entomopathogen biodiversity increases host mortality. *Biological Control* 59, 277–283.

Jaffuel, G., Mäder, P., Blanco-Perez, R., Chiriboga, X., Fliessbach, A. *et al.* (2016) Prevalence and activity of entomopathogenic nematodes and their antagonists in soils that are subject to different agricultural practices. *Agriculture, Ecosystems & Environment* 230, 329–340.

Jaffuel, G., Blanco-Pérez, R., Büchi, L., Mäder, P., Fließbach, A., *et al.* (2017) Effects of cover crops on the overwintering success of entomopathogenic nematodes and their antagonists. *Applied Soil Ecology* 114, 62–73.

Jaffuel, G., Blanco-Pérez, R., Hug, A.-S., Chiriboga, X., Meuli, R.G. *et al.* (2018) The evaluation of entomopathogenic nematode soil food web assemblages across Switzerland reveals major differences among agricultural, grassland and forest ecosystems. *Agriculture, Ecosystems & Environment* 262, 48–57.

Jagodič, A., Ipavec, N., Trdan, S. and Laznik, Ž. (2017) Attraction behaviors: are synthetic volatiles, typically emitted by insect-damaged *Brassica nigra* roots, navigation signals for entomopathogenic nematodes (*Steinernema* and *Heterorhabditis*)? *BioControl* 62, 515–524.

Kapranas, A., Maher, A.M.D. and Griffin, C.T. (2017) The influence of organic matter content and media compaction on the dispersal of entomopathogenic nematodes with different foraging strategies. *Parasitology* 144, 1956–1963.

Karp, D.S., Chaplin-Kramer, R., Meehan, T.D., Martin, E.A., DeClerck, F. *et al.* (2018) Crop pests and predators exhibit inconsistent responses to surrounding landscape composition. *Proceedings of the National Academy of Sciences of the USA* 115, E7863–E7870.

Karunakar, G., Easwaramoorthy, S. and David, H. (2002) Compatibility of *Steinernema feltiae*, *S. glaseri* and *Heterarhabditis indica* with certain fertilizers, herbicides and pesticides. *Sugar Technology* 4, 123–130.

Kladivko, E. (2001) Tillage systems and soil ecology. *Soil and Tillage Research* 61, 61–76.

Klemola, N., Klemola, T., Rantala, M.J. and Ruuhola, T. (2007) Natural host-plant quality affects immune defence of an insect herbivore. *Entomologia Experimentalis et Applicata* 123, 167–176.

Koppenhöfer, A.M., Shapiro-Ilan, D.I. and Hiltpold, I. (2020) Entomopathogenic nematodes in sustainable food production. *Frontiers in Sustainable Food Systems* 4: 125.

Kostenko, O., Duyts, H., Grootemaat, S., de Deyn, G.B. and Bezemer, T.M. (2015) Plant diversity and identity effects on predatory nematodes and their prey. *Ecology and Evolution* 5, 836–847.

Kung, S.P., Gaugler, R. and Kaya, H.K. (1990) Influence of soil pH and oxygen on persistence of *Steinernema* spp. *Journal of Nematology* 22, 440–445.

Kung, S.P., Gaugler, R. and Kaya, H.K. (1991) Effects of soil temperature, moisture, and relative humidity on entomopathogenic nematode persistence. *Journal of Invertebrate Pathology* 57, 242–249.

Labaude, S. and Griffin, C.T. (2018) Transmission success of entomopathogenic nematodes used in pest control. *Insects* 9: 72.

Lacey, L.A., Granatstein, D., Arthurs, S.P., Headrick, H. and Fritts, R. Jr. (2006) Use of entomopathogenic nematodes (Steinernematidae) in conjunction with mulches for control of overwintering codling moth (Lepidoptera: Tortricidae). *Journal of Entomological Science* 41, 107–119.

Laznik, Ž. and Trdan, S. (2014) The influence of insecticides on the viability of entomopathogenic nematodes (Rhabditida: Steinernematidae and Heterorhabditidae) under laboratory conditions. *Pest Management Science* 70, 784–789.

Lehmann, J., Bossio, D.A., Kögel-Knabner, I. and Rillig, M.C. (2020) The concept and future prospects of soil health. *Nature Reviews Earth and Environment* 1, 544–553.

Leite, L.G., Schmidt, F.S., Harakava, R.O., Filho, A.B., Giometti, F.H.C. *et al.* (2015) The influence of mulch on the persistence of *Steinernema brazilense* (Nematoda: Steinernematidae) in sugarcane fields. *Revista Colombiana de Entomología* 41, 176–179.

Letourneau, D.K. and Bothwell, S.G. (2008) Comparison of organic and conventional farms: challenging ecologists to make biodiversity functional. *Frontiers in Ecology and the Environment* 6, 430–438.

Lewis, E.E., Campbell, J.F. and Gaugler, R. (1998). A conservation approach to using entomopathogenic nematodes in turf and landscapes. In: Barbosa, P. (ed.) (1998) *Conservation Biological Control*. Academic Press, San Diego, pp. 235–254.

Losey, J.E. and Vaughan, M. (2006) The economic value of ecological services provided by insects. *BioScience* 56, 311–323.

Lu, Q., Liu, T., Wang, N., Dou, Z., Wang, K. and Zuo, Y. (2020) A review of soil nematodes as biological indicators for the assessment of soil health. *Frontiers of Agricultural Science and Engineering* 7, 275–281.

Millar, L.C. and Barbercheck, M.E. (2001) Interaction between endemic and introduced entomopathogenic nematodes in conventional-till and no-till corn. *Biological Control* 22, 235–245.

Millar, L.C. and Barbercheck, M.E. (2002) Effect of tillage practices on entomopathogenic nematodes in a corn agroecosystem. *Biological Control* 25, 1–11.

Mráček, Z., Bečvář, S., Kindlmann, P. and Jersáková, J. (2005) Habitat preference for entomopathogenic nematodes, their insect hosts and new faunistic records for the Czech Republic. *Biological Control* 34, 27–37.

Nádasy, E. and Nádasy, M. (2006) Some harmful or useful environmental effect of nitrogen fertilisers. *Cereal Research Communications* 34, 49–52.

Özdemir, E., İnak, E., Evlice, E. and Laznik, Z. (2020) Compatibility of entomopathogenic nematodes with pesticides registered in vegetable crops under laboratory conditions. *Journal of Plant Diseases and Protection* 127, 529–535.

Pace, M.L., Cole, J.J., Carpenter, S.R. and Kitchell, J.F. (1999) Trophic cascades revealed in diverse ecosystems. *Trends in Ecology and Evolution* 14, 483–488.

Perez, E.E., Lewis, E.E. and Shapiro-Ilan, D.I. (2003) Impact of the host cadaver on survival and infectivity of entomopathogenic nematodes (Rhabditida: Steinernematidae and Heterorhabditidae) under desiccating conditions. *Journal of Invertebrate Pathology* 82, 111–118.

Preisser, E.L. (2003) Field evidence for a rapidly cascading underground food web. *Ecology* 84, 869–874.

Quist, C.W., Gort, G., Mooijman, P., Brus, D.J., van den Elsen, S. *et al.* (2019) Spatial distribution of soil nematodes relates to soil organic matter and life strategy. *Soil Biology and Biochemistry* 136: 107542.

Ram, K., Gruner, D.S., McGlaughlin, J.P., Preisser, E.L. and Strong, D.R. (2008) Dynamics of a subterranean trophic cascade in space and time. *Journal of Nematology* 40, 85–92.

Rasmann, S., Köllner, T.G., Degenhardt, J., Hiltpold, I., Toepfer, S. *et al.* (2005) Recruitment of entomopathogenic nematodes by insect-damaged maize roots. *Nature* 434, 732–737.

Rasmann, S., Ali, J.G., Helder, J. and van der Putten, W.H. (2012) Ecology and evolution of soil nematode chemotaxis. *Journal of Chemical Ecology* 38, 615–628.

Robert, C.A.M., Zhang, X., Machado, R.A., Schirmer, S., Lori, M., *et al.* (2017) Sequestration and activation of plant toxins protect the western corn rootworm from enemies at multiple trophic levels. *eLife* 6: e29307.

Romig, D.E., Garlynd, M.J., Harris, R.F. and McSweeney, K.M. (1995) How farmers assess soil health and quality. *Journal of Soil and Water Conservation* 50, 229–236.

Root, R.B. (1973) Organization of a plant-arthropod association in simple and diverse habitats: the fauna of collards (*Brassica oleracea*). *Ecological Monographs* 43, 95–124.

Rosenheim, J.A., Kaya, H.K., Ehler, L.E., Marois, J.J. and Jaffee, B.A. (1995) Intraguild predation among biological-control agents: theory and evidence. *Biological Control* 5, 303–335.

Rowen, E., Tooker, J.F. and Blubaugh, C.K. (2019) Managing fertility with animal waste to promote arthropod pest suppression. *Biological Control* 134, 130–140.
Rowen, E., Regan, K., Barbercheck, M., and Tooker, J. (2020) Is tillage beneficial or detrimental for insect and slug management? A meta-analysis. *Agriculture, Ecosystems & Environment* 294: 106849.
Russell, E.P. (1989) Enemies hypothesis: a review of the effect of vegetational diversity on predatory insects and parasitoids. *Environmental Entomology* 18, 590–599.
Şahin, Y.S. and Susurluk, A. (2018) Effects of some inorganic fertilizers on entomopathogenic nematodes, *Steinernema feltiae* (Tur-s3) and *Heterorhabditis bacteriophora* (Hb.H). *Türkiye Biyolojik Mücadele Dergisi* 9, 102–109.
Salame, L. and Glazer, I. (2015) Stress avoidance: vertical movement of entomopathogenic nematodes in response to soil moisture gradient. *Phytoparasitica* 43, 647–655.
Settele, J. and Settle, W.H. (2018) Conservation biological control: improving the science base. *Proceedings of the National Academy of Sciences USA* 115, 8241–8243.
Shapiro, D.I., Tylka, G.L. and Lewis, L.C. (1996) Effects of fertilizers on virulence of *Steinernema carpocapsae*. *Applied Soil Ecology* 3, 27–34.
Shapiro, D.I., Lewis, L.C., Obrycki, J.J. and Abbas, M. (1999a) Effects of fertilizers on suppression of black cutworm (*Agrostis ipsilon*) damage with *Steinernema carpocapsae*. *Journal of Nematology* 31 (Suppl.), 690–693.
Shapiro, D.I., Obrycki, J.J., Lewis, L.C. and Jackson, J.J. (1999b) Effects of crop residue on the persistence of *Steinernema carpocapsae*. *Journal of Nematology* 31, 517–519.
Shapiro-Ilan, D.I., Lewis, E.E., Behle, R.W. and McGuire, M.R. (2001) Formulation of entomopathogenic nematode-infected cadavers. *Journal of Invertebrate Pathology* 78, 17–23.
Shehata, I.E., Hammam, M.M.A. and Abd-Elgawad, M.M.M. (2021) Effects of inorganic fertilizers on virulence of the entomopathogenic nematode *Steinernema glaseri* and peanut germination under field conditions. *Agronomy* 11: 945.
Shields, E.J., Testa, A.M. and O'Neil, W.J. (2018) Long-term persistence of native New York entomopathogenic nematode isolates across crop rotation, *Journal of Economic Entomology* 111, 2592–2598.
Shields, M.W., Johnson, A.C., Pandey, S., Cullen, R., González-Chang, M. *et al.* (2019) History, current situation and challenges for conservation biological control. *Biological Control* 131, 25–35.
Snyder, W.E., Snyder, G.B., Finke, D.L. and Straub, C.S. (2006) Predator biodiversity strengthens herbivore suppression. *Ecological Letters* 9, 789–796.
Stelinski, L.L., Willett, D., Rivera, M.J. and Ali, J.G. (2019) 'Tuning' communication among four trophic levels of the root biome to facilitate biological control. *Biological Control* 131, 49–53.
Stott, D.E. and Moebius-Clune, B.N. (2017) Soil health: challenges and opportunities. In: Field, D.J., Morgan, C.L.S. and McBratney, A.B. (eds.) *Global Soil Security*. Springer, Cham, Switzerland, pp. 109–122.
Straub, C.S. and Snyder, W.E. (2006) Species identity dominates the relationship between predator biodiversity and herbivore suppression. *Ecology* 87, 277–282.
Straub, C.S. and Snyder, W.E. (2008) Increasing enemy biodiversity strengthens herbivore suppression on two plant species. *Ecology* 89, 1605–1615.
Straub, C.S., Finke, D.L. and Snyder, W.E. (2008) Are the conservation of natural enemy biodiversity and biological control compatible goals? *Biological Control* 45, 225–237.
Strong, D.R. (2002) Population of entomopathogenic nematodes in foodwebs. In: Gaugler, R. (ed.) *Entomopathogenic Nematology*. CAB International, Wallingford, UK, pp. 225–240.
Stuart, R.J., El-Borai, F.E. and Duncan, L.W. (2008) From augmentation to conservation of entomopathogenic nematodes: trophic cascades, habitat manipulation and enhanced biological control of *Diaprepes abbreviatus* root weevils in Florida citrus groves. *Journal of Nematology* 40, 73–84.
Stuart R.J., Barbercheck M.E. and Grewal P.S. (2015) Entomopathogenic nematodes in the soil environment: distributions, interactions and the influence of biotic and abiotic factors. In: Campos-Herrera, R. (ed.) *Nematode Pathogenesis of Insects and Other Pests: Ecology and Applied Technologies for Sustainable Plant and Crop Protection*. Springer, Cham, Switzerland, pp. 97–137.
Susurluk, A. and Ehlers, R.U. (2008) Field persistence of the entomopathogenic nematode *Heterorhabditis bacteriophora* in different crops. *BioControl* 53, 627–641.
Tamburini, G., Bommarco, R., Wanger, T.C., Kremen, C., van der Heijden, M.G.A. *et al.* (2020). Agricultural diversification promotes multiple ecosystem services without compromising yield. *Science Advances* 6: eaba1715.
Thiele-Bruhn, S., Bloem, J., de Vries, F.T., Kalbitz, K. and Wagg, C. (2012) Linking soil biodiversity and agricultural soil management. *Current Opinion in Environmental Sustainability* 4, 523–528.

Tscharntke, T., Bommarco, R., Clough, Y., Crist, T.O., Kleijn, D. *et al.* (2007) Conservation biological control and enemy diversity on a landscape scale. *Biological Control* 43, 294–309.

Turlings, T., Hiltpold, I. and Rasmann, S. (2012) The importance of root produced volatiles as foraging cues for entomopathogenic nematodes. *Plant and Soil* 358, 51–60.

Ulug, D., Hazir, S., Kaya, H.K. and Lewis, E. (2014) Natural enemies of natural enemies: the potential top-down impact of predators on entomopathogenic nematode populations. *Ecological Entomology* 39, 462–469.

USDA (US Department of Agriculture) (2024) Soil Health. Natural Resources Conservation Service Washington, DC. Available at: www.nrcs.usda.gov/conservation-basics/natural-resource-concerns/soils/soil-health (accessed 27 March 2024.)

Wallace, H.R. (1968) The dynamics of nematode movement. *Annual Review of Phytopathology* 6, 91–114.

Wardle, D.A., Yeates, G.W., Williamson, W. and Bonner, K.I. (2003) The response of a three trophic level soil food web to the identity and diversity of plant species and functional groups. *Oikos* 102, 45–56.

Wardle, D.A., Bardgett, R.D., Klironomos, J.N., Setäla, H., van der Putten, W.H. and Wall, D.H. (2004) Ecological linkages between aboveground and belowground biota. *Science* 304, 1629–1633.

Zhang, X., Machado, R.A.R., Doan, C.V., Arce, C.C.M., Hu, L. and Robert, C.A.M. (2019) Entomopathogenic nematodes increase predation success by inducing cadaver volatiles that attract healthy herbivores. *eLife* 8: e46668.

Zhang, X., Li, L., Kesner, L. and Robert, C.A.M. (2021) Chemical host-seeking cues of entomopathogenic nematodes. *Current Opinion in Insect Science* 44, 72–81.

Zhou, X., Kaya, H.K., Heungens, K. and Goodrich-Blair, H. (2002) Response of ants to deterrent factor(s) produced by the symbiotic bacteria of entomopathogenic nematodes. *Applied Environmental Microbiology* 68, 6202–6209.

29 Conclusions

David I. Shapiro-Ilan[1]* and Edwin E. Lewis[2]
[1]USDA-ARS, Southeastern Fruit and Tree Nut Research Station, Byron, Georgia, USA; [2]Department of Entomology, Plant Pathology and Nematology, University of Idaho, Moscow, Idaho, USA

The discipline of entomopathogenic nematology has passed the 100 year mark since the first taxonomic description in this group of *Steinernema kraussei* (Steiner, 1923). Since that time, entomopathogenic nematodes (EPNs) have become commercially viable biocontrol agents that are used to control a wide array of economically important pests. It has been more than 18 years since the last volume of this book was published (Grewal *et al.*, 2005a). Great advances have been made during this interval as detailed within this volume's chapters. In this chapter, we highlight some of the key advances and present some ideas on how these accomplishments serve as the foundation for progress in the future.

One major advancement that stands out since the discovery of *S. kraussei* is how many more species have been described. Indeed, more than 125 species of *Steinernema* and *Heterorhabditis* spp. are listed in Chapter 1 (this volume). In the 18 years since the previous volume of this book, more than 60% of known EPN species have been described. Certainly, the number of EPN species described will continue to increase as more areas of the world are explored. Along with new discoveries have come increased regulation concerning international movement of endemic species and the release of non-native biocontrol agents in general. The advent of molecular diagnostic technology has increased the precision with which species can be rapidly and accurately differentiated, which is crucial to meeting regulatory requirements. None the less, taxonomic approaches for EPNs and advancements in technology and unification of techniques still have room for improvement, and we expect that further discoveries in this vein will be forthcoming.

Molecular advances in entomopathogenic nematology have not been limited to taxonomy and phylogenetics. Great advances have been made in genomics (Chapter 4, this volume) with a notable landmark being the first draft sequence of *Heterorhabditis bacteriophora* (TT01 strain) in 2007 (Ciche, 2007). Currently, at least nine EPN genomes have been sequenced, and certainly there will be more to come. Moreover, annotation translation of the genomes will prove to be a gold mine of knowledge on genes responsible for traits relevant to biocontrol efficacy.

In addition to genomic advances, substantial advancements have been made in our elucidation of EPN biology and ecology. Understanding and clarifying the full role of EPNs in the ecosystem and food web has been challenging. The potential role of EPNs as scavengers, rather than exclusively as parasites, has been established (San-Blas and Gowen, 2008; Chapter 2, this volume), yet it is still unclear to what extent the

*Email: David.Shapiro@usda.gov

DOI: 10.1079/9781800620322.0029

scavenger role occurs in nature, what kind of ecosystem services scavenging may provide, and what the factors are that affect the choice of scavenging versus parasitism.

Beyond the EPN trophic role (scavenger versus parasite/pathogen), in a broader sense, strides have been made in elucidating EPN population biology (Chapter 7, this volume). A key factor in pushing our understanding of EPN population ecology has been the use of molecular tools to track EPN species distribution, persistence and abundance across landscapes (e.g. Campos-Herrera *et al.*, 2011). Additional advances will be made as molecular tools continue to be developed and more laboratories become versed with using them.

On a smaller scale, many novel discoveries have been made in the realm of EPN behavioral ecology (Chapter 6, this volume). For example, EPNs were found to infect and disperse in aggregate, and to display follow-the-leader behavior. Novel cues that can impact EPN navigation have been discovered such as vibration and electromagnetic signals. EPN communication has been revealed via the discovery of ascaroside pheromones that regulate behaviors such as dispersal and infectivity. Moreover, fascinating levels of communication between EPNs and plants have been discovered such as call-for-help signals (Hiltpold *et al.*, 2011). Additional knowledge is needed to define the chemicals involved in EPN communication, the scope and overlap among species, and the hierarchies of signals and cues. Certainly, our understanding of EPN communication and signaling, along with the chemistry behind these interactions and how to exploit them for improved biocontrol efficacy will continue to expand in the years to come.

Going from the landscape/population level to the individual level, substantial progress has been made in EPN infective juvenile (IJ) survival biology (Chapter 5, this volume). EPNs are sensitive to various sources of abiotic stress, such as ultraviolet (UV) radiation, desiccation and temperature extremes. Therefore, understanding the nature of survival strategies is beneficial and can lead to improved biocontrol efficacy and perhaps improved product shelf life. Genes and transcriptome expression for aspects of EPN stress tolerance (e.g. heat and desiccation tolerance) have been elucidated. Recently, the process of EPN desiccation has been studied and described in detail at various relative humidity levels (Ramakrishnan *et al.*, 2022). However, a deeper understating of EPN survival biology has yet to be translated directly into major advances in commercial formulation and storage capacity. Such advances will likely come as the bases of survival biology at the micro- and macro-levels continue to be unraveled.

EPN applications have resulted in high levels of field efficacy in controlling a wide array of arthropod pests, as supported by many studies in recent years (Chapters 14–26, this volume). Many of the pests listed are targeted commercially. Additionally, by harnessing the EPN symbiotic bacteria or their metabolites, the potential for biocontrol application has been expanded to plant-parasitic nematodes and plant diseases (especially fungal diseases) (Chapter 9, this volume). The list of target pests for EPNs continues to increase, and this success is facilitated by advancements in EPN production, formulation and application technology.

Production technology has been substantially enhanced in the past decade (Chapter 10, this volume). Significant advances have been made in improving media quality for *in vitro* production in solid and liquid fermentation, and in bioreactor parameters for liquid culture. A novel biphasic production method was developed that combines liquid and solid production, and avoids separate harvest and formulation steps (because the sponge medium also serves as the formulation substrate) (Leite *et al.*, 2023). Innovative methods to improve *in vivo* EPN production have also been developed, including automated technology to reduce labor costs, and grower-based systems that enable farmers to sustainably produce their own EPNs (Oliveira-Hofman *et al.*, 2023). Farmer-based production systems may be especially amenable to niche markets or areas that do not have access to commercially produced nematodes. The economic feasibility of these approaches needs to be tested. As methodologies for production continue to increase and costs are reduced, we can expect EPNs to be used on a wider scale.

Sensitivity to environmental conditions such as UV radiation and desiccation has largely prevented EPNs from being applied to aboveground targets. Thus, formulations that protect EPNs from UV light and desiccation will lead to their expanded use (Chapter 11, this volume). For example, several gel formulations have been highly successful in facilitating aboveground formulations. Novel nanoparticle formulations have also shown recent progress (Wu *et al.*, 2022). Additional research is needed to develop low-cost protective formulations, especially those that are compatible with organic agriculture.

Enhancements have also been made in application technology (Chapter 11, this volume). It has been recognized that different spray technologies including sprayer type, nozzles and pressure can be optimized for particular target pests and cropping systems. Application approaches that are targeted and minimize the number of IJs used will be superior in terms of cost and efficacy. For example, specialized trunk sprayers only spray EPNs when they pass the trunk (avoiding waste between trees or rows). Using chemical lures, sentinel trap trees concentrate the pest to a drastically reduced number of trees within an orchard, thereby substantially reducing the treatment area (Chapter 17, this volume; Leskey *et al.*, 2008). In another example, a smart sprayer detects nilgai antelope (raised for meat) when they come to feeding stations and EPNs are automatically sprayed on to the animals to kill ticks (Goolsby *et al.*, 2019). More research is needed along these lines to optimize application methods and equipment, and to develop approaches that effectively minimize the amount of EPNs needed per unit area. Moreover, enhancing the persistence of EPNs via conservation methods (Chapters 11 and 28, this volume) will also lead to improved efficacy and potentially reduced costs.

In the first volume, Grewal *et al.*, 2005b indicated three categories of major hurdles to expanding EPN use: product availability and cost, ease of use and efficacy. As we conclude the current volume, 18 years later, the challenges remain the same. However, as indicated above and detailed in the chapters herein, major progress has been made in all areas since the printing of the previously volume. Product availability has been expanded based on a greater number of companies producing EPNs and enhanced production capacity. However, the costs of EPN applications still tend to be higher (sometimes substantially higher) than chemical pesticides. Thus, additional improvements to reduce costs are needed, such as in production and application approaches. Ease of use has been improved largely due to improved formulation, but formulations containing bulk EPNs that can be kept at room temperature remain elusive and formulation costs should also be reduced. The efficacy of EPNs in field applications has also improved based on increased knowledge, but there are still many instances where EPNs fail unexpectedly (indicating the need for additional fundamental and applied research).

Climate change, food insecurity and pesticide regulation (involving the removal of chemical insecticides) accentuate the growing need for safe pest-control alternatives. In this book, we have presented basic knowledge and diverse applications to indicate how EPNs continue to fill an important role as potent biocontrol solutions. We have also provided guidance on research needed to expand EPN utility in the future. Since the first volume of this book, EPNs have been used in the first biocontrol experiment conducted in space (Kaplan *et al.*, 2020). The experiment was a look toward the future when humans will be growing crops in space and perhaps using EPNs as biocontrol agents for control of space pests. It will indeed be interesting to see the advancements in entomopathogenic nematology in the years to come.

References

Campos-Herrera, R., Johnson, E.G., El-Borai, F.E., Stuart, R.J., Graham, J.H. and Duncan, L.W. (2011) Long-term stability of entomopathogenic nematode spatial patterns in soil as measured by sentinel insects and real-time PCR assays. *Annals of Applied Biology* 158, 55–68.

Ciche, T. (2007) The biology and genome of *Heterorhabditis bacteriophora*. In: *WormBook: The Online Review of C.* elegans *Biology*. Wormbook, Pasadena, California. Available at: www.ncbi.nlm.nih.gov/books/NBK19785/ (accessed 27 March 2024).

Goolsby, J.A., Cantu, D., Vasquez, A. and Racelis, A. (2019) Development of a remotely activated field sprayer and evaluation of temperature and aeration on the longevity of *Steinernema riobrave* entomopathogenic nematodes for treatment of cattle fever tick-infested nilgai. *Subtropical Agriculture and Environments* 70, 1–5.

Grewal, P.S., Ehlers, R.U. and Shapiro-Ilan, D.I. (eds.) (2005a) *Nematodes as Biocontrol Agents*. CAB International, Wallingford, UK
Grewal, P.S., Ehlers, R.-U. and Shapiro-Ilan, D.I. (2005b) Critical issues and research needs for expanding the use of nematodes in biocontrol. In: Grewal, P.S., Ehlers, R.U. and Shapiro-Ilan, D.I. (eds.) *Nematodes as Biocontrol Agents*. CAB International, Wallingford, UK, pp. 479–489.
Hiltpold, I., Erb, M., Robert, C.A.M. and Turlings, T.C.J. (2011). Systemic root signalling in a belowground, volatile-mediated tritrophic interaction. *Plant, Cell & Environment* 34, 1267–1275.
Kaplan, F., Shapiro-Ilan, D. and Schiller, K.C. (2020) Dynamics of entomopathogenic nematode foraging and infectivity in microgravity. *NPJ Microgravity* 6: 20.
Leite, L.G., Chacon-Orozcoa, J.G., Shapiro-Ilan, D.I., Baldo, F.B. and Cardoso, J.M. (2023) Effects of temperature for optimizing production and storage of *Steinernema rarum* in a novel biphasic process, and efficacy of the nematode against *Sphenophorus levis*. *Biological Control* 187: 105381.
Leskey, T.C., Pinero, J.C. and Prokopy, R.J. (2008) Oder-baited trap trees: a novel management tool for plum curculio (Coleoptera: Curculionidae). *Journal of Economic Entomology* 101, 1302–1309.
Oliveira-Hofman, C., Steffan, S. and Shapiro-Ilan, D. (2023) A sustainable grower-based method for entomopathogenic nematodes production. *Journal of Insect Science* 23: 4.
Ramakrishnan, J., Salame, L., Nasser, A., Glazer, I. and Ment, D. (2022) Survival and efficacy of entomopathogenic nematodes on exposed surfaces. *Scientific Reports* 12: 4629.
San-Blas, E. and Gowen, S.R. (2008) Facultative scavenging as a survival strategy of entomopathogenic nematodes. *International Journal for Parasitology* 38, 85–91.
Steiner, G. (1923) *Aplectana kraussei* n. sp. Eine in der Blattwespe *Lyda* sp. Parasitierende Nematodenform, nebst Bemerkungen über das Seitenorgan der parasitischen Nematoden. *Zentralbatt fur Bakteriologiel Parasitenkunde Infektionskrankheiten und Hygiene* Abteilung II, 59, 14–18.
Wu, S., Mechrez, G., Ment, M., Toews, M.D., Ananth, K. *et al.* (2022) Tolerance of *Steinernema carpocapsae* infective juveniles in novel nanoparticle formulations to ultraviolet radiation. *Journal of Invertebrate Pathology* 196: 107851.

Index

Printed and bound by CPI Group (UK) Ltd, Croydon, CR0 4YY

25/06/2025

14694444-0002